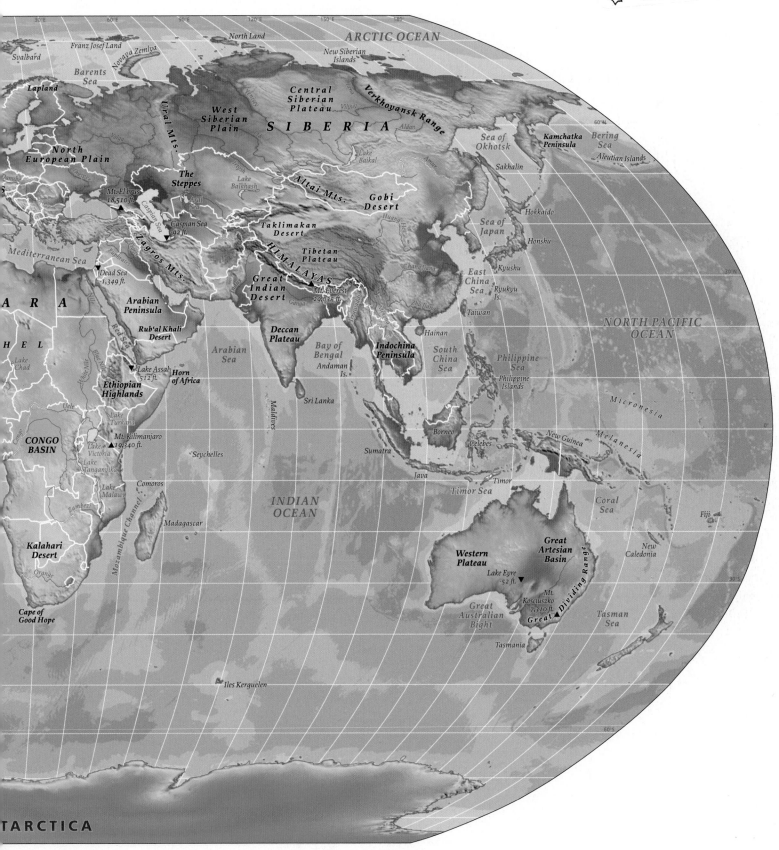

ARCTIC OCEAN

Svalbard

Franz Josef Land
Novaya Zemlya
North Land

New Siberian
Islands

Barents
Sea

Lapland

North
European Plain

Ural Mts.

West
Siberian
Plain

Central
Siberian
Plateau

Verkhoyansk Range

SIBERIA

Yenisey

Vilyuy

Aldan

Lake
Baikal

Amur

Sea of
Okhotsk

Kamchatka
Peninsula

Bering
Sea

Aleutian Islands

Sakhalin

Hokkaido

Sea of
Japan

Honshu

60 N

The
Steppes

Mt. Elbrus
18,510 ft.

Danube

Dnieper

Volga

Don

Lake
Balkhash

Altai Mts.

Gobi
Desert

Huang He

Caspian Sea
92 ft.

Aral
Sea

Caspian Sea

Zagros Mts.

Mediterranean Sea

Euphrates

Tigris

Dead Sea
1,349 ft.

Taklimakan
Desert

HIMALAYAS

Tibetan
Plateau

Great
Indian
Desert

Mt. Everest
29,035 ft.

Chang Jiang

Kyushu

Ryukyu
Is.

East
China
Sea

30 N

Taiwan

NORTH PACIFIC
OCEAN

A R A

H E L

Arabian
Peninsula

Red Sea

Rub'al Khali
Desert

Ganga

Deccan
Plateau

Bay of
Bengal

Indochina
Peninsula

Hainan

South
China
Sea

Philippine
Sea

Irrawaddy

Mekong

Zhu Jiang

Lake
Chad

White Nile

Blue Nile

Lake Assal
512 ft.

Horn
of Africa

Arabian
Sea

Andaman
Is.

Philippine
Islands

Ethiopian
Highlands

Uele

Lake
Turkana

Mt. Kilimanjaro
19,340 ft.

Sri Lanka

Maldives

Micronesia

0

CONGO
BASIN

Lake
Victoria

Seychelles

Lake
Tanganyika

Sumatra

Borneo

Celebes

New Guinea

Melanesia

Zambezi

Lake
Malawi

Comoros

Java

Timor

Mozambique Channel

Madagascar

INDIAN
OCEAN

Timor Sea

Coral
Sea

New
Caledonia

Fiji

Kalahari
Desert

Orange

30 S

Western
Plateau

Great
Artesian
Basin

Lake Eyre
-52 ft.

Great Dividing Range

Cape of
Good Hope

Great
Australian
Bight

Mt.
Kosciuszko
7,310 ft.

Great

Tasman
Sea

Tasmania

60 S

Iles Kerguelen

TARCTICA

# WORLD REGIONAL GEOGRAPHY

## Global Patterns, Local Lives

**FOURTH EDITION**

### LYDIA MIHELIČ PULSIPHER
*University of Tennessee*

### ALEX A. PULSIPHER
*Ph.D. Candidate*
*Graduate School of Geography*
*Clark University*

with the assistance of
**Conrad "Mac" Goodwin**
*University of Tennessee*

W. H. Freeman and Company
New York

**Publisher:** Sara Tenney
**Acquisitions Editor:** Marc Mazzoni
**Developmental Editor:** Susan Weisberg
**Project Manager:** Lisa Samols
**Senior Marketing Manager:** Scott Guile
**Senior Project Editor:** Mary Louise Byrd
**Cover and Text Designer:** Diana Blume
**Senior Illustration Coordinator:** Bill Page
**Maps:** University of Tennessee, Cartographic Services Laboratory, Will Fontanez, Director;
 Maps.com, Martha Bostwick, Lead Cartographer
**Photo Editors:** Patricia Marx and Ted Szczepanski
**Photo Researchers:** Elyse Rieder and Donna Ranieri
**Production Manager:** Julia De Rosa
**Editorial Assistant:** Maia Gil'Adi
**Supplements Editor:** Deepa Chungi
**Composition:** Sheridan Sellers, W. H. Freeman and Company, Electronic Publishing Center
**Printing and Binding:** RR Donnelly

FRONTLINE/World is a trademark of the WGBH Educational Foundation.
Frontline is a registered trademark of the WGBH Educational Foundation.

Library of Congress Control Number: 2007927108

ISBN-13: 978-0-7167-7792-2 (paperback); 978-1-4292-1729-3 (hardcover)
ISBN-10: 0-7167-7792-4 (paperback); 1-4292-1729-4 (hardcover)

Printed in the United States of America

First printing

W. H. Freeman and Company
41 Madison Avenue
New York, NY 10010
Houndmills, Basingstoke RG21 6XS, England
www.whfreeman.com

*To our three youngest sprouts, Anthony Louis, Samson Conrad, and Henry Vincent, and to the child depicted on the cover of this book; may their lives enrich the world we describe here.*

Lydia with Anthony Louis in Ljubljana, Slovenia, 2005.

**Lydia Mihelič Pulsipher** is a cultural-historical geographer who studies the landscapes of ordinary people through the lens of archaeology, historical geography, and ethnography. She has contributed to several geography-related exhibits at the Smithsonian Museum of Natural History in Washington, D.C., including "Seeds of Change," which featured her research in the eastern Caribbean. Lydia Pulsipher has ongoing research projects in the eastern Caribbean (historical archaeology) and in central Europe, where her graduate students are now studying issues of national identity in several countries. She has taught cultural, gender, European, North American, and Mesoamerican geography at the University of Tennessee at Knoxville since 1980; through her research, she has given many students their first experience in fieldwork abroad. Previously she taught at Hunter College and Dartmouth College. She received her B.A. from Macalester College, her M.A. from Tulane University, and her Ph.D. from Southern Illinois University.

**Alex A. Pulsipher** is a Ph.D. candidate in geography at Clark University, where he is studying real estate development and sustainability in the United States. In the early 1990s, Alex spent some time in South Asia working for a sustainable development research center and then went on to do an undergraduate thesis on the history of Hindu nationalism at Wesleyan University. Beginning in 1995, Alex worked full time on the research and writing of the first edition of this textbook. In 1999 and 2000, he traveled to South America, Southeast Asia, and South Asia, where he collected information for the second edition of the text and for the Web site. In 2000 and 2001, he returned to writing material and designing maps for the second edition. Alex participated fully in the writing of the fourth edition.

In the writing of *World Regional Geography*, Lydia and Alex Pulsipher were assisted in many ways by Lydia's husband, **Conrad "Mac" Goodwin**, a historical archaeologist who specializes in sites created during the European colonial era in North America, the Caribbean, and the Pacific. He has particular expertise in the archaeology of agricultural systems, gardens, domestic landscapes, and urban spaces. When not working on the textbook, Mac is a master organic gardener and slow-food chef. He holds a research appointment in the Department of Anthropology at the University of Tennessee.

(Left to right) Alex A. Pulsipher, Lydia Mihelič Pulsipher, Conrad "Mac" Goodwin.

# Brief Contents

*Preface*     xix

CHAPTER 1   Geography: An Exploration of Connections    1

CHAPTER 2   North America    53

CHAPTER 3   Middle and South America    115

CHAPTER 4   Europe    183

CHAPTER 5   Russia and the Newly Independent States    245

CHAPTER 6   North Africa and Southwest Asia    297

CHAPTER 7   Sub-Saharan Africa    353

CHAPTER 8   South Asia    413

CHAPTER 9   East Asia    473

CHAPTER 10   Southeast Asia    533

CHAPTER 11   Oceania: Australia, New Zealand, and the Pacific    583

*Glossary*    G-1

*Index*    I-1

# Contents

**FIGURE 1.9 (b)** Cultural patterns along the border of the United States and Mexico (page 8). [AP photo/David Muang.]

## CHAPTER 1  Geography: An Exploration of Connections  1

Where Is It? Why Is It There?  1
What Is Geography?  2
Geographers' Visual Tools  4
   *Longitude and Latitude*  5
   *Map Projections*  5
   *The Detective Work of Photo Interpretation*  6
The Region as a Concept  6
Globalization and Interregional Linkages  9

**WORLD** VIDEO VIGNETTE  INDIA **Hole in the Wall**  11

Cultural and Social Geographic Issues  12
Ethnicity and Culture—Slippery Concepts  12
Globalization and Cultural Change  13
Cultural Markers  14
   *Values*  14
   *Religion and Belief Systems*  15
   *Language*  16
   *Material Culture and Technology*  17
   *Gender Issues*  18
   *Race*  20
Physical Geography: Perspectives on the Earth  20
Landforms: The Sculpting of the Earth  20
   *Plate Tectonics*  20
   *Landscape Processes*  22
Climate  22
   *Temperature and Air Pressure*  23
   *Precipitation*  23
   *Climate Regions*  25
The Origins of Agriculture: Human Interaction with the Physical Environment  25

Economic Issues in Geography  29
Workers in the Global Economy  29
What Is the Economy?  30
What Is the Global Economy?  31
The Debate over Free Trade and Globalization  32

Measures of Development  33
Gross Domestic Product per Capita  33
Measures of Human Well-Being  34

Population Patterns  35
Global Patterns of Population Growth  35
Local Variations in Population Density and Growth  37
Age and Gender Structures  38
Population Growth Rates and Wealth  39

Humans and the Environment  41
Sustainable Development  41
   *Sustainable Agriculture*  41
   *Sustainability and Urbanization*  43
   *Changing Patterns of Resource Consumption*  45
Global Warming  45

Political Issues in Geography  48
Geopolitics  48
Nations and Borders  49
International Cooperation  49
**Chapter Key Terms**  50
**Some things to think about as you read this textbook**  51

## CHAPTER 2  North America  53

Global Patterns, Local Lives  53
Themes to Explore in North America  55
**I The Geographic Setting**  56

Terms to Be Aware Of   56
Physical Patterns   56
  Landforms   56
  Climate   57
Human Patterns over Time   57
  The Peopling of North America   57
  The European Transformation   59
  Expansion West of the Mississippi   61
  European Settlement and Native Americans   62
  The Changing Social Composition of North America   63
Population Patterns   64

II Current Geographic Issues   66
  The National and Global Geopolitical Repercussions
    of the Terrorist Attacks of September 11, 2001   66
  Relationships Between Canada and the United
    States   69
  Asymmetries   69
  Similarities   69
  Interdependencies   69
  Economic and Political Issues   70
    North America's Changing Agricultural Economy   70
    Changing Transport Networks and the North
      American Economy   72
    The New Service and Technology Economy   73
    Globalization and the Economy   74
  At the Global Scale  Wal-Mart   76
    Canadian and U.S. Responses to Economic Change   78
    Systems of Government: Shared Ideals, Different
      Trajectories   79
    Gender in National Politics and Economics   79
  Sociocultural Issues   80
    Urbanization   80
    Immigration   82
  At the Global Scale  Ethnicity in Toronto   85

WORLD VIDEO VIGNETTE  MEXICO  A Death in the Desert   86
    Race and Ethnicity in North America   86
    Religion   88
    Gender and the American Family   89
    Aging   91
  Environmental Issues   92
    Hazardous Waste   92
    Air Pollution   92
    Loss of Habitat for Plants and Animals   93

Water Resource Depletion   95
Measures of Human Well-Being   96

III Subregions of North America   97
  New England and the Atlantic Provinces   97
  Québec   99
  The Old Economic Core   100
  The American South (The Southeast)   102
  The Great Plains Breadbasket   104
  At the Local Scale  Meat Packing in the Great Plains   106
  The Continental Interior   107
  The Pacific Northwest   109
  Southern California and the Southwest   110
  Reflections on North America   112
  Chapter Key Terms   113
  Critical Thinking Questions   113

**FIGURE 3.29** Javanese Muslim women in Suriname (page 149).
[Robert Caputo/Aurora.]

CHAPTER 3   Middle and South America   115

Global Patterns, Local Lives   115
Themes to Explore in Middle and South America   115

I The Geographic Setting   118
  Terms to Be Aware Of   118
  Physical Patterns   118
    Landforms   119
    Climate   121
  Human Patterns over Time   122
    The Peopling of Middle and South America   122
    The Conquest   124
    A Global Exchange of Crops and Animals   124

*The Legacy of Underdevelopment*   125

Population Patterns   129

*Population Numbers and Distribution*   129

At the Local Scale  Mexico City   130

*Migration and Urbanization*   131

At the Local Scale  Curitiba, Brazil: A Model of Urban
Planning   132

**II Current Geographic Issues   134**

Economic and Political Issues   134

*Phases of Economic Development: Globalization
and Income Disparity*   134

*Phases of Economic Development: Structural
Adjustment*   139

At the Local Scale  The Informal Economy in Peru   140

*The Informal Economy*   140

*Regional Trade and Trade Agreements*   141

WORLD™ VIDEO VIGNETTE  GUATEMALA/MEXICO  Coffee Country   142

*Agriculture and Contested Space*   142

At the Global Scale  The Power to Map: An International
Movement by Indigenous People   145

*Is Democracy Rising?*   145

Sociocultural Issues   148

*Cultural Diversity*   149

*Race and the Social Significance of Skin Color*   149

*The Family and Gender Roles*   149

*Children in Poverty*   150

*Religion in Contemporary Life*   151

Environmental Issues   151

*Tropical Forestlands in the Global Economy*   153

*The Environment and Economic Development*   154

At the Local Scale  Ecotourism in the Ecuadorian Amazon   155

Measures of Human Well-Being   155

*A Rising Threat to Human Well-Being*   158

**III Subregions of Middle and South America   158**

The Caribbean   158

*Cuba and Puerto Rico Compared*   160

*Haiti and Barbados Compared*   161

Mexico   162

*The Geographic Distribution of Economic Sectors*   163

*NAFTA and the Maquiladora Phenomenon*   164

*Migration*   165

Central America   166

*Environmental Concerns*   168

*Civil Conflict*   168

The Northern Andes and Caribbean Coast   169

*The Guianas*   170

*Venezuela*   170

*Colombia*   170

The Central Andes   171

*Environments, Settlement Patterns, and Production
in the Central Andes*   172

At the Local Scale  Learning from the Incas: Agricultural
Restoration in the Peruvian Highlands   173

*Social Inequalities and Social Unrest*   174

The Southern Cone   174

*The Economies of the Southern Cone*   174

*Buenos Aires: A Primate City*   176

Brazil   176

*Urbanization*   178

At the Local Scale  African-Derived Religions Help
Harried Urban Dwellers   179

*Brasilia*   179

Reflections on Middle and South America   180

**Chapter Key Terms**   180

**Critical Thinking Questions**   181

**CHAPTER 4  Europe   183**

Global Patterns, Local Lives   183

Themes to Explore in Europe   184

**I The Geographic Setting   184**

*Terms to Be Aware Of*   184

Physical Patterns   186

*Landforms*   186

*Vegetation*   187

*Climate*   187

Human Patterns over Time   188

*Sources of European Culture*   189

*The Evolution of Feudalism as a Social, Economic,
and Political System*   189

*The Rise of Towns*   190

*Europe Looks Abroad*   190

*The Evolution of European Cities*   191

*An Age of Revolutions*   192

*Europe's Difficult Twentieth Century*   193

Population Patterns 195
  *Population Density and Access to Resources* 196
  *Modern Urbanization in Europe* 196
  *Europe's Aging Population* 197

**II Current Geographic Issues 198**
Economic and Political Issues 198
  *Goals of the European Union* 198
  *Central Europe and the European Union* 202
  Cultural Insight The Hungarian Sausage Experience
    and the European Union 203
  *Future Paths of EU Organizational
    Development* 203
  *Economic Change in Europe* 203
  *Agriculture in Europe* 206
Sociocultural Issues 208
  *Open Borders Bring Immigrants, Cultural Complexity,
    Needed Taxpayers, and Worries* 208
  At the Global Scale Soccer: The Most Popular Sport
    in the World 211
  *European Ideas About Gender* 212
  At the Regional Scale Muslims in Europe 213
  *Social Welfare Systems* 214
Environmental Issues 216
  *Air Pollution* 216
  At the Local Scale The Guerrilla Gardeners of London 218
  *Pollution of the Seas* 218

WORLD VIDEO VIGNETTE SPAIN The Lawless Sea 220

  Measures of Human Well-Being 221

**III Subregions of Europe 222**
West Europe 222
  *Benelux* 222
  At the Global Scale The Chocolate Standards War 225
  *France* 226
  *Germany* 227
  *The British Isles* 227
South Europe 230
  *Spain* 230
  *Italy* 232
North Europe 233
  *Scandinavia* 234
  *The Baltic States* 236
Central Europe 237
  *The Difficulties of Economic Transition* 238
  *Progress in Economic Transition* 239
  *New Experiences with Democracy* 240
  *Southeastern Europe: Understanding an Armed
    Conflict* 240
Reflections on Europe 242
**Chapter Key Terms** 242
**Critical Thinking Questions** 243

**FIGURE 5.4** Nalychevo (Nalitchevo) Nature Park in Kamchatka, Russia (page 248). [Christian Gluckman/http://www.edouardas.com.]

## CHAPTER 5 Russia and the Newly Independent States 245

Global Patterns, Local Lives 245
Themes to Explore in Russia and the Newly
  Independent States 247

**I The Geographic Setting 248**
  *Terms to Be Aware Of* 248
Physical Patterns 248
  *Landforms* 248
  *Climate and Vegetation* 250
Human Patterns over Time 251
  *The Rise of the Russian Empire* 251
  *The Communist Revolution and Its
    Aftermath* 254
  *World War II and the Cold War* 254
  *The Post-Soviet Years* 255
Population Patterns 256
  *Recent Population Changes* 256
  At the Local Scale Urban Gardens 257

II Current Geographic Issues   259

Economic and Political Issues   259

*The Former Command Economy*   259

At the Local Scale Norilsk: An Industrial City in Transition
to a Global, Free-Market Economy   261

*Reform in the Post-Soviet Era*   261

FRONTLINE WORLD™ VIDEO VIGNETTE   MOSCOW   Rich in Russia: The Brave
New World of Young Capitalists and
Tycoons   262

*Supplying Oil and Gas to the World*   266

*Greater Integration with Europe and the United
States*   266

*Political Reforms in the Post-Soviet Era*   266

*Russia's Internal Republics*   269

Sociocultural Issues   270

*Cultural Dominance of Russians and Other
Slavs*   270

*Cultural Revival in the Post-Soviet Era*   271

*The Social Effects of Unemployment and Loss
of the Safety Net*   272

*Gender and Opportunity in Free-Market
Russia*   272

At the Regional Scale The Trade in Women   273

*Corruption and Social Instability*   274

*Impatience with the Present, Nostalgia for
the Past*   274

Environmental Issues   274

*Resource Extraction and Environmental
Degradation*   275

*Urban and Industrial Pollution*   275

*Nuclear Pollution*   277

Measures of Human Well-Being   278

III Subregions of Russia and the Newly
Independent States   280

Russia   280

*European Russia*   280

*Siberian Russia*   283

*The Russian Far East*   284

Belarus, Moldova, and Ukraine   286

*Belarus*   286

*Moldova and Ukraine*   286

Caucasia: Georgia, Armenia, and
Azerbaijan   288

The Central Asian Republics   290

*The Physical Setting*   290

*Central Asia in World History*   291

*The Soviet and Post-Soviet Eras*   291

*The Free Market in Central Asia*   293

Reflections on Russia and the Newly Independent
States   294

Chapter Key Terms   294

Critical Thinking Questions   294

CHAPTER 6   North Africa and
Southwest Asia   297

Global Patterns, Local Lives   297

Themes to Explore in North Africa and Southwest
Asia   299

I The Geographic Setting   299

*Terms to Be Aware Of*   299

Physical Patterns   299

*Climate*   300

*Landforms and Vegetation*   300

Human Patterns over Time   302

*Sites of Early Cultivation*   302

*The Emergence of Gender Roles*   303

*The Coming of Judaism, Christianity, and
Islam*   303

*The Spread of Islam*   305

*Western Domination and State Formation*   306

*The Creation of the State of Israel on Palestinian
Lands*   308

Population Patterns   309

*Gender Roles and Population Growth*   311

*Migration and Urbanization*   312

*Refugees*   313

II Current Geographic Issues   314

Sociocultural Issues   314

*Religion in Daily Life*   314

*Role of Religion in Society*   315

Cultural Insight The Five Pillars of Islamic Practice   315

At the Global Scale The Diversity of Islam   316

*Family Values*   316

*Gender Roles and Gender Spaces*   317

Cultural Insight The Veil   319

*The Lives of Children*    320

*Language and Diversity*    320

*Islam in a Globalizing World*    321

At the Regional Scale  The Evolving Role of the Press    321

Economic and Political Issues    322

*The Oil Economy*    322

*The Traditional Economy: Agriculture and Herding*    323

*Economic Diversification and Growth*    325

*Side Effects of Development Efforts*    325

*The Economic and Political Legacy of Outside Influence*    326

 IRAQ  The Road to Kirkuk    327

*The Tension Between Religion and Democracy*    329

At the Global Scale  Terrorism as an Economic and Political Strategy    330

*Reform Efforts from Within the Arab Community*    330

Environmental Issues    331

*Water Availability*    332

At the Regional Scale  Turkey's Southeastern Anatolia Project (GAP)    334

*Desertification*    335

Measures of Human Well-Being    335

III  Subregions of North Africa and Southeast Asia    337

The Maghreb    337

*Violence in Algeria*    338

The Nile: Sudan and Egypt    339

*Sudan*    339

*Egypt*    341

The Arabian Peninsula    342

The Eastern Mediterranean    345

The Northeast: Turkey, Iran, and Iraq    346

*Turkey*    346

*Iran*    347

*Iraq*    349

Reflections on North Africa and Southwest Asia    350

**Chapter Key Terms**    350

**Critical Thinking Questions**    351

**FIGURE 7.12**  Soon to be elected president, Nelson Mandela votes for the first time in his life in 1994 (page 363). [David Turnley/CORBIS.]

# CHAPTER 7  Sub-Saharan Africa    353

Global Patterns, Local Lives    353

Themes to Explore in Sub-Saharan Africa    355

I  The Geographic Setting    355

*Terms to Be Aware Of*    355

Physical Patterns    356

*Landforms*    356

*Climate*    356

Human Patterns over Time    359

*The Peopling of Africa and Beyond*    359

*Early Agriculture, Industry, and Trade in Africa*    359

*Europeans and the Slave Trade*    360

*The Scramble to Colonize Africa*    361

*Case Study: The Colonization of South Africa*    362

*The Aftermath of Independence*    363

Population Patterns    364

*Africa's Carrying Capacity*    364

*Population Growth*    365

*Population and Public Health*    367

*HIV-AIDS in Africa*    367

II Current Geographic Issues    369

Economic and Political Issues    369

*Agricultural and Natural Resources*   370

At the Local Scale   Big Changes Come to Myeka High School   371

*The Current Economic Crisis*   374

*Alternative Pathways to Economic Development*   378

At the Local Scale   Cell Phones Change Lives and Societies
in Sub-Saharan Africa   380

*Political Issues: Colonial Legacies and African
Adaptations*   380

 NIGERIA   The Road North: What the
Miss World Riots Reveal About a
Divided Country   383

*Shifts in African Geopolitics*   384

At the Regional Scale   African Women Become Active
in Politics and Civil Society   385

Sociocultural Issues   385

*Settlement Patterns*   385

*Religion*   387

At the Regional Scale   The Gospel of Success   389

*Gender Relationships*   390

*Female Circumcision*   391

*Ethnicity and Language*   391

Environmental Issues   392

*Desertification*   394

*Forest Vegetation as a Resource*   394

*Wildlife and National Parks*   395

*Water*   396

Measures of Human Well-Being   397

III   Subregions of Sub-Saharan Africa   400

West Africa   400

At the Regional Scale   Water Management in the Niger
River Basin   401

*Conflict in Coastal West Africa*   401

Central Africa   402

*Congo (Kinshasa): A Case Study of Devolution*   404

East Africa   404

*The Interior Valleys and Uplands*   405

*The Coastal Lowlands*   406

*The Islands*   406

*Kenya and Tanzania: Case Studies in Contested
Space*   407

Southern Africa   408

*South Africa: A Model for Southern Africa*   409

Reflections on Sub-Saharan Africa   410

**Chapter Key Terms**   411

**Critical Thinking Questions**   411

CHAPTER 8   South Asia   413

Global Patterns, Local Lives   413

Themes to Explore in South Asia   414

I   The Geographic Setting   415

*Terms to Be Aware Of*   415

Physical Patterns   415

*Landforms*   415

*Climate*   416

Human Patterns over Time   418

*The Indus Valley Civilization*   418

*A Series of Invasions*   419

*The Legacies of Colonial Rule*   420

At the Global Scale   The Indian Diaspora   422

Population Patterns   424

*Population Growth Factors in South Asia*   425

At the Regional Scale   Should Children Work?   427

*HIV-AIDS in South Asia*   428

II   Current Geographic Issues   428

Sociocultural Issues   428

*Village Life*   428

*City Life*   429

*Language and Ethnicity*   430

*Religion*   431

*Caste*   434

*Geographic and Social Patterns in the Status
of Women*   435

Economic Issues   438

*Agriculture and the Green Revolution*   439

*Industry over Agriculture: A Vision of Self-Sufficiency*   440

*Economic Reform: Achieving Global Competitiveness*   440

At the Global Scale   Wall Street Moves to Dalal Street   442

 BHUTAN The Last Place   443

Political Issues   444

*Caste and Democracy*   444

*Religious Nationalism*   444

At the Regional Scale   Babar's Mosque or Ram's Temple:
The Geography of Religious Nationalism   445

*Regional Political Conflicts*   445

*War and Reconstruction in Afghanistan*   447

*The Future of Democracy*   448

Environmental Issues   448

*Deforestation*   450

At the Local Scale   A Visit to the Nilgiri Hills   450

      *Water Issues*   451

      *Industrial Pollution*   452

   Measures of Human Well-Being   452

**III Subregions of South Asia   454**

   Afghanistan and Pakistan   454

      *Afghanistan*   455

      *Pakistan*   456

   Himalayan Country   457

   At the Local Scale   Salt for Grain and Beans: A Geography of Trade   458

   Northwest India   459

   Northeastern South Asia   461

      *The Ganga-Brahmaputra Delta*   461

      *West Bengal*   462

   At the Local Scale   Kolkata's Building Boom   463

      *Far Eastern India*   463

      *Bangladesh*   464

   Central India   464

   Southern South Asia   467

      *Bangalore and Chennai*   467

      *Kerala*   468

      *Sri Lanka*   468

   At the Local Scale   Kerala's Fishing Industry   469

   Reflections on South Asia   470

   **Chapter Key Terms**   470

   **Critical Thinking Questions**   471

**FIGURE 9.30** The Qinghai–Tibet railway, the world's highest railway, connects Beijing and Lhasa (page 501). [Wu Hong/epa/CORBIS.]

## CHAPTER 9   East Asia   473

   Global Patterns, Local Lives   473

   Themes to Explore in East Asia   475

**I The Geographic Setting   475**

      *Terms to Be Aware Of*   475

   Physical Patterns   475

      *Landforms*   475

      *Climate*   477

   Human Patterns over Time   478

      *The Beginnings of Chinese Civilization*   478

      *Confucianism*   479

      *China's Preeminence*   480

      *European Imperialism in East Asia*   480

      *China's Turbulent Twentieth Century*   481

      *Japan Becomes a World Leader*   482

      *Conflict and Transfers of Power in East Asia*   484

   Population Patterns   485

   At the Local Scale   An Aging Population and the Immigration Debate in Japan   486

      *Declining Population Growth*   486

      *HIV-AIDS in East Asia*   486

**II Current Geographic Issues   487**

   Economic and Political Issues   487

      *The Japanese Model*   487

      *The Communist Command Economy*   489

      *Globalization and Market Reforms in China*   490

   At the Global Scale   China's "Button Town" Fights Itself for Global Market Share   492

   Sociocultural Issues   496

   At the Regional Scale   The Information Highway in China—It's Censored!   497

      *Population Policies and the East Asian Family*   497

      *Family and Work in Industrialized East Asia*   499

      *Indigenous Minorities*   500

   Environmental Issues   502

      *Air Pollution in China*   502

   At the Local Scale   The Three Gorges Dam   504

      *Water in China: Too Much, Too Little, and Polluted*   505

      *Environmental Problems Elsewhere in East Asia*   506

   Measures of Human Well-Being   507

**III Subregions of East Asia   509**

   China's Northeast   509

      *The Loess Plateau*   510

      *The North China Plain*   510

      *China's Far Northeast*   510

      *Beijing and Tianjin*   511

   Central China   511

      *Sichuan Province*   512

At the Local Scale  Silk and Sericulture  512

   *The Central and Coastal Plains*  513

At the Local Scale  Shanghai's Urban Environment  514

At the Global Scale  The Overseas Chinese  515

China's Far North and West  516

   *Xinjiang*  516

   *The Plateau of Tibet*  517

Southern China  519

   *The Yunnan-Guizhou Plateau*  519

   *The Southeastern Coast*  520

   *Hong Kong*  520

   *Macao*  520

Japan  520

   *Food Production and the Challenge of Limited
      Resources*  522

   *Living in Japan*  523

   *Working in Japan*  524

Taiwan  524

Korea, North and South  525

   *Contrasting Political Systems*  526

   *Contrasting Economies*  526

Mongolia  527

   *History*  528

   *The Communist Era in Mongolia*  529

   *Recent Economic Issues*  529

   *Gender Roles*  530

Reflections on East Asia  530

**Chapter Key Terms**  531

**Critical Thinking Questions**  531

**FIGURE 10.7**  Hmong women returning to their village (page 539).
[QT Luong/terragallery.com.]

## CHAPTER 10  Southeast Asia    533

Global Patterns, Local Lives  533

Themes to Explore in Southeast Asia  535

I **The Geographic Setting**  536

   *Terms to Be Aware Of*  536

Physical Patterns  536

   *Landforms*  536

   *Climate*  538

Human Patterns over Time  539

   *The Peopling of Southeast Asia*  539

   *Cultural Influences*  539

   *Colonization*  541

   *Struggles for Independence*  541

WORLD VIDEO VIGNETTE  CAMBODIA  Pol Pot's Shadow  543

Population Patterns  543

   *Population Dynamics*  544

   *Southeast Asia's HIV-AIDS Tragedy*  545

II **Current Geographic Issues**  546

Economic and Political Issues  546

   *Agriculture*  546

   *Patterns of Industrialization*  548

At the Regional Scale  Tourism Development in the
      Greater Mekong Basin: Economic Boom or
      Cultural Bust?  551

   *Economic Crisis and Recovery: The Perils
      of Globalization*  552

   *The Association of Southeast Asian Nations*  553

   *Pressures For and Against Democracy*  554

Cultural Insight  Pancasila: Indonesia's National
      Ideology  556

Sociocultural Issues  557

   *Cultural Pluralism*  557

   *Religious Pluralism*  558

   *Family, Work, and Gender*  560

   *Migration*  561

Environmental Issues  564

   *Deforestation*  566

   *Mining*  567

   *Air Pollution*  567

Measures of Human Well-Being  568

III **Subregions of Southeast Asia**  569

Mainland Southeast Asia: Burma and Thailand   569

   *Burma*   569

   *Is Thailand an Economic Tiger?*   569

At the Local Scale  The Yadana Project: A Joint Venture
   with Mixed Results   572

Mainland Southeast Asia: Vietnam, Laos, and
   Cambodia   572

   *Vietnam*   574

Island and Peninsular Southeast Asia: Malaysia,
   Singapore, and Brunei   574

   *Malaysia*   574

   *Singapore*   576

Indonesia and Timor-Leste   576

The Philippines   577

Reflections on Southeast Asia   580

**Chapter Key Terms**   580

**Critical Thinking Questions**   581


## CHAPTER 11 Oceania: Australia, New Zealand, and the Pacific    583

Global Patterns, Local Lives   583

Themes to Explore in Oceania   585

### I The Geographic Setting   585

   *Terms to Be Aware Of*   585

Physical Patterns   585

   *Continent Formation*   585

   *Island Formation*   587

   *Climate*   587

   *Flora and Fauna*   589

Human Patterns over Time   591

   *The Peopling of Oceania*   591

   *Arrival of the Europeans*   592

   *The Colonization of Australia and New Zealand*   593

   *Oceania's Shifting Ties with Other Countries*   594

Population Patterns   594

   *Population Growth and Distribution in Oceania*   595

   *Urbanization in the Pacific Islands*   596

### II Current Geographic Issues   597

Sociocultural Issues   598

   *Ethnic Roots Reexamined*   598

   *Forging Unity in Oceania*   601

   *Women's Roles in Oceania*   602

   *Being a Man: Persistence and Change*   603

Economic and Political Issues   604

   *From Export to Service Economies*   604

   *Tourism in Oceania*   605

   *New Asian Orientations*   606

Environmental Issues   609

   *Australia: Human Settlement in an Arid,*
     *Urbanized Land*   609

   *New Zealand: Losses of Forest and Wildlife*   610

   *The Pacific Islands: At Risk from Global Trends*   610

Measures of Human Well-Being   613

Reflections on Oceania   613

**Chapter Key Terms**   615

**Critical Thinking Questions**   615

GLOSSARY   G-1

INDEX   I-1

# Preface

Over the past three editions of this text, we have tried to portray the rich diversity of human culture throughout the world and to humanize geographic issues by representing the lives of women, men, and children in various regions of the globe. We wanted a book that made global patterns of trade and consumption meaningful for students by showing how they affect the regions of the world and the daily lives of ordinary people. In summary, we have tried to craft a text that explains and illustrates global patterns while helping readers engage with the way these patterns affect individuals. The fourth edition of text continues in this tradition with additional improvements to make the text as current, useful, and beautiful as possible.

## New to the Fourth Edition

**Enhanced Maps** In this fourth edition, we have focused our attention on improving what has often been cited as a strength of this text: high-quality and relevant maps. Our goal for this edition was to make sure that each map clearly highlights main concepts and patterns. This goal was approached on two fronts: we streamlined the data presented on the maps to bring the main ideas into sharper focus, and bolstered this focus with a completely new map style that uses bold colors, symbols, and lines to point out important information. The European economic activities map shown here exemplifies the clean, simple, bold kind of map that appears throughout the book. Looking at this map, one can clearly see the predominant industries of each country and which industries fall in the delineated industrial core. In this edition, we have worked diligently with some of the best cartographers in the country to update and refine the maps. We have used the most current cartographic technology and extensive reviewer input to create maps that are beautiful, accurate, and pedagogically sound.

**Consistent Base Maps** To help students make conceptual connections easily every chapter contains the following five maps:

- Physical
- Climate
- Population
- Political
- Human Impact

These maps will have the same projection and region depicted throughout the chapter, providing a framework to guide student learning.

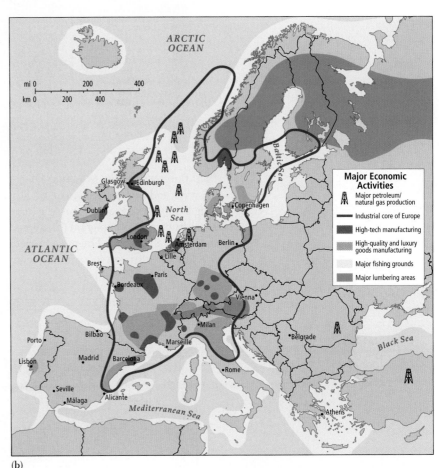

(b)

**FIGURE 4.20 (b)** Europe's principal industrial/manufacturing centers, 2000 (page 204). [Adapted from Terry G. Jordan-Bychkov and Bella Bychkova Jordan, *The European Culture Area: A Systematic Geography,* 4th Ed. (Lanham, Md.: Rowman & Littlefield, 2002), pp. 288 and 300.]

**FIGURE 8.13** Population density in South Asia (page 424). [Data courtesy of Deborah Balk, Gregory Yetman, et al., Center for International Earth Science Information Network, Columbia University; at http://www.ciesin.columbia.edu.]

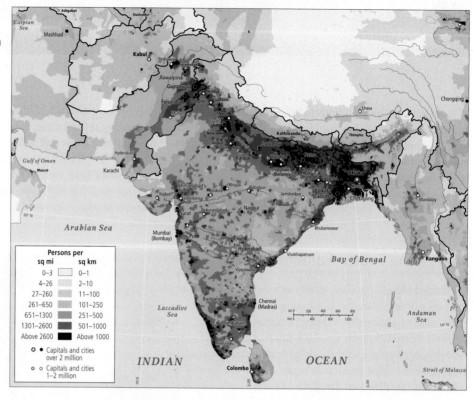

Map Builder: Custom Mapping Program  Instructors use maps in the classroom in many different ways, and we have tried to make it easy to use them by creating a flexible, friendly online mapping tool. W. H. Freeman and Maps.com have built a revolutionary, first-of-its-kind teaching tool for instructors. It is a program that allows instructors to create, display, and print *custom maps* using the data from our text. The program permits in-class display of thousands of different custom maps and allows instructors to zoom in on any portion of a map they want to enlarge. With just a few

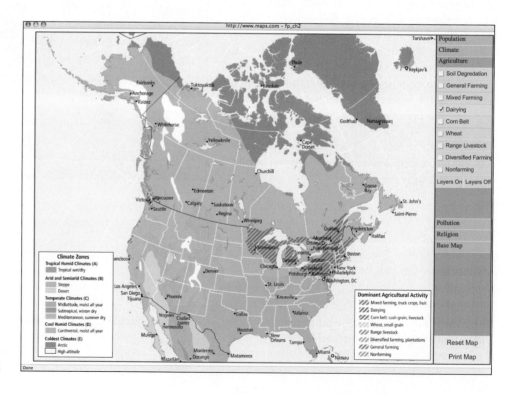

The custom mapping program allows instructors to choose data to layer together.

mouse clicks, instructors can choose the data from select maps that they would like to put together on a single map. For example, instructors can compare climate and agriculture in North America by choosing to display the continental climate zone and dairying together, as shown in the illustration at the bottom of the opposite page. These maps can then be printed in full color or displayed as slides in the classroom. For a demo of these maps, please ask your representative for a free sample, or visit www.whfreeman.com/pulsipher4e.

**Beautiful New Photos**  An aim of this text is to awaken the student to the circumstances of others around the world, and photos are a powerful way of accomplishing this objective. The fourth edition has nearly 50 percent more photos than the third edition. Each new photo has been carefully chosen to complement a concept or situation described in the text. All photos are now numbered, making it easier for students to relate them to the text as they read.

**FIGURE 3.5**  The Tigre River, a tributary of the Amazon, meanders through the Peruvian lowland rainforest (page 119). [Layne Kennedy/CORBIS.]

Not only have the photos been optimized, but they have also been given significantly more space and prominence in the page layout (as have all the book's graphics, including maps). The result is a more visually engaging and dynamic text.

**Current Content**  Because geography is far from static, one of the most important changes to this edition is updated content that is as close to current news as possible, including Hurricane Katrina; the elections of Evo Morales, the first indigenous leader of Bolivia, and of Ellen Johnson-Sirleaf, the first woman president of Liberia; the new influence of Arabic media outlet Al Jazeera around the world; and the consequences of global warming on Pacific Island nations.  Some of the major content areas of the book that have been updated include:

- Domestic and global implications of the U.S. political, economic, and military stances
- The role of terrorism in the realignment of power globally and locally
- Immigration and the ways it is changing countries economically and culturally
- The global phenomenon of labor outsourcing and its effect on countries importing and exporting jobs
- Continuing conflicts and growing turmoil in Southwest Asia and North Africa (Middle East)

- Expansion of the European Union and its consequences for the original EU members, new member states, and the global community
- Changing gender roles, particularly in developing countries
- The increasing role and influence of Islam around the world
- Climate change and its political as well as environmental implications

## THE ENDURING VISION: GLOBAL AND LOCAL PERSPECTIVES

**The Global View** In addition to these enhancements to the text, we retain the hallmark features that have made past editions of this text successful. For the fourth edition we continue to emphasize global trends and the interregional linkages that are changing lives throughout the world. The following linkages are explored in every chapter as appropriate:

- The **multifaceted economic linkages** among world regions. These include (1) the effects of colonialism; (2) trade; (3) the role of transnational corporations in the world economy; and (4) the influence exercised by the World Bank and the International Monetary Fund in the form of structural adjustment programs.
- **Migration.** Migrants are changing economic and social relationships in virtually every part of the globe. The text explores the local and global effects of the Indian diaspora; foreign workers in such places as Japan, Europe, the Americas, and Southwest Asia; and the increasing number of refugees resulting from conflicts around the world.
- Mass communications and marketing techniques are promoting **world popular culture** across regions. The text integrates coverage of popular culture and its effects throughout in discussions of topics such as TV viewing in North Africa and Southwest Asia and the blending of Western and traditional culture in Japan.

**The Local Level** Our approach pays special attention to the local scale—a town, a village, a household, an individual person. Our hope is, first, that stories of individual people and families will make geography interesting and real to students and, second, that seeing the effects of abstract processes and trends on ordinary lives will make the effects of these developments clearer to students. Reviewers have told us that students particularly appreciated the personal vignettes, which are often stories of real people (with names disguised). The following local responses are examined for each region as appropriate:

- **Cultural change:** Changes in the family, gender roles, and social organization in response to urbanization, modernization, and the global economy.
- **Impacts on well-being:** Ideas of what constitutes "well-being" differ from culture to culture, yet broadly speaking people everywhere try to provide a healthy life for themselves in a community of their choosing. Their success in doing so is affected by local conditions, global forces, and their own ingenuity.
- **Issues of identity:** Paradoxically, as the world becomes more tightly knit through global communications and media, ethnic and regional identities often become stronger. The text examines how modern developments such as the Internet are used to reinforce particular cultural identities.
- **Local attitudes toward globalization:** People often have ambivalent reactions to global forces: they are repelled by the seeming power of these forces, fearing effects on their own jobs and on local traditional cultural values, but they are also attracted by the opportunities that may emerge from greater global integration. The text looks at what a region's people say in favor of or against cultural or economic globalization.

**Personal Vignette** When Lee Xia and her sister went home to rural Sichuan the second time, Xia first tried to open a bar in the front room of her parents' house. She hoped to introduce the popular custom of karaoke singing she had enjoyed in Dongguan, but people couldn't stand the noise, and family tensions rose. Her sister, delighted to be home with her husband and baby, quickly found a job in a new small fruit processing factory. Her husband began farming again to fill the new demand for organic vegetables among the middle class in the Sichuan city of Chongqing. The government offered him a subsidy to learn new cultivation techniques if he would guarantee regular delivery of salad greens to a central purchasing depot.

Amid all this success and her bar's failure, news that training was now available in Dongguan for skilled electronics assembly convinced Xia to try again. Word was that the plant would be air-conditioned and that this time she could eventually earn U.S.$400 a month, enough to afford some serious shopping and perhaps one of the stylish new apartments built by the electronics firm for skilled workers.

*Sources: Paul Wiseman, "Chinese factories struggle to hire," USA Today, April 11, 2005, file:///Users/lydia/Desktop/%204th%20edition%201st%20drafts/East%20Asia/USATODAY.com%20-%20Chinese%20factories%20struggle%20to%20hire.webarchive; Louisa Lim, "The end of agriculture," "Reporter's Notebook," National Public Radio, May 19, 2006, http://www.npr.org/templates/story/story.php?storyId=5411325; Mei Fong, "A Chinese puzzle: Surprising shortage of workers," Wall Street Journal (August 16, 2004): B1; David Barboza, "Labor shortage in China may lead to a trade shift," New York Times (April 3, 2006), business section, 1.* ■

(page 495)

# ONE VISION, TWO VERSIONS: WITH OR WITHOUT SUBREGIONAL COVERAGE

The fourth edition is available in two versions to better serve the different needs of diverse faculty and curricula.

## World Regional Geography with Subregions (0-7167-7792-4)

The fourth edition continues to employ a consistent structure for each chapter. Each chapter is divided into three parts: I: The Geographic Setting; II: Current Geographic Issues; III: Subregions.

The subregions coverage provides a descriptive characterization of particular countries and places that expands on coverage in the main text. For example, the sub-Saharan Africa chapter considers the West, Central, East, and Southern Africa subregions, providing additional insights into the nations, issues, and people of this region of the globe.

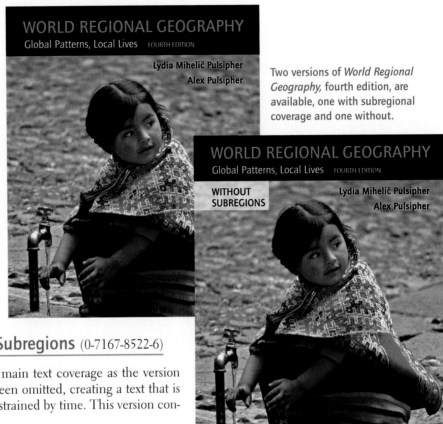

Two versions of *World Regional Geography,* fourth edition, are available, one with subregional coverage and one without.

## World Regional Geography without Subregions (0-7167-8522-6)

The briefer version provides essentially the same main text coverage as the version described above. The subregional sections have been omitted, creating a text that is much shorter, a useful alternative for teachers constrained by time. This version contains all the pedagogy found in the main version.

# FOR THE INSTRUCTOR

**Instructor's CD-ROM** (1-4292-0517-2)
To help instructors create their own Web sites and orchestrate dynamic lectures, the discs contain:

- **Map Builder** available in PowerPoint and Flash format. Built by W. H. Freeman and Company and Maps.com, this is a revolutionary, first-of-its-kind program that allows instructors to create custom maps and then print them or display them as slides in the classroom. This program allows for in-class display of thousands of different custom maps and for instructors to zoom in on any section they want to enlarge. For a free demo of these maps, please ask your representative or visit www.whfreeman.com/pulsipher4e.

- **All text images** in PowerPoint and JPEG formats with enlarged labels for better projection quality.

- **PowerPoint lecture outlines,** by Jason Dittmer, University College London. Main themes of each chapter are outlined and enhanced with images from the book, providing a pedagogically sound foundation on which to build personalized lecture presentations.

- **Instructor's Resource Manual,** by Jennifer Rogalsky, State University of New York, Geneseo, and Helen Ruth Aspaas, Virginia Commonwealth University, contains suggested lecture outlines, points to ponder for class discussion, and ideas for exercises and class projects; available as chapter-by-chapter Word files to facilitate editing and printing.

- **Test Bank,** by Jason Dittmer, University College London, and Andy Walter, West Georgia University. The *Test Bank* is designed to match the pedagogical intent of the text and offers more than 2000 test questions (multiple-choice, short answer, matching, true/false, and essay) in a Word format that makes it easy to edit, add, and resequence questions.

- **Clicker Questions,** by Jason Dittmer, University College London. Written in PowerPoint for easy integration into lecture presentations, *Clicker Questions* allow instructors to jump-start discussions, illuminate important points, and promote better conceptual understanding during lectures.

**Instructor's Web Site** (password-protected) www.whfreeman.com/pulsipher4e
In addition to all the resources available on the *Instructor's CD-ROM*, this password-protected Web site also includes:

- **Online Quizzing, powered by Questionmark,** by Jason Dittmer, University College London. Instructors can easily and securely quiz students using the online multiple-choice Sample Tests (25 questions per chapter). Students receive instant feedback and can take the quizzes multiple times. Instructors can go into a protected Web site to view results by quiz, student, or question, or can get weekly results via e-mail in a simple spreadsheet with all quizzes compiled and graded.

**Course Management**
All instructor and student resources are also available via WebCT/Blackboard to enhance your course. W. H. Freeman and Company offers a course cartridge that populates your site with content tied directly to the book.

**Overhead Transparency Set** (1-4292-0504-0)
Overhead transparencies showing every map from the text are available to adopters. All labels have been resized for easy readability.

The tiny kingdom of Bhutan, couched in the Himalayas between China and India, had for centuries been more or less secluded from the rest of the world. That all changed in June 1999, when a royal decree legalized television, making Bhutan the last country in the world to "plug in." Within a short time, several entrepreneurs, such as Rinzy Dorji, whom the Bhutanese call "The Cable Guy," were in business. For $5 a month, the price of a bag of chilis, Rinzy provides Bhutanese households with 45 cable TV channels—everything from the BBC to *Baywatch*.

Although Rinzy's business is booming, not everyone welcomes the new technology. As Kinley Dorji, editor of Bhutan's only newspaper, describes, "Soon after television started, we started getting letters . . . from children, children who seemed very [anxious]. The letters actually specifically asked about this World Wrestling Federation program, 'Why are these big men standing there and hitting each other? What is the purpose of it?' They didn't understand. . . . Now, a few months later my son jumps on me one morning and says, 'I am Triple H, and you can be Rock.' And, suddenly we are fighting. Suddenly these [TV wrestlers] are new heroes for our children."

Foreign Minister Lyonpo Jigma Thinley thoughtfully observes, "People have suddenly realized that there are so many things they desire, which they were not even aware of before. The truth is that most of these television channels are commercially driven. And some of the Bhutanese people are driven towards consumerism. And, that is inevitable. It's unfortunate, but inevitable." On the other hand, he has also heard people saying, "My God! We didn't know that we are living in a peaceful country. There seems to be violence and crime everywhere in the world." He concludes, "So, in a way, the positive thing is that people realize how good a life they are living in this country."

*To learn more about the impact of satellite and cable TV in Bhutan, see the FRONTLINE/World Video "Bhutan: The Last Place."* ■

(page 443)

## FRONTLINE/World™ Video Anthology

(Available in VHS or DVD format)

W. H. Freeman and Company and the authors of *World Regional Geography* are proud to continue their partnership with the groundbreaking PBS television series FRONTLINE/World. Drawn from the FRONTLINE/World series, these ten video segments are **free to qualified adopters.** Each video segment is between 10 and 20 minutes long and concerns matters both current and relevant. The videos naturally complement not only the subject matter of the text but also its unique approach to world regional geography.

The book builds on the themes in the FRONTLINE/World videos in nine "Video Vignettes." A variation of the text's pioneering "Personal Vignettes," the "Video Vignettes" tell the stories of individuals featured in the FRONTLINE/World video segments. This multimedia package builds upon the book's purpose of putting a face on geography by giving students and instructors access to the fascinating personal stories of people from all over the world.

## FRONTLINE/World™ Video Anthology Instructor Guide

The guide by Jason Dittmer, University College London, provides background information and offers ideas and resources for connecting the videos with classroom discussions and homework assignments.

FRONTLINE/World videos available on VHS or DVD include:

**Chapter 1**    Geography: An Exploration of Connections
*India* "Hole in the Wall: Opening the Door to Cyberspace"

**Chapter 2**    North America
*Mexico* "A Death in the Desert: The Fatal Journey of a Migrant Worker"

**Chapter 3**    Middle and South America
*Guatemala/Mexico* "Coffee Country: Can Fair Trade Save the Farm?"

**Chapter 4**    Europe
*Spain* "The Lawless Sea: Investigating a Notorious Shipwreck"

**Chapter 5**    Russia and the Newly Independent States
*Moscow* "Rich in Russia: The Brave New World of Young Capitalists and Tycoons"

**Chapter 6**    North Africa and Southwest Asia
*Iraq* "The Road to Kirkuk: After Saddam's Terror, Can Kurds and Arabs Live Together?"

**Chapter 7**    Sub-Saharan Africa
*Nigeria* "The Road North: What the Miss World Riots Reveal About a Divided Country"

**Chapter 8**    South Asia
*Bhutan* "The Last Place: Television Arrives in a Buddhist Kingdom"

**Chapter 9**    East Asia
*Hong Kong* "Chasing the Virus: Trying to Stop the Deadly SARS Epidemic"

**Chapter 10**   Southeast Asia
*Cambodia* "Pol Pot's Shadow: Searching for a Mysterious Executioner"

## STUDENT SUPPLEMENTS

**Mapping Workbook and Study Guide** (1-4292-0498-2)
Jennifer Rogalsky, State University of New York, Geneseo, and Helen Ruth Aspaas, Virginia Commonwealth University
The *Study Guide* retains the exercises that have made it a useful pedagogical tool over three editions. Mapping exercises help students understand and explain geographic patterns by employing the skills geographers themselves routinely use. The list of Important Places asks students to place country names, provinces, cities, and physical features on the blank maps that appear at the end of each study guide chapter.

**Mapping Workbook** (1-4292-0499-0)
Jennifer Rogalsky, State University of New York, Geneseo, and Helen Ruth Aspaas, Virginia Commonwealth University
The Important Places lists and blank maps of the student study guide are also available as a stand-alone resource.

**World Regional Geography Online** at **www.whfreeman.com/pulsipher4e**
Tim Oakes and Chris McMorran, University of Colorado, Boulder

- **Map Builder Assessment**—Ten questions per chapter ask students to explore the innovative Map Layering program and answer questions about the patterns they see (available Spring 2008).

- **Map Learning Exercises**—Students use this interactive feature to identify and locate countries, cities, and the major geographic features of each region. These exercises make learning place locations fun and are instructive for future work.

- **Thinking Geographically**—These activities allow students to explore a set of current issues, such as deforestation, human rights, or free trade, and experience how geography helps clarify our understanding of them. Linked Web sites are matched with a series of questions or brief activities or both that give students an opportunity to think about the ways they are connected to the places and people they read about in the text. This aid helps students focus on key geography concepts, such as scale, region, place, and interaction, by using these concepts to drive analysis of compelling issues.

- **Working with Maps**—This feature offers two sets of map-related exercises that develop student analytical abilities:

  *Thematic Maps:* Students can place various maps from the text side-by-side to compare and contrast data. Associated questions accompany each option.

  *Animated Population Maps:* Animated maps show how regional populations have changed or fluctuated with time. Related questions ask students how and why the changes may have occurred.

- **Blank Outline Maps:** Printable maps of the world, and of each region, for use in note-taking or exam review or both, as well as for preparing assigned exercises.

- **Online quizzing**—A self-quizzing feature (25 questions per chapter) enabling students to review key text concepts and sharpen their ability to analyze geographic material for exam preparation. Answers (correct or incorrect) prompt feedback referring students to the specific section in the text where the question is covered.

- **Flashcards**—Matching exercises to teach vocabulary and definitions.

- **Audio Pronunciation Guide**—Spoken guide of place names, regional terms, and names of historical figures.

- **World Recipes and Cuisines**—From *International Home Cooking*, the United Nations International School Cookbook.

**Rand McNally's Atlas of World Geography, 2007 Edition, paperback, 176 pages**
This atlas contains:

- 52 physical, political, and thematic maps of the world and continents; 49 regional, physical, political, and thematic maps; and dozens of metro area inset maps.

- Geographic facts and comparisons covering topics such as population, climate, and weather.

- A section on common geographic questions, a glossary of terms, and a comprehensive 25-page index.

## ACKNOWLEDGMENTS

**Fourth Edition**

Robert Acker
*University of California, Berkeley*

Joy Adams
*Humboldt State University*

John All
*Western Kentucky University*

Jeff Allender
*University of Central Arkansas*

David L. Anderson
*Louisiana State University, Shreveport*

Donna Arkowski
*Pikes Peak Community College*

Jeff Arnold
*Southwestern Illinois College*

Richard W. Benfield
*Central Connecticut University*

Sarah A. Blue
*Northern Illinois University*

Patricia Boudinot
*George Mason University*

Michael R. Busby
*Murray State College*

Norman Carter
*California State University, Long Beach*

Gabe Cherem
*Eastern Michigan University*

Brian L. Crawford
*West Liberty State College*

Phil Crossley
*Western State College of Colorado*

Gary Cummisk
*Dickinson State University*

Kevin M. Curtin
*University of Texas at Dallas*

Kenneth Dagel
*Missouri Western State University*

Jason Dittmer
*University College London*

Rupert Dobbin
*University of West Georgia*

James Doerner
*University of Northern Colorado*

Ralph Feese
*Elmhurst College*

Richard Grant
*University of Miami*

Ellen R. Hansen
*Emporia State University*

Holly Hapke
*Eastern Carolina University*

Mark L. Healy
*Harper College*

David Harms Holt
*Miami University*

Douglas A. Hurt
*University of Central Oklahoma*

Edward L. Jackiewicz
*California State University, Northridge*

Marti L. Klein
*Saddleback College*

Debra D. Kreitzer
*Western Kentucky University*

Jeff Lash
*University of Houston, Clear Lake*

Unna Lassiter
*California State University, Long Beach*

Max Lu
*Kansas State University*

Donald Lyons
*University of North Texas*

Shari L. MacLachlan
*Palm Beach Community College*

Chris Mayda
*Eastern Michigan University*

Armando V. Mendoza
*Cypress College*

Katherine Nashleanas
*University of Nebraska, Lincoln*

Joseph A. Naumann
*University of Missouri at St. Louis*

Jerry Nelson
*Casper College*

Michael G. Noll
*Valdosta State University*

Virginia Ochoa-Winemiller
*Auburn University*

Karl Offen
*University of Oklahoma*

Eileen O'Halloran
*Foothill College*

Ken Orvis
*University of Tennessee*

Manju Parikh
*College of Saint Benedict and Saint John's University*

Mark W. Patterson
*Kennesaw State University*

Paul E. Phillips
*Fort Hays State University*

Rosann T. Poltrone
*Arapahoe Community College, Littleton, Colorado*

Waverly Ray
*MiraCosta College*

Jennifer Rogalsky
*SUNY-Geneseo*

Gil Schmidt
*University of Northern Colorado*

Yda Schreuder
*University of Delaware*

Tim Schultz
*Green River Community College, Auburn, Washington*

Sinclair A. Sheers
*George Mason University*

D. James Siebert
*North Harris Montgomery Community College, Kingwood*

Dean Sinclair
*Northwestern State University*

Bonnie R. Sines
*University of Northern Iowa*

Vanessa Slinger-Friedman
*Kennesaw State University*

Andrew Sluyter
*Louisiana State University*

Kris Runberg Smith
*Lindenwood University*

Herschel Stern
*MiraCosta College*

William R. Strong
*University of North Alabama*

Ray Sumner
*Long Beach City College*

Rozemarijn Tarhule-Lips
*University of Oklahoma*

Alice L. Tym
*University of Tennessee, Chattanooga*

James A. Tyner
*Kent State University*

Robert Ulack
*University of Kentucky*

Jialing Wang
*Slippery Rock University of Pennsylvania*

Linda Q. Wang
*University of South Carolina, Aiken*

Keith Yearman
*College of DuPage*

Laura A. Zeeman
*Red Rocks Community College*

### First Edition
Helen Ruth Aspaas
*Virginia Commonwealth University*

Brad Bays
*Oklahoma State University*

Stanley Brunn
*University of Kentucky*

Altha Cravey
*University of North Carolina at Chapel Hill*

David Daniels
*Central Missouri State University*

Dydia DeLyser
*Louisiana State University*

James Doerner
*University of Northern Colorado*

Bryan Dorsey
*Weber State University*

Lorraine Dowler
*Penn State University*

Hari Garbharran
*Middle Tennessee State University*

Baher Ghosheh
*Edinboro University of Pennsylvania*

Janet Halpin
*Chicago State University*

Peter Halvorson
*University of Connecticut*

Michael Handley
*Emporia State University*

Robert Hoffpauir
*California State University, Northridge*

Glenn G. Hyman
*International Center for Tropical Agriculture*

David Keeling
*Western Kentucky University*

Thomas Klak
*Miami University of Ohio*

Darrell Kruger
*Northeast Louisiana University*

David Lanegran
*Macalester College*

David Lee
*Florida Atlantic University*

Calvin Masilela
*West Virginia University*

Janice Monk
*University of Arizona*

Heidi Nast
*De Paul University*

Katherine Nashleanas
*University of Nebraska*

Tim Oakes
*University of Colorado, Boulder*

Darren Purcell
*Florida State University*

Susan Roberts
*University of Kentucky*

Dennis Satterlee
*Northeast Louisiana University*

Kathleen Schroeder
*Appalachian State University*

Dona Stewart
*Georgia State University*

Ingolf Vogeler
*University of Wisconsin, Eau Claire*

Susan Walcott
*Georgia State University*

### Second Edition
Helen Ruth Aspaas
*Virginia Commonwealth University*

Cynthia F. Atkins
*Hopkinsville Community College*

Timothy Bailey
*Pittsburgh State University*

Robert Maxwell Beavers
*University of Northern Colorado*

James E. Bell
*University of Colorado, Boulder*

Richard W. Benfield
*Central Connecticut State University*

John T. Bowen, Jr.
*University of Wisconsin, Oshkosh*

Stanley Brunn
*University of Kentucky*

Donald W. Buckwalter
*Indiana University of Pennsylvania*

Gary Cummisk
*Dickinson State University*

Roman Cybriwsky
*Temple University*

Cary W. de Wit
*University of Alaska, Fairbanks*

Ramesh Dhussa
*Drake University*

David M. Diggs
*University of Northern Colorado*

Jane H. Ehemann
*Shippensburg University*

Kim Elmore
*University of North Carolina at Chapel Hill*

Thomas Fogarty
*University of Northern Iowa*

James F. Fryman
*University of Northern Iowa*

Heidi Glaesel
*Elon College*

Ellen R. Hansen
*Emporia State University*

John E. Harmon
*Central Connecticut State University*

Michael Harrison
*University of Southern Mississippi*

Douglas Heffington
*Middle Tennessee State University*

Robert Hoffpauir
*California State University, Northridge*

Catherine Hooey
*Pittsburgh State University*

Doc Horsley
*Southern Illinois University, Carbondale*

David J. Keeling
*Western Kentucky University*

James Keese
*California Polytechnic State University*

Debra D. Kreitzer
*Western Kentucky University*

Jim LeBeau
*Southern Illinois University, Carbondale*

Howell C. Lloyd
*Miami University of Ohio*

Judith L. Meyer
*Southwest Missouri State University*

Judith C. Mimbs
*University of Tennessee, Chattanooga*

Monica Nyamwange
*William Paterson University*

Thomas Paradis
*Northern Arizona University*

Firooza Pauri
*Emporia State University*

Timothy C. Pitts
*Edinboro University of Pennsylvania*

William Preston
*California Polytechnic State University*

Gordon M. Riedesel
*Syracuse University*

Joella Robinson
*Houston Community College*

Steven M. Schnell
*Northwest Missouri State University*

Kathleen Schroeder
*Appalachian State University*

Dean Sinclair
*Northwestern State University*

Robert A. Sirk
*Austin Peay State University*

William D. Solecki
*Montclair State University*

Wei Song
*University of Wisconsin, Parkside*

William Reese Strong
*University of North Alabama*

Selima Sultana
*Auburn University*

Suzanne Traub-Metlay
*Front Range Community College*

David J. Truly
*Central Connecticut State University*

Alice L. Tym
*University of Tennessee, Chattanooga*

## Third Edition

Kathryn Alftine
*California State University, Monterey Bay*

Donna Arkowski
*Pikes Peak Community College*

Tim Bailey
*Pittsburg State University*

Brad Baltensperger
*Michigan Technological University*

Michele Barnaby
*Pittsburg State University*

Daniel Bedford
*Weber State University*

Richard Benfield
*Central Connecticut State University*

Sarah Brooks
*University of Illinois at Chicago*

Jeffrey Bury
*University of Colorado, Boulder*

Michael Busby
*Murray State University*

Norman Carter
*California State University, Long Beach*

Gary Cummisk
*Dickinson State University*

Cyrus Dawsey
*Auburn University*

Elizabeth Dunn
*University of Colorado, Boulder*

Margaret Foraker
*Salisbury University*

Robert Goodrich
*University of Idaho*

Steve Graves
*California State University, Northridge*

Ellen Hansen
*Emporia State University*

Sophia Harmes
*Towson University*

Mary Hayden
*Pikes Peak Community College*

R. D. K. Herman
*Towson University*

Samantha Kadar
*California State University, Northridge*

James Keese
*California Polytechnic State University, San Luis Obispo*

Phil Klein
*University of Northern Colorado*

Debra D. Kreitzer
*Western Kentucky University*

Soren Larsen
*Georgia Southern University*

Unna Lassiter
*California State University, Long Beach*

David Lee
*Florida Atlantic University*

Anthony Paul Mannion
*Kansas State University*

Leah Manos
*Northwest Missouri State University*

Susan Martin
*Michigan Technological University*

Luke Marzen
*Auburn University*

Chris Mayda
*Eastern Michigan University*

Michael Modica
*San Jacinto College*

Heather Nicol
*State University of West Georgia*

Ken Orvis
*University of Tennessee*

Thomas Paradis
*Northern Arizona University*

Amanda Rees
*University of Wyoming*

Arlene Rengert
*West Chester University of Pennsylvania*

B. F. Richason
*St. Cloud State University*

Deborah Salazar
*Texas Tech University*

Steven Schnell
*Kutztown University*

Kathleen Schroeder
*Appalachian State University*

Roger Selya
*University of Cincinnati*

Dean Sinclair
*Northwestern State University*

Garrett Smith
*Kennesaw State University*

Jeffrey Smith
*Kansas State University*

Dean Stone
*Scott Community College*

Selima Sultana
*Auburn University*

Ray Sumner
*Long Beach City College*

Christopher Sutton
*Western Illinois University*

Harry Trendell
*Kennesaw State University*

Karen Trifonoff
*Bloomsburg University*

David Truly
*Central Connecticut State University*

Kelly Victor
*Eastern Michigan University*

Mark Welford
*Georgia Southern University*

Wendy Wolford
*University of North Carolina at Chapel Hill*

Laura Zeeman
*Red Rocks Community College*

This book has been a family project many years in the making. I came to the discipline of geography at the age of five, when my immigrant father, Joe Mihelič, hung a world map over the breakfast table in our home in Coal City, Illinois, where he was pastor of the New Hope Presbyterian Church. We soon moved to the Mississippi Valley of eastern Iowa, where my father, then a professor in the Presbyterian theological seminary in Dubuque, continued his geography lessons on the passing landscapes whenever I accompanied him on Sunday trips to small country churches. My sons, Anthony and Alex, got their first doses of geography in the bedtime stories I used to tell them. For plots and settings, I drew on Caribbean colonial documents I was then reading for my dissertation. They first traveled abroad and learned about the hard labor of field geography when as mid-sized children they were expected to help with the archaeological and ethnographic research my colleagues and I were conducting on the eastern Caribbean island of Montserrat. Alex, who was a researcher on the first edition and a co-author on the second, third, and fourth editions of this book, is now completing a Ph.D. in geography at Clark University in Massachusetts. Mac Goodwin, my husband, has given up several years of his career as a colonial sites archaeologist to help with the research, writing, and production of this book, with primary responsibility for all graphics. It was my brother John Mihelič who first suggested that we write a book like this one, after he too came to appreciate geography. He has been a loyal cheerleader during the process, as have our extended family and friends in Knoxville, Montserrat, San Francisco, Slovenia, and beyond.

Holly Hapke, who was a co-author on the third edition, has influenced the content, especially that on South Asia, for all editions.

My graduate students and faculty colleagues in the Geography Department at the University of Tennessee have been generous in their support, serving as helpful impromptu sounding boards for ideas. Ken Orvis, especially, has advised us on the physical geography sections of all editions.

Maps for this edition were conceived by Mac Goodwin with the help of Alex Pulsipher and produced by Will Fontanez and the University of Tennessee cartography shop staff: Joshua Calhoun, Tracy Pollock, and Steve Ahrens and by Martha Bostwick and her staff at Maps.com: Deane Plaister, Brandi Webber, Mike Powers, Lisa Basanese, Lucy Pendl, and Jesse Wickizer.

Sara Tenney and Liz Widdicombe at W. H. Freeman and Company were the first to persuade us that together we could develop a new direction for *World Regional Geography*, one that included the latest thinking in geography written in an accessible style. In accomplishing this goal, we are especially indebted to our first developmental editor, Susan Moran, and to the W. H. Freeman staff for all they have done to ensure that this book is well written, beautifully designed, and well presented to the public. Susan Moran was unfailingly wise, temperate, insightful, and probing, yet also kind and patient with us then novice authors. Susan Weisberg has ably succeeded her on this fourth edition, and has proved remarkably efficient and helpful in shortening the text and in reconciling all the many details that make for a useful and elegant book.

We would also like to gratefully acknowledge the efforts of the following people at W. H. Freeman: Susan Brennan, acquisitions editor for the first edition, Jason Noe, for the second and third editions, and Marc Mazzoni for the fourth; Mary Louise Byrd, senior project editor for all four editions; Lisa Samols, project manager; Scott Guile, senior marketing manager; Norma Roche, copy editor; Diana Blume, design manager; Bill Page, senior illustration coordinator; Julia De Rosa, production manager; Sheridan Sellers, W. H. Freeman and Company Electronic Publishing Center; Kathy Bendo, photo editor, and Inge King, photo researcher for the first edition; Meg Kuhta, photo editor, and Bianca Moscatelli and Julie Tesser, photo researchers for the second edition; Bianca Moscatelli, photo editor, and Elyse Rieder, photo researcher for the third edition; Trish Marx and Ted Szczepanski, photo editors, and Elyse Rieder and Donna Ranieri, photo researchers for the fourth edition; Deepa Chungi, supplements editor; Eleanor Wedge and Martha Solonche, proofreaders for the text and maps; and Maia Gil'Adi, editorial assistant. We are also grateful to the supplements authors, who have created what we think are unusually useful, up-to-date, and labor-saving materials for instructors who use our book: Jennifer Rogalsky and Helen Ruth Aspaas, authors of the *Mapping Workbook and Study Guide* and *Instructor's Resource Manual*; and Jason Dittmer, author of the Test Bank, PowerPoint lecture notes, Clicker questions, sample tests, and video guides.

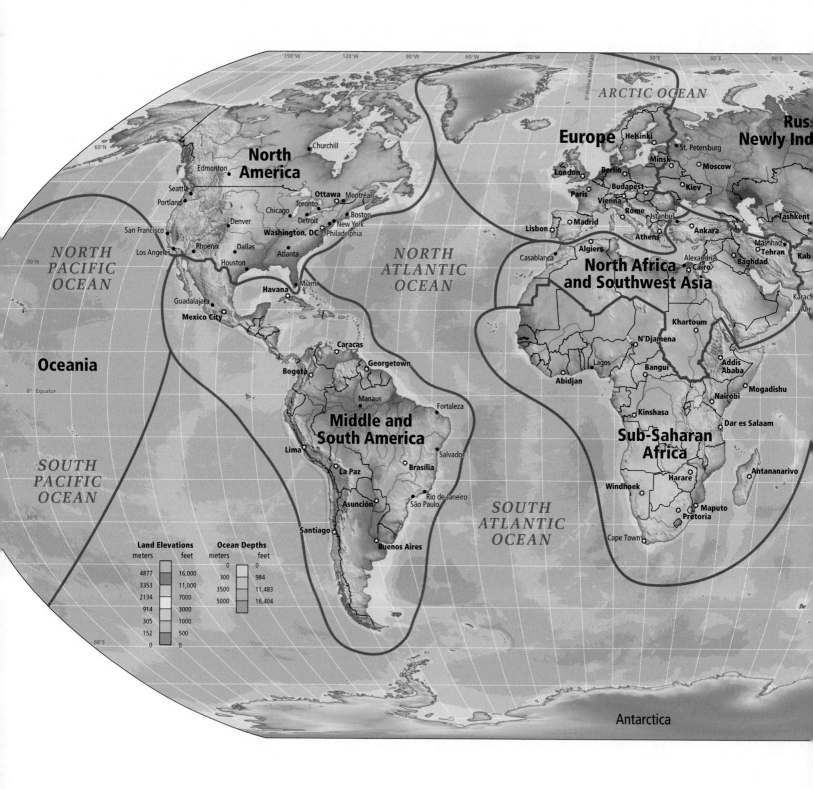

FIGURE 1.1 Regions of the world.

# CHAPTER 1

# GEOGRAPHY: An Exploration of Connections

## WHERE IS IT? WHY IS IT THERE?

Where are you? You may be in a house or a library or sitting under a tree on a fine fall afternoon. You are probably in a community (perhaps a college or university), and you are in a country (perhaps the United States) and a region of the world (perhaps North America, Southeast Asia, or the Pacific). Why are you where you are? There are immediate answers, such as "I have an assignment to read." But there are also larger explanations, such as your belief in the value of an education, your career plans, or someone's willingness to sacrifice to pay your tuition. Even past social movements in Europe and America that opened up higher education to more than a fortunate few may help to explain why you are where you are.

The questions *where* and *why* and *how* are central to geography. Like anyone who has had to find the site of a party on a Saturday night, the location of the best grocery store, or the fastest and safest route home, geographers are interested in location, spatial relationships, and connections between the environment and people.

Different places have different sights, sounds, smells, and arrangements of features. Understanding what has contributed to the look and feel of a place, to the standard of living and customs of its people, and to the way people in one place relate to people in other places, near and far, helps geographers answer the question of why places are as they are. Furthermore, geographers often think at several scales: for example, when choosing the best local situation for a new grocery store, a geographer might consider the socioeconomic circumstances of the local neighborhood as well as the best location from the perspective of the city at large. A potential site would also be considered in relation to national or even international transportation routes, possibly to determine cost-efficient connections to suppliers.

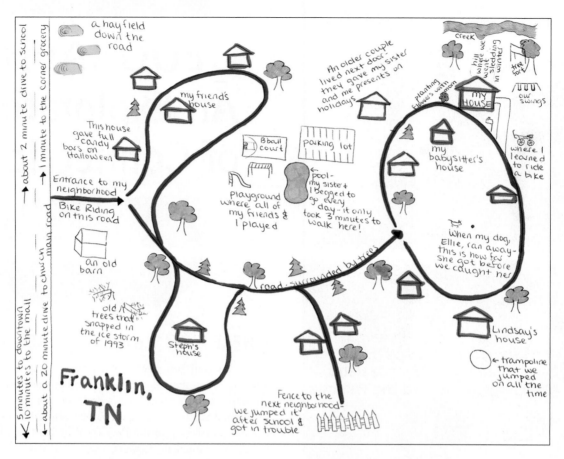

**FIGURE 1.2** A childhood landscape map. Julia Stump drew this map of her childhood landscape in Franklin, Tennessee, as an exercise in Dr. Pulsipher's geography class, January 2006. [Courtesy of Julia Stump.]

To make it easier to understand a geographer's many interests, please try this exercise. On a piece of blank paper, draw a map of your favorite childhood landscape. Relax, and let your mind recall the objects and experiences that were most important to you in that place. If the place was your neighborhood, you might start by drawing and labeling your home, then fill in other places you encountered regularly as you went about your life—such as your backyard and the objects in it, your best friend's home, your school. Don't worry about creating a work of art—just make a map that reflects your experiences as you remember them. For example, Figure 1.2 shows the childhood landscape of Julia Stump in Franklin, Tennessee.

When you are finished with your map, think about how the map reveals the ways in which your life was structured by space. What is the **scale** of your map? That is, how much space did you decide to illustrate on the paper? Were there places you were not supposed to go? The amount of space you covered with your map may represent the degree of freedom you had as a child to go about on your own, or how aware you were of the wider world around you. Does your map reveal, perhaps in a very subtle way, such emotions as fear, pleasure, or longing? Does it indicate your sex, your ethnicity, or the makeup of your family? Did you use symbols to show certain features? In making your map and analyzing it, you have engaged in several aspects of geography:

- Landscape observation

- Description of the earth's surface and consideration of the natural environment

- Historical reconstruction of bygone places

- **Spatial analysis** (the study of how people, objects, or ideas are, or are not, related to one another across space)

- The use of different scales of analysis (your map probably shows the spatial features of your childhood at a detailed local scale)

- Cartography (the making of maps)

As you progress through this book, you will acquire geographic information and skills. Whether you are planning where to travel, thinking about investing in East Asian timber stocks, searching for a good place to market an idea, or trying to understand current local events in the context of world events, knowing how to practice geography will make the task easier and more engaging.

## WHAT IS GEOGRAPHY?

Geography might be defined as the study of our planet's surface and the processes that shape it. Yet such a succinct definition does not begin to convey the fascinating interactions of physical and human forces that have given the earth its diversity of landscapes and ways of life. Geography, as an academic discipline, links the physical sciences—such as geology, physics, chemistry, biology, and botany—with the social sciences—such as anthropology, sociology, history, economics, and political science. **Physical geography** is the study of earth's physical processes to learn how they work, how they affect humans, and how they are affected by humans in return. **Human geography** is the study of various aspects of human life that create the distinctive landscapes and regions of the world. Physical and human geography are often tightly linked. For example, geographers might aim to understand

- how and why people came to occupy a particular place
- how they assess the physical aspects of that place (climate, landforms, resources) and then set about using and modifying them to suit their particular needs

- how people may create environmental problems by the way they use a place
- how people interact with other places, far and near

People are not always bound by their physical surroundings. Transport technology and communications media link people in one physical place to distant locations and resources, so people no longer need to design their lives around what is available locally. Thus the investigations of some geographers may seem to have little direct connection to the physical environment. For example, geographers may study such issues as political activism in certain neighborhoods, the spatial allocation of tax dollars, and socioeconomic variables in housing choice.

Geographers usually specialize in one or more fields of study, or *subdisciplines* (Figure 1.3), and you will see some of these particular types of geography mentioned as we go along. Despite their individual specialties, geographers often find it useful to

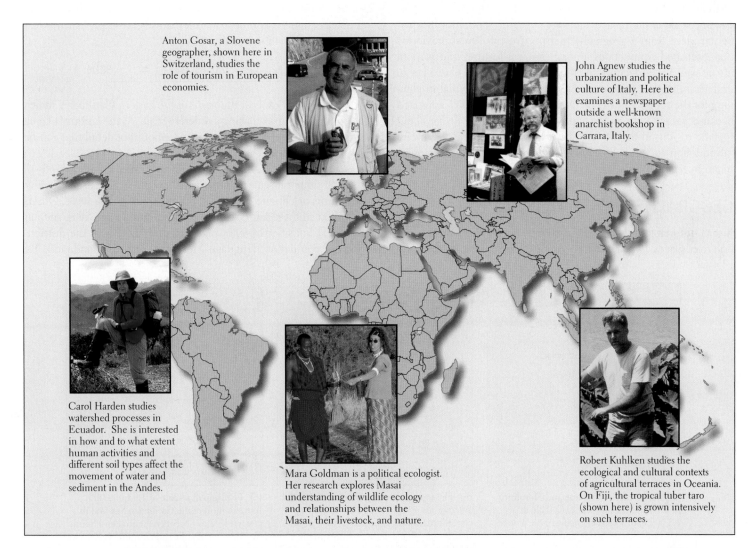

Anton Gosar, a Slovene geographer, shown here in Switzerland, studies the role of tourism in European economies.

John Agnew studies the urbanization and political culture of Italy. Here he examines a newspaper outside a well-known anarchist bookshop in Carrara, Italy.

Carol Harden studies watershed processes in Ecuador. She is interested in how and to what extent human activities and different soil types affect the movement of water and sediment in the Andes.

Mara Goldman is a political ecologist. Her research explores Masai understanding of wildlife ecology and relationships between the Masai, their livestock, and nature.

Robert Kuhlken studies the ecological and cultural contexts of agricultural terraces in Oceania. On Fiji, the tropical tuber taro (shown here) is grown intensively on such terraces.

**FIGURE 1.3 Geography by its nature is interdisciplinary.** Most geography departments contain professors who were trained in geography and in one or more other disciplines. As a result, the research that geographers do is wide-ranging. Yet the research almost always relates in some way to human interactions across space and with the environment. [Photos courtesy of scholars shown.]

cooperate in studying the interactions between people and places. For example, climate geographers, cultural geographers, and economic geographers are now cooperating to understand the spatial distribution of carbon dioxide emissions, which cultural practices might be changed to limit such emissions, and what the economic effects of such changes might be. In efforts to understand the roots of terrorism, historical geographers can shed light on how European colonialism set up situations that frustrated the development of democratizing customs, and economic geographers can explain how the inequitable distribution of wealth and well-being can lead to disaffection and even violence. Cultural geographers specializing in the study of religion can examine the extreme circumstances under which fundamentalist religious beliefs can lead to the rationalization of terrorist violence.

Many geographers are led by their interest in complex spatial relationships and by the difficulties of analyzing any phenomenon on a global scale to specialize in a particular region of the world, or even in a small place within that region. **Regional geography** is the analysis of the geographic characteristics of a particular place (the size and scale of such places can vary radically). The study of a particular place from a geographic perspective can reveal previously unappreciated connections among physical features and ways of life, as well as connections to other places. These links are a key to understanding the present (and the past) and are essential in planning for the future. This book follows a "world regional" approach because people new to geography find general knowledge about specific regions of the world to be the most interesting and useful introduction to the subject. (We will see just what geographers mean by *region* a little later in this chapter.)

## GEOGRAPHERS' VISUAL TOOLS

Just as historians study change over time, geographers study variation over spaces, large and small. Among geographers' most impor-

tant tools are maps, which they use to record and analyze spatial relationships, just as you recorded and analyzed the features that made up your favorite childhood landscape. Geographers who specialize in depicting geographic information graphically are called *cartographers.* The following discussion of map reading will help you to understand the maps in this book and elsewhere. Where appropriate, map captions will identify and explain what is being depicted.

When geographers want to show spatial aspects of particular topics on a map—say, the locations and relative sizes of islands in the Caribbean Sea—they can begin by deciding at what scale the information should be mapped. The scale of a map represents the relationship between the distances shown on the map and the actual distances on the earth's surface. Sometimes a numerical ratio shows what one unit of measure on the map equals in the same units on the face of the earth: for example, 1:1,000,000, or the fraction 1/1,000,000, means that 1 inch or 1 centimeter on the map equals 1 million inches or 1 million centimeters on the face of the earth. Other maps may express scale using a phrase such as "1 centimeter equals 10 kilometers." Alternatively, a simple bar may express the information visually:

| 0 | 10 | 20 | 30 | 40 | 50 km |

Notice that each of the maps of the eastern Caribbean shown in Figure 1.4 is drawn using a different scale. As the amount of area shown on a map becomes larger, the amount of detail that can be displayed decreases.

Throughout this book, you will encounter different kinds of maps at different scales. Some will show physical features such as landforms or climate patterns at the region or global scale. Others will show at these same scales aspects of human activities, such as historical transitions, routes taken by drug traders, or the distributions of population, languages, religions, or soil degradation. Yet

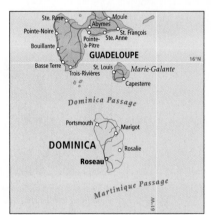

(a) **A map of Guadeloupe and Dominica,** in the eastern Caribbean, at a scale of 1:3,000,000. This scale makes it possible to show towns, a few roads, and a few landforms, but not much else.

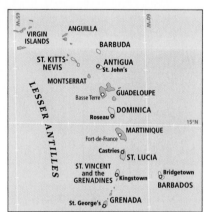

(b) **The same area at a scale of 1:15,000,000.** You can see much more of the eastern Caribbean, but the only detail that can be shown is the shape of the islands and the locations of some capital cities.

(c) **The map at a scale of 1:45,000,000.** It shows most of the Caribbean Sea and its general location between Central and South America, but now the eastern Caribbean islands are too small to identify clearly.

**FIGURE 1.4** Examples of scale. For comparison, Figure 1.1, on page 1, is at a scale of 1:95,000,000. [Adapted from *The Longman Atlas for Caribbean Examinations,* 2nd ed. (Essex, U.K.: Addison Wesley Longman, 1998), p. 4.]

other maps will show settlement or cultural features at the scale of cities or countries.

The *title*, *caption*, and *legend* give basic information about the map. The title tells you the subject of the map, and the caption usually points out some features of the map that the author wants you to notice. The legend is the box that explains what the symbols and colors on the map represent (see, for example, Figure 1.11 on page 10). The map's scale is usually placed in an open space or near the map's legend (see, for example, Figure 2.1 on pages 52–53). Often world maps will not have scale information, though Figure 1.1, a map of world regions, has a scale bar.

## Longitude and Latitude

Most maps contain *lines of latitude* and *longitude*, which enable us to establish a position on the map relative to other points on the globe. Lines of **longitude** (also called meridians) run from pole to pole; lines of **latitude** (also called parallels) run parallel to the equator (Figure 1.5).

Both latitude and longitude lines describe circles, so there are 360° in each circle of latitude and in each circle of longitude and 180° for each hemisphere. Each degree spans 60 minutes (designated with the symbol ′), and each minute has 60 seconds (designated with the symbol ″). Keep in mind that these are measures

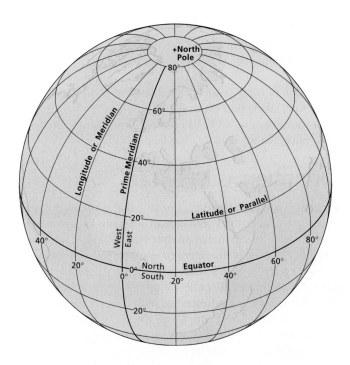

**FIGURE 1.5 Summary of longitude and latitude.** Lines of longitude (meridians) extend from pole to pole. The distance between them on the globe decreases steadily toward the poles, where they all meet. Lines of latitude (parallels) are equally spaced north and south of the equator and intersect the longitude lines at right angles. The only line of latitude that spans the complete circumference of the earth is the equator; all other lines of latitude describe ever smaller circles heading away from the equator. [Adapted from *The New Comparative World Atlas* (Maplewood, N.J.: Hammond, 1997), p. 6.]

of relative space on a circle, not time, nor even real distance (because the circles of latitude get successively smaller to the north and south of the equator, until they become a virtual dot at the poles).

The *prime meridian*, 0° longitude, runs from the North Pole through Greenwich, England, to the South Pole. The half of the globe's surface west of the prime meridian is called the Western Hemisphere; the half to the east, the Eastern Hemisphere. The longitude lines both east and west of the prime meridian are labeled by their direction and distance in degrees from the prime meridian, from 1° to 180°. For example, 20 degrees east longitude would be written as 20° E. The longitude line at 180° runs through the Pacific Ocean and is used as the *international date line*, where the calendar day officially begins. Latitude is measured from 0° at the equator to 90° north and south at the North Pole and South Pole, respectively.

Lines of longitude and latitude form a grid that can be used to designate the location of a place. In Figure 1.4a, you can see that the island of Marie-Galante lies just south of the parallel at 16° N and just about 18′ west of the 61st west meridian. Hence, the position of Marie-Galante's northernmost coast is 16° N, 61°18′ W.

## Map Projections

All maps must solve the problem of showing the spherical earth on a flat piece of paper. When reading a map, it is important to understand the limitations of the different strategies devised to do so. Imagine the problems of distortion if you drew a map of the earth on an orange, peeled the orange, and then tried to flatten out the orange-peel map and transfer it exactly to a flat piece of paper.

The various ways of showing the spherical surface of the earth on flat paper are called *projections*. All projections entail some distortion. For maps of small parts of the earth's surface, the distortion is minimal. Developing a projection for the whole surface of the earth that minimizes distortion is much more challenging. The *Mercator projection* (Figure 1.6a) is popular, but geographers rarely use it because of its gross distortion near the poles. To make his flat map, the Flemish cartographer Gerhardus Mercator (1512–1594) stretched out the poles, depicting them as lines equal in length to the equator! Greenland, for example, appears about as large as Africa, even though it is only about one-fourteenth Africa's size. Nevertheless, Mercator's projection is still useful for navigation because it portrays the shapes of landmasses more or less accurately, and because a straight line between two points on this map gives the compass direction between them. Furthermore, this type of projection can be safely used for parts of the globe that are within 15 degrees south or north of the equator because the distortion in this range is minimal.

*Goode's interrupted homolosine projection* (Figure 1.6b) flattens the earth rather like an orange peel, thus preserving some of the size and shape of the landmasses. In this projection, the oceans get snipped up. The *Robinson projection* (Figure 1.6c) shows the longitude lines curving toward the poles to give an impression of the earth's curvature, and it shows an uninterrupted view of land and ocean, but accuracy is sacrificed. In this book we often use the Robinson projection for world maps.

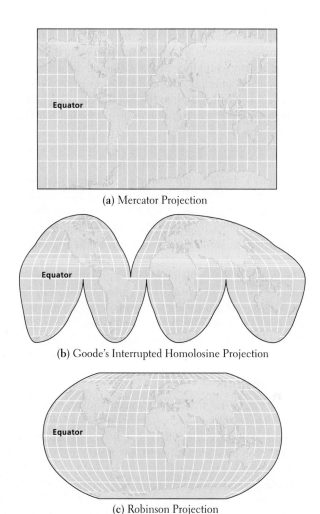

(a) Mercator Projection

(b) Goode's Interrupted Homolosine Projection

(c) Robinson Projection

**FIGURE 1.6 Three common map projections.** [Mercator and Robinson projections adapted from *The New Comparative World Atlas* (Maplewood, N.J.: Hammond, 1997), pp. 6–7. Goode's interrupted homolosine projection adapted from *Goode's World Atlas*, 19th ed. (Chicago: Rand McNally, 1995), p. x.]

Maps are not unbiased. Most currently popular world projections reflect the European origins of modern cartography. Europe or North America is usually placed near the center of the map, where distortion is minimal; other population centers, such as East Asia, are placed at the highly distorted periphery. For less biased study of the modern world, we need world maps that center on different parts of the globe. For example, much of the world's economic activity is now taking place in and around Japan, Korea, China, Taiwan, and Southeast Asia. Discussions of this activity require maps that focus on these regions but still include the rest of the world in the periphery.

## The Detective Work of Photo Interpretation

While maps are useful for translating vast distances or extensive quantities of information into a small, two-dimensional space, photos also can effectively catalog large amounts of physical and cul-

tural data. Most geographers make use of photographs to help them understand or explain a geographic issue or depict the character of a place. Interpreting a photo to extract its geographical information can sometimes be like detective work. Below are some points to keep in mind as you look at pictures; try them out with the photos in Figures 1.7 and 1.8.

- *Landforms:* Notice the lay of the land and the landform features. Speculate on any influences the landforms may have on climate or human activities. Is environmental stress visible?
- *Vegetation:* Notice whether the vegetation indicates a wet or dry, warm or cold environment. Can you recognize species, such as a palm tree or an evergreen? Does the vegetation appear to be natural or disturbed by human use?
- *Ambience:* Does the picture convey a mood? Is emotion apparent? What is the time of day, weather, and season?
- *Material culture:* What is indicated by the architecture, furnishings, or any tools you see? Is a particular level of technology or wealth, or an aesthetic sense, revealed?
- *Activities and their possible meaning:* If there are people in the photo, do their activities convey their level of education or technology, or their living standards?
- *Evidence about the economy:* Does the situation appear to be influenced only by local issues, or can you see evidence of the global economy?
- *Location:* From your observations, can you tell where the picture was taken or narrow down the possible locations that might be depicted? What is your rationale?

Now, think of every possible statement that you can make about the picture, taking note of any doubts you have, as they can be useful in your detective work. You can use this system to analyze any of the photos in the book or elsewhere.

## THE REGION AS A CONCEPT

A **region** is a unit of the earth's surface that contains distinct patterns of physical features or of human activities. We could speak of a desert region, a region that produces table wine, or a region that at a particular time is characterized by ethnic violence. It is rare for any two regions to be described by the same set of indicators. One assemblage of places might be considered a region because of its distinctive vegetative landscape and such cultural features as foods, dialects, and music; for example, see if you can list traits that characterize and distinguish the southern and southwestern regions of the United States. Another region might be defined primarily by its cold climate and sparse population; Siberia is one such region. Still another group of places might be considered a region primarily because of common historical experiences. Middle and South America are considered regions in large part because they experienced colonization by Spain and Portugal after 1492.

Another problem in defining regions is that their boundaries are rarely crisp. The more closely we look at the border zones, the

**FIGURE 1.7** Interpreting a photo: Ecuador. Orlando Ayme, his wife Ermelinda (carrying her baby and some groceries), and daughter Livia (with her schoolbooks) are returning home from their weekly trip to market. They live in Tingo, Ecuador, a village in the central Andes. Notice the well-used path, the sparse trees, the cultivated fields and grazing animals, the relative dryness, and the high mountain peaks. Also note the warm clothing that the family is wearing to ward off the cold winds and temperatures at high altitudes. [Peter Menzel/www.menzelphoto.com.]

fuzzier the divisions appear. Take the case of the boundary between the United States and Mexico (Figure 1.9). The clearly delineated political border is not really a marker of separation between cultures or economies. In a wide band extending over both sides of the border, one can find a blend of Native American, Spanish colonial, Mexican, and Anglo-American cultural features: languages, place-names, food customs, music, and family organization, to name but a few. And the economy of the border zone depends on interactions across a broad swath of territory. In fact, the Mexican national economy is becoming more closely connected to the economies of the United States and Canada than to those of its close neighbors in Middle America (the countries of Central America and the Caribbean).

Why, then, does this book place Mexico in Middle America? At this point in time, Mexico still has more in common overall with Middle America than with North America. Its colonial history and

**FIGURE 1.8** Interpreting a photo: Bhutan. The Namgay family of the village of Shingkhey, Bhutan, display a week's worth of food in the prayer room of their house. Take particular note of the number of people who are part of this family and of the kinds of foods that they eat regularly, especially the types of produce and the amount of rice. Notice also the chili peppers, potatoes, and tomatoes, all of which originated in Central and South America. [Peter Menzel/www.menzelphoto.com.]

**FIGURE 1.9** Cultural patterns along the border of the United States and Mexico. **(a)** Along the border there is a wide band of cultural blending in which language, food, religion, and architecture show influences of both U.S. and Mexican cultures. Sometimes a phenomenon new to both cultures emerges. **(b)** Because prescription drugs are expensive in the United States, increasingly U.S. citizens are crossing to Mexico to buy them at sharply lower prices, and then enjoying a short holiday. [(a) Adapted from *The World Book Atlas* (Chicago: World Book, 1996), p. 89. (b) AP Photo/David Maung.]

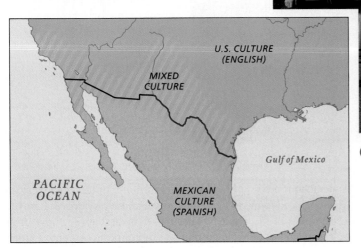

**(b)** U.S. citizens shopping in Mexico.

**(a)** Map of the Mexico–U.S. border region.

the resulting use of Spanish as its official language tie Mexico to Middle and South America, where the dominant language is Spanish (except in Brazil, where Portuguese is primary and several other small former British, French, or Dutch colonies where English, French, or Dutch is primary). The dominant language of the United States and Canada is English (except in Québec, where French is primary). These language patterns are symbolic of the larger cultural and historical differences between the two regions, which will be discussed in Chapters 2 and 3. Even though Mexico has much in common with parts of the southwestern United States, on the whole the similarities do not yet override the differences. However, in the future, if the economic connections among the United States, Canada, and Mexico grow stronger and as more citizens from all three countries emigrate across the borders, it might make sense to include Mexico in the North America world region.

If regions are so difficult to define and describe and are so changeable, why do geographers use them? The obvious answer is that it is impossible to discuss the whole world at once, and so we seek a reasonable way to divide it into manageable parts. There is nothing sacred about the criteria or the boundaries we use; they simply need to be practical. In defining the world regions for this book, we have considered such factors as physical features, political boundaries, cultural characteristics, and what the future may

hold. Geographers find that using multiple factors enables them to arrive at regional definitions with which most people can agree.

We have organized the material in this book into four levels, or scales: the global scale, the world regional scale, the subregional scale, and the local scale. At the **global scale,** explored in this chapter, the entire world is treated as a single area—a unity that is still new to all of us but is more and more relevant as we become used to thinking of our planet as a global system. We use the term **world region** for the largest divisions of the globe, such as East Asia, Southeast Asia, and North America (see Figure 1.1). We have defined ten world regions, each of which is covered in a separate chapter. Each regional chapter considers the interaction of human and physical geography in relation to cultural/social, economic, population, environmental, and political topics. The world regions are then divided into **subregions,** which may be independent countries, groups of countries, or even parts of a single country. For example, the world region of North America contains several subregions. In Canada, the province of Québec is considered a subregion because of its distinct French heritage (Figure 1.10). In the United States, no one state stands out so distinctly, and there subregions tend to be groups of states, such as the Southwest or the Northeast. In several cases, subregions include territory in both the United States and Canada.

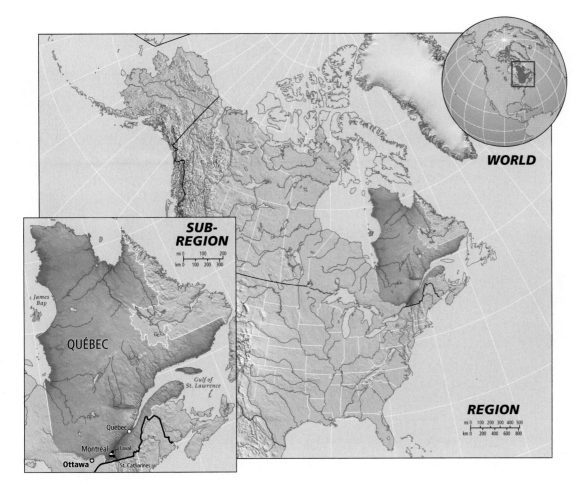

FIGURE 1.10 **Three geographic scales:** the world (upper right corner), a region (North America), and a subregion (Québec).

Most of us are familiar with geography at the **local scale**—where we live and work, whether in a city, town, or rural area. The options afforded by our local geography shape our lifestyles and the culture we create together. Throughout this book, life at the local scale is illustrated with vignettes about people in a wide range of places. Often the lives of individuals reflect trends that can be tracked across several world regions or may even occur at the global scale.

The regions, subregions, and local places discussed in this book vary dramatically in size and complexity. A region can be relatively small, such as Europe, or very large, as in the case of East Asia. Subregions can also vary greatly in size: both the continent of Australia, with an area of nearly 3 million square miles (7.8 million square kilometers), and the small group of tiny islands known as Micronesia, which together cover only 270 square miles (700 square kilometers), are here considered subregions of the world region of Oceania. Continuing with Oceania as an example, at the local scale, a place can be a backyard in Polynesia, a neighborhood in Honolulu, or a town in New Zealand.

In summary, regions have the following traits:

- A region is a unit of the earth's surface that contains distinct environmental or cultural patterns.

- Regions are constructed by people to help them define spaces for varying purposes; hence, regional definitions are fluid.

- No two regions are necessarily described by the same set of indicators.

- Regions can vary greatly in size (scale).

- The boundaries of regions are usually fuzzy and hard to agree upon.

## GLOBALIZATION AND INTERREGIONAL LINKAGES

Due to economic, technological, social, and political changes, regions widely separated in space can now have interdependent economic relationships that used to be possible only between close neighbors or **contiguous regions** (those adjacent to one another). For example, much of the clothing and tropical hardwoods used by North Americans come from Southeast Asia. Likewise, South Americans and Africans produce cash crops that end up on European tables. These connections between distant regions (**interregional linkages**) began to attain their present global reach during the early stages of **European colonialism,** the practice of

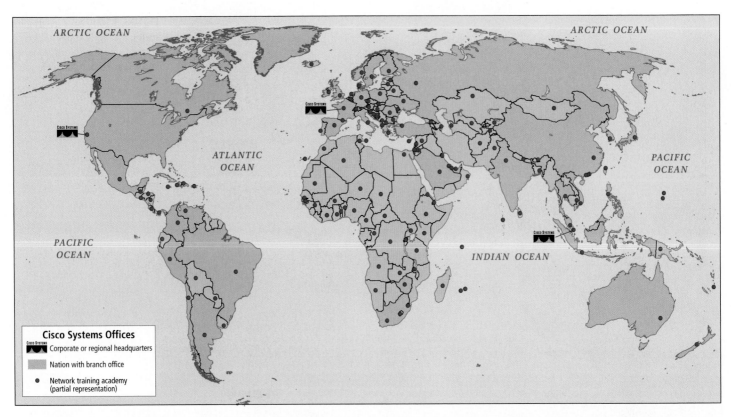

**FIGURE 1.11** Interregional linkages: Cisco Systems' global network. Cisco provides hardware and services for Internet networking. This map shows several levels of Cisco's activities: corporate and regional headquarters, countries with branch offices, and the locations of network training academies. Notice the uneven distribution of Cisco Systems offices.

taking over the human and natural resources of often distant places in order to produce wealth for Europe. Their voyages to America after 1492 led Spain and Portugal to establish colonies in Middle and South America. The British, Portuguese, Spanish, Dutch, and French founded colonies in North America and Asia in the sixteenth and seventeenth centuries and in the Pacific islands in the eighteenth century. By the end of the nineteenth century, most of Africa was divided into European colonies. Although most colonies are now independent countries, European colonization led to transglobal ties that still remain. For example, there are Muslims of Indonesian descent who have lived for generations in Suriname on the northern coast of South America. Suriname and Indonesia were both colonies of the Dutch.

Wherever it took hold, European colonization changed local economies and landscapes. In the Caribbean and the islands of Southeast Asia, for example, forest cover was lost through plantation agriculture, logging, and mining. Sometimes valuable cash crops and the laborers to grow them were transferred to entirely new places; for example, nutmeg trees were sent from the Molucca islands of Southeast Asia to the British colony of Grenada in the Caribbean. There, with the help of enslaved labor from Africa, nutmeg plantations became an important source of income for Britain. Many of the broad regional linkages we see today evolved from linkages that began hundreds of years ago, in large part under the influence of European colonialism. Similarly, many of today's violent conflicts have their roots in the colonial era.

Today, because of rapid transportation and the speedy flow of electronic information, widely separated places can be so intimately linked that the details of their relationships can change from day to day. For example, foreign construction workers in Saudi Arabia can be sending home substantial **remittances,** portions of their wages going to their families in Pakistan and Malaysia. Then, in a matter of days, hostilities in and around Saudi Arabia can send hundreds of thousands of workers scurrying home, not only disrupting construction in Saudi Arabia but also reducing tens of thousands of family budgets in Pakistan and Malaysia and limiting the fortunes of shopkeepers in their home villages.

The term **globalization** encompasses the many types of interregional linkages and flows and the changes they are bringing about. Here are some trends that have contributed to the globalization of societies around the world:

- As the volume of international trade grows, localities are becoming increasingly connected to this trade, but local producers and workers have little power over the circumstances of the new trade links.

- Businesses and corporations in rich countries have established manufacturing, sales, and service branches in poorer countries to be closer to their markets and to take advantage of low-cost resources, land, labor, and professional expertise (for example, see Figure 1.11). Some transnational or multinational organizations also seek lower taxes and less rigorous environmental and trade regulations than are found in their home countries.

- As manufacturing and some information processing jobs move to places where labor is cheaper, workers in the old locations are left unemployed and must retrain or move, while workers in the new location may be forced to work at very low wages and in inhumane conditions.

- International migration is surging as people seek opportunities in other locations.

- As people, goods, and capital increasingly flow across regional and national borders, national governments must vie with international and supranational organizations to control the new flows and relationships.

- Increased cross-cultural contact is leading to greater cultural homogenization, especially through the spread of modern popular culture. But, at the same time, rapid communication is reinforcing distinct cultural identities.

- Technology now facilitates the rapid transfer of money, making it easier for legitimate firms to do business and for migrant workers to send money to their families. Meanwhile, illegal activities, such as drug trafficking and theft, also benefit from rapid international money transactions, in part because identities can be disguised more easily.

- Access to technology, such as the Internet, is unevenly distributed, resulting in what is called the "digital divide" (see the Video Vignette below).

- Local environments increasingly show the effects of worldwide trade and resource consumption. Resources are moved from one region to serve the needs of another, environmental degradation in one place is often the result of demands for resources or goods in other places, and air and water pollution can flow across national borders.

Because globalization has multiple facets—economic, social, political, environmental—that influence one another, we will discuss globalization at many points in this book. Individuals can be harmed or empowered by its effects, and sometimes both at once. As you read the many references to globalization in the following sections, reflect on how you as an individual are part of the picture, perhaps through the inexpensive products made by low-wage labor that you buy at Wal-Mart, or through the factory or high-tech job lost by a relative when a company moved abroad.

Sugata Mitra: "The hole in the wall gives us a method to create a door, if you like, through which large numbers of children can rush into this new arena. When that happens it will have changed our society forever."

**FIGURE 1.12** **Hole-in-the-wall computing.** When Sugata Mitra first decided to address the "digital divide," he did so with one computer stuck in a wall facing a New Delhi slum. The experiment was immediately so successful that hundreds of such computers, like that shown here, are now available to poor children across India and beyond. [© Hole in the Wall Education, India/www.holeinthewall.com.]

Rajinder is an 8-year-old slum kid in New Delhi, India, who has a history of not doing well in school. Yet, in just a few minutes, he has leapt across the digital divide after encountering one of Sugata Mitra's kiosk computers. The **digital divide** is the gap between the small percentage of the world's population that has access to computers and the huge majority who do not.

Mitra is head of research and development for NIIT, a computer software and training company with offices around the world and annual sales of $300 million. NIIT was interested in developing kiosk computers for the global market—computers that give passersby quick access to the Internet.

Mitra cut a hole in the wall of his office compound in New Delhi and installed a high-speed computer with Internet access facing an adjacent slum. As he had guessed, within hours the local kids were browsing the Internet (Figure 1.12). There "was a spiral of self-instruction," says Mitra, with one kid making a discovery, three witnessing it, saying "Cool!" and then sharing three or four more discoveries while they explored together. In a matter of 5 hours, Rajinder had visited the Disney site, learned to use a drawing tool, and read news stories about the Taliban in Afghanistan. A young girl found a graphics program to help her father, a tailor, design the clothes he sews.

There are now hundreds of such computers in slum neighborhoods across India. This is important because although India is a leader in technological development, it is plagued by a digital divide. In the southern state of Karnataka, where Bangalore, known as India's "Silicon State" and ranked by the United Nations as the world's fourth-best hub of technological innovation, is located, 85 percent of the people still don't have access to a computer, and 100,000 school-age children don't go to school.

*Find out more about Rajinder and the hundreds of Indian children who love the Internet. Watch the FRONTLINE/World video "Hole in the Wall." Also visit the Web site http://www.hole-in-the-wall.com/.* ■

# CULTURAL AND SOCIAL GEOGRAPHIC ISSUES

**Culture** is an important distinguishing characteristic of human societies. It comprises everything we use to live on earth that is not directly part of our biological inheritance. Culture is represented by the ideas, materials, and institutions that people have invented and passed on to subsequent generations. Among other things, culture includes language, music, gender roles, belief systems, and moral codes (for example, those prescribed in Confucianism, Islam, and Christianity). **Material culture** comprises all the things that people use: clothing, houses and office buildings, axes, guns, computers, earthmoving equipment (from hoes to work animals to bulldozers), books, musical instruments, domesticated plants and animals, agricultural and food-processing equipment—the list is virtually endless. **Institutions** are formal or informal associations among people. The family, in its many different forms, is an **informal institution**. So is a community, which can be made up of the people who share a physical space (a village or a neighborhood) or those who share only a belief system. The community of Roman Catholic adherents, for example, stretches across continents (Figure 1.13).

**Formal institutions** include official religious organizations such as the Roman Catholic Church; local, state, and national governments; nongovernmental organizations that provide philanthropic services, such as the Red Cross and Red Crescent; and specific business corporations.

In this section we explore the concept of culture groups and some of the cultural attributes, such as value systems and languages, that help to define them. We also examine gender roles and perceptions about race as cultural phenomena that vary over space and play an important role in human relationships.

## ETHNICITY AND CULTURE: SLIPPERY CONCEPTS

A group of people who share a set of beliefs, a way of life, a technology, and usually a common ancestry and a place form an **ethnic group**. The term **culture group** is often used interchangeably

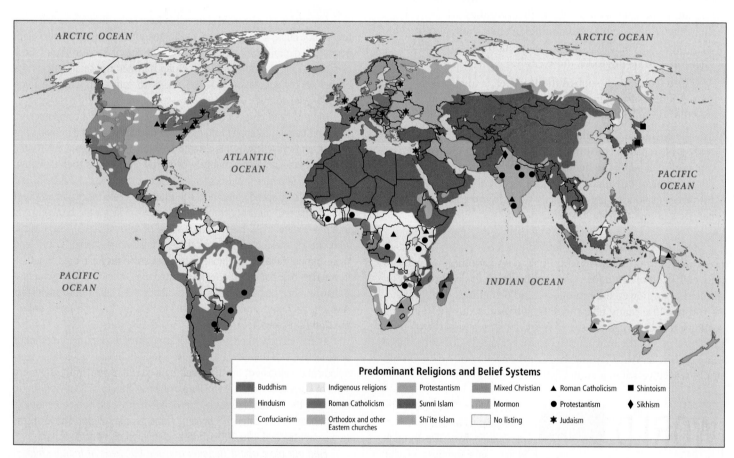

**FIGURE 1.13** Major religions around the world. The small symbols indicate a localized concentration of a particular religion within an area where another religion is predominant. [Adapted from *Oxford Atlas of the World* (New York: Oxford University Press, 1996), p. 27.]

**FIGURE 1.14** What does it mean to be Kurdish? **(a)** In rural areas away from the war zone in Iraq, Kurdish life is still peaceful and agriculturally based. Here, two young women carry sacks of grain home from the fields to feed their animals. [AP Photo/Brennan Linsley.]

**(b)** The Mazi supermarket in the Kurdish city of Duhok is Iraq's largest and draws shoppers from all over the country. The store and a mall were built on the site of a military facility where Kurds were imprisoned and tortured under Saddam Hussein. [Ed Kashi/National Geographic.]

with ethnic group, but these terms can become problematic. Like the concept of region, the concepts of culture and ethnicity are imprecise, especially as they are popularly used. For instance, as part of the modern globalization process, migrating people often move well beyond their customary cultural or ethnic boundaries to cities or to distant countries. In these new places, where they are often a minority, they take on new ways of life or even new beliefs, yet they still identify with their culture of origin. For example, long before the U.S. war in Iraq, the Kurds in Southwest Asia were asserting their right to create their own country in the territory where they traditionally lived as nomadic herders, but which is now claimed by Syria, Iraq, Iran, and Turkey (see Figure 6.49). Many Kurds who actively support the cause of the herders are now urban dwellers living and working in modern settings in Turkey, Iraq, or even London. Although these people think of themselves as ethnic Kurds and are so regarded in the larger society, they do not follow the traditional Kurdish way of life (Figure 1.14). Hence, we could argue that these urban Kurds both are and are not part of the Kurdish culture or ethnic group.

Another problem with the concept of culture is that it is often applied to a very large group that shares only the most general of characteristics. For example, one often hears the terms *American culture, African-American culture,* or *Asian culture.* In each case, the group referred to is far too large to share more than a few broad characteristics.

It might fairly be said, for example, that American culture (in both Canada and the United States) is characterized by beliefs that promote individual rights, autonomy, and responsibility; by high levels of consumption; by dependence on modern technology; by a market economy based on credit; and by widely accessible mid-

dle and higher education. If we try to be more specific, we quickly run into disagreements. Think for a moment about current debates over public school dress codes, whether the terminally ill have the right to "managed death," who should have the right to marry, or how much control women should have over their reproductive behavior. In fact, U.S. culture encompasses many subcultures that share some of the core set of beliefs, but disagree over parts of the core and over a host of other matters. The same is true, in varying degrees, for all other regions of the world.

## GLOBALIZATION AND CULTURAL CHANGE

There are indications that the diversity of culture is fading as trends and fads circle the globe via the many types of instant communication now available (Figure 1.15). American fast food, popular music, and clothing styles can now be found from Mongolia to Mozambique. At the same time, a wide variety of ethnic music, textiles, cuisines, and even modes of dress from distant places now graces the lives of consumers in the United States and across the world. Thus, as globalization proceeds, especially as people migrate and ideas are diffused, some measure of **cultural homogeneity**—more overall similarity between culture groups—will occur.

Are we all drifting toward a common repertoire of material culture, and perhaps even similar ways of thinking? Possibly, but there are also countervailing trends. It is now possible for people to reinforce their feeling of **cultural identity** with a particular group through the same channels that are encouraging homogenization. Consider the example of the people of Aceh in Southeast Asia.

Aceh is a province with a distinctive cultural identity located in northern Sumatra, an island within the country of Indonesia.

13

**FIGURE 1.15** Global culture. Mobile phones have spread to nearly every corner of the world, including cattle ranges in the Masai territory of Kenya. [Courtesy of Joseph Van Os/Getty Images.]

The people of Aceh desire greater **autonomy**—the right to control their own affairs and especially to retain control of their own resources—but the central Indonesian government views all resources as national, not local. The Indonesian government has sent military troops to enforce its interests, with lethal consequences to the Acehese. The Acehese established several Web sites and Internet chat groups to build awareness and a sense of identity among Acehese migrants worldwide. Then after a devastating tsunami struck Aceh and other sites around the Indian Ocean in December 2004, the cultural solidarity already built through the Internet helped the Acehese diaspora (those living outside Aceh) to raise funds efficiently and distribute assistance to victims.

The ability to communicate easily over the Internet and to travel quickly can reaffirm cultural identity, but it also enhances the conditions necessary for **multiculturalism**: the state of relating to, reflecting, or being adapted to several cultures. For example, the young Webmaster from Aceh who helped us to understand the perspective of the Acehese in Indonesia is now an artist in the United States, where he earns a living painting not only scenes from his tropical homeland, but also the mansions of U.S. corporate executives, as well as portraits of these executives.

## CULTURAL MARKERS

Members of a particular culture group share features, such as language and common values, that help to define the group. These shared features are called **cultural markers.** The following sections examine the roles of values, religion, language, and material culture as cultural markers. Notice how colonization and modern communications are causing some cultural markers to disappear and others to become more dominant.

## Values

Occasionally you will hear someone say, "After all is said and done, people are all alike," or "People ultimately all want the same thing." It is a heartwarming sentiment, but an oversimplification. Culturally, people are not all alike, and that is in large part what makes the study of geography interesting. We would be wise not to expect or even to want other people to be like us; it is often more fruitful to look for the reasons behind differences among people than to search hungrily for similarities. **Cultural diversity** is one reason that humans are such successful and adaptable animals. The various cultures serve as a bank of possible strategies for responding to the social and physical challenges faced by the human species. The reasons for differences in behavior from one culture to the next are usually complex, but they are often related to differences in values.

Let us look at an example that contrasts the values and **norms** (accepted patterns of behavior based on those values) held by modern urban individualistic culture with those held by rural community-oriented culture. In urban areas in many parts of the world, it is often acceptable, even desirable, for a woman to choose clothing that allows her to make a strong statement about who she is as an individual and how well off she is. One recent rainy afternoon, a beautiful forty-something Asian woman walked alone down a fashionable street in Honolulu, Hawaii. She wore high-heeled sandals, a flared skirt that showed off her long legs, and a cropped blouse that allowed a glimpse of her slim waistline. She carried a laptop case and a large fashionable handbag. Her long shiny black hair was tied back. Everyone noticed and admired her because she exemplified an ideal Honolulu businesswoman: beautiful, self-assured, and rich enough to keep herself outstandingly well dressed. But in her grandmother's village—whether it be in Japan, Korea, Taiwan, or rural Hawaii—people might regard such dress as outrageously immodest and dangerously antisocial. In a rural Asian community, her clothes would breach a widespread traditional ethic that no individual should stand out from the group. Furthermore, her costume exposed her body to open assessment and admiration by strangers of both sexes. In a village context, such behavior would signal that she lacked modesty. The fact that she walked alone down a public street—unaccompanied by her father, husband, or female relatives—might even indicate that she was not a respectable woman. Thus a particular behavior may be admired when judged by one set of values and norms, yet considered questionable or even despicable when judged by another.

If culture groups have different sets of values and standards, does that mean that there are no overarching human values or standards? This question increasingly worries geographers, who try to be sensitive both to the particularities of place and to larger issues of human rights. Those who lean too far in the direction of appreciating difference could be led to the tacit acceptance of inhumane behavior, such as the oppression of minorities and women or even torture and genocide. Acceptance of difference does not mean that we cannot make judgments about the value of certain extreme customs or points of view. Nonetheless, although it is important to take a stand against cruelty, deciding when and where to take that stand is rarely easy.

## Religion and Belief Systems

The **religions** of the world are formal and informal institutions that embody value systems. Most have roots deep in history, and many include a spiritual belief in a higher power (God, Yahweh, Allah) as the underpinning for their value systems. These days, religions often focus on reinterpreting age-old values for the modern world. Some formal religious institutions—such as Islam, Buddhism, and Christianity—*proselytize*; that is, they try to extend their influence by seeking converts. Others, such as Judaism and Hinduism, accept converts only reluctantly. Informal religions, often called *belief systems*, have no formal central doctrine and no firm policy on who may or may not be a practitioner.

Religious beliefs are often reflected in the landscape. For example, settlement patterns often demonstrate the central role of religion in community life: village buildings may be grouped around a mosque (Figure 1.16) or synagogue, or an urban neighborhood may be organized around a Catholic church. In some places, religious

**FIGURE 1.16 Religion at the center of community life.** This mosque, in Banja Luka, Serbia, is one of twenty-one mosques that were destroyed in the town during the 1992–1995 war. Rebuilding the mosque became a central goal of the Muslim community, as a symbol of their survival. [AFP/Getty Images.]

rivalry is a major feature of the landscape. Certain spaces may be clearly delineated for the use of one group or another, as in Northern Ireland's Protestant and Catholic neighborhoods.

Religion has also been used to wield power. For example, religion served as a political instrument during the era of European colonization—as a way to impose a quick change of attitude on conquered people, and the influence lingers. Figure 1.13 (page 12), which shows the distribution of the major religious traditions on earth today, demonstrates some of the religious consequences of colonization; note, for instance, the distribution of Roman Catholicism in the parts of the Americas, Africa, and Southeast Asia colonized by European Catholic countries. Religion can also spread through trade contacts. In the seventh and eighth centuries, Islamic people used a combination of trade and political power (and, less often, actual conquest) to extend their influence across North Africa, throughout Central Asia, and eventually into South and Southeast Asia.

The history and distribution of belief patterns throughout the world is complex. The distribution of major religions has changed many times over the course of history. Moreover, a world map is too small in scale to convey actual religious spatial patterns, such as where two or more religious traditions intersect at the local level. And as the world's cultural traditions become increasingly mixed and urban life spreads, **secularism**, a way of life informed by values that do not derive from any one religious tradition, is spreading.

## Language

Language is one of the most important criteria in delineating cultural regions. The modern global pattern of languages (Figure 1.17) reflects the complexities of human interaction and isolation over several hundred thousand years. But the map does not begin to depict the actual details of language distribution: between 2500 and 3500 languages are spoken on earth today—some by only a few dozen people in isolated places. Many languages have several **dialects**—regional variations in grammar, pronunciation, and vocabulary.

The geographic pattern of languages has continually shifted over time as people have interacted through trade and migration. The pattern changed most dramatically, however, after the age of European exploration and colonization began around 1500. From that point on, the languages of European colonists often replaced the languages of the colonized people. This is why we find large patches of English, Spanish, Portuguese, and French in the Americas, Africa, Asia, and Oceania. In North America, European languages largely replaced Native American languages, but in Middle and South America, Africa, and Asia, European and native tongues coexisted. Many people became bilingual or trilingual.

Today, with increasing trade and instantaneous global communication, a few languages have become dominant. At the same time, other languages are becoming extinct because children no longer learn them. Arabic is an important **lingua franca,** or language of international trade, as are English, Spanish, and Chinese. Among these, English dominates, largely because the British colonial empire introduced English as a second language to many places around the globe, including North America. U.S. economic influence is another factor that led to the dominance of English, but the language has now taken on a life of its own, being the second language of an estimated 1.5 billion people. As a native language, however, English ranks fourth, behind Chinese, Hindi, and

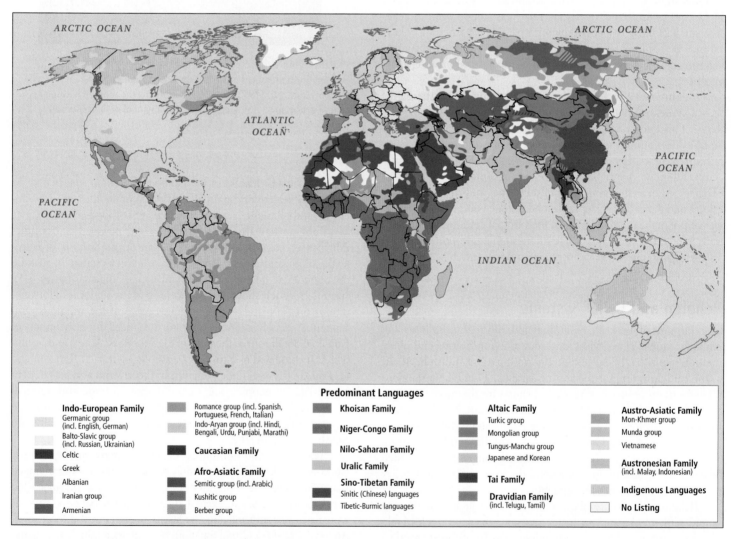

**FIGURE 1.17** The world's major language families. Distinct languages (Spanish and Portuguese, for example) are part of a larger language group (Romance languages), which in turn is part of a language family (Indo-European). [Adapted from *Oxford Atlas of the World* (New York: Oxford University Press, 1996), p. 27.]

**FIGURE 1.18** Material culture. **(a)** This home, in Canyon, California, belongs to the Cavin family. They are pictured (in the foreground) with all of their possessions. [Peter Menzel/Material World.]

**(b)** The six members of the Batsuur family live in a *ger*, the traditional tentlike Mongolian house that can easily be dismantled and moved to another place. The house has electricity (for the hot plate and television) and a coal-burning stove. [Leong Ka Tai and Peter Menzel/Material World.]

Spanish. Now the need for a common world language in the computerized information age is rapidly pushing the world community to accept English as the primary lingua franca.

For many people, a particular language represents a cultural and geographic identity that they wish to retain. For example, the recently independent small country of Slovenia (once part of Yugoslavia in southeastern Europe) jealously guards its language, Slovene, as one of only a few remaining cultural markers that give it identity. Slovenes worry that their language will vanish as their country becomes integrated into Europe and children study French, German, and English in school. As a result, Slovenes devote much energy to preserving their language. They have one of the highest rates of book publishing in the world and translate many foreign texts into Slovene. They even attempt to extend the use of Slovene beyond the borders of Slovenia by inviting the children and grandchildren of emigrants to return from the Americas, Western Europe, and Australia for heavily government-subsidized summer language courses. The Slovenes are an example of a culture group that is intensifying its distinct identity even as its connections to the wider world increase.

## Material Culture and Technology

A group's material culture reflects its **technology**, which is the integrated system of knowledge, skills, tools, and methods upon which a culture group bases its way of life. Material culture and technology help to define a culture, and in fact archaeologists use the interconnectedness of material artifacts, technology, and culture to deduce—from an excavated assemblage of potsherds (bits of broken pottery), tools, animal bones, and other material remains—what was going on long ago in a particular place.

Housing provides an example of how material culture reveals a particular culture group's way of life, especially its values, technology, and resource use. With its distinctive architecture, electrical and plumbing systems, and landscaping, the typical North American suburban ranch house silently reveals a great deal about American family values (Figure 1.18a). It reflects a nuclear family structure (mother, father, children)—a type that remains an ideal in American society, though it now constitutes less than 25 percent of American families—and often has to be modified to accommodate extended families. The ranch house reflects certain ideas about privacy (multiple bedrooms) and gender roles (Mom's special spaces may be the kitchen and laundry room; Dad's, the TV room, the garage, and the toolshed). This house bespeaks a certain level of affluence and leisure, equipped as it is with labor-saving devices and conveniences and set apart on a green lawn requiring constant maintenance. It also symbolizes American ideas about private property, polite neighborliness, and mobility.

Because we in North America move so often in our quest for a better job and standard of living, yet insist on privately owned homes, a vast system of interchangeable dwellings has evolved across America, supported by an institution: the American home real estate market. Thus a family moving to a distant city can find a home similar in floor plan and surroundings to the one they left.

An astute foreign observer could learn a great deal about Americans simply by "reading" the material culture of their homes and surrounding landscapes. Were you such an astute observer, you could closely observe housing in Japan, Mongolia (Figure 1.18b), France, or the West Indies and learn much about each particular culture's notions of proper family structure, gender roles, intimacy rules, aesthetic values, property rights, building technology, use of resources, and trading patterns.

## Gender Issues

Geographers have begun to pay more attention to gender roles in different culture groups. In virtually all parts of the world, and for at least tens of thousands of years, the biological fact of maleness and femaleness has been translated into specific roles for each sex. Although the activities assigned to men and to women can vary greatly from culture to culture and from era to era, there are some rather startling consistencies around the globe and over time.

Men are usually expected to fulfill public roles, working outside the home in such positions as traveling executives, animal herders, hunters, or government workers. Women are usually expected to fulfill private roles, keeping house, bearing and rearing children, caring for the elderly, and preparing the meals, among many other tasks. In nearly all cultures women are defined as dependent on men—their fathers, husbands, brothers, or adult sons—even when the women may produce most of the family sustenance. Because their activities are focused on the home, women typically have less access to education and paid employment, and hence less access to wealth and political power. When they do work outside the home (as is the case increasingly in every region), women tend to fill lower-paid positions—whether as laborers, service workers, or professionals—and to retain their household duties as well.

**Gender**—the sexual category of a person—is both a biological and a cultural phenomenon. Men and women have different reproductive roles, and there are certain other physical differences as well. Men have larger muscles, can lift heavier weights, and can run faster than women (but not necessarily for longer periods). In some physical exercises, average women have more endurance and are capable of more precise movements than average men, and in populations that enjoy overall good health, women tend to outlive men by an average of 3 to 5 years. Women's physical capabilities are somewhat limited during pregnancy and nursing—and, for some, during menstruation—but their susceptibility to pregnancy is limited to about 30 years. Beyond about age 45, women are no longer subject to the physical limits imposed by reproduction, and most contribute in some significant way to the well-being of their adult children and grandchildren. In societies that seclude women, postmenopausal women have more freedom to move about outside in public spaces. A growing number of evolutionary biologists postulate that the evolutionary advantage of menopause in midlife is that it gives women the time and energy and freedom to help succeeding generations thrive—an idea sometimes labeled the *grandmother hypothesis*.

Although average physical gender differences exist, in most cultures they are amplified to carry greater social significance than the biological facts would warrant. Customary ideas about masculinity and femininity, proper gender roles, and sexual orientation are handed down from generation to generation within a particular culture group or society, and have enormous effects on the everyday lives of men, women, and children. Perhaps more than for any other culturally defined human characteristic, there is significant agreement from place to place and over time that gender is important.

The historical and modern global gender picture is a puzzlingly negative one for women. In nearly every culture, in every region of the world, and for a great deal of recorded history, women have had

**TABLE 1.1** Comparisons of male and female income in countries where average education levels are higher for females than for males

| Country | Female income (PPP[a] U.S.$, 2003) | Male income (PPP U.S.$, 2003) | Female income as percent of male income |
|---|---|---|---|
| Austria | 15,878 | 45,174 | 35 |
| Barbados | 11,976 | 19,687 | 61 |
| Canada | 23,922 | 37,572 | 64 |
| Japan | 17,795 | 38,612 | 46 |
| Jordan | 2,004 | 6,491 | 31 |
| Kuwait | 8,448 | 24,204 | 35 |
| Poland | 8,769 | 14,147 | 62 |
| Russia | 7,302 | 11,429 | 62 |
| Saudi Arabia | 4,440 | 20,717 | 21 |
| Sweden | 21,842 | 31,722 | 69 |
| United Kingdom | 20,790 | 33,713 | 62 |
| United States | 29,017 | 46,456 | 62 |

[a]PPP, purchasing power parity, is the amount that the local currency equivalent of U.S.$1 will purchase in a given country.

*Source: United Nations Human Development Report 2005* (New York: United Nations Development Programme), Table 25, "Gender-related development index," pp. 299–302, and Table 27, "Gender inequality in education," pp. 307–310.

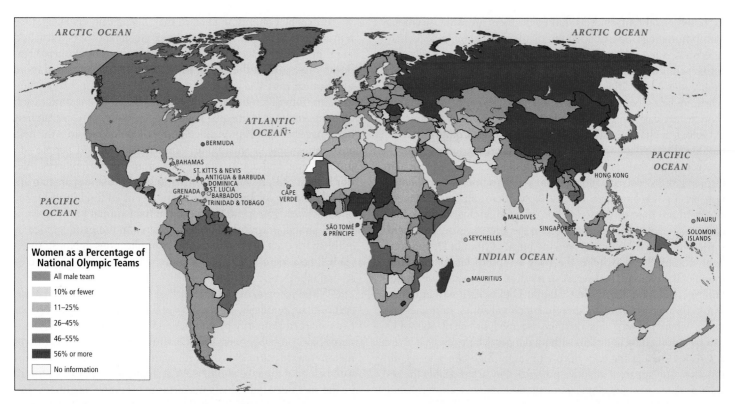

**FIGURE 1.19** Women's participation in the 2004 Olympics. Traditional notions of femininity, female roles, and female strength, speed, and endurance are being strongly challenged by Olympic athletes. In the first modern Olympic Games, in 1896, women were barred from participating. In 2004, women made up 44 percent of the Olympic competitors, and in some countries (China, for instance), more than 50 percent of the athletes were women. [Adapted from Joni Seager, *The State of Women in the World Atlas* (New York: Penguin, 2003), pp. 50–51; updated from http://multimedia.olympic.org/pdf/en_report_1000.pdf.]

(and still have) an inferior status. It is hard to find exceptions, although the intensity of this second-class designation varies considerably. People of both sexes routinely accept the idea that males are more productive and intelligent than females. In nearly all cultures, families prefer boys over girls because, as adults, boys have greater earning capacity (in large part as a result of discrimination against females; Table 1.1), have more power in society, and will perpetuate the family name (because of patrilineal naming customs). Around the globe, females have less access to medical care, are more likely to die in infancy (see Figure 1.34 on page 39), start work at a younger age than males, work longer hours, and eat less well. In the United States, when domestic violence occurs, females are the victims 96 percent of the time. The question of how and why women became subordinate to men has not yet been well explored because, oddly enough, few thought the question significant until recently.

There are growing challenges to traditional notions of femininity and ideas of female capabilities. In many countries around the world, females are now acquiring education at higher rates than males, a fact that should eventually make women competitive with men for jobs. Unless discrimination persists, women should also begin to earn pay equal to men. Between 2000 and 2004 female athletes increased their participation in the summer Olympic games from 30 to 40 percent of the competitors, and in some countries (China, for example), more than half were female (Figure 1.19).

Yet looking only at women's woes misses half the story. Men are also disadvantaged by strict gender expectations. Probably for most of human history, young men have borne the lion's share of onerous physical tasks and dangerous undertakings. Until recently it was usually only young men who left home to migrate to distant, low-paying jobs. Overwhelmingly, it is young men who die in wars. It is boys who are taught to repress their feelings, although some cultures emphasize this more than others. Men who grow up with negative attitudes toward women are taught those attitudes by their families (often female kin) and the larger society. This book will return repeatedly to the question of gender disparities in an effort to investigate this most perplexing cross-cultural phenomenon. Our examination will also reveal that many societies are addressing gender inequalities and that, in every country on earth, what it means to be a man or a woman is being renegotiated, at least in small ways.

## Race

Just as ideas about gender roles affect life in all world regions, ideas about race affect human relationships everywhere on earth. However, according to the science of biology, all people now alive on earth are members of one species, *Homo sapiens sapiens*, and the popular markers of what we call **race** (skin color, hair texture, face and body shapes) have no biological significance. For any supposed racial trait, such as skin color, there are wide variations within human groups. In addition, many invisible biological characteristics, such as blood type and DNA patterns, actually cut across skin color distributions and are shared by what are commonly viewed as different races. Over the last several thousand years, in fact, there has been such massive gene flow among human populations that no modern group presents a discrete set of biological characteristics. Although we may look quite different, biologically speaking we are all closely related.

It is likely that some of the easily visible features of particular human groups evolved to adapt them to environmental conditions, but precise explanations of just how these physical changes occurred are hard to come by. For example, biologists think that darker skin evolved in regions close to the equator, where sunlight is most intense. All humans need the nutrient vitamin D, and sunlight striking the skin helps the body to absorb vitamin D. Too much of the vitamin, however, can result in improper kidney functioning. Dark skin absorbs less vitamin D than light skin and thus would be a protective adaptation in equatorial zones. In higher latitudes, where the sun's rays are more dispersed, sufficient vitamin D can be absorbed by light skin without danger. Many physical characteristics, such as big ears, deep-set eyes, or high cheekbones, do not serve any apparent adaptive purpose; they are probably the result of random chance and ancient inbreeding within isolated groups. Similarly, there is no evidence that any so-called "race" has particularly high math ability or athletic ability; such characteristics are present in individuals in all human populations and may become enhanced by cultural practices.

It may be comforting to learn that, biologically speaking, race is meaningless. But we cannot overlook the significant political and social import that race (often paired with culture or ethnicity) carries in most parts of the world. Race seems to have acquired more significance over the last several thousand years, especially when humans from different parts of the world began to encounter each other in situations of unequal power. For example, some researchers have suggested that European colonizers adopted **racism**—the negative assessment of unfamiliar, often darker-skinned people—to justify taking land and resources away from those supposedly inferior beings. Still, it would be wrong to suggest that disparaging appraisals and exploitation of others are recent cultural innovations, or particularly European traits. The human animal has long committed atrocities against its own kind, often in the name of race or ethnicity, or even gender. Race and its implications in North America will be a focus in Chapter 2, and the topic will be discussed in several other world regions as well.

While recognizing all the ills that have emerged from racism and similar prejudices, we should not infer that human history has been marked primarily by conflict and exploitation. Actually, humans have probably been so successful because of a strong inclination toward *altruism*, the willingness to sacrifice one's own well-being for the sake of others. Writ small, altruism can be found in the sacrifices individuals make to help family, neighbors, and community. Writ large, it includes charitable giving to help anonymous people in need. It is probably our capacity for altruism that causes us such deep consternation over the relatively infrequent occurrences of inhumane behavior.

# PHYSICAL GEOGRAPHY: PERSPECTIVES ON THE EARTH

Humans have always had to adapt to the physical environment, although the nature of the interactions between humans and their environments has changed over time and has varied from culture to culture. In this section we look at two components of the physical environment that are of particular interest to physical geographers: landforms and climate. We finish by examining the origins of agriculture as an example of how the interactions between humans and the environment can alter both the physical environment and human society.

## LANDFORMS: THE SCULPTING OF THE EARTH

The processes that create the world's varied **landforms**—mountain ranges, continents, and the deep ocean floor—are some of the most powerful and slow-moving forces on earth. Originating deep beneath the earth's surface, these forces (**internal processes**) can move entire continents and often take hundreds of millions of years to do their work. Many of the earth's features, however, such as a beauti-

ful waterfall or a dramatic rock formation, are formed by more rapid and delicate processes that take place on the surface of the earth (**external processes**). All of these forces are studied by *geomorphologists*, who focus on the processes that constantly shape and reshape the earth's surface.

## Plate Tectonics

Two key ideas related to internal processes in physical geography are the Pangaea hypothesis and plate tectonics (Figure 1.20). The **Pangaea hypothesis**, first suggested by geophysicist Alfred Wegener in 1912, proposes that all the continents were once joined in a single vast continent called *Pangaea* (meaning "all lands"), which fragmented over time into the continents we know today. As one piece of evidence for his theory, Wegener pointed to the neat fit between the west coast of Africa and the east coast of South America. For decades, most scientists rejected Wegner's hypothesis, but we now know that the earth's continents have been assembled into supercontinents at least three different times, only to break apart again. Pangaea, which existed from about 237 million years ago, was only

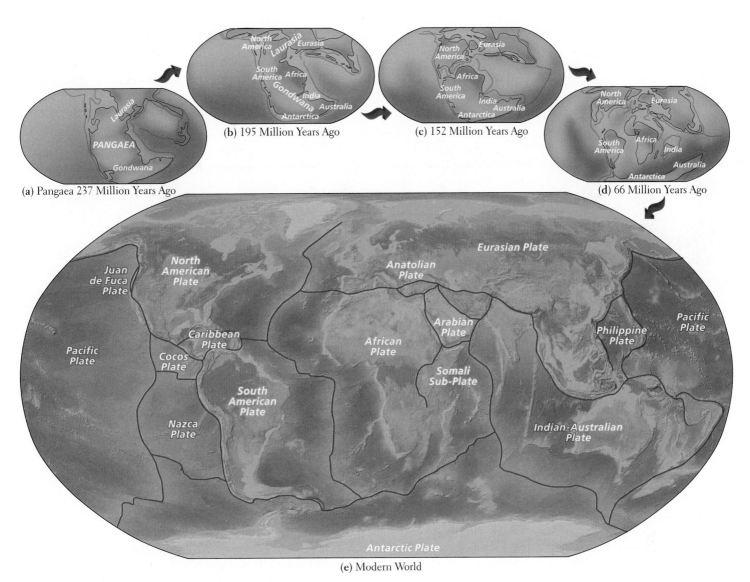

(a) Pangaea 237 Million Years Ago

(b) 195 Million Years Ago

(c) 152 Million Years Ago

(d) 66 Million Years Ago

(e) Modern World

**FIGURE 1.20** **The breakup of Pangaea and resulting continental drift.**
The Modern World map (e) depicts the current boundaries of the major
tectonic plates. Pangaea is only the latest of several global configurations that
have coalesced and then fragmented over the last billion years. [Adapted
from Frank Press, Raymond Siever, John Grotzinger, and Thomas H. Jordan,
*Understanding Earth,* 4th ed. (New York: W. H. Freeman, 2004), pp. 42–43.]

the latest such supercontinent; it was preceded by Rodinia and Pan-
notia. In the distant future, our still-moving continents are predicted
to coalesce once again into the supercontinent Pangaea Ultima. All
of this activity is made possible by plate tectonics, a process of continental motion discovered in the 1960s, long after Wegener's time.

The premise of **plate tectonics** is that the earth's surface is
composed of large plates that float on top of an underlying layer of
molten rock. The plates are of two types. *Oceanic plates* are dense
and relatively thin, and they form the floor beneath the oceans.
*Continental plates* are thicker and less dense. Much of their surface
rises above the oceans, forming continents. These massive plates
move slowly, driven by the circulation of the underlying molten
rock flowing from hot regions deep inside the earth to cooler surface regions, and back. The creeping movement of tectonic plates

created the continents we know today by fragmenting and separating Pangaea (see Figure 1.20).

Plate movements influence the shapes of major landforms,
such as continental shorelines and mountain ranges. The continents have piled up huge mountains on their leading edges as the
plates carrying them collided with other plates, folding and warp-
ing in the process. Hence, the theory of plate tectonics accounts for
the long, linear mountain ranges extending from Alaska to Chile in
the Western Hemisphere and from Southeast Asia to the European
Alps in the Eastern Hemisphere. The highest mountain range in
the world, the Himalayas of South Asia, was created when what is
now India, at the northern end of the Indian-Australian Plate,
ground into Eurasia. The only continent that lacks these long, linear mountain ranges is Africa. Often called the "plateau continent,"

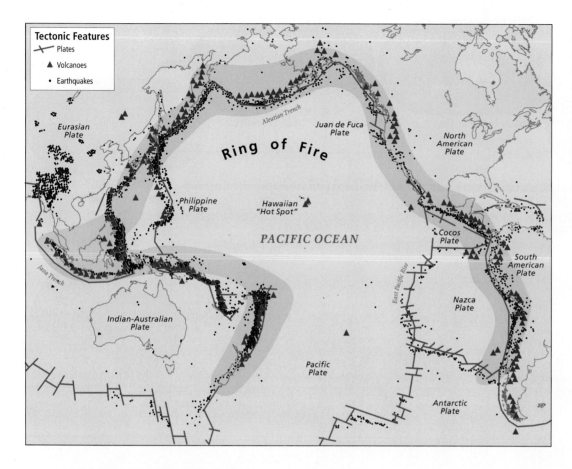

**FIGURE 1.21** The Ring of Fire. Volcanic formations encircling the Pacific Basin form the so-called Ring of Fire, a zone of frequent earthquakes and volcanic eruptions. [Adapted from http://vulcan.wr.usgs.gov/Glossary/PlateTectonics/Maps/map_plate_tectonics_world.html; and Frank Press, Raymond Siever, John Grotzinger, and Thomas H. Jordan, *Understanding Earth*, 4th ed. (New York: W. H. Freeman, 2004), p. 27.]

Africa is believed to have been at the center of Pangaea and to have moved relatively little since the breakup.

Humans encounter tectonic forces most directly as earthquakes and volcanoes. Plates slipping past each other create the catastrophic shaking of the landscape we know as an **earthquake.** When plates collide and one slips under the other, this is known as **subduction. Volcanoes** arise at zones of subduction or sometimes in the middle of a plate, where gases and molten rock (called magma) can rise to the earth's surface through fissures and holes in the plate. Volcanoes and earthquakes are particularly common around the edges of the Pacific Ocean, an area known as the **Ring of Fire** (Figure 1.21).

## Landscape Processes

The landforms created by plate tectonics have been further shaped by external processes, which are more familiar to us because we can observe them daily. One such process is **weathering.** Rock, exposed to the onslaught of sun, wind, rain, snow, ice, and the effects of life-forms, fractures and decomposes into tiny pieces. These particles then become subject to another external process, **erosion.** In erosion, wind and water carry rock particles away and deposit them in new locations. The **deposition** of eroded material can raise and flatten the land around a river, where periodic flooding spreads huge quantities of silt. As small valleys between hills are filled in by silt, a **floodplain** is created. Where rivers meet the sea, floodplains often fan out roughly in the shape of a triangle, creating a **delta.** External processes tend to smooth out the dramatic mountains and valleys created by internal processes.

Human activity often contributes to external landscape processes. By altering the vegetative cover, agriculture and forestry expose the earth's surface to sunlight, wind, and rain—agents that increase weathering and erosion. Flooding becomes more common because the removal of vegetation limits the ability of the earth's surface to absorb rainwater. As erosion increases, rivers may fill with silt and deltas extend into the oceans. Building with concrete, asphalt, and steel often covers formerly wooded land with impervious surfaces. Again, flooding is the result, because rainwater runs over the surface to the lowest point instead of being absorbed into the ground. The physical effects of human activities vary in degree from one culture to another depending in part on the tools used: mechanized earthmovers used to build roads change the earth's surface more rapidly and profoundly than machetes used to clear a path.

## CLIMATE

The processes associated with climate are generally more rapid than those that shape landforms. **Weather,** the short-term expression of climate, can change in a matter of minutes. **Climate** is the long-term balance of temperature and precipitation that keeps weather patterns fairly consistent from year to year. By this definition, the last major global climate change took place 15,000 years ago, when the glaciers of the last ice age began to melt.

Energy from the sun gives the earth a temperature range hospitable to life. The earth's atmosphere, oceans, and land surfaces absorb huge amounts of solar energy, and the atmosphere traps much of that energy at the earth's surface, insulating the earth from the deep cold of space. Solar energy is also the engine of climate. The most intense, direct sunlight falls in a broad band stretching about 30° north and south of the equator. The highest average temperatures on earth occur within this band. Moving away from the equator, sunlight becomes less intense, and average temperatures drop.

## Temperature and Air Pressure

The wind and weather patterns we experience daily are largely a product of complex patterns of air temperature and air pressure. **Air pressure** can best be understood by thinking of air as existing in a particular unit of space—for example, a column of air above a square foot of the earth's surface. Air pressure is the amount of force exerted by that column on that square foot of surface. Air pressure and temperature are related: the gas molecules in warm air are relatively far apart and are associated with low air pressure, whereas in cool air the gas molecules are relatively close together (dense) and are associated with high air pressure.

As a unit of cool air is warmed by the sun, the molecules move farther apart. The air becomes less dense and exerts less pressure. Air tends to move from areas of higher pressure to areas of lower pressure, creating wind. If you have been to the beach on a hot day, you may have noticed a cool breeze blowing in off the water. Land heats up (and cools down) faster than water, so on a hot day, the air over the land warms, rises, and becomes less dense than the air over the water, causing that cooler, denser air to flow inland. At night the breeze often reverses direction, blowing from the now cooling land onto the now relatively warmer water.

These air movements have a continuous and important influence on global weather patterns. Over the course of a year, continents heat up and cool off much more rapidly than the oceans that surround them. Hence, the wind tends to blow from the ocean to the land during summer and from the land to the ocean during winter. It is almost as if the continents were breathing once a year, inhaling in summer and exhaling in winter.

## Precipitation

Perhaps the most tangible way we experience changes in air temperature and density is through the falling of rain or snow. Precipitation occurs primarily because warm air holds more moisture than cool air. When this moist air rises to a higher altitude, its temperature drops, which reduces its ability to hold moisture. The moisture condenses into drops to form clouds and may eventually fall as rain or snow.

Several conditions that encourage moisture-laden air to rise influence the pattern of precipitation observed around the globe (Figure 1.22). When moisture-bearing air is forced to rise as

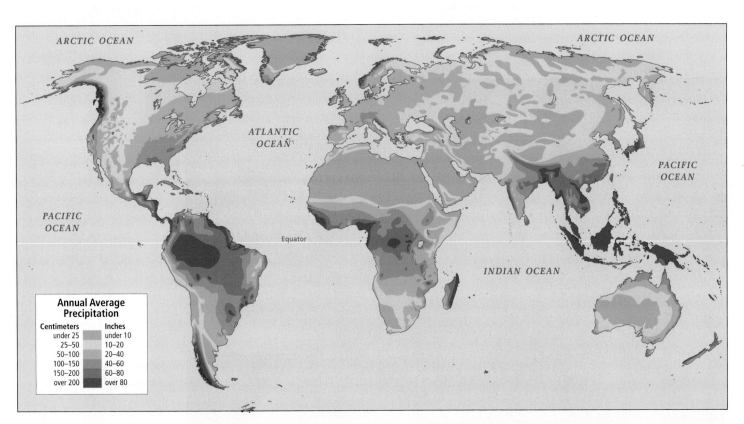

**FIGURE 1.22** World average annual precipitation. Notice the concentration of the heaviest precipitation primarily in a wide irregular band on either side of the equator. [Adapted from *Goode's World Atlas*, 21st ed. (Chicago: Rand McNally, 2005), pp. 20–21.]

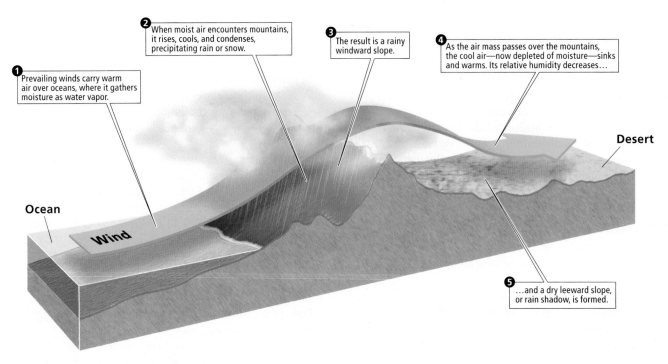

① Prevailing winds carry warm air over oceans, where it gathers moisture as water vapor.

② When moist air encounters mountains, it rises, cools, and condenses, precipitating rain or snow.

③ The result is a rainy windward slope.

④ As the air mass passes over the mountains, the cool air—now depleted of moisture—sinks and warms. Its relative humidity decreases...

⑤ ...and a dry leeward slope, or rain shadow, is formed.

Ocean

Wind

Desert

**FIGURE 1.23** Orographic rainfall (rain shadow). [Adapted from Frank Press, Raymond Siever, John Grotzinger, and Thomas H. Jordan, *Understanding Earth*, 4th ed. (New York: W. H. Freeman, 2004), p. 281.]

it passes over mountain ranges, the air cools and the moisture condenses to produce rainfall. This process, known as **orographic rainfall**, is most common in coastal areas where wind blows moist air from above the ocean onto the land and up the side of a coastal mountain range (Figure 1.23). Most of the moisture falls as rain as the air is rising along the coastal side of the range. On the inland side, the descending air warms and ceases to drop its moisture. The drier side of a mountain range is said to be in the **rain shadow.** Rain shadows may extend for hundreds of miles across the interiors of continents, as they do on the Mexican Plateau, east of California's Pacific coast, or north of the Himalayas of Eurasia.

**FIGURE 1.24** An Asian monsoon. This picture shows a summer monsoon rainstorm approaching Sri Lanka off the southeast coast of India. [Mark Henley/Panos.]

A rain belt associated with the equator is primarily the result of moisture-laden tropical air being heated by the strong equatorial sunlight and rising to the point where it releases its moisture as rain. Equatorial areas in Africa, Southeast Asia, and South America are watered in this way (see Figure 1.22 on page 23). Neighboring non-equatorial areas also receive some of this moisture when seasonally shifting winds blow the rain belt north and south of the equator. The huge downpours of the Asian summer **monsoon** are an example (Figure 1.24). The Eurasian continental landmass heats up during the summer, causing the overlying air to expand, become less dense, and rise. The somewhat cooler, yet moist, air of the Indian Ocean is drawn inland. The effect is so powerful that the equatorial rain belt is sucked onto the land, resulting in tremendous, sometimes catastrophic, rains throughout virtually all of South and Southeast Asia and much of coastal and interior East Asia (see Figures 8.4 and 8.5 on pages 416 and 417). Similar forces pull the equatorial rain belt south during the Southern Hemisphere's summer.

Much of the moisture that falls on North America and Eurasia is **frontal precipitation** caused by the interaction of large air masses of different temperatures and densities. These masses develop when air stays over a particular area long enough to take on the temperature of the land or sea beneath it. Often when we listen to a weather forecast we hear about warm fronts or cold fronts. A front is the zone where warm and cold air masses come into contact, and it is always named after the air mass whose leading edge is moving into an area. At a front, the warm air tends to rise over the cold air, carrying its clouds to a higher altitude, and rain or snow may follow. Much of the rain that falls along the outer edges of a hurricane is the result of frontal precipitation.

## Climate Regions

Geographers have several systems for classifying the world's climates that are based on the patterns of temperature and precipitation just described. This book uses a modification of the widely known Köppen classification system, which divides the world into several types of climate regions, labeled A, B, C, D, and E on the climate map in Figure 1.25 (pages 26–27). As you look at the regions on this map and read the accompanying climate descriptions, the importance of climate to vegetation becomes evident. Each regional chapter will contain a climate map; when reading these maps, you can refer to the verbal descriptions in Figure 1.25 if necessary. Keep in mind that the sharp boundaries shown on climate maps are in reality much more gradual transitions.

## THE ORIGINS OF AGRICULTURE: HUMAN INTERACTION WITH THE PHYSICAL ENVIRONMENT

The development of agriculture provides a compelling illustration of how human interactions with the physical environment can transform human society and ways of life. Agriculture includes animal husbandry, or the raising of animals, as well as the cultivation of plants. The practice of agriculture has had long-term effects on human population growth, rates of natural resource use, the development of towns and cities, and ultimately on the development of civilization.

Where and when did plant cultivation and animal husbandry first develop? Very early humans hunted animals and gathered plants and plant products (seeds, fruits, roots, fibers) for their food, shelter, and clothing. Early assumptions by prehistorians that males were the hunters and females the gatherers are now disputed; very likely both genders engaged in both activities. During the hunting and gathering stage, humans were manipulating the environment of wild plants and animals by reducing their numbers and changing their habitats. It is very likely that the transition from hunting and gathering to tending pastures and gardens was a gradual process arising from a long familiarity with the growth cycles and reproductive mechanisms of the plants and animals that humans liked to use.

Genetic studies support the view that at varying times between 8000 and 20,000 years ago, people in many different places around the globe independently learned to develop especially useful plants and animals through selective breeding, a process known as **domestication.** This time of change in the economic base of human society is sometimes known as the **Neolithic Revolution,** a period characterized by the expansion of agriculture and the making of polished stone tools. The map in Figure 1.26 (page 28) shows several well-known centers of domestication. A continuing process of agricultural innovation also occurred in many places outside these centers. Sometimes diffusion was the mechanism that spread agriculture as farmers or herders moved into new areas, taking their tools, methods, or plants and animals with them. The geographer Carl Sauer posited in the 1950s that women were probably as active as men in plant domestication and the development of agriculture, and many other scholars since then have noted the dominance of women in Native American agriculture in prehistoric times. In Greek and Roman agricultural mythology, deities in charge of agriculture and harvests were both female—Demeter, Ceres, Egesta, and Horta—and male—Saturn and Cronus.

Why did agriculture and animal husbandry develop in the first place? Certainly the desire for more secure food resources played a role. But Richard MacNeish, an archaeologist who has studied plant domestication in Mexico and Central America, suggests that the opportunity to trade was also a major motivation and may have been just as important as the simple desire to increase or improve food supplies. Many of the known locations of agricultural innovation lie near early trade centers, where people would have had access to new information and new plants and animals brought by traders, and would have needed products to trade with the people passing through. This link of agriculture to trade also provides a glimpse of how cities may have emerged at trading crossroads: people were attracted to these centers, and some developed occupations that served the needs of others who gathered to trade. Through trade they were able to access food they had not produced themselves. Perhaps, then, agriculture was at first a profitable hobby for hunters and gatherers that eventually, because of food security *and* market demand, grew into a "day job" for some—their primary source of sustenance.

(*Text continues on page 28.*)

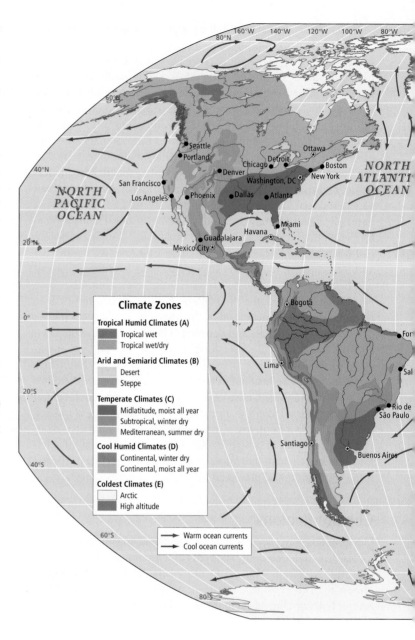

**FIGURE 1.25** Climate regions of the world.

Tropical Humid Climates (A). These climates occupy a wide band reaching 15° to 20° north and south of the equator, extending to higher latitudes when moderated by marine influences. Here we have simplified the variations to just two distinct climates: tropical wet and tropical wet/dry.

In **tropical wet climates,** rain falls predictably every afternoon and usually just before dawn. The natural vegetation is the tropical rain forest, consisting of hundreds of species of broad-leafed evergreen trees that form a several-layered canopy above the forest floor.

The **tropical wet/dry climate,** also called **tropical savanna,** experiences a wider range of temperatures than the tropical wet climate and may actually receive more total rainfall, but the rain comes in great downpours during the heat of the summer. The vegetation is mixed grassland and tropical forest. All species may show signs of having to survive long dry periods that occur unpredictably, during which they may drop their leaves to conserve moisture.

Arid and Semiarid Climates (B). Arid and semiarid climates may be either deserts or steppes.

**Deserts** generally receive very little rainfall (two inches or less per year), and most of that comes in downpours that are extremely rare and unpredictable, but are capable of bringing a brief, beautiful flourishing of desert life. Usually, deserts have sparse vegetation and almost no cloud cover, which leads to wide swings in temperature between day and night. Life is a struggle for both plants and animals because they must be able to survive heat stress during the day and freezing at night.

**Steppes,** such as the pampas of Argentina or the Great Basin of the American West, have climates similar to those of deserts, but more moderate. They usually receive about 10 inches more rain per year than deserts and are covered with grass and/or scrub.

Arid and semiarid climates are found primarily in two locations: the subtropics (slightly poleward of the tropics) and the midlatitudes. Subtropical deserts and steppes are found between 20° and 30° north and south latitudes, where high pressure air descends in a belt around the planet. They are generally much warmer than midlatitude deserts and steppes, which are found farther toward the poles in the interiors of continents, often in the rain shadows of high mountains. Although soils are generally thin and unproductive in most

deserts and steppes, some midlatitude steppes, such as the Great Plains of North America, have some of the thickest and richest soils in the world. The slightly colder temperatures and pronounced seasonality of these steppes keep down rates of decay and hence encourage the accumulation of organic matter in the soil over time.

Temperate Climates (C). In this book, we distinguish just three temperate climates.

**Midlatitude climates,** such as those in southeastern North America and China, are moist all year and have short, mild winters and long, hot summers. A variant of the midlatitude climate is the **marine west coast climate,** such as that of western Europe, which is noted for fine drizzling rains.

**Subtropical climates** differ from midlatitude climates in that winters are dry.

**Mediterranean climates** have moderate temperatures but are dry in summer and wet in winter. Plants do not get moisture when temperature and

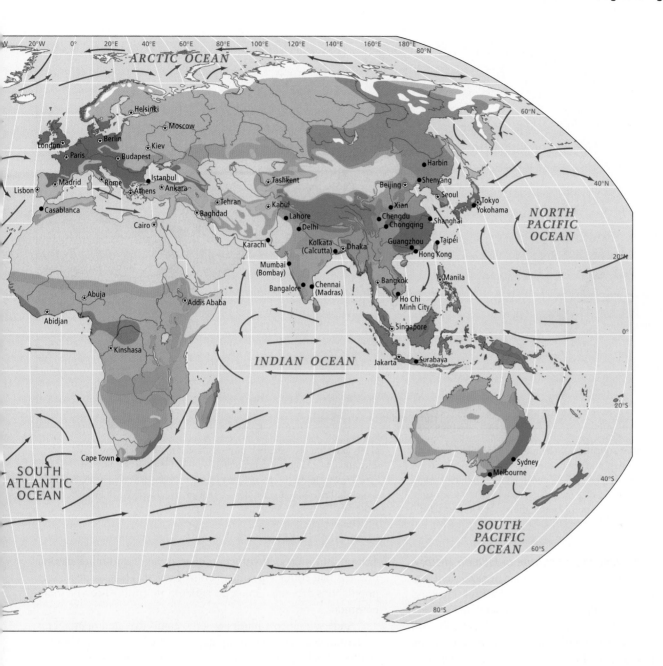

evaporation rates are highest, so the plant species that live in this climate tend to be adapted to dry conditions, with scrubby, waxy leaves capable of storing moisture. California, Portugal, northwestern Africa, southern Italy, Greece, and Turkey are examples of places with this climate type.

Cool Humid Climates (D). Stretching across the broad interiors of Eurasia and North America are continental climates, either **dry winters** (northeastern Eurasia) or **moist all year** (North America and north-central Eurasia). Summers in cool humid climates are short but can have very warm days. The natural vegetation of southern cool humid climates is broad-leafed deciduous and evergreen forest. Here the soil is deep and rich as a result of seasonally low temperatures that inhibit decay, as in the midlatitude steppes. In the more northerly areas, winters are long and cold. Vast needle-leafed evergreen forests called *taiga* stretch across the cold interior. In the taiga, soils can be deep, but they are not as rich as they are farther south; growing seasons are so short that cultivation is minimal.

Coldest Climates (E). Arctic and high-altitude climates are by far the coldest and are also among the driest. Although moisture is present, there is little evaporation because of the low temperatures.

The **Arctic climate** is often called *tundra*, after the low-lying vegetation that covers the ground. This dwarfed vegetation is a response to the 7 to 11 months of below-freezing temperatures. What little precipitation there is usually comes during the warmer months, and even this may fall as snow.

The **high-altitude** version of this climate, which may occur far from the Arctic, is more widespread and subject to greater daily fluctuations in temperature. High-altitude microclimates, such as those in the Andes and the Himalayas, can vary tremendously depending on factors such as available moisture, orientation to the sun, and vegetation cover. As one ascends in altitude, the climate changes loosely mimic those found as one moves from lower to higher latitudes. These changes are known as temperature-altitude zones (see Figure 3.7 on page 121).

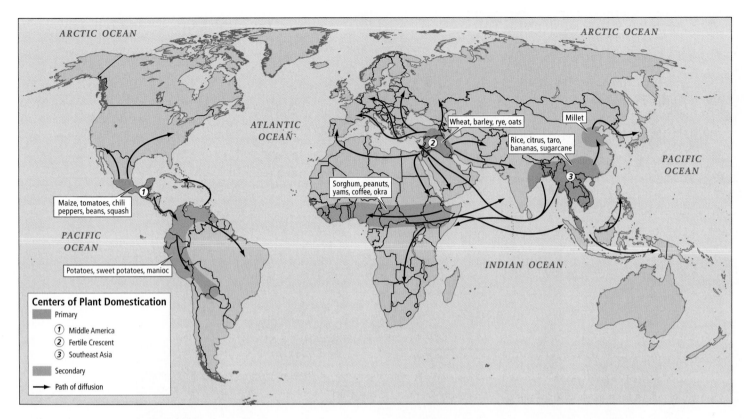

**FIGURE 1.26** The origins of agriculture. Scientists have identified six main areas of the world (three primary and three secondary) where agriculture emerged independently. For lengthy periods, people in these different places tended plants and animals and selected for the genetic characteristics they valued. This knowledge eventually spread around the world. Domesticated plants and animals were then further adapted to new locations. This selection and adaptation process continues in the present. [Adapted from Terry G. Jordan-Bychkov, Mona Domosh, Roderick P. Neumann, and Patricia L. Price, *The Human Mosaic*, 10th ed. (New York: W. H. Freeman, 2006), pp. 274–275.]

### Recent Insights on the Role of Agriculture in Human History.

Agriculture is usually viewed as a major human achievement that made it possible to amass surplus stores of food for lean times and allowed some people to specialize in activities other than food procurement. While agriculture did produce these effects, the full picture is more complex. Anthropologist E. N. Anderson, writing about the beginnings of agriculture in China, suggests that agricultural production may have led to several developments now regarded as problems: rapid population growth, extreme social inequalities, environmental degradation, and even famine. He suggests that as groups turned to raising animals and plants for their own use or for trade, more labor was needed. As the population expanded to meet this need and more resources were used to produce food, natural habitats were destroyed, and hunting and gathering were gradually abandoned. One consequence was that the quality of human diets may have declined, as people abandoned foraging among diverse wild plants and began to eat primarily corn, wheat, or rice. Moreover, everyone did not share the wealth provided by trade equally, and some of those who specialized in nonagricultural activities began to amass wealth and power. Gradually, a small elite emerged atop a mass of much poorer people. Further, land clearing for agriculture led to soil erosion and increased vulnerability to drought and other natural disasters that could wipe out an entire harvest. Thus, as ever-larger populations depended solely on agriculture, famine became more common.

Anderson's theory illustrates some points that are important for understanding modern geographic issues, especially general human well-being. The potential for human impact on the world's environments has increased markedly as fields of cultivated plants and pastures for cattle, sheep, and goats have replaced forests and grasslands. In addition, we apply ever-larger amounts of chemicals and irrigation water to keep our fields and pastures productive. The trend of increasingly intense human impact has become even more pronounced over the past few centuries as the human population has doubled and redoubled. Today, housing tracts are replacing cultivated fields.

The invention of agriculture was an early step in the development of economic systems, all of which are based in some way on interactions between humans and their environments. Over the course of history, local economic systems became linked to increasingly distant systems. We next consider some current economic issues that have grown out of the resulting complex linkages.

# ECONOMIC ISSUES IN GEOGRAPHY

Geographers have long been interested in the interactions of economies in different parts of the world and in how economies affect the resources that people use and the ways that people arrange themselves across the land. In recent decades, those who study economic geography have focused increasingly on the economic aspects of globalization, or the **global economy**—the ways in which goods, capital, labor, and resources are exchanged among distant and very different places. We begin by looking at a few examples of people who are part of that exchange and whose day-to-day lives are strongly affected by the global economy.

## WORKERS IN THE GLOBAL ECONOMY

**Personal Vignette**    Olivia lives in Soufrière on St. Lucia, an island in the Caribbean (Figure 1.27). Soufrière was once a quiet fishing village, but it now hosts cruise ship passengers several times a week. Olivia is 60. She, her daughter Anna, and three grandchildren live in a wooden house surrounded by a leafy green garden dotted with fruit trees. Anna has a tiny shop at the side of the house, from which she sells various small everyday items and preserves that she and her mother make from the garden fruits. On days when the cruise ships dock, Olivia strolls down to the market shed on the beach with a basket of papaya, roasted peanuts packed up in tiny paper bags, and rolls of cocoa paste made from cacao beans picked in a neighbor's yard, local nutmeg, cloves, and cinnamon. She calls out to the passengers as they are rowed to shore, offering her spices and snacks for sale. In a good week she makes U.S.$50. Her daughter makes about

U.S.$100 a week in the shop and is constantly looking for other ways to earn a few dollars, making necklaces for tourists or taking in washing. Usually the family of five makes do with about U.S.$170 a week (U.S.$8840 per year). From this income they pay rent on the house, the electric bill, and school fees for the granddaughter who will go to high school in the capital next year and perhaps college if she succeeds. They also buy clothes for the children and whatever food (chiefly flour and sugar) they can't grow themselves, which puts them at or above the standard of living of most of their neighbors.

Thirty-year-old Reyhan sits on the dock in Riau, Indonesia, anxiously waiting with several other male and female Indonesians for the boat that will take them across the Strait of Malacca to Malaysia, where they plan to work illegally (Figure 1.28). They will join more than a million fellow Indonesians in Malaysia, some working legally, some not. All are attracted by Malaysia's booming economy, where average wages are four times higher than at home. This is Reyhan's second trip to Malaysia. His first was in 1996, when he signed up for an overseas employment program of the Malaysian government's Ministry of Manpower. Although he was promised a 2-year work visa and a contract to work legally on a Malaysian oil palm plantation, upon arrival Reyhan found that his first 3 months' wages would go toward paying off his boat fare and that his visa was valid for only 2 months. Not one to give in easily, he soon escaped to the city of Malacca, where he secured a construction job earning U.S.$10 a day (about U.S.$2600 a year). On this wage, he was able to send enough money to his wife and two children in Indonesia to pay for their food, school fees, and a new roof for their house. In 1998, however, Reyhan was one of thousands of Indonesians deported by Malaysia, which was suffering growing unemployment due to the Asian financial crisis and wanted

**FIGURE 1.27** Soufrière. From a distance, cruise ship passengers see the village of Soufrière, St. Lucia, as a quaint place. Once ashore, they encounter a small but vibrant entrepreneurial community. [Lydia Pulsipher.]

**FIGURE 1.28** Illegal migrants. These construction workers, male and female, have fled the economic crisis in Indonesia to work illegally in Malaysia. They are under arrest on a Malaysian construction site and will be questioned by immigration officials there before being deported. Notice their faces and try to imagine their worries at this point in their lives. [AP Photo/S. Thinakaran.]

to create more jobs for locals by expelling foreign workers. The crisis affected Indonesia even more severely, and after several years of only part-time farm work, Reyhan is willing to give Malaysia another try.

Tanya is a 50-year-old grandmother with a son in high school and a married daughter. Tanya works at a fast-food store in North Carolina, making less than U.S.$6 an hour. She had been earning U.S.$8 an hour sewing shirts at a textile plant until it closed and moved to Indonesia. Her husband is a delivery truck driver for a snack-food company. Between them, they make $27,000 a year; but from this income they must cover their mortgage and car loan, meet regular monthly expenses for food and utilities, and help their daughter, Rayna, who quit school after eleventh grade and married a man who is now out of work. They and a baby live in the old mobile home at the back of the lot that Tanya and her family once lived in. With Tanya's lower wage ($5000 less a year), there will not be enough money to pay the college tuition for her son, who had hoped to be an engineer and would be the first in the family to go past high school. For now, he is working at the local gas station.

*Adapted from Lydia Pulsipher's field notes, 1992–2000 (Olivia and Tanya), and Alex Pulsipher's field notes, 2000 (Reyhan).* ∎

These people, living worlds apart, are all part of the global economy. Workers around the world are paid at startlingly different rates for jobs that require about the same skill level. But varying costs of living and varying local standards of wealth make the difference between the near poverty and threatened hopes felt by Tanya's family, which actually has the highest income by far, and Olivia's family who, though not well off, do not think of themselves as poor because they have what they need and because others around them live in similar circumstances. However, their livelihoods are similarly precarious. While the globally growing tourist trade promises increased income, it also means dependence on circumstances

beyond their control—in an instant, the cruise-line companies can choose another port of call. Reyhan, by far the poorest, seems trapped by his status as an illegal worker, which robs him of many of his rights. Still, the higher pay that he can earn in Malaysia offers him a possible way out of poverty.

## WHAT IS THE ECONOMY?

Economic geography focuses on how people interact with their environment as they go about earning a living. The **economy** is the forum in which people make their living, and **resources** are what they use to do so. Some resources are tangible materials, such as mineral ores, timber, plants, and soil. Because they must be mined from the earth's surface or grown from its soil, they are called **extractive resources.** There are also **nonmaterial resources,** such as skills and brainpower. Often resources must be transformed to produce new commodities (such as refrigerators or sugar) or bodies of knowledge (such as books or computer software). **Extraction** (mining and agriculture) and **industrial production** are two types of economic activities, or *sectors* of the economy. A third is the **exchange** or **service sector:** the bartering and trading of resources, products, and services. Generally speaking, as people in a society shift from extractive activities, such as farming, to industrial and service activities, their material standards of living tend to rise; but their actual well-being may be compromised by dependence on cash, the need to migrate, urban crowding, or poor working conditions.

The **formal economy** includes all the activities that are recorded as part of a country's official production. Examples from the vignettes at the beginning of this section include Olivia's daughter in her shopkeeper's role, and Tanya, her husband, and their son. All are registered workers who earn recorded wages and pay taxes to their governments. The activity of formal economies is measured by the **gross domestic product (GDP),** a number that gives the total value, in monetary terms, of all goods and services officially recognized as produced in a country during a given year.

Many goods and services are produced outside formal markets, in the **informal economy.** Here, work is often traded for payment other than cash or for cash payment that is not reported to the government as taxable income. Only recently have economists begun paying attention to informal economies, yet it is estimated that one-third or more of the world's work falls into this category. Examples of workers in the informal economy include Olivia when she sells her goods to the tourists; Reyhan when he works illegally in Malaysia; and any members of the three families when they contribute to their own or someone else's well-being through such unpaid services as housework, gardening, and elder and child care. If remittances are sent home through banks or similar financial institutions, they become part of the formal economy of the receiving society. If they are transmitted in pockets or via the mail as cash, or some other "off-the-books" (perhaps illegal) transaction, they are part of the informal economy.

## WHAT IS THE GLOBAL ECONOMY?

The global economy includes the parts of any country's economy that are involved in global flows of resources: extracted materials, manufactured products, money, and people. Most of us participate in the global economy every day. For example, this book was manufactured using paper made from trees cut down in Southeast Asia, North America, or Siberia and shipped to a paper mill in Oregon. It was printed in the United States, but many books are now printed in Asia because labor costs are lower there. Such long-distance

movement of resources and products has grown tremendously in the past 200 years, but it existed at least 2500 years ago, when silk and other goods were traded along Central Asian land routes that connected Rome with China.

Starting in about 1500, long-distance trade entered a period of expansion and change as Europeans began extracting resources from their colonies and organizing systems that processed those resources into higher-value goods to be traded back in Europe or wherever else there was a market. Sugarcane, for example, was grown on Caribbean, Brazilian, and Asian plantations (Figure 1.29), made into crude sugar and rum locally, and then sold in Europe and North America, where further refining took place. The global economy grew as each region produced goods for export, rather than just for local consumption, and became increasingly dependent on imported food, clothing, machinery, energy, and knowledge.

New wealth and ready access to global resources led to Europe's **Industrial Revolution,** a series of innovations and ideas that changed the way goods were produced. One such innovation was the orderly timing and spacing of production steps and the use of specialized labor. Instead of one woman producing the cotton or wool for cloth, spinning thread, weaving the thread into cloth, and sewing a shirt, tasks were spread out among workers, often in distant places, with some specializing in producing the fiber, others in spinning, weaving, or sewing. These innovations were followed by labor-saving improvements such as mechanized reaping, spinning, weaving, and sewing.

This larger-scale mechanized production created a demand for raw materials and a need for markets in which to sell finished goods. European colonies in the Americas, Africa, and Asia provided both. For example, in the British Caribbean colonies, hundreds of thousands of African slaves wore garments made of cotton cloth woven

**FIGURE 1.29** European use of colonial resources. Among the first global economic institutions were Caribbean plantations like Old North Sound on Antigua, shown here in an old painting. In the eighteenth century, thousands of sugar plantations in the British West Indies, subsidized by the labor of slaves, provided huge sums of money to England and helped fund the Industrial Revolution. [Museum of Antigua and Barbuda.]

in England from cotton grown in other colonies in America, Africa, and Asia. The sugar they produced on British-owned plantations was processed in British-made cast-iron industrial equipment and transported to European markets in ships made in the British Isles of trees and resources from various parts of the world.

Until the early twentieth century, much of the activity of the global economy took place within the huge colonial empires ruled by a few European nations. Global economic and political changes brought an end to these empires by the 1960s, and almost all colonial territories are now independent countries. Nevertheless, the global economy continues to grow as the flows of resources and manufactured goods are sustained by private companies, many of which first developed during colonial times. **Multinational corporations** such as De Beers, Shell, Wal-Mart, Bechtel, and Cisco (see Figure 1.11 on page 10) operate across international borders, extracting resources from many places, producing products in factories carefully located to take advantage of cheap labor and transport facilities, and marketing their products wherever they can make the most profit. One of the key characteristics of multinationals is that their global influence, wealth, and importance to local economies enable them to influence the economic and political affairs of the countries in which they operate.

Multinationals are important conduits for the flow of **capital**, or investment money. For example, consider what was once a family-owned Mexico City cement company we will call MEXCRETE. It recently borrowed heavily from international banks to acquire controlling interests in cement industries in Texas, Mexico, the Philippines, and Thailand. The focus of the company is increasingly global, and MEXCRETE hopes to outmaneuver its rivals by using borrowed money to build new cement plants that use cutting-edge technology in order to expand sales to Europe and especially to China, where a building boom has increased the demand for cement.

Such international investment has advantages and disadvantages. In this case, high-quality cement may be delivered more efficiently and cheaply to all the markets served. Competition may spur technological advancement, and jobs will be created in Southeast Asia, Mexico, and Texas and in the markets the company serves. But to maximize its profits, MEXCRETE may not pay its workers enough to live on or provide a healthy workplace, and its cement plants may cause pollution. Competing smaller cement providers may fail, creating local unemployment. By taking the profits home to Mexico, MEXCRETE deprives local economies of capital needed for investment. Moreover, if, because of rising oil prices or other disturbances, demand for concrete slows and MEXCRETE misses payments on the huge debt that it amassed while expanding, its creditors could foreclose, putting thousands of jobs from Texas to Southeast Asia at risk.

## THE DEBATE OVER FREE TRADE AND GLOBALIZATION

**Free trade** is the unrestricted international exchange of goods, services, and capital. Currently, all governments impose some restrictions on trade to protect their own national economies from foreign competition. Restrictions take two main forms: tariffs and import quotas. **Tariffs** are taxes imposed on imported goods that have the effect of increasing the cost of those goods to the consumer. **Import quotas** set limits on the amount of a given good that may be imported over a period of time, thereby increasing the price of the imported good relative to domestic goods.

These and other forms of trade protection are a subject of contention, and views on the value of free trade and the globalization of markets vary widely. Proponents of free trade argue that it encourages efficiency and the production of higher-quality goods and services and gives consumers more product choices at lower prices. With free trade, companies can sell to larger markets and take advantage of mass-production systems that lower costs further. As a consequence, they can grow faster, thereby providing people with jobs and opportunities to raise their standard of living.

Proponents of free trade have been quite successful in recent years, and restrictions on trade imposed by individual countries are being reduced through the formation of regional trade blocs and the support of global free trade institutions. Several **regional trade blocs**—associations of neighboring countries that agree to lower trade barriers for one another—have been formed, including the North American Free Trade Agreement (NAFTA), the European Union (EU), the Southern Common Market (Mercosur) in South America, and the Association of Southeast Asian Nations (ASEAN).

One of the main global institutions that support free trade is the **World Trade Organization (WTO)**, whose stated mission is to lower trade barriers and establish ground rules for international trade. Two other global institutions are the **World Bank** (officially named the International Bank for Reconstruction and Development) and the **International Monetary Fund (IMF)**, both of which make loans to countries that need money to pay for development projects. Before approving a loan, the World Bank or the IMF may require a borrowing country to reorganize its national economy to achieve freer trade. For example, the country may have to reduce and eventually remove tariffs and import quotas and reduce taxes and revenue deficits by cutting government services such as schools, health care, and social services. Such economic reorganization requirements are called **structural adjustment policies (SAPs)**. SAPs have become highly influential and controversial in virtually every region of world. The most detailed explanation of SAPs is given in Chapter 3 (see pages 139–140); please refer to that discussion as you go through the book.

Groups opposed to free trade argue that its gains are offset by the instability—even chaos—that comes with a less-regulated global economy, which is more prone to rapid cycles of growth and decline that can wreak havoc on smaller national economies (Figure 1.30). Labor unions point out that as corporations relocate factories and services to poorer countries where wages are lower, jobs are lost in richer countries. In the poorer countries, multinational corporations often work with governments to prevent workers from organizing themselves into unions that could bargain for **living wages** (wages that support a minimum healthy life). Environmentalists argue that in newly industrializing countries, which often don't have effective environmental protection laws, multinational

**FIGURE 1.30** A demonstration against the World Trade Organization. Korean farmers demonstrate at the December 2005 WTO meeting in Hong Kong. When they were arrested, a hunger strike in their support ensued. [AP Photo/Kin Cheung.]

corporations tend to use highly polluting or unsafe, unsustainable production methods because such methods lower production costs and raise profits. Many fear that a "race to the bottom" in wages, working conditions, government services, and environmental quality is under way as countries compete for profits and for the attention of potential investors.

**Fair trade,** defined as trade that values equity throughout the international trade system, is now proposed as an alternative to free trade. Some profits are sacrificed in order to provide markets for producers from developing countries. For example, "fair trade" coffee and chocolate are now marketed to North American and European consumers at somewhat higher prices with the agreement that the extreme profits of middlemen will be eliminated and growers of coffee and cocoa beans will receive living wages and improved working conditions.

In developing an opinion about free trade and globalization, consider how many of the things you own or consume—computer, clothes, furniture, appliances, car, and foods—were produced in the global economy. These products are cheaper for you to buy, and your standard of living is higher, as a result of lower production costs and competition among many global producers. Also, as a result of free trade within the United States, you can travel unencumbered by border crossings at state lines, and you can hunt for a job and live in any part of the country. On the other hand, you or a relative may have lost a job because a company moved to another location where labor and resources are cheaper. You may be concerned that the products you buy so cheaply were made under harsh conditions by underpaid workers (even children), or that resources were used unsustainably. High levels of pollution may have occurred in the manufacturing and transport processes. Given all these considerations, what might be the advantages and drawbacks of free or fair trade?

## MEASURES OF DEVELOPMENT

Until recently, the term **development** was used to describe only economic changes that lead to better standards of living. These changes often accompany the greater productivity in agriculture and industry that comes from such technological advances as mechanization and computerization. Increasingly, however, the question **"Development for whom?"** is being asked, as it becomes clear that merely raising average national productivity may benefit primarily those who are already economically well off, leaving the majority in circumstances that are little improved, or even worsened. Furthermore, development as measured by average economic gains often results in environmental side effects, such as air and water pollution, that

reduce the quality of life for everyone. Some development experts (for example, the Nobel Prize–winning economist Amartya Sen) are advocating a broader definition of development that includes measures of **human well-being** (a healthy and socially rewarding standard of living), as well as measures of environmental quality.

### GROSS DOMESTIC PRODUCT PER CAPITA

The most popular economic measure of development is **gross domestic product (GDP) per capita,** which is the total value of all goods and services produced in a country in a given year,

TABLE 1.2   Sample human well-being table[a]

| Selected countries (I) | GDP per capita, adjusted for PPP[b] ranking among 177 countries), 2005 (II) | Human Development Index (HDI) ranking among 177 countries,[c] 2005 (III) | Gender Development Index (GDI) ranking among 140 countries, 2005 (IV) | Gender Empowerment Measure (GEM) ranking among 80 countries, 2005 (V) |
|---|---|---|---|---|
| Barbados | 15,720  (36) | 30  (high) | 29 | 25 |
| Japan | 27,967  (13) | 11  (high) | 14 | 43 |
| Kenya | 1,037  (155) | 154  (low) | 117 | ND |
| Kuwait | 18,047  (30) | 44  (high) | 39 | ND |
| Malaysia | 9,512  (54) | 61  (medium) | 61 | 51 |
| United States | 37,562  (4) | 10  (high) | 8 | 12 |

[a] Rankings are in descending order; i.e., low numbers indicate high rank.

[b] PPP = purchasing power parity, figured in 2003 U.S. dollars.

[c] The high, medium, and low designations indicate where the country ranks among the 177 countries classified into three categories by the United Nations.

ND = No data available.

*Sources: United Nations Human Development Report 2005* (New York: United Nations Development Programme), Table 1, "Human development index," Table 25, "Gender-related development index," and Table 26, "Gender empowerment measure," at http:/hdr.undp.org/reports/global/2005/pdf/HDR05_complete.pdf.

divided by the number of people in the country. This figure (Table 1.2, column II) is often used as a crude indicator of how well people are living in a given country. There are, however, several problems with GDP per capita as a measure of overall well-being. First is the matter of wealth distribution. Because GDP per capita is an average figure, it can hide the fact that a country has a few fabulously rich people and a mass of abjectly poor people. For example, a GDP per capita of U.S.$20,000 would be meaningless if a few lived on millions per year and most lived on $5000 per year. Second, the purchasing power of currency varies widely around the globe, so a GDP of U.S.$15,000 per capita in Barbados might represent a middle-class standard of living, whereas that same amount in New York City could not buy even basic food and shelter. Because of these purchasing power variations, in this book we use GDP per capita figures that have been adjusted for **purchasing power parity (PPP)**. PPP is the amount that the local currency equivalent of U.S.$1 will purchase in a given country. For example, according to *The Economist*, on January 12, 2006, a Big Mac at McDonald's in the United States cost U.S.$3.15, in Australia it cost the equivalent of U.S.$2.44, and in China it cost just U.S.$1.30. Of course, for the consumer in China, where per capita incomes average about $4000, this would be a very expensive meal.

A third problem with GDP per capita is that it measures only what goes on in the formal economy, whereas in some places the informal economy actually accounts for more activity. Recently, researchers who examined all types of societies and cultures have shown that, on average, women perform about 60 percent of all the work done on a daily basis, much of it unpaid. Nonetheless, only their paid work performed in the formal economy appears in the statistics. Statistics also neglect the contributions of millions of men and children who work in the informal economy as subsistence farmers, traders, service people, or seasonal laborers.

The most important failing of GDP per capita as a measure for comparing countries is that it ignores all aspects of development other than economic ones. For example, there is no way to tell from GDP per capita figures how fast a country is consuming its natural resources, or how well it is educating its young or maintaining its environment. There are a number of movements to refine the definition of development to include these factors and others. Therefore, along with the traditional GDP per capita figure, we have used several other measures of development in this book: the United Nations Human Development Index (HDI), the United Nations Gender Development Index (GDI), and the United Nations Gender Empowerment Measure (GEM). Together, these measures reveal some of the subtleties and nuances of well-being and make comparisons between countries somewhat more valid. Because these more sensitive indexes are also more complex than the purely economic GDP per capita, they are all still being refined by the United Nations.

## MEASURES OF HUMAN WELL-BEING

***The United Nations Human Development Index (HDI).*** The Human Development Index considers adjusted real income, which takes into account what people can buy with what they earn, as well as data on life expectancy at birth and on educational attainment. Countries are ranked on the HDI factors from highest

(1) to lowest (177). The HDI provides no way to score a country directly on the equality of its distribution of income or purchasing power. These factors are indicated only indirectly by the information about health and education. The assumption is that a country providing widely available health and education services has a more equitable distribution of wealth than a country that provides low levels of access to health care and education. Also, notice that the ranks of some countries based on HDI (see Table 1.2, column III) are quite different from those ranks based solely on GDP per capita (the figures in parentheses in column II). The United States, for example, ranks fourth globally in GDP per capita, but because it falls short in providing equitably for its citizens, it falls to tenth on the HDI. The HDI also classifies all countries into one of three clusters of achievement on the HDI scale: high, medium, or low.

### The United Nations Gender Development Index (GDI).
The Gender Development Index looks at whether countries make basic literacy, health care, and access to income available to both women and men (see Table 1.2, column IV). Because of a lack of data, only 140 countries of the 177 ranked by HDI are in this index; lower numbers represent higher rankings. GDI does not measure general social acceptance of the idea of gender equality, which is better measured by GEM.

### The United Nations Gender Empowerment Measure (GEM).
The United Nations devised the Gender Empowerment Measure to score and rank countries according to how well they enable participation by women in the political and economic life of the country (see Table 1.2, column V). Lower numbers indicate greater empowerment of women. The indicators used include percentage of women holding parliamentary seats; percentage of administrators, managers, and professional and technical workers who are women; and women's GDP per capita. Although the GEM tells us something about the relative power of men and women in a soci-

ety, the indicators now available are unsatisfactory. In most countries, very little data are collected separately on men and women, and in many cases data on women are missing altogether, making comparisons impossible. Because sufficient data were available to rank only 80 of the possible 177 countries in the year 2005, a GEM rank higher than an HDI rank may be merely the result of the small number of countries ranked on GEM relative to HDI. Also, a high rank does not indicate that a country is treating women and men equally, but only that the country is doing better than those ranked lower. Although women are half the world's population, nowhere on earth do women hold 50 percent of the parliamentary seats or have an average earning power even close to that of men (see Table 1.1 on page 18).

### A Sample Human Well-Being Table.
Table 1.2 shows GDP per capita, with rankings, and also HDI, GDI, and GEM rankings for six selected countries: Barbados, Japan, Kenya, Kuwait, Malaysia, and the United States. Notice how the rankings change in the four categories. Kuwait's rank drops significantly from 30th in the GDP per capita column (II) to 44th in the HDI column (III); Japan (and the United States, though to a lesser extent) shows a drastic drop from GDI to GEM rank; Kenya's ranks are consistently low; meanwhile, Barbados, with a relatively low GDP per capita, rises in rank from the GDP (36) to the HDI (30). The higher GDI (29) and GEM (25) ranks for Barbados are less significant because so few countries are ranked. Perhaps you can already suggest some explanations for why some country rankings are similar across categories while others differ significantly.

Each chapter in this book is designed to provide historical, demographic, cultural, social, economic, and political information that will help you to interpret the rankings for the countries in each world region. A human well-being table similar to Table 1.2, with a short discussion of the rankings, will be included in each chapter.

## POPULATION PATTERNS

The changing levels of well-being associated with economic development have had dramatic effects on population patterns, such as growth, decline, and distribution. Because geographers are concerned with the interaction between people and their environments, **demography,** the study of population patterns and changes, is an important part of geographic analysis.

## GLOBAL PATTERNS OF POPULATION GROWTH

It took between 1 million and 2 million years, or at least 40,000 generations, for humans to evolve and to reach a population of

2 billion, which happened in about 1945. Then, in just 55 years, by the year 2000, the world's population more than tripled to 6.1 billion (Figure 1.31). What happened to make the population grow so very quickly in such a short time?

The explanation lies in changing relationships between humans and the environment. For most of human history, fluctuating food availability, natural hazards, and disease kept human death rates high, especially for infants. Periodically, there would even be crashes in human population—as happened in the 1300s throughout Europe and Asia during the pandemic known as the Black Death. An astonishing upsurge in human population began about 1500, at a time when the technological, industrial, and scientific revolutions were beginning in some parts of the world. Human life

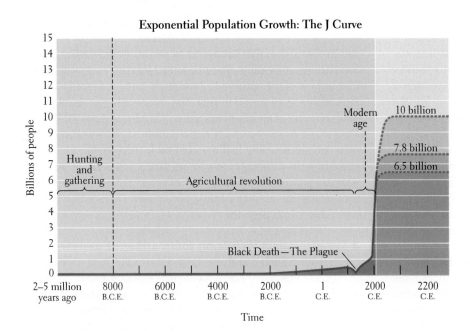

**Exponential Population Growth: The J Curve**

**FIGURE 1.31** Exponential growth of the human population. The curve's J shape is a result of successive doublings of the population: it starts out nearly flat, but as doubling time shortens, the curve bends ever more sharply upward. Note that B.C.E. (before the common era) is equivalent to B.C. (before Christ); C.E. (common era) is equivalent to A.D. (anno domini). [Adapted from G. Tyler Miller, Jr., *Living in the Environment*, 8th ed. (Belmont, Calif.: Wadsworth, 1994), p. 4.]

expectancy increased dramatically, and more and more people lived long enough to reproduce successfully, often many times over. The result was an exponential pattern of growth that is often called a J curve (see Figure 1.31), because the ever-shorter periods between doubling and redoubling of the population cause an abrupt upward swing in the growth line when depicted on a graph.

Today, the human population is growing in virtually all regions of the world, but more rapidly in some places than in others. Even if all couples agreed to have only one child, world population would probably grow to at least 7.8 billion before zero growth would set in simply because there are currently so many people who are about to reach the age of reproduction. Nevertheless, there are indications that the rate of growth is slowing globally. In 1993, the world's population was 5.5 billion and the world growth rate was 1.7 percent; by 2005, the world's population was 6.5 billion, but the growth rate had decreased to 1.2 percent. If present slower growth trends continue, world population may level off at about 7.8 billion before 2050, but this projection is contingent on couples in less-developed countries having access to birth control technology and information on why smaller families might be advisable for them. Nearly 8 billion people will tax the earth's resources beyond imagining, especially if more and more people have lifestyles based on mass consumption, as is increasingly the case.

In a few countries, especially in Central Europe, population is actually declining and aging, due to low birth and death rates. This situation could prove problematic as younger workers become scarce and those who are elderly and dependent become more numerous and pose a financial burden to working-age people. HIV–AIDS is affecting population patterns to varying extents in all world regions. In Africa, the epidemic is severe; several countries are beginning to show sharply lowered life expectancies, and some countries are beginning to show declines in population. The effects of HIV–AIDS are discussed in several world region chapters.

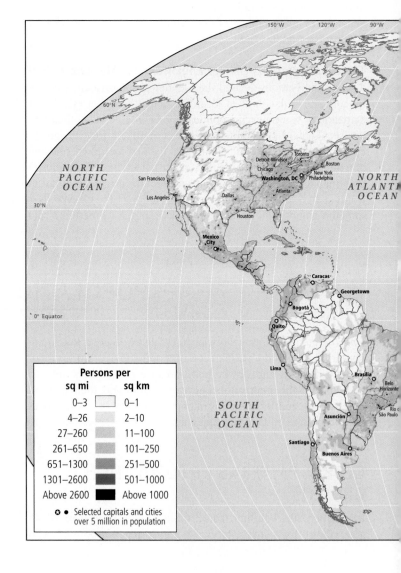

| Persons per | |
|---|---|
| **sq mi** | **sq km** |
| 0–3 | 0–1 |
| 4–26 | 2–10 |
| 27–260 | 11–100 |
| 261–650 | 101–250 |
| 651–1300 | 251–500 |
| 1301–2600 | 501–1000 |
| Above 2600 | Above 1000 |

⊙ ⊗ ● Selected capitals and cities over 5 million in population

# LOCAL VARIATIONS IN POPULATION DENSITY AND GROWTH

If the more than 6.5 billion people on earth today were evenly distributed across the land surface, they would produce an **average population density** of about 113 people per square mile (43.6 per square kilometer). People are not evenly distributed across the face of the earth, however (Figure 1.32). Nearly 90 percent of all people live north of the equator, and most of them live between 20° N and 60° N. Even within that limited territory, people are concentrated on about 20 percent of the available land, mainly in zones that have climates warm and wet enough to support agriculture, along rivers, in lowland regions, or fairly close to the sea. In general, people are located where resources are available.

Nevertheless, many places with meager resources contain a great many people because the resources to support them can be garnered from elsewhere. If people have the means to pay for them, then food, water, and clothing; raw materials for building, manufacturing, and producing electricity; and other material support can all be imported. An extreme example is Macao, the former Portuguese trading enclave in South China. The population density in Macao is the highest on earth: 61,000 per square mile (23,800 per square kilometer). Life there is sustained at a high standard not by local resources, but by global trading connections enhanced by Macao's location on the edge of the huge and populous Chinese mainland. For many centuries, residents of cities around the world have relied on distant resources, and this is increasingly true for more and more of the world's people, no matter where they live.

Just as there is no easy correlation between population density and richness of resources, there is no easy correlation between density and poverty or wealth. Some densely populated places, such as parts of Europe and Japan, are very wealthy, while others, such as parts of India and Bangladesh, are desperately poor. To explain population density patterns today, we must look to cultural, social, and economic factors and to such historical events as experiences with colonialism. We also must identify present

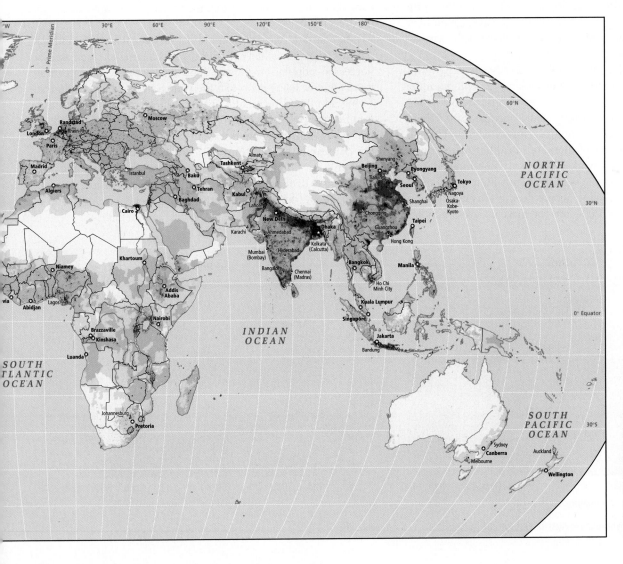

**FIGURE 1.32** World population density. [Data courtesy of Deborah Balk, Gregory Yetman, et al., Center for International Earth Science Information Network, Columbia University; at http://www.ciesin.columbia.edu.]

circumstances that attract migrants to some places and cause them to flee others.

Usually, the variable that is most important for understanding population growth in a region is the **rate of natural increase** (often shortened to **growth rate**). The rate of natural increase is the relationship between the number of people being born (**birth rate**) and the number dying (**death rate**) in a given population, without regard to the effects of migration. The rate of natural increase is usually expressed as a percentage. For example, in Austria (in Europe), which has 8,200,000 people, the annual birth rate in 2005 was 10 per 1000 people, and the death rate was 9 per 1000 people. Therefore, the annual rate of natural increase is 1 per 1000 (10 − 9 = 1), or 0.1 percent. Just 16 percent of Austrians are under 15 years of age, and 15 percent are over 65 years of age. For comparison, consider Jordan (in Southwest Asia), which has 5,800,000 people. The birth rate is 29 per 1000 and the death rate is 5 per 1000. The annual rate of natural increase is thus 24 per 1000 (29 − 5 = 24), or 2.4 percent per year. At this rate, Jordan will double its population in just 29 years. As you might expect, the population of Jordan is very young: 37 percent are under 15 years of age, and just 3 percent are over 65. **Total fertility rate (TFR)** is another term used to indicate trends in population. TFR is the average number of children women in a country are likely to have at the present rate of natural increase.

Migration by refugees and workers can also be a powerful contributor to population growth, and Jordan is a good example. In 1948, when the state of Israel was created from land belonging to the Palestinians, Jordan had just 500,000 people. Since then, Jordan has received at least 1.7 million refugees who fled Israeli-controlled territory and tens of thousands of others who voluntarily moved to Jordan to work and live. It is estimated that Palestinians and their descendants now constitute one half of the Jordanian population, while people fleeing wars in Lebanon (1975–1991) and Iraq (1991; 2003–present) make up at least another

1.5 million. The result is that Jordanians of native descent, now about 1.4 million, are a minority in their own land—76 percent are foreign born. For comparison, less than 9 percent of Austria's population are foreign born.

## AGE AND GENDER STRUCTURES

The **age distribution** (also known as the **age structure**) of a population is the proportion of the total population in each age-group, and the **gender structure** is the proportion of males and females in each age-group. The age and gender structures reflect past and present social conditions, and knowing these structures helps us to predict future population trends.

The **population pyramid** is a graph that depicts age and gender structures. Careful study of the population pyramids of places such as Austria and Jordan (Figure 1.33) reveals the age and gender distribution of the populations of those countries. As we have noted, nearly 40 percent in Jordan are very young, so they are clustered toward the bottom of the pyramid, with the largest groups in the age categories 0 through 9. The pyramid tapers off quickly as it rises above age 34, showing that in Jordan most people die before they reach old age.

In contrast, Austria's pyramid has an irregular shape, indicating that Austria has had experiences that alternately increased or decreased population growth. The narrow base indicates that there are now fewer people in the younger age categories than in young adulthood or middle age, and those over 70 greatly outnumber the youngest (ages 0 to 4). This age distribution tells us that many Austrians now live to an old age and illustrates that in the last several decades Austrian couples have been choosing to have only one child, or none. Austrians worry that their population will begin to decline and be weighted with elderly people who will need care and financial support from an ever-declining group of working-age people.

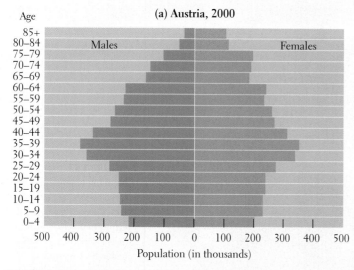

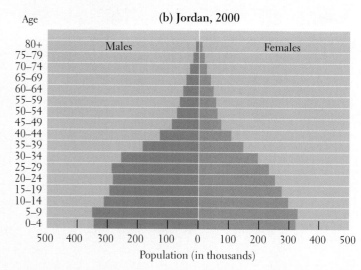

**FIGURE 1.33** Population pyramids for Austria and Jordan. [Adapted from U.S. Bureau of the Census, International Data Base (IDB), at http://www.census.gov/ipc/www/idbpyr.html.]

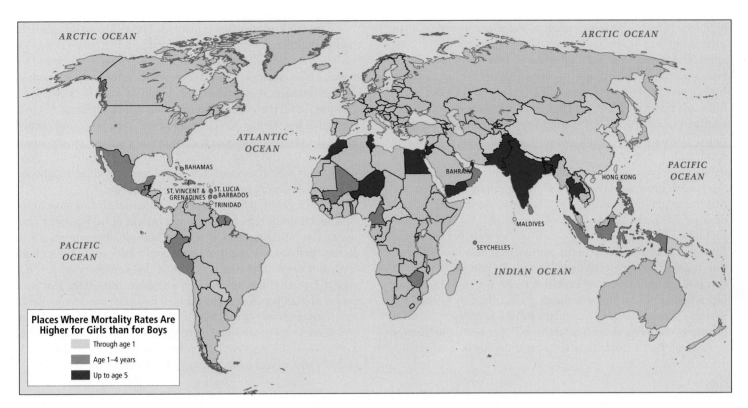

**FIGURE 1.34 Young female mortality rates.** In the countries shown in shades of purple, the mortality rate for girls is abnormally high. The darker the color, the longer the risk to girls lasts. [Adapted from Joni Seager, *The State of Women in the World Atlas* (New York: Penguin, 1997), p. 35.]

Population pyramids also reveal subtle gender differences within populations. Look closely at the right (female) and left (male) halves of the pyramids in Figure 1.33. In several age categories the sexes are not evenly balanced on either side of the line. In the Austrian pyramid, there are more women than men near the top (age 70 and older), reflecting the deaths among male soldiers in World War II and the as yet poorly understood trend in wealthy countries for women to live about 5 years longer than men. In Jordan, the gender imbalances occur at younger ages; for instance, there are about 230,000 women and about 290,000 men age 25 to 29.

Because gender-based research is so new, explanations for the imbalances are only now being proposed. One possible explanation is the worldwide cultural preference for sons that we noted earlier in this chapter. The normal ratio is for about 95 females to be born for every 100 males. Because boy babies are somewhat weaker than girls, the ratio evens out naturally within the first 5 years. Over the last 100 years, the ratio of females to males at birth has declined further to 92.6 females to 100 males globally and as low as 80 females to 100 males in parts of South and East Asia. The widespread preference for boys over girls appears to increase as couples choose to have fewer children. Some female fetuses are never born because they are aborted. Other girl babies die because, in societies afflicted by poverty, girls are sometimes fed less well than boys and receive less health care; hence they are more likely to die in early childhood (Figure 1.34).

Even surviving females may be invisible in statistics. For example, research has shown that around the world, in every country from Sweden to Swaziland, females benefit less from development than do males. But development statistics rarely show gender differences, so the lesser well-being of females is not apparent. Another aspect of the "missing female" problem is found in global statistics related to work. Almost universally, women's contributions as subsistence farmers, as homemakers, as domestic servants, as child-care providers, as volunteers, and as workers elsewhere in the informal economy have not been considered productive labor, so much of women's work has not been calculated into national income statistics. Even though the oversight has been documented for more than 25 years, the customs of statistics gathering are only beginning to change, and the missing female phenomenon continues.

## POPULATION GROWTH RATES AND WEALTH

Although there is a wide range of variation, regions with slow population growth rates usually tend to be affluent, and regions with fast growth rates tend to have widespread poverty. The reasons are complicated; again, Austria and Jordan serve as examples.

Austria has a GDP per capita (PPP) of $30,094. Its highly educated population is 100 percent literate and is employed largely in high-tech industry, business, research, teaching and writing, manufacturing, and upscale tourism. The large amounts of time, effort, and money required to prepare a child to compete in this economy cause many couples to choose not to have any children. Jordan, on the other hand, has a GDP per capita (PPP) of $4320. Although many people in Jordan live in urban areas (and 97 percent of the recorded GDP comes from industry and services), much everyday work is still done in the informal economy by hand, and each new child is seen as a potential contributor to the family well-being and income and as an eventual caregiver when the parents grow old. Also, not producing enough children who will live to adulthood is still a significant worry, because 22 children per 1000 die before they are 5 years old. (In Austria, only 5 children per 1000 die before they are 5.) It is not surprising, then, that only 56 percent of the women in Jordan use birth control (more than 67 percent do in Austria), in part because in Jordan birth control is less accessible, but also because the chance of losing a child to illness is much greater than in Austria. But the situation is changing in Jordan. Just 25 years ago, the GDP per capita was $993, infant mortality was 77 per 1000, and the average woman had 8 children, whereas now she has on average only 3.7 children.

The kinds of changes evident in Jordan are taking place in many countries. As agricultural production is mechanized, there is less need for agricultural labor, and, instead, there is a demand for fewer, well-trained specialists. Furthermore, **subsistence economies**—circumstances in which a family produces most of its own food, clothing, and shelter—are losing their appeal because cash is needed to buy such goods as television sets, blue jeans, and canned goods. In **cash economies,** children become a drain on the family income. Each child must be educated in order to qualify for a good cash-paying job. There are often school fees, and children who are in school all day are not available to earn money.

Until recently, it took people one full generation or more to see that, when death rates fall, there is a benefit to having fewer children. Now birth rates as well as death rates are dropping quickly in Jordan, perhaps as a result of better public education. When both birth and death rates in a region slow, demographers say that the region has gone through the **demographic transition:** that is, a period of high population growth rates has given way to a period of much lower (or no) growth rates, at the same time that major social and economic changes are taking place within the society (Figure 1.35). Given the information you have here, would you say that, as of 2005, Jordan had passed through the demographic transition?

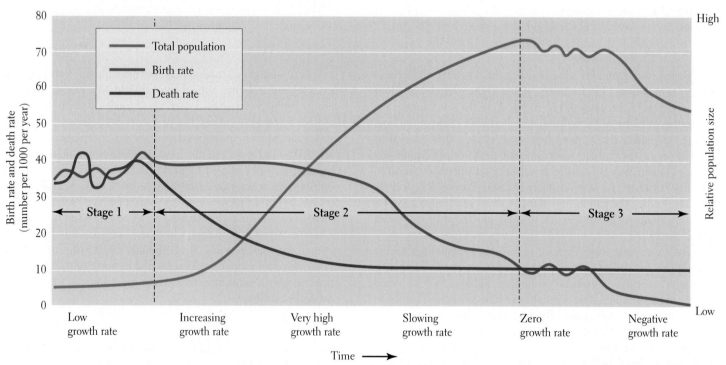

**FIGURE 1.35 The demographic transition.** In traditional societies (Stage 1), both birth rates and death rates are usually high (left vertical axis), and population numbers (right vertical axis) remain low and stable. With advances in food production, education, and health care (Stage 2), death rates usually drop rapidly, but strong cultural values regarding reproduction remain, often for generations, so birth rates drop much more slowly, with the result that for decades or longer, the population continues to grow significantly. When changed social and economic circumstances enable most children to survive to adulthood and it is no longer necessary to produce a cadre of family labor (Stage 3), population growth rates slow and may eventually drift into negative growth. At this point, demographers say that the society has gone through the demographic transition. [Adapted from G. Tyler Miller, Jr., *Living in the Environment*, 8th ed. (Belmont, Calif.: Wadsworth, 1994), p. 218.]

# HUMANS AND THE ENVIRONMENT

The long, continuous give-and-take between humans and their physical environment has resulted in drastic changes in the circumstances of human life and enormous impacts on physical environments. Some of the human effects on the environment are now regarded as harmful (and perhaps irreversible), including ozone depletion, soil erosion, global warming, acid rain, deforestation, and water pollution. Mass consumption of resources tends to alter the environment most profoundly, but all human ways of life—whether on small cultivation plots in the tropics or in huge industrial cities—have environmental effects. Figure 1.36 shows the impact of humans on the planet's land surface. The intensity and nature of the impact varies greatly, but human impact can be found virtually everywhere.

Mounting awareness of environmental impacts has prompted numerous proposals to limit the damage, from relying on technological advances to reducing resource consumption. Halting and reversing environmental damage may be the greatest challenge our species has yet faced.

## SUSTAINABLE DEVELOPMENT

The United Nations defines **sustainable development** as the effort to improve present living standards in ways that will not jeopardize those of future generations. By destroying resources (as in deforestation) or poisoning them (as in pollution of water and air), we may be depriving future generations of the resources they will need for their well-being. Those who support the idea of sustainability hope to promote more environmentally and socially friendly development for all, but especially for the vast majority of earth's people who do not yet enjoy an acceptable level of well-being (Figure 1.37 on page 44). Strategies are being devised in all parts of the world to create more sustainable ways of life, but no country or organization has yet devised a comprehensive workable policy.

Geographers who study the interactions among development, human well-being, and the environment are called **political ecologists.** They often apply a question that we introduced earlier—"Development for whom?"—as they examine how the power relationships in a society affect the ways in which development proceeds, whose needs it addresses, and how success is measured. For instance, in a Southeast Asian country, the clearing of forests to grow oil palm trees might earn profits for the growers and raise tax revenue for the government through the sale of palm oil. However, this must be balanced against the resulting loss of forest resources and soil fertility when a single plant species replaces a multispecies forest, or the destruction of a way of life when forest dwellers are forced to migrate to cities, where their woodland skills are useless. If this same country measured its development by improvements in average human well-being and in environmental quality, and in the potential for sustaining those improvements

into the future, it would soon see that oil palm plantation development would fail to meet both its development and sustainability goals. The answer to "Development for whom?" would be that only a few would benefit, whereas the majority of citizens, the environment, and future generations would lose.

Developing sustainably is not easy. Robert Prescott-Allen, a Canadian specialist in sustainability, found that the countries that provide the highest overall standard of living for their citizens (Sweden, Finland, Norway, and Iceland) achieve this standard only at the expense of the environment. The following sections examine some of the issues of sustainability in agriculture, industry, and urban growth, and the issue of global warming.

## Sustainable Agriculture

Farming that meets human needs without poisoning the environment or using up water and soil resources is called **sustainable agriculture.** This term is related to **carrying capacity,** which refers to the maximum number of people a given place can support sustainably. Technology has increased food production on earth remarkably, especially over the last several decades. In the 25 years between 1965 and 1990, total food production rose between 70 and 135 percent, depending on the region. But population also rose quickly during this period, so the gains were much less per capita, and in Africa, per capita food production actually decreased. By 2004, it appeared that the pattern of growth in global agricultural production was slowing, but overall the global system was probably capable of producing more than enough food for all. However, according to the United Nations Food and Agriculture Organization (UNFAO), while undernutrition had fallen in Asia and Latin America by 2004, it rose in Southwest Asia and North Africa and in sub-Saharan Africa (Figure 1.38 on page 44). For a variety of logistical and financial reasons, the food was not getting to the hungry.

Sustainable agriculture requires that food production meet the needs of everyone on the planet. However, one-fifth of humanity subsists on a diet too low in total calories and vital nutrients to sustain adequate health and normal physical and mental development. At the same time, wealthy countries have experienced surplus food production that, when sold on the world market, has depressed global prices and contributed to farm failures even in poor countries where food is scarce. Hence, the problem of hunger, so far, is not that the world's agricultural systems cannot produce enough, but that farmers in many places have been forced out of production, often by "free market" forces. In addition, political, economic, and environmental crises can prevent existing food from reaching hungry people. Many experts now agree that the most promising solution to hunger seems to be for individual countries and regions to develop their own plans for sustainable agricultural development. Wealthier countries should help when asked, but in ways that will encourage sustainability, not dependence. In major food-exporting

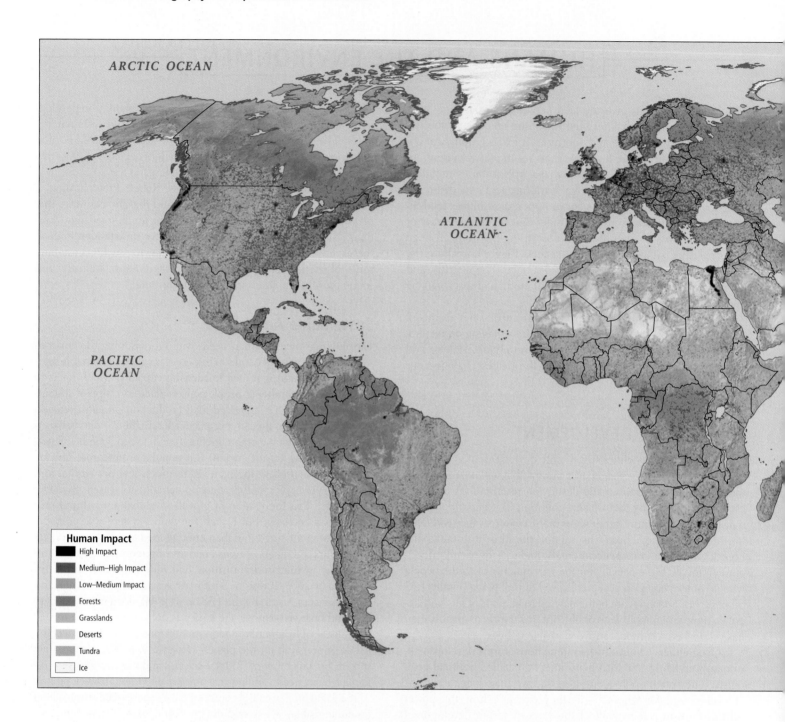

**Human Impact**

- High Impact
- Medium–High Impact
- Low–Medium Impact
- Forests
- Grasslands
- Deserts
- Tundra
- Ice

countries such as the United States, that may mean cutting production to avoid glutting the world market and hurting small producers.

Just how sustainable are the world's present agricultural systems? The answer is unclear, but scientists from many disciplines think the world will reach the limit of its carrying capacity within the next 50 years because, as populations grow, environmental problems proliferate. There is some hope that technological advances, such as advanced irrigation techniques, will make present agricultural land more productive and unused land useful. However, previously unrecognized side effects of agricultural development are

just coming to light. Many of the most agriculturally productive parts of North America, Europe, and Asia have already suffered moderate to serious losses of soil through erosion. Globally, soil degradation and other problems related to food production affect about 7 million square miles (2000 million hectares), putting the livelihoods of a billion people at risk.

The main causes of soil degradation are overgrazing, deforestation, and mismanagement of farmland through the overuse of irrigation and chemicals. Irrigation often makes soil salty and infertile over time, and it can deplete water resources by using them at

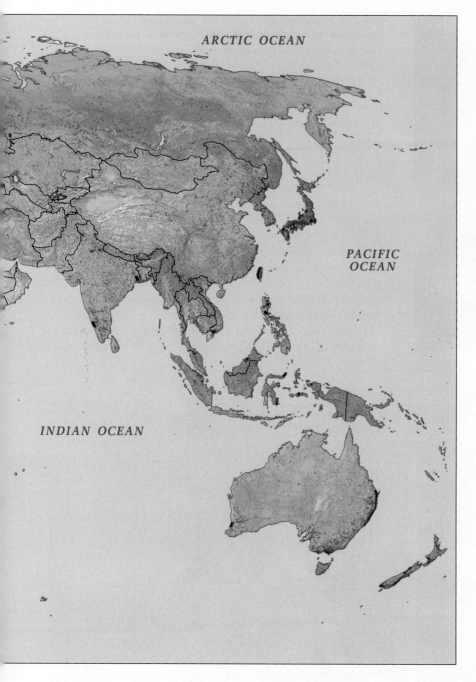

ARCTIC OCEAN

PACIFIC
OCEAN

INDIAN OCEAN

**FIGURE 1.36** Human impact on the earth, 2002. Variables used to assess human impact on the planet include land areas that have been urbanized; land areas that have been converted to other uses (agriculture, mining, forestry, transportation, communication, and energy flow systems, and so forth); land areas that are fragmented and undergoing conversion; and areas with relatively intact ecosystems. The impacts depicted here are derived from a synthesis of hundreds of studies and adjusted according to geographical region, land cover, and climate. To see details of human impact, go to the Web site below, download the poster of World, Robinson projection pdf, and zoom in on the place of interest. [Adapted from *United Nations Environment Programme, 2002, 2003, 2004, 2005, 2006* (New York: United Nations Development Programme), at http://maps.grida .no/go/graphic/human_impact_year_1700_approximately and http://maps.grida.no/go/graphic/human_impact_year_2002.]

too fast a rate. Modern agricultural techniques pioneered in the United States and now spreading throughout the world—such as the use of fertilizers, pesticides, and herbicides—have caused massive die-offs of birds, insect pollinators, fish, and other life-forms.

Few truly sustainable agricultural solutions have been developed so far. On the one hand, mass production almost always results in environmental problems and in misallocations. On the other hand, small-scale production that may be more environmentally sound and able to meet local needs cannot meet the huge food demands of growing urban populations. The agricultural suc-

cesses and failures of various countries and agencies will be examined further in the chapters on world regions.

## Sustainability and Urbanization

The world is fast becoming urbanized. In 1700, less than 10 percent of the world's total population—about 7 million people— lived in cities, and only 5 of those cities had populations as high as several hundred thousand people. By 2005, 47 percent of the world's population lived in cities, and there were more than

**FIGURE 1.37** Sustainable development. The Rainforest Alliance has certified hundreds of Colombian coffee growers like this one for their environmentally sustainable and socially responsible farming practices. Such practices include providing good working conditions and sustainable livelihoods for workers. [www.rainforest-alliance.org.]

400 cities of over 1 million and about 25 cities of over 10 million. Although many people anticipate enjoying city life, it is the elusive ideal of life in a modern, wealthy city that they cherish. Most urban dwellers today cope with far more difficult realities.

Water and sanitation are cases in point. In rapidly growing cities in Asia, Africa, and Latin America, widespread poverty means that water is acquired with a pail from a communal faucet (Figure 1.39). Usually this water should be boiled before use, even for bathing, and truly safe drinking water must be purchased. But the poorest urban inhabitants, who are often a majority of the population, lack the money to buy clean water or the fuel to boil it. As a result, adults are often chronically ill, and waterborne diseases cause many children to die before the age of 5.

Urban housing is often self-built of scavenged materials, and sanitation systems are absent. Sewage and other wastewater are

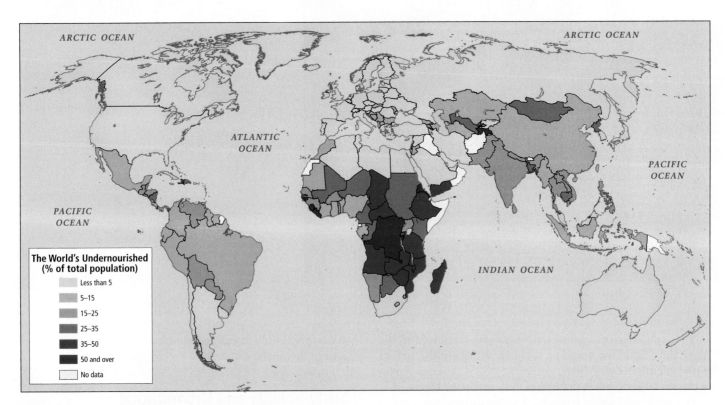

**FIGURE 1.38** Undernourishment in the world. The proportion of people suffering from undernourishment—lack of adequate nutrition to meet their daily needs—has declined in the developing world over the past several years. However, an estimated 824 million people were still affected by chronic hunger in 2003. As you can see from the map, people in much of Africa, parts of Central Asia (probably including Afghanistan, if data were available), Mongolia, Bangladesh, North Korea, Cambodia, Nicaragua, and parts of the Caribbean suffer the worst. To see a country-by-country animated map of undernourished populations between 1970 and 2003, go to http://www.fao.org/es/ess/faostat/foodsecurity/FSMap/flash_map.htm. [Adapted from the Food and Agriculture Organization of the United Nations at http://www.fao.org/es/ess/faostat/foodsecurity/.]

**FIGURE 1.39** Lack of safe water. In sections of Kolkata (Calcutta), India, the only available public water supply comes from small pipes located at curbsides. Here a woman (left center) fills buckets for washing dishes and clothes from a public spigot (under her right hand). Notice the entire context of the photo, including what other people are doing. [Dilip Mehta, Contact/The Stock Market.]

often pumped into a nearby river, swamp, or ocean, even from modern high-rise apartments and luxury hotels. This method of waste disposal causes serious health hazards and widespread ecological damage.

Building adequate wastewater collection and water purification systems in cities already housing several million inhabitants is so costly that this option is rarely considered, yet new affordable sources of clean water are not available. As we will see in later chapters, clean water is only one of several problems faced by the world's urban dwellers. Technological and other advances will no doubt help alleviate some problems, but we are now far short of the sustainable ideal for the world's cities.

## Changing Patterns of Resource Consumption

As people move from rural agricultural work to industry or service sector jobs in cities, they begin to use more resources per capita, and they draw their resources from a wider and wider area. Water once fetched from nearby village wells may now be piped into urban apartments or shantytowns from hundreds of miles away. Clothing once made laboriously by hand at home is now purchased from manufacturers half a world away. Consumers in wealthy countries may have access to a variety of imported products at lower prices than they would pay for locally produced items. But there is a downside to accessing a wide net of resources: the tendency to overconsume. When water is piped into the house or yard, its very convenience encourages waste. When clothes can be purchased at very low prices, closets become crowded. In fact, resource use has become so skewed toward the affluent with cash to pay that, in any given year, the relatively rich minority (about 20 percent of the world's population) consumes more than 80 percent of the available world resources, leaving the poorest 80 percent of the population with less than 20 percent.

In 1973 the economist E. F. Schumacher commented in his book *Small Is Beautiful* that "the problem passengers on spaceship Earth are the first class passengers." The point is still valid: the wealthiest 20 percent of the world's population produces close to 90 percent of the world's hazardous wastes, and this same group also consumes well over 50 percent of the world's fossil fuel, metal, and paper resources. But this observation does not recognize the full complexity of the situation. Although the people of the developing world currently consume much less as individuals than do those of the developed world, the developing world is already suffering from environmental deterioration. As populations in these countries increase, and as their consumption per capita grows closer to consumption rates in the developed world, negative environmental effects will also increase proportionately. A case in point is global warming.

## GLOBAL WARMING

The term **global warming** refers to the observed warming of the earth's surface and climate. The burning of coal and oil, and other human activities, have increased atmospheric levels of carbon dioxide ($CO_2$), methane, water vapor, and other gases above natural levels. These gases are collectively known as "greenhouse gases" because their presence allows large amounts of heat from sunlight to be trapped in the earth's atmosphere in much the same way that heat is trapped in a greenhouse or a car parked in the sun.

Greenhouse gases exist naturally in the atmosphere. In fact, it is their heat-trapping ability that makes the earth warm enough for life to exist. Increase their levels, as humans are doing now, and the earth becomes warmer still.

Over the last several hundred years, humans have greatly intensified the release of greenhouse gases. Industrial processes and transport vehicles burn large amounts of $CO_2$-producing fossil fuels such as coal and oil, and cash-crop agriculture uses nitrogen-based fertilizers. Even the large-scale raising of grazing animals contributes methane through flatulence. Unusually large quantities of greenhouse gases from these sources are accumulating in earth's atmosphere, and their presence has already led to significant warming of the planet's climate.

At the same time, deforestation is removing growing plants (particularly trees) from many landscapes around the world, especially in developing countries. Living forests are a reservoir of carbon, which is continuously taken up, processed, and released back into the atmosphere by the plants. When trees are cut down and their biomass used for fuel or other purposes, the carbon reservoir is converted, either immediately or eventually, into carbon dioxide. Scientists at the World Resources Institute in Washington, D.C., say that as much as 30 percent of the buildup of $CO_2$ in the atmosphere results from the loss of trees and other forest organisms.

Climatologists and other scientists are documenting long-term global warming and cooling trends by examining evidence in tree rings, fossilized pollen and marine creatures, and glacial ice. These data indicate that the twentieth century was the warmest in 600 years and that the decade of the 1990s was the hottest since the late nineteenth century. It is estimated that, at present rates of emissions, average global temperatures could rise between 2.5°F and 10°F (about 2°C to 5°C) by 2100, but it is not clear just what the consequences of such a rise in temperature will be. Certainly the effects will not be uniform across the globe.

One prediction is that glaciers and the polar ice caps will melt, causing a corresponding rise in sea level. In fact, this phenomenon is already observable (Figure 1.40). Satellite imagery analyzed by scientists at the National Aeronautics and Space Administration (NASA) shows that, in just 26 years, between 1979 and 2005, the polar ice caps shrank by about 23 percent, releasing meltwater into the oceans. If this trend continues, at least 60 million people in coastal areas and on low-lying islands could be displaced by rising sea levels. Scientists also forecast a shift of warmer climate zones northward in the Northern Hemisphere and southward in the Southern Hemisphere, a pattern that is already discernible: there have been sightings of robins in Alaska, and the range of the mosquito that carries the West Nile virus is spreading to the north and south. Such climate shifts might lead to the displacement of huge numbers of people, because the zones where specific crops can grow would change dramatically. Animal and plant species that cannot adapt rapidly to the change will disappear. Another effect of global warming could be a shift in ocean currents, resulting in more chaotic and severe weather, such as hurricanes, and possible changes in climate for places such as western Europe.

Most of the responsibility for $CO_2$ emissions measured in tons produced rests with the industrialized countries of the United States, parts of Europe, and Russia; China and India are also major emitters (Figure 1.41a). But notice that when tons *per capita* is the measurement (Figure 1.41b), the geographic patterns change in important ways: North America, Australia, New Zealand, and Russia, plus some of the Arab states, produce the most per capita. For the

(a) The Muir Glacier in August 1941

(b) The Muir Glacier in August 2004

**FIGURE 1.40** Effects of global warming: The Muir Glacier in Alaska. Both photos were taken from the same vantage point. Geologist Bruce Molnia, with the U.S. Geological Survey, reports that the glacier retreated 7 miles (12 km) and thinned more than 875 yards (800 m). [National Snow and Ice Data Center: (a) W. O. Field; (b) B. F. Molnia.]

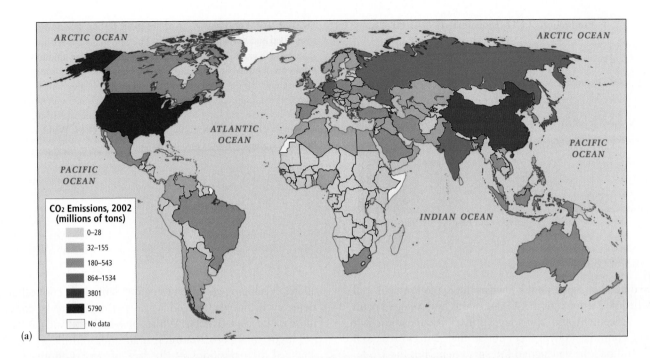

(a)

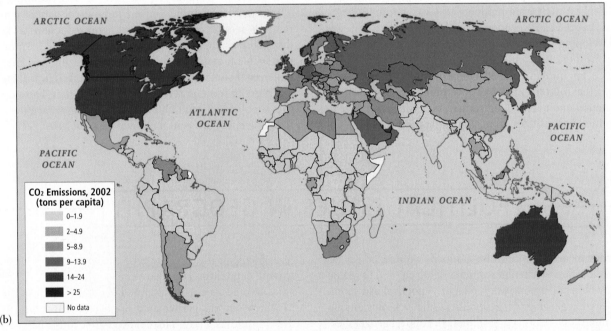

(b)

**FIGURE 1.41** Carbon dioxide emissions around the world. (a) Total emissions by tons, 2002. The United States leads the entire world (23.32 percent of the world total), with China second (15.28 percent). (b) Tons of emissions per capita. Rates are very high for Arab states (for example, Qatar emits 41 tons per capita). Some other high-emitting countries are the United States, 20 tons per capita; Australia, 17.2 tons per capita; and Canada, 16.5 tons per capita. Note that by this measurement, China (3 tons per capita) is not a leading emitter. [Adapted from World Resources Institute at http://cait.wri.org/cait .php?page=yearly&mode=view&sort=val-desc&pHints=shut&url=form&year= 2002&sector=natl&co2=1.]

period 1859–1995, developed countries produced roughly 80 percent of the greenhouse gases from industrial sources, and developing countries produced 20 percent. But by 2000, the developing countries were catching up, accounting for nearly 30 percent of total $CO_2$ emissions. As developing nations industrialize over the next century and continue to cut down their forests, they will release more and more greenhouse gases every year (Figure 1.42). If present patterns hold, greenhouse gas contributions by the developing countries will *exceed* those of the developed world by 2040.

In 1992, members of the **Organization for Economic Cooperation and Development (OECD)**, which includes the highly industrialized countries of North America, Europe, East Asia, and

Production of $CO_2$ Emissions, 1990–2100

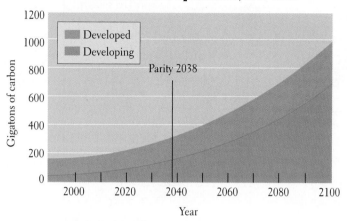

**FIGURE 1.42** Contributions to atmospheric carbon dioxide by developed and developing countries. The carbon dioxide releases shown in this graph include both industrial emissions and amounts released as a result of deforestation. When both these sources are taken into account, the developing countries will exceed the developed countries in $CO_2$ production after 2038. [Adapted from Duncan Austin, José Goldemberg, and Gwen Parker, "Contributions to climate change: Are conventional metrics misleading the debate?" *Climate Notes* (World Resource Institute Climate Protection Initiative) (October 1998).]

Oceania, drafted an agreement known as the Kyoto Protocol, calling for scheduled reductions in $CO_2$ emissions by developed countries. The agreement also encourages OECD cooperation with developing countries to help them curtail their emissions as well. By 1997, 160 nations had signed the agreement, but the United States, the highest per capita producer of $CO_2$, refused to sign, arguing that it would cause economic hardship for U.S. businesses and that developing countries should first be required to reduce their emissions more stringently.

Evidence for human-induced global warming is now sufficiently strong that most scientists and government leaders acknowledge the phenomenon. The disagreement now hinges on how to respond. In the United States, some argue that the Kyoto Protocol allows developing countries too much latitude in controlling emissions, even though their emissions per capita are far lower than those in the United States. While some in the oil and gas industries still argue that global warming is an unproved theory, most have changed their positions in the face of overwhelming evidence. Those in the scientific and environmental activist communities not only see the currently agreed-upon reductions as far too low, they call for stepped-up energy conservation and for more research into such alternatives as solar, wind, and geothermal energy, noting that even if comprehensive measures are taken, it will take perhaps a century or more to reverse the present trends. Developing countries such as India and China are beginning to face the reality that reducing emissions is in their own best interests.

# POLITICAL ISSUES IN GEOGRAPHY

Political issues in geography revolve primarily around power: its exercise, its allocation to different segments of society, and its spatial distribution within and among world regions. A political issue commonly discussed in geography is **geopolitics**—the jockeying among countries for territory, resources, or influence. Political geographers also look at how these issues are addressed at smaller scales such as local and state governments, and how international organizations and social movements play a role in the allocation of power.

## GEOPOLITICS

Geopolitics encompasses the strategies that countries use to ensure that their own interests are served in their relations with other countries. Geopolitics typified the **cold war era,** the period from 1946 to the early 1990s when the United States and its allies in Western Europe faced off against the Union of Soviet Socialist Republics (USSR) and its allies in Eastern Europe and Central Asia. Ideologically, the United States promoted a version of free market capitalism and democracy, whereas the USSR and its allies favored a centrally planned economy and a socialist system in which citizens participated in government indirectly through the Communist Party. The cold war grew into a race to attract the loyalties of unallied countries and to arm them. Sometimes the result was that very unsavory dictators were embraced as allies by one side or the other. Eventually, the cold war influenced the internal and external policies of virtually every country on earth, often oversimplifying complex local issues into a contest of democracy versus communism.

In the post–cold war period of the 1990s, geopolitics shifted: the Soviet Union dissolved, creating many independent states, nearly all of which began to implement some democratic and free market reforms. Globally, countries jockeyed for a better position in what appeared to some to be a new era of trade and amicable prosperity, rather than war. But throughout the 1990s, while the developed countries enjoyed unprecedented prosperity, there were many unresolved political conflicts in Asia, Africa, the Middle East, and Europe, which too often erupted into bloodshed and **genocide,** the systematic attempt to kill all members of an ethnic or religious group. The most worrisome of these conflicts was the long-standing dispute between Israel and the Palestinian people, which had raged with varying intensity since 1948, when Israel was created on Palestinian lands. The terrorist attacks on the United States on September 11, 2001, ushered in a new geopolitical era that is still evolving. Because of the size and the geopolitical power of the United States, the attacks and the U.S. reactions to them affected virtually all international relationships, public and private, creating adjustments that are directly or indirectly affecting the daily lives of billions of people around the world. We will discuss the ramifications of these various geopolitical changes in the regional chapters, particularly Chapter 2 and Chapter 6.

## NATIONS AND BORDERS

In this book, you will see regions defined primarily as a set of countries. A **country** is a common unit of geographic analysis, and the boundaries of countries are a major feature of the maps in this book. Why a country has its particular boundaries is a complex subject; the following discussion addresses some of the determining factors.

The eighteenth-century French writer Jean-Jacques Rousseau believed that people realize their true potential by joining together to create a cohesive nation-state anchored to a particular territory. According to this view, a **nation,** per se, is not a political unit or an official country, but a group of people who share language, culture, and political identity: the Cherokee Nation, for example, or the Palestinians. Once those people formally establish themselves as occupying a particular territory and become a country, they are a **nation-state.** Originally the concept of the nation-state was closely linked to homogenous culture (like Japan or Slovenia, which have only tiny minorities), but now the term is more commonly synonymous with country.

Ideas about the nature of **national identity**—the feeling of strong allegiance to a particular country, often expressed as nationalism—spread from France to the rest of Europe, the Americas, and beyond. Cultural diversity usually prevented the establishment of culturally homogenous nation-states in the former colonies of Europe, but eventually nearly all became independent countries. Some of these new countries that emerged from colonies were **pluralistic states,** such as Indonesia or Trinidad and Tobago, in which power was shared (not always equally) among several distinct culture groups. Often one group managed to monopolize power, leaving others at a disadvantage. As countries were created where none had existed before, indigenous peoples often found that the territory they had traditionally occupied was claimed by the new countries, leaving them without a homeland, relocated to a tiny section of their original territory, or forced to migrate to entirely new territory. Such was the case with almost all the indigenous peoples of North, South, and Middle America and Australia and New Zealand. In many of these cases, and in others around the world, indigenous groups remain in conflict with national governments over rights to territory.

The emergence of the idea of the nation-state was linked with the concept of **sovereignty,** which means that a country is self-governing (though not necessarily according to democratic principles) and can conduct its internal affairs as it sees fit without interference from outside. Sovereignty increased the importance of precise legal boundaries as demarcations of power and control. A country's borders mark the extent of its territory and hence its sovereignty. When we look at a map, we often are viewing a spatial representation of the outcomes of struggles over control of territory; this is why political maps can change over time.

## INTERNATIONAL COOPERATION

The idea of the nation-state has been recognized as deeply flawed, and there is reason to think that the idea of national sovereignty is also becoming less viable. Increasingly, people, goods, and capital are flowing freely across national borders, and this trend favors interdependence over national sovereignty. It also creates a need for some way to enforce laws governing business, trade, and human rights at the international level. When extreme human rights abuses became painfully obvious in the former Yugoslavia, first in Bosnia and then in Serbia and Kosovo, the United States and Europe reluctantly decided that it was appropriate to intervene and thus breach the concept of sovereignty. The U.S. invasion of Iraq in 2003 was a geopolitical move that was also a breach of sovereignty.

There is some evidence that the role of the individual state may decline as international or supranational organizations play an increasing role in world affairs. The prime example of government-to-government supranational cooperation is the **United Nations (UN),** an assembly of 185 member states that sponsors programs and agencies focusing on scientific research, humanitarian aid, economic development, general health and well-being, and peacekeeping assistance in hot spots around the world. Thus far countries have been unwilling to relinquish sovereignty; hence, the United Nations has limited legal power and can often enforce its rulings only through the power of persuasion. Even in its peacekeeping mission, there are no true UN forces, but rather troops from member states who wear UN designations on their uniforms and take orders from temporary UN commanders.

The World Bank, the International Monetary Fund, and the World Trade Organization, discussed earlier (see page 32), are important international organizations that affect economies and trade practices throughout the world.

**Nongovernmental organizations (NGOs)** are an increasingly important embodiment of globalization. In such associations, individuals, often from widely differing backgrounds and locations, agree on a political, economic, social or environmental goal, such as protecting the environment (for example, the World Wildlife

**FIGURE 1.43** Help from a nongovernmental organization (NGO). *Médecins Sans Frontières* (Doctors Without Borders), an international humanitarian aid organization, provides emergency medical assistance to populations in danger in more than 70 countries. Here, a doctor holds a malnourished and dehydrated child in eastern Democratic Republic of Congo in 2005. [AP Photo/Ron Haviv.]

Fund, the Audubon Society) or dispensing medical care to those who need it most (Médecins Sans Frontières, or Doctors Without Borders [Figure 1.43]). The educational efforts of an NGO, such as Rotary International, raises awareness of important issues among the global public. There is some concern that huge international NGOs wield such power that they could undermine democratic processes, especially in small countries, and that NGO officials may constitute a powerful do-gooder elite that does not interact sufficiently with local people. On the other hand, NGOs such as OXFAM International (a group of NGOs) have provided relief and rebuilding services to the victims of the Indian Ocean tsunami (2004), to those affected by Hurricane Katrina in 2005, and to victims of war in Lebanon, Israel, and Gaza in 2006. Although these organizations may operate on a global scale (especially since the advent of the Internet), at the local level they often resemble community action groups in that they solicit input from a wide range of individuals—a feature that political scientists consider essential to participatory democracy.

## Chapter Key Terms

age distribution or age structure 38
air pressure 23
autonomy 14
average population density 37
birth rate 38
capital 32
carrying capacity 41
cash economy 40
climate 22
cold war era 48
colonialism 9
contiguous region 9
country 49
cultural diversity 14
cultural homogeneity 13
cultural identity 13
cultural marker 14
culture 12
culture group 12

death rate 38
delta 22
demographic transition 40
demography 35
deposition 22
development 33
  "Development for whom?" 33
dialect 16
digital divide 11
domestication 25
earthquake 22
economy 30
erosion 22
ethnic group 12
European colonialism 9
exchange or service sector 30
external process (geophysical) 20
extraction 30
extractive resource 30

fair trade 33
floodplain 22
formal economy 30
formal institution 12
free trade 32
frontal precipitation 25
gender 18
gender structure 38
genocide 49
geopolitics 48
global economy 29
global scale 8
global warming 45
globalization 10
gross domestic product
  (GDP) 30
gross domestic product (GDP)
  per capita 33
growth rate 38

human geography 3
human well-being 33
import quota 32
industrial production 30
Industrial Revolution 31
informal economy 31
informal institution 12
institution 12
internal process (geophysical) 20
International Monetary Fund (IMF) 32
interregional linkage 9
landform 20
latitude 5
lingua franca 16
living wage 32
local scale 9
longitude 5
material culture 12
monsoon 25
multiculturalism 14
multinational corporation 32
nation 49
national identity 49

nation-state 49
Neolithic Revolution 25
nongovernmental organization (NGO) 49
nonmaterial resource 30
norm 14
Organzation for Economic Co-operation and Development (OECD) 47
orographic rainfall 24
Pangaea hypothesis 20
physical geography 3
plate tectonics 21
pluralistic state 49
political ecologist 41
population pyramid 38
purchasing power parity (PPP) 34
race 19
racism 20
rain shadow 24
rate of natural increase (growth rate) 38
region 6
regional geography 4
regional trade bloc 32
religion 15
remittance 10

resource 30
Ring of Fire 22
scale 2
secularism 15
sovereignty 49
spatial analysis 2
structural adjustment policies (SAPs) 32
subduction 22
subregion 8
subsistence economy 40
sustainable agriculture 41
sustainable development 41
tariff 32
technology 17
total fertility rate (TFR) 48
United Nations (UN) 49
volcano 22
weather 22
weathering 22
World Bank 32
world region 8
World Trade Organization (WTO) 32

## Some things to think about as you read this textbook

1. What are the fallacies inherent in the idea that people in the United States should consume at high levels because this is what keeps the world economy going?

2. What are the real causes of the huge increases in migration, legal and illegal, that have taken place over the last 25 years?

3. What would happen in the global marketplace if all people had a living wage?

4. As people live longer and also decide not to raise large families, how can they beneficially spend the last 40 to 50 years of their lives?

5. Given that global warming exists and seems to threaten deleterious results, what are the most important steps to take now?

6. As other ways of life, other values, and other perspectives on the world come up in the reading, reflect on the appropriateness of force as a way of resolving conflicts.

7. Reflect on the various careers one could take up that would address one or more of the issues raised in particular chapters.

8. Reflect on why some people have so much and some so little.

9. What would be some of the disadvantages and advantages of abolishing gender roles in any given culture?

10. Reflect on the ways that access to the Internet enhances prospects for world peace and the ways that it contributes to discord.

# CHAPTER 2

# NORTH AMERICA

**Global Patterns, Local Lives** Fifteen women friends from Vera Cruz, Mexico, are settling down in front of the television in the living room of their neat and attractive group quarters in Terrebone Parish, Louisiana. They all have H1B temporary visas arranged for them by Motivatit Seafood, one of the state's largest harvesters and processors of oysters. Motivatit Seafood, owned by the Voisin family, was founded by a French immigrant many generations ago. In two hours the women will join a few local workers to start their shift in the chilly workroom where they shuck, wash, and package oysters for between $5.50 and $6.00 an hour.

In the months after hurricanes Katrina and Rita hit the U.S. Gulf Coast (Figure 2.1) in August and September of 2005, business owners struggled to find the resources to reopen. Building materials were in short supply, but most important, a shortage of workers kept some of the region's most viable industries from getting back into production. Responding to the massive destruction of the storms and the inability of local, state and national governments to meet their immediate needs, hundreds of thousands of coastal residents fled to other parts of the United States. With no prospect of quickly reclaiming their homes or jobs, many decided to stay wherever they had relocated. Soon, some of the demand for workers along the Gulf Coast was being filled by workers from Mexico, documented and undocumented. The influx of immigrants was so great that the mayor of New Orleans, which had been 62 percent African-American before the storms, worried that the culture of his city would be changed forever.

Forty percent of the nation's oysters come from Louisiana, and the oyster beds suffered serious storm damage; but by February of 2006, nature was repairing itself. Motivatit Seafood, with the help of the Mexican women, was getting back to normal. Still, the company could not fill the demand for oysters because it needed at least thirty more workers. Mike Voisin, the present owner, says that even before the storms it was nearly impossible to keep local employees because they spurned the hard labor of oyster work. He is hoping to get permission for yet more H1B visas for Mexican workers.

**FIGURE 2.1** Regional map of North America.

Mike's grandson, Kevin Voisin, says that he was originally against the idea of Mexican workers, but he changed his mind when he met the women. He visits them often with his wife. "These are great people. This is what America is all about. [Many of] my people came from Ireland. People worried then that the Irish would take over America; but this [hasn't become] the Irish Republic of America." Kevin Voisin went on to express his admiration for the Mexican women, their dedication to hard work, and their frugality. In their home village, only field work at the equivalent of a few dollars a day was available. Now, even at the low Louisiana wages, all are managing to save enough to send both money and clothes from Wal-Mart to their families in Vera Cruz.

*Adapted from the National Public Radio's "Sunday Morning Edition" series on Motivatit Seafood, by Lianne Hanson et al., September 11 and 25 and October 2, 2005, and February 12 and 19, 2006.* ∎

The influx of Mexican and other Hispanic workers into Louisiana and Mississippi in the aftermath of hurricanes Katrina and Rita is only one circumstance of many in which North American employers have welcomed the availability of immigrant workers—some highly skilled technical specialists such as computer engineers and medical doctors, some ordinary laborers. The U.S. Chamber of Commerce says that there is already a large unmet need for workers, and, with the retirement of the baby boomers, by 2010 there will be a deficit of perhaps 15 million workers of all types.

Given the demand, it is not surprising that undocumented workers fill some of these positions, particularly those that are lower paid. As of September 2006, estimates on the number of those currently working without the proper papers run between 11 million and 12 million.

Figure 2.2 shows recent patterns of immigration to North America. Immigrants from Asia, Africa, and Middle and South America are bringing new customs and new international sensibilities to even the most remote communities of North America. Rural villages in the mountains of North Carolina have Mexican grocery stores catering to farm laborers from Michoacan in central Mexico. Physicians from India live in eastern Kentucky hill-country towns and treat predominantly Appalachian patients (Figure 2.3). In Minneapolis, Somali women in Muslim veils lead labor protests for better working conditions for everyone in Minnesota's workplaces (Figure 2.4).

Americans are ambivalent about these new immigrants and guest workers and the major social and economic changes that they represent. Farming and manufacturing, long the mainstay of North America's economy, are being replaced by a new and still unstable economy based on sophisticated technology and the high-speed exchange of information, goods, and money on a global scale. There are rapid changes in employment levels and in the locations of jobs across the continent. Immigrants have a central role in these

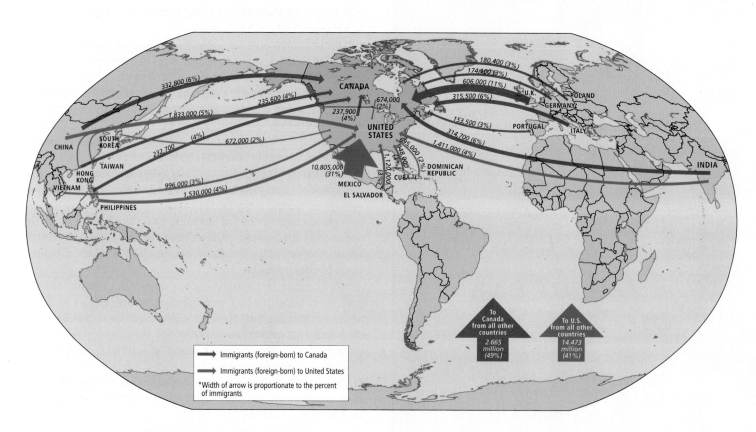

**FIGURE 2.2 Total foreign-born residents of Canada and the United States by country of birth.** Foreign-born residents include legal and illegal immigrants, temporary workers, students, and refugees. The numbers for Canada are as of the 2001 census (the latest available); those for the United States are as of 2005. Percentages refer to proportion of total immigrants to a particular country.

**FIGURE 2.3** Indian immigrants bringing medical care to the rural United States. Doctors Rakesh Sachdeva and Seema Sachdeva, both from India but now U.S. citizens, examine Connor Mayhorn at their medical office, in Pikeville, Kentucky. Pikeville is a small city in eastern Kentucky in a county that for years was underserved by doctors. [Dena Potter/*Appalachian News-Express*.]

**FIGURE 2.4** Somali women leading the labor movement in Minnesota. Here, as members of the Hotel and Restaurant Employees Union, Somali women urge a taxi driver not to cross their picket line. [Craig Borck/Pioneer Press.]

changes, some entering the job market at the low-skill end to take jobs that Americans no longer wish to fill, others entering at the high-skill end to take jobs for which there are too few qualified Americans. Those with medium-level skills, such as construction workers, may well be taking jobs away from American workers. All of these immigrants are changing the face of America, increasing its multiethnic hue, adding a youthful vigor to its body, and contributing significantly to its tax coffers.

It is likely that you are a resident of North America. You already have some impressions of this land and the changes it is undergoing. Please bring your own experience to the task of defining the evolving character of North America. If you grew up in North America, you might want to glance at the map of your childhood landscape that you drew at the beginning of Chapter 1 and place it in its broader regional context. Think of this chapter as a guide to enhance your understanding of what you already know and to help you construct, over a lifetime, an ever more complex understanding of the geography of North America.

## Themes to Explore in North America

Some central patterns in North American life emerge in this chapter:

1. Mobility. The ability to move around the continent is related to the individualism that characterizes North American lifestyles. Migrating to a new geographic location is accepted as a way to achieve individual fulfillment and financial success. About one out of every three citizens of the United States between ages 5 and 64 moved in the period from 1995 to 2000, and that pattern continues.

2. Changing population composition. North America is becoming less European and African as new residents arrive from Middle and South America and from various parts of Asia. According to the 2000 U.S. census, Euro-Americans (whites) are now minorities in 50 of the 100 largest U.S. cities, and African-Americans, long the largest minority, are now outnumbered by Hispanic Americans.

3. Changing population distribution. Overall, there has been an internal shift in the distribution of the population from the northeast and middle of North America to the south and west.

4. Similar landscapes; regional differences. All major cities in Canada and the United States have similar suburbs, office towers, malls, ethnic restaurants, and home improvement centers. Yet, despite this sameness in landscape, over the past few decades people within both countries have become more, rather than less, culturally and economically diverse.

5. Wealth and poverty. Although Americans as a whole are prospering more than ever, the gap between rich and poor is widening. According to the 2000 U.S. census, there were 4 million millionaires (1000 times more than in 1900), but 37 million people (13 percent of the population) remained poor. For the same period, 11 percent of Canadians remained poor.

6. Increasing global awareness and interaction. On a personal scale, Americans can now maintain daily relationships with individuals in virtually any country on earth by using the Internet. And Americans must be agile in their response to global events, such as terrorism or economic changes, because their reactions can cause reverberations both locally and around the world.

# | THE GEOGRAPHIC SETTING |

## Terms to Be Aware Of

The world region discussed in this chapter consists of Canada and the United States. The term *North America* is used to refer to both countries. Even though it is common on both sides of the border to call people in the United States "Americans," in this text the term *American* describes citizens of, or patterns in, both countries, not just the United States.

Other terms relate to the growing cultural diversity in this region. For example, the text uses both **Hispanic** and **Latino.** Both are loose ethnic (not racial) terms that refer to all Spanish-speaking people from Latin America and Spain, whose ancestors may have been black, white, Asian, or Native American. In Canada, the **Québecois,** or those French Canadians living in Québec, are an ethnic group that is distinct from the rest of Canada. They are the largest of an increasingly complex mix of minorities in that country, most of whom are still content to be called simply Canadians.

## PHYSICAL PATTERNS

The continent of North America is a magnificent display of mountain peaks, ridges and valleys, expansive plains, long winding rivers, sparkling lakes, and extraordinarily long coastlines. You will find it easier to learn the physical geography of this large and complex territory if you break it into a few smaller segments, as described in the following section on landforms.

## Landforms

The most dramatic and complex North American landform is the wide mass of mountains and basins in the west, known as the Rocky Mountain zone of Canada and the United States (see Figure 2.1). It sweeps down from the Bering Strait in the far north, through Alaska, and continues on into Mexico, where it is known as the Sierra Madre. This zone formed about 200 million years ago, when, as part of the breakup of the supercontinent Pangaea (see Figure 1.20 on page 21), the Pacific Plate pushed against the North American Plate, thrusting up the broad band of mountains along the western edge of the continent. The Pacific Plate continues to slip to the northwest, and the North American Plate to the southeast, causing the earthquakes that are common along the Pacific coast of North America. The much older and hence more eroded Appalachian Mountains, which stretch along the eastern edge of North America from New Brunswick and Maine to Georgia, resulted from earlier collisions of the North American Plate with North Africa.

Between these two mountain ranges lies a huge, irregularly shaped *central lowland* of undulating plains that stretch from the Arctic to the Gulf of Mexico. The central lowland was created by the deposition of material eroded from the mountains and carried to the region by wind and rain and by the rivers flowing east and west into what is now the Mississippi drainage basin. These deposits can be several kilometers deep.

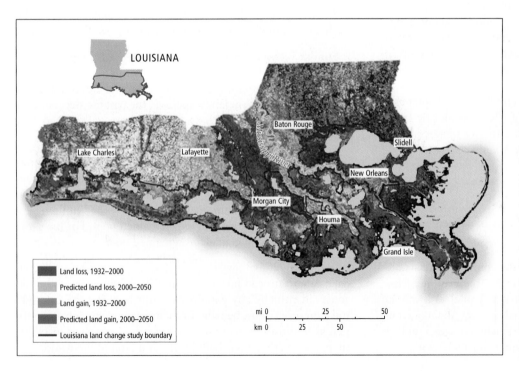

**FIGURE 2.5** 100+ years of land change for coastal Louisiana. The Louisiana coastline and the lower Mississippi River basin are vital to the nation's interests, providing coastal wildlife habitat, recreational opportunities, and transport lanes that connect the vast interior of the country with the world ocean and access to offshore oil and gas. Most important, the wetlands provide a buffer against damage from hurricanes. But largely due to human impacts on natural systems, Louisiana has lost one-quarter of its total wetlands over the last century. The remaining 3.67 million acres constitute 14 percent of the total wetland area in the lower 48 states. [USGS/National Wetlands Research Center.]

Glaciers have covered the northern portion of the lowlands, as well as adjoining mountain ranges to the west and east, during periodic ice ages occurring over the last 2 million years. During the height of the most recent ice age (between 25,000 and 10,000 years ago) glaciers, sometimes as much as 2 miles (about 3 kilometers) thick, scoured the very old, exposed rock surface of the Canadian Shield in the far north and picked up sediment from lands to the south. As the glaciers melted, the sediment was dumped. The Great Lakes are depressions left by glacial scouring, as are the smaller lakes, ponds, and wetlands that dot central Canada and the United States from Minnesota to Massachusetts. In the upper Mississippi drainage basin, some of the sediment dumped by the melting glaciers has been picked up and redeposited by the wind. This wind-deposited layer of soil, called **loess,** is often many meters deep; it has proved particularly suitable for large-scale mechanized agriculture, but it remains susceptible to wind and water erosion.

East of the Appalachians, *a coastal lowland,* which is intermittent and narrow in the north but wider in the south, stretches from New Brunswick to Florida and then sweeps west to the southern reaches of the central lowland along the Gulf of Mexico. In Louisiana and Mississippi is the Mississippi delta, which was formed by huge loads of silt deposited by North America's largest river system during floods over the past 150 million years. The delta originally began at what is now the junction of the Mississippi and Ohio rivers and then slowly advanced 1000 miles (1600 kilometers) into the Gulf of Mexico, filling in much of the southern central lowland. Over the past centuries, human activities such as deforestation, deep plowing, and heavy grazing have added to the silt load of the rivers (Figure 2.5). At the same time, the construction of levees along riverbanks has drastically reduced flooding. As a result, much of the southern part of the Mississippi delta, which is a low, flat transition zone between land and sea characterized by swamps, lagoons, and sandbars, is now sinking into the Gulf of Mexico. Meanwhile, the huge loads of silt, which used to spread widely across the delta region, are now contained by levees until they reach the deep waters of the Gulf of Mexico.

## Climate

Every major type of climate on earth, except tropical wet, exists in North America. The landforms over this continental expanse contribute to its enormous climate variety by influencing the movement and interaction of air masses.

Look at the climate map of North America (Figure 2.6) and observe the thin band of light green along the Pacific coast. The mild climate (Mediterranean) here is cool and moist in the winter and warm and dry in the summer. Conditions in the north (around Seattle) are generally cooler and wetter than in the south (near Los Angeles). The west coast of North America, especially north of San Francisco, receives moderate to heavy rainfall (see Figure 1.22 on page 23) because moist air passes onto land from the Pacific Ocean. When this moist air encounters the coastal mountain ranges, it rises, cools, and dumps rain along the windward (western) slopes (see the discussion of orographic rainfall on pages 23–24). Climates are much drier to the east of the coastal moun-

tain ranges because air masses tend to hold what little moisture they have left as they move eastward across the Great Basin and Rocky Mountains, resulting in a rain shadow. In this arid region, many expensive dams and reservoirs for irrigation projects have been built to make agriculture and large-scale human habitation possible.

On the eastern side of the Rocky Mountains, the main source of moisture is the Gulf of Mexico. When the continent is warming in the spring and summer, the air masses above it rise, pulling in moist, buoyant air from the Gulf. This air interacts with cooler, drier, heavier air masses moving into the central lowland from the north and west, often creating violent thunderstorms and associated tornadoes. Generally, the central part of North America is wettest in the east and south and driest in the north and west. Along the Atlantic coast, moisture is supplied by warm wet air above the Gulf Stream, a warm ocean current that flows up the eastern seaboard from the tropics.

The large size of the North American continent creates wide temperature variations in the continental interior, with hot summers and cold winters. Because land heats up and cools off more rapidly than water, temperatures in the continental interior are hotter in the summer and colder in the winter than areas along the coast, where temperatures are moderated by the oceans.

The geographic distribution of the landforms and climates of North America have influenced the settlement patterns of the continent and the behavior, occupations, and problems of the people living there. We now discuss the historical geography of North America.

## HUMAN PATTERNS OVER TIME

The human history of North America is a series of successive arrivals and dispersals of people across the vast continent, beginning in prehistoric times with the original peopling of the continent from Eurasia via Alaska (and possibly across the Pacific), and continuing with waves of European immigrants, enslaved Africans, and their descendants, who settled the country primarily from east to west from the 1600s on. Today, immigrants from Asia and Latin America are arriving primarily in the West. The patterns of these movements have contributed to the formation of distinct subregions. Internal migration is still a defining characteristic of life for North Americans, who are among the most mobile people in the world.

### The Peopling of North America

Recent evidence suggests that humans first came to North America from northeastern Asia between 25,000 and 14,000 years ago, during an ice age. At that time, the global climate was cooler, polar ice caps were thicker, sea levels were lower, and the Bering land bridge, a huge, low landmass more than 1000 miles (1600 kilometers) wide, connected Siberia to Alaska. Small bands of hunters crossed by foot or small boats over into Alaska and traveled down the west coast of North America. By 15,000 years ago, humans had reached nearly to the tip of South America and had spread deep into the continent. By 10,000 years ago, global temperatures began to rise, and, as the

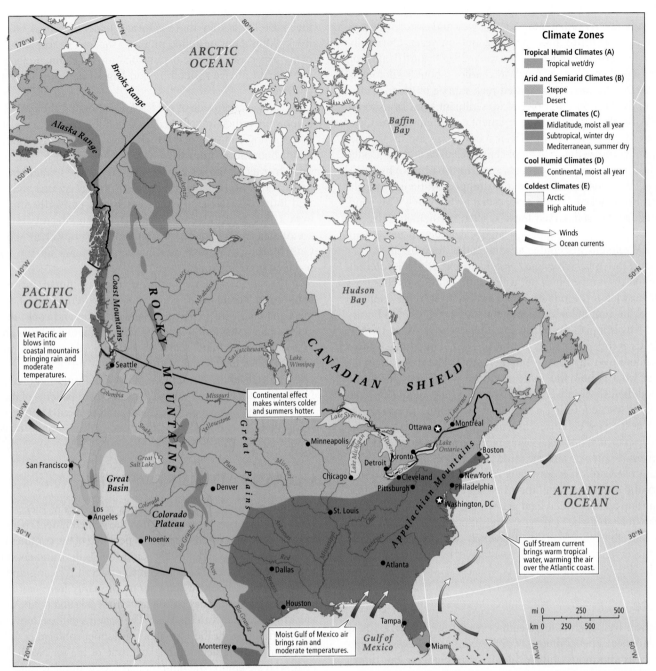

**FIGURE 2.6 Climates of North America.** Climate zones, like those shown on this map, do not have sharp boundaries but, rather, grade into each other over broad transition zones. For example, the boundary between the midlatitude climate and the continental climate is a wide band of several hundred miles that shifts north or south from year to year.

ice caps melted, the Bering land bridge sank beneath the sea. In North America a mid-continent corridor opened up through glaciers, allowing more people to pass to the south; soon virtually every climate region throughout the Americas was occupied.

Over thousands of years, the people settling in the Americas domesticated plants, created paths and roads, cleared forests, built permanent shelters, and sometimes created elaborate social systems. They developed a wide range of cultural traditions attuned to the circumstances in which they lived. The domestication of corn

is thought to be closely linked to settled life and population growth in North America. Corn was introduced from Mexico into what is now the southwestern desert about 3000 years ago along with other Mexican domesticates, particularly squash and beans.

These foods provided surpluses that allowed some community members to specialize in activities other than agriculture, hunting, and gathering. The impressive harvests made possible large, city-like regional settlements. For example, by about 1800 years ago, such distinctive culture groups as the Anasazi, Hohokam, and Mongol-

**FIGURE 2.7** Cahokia in 1150, an artist's interpretation. Long before European settlements in North America, 30,000 people lived in this urban area in what became central Illinois. [Painting by William R. Iseminger.]

lon had emerged in the southwest of North America. By 1000 years ago, the urban settlement of Cahokia (in what is now central Illinois) covered 5 square miles (12 square kilometers) and was home to an estimated 30,000 people (Figure 2.7).

***The Arrival of the Europeans.*** North America today bears little resemblance to the land that the Native Americans had created by the 1500s. The vegetation, land use, technology, and cultural orientation of the population have been transformed by the sweeping occupation of North America by Europeans and subsequently by other immigrants.

From the early 1600s to the present, immigrant settlements have rolled over North America like an unstoppable tide. In the early seventeenth century, the British established colonies along the Atlantic coast in what is now Virginia (1607) and Massachusetts (1620). Over the next two centuries, colonists from northern Europe built villages, towns, port cities, and plantations up and down the east coast of North America. By the mid-1800s they had occupied most Native American lands throughout the Appalachian Mountains and into the central part of the continent.

***The Disappearance of Native Americans.*** The rapid expansion of European settlement was assisted by the fact that Native American populations had long been isolated from the rest of the world and hence had no immunity to diseases such as measles and smallpox. These diseases, which were transmitted by Europeans and Africans who had built up immunity to them, were particularly deadly to Native Americans, killing an average of 90 percent of any given group within 100 years of contact. The encroaching Europeans, with their technologically advanced weapons, trained dogs,

and horses, also took a large toll. Often the Native Americans had only stones, bows, and poisoned arrows, although by the late 1500s Native Americans in the Southwest had acquired horses from the Spanish and learned to breed and train them for use in warfare against the Europeans. Simple numbers reveal the effect of European settlement on Native American populations. There were roughly 18 million Native Americans in North America in 1492. By 1542, after only a few Spanish expeditions such as that led by Hernando de Soto, there were half that many. By 1907, slightly more than 400,000, or just 0.02 percent, remained.

## The European Transformation

European settlement erased many of the landscapes familiar to Native Americans and imposed new ones that fit the varied physical and cultural needs of the new occupants. Hence distinct subregions, which still exist today, developed as settlement proceeded.

***The Southern Settlements.*** European settlement of eastern North America began with the Spanish in Florida in 1565 and with the establishment of the British colony of Jamestown in Virginia in 1607. By the late 1600s, the colonies of Virginia, the Carolinas, and Georgia were cultivating cash crops such as tobacco and rice on large holdings known as *plantations*. To secure a large, stable labor force, enslaved African workers were brought into North America, beginning in 1619. Slavery was not widely accepted at first, but within 50 years enslaved Africans were the dominant labor force on some of the larger southern plantations. By the start of the Civil War in 1861—a conflict that grew largely out of competition among the northern, mid-Atlantic, and southern economies—slaves made up about one-third of the population throughout the southern states and were concentrated where plantations were located.

The plantation system was detrimental to the economic and social development of the South because it encouraged wide disparities of wealth and stark race and class divisions. Much of the area's wealth was in the hands of a small class of plantation owners, who made up just 12 percent of southerners in 1860. Planter elites kept taxes low and invested their money in Europe or the more prosperous northern colonies, instead of in industry and social services at home. More than half of southerners were poor white subsistence and small cash-crop farmers who consumed most of what they produced and therefore had little to sell and few earnings to invest. Because both slaves and poor whites were forced to live simply, their meager consumption did not provide a demand for goods, so there were few market towns. Plantations tended to be largely self-sufficient and generated few of the independent spin-off enterprises—such as transport and repair services, shops, garment making, and small manufacturing—that could have invigorated the regional economy. After the Civil War (1861–1865), the plantation economy declined, and the South sank deep into poverty, remaining economically and socially underdeveloped well into the twentieth century.

We can see lingering effects of the plantation era in southern place names and architecture, and in North America's largest concentrations of African-Americans, which are still in the southeastern states, where the slave-based plantation economy was centered (Figure 2.8).

**The Northern Settlements.** Throughout the seventeenth and eighteenth centuries, relatively poor subsistence farming communities dominated agriculture in the colonies of New England and southeastern Canada. There were no plantations and few slaves, and not many cash crops were exported. Farmers lived in interdependent communities that prized education, ingenuity, and thrift. At first, incomes were augmented with exports of timber and animal pelts, and some communities depended heavily on the rich fishing grounds of the Grand Banks, located off the shores of Newfoundland and Maine. However, by the late 1600s, New England was implementing ideas and technology from Europe that led to the first industries. By the 1700s, metalworks and pottery, glass, and textile factories in Massachusetts, Connecticut, and Rhode Island were supplying markets in North America and also exporting to British plantations in the Caribbean.

Industry began to flourish in the early 1800s, when Francis Lowell designed a water-powered loom, opening up opportunities for women as a side effect. By 1822, "factory girls" were working in the textile mills of Lowell, Massachusetts, and living as single women in company-owned housing. Southern New England, especially the region around Boston, became the center of manufacturing in North America, drawing on male and female immigrant labor from French Canada and Europe.

**The Economic Core.** New England and southeastern Canada were eventually surpassed in numbers and in wealth by the mid-Atlantic colonies of New York, New Jersey, Pennsylvania, and Maryland. This region benefited from more fertile soils, a slightly warmer climate, multiple deepwater harbors, and better access to the resources of the interior. By the end of the Revolutionary War in 1783, the mid-Atlantic region was on its way to becoming the **economic core**, or the dominant economic region, of North America.

Both agriculture and manufacturing in the region boomed in the early nineteenth century, drawing immigrants from much of

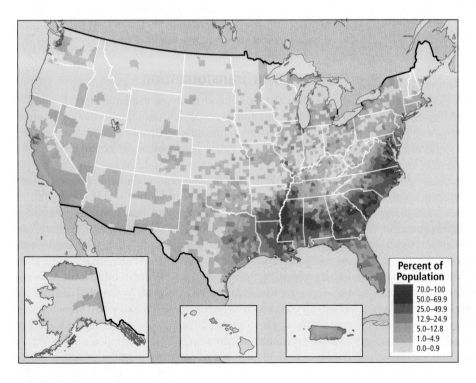

**FIGURE 2.8** African-American population, 2000. While many African-Americans live in cities across the United States, the majority continue to live in the southeastern parts of the country, where they pursue urban professional careers or work in blue- and white-collar jobs. As of 2000, African-Americans were returning to the Southeast in greater numbers than were leaving, many retiring on pensions earned elsewhere. [U.S. Census Bureau, Mapping Census 2000: The Geography of Diversity.]

**Percent of Population**
70.0–100
50.0–69.9
25.0–49.9
12.9–24.9
5.0–12.8
1.0–4.9
0.0–0.9

**(a) Travel Times from New York City, 1800.**
It took a day to travel by wagon from New York City to Philadelphia and a week to go to Pittsburgh.

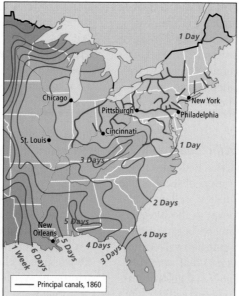

**(b) Travel Times from New York City, 1857.**
The travel time from New York to Philadelphia was now only 2 or 3 hours and to Pittsburgh less than a day because people could go part of the way via canals (dark blue). On the canals they could easily reach principal cities along the Mississippi River, and travel was less expensive and onerous.

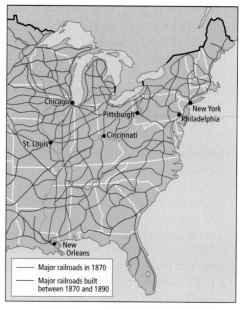

**(c) Railroad Expansion by 1890.**
With the building of railroads, which began in the decade before the Civil War, the mobility of people and goods increased dramatically. By 1890, railroads crossed the continent, though the network was most dense in the eastern half.

**FIGURE 2.9** Nineteenth-century transportation. [Adapted from James A. Henretta, W. Elliot Brownlee, David Brody, and Susan Ware, *America's History*, 2nd ed. (New York: Worth, 1993), pp. 400–401; and James L. Roark, Michael P. Johnson, Patricia Cline Cohen, Sarah Stage, Alan Lawson, and Suan M. Hartmann, *The American Promise: A History of the United States*, 3rd ed. (Boston: Bedford/St. Martin's, 2005), p. 601.]

northwestern Europe. The tremendous success of agriculture, particularly grain, laid the groundwork for investment in related industries such as baking, food processing, and meatpacking. As farmers prospered, they bought mechanized equipment, appliances, and consumer goods, much of it made in nearby cities. Banks catering to the richest farmers put up money to build the fac_____ cities such as New York, Philadelphia, _____ because they were well positioned to t_____ trial products to both Europe, via ocea____ nental interior and Great Lakes regi____ canals, and eventually railways. This ___ region yielded almost a third of U.S. i_____ of the Civil War (1860).

Throughout the nineteenth cen_____ panded as connections with interior a____ include what became the states of O____ nois, and Wisconsin (Figure 2.9). Th____ ing economic core were farmers wh____ urban populations, but by the mid-n____ omy of the core was increasingly base____ diffused westward to the Great Lakes i____ Detroit, and Chicago. The steel indu____ mining of local deposits of coal and iro____ tion and manufacture of a wide rang____ household tools and appliances, such a____

and iron stoves. As mechanization proceeded, heavy farm and railroad equipment, including steam engines, became important products of this region.

By the late nineteenth century, the economic core stretched from the Atlantic to Chicago and from Toronto to Cincinnati. This _____idly industrializing region dominated the other parts of N___ arid character of this land eventually created an ecological disaster for Great Plains farmers. In the 1930s, after ten especially dry years, a series of devastating dust storms blew away topsoil by the ton; animals died, and entire crops were lost (Figure 2.10). This hardship was made worse by the widespread economic depression of the 1930s, and many Great Plains farm families packed up what they could and headed west to California and other states on the Pacific Coast.

### Settlement of the Mountain West and Pacific Coast.

While the Great Plains filled up, other settlers were alerted to the possibilities farther west. Settlers were drawn to the valleys of the Rocky Mountains (especially the Great Basin) and to the well-watered and fertile coastal zones of what was then known as the Oregon Territory and California. The great rush to Oregon began in the 1840s, and by the 1860s, more than 350,000 people had walked the Oregon Trail, carrying their possessions in wagons that wore ruts 3 feet deep that are still visible today. News of gold in California in 1849 diverted thousands from farming to the idea of getting rich quick. The vast majority of gold seekers were unsuccessful and, by 1852, had to look for employment elsewhere. Although farming was at first the major occupation in the Oregon Territory, logging eventually became the dominant industry in the Pacific Northwest as vast stands of enormous redwoods, Douglas firs, spruces, and many other species were clear-cut throughout the region. (**Clearcutting** is the cutting down of all trees on a given plot of land, regardless of age, health, or species.) The extension of the railroads across the continent in the nineteenth century facilitated the movement of manufactured goods to the West and raw materials to the East (see Figure 2.9).

Among the earliest and most successful new settlers in the West were a group of utopian Christians known as the Mormons, who demonstrated that the semiarid Great Basin region could be farmed with the use of irrigation. The Mormons (members of the Church of Jesus Christ of Latter-day Saints) remain an important cultural influence in the Great Basin states to this day.

### Settlement in the Southwest.

The Southwest was first colonized by people of Spanish extraction who came from the Spanish colony of Mexico at the end of the sixteenth century. They established sheep ranches and missions in central and southern California Territory, and in the southern reaches of what eventually became the states of Arizona, New Mexico, and Texas. These early settlements were sparse, and as settlers from the United States expanded into the region, Mexico found it increasingly difficult to retain its claimed territory. By 1850, nearly all of the Southwest was under U.S. control, and the new state of California in particular was attracting farmers and laborers.

By the twentieth century, a vibrant agricultural economy was developing in central and southern California, where the mild Mediterranean climate made it possible to grow vegetables almost year-round. With the advent of refrigerated railroad cars, fresh California vegetables could be sent to the major population centers of the East. Massive government-sponsored irrigation schemes greatly increased the amount of arable land. These projects diverted whole rivers to supply fields and urban developments hundreds of miles away, totally transforming the natural course of water drainage and dramatically changing southern California and much of the West.

## European Settlement and Native Americans

Native Americans living in the eastern part of the continent who survived early encounters with Europeans occupied land that Europeans wished to use. Almost all were forcibly relocated to relatively small reservations with few resources. The best known of these relocations occurred in the late 1830s, when the Creek and Cherokee of the southeastern states, who had already adopted many Euro-

**FIGURE 2.10**  Dust storm approaching Stratford, Texas, April 18, 1935. [NOAA/ George E. Marsh Album.]

**FIGURE 2.11** Choctaw manufacturing plant. Some tribes have been successful in taking creative approaches to development. The Choctaw in Mississippi have developed factories that produce plastic utensils for McDonald's restaurants, electrical wiring for automobiles, and greeting cards. They employ not only their own people but also 1000 non–Native Americans who come onto the reservation to work. [Courtesy of Mississippi Band of Choctaw Indians.]

pean methods of farming, building, education, government, and religion, were rounded up by the U.S. Army and marched to Oklahoma. The route they took became known as the Trail of Tears because of the many who died along the way.

Reservations were created primarily west of the Mississippi, usually with insufficient attention paid to the needs and customs of Native Americans. As European settlers occupied the Great Plains and prairies, many of the reservations were further shrunk or relocated on ever less desirable land. Today, reservations cover just over 2 percent of the land area of the United States but more than 20 percent of Canada since the creation of the Nunavut territory in 1999 and of the Dogrib territory in 2003, both in Canada's far north (see Figure 2.1). The Nunavut and Dogrib stand out as having won the right to control their own resources and the right to home rule, including an elected legislative assembly, a public service system, and a court. (In contrast to the United States, it is unusual for native groups in Canada to have legal control over their territory.) Most reservations in North America have insufficient resources to support their populations at the standard of living enjoyed by other citizens, and, after centuries of mistreatment, many Native Americans are more familiar with poverty than with true native ways; many are totally dependent on meager welfare money. Severely demoralized by the lack of opportunity, some have turned to alcohol, which often leads to family violence. In the United States, Native American suicide and homicide rates are almost twice the national average.

Economic development is now transforming the lives of some Native Americans. For years, native young people faced a difficult choice: they could remain in poverty with family and community on the reservation, or they could seek a more affluent way of life far from home and kin. Outside their communities, many young people lose touch with traditional ways and encounter considerable prejudice. Beginning in the 1990s, several tribes found avenues to greater affluence on the reservations through monetary compensation from the government for past losses, or by establishing manufacturing industries (Figure 2.11), developing fossil fuel and mineral deposits underneath reservations, or opening gambling casinos designed to lure nonnative gamblers. Although these few roads to prosperity can produce enormous income (especially casinos), the stress on weakened tribal structure can be intense. In some cases, graft has diverted funds to a few elites and Washington lobbyists. In others, economic development has been accompanied by environmental pollution, as in the case of the many oil and gas wells and coal and uranium mines located on reservations. In Canada, Native Americans have a higher standard of living than their U.S. counterparts, and the government spends much more per capita on them. But with the exception of the new Nunavut and Dogrib territories in northern Canada, Canadian tribes have less control over the resources on their lands, which are often developed for use by non–Native Americans.

Native American populations overall are now expanding, after plunging from 18 million to less than 400,000 between 1500 and 1900. In 2000, there were almost 2.5 million Native Americans in North America, most living in the United States, where they constitute less than 1 percent of the total population.

## The Changing Social Composition of North America

The subregions of European-led settlement still remain in North America, but their boundaries and distinctive characteristics are becoming increasingly blurred. The economic core no longer dominates North America, as industry has spread to other parts of the continent. Some regions that were once dependent on agriculture, logging, or mineral extraction now have high-tech industries as well. In cultural terms, once-distinct ethnic groups have become assimilated as succeeding generations have married outside their groups and moved to new locations. Before we look more closely at these emerging patterns, however, some information about the changing population of North America will be useful.

## POPULATION PATTERNS

The population map of North America (Figure 2.12) shows the uneven distribution of the more than 332 million people who live on the continent. Canadians, who account for just under one-tenth (32.2 million) of the population of North America, live primarily in southeastern Canada, close to the border with the United States. Sixty percent of Canadians live in the Great Lakes region of southern Ontario and in Québec along the St. Lawrence River. In October 2006, the population of the United States reached 300 million, with the greatest concentration of people in the northeastern part of the country not far south of the border with Canada. A quadrant

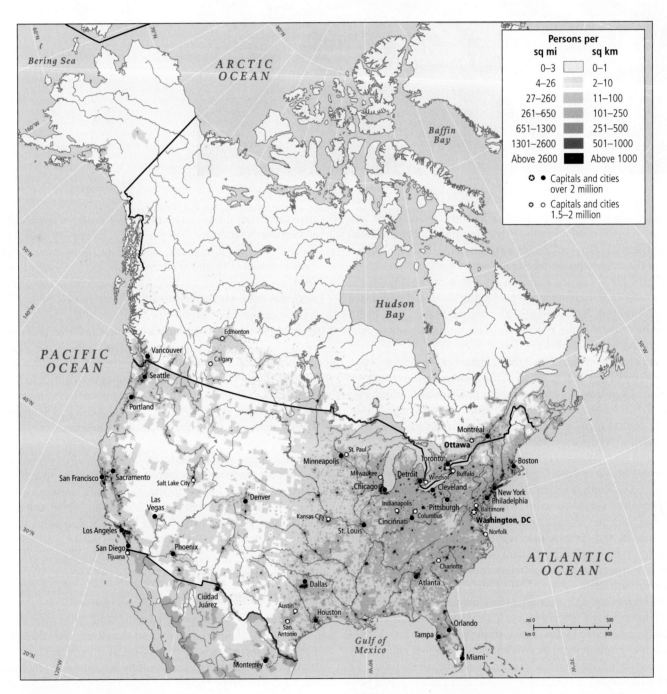

**FIGURE 2.12** Population density in North America. [Data courtesy of Deborah Balk, Gregory Yetman, et al., Center for International Earth Science Information Network, Columbia University, at http://www.ciesin.columbia.edu.]

marked by Boston and Washington, D.C., along the Atlantic seaboard, St. Louis on the Mississippi River, and Chicago on Lake Michigan contains 7 of the 12 most populous states in the country. Historically, this was the economic core; however, other regions of the country are now growing much faster (Figure 2.13). The 2000 U.S. census shows that the Northeast grew by just 5.5 percent in the 1990s, while the West grew by 20 percent and the South, east of the Mississippi, by 17 percent.

The population core of the Northeast is flanked to the south and west by a zone of less dense settlement on rich agricultural land dotted with many medium-size cities, most with populations under 2 million (see Figure 2.12). On average, rural densities here are 3 to 129 people per square mile (1 to 50 per square kilometer). Many farm towns in the Middle West (the large central farming region of North America), once magnets for European immigrants, are now losing population as farms consolidate under corporate ownership, labor needs decrease, and young people choose better-paying careers in cities (notice the changes in the area labeled Middle West in Figure 2.13). Middle Western cities are growing modestly and becoming more ethnically diverse, with rising populations of Hispanics and Asians in such places as Indianapolis, St. Louis, and Chicago. Meanwhile, many cities to the south—Atlanta, Birmingham, Nashville, and the Raleigh–Durham area, for example—are growing rapidly, and are also attracting Hispanic residents and others, especially from East Asia and South Asia.

To the west and north of the Middle West (see Figure 2.12), in the continental interior, settlement is very light. In fact, roughly two-thirds of this subregion, including northern Canada and Alaska, has fewer than 3 people per square mile (1 per square kilometer). This low population density is correlated with mountainous topography, lack of rain, and, in northern or high-altitude zones, a growing season that is too short to sustain agriculture. Some population clusters exist in irrigated agricultural areas, such as in the Utah Valley, and near rich mineral deposits and resort areas. Many states within this region grew only very slowly in the 1990s. A major exception is Las Vegas in Nevada, at the southern end of the region, which grew 66 percent, making this the fastest growing area in the United States.

Along the Pacific coast, a band of population centers stretches north from San Diego to Vancouver and includes Los Angeles, San Francisco, Portland, and Seattle. These are all port cities engaged in trade around the **Pacific Rim** or **Pacific Basin** (all the countries that border the Pacific Ocean on the west and east) and with countries of the Atlantic community via the Panama Canal. These Pacific cities are flanked by irrigated agricultural zones that supply a major portion of the fresh produce for North America. Over the past several decades, these cities also have become centers of innovation in computer technology.

During the twentieth century, there have been clear shifts in the U.S. population from the Northeast to the South and West (see Figure 2.13). Because Americans are among the most mobile people on earth, this kind of redistribution is common. Every year, in the United States, almost one-fifth of the population relocates. Some are changing jobs, others are attending school or retiring to a warmer climate, others are merely moving across town or to the suburbs or countryside. Still other people are arriving from outside the region: immigrants enter North America at the rate of about 5000 per day, leading to major economic and political consequences, as we shall see later in this chapter.

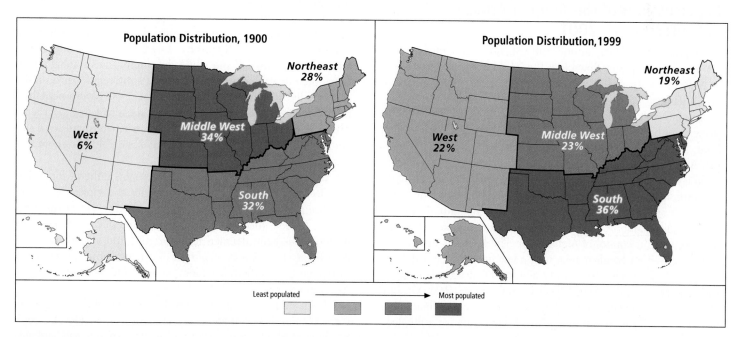

**FIGURE 2.13** Percentage of population by region in the United States, 1900 and 1999. Notice how much the percentage has increased in the West and decreased in the Northeast and Middle West over the past 100 years.

[Adapted from "America's diversity and growth," *Population Bulletin* (June 2000): 12. See http://www.statcan.ca/english/research/11F0019MIE/11F0019MIE2005254.pdf.]

People's perceptions of which regions are desirable places to live change over time, so the movement of people ebbs and flows. Many factors are at work. Internal migrants are often lured by opportunities for employment. Immigrants from Middle and South America and Central Europe are recruited by towns in the Middle West to work in farm-related industries, just as immigrant workers from Europe were sought in the nineteenth century. Southern cities such as Atlanta, Georgia, and Birmingham, Alabama, featuring warm climates and lower energy and living costs, are drawing workers and retirees from the the old economic core of the Middle West, as well as East Asian and Hispanic migrants. These cities and others farther west have sprouted satellite or "edge" cities around their peripheries, often based on businesses dealing with technology and international trade. You will find additional explanations for these population movements in other sections of this chapter.

The rate of natural increase in North America (0.6 percent per year) is low, less than one-half the rate of the rest of the Americas (1.6 percent). Still, according to the Population Reference Bureau, North Americans are adding to their numbers through births and immigration fast enough to reach 420 million by 2050. Many of the important social issues now being debated in North America, and discussed below, reflect changing population patterns—issues such as legal and illegal immigration, voting patterns among new immigrants, the language of instruction in public schools, the aging of the population, availability and use of birth control, the social costs and benefits of low birth rates, changing family structures, and gender and ethnic diversity.

# II CURRENT GEOGRAPHIC ISSUES

While North America is privileged by its huge economy and still-plentiful resources, the region faces complex problems posed by rapidly changing regional and global economies, an increasingly diverse population, and rising environmental concerns. These long-standing challenges must now be understood in the context of the terrorist attacks of 9/11/2001, which have had repercussions in almost every aspect of life in North America and indeed the entire world.

## The National and Global Geopolitical Repercussions of the Terrorist Attacks of September 11, 2001

On September 11, 2001, around 9:00 A.M., two fully loaded passenger jets were purposely flown into the World Trade Center towers in New York City. A third passenger jet struck the Pentagon, and a fourth plane crashed in rural Pennsylvania, forced down by its passengers before it could reach its target in Washington, D.C. All four planes had been commandeered by suicide hijackers. Within the hour, the World Trade Center towers collapsed, killing 2792 people. Altogether, 3016 people died as a consequence of the attacks. Soon it was learned that the attacks were masterminded by a radical Muslim organization known as Al Qaeda, led by a dissident Saudi Arabian millionaire, Osama bin Laden, who had allied himself with, and lived among, a faction of Muslim extremists known as the Taliban, then in control of Afghanistan. The hijackers themselves were mostly from Saudi Arabia, with a few from the United Arab Emirates, Egypt, and Lebanon. The attacks were designed to undermine the power of the West. Especially in Southwest Asia, Muslim societies have long been strained by the simultaneous embrace and resentment of Western cultural, political, and economic influence.

***Security Issues Become Paramount.*** A wave of fear swept the United States during and after the attacks with the sudden realization that the country was more vulnerable than previously thought. In the nearly 60 years since the Japanese had bombed Pearl Harbor in December 1941, U.S. citizens had come to believe that on the North American continent they were safe from foreign attack. On September 11, it became clear that the very openness of U.S. society and its physically unguarded borders allowed multiple opportunities for intrusion by those who might do harm.

The motivation for the terrorist attacks was difficult for most people in the United States to understand because they had long assumed that most people around the world sought to emulate their way of living and thinking. Many were horrified by the sudden realization that there are many people around the world who resent U.S. economic, military, political, and cultural power, and that the 9/11 hijackers themselves had lived, worked, and studied among Americans, all the while developing harsh antagonisms. One outcome was an anti-Muslim (and antiforeign) backlash that at times led to violent vigilantism. The U.S. government itself used the Patriot Act (hastily passed after the attacks) to incarcerate without due process hundreds of Muslim citizens and legal residents as well as illegal aliens. Some have been held for more than five years, without due process.

Several years after the attack, it remains difficult for Americans to agree on which security strategies are appropriate, given the strong value placed on personal freedom in North American society. Indeed, the Patriot Act (slightly amended and reenacted in 2006) has been deemed too restrictive and abusive of individual liberty by a diverse coalition of interest groups, such as the American Civil Liberties Union (ACLU) and the National Rifle Association (NRA).

Even more divisive has been the issue of what strategies the United States should implement on the international level. Should the United States focus primarily on its own interests and take aggressive stands against perceived enemies and possible terrorists, or should it cooperate with the rest of the world to alleviate the festering resentment and outrage (justified or not) that had led to the 9/11 events?

National debate was cut short when, in the fall of 2001, the administration of George W. Bush, with the advice and consent of Congress, launched what was called the War on Terror. The first target was Afghanistan, thought to be host to the elusive Al Qaeda network; the aim was to capture bin Laden (as of mid-2007 his whereabouts were still unknown). Then, in the spring of 2003, based on unsubstantiated claims that President Saddam Hussein of Iraq may have been involved in the 9/11 events, had chemical weapons, and was seeking to acquire materials for nuclear weapons, President Bush brought the War on Terror to Iraq.

***The World Is Outraged.*** The international community extended warm sympathy to the United States after 9/11 and generally supported the war in Afghanistan, but the manner in which the war in Iraq was conceived and carried out led to the most severe diplomatic isolation in U.S. history. The United Nations and all major U.S. allies, with the exception of the United Kingdom, advised against attacking Iraq and did not cooperate in the attack. World opinion was overwhelmingly against U.S. policy in Iraq, and the failure of the United States to substantiate any of its prewar justifications for the invasion of Iraq brought about high levels of opposition, in the United States as well as internationally (Figure 2.14). World opposition solidified when, after its initial success in removing Saddam Hussein from power, the United States was unable to establish order and democracy. Iraqi resistance to the U.S. occupation has grown, and tension among Iraq's different ethnic and religious groups has brought the country to the brink of full-scale civil war. A 2006 report from the Johns Hopkins Bloomberg School of Public Health revealed that over 600,000 Iraqi civilians had died as a result of the war. More than 3000 U.S. armed service people had been killed by February 2007.

**FIGURE 2.14** **Antiwar protest in New York City.** Large protests against the Iraq war have been regular events in many communities since the war began in 2003. This picture was taken on the third anniversary of the war, March 18, 2006, when thousands of protesters moved along 42nd Street in New York. Similar marches occurred around the world. [Ashley Gilbertson/Aurora Photos.]

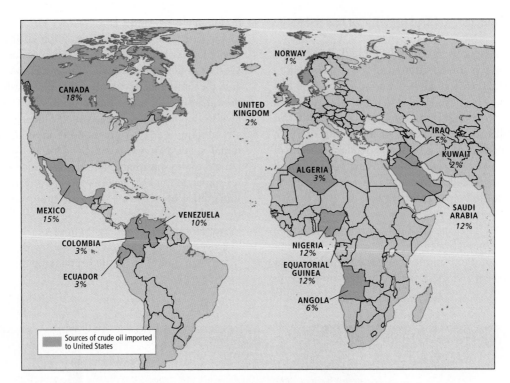

**FIGURE 2.15** Sources of crude oil imported to the United States. The United States is dependent on crude oil from many locations around the world. This map shows the degree of dependence on specific suppliers as of November 2005. [Adapted from http://www.gravmag.com/oil.html.]

Equally damaging has been the U.S. decision not to treat captured prisoners in the War on Terror according to the **Geneva Conventions** (treaties that protect the rights of prisoners of war). Repeated discoveries of physical and psychological torture of prisoners by the U.S. military have sparked outrage both at home and abroad. Indeed, many critics of the U.S. responses to 9/11 point out that the various actions may have made the country more vulnerable to terror, not less, because of the loss of international support that resulted. By 2006, opinion polls around the world revealed little support for the U.S. War on Terror and increasing criticism of the United States in general, though most of those polled distinguished between the U.S. government and the people of the United States.

***The Economy Is Shaken.*** The 9/11 attacks had immediate negative effects on the economy. The recession that was already looming in the spring of 2001 deepened as investors lost confidence, and hundreds of thousands of people lost jobs. Just in New York City, 79,000 people lost their jobs as a direct result of the destruction of the World Trade Center. Perhaps the most important impact was a reduction in air travel. Because the hijackers had been able to breach airport security easily and had used commercial jets as missiles, Americans suddenly canceled planned trips and reevaluated the role air travel would play for them in the future. Cities and businesses dependent on tourism or business travel suffered huge losses. It took until the end of 2005 for U.S. air traffic to reach the level it had been at in the summer of 2001. By that time, stringent and costly security measures were in place, not only at all U.S. airports, but in every airport in the world with connections to the United States. The annual budget of the Federal Transportation Security Agency grew to $5.5 billion (equal to the annual FBI budget).

The economic impacts of the 9/11 attacks are now felt mainly through the cost of the wars in Afghanistan ($1 billion a month as of March 2006) and Iraq ($8.4 billion a month, or $200 billion per year by January 2007), which were 14 times higher than estimated by the Bush administration in 2003. For comparison purposes, just $35 billion would fund preschool training for every 3- and 4-year-old in the United States for a year.

***Dependence on Oil Is Questioned.*** The 9/11 attacks, connected as they were to Muslim extremists, led to alarms about U.S. dependence on oil produced in predominantly Muslim countries. In 2003, roughly 30 percent of the oil used in the United States came from countries with large Muslim populations, and by 2005, it was up to 34 percent (Figure 2.15). It is, doubtful, however, that all Muslim countries would participate in disrupting flows of oil to the United States, because they are too dependent on the oil income. Most oil price analysts argue that, over the long term, global oil prices will be most affected not by the aftermath of 9/11, but by spiraling demand for oil in growing Asian economies and by worries over the effects of oil-based emissions on global warming. Nevertheless, the price rises immediately after 9/11 and during the Iraq war and the ever-increasing demand for energy have sparked greater interest in alternatives to oil, especially renewable energy sources.

# RELATIONSHIPS BETWEEN CANADA AND THE UNITED STATES

Citizens of Canada and the United States share many characteristics and concerns. Indeed, in the minds of many people—especially those in the United States—the two countries are one. Yet that is hardly the case. In fact, Canadians often make the point that their national identity is focused on noting how different Canada is from the United States. Three key words characterize the interaction between Canada and the United States: *asymmetries*, *similarities*, and *interdependencies*.

## Asymmetries

*Asymmetry* means "lack of balance." Although the United States and Canada occupy about the same amount of space (Figure 2.16), much of Canada's territory is sparsely inhabited, cold country. The U.S. population is nearly ten times the Canadian population. And although Canada's economy is one of the largest and most productive in the world (producing U.S.$1.11 trillion purchasing power parity (PPP) in goods and services in 2005), it is dwarfed by the U.S. economy, which is more than ten times larger ($12.37 trillion PPP in 2005).

There is also asymmetry in international affairs. The United States is an economic, military, and political superpower with a world leadership role that preoccupies it. Since 2000, U.S. foreign policy has focused primarily on relations with Southwest Asia, Russia, Europe, China, Japan, Southeast Asia, and Middle and South America. Canada is only an afterthought in U.S. foreign policy, in part because the country is so secure an ally. But for Canada, managing its relationship with the United States is the foreign policy priority. As a result, Canadians focus much more on events and circumstances in the United States and how they should react to them than vice versa. These asymmetries are cause for discontent in Canada, and, as the importance of transborder issues such as migration, free trade, environmental pollution, and security against terrorism increases, people in the United States may have to pay closer attention to their northern neighbors.

## Similarities

The United States and Canada do have much in common. Both are former British colonies, and from this experience they have developed comparable political traditions. Both are federations (of states or provinces), and both are representative democracies; their legal systems are also alike. Not the least of the features they share is a 4200-mile (6720-kilometer) border, which, despite heightened security against terrorism since 9/11, still contains the longest sections of unfortified border in the world. Official checkpoints on highways have been increased, and there are now 1000 U.S. border guards, rather than the 300 deployed in 2001 (but far fewer than the nearly 10,000 agents along the border with Mexico, which is half as long). Canada has only several hundred border guards, and they are unarmed. For thousands of miles along the U.S.–Canada border, there is not even a fence, and rural residents pass out of one country and into the other many times in the course of a day simply by going about their usual business.

Well beyond the border country, Canada and the United States share many other landscape similarities. Their cities and suburbs look much the same. The billboards that line their highways and freeways advertise the same brand names. Shopping malls have followed suburbia into the countryside, encouraging similar patterns of mass consumption. And the two countries share similar patterns of ethnic diversity that developed, as we shall see shortly, in nearly identical stages of immigration from abroad.

## Interdependencies

Canada and the United States are perhaps most intimately connected by their long-standing economic relationship, which includes mutual tourism, direct investment, migration, and, most of all, trade. By 2005, that trade relationship had evolved into a two-way flow of U.S.$1 trillion annually (Figure 2.17). Each country is the other's largest trading partner. Canada sells 85 percent of its exports to the United States and buys 59 percent of its imports from the United States. The United States, in turn, sells 23 percent of its exports to Canada and buys 17 percent of its imports from Canada. Notice that asymmetry exists even in the realm of interdependencies: Canada's smaller economy is much more dependent on the United States than the reverse. Nonetheless, if Canada were to disappear tomorrow, as many as a million American jobs would be threatened. The asymmetry of this relationship will undoubtedly persist; because of its size, large population, and giant economy, the United States is likely to remain the dominant and somewhat insensitive partner. As former Canadian Prime Minister Pierre Trudeau once told the U.S. Congress, "Living next to you is in some ways like sleeping with an elephant: No matter how friendly and even-tempered the beast, one is affected by every twitch and grunt."

**FIGURE 2.16** Political map of North America.

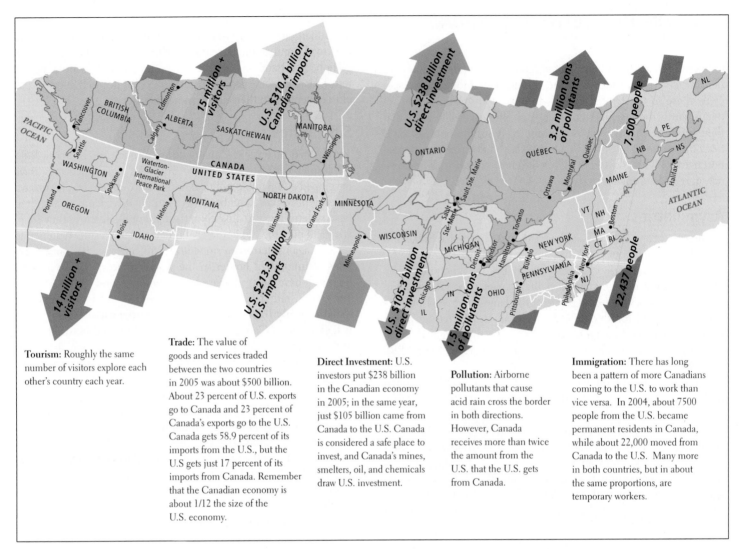

**Tourism:** Roughly the same number of visitors explore each other's country each year.

**Trade:** The value of goods and services traded between the two countries in 2005 was about $500 billion. About 23 percent of U.S. exports go to Canada and 23 percent of Canada's exports go to the U.S. Canada gets 58.9 percent of its imports from the U.S., but the U.S gets just 17 percent of its imports from Canada. Remember that the Canadian economy is about 1/12 the size of the U.S. economy.

**Direct Investment:** U.S. investors put $238 billion in the Canadian economy in 2005; in the same year, just $105 billion came from Canada to the U.S. Canada is considered a safe place to invest, and Canada's mines, smelters, oil, and chemicals draw U.S. investment.

**Pollution:** Airborne pollutants that cause acid rain cross the border in both directions. However, Canada receives more than twice the amount from the U.S. that the U.S. gets from Canada.

**Immigration:** There has long been a pattern of more Canadians coming to the U.S. to work than vice versa. In 2004, about 7500 people from the U.S. became permanent residents in Canada, while about 22,000 moved from Canada to the U.S. Many more in both countries, but in about the same proportions, are temporary workers.

**FIGURE 2.17** Transfers of tourists, goods, investment, pollution, and immigrants between the United States and Canada. Canada and the United States have the world's largest trading relationship. The flows of goods, money, and people across the long Canada–U.S. border are essential to both countries, but because of its relatively small population and economy, Canada is more reliant on the United States than the reverse. Restrictions on travel and residency permits since September 11, 2001, have affected Canadians more than U.S. citizens. Pollution that moves across the border also affects Canada to a much greater extent. All amounts shown are in U.S. dollars. [Adapted from *National Geographic* (February 1990): 106–107, and augmented with data from the Office of Travel and Tourism Industries, at http://tinet.ita.doc.gov/view/f-2000-04-001/index.html?ti_cart_cookie= 20030901.194057.17337; International Travel Forecasts, Fourth Quarter Update, 2003, at http://ftp.canadatourism.com/ctxUploads/en_publications/ InternationalTravelForecastsQ4.pdf; the Canadian Embassy, Washington, D.C., at http://www.canadianembassy.org/trade/wltr-en.asp; *CIA Factbook*, 2002, 2003; The Green Lane, Acid Rain, at http://www.ec.gc.ca/acidrain/acidhealth.html; and *Population Today* 30(2) (February–March 2002).]

## ECONOMIC AND POLITICAL ISSUES

The economic and political systems of Canada and the United States have much in common. Both countries evolved from societies based primarily on family farms. After an era of industrialization, both now have primarily service-based economies with important technology sectors and economic influence that reaches worldwide. Politically, the two countries have similar governments and face similar issues. Yet they often take very different approaches to issues such as unemployment, health care, and international relations.

## North America's Changing Agricultural Economy

North America benefits from an abundant supply of food, and it is an important producer of food for both domestic and foreign consumers (Figure 2.18). Exports of agricultural products were once the backbone of the American economy; today, most American farm income derives from domestic sales, with exports accounting for just 20 to 30 percent. The overall economic role of agriculture within North America has been shrinking for many years. It now

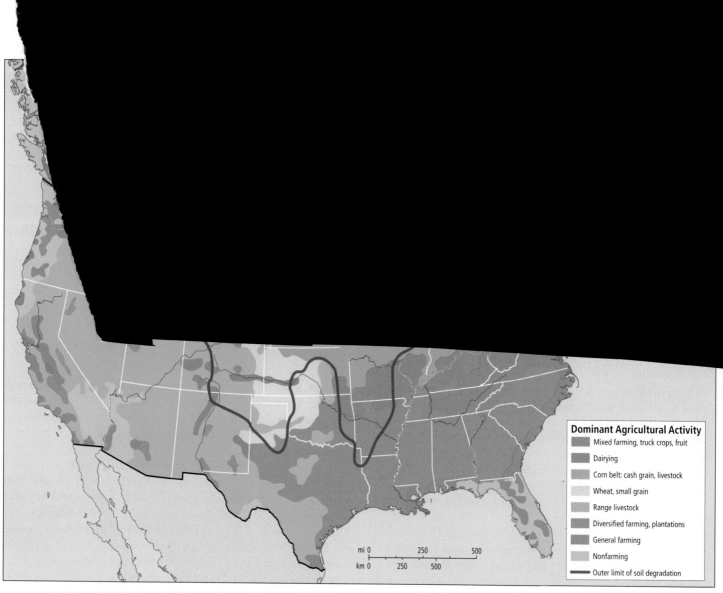

**Dominant Agricultural Activity**

- Mixed farming, truck crops, fruit
- Dairying
- Corn belt: cash grain, livestock
- Wheat, small grain
- Range livestock
- Diversified farming, plantations
- General farming
- Nonfarming
- Outer limit of soil degradation

mi 0     250     500
km 0   250   500

**FIGURE 2.18  Agriculture in North America.** North America is remarkable in that some type of agriculture is possible throughout much of the continent. The major exceptions are the northern reaches of Canada and Alaska and the dry mountain and basin region (the continental interior) lying between the Great Plains and the Pacific coastal zone. However, even some marginal areas, such as southern California, southern Arizona, and the Utah Valley, are cultivated with the help of irrigation. (Hawaii is not included here because it is covered in Oceania.) [Adapted from Arthur Getis and Judith Getis, eds., *The United States and Canada: The Land and the People* (Dubuque, Iowa: William C. Brown, 1995), p. 165.]

accounts for just 1.7 percent of the region's gross domestic product—this despite the fact that American agriculture is highly productive per worker and per acre.

The shift to mechanized agriculture in North America has brought about sweeping changes in employment and farm ownership. Agriculture employed 90 percent of the American workforce in 1790 and 50 percent in 1880. Thousands of highly productive family-owned farms, spread over much of the United States and southern Canada, provided the majority of food and fiber for domestic consumption and the majority of all exports until 1910. Currently, agriculture employs less than 2 percent of the North American workforce because today's highly mechanized farms need few workers. A successful farm now requires so much capital investment in land and machinery that large **agribusiness** corporations,

with their ready access to cash, have an advantage over individual farmers, especially because they can engage in all aspects of food production: farming, marketing, research, processing, transport, and delivery to customers. Corporate agriculture provides a wide variety of food at low prices for North Americans but makes it difficult for small farms to compete; in many rural areas, corporate farms have depressed local economies and created social problems. Communities in such places as rural Iowa, Nebraska, and Kansas were once made up of farming families with similar incomes, similar social standing, and a commitment to the region. By 2000, farm communities were increasingly composed of a few wealthy farmer-managers amid a majority of poor, often migrant, Hispanic or Asian laborers working for low wages on corporate farms and in food-processing plants (see Box 2.3 on page 106). The result is increasing class disparity, with low levels of education and well-being for many people, and local governments that no longer are responsive to many residents, who may not even be citizens. Under the assumption that the family farm has inherent value, the U.S. federal government and some states are increasing efforts to protect family farms and the rural communities of which they are a part. The states of Iowa, Kansas, Minnesota, Missouri, Nebraska, North and South Dakota, Oklahoma, and Wisconsin all have laws that restrict corporate involvement in agriculture.

Corporate agriculture relies heavily on the use of highly modified seeds and chemical fertilizers, pesticides, and herbicides to increase crop yields. These farming methods have the potential to contaminate food and pollute environments. Compared with owner-operators who have a personal stake in the long-term sustainability of a farm (Figure 2.19), agribusiness can more easily ignore such environmental effects as erosion and stream pollution. Many North American farming areas had lost as much as one-third of their topsoil even before corporate farms took over, and many fear that corporate agriculture will hasten the demise of America's productive topsoil. Perhaps an even more controversial aspect of corporate agriculture is **genetic engineering,** the modification of the DNA of animals and crops to create varieties that are bigger, more productive, or resistant to pests and diseases. In Europe and other parts of the world to which North American farms have traditionally exported, genetic manipulation of food crops and animals is seen as potentially so unsafe that, in a number of cases, Europe has refused to import U.S. agricultural products.

## Changing Transport Networks and the North American Economy

North America is going through dramatic transformations, many of them related to changes in technology and trade. As manufacturing fades in economic importance, just as agriculture did

earlier, a vibrant new service economy, based increasingly on technology and information exchange, is emerging. The old economic core no longer dominates, as new centers of production spring up across the continent, facilitated by ground and air transportation networks.

The development of the inexpensive mass-produced automobile in the 1920s changed North American transport radically. Soon trucks were delivering cargo more quickly and conveniently than the railroads could, and the number of miles of railroad track decreased between 1930 and 1980. The **Interstate Highway System,** a 45,000-mile (72,000-kilometer) network of high-speed, multilane roads, was begun in the 1950s and completed in 1990. Because this network was connected to the vast system of local roads, it introduced previously unimagined flexibility, speed, and low cost to the delivery of manufactured products. Thus the highway system made possible the dispersal of industry and related services into nonurban locales across the country, where labor, land, and living costs were cheaper.

After World War II, air transport also served economic growth in North America. Although air transport is used to move goods, its primary niche is business travel. Despite the increasing use of instant communication via the Internet, face-to-face contact is still seen as essential to American business culture. To address North America's pattern of widely dispersed industries in numerous medium-size cities and to facilitate one-day travel between any two points, air service is organized as a **hub-and-spoke network.** Hubs are strategically located airports used as collection and transfer points for passengers and cargo that are then shifted to flights departing to other hubs or final destinations. The hub system has simplified the transport of passengers and cargo, but for passengers, travel time and distance may be lengthened, and delays can affect the synchronization of the system.

Flying is a way of life in North America, used for much personal as well as business travel. In 1999, North American airlines carried nearly 640 million passengers, more than double the continental population. A Gallup survey in 2000 revealed that 80 percent of the U.S. adult population had flown at least once, more than one-third of them in the previous 12 months. The air transport industry was deeply affected by the aftermath of 9/11 and by rising oil prices. Several airlines went into bankruptcy after 2001, and the industry has reorganized to cut costs by reducing flights and services. Nonetheless, by 2005, combined Canadian and U.S. air traffic had more than rebounded to about 700 million passengers.

## The New Service and Technology Economy

By 2002, the employment picture in North America had changed in two important ways. First, occupational structure had changed: routine, low-skill, mass-production industrial jobs were less available due to mechanization and firms moving such operations abroad. Second, education requirements had increased: knowledge-intensive jobs requiring specialized professional training in technology and management were growing fast.

*Decline in Manufacturing Employment.* By the 1960s, the geography of manufacturing was changing. Because of increased costs of production resulting from higher pay and benefits and better working conditions won by labor unions, as well as the cost of keeping equipment technologically current, the old economic core began to lose its advantageous position. Many companies began moving their factories elsewhere. At first, the southeastern United States was a major destination for these industries because the absence of labor unions there resulted in lower wages, and the warm climate meant lower energy costs than elsewhere in the country. However, by the 1980s, and especially after the passage of the North American Free Trade Agreement (NAFTA) in 1994, certain industries, such as clothing, electronic assembly, and auto parts manufacturing, were moving farther south to Mexico or overseas. In these places labor was vastly cheaper, and laws mandating environmental protection and safe and healthy workplaces were absent or less strictly enforced—a further source of savings.

Another factor in the decline of manufacturing employment is automation. The steel industry provides an illustration. In 1980, huge steel plants, most of them in the economic core, employed more than 500,000 workers. At that time, it took about 10 person-hours and cost about $1000 to produce 1 ton of steel. Spurred by more efficient foreign competitors, the North American steel industry applied new technology in the 1980s and 1990s to lower production costs, improve efficiency, and increase production. Revamped plants in the late 1990s produced 1 ton of steel with just 3 person-hours of labor (a 70 percent reduction) at a cost of just $200 per ton. In addition, new technology has allowed the use of cheap scrap metal from many sources in smaller operations (minimills) that are located in less urbanized areas (principally the Southeast). By 2006, these minimills produced steel at the rate of just 0.44 person-hour per ton (a more than 90 percent reduction since 1980) and at a cost of about $165 per ton. The steel industry in the United States now employs fewer than half the workers it did in 1980.

Throughout North America, far fewer people are now producing more of a given product at far lower cost than was the case 20 years ago. So, although the share of the gross domestic product (GDP) produced by manufacturing has declined, the actual level of industrial production often has not.

*Growth of the Service Sector.* As factories moved outside the United States and manufacturing employment declined, the economic base of North America shifted increasingly to a broad and varied **service sector** (economic activity that involves the sale of services). As of 2000, in both Canada and the United States, about three-fourths of the GDP and a majority of the jobs were in the service sector. In Canada, 70 percent of workers, and in the United States, 80 percent, now work in services: transportation, utilities, wholesale and retail trade, health, leisure, maintenance, government, and education. High-paying jobs exist in all of these categories, but low-paying jobs are common. Wal-Mart and Manpower Inc. (an agency that arranges temporary contracts between workers and employers, often with no health care or retirement benefits), are the largest employers in the United States (see Box 2.1 on page 76). Service jobs are often connected in some way to international trade, entailing the processing, transport, and trading of agricultural and manufactured products and information that are either imported or exported from North America; hence international events can shrink or expand the numbers of these jobs.

An important subcategory of the service sector involves the creation, processing, and communication of information—what is often labeled the **knowledge economy.** The knowledge economy includes workers who manage information, such as those employed in finance, publishing, print and electronic media, higher education, research and development, and professional aspects of health care. Industries that rely on the use of computers and the Internet to process and transport information—banks, software companies, medical technology companies, publishing houses—are increasingly called **information technology (IT)** industries. They are freer to locate where they wish than were the manufacturing industries of the old economic core, which depended on massive amounts of steel and energy, especially from coal. By contrast, IT industries depend on highly skilled thinkers and technicians and hence they are often located near major universities and research institutions.

By the late 1990s, yet another technological change—the Internet—was expanding the networks of individuals, communities, corporate organizations, and governments that underlie the IT-based economy. The **Internet** is a computer network that allows for the electronic transfer of all kinds of information virtually instantaneously to any place in the world—information that used to be carried via mail over roads, rails, water, and air. The Internet is emerging as an economic force more rapidly in North America than in any other world region. It was here that the Internet was first widely available, and, though it has only 5 percent of the world's population, North America accounts for 22 percent of the world's Internet users. As of 2006 roughly 68 percent of the population of Canada and the United States used the Internet, compared to 50 percent in the European Union and 16 percent for the world as a whole.

The geography of the "new economy" of information technology bears some striking resemblances to the hierarchies of the "old economy" of industrial production. In the old geographic hierarchy of places connected by transport and communications systems, the large cities and highly productive industries were usually connected early and well. Less influential outlying towns and enterprises were connected later, and rural areas later still. Today, those places first "wired" for information technology tend to be large and wealthy cities, major government research laboratories, and universities on the cutting edge of technology innovations. A spatial **digital divide** has developed. Though rural middle-class homes are increasingly online, important portions of the population are left out: the poor, the elderly, women, and minorities are less likely to have a computer and to be able to afford Internet connections. For them, the public library or the county courthouse may be the only access. Once the information technology infrastructure is in place, however, distance from centers is far less of a hindrance to participation in the new IT economy than it used to be in the industrial economy.

## Globalization and the Economy

Throughout the previous discussion, globalization—the forging of worldwide economic linkages, sometimes by country-to-country agreements, sometimes by general reductions in barriers to trade—has been recognized as a force driving economic change in North America. North America has a wide trading network (Figures 2.20 and 2.21) and, because of its size and wealth, wields great power in the international organizations that affect global economic relations, such as the United Nations, the World Bank, the International Monetary Fund (IMF), and the World Trade Organization (WTO).

In the past, trade barriers have been important aids to North American development. For example, upon achieving independence from Britain in 1776, the new U.S. government imposed tariffs and quotas on imports and gave subsidies to domestic producers to protect fledgling domestic industries and commercial agriculture. As a result, the economic core region flourished. But if tariffs and quotas are kept in place too long, they can result in stagnation. Without competition, quality may suffer and prices will remain high.

The United States is now an active partner in the World Trade Organization, which seeks to lower trade barriers and increase competition. Nonetheless, some tariffs and quotas against low-priced manufactured goods from developing countries (such as textiles, shoes, and agricultural products) remain in effect in the United States. The stated role of these tariffs is to prevent less-expensive foreign-produced versions of these goods from driving North American producers out of business, but they have the

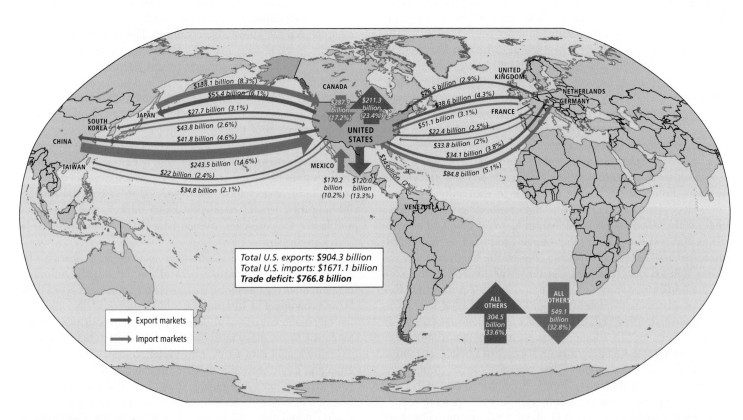

**FIGURE 2.20** Economic issues: Top ten U.S. import and export trading partners (goods only). As you would expect because of geography and NAFTA, Canada and Mexico are the top trading partners of the United States. (The arrow size represents percentage of trade, not dollar amount.)

[Adapted from FTDWebMaster, Foreign Trade Division, U.S. Census Bureau, Washington, D.C., at http://www.census.gov/foreign-trade/statistics/highlights/top/top0512.html#exports.]

added effect of shutting poor countries out of lucrative markets just when these fledgling economies need market opportunities, as the United States once did. Tariffs aimed at developing economies are one thing that spurs the public protests against the United States and the European Union that often accompany WTO meetings.

**Government subsidies**—payments and tax breaks that cover part of the production costs of some farm products—also help North American producers to sell their goods on the global market for less than their foreign competitors, even those in very poor countries, who may thereby be driven out of production. Nonetheless, in the highly competitive global market, less-developed countries and world regions have some advantages, such as lower wages, fewer financial and environmental controls, and lower taxes. Increasingly, workers in North America, like those in developing countries, are noting that globalization is a mixed bag, offering both opportunities—low-cost goods at Wal-Mart—and risks—loss of job security—for the individual worker (Box 2.1).

### The Role of Regional Trade Agreements in Global Trade: NAFTA.

The United States and Canada have long had internal free trade unrestricted by tariffs or quotas between the states and provinces. In recent years, Canada and the United States have promoted the idea of free trade in the whole of North America based on the idea that tariffs and quotas are constraints on economic growth that raise prices for consumers, discourage competition

among producers, and restrict the movement of workers. The process formally began with the Canada–U.S. Free Trade Agreement of 1989. The creation in 1994 of the **North American Free Trade Agreement (NAFTA)** brought in Mexico as well. Today, NAFTA is the world's second-largest trading bloc, both in population size and in dollar value of trade (the European Union, which includes many more countries, is the largest). A major long-term goal of NAFTA is to increase the level of trade among Canada, the United States, and Mexico, though not necessarily to balance that trade (the Canadian and Mexican economies, at $1.08 trillion and $1.06 trillion per year, respectively, are together only about 17 percent the size of the U.S. economy of $12.37 trillion per year).

By the time of its full implementation in 2009, NAFTA will have largely integrated the economies of Canada, the United States, and Mexico by removing or reducing most trade barriers, except that against the free movement of labor across borders. Establishing uniform standards is part of this process. For example, all trucks eventually will be equipped with brakes and pollution control devices that meet agreed-upon standards so that they can move freely through all three countries.

It is not possible at this time to assess the overall effects of NAFTA with accuracy, partly because data have not been collected for a long enough time and partly because it is difficult to know whether perceived changes are due to NAFTA or to other changes in world trade. However, a few things are now clear. NAFTA seems

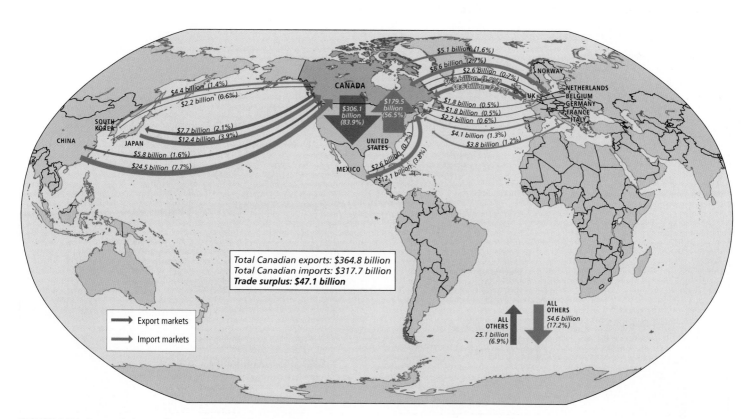

**FIGURE 2.21** Economic issues: Top ten Canadian import and export trading partners (goods only). The United States is Canada's top trading partner, with more than 71 percent of Canada's total trade. Mexico is among the country's top ten trading partners, but Japan and China both exceed Mexico. (The arrow size represents percentage of trade, not dollar amount.) [Adapted from Strategis, Industry Canada, September 2006, at http://www.2ontario.com/welcome/coca_401.asp.]

## BOX 2.1 | AT THE GLOBAL SCALE Wal-Mart

One of the most common places where North Americans encounter the global economy is at the local Wal-Mart store, where they can buy a wide range of items, including groceries, produced abroad. The selection is so good and the prices are so low that in nearly every market, Wal-Mart is driving local merchants and manufacturers, and even large chain stores, out of business.

Based in Bentonville, Arkansas, Wal-Mart is the world's largest retailer (Figure 2.22) and the world's largest company in terms of annual sales ($312.4 billion globally as of January 31, 2006). It is the largest private sector employer in the United States (1.3 million employees) as well as in Canada and Mexico; the company also has stores in Brazil, Germany, Japan, Argentina, Puerto Rico, South Korea, and Britain, and recently opened 68 stores in China, where it employs 36,000 people. More than 176 million people visit Wal-Mart stores worldwide every week. In the United States, eight out of ten households shop at a Wal-Mart store weekly.

Wal-Mart has perfected the art of obtaining goods at low prices, and because it buys and sells in such volume, it can influence the ways people shop, what kinds of products they buy, even how products are packaged. Each year Wal-Mart buys nearly $18 billion of goods directly from China, and the company's suppliers purchase an equal amount from China. Criticism of large "big box" retailers such as Wal-Mart is growing, however. First of all, the jobs associated with them provide low wages and few or no health benefits, and unionization is discouraged. Taxpayers in local communities and states have to absorb most of the health-care costs of Wal-Mart employees. Stores are usually located on relatively inexpensive land at the edges of suburban communities, where they take up open space and farmland, cause environmental disruption, and contribute to sprawl and automobile dependence. Some cities, such as Phoenix, Arizona, welcome the stores. By contrast, in May 2004, the National Trust for Historic Preservation designated the state of Vermont one of America's 11 most endangered historic places because huge retailers like Wal-Mart are putting out of business the hundreds of small stores that for years have been crucial to the special lifestyle of Vermont's many villages.

*Sources:* http://www.walmartstores.com/wmstore/wmstores/Home Page.jsp; "Wal-Mart and the World," PBS broadcast on "Now with Bill Moyers," December 19, 2003; "National Trust Names the State of Vermont One of America's 11 Most Endangered Historic Places," http://www.ptvermont.org/endangered2004.htm; Sana Siwolop, "Wal-Mart's Mixed Success Where Land Is Costly," NYTimes.com, May 26, 2004; Charles Fishman, *The Wal-Mart Effect: How the World's Most Powerful Company Works and How It Is Transforming the American Economy* (New York: Penguin, 2006).

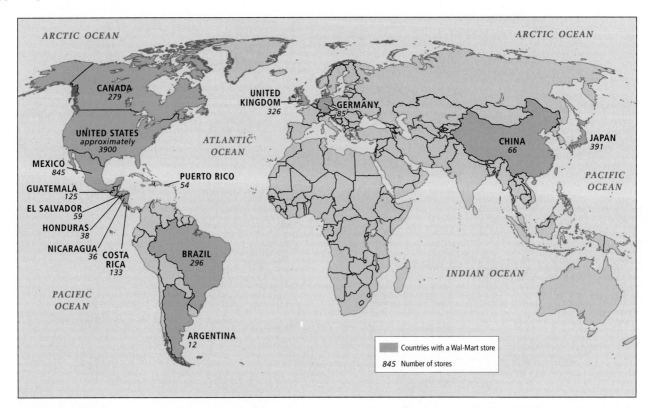

**FIGURE 2.22** Wal-Mart at the global scale. In late 2006, Wal-Mart had 3944 store operations in the United States and 2745 in 14 other countries. [Adapted from http://www.wal-martchina.com/english/walmart/wm_world.htm.]

**FIGURE 2.23** Toyota plant in Georgetown, Kentucky. Donna Jones puts control panels into Camrys as they move along the assembly line. Toyota leads foreign automakers in U.S. sales, and in 2006 had about 13 percent of the U.S. auto market. Eleven Toyota plants are located in the United States and one in Canada. [AP Photo/Al Behrman.]

to have worsened the long-standing U.S. **trade deficit,** the extent to which the amount of money earned by exports is exceeded by the amount of money spent on imports. Between 1993 and 2004, the value of U.S. exports to Canada and Mexico increased 77 percent, while the value of imports increased 137 percent. NAFTA also seems to have resulted in a net loss of 1 million jobs in the United States, with increased imports from Canada and Mexico displacing about 2 million U.S. jobs, and increased exports to these countries creating only about 1 million. While Canada's trade with the United States and Mexico has boomed, both internal trade and trade with the rest of the world have decreased, eliminating jobs in Canada. Moreover, although some new NAFTA-related jobs pay up to 18 percent more than the average American wage, those jobs are usually in different locations from the ones that were lost, and the people who take them tend to be younger and more highly skilled than those who lost jobs. Often the new jobs taken by former factory workers are short-term contract jobs or low-skill jobs that pay minimum wage and carry no benefits.

In Mexico, NAFTA appears to have increased exports and levels of foreign investment, but these gains have been concentrated in only a few firms along the U.S. border and have not increased growth in the Mexican economy. This may be due in part to stiff competition from U.S. agribusiness, which has resulted in about 1.3 million job losses in the Mexican agricultural sector. These losses, in turn, have fueled legal and illegal immigration from Mexico to the United States, which increased from 350,000 per year just before the start of NAFTA in 1992 to 500,000 to 600,000 per year in 2005.

Even as the benefits and drawbacks of NAFTA are being assessed, there is talk of extending NAFTA to the entire Western Hemisphere under what would be called the Free Trade Area of the Americas (FTAA). Such an agreement would have its own drawbacks and benefits, and a number of countries, such as Brazil, Bolivia, Ecuador, and Venezuela, are wary of being overwhelmed by the U.S. economy. This emerging trade agreement is discussed in Chapter 3.

***The Asian Link.*** NAFTA is only one way in which the North American economy is becoming globalized. The lowering of trade barriers has encouraged the growth of trade between North America and Asia, which, during the early 1990s, surpassed growth in trade with Europe. Although U.S.–Asian trade slowed during the Asian recession of the late 1990s, the growing importance of this region to the U.S. economy is obvious.

Examples of U.S.–Asian trade are the Japanese and Korean automotive companies that are locating plants in North America to be near their most important pool of car buyers—commuting North Americans. In rural locations that are also close to the Interstate Highway System, particularly east of the Mississippi River, the companies have found a ready, inexpensive labor force among people who value life in rural America yet do not wish to farm. Like those who will eventually buy the cars, these automotive workers are willing to commute 20 miles (32 kilometers) or more for secure jobs that pay reasonably well and include health and retirement benefit packages.

The Toyota Camry plant constructed during the 1980s in Georgetown, Kentucky, is one such enterprise. The Kentucky plant is Toyota's largest outside Japan, employing 7500 people who turn out more than 500,000 vehicles and engines per year (Figure 2.23). Several dozen smaller Japanese and U.S. plants in central and western Kentucky produce parts for the main plant. The Toyota plant has brought social and physical change to Georgetown. The arrival of Japanese executives and their families brought new ethnic diversity and required adjustments in the public schools, such as the teaching of Japanese and of English as a second language. New religious institutions were built for those Japanese who practice Buddhism. Water and electricity supplies had to be increased greatly to accommodate the new plant and labor force, and new businesses were established to cater to those workers. In 2006, as the venerable U.S. auto manufacturer General Motors announced it was downsizing, Toyota announced a new Camry installation in Indiana with 1000 new jobs. The Camry is the biggest-selling sedan in North America, and Toyota employs 38,000 North Americans and sells 2.5 million vehicles (of all models) a year.

***New Competition from Developing Economies for IT Jobs.*** By the early 2000s, the North American economy was affected by yet another aspect of globalization. Information technology (IT) jobs were being **outsourced,** meaning that a range of jobs, from software programming to telephone-based customer support services, were shifted to lower-cost zones outside North America. By mid-2003, 500,000 jobs had been outsourced, and forecasts are that by 2020, 3.3 million more IT jobs will follow. Newly developed IT centers in India, China, Southeast Asia, Central Europe, and Russia, where large pools of highly trained, English-speaking young people will work for wages that are less than 15 percent of their American counterparts', won contracts from North American firms such as Microsoft and Oracle. Some advocates of outsourcing argue that rather than depleting jobs, the trend will actually help job creation in North America by saving corporations money, which will then be reinvested in new ventures. The truth of this argument remains to be seen. One North American beneficiary of U.S. outsourcing is Canada, where the U.S. firm Allmerica saved 20 percent on software development expenses. The firm reports that outsourcing to Canada has four advantages: the work is done more cheaply than in the United States; travel and communications costs are lower than in more distant locations such as India or Europe; few cultural adjustments are necessary for workers; and there are no disruptive time zone changes. The reaction among American consumers to outsourcing of call centers and other services to places outside North America has been more negative than positive, with many complaining about having to deal with workers whose English is less than perfect.

## Canadian and U.S. Responses to Economic Change

The U.S. and Canadian governments have responded differently to the displacement of workers by economic change. For many decades, Canada has protected its working population from job losses or declines in income by enacting strict unemployment insurance laws. It also has spent more than the United States on social programs that lessen the financial burdens of working people in times of economic crisis. These policies have made the financial lives of working people more secure, but, until recently, the higher taxes and greater economic regulation they entail have also made Canada slightly less attractive than the United States to new businesses. This has contributed to Canada's unemployment rate (usually several percentage points higher than that in the United States) and its slightly lower rate of economic growth. By contrast, the U.S. approach provides little job protection or government unemployment assistance. Many poor rural and urban areas in the United States that have lost agricultural and manufacturing jobs have experienced increases in ill health, violent crime, drug abuse, and family disintegration—all of which are social problems connected with declining incomes and the lack of a social safety net. Nonetheless, for years the prevailing political position has been that the U.S. system is better for the economy because a spare social safety net means lower taxes for businesses and attracts new investment and new jobs, the benefits of which will trickle down to those most in need.

***The Link Between Health Care and Job Creation.*** Some U.S. firms are beginning to eye Canada as a desireable place to relocate precisely because of the country's social safety net, especially its government-sponsored health-care system. At present, among those U.S. firms that provide benefits, about 53 percent of the cost of every new employee is health insurance—a cost that places U.S. firms at a disadvantage in the global economy. As concerns grow

**TABLE 2.1   Health-related indexes for Canada and the United States**

| Country | Health-care costs as percentage of GDP[a] | Percentage of population fully insured | Deaths per 1000 | Infant mortality per 1000 live births | Maternal mortality per 100,000 live births | Life expectancy at birth (years) | Health expenditures per capita (PPP U.S.$)[a] |
|---|---|---|---|---|---|---|---|
| Canada | 9.6 | 100 | 7.5 | 5 | 6 | 79 | 2931 |
| United States | 14.6 | 66 | 8.3 | 7.3 | 17 | 77 | 5274 |

[a]Data from 2002; PPP = purchasing power parity.

*Sources: United Nations Human Development Report 2005* (New York: United Nations Development Programme), Table 6 and Table 10; World Resources Institute, "Population, health, and human well-being," at http://earthtrends.wri.org/pdf_library/data_tables/pop2_2005.pdf, in *Health Care Spending in 23 Countries,* at http://www.thirdworldtraveler.com and http://www.thirdworldtraveler.com/Health/O_Canada.html.

about the rising cost of private health insurance in the United States, interest in Canada's system has also grown.

The contrasts between the two systems are striking. The Canadian health-care system covers 100 percent of the population; in the United States, about 40 percent of the population has no health-care coverage. In Canada, health care is heavily subsidized by the government and is less expensive per capita and more effective in preventing illness than the largely private health-care system in the United States. The United States spends 14.6 percent of its gross domestic product on health care; Canada spends only 9.6 percent. Total health expenditures per capita in Canada are just 56 percent of those in the United States. Nevertheless, Canada does better than the United States on most indicators of overall health (Table 2.1). Medicines are also notably less expensive in Canada. Having guaranteed health care removes considerable pressure from those who work for low wages or who lose their jobs, even temporarily. While the taxes to cover a government health-care system are substantial, when spread over the population these taxes are more than offset by the lower costs to employers who don't have to pay for private care plans, hence the cost of creating new jobs is lessened.

## Systems of Government: Shared Ideals, Different Trajectories

Canada and the United States have similar democratic systems of government, but there are differences in the way power is divided between the federal government and provincial or state governments. There are also differences in the way the division of power has changed since each country achieved independence.

Both countries have a federal government, in which a union of states recognizes the sovereignty of a central authority while retaining certain residual governing powers. In both Canada and the United States, the federal government has an executive branch, a legislature, and a judiciary. In Canada, however, the executive branch is more closely bound to follow the will of the legislature, and the Canadian federal government has more and stronger powers (at least constitutionally) than does the U.S. federal government. Over the years, both the Canadian and U.S. federal governments have moved away from the original intentions of their constitutions. Canada's originally strong federal government has become somewhat weaker, largely in response to demands by provinces for greater autonomy over local affairs. The most assertive effort for provincial control has come from the French-speaking province of Québec, where many people feel overwhelmed by the rest of Canada. In the past decades, several referenda on political independence for Québec have narrowly failed.

In contrast to Canada, the initially more limited federal government in the United States has extended its powers. The U.S. federal government's original source of power was its mandate to regulate trade between states. Over time, this mandate has been interpreted ever more broadly. By the late twentieth century, the federal government was affecting life at the state and local levels primarily through its ability to dispense federal tax monies through programs such as grants for school systems, federally assisted housing, and health care (medicare, medicaid), as well as through the establishment of military bases, anticrime legislation, and the building of interstate highways. Money for these programs is withheld if state and local governments do not conform to federal standards. This practice has made some poorer states dependent on the federal government, but it also has encouraged some state and local governments to enact more enlightened laws than they might have done otherwise. In the 1960s, the federal government promoted civil rights for African-American citizens by providing enriched educational programs for minority children. Eventually, federal money was made available for programs promoting adult literacy, employment training, job opportunities, school vouchers, voter registration, and community development for disadvantaged African-American and Hispanic communities.

In both countries the courts at the federal level have ruled in ways that have enhanced the rights of minorities. Several U.S. Supreme Court decisions promoted racial integration in the schools and the workplace. In Canada, courts have ruled in favor of native language use by immigrants, have given large swatches of Canadian territory back to native groups, and have awarded reparations to Chinese immigrants who in the early twentieth century were charged a head tax when immigrants from Europe were not.

## Gender in National Politics and Economics

There are some powerful political contradictions in North America with regard to gender. Women apparently decided the U.S. presidential election of 1996, voting overwhelmingly for Bill Clinton, and in the 2006 interim elections 55 percent of women voted Democratic, registering strong anti-war sentiment. Such women's issues as family leave, health care, equal pay, day care, and reproductive rights were forced into the national debate. However, of the 538 people in the U.S. Congress as of 2006, just 85, or 16 percent, are women (up from 12.5 percent in 1998). At the state level, 23 percent of legislators are women. In the Canadian Houses of Parliament, 21 percent of the members are women and 20 percent of provincial elected officials are women. Both the United States and Canada are well behind several other countries in their percentages of women in national legislatures: Cuba (36), South Africa (30), Mozambique (35), Norway (38.2), and Sweden (45.3). Canada's governor-general (a largely ceremonial post appointed by the United Kingdom's Queen Elizabeth) is a woman (Figure 2.24).

On the economic side, in the United States men still hold 70 percent of the top executive positions in business and government, and female workers on average earn only about 62 cents for every dollar that male workers earn. This is up only 3 percent from 1970. Canadian female workers earn slightly more relative to men than do their U.S. counterparts. In both countries, however, women tend to work more hours than men, so their hourly earnings are even lower, relative to men's. It is estimated that if women and men earned equal wages, the effect would be to reduce the poverty rate by half. North American women are now about half the labor

**FIGURE 2.24** Canada's governor-general, Michaëlle Jean. Canada is both a parliamentary democracy and a constitutional monarchy that recognizes the sovereign of the United Kingdom as head of state, though not as head of government. Canada's governor-general is appointed by the Queen to "carry out Her Majesty's duties." Here, Michaëlle Jean (in beige suit) is being greeted by the Honorable Iona Campagnolo, lieutenant governor of British Columbia, in March 2006. [Sgt. Eric Jolin, Rideau Hall/Government General of Canada.]

force, with most working for male managers. In both countries women entrepreneurs are increasingly active, starting close to half of all new businesses. Such businesses tend to be small, however, and less financially secure than those in which the dominant control is held by men, in part because it is harder for women business owners to obtain loans and to land large contracts.

In education, Canadian women have attained virtually the same level as men in most categories, and this is nearly the case in the United States, where 25 percent of women age 25 and over hold an undergraduate degree, while 29 percent of men do. This imbalance is likely to disappear, because by 2000, U.S. women were earning slightly more undergraduate degrees than men. Nevertheless, the disparity in earnings between men and women persists, in both countries. In the United States there are also disparities in workers' compensation and retirement benefits; this is less the case in Canada.

## SOCIOCULTURAL ISSUES

North America is undergoing rapid sociocultural change, and much of this change has geographic aspects. In this section we discuss some especially widespread changes that are likely to persist into the future. First we examine a set of issues—urbanization, immigration, race and ethnicity, and religion—that stem from the fact that North Americans are an increasingly diverse people living in concentrated settlements where their differences can make it difficult to find common ground. Then we discuss two issues—the American family and aging—that illustrate changing social organization in this large, complex, and prosperous region.

## Urbanization

Urbanization is an old and important geographic phenomenon that has greatly affected life in North America and contributed to the region's prominent role in the global economy. Two of the world's most economically influential cities are in this region: New York and Los Angeles. (Other world-class cities are London, Tokyo, Paris, Hong Kong, and Singapore.) It is not population that sets such cities apart, but rather their innovative, trendsetting, knowledge-based economies.

*Metropolitan Areas.* Close to 80 percent of North Americans now live in **metropolitan areas,** cities of 50,000 or more plus their surrounding suburbs and towns. In the nineteenth and early twentieth centuries, cities in Canada and the United States consisted of dense inner cores and less dense urban peripheries. Starting in the early 1900s, central cities began losing population and investment while urban peripheries began growing. Workers were drawn to the peripheries of crowded cities (the suburbs) by the opportunity to raise their families in single-family homes with large lots in secure and pleasant surroundings. Some continued to travel to the inner city to work, usually on streetcars, then increasingly in private cars. After World War II, suburban growth accelerated as factories and firms seeking less expensive land on which to develop, as well as better access to highways and airports, followed their workers. They have been joined in the suburbs in the past two or three decades by the high-tech businesses of the new IT economy. Eventually, huge tracts of inner-city land that once held manufacturing industries were abandoned, and cities lost the tax revenue once generated there. These old industrial sites are called **brownfields** because their degraded conditions—often contaminated with chemicals and covered with obsolete structures—pose obstacles to redevelopment.

Also left behind in the inner cities were the least-skilled and least-educated citizens, many of whom are members of racial and ethnic minorities. In the early 1990s, 70 of the 100 largest U.S. cities had white majorities, but by 2000 almost half of the largest U.S. cities had nonwhite majorities. The majority population in these cities is a mixture of African-Americans, Asians, Hispanics who are nonwhite, and other groups who identify themselves as nonwhite. As many whites and some African-Americans moved to the suburbs and urban fringe, immigrants (especially Hispanics) took their places in the inner cities. Often those who remain in the inner cities are in great need of the very services—health care, social support (including churches, synagogues, and mosques), and schools—that have moved with more affluent people to the suburbs.

In Canada, the basic patterns are similar. For example, in Toronto in 1971, 95 percent of the population identified themselves as of European origin. By the mid-1990s, self-identified Euro-Canadians accounted for only 50 percent of Toronto's population. The change was not due to a reduction in this population, but to the fact that new arrivals from Asia, Africa, the Caribbean, and Middle and South America reduced the relative proportion of those of European origin.

Geographers James Fonseca and David Wong have observed that urban and suburban densities continually fluctuate. **Dense nodes** (small regions with extremely dense populations) may appear in urban peripheries where high-tech industries attract thousands of skilled workers to live and work. Nodes can also reappear in old urban centers and then grow into larger dense settlements when new businesses and industries locate on old industrial brownfields that have been rehabilitated. In this way, portions of the old industrial Northeast (Baltimore, Maryland, and Hackensack, New Jersey, for example) are becoming densely populated again, not so much by population growth as by population redistribution. In what has been labeled the **New Urbanism** movement, people are moving back to the city from the suburbs to take newly available jobs and because of their appreciation for the many advantages of urban life such as less dependence on automobiles, the convenient location of housing and jobs, and the cultural diversity of residents. **Gentrification** of old urban residential districts is increasingly common as primarily well-educated people invest substantial sums of money in renovating old houses and apartments, often displacing poor residents in the process. The effect of gentrification on the displaced poor in Canada appears to be somewhat less harsh than in the United States, primarily because of Canada's stronger social safety net that better ensures housing and social services.

In yet another fluctuation, during the 1990s, a significant number of people moved even farther out than the suburbs. Most of these new rural residents remain employed in urban areas, commuting long distances; others work at home via computers or as skilled craftspeople.

***The Megalopolis and Urban Sprawl.*** The growth of cities and the spread of their suburbs have resulted in two common urban phenomena: the megalopolis and urban sprawl. A **megalopolis** is created when several cities grow to the extent that their edges meet and coalesce. "Megalopolis" is the name chosen by French geographer Jean Gottman in the 1960s to describe the 500-mile (800-kilometer) band of urbanization stretching from Boston through New York City, Philadelphia, and Baltimore, to south of Washington, D.C. (Figure 2.25). Other megalopolis formations in North

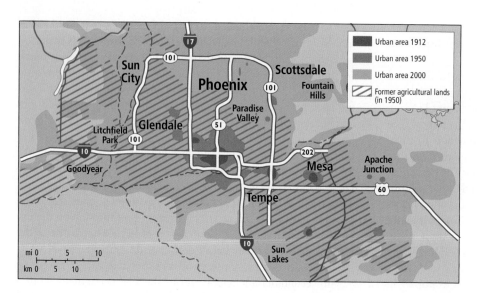

**FIGURE 2.26** Urban sprawl in Phoenix, Arizona, 2004. To see an informative animation of the rate and extent of urban sprawl in Phoenix, go to http://sciencebulletins .amnh.org/bio/v/sprawl.20050218/. [American Museum of Natural History 2004.]

America include the San Francisco Bay area, Los Angeles and its environs, the region around Chicago, and the stretch of urban development from Eugene, Oregon, to Vancouver, British Columbia.

Among the many consequences of industrialization and urbanization in the economic core and elsewhere across North America is the phenomenon known as **urban sprawl:** the invasion of farmland, forest, grassland, and desert by bulldozers preparing the way for suburban development—residences, malls, and discount outlets. Typically, real estate developers find rural land cheaper to develop than deteriorated city properties, because it is taxed at a lower rate (at least at the beginning of the development process), is less polluted, and requires less preparation for development. Residents of sprawl areas require automobiles to get through all aspects of daily life, and long commutes

helps the nation as a whole by reducing the need for food imports and by protecting against volatility in food prices. The town of Pittsford, New York, a suburb of Rochester, decided in 1997 that farms were a positive influence on the community. Mark Greene's 400-acre 200-year-old farm lay at the edge of town and, as the population grew, the chances of the farm remaining in business for another generation looked dim. As land prices rose, so did property taxes, and the Greene family could not meet its tax payments. Residents in the new suburban homes sprouting up on what had been neighboring farms pushed local officials to halt normal farm practices, such as noisy nighttime harvesting or planting, spreading smelly manure, or importing bees to pollinate fruit trees. Pittsford decided to stand by the farmers by issuing $410

## Regions and Countries of Origin for U.S. Immigrants

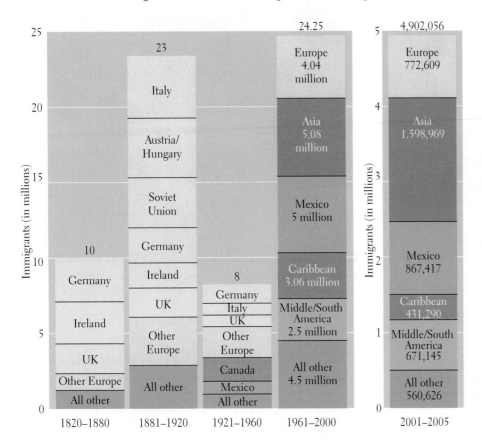

**FIGURE 2.27** Changing national origins of legal U.S. immigrants. The graph for 2001–2005 is at a different scale from those for 1820–2000 because the time span is only five years. Note that the number of legal immigrants from Asia, Mexico, and other Middle and South American countries continues to outnumber European and other immigrants. [Adapted from Philip Martin and Elizabeth Midgley, "Immigration to the United States: Journey to an uncertain destination," *Population Bulletin* 49 (September 1994): 25; Martha Farnsworth Riche, "America's diversity and growth," *Population Bulletin* 55 (June 2000): 11; Table 2, "Immigration by region and selected country of last residence: fiscal years 1820–1998," *1998 Statistical Yearbook of the Immigration and Naturalization Service,* p. 10; and *Yearbook of Immigration Statistics: 2005,* Table 2, at http://www.uscis.gov/graphics/shared/statistics /yearbook/LPR05.htm.]

Minnesota and the surrounding states and provinces. Many Germans settled in a wedge across the central Middle West, from Ohio to Missouri and in Nebraska and the Dakotas. Other Germans went to southern Appalachia and to south Texas. When immigrants send for friends and relatives, word of an attractive American location spreads to adjacent villages in their homeland. As a result of this phenomenon, which geographers have dubbed **chain migration,** people from particular places tend to be concentrated in certain urban neighborhoods or rural communities. Examples are the women from Vera Cruz, Mexico, who have come to work in post-Katrina New Orleans (see the chapter opening on page 53) or the shrimp fishermen from coastal Vietnam who have, since the 1960s, congregated along the Gulf Coast in Louisiana and Texas, where they continue to ply their trade.

Canada's general immigration pattern is similar to that of the United States, but specific settlement patterns differ. The English, Irish, and Scots tended to settle in the Maritime Provinces bordering the Atlantic coast, around the Great Lakes, and on farms in the Prairie Provinces; they were joined by Germans and Scandinavians. The French settled primarily along the lower St. Lawrence River, and their descendants remain concentrated in and near Québec. Canada never had a slave-based economy, so Canadians with some African ancestry have roots among those who escaped to Canada

from the United States during slavery or who came later from the United States, the Caribbean, or Africa. They tend to live in the cities of Ontario and Québec. During most of the twentieth century, Canada encouraged immigration to fill up its empty lands. After farmers from the British Isles arrived, others from Eastern Europe, Russia, and Ukraine settled mostly on the prairies. By the twenty-first century, as in the United States, the immigrants' places of origin had shifted away from Europe to Middle America and Asia. By the 1990s, 55 percent of Canada's new immigrants came from Asia. Many of these Asian immigrants are educated and affluent, and they are particularly attracted to Canada's Pacific coast urban areas.

***Diversity and Immigration.*** In both Canada and the United States, the relatively recent influx of immigrants from areas other than Europe has challenged the long-held assumption that the dominant culture of North America would forever be derived from Europe, with only a few small minorities of Native Americans, African-Americans, Hispanics, and Asians. Especially in cities, North America is increasingly characterized by wide ethnic diversity; and in Canada, with its relatively small population, the recent surge in immigration has led to near majorities of foreign-born residents in a number of leading cities. Furthermore, according to census data, although until 2000 the total number of immigrants tended to

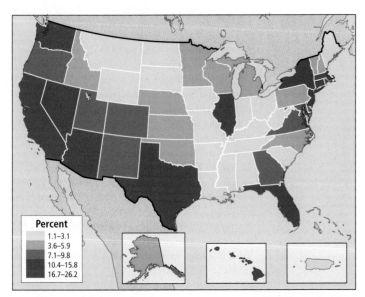

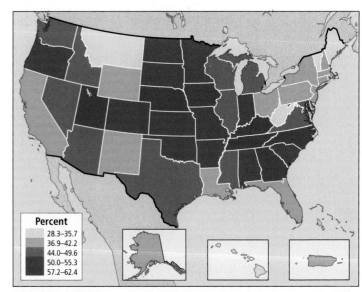

(a) Percent of total foreign-born within each state as of 2000. Includes all living foreign-born as of that year.

(b) Percent of foreign-born who entered each state during the 1990–2000 decade.

**FIGURE 2.28** Changing patterns of foreign-born settlements in the United States. [U.S. Census Bureau.]

settle in states around the periphery of the United States (Figure 2.28a), a change began to emerge in the 1990s. After 1990, immigrants chose to settle primarily in the American South, the Great Plains, and Utah in the Great Basin (Figure 2.28b).

Of the two countries, Canada seems to be the more receptive to newcomers. In recent decades, Canadians have accepted with equanimity many refugees from Asia and Africa. More than in the United States, emigrants to Canada are encouraged to retain their culture, use their native languages in schools, and maintain (usually voluntarily) residential segregation. The idea, first articulated by the Québecois and now openly accepted by most Canadians, is that the variety of immigrant cultures enriches Canada. Toronto, Ontario, provides a model of how Canadians accommodate ethnicity (Box 2.2).

In the United States, immigration and cultural diversity are topics of increasing public debate. By 2006, polls showed that a majority of U.S. residents felt future immigration should be controlled, but a clear majority also felt that immigration is the strength of the country and 76 percent supported the idea that illegal immigrants should have a chance to become citizens.

In a number of smaller U.S. cities that have been losing population, an influx of immigrants has proved to be an economic boost. For example, in Lewiston, Maine, a town of 36,000, 1200 Muslim Somalis who had spent a decade or more in refugee camps have found a home and started businesses, helping to revitalize main street. There were some nasty clashes, but in time older residents began to realize that the Somalis liked Lewiston for the same reasons they did: clean, quiet, safe neighborhoods, good schools, and strong families. Every year since 1960, Lewiston had lost population; now it is one of the few Maine towns that is growing. The Somalis are shopping, paying taxes, and buying property, and because there are now more students, federal and state funding has increased.

Below we examine some of the main aspects of the immigration debate in the United States.

### Do New Immigrants Cost U.S. Taxpayers Too Much Money?
Many people believe that immigrants use public services such as food stamps and welfare and thus are a liability to taxpayers. But repeated studies have shown that, over the long run, immigrants contribute more to the U.S. economy than they cost. Most immigrants start to work and pay taxes within a week or two of arrival. Even immigrants who draw on public services tend to do so only in the very first few years after they arrive. In fact, immigrants play important roles as taxpayers: payroll taxes, sales taxes, excise taxes, and indirect property taxes through rent. (Even illegal immigrants make many of these tax payments.) Perhaps most noteworthy is the role of immigrants in supporting the elderly. As the U.S. population ages, more people will be drawing Social Security, and the money to pay their benefits will come out of the pockets of young and middle-age workers. Yet the base of native-born young workers is shrinking because those people now reaching retirement age had relatively few children. Therefore, Social Security contributions of new immigrant workers, who tend to be young adults, are increasingly essential to older U.S. residents. Interestingly, a 2004 study by the National Institutes of Health (NIH) reports that immigrants are healthier and live longer than native U.S. residents and hence are themselves less of a drain on the health-care and social service systems. The NIH attributes this difference to a stronger work ethic, a healthier lifestyle, and more nutritious eating patterns of new residents when compared to U.S. society at large.

### Do Immigrants Take Jobs Away from U.S. Citizens?
Professionals in the United States are occasionally in competition with highly trained immigrants for jobs. This competition is not usually a big

## BOX 2.2 | AT THE GLOBAL SCALE   Ethnicity in Toronto

The lines dividing Toronto's ethnic neighborhoods are neither strict nor exclusionary, but when a city bus moves through town, Italians get off at particular stops, Chinese at others, and Greeks at yet others. Many of the neighborhoods have been rehabilitated and upgraded, or gentrified, to attract the middle-class children and grandchildren of the much poorer laborers who arrived earlier in the twentieth century. There are about 400,000 Torontonians of Italian descent, many of whom live in Corso Italia (the Little Italy of Toronto), an area of shops, sidewalk cafés, and elegant homes and apartments. The Chinese community of 350,000 is the fastest growing. Many Chinese who arrived in the 1990s are prosperous businesspeople from Hong Kong who feared the consequences of China's resumption of control of Hong Kong in June 1997. Throughout the city, Koreans and Vietnamese own many of the fresh produce and grocery stores, and Koreatown is home to herb shops and import emporiums. The Greek community revolves around the National Bank of Greece and the Greek Orthodox Church. Jews and people from the Caribbean are sprinkled in small enclaves throughout the city, while Polish people often settle together and specialize in owning bakeries and butcher shops. All in all, there are said to be at least 80 different ethnic neighborhoods in Toronto.

*Source:* Adapted from Richard Conniff, "Toronto," *National Geographic* 189(6) (June 1996): 120–139.

**FIGURE 2.29** Multilingual Toronto. Throughout Toronto, signs on city streets are written in English, the lingua franca, as well as in the predominant language of the neighborhood—in this case, Vietnamese and Chinese. [David R. Frazier/Tony Stone Images.]

problem, because there are not enough trained native-born people to fill these positions. The computer technology industry, for example, regularly recruits abroad in such places as India, where there is a surplus of high-tech workers. Certainly, the least-educated, least-skilled American workers may find themselves competing with immigrants for jobs, and in a local area a large pool of immigrant labor can drive down the wages of some Americans. Usually, however, immigrants with little education fill the very lowest paid service and agricultural jobs, which U.S. workers have already rejected.

### Are Too Many Immigrants Being Admitted to the United States?

Several circumstances have caused some to call for curtailing immigration into the United States. First, after a lull at midcentury, the immigration rate has picked up again (see Figure 2.27). In the 1990s, the rate of immigration appeared to be increasing faster than ever before (although this increase may be partly a result of changes in census coverage): 10.6 million legal immigrants entered the United States, and 6.4 million children were born to them. Immigrants and their children accounted for 78 percent of

U.S. population growth in the 1990s. At present rates, by the year 2050, the U.S. population will be over 400 million, nearly one-fourth larger than it would be if all immigration stopped now.

Second, undocumented (illegal) immigration reached unprecedented levels during the 1980s, 1990s, and early 2000s. The Census Bureau estimates that the total undocumented immigrant population in the United States is about 11 million, with about 700,000 more entering each year. This trend is worrisome because undocumented immigrants tend to lack skills, their numbers are uncontrolled, and they are not screened for criminal background or terrorist intentions. However, the vast majority do not engage in criminal activity. Since September 2001, border controls have tightened significantly; hence, many fewer illegal immigrants are entering. Furthermore, because they must hide from the authorities, illegal immigrants do not usually partake of public programs, so their drain on government services is negligible. In reality, illegal immigrants contribute to overall productivity by subsidizing the cost of products and services with their very low wages. Moreover, a great deal of essential work, such as farm labor, construction, and maintenance, is done by undocumented workers. The Inter-American Development Bank estimates that $418 billion is contributed annually to the U.S. economy by migrants from Middle and South America; meanwhile the immigrants send $32 billion in remittances to improve conditions in their home countries, a topic that is discussed in several of the chapters to come.

The debate regarding immigration became harsher in 2006, when federal and state laws were proposed to criminalize any interaction with undocumented immigrants. In the spring of that year, major street demonstrations were mounted in large cities across the United States (100,000 people demonstrated in Chicago, 500,000 in Los Angeles) by those opposed to this legislation. Some states, such as Illinois, Washington, Idaho, and New Mexico, have rallied to support immigrants by reinforcing programs that help them and their children. On the other hand, Virginia, Kentucky, South Carolina, and Arizona have passed laws that crack down on immigrants. Along the U.S.–Mexico border, the work of some 10,000 border guards was judged insufficient by conservative private militia groups, such as the Minutemen, who placed themselves along the border as vigilante enforcers. Immigration had become a central political issue at the national and state levels.

Sheriff's deputy Michael Walsh works along the Arizona-Mexico border. Talking with a reporter about his recent encounter with two bereaved Mexican young men holding the body of their relative, Walsh said, "Mostly you just find skeletons in the desert. This time there was a lot more emotion, family emotion. . . . Obviously they were close, they were crying. You have a name, you know that he had a brother, a cousin, a family in Mexico. . . ." Deputy Walsh's voice trails off.

Matias Garcia, age 29, died after walking 32 miles through the desert. He was a Zapotec Indian who lived near Oaxaca (southern Mexico) with his wife and three children, as well as his parents, younger brothers, and several cousins. Matias's cash crop of chili peppers was ruined by a spring frost just as they were ripening, leaving him in debt. So he reluctantly decided to risk a trip to the United States to work in some vineyards where he had worked on and off since he was a teenager. From there he could send money home to his family to keep the house in repair and to send his children to school. But it takes money just to cross the border, and it took until May for Matias, his younger brother, and a cousin to save the necessary amount.

May is one of the hottest and driest months in this part of the Americas, and since the NAFTA agreement of 1994, the border has been more carefully patrolled. This meant that Matias and the two younger men would attempt to cross the less patrolled but more perilous Arizona desert on a route to the United States that has come to be known as the "Devil's Highway." The men tried to avoid the worst of the heat by walking at night, but they didn't reach the highway on the Arizona side by dawn. Their water ran out, the sun turned especially hot, and Matias began having seizures. His brother and cousin carried him, desperately looking for the highway, but he died shortly before they found it and could flag down someone to call for help. That's where Deputy Walsh found the two men grieving over Matias's body.

*Find out more about Matias Garcia, why he migrated, how he died, and what his death has meant back in his village. Watch the FRONTLINE/World video "Mexico: A Death in the Desert."* ■

## Race and Ethnicity in North America

In any discussion of race and ethnicity in North America, it is important to remember that "race" has no biological meaning and is a socially constructed concept (see Chapter 1, pages 19–20). That is, to greater and lesser extents around the world, people have decided to invest meaning in skin color and other superficial features. The same may be said for **ethnicity,** which is the cultural counterpart to race in that people may also ascribe overwhelming (and unwarranted) significance to cultural characteristics, such as religion or family structure or gender customs. Because race and ethnicity are social constructs, people can also decide that those concepts are obsolete and irrelevant. Unfortunately, dispensing with racial and ethnic prejudice and its social consequences can take time.

Although North America is ethnically and racially diverse, in this region the term *race* usually has been used in relation to deeply embedded discrimination against African-Americans. Prejudice and lack of opportunity have clearly hampered the ability of African-Americans to reach social and economic parity with Americans of other ethnic backgrounds. Unfortunately, in the United States and somewhat less so in Canada, despite the removal of legal barriers to equality, African-Americans as a group still experience lower life expectancies, higher infant mortality rates, lower levels of academic achievement, higher poverty rates, and greater unemployment than do other Americans. (Hispanics of all racial backgrounds, Chinese, and South Asians have similar difficulties, though not as extreme; and many Native Americans have especially low living standards.)

In recent years, there has been a tendency to include Hispanics, Asians, Native Americans, and Pacific Islanders in discussions of race and ethnicity in both the United States and Canada, because these other minority groups are growing as a proportion of the population. In the United States, in June 2001, African-Americans, whose numbers are growing only slowly, were overtaken by Hispanics as the largest minority group in the United States. Because of a higher birth rate and a high immigration rate, Hispanics (of all racial identities) increased by 58 percent in the 1990s. Asians, who in 2000 made up only 3.6 percent of the U.S. population, nonetheless increased by 48 percent. Figure 2.30 shows the changes in the ethnic composition of the U.S. population in 1950, 2004, and projected to 2050.

Some people in the United States are concerned that so many recent immigrants are Hispanic and Asian. A Kaiser Family Foundation study in 2004 showed that two out of three Euro- and African-Americans, despite liking the idea of a multicultural America, worry about losing their sense of home in a United States of America where the largest minority is Hispanic, where Spanish is a prominent second language, and where another growing minority is from Asia. Will the traditional U.S. national identity and character be diluted by so many newcomers from different backgrounds? Or can a new national identity be constructed that will be more relevant in the emerging global society?

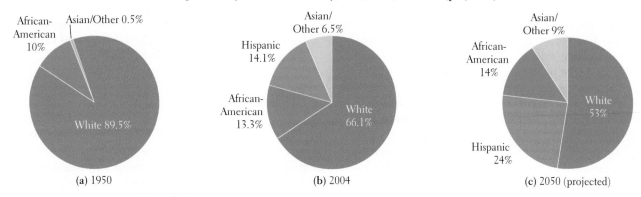

U.S. Population by Race and Ethnicity, 1950, 2004, and 2050 (projected)

(a) 1950

(b) 2004

(c) 2050 (projected)

**FIGURE 2.30** **The changing U.S. ethnic composition.** (The U.S. Census did not begin counting Hispanics as a separate category until 1970.) [Adapted from Jorge del Pinal and Audrey Singer, "Generations of diversity: Latinos in the United States," *Population Bulletin* 52 (October 1997): 14; "U.S. Census Bureau, 2004 Updated Report," at http://www.census.gov/Press-Release/www/releases/archives/population/005164.html; and "Census of Population," volume 2, part 1, "United States Summary," Table 36, at http://www2.census.gov/prod2/decennial/documents/21983999v2p1ch3.pdf.]

One answer to these questions is that new immigrants, whose background is distinctly different from the long-standing Euro-American and African-American heritage of the country, are adding talent, entrepreneurial vigor, and creativity to the U.S. cultural mix. Increasingly, they are opening small businesses, and they work long hours to make them succeed. In addition, according to Harvard's Joint Center for Housing Studies, in 2003, 50 percent of all immigrants owned their own homes, and immigrants were acquiring homes at a faster rate than native-born Americans. (In both the United States and Canada, homeownership has long been thought to indicate responsible citizenship.) Moreover, ethnic diversity contributes significantly to U.S. culture by enriching friendships, entertainment, business opportunities at home and abroad, and political perspectives. At the global scale, U.S. leadership in international affairs is enhanced by a less monolithic Euro-American presence at conferences and negotiating sessions.

Another source of worry in the United States is that there are increasing discrepancies in income among Asians, Euro-Americans, Native Americans, Hispanics, and African-Americans (Figure 2.31). There is considerable public debate over the causes and significance of these racial and ethnic differences in income. Over the past few decades, many African-Americans and Hispanics have joined the middle class and achieved success in the highest ranks of government and business. In particular, African-Caribbean emigrants to North America have been outstandingly successful, yet their roots in slavery and past discrimination are very similar to those of other African-Americans. A similar observation can be made about North Americans of Chinese and South Asian background; though often from humble backgrounds, they tend to seek advanced education, which has contributed to their prosperity. Numerous surveys show that Americans of all backgrounds favor equal opportunities for groups that have experienced past discrimination. Nonetheless, in both the United States and Canada, many middle-class African-Americans, Native Americans, and Hispanics (and, to a lesser extent, Asians) report experiencing both overt and covert discrimination that affects them economically.

Is anything besides the prejudice of others holding back African-Americans, Hispanics, and Native Americans? Experts disagree, but there is a consensus that poverty and low social status breed the perception among their victims that there is no hope for them and hence no point in trying to succeed. Furthermore, the poorest Americans, no matter what their background, have little opportunity to learn the basic skills for success at home, in their neighborhoods, or in school. The increasing number of poor single-parent families is also thought to play a role. For example, 50 years ago, African-American families were primarily two-parent units or extended families, and fathers, uncles, and grandfathers influenced the development of their children. In the 1990s, nearly 75 percent of African-American children were born to single mothers, and often the fathers of those children were not active in

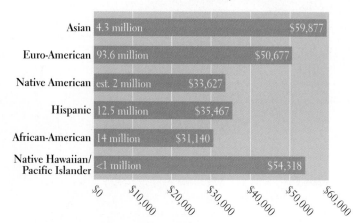

**FIGURE 2.31** U.S. median household income by race and ethnicity, average for 2003-2005. [Adapted from "Income of households by race and Hispanic origin using 3-year-average medians: 2003 to 2005," *Income, Poverty, and Health Insurance Coverage in the United States: 2005*, Table 2, U.S. Census Bureau, August 2006, at http://www.census.gov/prod/2006pubs/p60-231.pdf.]

their support and upbringing. The enormous responsibilities of both child rearing and breadwinning are left in the hands of often undereducated young mothers who are unable to help their children advance.

There is growing evidence that patterns of poverty, underachievement, and poor health in all groups are part of a larger class problem—the growing economic disparity and spatial separation between poor and rich Americans. In both the United States and Canada, the increasingly prosperous middle class, of whatever race or ethnicity, has fled to the suburbs. Geographer John B. Strait found in his study of neighborhoods of Atlanta, Georgia, that isolation of the poor into primarily inner-city slum neighborhoods makes it difficult for the affluent of all racial and ethnic backgrounds to see how they might help individuals move out of poverty, and the very poor rarely have the chance to associate with models of success. Evidence for this class-based explanation is that privileged Americans of all racial backgrounds are beginning to share middle- and upper-class

neighborhoods, workplaces, places of worship, and marriages, with markedly decreasing attention paid to matters of skin color. Meanwhile, the poor of all races see the material evidence of the success of others all around them, but have little access to the life choices that made that success possible for others.

## Religion

In both the United States and Canada, one is legally free to accept whatever creed one likes, or none at all. Because so many early immigrants to North America were Christian in their home countries, Christianity is currently the predominant religious affiliation claimed by North Americans. Nonetheless, virtually every medium-size city has at least one synagogue, mosque, and Buddhist temple; in some localities, adherents of Judaism, Islam, or Buddhism are numerous enough to constitute a prominent cultural influence.

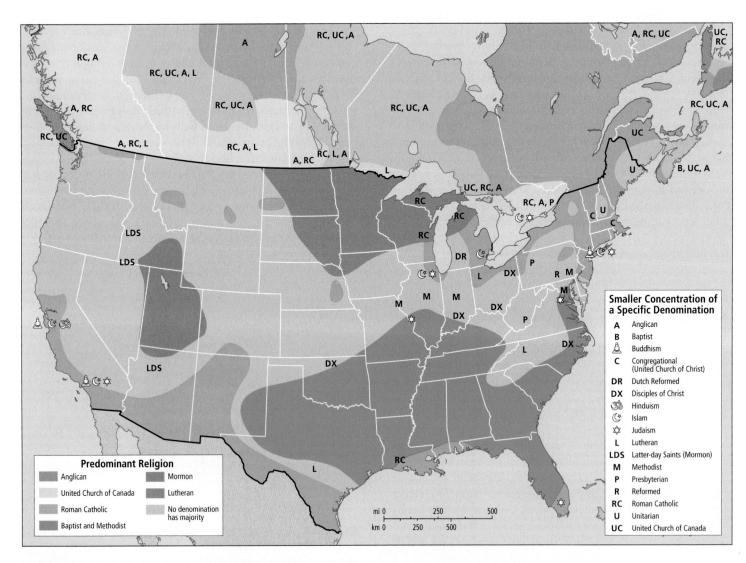

**FIGURE 2.32** Religious affiliations across North America. [Adapted from Jerome Fellmann, Arthur Getis, and Judith Getis, *Human Geography* (Dubuque, Iowa: Brown & Benchmark, 1997), p. 164.]

There are many versions of Christianity in North America, and their geographic distributions are closely linked to the settlement patterns of the immigrants who brought them (Figure 2.32). Roman Catholicism dominates in regions where Hispanic, French, Irish, and Italian people have settled—in southern Louisiana, the Southwest, and the far Northeast in the United States, and in Québec and other parts of Canada. Lutheranism is dominant where Scandinavian people have settled, primarily in Minnesota and the eastern Dakotas. Baptists are prominent in the religious landscapes of the South. Generally speaking, Southern Baptists and other evangelical Christians tend to take the teachings of the Bible more literally than many other Christian denominations do, so this part of the United States has come to be called the Bible Belt. Christianity—especially the Baptist version—is such an important part of community life in the South that newcomers to the region are asked almost daily what church they attend. (This question is rarely asked in most other parts of the continent.) Those who reply "none" or something other than "Baptist" may be left with the feeling that they have disappointed the questioner. Especially in the southern parts of the United States (much less so in Canada), Mormons and evangelical Christians are the fastest-growing Christian groups.

The proper relationship of religion and politics has long been a controversial issue in the United States, in large part because the framers of its Constitution, in an effort to ensure religious freedom, supported the idea that church and state should remain separate. Even now, most people try to avoid mixing the two in daily life and conversation. In the past decade or two, however, some of the more conservative Protestants have pushed for a closer integration of religion and public life through, for example, prayer in the public schools, teaching the biblical version of creation instead of evolution, and banning abortion. The policies of conservative Christians have met with the most success in the South, but their goals are shared by a minority scattered across the country.

New immigrants have brought their own faiths and belief systems, and they are contributing to the debate about religion and public life. The long-term outcome of these conflicts in American religious and political life is not yet apparent, but recent national surveys indicate that a substantial majority of Americans favor the continued separation of church and state and favor personal choice in belief and in reproductive and mating matters.

## Gender and the American Family

The family has repeatedly been identified as the institution most in need of shoring up in today's fast-changing and ever more impersonal America. Humans have long resided in communities of kinfolk, and a century ago, most Americans lived in extended families of several generations. Families pooled their incomes and shared chores. Aunts, uncles, cousins, siblings, and grandparents were almost as likely to provide daily care for a child as its mother and father were. The **nuclear family**, consisting of a married father and mother and their children, is a rather recent invention of the industrial age.

Beginning after World War I, and especially after World War II, many young people left their large kin groups on the farm and migrated to distant cities, where they established new nuclear families. Soon suburbia, with its many similar single-family homes, seemed to provide the perfect domestic space for the emerging nuclear family. This small, compact family suited industry and business, too, because it had no firm ties to other relatives and hence was portable. Many North Americans born since 1950 moved as often as five to ten times before reaching adulthood. The grandparents, aunts, and uncles who were left behind missed helping to raise the younger generation, and they had no one to look after them in old age. Institutional care for the elderly proliferated.

In the 1970s, the whole system began to come apart. It was a hardship to move so often. Suburban sprawl meant onerous commutes to jobs for men and long, lonely days at home for women. Women began to want their own careers, and rising consumption patterns made their incomes increasingly useful to family economies. By the 1980s, 70 percent of the females born between 1947 and 1964 were in the workforce, compared with 30 percent of their mothers' generation. But once employed, women could not easily move to a new location with an upwardly mobile husband. Nor could working women manage all of the family's housework and child care as well as a job. Some married men began to handle part of the household management and child care, but the demand for commercial child care grew sharply. With husband and wife both spending long hours at work, many people of both sexes found that their social life increasingly revolved around work, while family life receded in importance. With kinfolk no longer around to shore up the marital bond, and with the new possibility of self-support available to women in unhappy marriages, divorce rates rose drastically. By 2006, the U.S. divorce rate had reached nearly 50 percent; in Canada, the divorce rate was about 40 percent.

By the late 1980s, the nuclear family, which had been dominant for only a few decades, was less and less representative of the American family. As Figure 2.33 shows, the percentage of nuclear

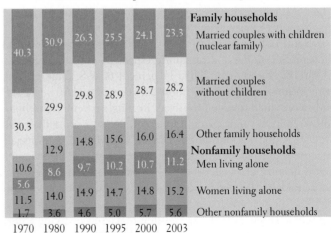

**Household Composition, 1970–2003 (percent)**

**Family households**
Married couples with children (nuclear family): 40.3, 30.9, 26.3, 25.5, 24.1, 23.3
Married couples without children: 30.3, 29.9, 29.8, 28.9, 28.7, 28.2
Other family households: 10.6, 12.9, 14.8, 15.6, 16.0, 16.4
**Nonfamily households**
Men living alone: 5.6, 8.6, 9.7, 10.2, 10.7, 11.2
Women living alone: 11.5, 14.0, 14.9, 14.7, 14.8, 15.2
Other nonfamily households: 1.7, 3.6, 4.6, 5.0, 5.7, 5.6

1970   1980   1990   1995   2000   2003

**FIGURE 2.33** U.S. households by type, 1970–2003. [Adapted from Jason Fields, "America's family and living arrangements: 2003," *Current Population Reports*, November 2004, U.S. Census Bureau, at http://www.census.gov/prod/2004pubs/p20-553.pdf.]

**FIGURE 2.34** An American family.
[© Elisabeth Carecchio/CORBIS SYGMA.]

families has continually shrunk. By 2003, less than 24 percent of U.S. households were nuclear families; the most common household type—28.2 percent—was a married couple *without* children, while the fastest-growing household type, a single person living alone, had reached 26.4 percent. More Americans than ever before are living alone; the majority are over the age of 45 and were once part of a nuclear family that dissolved due to divorce or death. There is no longer a typical American household, only an increasing diversity of forms (Figure 2.34).

Some of these new forms do not necessarily represent an improvement over the nuclear family, especially in providing for the welfare of children. In 2005, 35 percent of U.S. children were born to unmarried women (most between 20 and 30 years old), and more than one-fourth lived in single-parent households. Although most single parents are commited to rearing their children well, the responsibilities can be overwhelming, especially because single-parent families tend to be hampered by economic hardship and lack of education. The vast majority of single-parent households are headed by young women, whose incomes are, on average, 35 percent lower than those of single male heads of household (Figure 2.35). One result is that children in the United States are disproportionately poor. In 2005, the Annie E. Casey Foundation reported that 18 percent of U.S. children lived in poverty, whereas only 11.4 percent of adults did. In Canada, 14.7 percent of children were poor. In Sweden, by comparison,

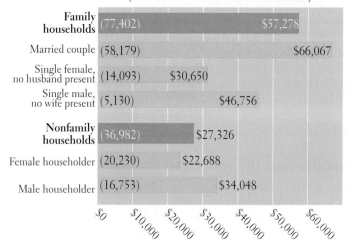

Median U.S. Income
by Type of Household, 2005
(number of households in thousands)

Family households (77,402) — $57,278
Married couple (58,179) — $66,067
Single female, no husband present (14,093) — $30,650
Single male, no wife present (5,130) — $46,756
Nonfamily households (36,982) — $27,326
Female householder (20,230) — $22,688
Male householder (16,753) — $34,048

**FIGURE 2.35** U.S. median income by household type, 2005. Notice the gender differences in income. Nonfamily households can mean a single person living alone, an unmarried same-sex couple, or unrelated acquaintances living together. [Adapted from "Income of households by race and Hispanic origin using 3-year-average medians: 2003 to 2005," *Income, Poverty, and Health Insurance Coverage in the United States: 2005*, Table 1, U.S. Census Bureau, August 2006, at http://www.census.gov/prod/2006pubs/p60-231.pdf.]

2.4 percent of children lived in poverty; in Ireland, 12.4 percent; in Spain, 12.4 percent; and in Poland, 12.7 percent.

## Aging

In most societies, people between youth and old age (15 and 65 years) must support those who are younger or older. The age structure of a population tells us how great their burden is likely to be. Wealthier, developed societies tend to have lower birth rates, and their members tend to have longer life spans. Thus, over time, the proportion of those younger than 15 becomes smaller, while the proportion of those over 65 becomes larger. The populations of Canada and the United States are aging, meaning that the number of people over 65 is growing more rapidly than the number of those under 15. During the twentieth century, the number of older Americans grew rapidly. In 1900, 1 in 25 individuals was over the age of 65; by 1994, 1 in 8 was. By 2050, when most of the current readers of this book will be over 65, 1 in 5 Americans will be elderly.

In North America, the number of elderly people will shoot up especially fast between the years 2010 and 2030. This predicted spurt is the result of a marked jump in birth rate that took place in the years after World War II, from 1947 to 1964. The so-called **baby boomers** born in those years constitute the largest age-group in North America, as indicated by the wide bulge through the middle of the population pyramids shown in Figure 2.36. As this group reaches age 65 and retires, payments from Social Security and private pensions will be high, and medical costs will leap upward. The boomers had fewer children than their parents had, so there will be fewer people of working age to pay the taxes and pension fund contributions necessary to meet these costs (although, as mentioned earlier, the contri-

butions of new young adult immigrants will help). Moreover, people in this smaller population will have fewer brothers and sisters with whom to share responsibility for the daily personal care and companionship needed by their elderly kin. Most families will not be able to afford assisted living and residential care, which already (in 2006) costs from $30,000 to $70,000 per year for one person. We might expect that once the boomers begin to retire in large numbers, the makeup of many households and the spaces in which they live will change to reflect people's efforts to find humane and economical means to care for elderly family members at home.

The maps in Figure 2.37 show where the elderly were concentrated in the 1990s and how these patterns are expected to look by 2020. In rural areas (through the Middle West and Northeast), the young have departed in large numbers for other parts of the country, leaving aging parents behind. In other areas—Florida, especially, but also all across the southern United States—the elderly are the new residents, attracted by the warm, pleasant climate of the Sunbelt and gated communities that appeal to the security-conscious affluent elderly. California has the largest number of elderly people, but they make up only a small percentage of the total population because there are so many young people in that state.

The problems presented by aging populations reveal a paradox. On the one hand, it is widely agreed that population growth should be reduced to lessen the environmental impact of human life on earth, especially that of the societies that consume the most. On the other hand, slower population growth means that as more of the world's citizens grow old, there will be fewer young, working-age people to keep the economy going and to provide the financial and physical help the elderly require (unless immigrants assist).

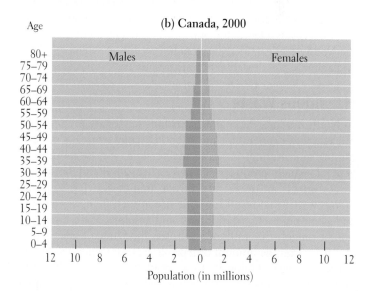

**FIGURE 2.36** Population pyramids for the United States and Canada, 2000. The "baby boomers," born between 1947 and 1964, constitute the largest age group in North America, as indicated by the wider middle portion of these population pyramids. [Adapted from "Population pyramids of the United States, 2000" and "Population pyramids of Canada, 2000" (Washington, D.C.: U.S. Bureau of the Census), International Data Base, at http://www.census.gov/ipc/www/idbpyr.html.]

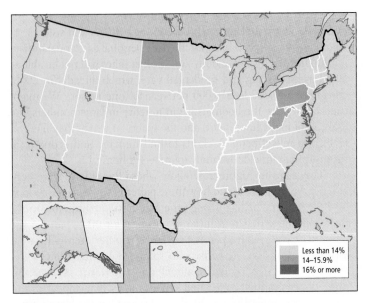

(a) Percent of total state population 65 years and over, 2004

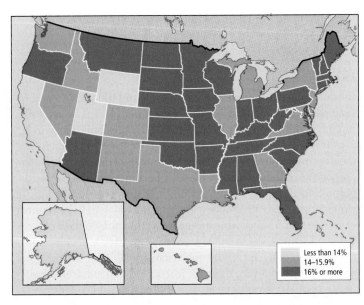

(b) Percent of total state population 65 years and over, 2020 (projected)

**FIGURE 2.37** Changing distribution of the elderly in the United States. The proportion of elderly people in the population is expected to increase across the country in the coming decades. The large numbers of elderly in California are masked on these maps by the fact that this state also has many young people. [Data from U.S. Census Bureau, 2000 census and 2004, at http://factfinder.census.gov/servlet/ThematicMapFramesetServlet?_bm=y&-geo_id=01000US&-tm_name=ACS_2004_EST_G00_M00623&-ds_name=ACS_2004_EST_G00_&-_dBy=040.]

## ENVIRONMENTAL ISSUES

North America's wide range of resources and seemingly limitless stretches of forest and grasslands long diverted attention from the environmental impacts of settlement and development. This section opens with a discussion of one of the most significant and unsustainable outcomes of human activities—hazardous waste—and then covers a few of the other environmental consequences of the North American lifestyle: air pollution, loss of habitat for plants and animals, and depletion of water resources.

### Hazardous Waste

In North America the major **hazardous wastes** are produced by nuclear power generation and weapons manufacture, by mineral mining and drilling, by waste incinerators, and by industry in general, which produces 80 percent of all liquid hazardous waste. Small businesses and private homes also pass on hazardous wastes that result from the use of such items as cleaning and paint products, weed killers, and gasoline. With a population of roughly 300 million, each year the United States generates five times the amount of hazardous waste generated by the entire European Union (with a population of about 460 million). Canada generates much less hazardous waste per capita each year than the United States, although its citizens generate several times the global average of about 2.2 pounds (1 kilogram) per person, per day.

The disposal of hazardous waste within the United States has a geographic pattern. Sociologist Robert D. Bullard has shown that a disproportionate amount of hazardous waste is disposed of in the South and in locations inhabited by poor Native American, African-American, and Hispanic people. Bullard writes that "nationally, 60 percent of African-Americans and 50 percent of Hispanics live in communities with at least one uncontrolled toxic-waste site." Military bases are also major sources of hazardous waste, generating more wastes than the top five U.S. chemical companies combined.

### Air Pollution

With only 5 percent of the world's population, North America produces 26 percent of the greenhouse gases released globally by human activity. This large share can be traced to North America's high consumption of fossil fuels, which is in turn related to its oil-dependent industrial and agricultural processes, the heating of its homes and offices, and its dependency on automobiles. As we discussed in Chapter 1, greenhouse gases lead to global warming (see pages 45–48). In addition, most greenhouse gases contribute to various forms of air pollution.

**Smog** is a combination of industrial emissions and car exhaust that frequently hovers over many North American cities, causing a variety of health problems for their inhabitants (Figure 2.38). In Los Angeles the intensity of the smog is due in large part to the city's warm land temperatures and West Coast seaside location, which often results in a **thermal inversion**—a warm mass of stagnant air that is trapped beneath cooler air blowing in off the ocean. The inversion is held in place, often for days, by the mountains that surround the city.

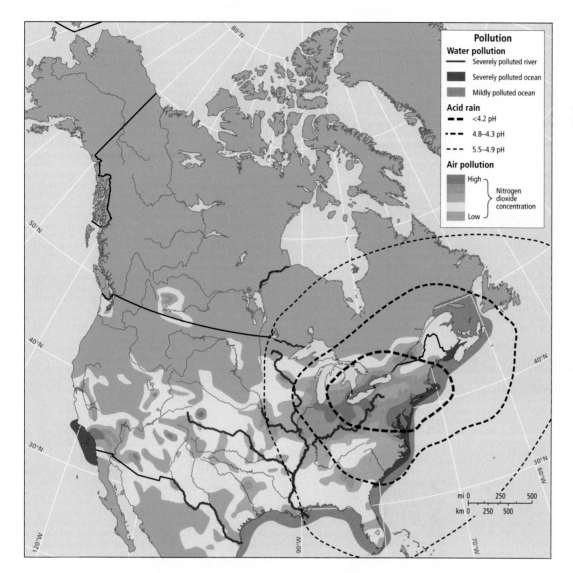

**FIGURE 2.38** Air and water pollution in North America. Nitrogen dioxide (NO$_2$), a toxic gas that comes primarily from the combustion of fossil fuels by motor vehicles and power plants, transforms in air to gaseous nitric acid and toxic organic nitrates.

The burning of fossil fuels also contributes to **acid rain** (and acid snow) because it releases sulfur dioxide and nitrogen oxides into the air. Acid rain is created when these gases dissolve in falling rain and make it acidic. Acid rain can kill certain trees; when concentrated in lakes and streams and snow cover, can destroy fish and wildlife; and can corrode buildings and other structures. The eastern half of the continent, from the Gulf Coast to Newfoundland and including the entire eastern seaboard, is greatly affected by acid rain. The United States, with its larger population and more extensive industry, is responsible for the vast majority of this acid rain. Due to continental weather patterns, however, the area most affected by acid rain encompasses a wide area on both sides of the eastern U.S.–Canada border (see Figure 2.38).

## Loss of Habitat for Plants and Animals

Prior to 1500, North America was not severely impacted by humans (Figure 2.39a). Today, throughout the continent, many plants and animals are in danger of becoming extinct, primarily because humans are destroying the environments in which they live. Of all countries in the world, as of 2003, the United States had the *highest percentage* (33 percent) of the world's plant and animal species threatened by extinction. As North America was colonized and developed, millions of acres of forests and grasslands were cleared to make way for farms. Now the last bits of natural land near cities are disappearing to make way for residential developments, highways, golf courses, office complexes, and shopping centers (Figure 2.39b). North American plants and animals have been forced into ever smaller territories, and many have died out entirely and replaced by alien species. Biogeographers warn that because of the complex interdependence of the biological world, the depletion or extinction of species will have long-term negative effects for life on earth, though the specific effects are not yet well understood. The destruction of wetlands is especially significant because they are important reproductive zones for many bird and aquatic species. Overuse of ocean resources has reached

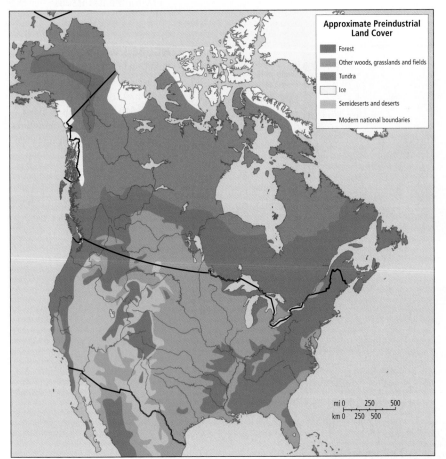

**Approximate Preindustrial Land Cover**

- Forest
- Other woods, grasslands and fields
- Tundra
- Ice
- Semideserts and deserts
- —— Modern national boundaries

mi 0   250   500
km 0   250   500

(a) Preindustrial

**FIGURE 2.39** Land cover (preindustrial) and human impact (2002). In the preindustrial era (**a**), humans had cleared forests, extended grasslands (by burning), and planted agricultural fields. Three hundred years later (**b**), some impact of human agency was observable virtually everywhere in North America. [Adapted from *United Nations Environment Programme*, 2002, 2003, 2004, 2005, 2006 (New York: United Nations Development Programme), at http://maps.grida.no/go/graphic/human_impact_year_1700 _approximately and http://maps.grida.no/go/graphic/human _impact_year_2002.]

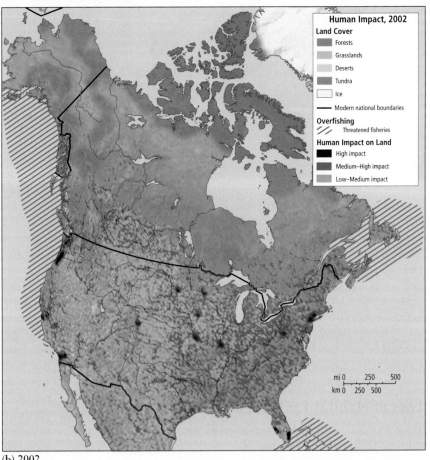

**Human Impact, 2002**

**Land Cover**
- Forests
- Grasslands
- Deserts
- Tundra
- Ice
- —— Modern national boundaries

**Overfishing**
- Threatened fisheries

**Human Impact on Land**
- High impact
- Medium–High impact
- Low–Medium impact

mi 0   250   500
km 0   250   500

(b) 2002

such a critical state that scientists now think it is likely that humans will force even the great whales into extinction.

## Water Resource Depletion

People who live in the humid eastern part of North America find it difficult to believe that water is becoming scarce even there. But as populations grow and per capita water usage increases, it has become necessary to look farther and farther afield for sufficient water resources. New York City, for example, obtains most of its water from the distant Catskill–Delaware watershed in upstate New York. Atlanta, Georgia, with over 5 million residents, is located in the usually moist Southeast, but absorbs water so insatiably that downstream users in Alabama and Florida have sued Atlanta for depriving them of their rights to water. Disputes over water rights exist around the world and are difficult to resolve, as we will see in our discussion of many world regions.

In North America, water becomes increasingly precious the farther west one goes. On the Great Plains, rainfall in any given year may be too sparse to support healthy crops and animals. To make farming more secure and predictable, taxpayers across the continent have subsidized the building of stock tanks and reservoirs. Irrigation has also increased in recent decades, drawing on "fossil water" that has been stored over the millennia in natural underground reservoirs called **aquifers.** The **Ogallala aquifer** (Figure 2.40) underlying the Great Plains is the largest such body of water in North America. In parts of the aquifer, water is being pumped out at rates that exceed natural replenishment by 10 to 40 times. Although local precipitation plays a small role in recharging the aquifer, the Ogallala received most of its water from

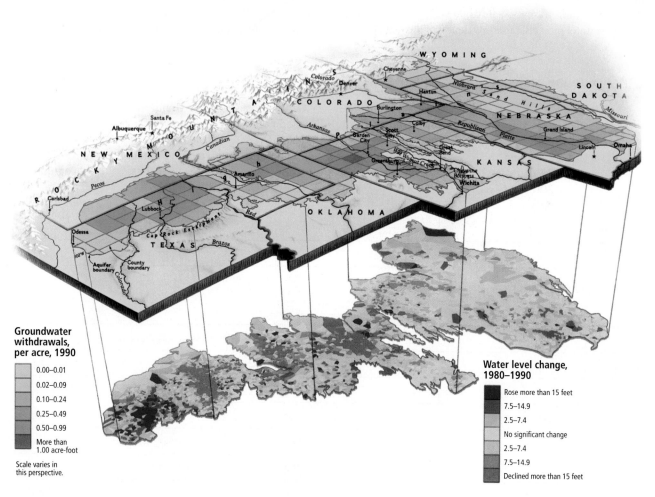

**Groundwater withdrawals, per acre, 1990**

- 0.00–0.01
- 0.02–0.09
- 0.10–0.24
- 0.25–0.49
- 0.50–0.99
- More than 1.00 acre-foot

Scale varies in this perspective.

**Water level change, 1980–1990**

- Rose more than 15 feet
- 7.5–14.9
- 2.5–7.4
- No significant change
- 2.5–7.4
- 7.5–14.9
- Declined more than 15 feet

**FIGURE 2.40  The Ogallala aquifer.** From the 1940s through the 1980s, the aquifer lost an average of 10 feet (3 meters) of water overall, and more than 100 feet (30 meters) of water in some parts of Texas. During the 1980s, the aquifer declined less because of abundant rain and snow, but this is an area where the climate fluctuates from moderately moist to very dry. A drought began in mid-1992 and continued until late 1996. Large agribusiness firms pumped water from the aquifer for irrigation to supplement precipitation. As a result, water levels declined an average of 1.35 feet per year during the mid-1990s. [Graphic adapted from *National Geographic* (March 1993): 84–85, with supplemental information from High Plains Underground Water Conservation District 1, Lubbock, Texas, at http://www.hpwd.com; and Erin O'Brian, Biological and Agricultural Engineering, National Science Foundation Research Experience for Undergraduates, Kansas State University, 2001.]

highland streams to the west. Because of geologic change, access to much of this stream flow has been cut off.

Irrigation is also a major issue in California, which supplies much of the fresh fruit and vegetables consumed in the United States, but at a high cost to taxpayers. Billions of dollars of federal and state funds have paid for massive water engineering projects that bring enormous quantities of water to the southern part of the state from hundreds of miles away in Washington and Oregon, northern California, Colorado, and Arizona, at times pumping it up and over mountain ranges. Irrigation in Southern California also deprives Mexico of this much-needed resource. The mouth of the Colorado River, which used to be navigable, is now dry and sandy. Only a mere trickle gets to Mexico, which also would like to use more water for irrigation.

Increasingly, citizens in western North America are recognizing that the use of scarce water for irrigated agriculture is unsustainable and uneconomical. Conflicts over moving water from wet regions to dry ones, or from sparsely inhabited to urban areas, have halted some new water projects, raised awareness of the need to con-serve water, and resulted in some reduction in the use of water for irrigation. Nevertheless, government subsidies keep water artificially low in price, and past successes in harnessing new water supplies provide a disincentive to change.

## MEASURES OF HUMAN WELL-BEING

Both Canada and the United States consistently rank high on global scales of well-being, yet a comparison of these two wealthy countries makes it clear why old ways of making such comparisons are misleading. If we looked only at gross domestic product (GDP) per capita for the two countries, we would see that Canada's GDP per capita is U.S.$30,667 (seventh highest in the world) and that of the United States is U.S.$37,562 (fourth highest). From these numbers, we might conclude that Canadians are doing notably less well than people in the United States. But remember that GDP per capita ignores all measures of well-being other than income and is only an average for the entire country. In fact, despite its lower GDP per capita, Canada ranks nearly the same as the United States on many

**TABLE 2.2**  Human well-being rankings of Canada, the United States, and selected developed countries[a]

| Country (I) | GDP per capita, adjusted for PPP[b] (GDP ranking among 177 countries), 2005 (II) | Human Development Index (HDI) ranking among 177 countries, 2005 (III) | Gender Development Index (GDI) ranking among 140 countries, 2005 (IV) | Gender Empowerment Measure (GEM) ranking among 80 countries, 2005 (V) | Women's average GDP per capita as a percentage of men's average GDP per capita,[b, c] 2003 (VI) |
|---|---|---|---|---|---|
| Canada | 30,677  (7) | 5 | 5 | 10 | 64 |
| United States | 37,562  (4) | 10 | 8 | 12 | 62 |
| **Selected Developed Countries** | | | | | |
| Australia | 29,632  (10) | 3 | 2 | 7 | 72 |
| Barbados | 15,720  (36) | 30 | 29 | 25 | 61 |
| Germany | 27,756  (14) | 20 | 20 | 9 | 54 |
| Japan | 27,967  (13) | 11 | 14 | 43 | 46 |
| Kuwait | 18,047  (30) | 44 | 39 | ND | 35 |
| Mexico | 9,168  (56) | 53 | 46 | 38 | 38 |
| Sweden | 26,750  (20) | 6 | 4 | 3 | 69 |
| World | 8,229 | — | — | — | 52 |

[a] Rankings are in descending order; i.e., low numbers indicate high rank.

[b] PPP = purchasing power parity, figured in 2003 U.S. dollars.

[c] Male/female percentage of GDP not available in 2005; 2003 figures used.

ND = No data available.

*Source: United Nations Human Development Report 2005* (New York: United Nations Development Programme), Table 14, Table 25, and Table 26, at http:/hdr.undp.org/reports/global/2005/pdf/HDR05_complete.pdf.

indicators of well-being and, as we have seen, exceeds the United States in gender empowerment, life expectancy, and infant and maternal survival rates (see Table 2.1, page 78, and the discussion on pages 78–79). Table 2.2 compares Canada, the United States, and some selected developed countries in terms of GDP per capita and the United Nations measures we introduced in Chapter 1 (see pages 33–35). On all the UN measures included in Table 2.2, Canada actually outranks the United States.

# III SUBREGIONS OF NORTH AMERICA

When geographers try to understand patterns of human geography in a region as large and varied as North America, they usually impose some sort of subregional order on the whole (Figure 2.41). In this section we divide North America into subregions, and then sketch in the features that give each subregion its "character of place." Keep in mind, however, that, as we observed in Chapter 1, geographers often have trouble reaching consensus on just where regional boundaries should be drawn.

## NEW ENGLAND AND THE ATLANTIC PROVINCES

Among the earliest parts of North America to be settled by Europeans were New England and Canada's Atlantic Provinces (Figure 2.42). Of all the regions of North America, this one may maintain the strongest connection with the past, and it holds a certain cultural prestige and reputation as North America's cultural

**FIGURE 2.41** Subregions of North America. There are many schemes for dividing North America into subregions. The scheme we use here is partly indebted to Joel Garreau and his book *The Nine Nations of North America* (1981), in which he proposed a set of regions that cut not only across state boundaries but also across national boundaries. Though Garreau includes parts of Mexico and the Caribbean, our examination is limited to the United States and Canada. [Adapted from Joel Garreau, *The Nine Nations of North America* (Boston: Houghton Mifflin, 1981).]

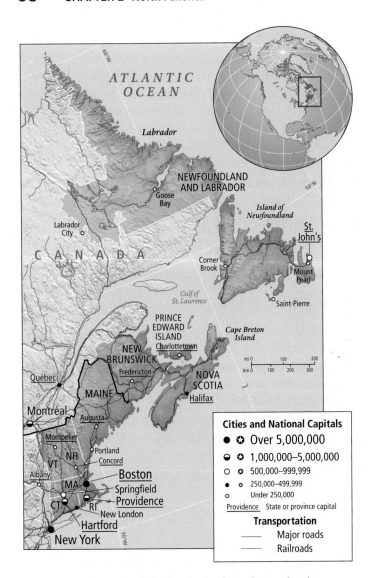

**FIGURE 2.42** The New England and Atlantic Provinces subregion. [Map adapted from Joel Garreau, *The Nine Nations of North America* (Boston: Houghton Mifflin, 1981).]

Although farmers settled here, they struggled to survive. Today, many areas try to capitalize on their rural and village ambiance and historic heritage by enticing tourists and retirees, but economically, northern New England and the Atlantic Provinces of Canada remain relatively poor.

After being cleared in the days of early settlement, New England's evergreen and hardwood deciduous forests have slowly returned to fill in the fields abandoned by farmers. These second-growth forests are now being clear-cut by logging companies, and, in rural areas, paper milling from wood is still supplying jobs as other blue-collar occupations die out.

Abundant fish were the major attraction that drew the first wave of Europeans to New England. In the 1500s, hundreds of fishermen from Europe's Atlantic coast came to the Grand Banks, offshore of Newfoundland and Maine, to take huge catches of cod and other fish. The fishing lasted for over 400 years, but now the fish stocks of the Grand Banks are badly depleted by modern fishing vessels (some from outside North America) using equipment that destroys the marine ecosystem (see Figure 2.39b on page 94). The fishing industry has also been hurt by competition from fish farming, a rapidly growing industry in many parts of North America. There is a global connection here: farmed fish are fed with millions of tons per year of wild ocean fish. The harvesting of these wild fish throughout the world ocean to feed farmed fish for markets in North America poses a threat to oceanic resources not only off the shores of New England and the Atlantic Provinces, but also in countless places where the poor are dependent on fishing for their basic nutrition.

Many parts of southern New England now thrive on service and knowledge-based industries that require skilled and educated

hearth. Philosophically, New England laid the foundation for religious freedom in North America with a strong conviction that eventually became part of the U.S. Constitution: that there should be no established church and no requirement that citizens or public officials hold any particular religious beliefs. New England is especially noted as the source of the classic American village style, arranged around a town square, and of such early American house styles as Cape Cod and other wood-framed houses (Figure 2.43). Many domestic material culture styles also originated in New England; in all classes of American homes, popular interior furnishings that are copies of New England–designed furniture and accessories can be found.

Many of the continuities between the present and past economies of New England and the Atlantic Provinces derive from the region's geography. During the last ice age, glaciers scraped away much of the topsoil, leaving behind some of the finest building stone in North America but only marginally productive land.

**FIGURE 2.43** New England style. This historic wood-framed house is in the town of Eastham, Massachusetts, on Cape Cod. [© Stephen Saks Photography/ Alamy.]

workers: insurance, banking, high-tech engineering, and genetic and medical research, to name but a few. New England's considerable human resources derive in large part from the strong emphasis placed on education, hard work, and philanthropy by the earliest Puritan settlers, who established many high-quality schools, colleges, and universities. The cities of Boston and Cambridge have some of the nation's foremost institutions of higher learning (Harvard, the Massachusetts Institute of Technology, Boston University). The Boston area has capitalized on its supply of university graduates to become North America's second most important high-technology center, after California's Silicon Valley.

At present, New England finds itself confronted by some startling discontinuities with its past. This has long been a region of relatively large cities (Boston, Providence, Hartford), but it is now becoming both more urban and more ethnically diverse. The remaining agricultural enterprises and rural industries have become increasingly mechanized and larger in scale, so rural New Englanders are flocking to the cities for work. There, they discover that New England is not the Anglo-American stronghold they had thought it to be. Native Americans, African-Americans, and immigrants from Portugal and southern Europe have long shared New England with those from the British Isles and elsewhere. But today, in the suburbs of Boston, corner groceries are owned by Koreans and Mexicans; Jamaicans sell meat patties and hot wings; restaurants serve food from Thailand; and schoolteachers are Filipino, Brazilian, and West Indian. The blossoming cultural diversity of New England and the entrepreneurial skills of new migrants are helping New England to keep pace with change across the continent. These trends are less obvious in the Atlantic Provinces of Canada, but there, too, immigrants from Asia and elsewhere are influencing ways of life.

## QUÉBEC

Québec is the most culturally distinct formal region in Canada or the United States (Figure 2.44). For more than 300 years, a substantial portion of the population has been French-speaking. French Canadians are now in the majority in Québec province and are struggling to resolve Québec's relationship with the rest of Canada.

In the seventeenth century, France encouraged its citizens to settle in Canada. By 1760, there were 65,000 French settlers in Canada. Most lived along the St. Lawrence River, on long, narrow strips of land stretching back from the river's edge. This **long-lot system** gave the settlers access to resources from both the river and the land: they fished and traded on the river, they farmed the fertile soil of the floodplain, and, on higher forested ground, they hunted. Because of the orientation of the long lots to the riverside, early French colonists joked that one could travel along the St. Lawrence and see every house in Canada. Later, the long-lot system was repeated inland, so that today narrow farms also stretch back from roads that parallel the river, forming a second tier of long lots.

Through the first half of the twentieth century, Québec remained a land of farmers eking out a living on poor soils similar to those of New England and the Atlantic Provinces, growing only enough food to feed their families. After World War II, Québec's economy grew steadily, propelled by increasing demand for the

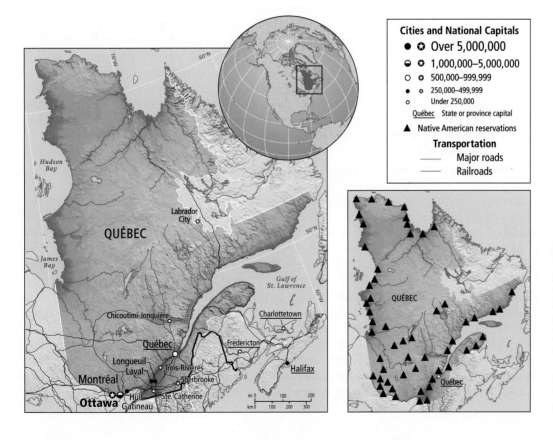

### Cities and National Capitals

| | | |
|---|---|---|
| ● ✪ | Over 5,000,000 | |
| ◒ ✪ | 1,000,000–5,000,000 | |
| ○ ✪ | 500,000–999,999 | |
| • ○ | 250,000–499,999 | |
| ○ | Under 250,000 | |

Québec  State or province capital

▲  Native American reservations

### Transportation

—— Major roads

—— Railroads

**FIGURE 2.44** The Québec subregion. Native Americans from various ethnic groups live throughout Québec, but one-fourth of them live in far northern and eastern rural Québec, as indicated in the map at the right. [Adapted from Joel Garreau, *The Nine Nations of North America* (Boston: Houghton Mifflin, 1981); and "Map of Indian Nations of Québec," Indian and Northern Affairs Canada, at http://www.ainc-inac.gc.ca/qc/map/index_e.html.]

**FIGURE 2.45** Québec City with landmark Chataeu Frontenac hotel (upper left). Québec City's location on the St. Lawrence and its deep, year-round seaport helped it to become the economic engine of eastern Québec. The Chateau Frontenac hotel was built by a railroad mogul for travelers transferring from land to sea. The city is the provincial capital and the seat of Québec's parliament and is increasingly a biomedical high-tech center featuring leading-edge software development, especially in medicine. [Andrea Pistolesi/Tips Images.]

natural resources of northern Québec, such as timber, iron ore, and hydroelectric power. In the cities of the St. Lawrence River valley, some prospered from the transport and processing of these resources and built fine buildings (Figure 2.45), although most of these enterprises were in the hands of Anglo-Canadians (those whose ancestors came from the British Isles). Québec's new prosperity was most visible in the rapid growth of Montréal. This city's interior location, near the confluence of several rivers, had always made it an attractive site for British entrepreneurs interested in exporting the natural resources of Québec.

As rural French Canadians moved into Québec's cities, the so-called Quiet Revolution began. With increasingly better access to education and training, the Québecois were for the first time able to challenge English-speaking residents of Québec for higher-paying jobs and power. Gradually, their conservative Catholic agricultural society gave way to a more cosmopolitan one that was increasingly resentful of discrimination at the hands of English speakers. This discrimination had taken the form of social denigration of French culture as well as blocked access to education and economic participation. In the 1970s, the Québecois pushed for increased autonomy and then outright national independence. At that time, the province passed laws that heavily favored the French language in education, government, and business. In response, many English-speaking natives of Québec relocated to Ontario and elsewhere. Some Québecois themselves began to fear for Québec's economic survival if the province left Canada, and a referendum on independence failed narrowly in 1996. By 2006, the issue of Québec national independence was again gaining momentum within the province, but Canadians generally did not favor such a move.

Québec's northern portions extend into areas around Hudson Bay that are rich in timber and mineral deposits (iron ore, copper, and oil). These resources are hard to reach in the remote, difficult terrain of the Canadian Shield (see Figure 2.1 on pages 52–53). Although these are the native lands of the Algonquin-speaking Cree people, the Québec provincial government has legal control of the resources. In the 1960s, that government became interested in developing hydroelectric power in the vicinity of James Bay (part of Hudson Bay) in order to run mineral-processing plants, sawmills, and paper mills. The Cree protested the clear-cutting of their forests and the changes to community life brought by outsiders. In the 1990s, further hydroelectric projects were put on hold in response to Cree protests that the enormous shallow lakes created by the dams flooded sacred ancestral hunting and burial grounds. Nonetheless, development of resources in Cree lands by Euro-Canadians continues, and some electric power from Cree country is sold to the northeastern United States.

## THE OLD ECONOMIC CORE

Southern Ontario and the north-central part of the United States, from Illinois to New York (Figure 2.46), represents less than 5 percent of the total land area of the United States and Canada, but it was once the economic core of North America. As recently as 1975, its industries produced more than 70 percent of the continent's steel and a similar percentage of its motor vehicles and parts, an output made possible by the availability of energy and mineral resources in the region, or just beyond its boundaries. Today, the industrial economy of this region has gone into severe decline, and large cities dependent on manufacturing, such as Detroit, and hundreds of smaller cities and towns have suffered near economic collapse. Meanwhile, some of North America's largest cities, such as New York, Toronto, and Chicago, continue to prosper here largely be-

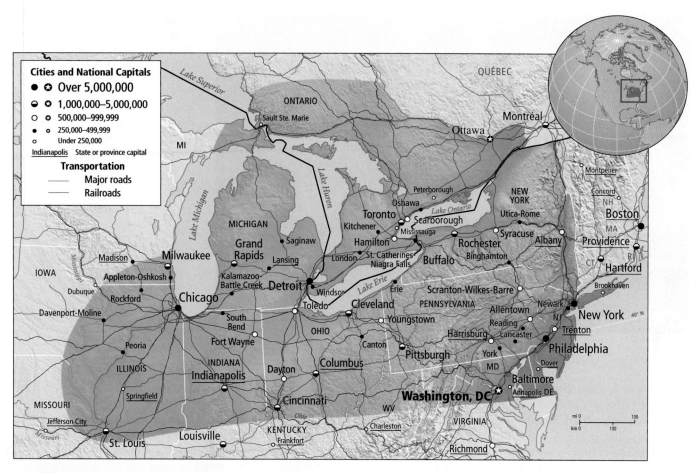

**FIGURE 2.46** The old economic core subregion. [Adapted from Joel Garreau, *The Nine Nations of North America* (Boston: Houghton Mifflin, 1981).]

cause of their strong service industries that connect them to regional and global economies.

Plants started closing in the 1970s, and by the 1990s, most manufacturing jobs had moved to the South, Middle West, and Pacific Northwest, well beyond the old industrial heartland. Industrial resources that once supplied factories now come from elsewhere: coal from Wyoming, British Columbia, Alberta, and Appalachia; steel, the mainstay of the automotive and construction industries, can be more cheaply derived from scrap metal or purchased abroad from Brazil, Japan, or Europe. Hence, large, outdated factories sit empty or underused throughout much of the economic core, and thousands of aging factory workers have retrained or entered an early, impoverished retirement. Some people have taken to calling this obsolete industrialized region the **Rust Belt** (Figure 2.47). While some of the food-related industries that grew as offshoots of the farming industry remain in the old economic core, many others, such as the meatpacking industry, have consolidated or moved to the South, where labor unions scarcely exist and safety and environmental regulations are more lax.

The loss of millions of manufacturing jobs in the old industrial core has had serious social consequences, as sociologist William

Julius Wilson notes in his book *When Work Disappears: The World of the New Urban Poor.* "Neighborhoods that are poor and jobless are entirely different from neighborhoods that are poor and working," Wilson writes. "Work is not simply a way to make a living and support a family. It also constitutes a framework for daily behavior because it imposes discipline."

Industrial jobs drew millions of rural men and women, both black and white, from the South to the industrial core after World War II. Many of their sons and daughters, now without work and without funds to retrain or relocate, constitute some of the more than 35 million U.S. citizens living in poverty. Consider the case of one inner-city neighborhood on the west side of Chicago, called North Lawndale. In 1960, 125,000 people lived there with two large factories employing 57,000 workers, and a large retail chain's corporate headquarters supporting thousands of secretaries and office workers. One factory closed in the late 1960s, removing 14,000 jobs; in 1974, the retail headquarters had moved downtown; and in 1984, the other large factory had closed, eliminating a breathtaking 43,000 jobs. North Lawndale began to disintegrate with the first loss of jobs. Because no one had money to spend anymore, thousands of service jobs disappeared and, with them, many

**FIGURE 2.47 The Rust Belt.** The foreground of this photo of Youngstown, Ohio, is dominated by buildings of the Republic Steel plant, which once employed 10,000 people. The Republic plant is now closed, but several foreign-held firms operate minimill facilities on the site and employ 250 workers. Early twentieth-century frame houses surround a large church in the nearby working-class neighborhood. [Richard Kalvar/Magnum Photos.]

middle-class families. By the early 1990s, the housing stock had deteriorated. As families and businesses left to look for opportunities elsewhere, landlords abandoned buildings, and financial institutions saw little reason to support reinvestment. By 2003, North Lawndale had shrunk to just 50,000 people. However, in many economically depressed communities in the old economic core, people have not given up hope of rebuilding. Some cities have turned defunct factories to other uses, such as condominiums and museums. The North Lawndale Industrial Development Team, with the help of a family foundation and local civic groups, actively recruits industry and offers to provide a custom-trained labor force.

Despite the kinds of situations we have been describing, it would be wrong to think of the economic core, even during its heyday, as solely a region of factories and mines. The thousands of miles of riverfront, shoreland around the Great Lakes, and ocean front along the eastern seaboard provide both winter and summer recreational landscapes, as do the mountains of New York State and Pennsylvania. In between the great industrial cities of this region are thousands of acres of some of the best temperate climate farmland in North America.

## THE AMERICAN SOUTH (THE SOUTHEAST)

The regional boundaries of the American South are perhaps fuzzier and based more on a perceived state of mind and way of life than is the case for most other U.S. regions. This region, in fact, covers only the southeastern part of the country, not the whole of the southern United States (Figure 2.48). Somewhere east of Austin, Texas, the American South grades into the Southwest, a region with noticeably different environmental and cul-

tural features. But what characteristics best define the South? And where should we draw its borders?

Within this region there is a complex of features that many people would identify as southern: certain dialects and comfort food; stock car racing, bluegrass, jazz, and blues music; the open friendliness of the people; conservative stances on gun ownership, environmental protection, and the role of religion in public life; the early onset of spring; and rural settlement patterns, such as those that keep large kin groups together in clusters of old wooden cabins, mobile homes, or fine newly built houses. But southern cultural features are hard to measure: any given southern place has only a few of the features listed here; there are few clear spatial distributions, and the patterns are not contiguous. Furthermore, arguments ensue if one tries to define just which accents, or what recipes, or which styles of music are "southern." Some places located in the South have few recognizable southern qualities. Parts of Missouri, Oklahoma, and Texas have strong ties to the Great Plains; Miami, on the far southern tip of Florida, with its cosmopolitan Latino culture and its trade and immigration ties to the Caribbean and South America, hardly seems southern at all. New Orleans, also in the heart of the South, has a unique set of cultural roots—French, African, Cajun, Creole—that set it apart from other southern cities.

North Americans and foreign visitors to the South tend to hold many outdated images from the civil rights efforts of the 1960s, and even the Civil War era. It is true that significant racial segregation still persists in that black people and white people tend to live and worship separately, but, in fact, many more whites and blacks share neighborhoods in the South than in the old economic core. And in many rural and urban settings, schools are integrated as a result of residential patterns as well as busing plans and magnet schools.

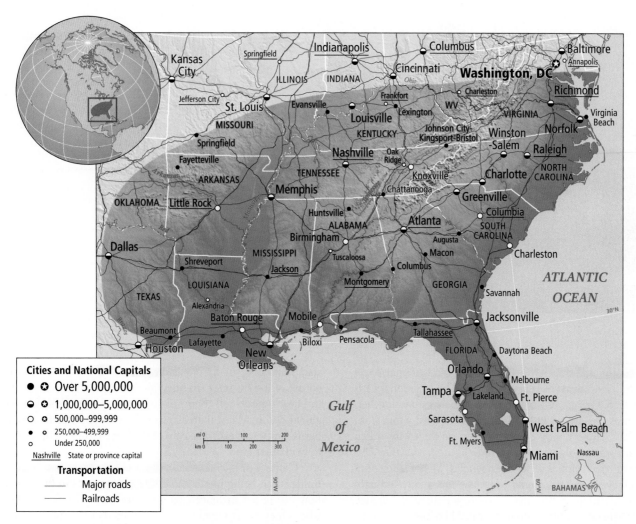

**FIGURE 2.48** The American South subregion. [Adapted from Joel Garreau, *The Nine Nations of North America* (Boston: Houghton Mifflin, 1981).]

The workplace is now integrated, and it is not uncommon to find African-Americans in supervisory and administrative positions, especially in business, government, and educational institutions. In the 1990s, Hispanic immigrants added another dimension to the South. Now in most southern cities much of the restaurant service, road maintenance, and construction work is done by Spanish-speaking workers from Mexico and Central America, many of whom are sinking roots in the region and saving to establish businesses and buy homes (Figure 2.49).

Poverty is still a problem: the South has the nation's highest concentration of families living below the poverty line. On the other hand, most southerners—black, white, or Hispanic—are able to maintain a substantially better standard of living than their parents did. In the first decade of the 2000s, the vast majority of southerners work in the service sector of the economy. Many who left for factory jobs in the old economic core are being attracted back to the region by jobs, business opportunities, lower costs of living, a milder climate, and safer, more spacious, and friendlier neighborhoods than they could afford in such places as New York, Illinois,

or California. In the decade of the 1990s, black migration to the South grew dramatically: 3.5 million people who identified themselves as black moved to the South from other parts of the United States. Seven of the ten metropolitan areas that gained the most black migrants during the 1990s were in the South. For instance, of the more than 600,000 new residents who came to Atlanta between 1995 and 2003, at least 200,000 of them were black, many of these young educated professionals.

The South is steadily improving its position as a growth region. The federally funded Interstate Highway System opened the region to auto and truck transport. Inexpensive industrial locations, close to arterial highways, have drawn many businesses to the South, including automobile and modular home manufacturing, light-metals processing, high-tech electronics assembly, and high-end crafts production. For several decades, tourists and retirees by the hundreds of thousands have been driving south on the Interstate Highway System, many attracted by the bucolic rural landscapes and warm temperatures, the scenery of the national parks, outlet mall shopping, and recreational theme parks.

**FIGURE 2.49** A Latino-theme shopping center in Hoover, Alabama. Owned by two Mexican brothers, the center includes a Mexican restaurant, a market, a laundromat, a western-wear store—and space for more. [Paul McDaniel.]

Southern agriculture is now mechanized, and is at once more diversified and specialized than ever before. The traditional cash crops of tobacco, cotton, and rice are being replaced by strawberries, blueberries, peaches, tomatoes, mushrooms, and wines from local vineyards. These luxury crops are often produced for urban consumers on small holdings by part-time farmers who may also have jobs in nearby factories or service agencies. Most of the country's broiler chickens are now produced by large operations throughout the South. Interestingly, the laborers on these factorylike chicken farms are often immigrants from places like Russia, Ukraine, Vietnam, Haiti, or Honduras who are willing to take these low-wage jobs, at least for a while.

## THE GREAT PLAINS BREADBASKET

The Great Plains (Figure 2.50) receives its nickname, the Breadbasket, from the immense quantities of grain—wheat, corn, sorghum, barley, and oats—it produces. Other crops include soybeans, sugar beets, and sunflowers, and much of this production is exported. The gently undulating prairies give the region a certain visual regularity; its weather and climate, in contrast, can be extremely unpredictable, making life precarious at times. More than a hundred tornadoes may strike across the region in a single night, taking lives and demolishing whole towns. Summers in the middle of the continent, even as far north as the Dakotas, can be oppressively hot and humid, while winters can be terribly cold and dry, or extremely snowy. Rainfall is also unpredictable from year to year.

Nevertheless, the people of the plains have learned to adapt ever since settlers began to stake claims there in the 1860s. As in yet drier regions to the west, Great Plains farmers often irrigate their crops with water pumped from deep aquifers. The primary crop remains wheat, much of it genetically selected to resist frost damage. Most wheat is harvested by traveling teams of combines that start harvesting in the south in June and move north over the course of the summer. To select the most profitable crops and find the best time to sell them, plains farmers keep close tabs on the global commodities markets, usually with computers installed in barns, homes, or even the cabs of the huge farm machines they use to work their land.

Cattle raising is another important activity on the Great Plains. Rather than being herded on the open range, as in the past, cattle now are raised in fenced pastures and then shipped to feedlots, where they are fattened for market on a diet rich in sorghum and corn. Cattle (and other livestock such as hogs and turkeys) are slaughtered and processed for market in small plants across the plains (Box 2.3).

Erosion is a serious problem on the plains: soil is disappearing more than 16 times faster than it can form. Grain fields and pasture grasses do not hold the soil as well as the original dense prairie grasses once did. The sharp hooves of the cattle loosen the soil so that it is more easily carried away by wind and water erosion. Experts estimate that each pound of steak produced in a feedlot results in 35 pounds of eroded soil, so cattle raising is by no means a sustainable economic activity.

Women and men have worked equally hard on the Great Plains since the first settlers arrived. For generations, women have borne and raised large families; grown, preserved, and cooked most of the food; and often managed the bookkeeping. In addition, they have cared for farm animals, driven farm equipment, taught in the country schools, and organized the church-related social functions that reinforced community among widely dispersed farm families.

It is common now to find farm families in which several members of both sexes have college degrees. Increasingly, farms are jointly managed by several family members, while others work in a nearby town, not only to earn salaries but also to gain health and retirement benefits for their families.

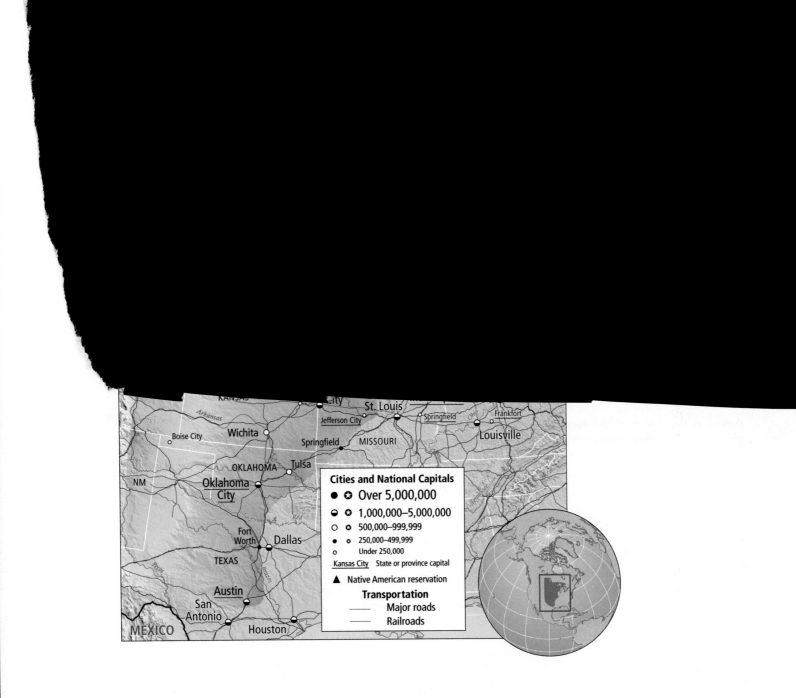

Population patterns on the Great Plains are changing. The few cities around the periphery of the region (Denver, Minneapolis, St. Louis, Dallas, Kansas City) are growing as young people leave the small towns on the prairies. Mechanization has reduced the number of jobs in agriculture and encouraged the consolidation of ownership. Increasingly, the farms of the Great Plains are owned by corporations. Individuals who still farm may own several large farms in different locales. They often choose to live in cities, traveling to their farms seasonally. As a result of this rural depopulation, thousands of small prairie towns are dying out (Figure 2.51). The 2000 census revealed that 60 percent of the U.S. counties in the Great Plains had lost population; the area now has fewer than 6 people per square mile (2 per square kilometer). Those left behind are struggling to provide schools and opportunities for their own children and, in some towns, for the children of poor Hispanic or Asian immigrants attracted by low-paying meatpacking or crop-processing jobs.

A countertrend in some parts of the Great Plains is that Native Americans are coming home, often to work in newly established gambling casinos. Counties with significant numbers of Native Americans are the only counties growing in the area. Although still only a fraction of the total plains population, the overall Native American population grew 12 to 23 percent between 1990 and 2000 in North and South Dakota, Montana, Nebraska, and Kansas. Myron Gutmann, a University of Texas professor, says these and other statistics suggest that European agricultural settlement on the Great Plains may have been an experiment that is ending. In hundreds of counties, native grasses and wildflowers, prairie dogs, and bison have made comebacks. Cattle ranches and grain farms still prevail in the middle and southern plains, but in the northern plains, there are now 300,000 bison, which, unlike cattle, require little management. Mike Faith, a Sioux trained at an accredited Native American college, manages a bison herd at Standing Rock Reservation, not far from where Sitting Bull was killed in 1890. Bison meat, which is low in cholesterol, may help to control excessive weight gain, a recognized problem among Native Americans. Faith adds, "Just having these animals around, knowing what they meant to our ancestors, and bringing kids out to connect with them has been a big plus."

---

## BOX 2.3 | AT THE LOCAL SCALE Meatpacking in the Great Plains

In the 1970s, a meatpacking job in the Great Plains provided a stable annual income of $30,000 or more, relatively high for the times. The workforce was unionized, and virtually all workers were descendants of German, Slavic, or Scandinavian immigrants. But in the 1980s, a number of highly unionized meatpacking companies closed their doors. Other nonunionized plants have opened, often in isolated small towns in Iowa, Nebraska, and Minnesota. The labor is supplied not by local residents, but by immigrants from Mexico, Central America, Laos, and Vietnam. The meatpacking companies usually pay $6.00 an hour, which yields an income, after taxes, of less than $12,000 per year. And union-won work rules are gone. Working hours are long or short at the convenience of the packing house manager, and those who protest may be summarily fired.

Many of the Latino and Asian workers are refugees from war in their home countries, and most of the Asians spent a decade or more in refugee camps in Southeast Asia before coming to the United States. These immigrant workers have difficulty affording housing on their wages, and midwesterners are often reluctant to rent to them. Hence, many live in makeshift housing, like the Laotian families in Storm Lake, Iowa, who occupy a series of old railroad cars and shanties. Down the road, Latino workers live in two dilapidated trailer parks.

The Reverend Tom Lo Van, a Laotian Lutheran pastor, sees little chance that Laotian youth will prosper from their parents' toil. "This new generation is worse off," he says. "Our kids have no self-identity, no sense of belonging . . . no role models. Eighty percent of [them] drop out of high school."

On June 22, 2003, an article in the *Des Moines Register* documented further how globalization is complicating the raising and processing of meat in the North American Great Plains. A Taiwanese family firm recently bought the Hospers, Iowa, meatpacking plant (near Storm Lake) and joined with a northwestern Iowa meat producers' confederation to supply meat to customers in Taiwan, Korea, and Japan. The Iowa meat producers were happy for the deal because they can no longer afford to independently grow, process, and market their meat in the global marketplace. Now, with a guaranteed market in Asia, there is suddenly a demand for much more Iowa meat. But the laborers in the packing plants are still poor migrants, primarily from Latin America and Asia, working now for about $7.00 an hour, $14,560 a year.

*Source:* Adapted from Marc Cooper, "The heartland's raw deal: How meatpacking is creating a new immigrant underclass," *The Nation* (February 3, 1997): 1–18; updated in 2003, http://rwor.org/a/v19/920–29/920/storm.htm; http://lombardiwctc.net/~hchao/; http://desmoinesregister.com/business/stories/c4789013/21538112.html.

# THE CONTINENTAL INTERIOR

Among the most striking features of the continental interior (Figure 2.52) are its huge size, its physical diversity, and its very low population density. This is a land of extreme physical environments, characterized by rugged terrain, frigid temperatures, and lack of water (compare the maps in Figures 2.1, 2.6, and 2.12 with that in Figure 2.52 to gain an appreciation of these characteristics). These physical features restrict many economic enterprises and account for the low population density: fewer than 2 people per square mile (1 person per square kilometer) in most parts of the region.

Physically, there are four distinct zones within the continental interior: the Canadian Shield, the frigid and rugged lands of Alaska,

**FIGURE 2.52** **The continental interior subregion.** [Adapted from Joel Garreau, *The Nine Nations of North America* (Boston: Houghton Mifflin, 1981).]

the Rocky Mountains, and the Great Basin. The **Canadian Shield** is a vast glaciated territory lying north of the Great Plains that is characterized by thin or nonexistent soils, innumerable lakes, and large meandering rivers. The rugged lands of Alaska lie to the northwest of the shield. The shield and Alaska have northern coniferous (**boreal**) and subarctic (**taiga**) forests along their southern portions. Farther north, the forests give way to the **tundra,** a region of winters so long and cold that the ground is permanently frozen several feet below the surface. Shallow-rooted, ground-hugging plants such as mosses, lichens, dwarf trees, and some grasses are the only vegetation. The Rocky Mountains stretch in a wide belt from southeastern Alaska to New Mexico. The highest areas are generally treeless, with glaciers or tundralike vegetation, while forests line the rock-strewn slopes on the lower elevations. Between the Rockies and the Pacific coastal zone is the **Great Basin** (see Figure 2.52), a dry region of widely spaced mountains covered mainly by desert scrub and some woodlands.

The continental interior has the greatest concentration of Native Americans in North America. Most of them live on reservations. In the United States and southern Canada, the reservations are usually not part of their original native lands. Many of those who live in the vast tundra and northern forests of the Canadian Shield, however, still occupy their original territory. The Nunavut, for example, recently won rights to their territory after 30 years of negotiation. They are now able to hunt and fish and generally maintain the ways of their ancestors, although most of them now use snowmobiles and modern rifles. In September 2003, the Dogrib people, a Canadian Native American group of 3500, reclaimed 15,000 square miles of their land just south of the Arctic Circle. The Dogrib, who have a strong sense of cultural heritage and a strong entrepreneurial streak as well, will manage the resources of their ancestral lands, which may include oil, gas, gold, and diamonds.

Although much of the continental interior remains sparsely inhabited, nonindigenous people have settled in considerable density in a few places. Where irrigation is possible, agriculture has been expanding—in Utah and Nevada (Figure 2.53), in the lowlands along the Snake and Columbia rivers of Idaho, Oregon, and Washington, and as far north as the Peace River district in the Canadian province of Alberta. There are many cattle and sheep ranches throughout much of the Great Basin. Overgrazing and erosion are problems there; more serious problems include groundwater pollution from chemical fertilizers and the malodorous effluent from feedlots where large numbers of beef cattle are fattened for market. Efforts by the U.S. government to curb abuses on federal land, which is often leased to ranchers in extensive holdings, have not been very successful.

More than half the land in the continental interior is federally owned. Mining and oil drilling, often on leased federal land, are by far the largest industries. Since the mid-nineteenth century, most of the region's major permanent settlements have been supported by its wide range of mineral resources, which links them to the global economy; even in remote towns, alternating booms and busts are the result of fluctuating world market prices for minerals. In recent times, the most stable mineral enterprises have been oil-drilling

**FIGURE 2.53** Irrigation in the Great Plains and continental interior. Central pivot irrigators create these great circles in the landscape near Tuscarora, Nevada. The water supports crops that would not otherwise grow in this semiarid environment. Often a well at the center of each irrigator taps an underground aquifer, which is then depleted at an unsustainable rate. When the aquifer is depleted, the circles fade to a brownish green. [Alex S. MacLean/Landslides.]

operations along the northern coast of Alaska. From here the Trans-Alaska Pipeline runs southward for 800 miles to the port of Valdez. Often running above ground to avoid shifting as the earth above the permafrost freezes and thaws, the pipeline constitutes a major ecological disruption, interfering with caribou migrations and always posing the threat of oil spills. Oil spills are also a danger offshore; the giant spill from the oil tanker *Exxon Valdez* in 1989 devastated 1100 miles of Alaskan shoreline, killed much wildlife, and ruined native livelihoods and commercial fishing.

Increasing numbers of people are migrating to the continental interior. The many national parks in the region attract millions of tourists as well as both seasonal and permanent workers. Other people come to exploit the area's natural resources. Towns such as Laramie, Wyoming, and Calgary, Alberta, have swelled with workers in the mining industry. These various groups often find themselves in conflict with each other and with the indigenous people of the continental interior. In the United States, environmental groups have pressured the federal government to set aside more land for parks and wilderness preserves. Environmentalists also want to limit or eliminate activities such as mining and logging, which they see as damaging to the environment and the scenery. A switch to more recreational and preservation-oriented uses would change, and probably lessen, employment opportunities throughout the region, while the increasing numbers of visitors would place stress on natural areas, particularly on water resources. The continental interior is one of the most intense battlegrounds in North America between environmentalists and resource developers.

# THE PACIFIC NORTHWEST

Once a fairly isolated region, the Pacific Northwest (Figure 2.54) is now at the center of debates about how North America should deal with environmental and development issues. The economy in much of this region is shifting from logging, fishing, and farming to information technology industries. As this happens, its residents' attitudes about their environment are also changing. Forests that were once valued primarily for their timber are now also prized for their recreational value, especially their natural beauty and their wildlife, both of which are threatened by heavy logging. Energy, needed to run technology and other industries, is now more than ever a crucial resource for the region.

The physical geography of this long coastal strip consists of fiords and islands, mountains and valleys. Most of the agriculture, as well as the largest cities, are located in the southern part, in a series of lowlands lying along long, rugged coastal mountain ranges extending north and south. Throughout the region, the climate is wet. Winds blowing in from the Pacific Ocean bring moist and relatively warm air inland, where it is pushed up over the mountains, resulting in copious orographic rainfall and snowfall. The close proximity of the ocean gives this region a milder climate than is found at similar latitudes farther inland. In this rainy, temperate zone, enormous trees and vast forests have flourished for eons. The balmy climate in much of the region, along with the spectacular scenery, attracts many vacationing or relocating Canadians and U.S. citizens.

Pacific Northwest logging provides most of the construction lumber and an increasing amount of the paper used in North America. Lumber and wood products are also important exports to Asia, and the lumber industry is responsible directly and indirectly for hundreds of thousands of jobs in the region. As the forests have shrunk, environmentalists have harshly criticized the logging industry for such practices as clear-cutting (Figure 2.55), which destroys wild animal and plant habitats, thereby reducing species diversity, and leaves the land susceptible to erosion and the adjacent forest susceptible to diseases and pests. There are many reasons to criticize clear-cutting, but an alternative logging method that will preserve forest diversity and not be unduly costly in time and money is not easy to come by. The battle lines have been drawn between those who make a living from logging or related activities and those who make a living from occupations that tout the beauty of Pacific Northwest forests—urban and rural people who are often advocates of strict environmental protection. But there are also battle lines between those who favor different environmental activist tactics.

**Personal Vignette** Tim Hermack lives in Eugene, Oregon, home of the University of Oregon, long a center for environmental activism. He joined the movement when he returned from the Vietnam War to find that his favorite camping places had been clearcut. But Hermack became conflicted about his role as an activist when, in the mid-1990s, the Earth Liberation Front (ELF) began firebombing lumber

**FIGURE 2.54** The Pacific Northwest subregion. [Adapted from Joel Garreau, *The Nine Nations of North America* (Boston: Houghton Mifflin, 1981).]

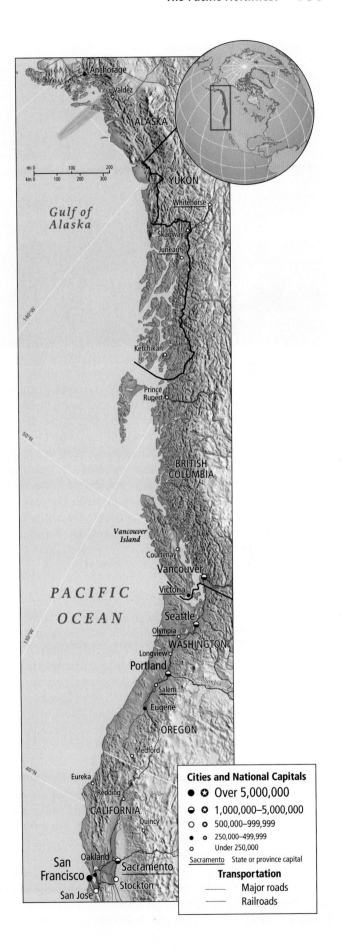

**Cities and National Capitals**
- ● ✪ Over 5,000,000
- ◓ ✪ 1,000,000–5,000,000
- ○ ✪ 500,000–999,999
- • ◦ 250,000–499,999
- ∘ Under 250,000
- Sacramento  State or province capital

**Transportation**
- —— Major roads
- —— Railroads

**FIGURE 2.55** Clear-cutting in Washington. Near Cathlamet, in central Washington State, this landscape shows various stages of clear-cutting. The dark green is uncut forest, the bright green is regrowing forest, and the gray is newly clear-cut forest. Logging roads are also visible. In the Pacific Northwest, logging companies have been known to leave wide strips of forest along public highways so that the citizenry remains unaware of the extent of the forest removal. [Jim Wark/Airphoto.]

companies and ski resorts to draw attention to the devastation caused by clear-cutting. Hermack was outraged by these tactics, saying that by using violence ELF was not protecting the liberty and justice he held dear. Now, he is further outraged by recent activities of the U.S. Justice Department, which in 2005 charged ELF activists with ecoterrorism and sentenced one to 22 years in prison for arson. Hermack notes that a far more flagrant arsonist, an employee of the U.S. Forest Service who set 35 fires because she was upset about overtime pay, got only probation. He sees the U.S. Justice Department action as a witch hunt exploiting the post-9/11 fears of terrorism to discredit the worthy cause of opposition to clear-cutting.

Source: "Harsh Sentences Silence Radical Environmentalists," National Public Radio, "All Things Considered," March 8, 2006. ■

There are also disputes over how to acquire sufficient energy for the region. The presence of several large hydroelectric dams in the Pacific Northwest has attracted industries in need of cheap electricity, particularly aluminum smelting and associated manufacturing industries in aerospace and defense. These industries are major employers, but their demand for labor is erratic and periodic layoffs are common. In addition, the hydroelectric dams have been criticized by another major employer in the region, the fishing industry. The dams block the seasonal migrations of salmon to and from their spawning grounds, so some of the region's most valuable fish cannot produce young.

Given the environmental impacts of the logging and hydroelectric industries, it is not surprising that many people are enthusiastic about the increasing role of information technology within the major urban areas of the Pacific Northwest: San Francisco, Portland, Seattle, and Vancouver. With Silicon Valley just outside San Francisco and with a growing number of IT companies around Seattle, the Pacific Northwest is a world leader in information technology. However, IT firms produce hazardous waste such as lead and other heavy metals used in circuit boards, and they are particularly dependent on secure and steady sources of electricity—

whether from waterpower, gas, oil, coal, or nuclear fuels—that have significant negative environmental impacts.

The Pacific Northwest, and particularly its major cities, often lead the rest of the United States and Canada in social and economic innovation. This role is illustrated by the region's growing connections to Asia. The adjustment of a predominantly Euro-American society to large numbers of Asian immigrants has been well under way along the Pacific coast for many years. This transition has been accompanied by the growing importance of the countries around the Pacific Rim as trading partners. Vancouver, Seattle, Portland, and San Francisco are increasingly oriented toward Asia, not just in the trade of forest products, but in banking and tourism, and in the design and manufacture of information technology equipment, some of which is now being outsourced to Asia.

## SOUTHERN CALIFORNIA AND THE SOUTHWEST

Southern California and the Southwest (Figure 2.56) are united primarily by their warm, dry climate, which has attracted many migrants from across America; by their long and deep ties to Mexico; and by the promise that the North American Free Trade Agreement will make the subregion a nexus of development.

The varied landscapes of the Southwest include the Pacific coastal zone, the hills and interior valleys of Southern California, the widely spaced mountains and dramatic mesas and canyons of southern Arizona and New Mexico, and the gentle coastal plain of south Texas. The common physical characteristics of these landscapes are their warm average year-round temperatures and arid climate. Most areas receive less than 20 inches of rainfall annually, and the dominant vegetation is scrub, bunchgrass, and widely spaced trees. Ranching is the most widespread form of land use, but other forms are more important economically and in numbers of employees.

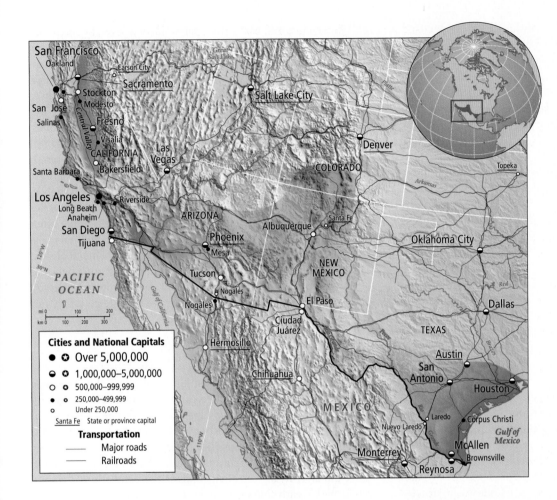

**FIGURE 2.56** The Southern California and Southwest subregion. [Adapted from Joel Garreau, *The Nine Nations of North America* (Boston: Houghton Mifflin, 1981).]

Where irrigation is possible, farmers take advantage of the nearly year-round growing season to cultivate fruits and vegetables. California's Central Valley is a leading producer of fruits and vegetables in the United States and is one of the most valuable agricultural districts in the world. Many of the crops are produced on vast, plantation-like farms. Migrant Mexican workers, often illegal immigrants and sometimes mere children, supply most of the labor and lower the cost of North American food by working for very low wages (Figure 2.57).

Although agriculture is important, the bulk of the region's economy is nonagricultural. The coastal zones of Southern California and Texas are home to oil drilling and refining and associated chemical industries, as well as other industries that need the cheap transport provided by the ocean. Access to both the Pacific Ocean and the Gulf of Mexico, mild climates, sunny weather, and spectacular scenery have made this region a popular location for a wide range of service industries, which employ the vast majority of the region's workers. Los Angeles—with its major trade, transport, media, entertainment, and finance industries, as well as important research facilities—is now the second most populous city in the United States. Like San Francisco, Seattle, and other Pacific coastal cities, Los Angeles is strategically located for trade with Pacific Rim countries, and it recently replaced New York as the largest port in the United States.

**FIGURE 2.57** Migrant farmworkers in the Southwest. José Madrid, 11, picks green chilis in New Mexico. "I'm not good at math, but I'm good at money," he says. Like many child migrant workers in the United States, he goes to school only intermittently. [Eric Draper/AP/Wide World Photos.]

On the eastern flank of this region, Austin, Texas, has attracted information technology industries, lured by research activities connected to the University of Texas, by the warm climate, and by the laid-back, folksy lifestyle of central Texas. Austin has close to 3 million people and grew by more than 30 percent from 1990 to 2000. Across the region, the climate draws large numbers of vacationers and retirees; entire planned communities of 10,000 or 20,000 people have grown up in the deserts of New Mexico and Arizona within just a few years.

In the first decade of the 2000s, residents of Southern California and the Southwest must worry about energy costs, congestion, smog, and increasingly scarce water. In the case of energy, a number of factors, including high usage based on originally low prices, deregulation of energy providers leading to escalating prices, corruption among some private generators and distributors, and decreased supplies led to an energy crisis in early 2001 that provoked questions about the region's economic future. In the short term, the electricity supply problem has been alleviated by controls on prices and reduction of nonessential use. But like the rest of the United States, Southern California and the Southwest have been consuming energy at rates that are not sustainable, affordable, or safe for the environment.

Much of this region was originally a colony of Spain, and the area has maintained the Spanish language, a distinctive Hispanic culture, and other connections with Mexico (see Figure 1.9 on page 8). Today, Hispanic culture is gaining prominence in the Southwest and spreading beyond it. As we have seen, large numbers of immigrants are arriving from Mexico (see Figure 2.27. on page 83), and the economies of the United States and Mexico are becoming increasingly interdependent as NAFTA fosters increased trade (see the discussion on pages 75, 77).

Among aspects of this interdependence are the factories, known as **maquiladoras,** set up by U.S., Canadian, European, and Asian companies in Mexican towns just across the border from U.S. towns. Maquiladoras in such places as Ciudad Juárez (El Paso), Nuevo Laredo (Laredo), Nogales (Nogales), and Tijuana (San Diego) produce manufactured goods for sale primarily in the United States and Canada, taking advantage of cheaper Mexican wages, cheaper land and resources, lower taxes, and weaker environmental regulations. (The maquiladora phenomenon is discussed further in Chapter 3.) These factories are a key part of a larger transborder economic network that stretches across North America. In fact, the U.S.–Mexico border has been one of the world's most **permeable national borders,** meaning that people and goods flow across it easily (though not without controversy). This permeability was greatly increased by the North American Free Trade Agreement of 1994. There are now as many as 200 million total legal border crossings each year. According to a U.S. Immigration and Naturalization Service study in 1998, illegal border crossings then were less than 10 percent of the total, at about 150,000 per year. Since the terrorist attacks of 2001, the border has become much less permeable and the trip more dangerous. Smugglers who bring workers across the more heavily guarded border charge high fees, often cheat their passengers, and in a number of cases have left whole truckloads of people to die in the desert heat, without food or water.

As we discussed earlier (see pages 82–86), several contentious issues surround the estimated 11 million undocumented immigrants currently residing in the United States, of whom perhaps 6 million are Mexicans. Many Mexican migrants remain in California and the Southwest, where, although they contribute vitality to the local and national economy, their willingness to work for low wages keeps down the wages of other low-skilled workers. The presence of these migrants also increases tension over cultural change. Although English speakers in the Southwest far outnumber Spanish speakers, some fear that English could be challenged in much the same way it has been challenged by French Canadians in Québec. Accordingly, Arizona and California recently became the first states to make English the official language of government. Some bilingual programs, installed in the 1970s to help migrant children make the transition slowly from Spanish to English, are being abolished by those who feel it is better for those children to learn English as quickly as possible. Other educators think that migrant children do better in math and science and enjoy higher self-esteem if they can study for a period in their native language.

## Reflections on North America

It is not hard to rhapsodize about North America: the sheer size of the continent, its incredible wealth in natural and human resources, its superior productive capacities, and its powerful political position in the world are attributes possessed by no other world region. North America enjoys this prosperity, privilege, and power as a result of fortunate circumstances, the hard work of its inhabitants, the diversity and creativity of its settlers, and astute planning on the part of early founders, especially the writers of the U.S. Constitution—but also because it is able to access resources and labor worldwide at favorable prices. Perhaps the most important factor in North America's success has been its democratic governments and supporting institutions, which allow for individual freedom and flexibility as times and circumstances change.

Yet life has not been good to everyone in North America, nor has the influence of North America on other parts of the world always been benign. Native Americans, enslaved Africans and their descendants, and other ethnic minorities suffered as Canada and the United States were being created out of lands long occupied by indigenous peoples. Settlers of all origins and their descendants had, and continue to have, a significant negative impact on the continent's environments, and the standard of living expected by all North Americans promises to increase the strain on the environment. Furthermore, as Canada and the United States have developed into wealthy world powers, their impact on environments and people elsewhere around the world, through trade and cultural diffusion, has increased. Canada's influence has thus far been judged largely benign; but the influence of the United States has been increasingly viewed as problematic, even threatening.

There is no guarantee that North America will continue its leadership role into the future; in fact, in the post-9/11 world, challenges to that leadership are common in every corner of the globe. Some critics say that the recent U.S. tendency to use military force before adequate consultation with allies is the major cause for concern.

Others cite insensitive trading policies or mistreatment of employees in the overseas workplaces of U.S.-based corporations such as Wal-Mart or Nike. North American models for development—based on democracy and on assumptions of rich and inexhaustible resources—are being challenged as inappropriate for much of the rest of the world. As we shall see in subsequent chapters, societies elsewhere are beginning to prosper without following North American examples, and sometimes without first installing democratic institutions. Environmental concerns, increasingly a part of the North American consciousness (if not always practice), are not central in many developing countries. Rather, material prosperity is often the chief goal, just as it still is for many in North America. Many people around the world are eager to bring the North American material miracle to their lands, regardless of negative environmental impacts.

The closer formal economic association of Canada, the United States, and Mexico through NAFTA has strengthened the already strong economic relationships among these three countries. Will NAFTA be the forerunner of new alignments between North America and countries in Middle and South America, or will its difficulties dampen further trade unity in the hemisphere? As we shall see in Chapter 3, over the past few centuries Middle and South America have had experiences very different from those of North America, and now social and economic changes in that region seem to suggest a new era of greater autonomy and prosperity for a number of countries. Such changes could actually enhance the possibility that all of the Americas will eventually be more integrated economically and, possibly, socially and politically as well.

## Chapter Key Terms

acid rain 93
agribusiness 71
aquifer 95
baby boomer 91
boreal forest 108
brownfield 80
Canadian Shield 108
chain migration 83
clear-cutting 62
dense node 81
digital divide 74
economic core 60
ethnicity 86
genetic engineering 72
Geneva Conventions 68
gentrification 81
government subsidy 75

Great Basin 108
hazardous waste 92
Hispanic 56
hub-and-spoke network 72
information technology (IT) 73
Internet 73
Interstate Highway System 72
knowledge economy 73
Latino 56
loess 57
long-lot system 99
maquiladora 112
megalopolis 81
metropolitan area 80
New Urbanism 81
North American Free Trade Agreement (NAFTA) 75

nuclear family 89
Ogallala aquifer 95
outsourced 78
Pacific Rim (Basin) 65
permeable national border 112
Québecois 56
Rust Belt 101
service sector 73
smog 92
taiga 108
thermal inversion 92
trade deficit 77
tundra 108
urban sprawl 82

## Critical Thinking Questions

1. Discuss the ways in which North American culture is adapted to the high rate of spatial mobility experienced by Americans.

2. North America's population is changing in many ways. Explain why the aging of America is of interest and discuss how you personally are likely to be affected.

3. North American family types are changing. Explain the general patterns and discuss whether or not the nuclear family is or should be the sought-after norm.

4. The influence of globalization is now felt in small, even isolated, places in North America. Pick an example from the text or from your own experience and explain at least four ways in which this place is now connected to the global economy or global political patterns.

5. North Americans profit from having undocumented workers produce goods and services. Explain why this is the case and discuss if and how the situation should be changed. What is the argument for doing nothing or very little?

6. People like to say that the attacks on 9/11 changed everything for North Americans (especially for those living in the United States). What do they mean by this? How were U.S. attitudes toward the rest of the world modified? Do you think these modifications will be useful in the long run?

7. When you compare the economies of the United States and Canada and the ways they are related, what are the most important factors to mention?

8. When U.S. farmers or producers get subsidies from the federal government, what is intended? Why do producers and farmers in poor countries say that this practice hurts them?

9. Examine the connections between urban sprawl, rising consumption of gasoline, and rising air pollution in North American urban areas, and then propose a solution that you could abide by yourself.

10. The United States and Canada have quite different approaches to treatment of citizens who experience difficulties in life. Explain those different approaches and the apparent philosophies behind them.

ATLANTIC OCEAN

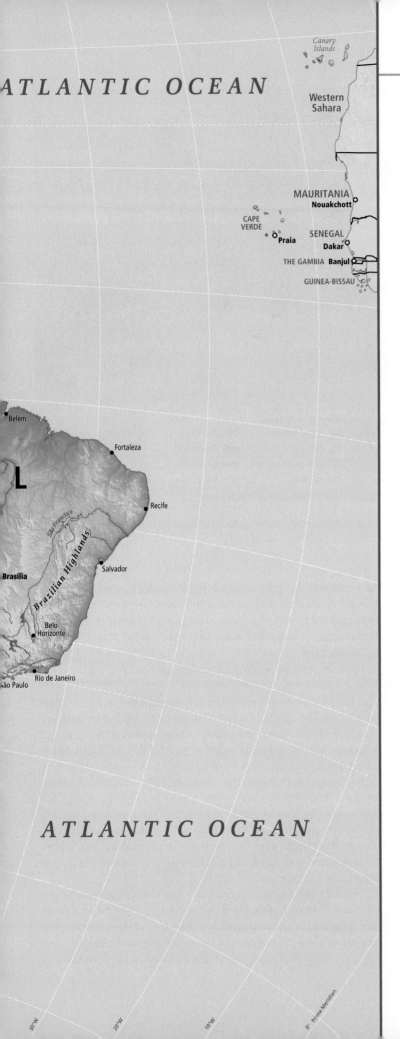

ATLANTIC OCEAN

# CHAPTER 3

# MIDDLE AND SOUTH AMERICA

On a trip through the Ecuadorian Amazon several years ago, Alex Pulsipher looked into the geography of the region's rapidly expanding oil industry and its effects on local populations.

**Global Patterns, Local Lives** The boat trip down the Aguarico River in Ecuador took me into a world of magnificent trees, river canoes, and houses built high up on stilts to avoid floods and pests and to catch the breeze. I was there to visit the Secoya, a group of 350 indigenous people in the Ecuadorian Amazon who were negotiating with the U.S. oil company Occidental Petroleum over its plans to drill for oil on Secoya lands. Oil revenues supply 40 percent of the Ecuadorian government's budget and are essential to paying off its national debt. Hence the government had threatened to use military force to compel the Secoya to allow drilling. The Secoya wanted to protect themselves from the pollution and cultural disruption that come with oil development. As Colon Piaguaje, chief of the Secoya, put it to me, "A slow death will occur. Water will be poorer. Trees will be cut. We will lose our culture, our language, alcoholism will increase, as will marriages to outsiders, and eventually we will disperse to other areas. There is no other way. We will negotiate, but these things will happen." Given all the impending changes, Chief Piaguaje asked Occidental to use the highest environmental standards in the industry and to establish a fund to pay for the educational and health needs of the Secoya people.

Like the Secoya, indigenous peoples around the world are bearing the consequences of development (Figure 3.2). Chief Piaguaje based his predictions for the future on what has happened in parts of the Ecuadorian Amazon that have already experienced several decades of oil development. The U.S. company Texaco was the first major oil developer to establish operations in Ecuador. From 1964 to 1992, its pipelines and waste ponds leaked almost 17 million gallons of oil into the Amazon

**FIGURE 3.1** Regional map of Middle and South America.

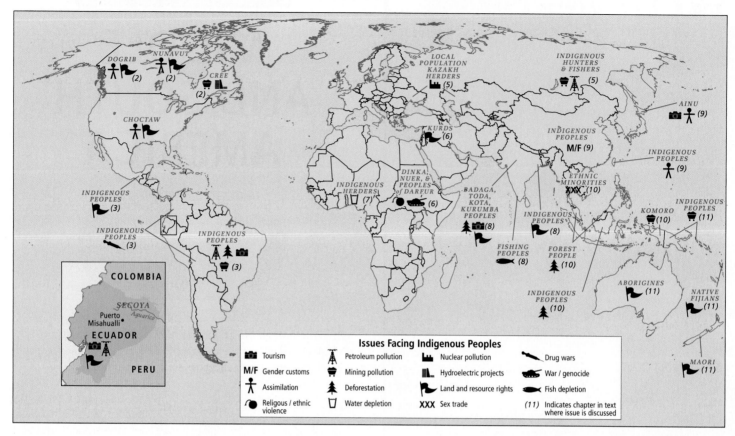

**FIGURE 3.2** Indigenous peoples and environmental issues in this text. Issues relating to indigenous peoples are mentioned in many places in this book. In all cases, the issues are in some way related to indigeous peoples' interactions with the outside world.

Basin—50 percent more than the 11 million gallons the oil tanker *Exxon Valdez* spilled into Prince William Sound in Alaska. Although Texaco sold its operations to the government and left Ecuador in 1992, its oil wastes continue to leak into the environment from hundreds of open pits (Figure 3.3). In 1993, some 30,000 people—both indigenous people and settlers from the densely populated highlands who had established farms along Texaco's service roads—sued Texaco in New York State, where the company (now called ChevronTexaco) is headquartered, for damages from the pollution. Several epidemiological studies concluded that oil contamination has contributed to higher rates of childhood leukemia, cancer, and spontaneous abortions among people who live near pollution created by Texaco.

Six years after my visit to the Secoya, many of Chief Piaguaje's worries have been borne out. Occidental did establish a fund to help the Secoya deal with the disruption of oil development, but part of this fund was distributed to individual households. Some people invested their money in ecotourism and other commercial enterprises and have prospered, while others simply spent their money and are now working for others in the community. Such employer-employee relationships are new to the Secoya, and they are changing, and some would argue degrading, what was once an especially egalitarian culture.

The oil development has had many negative effects on the environment. Air and water pollution have increased rates of illness, and the animals that the Secoya used to depend on have disappeared almost entirely due to overhunting by new settlers from the highlands. In 2002,

the Ecuadorian suit against Texaco was dismissed by the U.S. Court of Appeals. It was refiled in Ecuador in 2003 with the proviso that it could be re-appealed in the United States if the plaintiffs believed they had not received justice. A judgment had not yet been handed down in Ecuador by January 2007.

*Sources: Amazon Watch 2006; Oxfam America 2005.* ∎

The rich resources of the Americas have attracted outsiders since the first voyage of Christopher Columbus in 1492. Europe's encounter with the Americas marked a major expansion of the global economy, but for several hundred years Middle and South America occupied a peripheral position in global trade, supplying primarily raw materials that aided the Industrial Revolution in Europe. Recently, however, this region has begun to forge stronger global connections by forming trade blocs within the region, by sending abroad migrants who retain attachments to home, by attracting outside investment, and by developing trade partnerships within the region and with Japan, China, European countries, the United States, and Canada.

Middle and South America differ from North America in several ways. Physically, this world region is larger and exhibits a more varied mosaic of environments (see Figure 3.1). Its cultural complexity is different from that of the continent to the north; here, there are both large and varied **indigenous** (native) populations

**FIGURE 3.3** Pollution from oil development in Ecuador. A worker samples one of the several hundred open waste pits that Texaco left behind in the Ecuadorian Amazon. Wildlife and livestock trying to drink from these pits are often poisoned or drowned. After heavy rains, the pits overflow, polluting nearby streams and wells. [Alex Pulsipher.]

and immigrant groups from Europe, Asia, and Africa, all of which have a strong influence on daily life. This cultural variety is made yet more complex by highly stratified social systems based on class, race, and gender. Politically, too, the region is more complex: it includes more than three dozen independent countries that implement different models of self-government with different degrees of success. Economically, levels of well-being vary more markedly than they do in North America.

Despite the distinct features of the many parts of Middle and South America, there are commonalities within the region: widespread use of Spanish, cohesive multigenerational families, and a strong social and political role for the Roman Catholic Church are just a few examples. Most of these commonalities originate from many of the countries' shared experience as colonies of Spain or Portugal. The focus on production for export that began with colonialism greatly altered local landscapes, just as oil extraction is altering the Ecuadorian Amazon today. Indigenous lands once devoted to multiple uses were transformed into vast stretches of a single crop, and the surface of the land was sometimes stripped away for mining. In both colonial and postcolonial times, as local elites allied first with European and then with U.S. private investors to exploit local lands and extract resources, indigenous people have been denied equal rights, held in subservient class positions, and discriminated against.

Nonetheless, in every country of the region, people like Chief Colon Piaguaje in Ecuador are making efforts to change the patterns of the colonial and postcolonial eras, and successes are not hard to find. Self-help projects and efforts to democratize everyday life are evident in countless villages and city neighborhoods. The military, which in the past was often called on to maintain order (for good or ill), is now much less in evidence. The economies of

many countries are being reorganized to encourage economic growth, and government controls on markets are being removed. While the poor have been hit hard by such **economic restructuring** and the lost jobs and higher prices it brings, many have managed to survive using informal networks and communal self-help strategies. In confronting the hardships wrought by restructuring, many people of the region are realizing that development is not just a matter of economic growth; to be judged successful, development must change the lives of the majority for the better.

## Themes to Explore in Middle and South America

You will encounter a few major themes repeatedly in this chapter:

1. **Cultural and physical diversity.** The region of Middle and South America is notable for its cultural variety and richness, but also for its wide array of physical environments.

2. **Increasing regional integration of trade.** The countries of the region are forming economic links with one another by lowering old colonial trade barriers and establishing free trade blocs within the Americas as well as with the European Union and several countries on the Pacific Rim.

3. **Production of raw materials.** Many countries still rely on income from the export of agricultural products and extracted resources, though manufacturing for export is a rapidly growing economic sector.

4. **Beginnings of change in highly stratified social systems.** Although in many countries political and economic power remains concentrated in the hands of a small, rich minority, often of

European descent, the poor majority (which includes many people of indigenous, mestizo, and African descent) is beginning to exercise more political influence. The disparities in wealth are wider than in any other global region, yet in some areas these disparities have decreased slightly.

5. **Rural-to-urban and international migration.** In all parts of the region, people, especially the young, are migrating from the countryside to cities and from the region to North America and Europe. The money and goods they send home form a significant portion of national incomes.

6. **The extended family.** Despite declining birth rates, large multigenerational families, with defined roles for men and women, continue to predominate in the region and serve as a source of mutual support during hard times.

# I THE GEOGRAPHIC SETTING

## Terms to Be Aware Of

In this book, **Middle America** refers to Mexico, the narrow ribbon of land south of Mexico that makes up Central America, and the islands of the Caribbean. **South America** refers to the continent south of Central America. For several reasons, we usually don't use the term *Latin America* in this book, even though it is the term most often used by others for what we call Middle and South America. Latin America is so called because it was colonized for the most part by Spain and Portugal, whose cultures are thought of as having Latin (Roman) origins. Therefore, when *Latin* is used to refer to the Americas, it serves as a permanent reminder of the region's former colonial status. Furthermore, the designation ignores not only the other cultures present in the region, most notably those of the various indigenous groups, but also those of non-Latin people who arrived during and after the colonial period: Africans, Dutch, Germans, British, Chinese, Japanese, and others from elsewhere in Asia. Nor does it acknowledge the new, distinctly mixed cultures—often called **Creole cultures**—that have been created from these many strands. In this chapter, we use the term *indigenous groups* or *peoples* to refer to the native inhabitants of the region, rather than *Native Americans*.

The terms *New World* and *Old World* are also problematic. Is the New World so designated because it is actually newer in some way than the Old World, or only because it was new to Europeans in 1492? In this book, we occasionally use the term *Old World* to mean an entity other than the Americas, which includes virtually the entire rest of the world.

## PHYSICAL PATTERNS

You can see from the map in Figure 3.1 that the region of Middle and South America extends south from the midlatitudes of the Northern Hemisphere all the way across the equator through the Southern Hemisphere, nearly to Antarctica. This expanse is part of the reason for the wide range of climates in the region. Another contributing factor is the variation in altitude. Tectonic forces have shaped the primary landforms of this huge territory to form an overall pattern of highlands to the west and lowlands to the east.

**FIGURE 3.4** Soufrière volcano on Montserrat. The volcano became active in 1995 and remained so in 2007. Volcanologists expect its activity to persist for decades. In this photograph, taken on December 30, 2006, the volcano emits an ash cloud as a portion of the lava dome collapses and pyroclastic flows fall down the slope on the left. A week later, a major pyroclastic flow sent ash 5 miles into the air. One-third of the island remains safe from the volcanic activity. [Courtesy of Mac Goodwin.]

# Landforms

***Highlands.*** A grandly curving and nearly continuous chain of mountains stretches along the western edge of the American continents for more than 10,000 miles (16,000 kilometers) from Alaska to Tierra del Fuego, at the southern tip of South America (see Figures 1.1 on page 1, 2.1 on page 53, and 3.1). The southern part of this long mountain chain is known as the Sierra Madre in Mexico, by various local names in Central America, and as the Andes in South America. It was formed by the eastward-moving oceanic plates—the Cocos Plate and the Nazca Plate—plunging beneath the three continental plates—the North American Plate, the Caribbean Plate, and the South American Plate—at a lengthy **subduction zone** (the zone where one tectonic plate slides under another) running thousands of miles along the western coast of the continents (see Figure 1.20 on page 21). In a process that continues today, the overriding plates crumple to create mountain chains, often developing fissures that allow molten rock from beneath the earth's crust to ascend to the surface and form volcanoes (see Figure 1.21 on page 22). These volcanic highlands constitute a major barrier to transportation and communication throughout their length, but in northern and central South America, where the population is dense, they pose a special challenge.

The chain of low and high mountainous islands in the eastern Caribbean is similarly volcanic in origin, created as the Atlantic Plate thrusts under the eastern edge of the Caribbean Plate. It is not unusual for volcanoes to erupt in this active tectonic zone. On the island of Montserrat, for example, people have been coping with an active volcano for more than a decade. The eruptions have been violent **pyroclastic flows**—blasts of superheated rock, ash, and gas that move with great speed and force (Figure 3.4). In the blast of July 1995, two-thirds of the island was buried in ash, 20 people were killed, and two-thirds of the population was evacuated. About 4000 others, who relocated to the north of the island, are adapting to life in a still beautiful but greatly changed place.

***Lowlands.*** A look at the regional map in Figure 3.1 will show that lowlands generally extend over most of the land to the east of the western mountains. In Mexico, however, the Sierra Madre is divided into a broad **U**. Within the **U** sits a high plateau, and to the east of the Sierra Madre, a coastal plain borders the Gulf of Mexico. Farther south, in Central America, wide aprons of sloping land descend to the Caribbean coast. In South America, a huge wedge of lowlands, widest in the north, stretches from the Andes east to the Atlantic Ocean. These South American lowlands are interrupted in the northeast and the southeast by two modest highland zones: the Guiana Highlands and the Brazilian Highlands. Elsewhere in the lowlands, grasslands cover huge, flat expanses, including the llanos of Venezuela and Colombia and the pampas of Argentina.

The largest feature of the South American lowlands is the Amazon Basin, drained by the Amazon River and its tributaries (Figure 3.5). This basin lies within Brazil and the neighboring countries to its west, but it has global biological significance: it contains the earth's largest expanse of tropical rain forest, 20 percent of the earth's fresh water, and more than 100,000 species of plants and animals. The basin's rivers are so deep that ocean liners can steam 2300 miles (3700 kilometers) upriver from the Atlantic Ocean all the way to Iquitos, jokingly referred to as Peru's "Atlantic seaport." The vast Amazon River system starts as streams high in the Andes. These streams eventually unite as rivers that flow eastward toward the Atlantic, joined along the way by rivers flowing from the Guiana Highlands to the northeast and the Brazilian Highlands to the southeast. Once these rivers reach the flat land of the Amazon Plain, their velocity slows abruptly, and fine soil particles, or **silt,** sink to the riverbed. When the rivers flood, silt and organic material transported by the floodwaters renew the soil of the surrounding

**FIGURE 3.5** The Amazon lowlands. The Tigre River, a tributary of the Amazon, meanders through the Peruvian lowland rain forest. This scene is typical of those parts of the Amazon Basin that have not been disturbed by roads or forest removal. [Layne Kennedy/CORBIS.]

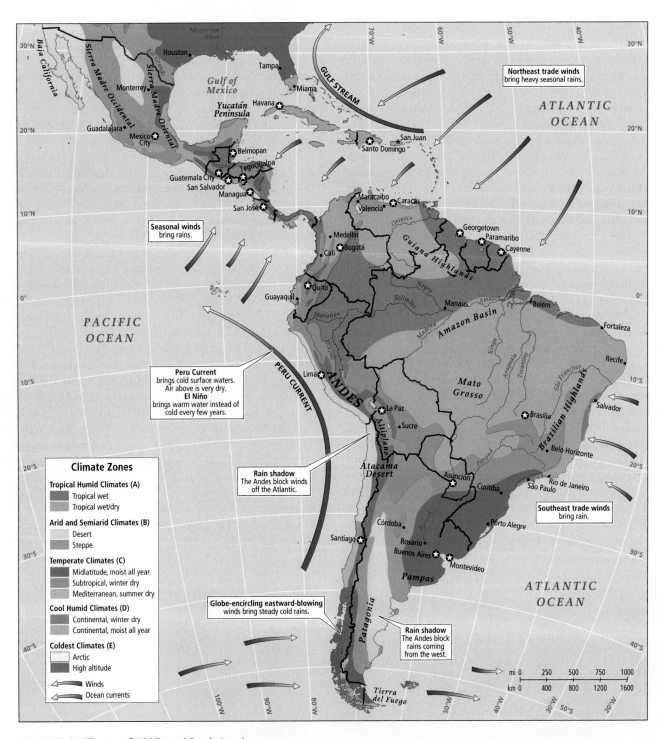

**FIGURE 3.6** Climates of Middle and South America.

areas, nourishing millions of acres of tropical forest. Not all of the Amazon Basin is rain forest, however. Variations in rainfall, cloud cover, wind patterns, and soil types, as well as human activity, can create the conditions for seasonally dry deciduous tropical forest or even grassland.

The interior reaches of the Amazon Basin are home to some of the last relatively undisturbed indigenous cultures. When Europeans first came to these tropical wetlands, there were few ways to

exploit them for profit, so for several centuries they remained largely unexplored by Europeans. The Amazon Basin was home to perhaps 2 million hunters and gatherers and subsistence cultivators. In the nineteenth century, European settlement and commercial exploitation (mining, logging, and some ranching and plantation cultivation) began in earnest, and soon the native peoples were in decline. In 1900, there were 230 known ethnic groups (often referred to as tribes) in the Amazon Basin. Since then, 87 entire tribes have become extinct, and in some isolated tribes, such as the Secoya of Ecuador, only a few hundred members survive. Many people died as a result of introduced diseases or mistreatment while laboring on plantations or in mines. Recently, others have been forced to leave by losses of habitat and environmental quality caused by logging, ranching, oil exploitation, commercial agriculture, and new settlements.

## Climate

From the jungles of the Amazon and the Caribbean to the high, glacier-capped peaks of the Andes to the parched moonscape of the Atacama Desert, the climate variety of Middle and South America is astounding (Figure 3.6). This variety results from several factors. The wide range of temperatures reflects the great distance the landmass spans on either side of the equator: northern Mexico lies at 33° N latitude, while Tierra del Fuego, at the southern tip of South America, lies at 55° S latitude. The region's long mountainous spine creates tremendous changes in altitude that also contribute to the wide range of temperatures. In addition, global patterns of wind and ocean currents result in a distinct pattern of precipitation.

***Temperature-Altitude Zones.*** Four main **temperature-altitude zones** are commonly recognized in the region (Figure 3.7). As altitude increases, the temperature of the air decreases by about 1°F per

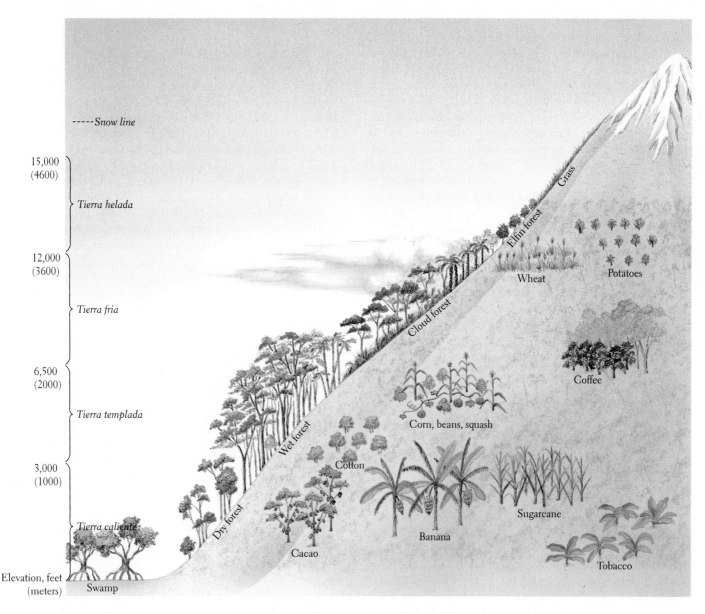

**FIGURE 3.7** Temperature-altitude zones of Middle and South America. As the temperature changes with altitude, the natural vegetation on mountain slopes also changes, as shown along the edge of the mountain here. The same is true for crops, some of which are suited to lower, warmer elevations and some to higher, cooler elevations. [Illustration by Tomo Narashima, based on fieldwork and a drawing by Lydia Pulsipher.]

300 feet (1°C per 165 meters) of elevation. Thus temperatures are warmest in the lowlands, which are known in Spanish as the *tierra caliente,* or "hot land." The *tierra caliente* extends up to about 3000 feet (1000 meters) in elevation, and in some parts of the region these lowlands cover wide expanses. Where moisture is adequate, tropical rain forests thrive, as do a wide range of tropical crops, such as bananas, sugarcane, cacao, and pineapples. Many coastal areas of the *tierra caliente,* such as northeastern Brazil, have become zones of plantation agriculture that support populations of considerable size.

Between 3000 and 6500 feet (1000 to 2000 meters) of elevation are the cooler *tierra templada* ("temperate lands"). The year-round springlike climate of this zone drew large numbers of indigenous people in the distant past and, later, Europeans. Here such crops as corn, beans, squash, various green vegetables, wheat, and coffee are grown. Between 6500 and 12,000 feet (2000 to 3600 meters) are the *tierra fria* ("cool lands"). Many midlatitude crops—such as wheat, fruit trees, and root vegetables (potatoes, onions, and carrots)—and cool-weather vegetables—such as cabbage and broccoli—do very well at this altitude. Several modern population centers are in this zone, including Mexico City and Quito, Ecuador. Above 12,000 feet (3600 meters) are the *tierra helada* ("frozen lands"). At the lowest reaches of this zone, some grains and root vegetables are cultivated, and some animals—such as llamas, sheep, and guinea pigs—are raised for food and fiber. Higher up, vegetation is almost absent, and mountaintops emerge from under snow and glaciers. The remarkable feature of such tropical mountain zones is that in a single day of strenuous hiking, one can encounter most of the climate types known on earth.

***Precipitation.*** The pattern of precipitation throughout the region (see Figure 1.22 on page 23) is influenced by the interaction of global wind patterns, topographic barriers such as mountains, and nearby ocean currents. The **trade winds** (tropical winds that blow from the northeast and the southeast toward the equator) sweep off the Atlantic, bringing heavy seasonal rains (see Figure 3.6). Winds blowing in from the Pacific bring seasonal rain to the west coast of Central America, but mountains block the rain from reaching the Caribbean side.

The Andes are a major influence on precipitation in South America. Rains borne by the northeast and southeast trade winds off the Atlantic are blocked, creating a rain shadow on the western side of the Andes in northern Chile and western Peru. Southern Chile is in the path of eastward-trending winds that sweep north from Antarctica, bringing steady cold rains that support forests similar to those of the Pacific Northwest of North America. The Andes block this flow of wet, cool air and divert it to the north, thus creating an extensive rain shadow on the eastern side of the mountains along the southeastern coast of Argentina (Patagonia).

The pattern of precipitation is also influenced by the adjacent oceans and their currents. Along the west coasts of Peru and Chile, the cold surface waters of the Peru Current bring cold air that is unable to carry much moisture. The combined effects of the Peru Current and the central Andes rain shadow have created what is possibly the world's driest desert, the Atacama of northern Chile.

***El Niño.*** An interesting and only partly understood aspect of the Peru Current is its tendency to change its direction every few years. When this happens, warm water flows eastward from the western Pacific, bringing warm water and torrential rains, instead of cold water and drought, to parts of the west coast of South America. The phenomenon was named **El Niño,** or "the Christ Child," by Peruvian fishermen, who noticed that when it occurs (on an irregular cycle), it reaches its peak sometime in December, when Hispanic cultures celebrate Christmas. El Niño has effects not only in South America, but worldwide, periodically bringing cold air and drought to normally warm and humid western Oceania, unpredictable weather patterns to Mexico and the southwestern United States, droughts to the Amazon, and perhaps fewer hurricanes to the Caribbean. The El Niño phenomenon in the western Pacific is discussed in Chapter 11, where Figure 11.9 (page 589) illustrates its transpacific effects.

***Hurricanes.*** Large tropical low-pressure storm systems form annually in the Atlantic Ocean, primarily north of the equator, close to Africa (and north and south of the equator in the Pacific and Indian oceans as well). A tropical storm begins as a group of disorganized thunderstorms that, when warming wet air reaches a critical stage, organize themselves into a swirl of wind that moves across the earth's surface from east to west. The lower the air pressure, the higher the speed of the winds. The highest wind speeds are found at the edge of the eye, or center, of the storm. Once wind speeds reach 75 miles (120.7 kilometers) per hour, such a storm is officially called a hurricane. Hurricanes usually last about a week and will slow down and eventually dissipate as they move over cooler water or over land. Tropical storms and hurricanes are the most destructive of all storms, causing major wind and flood damage to human habitations and crops. The effects of these storms have always been significant, but their consequences are increasing as coastal population densities increase. Some scientists also think that global climate change is leading to an increase in the number and intensity of hurricanes.

## HUMAN PATTERNS OVER TIME

The conquest of the Americas by Europeans set in motion a series of changes that helped create the ways of life found in Middle and South America today. That conquest wiped out much of indigenous civilization and set up colonial regimes in its place. It introduced many new cultural influences to the Americas and led to the lopsided distribution of power and wealth that continues to this day.

### The Peopling of Middle and South America

Between perhaps 30,000 and 14,000 years ago, groups of hunters and gatherers from northeastern Asia spread throughout North America after crossing the Bering land bridge on foot, moving along shorelines in small boats, or both. Some of these groups remained in North America, while others ventured south over the Central American land bridge between North and South America. Anthropologists now think that these people managed to adapt to a wide range of ecosystems in only a few thousand years, reaching the tip of South America by about 13,000 years ago. But scientists differ over how long South America has been settled. Recent evidence from the archaeological site Monte Verde in the southern Andes indicates occupation perhaps as long as 30,000 years ago.

By 1492, there were 50 million to 100 million indigenous people in Middle and South America. In some places, even rural population densities were high enough to threaten sustainability. These people altered the landscape in many ways. They modified drainage to irrigate crops, constructed raised fields in lowlands, terraced hillsides, built paved walkways across swamps and even mountains, constructed cities with sewer systems and freshwater aqueducts, and raised huge earthen and stone ceremonial structures that rivaled the Pyramids of Egypt. They also perfected the system of **shifting cultivation** that is still common in wet, hot regions in Central America and the Amazon Basin. Small plots are cleared in forestlands, the brush is dried and burned to release nutrients to the soil, and the clearings are planted with multiple crop species. Each plot is used for only 2 or 3 years and then abandoned for many years of regrowth. This system is highly productive per unit of land and labor if there is sufficient land for long periods of regrowth. If population pressure increases to the point that a plot must be used before it has fully regrown, yields will decrease drastically.

These various preconquest activities significantly changed the habitats of land and water species, some of which became extinct; in addition, some animals were hunted to extinction. Then the trauma of the Spanish conquest and the ensuing 500 years of European dominance further affected physical environments and obliterated many of the remarkable accomplishments of indigenous cultures, some of which are only now coming to light through archaeological research.

Although they lacked the wheel and gunpowder, the **Aztecs** of the high central valley of Mexico had some technologies (such as urban water and sewage systems) and levels of social organization (such as highly organized marketing systems) that rivaled or surpassed those of contemporaneous Asian or European civilizations. Historians have recently concluded that, on the whole, all social classes of Aztecs lived better and more comfortably than did their contemporaries in Europe.

In 1492, the largest state in the Americas was that of the **Incas,** whose domain stretched from southern Colombia to northern Chile and Argentina, with the main population clusters in the Andes highlands. The cooler temperatures at these high altitudes eliminated the diseases of the tropical lowlands, yet proximity to the equator guaranteed mild winters and long growing seasons. For several hundred years, the Inca empire was one of the most efficiently managed in the history of the world. The ruling class divided the population into a hierarchy of family groups, each unit strictly controlled by a leader who reported to his superior. Highly organized systems of cooperative and reciprocal labor were used to construct paved road systems linking high Andean communities and to build great stone cities in the highlands. Incan agriculture was advanced, particularly in the development of staple crops such as numerous varieties of potatoes and grains (Figure 3.8).

**FIGURE 3.8** Ancient Incan terraces rediscovered and put to use. Built well before the Spanish arrived, these circular sunken terraces at 12,000 feet in Moray, Peru, were used to grow experimental crops of wheat, quinoa, and other plants. The terraces had a system of canals that delivered rainwater for irrigation. Because they were below grade level, the terraces protected the plants from the wind and cold temperatures. Each descending level had a different microclimate so that elaborate experimental agriculture was possible. [Tom Dempsey, www.photoseek.com.]

## The Conquest

The European conquest of Middle and South America was one of the most significant events in human history, rapidly altering landscapes and cultures and ending the lives of millions of indigenous people through slavery and cruelty (Figure 3.9). Columbus established the first Spanish colony in 1492 on the Caribbean island of Hispaniola, now occupied by Haiti and the Dominican Republic. This initial seat of the Spanish empire expanded to include the rest of the Greater Antilles—Cuba, Puerto Rico, and Jamaica. After learning of the Americas from Columbus, other Europeans, mainly from Spain and Portugal on Europe's Iberian Peninsula, conquered Middle and South America. By the 1530s, a mere 40 years after Columbus's arrival, all major population centers in the Americas had been conquered and were rapidly being transformed by Iberian colonial policies. The colonies soon became part of extensive trade networks within the region and with Europe, Africa, and Asia (Figure 3.10). By 1570, Spain was trading from its Mexican colony across the Pacific with its colony in the Philippine Islands.

The superior military technology of the Iberians speeded the process of conquest and control considerably, but the major factor that explains the swiftness and completeness of the conquest was the vulnerability of the indigenous people to diseases such as smallpox and measles carried by the Europeans. In about 150 years, the total population of the Americas was reduced by more than 90 percent, to just 5.6 million. To obtain a new supply of labor to replace the indigenous people, the Spanish initiated the first shipments of enslaved Africans to the Americas in the early 1500s; soon the Portuguese became important suppliers of African slaves.

Roman Catholic diplomacy prevented conflict between Spain and Portugal over the lands of the Americas. The Treaty of Tordesillas of 1494 divided the Americas at approximately 46° W longitude (see Figure 3.10). Portugal took all lands to the east and eventually acquired much of what is today Brazil; Spain took all lands to the west.

The first part of the mainland to be conquered was Mexico, home to several advanced indigenous civilizations, most notably the Aztecs. The Spanish, unsuccessful in their first attempt to capture the Aztec capital of Tenochtitlán, succeeded a few months later after a smallpox epidemic they inadvertently brought with them had decimated the native population. The Spanish demolished the Aztec capital in 1521, including its grand temples, public spaces, causeways, residences, and aqueducts. Mexico City was built on its ruins, becoming one of two main seats of the Spanish empire in the Americas. Called the Viceroyalty of New Spain, this part of the empire extended from Panama in the south all the way to what is now San Francisco on the northern California coast. Wealth from the gold and silver mines of Mexico flowed through the port of Veracruz on the Caribbean and from there on to Spain.

The conquest of the Incas in South America bore a remarkable resemblance to that of the Aztecs in that a tiny band of Spaniards was able to capture the capital city of an extensive empire. Their leader, Francisco Pizarro, himself admitted that his campaign received its greatest assistance from a smallpox epidemic (brought unintentionally by earlier Spanish scouts) that preceded his arrival. Out of the ruins of the Inca empire the Spanish created the Viceroyalty of Peru, which originally encompassed all of South America except Portuguese Brazil. The newly constructed capital of Lima flourished on the profits of the rich silver mines established in the highlands of Bolivia at Potosí.

The conquest of Brazil by the Portuguese was similar to the Spanish conquest of other areas in that land was seized and people were killed or enslaved, but it differed in some key respects. Brazil was apparently only sparsely populated by indigenous people who lived in small, impermanent villages, and there were no highly organized urban cultures. Most Atlantic coastal cultures were annihilated early on, and the populations of the huge Amazon Basin declined sharply as contagious diseases spread through trading. Because it was difficult to extract quick wealth from lowland tropical forests, the Portuguese focused on extracting mineral wealth, especially gold and precious gems, from the Brazilian Highlands and on establishing plantations along the Atlantic coast.

## A Global Exchange of Crops and Animals

Beginning in the earliest days of conquest, a number of plants and animals were exchanged among the Americas, Europe, Africa, and Asia via the trade routes illustrated in Figure 3.10. Many plants from the Old World are essential to agriculture in Middle and South America today: rice, sugarcane, bananas, citrus, melons, onions, apples, wheat, barley, and oats are just a few examples. When disease decimated the native populations of the Americas,

**FIGURE 3.9  The European conquest of Middle and South America.**
The Franciscan priest Bernardino de Sahagún interviewed Aztecs who had survived the conquest of 1519–1521. They reported thinking at the time that the armored Spanish soldiers on horseback were actually single, huge, otherworldly creatures. This is one of the pictures they drew for Sahagún as illustration. [Arthur J. O. Anderson and Charles E. Dibble, *The War of Conquest* (Salt Lake City: University of Utah Press, 1978), p. 31.]

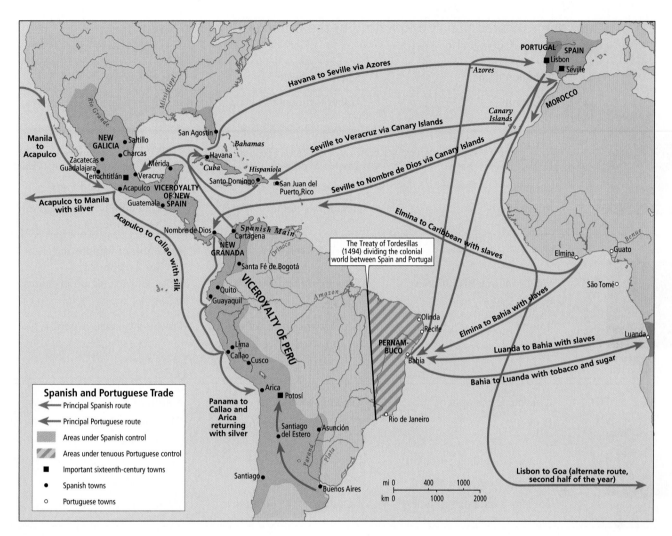

**FIGURE 3.10** Spanish and Portuguese trade routes and territories in the Americas, circa 1600. The major trade routes from Spain to its colonies led to the two major centers of its empire, Mexico and Peru. The Spanish colonies could trade only with Spain, not directly with one another. By contrast, there were direct trade routes from Portuguese colonies in Brazil to Portuguese outposts in Africa, as well as to Portugal. Many millions of Africans were enslaved and traded to Brazilian plantation and mine owners (as well as to Spanish, British, French, and Dutch colonies in the Caribbean and Middle and South America). [Adapted from *Hammond Times Concise Atlas of World History* (Maplewood, N.J.: Hammond, 1994), pp. 66–67.]

the colonists turned the abandoned land into pasture for herd animals imported from Europe, including sheep, goats, oxen, cattle, donkeys, horses, and mules. European draft animals helped fill in indigenous irrigation canals, drain lakes, and plow the relics of complex gardens into huge one-crop fields of sugarcane or wheat. Surviving native people living on the plains of Mexico and Argentina adopted European-introduced horses, using them to hunt large game. Others learned to herd sheep and adapted their fleece to ancient spinning and weaving technologies.

Plants first domesticated by indigenous people of Middle and South America have changed diets everywhere and have become essential components of agricultural economies around the globe. The potato, for example, had so improved the diet of the European poor by 1750 that it fueled a population explosion. Manioc (cassava), though less nutritious, played a similar role in West Africa. Corn is a widely grown garden crop in Africa, and peanuts and cacao (chocolate) are essential cash crops to many African farmers. Peppers are a cash crop in China, as are pineapples in the Pacific, and tomatoes have transformed Mediterranean cuisine. Table 3.1 lists some of the more common globally used plants originally from Middle and South America and their sites of probable domestication.

## The Legacy of Underdevelopment

Today, despite Middle and South America's vast human and natural resources, perhaps 30 percent of the people are poor and lack access to land, adequate food and shelter, and basic education. Meanwhile, a small elite class enjoys levels of affluence equivalent to those of the wealthy in the United States. In part, these inequalities result from

**TABLE 3.1   Major domesticated plants originating in the Americas and now used commercially around the world**

| Type | Common name | Scientific name | Place of origin |
|---|---|---|---|
| **Seeds** | Amaranth | *Amaranthus cruentus* | Southern Mexico, Guatemala |
| | Beans | *Phaseolus* (4 species) | Southern Mexico |
| | Maize (corn) | *Zea mays* | Valleys of Mexico |
| | Peanut | *Arachis hypogaea* | Central lowlands of South America |
| | Quinoa | *Chenopodium quinoa* | Andes of Chile and Peru |
| | Sunflower | *Helianthus annuus* | Southwestern and southeastern North America |
| **Tubers** | Manioc (cassava) | *Manihot esculenta* | Lowlands of Middle and South America |
| | Potato (numerous varieties) | *Solanum tuberosum* | Lake Titicaca region of Andes |
| | Sweet potato | *Ipomoea batatas* | South America |
| | Tannia | *Xanthosoma sagittifolium* | Lowland tropical America |
| **Vegetables** | Chayote (christophene) | *Sechium edule* | Southern Mexico, Guatemala |
| | Peppers (sweet and hot) | *Capsicum* (various species) | Many parts of Middle and South America |
| | Squash (including pumpkin) | *Cucurbita* (4 species) | Tropical and subtropical America |
| | Tomatillo (husk tomato) | *Physalis ixocarpa* | Mexico, Guatemala |
| | Tomato (numerous varieties) | *Lycopersicon esculentum* | Highland South America |
| **Fruit** | Avocado | *Persea americana* | Southern Mexico, Guatemala |
| | Cacao (chocolate) | *Theobroma cacao* | Southern Mexico, Guatemala |
| | Papaya | *Carica papaya* | Southern Mexico, Guatemala |
| | Passion fruit | *Passiflora edulis* | Central South America |
| | Pineapple | *Ananas comosus* | Central South America |
| | Prickly pear cactus (tuna) | *Opuntia* (several species) | Tropical and subtropical America |
| | Strawberry (commercial berry) | *Fragaria* (various species) | Genetic cross of Chile berry + wild berry from North America |
| | Vanilla | *Vanilla planifolia* | Southern Mexico, Guatemala, perhaps Caribbean |
| **Ceremonial and drug plants** | Coca (cocaine) | *Erythroxylon coca* | Eastern Andes of Ecuador, Peru, and Bolivia |
| | Tobacco | *Nicotiana tabacum* | Tropical America |

Quinoa   [Alamy.]

Tannia (foreground); maize (background)   [Jacques Jangoux / Peter Arnold, Inc.]

Chayotes   [CORBIS.]

Cacao   [Alamy.]

Coca   [Carlos Cazalis / CORBIS.]

*Source:* B. Kermath and L. Pulsipher, "Guide to Food Plants Now Used in the Americas," unpublished manuscript.

**FIGURE 3.11** The colonial heritage of Middle and South America. Most of Middle and South America was colonized by Spain and Portugal, but important and influential small colonies were held by Britain, France, and the Netherlands. Nearly all the colonies had achieved independence by the late twentieth century; those for which no date appears on the map are still linked in some way to the colonizing country. [Adapted from *Hammond Times Concise Atlas of World History* (Maplewood, N.J.: Hammond, 1994), p. 69.]

the lingering effects of European colonization (Figure 3.11): economic policies that still favor the export of raw materials and foster the dominance of outside investors who spend their profits elsewhere rather than reinvesting them within the region. Other factors leading to inequality are official corruption at all levels of government and social attitudes and policies that stifle local entrepreneurial development and upward social mobility. All these factors have resulted in an undereducated and underskilled populace and an infrastructure that is inadequate by modern standards. For these reasons, it is difficult for the region to compete in the global marketplace.

The first colonial enterprises to be established were based on the extraction of raw materials, including gold, silver, and other minerals, timber, and various agricultural products such as cotton, indigo, tobacco, and sugar. These activities were part of an emerging policy of **mercantilism** by which European rulers sought to increase the power and wealth of their realms by managing all aspects of production, transport, and trade. Resources were mined or grown in the colonies at the lowest possible cost, brought home to Europe, and made into trade goods (cotton into cloth, indigo into dye, sugar into rum). These manufactured products were then sold to the colonies and in other world markets. Money flowed into Europe as payment for the manufactured goods, enriching the merchants, and taxes on the trade brought revenues to the national treasuries. Each European country maintained a *merchant marine* (a fleet of privately owned trading ships) to transport resources, man-

ufactured goods, and, once the slave trade began in earnest in the 1600s, people. Nearly all commercial activity was focused solely on enriching the colonizing country, or **mother country,** and little or no effort was made to build stable economies in the Americas. For example, Spain's colonies were allowed to trade only with Spain—not with one another. If businesspeople in Peru wanted to trade legally with their counterparts in Mexico, the goods first had to cross the Atlantic in Spanish ships, be taxed in Spain, and then be reshipped back to Mexico in Spanish ships. These cumbersome and expensive restrictions nurtured a huge underground economy and institutionalized dishonesty as merchants and customers often bribed their way around the trade laws, and they led to considerable resentment of the Spanish authorities.

In the early nineteenth century, wars of independence transformed the region. The modern countries of Middle and South America emerged, and Spain was left with only a few colonies in the Caribbean. Figure 3.11 gives the dates of independence for the various countries in the region (in parts of the Caribbean, independence was not achieved until the mid-twentieth century). Many supporters of these nineteenth-century revolutions were not true reformists, but rather **Creoles** (people of usually European descent born in the Americas) who wished to consolidate their control of economic assets in the colonies and relatively wealthy **mestizos** (people of mixed European, African, and indigenous descent) whose access to the profits of the colonial system had been restricted due to their

**FIGURE 3.12** Population density in Middle and South America. [Data courtesy of Deborah Balk, Gregory Yetman, et al., Center for International Earth Science Information Network, Columbia University; at http://www.ciesin.columbia.edu.]

race. Once these pseudo-revolutionary leaders came to power, they became a new elite that controlled the state and monopolized economic opportunity, doing little to expand economic development or the majority's access to political power.

During the twentieth century, some countries in the region began to experiment with radical ways of fostering development, as we shall see in the section on economic and political issues. Today, the economies of Middle and South America are much more complex and technologically sophisticated than they once were. Nevertheless, the colonial pattern of dominance by an elite and dependence on extraction of raw materials remains, contributing to a persistent pattern of underdevelopment, unsustainable use of resources, and unequal distribution of wealth.

## POPULATION PATTERNS

It would be difficult to find a case elsewhere in human history to rival the massive shift in human population that occurred in Middle and South America over the past 500 years. Today, the patterns of human settlement continue to change, but now these changes are being generated from within the region, rather than coming from outside. The population continues to climb, but primarily because of high birth rates rather than immigration. At present, the major migration trend is internal: rural-to-urban migration is taking place everywhere in the region and transforming traditional ways of life. A second, international migration trend is growing, however: many people from the Caribbean, Mexico, and Central America are leaving their countries, temporarily or permanently, to seek opportunities elsewhere, primarily in the United States and Europe. In an interesting historical reversal, since 1996, several hundred thousand Peruvians and Ecuadorians have migrated to Spain and Portugal in search of jobs within the European Union.

### Population Numbers and Distribution

As of 2005, 560 million people were living in Middle and South America—close to ten times the population of the region in 1492.

This number is about 230 million more than presently live in North America.

***Population Distribution.*** The population density map in Figure 3.12 reveals a very unequal distribution of people in the region. If you compare Figure 3.12 with the regional map (Figure 3.1), you will see that there is no obvious, consistent relationship between population density patterns and landforms. Some of the highest densities, such as those around Mexico City and in Colombia and Ecuador, are in highland areas. Elsewhere, high concentrations are found in lowland zones along the Pacific coast of Central America and especially along the Atlantic coast of South America. In tropical and subtropical zones, the cool uplands (*tierra templada*) are particularly pleasant and healthful and were comparatively densely occupied even before the European conquest. Most of the coastal lowland concentrations, in *tierra caliente*, are near important seaports that attract people with their vibrant and varied economies and interesting social life. Seaside living is also attractive because the ocean water moderates the heat and humidity, making the coast relatively cooler and breezier than lowlands in the interior.

Figure 3.12 also reveals lands that are relatively empty of human occupants. For example, cold and windy Patagonia at the far southern end of South America has very few people, as do the desert regions of Chile and Peru along the Pacific coast. The vast Amazon Basin is only lightly settled. Despite recent efforts to use advancements in technology to develop and populate this wet tropical zone, no such strategies have so far proved sustainable for more than a few years. It seems able to truly sustain only hunting and gathering, shifting cultivation, and light agroforestry (the growing of tree crops).

***Population Growth.*** Rates of natural population increase were high in Middle and South America in the twentieth century. (Recall from Chapter 1 that the rate of natural increase is population growth resulting from births alone; it does not count immigration.) Although rates of natural increase are now declining, as shown in Figure 3.13, at present rates the region's population could double in just 43 years. (For comparison, in North America,

**FIGURE 3.13** Trends in natural population increases, 1975–2015. Comparison of the orange and blue columns indicates that rates of natural increase have declined steadily throughout the region and are projected to continue to do so into the future. Nevertheless, in many countries natural population increase remains high enough to outstrip efforts at improving standards of living. Note that the rates of natural increase given here are for ranges of time. [Adapted from *United Nations Human Development Report 2005* (New York: United Nations Development Programme), Table 5.]

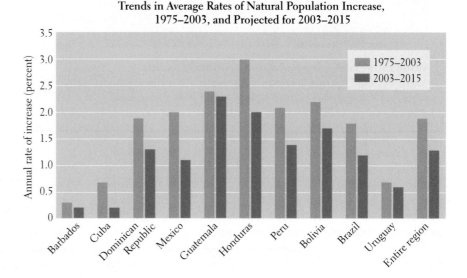

Trends in Average Rates of Natural Population Increase, 1975–2003, and Projected for 2003–2015

## BOX 3.1 | AT THE LOCAL SCALE Mexico City

Mexico City is one of the world's largest primate cities (Figure 3.14). It has drawn so many migrants because it is a vibrant center of cultural heritage, with several large universities and museums, as well as a world financial center, home to many international banking and financial institutions and the headquarters of several multinational corporations. Mexico City and the adjacent cities such as Cuernavaca, Toluca, and Puebla are also home to thousands of smaller formal and informal enterprises that employ millions of people from the continuous stream of migrants arriving from the countryside.

The city's physical situation, its high population density, and its rapid modernization have all contributed to major environmental crises. Automobiles and trucks crowd the streets, often creating gridlock traffic jams that take hours to clear. The snarl of motor vehicles contributes tons of pollution to the air, which is prevented from dispersing by the surrounding moun-

**FIGURE 3.15** Low-income housing in Ixtapaluca, a suburb in the Mexico City metropolitan area. There are more than 10,000 houses in this development, built since 2000 during a major housing initiative by the Vicente Fox administration. Although many conclude that this picture is not real (perhaps a SimCity image), the photographer, Carlos Ruiz, a helicopter pilot, and many others have confirmed its authenticity. [Carlos Oscar Ruiz.]

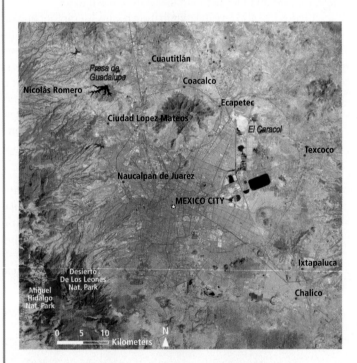

**FIGURE 3.14** Satellite image of Mexico City metropolitan area, March 21, 2000. The natural vegetation is green, while the urban infrastructure is depicted in shades of gray. [United Nations Environment Programme.]

tains. Although the *tierra fria* climate of Mexico City once made it a place of sparkling clear vistas, heavy brown smog now obstructs views on most days and causes asthma and related respiratory maladies in the citizenry.

The millions of poor migrants coming to the city have had to build their own housing, which may be unsafely constructed and usually lacks adequate plumbing—another source of pollution. Furthermore, the location of the Mexico City conurbation on a drained lake bed in a tectonically active zone makes it vulnerable to **subsidence** (sinking land), disastrous earthquake damage, and even volcanic hazards. But still the people come, because, for all its faults, the city promises opportunities and rich life experiences.

In recent years, as an answer to chaotic urban sprawl in Mexico City, the Vicente Fox administration built thousands of homes for low-income families. Figure 3.15 shows one such project, the Ixtapaluca subdivision, where family income is reported to be $4300 per year, less than one-quarter of the national average

doubling is expected to take 116 years.) Rapid population doubling rates are a disturbing prospect for a region in which the majority of people already suffer from a low standard of living. Any gains in well-being might be checked by the costs of supplying more and more people with houses, schools, and hospitals.

There are a number of reasons why rates of natural increase in this region have remained relatively high. One is that until re-

cently, most people in Middle and South America lived in agricultural areas. Children were seen as sources of wealth because they could do useful work at a young age and eventually would care for their aging elders. High infant death rates encouraged parents to have many children to be sure of raising at least a few to adulthood. In addition, the Catholic Church has discouraged systematic family planning. Further, as will be discussed later, the cultural mores of *machismo* and *marianismo* reinforced the idea that both men and women validate themselves as adults by reproducing prolifically. When medical care in the region began improving in the 1930s, death rates began to decline rapidly. By 1975, death rates were about one-third what they had been in 1900. Because birth rates typically do not decrease as quickly as death rates, population growth was especially rapid between 1940 and 1975.

By the 1980s, for a number of reasons, Middle and South Americans were beginning to limit their family sizes, and population growth rates started to fall. Now the region is beginning to undergo the demographic transition (see Figure 1.35 on page 40). Between 1975 and 2005, the annual rate of natural increase for the entire region fell from about 1.9 percent to 1.3 percent—still a higher rate of growth than the world average of 1.1 percent.

In 2005, Caribbean countries generally had the lowest annual rates of natural increase in the region (averaging 1.1 percent). Several factors explain these low rates. Cuba, Trinidad and Tobago, and Barbados, all of which had rates about the same as North America (0.6 percent), provide both women and men with education, meaningful work, and basic health care. Infant mortality rates are low, so people can expect to see their one or two children grow to adulthood. Barbados prospers from a diversified economy of manufacturing, computerized information processing, tourism-related services, remittances (money sent home by migrants), and agriculture. Its skilled citizens have many economic, civic, and social options for constructing interesting lives that make large families less attractive. Trinidad and Tobago, while less prosperous, has oil-related industries and a thriving business community. Cuba has made a strong effort to improve social welfare and encourage smaller families over the past 35 years.

In Mexico and Central America, poverty is more widespread and women have less access to education and employment than in the Caribbean, both factors associated with high rates of natural increase (2.0 percent for these subregions). South America, also afflicted with high rates of poverty, has a rate of just 1.5 percent. Within South America, however, there is considerable variation, explained partially by differing standards of living and access to education and jobs, especially for women: in the year 2005, the rate of natural increase in South America varied from a high of 2.1 percent in Bolivia to a low of 0.6 percent in Uruguay. The rising incidence of HIV-AIDS will affect the population projections for this region.

## Migration and Urbanization

Since the early 1970s, Middle and South America have led the world in rates of migration from rural to urban communities. As a result, cities throughout the region have grown remarkably quickly; more than 75 percent of the people in the region now live in towns with populations of at least 2000. Increasingly, in Middle and South America as in other underindustrialized countries, one city or one metropolitan area of several contiguous cities (called a **conurbation**) is vastly larger than all the others, sometimes accounting for one-fourth or more of the country's total population. Such cities are called **primate cities;** examples are Mexico City (with

21 million people and about 20 percent of Mexico's population; Box 3.1); Managua, Nicaragua (20 percent of that country's population); Lima, Peru (30 percent); Santiago, Chile (34 percent); and Buenos Aires, Argentina (38 percent).

The concentration of people into just one or two primate cities in a country leads to uneven spatial development and urban bias (government policies and social values that favor urban areas). Wealth and power are concentrated in one place, while distant rural areas, and even other towns and cities, have difficulty competing for talent, investment funds, industries, and government services. Many provincial cities languish as their most educated youth leave for the primate city. Because the rush to the cities has not been accompanied by a general rise in productivity and prosperity, no city in the region is prepared for the influx. Spending on urban infrastructure (such as roads and sanitation) and on social services (such as schools and hospitals) has failed to keep pace with the inflow of people, and the signs of urban decay are everywhere.

Although there are wealthy neighborhoods in the cities, they are often walled and heavily guarded because of their proximity to unsightly, uncontrolled shantytowns built by the poor (Figure 3.16).

**FIGURE 3.16** A favela in Rio de Janeiro. These people are crossing a bridge in the Rocinha favela in Rio de Janeiro to cast their votes in Brazil's national election of October 29, 2006. [AP Photo/Silvia Izquierdo.]

## BOX 3.2 | AT THE LOCAL SCALE  Curitiba, Brazil: A Model of Urban Planning

The southern Brazilian city of Curitiba, capital of the grain-producing state of Paraná, has gained international renown for its environmentally friendly urban planning. Like most other Brazilian cities, Curitiba has mushroomed, doubling its population in just 20 years to more than 1.8 million people. But Curitiba is unusual in that it has carefully oriented its expansion around a master plan (dating from 1968), which included an integrated transport system funded by a public-private collaboration. Eleven hundred minibuses making 12,000 trips a day bring 1.3 million passengers from their neighborhoods to terminals, where they meet express buses to all parts of the city (Figure 3.17). The fifty-square-block pedestrian-only inner city has proved to be a boon for merchants. People spend more time shopping because they do not need to look for parking. Being able to get to work quickly and cheaply has helped Curitiba's poor to find and keep jobs. Car ownership is common in Curitiba, but people use cars less than in other cities; hence emissions are negligible, congestion low, and urban living pleasant. The city's streets and its many parks and green spaces are kept spotlessly clean and decorated with flowers. The city also has a decentralized public health serv-

ice. A trash recycling program encourages people in the informal economy to collect specific kinds of trash and sell them to recycling companies.

One goal is to keep migrants from swamping Curitiba's well-designed, environmentally sensitive urban environment with shantytowns. The city first tries to stem the flow by offering free bus tickets back home to new migrants; 25,000 people have used the tickets. To accommodate those migrants who come to take what are often short-term jobs in Curitiba's industrial sector, the city is building rural satellite towns, called *vilas rural*, where people can live and maintain their agricultural skills by farming small plots when they are between industrial jobs. With financing from the World Bank and the Inter-American Development Bank, 5 *vilas rural* have been built, 15 are under construction, and 60 more are planned. The strategy is to accommodate a significant proportion of Paraná's landless farmers in these urban fringe communities, where the advantages of both rural and urban life can be enjoyed by those of meager means.

*Source:* http://www.dismantle.org/curitiba.htm; http://www.pbs.org/frontlineworld/fellows/brazil1203.

**FIGURE 3.17**  Public transport in Curitiba. Express buses and tube stations serve patrons with fast, safe, and efficient transport throughout the city. These buses are part of an integrated transport system that uses five different bus types and eight different lines extending from the city center to outlying districts. [Alamy.]

These self-help settlements—called **colonias, barrios, favelas,** or **barriadas**—can arise spontaneously. Because most recent urban migrants can't afford to buy or lease land, they become squatters who invade a piece of property after dark and set up simple dwellings by morning. Once established, these communities are extremely difficult to dislodge, because the impoverished are such a huge portion of the urban population that even those in positions of power will not challenge them directly. The result is a city landscape that is remarkably different from the common U.S. pattern of a poor inner city surrounded by affluent suburbs, with relatively clear spatial separation of classes, races, and family styles. Rather,

various housing types are intermingled with factories, warehouses, and commercial enterprises (Figure 3.18). Inner-city neighborhoods of vastly different socioeconomic classes nestle close to one another, and the suburban fringe is also a mixture of wealthy and poor neighborhoods. The poor neighborhoods, which predominate, are often built of unconventional materials and laid out in an uncontrolled and seemingly chaotic pattern. (Box 3.2 describes an exception to this pattern.)

The squatters are often enterprising and admirable people who have simply made the most of the bad hand dealt them. Such settlers sometimes organize themselves to press for social services, and some farsighted urban governments, such as that of Fortaleza (Ceará) in northeastern Brazil, contribute building materials so that the squatters can build more permanent structures with basic indoor plumbing. Expanses of shacks and lean-tos have thereby been transformed through self-help into crude but livable suburbs (see the personal vignette on this page). Moreover, these communities can be centers of pride and support for their residents, where community work, music, folk belief systems, and crafts flourish. For example, many of the most prestigious steel bands of Port of Spain, Trinidad, have their homes in such shantytowns. In Bahia, Brazil, the favelas are home to samba clubs (musical ensembles) and folk religious movements that focus on helping migrants cope with urban life.

Interestingly, rural women are just as likely to migrate to the city as rural men. This is especially true when employment is available in foreign-owned factories that produce goods for export, where women are preferred because they are a low-cost, usually passive labor force. Ironically, rural development projects may encourage female migration because, due to mechanization, these projects often end up decreasing available jobs, and women are rarely considered for training as farm equipment drivers and

mechanics. In urban areas, unskilled women migrants can usually find work as domestic servants. Low wages, however, force them to live in the households where they work, where they are often subject to sexual overtures by their employers, yet are themselves blamed if they fail to adhere to rigid standards of behavior. Male urban migrants tend to depend on short-term day work, most often in low-skill jobs in construction, maintenance, small-scale manufacturing, and petty commerce; many make some cash in the informal economy as street vendors, errand runners, car washers, and trash recyclers, and some may engage in crime. The loss of family ties and village life is sorely felt by both men and women, and the chances for recreating normal family life in the urban context are extremely low.

**Personal Vignette**   Favelas are everywhere in Fortaleza, Brazil. The city grew from 30,000 to 300,000 during the 1980s. By 2006, there were more than 3 million residents, most of whom had fled drought and rural poverty in the interior. City parks of just a square block or two in middle-class residential areas were suddenly invaded, and within a year 10,000 or more people were occupying crude stacked concrete dwellings in a single park space, completely changing the ambience of the neighborhood. In the early days of the migration, for lack of water and sanitation, the migrants often had to relieve themselves on the street.

One day, while strolling on the waterfront, Lydia Pulsipher chanced to meet a resident of a beachfront favela who invited her to join him on his porch. There he explained just how one becomes an urban favela dweller. He maintained a small refreshment stand, and his wife ran a beauty parlor that catered to women from the area. They had come to the city 5 years before, after being forced to leave the drought-plagued interior when a newly built irrigation reservoir flooded the rented land their families had cultivated for generations. With no way to make a living, they set out on foot for the city. Here in Fortaleza, they constructed the building they used for home and work from resources they collected along the beach, and eventually they were able to purchase the roofing tiles that gave it an air of permanency.

*Source: Lydia Pulsipher's field notes in Brazil, updated with the help of John Mueller, Fortaleza, 2006.* ■

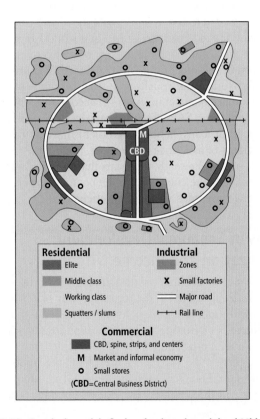

**FIGURE 3.18** Crowley's model of urban land use in mainland Middle and South America. William Crowley, an urban geographer specializing in Middle and South America, developed this model to depict how residential, industrial, and commercial uses are mixed together, with people of widely varying incomes living in close proximity to one another and to industries. Squatters and slum dwellers ring the city in an irregular pattern. [From *Yearbook of the Association of Pacific Coast Geographers* 57 (1995): 28; printed with permission.]

It always takes some resourcefulness to move from one place to another. Usually those people who already have some advantages —some years of education and unusual ambition—are the ones who migrate to cities (where their talents may, however, go to waste). When young adults leave rural communities that have invested years in nurturing and educating them, the loss is often referred to as **brain drain.** Brain drain happens at several scales in Middle and South America: there is rural-to-urban migration from villages to regional towns and from towns and small cities to primate cities, as well as international migration from the many countries in the region to North America and Europe. On the other hand, families often encourage their youth to migrate so that they can benefit from the remittances, goods, and services that migrants provide to their home communities. These contributions somewhat offset brain drain and can bring considerable prestige to returning migrants, who, with their skills enhanced by their

Within the figure legend:

**Residential**
- Elite
- Middle class
- Working class
- Squatters / slums

**Industrial**
- Zones
- X Small factories
- Major road
- Rail line

**Commercial**
- CBD, spine, strips, and centers
- M Market and informal economy
- o Small stores

(**CBD**=Central Business District)

overseas sojourn, are in a good position to become local leaders. Thus migration has a mix of consequences for sending communities—some positive, some negative.

The receiving societies, whether cities such as Rio de Janeiro or Mexico City or countries such as the United States, garner considerable benefit from immigration: a huge, inexpensive labor pool and a concentration of skilled professionals. For example, in 2000, the American Medical Association reported that one-fourth of the doctors practicing in the United States were immigrants, as were about 50 percent of medical students. Of the practicing doctors who were immigrants, more than 5 percent were from Middle and South America. The United States saves money when other countries supply the early education for these physicians, who may then spend their most productive years treating patients in the United States.

# II CURRENT GEOGRAPHIC ISSUES

Historically, power and wealth in the countries of Middle and South America have been concentrated in the hands of a few. The economic modernization of most countries and their transformation from rural to urban societies over the last century have not changed that reality. The rewards of urbanization and industrialization have spread only thinly to the masses of poor and not-so-poor, who instead have had to absorb the effects of large government debts incurred in the 1970s and 1980s.

## ECONOMIC AND POLITICAL ISSUES

Although Middle and South America is not as poor on average as sub-Saharan Africa, South Asia, or Southeast Asia, it has the widest gap between rich and poor. Except for a few relatively well-off Caribbean countries, such as Barbados and Cuba, by 2000 the richest 20 percent of the population was between 12 and 26 times richer than the poorest 20 percent (Table 3.2 [see page 136]). In 2005, though disparities were shrinking slightly in Brazil and Venezuela due to newly socially active governments, nearly 40 percent of the regional population was living in poverty. This **income disparity** is troubling in many ways. From a moral standpoint, many people argue that it is wrong for wealth and resources to be concentrated among a tiny number of people while so many have so little. From a social viewpoint, poverty can physically inhibit the development of children, leaving them dependent rather than productive citizens. From an economic perspective, poverty prevents people from contributing to economies with their own purchasing power or with their skills, which tend to remain at a low level. Thus, wherever large numbers of people are poor, the potential for a nation's development diminishes. Political instability also increases because impoverished people may rebel violently and be violently repressed by governments. Hundreds of thousands of people in Bolivia, Peru, Chile, Brazil, Argentina, Colombia, Guatemala, Nicaragua, Honduras, El Salvador, Mexico, and Haiti have died in repeated waves of government repression, revolution, counterrevolution, and coups d'état.

While many hope that globalization will reduce income disparity in the region as increases in international trade provide more jobs for the poorest, results thus far have been disappointing. Lowered trade barriers have expanded opportunities for local entrepreneurs and have encouraged multinational corporations to invest in the region (an activity called **foreign direct investment,** or **FDI**), building factories and hiring semiskilled workers and some local managers. Moreover, markets for products from the region, such as meat and building materials, are expanding in Asia and Europe. The clearest beneficiaries of globalization, however, have so far been highly educated, urban, technologically proficient businesspeople (sometimes called *technocrats*). While significant numbers of the less educated poor have also benefited from jobs created by globalization, for them, job security has proved to be precarious as businesses open and close and expand and contract in response to global economic conditions. Furthermore, as private foreign direct investment has increased, government-funded safety-net programs for the poor and aid from the United States and Europe have decreased, on the assumption that market economies will eventually make everyone self-supporting. Multilateral development agencies, such as the World Bank, usually require governments to streamline their economies to attract investors (in accordance with structural adjustment policies, SAPs, discussed on pages 139–140) before giving them aid or loans; hence governments have dramatically cut their programs of assistance to the poor, hoping the trickle-down effect will suffice. Presently it appears that the trickle-down effect has not balanced the cuts in aid to the poor, and levels of income disparity have remained more or less unchanged.

## Phases of Economic Development: Globalization and Income Disparity

The current economic and political situation in Middle and South America is deeply rooted in the region's history of dependent relationships with external forces: colonial mother countries, the economic interests of Europe and America, and foreign and multinational corporations. This history can be divided into three major economic phases: the early extractive phase, the import substitution industrialization phase, and the structural adjustment phase of the present. All three phases have helped entrench wide income disparities.

***The Early Extractive Phase.*** The **early extractive phase,** or *mercantilist phase,* began with the European conquest and continued until the early twentieth century. It was characterized by lopsided trade: a small flow of foreign investment and manufactured

**FIGURE 3.19** Political map of Middle and South America.

goods came into the region, and a vast flow of resources left for Europe and beyond. Foreign money to fund resource extraction for export (farms, plantations, mines, sawmills, roads and railways to ports) came first from Europeans and later from North Americans and other international sources. The profits were usually banked abroad.

The consequences of this pattern are highlighted by comparing it with the economic development of North America at the time. After the American Revolution, the profits from U.S. extractive industries went to owners who tended to live in the region. They invested their money locally in industries that processed North America's raw materials into more valuable finished goods, which were then bought by American workers or exported, provid-

ing the foundation for North America's history of economic stability and high living standards. In contrast, the colonizers of Middle and South America did not reinvest in the region; industries did not develop, and hence even essential items, such as farm tools and household utensils, had to be purchased from Europe and North America at relatively high prices. Many people simply did without.

A number of economic institutions arose in Middle and South America to feed raw materials to Europe and North America. Large rural estates called **haciendas** were granted to colonists as a reward for conquering territory and people for Spain, then passed down through the families of those colonists. Generally, haciendas produced several types of crops for local consumption and limited

## TABLE 3.2  Income disparity in selected countries[a]

| Country | Ratio of wealth of richest 20% to wealth of poorest 20%[b] | | |
|---|---|---|---|
| | 1987 | 1998–2000[c] | 2005 |
| **Middle and South America** | | | |
| Bolivia | 9:1 | 12:1 | 12:1 |
| Brazil | 26:1 | 29:1 | 26:1 |
| Chile | 17:1 | 19:1 | 19:1 |
| Colombia | 20:1 | 20:1 | 23:1 |
| Guatemala | 30:1 | 16:1 | 24:1 |
| Mexico | 16:1 | 17:1 | 19:1 |
| Peru | 12:1 | 12:1 | 18:1 |
| **Other selected countries** | | | |
| China | 8:1 | 8:1 | 11:1 |
| France | 6:1 | 6:1 | 6:1 |
| Jordan | 6:1 | 6:1 | 6:1 |
| Philippines | 10:1 | 10:1 | 10:1 |
| South Africa | 33:1 | 18:1 | 18:1 |
| Thailand | 8:1 | 8:1 | 8:1 |
| Turkey | 8:1 | 8:1 | 8:1 |
| United States | 9:1 | 9:1 | 8:1 |

[a]The UN used data from 1987, 1998 to 2000, and 2005 on either income or consumption to calculate an approximate representation of how much richer the wealthiest 20 percent of the population is than the poorest 20 percent in selected countries. The lower the ratio, the more equitable the distribution of wealth in the country.

[b]Decimals rounded up or down.

[c]Survey years fall within this range.

*Source: United Nations Human Development Report 2000, Table 13; 2003, Table 13; 2005, Table 15.*

exports, but used only a fraction of their potential agricultural land. Hacienda laborers were virtually feudal serfs, while the owners, who often lived in a distant city or in Europe, usually had little interest in the day-to-day operations.

**Plantations,** large factory farms growing a single crop such as sugar, coffee, cotton, or (more recently) bananas, were first developed by the European colonizers of the Caribbean in the 1600s. Northeastern Brazil had some early sugar plantations, and they became more common in South America by the late nineteenth century. Owners of plantations made larger investments in equipment than did hacienda owners, and instead of depending on local populations, they imported slave labor from Africa. Unlike haciendas, which were often established in the continental interior in a variety of climates, plantations were for the most part situated in tropical coastal areas with year-round growing seasons. Their coastal

location also gave them easier access to global markets via ocean transport.

As markets for meat, hides, and wool grew in Europe and North America, the livestock ranch emerged, specializing in raising cattle and sheep. Today, commercial ranches are found in the drier grasslands and savannas of South America, Central America, and northern Mexico, and even in the wet tropics on cleared rainforest lands.

Mining was another early extractive industry. Important mines (at first primarily gold and silver) were located on the island of Hispaniola, in north-central Mexico, in the Andes, in the Brazilian Highlands, and in many other locations as well. Today, oil and gas have been added to the mineral extraction industry, but rich mines throughout the region continue to produce gold, silver, copper, tin, precious gems, titanium, bauxite, and tungsten (Figure 3.20).

**FIGURE 3.20** The Yanacocha gold mine in Cajamarca, Peru. This mine, the largest in the Americas, is run by a subsidiary of the U.S. mining group Newmont. One of the most polluting of all extractive industries, the mine has been the scene of widespread protests by local people who say it is polluting their water supply. The protests caused the suspension of mining in August 2006. [STR/AFP/Getty Images.]

The mines, ranches, plantations, and haciendas of the early extractive phase did not contribute much to the development of the region because they paid low wages and because the owners generally invested their earnings outside the region. By the late nineteenth and early twentieth centuries, when nearly all the countries of Middle and South America had gained independence from their mother countries, wealthy European and North American private investors had purchased many of the extractive industries in the region, thus ensuring that the profits would keep leaving the region. These investments came to be known as *neocolonial* because the power these foreign investors wielded resembled that of colonial officials. (**Neocolonialism** refers to modern efforts by dominant countries to further their own aims by controlling economic and political affairs in other countries.)

***The Import Substitution Industrialization Phase.*** In the early and mid-twentieth century, there were waves of protest against the domination of the economy and society by local elites and foreign businesses that was characteristic of the extractive phase. Many governments—Mexico and Argentina most prominent among them—proclaimed themselves socialist democracies and enacted a set of policies that came to be known as **import substitution industrialization (ISI)**. The goal of ISI policies was to keep money and resources within the region in order to foster economic self-sufficiency. First, national governments seized the most profitable extractive industries from foreign owners (usually with some payment). The intention was to use any profits to create local manufacturing industries that could supply the goods once purchased from

Europe and North America. To encourage people to buy manufactured goods from local suppliers, governments placed high tariffs on imported manufactured goods. The money and resources kept within each country were expected to provide the basis for further industrial development that would create well-paying jobs, raise living standards for the majority of people, and ultimately replace the extractive industries as the backbone of the economy.

Although ISI strategies still survive in some countries and are being reconsidered for reimplementation in a few others (Bolivia, Venezuela), they failed to bring about a thorough transformation of the industrial sector. Income disparities were reduced as governments increased spending on public health, education, and infrastructure (electrification, roads, housing). **Land reform** broke up some large landholdings and distributed them among landless farmers. But the goal of lifting large numbers of people out of poverty through jobs and general economic growth was never realized. The state-owned manufacturing sectors on which the success of ISI depended were never able to expand sufficiently to increase exports, provide tax revenues for social programs, or provide jobs for the migrants flooding into the cities; hence the number of people able to afford locally manufactured goods did not increase significantly.

Not all state-owned corporations were losing propositions, however. Brazil—with its aircraft, armament, and auto industries—and Mexico—with its oil and gas industries—both experienced successes in their ISI development. Venezuela has a successful joint venture with CVG Venalum Company, the largest aluminum smelter in Middle and South America, and until recently had also

**FIGURE 3.21** **Economic issues for Middle and South America: Debt, maquiladoras, and trade blocs.** The high rate of debt in the region stems from countries borrowing money to finance industrial and agricultural development projects intended to replace imports and reduce poverty by providing jobs. Debt is presented as total debt service (repayment of public and private loans) as a percentage of a country's exports of goods and services. Maquiladoras, usually foreign owned but staffed by a poor local population, have played a major role in efforts to reduce government debts. The creation of trade blocs such as NAFTA and Mercosur is another means countries in the region are using to reduce debt by lowering tariffs and reducing dependence on higher-cost imports from outside the region. [Adapted from *United Nations Human Development Report 2005* (New York: United Nations Development Programme), Table 20.]

cooperated with foreign oil and gas marketers. Interest in state-supported manufacturing industries continues, but the general trend is now toward more market-oriented management and global competitiveness. Dependence on the export of raw materials persists.

***The Debt Crisis.*** Ultimately, a global financial crisis diminished the ISI phase. A period of global prosperity that began in the early 1950s ended in the 1970s due to increases in oil prices and decreases in the prices of raw materials on global markets. While earnings from exports were falling, governments and private interests continued to pursue expansive plans to modernize and industrialize their national economies. They paid for these projects by borrowing millions of dollars from major international banks, most of which were in North America or Europe. The beginning of a global recession in 1980 put a halt to such ambitious plans. Dragged down by the recession, the Middle and South American economies could not meet their targets for growth, and thus the governments were unable to repay their loans. The damage to the region was made worse by the fact that the biggest borrowers—such as Mexico, Brazil, and Argentina—also had the largest economies in the region. Hence huge **external debts** (debts owed to foreign banks or governments and repayable only in foreign currency) now burdened the very countries that had been the most likely to grow (Figure 3.21).

## Phases of Economic Development: Structural Adjustment

***Structural Adjustment Policies.*** In the 1980s, at the urging of the foreign banks that had made the loans to governments in the region, the International Monetary Fund (see Chapter 1, page 32) developed and enforced the implementation of structural adjustment policies (SAPs) that mandated profound changes in the organization of economies. Free trade and privatization, rather than government-funded industries, were considered the soundest ways to achieve economic expansion and hence repayment of debts. In order to obtain further loans, governments were required to begin removing tariffs on imported goods of all types and to sell state-owned raw materials and manufacturing industries to private investors—often multinational corporations located in Asia, North America, and Europe. SAPs also reversed the ISI-era trend toward the expansion of government social programs. Again, to generate funds for debt repayment, governments were required to fire many civil servants and drastically reduce spending on public health, education, and infrastructure. SAPs encouraged the (preferably unregulated) expansion of industries that produce goods for export. In most developing countries, these export industries are still based on the extraction of raw materials because this is seen as a reliable way to earn the money necessary to first repay national debts quickly and then invest in job-creating manufacturing and service industries.

***Outcomes of SAPs.*** SAPs have increased some kinds of economic activity in Middle and South America, resulting in modest rates of national GDP growth (averaging about 5 percent per year). They have aided the expansion of U.S.-, European-, and Asian-owned multinational corporations in the mining, agriculture, forestry, and fishing industries. Partnerships between foreign companies and local businesspeople operate the largely unregulated gold mining, forestry, and other extractive industries of the Amazon Basin. Not surprisingly, because relaxing regulations of all types is an important part of structural adjustment, the very industries that have expanded as a result of SAPs are those that raise worries about environmental impacts, worker safety, and sustainability, as we will see in our discussion of environmental issues in the region.

SAPs encouraged the expansion of manufacturing industries in **Export Processing Zones (EPZs)**, also known as *free trade zones*. EPZs are specially created areas within a country where duties and taxes are not charged in order to attract foreign-owned factories that assemble products strictly for export back to the owner's country. EPZs exist in nearly all countries on the Middle and South American mainland and on some Caribbean islands, but by far the largest of the EPZs is the conglomeration of assembly factories, called **maquiladoras**, located along the Mexican side of the U.S.–Mexico border. (The word *maquiladora* comes from the Arabic *makilah*, meaning the share of meal kept by the miller as payment.) Because the taxes paid by these industries are low, the main benefit to the host country is the employment of local labor, which eases unemployment and brings money into the economy. EPZs now compete rather fiercely with one another.

Evidence is accumulating that SAPs have failed to produce the economic growth that was expected to relieve the debt crisis and help the people of the region achieve greater prosperity. Wealth disparities in Mexico jumped more than 30 percent between 2001 and 2005, and while the burden of debt has fallen throughout the region, it is still high, and it is likely to increase again as countries return to spending for badly needed social programs. Basic economic indicators, such as per capita growth in income, show a marked decline during the SAP phase as compared with the ISI phase. During the last two decades of the ISI phase (1960 to 1980), per capita GDP grew by 82 percent for the region as a whole. When SAPs were implemented across the region (1980 to 2000), growth in per capita GDP slowed to just 12.6 percent, and between 2000 and 2005 it slowed to only 1 percent.

SAPs were a disappointment not only because of the hardship they inflicted on the poor, but also because they encouraged greater dependence on exports of raw materials just when prices were falling for those materials on the global market because of increased competition.

***The Backlash Against SAPs.*** A political and economic backlash against the IMF and its SAPs has been gaining strength in recent years. Since 1999, presidents that explicitly oppose SAPs have been elected in eight countries in the region: Argentina, Bolivia, Brazil, Chile, Ecuador, Nicaragua, Uruguay, and Venezuela. Both Brazil and Argentina have made considerable sacrifices in order to pay off their debts and hence be liberated from the restrictions of SAPs. Under President Hugo Chavez, Venezuela has emerged as a leader

## BOX 3.3 | AT THE LOCAL SCALE The Informal Economy in Peru

The informal economy is a lifesaver in Peru, which has been hit by unprecedented economic recession, losses in real wages, and underemployment for up to 70 percent of urban workers. By the year 2000, as many as 68 percent of urban workers were in the informal sector, and they generated an estimated 42 percent of the country's total gross domestic product. Most of them work as street vendors; up to a fourth of the working population is so employed (Figure 3.22).

Geographer Maureen Hays-Mitchell, who studies street vending in Peruvian cities, found street vending in Peru to be a vibrant, organized, highly rational sector of the informal economy. Vendors assess the market and the risks entailed in this semi-illegal activity, compete intensely for particular urban spaces, and strategize for market advantages. Hays-Mitchell also found that street vending not only was an important source of sustenance for the vendors, but also improved the overall quality of urban life by making goods available in convenient places and at affordable prices, something the inefficient formal retailing system seemed unable to do.

Street vendors specialize in particular products and carefully place their stands where they can attract the most customers for the products they sell. Vendors sell food and small gifts in front of hospitals during visiting hours; candy, games, and toys in front of schools; shoeshines and newspapers near hotels and restaurants; lotion, hats, and bikinis near beaches. Few opportunities are left unexploited: movie patrons can buy comics and magazines while waiting in line; outside jails, visitors, guards, and even prisoners can buy food, souvenirs, and handicrafts produced by inmates. The most popular locations, though, are along the edges of the streets surrounding a city's central retailing district, near the entrances to specific buildings and stores, and at the intersections of

**FIGURE 3.22** A vegetable vendor on a street corner in Lima, Peru. [© Lin MacDonald.]

major streets. A single city block may contain, on average, from 15 to 30 street vendors.

A desirable location has itself become a commodity for sale, though the vendors have no legal right to any particular space. Hays-Mitchell observed vendors painting the outlines of their stands on the street itself. The operators have sometimes sold the rights to use such spaces to new vendors for several hundred dollars.

of the SAP backlash, forcing more favorable contracts on foreign oil companies and using its considerable oil wealth to help other countries, such as Bolivia, pay off their debt. Domestically, Venezuela appears to have violated the logic of SAPs by implementing dramatic expansions of social services to the poor while at the same time reversing a 30-year economic decline. Spiking global oil prices have helped, since most of Venezuela's foreign exchange income comes from oil. Nonetheless, the effects of more oil income have been slow to reach the poor majority; wealth disparity in Venezuela actually increased by 43 percent between 2001 and 2003.

## The Informal Economy

For centuries, low-profile businesspeople throughout the Americas have operated outside the law in the informal economy (described in greater detail in Chapter 1 on page 31), paying no taxes but supporting their families through inventive entrepreneurship. Most people working in the informal economy are small-scale operators who see an opportunity to earn some money in the market economy (Box 3.3). This entrepreneurship frequently entails recycling used items, such as clothing, glass, or wastes. In some instances it can serve as an incubator for businesses that might eventually provide legitimate jobs for family and friends. Throughout Middle and South America, the informal economy is huge, and there are probably few middle- and lower-income families who do not depend on it in some way. Some economists argue, however, that too often informal workers are just treading water, making too little to ever expand their businesses significantly. Moreover, the bribes they have to pay (sometimes labeled "informal taxes") to avoid prosecution are much less beneficial to the economy as a whole than the taxes paid by legitimate businesses. Work in the informal economy is also risky because there is no protection of workers' health and safety and no retirement or disability benefits. In many ways, the growth of the informal economy is a symptom of the increasing disparity between those with substantial access to wealth and power and those with very little.

## Regional Trade and Trade Agreements

The growth in international trade and foreign investment in the region, encouraged in part by SAPs, has been joined by growth in regional free trade agreements, which reduce tariffs and other barriers to trade among a group of neighboring countries. The two largest free trade agreements within the region are NAFTA and Mercosur (see Figure 3.21). The **North American Free Trade Agreement (NAFTA)**, created in 1994, is by far the larger of the two. It links the economies of Mexico, the United States, and Canada in a free trade bloc containing more than 435 million people, with a total annual economy worth at least $14 trillion. **Mercosur** is older, having been created in 1991, and smaller. It links the economies of Argentina, Brazil, Paraguay, Uruguay, and Venezuela to create a common market with 250 million potential consumers and an economy worth nearly $2 trillion per year. (Bolivia, Chile, Colombia, Ecuador, and Peru are associate members and are not included in these figures.) There is a growing number of other free trade agreements within the region, some of them spurred by competition between the European Union and the United States to expand their influence in the region. Traditionally, the United States and Canada have traded and invested more heavily in the northern parts of Middle and South America—as reflected by Mexico's position within NAFTA and by U.S. economic ties to Central America and the Caribbean—because these countries are the closest, so transport costs are lower, and because this part of the region is seen as strategically important, especially to the United States. Europe has been more heavily involved in South America, as reflected by the European Union's growing interest in Mercosur and the Andean Community (CAN—a free trade bloc consisting of Bolivia, Peru, Ecuador, Chile, and Colombia).

In the 1990s, hoping to check, at least partially, European trading power in the region, the United States began pushing for the creation of a NAFTA-like trade bloc for all of the Americas called the Free Trade Agreement of the Americas (FTAA). Rising opposition to SAPs, which many in the region see as working hand in hand with U.S.-dominated free trade agreements such as NAFTA, has stalled the FTAA in recent years. Mercosur is increasingly seen as a more equitable alternative to NAFTA and the FTAA.

More global efforts to promote free trade, led by the World Trade Organization (WTO), were dealt a setback in 2003 at the Cancún meeting of the WTO. The G22, a global coalition of 22 developing countries, led in part by Argentina, Brazil, Cuba, and Venezuela, challenged the rich and powerful developed countries of the G8 (Group of Eight: the United States, Canada, France, Germany, Italy, Japan, Russia, and the United Kingdom) to stop the hypocrisy of promoting free trade for others while practicing protectionist policies themselves. The G22 protested tariffs imposed by the G8 that make farm products from the G22 uncompetitive within G8 countries. The G22 also protested subsidies to farmers in G8 countries that help them sell their products on the global market at less than cost, putting farmers from poor countries out of business in their own homelands. One consequence of these policies has been increasing dependence on imported food and rising food prices for people who are already malnourished. The G22 protest in Cancún stopped the WTO meeting and has subsequently dampened further efforts by the WTO to expand free trade agreements elsewhere around the world (Figure 3.23).

The record of regional free trade agreements so far is mixed. While they have increased trade, the benefits of that trade usually are not spread evenly among regions or among sectors of society. In Mexico, the benefits of NAFTA have been concentrated in the

**FIGURE 3.23** Protest in Cancún, Mexico. A Mexican farmer sits before a banner reading "Resistance ... Rebellion" at a gathering of opponents to the World Trade Organization in Cancún on September 9, 2003. The WTO meeting, intended to finalize a comprehensive world trade pact, was cut short by a large protest by workers and farmers from more than 22 countries. [Reuters/Juan Carlos Ulate AW/GN.]

northern states that border the United States, where the agreement facilitated the growth of maquiladoras by easing cross-border finances and transit. Still, most of those who work in the maquiladoras are paid only marginally above the going national wage rate, and living conditions are so difficult (poor housing, schools, and sanitation) that the actual standard of living of these workers has risen only slightly, if at all. Meanwhile, as many as one-third of small-scale Mexican farmers have lost their jobs as a result of increased competition from U.S. corporate agriculture, which now has unfettered access to Mexican markets. Research is beginning to show that these now-unemployed agricultural workers make up a significant number of the undocumented workers coming into the United States.

Producers within the Mercosur sphere differ in their views of that trade agreement. Chilean grain farmers fear competition from cheaper Argentine and Uruguayan wheat and corn, while Chilean fruit growers and winemakers see greater access to buyers in neighboring Brazil as beneficial. Brazil favors a customs union with no tariffs between Mercosur countries, but high external tariffs on goods entering from outside Mercosur, in order to strengthen regional markets for goods produced within Mercosur. However, such tariffs would hamper the ability of Argentina and Uruguay to attract foreign industrial capital. Benefits to all Mercosur members would grow if roads, railroads, and air service between them, which are currently in poor condition, could be improved. So far, trade within Mercosur accounts for less than one-fifth of its members' total trade.

Reporter Sam Quinones speaks to his audience: "In the current crisis peasant coffee growers have to learn the Starbucks lesson and focus on quality. Consumers meanwhile have to be willing to pay extra for the best coffee, searching out regional coffees, the way they do with wine."

The Guatemalan coffee farmer listens to the speaker and frowns down at a slick package of Green Mountain organic coffee. It has his picture on the front. It's the first time he has ever seen how the coffee he grows is marketed to upscale consumers in North America. He has never tasted coffee himself, preferring sweet soda drinks. Only a harsh, low-quality coffee is sold in his village, and virtually no one drinks it.

Coffee used to be a primary commercial crop in the Pacific coastal uplands of Guatemala. Now, many of the coffee plantations (*fincas*) are abandoned; the workers have migrated to Mexico and the United States seeking some other way to support their families. A glut in the global coffee market caused by too many poor tropical farmers on 10 or 12 acres trying to make a living in coffee means that most are paid only a tiny amount per pound by the itinerant coffee trader (called a coyote) who is their only access to the market. When you buy a $1.00 cup of coffee, the grower gets just 10 to 12 cents, the trader 3 cents, and the shipper 4 cents, but the roaster—usually a large multinational company—gets 65 to 70 cents. The retailer typically gets 10 to 15 cents.

Green Mountain and other *fair trade* (as opposed to free trade) coffee marketers are trying to change that lopsided allocation of profits by buying directly from the growers (Figure 3.24). This direct-to-market ap-

**FIGURE 3.24** Fair trade coffee in Guatemala. These people are sifting coffee beans. [Photo from FRONTLINE/World, "Coffee Country"; courtesy of Bill Kinzie, Green Mountain Coffee Roasters Foundation.]

proach means that growers can get as much as $1.26 a pound for their coffee. But it also means that they have to produce high-quality coffee without chemical pesticides and fertilizers—coffee that passes the high standards of professional coffee tasters. So the farmers themselves must learn to judge coffee by its taste. The goal of the fair trade coffee movement is to teach them to be coffee connoisseurs (as vintners are connoisseurs of wine), no longer just producers of beans.

*For more on how fair trade affects the coffee market and the lives of growers, see the FRONTLINE/World video "Coffee Country."* ■

## Agriculture and Contested Space

The agricultural lands of Middle and South America (Figure 3.25) are an example of what geographers call **contested space**: various groups are in conflict over the rights to use a specific territory as they see fit.

***At the Personal Scale.*** Throughout Middle and South America, increasing numbers of rural people find themselves displaced from lands they once cultivated and forced to work as itinerant laborers for low wages, as this vignette illustrates.

**Personal Vignette**   Aguilar Busto Rosalino used to work on a Costa Rican hacienda, where he had a plot for his own subsistence and in return worked 3 days a week for the hacienda. Since a banana plantation took over the hacienda, he rises well before dawn and works 5 days a week from 5:00 A.M. to 6:00 P.M.., stopping only for a half-hour lunch break. Aguilar places plastic bags containing pesticide around bunches of young bananas, work he prefers to his last assignment of spraying a more powerful pesticide, which left him and 10,000 other plantation workers sterile. He works very hard because he is paid according to how many bananas he treats. Usually he earns between $5.00 and $14.50 a day. Right now he is working for a plantation that supplies bananas to Del Monte, but he thinks that in a few

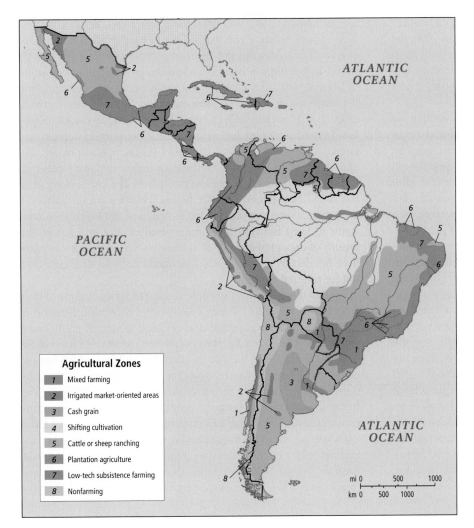

**FIGURE 3.25** Agricultural zones in Middle and South America. Subsistence farming (7) remains widespread, but this traditional mode of agriculture is losing ground to modern, mechanized, export-oriented agriculture (2, 3, 5, 6). The displaced farmers often join the stream of migration to urban areas and North America. [Adapted from Edward F. Bergman, *Human Geography--Cultures, Connections, and Landscapes* (Englewood Cliffs, N.J.: Prentice Hall, 1995), p. 194.]

**Agricultural Zones**

| | |
|---|---|
| 1 | Mixed farming |
| 2 | Irrigated market-oriented areas |
| 3 | Cash grain |
| 4 | Shifting cultivation |
| 5 | Cattle or sheep ranching |
| 6 | Plantation agriculture |
| 7 | Low-tech subsistence farming |
| 8 | Nonfarming |

months he will be working for another plantation nearby. It is common practice for these banana operations to fire their workers every 3 months so that they can avoid paying the employee benefits that Costa Rican law mandates. Although Aguilar makes barely enough to live on, he has no plans to press for higher wages, because he knows that he would be put on a blacklist of people that the plantations agree they will not hire.

*Source: Adapted from Andrew Wheat, "Toxic bananas," Multinational Monitor 17 (9) (September 1996): 6–7.* ■

For centuries, while people labored for very low wages on haciendas, the owners would at least allow them a bit of land to grow a small garden or some cash crops. In recent years, however, governments have encouraged a shift to large-scale, commercial, export-oriented farms and plantations because these operations have the potential to earn large amounts of the foreign currency their countries need to pay off debts. SAPs forced the relaxation of restrictions on foreign ownership, so it is easier for foreign multinationals such as Del Monte—highly efficient, state-of-the-art operations—to dominate the most profitable agribusinesses. Meanwhile, smaller farmers are squeezed out because they cannot afford the latest production techniques that would allow them to sell their produce at competitive prices. Farmers who lose out often must migrate to cities or work as wage laborers on plantations under conditions similar to those experienced by Aguilar Busto Rosalino in Costa Rica.

***At the Provincial Scale.*** In the southern Mexican state of Chiapas, agricultural activists and indigenous leaders have mobilized armed opposition to the entrenched systems that have left them poor and powerless. The Mexican government redistributed some hacienda lands to poor farmers early in the twentieth century, but most land in Chiapas is still held by a wealthy few who grow cash crops for export. The majority farms tiny plots on unfertile hillsides. In 2000, about three-fourths of the rural population was malnourished, and one-third of the children did not attend school. The Zapatista rebellion (named for the Mexican revolutionary hero Emiliano Zapata) began on the day the North American Free Trade Agreement took effect in 1994. The Zapatistas view NAFTA as a threat because it has diverted the support of the Mexican federal government from land reform to large-scale export agriculture. The Mexican government used the army to repress the rebellion. In 2003, after 12 years of armed resistance, the Zapatista movement redirected part of its energies toward a nonviolent political campaign to democratize local communities by setting up people's governing bodies parallel to local official governments. In 2006, the Zapatistas toured the country prior to national elections in an effort to turn the political climate in Mexico against globalization and toward greater support for indigenous people and the poor.

***At the National Scale.*** The trends in rural agriculture that we have been describing have sparked resistance in a number of countries. In Brazil, 65 percent of arable and pasture land is owned by wealthy farmers who make up just 2 percent of the population. Since 1985, due to privatization (the selling of industries and other firms, formerly owned and operated by governments, to private companies or individuals) and structural adjustment policies, more than 2 million small-scale farmers have been pressed to sell their land to large corporate farms specializing in the production of cattle and other agricultural products for export, and many have been forced to migrate. To help these farmers, organizations such as the Movement of Landless Rural Workers (MST) began taking over unused portions of some large farms, arguing that it is wrong for agricultural land to lie unused while there are people living in extreme poverty. Since the mid-1980s, the landless movement has coordinated the occupation of more than 51 million acres (21 million hectares) of Brazilian land (an area about the size of Kansas). Some 250,000 families have gained land titles, while the elite owners have been paid off by the Brazilian government and have moved elsewhere. MST played a major role in the political success of Brazilian President Lula da Silva in 2002 and 2006. Public opinion polls show that 77 percent of Brazilians support the landless movement. Movements with goals similar to those of MST now exist in Ecuador, Venezuela, Colombia, Peru, Paraguay, Mexico, and Bolivia.

In January 2006, Bolivians inaugurated Evo Morales, the first indigenous (Aymara) national leader since the Spanish conquest (Figure 3.26). Morales got his start in politics as the leader of a movement of impoverished coca farmers, protesting the acquisition of Bolivian agricultural land by corporate export-oriented firms and the stigmatization and criminalization of Bolivia's time-honored crop, coca. Coca has been grown in Bolivia for local use for millennia. Its mild stimulant and palliative qualities provide strength and energy for those doing hard labor and relieve upset stomachs and altitude sickness. Because of these qualities, the plant plays an important ceremonial role as well. Only in the last century or so has coca been refined into the highly addictive cocaine, and the primary market for cocaine is North America, not Bolivia. Morales wants to keep Bolivian land under local control and restore the legality of coca growing so that the farmers may once again support their families by supplying coca for traditional uses within Bolivia, not for external markets.

All of these instances of contested use of agricultural land have arisen from inequities within globalizing economies. Because of the money-making potential and political power of commercial agriculture, only rarely has the contest been resolved in favor of the workers (Box 3.4). Yet the story of contested agricultural land would not be complete without recognizing that commercial agriculture is sometimes essential to the economy and well-being of the region. In rapidly urbanizing countries, only commercial-scale agriculture can supply sufficient food for city people. Some commercial farms have been around for a long time. For example, a wide belt of commercial (cash) grain production, similar to that found in the midwestern United States, stretches through the Argentine pampas (see Figure 3.25). Zones of modern mixed farming, including the

**FIGURE 3.26** Bolivian President Evo Morales, the first elected indigenous head of state in South America. Morales (*right foreground*) shakes hands with Bolivia's energy minister after reaching a compromise over control of the country's energy industries. Morales has promised to reform the ways in which foreign-owned extractive industries pay Bolivia for the privilege of operating there. [Reuters/David Mercado.]

## BOX 3.4 | AT THE GLOBAL SCALE  The Power to Map: An International Movement by Indigenous People

Indigenous people in Middle and South America have become leaders of an international movement to map native lands. They were introduced to this idea by ecologists and environmentalists seeking to better understand just what territory a given group used in their day-to-day livelihood strategies. Indigenous people quickly recognized that such maps could help them get official acknowledgement of their claims to land and protection for their spiritual and economic (hunting, fishing, farming) practices on that land. Soon indigenous people were learning to use modern mapping technology such as global positioning systems and computer cartography (see the discussion in Chapter 10), and the maps they have produced are now used to educate their communities about their own cultural heritage. At the national and global scales, the maps are serving as a powerful tool for getting constitutional changes and international agreements that acknowledge indigenous rights. The World Bank and related lending institutions, as a result of international public pressure, now fund indigenous mapping projects in an effort to stabilize and regularize property claims and as an aid in preserving areas of biodiversity from development. Most important, the spread of indigenous mapping to many parts of the world has led to transnational alliances of indigenous peoples who formerly had no contact. This network now connects indigenous peoples in such places as Indonesia, Malaysia, Australia, Afghanistan, Bangladesh, Canada, the Congo Basin in Africa, and many places in Middle and South America.

These are important accomplishments, but there are some drawbacks to the mapping movement. Among them is the fact that nomadic people who lay claim not to land, but rather to the right to pass through land and use its resources in a transitory fashion, may lose that right if the land is formally claimed by others. Further, when formal banking institutions seek to stabilize and regularize the property regimes of indigenous people, they may later use the maps to justify environmentally destructive development on nearby unprotected lands. Linking people to land through maps can also foster ethnic conflict in places where informal negotiation has averted such tensions in the past.

*Source: Karl Offen,* "Map or be mapped: Indigenous and black territorial politics of identity in Latin America," Fulbright Lecture Series, Barranquilla, Colombia, 2004.

---

production of meat, vegetables, and specialty foods for sale in urban areas, are located around the major urban centers of southern Brazil, Argentina, and Uruguay, as well as in south central Chile. Irrigated agriculture is practiced along the dry, narrow coastal desert plain of Peru and northern Chile, as well as on the other side of the Andes in the Argentine pampas and in northwestern Mexico. These various types of market-based agriculture on large holdings have long been essential to the trade of many countries, and because they help to feed local urbanites, they decrease the amount of food that must be imported.

## Is Democracy Rising?

For the first time in a long while, most countries in the region have multiparty political systems and democratically elected governments. Cuba is the main exception. In the last 25 years, there have been repeated peaceful democratic transfers of power in countries once dominated by monolithic alliances composed of the military, wealthy rural landowners, wealthy urban entrepreneurs, foreign corporations, and even foreign governments such as the United States. Nonetheless, election results are frequently contested. Some elected governments are repeatedly threatened with coups d'état because of policies that are unpopular with the masses, powerful elites, or the United States—Venezuela, Ecuador, Colombia, and Bolivia were so threatened in the last decade. Change has been especially dramatic (and not necessarily peaceful) in Mexico, where the entrenched political machine of the Institutional Revolutionary Party (PRI), which had ruled the country for 71 years, was unseated in 2000 by a coalition led by the maverick businessman Vicente Fox. Despite a PRI majority in the Mexican Congress, President Fox managed to pass a number of reform-oriented laws in his 6-year term. In the 2006 presidential election, the conservative Felipe Calderon narrowly won over the populist Andres Lopez Obrador, a popular mayor of Mexico City. Lopez Obrador, with the public support of many factions, created a constitutional crisis by refusing to accept the election results. He organized a so-called parallel government, declaring himself president and appointing ministers. The goal seemed to be to force President Calderon to resign, which he did not do.

In Venezuela, where the profits from the country's rich oil deposits have not trickled down to the poor—one-third of the population lives on $2.00 (PPP) a day, and per capita GDP (PPP) was lower in 1997 than it was in 1977—the landslide election of Hugo Chavez as president in 1998 appeared to advance the cause of democracy. Chavez called for fundamental restructuring of the system of government along more egalitarian lines. Although President Chavez originally had the support of the middle class, that support waned when his administration focused on providing jobs, community health care, and subsidized food for the large underclass. Chavez was reelected in 2000, then was briefly deposed in a coup d'état in 2001, but was quickly reinstated by a groundswell of popular support among workers. His position was sustained in a referendum in 2004 and again in the landslide election of 2006. The Chavez-led government has tried unorthodox strategies to alleviate the effects of poverty. For example, it has contracted with Cuba, which has a surplus of medical personnel, to supply qualified doctors to community clinics. By 2005, infant mortality rates and some other health indicators were improving.

Socialist-leaning governments have also been elected elsewhere in the region, apparently in a popular backlash against rightist governments that had acceded to the structural adjustment and free market policies of the IMF, World Bank, and WTO—policies that were especially hard on the poor and were therefore considered by some analysts to be antidemocratic. In Brazil, the 2002 election of populist Luiz Inácio Lula da Silva was expected to diminish privatization, but in fact, the disruption of privatization policies under President da Silva has been only slight. Moderately leftist governments in Argentina and Chile are not likely to abandon privatization, but rather to pay more attention to relieving the hardships such policies work on the poor. The recently elected governments in Peru and Bolivia are likely to take somewhat stronger stands against privatization, but even there, strong free market factions will continue to wield influence.

Some pro-market policies, such as expansion of Internet access to the general public, clearly further public participation in civil society. However, in much of the region, democracy is still fragile and not particularly *transparent*—meaning that official decisions are not open to public input and review. Policy is formulated without sufficient attention to its potential effects on the poor or those who lack strong political voices. Officials may remain in power not because they have broad public support, but because they are backed by powerful people, who may even be drug czars or otherwise corrupt.

***Political Corruption.*** High-level corruption undermines democratic institutions, in part by setting an example for the rest of society that encourages the diversion of funds from needed public projects, facilitates all types of crime, and even encourages election fraud. Virtually every country in the region has had a serious scandal in the past few years involving high-level officials performing million-dollar favors for friends and families or stealing money from taxpayers outright. The international banks in Miami, Florida, are well-known as depositories for such stolen funds, so North American financial institutions have benefited indirectly from the corruption, while Middle and South American countries have been robbed of much-needed public funds and private investment capital.

***The Drug Trade.*** The international drug trade is a major factor contributing to corruption, violence, and the subversion of the democratic process throughout Middle and South America. Central America and northwestern South America are the primary sources for the raw materials of the drug trade—coca for cocaine and marijuana. Figure 3.27 graphically illustrates the geographic

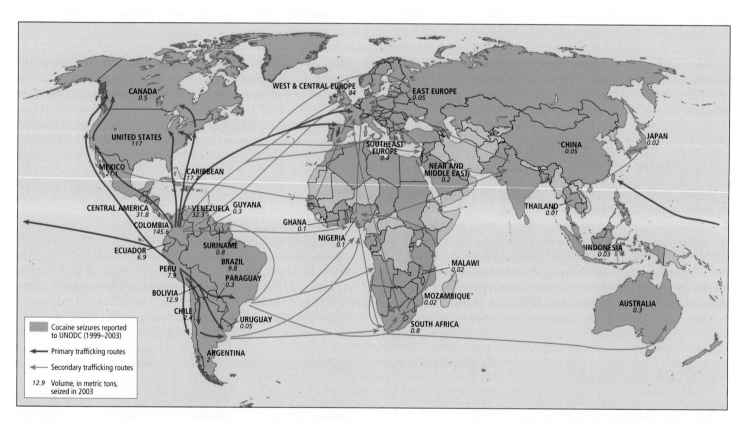

**FIGURE 3.27** Interregional linkages: Cocaine sources, trafficking routes, and seizures worldwide. Colombia, Peru, and Bolivia are the most important sources for the cultivation and production of cocaine in the world. The big cocaine markets are in North America (primarily the United States) and in Europe (primarily in Spain and the United Kingdom), but the drug is also widely used in western South America. The red lines show main trafficking routes. The green color shows that cocaine was seized in almost every country in the world in 2002–2003. [Adapted from *2005 World Drug Report*, Volume 1: *Analysis* (Vienna: United Nations Office on Drugs and Crime, 2005), pp. 65, 75, and 78; available at http://www.unodc.org/pdf/WDR_2005/volume_1_web.pdf.]

distribution of cocaine seizures during 2002–2003. (Such seizures are the most accurate, yet still a flawed, way to measure the cocaine trade.) The map shows little action in Asia and Africa, but this may be simply a result of lax law enforcement there. Judging by the amount seized, Colombia is probably the largest global producer and the United States the largest market, with Europe close behind. Mexico, Central America, and the Caribbean are major conduits to North America, but the trade connections to the rest of the world are widely dispersed throughout the Americas. Street prices for cocaine dropped in 2003 and have not risen substantially since then, indicating that surplus stockpiles stored in the Andean highlands might be entering the market, increasing supplies.

Cocaine is derived from the coca plant, the leaves of which have been chewed by Andean people as a mild stimulant for thousands of years. Today, most coca growers are small-scale farmers of indigenous or mestizo origin in remote locations, who can make a better income for their families from these plants than from other cash crops. Although production of addictive drugs is illegal in all of Middle and South America, public figures, from the local police on up to high officials, are paid to turn a blind eye to the industry. Because law enforcement is lax in drug-producing areas, drug lords can fund their own private armies to protect drug transport routes and force the cooperation of local citizens.

The U.S. "war on drugs" is controversial in the region because it focuses on halting production and destroying enough crops to stop the flow of drugs into the United States. A more direct approach would be to significantly reduce demand by acting on U.S.-sponsored research showing that social programs providing emotional and physical support for youth are most likely to reduce drug usage. However, funding for such programs in the United States has been cut in recent years.

The effort to stop the production of drugs in Middle and South America has led to the largest U.S. military presence in the region in history. The U.S. effort is focused on supplying intelligence, eradication chemicals, military equipment, and training to military forces in the region. U.S. military aid to Middle and South America is now about equal to U.S. aid for education and other social programs in the region. Meanwhile, drug production in Middle and South America is now so great that it exceeds demand in the U.S. market, so much of the production is traded within the region, where drug use is increasing.

### Foreign Involvement in the Region's Politics.

Interventions by powers outside the region have compromised democracy, even though their stated aim may have been to enhance it. Although the former Soviet Union, Britain, France, and other European countries have wielded influence in the region during the twentieth century, by far the most active foreign power has been the United States. That country proclaimed the Monroe Doctrine in 1823 to warn Europeans that the United States would allow no further colonization in the Americas. Successive U.S. administrations have interpreted the Monroe Doctrine more broadly to mean that the United States itself had the right to intervene in the affairs of other countries in the Americas. And the United States

has done so a number of times, ostensibly to make those countries safe for democracy, but in reality to protect U.S. political and economic interests. One result has been the installation of U.S.-backed unelected political leaders, many of them military dictators, in almost every country in the region at some point during the past 150 years. In recent decades, the United States supported armed insurgencies for the purpose of preventing or reversing Communist takeovers in Cuba (1961), the Dominican Republic (1965), and Nicaragua (1980s). In Chile, the elected government of Salvador Allende was overthrown with the complicity of the U.S. Central Intelligence Agency (CIA) on September 11, 1973, and General Augusto Pinochet was installed as a military dictator with U.S. approval. Over its 17-year rule, the Pinochet regime imprisoned and killed thousands of Chileans who protested the loss of democracy.

The United States invaded Panama in 1989 and arrested the Panamanian general, and then president, Manuel Noriega. Noriega had shadowy connections with both drug dealing and CIA covert operations in the region. He is now in jail in Florida. Since the end of the cold war, the United States has curtailed direct military intervention in the region's politics, instead funding local opposition groups that support U.S. interests and lead attempted coups d'état, as was the case in Venezuela in 2002. Although resentment of the United States has grown dramatically throughout the region in recent years, it still wields considerable influence.

### The Political Impacts of Information Technology.

The Internet already plays a role in the politics of Middle and South America by helping activists get their message out to the rest of the world. Broader impacts will depend on how many people have access to the Internet. A serious digital divide exists in the region (Figure 3.28), but more and more countries are adopting policies that will increase public access to the Internet. A recent United Nations study of global Internet use showed that Internet users made up just 3.2 percent of the region's population in 2000, but by 2006, the rate for the region was 14.4 percent, and Chile's rate was 35.7 percent. The highest rate of use was in the Caribbean: in Barbados, 56.2 percent of the population used the Internet in that year.

Overall, Middle and South America is more advanced in technological achievement than other developing regions of the world, and a 2000 study by *Wired* magazine considered Mexico, Argentina, Costa Rica, Chile, and Brazil to have excellent potential for leadership in global technological innovation. Brazil, which ranked 10th in the world in 2006 for number of Internet users, has two world-class technology hubs in the environs of São Paulo and is attempting to cover the whole country with broadband fiber-optic cable networks for telecommunications and Internet service. In 2003, the government of newly elected President Lula da Silva decided to liberalize the Internet governance structure for Brazil to increase access for all citizens. Mexico (which ranks 15th in number of Internet users worldwide) has recently launched a program to give its citizens access to training and higher education via the Internet as part of an effort to stem the flow of migration to the United States.

**FIGURE 3.28** Internet use in Middle and South America. The numbers on the map indicate the number of Internet users in each country; percents indicate the percentage of the country's population that uses the Internet. By March 2006, 14.4 percent of the people (nearly 80 million) in Middle and South America were using the Internet, a 342 percent increase since 2000. This region has the highest percentage of Internet users in the developing world. Chile has the highest percentage of users (35.7) in South America, Barbados (56.2 percent) the highest in the Caribbean, and Costa Rica (22.7) the highest in Middle America. [Data from "Internet usage statistics—the big picture" and "Internet usage statistics for the Americas," from Internet World Stats at http://www.internetworldstats.com/stats.htm and http://www.internetworldstats.com/stats2.htm.]

## SOCIOCULTURAL ISSUES

Under colonialism, a series of social structures evolved that guided daily life—standard ways of organizing the family, the community, and the economy. They included rules for gender roles, rules for race relations, and ways of religious observance. These social structures are still widely in place in the region, but they are changing in response to urbanization, economic development, and the diffusion of ideas from outside. The results are varied. In the best cases, change is leading to a new sense of initiative on the part of women, men, and the poor; in the worst cases, to the breakdown of family life and the abandonment of children.

## Cultural Diversity

The region of Middle and South America is one of the most culturally complex on earth. Contributing to this diversity are the many distinct indigenous groups that were present when the Europeans arrived and the many cultures that were introduced during and after the colonial period. In the Caribbean, the Guianas (Guyana, Suriname, and French Guiana), and Brazil, the arrival of a relatively small number of European newcomers resulted in the annihilation of indigenous cultures, and these regions then became populated almost entirely by various people from the Old World. From 1500 to the early 1800s, some 10 million Africans were brought to plantations in the islands and coastal zones of Middle and South America. After the emancipation of African slaves in the Caribbean in the 1830s, more than half a million Asians were brought there from India, Pakistan, and China as indentured agricultural workers; their cultural impact remains most visible in Trinidad and Tobago, Jamaica, and the Guianas (Figure 3.29). In some parts of Mexico and in Guatemala, Ecuador, and other Andean zones, indigenous people have remained numerous and, to the unpracticed eye, may appear little affected by colonization.

Mestizos are now the majority in Mexico, Central America, and much of South America. In some places, such as Argentina, Chile, and southern Brazil, Euro-Americans are now so numerous that they dominate the landscapes and ways of life. The Japanese, though a tiny minority everywhere in the region, are increasingly influential in innovative agriculture and industry in Brazil, the Caribbean, and Peru. In some ways, diversity is increasing as the media and trade introduce new influences into the region. Yet culture groups are becoming less distinct in urban areas, where various groups live in close proximity. The increasing contact between people of widely different backgrounds is accelerating the rate of **acculturation** (cultural borrowing) and **assimilation** (loss of old ways and adoption of new ones). In the big cities, such as Mexico City, Lima, and Rio de Janeiro, there is a blend of many different strains, so that no one cultural component remains unaffected by the others.

## Race and the Social Significance of Skin Color

People from Middle and South America, and especially those from Brazil, often proudly claim that race and color are of less consequence there than in North America, and they are right in certain ways. Skin color is less associated with status than in the colonial past. In Middle and South America, a person of any skin color, by acquiring an education, a good job, a substantial income, the right accent, and a high-status mate, may become recognized as upper class. Today, the terms *white, clear-skinned* for "light," and *black, mulatto,* or *pardo* for "brown" tend to be merely descriptive and do not necessarily carry social meaning. The dark skin of an upper-class person would not be thought of as racially significant; on the other hand, that person's family, wealth, education, place of residence, and occupation would be very significant.

Nevertheless, the ability to erase the significance of skin color through accomplishments is not quite the same as its having no significance at all. This point is illustrated by the fact that overall, those who are poor, less well educated, and of lower social standing tend to have darker skins than those who are educated and well-off. And while there are poor light-skinned people throughout the region (often the descendants of migrants from Central Europe over the last century), most light-skinned people are middle and upper class. These observations seem to indicate that race and color have not yet disappeared as social factors in Middle and South America. In some countries—Cuba, for example, where overt racist comments are socially unacceptable—it is common for a speaker to silently indicate to a listener by tapping his forearm with two fingers that the person referred to is African-Caribbean.

## The Family and Gender Roles

The **extended family** is the basic social institution in all the societies of Middle and South America. Throughout this region, it is generally accepted that the individual should subvert his or her interests to those of the family and local community and that individual well-being is best secured by doing so.

The spatial arrangement of domestic life illustrates these strong family ties. Families of adult siblings, their mates and children, and their elderly parents frequently live together in domestic compounds of several houses surrounded by walls. Social groups out together in public are most likely to be family members of several generations, rather than unrelated groups of single young adults or of married couples, as would be the case in Europe or the United States. A woman's best friends are likely to be her female relatives. A man's business or social circle will almost certainly include male family members or long-standing family friends.

Gender roles have their roots in the extended family as well. Throughout Middle and South America, the Catholic Church

**FIGURE 3.29 Javanese Muslim women in Suriname.** These women, descendants of indentured servants brought by the Dutch from their colony in Indonesia in the 1850s, are marking the end of Ramadan, a major religious observance. [Robert Caputo/Aurora.]

was instrumental in the colonizing process and in the establishment of official and popular mores. The church remains influential in daily life. The Virgin Mary, the mother of Jesus, is held up as the model for women to follow through a set of values, known as *marianismo,* that puts a priority on chastity, motherhood, and service to the family. The ideal woman is the day-to-day manager of the house and of the family's well-being, training her sons to enter the wider world and her daughters to serve within the home. Over the course of her life, a woman's power increases as her skills and sacrifices for the good of all are recognized and enshrined in family lore.

Her husband, the titular family head, is expected to work and provide most of his income to his family. The rules for how men are to contribute to family stability and well-being are less clearly spelled out than the rules for women. Men have a good bit more autonomy and freedom to shape their lives than women do simply because they are expected to move about the community and establish relationships, both economic and personal. A man's social network is deemed just as essential to the family's prosperity and status in the community as his work. In addition, as is the case in most societies, there is an overt double sexual standard for males and females. While expecting strict fidelity in mind and body from his wife, a man is freer to associate with the opposite sex. Males measure themselves by the model of *machismo,* which considers manliness to consist of honor; respectability; and the ability to father children and be attractive to women, to be an engaging raconteur in social situations, and to be the master of the household. Under traditional rules of machismo, the ability to acquire money was secondary to the other symbols of maleness. Now a new market-oriented culture also prizes visible affluence as a desirable male attribute.

Many factors are transforming these family and gender roles. For one thing, with infant mortality declining steeply, couples are now having only two or three children, instead of five or more. For men, this change may require an adjustment of the machismo ideal that defines manliness in part by high fertility. Moreover, because most people still marry early, most parents are free of child-raising responsibilities by the time they are 40, leaving women with 30 or more years of active life to fill in some other way. As it happens, economic change throughout the region has provided a solution. Increasingly, many women, despite their lack of formal education, have been able to find factory jobs or other employment that puts to use the organizational and problem-solving skills they perfected while supervising a family. For more and more women of all ages, employment outside the home (often in a distant city) is a way to gain a measure of independence and also contribute to the needs of the extended family. Many families simply bend to accommodate these new situations; others lose many members to migration, and some disintegrate.

## Children in Poverty

Despite often difficult conditions, the vast majority of parents in Middle and South America, even those in the worst urban slums, heap loving care on their children, providing all they possibly can and sacrificing so that their children might have a better life. Still, many families need the income that even young children can gen-

erate. It is estimated that one-third of the children in Middle and South America are economically active. In the countryside, children work in agriculture, logging, and mining. But city children also work, often on the street selling items such as chewing gum, flowers, and flip-flops (rubber sandals), cleaning cars, and performing small services.

It is understandable that children are asked to help alleviate a family's poverty by doing some work, paid or unpaid. Nonetheless, in societies that traditionally place such high value on the family and children, it is difficult to understand recent news stories about homeless and abandoned children in such countries as Mexico, Guatemala, Colombia, and Brazil. Some of these reports may be overblown: a study in the year 2000 in São Paulo, Brazil, South America's largest city, revealed that just 609 children were homeless there, one-tenth of what had been thought to be the case. Nonetheless, throughout the region, many children live on the streets of cities, and thousands have been pushed out of the home at a tender age by overburdened parents. These children must fend for themselves, sleeping in doorways, hustling for money or food. Reports of abandoned children are increasing around the world, and the explanations for this phenomenon are numerous.

In Middle and South America, structural factors are largely responsible for the difficult conditions under which families must operate. As discussed previously, rural-to-urban migration by young adults is common. Many inexperienced young women migrants are equipped for only the most menial jobs and rarely earn a living wage, so they may ally themselves with a man to gain some measure of security, or even just a place to live. Often, children result because birth control is unavailable, too costly, or rejected for religious reasons. The man may already have several informal mating relationships, so his unpredictable income is stretched too thin to support several households adequately. Women are commonly

**FIGURE 3.30 Glue sniffing.** A Nicaraguan boy sniffs glue from a baby-food jar in the Oriental Market in Managua, while his friend pulls his own jar out of his shirt. [Richard Sennott/*Minneapolis–St. Paul Star Tribune.*]

expected to undertake the daily care of children, and in the case of young migrant mothers, they must also keep working. With no extended families to help with child care, and with no older, solid role models—aunts, grandmothers—to enforce traditional values, the children may grow up neglected, malnourished, and unruly, with lonely, dysfunctional mothers who may turn to drugs or alcohol for solace. Some of these children turn to brain-damaging glue sniffing to ease their anxiety (Figure 3.30).

## Religion in Contemporary Life

The Roman Catholic Church remains one of the most influential institutions throughout Middle and South America. From the beginning of the colonial era, the church was the major partner of the Spanish and Portuguese colonial governments. It received extensive lands and resources from those governments, and it sent thousands of missionary priests to convert indigenous people. For many centuries, the Catholic Church encouraged working people to accept their low status, be obedient to authority, and postpone their rewards until heaven. Thus, while serving to unify the region and spread European culture, the church also reinforced class differences. Furthermore, the church ignored those teachings of Christ that admonish the privileged to share their wealth and attend to the needs of the poor. Nonetheless, poor people throughout Spanish and Portuguese America embraced the faith and still make up the majority of church members. Indigenous people throughout the region have put their own ethnic spin on Catholicism, creating multiple folk versions of the religion with music, liturgy, and interpretations of Scripture that vary greatly from European ones.

The Catholic Church began to see its power erode in the nineteenth century in places such as Mexico. **Populist movements**—popularly based efforts seeking relief for the poor—seized and redistributed church lands and canceled the high fees the clergy had been charging for simple rites of passage such as baptisms, weddings, and funerals. Over the years, the church became less obviously connected to the elite and more attentive to the needs of poor and non-European people. In the mid-twentieth century, the church began ordaining indigenous and African-American clergy as a matter of course. After the Second Vatican Council in the 1960s, women were allowed to perform certain tasks during Mass.

In the 1970s, a more radical movement known as **liberation theology,** which sought to reform the church from the ground up, emerged. Begun by a small group of priests and activists, the movement (not the official church) portrayed Jesus Christ as a social revolutionary who symbolically spoke out for the redistribution of wealth when he divided the loaves and fishes among the multitude. It viewed the perpetuation of gross inequality and political repression as sinful and promoted social and economic reform as liberation from evil.

At its height, the liberation theology movement was the most articulate movement for region-wide social change and had more than 3 million adherents in Brazil alone. Today, the movement is reduced in strength, and its positions have been severely criticized by the Vatican. In such places as Guatemala and El Salvador, liberation theology became the target of state-sponsored attacks that vilified its participants as Communist conspirators. In addition, it has had to compete with newly emerging evangelical Protestant movements that also have many attractions for the poor.

**Evangelical Protestantism** diffused into Middle and South America from North America. It is now the fastest-growing religious movement in the region; perhaps as many as 10 percent of the population, or about 50 million people, are adherents. Like liberation theology, it appeals to the rural and urban poor—that segment of society most in need of hope for a better future. Some evangelical Protestants teach a "gospel of success," stressing that those who are true believers and give themselves to Christ and to a new life of hard work and clean living will experience prosperity of the body (wealth) as well as of the soul. The movement is charismatic, meaning that it focuses on personal salvation and empowerment of the individual through miraculous healing and transformation, rather than on general social change aimed at benefiting all citizens. The movement has no central authority, consisting of a host of small, independent congregations whose leadership is often passed around to both men and women. There is considerable evidence that the gospel of success has been an important impetus to the emergence of a middle class, and some theologians have noted that this version of Christianity meshes nicely with the idea that social justice can be achieved in a trickle-down fashion through the success of the market economy. Others believe that general societal improvements are unlikely to spring from a religious movement that encourages individual well-being but does not mobilize the faithful to reform society on behalf of the poor and the marginalized.

There are a considerable number of Protestants throughout Middle and South America, but the extent to which these geographic patterns reflect the spread of the evangelical version of Protestantism is not yet well understood. Evangelical Protestantism is growing in such formerly Catholic countries as Brazil and Chile, and is also strong in the Caribbean, Mexico, and Middle America, among both the poor and middle classes. The revival tents of North American evangelical pastors are a not uncommon sight in the urban fringe of cities throughout the region.

Other religions found across the region include Judaism, Islam, Hinduism, and indigenous beliefs. A range of African-based belief systems (Condomble, Umbanda, Santeria, Obeah, Voodoo) combined with Christian beliefs are found in Brazil, the Caribbean, and wherever the descendants of Africans have settled. These African-based religions have attracted adherents of European or indigenous backgrounds as well, especially in urban areas.

## ENVIRONMENTAL ISSUES

Environments in Middle and South America were among the first to inspire concern about the use and misuse of the earth's resources. Beginning in the 1970s, construction of the Trans-Amazon Highway provided migrant farmers with access to the Amazon Basin. They followed the new road into the rain forest and began clearing it to grow crops. Scholars such as anthropologist Emilio Moran warned of an impending crisis in the tropical rain forests of the Americas. Scientists then began to compile records to show the

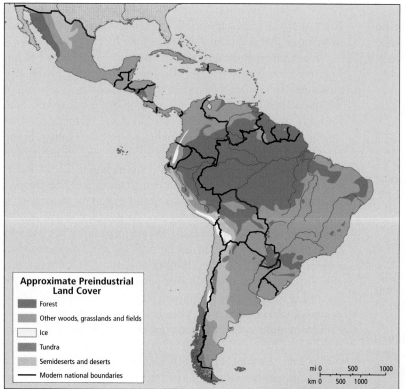

(a) Preindustrial

**FIGURE 3.31** Land cover (preindustrial) and human impact (2002). In the preindustrial era (**a**), humans had cleared some forest, planted plantation crops, and built some cities; but the overall human impact, at least as shown at this scale, was minimal. By 2002 (**b**), human impact was observable in most of the region, with the possible exception of some parts of the Amazon. [Adapted from *United Nations Environment Programme, 2002, 2003, 2004, 2005, 2006* (New York: United Nations Development Programme), at http:////maps.grida.no/go/graphic/human_impact _year_1700_approximately and http://maps.grida.no/go/graphic /human_impact_year_2002.]

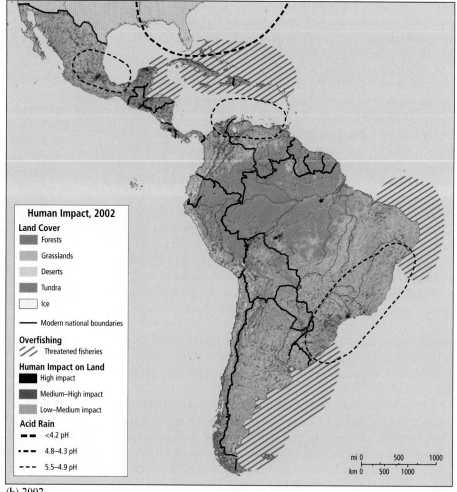

(b) 2002

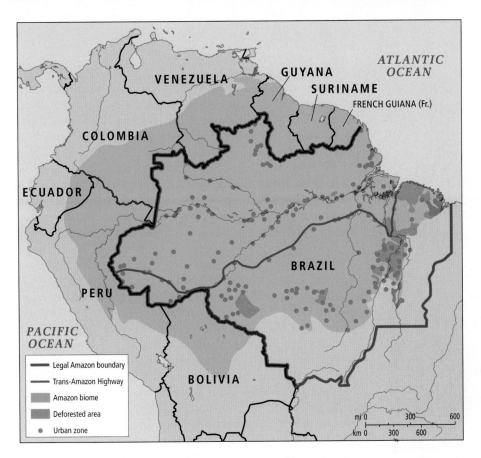

**FIGURE 3.32** Effects of development in the Amazon Basin. The Amazon biome is huge and extends into many countries, as shown in green. Brazil legally defines its part of the Amazon somewhat differently, as shown by the dark boundary, all of which is called Amazonia. [Adapted from "Human pressure in the Brazilian Amazon, all indicators," World Resources Institute, at http://pubs.wri.org/pubs_maps_description .cfm?image ID=2475.]

extent of human impacts on environments in this region over time (Figure 3.31). Scholars now know that even in prehistory, every human settlement has had consequences for environments across the Americas. Today's impacts, however, are particularly severe because the population has doubled and redoubled at the same time that per capita consumption of resources within the region and exports out of the region have increased dramatically.

## Tropical Forestlands in the Global Economy

Amazon forest environments, though vast, face multiple threats (Figure 3.32). Primary among them are the logging of valuable hardwoods (especially mahogany), the clearing of land to raise cattle and to grow cash crops (especially soybeans), and the extraction of underlying minerals, including oil and gas. All of these activities produce raw materials for the global export market and connect investors, producers, and consumers across several world regions. The logging of Amazon hardwoods is often carried out by local entrepreneurs and landless loggers working for large landowners, who are encouraged to log by government policies. Investment capital comes from Asian multinational companies that have turned to the Amazon forests after having logged up to 50 percent of the tropical forests in Southeast Asia. Logging firms from Burma, Indonesia, Malaysia, the Philippines, and Korea are now purchasing forestlands in Brazil after already having secured logging rights to Amazon Basin forests in the neighboring countries of Peru, Colombia, Venezuela, and the Guianas. A single tree can be worth $6000 to the barefoot men who fell the tree, cut it up, and raft it to a transshipment point, but the same tree will be worth $300,000 once it is turned into furniture or paneling in the United States or

Europe. The World Wildlife Fund estimates that 50,000 such trees ended up in the U.S. market in 2002. According to an industry spokesperson, U.S. consumers typically do not inquire about the origins of the wood in the furniture they buy.

The governments of Peru and Brazil encourage surplus urban populations to occupy cheap land along the newly built logging roads. As a result, landless settlers from the highlands of the Andes and from southern and eastern Brazil clear yet more land for use as subsistence cultivation plots (Figure 3.33). After a few years of farming the poor rain-forest soils, the settlers often abandon the land, now eroded and depleted of nutrients, and move on to new plots. Ranchers may buy the worn-out land from these failed small farmers to use as cattle pastures. As discussed at the opening of this chapter, oil and gas extraction by multinational petroleum corporations is polluting the water and soil on remote indigenous lands. The access roads built for oil extraction also encourage migrants, who inadvertently bring in diseases that are deadly to indigenous forest dwellers who rarely encounter people from the outside world.

The days of unopposed exploitation of the Amazon Basin may be waning. Until the mid-1980s, Brazilian government policies encouraged forest clearing, considering it development of unused land, and those who developed forestland did not have to pay taxes on it. Eventually, the Brazilian government changed its taxation policies in response to environmental concerns. Although they come late, environmental regulations are now being developed in all the countries of the Amazon. Nevertheless, the area is not free of environmental threats. While Brazil now strictly controls the mahogany trade, illegal logging is on the rise. The pace of clear-cutting for pasture and small farming is slowing, but the rapid growth of urban settlements and suburbs, the increased use of forest timber as fuelwood, and the

**FIGURE 3.33 Logging in the Amazon Basin.** In the Brazilian state of Rondônia, in the Amazon Basin, colonists from other areas clear the rain forest for settlement and large-scale agriculture. New roads, like the one at the upper left, provide the settlers with access to the forest and become the conduit for many other changes. Some of the wood is harvested for local use or sold, but much is wasted. [Randall Hyman.]

clearing of forest tracts for the large-scale production of cash crops such as soybeans all continue to threaten Amazon forestlands.

The destruction of the Amazon rain forest has global implications. Scientists now understand that the absorption of carbon dioxide and the release of free oxygen are some of the most important functions of the world's forests. Thus the loss of huge tracts of forested land in the Amazon Basin contributes to global warming (see the discussion on pages 45–48 in Chapter 1). One positive note is that reforestation is more widespread than previously thought. The tropical climate encourages fairly rapid rates of regrowth: abandoned pastures and farmland can regain a mature forest in 15 to 20 years. Planting fast-growing trees in forest plantations is one way to maintain the exchange of carbon dioxide and oxygen, and the use of wood from such plantations may save some natural forests from cutting in the future. However, forest plantations fall far short of replicating the complex biodiversity of the rain forest, which may have as many as 300 different species of trees on a single hectare

(2.5 acres) of land, compared with the one or two species that would cover a plantation of hundreds of hectares. It is estimated that tropical rain forests contain 60 percent of all species found on earth, so losing these forests at the present rates means losing an enormous part of the planet's genetic inheritance.

The forests in other parts of Middle and South America are also at risk. About 65 percent of the forest clearing in Central America is intended to create pastures for beef cattle grown primarily for the U.S. fast-food industry. The forests of Central America and the Caribbean have also been cleared for export crops, such as sugar, cotton, and tobacco in the past and bananas today. If deforestation continues in these areas at the present rates, the natural forest cover in these areas will be entirely gone in 20 years.

The forests of Middle and South America are another example of contested space (see the discussion on pages 142–145). For example, one well-financed government research agency in Brazil promotes the use of forest products, while another less well financed government agency works to undo the resulting damage and protect indigenous tribes from the encroachment of development. Meanwhile, ecotourism entrepreneurs take their clients to resorts in the rain forest, where luxury and spectacle often have priority over environmental sustainability (Box 3.5).

## The Environment and Economic Development

In the past, governments in the region argued that economic development was so desperately needed that environmental regulations were an unaffordable luxury. Now there are increasing attempts to take a middle ground by embracing economic development as necessary to raise standards of living, yet trying to minimize its negative effects on the environment. A chief worry is access to sufficient clean water, especially for poor urban dwellers. In Bolivia, Agua para Todos (Water for All), a UN/World Conservation Union–funded agency, has helped local people to create their own sustainable water distribution system, which has reduced the cost of water and its wastage. Elsewhere, the text describes two locally designed projects that are sensitive to environmental concerns: one fostering urban quality of life in Curitiba, Brazil (see Box 3.2 on page 132), and one reviving ancient cultivation systems in the Peruvian Andes (see Box 3.6 on page 173). Across the region, grassroots environmental movements, often assisted by international nongovernmental organizations (NGOs), are active from the neighborhood level on up to the national level.

***Ecotourism.*** One effort at sustainable economic development that has been tried in tropical zones around the world is **ecotourism.** Ideally, ecotourism offers both natural and cultural travel experiences in unfamiliar environments—experiences that sensitize both travelers and hosts to the complexity of environmental issues. It encourages sustainable use and conservation of resources while providing a livelihood to local people and the broader host community. Ecotourism is now the most rapidly growing segment of the global tourism and travel industry, which itself is the world's largest industry, with $3.5 trillion spent annually. We explore some of the issues that are raised by ecotourism efforts in Box 3.5.

## BOX 3.5 | AT THE LOCAL SCALE   Ecotourism in the Ecuadorian Amazon

In a world where human efforts to prosper, or merely survive, continually threaten natural environments, one new strategy for both using and respecting nature is ecotourism (Figure 3.34). Unfortunately, ecotourism doesn't always live up to its ideals, and those projects that do often fail financially.

**Personal Vignette** Puerto Misahualli is a small river boom-town that is currently enjoying significant economic growth due to the many European, North American, and other foreign travelers who come for experiences that will bring them closer to the now legendary rain forests of the Amazon. The array of ecotourism offerings can be perplexing. One indigenous man offers to be a visitor's guide for as long as desired, traveling by boat and on foot, camping out in "untouched forest teeming with wildlife." His guarantee that they will eat monkeys and birds does not seem to promise the non-intrusive, sustainable experience the visitor might be seeking. At a well-known "eco-lodge," visitors are offered a plush room with a river view, a chlorinated swimming pool, and a fancy restaurant serving "international cuisine," all on a private 740-acre (300-hectare) nature reserve separated from the surrounding community by a wall topped with broken glass. It seems more like a fortified resort than an eco-lodge.

By contrast, the solar-powered Yachana Lodge (Figure 3.34) has simple rooms, local cuisine, and a knowledgeable resident naturalist who is a veteran of many campaigns to preserve Ecuador's wilderness. Profits from the lodge fund a local clinic and various programs that teach sustainable agricultural methods that protect the fragile Amazon soils while increasing farmers' earnings from surplus produce. The "local" cultivators are actually poor migrants from Ecuador's densely populated cool highlands who have been given free land in the Amazon by the government—but no training in how to farm in the lowland rain forest. Many of their attempts at cultivation have resulted in extensive soil erosion and deforestation. The nonprofit group running the Yachana Lodge—the Foundation for Integrated Education and Development—is earning just barely enough to sustain the clinic and agricultural programs. [*Source: Alex Pulsipher's field notes in Ecuador.*] ∎

This story touches on two themes related to environmental issues in Middle and South America: the tensions between economic development and environmental preservation, and the ways in which poverty is linked to environmental degradation. Efforts labeled "ecotourism" can be little different from other kinds of tourism that damage the environment and give little back to the surrounding community. There is the potential to use the profits of ecotourism to benefit local communities and the environment, but the margins of profit may be small.

**FIGURE 3.34** A typical ecotourism lodge in Ecuador.
[Gary Irvine/Alamy.]

## MEASURES OF HUMAN WELL-BEING

The wide disparities in wealth that are a feature of life in Middle and South America have been discussed above. Too often, development efforts have only increased the gap between rich and poor. The vignette below reminds us of the situations that can create such disparities.

**Personal Vignette** In the 1970s, Cancún was a sparsely populated coastal wetland on the Yucatán coast of Mexico. Today, it is a tourist town of 400,000, where thousands of American college students enjoy spring break party time. Local people once fished and swam along the beach, but now a hundred hotels tower over the blue water, and the beach is reserved for the tourists. Although the new town of Cancún has tree-lined streets with modest homes, low-wage tourism workers live in stark poverty on the edge of town.

Miguel Elbaje lives in Zone 288, a 15-year-old settlement of ramshackle homes on dirt tracks. A construction worker who has helped to build several hotels, Elbaje has a simple hut of planks and branches and gets water and electricity only a few hours of the day. He feels cheated by the tourism economy. His neighbor, Elsie Carerra, disagrees. She thinks tourism has made their lives better. Her husband is a waiter in one of

**TABLE 3.3** Human well-being rankings of countries in Middle and South America and other selected countries[a]

| Country (I) | GDP per capita, adjusted for PPP[b] (GDP ranking among 177 countries), 2005 (II) | Human Development Index (HDI) ranking among 177 countries,[c] 2005 (III) | Gender Empowerment Measure (GEM) ranking among 80 countries, 2005 (IV) | Female literacy (percent), 2003 (V) | Male literacy (percent), 2003 (VI) |
|---|---|---|---|---|---|
| **Selected countries for comparison** | | | | | |
| Japan | 27,967 (13) | 11 (high) | 43 | 99 | 99 |
| Kuwait | 18,047 (33) | 44 (high) | ND | 81 | 85 |
| United States | 37,562 (4) | 10 (high) | 12 | 99 | 99 |
| **Caribbean** | | | | | |
| Antigua and Barbuda | 10,294 (53) | 60 (medium) | ND | 88[d] | 90[d] |
| Bahamas | 17,159 (37) | 50 (high) | 17 | 96 | 95 |
| Barbados | 15,720 (39) | 30 (high) | 25 | 99.7 | 99.7 |
| Cuba | 3,300[d] (112) | 52 (high) | 25 | 97 | 97 |
| Dominica | 5,448 (89) | 70 (medium) | ND | 94[d] | 94[d] |
| Dominican Republic | 6,823 (73) | 95 (medium) | 45 | 87 | 88 |
| Grenada | 7,959 (63) | 66 (medium) | ND | 98[d] | 98[d] |
| Guadeloupe | 7,900[d] (64) | ND | ND | 90 | 90 |
| Haiti | 1,742 (139) | 153 (low) | 71 | 50 | 54 |
| Jamaica | 4,104 (99) | 98 (medium) | ND | 91 | 84 |
| Martinique | 14,400[d] (41) | ND | ND | 98 | 97 |
| Netherlands Antilles | 11,400[d] (47) | ND | ND | 97 | 97 |
| Puerto Rico | 18,500[d] (32) | ND | ND | 94 | 94 |
| St. Kitts and Nevis | 12,404 (45) | 49 (high) | ND | 99 | 99 |
| St. Lucia | 5,709 (87) | 76 (medium) | ND | 91 | 90 |
| St. Vincent and the Grenadines | 6,123 (80) | 87 (medium) | ND | 96 | 96 |
| Trinidad and Tobago | 10,766 (51) | 57 (high) | 23 | 98 | 99 |

[a]Rankings are in descending order; that is, low numbers indicate high rank.

[b]PPP = purchasing power parity, figured in 2003 U.S. dollars.

[c]The high, medium, and low designations indicate where the country ranks among the 177 countries classified into three categories by the United Nations.

[d]Data from *CIA World Factbook*, 2005; dates of CIA estimates vary.

the restaurants, and even though in the high season he makes just U.S. $500.00 a month, their oldest son is attending college. Another neighbor, fisherman Luis Fuentes, resents being displaced by tourism from his favorite fishing sites. With his résumé in hand, he is setting out to apply for a security guard job at one of the hotels.

*Source: Adapted from Gerry Hadden, "Cancún prospers, but most Mexicans don't share the wealth," "Morning Edition," National Public Radio, September 15, 2003.* ■

This vignette helps us to understand how development can create prosperity for some and at the same time leave others impoverished and detached from their former environments. It is important to keep in mind that national-level statistics mask the marginalized circumstances of people like those in Zone 288. Table 3.3 gives the human well-being statistics for Middle and South America. You can review the details of the different measures used in Chapter 1 (pages 34–35).

## TABLE 3.3 *(Continued)*

| Country (I) | GDP per capita, adjusted for PPP[b] (GDP ranking among 177 countries), 2005 (II) | Human Development Index (HDI) ranking among 177 countries,[c] 2005 (III) | Gender Empowerment Measure (GEM) ranking among 80 countries, 2005 (IV) | Female literacy (percent), 2003 (V) | Male literacy (percent), 2003 (VI) |
|---|---|---|---|---|---|
| **Mexico and Central America** | | | | | |
| Mexico | 9,168 (60) | 53 (high) | 38 | 89 | 92 |
| Belize | 6,950 (71) | 91 (medium) | 57 | 77 | 77 |
| Costa Rica | 9,606 (57) | 47 (high) | 19 | 96 | 96 |
| El Salvador | 4,781 (95) | 104 (medium) | 62 | 77 | 82 |
| Guatemala | 4,148 (102) | 117 (medium) | 35[d] | 63 | 75 |
| Honduras | 2,665 (115) | 116 (medium) | 74 | 80 | 80 |
| Nicaragua | 3,262 (112) | 112 (medium) | ND | 77 | 77 |
| Panama | 6,854 (72) | 56 (high) | 40 | 91 | 93 |
| **South America** | | | | | |
| Argentina | 12,106 (46) | 34 (high) | 20 | 97 | 97 |
| Bolivia | 2,587 (118) | 113 (medium) | 47 | 80 | 93 |
| Brazil | 7,790 (64) | 63 (medium) | 68 | 87 | 88 |
| Chile | 10,274 (54) | 37 (high) | 61 | 96 | 96 |
| Colombia | 6,702 (76) | 69 (medium) | 52 | 95 | 94 |
| Ecuador | 3,641 (108) | 82 (medium) | 55 | 90 | 92 |
| French Guiana | 8,300[d] (61) | ND | ND | 82[d] | 84[d] |
| Guyana | 4,230 (101) | 107 (medium) | ND | 98 | 99 |
| Paraguay | 4,684 (97) | 88 (medium) | 65 | 90 | 93 |
| Peru | 5,260 (90) | 79 (medium) | 48 | 82 | 94 |
| Suriname | 6,552[d] (77) | 86 (medium) | ND | 84 | 92 |
| Uruguay | 8,280 (62) | 46 (high) | 50 | 98 | 97 |
| Venezuela | 4,919 (94) | 75 (medium) | 64 | 93 | 93 |

ND = No data available.

*Sources: United Nations Human Development Report 2005* (New York: United Nations Development Programme), Table 1, Table 25, and Table 26, at http://hdr.undp.org/reports/global/2005/pdf/HDR05_complete.pdf; *CIA World Factbook*, 2005, at https://www.cia.gov/cia/publications/factbook/index.html.

As you can see in column II of Table 3.3, the countries in Middle and South America lag far behind Japan, the United States, and Kuwait in GDP per capita (adjusted for purchasing power parity—PPP—in U.S. dollars). Even these adjusted figures can be misleading because they are averages and hence do not show the very wide disparity in wealth in Middle and South America, but they do indicate that purchasing power is limited throughout the region. Nonetheless, purchasing power is not

the best measure of well-being, as the following comparisons demonstrate.

In column III of Table 3.3, you can see that some countries in Middle and South America rank close to Kuwait (44) on the United Nations Human Development Index (HDI) despite having much lower GDP per capita figures, such as Barbados (30), Cuba (52), Costa Rica (47), Chile (37), and Uruguay (46). These countries' rankings are close to or higher than Kuwait's partly because education is more available to both sexes and across classes in Middle and South America than in Southwest Asia, where women are secluded. Nonetheless, investment in basic and secondary education in Middle and South America is not sufficient to prepare most people for skilled jobs, and poor health care is holding down life expectancy to age 72 for the region as a whole. The HDI rankings also indicate that poverty is especially deep in Haiti (153), Guatemala (117), Honduras (116), Nicaragua (112), and Bolivia (113). In these countries, the burden of a low income is not eased by government social programs, as it is in most Caribbean countries.

The United Nations Gender Empowerment Measure (GEM) (Table 3.3, column IV) measures the extent to which females have opportunities to participate in economic and political life and takes female per capita GDP into account. Available figures show that five countries in the region rank relatively high on this ranking of 80 countries: the Bahamas (17), Barbados (25), Trinidad and Tobago (23), Costa Rica (19), and Argentina (20). All five of these countries rank higher than Japan (43). Moreover, their GEM ranks are higher than their HDI ranks, which can mean that despite economic problems, the society is comparatively open to female participation in education, jobs outside the home, entrepreneurship, leadership in business, and government. In the Caribbean, women have long been powerful family members while working outside the home and leading community organizations. They are increasingly serving in government and usually have higher educational attainments than men. Many other countries in Middle and South America, however, rank relatively low on the GEM, such as Brazil (68), Chile (61), Paraguay (65), and Venezuela (64).

Returning to the case of the people in Zone 288 in Cancún, we can now see that the income of the Carerra family of four (at most $6000 a year, $1500 per person) is markedly lower than the average per capita income for Mexico ($9168). Yet their circumstances are not unusual; wealth disparity in Mexico is wide, with the richest 20 percent receiving 19 times the income of the poorest 20 percent. The lack of water and electricity service and the building of tourism facilities that marginalize local people illustrate the side effects of free market and structural adjustment policies. The hotel developers were not asked to consider the social costs to local people.

## A Rising Threat to Human Well-Being

The global epidemic of HIV-AIDS is now taking a serious toll throughout Middle and South America and can be expected to affect the region's well-being statistics negatively. The UN reported in 2005 that more than 1.8 million people in the region were HIV-positive; of these at least 60,000 were children. The Caribbean, where transmission is primarily through heterosexual contact, has been widely reported as having one of the highest HIV rates on earth, but this figure is misleading because Haiti accounts for about two-thirds of the infections and deaths. In the rest of the Caribbean, aggressive public education programs, high education rates, and the relatively high status of women are resulting in the quick adoption of safe-sex measures, and the numbers of new cases are decreasing.

The Venezuelan journalist Silvana Paternostro, in her book *In the Land of God and Man: Confronting Our Sexual Culture* (1998), has written that cultural practices in mainland Middle and South America probably contribute to the rate of infection there. Cultural mores discouraging the discussion of sex, as well as the low status of women in much of the region, militate against safe-sex practices. Men are customarily not expected to be monogamous, and there are many sex workers, male and female. The participation of husbands as customers in the sex trade means that married women are vulnerable to being infected and to infecting any children they bear. Condom use is only beginning to be openly accepted, and a wife would be loath to ask her husband to use a condom. Drug use is also a significant factor in the spread of HIV.

Brazil is now producing antiretroviral drugs at a fraction of the cost of buying them from multinational pharmaceutical corporations, and drug therapy is also available in the richer countries of Argentina, Chile, Mexico, Uruguay, and Venezuela, as well as in some of the Caribbean countries. In the poorer countries, especially in Middle America, the drugs necessary for treating HIV-AIDS are much less available.

# III SUBREGIONS OF MIDDLE AND SOUTH AMERICA

This tour of the subregions of Middle and South America focuses primarily on the themes of cultural diversity, economic disparity, and environmental deterioration. These themes are reflected somewhat differently from place to place. For every subregion, examples of connections to the global economy are given.

## THE CARIBBEAN

Much of the Caribbean has a strong record of fostering human well-being, but visitors are often unaware of this and tend to be struck by the failure of the ramshackle lived-in landscapes of the islands to match the glamorous expectations inspired by the tourist brochures and television ads. Because Caribbean tourists are usually (and unnecessarily) isolated in hotel enclaves or on cruise ships, glimpsing only tiny swatches of island settlements through green foliage, they rarely learn that the humble houses, garden plots, and quaint, rutted, narrow streets mask social and economic conditions that are supportive of a modest but healthy and productive way of life.

Over the past half-century, most of the island societies have emerged from colonial status to become independent, self-governing states (Figure 3.35). These islands, with the exception of Haiti and parts of the Dominican Republic, are no longer the poverty-stricken places they were 30 years ago. Rather, they are managing to provide a reasonably high quality of life for their people. Children go to school, and literacy rates for all but the elderly average close to 95 percent. There is basic health care: mothers receive prenatal care, nearly all babies are born in hospitals, infant mortality rates are low, and diseases of aging are competently treated. Life expectancy is in the 70s, and people are choosing to have fewer children; the overall rate of population increase for the Caribbean is the lowest in Middle and South America. A number of Caribbean islands rank high on the Human Development Index, and some do particularly well on the Gender Empowerment Measure (see Table 3.3). Returned emigrants often say that the quality of life on their home Caribbean islands actually exceeds that of far more materially endowed societies because life is enhanced by strong community and family support. And, of course, there is the healthful and beautiful natural environment.

Beyond the beaches and quaint villages, there are local civic and social organizations such as the Rotary and Lions clubs, libraries, gourmet cooking clubs, garden societies, and active churches. These organizations practice participatory democracy daily: citizens meet to educate one another about social and environmental issues, and they continually design and implement solutions to local problems. The progress indicated by the demographic statistics is the result of hard work and civic responsibility, as well as aid from the old colonial powers and from Canada (and, in Puerto Rico, from the United States).

Island governments continually search for ways to turn former plantation economies, once managed from Europe and North America, into more self-directed, self-sufficient, and flexible entities that can adapt quickly to the perpetually changing markets of the global economy. On the one hand, they are tempted to specialize; on the other, they are wary of the dependence and vulnerability that too much specialization can bring. When plantation cultivation of sugar, cotton, and copra (dried coconut meat) died out in the 1950s and 1960s, some islands turned to producing "dessert" crops, such as bananas and coffee, which they sold for high prices under special agreements with the countries that once held them as colonies.

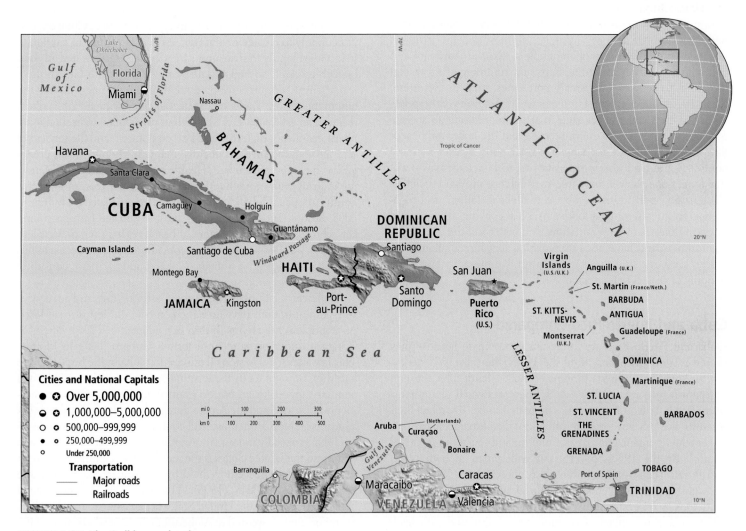

**FIGURE 3.35** The Caribbean subregion.

Now these protections are disappearing as the European Union makes such agreements illegal in Europe. Other island countries turned to the processing of their special resources (petroleum in Trinidad and Tobago, bauxite in Jamaica), the assembly and finishing of such high-tech products as computer chips and pharmaceuticals (in St. Kitts and Nevis), or the processing of computerized data (in Barbados). Most islands combine one or more of these strategies with tourism development.

For some islands, tourism and related activities contribute as much as 80 percent of the gross national product, as was the case in Antigua and Barbuda, Barbados, the Bahamas, and St. Martin in the 1990s. In most years, Jamaica (population 2.7 million) receives visitors numbering more than its population; Antigua and Barbuda (population 66,000) hosts more than 4 times its population, and St. Martin (population 20,000), 20 times its population. Heavy borrowing to build hotels, airports, water systems, and shopping centers for tourists has left some islands with huge debts to pay off. Furthermore, there is a special stress that comes from dealing perpetually with hordes of strangers, especially when they act in ways that violate local mores.

Host countries consider tourism difficult to regulate, in part because it is a complex multinational industry controlled by external interests such as travel agencies, airlines, and the investors needed to fund the construction of hotels, resorts, restaurants, and golf courses. Caribbean cruise ships now routinely bring several thousand passengers into small port cities such as Soufrière, St. Lucia, or Georgetown, Grenada. In 2006, Disney Enterprises launched even larger cruise ships that are troublesome to accommodate because of their enormous size and the large numbers of day visitors (3500) they discharge into small ports. To exercise some control over the industry, some islands are seeking tourists who will stay longer and show a more informed interest in island societies. Ecotourism, sport tourism (such as small-boat sailing), and high-fee special-interest seminar tourism (such as cooking, art, yoga, and physical fitness) are three variations that better suit the Caribbean environment and have been promoted as low-density alternatives to the "sand, sea, and sun" mass tourism that presently brings the majority of visitors to the region.

## Cuba and Puerto Rico Compared

Cuba and Puerto Rico are interesting to compare because they shared a common history until the 1950s, then diverged starkly. North American interests dominated both islands after the end of Spanish rule around 1900. U.S. investors owned plantations, resort hotels, and other businesses on both islands. To protect their interests, the U.S. government influenced dictatorial local regimes to keep labor organization at a minimum and social unrest under control. By the 1950s, poverty was widespread in both Cuba and Puerto Rico. Infant mortality was above 50 per 1000 births, and most people worked as agricultural laborers. Then each island took a different course toward social transformation. In 1959, Cuba experienced a Communist revolution under the leadership of Fidel Castro; meanwhile, Puerto Rico experienced a more gradual capitalist metamorphosis into a high-tech manufacturing center.

*Cuba.* Since Castro seized control, Cuba has dramatically improved the well-being of its general population in all physical categories of measurement (life expectancy, literacy, and infant mortality). By 1990, it had a solid human well-being record despite its persistently rather low GDP per capita (see Table 3.3). Cuba thus proved that, once the requisite investment in human capital is made, poverty and social problems are not as intractable as some people had thought. Unfortunately, Cuba's successes are not replicable elsewhere in the region, first, because it has been politically repressive in the extreme, forcing out the aristocracy and jailing and executing dissidents. Second, Cuba has been inordinately dependent on outside help to achieve its social revolution. Until 1991, the Soviet Union was Cuba's chief sponsor. It provided cheap fuel and generous foreign aid and bought Cuba's main export crop, sugar, at artificially high prices.

With the demise of the Soviet Union, Cuba's economy declined sharply during what became known as the Special Period, when all Cubans were asked to make sacrifices and cut back their consumer spending. To survive this economic crisis, Cuba opened its economy to outside investment, especially European capital, to help redevelop its tourism industry. Canada has invested in joint biotech research on hepatitis and cancer drugs with Cuban medical scientists, who receive particularly good training—among the best in the region. Trade with Venezuela is growing; in 2005, Cuba received an estimated U.S.$1.1 billion worth of oil in exchange for pharmaceuticals, high-tech medical equipment, and the services of several thousand medical doctors, many of whom serve in clinics for low-income Venezuelans. Health and spa tourism is also a growing industry with more than 5000 customers per year, mostly from South America. Cuban tourism grossed an estimated $2 billion in 2005, hosting 2.3 million visitors, mostly from the European market, who spend a week or more at beach resorts such as Varadero (Figure 3.36). Despite growing income from tourism, access to this income is restricted for ordinary Cubans, and for them, consumer goods remain in short supply.

Political relations between the Castro regime and successive U.S. administrations have remained on cold war terms since 1961. An example is the Helms-Burton Act, adopted in 1996 by the U.S. Congress under pressure from former Cubans living in Florida. This law attempted to stop all trade with Cuba in the hope that the hardships brought on by the ensuing economic crisis would lead to popular support for an end to Castro's leadership. Other governments in Europe and the Americas regard the Helms-Burton Act as a breach of international law and continue to trade with and invest in Cuba, especially in tourism. Another contentious issue between Cuba and the United States is the naval base at Guantánamo in far southern Cuba, on territory held by the United States since the Spanish-American War (1898). This base gained international fame in 2002, when the United States began using it to indefinitely detain prisoners captured in Afghanistan and elsewhere around the world suspected of vaguely defined links to terrorism. The Bush

**FIGURE 3.36** Tourism in Cuba. Varadero, one of Cuba's new tourism areas, is on the island's north coast, about 100 miles (160 kilometers) east of Havana. [David Alan Harvey/Magnum Photos.]

administration contended that the prisoners held there incommunicado (435 remained as of November 2006) were not protected by the Geneva Accords on prisoners of war.

***Puerto Rico.*** In the 1950s, Puerto Rico began Operation Bootstrap, a government-aided program to transform the island's economy from its traditional sugar plantation base to modern industrialism. Many international industries took advantage of generous tax-free guarantees and subsidies to locate plants on the island. Since 1965, this industrial sector has shifted from light to heavy manufacturing, from assembly plants to petroleum processing and pharmaceutical manufacturing. The Puerto Rican seaboard is heavily polluted, probably as a consequence of the chemicals released by these more recent industries.

Because Puerto Rico is a commonwealth within the United States and its people are U.S. citizens, many Puerto Ricans migrate to work in the United States. Their remittances, as well as the manufacturing jobs in Puerto Rico, have greatly improved living conditions on the island, but social investment by the U.S. government has also upgraded the standard of living. Many Puerto Ricans receive some sort of support from the federal government, including retirement benefits and health care. The infant mortality rate in 2005 was 9.8 (Cuba's was just 5.8), and life expectancy is 77 on both islands. The connection with the United States helps Puerto Rico's tourism and light industry economy, but outside of San Juan, with its skyscraper tourist hotels, Puerto Rico's landscape reflects stagnation: few interesting or well-paying jobs, an inadequate transport system, little development at the community level, mediocre schools, and few opportunities for advanced training.

Statehood for Puerto Rico, which some people desire, would mean the end of tax holidays for U.S. companies based there, and hence the end of Puerto Rico's ability to attract assembly plants, some of which are already moving to cheaper labor markets in Asia. In addition, many fear that statehood would eventually bring cultural assimilation and the loss of the island's Spanish linguistic and cultural heritage.

## Haiti and Barbados Compared

Haiti and Barbados present another study in contrasts (Figure 3.37). During the colonial era, both were European possessions with plantation economies, Haiti a colony of France and Barbados a colony of Britain. Yet they have had very different experiences, and today they are far apart in economic and social well-being. Haiti, though not without useful resources, is the poorest nation in the Americas, with a ranking on the United Nations Human Development Index of 153 out of 177. Barbados has one of the highest HDI rankings in the entire region (30), even though that island has much less space, a higher population density, and few resources other than its limestone soil, its beaches, and its people.

***Haiti.*** Haiti had the richest plantation economy in the Caribbean by the end of the eighteenth century. When Haitian slaves revolted against the brutality of the French planters in 1804, Haiti became the first colony in Middle and South America to achieve independence. Haiti's early promise was lost, however, when the former-slave reformist leaders were overthrown by other former slaves who were violent and corrupt militarists. They neither reformed the exploitative plantation economy nor sought a new economic base. Under a long series of incompetent authoritarian governments, the people sank into abject poverty, while the land was badly damaged by particularly wasteful and unprofitable plantation cultivation. In

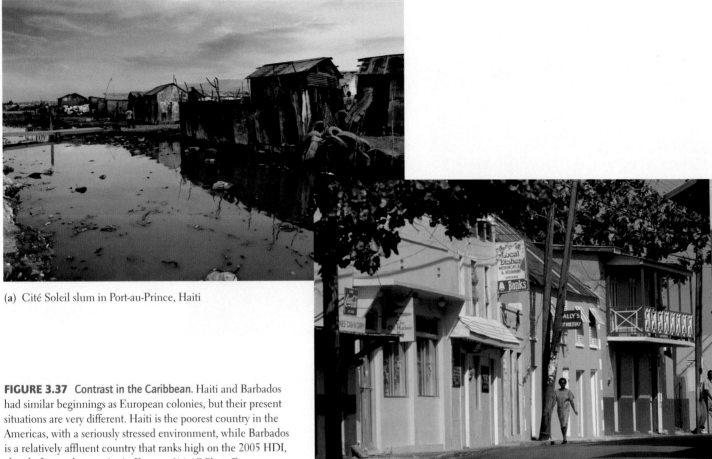

(a) Cité Soleil slum in Port-au-Prince, Haiti

**FIGURE 3.37 Contrast in the Caribbean.** Haiti and Barbados had similar beginnings as European colonies, but their present situations are very different. Haiti is the poorest country in the Americas, with a seriously stressed environment, while Barbados is a relatively affluent country that ranks high on the 2005 HDI, ahead of several countries in Europe. [(a) AP Photo/Brennan Linsley. (b) Bob Sacha/CORBIS.]

(b) Speightstown, Barbados

the mid-twentieth century, a class-based reign of terror under François "Papa Doc" Duvalier (1957–1986) pitted the mulatto elite against the black poor.

Today, Haiti remains overwhelmingly rural, with widespread illiteracy and an infant mortality rate of 80 per 1000 births (Barbados's rate is 13). Haiti's lands are deforested and eroded and are subject to disastrous flooding. Efforts to establish democracy have repeatedly devolved into violence. Since the early 1990s, the United Nations has maintained peacekeeping troops in Haiti, and several humanitarian aid organizations run programs there. Multinational corporations have opened maquiladora-like assembly plants, employing primarily young women, but the plants are not yet flourishing. Although minerals such as bauxite, copper, and tin exist in Haiti, cost-effective development of these resources is not possible.

***Barbados.*** Far more prosperous but tiny Barbados has fewer natural resources and is more than twice as crowded as Haiti. With 166 square miles (430 square kilometers), Barbados has 1500 people per square mile (618 per square kilometer); by contrast, Haiti, with 10,700 square miles (27,800 square kilometers) has just 700 people per square mile (268 per square kilometer). Although both Haiti and Barbados began the twentieth century with large, illiterate,

agricultural populations, their development paths have diverged sharply. In 2005, two-thirds of Haitians were still agricultural workers, and only half were able to read and write. Barbados, on the other hand, now has a diversified economy that includes tourism, sugar production, remittances from migrants, information processing, offshore financial services, and modern industries that sell products throughout the Caribbean. Barbados's present prosperity is explained by the fact that its citizens successfully pressured the British government to invest in the people and infrastructure of its colony before giving it independence in 1966. Barbadians hold jobs that demand skilled, literate employees; they are well educated and well fed, and most are homeowners. Meanwhile, the Barbadian government and private businesspeople constantly seek new employment options for the citizens and occupy a central role in Caribbean economic and social development.

## MEXICO

Mexico today (Figure 3.38) is working toward becoming a reasonably well-managed, middle-income democracy (see the discussion on page 145). Most of the efforts to achieve this goal, both private and governmental, are focused in some way on Mexico's relation-

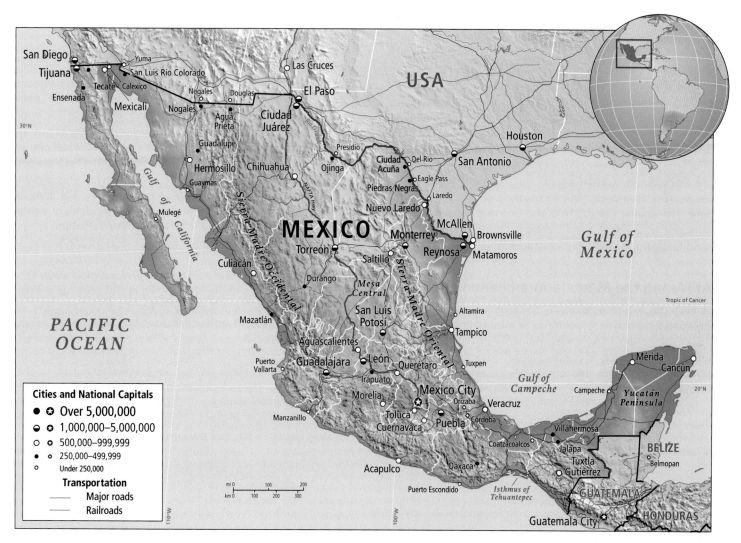

**FIGURE 3.38** The Mexico subregion.

ship with the United States, its main trade partner. The integration of the Mexican, U.S., and Canadian economies has proceeded rapidly since NAFTA became official in the mid-1990s. Two of the most obvious geographic markers of this integration are interconnecting highway systems in the three countries and, in Mexico, the phenomenon of the maquiladoras.

Mexico's formal economy has modernized rapidly in recent decades. Its main components are mechanized, export-oriented agriculture, a vastly expanded manufacturing sector, petroleum extraction and refining, tourism, a service and information sector, and the remittances of millions of migrants working abroad. Mexico's large and varied informal economy employs an estimated 16 million people at least part time—about half of the total working population of 33 million. These workers contribute at least 13 percent of the GDP through such activities as subsistence agriculture, fine craft making, and many types of services, ranging from street vending and tutoring to trash recycling. The informal economy also includes any remittances smuggled home, a flourishing black market, and the crime sector.

## The Geographic Distribution of Economic Sectors

Mexico's economic sectors have a distinct geographic distribution pattern. Just 12 percent of Mexico is good agricultural land, and much of that is on hillsides. The largest agricultural units are corporate farms, which, with the aid of government marketing, research, and irrigation programs, have largely replaced the old, inefficient haciendas and ranches. These farms produce high-quality meat and produce such as asparagus, peppers, winter vegetables, and raspberries, as well as sorghum for animal feed. Large corporate farms located in places such as the state of Sonora, in the arid north close to the U.S. border, send produce to the North American market. In the tropical coastal lowlands farther south, subsistence cultivation on small plots is much more common, but here, too, tropical fruits and products such as jute, sisal, and tequila are produced for export as well as for internal consumption. Mexico's convoluted upland landscapes provide many small and specialized niches where a wide variety of crops are grown, mostly on small farms, including coffee, corn, tomatoes, sesame seeds, hot peppers, and flowers. About 18 percent of Mexico's population is employed in agriculture, though agriculture produces only 4 percent of Mexico's GDP.

The petroleum refining industry is located along the Gulf coastal plain, and tourism facilities are found primarily along the Caribbean and Pacific coastal plains. The service sector employs 58 percent of the population and produces 70 percent of the GDP. Although it includes restaurants, hotels, financial institutions, consulting firms, and museums, this sector is dominated by small operations that cater to everyday needs: shoe repair services, Internet cafés, cleaning businesses. Most such service jobs are concentrated in cities throughout the country. The informal economy, which can include agricultural, manufacturing, and service activities, is everywhere, but it is most varied and noticeable in the big cities.

## NAFTA and the Maquiladora Phenomenon

Manufacturing, which employs about 24 percent of Mexico's registered workforce and accounts for 27 percent of the GDP, is concentrated along the U.S. border, primarily in maquiladoras, in cities such as Tijuana, Mexicali, Ciudad Juárez, and Reynosa. Maquiladoras are plants that hire people to assemble manufactured goods, such as clothing or electronic devices, usually from duty-free imported parts. The finished goods are returned to the country of origin—90 percent to the United States—for sale there or elsewhere. There are thousands of maquiladoras along the Mexican side of the U.S.–Mexico border as well as elsewhere in the country. Many are branches of American, European, or Asian (especially Japanese) corporations, though Mexicans hold the majority interest in about half of them.

The Mexican side of the border is a desirable location for the maquiladoras because it provides an inexpensive labor pool within easy reach of the United States. In 2006, Mexican factory workers earned an average of about $9 a day ($1.10 an hour). Furthermore, Mexico has far fewer regulations covering worker safety, fringe benefits, and environmental protection than the United States, and it charges much lower taxes on industries than the United States. These conditions attracted a considerable number of maquiladoras

even before NAFTA took effect in 1994, and thereafter, the number of maquiladoras increased dramatically. As the graph in Figure 3.39 shows, total maquiladora employment grew until 2000, but then wavered for several years as multinational firms found even less expensive labor in Asia (see the personal vignette on the next page). Recent Mexican business reforms are aimed at attracting new firms.

Studies in the late 1990s showed that more than half of the maquiladora laborers are women, many of them young and unmarried (Figure 3.40). Often living for the first time without the protection of their families, these young women have been subjected to high rates of criminal assault and even murder. The film *La Señorita Extraviada* (2002) documents the failure of Mexican officials to investigate a series of such killings in Ciudad Juárez and chronicles the efforts of women on both sides of the border to demand an accounting.

Women often can find work in maquiladoras more easily than men because they are more compliant and will work for less pay. But men, displaced from agriculture and local factory work, are increasingly employed in maquiladoras, filling traditionally female jobs such as sewing or assembling electronic devices. Maquiladora laborers, male and female, migrate from rural locations across Mexico, and the remittances they send home are crucial to their families' welfare. Because of their financial contributions, female workers enjoy status in the family that earlier generations of women did not have. Recent studies have shown that the majority do not use maquiladora work as a precursor to migration to the United States, as has often been claimed; if these migrants fail to find maquiladora work, they tend to return to their home villages.

Although many Mexicans are pleased to have the maquiladoras and the jobs they bring, the country is increasingly facing negative side effects. The border area has developed serious groundwater and air pollution problems because of the high concentration of poorly regulated factories and the unplanned worker shantytowns that have sprung up around them. Social problems are also developing among those crowded into unsanitary living conditions, and family violence is high. The schools in the border cities are inadequate for the increasing numbers of children. In Ciudad Juárez in 2001, for example, there were just ten day-care

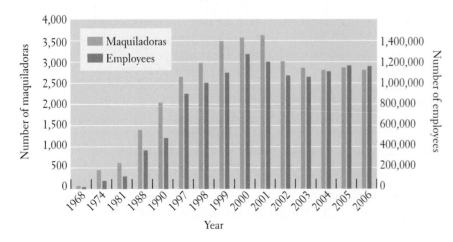

**FIGURE 3.39** Growth of the maquiladora sector. Employment in maquiladoras has shrunk some since a high in 2000, due in part to increased mechanization, as well as the movement of some factories to Asia, especially to China and Southeast Asia.

**FIGURE 3.40** Maquiladora workers. In March 2006, these young Mexican women were assembling car radios for Delphi Delco Electronics de Mexico in Matamoros, Mexico. The plant, a maquiladora across the border from Brownsville, Texas, is one of seven Delphi plants in the vicinity employing a total of 11,000 Mexican workers. [Bob Daemmrich/ CORBIS.]

centers serving 180,000 women workers. A further problem has to do with fragmented government authority in the long border zone where maquiladoras dominate employment. There are more than 13 paired U.S.–Mexican cities along the border, and the multiplicity of governments makes it difficult to address common problems. Because NAFTA purports to be a trade agreement among equals, cooperation along the border to improve the lives of maquiladora workers is now gaining emphasis. The U.S.–Mexico Border XXI Project has joint task forces focusing on infant immunization, water and air pollution mitigation, hazardous waste cleanup and prevention, environmental studies, and emergency preparedness. In the next few years, Mexico hopes to provide 93 percent of the border population with clean drinking water, and 16 water-related projects are currently under way along the border, some in Mexico, some in the United States.

**Personal Vignette**  Orbalin Hernandez has just returned to his self-built shelter in Mexicali. He smiles as he enters the yard, and his wife breathes a sigh of relief. Recently fired for taking off his safety goggles while loading TV screens onto trucks at Thompson Electronics, he went to the personnel office to ask for his job back. Because there were no previous problems with him, he was rehired at his old salary of $300 a month ($1.88 an hour). Thompson, a French-owned electronics firm, took advantage of NAFTA and moved in 2001 to Mexicali from Scranton, Pennsylvania, where its 1100 workers had been paid an average of $20.00 an hour. Now 20 percent of the Scranton workers are unemployed, and many feel bitter toward the Mexicali workers.

Orbalin's salary at Thompson-Mexicali is barely enough for him to support his wife, Mariestelle, and four kids. They are from a farming community in Tabasco State, where, for people with no high school education, wages averaged $60 a month, so the Hernandez family is grateful for the job. Thompson is known for treating its Mexican workers well. It pays into a housing fund, supports local educational facilities, and provides bus transport for workers to and from home. But a recent study at the Autonomous University in Mexico City found that over the last 20 years of the maquiladora boom, Mexican wages have lost 81 percent of their buying power. A typical maquiladora worker has to work more than an hour to buy a kilo (2.2 pounds) of rice, whereas a dockworker in Los Angeles can buy that rice after just 3 minutes of work, and an undocumented minimum-wage worker in L.A. after 12 minutes of work.

Nonetheless, Orbalin and his fellow Mexicali workers are now being told that their wages are too high. Maquiladora firms claim that to compete globally they must cut costs even further. Fourteen Mexicali plants have closed and moved to Asia, where, in 2005, workers with the same skills as Orbalin earned just $0.35 an hour. Those firms remaining in Mexicali have cut wages and reduced benefits in response. But the tide may be turning: because of a growing labor shortage in China, wages there are increasing by 10 to 40 percent per year, and soon it will make little sense to move a Mexican factory to China.

*Source: NPR reports by John Idste, August 14, 2001, August 25, 2003, August 16, 2003, and by Gary Hadden, August 27, 2003; David Bacon, "Anti-China campaign hides maquiladora wage cuts," February 2, 2003, http://csf.colorado.edu/forums/labor-rap/current-discussion/ msg01050.html;April, 2006; http://www.maquilaportal.com/cgi-bin/ public/board.pl?klie=10; http://www.businessweek.com/magazine/ content/06_13/b3977049.htm.* ∎

## Migration

Migration is a strategy often used to alleviate the problems of poverty. Rapid rural-to-urban migration in Mexico is legendary: 75 percent of Mexicans (79 million people) live in cities, the majority of them in large cities. The emigration of Mexicans to North America also has an enormous effect on Mexico as a nation. On the one hand, the skills of some of the best and brightest are lost, at least temporarily; on the other, some benefits are gained from their industry.

Unlike the many Europeans who cut their ties with family and native land when they came to North America, Mexicans often undertake their migrations with the express purpose of helping out their families and communities. In 2005, an estimated $20 billion was remitted to Mexican communities by workers in the United States. This sum was less than the $105 billion in foreign exchange earned for Mexico by maquiladoras, but more than the estimated $13 billion earned by the tourism sector. Most Mexican households receiving remittances are located in relatively better-off states such as Michoacan and Zacatecas, as residents of the poorest states find it difficult to finance a migration to the United States.

A paper by geographer Dennis Conway and anthropologist Jeffrey H. Cohen reports that couples who migrate from indigenous villages in Mexico often work just long enough at menial jobs in the United States to save a substantial nest egg. Then they return home to build a house (usually a family self-help project) and buy appliances. Because a few thousand dollars saved in the United States will accomplish a great deal in rural Mexico, the family may live off their nest egg for several years while one or both members of the couple renders volunteer community service to the home village. When the money runs out, the couple, or just the husband, may migrate again to save another nest egg. Often these civic-minded migrants cycle other members of their family through a menial U.S. job, such as dishwashing, thus keeping the job open for themselves when they need it again. Their bosses and co-workers in the United States never guess that such unprepossessing minimum-wage workers are actually influential and public-spirited citizens in their home villages, who use migration and hard work to enhance community living standards and participatory democracy.

The migration of Mexicans to the United States to find work is controversial in Mexico, no less than in the United States. There is worry over the safety of such workers while crossing the border and on the job, as well as the feeling that Mexico is losing its most pro-ductive citizens. Does migration have to be the main route to advancement for Mexico's youth? Maybe not. One month after he was elected president in January 2001, Vicente Fox founded "e-Mexico," designed to bring the Internet to 10,000 communities by the end of his term in 2006 (eMexico online: http://www.e-mexico.gob.mx/). Fox thinks the Internet will be a way for Mexicans to leap out of poverty by gaining access to training and higher education. One example of its effectiveness can be seen in the tiny mountain town of Santa Ana de Allende, where there is one telephone for 1400 people. There, Oracio Covarrubias uses the Internet to study at Tec de Monterrey Virtual University. Already students in Santa Ana are opting to stay in town, use their Internet training to establish businesses, and work for local change, rather than migrating to the United States. By 2005, Mexicans with more than a high school education were much less likely to migrate than those with 9 to 12 years of schooling. This Internet initiative in Mexico resembles thousands of others around the world.

## CENTRAL AMERICA

Central America's wealth is in its soil. Although industry and services account for increasing proportions of the gross domestic product of all countries, fully one-third of the people of the seven countries of Central America (Figure 3.41) remain dependent on the production of their plantations, ranches, and small farms (Figure 3.42). The disparity in income seen in the region is due primar-

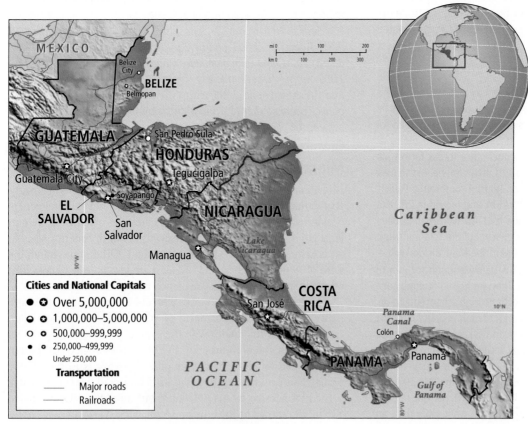

**FIGURE 3.41** The Central America subregion.

ily to the fact that most of the land is controlled by a tiny minority of wealthy individuals and companies.

Most of the Central American **isthmus** (a narrow strip of land joining two larger land areas) consists of three physical zones that are not well connected with one another: the narrow Pacific coast, the highland interior, and the long, sloping, rainwashed Caribbean coastal region. Along the Pacific coast, mestizo (**ladino** is the local term) laborers work on large plantations that grow sugar, cotton, and bananas and other tropical fruits; coffee is grown in the hills behind the coast. In the highland interior of Guatemala, Honduras, and Nicaragua, cattle ranching and commercial agriculture have recently displaced indigenous subsistence farmers. Similarly, the humid Caribbean coastal region, long sparsely populated with indigenous and African-Caribbean subsistence farmers, is increasingly dominated by commercial agriculture, forestry, tourism development, and resettlement projects for small farmers displaced from the highlands.

In Central America, the majority of people are either indigenous or ladino, and about half still live in small villages. In these villages, most people make a sparse living by cultivating their own food and cash crops on land they own or rent, as sharecroppers, or as seasonal laborers on large farms and plantations (see the story of Aguilar Busto Rosalino in the personal vignette on pages 142–143). The people of this region have experienced centuries of hardship, including long hours of labor at low wages and the loss of most of their farmlands to large landholders. In both rural and urban areas, infrastructure development has lagged. Roads are primitive and few. Most people lack clean water, sanitation, health care, protection from poisoning by agricultural chemicals, and basic education. Often they do not have access to enough arable land to meet their basic needs. The majority of the land is held in huge tracts—ranches, plantations, and haciendas—owned by a few families. As a result, on the small bits of farmland available to the poor for grow-

**FIGURE 3.42 Minifundios.** The pattern of tiny farm plots (minifundios) in El Salvador is visible in this photo. Some are the result of land redistribution efforts, but because of high population density, the minifundios are often too small to support a family. [Tomasz Tomaszewski/National Geographic Image Collection.]

ing their own food and cash crops, local densities may be 1000 people or more per square mile.

It is this control of large tracts by a few rich landholders, not overpopulation, that accounts for the poverty of the majority. Only tiny El Salvador is truly densely populated, with an average of 847 people per square mile (325 per square kilometer) (see Table 3.4, column III). In fact, arable land there is so scarce that people have slipped over the border to cultivate land in Honduras.

## TABLE 3.4 Population data on Mexico and Central America, 2005

| Country (ordered by size) (I) | Population (II) | Population density per square mile (III) | Rate of natural increase (percent) (IV) | Literacy rate (percent) (V) |
|---|---|---|---|---|
| Mexico | 107,000,000 | 142 | 1.9 | 90 |
| Guatemala | 12,700,000 | 302 | 2.8 | 69 |
| Honduras | 7,200,000 | 167 | 2.8 | 80 |
| El Salvador | 6,900,000 | 847 | 2.0 | 80 |
| Nicaragua | 5,800,000 | 115 | 2.7 | 77 |
| Costa Rica | 4,300,000 | 220 | 1.3 | 96 |
| Panama | 3,200,000 | 111 | 1.8 | 92 |
| Belize | 300,000 | 33 | 2.3 | 77 |

*Source:* Population Reference Bureau, *2005 World Population Data Sheet,* 2005, http://www.prb.org/Content/NavigationMenu/PRB/PRB_Library/Data_Sheets/Data_Sheets.htm; *United Nations Human Development Report 2005* (New York: United Nations Development Programme), Table 1, http://hdr.undp.org/reports/global/2005/pdf/HDR05_complete.pdfxs.

Costa Rica is an exception to these extreme patterns of elite monopoly, mass poverty, and rapid population growth (as is, to some extent, Panama) (see Tables 3.3 and 3.4). The huge disparities in wealth between colonists and laborers did not develop in Costa Rica, chiefly because the fairly small native population died out soon after the conquest. Without a captive labor supply, the European immigrants to Costa Rica set up small but productive family farms that they worked themselves, not unlike early North American family farms. Costa Rica's democratic traditions stretch back to the nineteenth century, and it has unusually enlightened elected officials as well as one of the region's soundest economies. Population growth is low for the region at 1.3 percent per year, and investment in schools and education has been high. As a result, Costa Rica has Central America's highest literacy rates, and on many scales of comparison, including GDP per capita and HDI, the country stands out for its high living standards (see Table 3.3). Thus Costa Rica has often been hailed as a beacon for the more troubled nations of Central America.

Panama is known primarily for the canal that joins the Atlantic Ocean and Caribbean Sea with the Pacific, shortening the sea voyage around the tip of South America by many days. But the canal is no longer large enough to accommodate the huge cargo, tanker, and cruise ships of the modern era, and it has become a bottleneck for world trade. Opened in 1914, the Panama Canal was built primarily with money from the United States and labor from the British Caribbean. The United States managed the canal and maintained a large military presence there. Panama remained a virtual colony of the United States until 1997, when the United States agreed to turn the canal over to Panama and remove itself as a dominant presence. Interestingly, the turnover of the canal came at a time when it was becoming obsolete, experiencing competition from nearby pipelines and possibly from another potential canal route through Nicaragua. In October 2006, Panamanians overwhelmingly approved a U.S.$5.2 billion project to widen the canal and make it usable by mega-ships of the future.

## Environmental Concerns

Although poor farmers may use unwise practices in wrenching enough to feed their families from the tiny plots they are allowed, and are often blamed by the news media for environmental degradation, most environmental problems in Central America are caused by large-scale agriculture and cattle ranching. Each year, hundreds of children die from allergies and other effects of agricultural chemical poisoning. Land management by the large landowners is also a problem. Clearing enough land for large-scale agriculture and cattle grazing leads to tremendous amounts of erosion. In Honduras, the reservoir for a large hydroelectric dam built only a few years ago has been nearly filled with silt eroded from surrounding cleared land. Its electricity output must now be supplemented with generators run on imported oil.

Costa Rica has been a leader in the environmental movement in Central America. In the 1980s, the country established wetland parks along the Caribbean coast and encouraged ecotourism at a number of nature preserves, while at the same time acknowledging the potentially negative environmental side effects of tourism. Costa Rica supports scientific research through several international study centers in its central highlands and lowland rain forests, where students from throughout the hemisphere study tropical environments. Elsewhere in Central America, support for national parks is growing. With the help of the U.S. National Park Service and international NGOs such as the Audubon Society, 175 small parks have been established.

## Civil Conflict

Frustrated with governments unresponsive to their plight, indigenous people and other rural people in Guatemala, Honduras, Nicaragua, and El Salvador began organizing protest movements in the 1970s. In some cases, they had the help of liberation theology advocates and Marxist revolutionaries from outside the region. Despite reasonable requests, they met with stiff resistance from wealthy elites assisted by national military forces. Destructive civil wars have plagued these countries over recent decades.

In the 1980s, protests were particularly strong in Guatemala, which has the largest indigenous population in Central America. During those dying days of the cold war, the United States backed and armed military dictatorships because it was convinced that the revolutionaries posed a Communist-inspired threat to the United States. U.S.-equipped national armies then killed many thousands of native protesters and drove 150,000 into exile in Mexico. In time, the protesters responded with guerrilla tactics. Rigoberta Menchu, an indigenous Guatemalan woman, won the 1992 Nobel Peace Prize for her efforts to stop government violence against her people. Her autobiographical account attracted public attention to the carnage and was important in awakening worldwide concern. Eventually, after a number of regional peacemakers joined Menchu in bringing international pressure to bear on the Guatemalan government, the Guatemalan Peace Accord was signed in September 1996.

Since 2002, Central American rural grassroots resistance has been redirected against the globalization of markets across the region. The proposed Free Trade Area of the Americas (FTAA), which would extend many of the provisions of NAFTA to the entire hemisphere, and the Central American Free Trade Association (CAFTA) are seen by Central American farmers as a threat to their survival. For example, Oxfam, an international NGO concerned with issues of hunger and human rights, estimates that if the United States gained the right to trade its cheaply produced surplus corn in the region, U.S. exports of corn to Central America would increase 1000 percent in just one year. Such a volume of corn would lower the market price to the point where local farmers would be squeezed out of corn production and would probably have to migrate to find work.

***A Case Study of Civil Conflict: Nicaragua.*** Until the late twentieth century, Nicaragua showed land ownership patterns characteristic of the region: a tiny elite held the usual monopoly on land, while the mass of laborers lived in poverty. By 1910, North Americans had invested in the coffee and fruit plantations on the

Nicaraguan Pacific uplands and coastal plain. Between 1912 and 1933, U.S. Marines were kept in Nicaragua to quell labor protests that threatened U.S. interests in Nicaragua's export economy. This interference helped the wealthy Somoza family to establish its members as brutal dictators in Nicaragua in the 1930s.

The Somoza regime was finally ousted by the Marxist-leaning Sandinista revolution of 1979. The Sandinistas, who eventually won several national elections, embarked on a program of land and agricultural reform and improved basic education and health services. Soon, however, a debilitating war with the "contras," right-wing counterinsurgents backed by local elites, undid most of the social progress. Again, the United States became involved as the Communist-wary Reagan administration supported the contras (something Congress disapproved) with funds from secret arms sales to Iran. A trade embargo imposed by the United States further contributed to the ruin of the Nicaraguan economy. By the end of the 1980s, Nicaragua was one of the poorest nations in the Western Hemisphere.

In national elections in 1990, the Sandinistas were defeated by an electorate weary of violence. Since 1997, several free elections, in which 75 percent of the eligible citizens voted, have resulted in moderate governments that have continued to find it difficult to bring Nicaragua any measure of prosperity. In 2006, Daniel Ortega, once a Sandinista leader and president in the late 1980s, was elected president on a moderate platform.

In 2005, the Nicaraguan government ratified the CAFTA agreement over the mass protests of farmers who feared competition with large U.S. agribusiness firms. The government says CAFTA will bring jobs and expanded investment to Nicaragua, but civil society organizations, NGOs, and labor unions note that in Mexico, free trade agreements have hurt small farmers and laborers and have contributed to environmental deterioration. In the months leading up to the October 2005 signing, Nicaraguan farmers were joined in their protests by farmers in Costa Rica, El Salvador, Guatemala, and southern Mexico.

# THE NORTHERN ANDES AND CARIBBEAN COAST

The five countries in the northernmost part of South America share a Caribbean coastline and extend into a remote interior of wide river basins and humid uplands (Figure 3.43). The Guianas resemble Caribbean countries in that they were once traditional plantation colonies worked by slave and indentured labor, and today their multicultural societies are made up of descendants of African, East Indian, Pakistani, Southeast Asian, Dutch, French, and English settlers. Venezuela and Colombia share a Spanish colonial past and are predominantly mestizo, with a small upper class of primarily European heritage and a small population of African derivation in the western and Caribbean lowlands. In all the countries of this subregion, small indigenous populations survive, mainly in the interior lowland Orinoco and Amazon basins, where they hunt, gather, and grow subsistence crops.

**FIGURE 3.43** The northern Andes and Caribbean coast subregion.

## The Guianas

To the north of Brazil lie the three small countries known collectively as the Guianas: Guyana, Suriname, and French Guiana. Guyana gained independence from Britain in 1966 and Suriname from the Netherlands in 1975. French Guiana, on the other hand, is not independent, but rather is considered part of France. Today, the common colonial heritage of these three countries is still visible in both their economies and their people. Sugar, rice, and banana plantations established by the Europeans in the coastal areas continue to be economically important, but endeavors related to logging and the mining of gold, diamonds, and bauxite in the resource-rich highlands are now the leading economic activities.

The population descends mainly from laborers who once worked the plantations. These laborers formed two major culture groups: Africans, brought in as slaves from 1620 to the early 1800s, and South and Southeast Asians, brought in as indentured servants after the abolition of slavery. The descendants of Asian indentured servants are mostly Hindus and Muslims. Many of them became small-plot rice farmers or owners of small businesses. Those of African descent are Christian and are both agricultural and urban workers. Politics in the Guianas is complicated by the social and cultural differences between citizens of Asian and African descent. These differences have also slowed economic and human development to the point that these three countries lag a bit behind several Caribbean island countries that experienced similar colonial histories, such as Trinidad and Tobago and Barbados.

## Venezuela

Venezuela has long had the potential to become one of the wealthier countries in South America, primarily because it holds large oil deposits and is an active member of the Organization of Petroleum Exporting Countries (OPEC). Oil has been the backbone of the country's economy since the mid-twentieth century. Venezuela not only is among the top suppliers of oil to the United States, but also supplies oil to Cuba, Canada, and South America and is seeking deals with China, Japan, and Europe. The U.S. invasion of Iraq in 2003 illustrates how geopolitics can link distant regions: Venezuelans feared that the invasion would result in the takeover of Iraqi oil management by the United States and Britain. This could have caused Venezuela to lose market share in the United States, and the result in Venezuela, already economically strapped, could have been civil disorder. That scenario did not play out; instead, a drawn-out war in Iraq allowed Venezuela to gain power globally as oil prices rose sharply.

Venezuela's oil resource was nationalized in the 1970s, with the idea that the profits would fund public programs to help the poor majority. Market-oriented governments relaxed this policy, however, and for a while foreign firms, such as Chevron Oil (now called ChevronTexaco), controlled 25 percent of production. Oil did not generate widespread prosperity; instead, oil profits stayed with the elite and the small middle class. By the late 1990s, 63 percent of the country's wealth was controlled by 10 percent of the population, mainly those of European descent, leaving those of mixed native and African descent at the bottom of the income pyramid. Taxes on the wealthy and on foreign investors were kept low, so oil and other assets did not generate enough government revenue to fund badly needed improvements in education, health care, transport, and communications. As a result, Venezuela failed to build a base for general economic and social advancement. Although the capital city of Caracas is bedecked with gleaming skyscrapers, modern freeways, and universities, it is surrounded by poor shantytowns that lack access to clean drinking water and sanitation and are served by substandard schools and inadequate roads. A third of the population lives in deep poverty, on about $2.00 a day per capita.

Despite periodically high oil prices in the 1980s, Venezuela accumulated a large debt through corruption and poor management of development programs such as land reform. A slump in oil prices during the early 1990s left the country still deeper in debt. Structural adjustment policies, intended to free money for debt repayment, reduced the role of government in industry and in social programs. Many government jobs disappeared, and the poor lost rent and food subsidies and access to education. Living standards fell even further, and violent riots were one result. Since the late 1990s, the populist policies of President Hugo Chavez have discouraged outsiders contemplating investment in Venezuela's industries and mechanized agriculture, thereby hindering economic expansion (see pages 145–146). Nevertheless, Chavez survived, partly because a rebound in oil prices brought much-needed capital back into the Venezuelan government coffers, making it possible to pay Venezuela's national debt and to create the promise of significantly higher income for the nation. In 2005, taxes on the profits of ChevronTexaco were increased by 50 percent retroactive to 2001, and in 2006 Venezuela converted all oil operations into state-controlled joint ventures. Global oil prices are not yet high enough for profitable extraction of oil in the Orinoco Basin, where deposits may exceed the declared oil reserves of the rest of the world combined. By reinstating many socialist ideas that predated the recent era of capitalist structural adjustment policies, Chavez is restoring a variety of social programs, and his popularity with the masses is high. He has expanded his efforts at regional leadership by advising and giving donations to a wide range of politicians and by donating funds or subsidized oil to people he defines as needy, but this approach is not always welcome, and his leadership role in the region is still ambiguous.

## Colombia

An ongoing civil war in Colombia has displaced more than 1.5 million people over the past decade, leaving Colombia with one of the highest totals of internally displaced people in the world. The current wave of violence is part of a long string of conflicts that have arisen out of inequalities in the country's social order. Although Colombia is today the world's second largest exporter of coffee and a major exporter of oil and coal, a small proportion of the population, mostly of European descent, has monopolized most of the income by keeping wages low and resisting the pay-

**FIGURE 3.44** Cartagena: Chess on the beach. Entertainments for tourists along the beach in Cartagena, Colombia, include an ongoing chess game with chess pieces made of recycled plastic bottles painted and filled with sand, and a ride in a *chiva*, one of the ubiquitous Colombian party buses that tour around while the riders enjoy music, drink, and dancing. [Christopher Kirk.]

ment of taxes. In addition, the wealthy have resisted efforts to redistribute some of their extensive landholdings. On one side of the civil war are revolutionary guerrilla bands seeking government-sponsored reforms for the poor; a group known as FARC controls a large part of the Colombian interior south of Bogotá. On the other side are the private armies of the wealthy, who have little faith in the will and ability of the ill-equipped Colombian military to defeat the guerrillas.

The hostilities are complicated by the fact that, to raise money, all the warring parties participate in the cocaine trade to varying degrees. This trade, as we have seen, depends on the leaves of the ancient Andean coca plant, traditionally chewed as a fatigue and hunger suppressant, but now processed into the much stronger cocaine and sold internationally (see pages 146–147). Although coca growers make a somewhat better income from coca than they would from other crops, most of the profits are reaped by competing drug-smuggling rings whose members bribe, intimidate, and murder Colombian government officials who seek to reduce or eradicate the drug trade.

Are there no positive aspects to life in Colombia? In fact, there are signs that the drug-related violence is beginning to decrease, due perhaps in part to training and equipment supplied by the United States, but primarily to local initiatives such as the *No Mas* (No More) movement, a civic effort to stop the killing. Millions have demonstrated against the drug cartels, and President Uribe has succeeded in abating the violence on the right and the left, in part by

rehabilitating drug trade operatives. There are many efforts to find sources of income for rural people other than coca; in the mountains of southwestern Colombia, 1200 farm families have formed a cooperative that produces a line of 20 products—preserved fruits, sauces, and candies—marketed especially to Hispanics in North America and Europe. The cooperative also focuses on increasing child and adult education and on enhancing marketable skills. Elsewhere in the region, rural people working with the food scientists at the International Center for Tropical Agriculture have developed new varieties of corn that will be more productive and nutritious.

Through all the news about drug violence, Colombia has maintained a vibrant tourist economy. Cartagena, for example, is known for its elegant colonial buildings, beautiful beaches, and lively entertainments for affluent and budget-minded tourists alike (Figure 3.44).

## THE CENTRAL ANDES

The central Andes, which includes the countries of Ecuador, Peru, and Bolivia, is the poorest subregion in Middle and South America (Figure 3.45). On the eve of the Spanish conquest, it was the home of the Inca empire. The legacy of the Incas is reflected in numerous ruins, such as those at Machu Picchu in Peru, in thousands of miles of trade routes, and especially in the roughly one-half of the population that is indigenous—the largest proportion in South America.

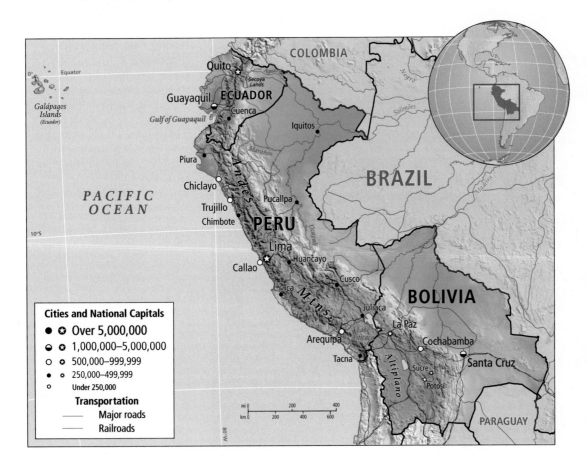

**FIGURE 3.45** The central Andes subregion. Bolivia has two capitals: La Paz is the seat of the government, and Sucre is the seat of the judiciary.

After the fall of this subregion to the Spanish, as in other parts of the Americas, a tiny group of landowners prospered while the vast majority lived in poverty. Most of the twentieth century was marked by failed efforts at social change and the violence resulting from those failures, but the growing political involvement of the large indigenous population may yet help achieve greater social equity.

## Environments, Settlement Patterns, and Production in the Central Andes

Only Ecuador and Peru have coastal lowlands; Bolivia has been landlocked since its mineral-rich Atacama Desert coastline was annexed by Chile in 1884. The coastal populations of Ecuador and Peru are mainly mestizo and African. The Pacific coast in this subregion is a zone of productive agricultural land and is also home to large and modern cosmopolitan cities, including Peru's capital and commercial center, Lima, and Ecuador's leading industrial center, Guayaquil. Although the climate is often dry, especially in Peru, plantations and other agricultural enterprises that produce crops for export thrive along the coast with the help of irrigation. The production of export crops such as bananas, cotton, tobacco, grapes, citrus, apples, and sugarcane has increased dramatically. Shrimp farms and ocean fishing grounds nourished by the nutrient-rich Peru Current sustain a vital export-oriented fishing industry. Food production for local consumption can be precarious, so reliance on traditional foods is now encouraged (Figure 3.46). Overall, dependence on food imports (chiefly cereals) is increasing, reflecting the fact that consolidated farms are producing for the export market rather than for local consumption, as small farms used to do.

Like other parts of Middle and South America, for most of the 1990s and up to the present, this area has tightened its belt in response to government debts acquired through borrowing to fund development. Structural adjustment policies mandated by the IMF forced the privatization of state-run industries and the streamlining of government. The result was job loss and social turmoil along with only modest economic growth. Governments dramatically increased prices for gasoline, electricity, and transport in an effort to raise funds to pay off debts. Such policies hurt the urban poor, whose low wages could not cover the increases. In rural coastal areas, SAPs removed government assistance to small-scale farmers. Small farms are now being replaced by export-oriented corporate farms and high-tech operations such as shrimp farms, most of which cause environmental degradation and yet receive government assistance. In 1997, protests by working people brought down the government of Ecuador, and the same thing might have happened in Peru had it not been for suppression by the military. Eventually, popular uprisings forced the resignation of both the president of Peru in 2000 and the president of Bolivia in 2003. Ecuador declared bankruptcy in 1998 because it couldn't repay its debts to the international financial community, and its economy and society are still in a state of crisis and uncertainty.

**FIGURE 3.46** Traditional cuisine in Ecuador. The chef is barbecuing a *cuy*, or guinea pig (*Cavia porcellus*), a traditional staple of Andean cuisine, often served with potatoes and a spicy sauce. Raised commercially today, the *cuy* reportedly tastes similar to rabbit. [© 2004 Julia Ng.]

The highlands, known as the **altiplano,** are the home of the bulk of the region's indigenous people, who for centuries worked on large haciendas and in rich mines (copper, lead, and zinc in Peru; tin, bauxite, lead, and zinc in Bolivia) owned by a small group of people of European descent. Most of the indigenous people today obtain their incomes from subsistence cultivation of a wide array of vegetables acclimated to the cool climate: more than 50 varieties of potatoes, for example, are cultivated in the altiplano. Surpluses are traded on market days in towns throughout the altiplano. Sheep herding, wool processing and weaving, and (for some men) mining also contribute to family incomes. Recently, foreign and local scientists have explored imaginative cultivation techniques and crops—both new and ancient—that may improve standards of living (Box 3.6). In Peru, for example, agronomist Angel Valledolid oversees 7500 acres (3000 hectares) of nuña, or "popping beans," an ancient crop plant that grows without pesticides. Marketed as a cocktail snack in Europe and North America, the popping beans could improve the incomes of 200,000 Peruvian cultivators and their families. In the last few decades, highland Peruvians who cannot get access to land or other means of support have been moving east into Amazon lowland regions.

The Amazon lowlands of Ecuador, Peru, and Bolivia have traditionally been the home of scattered groups of indigenous people. In recent years, this area has seen rapid and often destructive development of natural resources. National governments have encouraged private extraction, first of trees and then of minerals and agricultural products, and these activities have severely damaged the home territories of indigenous people (see the description that opens this chapter). Roads—such as the one cut across the Andes in Peru from Lima to Pucallpa at the headwaters of the Amazon—have been built to gain access to timber and minerals, but they have also opened up this region to waves of migrants from the highlands and coastal lowlands. Accompanying these migrants are diseases that are decimating the indigenous people of the Amazon Basin, who lack resistance to them.

## BOX 3.6 | AT THE LOCAL SCALE Learning from the Incas: Agricultural Restoration in the Peruvian Highlands

Around the ancient Incan capital of Cusco, in the Peruvian highlands, terraces and irrigation canals once supported crops that fed hundreds of thousands of people on what would otherwise have been barren mountain slopes. Much of this ancient infrastructure still remains, but it was not being used until recently. Now some of it is being restored, thanks mainly to more than two decades of careful research by the British rural development specialist and archaeologist Ann Kendall. Kendall recognized that Incan agricultural technologies were more advanced than archaeologists had thought. For example, on steep slopes, the construction of terraces makes the recycling of nutrients more efficient by inhibiting erosion and trapping moisture, thus making high-yield organic cultivation possible. In particular, Kendall found that the Incan use of clay as a flexible semiwet sealant on both terraces and canals was extremely well adapted to the earthquake-prone environment of the Peruvian highlands because it doesn't break the way concrete does and because cracks tend to reseal themselves. So far, reviving these traditional methods has proved highly productive, allowing some highland people who had migrated to lowland cities to return home and take up farming. One canal and its associated terrace system can irrigate enough land to support more than 2000 people.

*Source:* Kevin Krajick, "Archaeology: Green farming by the Incas?" *Science* 281(5375) (July 17, 1998): 322.

## Social Inequalities and Social Unrest

In the twentieth century, various strategies for lessening the subregion's gross social inequalities were tried, with varying degrees of success. In the 1950s, Bolivia became the first country to attempt reform when the government took over the tin mines in the altiplano after a revolution in which miners' unions played a key role. Wages were raised and conditions improved for several decades, but then world prices for tin declined because plastic packaging was replacing tin cans just as tin mining was intensifying around the globe. As tin prices fell, illegal drugs replaced tin as Bolivia's chief export. As in Colombia, the military and the government publicly condemn the drug trade, yet members of both are extensively involved, as are local cultivators who grow poppies for heroin or coca for cocaine.

In the 1970s, Peru's military dictatorship sought to improve the lives of indigenous people by transforming the haciendas of the elite into state-run communes. The experiment was not popular, and it soon inspired a resistance movement among indigenous small farmers, which was taken over in 1980 by a Communist guerrilla movement called the Sendero Luminoso (Shining Path). Followers of the Sendero Luminoso forced cooperation by killing farmers if they participated in the capitalist economy in any way or were associated with the central government, whether as traders, community officials, participants in development projects, or even just voters. Eventually, the farmers banded together to fight the Sendero Luminoso with the aid of the military. The level of violence has decreased for several years now, but the issues of affordable access to good agricultural land and democratic participation for the majority remain. The Peruvian government, hoping to avoid reigniting past grievances, supplies highland communities with subsidies and protections from the harsher aspects of the SAP reforms that are straining the lowlands.

The indigenous people of South America have visibly increased their participation in national politics. They played a part in forcing the resignation of Peru's president, Alberto Fujimori, in the autumn of 2000 after a series of corruption scandals. In Ecuador, CONAIE, a federation representing all of Ecuador's indigenous groups, has been a strong force fighting for the land rights of indigenous people. CONAIE's support was central to the success of the protests against SAPs that brought down Ecuador's government in 1997, and indigenous demands for an increased voice in national policy have continued.

In the 1980s, Bolivia embraced the free market model and sold off state-owned industries to largely foreign interests. These policies, urged on Bolivia by the World Bank and IMF to reduce its debt, enriched a few, such as President Gonzalo Sanchez de Lozada, but also brought high unemployment and a drop in real per capita income for Bolivia's majority indigenous population. Native people mounted massive protests, spurred by government plans to export natural gas from Bolivia to the United States via a pipeline to a port in Chile. Protesters interviewed by *New York Times* reporter Larry Rohter said they saw an "unbroken line" between the rapacious Spanish colonial policies of the past and modern movements linked to globalization and free trade. In one weekend, the Bolivian military shot to death 50 unarmed protesters who were demanding the president's resignation. President Lozada eventually resigned in October 2003, and in 2005, Bolivians elected in a landslide the first indigenous head of state in the Americas: Evo Morales, a socialist and former coca farmer.

## THE SOUTHERN CONE

The countries of Chile, Paraguay, Uruguay, and Argentina (Figure 3.47) have diverse physical environments but remarkably similar histories. The so-called southern cone of South America experienced little European settlement during the Spanish empire, but in the late nineteenth and early twentieth centuries, European immigrants—mainly Germans, Italians, and Irish—were drawn there by temperate climates, economic opportunities, and the prospect of landownership. These immigrants swamped the surviving indigenous populations throughout most of the region. Paraguay is the only one of these countries that has a predominantly mestizo population, and it is the poorest country of the four (see Table 3.3).

## The Economies of the Southern Cone

Agriculture was once the leading economic sector in the southern cone, but it is no longer the main source of income or the main employer, having been replaced by the service sector. Nonetheless, agriculture remains prominent in the identity of the region and in earning foreign exchange through exports. The primary agricultural zone, the pampas, is an area of extensive grasslands and highly fertile soils in northern Argentina and Uruguay (and southern Brazil). The region is famous for its grain and cattle. Sheep raising dominates in Argentina's drier, less fertile southern zone, Patagonia. On the Pacific side of the Andes, in Chile's central zone, a Mediterranean climate supports large-scale fruit production that caters to the winter markets of the Northern Hemisphere; some fruit also goes into wine production. Chile also benefits from considerable mineral wealth, especially copper.

Although the agricultural and mineral exports of the southern cone created considerable wealth in the past, fluctuating prices for raw materials on the global market have periodically sent the economies of this subregion into a sudden downturn. The desire for economic stability was a major impetus for industrialization and urbanization in the mid-twentieth century. At first, the new industries were based on processing agricultural and mineral raw materials. Later, state policies supported diversification into import substitution industries (see pages 137, 139). However, inefficiencies, corruption, and the small size of local markets prevented industry from becoming a leading sector for the region. Although the service sector has now surpassed agriculture and industry in the southern cone countries, it too is lackluster, performing as only a modest leader in employment and contributions to GDP.

Economic policy has been a source of conflict within the subregion for decades (see the discussion of the Mercosur trade agreement on pages 141–142). Despite their considerable resources, these countries have always had substantial impoverished populations that were nonetheless well enough educated to demand attention to their grievances. In response, each country developed

**FIGURE 3.47** The southern cone subregion.

mechanisms for redistributing wealth—government subsidies for jobs, food, housing, basic health care, and transport, for example—that did alleviate poverty to some degree, though they did not contribute to viable development or the spread of democratic institutions. When global prices for raw materials fell in the 1970s and produced an economic downturn with accompanying depriva-tion, people clamored for more fundamental changes. In response, throughout the 1970s and into the early 1980s, military leaders fearful of civil unrest, and with the support of the elite, took control of governments (especially in Chile and Argentina) and waged bloody campaigns of repression against striking workers and the socialist and Communist opposition. In the years of state-sanctioned violence that followed, known as the "dirty wars," at least 3000 people were killed in Chile and 14,000 to 18,000 in Argentina, and tens of thousands more were jailed and tortured for their political beliefs (see page 147 for a discussion of the overthrow of President Salvador Allende in Chile in 1973).

The International Monetary Fund has consistently applied pressure for repayment of the subregion's massive debts, effectively shifting countries away from poverty alleviation policies and toward free market economic reform. The effects of these policies in Chile have been mixed. Chile's export-oriented industrial, mining, and agricultural sectors have all grown, but not sufficiently to alleviate social discontent. In 2006, Chileans elected as president a moderate socialist woman, Michelle Bachelet, who may downplay future free market reforms. Argentina, after suffering a long recession, continued job losses, and a crushing debt burden, defaulted on its debt payments in December 2001, causing a precipitous drop in foreign investment. An IMF package of structural reforms and loans to the Argentine government held off the crisis, but did not resolve it. In 2006, Argentina also elected a moderate socialist president, Néstor Kirchner, who is likely to increase, not restrict, government spending.

## Buenos Aires: A Primate City

The primary urban center in the southern cone is Buenos Aires, the capital of Argentina and one of the world's largest cities (Figure 3.48). Forty percent of the country's people live in this primate city. Buenos Aires boasts premier shopping streets, elegant urban landscapes, and dozens of international banks, yet six decades of decline have left it with environmental degradation, empty factories, social conflict, severe poverty, and declining human well-being. Some people argue that the downward slide in quality of life in Buenos Aires is an unavoidable result of the restructuring required to create an economy that will be competitive in the global arena. Past Argentine leaders contended that greater integration with the global economy would help to reverse decades of malaise. They wanted Buenos Aires to be seen as a world city—a center with pools of skilled labor that attracted major investment capital, a sophisticated city with a beautiful skyline and a powerful sense of place. While it is unlikely that the new socialist president has any lower ambitions for Buenos Aires, he will have to deal with an impatient local population living in rundown apartments and on wages too low to afford basic nutrition. He, as well as the city's poor residents, is likely to give more importance to obtaining such basics as decent housing, better food, and modernized transportation, goals that will not necessarily draw profit-seeking free market investors.

## BRAZIL

The observant visitor to Brazil (Figure 3.49) is quickly caught up in the country's physical complexity and in the richly exuberant, multicultural quality of its society. But its landscapes also plainly show the environmental effects of both colonialism and recent under-planned economic development. Brazil's 184 million people live in a highly stratified society made up of a small, very wealthy elite, a modest but rising middle class, and a majority that lives below, or just barely above, the poverty line. In Brazil's megacities of São Paulo and Rio de Janeiro, vast shantytowns, high crime rates, and homeless street children are signs of the gross inequities in opportunity and well-being. The richest 20 percent of Brazil's population has 26 times the wealth of the poorest 20 percent—one of the widest disparities on earth (see Table 3.2). Travelers find themselves delighted by the flamboyant creativity and elegance of the Brazilian people, yet sobered by the obvious hardships under which they labor.

Brazil has about the same area as the United States. Its three distinctive physical features are the Amazon Basin, the Mato

**FIGURE 3.48** Buenos Aires skyline, with a shantytown in the foreground. Taken in 2002, this picture is another illustration of Crowley's urban model (see Figure 3.18 on page 133). As many as half of Argentina's 36 million people were impoverished by a grueling economic crisis that began in the 1990s. Although the situation was improving by 2005, joblessness remains high, 38 percent are below the poverty line, and many now live in self-built housing projects, like this one near the city center. [AP Photo/ Pilar Capurro.]

**FIGURE 3.49** The Brazil subregion.

Grosso, and the Brazilian Highlands. The Amazon Basin, which covers the northern two-thirds of the country, is described on pages 119–121. The Mato Grosso is a seasonally wet/dry interior lowland south of the Amazon with a convoluted surface. Once covered with grasses and scrubby trees adapted to long dry periods, in the twentieth century it was extensively cleared for subsistence and commercial agriculture. The southern third of Brazil is occu-

pied mostly by the Brazilian Highlands, a variegated plateau that rises abruptly just behind the Atlantic seaboard 500 miles (800 kilometers) south of the mouth of the Amazon. The northern portion of the plateau is arid; the southern part receives considerably more rainfall. Settlement in northeastern Brazil is concentrated in a narrow band along the Atlantic seaboard and south through Salvador (see Figure 3.12). In Brazil's temperate zone, near Rio de

Janeiro, São Paulo, Curitiba, and Pôrto Alegre, settlement extends deeper inland.

The Brazilian economy is the largest in Middle and South America and the eighth largest in the world. The resources available for development in Brazil are the envy of most nations, but management of those resources in Brazil's best interests has been a challenge. Gold, silver, and precious gems have been important resources since colonial days, but it is industrial minerals—titanium, manganese, chromite, tungsten, and especially iron ore—that are most valuable today. These minerals are found in many parts of the Brazilian Highlands. Hydroelectric power is widely available because of the many rivers and natural waterfalls that descend from the highlands.

Local oil now provides for 60 percent of Brazil's needs, and more has recently been discovered west of Manaus in the Amazon. A large part of the remainder of Brazil's energy needs are supplied by ethanol, which is made from commercially grown sugarcane in a system that is precisely engineered for efficiency in terms of energy used to energy produced. Brazil now produces cars that can instantly switch from gas to ethanol and achieve 40 miles to the gallon.

There has been large-scale labor-intensive agriculture in Brazil for 400 years. Today, there is horticultural farming of fruits and vegetables in the south; large, increasingly mechanized, commercial farms growing primarily sugarcane, tobacco, cotton, cassava, and oil palm in the northeast; and ranching in the interior dry zones. Production of soybeans, pork, and beef has some potential to expand in the Mato Grosso, but most of Brazil's available land is in the tropical Amazon, where soils are fragile. There, agriculture and timber extraction are expanding rapidly at the expense of tropical rain forests. Thousands of square miles of forest fall yearly for the cultivation of such crops as soybeans. Overall, agricultural exports from Brazil are increasing and exceed imports, but as a proportion of Brazil's GDP (10 percent), agriculture is rapidly losing out to industry (40 percent) and services (50 percent).

Brazil is the most highly industrialized country in South America. Most of its industries—steel, motor vehicles, aeronautics, appliances, chemicals, textiles, and shoes—are concentrated in a triangle formed by the huge southeastern cities of São Paulo, Rio de Janeiro, and Belo Horizonte. Until the 1990s, the vast majority of Brazil's mining and industrial operations were developed using government funding, and many are still owned and run by the government. In the 1980s and 1990s, elected governments adopted structural adjustment policies and privatized many formerly government-held industries and businesses in an effort to make them more efficient and profitable. Many of these firms were sold to foreign investors at bargain-basement prices. In 2000 alone, direct foreign investors spent $32.8 billion on Brazilian properties. Privatizing industry often results in greater productivity, and indeed, beginning in 2001, exports began to exceed imports. By 2005, exports were $118 billion and imports just $74 billion. The sale of state-owned industries may be hurting Brazil in the long run, however, because environmental regulations on these firms are lax and because any profits that might materialize will go not to the public, who paid for the development of these industries, but to the new private owners, who may well invest those profits outside Brazil.

## Urbanization

Brazil has a number of large and well-known cities: Rio de Janeiro, Curitiba, São Paulo, Salvador, Recife, Fortaleza, Belém, Manaus, and Brasília, all but the last two on the Atlantic perimeter of the country. During the global economic depression of the 1930s, farmworkers throughout the country began migrating into urban areas as world prices for Brazil's agricultural exports fell; subsequent agricultural change pushed even more people off the land. During and after the 1960s, the chance of employment in the factories being built with government money pulled them into the cities. Brazil's competitive edge in the global market was its cheap labor. The military governments of the time thought that the mostly government-owned industries could continue to pay very low wages because there was such a surplus of workers, but they had to quell many protests by workers who could not live decently on their wages.

By 1995, 77 percent of Brazil's population was urban, and at least one-third of the urban dwellers, many of them unemployed, were living in favelas, the Brazilian urban shantytowns (see Figure 3.16). The poverty in the hardest-hit cities in the northeast rivals that of Haiti, the poorest country in the Americas. Brazilians, however, are intrepid at finding ways to make life worth living. Favela dwellers are famous for their efforts to create strong community life and support for those in distress. Many turn to religion of one sort or another as a source of strength, as described in Box 3.7.

Rio de Janeiro and other cities in Brazil's southeastern industrial heartland have districts that are elegant and futuristic showplaces, resplendent with the very latest technology and graced with buildings and high-end shops that would put New York, Singapore, Kuala Lumpur, or even Tokyo to shame. Brazilian planners now acknowledge, however, that in the rush to develop these modern urban landscapes, developers neglected to underwrite the parallel development of a sufficient urban infrastructure (for an exception, see the discussion of Curitiba in Box 3.2). In short supply are sanitation and water systems, up-to-date electrical wiring, transportation facilities, schools, housing, and medical facilities—all necessary to sustain modern business, industry, and a socially healthy urban population. But the Brazilians should not be criticized too severely for what in hindsight seems like an obvious error. During the 1970s and 1980s, development theory set forth by the World Bank and other financial institutions held that social transformation and infrastructure development would be naturally occurring side effects of investment in what were called **urban growth poles**—cities that would attract investment, innovative immigrants, and trade, hence stimulating further development. The debt crisis of the 1980s, which ate up profits with skyrocketing *inflation* (a rapid rise in consumer prices due to the falling value of the country's currency on the international market), put an end to such optimistic forecasts. Only very recently have all parties begun to recognize that planned development of human capital through education, health services,

## BOX 3.7 | AT THE LOCAL SCALE  African-Derived Religions Help Harried Urban Dwellers

It's Friday afternoon, and a crowd of white-clad women is gathering outside a house in the Felicidad favela. Like the surrounding houses, this one is modest, but it gleams white; all its surfaces are swathed in marble. Potted palms decorate the porch, and beside them, welcoming the women, is a tall and elegant middle-aged man, a religious leader in the movement known as Umbanda. Umbanda, Batuque, and related belief systems thrive in all the Atlantic coastal cities of Brazil from Belém to São Paulo (Figure 3.50). Each group is centered around a male or female spiritual leader, who invokes the spirits to help people cope with health problems and the ordinary stresses and strains

of a life of urban poverty. During the ceremonies, which can last as long as 8 hours, drumming, dancing, spirit possession, and psychological support are the central focus. Umbanda grew out of an older African-Brazilian belief system called Condomble, and similar movements (Voodoo, Santeria, Obeah) are found elsewhere in the Americas, including the Caribbean, the United States, and Canada. In Brazil, Umbanda appeals to an increasingly wide spectrum of the populace from African, European, Asian, and indigenous backgrounds.

*Source:* Lydia Pulsipher's field notes.

**FIGURE 3.50**  An Umbanda ceremony in Rio de Janeiro, 2000. Many urban Brazilians meet weekly to enjoy the exercise and healing effects of Umbanda, an African-Brazilian religion that encourages spiritual trances and dancing. Hours are spent in centers where the participants enjoy a sense of community. [Ricardo Azoury/CORBIS.]

and community development, whether privately or publicly funded, is a primary part of building strong economies.

## Brasília

Brasília, the modern capital of Brazil, is an intriguing example of the effort to lead development with urban growth poles. Built in the state of Goiás in just 3 years, beginning in 1957, Brasília lies about 600 miles (1000 kilometers) inland from the glamorous old administrative center of Rio de Janeiro. The official explanation for this move was that the city, built in this remote location, would serve as a **forward capital**—it would draw migrants and investment for the development of the western highland territories, the Mato Grosso, and, eventually, the Amazon Basin. The scholar William Schurz suggested an alternative explanation for Brasília's location: moving the capital so far away from the centers of Brazilian society was the most efficient way to trim the badly swollen and highly

inefficient government bureaucracy. In any case, symbolism figured more prominently than practicality in the design. The city was laid out to look from the air like a swept-wing jet plane. There was to be no central business district, but rather shopping zones in each residential area and one large mall. Pedestrian traffic was limited to a few grand promenades; people were expected to move even short distances in cars and taxis. Public buildings were designed for maximum visual and ceremonial drama, but safety was an afterthought.

Over five decades of actually using this urban landscape, people have made all sorts of interesting changes to the formal design. At the Parliament, legislative staff and messengers created footpaths where they needed them: through flower beds and—with little steps notched in the dirt—up and over landscaped banks. Thus they efficiently connected the administration buildings, bypassing the sweeping promenades. Commercial districts were retrofitted in and around hotel complexes. Urban workers who couldn't afford taxi fares wore direct-route footpaths across great green swards of grass. Little hints of the informal economy that characterizes life in the old cities of Brazil began to show up—a fruit vendor here, a manicurist there. And the shantytowns that the planners had tried hard to eliminate began to rise relentlessly around the perimeter. Today, life in the somewhat spiffed-up shantytowns is so much more interesting than life in the sterile environment of Brasília proper that tour buses take visitors to see them and shop there. Overall, in the 50 years since its construction, Brasília's success as a forward capital has been limited. Although the city has drawn poor laborers, the entire province of Goiás still has only 4.5 million people, just 6 percent of the country's total, and their average income is only half that of the country as a whole. Population and investment remain centered in the Atlantic coastal cities.

## Reflections on Middle and South America

In a number of ways, European colonialism in Middle and South America launched the modern global economic system. It was in this region that large-scale extractive practices were inaugurated. Raw materials were shipped at low prices to distant locales, where they were turned into high-priced products, the profits on which went to Europe. Local people were diverted from producing food and other necessities for themselves to working as low-paid or enslaved labor in the fields and mines. Rules governing private property were set down in societies that had long recognized communal land rights. In the process, the landscapes and settlement patterns of the Americas were reorganized as the focus shifted from producing for local and regional economies to producing for global markets.

After the colonial era, this region continued to serve as a testing ground for economic theories, both capitalist and socialist. Middle and South America tried out such capitalist ideas as mechanized agriculture for export, rural-to-urban migration as a solution to rural poverty, and government borrowing and financing for large-scale development projects. Socialist experiments included government-sponsored industrialization to create jobs and produce substitutes for imports as well as broad government subsidies to address the basic needs of the poor. External factors, however, intervened to thwart any such optimistic plans for economic expansion and societal transformation. The global recession of the early 1980s left the region's governments in debt, and their indebtedness forced free market structural transformations on them, which widened wealth inequities and increased social disparities. The rapidly growing numbers of urban poor had to resort to self-help solutions for their shelter; they sought meager livings in the informal economy, and with the reduction of public health and welfare services, many resorted to community and political activism.

Despite the hardships of the last 20 years—the increased harshness of everyday life for the rural and urban poor, especially women, children, and the elderly—there are signs that the situation is not beyond retrieval. There is increasing recognition that to be judged successful, development must first change the lives of the majority for the better. Middle and South Americans are beginning to build on their strengths and to invent solutions to problems common in many places around the world. Curitiba in Brazil is seeking simple but humane solutions to rapid urbanization and the pollution and social disruption it brings. Degraded agricultural lands are being rehabilitated in the Peruvian highlands by the revival of ancient indigenous practices. Environmental groups are addressing the root causes of deforestation: inequitable domestic distribution of lands and resources, unsustainable uses of forestlands such as ranching and cash-crop agriculture, and demand for forest products in distant markets. Increasingly, regional trade organizations are emerging with sufficient strength and motivation to negotiate for the good of the region rather than for parochial interests or those of distant global powers. Finally, in the first decade of the twenty-first century, Middle and South Americans have elected moderate but outspoken leaders who are advocating for the poor majority while also seeking balanced economic structures capable of competing in the global economy.

As you read other chapters of this book, it might be useful to reflect again on the geographic issues we have discussed in this chapter. For example, notice how Europe's situation today—its wealth, its position as a world leader, and its emerging commitment to help its former colonies—is related in part to its colonizing experiences, which began in the Americas. You will see that Africa and South Asia under colonial rule experienced some conditions similar to those in Middle and South America; more recently, they too have felt the sting of SAPs. Southeast Asia and Oceania have had comparable experiences as colonies. In the former Soviet Union and Central Asia, outside investors intent on exploiting oil and forest resources are reminiscent of the conquistadors and their successors in the Americas.

## Chapter Key Terms

acculturation 149
altiplano 173
assimilation 149
Aztecs 123
barriadas 132
barrios 132
brain drain 133

colonias 132
contested space 142
conurbation 131
Creole cultures 118
Creoles 127
early extractive phase 134
economic restructuring 117

ecotourism 154
El Niño 122
evangelical Protestantism 151
**Export Processing Zones (EPZs)** 139
extended family 149
external debts 139

favelas  132

foreign direct investment (FDI)  134

forward capital  179

hacienda  135

import substitution industrialization (ISI)  137

Incas  123

income disparity  134

indigenous  116

isthmus  167

ladino  167

land reform  137

liberation theology  151

*machismo*  150

maquiladoras  139

*marianismo*  150

mercantilism  127

Mercosur  141

mestizos  127

Middle America  118

mother country  127

neocolonialism  137

North American Free Trade Agreement (NAFTA)  141

plantation  136

populist movements  151

primate city  131

pyroclastic flow  119

shifting cultivation  123

silt  119

South America  118

subduction zone  119

subsidence  130

temperature-altitude zones  121

*tierra caliente*  122

*tierra fria*  122

*tierra helada*  122

*tierra templada*  122

trade winds  122

urban growth poles  178

## Critical Thinking Questions

1. If the European colonists had come to Middle and South America in a different frame of mind—say, they were simply looking for a new place to settle and live quietly—how do you think the human and physical geography of the region would be different today?

2. Explain two main ways in which tectonic processes account for the formation of mountains in Middle and South America.

3. Reflecting on the whole chapter, pick some locations where you were impressed with the ways in which people are presently dealing with either environmental issues or issues of wealth disparity or human well-being. Explain your selections.

4. Discuss the ways in which you see the historical circumstances of colonization affecting modern approaches to economic problems in Mexico, Bolivia, Brazil, Venezuela, or Cuba.

5. Describe the main patterns of migration in this region and discuss the effects of migration on both the sending and receiving societies.

6. Discuss how gender roles in this region have been affected by migration patterns.

7. Name three factors you see as important in increasing democratic participation in this region. In which countries would you say these factors are making the biggest difference?

8. Explain how the Amazon Basin and its resources constitute an example of contested space.

9. Argue for or against the proposition that free trade blocs, such as NAFTA and Mercosur, assist upward mobility for the lowest-paid workers.

10. How would you respond to someone who suggested that Middle and South America were helped toward development and modernization by the experience of European colonization?

# CHAPTER 4

# EUROPE

In May of 2004, ten new countries joined the European Union (EU). In January of 2007, two more joined. Ten of the twelve were formerly Communist countries in central Europe, most with lower standards of living and higher unemployment rates than western Europe. Soon hundreds of thousands of workers from the eastern fringes of Europe began taking advantage of the EU principle that, in theory, citizens of all member states have the right to move to any other EU state. In fact, only Ireland, with a newly booming economy based on high-tech and light manufacturing, Britain, and Sweden actually have open borders. Ireland is a favorite destination for these migrants.

**Global Patterns, Local Lives** It is a chilly, overcast late October day in Riga, Latvia. Janis Neulans sits on a creaky bed in a smelly fifth-rate hotel. He is about to take the midnight flight from Riga to Dublin, Ireland, where he hopes to find work. His last job was sandblasting the hulls of freighters in the Riga shipyard. Well-built, with powder blue eyes and a thatch of blond hair (Figure 4.2), Neulans comes from a distant village of eight houses near the Russian border. Although he is 39, he is not yet married. His schooling is minimal; his entire net worth is $420, including cash from the recent sale of his 1984 Volvo.

Neulans and Vladimir, whom he meets on the plane, arrive in Dublin in the middle of the night. Vladimir's friend, also a Latvian, meets them at the airport and takes them to two bare mattresses in spare quarters in a vast new housing development in a west Dublin suburb. By mid-morning, Neulans has paid the rent on the room he shares with Vladimir and begun his job search with the help of other Latvians who arrived some months ago. They help him write job ads and a curriculum vitae in English. For the first two weeks he subsists on bread, cheddar cheese, and diet cola while visiting dozens of newly built warehouses and factories seeking work. On the 11th day, down to his last $8, he agrees to work on a farm 40 miles (64.4 kilometers) from Dublin, milking cows and doing other chores for 14 hours a day. But soon Neulans receives a call from a door-frame factory, where he applied before taking the farm job. The farmer pays him just half of what he is owed.

Just two weeks after arriving in Ireland, Neulans has his dream job. He works for 10 hours a day at a drill press, making prefabricated door frames,

**FIGURE 4.1** Regional map of Europe.

183

**FIGURE 4.2** Janis Neulans in Ireland. Neulans, seen here on the subway in Dublin, polished his job-search skills in that city. Within two weeks he had landed a job at a door-frame factory in southeastern Ireland. [Kevin Sullivan.]

for the Irish minimum wage of $9.20 an hour (three times the wage in Latvia), plus overtime pay. In addition, he serves as the factory's night watchman, for which he gets the use of an on-site three-bedroom mobile home. There, he outlines his plans: he will work here for several years while saving to return to his Latvian village, where he will buy some calves to raise and enjoy the company of his family. "It's where my heart is," he says.

*Source: Kevin Sullivan, "East-to-west migration remaking Europe," Washington Post Foreign Service, November 28, 2005. A video of Kevin Sullivan accompanying Janis Neulans on his journey is available at http://www.washingtonpost.com/video.* ∎

Janis Neulans was able to move easily between Latvia and Ireland because both countries are members of the **European Union (EU)**, a supranational organization that unites most of the countries of West, South, North, and now Central Europe in a single economy within which people, goods, and money can move freely (Figure 4.3). Some western Europeans fear that expansion of the EU into central Europe, and eventually into southeastern Europe, will bring new political tensions and a throng of alien immigrants requiring social and educational services. At the same time, some who support the idea of European unity harbor the fear that if the EU succeeds, **cultural homogenization** will result: Latvian village culture, and other distinctive local ways of life, will disappear. Meanwhile, employers across Europe are saying that without the

immigrant workers, economic growth and competitiveness in the global market is impossible. The European Union will be discussed in more detail later in this chapter.

## Themes to Explore in Europe

The following themes are developed in this chapter:

1. **Economic union.** The lure of greater prosperity first drew the countries of western Europe into economic union. Twenty-seven countries now form the European Union, the world's strongest and most elaborate trading alliance.

2. **Countertrends that resist change.** Despite, or even because of, increasing economic integration, some Europeans feel a desire to establish a sense of national identity and to stem the tide of rapid change that may lead to cultural homogenization. Worried that the EU may change their way of life, some are reacting against a wave of immigrants; others are opposed to the regulatory aspects of the European Union.

3. **A large role for government.** European governments play an active role in shaping urban and rural environments by providing housing, education, health care, transportation, and other services, though the degree to which national governments provide these services varies.

4. **Regional disparities in economic development and well-being.** One long-term goal of the EU has been to foster economic growth in those subregions that are poorer and less developed, primarily those in South Europe and now Central Europe, where many countries are emerging from a half-century of control by the Soviet Union. All are trying, with differing levels of success, to enter the free market economy and to improve the well-being of their citizens. Varying degrees of government-provided services are reflected in the countries' differing indices of human well-being.

5. **Effects of globalization.** Europe's overall prosperity arises in part from the many benefits it has obtained—and still obtains—from the cheap labor and resources of its former colonies. At the same time, the global economy has placed some of Europe's workers and products in stiff competition with those in developing parts of the world. Like North America, Europe is losing jobs to other world regions.

6. **Environmental activism.** Europe's dense population and its high rates of consumption are contributing to air, soil, and water pollution, which have alarmed the European public. Environmental awareness and activism are growing in Europe.

# I THE GEOGRAPHIC SETTING

## Terms to Be Aware Of

The end of the cold war (as discussed on pages 48–49) changed political alignments across Europe and raised questions about just where the borders of Europe lie; hence special care is needed in

designating the various parts of the region. Remember that regions are just a convenient tool in geographic analysis and that their borders frequently change.

In this book, Europe is divided into four subregions (see Figure 4.14 on page 199). *North Europe* comprises **Scandinavia**—

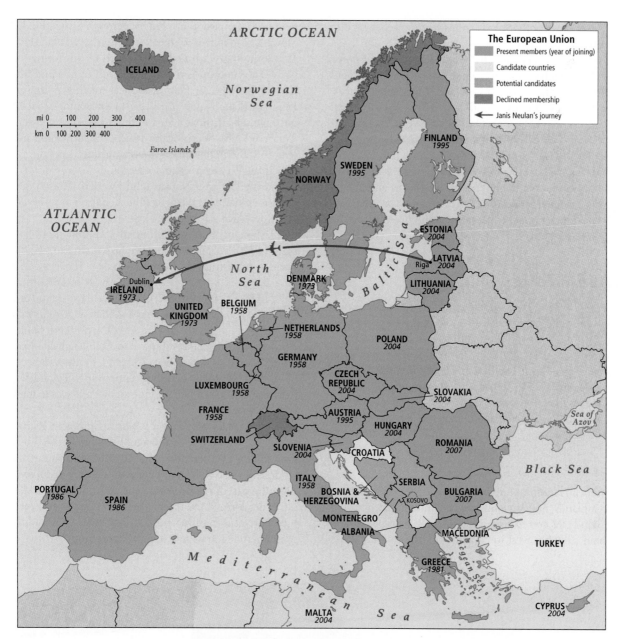

**FIGURE 4.3** The European Union (EU). The EU, formed with the initial goal of economic integration, has led the global movement toward greater regional cooperation. It is older and more deeply integrated than its closest competitor, the North American Free Trade Agreement (NAFTA), which was largely a response to the EU. Important issues in the EU include national sovereignty, challenges to national authority, and the consequences of eastward expansion. For example, since 1999 Kosovo, Serbia's southern province, has been a UN protectorate. Eventual independence is supported by the EU but opposed by Serbia and Russia.

Iceland, Denmark (including Greenland—not shown on most maps in this chapter—and the Faroe Islands), Sweden, Norway, and Finland—as well as Estonia, Latvia, and Lithuania. The last three countries, formerly part of the Soviet Union, are now considered part of North Europe because they share many economic and some cultural characteristics with that region. *West Europe* comprises the United Kingdom, the Republic of Ireland, France, Belgium, Luxembourg, the Netherlands, Germany, Switzerland, and Austria. *South Europe* comprises Portugal, Spain, Italy, Greece, Malta, and Cyprus. *Central Europe*, the largest subregion, contains (1) those Central European countries that were formerly within the Soviet Russian sphere of influence and practiced a form of Communism: Poland, the Czech Republic, Slovakia, Hungary, Albania, Romania, and Bulgaria; and (2) all the countries once known as Yugoslavia, also Communist but independent of Soviet control: Slovenia, Croatia, Bosnia and Herzegovina, Serbia (including Kosovo), Montenegro, and Macedonia. As we shall see, some of the Central European countries have recently joined the EU, while others have not. Altogether, 27 countries from all four European subregions were members of the EU as of January 2007, with others in Central Europe scheduled to join in the next several years.

For convenience, we occasionally use the term *western Europe* to refer to all the countries that were not part of the experiment with Communism in the Soviet sphere and Yugoslavia. That is, *western Europe* is used for the combined subregions of North Europe (except Estonia, Latvia, and Lithuania), West Europe (except the former East Germany), and South Europe. When we refer to the countries that were part of the Soviet sphere up to 1989, we use the pre-1989 label *eastern Europe*. When it is necessary to refer to the group of countries that includes Albania, Bosnia and Herzegovina, Bulgaria, Croatia, Macedonia, Montenegro, Romania, Serbia, and Slovenia, once known collectively as the *Balkans*, we use the term *southeastern Europe*.

## PHYSICAL PATTERNS

Europe is a region of peninsulas upon peninsulas. The entire European region is one giant peninsula extending off the Eurasian continent, and the whole of its very long coastline is itself festooned with peninsular appendages, large and small. Norway and Sweden share one of the larger appendages. The Iberian Peninsula (shared by Portugal and Spain), Italy, and Greece are other examples of large peninsulas. One result of these many fingers jutting into oceans and seas is that much of Europe feels the climate-moderating effect of the large bodies of water that surround it.

## Landforms

Although European landforms are fairly complex, the basic pattern is mountains, uplands, and lowlands, all stretching roughly east to west in wide bands. As you can see in Figure 4.1, Europe's largest mountain chain stretches west to east through the middle of the continent, from southern France through Switzerland, Austria, and Slovakia, and curves southeast into Romania. The Alps (Figure 4.4) are the highest and most central part of this formation. This network of mountains is mainly the result of pressure from the collision of the African Plate, which is moving northward, with the Eurasian Plate, which is moving to the southeast (see Figure 1.20 on page 21). Europe lies on the westernmost extension of the Eurasian Plate. South of the main Alps formation, mountains extend into the peninsulas of Iberia and Italy and along the Adriatic Sea through Greece. The northernmost mountainous formation is shared by Scotland, Norway, and Sweden. These northern mountains are old (about the age of the Appalachians in North America) and worn down by glaciers.

Extending northward from the central mountain zone is a band of low-lying hills and plateaus curving from Dijon (France) through Frankfurt (Germany) to Krakow (Poland). These uplands form a transition zone between the high mountains and the lowlands of the North European Plain, the most extensive landform in Europe. The plain begins along the Atlantic coast in western France and stretches in a wide band around the northern flank of the main European peninsula, reaching across the English Channel and the North Sea to take in southern England, southern Sweden, and most of Finland. The plain continues east through Poland, then broadens to the south and north to include all the land east to the Ural Mountains (in Russia).

The coastal zones of the North European Plain are densely populated all the way east through Poland. Crossed by many rivers and holding considerable mineral deposits, this coastal lowland is an area of large industrial cities and densely occupied rural areas. Over the past thousand years, especially in the Netherlands, people have transformed the natural seaside marshes and vast river deltas into farmland, pastures, and urban areas by building dikes and draining the land with wind-powered pumps.

**FIGURE 4.4** An Alpine village near Innsbruck, Austria. The rugged Alpine landscape harbors numerous villages like this one. Elaborately constructed roads and railroads weave through steep-sided valleys. [Mac Goodwin.]

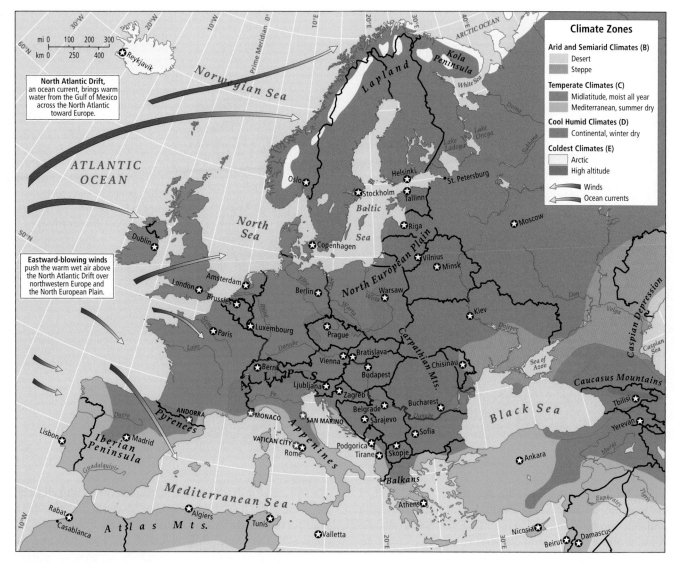

**FIGURE 4.5** Climates of Europe.

The rivers of Europe link its interior to the surrounding seas. Several of these rivers are navigable well into the upland zone, and Europeans have built large industrial cities on their banks. The Rhine carries more traffic than any other European river, and the course it has cut through the Alps and uplands to the North Sea also serves as the route for railways and motorways. The area where the Rhine flows into the North Sea is considered the economic core of Europe, and it is here that Rotterdam, Europe's largest port, is located. The larger and much longer Danube flows from Germany to the southeast, connecting the center of Europe with the Black Sea and passing the important and ancient cities of Vienna, Budapest, and Belgrade. As the EU expands to the east, the economic and environmental roles of the Danube River basin will be getting increased attention.

## Vegetation

Europe was once covered by forests and wild grasslands, most of which were cleared at one time or another in the distant past for farmland, pasture, towns, and cities. Today, primary forests exist only in scattered areas, especially on the more rugged mountain slopes and in the most northern parts of Scandinavia. In parts of central and southeastern Europe, forests have been sustainably managed for generations, and more are regenerating where small farms have been abandoned in favor of concentrated commercial farms. Today, forests of all sorts cover about one-third of Europe, but elsewhere the dominant vegetation is now crops and pasture grass. Vast areas are covered with industrial sites, railways, roadways, parking lots, canals, cities, suburbs, and parks.

## Climate

Europe has three main climate types: temperate midlatitude, Mediterranean, and humid continental (Figure 4.5). The **temperate midlatitude climate** dominates in northwestern Europe, where the influence of the Atlantic Ocean is very strong. A broad warm-water current called the North Atlantic Drift, which is really just the easternmost end of the Gulf Stream (see Figure 2.6 on page 58), brings

**FIGURE 4.6** Flooding on the Elbe River, 2006. Heavy snowmelt in the mountains of the Czech Republic and lots of rain across central Europe in April brought floodwaters to the Elbe River in eastern Germany. The village of Gohlis, near Dresden, shown here, remained flooded for several days. Most experts believe climate change (global warming) lies behind the periods of flood and drought that have taken hundreds of lives and caused billions of dollars in damage in Europe in the past few years. [AP Photo/Fabian Bimmer.]

large amounts of warm water that has traveled from the Gulf of Mexico along the eastern coast of North America and across the North Atlantic toward Europe. The air above the North Atlantic Drift is warm and wet, and eastward-blowing winds push this air over northwestern Europe and the North European Plain, bringing moderate temperatures and rain deep into the Eurasian continent. These factors create a climate that, though still fairly cool, is much warmer than elsewhere in the world at similar latitudes. To minimize the effects of precipitation runoff, people in these areas have developed elaborate drainage systems for their houses and communities (steep roofs, rain gutters, storm sewers, drain fields, and canals), and they grow crops, such as potatoes, beets, turnips, and cabbages, that thrive in cool, wet conditions.

Farther to the south, the **Mediterranean climate** of warm, dry summers and mild, rainy winters prevails. In the summer, warm, dry air from North Africa shifts north over the Mediterranean Sea as far north as the Alps, bringing high temperatures, clear skies, and dry periods. Crops grown in this climate, such as olives, citrus fruits, and wheat, must be drought-resistant or irrigated. In the fall, this warm, dry air shifts to the south and is replaced by cooler temperatures and thunderstorms sweeping in off the Atlantic. Overall, the climate here is mild, and houses along the Mediterranean coast are often open and airy to afford comfort in the hot, sunny, rainless Mediterranean summers.

In eastern Europe, the moderating influences of the Atlantic Ocean and the Mediterranean Sea are less or absent, and the climate is more extreme. In this region of **humid continental climate,** summers are fairly hot, and winters are longer and colder the farther north or the deeper into the interior of the continent one goes. Here, houses tend to be well insulated, with small windows and low ceilings. Crops must be adapted to much shorter growing seasons.

The European Union Environmental Agency reports that global warming may already be having recognizable effects on European climates. The 1990s were the warmest decade in 150 years, with the effects most obvious in central and southern Europe. Some recent extremes in temperature and precipitation may be linked to global warming: in 2003, the normal hot dry air mass from North Africa that drifts toward Europe in the summer hovered over western Europe through July and August. Temperatures in England exceeded 100°F (38°C), breaking all records. Europeans were unprepared for such extremely hot conditions. Crops failed, barges were stranded, forests burned, and deaths soared; in France alone, 3000 people died. In 2002 and 2006, rainfall and snowfall in central Europe reached record amounts, and in the spring of 2006, the rivers of central Europe—the Elbe, the Danube, the Morava—flooded for weeks (Figure 4.6).

## HUMAN PATTERNS OVER TIME

Some grand claims have been made in the name of Europe, and geographers have been among the most vociferous supporters of these ideas, as the following quotation attests.

*Europe has been a great teacher of the world. Almost every vital political principle active in the world today had its origin in Europe or its offspring, European North America. . . . [T]he same is true of the arts. Even though other parts of the world have produced rich folk arts, the culture of the West has become dominant.*

George F. Hepner and Jesse O. McKee, eds.,
*World Regional Geography: A Global Approach*
(Eagan, Minn.: West Publishing Co., 1992), p. 144.

We are so used to hearing the praises of Western (European) culture that for many people it is hard to spot the realities that they omit. The quotation above fails to recognize that Europe has been as much a learner from the world as a teacher of it, and that many "European" ideas and technologies were adopted from non-European sources. For example, the concept of the peace treaty, so vital to current global stability, was first documented not in Europe, but in ancient Egypt. Nevertheless, it is true that over the last 500 years (in fact, not a very long time), Europe has influenced how much of the rest of the world trades, fights, thinks, and governs itself. Attempts to explain this influence have ranged from arguments that Europeans are somehow a superior breed of humans to assertions that Europe's many bays, peninsulas, and navigable rivers have promoted commerce to a greater extent there than elsewhere. European capitalism is often cited as the crucial development that fueled Europe's geopolitical prominence. We may never have a single satisfying answer to the question of why Europe gained the leading role it continues to play. Still, it is worth taking a look at the broad history of this area and considering a few of the developments that have made Europe so influential over the past five centuries.

## Sources of European Culture

Starting about 10,000 years ago, the practice of agriculture and animal husbandry gradually spread into Europe from the Tigris and Euphrates river valleys in Southwest Asia (also known as Mesopotamia) and from farther east in Central Asia and beyond (see Figure 1.26 on page 28). The cultivation of wheat, barley, and numerous vegetables and fruits and the keeping of cattle, pigs, sheep, and goats came to Europe from the east and south (from Central and Southwest Asia and North Africa), as did pottery making, weaving, mining, metalworking, and mathematics. All these innovations opened the way for a wider range of economic activity, especially trade.

The first European civilizations were ancient Greece (800–86 B.C.E.) and Rome (100 B.C.E.–450 C.E.). The innovations of these societies and their extensive borrowings from other culture groups formed some of the most important cultural legacies of Europe. Located in southern Europe, Greece and Rome interacted more with the Mediterranean rim, Southwest Asia, and North Africa than with the rest of Europe, which was then relatively poor and thinly populated. Greek philosophers (Herodotus, Socrates, Aristotle), and mathematicians (Pythagoras, Plato, Euclid) were fascinated with the workings of both the natural world and human societies. Later European traditions of science, art, and literature

were heavily based on Greek ideas derived from yet earlier systems of thought. The Romans, perhaps the greatest borrowers of Greek culture, also left important legacies in Europe. Many Europeans (and Middle and South Americans) today speak Romance languages, such as Spanish, Portuguese, Italian, and French, which are largely derived from Latin, the language of the Roman Empire. The Roman notion of individual ownership of private property strongly influenced the development of Europe, as did the Roman practices used in colonizing new lands. After a military conquest, the Romans would secure control by establishing large plantation-like farms and communities of settlers transplanted from the homeland. By the second century C.E., hundreds of Roman towns dotted Europe. Roman systems of colonization were later used when European states laid claim to territory in the Americas, Asia, and Africa.

The influence of Islamic civilization on Europe is often overlooked. North African Muslims, called Moors in Europe, had a profound influence on language, music, food customs, architecture, and belief systems in Spain, which they ruled for over 700 years, starting in 711. Similarly, the Muslims of the Ottoman Empire left a deep imprint on various parts of southeastern Europe and Greece, which they dominated from the 1400s through the early 1900s. These Muslim rulers brought many textile and metal trade goods, food crops, architectural principles, and technologies to Europe from Arabia, China, India, and Africa. After the fall of Rome, while Europe slumbered during the *Dark Ages* (roughly 450–1000), pre-Muslim (Arab and Persian) and Muslim scholars preserved learning from Rome, Greece, Egypt, Persia, and other ancient civilizations. Eventually they brought Europe the Arabic numbering system as well as significant advances in medicine. Arab mathematicians, building on ideas picked up in India, introduced algebra and algorithms to Europe—both essential elements of modern engineering and architecture.

## The Evolution of Feudalism as a Social, Economic, and Political System

As the Roman Empire declined, a social system known as **feudalism** evolved during the **medieval period** (450–1500 C.E.). This system originated from the need to provide some stability and order and to defend rural areas against raiders: rival elites, local bandits, Vikings from Scandinavia, and nomads from the Eurasian interior. The objective of feudalism was to have a sufficient number of heavily armed, professional fighting men, or knights, ready to defend the farmers, or serfs, who cultivated plots of land for them. Over time, some of these knights became an elite class of warrior-aristocrats, called the nobility, who controlled, but did not actually own, certain territories, collected revenues from the serfs, and were bound together by a complex web of allegiances obligating them to assist one another in times of war. The often lavish lifestyles and elaborate fortifications of the wealthier nobility were supported by the labors of the serfs, who were legally barred from leaving the lands they cultivated for their protectors. The concept of a monarch or king, in whom authority was centralized, developed from that of the overlord who controlled the land and laid claim to its revenues

**FIGURE 4.7** Podsreda Castle, Slovenia. Begun in the twelfth century, modified over the centuries since, and now renovated and preserved as a national landmark in Kozjanski Park in eastern Slovenia, Podsreda Castle exemplifies the feudalism out of which modern Europe eventually emerged. [Mac Goodwin.]

and to the services and allegiances of those who lived on it—both the nobility and the serfs who served them. Most serfs lived in poverty outside castle walls (Figure 4.7). During later colonial times, aspects of feudalism were brought to the Americas, where the Spanish crown expropriated land from Native Americans and granted its use to colonists, along with the right to treat the native inhabitants as serfs.

## The Rise of Towns

While rural life followed established feudal patterns, new political and economic institutions were developing in Europe's towns, where thick walls provided defense against raiders and thus independence from feudal knights and later from kings. Located along trade routes, these towns sheltered artisans and merchants who modified ideas borrowed from China, India, and Southwest Asia to develop the precursors of some primary institutions of modern European capitalism: markets, banks, insurance companies, and corporations.

The rights of town citizens, set forth in documents called town charters, were eventually extended to all citizens. People with little power could make complaints against those more powerful and prevail, providing the basis for the European notion of civil rights. This movement toward allotting power to the people led eventually to the Magna Carta (agreed to by the Catholic Pope and King John of England in 1215), which limited the power of the king, especially the power to tax without consulting with those taxed. These social institutions allowed Europe's townsfolk to moderate the extreme divisions of status and wealth of the feudal system and to establish a pace of technological and social change, as well as contacts with the

outside world, that left the feudal hinterland literally in the Dark Ages. This division in European thought, between the "exciting, creative, cultured, urban" and the "behind-the-times, rude, rural," still shapes many of our attitudes toward economic development.

An outgrowth of European town life was the **Renaissance** (French for "rebirth"), a broad cultural movement that began in Italy in the fourteenth century, inspired by the older Greek, Roman, and Islamic civilizations. Renaissance thinkers turned their attention to science, politics, commerce, and the arts. By developing **humanism,** a philosophy that emphasizes the dignity and worth of the individual, they provided the foundation of modern European culture.

One aspect of life affected by the liberating influences of the new urban institutions and the Renaissance was religion. Since late Roman times, the Catholic Church had dominated not only religion but also politics and daily life throughout much of Europe. In the 1500s, however, a movement known as the **Protestant Reformation** arose in the towns of the North European Plain, Scandinavia, and the British Isles. Reformers such as Martin Luther challenged Catholic practices that stifled public participation in religious discussions, such as printing the Bible and holding church services only in Latin, a language that none but a tiny educated minority understood. The Protestant Reformation coincided with the invention of the European version of the printing press (by Johannes Gutenberg), which facilitated widespread literacy and the diffusion of reformist ideas and stimulated the development of written versions of local languages. Protestants also promoted individual responsibility and more open public debate of social issues.

## Europe Looks Abroad

Europe's **Age of Exploration,** a direct outgrowth of the greater openness of the Renaissance and the Protestant Reformation, began a period of accelerated global commerce and cultural exchange. In the fifteenth and sixteenth centuries, Portugal took advantage of Renaissance advances in navigation, shipbuilding, and commerce to set up a trading empire in Asia and a colony in Brazil. Spain, beginning with the first voyage of Christopher Columbus in 1492, founded a vast and profitable American-Pacific mercantile empire. By the seventeenth century, however, England and Holland (now called the Netherlands, and its people, the Dutch) had seized the initiative from Spain and Portugal. They perfected **mercantilism,** a strategy for increasing a country's power and wealth not only by acquiring colonies with their human and natural resources, but also by managing all aspects of production, transport, and trade for the colonizer's benefit. Mercantilism supported the Industrial Revolution in Europe by supplying cheap resources from around the globe for Europe's new factories as well as colonial markets for European manufactured goods (Figure 4.8). By the nineteenth century, the Spanish and Portuguese colonial empires were overshadowed by those of England and Holland. These two countries, influenced by the Protestant emphasis on individualism and innovation, improved on Iberian methods of colonial expansion and control, and they soon

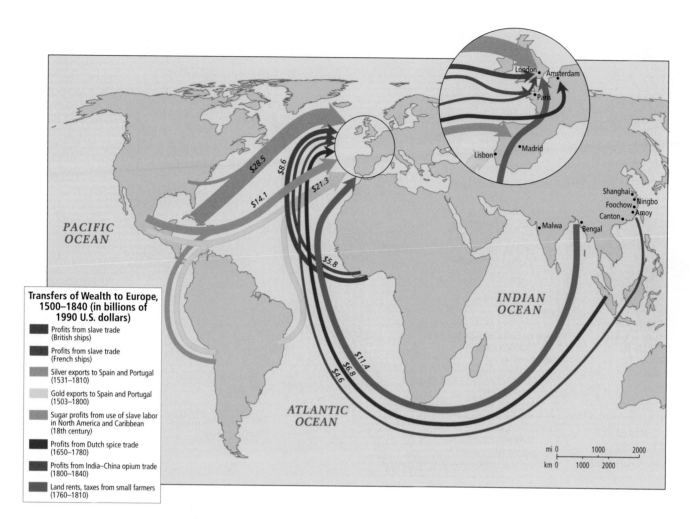

**FIGURE 4.8** Transfers of wealth from the colonies to Europe, 1500–1840. Europe received billions of dollars of income from its overseas colonies during this period of mercantilism. (For European colonial holdings in various regions of the world, see Figures 3.11, 6.11, 6.12 and 7.9.) [Adapted from Alan Thomas, *Third World Atlas* (Washington, D.C.: Taylor & Francis, 1994), p. 29.]

Transfers of Wealth to Europe, 1500–1840 (in billions of 1990 U.S. dollars)
- Profits from slave trade (British ships)
- Profits from slave trade (French ships)
- Silver exports to Spain and Portugal (1531–1810)
- Gold exports to Spain and Portugal (1503–1800)
- Sugar profits from use of slave labor in North America and Caribbean (18th century)
- Profits from Dutch spice trade (1650–1780)
- Profits from India–China opium trade (1800–1840)
- Land rents, taxes from small farmers (1760–1810)

extended their influence into Asia and Africa. By the twentieth century, European colonial systems had strongly influenced nearly every part of the world.

## The Evolution of European Cities

Roman towns—some quite large, but most with only 2000 or so inhabitants—were sprinkled across Europe by 200 C.E. During the medieval period (450–1500 C.E.), many of these towns fell into ruin, but some cities in northern Italy developed important trading links, by land and by sea. Figure 4.9a shows the trading links between Florence, Genoa, Venice, and Brugge, as well as coastal sea trading links, such as the Hanseatic League along the Baltic and North seas.

After 1500, Europe's increasing wealth was particularly evident in the flourishing cities of north-central Europe, to which the overseas colonies ruled by England and Holland brought enormous infusions of cash to be invested in industrialization. Wealth and investment capital were shifted from Mediterranean Europe to Augsburg, and to ancillary centers across Europe, such as Antwerp, Cologne, Madrid, and Seville (Figure 4.9b).

By the mid-1700s, industrial activities such as mining, milling, and manufacturing led to substantial migration of workers from rural areas to factory towns in England, Holland, Belgium, France, and Germany (Figure 4.9c). Large trading companies plying ports in the Americas, Asia, and Africa generated profits from the movement of raw materials and the sale of new products, bringing yet more wealth to these parts of Europe. Some cities, such as Paris, London, and Vienna, were elaborately rebuilt in the 1800s to reflect their roles as centers of empires.

By 1800, London and Paris, each of which had a million people, were Europe's largest cities, a status that eventually brought them to their present standing as **world cities** (cities of worldwide economic or cultural influence). London is a main global center of finance, and Paris a cultural center that has influence over global consumption patterns, from food to fashion to tourism. (For a discussion of modern patterns of urbanization, see page 196.)

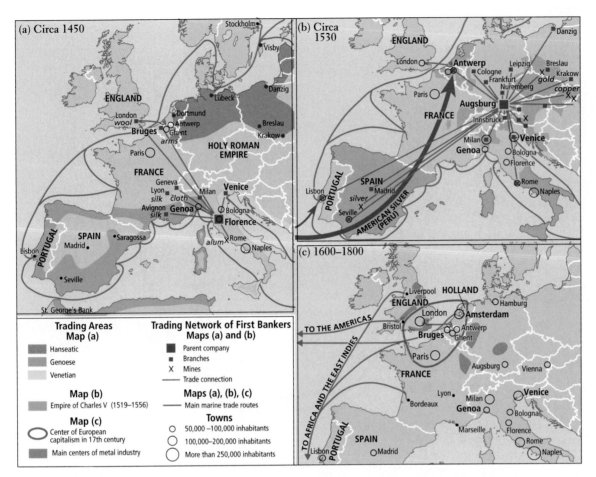

**FIGURE 4.9** Shifts of power among urban centers, 1450–1800. See page 191 for discussion. [This map was prepared for this text with the assistance of geographer John Agnew.]

## An Age of Revolutions

The wealth derived from Europe's colonialism helped fund two of the most dramatic transformations in a region already characterized by rebirth and innovation: the industrial and democratic revolutions.

*The Industrial Revolution.* Europe's Industrial Revolution was intimately connected with the Age of Exploration and colonial expansion. In particular, Britain's ascendancy as the leading industrial power of the nineteenth century had its origins partly in the expansion of its empire. In the sixteenth century, Britain was an island of only modest wealth and resources. In the seventeenth century, however, Britain developed a small but growing trading empire in the Caribbean, North America, and South Asia, which provided it with access to a wide range of raw materials and to markets for British goods. Sugar, produced by British colonies in the Caribbean, was an especially important trade crop (see Figure 4.8). Sugar production was a complex process requiring major investments in equipment and labor as well as long-term management of planting, harvesting, processing, storage, transport, and marketing activities, all of which contributed to the development of skills needed for industrialization. So, as the availability of sugar stimulated the demand for sweet foods, the mass production of sugar not only generated enormous wealth that helped fund industrialization in Britain, but also provided a production model for the Industrial Revolution.

By the late eighteenth century, Britain was straining to meet a demand for more goods than it could produce. The country met the challenge by introducing mechanization into its industries, first in textile weaving and then in the production of coal and steel. By the nineteenth century, Britain was a global economic power with a huge and growing empire, expanding industrial capabilities, and the world's most powerful navy, and its industrial technologies were spreading throughout Europe, North America, and elsewhere.

*The Democratic Revolution.* As Europe was industrializing, it experienced political and social transformations that redistributed power, and eventually wealth, more evenly throughout society. The road to democracy in Europe was rocky and violent, however, just as it is now in many parts of the world. For centuries, Europe's power structure had been feudal. But by the eighteenth century, especially in the cities of western Europe, the political elite was expanding to include merchants and industrialists as well as nobles. While kings and churches were forced by the expanding financial power of these elites to accept constitutions that restricted their power, the impoverished working-class majority, both rural and urban, still had no vote or any other formal role in the political system.

In 1789, the French Revolution led to the first major inclusion of the common people in the political process in Europe. Angered by the extreme disparities of wealth in French society, and inspired by the news of the popular revolution in North America, the poor rose up against the monarchy and elite-dominated power structure. As a result, the populace became involved in governing through elected representatives.

Throughout the nineteenth century, the political structure of Europe was transformed by the idea of **nationalism,** or allegiance to the state (see Chapter 1, page 49). In individual kingdoms, in the case of France, or collections of kingdoms, in the case of Germany and Italy, the notion spread that all the people together formed a nation and that loyalty to that nation should supersede all other loyalties. Elites found this rallying call to nationalism an effective way to deflect attention away from the poor living conditions the majority of people experienced, and eventually the whole map of Europe was configured as a mosaic of small nation-states. However, virtually none of these political units was made up of a single culture group; all contained numerous minorities that did not fit into the idealized nation-state model. France, for example, had the Bretons and Basques; Spain the Basques, Galicians, and Catalans; Italy the Slovenes and the Sicilians; Austria the Tyroleans; and all of Europe had **Roma** (Gypsy) and Jewish minorities. So loyalty within the nation-states was tenuous, and there were numerous efforts to expel, or even annihilate, those who were not considered sufficiently French or German or Italian or Spanish.

The failure of the Industrial Revolution to raise living standards for the vast majority of Europeans also led to political transformations. The demand for skilled and unskilled factory workers triggered by the Industrial Revolution led to steady migration from the countryside to the cities, where low wages and unsafe, unsanitary conditions prevailed, but where political organization was more possible. Periodically, popular discontent erupted in violent protests and revolutionary movements that threatened the established civic order. The political philosopher and social revolutionary Karl Marx, who grew up in an industrializing Europe, witnessed the suffering of the impoverished working class (the *proletariat*). He framed the mounting social unrest in terms of a class struggle in his treatises *The Communist Manifesto* (1848) and *Das Kapital* (1867). His ideas helped social reformers across Europe articulate socialist ideas on how wealth could be more equitably apportioned.

In Russia, the ideas of Marx and his colleague Vladimir Lenin lent inspiration to the Bolshevik revolution of 1917. In western Europe, political pressure from workers, combined with drastically increased industrial productivity, convinced some governments to become **welfare states.** Such governments accept responsibility for the well-being of their people, guaranteeing such basic necessities of life as education, employment, and health care for all citizens. In time, government regulations on wages, hours, safety, and vacations established more harmonious relations between workers and employers, reduced wealth disparities, and increased overall civic peace and prosperity. Although Europeans today generally enjoy high levels of well-being and a state-supported safety net that meets their basic needs, just how much support the welfare state should provide is still a hot topic of debate, and one that has been resolved in different ways across Europe (discussed further on pages 214–216).

**Democratic institutions** in Europe, such as constitutions, elected parliaments, and impartial courts, developed only slowly and unevenly. The Magna Carta gave the English (though only a privileged few of them) the right to challenge the authority of the king, and this right eventually developed into the British Parliament. But the right to actually elect leaders came first to some European men only in the early nineteenth century, and to women only after considerable delay and political agitation; in Switzerland, women did not obtain the vote until 1971! In Central Europe, democratic institutions were not adopted until the 1990s. When assessing progress toward democracy in other world regions, it is helpful to remember this long and difficult path to democracy in Europe.

## Europe's Difficult Twentieth Century

*World Wars, Cold War, and Decolonization.* Despite Europe's many advances in industry and politics, the region still lacked a system of collective security that could prevent war between its rival nations. Between 1914 and 1945, two horribly destructive world wars removed Europe from its position as the dominant region of the world. By the mid-twentieth century, Europe lay in ruins, and millions had died in the **Holocaust,** a massive ethnic cleansing of 6 million Jews (and several million others: Roma, Slavs, the infirm, and political dissidents) perpetrated primarily by the Nazi government in Germany and the Fascist government in Italy (Figure 4.10). The defeat of Germany, seen as the instigator of both world wars, resulted in a number of enduring changes in Europe. After the end of World War II, in 1945, Germany was divided into two parts. West Germany became an independent democracy allied with the rest of western Europe, especially Britain and France, and the United States. Through the Marshall Plan, the United States provided financial assistance to rebuild western Europe's basic facilities, such as roads, housing, and schools. Western European countries continued their free market economic system of privately owned businesses that adjusted prices and output to match the demands of the market.

East Germany and the rest of what was then called eastern Europe (Latvia, Lithuania, Estonia, Poland, Czechoslovakia, Hungary, Romania, Bulgaria, Ukraine, Moldova, and Belarus) fell under the control of the Soviet Union, a revolutionary Communist state that had emerged out of the Russian Bolshevik revolution in 1917. The line between East and West Germany was part of what was called the **iron curtain,** a long, fortified border zone that separated western Europe from (then) eastern Europe. The Soviet Union integrated much of eastern Europe into its sphere of so-called Communist states, which employed a form of socialism in which the state owned all farms, industry, land, and buildings. In contrast to the market economies of western Europe, these economies were **centrally planned:** a central bureaucracy dictated prices and output with the stated aim of allocating goods equitably across society according to need.

The division of Europe laid the foundation for the **cold war** between the United States and the Soviet Union, an era lasting from 1945 to 1991, during which the entire world became a stage on which these two superpowers competed for dominance. Once-powerful Europe became subject to the geopolitical manipulations of the two superpowers. Yet another manifestation of Europe's decline was its loss of control over its colonial empires. By the 1960s, most former European colonies were independent and had

**FIGURE 4.10** The Holocaust. On April 16, 1945, at the end of World War II, the Allies liberated the Buchenwald (Germany) concentration camp and found many scenes like this. Many of the Jews in the camp were on the edge of starvation; the soldiers found the bodies of many others who had died shortly before the liberation. Some prisoners in Buchenwald were not Jews, but rather resisters of Nazi domination from across Europe, or other ethnic minorities, such as Roma. [Pvt. H. Miller/CORBIS.]

entered a difficult period—which still persists—of finding ways to thrive on their own.

***Europe's Rebirth and Integration.*** In the decades after World War II, economic reconstruction proceeded rapidly in the free market democracies of western Europe. But progress was much slower in socialist eastern Europe, where the Soviet Union wielded a strong influence in much the same way that colonizers such as the British, French, and Dutch once controlled the resources and people of the Americas, Africa, and Asia. Eastern European state-run industries were inefficient and highly polluting, and public debate and citizen participation were discouraged. Local attempts at reform were squelched by the Soviets in 1953 and 1956 in Hungary and in 1967 in Czechoslovakia. In the early 1980s, however, political and economic reforms in eastern Europe began with labor protests in Poland (led by the federation of trade unions known as *Solidarity*), and by the end of that decade, much of eastern Europe had abandoned socialism. East Germany, which had been the wealthiest and most technologically advanced area under Soviet control, was politically reunified with West Germany, which undertook hugely expensive and still only partially successful efforts toward its economic reintegration. These changes hastened the economic and political collapse of the Soviet Union in 1991 because the quasi-colonial system of the Soviet Union had depended greatly on the resources and skilled labor of eastern Europe.

Meanwhile, beginning in the 1950s, some of the free market democracies of western Europe began lowering barriers to trade among themselves, a process that led eventually to the establish-

ment of the European Union (see Figure 4.3 on page 185). Today, the EU encourages **economic integration**: the free movement of people, goods, money, and ideas among member countries. EU nations also exhibit some degree of policy coordination in civil, judicial, economic, military, environmental, and foreign affairs. (The structure and activities of the European Union are explained on pages 198–203.) Because Europe increasingly acts as a single large unit, encompasses a large and affluent population, and is pioneering innovative social programs, it is now challenging the dominance of the United States in world affairs. Later in this chapter, however, we shall see that as the EU expands, primarily into the poorer formerly Communist countries of Central Europe, and takes on a larger global role, it is encountering many challenges of its own.

***Ethnic Cleansing in Southeastern Europe.*** The peaceful absorption of former Communist states into the European Union was arrested by bloody conflict in southeastern Europe in the 1990s. After World War II, the country of Yugoslavia was formed from six territories. In five of them—Serbia, Croatia, Slovenia, Macedonia, and Montenegro—a single ethnic group was dominant, although other ethnic minorities were present in most of the five. The sixth territory, Bosnia and Herzegovina, did not have a majority ethnic group. Rather, most of its people had been in Bosnia long enough to think of themselves as simply Bosnian, though they were aware of having historical roots as Croatians or Serbians. Many of these people were either Catholic or Orthodox Christians. There was also a third "ethnic" (culture) group, consisting of Muslims who had converted from Christianity generations ago and were largely secular in their religious practices.

In 1991, the first free elections ever held in Yugoslavia resulted in declarations of independence, first by Slovenia, then by Croatia and Macedonia, and finally by Bosnia and Herzegovina. Although Slovenia and Macedonia were allowed to separate relatively peacefully, the resulting much smaller Yugoslavia, then composed only of Serbia and Montenegro, fought protracted wars to retain Croatia and Bosnia and Herzegovina. Both countries had Serb-populated provinces, which Serbia argued should be part of one Serbian-led country. To rid the coveted provinces of non-Serbs, some Serbs instigated genocidal **ethnic cleansing** campaigns against civilians in Bosnia and Herzegovina, in Croatia, and later in Kosovo, a province of Serbia. Their goal was to create ethnically "pure" nation-states, or independent countries consisting of just one nationality.

Only after agonizing delays did this systematic violence gain the attention of the EU countries and the United States, both of which sent military peacekeepers. Peace accords were signed in Dayton, Ohio, in 1995. By this time, the results of the war were devastating, and Serbia itself was a shambles. It lost the territory it had illegally claimed, and its economy was ruined. In Bosnia and Herzegovina alone, 200,000 people—5 percent of the population—had died in the war; many were the victims of Serbian war crimes. The cosmopolitan city of Sarajevo, site of the Winter Olympics in 1984, was destroyed, as were countless villages. Montenegro and Macedonia were independent, and Kosovo was seeking to separate from Serbia.

## POPULATION PATTERNS

Enumerating just how many people live in Europe is complex because, as a result of recent political changes, the geographic region has no universally agreed-upon eastern border. According to the way we define Europe in this book (see Figure 4.1 on pages 182–183), there are about 525 million Europeans, and they are distributed unevenly. A population map of Europe (Figure 4.11) shows its

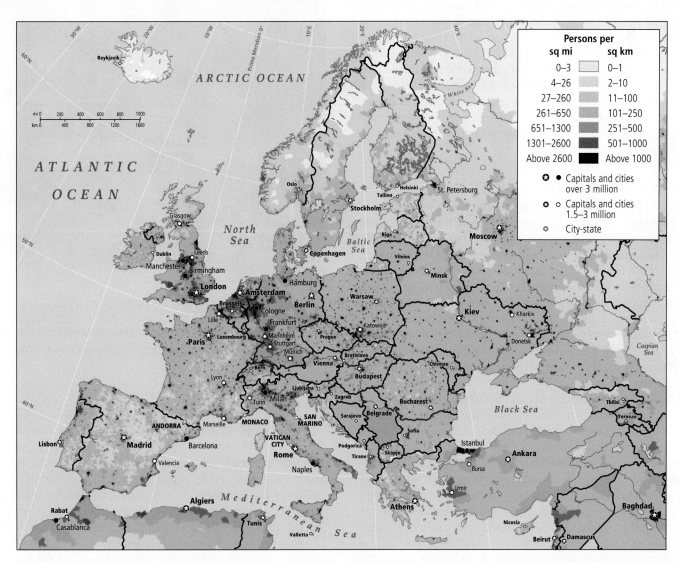

**FIGURE 4.11** Population density in Europe. [Data courtesy of Deborah Balk, Gregory Yetman, et al., Center for International Earth Science Information Network, Columbia University; at http://www.ciesin.columbia.edu.]

densest settlement stretching in a disconnected band from the United Kingdom and northern coastal France, east through the Netherlands and central Germany, all the way to Warsaw and Bucharest, and continuing into Ukraine (see Chapter 5). Northern Italy is another zone of density, and pockets along the coasts of Portugal, Spain, southern France, Sicily, and southern Italy, as well as in southeastern Europe, are also densely populated. A glance at the global population map in Figure 1.32 (pages 36–37) will show that, overall, Europe is one of the most densely occupied regions on earth.

## Population Density and Access to Resources

Population density alone does not tell us whether or not people have access to adequate resources or whether those resources are allocated fairly among the population. A region's **resource base** is the selection of raw materials, such as coal, petroleum, iron ore, cotton and wool fiber, food, soil, and water, that are available for domestic use and industrial development. Europe depleted many of its own resources, especially forests and minerals, early in the Industrial Revolution. Nonetheless, Europe is now both densely populated and wealthy because for hundreds of years it has had access to a resource base that reaches far beyond its boundaries (see Figure 4.8). Since about 1500, wealth derived from its colonies has added considerably to the overall standard of living of Europeans and to the ability of Europe to develop an industrial economy. Even ordinary laborers in Europe have benefited. In contrast, those far away in former colonies who have worked to create wealth for Europeans have often lived in poverty.

## Modern Urbanization in Europe

Today, Europe is a region of cities surrounded by well-developed rural hinterlands. These cities are the focus of the modern European economy, which, though long grounded in agriculture, trade, and manufacturing, is increasingly service based. Even in sparsely settled North Europe, 72 percent of the people live in urban areas. Central Europe is least urbanized, with 62 percent living in cities. Many European cities began as trading centers more than a thousand years ago and still bear the architectural marks of medieval life in their historic centers. These old cities are located either on navigable rivers in the interior or along the coasts because water transport figured prominently (and still does) in Europe's trading patterns (see Figures 4.9a and b).

Since World War II, nearly all the cities in Europe have expanded around their perimeters in concentric circles of apartment blocks. Well-developed rail and bus lines link these blocks to one another and to the old central city. Land is scarce and expensive in Europe, so only a small percentage of Europeans live in single-family homes, although the number is growing. Even single-family homes tend to be attached or densely arranged on small lots; rarely does one see the sweeping lawns that North Americans spend so much effort grooming. Publicly funded transportation is widely available, so many people prefer to live in apartments near city centers, though many others commute daily from suburbs or ancestral villages to work in nearby cities.

**FIGURE 4.12** Cosmopolitan urban life in Europe. Sidewalk cafés like this one in Paris are found in central cities and residential neighborhoods across Europe. City dwellers tend to entertain their friends in cafés, in part because their apartments are relatively small. [Peter Adams/Agency Jon Arnold Images/Agefotostock.]

Despite the size of many of Europe's large cities—London has 12 million people in its metropolitan area, Paris 9.9 million, Madrid 5.6 million, and Berlin 4.2 million—life can be quite intimate and personal as friends congregate in sidewalk restaurants, ancient pubs, and elegant public squares (Figure 4.12). Although deteriorating housing and slums do exist, substantial public spending on social welfare systems, sanitation, water, utilities, education, housing, and public transportation maintains a generally high standard of urban living in Europe and a quality of life that many feel exceeds that in the United States. Tourists from around the world are drawn to Europe's large cities by their art, music, museums, architectural heritage, outdoor spaces, and generally pleasant walkable ambience.

## Europe's Aging Population

Although Europe is densely occupied, its population is aging, as European families are choosing to have ever fewer children. In 2006, birth rates in Europe were the lowest on earth, and for the region as a whole, there is actually a negative rate of natural increase (−0.1). The relationships are circular here: death rates are higher than birth rates in part because the average age of the population is high, but low birth rates contribute to this high average age. This situation is most marked in the countries that were part of the former Soviet Union or Yugoslavia: because birth rates are low, there are many elderly people, and the quality of health care has deteriorated, so deaths are more common. The one-child family is increasingly common throughout Europe. In western Europe, immigrants are a major, and needed, source of population growth; however, once they have assimilated to European life, immigrants, too, choose to have small families.

The declining birth rate is illustrated in the population pyramids of European countries, which look more like lumpy towers than pyramids. The population pyramid of Germany (East and West, reunited in 1990, and are combined here) is an example (Figure 4.13a). The pyramid's narrowing base indicates that for the last 35 years, far fewer babies have been born than in the 1950s and 1960s, when there was a baby boom across Europe. By 2000, 35 to 40 percent of Germans were choosing to have no children at all.

The reasons for these trends are complex. For one thing, more and more women desire professional careers. The need for advanced education alone could account for late marriages (25 percent of Germans were choosing to remain unmarried well into their thirties) and low birth rates. In addition, many governments make very few provisions for working mothers beyond giving them some months off with pay after giving birth. In Germany, for example, there is little day care available for children under 3. In addition, in places such as Austria, Italy, and Slovenia, school days are short, ending at 2 P.M., or no lunches are provided, so mothers must be home by early afternoon. Hence many women are choosing not to become mothers because they would have to settle for part-time jobs. To encourage higher birth rates, there is a move within the European Union to give one parent (mother or father) a full year off with reduced pay after a child is born or adopted, but this provision is not yet fully operational.

Two results of the small number of births in Germany are that the population pyramid continues to narrow at the base, and that the vast majority of Germans are over the age of 30. If these trends continue, then as the older generations die, the population will eventually settle into a stable age structure, as Sweden's population has already done (Figure 4.13b), in which each age category has roughly similar numbers of people, tapering only at the top after age 60.

A stable population with a low birth rate has several consequences. Although families with few children have extra money for luxury spending, new consumers are not being produced at the rate at which affluent elderly consumers are dying. Hence, in time, markets will contract unless immigrants can be attracted. Demand for new workers, especially highly skilled ones, may go unmet, again unless immigration supplies a solution, as is increasingly the case in Germany. The number of younger people available to provide expensive and time-consuming health care for the elderly, either personally or through tax payments, is small;

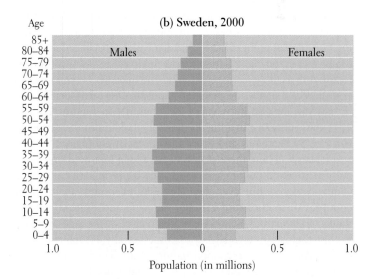

**FIGURE 4.13** Population pyramids for Germany and Sweden, 2000. These two population pyramids have quite different shapes, but both exhibit a narrow base, indicating low birth rates. Observe that the scales along the bottom of the pyramids are different for the two countries, indicating that Sweden's total population is much smaller than Germany's. [Adapted from "Population pyramids of Germany" and "Population pyramids of Sweden" (Washington, D.C.: U.S. Census Bureau, International Data Base, 2000).]

currently, for example, there are just two German workers for every retiree.

For centuries, economic growth has been the hallmark of progress, and the success of states has been measured by the ever-increasing wealth of their economies. But with declining and aging populations, economic growth in Europe may be difficult to maintain. Immigration provides one solution, but Europeans are reluctant to absorb large numbers of immigrants, especially from distant parts of the world whose cultures are very different from their own (see the discussion on pages 208–212).

# II CURRENT GEOGRAPHIC ISSUES

At the start of the twenty-first century, the social, economic, and political geography of Europe (Figure 4.14) is in a state of flux. This situation is the result of three major changes that occurred during the 1990s: the demise of the Soviet Union, the end of the cold war, and the effort toward economic and political integration that has produced the European Union. These developments, especially the emergence of the EU, could ultimately bring greater peace, prosperity, and world leadership to Europe, but many problems and tensions still remain to be resolved.

## ECONOMIC AND POLITICAL ISSUES

Almost all economic and political issues in Europe today are linked to the European Union in one way or another. At the end of World War II, European leaders felt that closer economic ties would prevent the hostilities that had led to two world wars. The first major step in achieving this economic unity took place in 1958, when Belgium, Luxembourg, the Netherlands, France, Italy, and West Germany formed the **European Economic Community (EEC)**. The members of the EEC agreed to eliminate certain tariffs against one another and to promote mutual trade and cooperation. Five episodes of expansion followed, including, in 1992, a treaty enlarging the concept of the EEC to that of a European Union, signed at Maastricht in the Netherlands. Figure 4.3 (page 185) shows the current members of the European Union and the dates of their joining, as well as candidates and potential candidates for membership.

As you look at the maps of EU countries in Figures 4.3 and 4.16, you will notice that there is one lone holdout in the center of Europe (Switzerland) and two on its northern periphery (Norway and Iceland). These three countries have long treasured their neutral role in world politics and, as wealthy countries, are concerned about losing control over their domestic economic affairs. Two countries in southeastern Europe, Romania and Bulgaria, joined the EU in January 2007 and are undergoing many changes in order to align themselves with EU standards, especially its financial, judiciary, legal, and anticorruption requirements. Turkey and the former Yugoslav states—especially Croatia and Macedonia—are also being prepared to become part of the EU in several years. Eventually, countries on the perimeter of Europe, such as Russia and Ukraine, and perhaps even the Caucasian republics (Armenia, Azerbaijan, and Georgia), may be invited to join.

Turkey, which in this book continues to be treated as part of the region of Southwest Asia and North Africa, is being considered for many reasons. It became an important ally of western Europe and a member of the North Atlantic Treaty Organization (NATO) during the cold war. It has had a long history of economic and social interaction with Europe, and millions of Turks are now living in Europe. While some aspects of Turkey's culture are markedly different from Europe's, many Turks identify themselves as European. The present Turkish government is interested in joining the European Union for economic and strategic reasons, but there is opposition to this idea in both Turkey and Europe. One point of contention is the island of Cyprus, which has residents of both Greek and Turkish heritage. Cyprus was admitted to the EU as an independent country in 2004, but the culturally Turkish part of the island (the northeast) is still administered by Turkish Cypriots.

## Goals of the European Union

The original plan at the founding of the EEC, which remains the core idea of the European Union, was simply to work toward a level of economic and social integration that would make possible the free flow of goods and people across national borders. But two other roles for the EU have emerged. With the increasing importance of the global economy and the geopolitical maneuvering that followed the terrorist attacks on the United States on September 11, 2001 (first the war in Afghanistan and, more important, the U.S. attack on Iraq), some Europeans feel that the EU should become both a global economic power competing with the United States and Japan and a politically and militarily independent counterforce to the United States.

***European Union Governing Institutions.*** The European Union has three administrative components, each with a different constituency. The **European Parliament** is directly elected and represents EU citizens. The number of seats each country has is based on its population, much like the U.S. House of Representatives. The **Council of the European Union** represents the individual member states and consists of ministers from the national governments of those states, who attend meetings according to the topics under discussion: economics, finance, agriculture, foreign affairs, environment, and so forth. The **European Commission** repre-

**FIGURE 4.14** Political map of Europe.

ICELAND

Faroe Islands
(Denmark)

SWEDEN

FINLAND

NORWAY

ESTONIA

DENMARK

LATVIA

UNITED
KINGDOM

LITHUANIA

RUSSIA

REPUBLIC of
IRELAND

NETHERLANDS

POLAND

GERMANY

BELGIUM

LUXEMBOURG

CZECH
REPUBLIC

SLOVAKIA

FRANCE

LIECHTENSTEIN

AUSTRIA

HUNGARY

SWITZERLAND

SLOVENIA

ROMANIA

CROATIA

SERBIA

BULGARIA

PORTUGAL

SPAIN

Corsica
(France)

ITALY

ALBANIA

Sardinia
(Italy)

GREECE

North Europe
West Europe
Central Europe
South Europe

Sicily
(Italy)

MALTA

Crete
(Greece)

1 BOSNIA & HERZEGOVINA
2 MONTENEGRO
3 MACEDONIA
4 KOSOVO

CYPRUS

sents the interests of the EU as a whole and is independent of national governments. There are 27 members, one from each country, nominated and approved by the European Parliament. They do not represent the governments of their home countries. Instead, each is responsible for a particular policy area, such as environment, justice, foreign affairs, or immigration. The Commission proposes new laws, and the Parliament and the Council adopt or reject them. Two other institutions play a vital part in the EU: the Court of Justice of the European Communities upholds the rule of EU law, and the European Court of Auditors checks the financing of EU activities. These powers and responsibilities are laid down in treaties agreed to by country presidents and ratified by country parliaments. A formal constitution for the EU has

been developed, but not yet adopted (ratification by all members is required). The Council of the European Union meets in Brussels, Belgium, as does the European Commission (Figure 4.15). The European Parliament meets primarily in Strasbourg, France. The Court of Justice and Court of Auditors meet in Luxembourg.

***The Economic and Social Integrative Role of the European Union.*** Individual European countries have far smaller populations than their competitors in North America and Asia. With smaller markets for their products, companies in small countries earn lower profits. Before the European Union, European businesses could sell their products to neighboring countries, as is common in the global economy, but the extra costs imposed by tariffs, currency exchanges,

**FIGURE 4.15** Headquarters of the European Union in Brussels, Belgium.
[Jan Rietz/Getty Images/Nordic Photos.]

and border regulations sapped their earnings. The EU solved this problem by joining European national economies into a common market, thus providing the firms in any one country with access to a much larger market and the potential for larger profits through **economies of scale** — reductions in the unit costs of production that occur when goods or services are produced in large amounts, resulting in a rise in profits per unit.

The EU economy now encompasses close to 489 million people (out of a total of 525 million in the whole of Europe) — roughly 200 million more than live in the United States. Collectively, the EU countries are wealthy, and their joint economy was more than $12 trillion (PPP) in 2005, nearly equal to that of the United States. The combined total external trade (imports and exports) of the EU countries was 19 percent of the world's total, also equal to that of the United States. But whereas the United States imports far more than it exports, resulting in a trade deficit of $8 billion in 2005, the EU enjoys a trade balance (the values of imports and exports are nearly equal). On the other hand, economic growth in the EU is sluggish at 1.5 percent per year (the United States does somewhat better at 3.5 percent per year), and the average gross domestic product (GDP, PPP) per capita for the EU ($28,100) is significantly less than that of the United States ($37,562).

Some EU countries are notably wealthier than others (Figure 4.16; see also Table 4.1 on pages 220–221). One aim of the EU is to promote the equitable distribution of economic activity, opportunity, human well-being, and environmental quality across Europe while at the same time building institutions that respect the many different regional identities within Europe. EU funds are raised through an annual 1.27 percent tax on the gross national product (GNP) of all members. Although financial allotments change from year to year, most member countries in North and West Europe contribute more than they receive back in grants, while most in South and Central Europe tend to receive more than they contribute.

Since 1993, the EU's agenda has expanded to include the creation of a common European currency (the euro), the defense of Europe's interests in international forums, and negotiation of EU-wide agreements on human rights and social justice. The EU is also experimenting with various forms of political unification (including the creation of a common European military). One day, the people of Europe may live under a common EU constitution, but so far this idea has not received the unanimous acceptance required by law.

*A Common European Currency.* On January 1, 2002, the **euro** (€) became the official currency in 12 of the then 15 EU countries: Austria, Belgium, Finland, France, Germany, Greece, Ireland, Italy, Luxembourg, the Netherlands, Portugal, and Spain. Slovenia became the thirteenth EU country to adopt the euro in January 2007. Three EU members — the United Kingdom, Denmark, and Sweden — had earlier elected not to use the euro because they feared losing control over their own national economies and over trade relations beyond Europe. In addition, they preferred to keep their national currencies alive as a patriotic symbol. It is thought that a single regional currency will bring gains in efficiency; for example, by eliminating currency exchange fees charged to international travelers and by simplifying auditing procedures. Countries that use the euro have a greater voice in the creation of EU economic policies. Furthermore, those that do not use it face the economic uncertainty of using a currency whose value fluctuates relative to the value of the euro. For example, when the pound rises in value relative to the euro, British exporters are at a disadvantage when selling in Europe because their goods become more expensive to consumers using the euro.

*The European Union in the Global Economy.* To be more competitive with the United States and Japan, the EU is using fund pools such as the European Regional Development Fund (ERDF) to update its members' technologies and infrastructure. Now, however, Europe's firms must face the much more difficult problem of competing with developing economies in Asia, Africa, and South America, where production costs, including labor and raw materials, are much cheaper and environmental regulations are less stringent and hence less costly to meet.

Managing the locations of industries within Europe is one way in which the EU tries to remain competitive at the global scale. If factories can be relocated from the wealthiest countries to the EU's relatively poorer, lower-wage countries (see Figure 4.16), the costs of production will fall. The EU's main economic leaders are currently clustered in West and North Europe; Italy, the only such

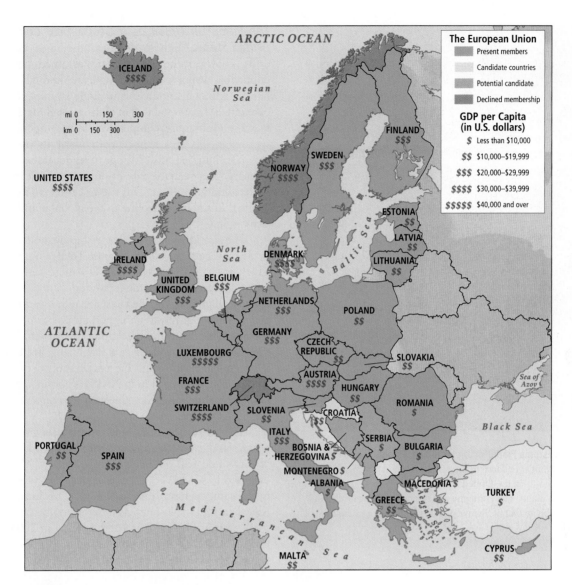

**FIGURE 4.16** The EU economy: GDP per capita. GDP per capita varies considerably among the 27 countries in the EU.

leader in South Europe, has experienced declining growth since 2001 and chronic low employment. The relatively high wages, worker benefits, and taxes in these countries drive up production costs. The EU has encouraged its economic leaders to locate new facilities in the poorer, lower-wage countries of Portugal, Spain, Ireland, Greece, and the newly independent countries of Central and North Europe (Figure 4.17). It is hoped that these efforts will help poorer European countries to prosper and keep the costs of doing business in the EU low enough to restrain European companies from moving to Mexico, Southeast Asia, or China, where costs are lower still. This effort will undoubtedly mean that fewer jobs will be created in Europe's wealthiest countries.

Like the United States, the EU exerts a powerful influence in the global trading system, especially in the World Trade Organiza-

tion (WTO), where it often negotiates privileged access to world markets for European firms and farmers and for former European colonies (Figure 4.18). Generally, the EU employs protectionist measures that favor European producers at the expense of both European consumers, who must then pay higher prices for goods, and non-European producers, who lose access to EU markets. The unwillingness of the EU, along with most of the world's rich countries, to reduce tariffs and subsidies on a variety of goods is currently a major threat to the credibility of the WTO, which is predicated on the notion that trade should be free of regulations. In recent years, poor countries throughout the world have united to protest the failure of the EU and other rich countries to open their economies to foreign competition. These protests are especially boisterous at WTO meetings (see Figure 1.30 on page 33 and Figure 3.23 on page 141).

**FIGURE 4.17** The Estonia Piano Factory in Tallinn, Estonia. This factory has been revived from near-collapse in the last decade and now markets its pianos in the West, mostly in the United States. Here a worker installs the soundboard in a new piano. The company says its craftsmen create a sound that compares favorably to that of Steinways but at half the cost. [Courtesy of Estonia Pianos.]

**EU's Share of World Trade, 2003**

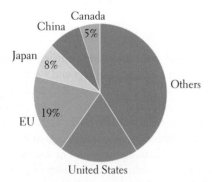

**FIGURE 4.18** The European Union's share of world trade, 2003. At 19 percent, the EU's share of world trade (imports and exports combined) was equal to that of the United States and more than those of China and Japan combined. [Adapted from *Europe in Figures: Eurostat Yearbook 2005* (Luxembourg: European Commission, 2005), p. 187, at http://bookshop.europa.eu /eubookshop/FileCache/PUBPDF/KSCD05001ENC/KSCD05001ENC_002.pdf.]

***The European Union as a Geopolitical Counterforce to the United States.*** With the demise of the Soviet Union and the end of the cold war, the United States assumed global leadership in political as well as economic matters. Although the United States has the largest economy of any single country, the combined EU economy is now equal to that of the United States, and it promises to grow even more competitive. More important, the U.S. responses to the 9/11 attacks, particularly its preemptive military action in Iraq, led most European leaders to adopt a global intermediary, peacemaking role in order to curtail future U.S. use of such military solutions. This rise in European leadership has contributed to a general decline in the global political prestige of the United States.

The rise in geopolitical leadership by the EU has contributed to the prestige of the **North Atlantic Treaty Organization (NATO)**, which is based in Europe. NATO nations (originally the United States, Canada, the countries of western Europe, and Turkey) cooperated militarily and learned to share authority during the cold war, and this experience was an important precursor to European economic union. With the breakup of the Soviet Union, NATO has tried to reconfigure itself as a stabilizing institution that fosters international security and the cooperation needed to expand the EU. Indeed, membership in NATO is considered a stepping-stone to membership in the EU, especially for countries once allied with the Soviet Union.

As cold war tensions fade, and if strengthening global economic and political forces bring the world closer together, NATO's role will doubtless change further. Most observers had expected to see NATO expand its peacekeeping role to hot spots in adjacent regions, such as Russia (Chechnya) and Southwest Asia (Israel and Palestine), as it did in southeastern Europe in the 1990s and in Central Asia (Afghanistan) in 2003. But when the United States and Britain attacked Iraq in 2003, without the assistance of NATO (and despite strong European dissent), the future viability of this once-powerful alliance was called into question. However, because the war in Iraq has not gone smoothly for the U.S.–British alliance, because the threat of terrorism remains strong, and because NATO is providing the majority of the troops in Afghanistan, it is possible that NATO, or a successor alliance with Europe playing a leading role, will take on the very difficult job of addressing global security issues.

## Central Europe and the European Union

Membership in the EU became especially attractive to countries in Central Europe after the demise of the Soviet Union, when their economies and socialist safety nets began to deteriorate. The general move toward more open and competitive market systems has placed many of these countries in jeopardy. Former state-owned industries are now forced to survive on their own revenues; hence operations have been streamlined, and many workers have lost their jobs. In some of the poorest countries, such as Romania, Bulgaria, and Serbia, social turmoil and organized crime threaten stability. Although the social programs and funds that benefit EU

## BOX 4.1 | CULTURAL INSIGHT The Hungarian Sausage Experience and the European Union

The Hungarian writer, humorist, and self-professed anthropologist Dork Zygotian (probably a pseudonym) assesses the prospects of European union from the perspective of a sausage lover. Zygotian writes that when you bite into a Hungarian sausage (*kolbasz*), "great torrents of paprika colored grease and juice should explode into the atmosphere around you. If you eat more than two, you should expect to bite on some piece of bone or possibly find a tooth or hair sometime during your meal. There will be a large yellow gelatinous bit somewhere in your sausage that you should not be able to identify." This is all part of the tasty Hungarian kolbasz experience (Figure 4.19).

But now Hungary, long a part of Eastern Europe, has joined the EU, and Zygotian worries about the effect of membership on his country's sausages, which he rates as Europe's best. He fears that overzealous EU standards of cleanliness and purity will kill the special flavors of Hungarian sausages. Worse, "Eurofication" (his term) may well leave Hungarians unable to afford their own beloved kolbasz because prices in the small, poorer countries will rise to match those in wealthy Germany, France, the Netherlands, and Switzerland.

**FIGURE 4.19** Sizzling kolbasz. Sausages cook at a night market in Budapest. [Suzanne Long/Alamy.]

members are appealing to these countries, it is the possibility of attracting foreign investment as EU members that they see as a main source of economic salvation. Existing and proposed investments by neighboring EU member countries in West and North Europe have smoothed the path to EU membership for all the Central European EU members.

Standards for EU membership, however, are exacting. To be considered for membership, a country must achieve political stability and a democratically elected government, at least to the satisfaction of all EU members at the time. Each country has to adjust its constitution to EU standards that guarantee the rule of law, human rights, and respect for minorities. Each also has to have a functioning market economy open to investment by foreign-owned companies, the capacity to cope with competition from within the EU, and the ability to take on the financial and administrative obligations of membership.

Meeting these standards can be both annoying and painful. For example, in the summer of 2004, Slovenia was abuzz with the news that the country's cucumbers would henceforth have to be a precise size and shape to be accepted for market. And, sure enough, in short order, the "EU Protocol for Distinctness, Uniformity and Stability for *Cucumis sativus* L. (cucumber)" was adopted and published. Regulated cucumbers may be only the beginning, as Box 4.1 suggests. The idea that European unification will erase distinguishing cultural features, encourage boring homogenization, and even diminish well-being comes up again and again throughout Europe.

## Future Paths of EU Organizational Development

At present, the EU governing body in Brussels makes decisions on a consensus basis: issues are discussed and proposals are adjusted until all members agree. A single country has the power to block projects or policies that all others have approved. In 2005, for example, the Netherlands and France blocked ratification of the EU constitution. Supporters of a stronger EU believe it can be strengthened by moving to majority rule, thereby giving the central governing bodies greater power.

Many citizens of EU member countries, especially the smaller ones, are wary of this proposed change because, having only a few votes, they fear losing control of decisions important to their own countries. They prefer a more flexible arrangement in which groups of countries with common interests could band together to work out individual agreements. A bloc of EU countries, for example, could vote to form a European army that would be staffed by soldiers only from the countries that agreed to this program. The other member countries would not have to participate, but they could not block the arrangement and might lose control over how that army would be used.

## Economic Change in Europe

The European Union powerfully influences all aspects of life in Europe as it works for progressive changes in European societies.

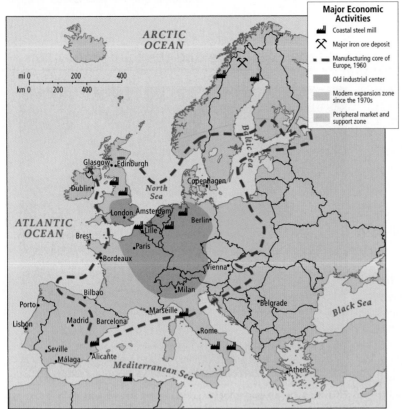

(a)

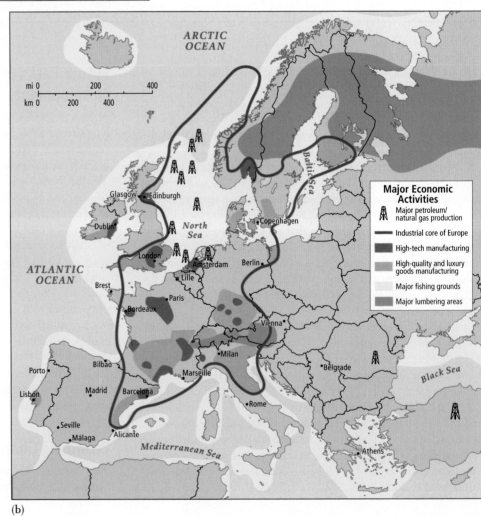

(b)

**FIGURE 4.20** Europe's principal industrial/manufacturing centers: Shifts from 1960 (a) to 2000 (b). [Adapted from Terry G. Jordan-Bychkov and Bella Bychova Jordan, *The European Culture Area: A Systematic Geography*, 4th ed. (Lanham, Md.: Rowman & Littlefield, 2002), pp. 288 and 300.]

Here we discuss the changes it has brought to the manufacturing, service, and agricultural sectors of European economies.

*Deindustrialization.* Over the past several decades, Europe's industries have modernized, diversified, and become distinctly more efficient, so although production has been increasing, demand for labor has been declining. *Deindustrialization* has also resulted from the relocation of industries from the nineteenth-century coal, steel, and manufacturing centers of Britain and the North European Plain to less expensive areas in South and Central Europe, as shown in Figure 4.20. Several factors made this move easier: the EU's easing of trade and migration restrictions; the switch from coal to oil, gas, and nuclear power as sources of energy; and the expansion of transport systems, especially highways. In the original 15 EU countries in 1970, 30 percent of the workers were in manufacturing; by 1994, only 20 percent were. With the addition of the countries in Central Europe, the percentage in industry was back up to 27 in 2006, but it is likely to decline once again as global competition increases and inefficient factories close.

*Europe's Growing Service Economies.* As industrial jobs decrease, most Europeans (about 70 percent) find jobs in the service economy, which meets the many needs of governments and businesses for **producer services** (advice, information, training, testing, licensing, strategic planning), communication, and finance. As Europe's economies become better integrated region-wide, new jobs are created. Growth in services to the private sector has drawn hundreds of thousands of new employees to the main European cities. For example, financial services located in London and serving the entire world play a huge role in the British economy, and many transnational companies are headquartered in London. London's prominence in the global economy is a main reason that Britain has elected to not adopt the euro as its currency, because this step would place the British economy under the control of the European Central Bank, which is located in Frankfurt, Germany.

Service jobs in the government sector are also numerous because European countries provide many tax-supported social services to their citizens. In addition, hundreds of thousands of clerical, administrative, and advisory jobs support the European Union itself, as well as the European offices of international organizations associated with the United Nations, the Organization of Petroleum Exporting Countries (OPEC), and the International Atomic Energy Agency (IAEA).

Another important component of Europe's service economy is tourism. Europe is the most popular tourism destination on earth, and one job in eight in the European Union is related to tourism. The tourism industry generates 13.5 percent of the EU's gross domestic product and 15 percent of its taxes. Europeans are themselves enthusiastic travelers, visiting one another's countries as well as many exotic world locations frequently. This travel is made possible by the long vacations Europeans are granted by employers: most workers get at least 4 weeks off from the job every year (a cost to employers that makes them less competitive globally). In the last several decades, taking several weeklong trips, rather than a single long vacation, has become popular, and this custom has increased the demand within Europe for urban hotels, rural guest houses, folk festivals, car rentals, and airline service.

*Energy Resources.* At the same time as Europe's economy has become more oriented toward services, its energy use has steadily moved away from coal toward petroleum and natural gas. Europe imports much of its oil and gas from outside the region, buying natural gas from Algeria and Russia and much of its oil from the Middle East. Within Europe itself, there are large oil and gas deposits in the North Sea and under the Netherlands (see Figure 4.20), which supply much of the energy used in West Europe. The use of nuclear power to generate electricity has been more common in Europe than in North America, especially in France, where it accounts for 78 percent of the electricity generated (compared with only 20 percent in the United States). Although some Central European countries looked to nuclear power as a way to achieve future energy self-sufficiency, most West European countries have been phasing out their use of nuclear energy. The decline in support for nuclear power was partly a response to the 1986 accident in Chernobyl, Ukraine, in which a nuclear plant explosion sent a cloud of radiation drifting across western Europe. Now, as petroleum prices soar and consistent supplies become more doubtful, alternative energy projects are being pursued across Europe and beyond, with various countries expressing the intention to get 20 to 30 percent of their energy from renewable sources by 2010. Projects attracting investment include emission-free solar power generated in the North African desert and transported to Europe, wind farms on the Scottish shore, high-energy fuel pellets made from biomass, fuel-efficient wood stoves, and energy from nuclear fusion.

*The Transportation Infrastructure.* One goal of the EU has been to extend transportation networks to draw all of Europe together. Europeans have typically favored fast rail networks for both passengers and cargo, rather than multilane highways for cars and trucks (Figure 4.21). Yet, despite high and rising gasoline prices over the last two decades, there has been a noticeable trend toward less energy-efficient but more flexible motorized road transport. Now, however, fuel costs are increasingly being considered in the design of *multimodal transport* that links high-speed rail to road, air, and water transport. A 2005 EU transport report notes that 1 kilogram of gasoline can move 50 tons of cargo 1 kilometer by truck, but it can move 90 tons by rail and 127 tons by waterway.

*The Information Economy.* Although Europe has lagged behind North America in the development and use of personal computers and the Internet, the information economy is well advanced in Europe generally, with West Europe leading the way (Figure 4.22). In fact, Europe leads the world in cell phone use. In South Europe and Central Europe, where personal computer ownership is lowest, public computer facilities in cafés and libraries are common, and surfing the Internet is popular, especially among schoolchildren.

**FIGURE 4.21** The Trans-European transport network (TEN-T): Priority axes and projects. As of 2007, Europeans were reevaluating transport policies in light of rising fuel costs. Cheaper rail and water transport were getting renewed attention over cars and trucks, which, for a time, were favored for flexibility. [Adapted from "Transport, a driving force for regional development," *Inforegio panorama*, No. 18 (December 2005), p. 11, at http://ec.europa.eu/comm/regional _policy/sources/docgener/panora _en.htm.]

## Agriculture in Europe

Although, on average, only about 2 percent of Europeans are now engaged in full-time farming, Europeans like the idea of being self-sufficient in food. Toward this end, the European Common Agricultural Policy (CAP) aids farmers with tariffs on imported agricultural goods and with **subsidies** (payments to farmers to lower their costs of production). Subsidies are also paid to some farmers (especially in Central Europe) to retain quaint, traditional (and less efficient) farm buildings, equipment, and cultivation practices so as to maintain attractive rural landscapes for tourism. Such measures are expensive—payments to farmers are the largest expense category in the EU budget—and they raise food costs for the consumer. However, the tariffs and subsidies do help Europe to be self-sufficient in food, and they provide a decent living standard for farmers, thus keeping them on the land. Such policies, which are also found in the United States, Canada, and Japan, are especially unpopular in the developing world, whose farmers are locked out of major potential markets by the tariffs. Subsidies also encourage farmers to overproduce (to collect more payments), thereby causing a glut of farm products that are then sold cheaply on the world market (Figure 4.23). This practice, called **dumping**, lowers global prices and thus hurts farmers in developing countries more significantly than those in developed countries.

Increasingly, Europeans are interested in knowing that their food is pure and wholesome (and preferably organically produced). Such concerns have led to a movement against any component in imported food deriving from genetically modified (GMO) plant or animal material. Since the U.S. public has not pushed for government restrictions on the use of GMO plant material, Europe has restricted food imports from the United States, a policy that hurts U.S. farmers.

The average European farm is 45 acres (18 hectares), less than one-tenth the size of the average U.S. farm. Geographer Ingolf Vogeler points out that small family farms are disappearing in Europe, just as they did several decades ago in the United States, and that the trend is toward larger, more profitable farms with very few laborers. In part, this consolidation of farms is driven by the use

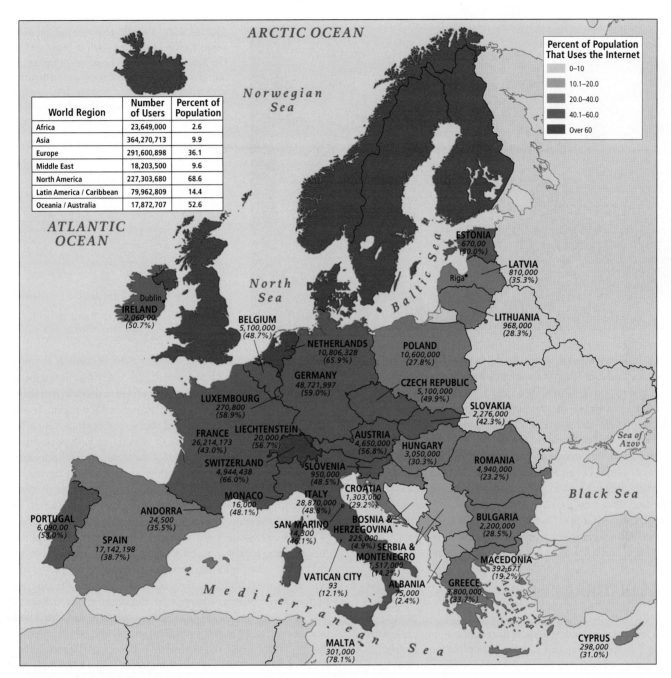

| World Region | Number of Users | Percent of Population |
|---|---|---|
| Africa | 23,649,000 | 2.6 |
| Asia | 364,270,713 | 9.9 |
| Europe | 291,600,898 | 36.1 |
| Middle East | 18,203,500 | 9.6 |
| North America | 227,303,680 | 68.6 |
| Latin America / Caribbean | 79,962,809 | 14.4 |
| Oceania / Australia | 17,872,707 | 52.6 |

**Percent of Population That Uses the Internet**
- 0–10
- 10.1–20.0
- 20.0–40.0
- 40.1–60.0
- Over 60

**FIGURE 4.22** Internet use in Europe. The numbers on the map indicate the number of Internet users in the country as of 2006; the percents indicate the percentage of the country's population that uses the Internet.

of expensive agricultural machinery, a practice that requires more acres in production to be profitable. While mechanization and consolidation have forced many farmers off their land, they have also raised living standards for the farm families who remain. Along with mechanization, the use of chemical fertilizers, pesticides, and herbicides has greatly increased production, though this has come at the cost of ever-increasing environmental degradation as these chemicals are washed into rivers and streams.

In Central Europe, large corporate farms are more common than in other parts of Europe. When Communist governments gained power in the mid-twentieth century, they consolidated many

small, privately owned farms into large collectives worked and managed by a large labor force under the supervision of the state. After the breakup of the Soviet Union, these farms, many reclaimed by their original owners, were rented to large corporations, which in turn further mechanized the farms and laid off all but a few laborers. Rural poverty rose and small towns declined as farmworkers and young people left for the cities. In a few places, however, farms remained small and less modernized. In Romania, for example, where close to one-third of the population is still employed in agriculture, farmers continue to use draft animals and other outdated techniques to produce crops that do not meet the standards of the

**FIGURE 4.23** French farmers protest lowered prices for their produce as well as high fuel costs. On September 22, 2005, hundreds of farmers dumped tons of fresh produce (here, butternut squash) on roads south of Avignon, blocking traffic for days. This literal dumping is not the same as the "dumping" of cheap farm products on the world market, which led to the protests. [AP Photo/ Claude Paris.]

EU. In wealthier Slovenia, where only 6 percent of workers are still in agriculture and even the tiniest farms have a tractor, production is still inefficient and farm incomes comparatively low.

In the open markets of present-day Europe, the larger, more profitable corporate farms could give the smaller family farms of western Europe stiff competition, especially as corporate farms in Central European countries join the market. The quandary is how the farmers of Central Europe should be treated now that they are in the EU. Were they to receive CAP subsidies like those typically paid to EU farmers, they would quickly bankrupt the EU budget. On the other hand, failure to support farmers in the new EU member states could leave entire regions in a second-class, impoverished state, fueling resentment and a flood of ex-farmers into cities across Europe.

## SOCIOCULTURAL ISSUES

Although the European Union was conceived primarily to promote economic cooperation and free trade, the economic integration that has resulted from its programs has social implications. In this section, we examine how the European Union is affecting and being affected by attitudes toward immigrants and gender roles and the evolution of social welfare programs. At the same time, though religion and language used to be divisive issues in the region, within the European Union they are now largely fading as a focus of disputes.

### Open Borders Bring Immigrants, Cultural Complexity, Needed Taxpayers, and Worries

Few Europeans mourn the demise of strict regulations at the borders between countries. Not long ago, border crossings meant tension-filled moments for travelers as armed border police closely inspected passports, vehicles, and luggage, often with a rude edginess. Then, in the 1990s, the EU and many of its neighbors approved the **Schengen Accord,** an agreement allowing free movement across common borders. Now, even at crossings between new EU member countries in Central Europe, border police are usually scrupulously polite, uttering mild pleasantries while barely examining passports. One has to request a country stamp on one's passport as a souvenir.

Open borders have eased life throughout the region, facilitating trade, employment, tourism, and social relationships. Open borders were also intended to bring new workers to European cities and towns. There has been ambivalence, however, in all European countries, whose people have long prided themselves on having distinct cultural identities, about freely admitting so many new people, even those from other parts of Europe. And now, in addition to people from Central Europe, immigrants are arriving from around the globe (Figure 4.24). Some, such as many of those from Turkey and North Africa, come legally as **guest workers,** who are expected to stay for only a few years, fulfilling Europe's need for temporary workers in certain sectors. Others are refugees from the world's trouble spots—Afghanistan, Albania, Democratic Republic of the Congo, Haiti, Iraq, Kosovo, Rwanda, Serbia, Sudan—who are supported by European governments and community groups. Still others enter Europe illegally. Here is the dilemma: while Europeans fear that the presence of large numbers of immigrants, especially from outside Europe, will make them feel like strangers in their own lands, most Europeans want to think of themselves as generous to those in need. Furthermore, Europeans desperately need these mostly young new residents to fill jobs at all levels of the economy and to lend vibrancy, creativity, and new young families to an aging Europe. In particular, they are needed to contribute to the tax coffers that will pay the pensions of the many Europeans about to retire.

The new arrivals are evident in schools, the workplace, sports arenas, and religious institutions. Schools in Switzerland may have students from Sri Lanka, Bosnia, Algeria, Russia, China, and Iran. Workplaces in Austria and Germany may have workers from Turkey, Iran, Ukraine, and Kazakhstan. Soccer teams may be made up of players originally from South Africa, Argentina, Zimbabwe, and Trinidad (Box 4.2). The presence of so many culture groups raises questions of national identity: Is Germany no longer a German

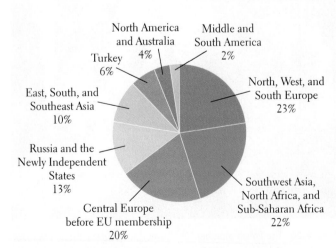

**Migration into the European Region, 2005**
- Original 15 EU countries
- 10 newest EU countries
- Other European countries
- No data

ICELAND
23,100
(7.8%)

ARCTIC OCEAN

*Norwegian Sea*

FINLAND
156,200
(3.0%)

SWEDEN
1.1 mil
(12.4%)

NORWAY
343,900
(7.4%)

ESTONIA
201,700
(15.2%)

LATVIA
449,200
(19.5%)

DENMARK
388,500
(7.2%)

*North Sea*

*Baltic Sea*

LITHUANIA
165,200
(4.8%)

ATLANTIC OCEAN

IRELAND
585,400
(14.1%)

UNITED KINGDOM
5.4 mil
(9.1%)

BELGIUM
719,300
(6.9%)

NETHERLANDS
1.6 mil
(10.1%)

POLAND
702,800
(1.8%)

GERMANY
10.1 mil
(12.3%)

CZECH REPUBLIC
453,300
(4.4%)

SLOVAKIA
124,500
(2.3%)

LUXEMBOURG
173,700
(37.4%)

FRANCE
6.5 mil
(10.7%)

AUSTRIA
1.2 mil
(15.1%)

HUNGARY
316,200
(3.1%)

ROMANIA
133,400
(0.6%)

SWITZERLAND
1.7 mil
(22.9%)

SLOVENIA
167,300
(8.5%)

CROATIA
661,400
(14.5%)

*Sea of Azov*

PORTUGAL
763,700
(7.3%)

SPAIN
4.8 mil
(11.1%)

ITALY
2.5 mil
(4.3%)

BOSNIA & HERZEGOVINA
40,800
(1.0%)

SERBIA & MONTENEGRO
512,300
(4.9%)

BULGARIA
104,800
(1.3%)

MACEDONIA

*Black Sea*

ALBANIA
82,700
(2.6%)

MALTA
10,700
(2.7%)

GREECE
973,700
(8.8%)

CYPRUS
116,100
13.9%

*Mediterranean Sea*

mi 0 100 200 300
km 0 250 500

(a) Immigrants in Europe, 2005

**FIGURE 4.24** Interregional linkages: Migration to western Europe. Migration to western Europe increased in the 1990s and continued to increase into the 2000s, becoming a crucial issue in EU debates. The numbers in part (a) indicate the numbers of immigrants living in western European countries in 2005 and their percentage of the total population of each country. [Data from (a) "World migrant stock: The 2005 revision population database," United Nations Department of Economic and Social Affairs, Population Division, at http://esa.un.org/migration/p2k0data.asp; (b) Christina Boswell, "Migration in Europe," a paper prepared for the Policy Analysis and Research Programme of the Global Commission on International Migration, September 2005, p. 4, at http://www.gcim.org/attachments/RS4.pdf.]

North America and Australia 4%

Middle and South America 2%

Turkey 6%

North, West, and South Europe 23%

East, South, and Southeast Asia 10%

Russia and the Newly Independent States 13%

Southwest Asia, North Africa, and Sub-Saharan Africa 22%

Central Europe before EU membership 20%

(b) Immigrant population in the 15 original EU countries, by country of origin, 2004

place (Figure 4.25)? Will France's self-image suffer if children of French–North African descent read textbooks that deal frankly with France's long repressive colonial regime in North Africa? Should the predominant religion of these children—Islam—have equal footing with Catholicism, the traditional religion of choice in France? Questions of public welfare also arise as countries try to accommodate unfamiliar value systems and family types.

European host countries have very different policies regarding immigrant workers. Some, such as the Netherlands, have lenient laws that allow immigrants from former colonial territories to stay indefinitely and to receive subsidized housing and other social welfare services equivalent to those available to citizens. France provides its immigrants with less adequate housing blocks in informally segregated working-class neighborhoods inconveniently located on the urban fringe, where the atmosphere is often hostile. In general, sentiment favors accommodation and acceptance, but extreme nationalism is not new to Europe and has become more active recently in response to the new wave of immigrants. In widely dispersed locations, nationalist feelings have produced violence against non-European immigrants, and in Germany, Austria,

**FIGURE 4.25** Who is German? Turkish migrants relax and play soccer in Berlin's Tiergarten, the city's central park. In Berlin, by the mid-1990s, one of every eight residents was foreign born. [Gerd Ludwig/National Geographic Image Collection.]

Italy, and France, minority right-wing parties now openly speak of forcing immigrants to leave. Officials at the community, country, and EU levels must be continually cognizant of the need to educate the public about the advantages of bringing in new young workers who will stay and contribute their talents and money.

An incident in France in 2005 shows how a sense of humor can diffuse malice. In June of that year, as a way of illustrating that workers from Central Europe were overrunning western Europe, French nationalists complained in the press that Polish plumbers were taking French jobs. (In fact, there was a significant shortage of plumbers in France.) Rather than reacting defensively, Poland responded by featuring on its tourism posters a "hunky" male model wielding plumbing tools (Figure 4.26) and saying seductively, "I'm staying in Poland, won't you come over?" The "Polish Plumber" is now shorthand across Europe for issues related to immigration and jobs.

***Citizenship.*** Until recently, achieving legal citizenship in a European country was very difficult for outsiders, especially those of non-European heritage. Germany, a country with a particularly large group of immigrants (in 2004, 8.8 percent of its 82.5 million people were foreign born), once had especially stringent rules. Even if they were born in Germany, the children and grandchildren of Turkish or North African immigrant workers were not considered citizens, and most were rejected if they applied for citizenship. But attitudes and policies have changed. As of January 2000, all children born in Germany are citizens, and people born elsewhere who have lived in Germany for 8 years can apply to become citizens. The law is retroactive to 1975.

France's citizenship laws are somewhat less generous than Germany's, although France has long prided itself on being a culturally diverse country, and it has turned a number of its former colonies in the Caribbean, Indian Ocean, and the Pacific into overseas *départements* (states). The United Kingdom, probably the most multicultural of all European countries, has absorbed many hun-

dreds of thousands of West Indians, Indians, Pakistanis, and Africans from its former colonies. The United Kingdom recently decided to give the several hundred thousand citizens of its few remaining colonies full citizenship and rights to social welfare benefits and subsidized higher education. As a result of all these new policies, an increasing number of Europeans can now trace their ancestry to such places as China, Nigeria, Thailand, Argentina, Brazil, the Caribbean, the Pacific, or Turkey, and these demographic changes are bringing about a redefinition of what it means to be a European.

***Rules for Assimilation.*** In Europe, race and skin color play less of a role in defining differences between people than does culture. An immigrant from Asia or Africa may be fully accepted into the community if he or she has wholeheartedly adopted European ways. This is especially true for immigrants skilled in the host country's language and for those who have mastered the finer points of its manners and decorum. Yet even certain minorities that have been in Europe for thousands of years, such as the Basques in Spain and the Roma (Gypsies) who reside in many countries, find it nearly impossible to blend into society if they retain their traditional ways. In Europe, **assimilation** usually means a comprehensive change of lifestyle. If immigrants give up the culture of home —language, dress, family relationships, food, customs, mores, and even religion—and take up instead the ways of their adopted country, they are more likely to be accepted.

Geographer Eva Humbeck studied how Thai women who come to Germany explicitly to marry German men grapple with

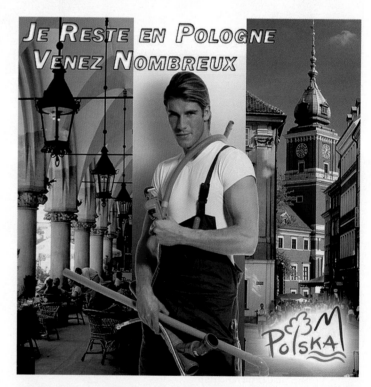

**FIGURE 4.26** The Polish plumber. This poster, inviting the French to meet the plumber in Poland, became an emblematic reaction to anti-immigrant sentiment. [SIPA.]

## BOX 4.2 | AT THE GLOBAL SCALE   Soccer: The Most Popular Sport in the World

In Europe, soccer teams are a major component of national identity. They bring together citizens of vastly different cultural, racial, and class backgrounds to enjoy a common experience. Countries go to great lengths to recruit the best players for their teams, often importing them from South America, South Asia, and Africa and granting them instant citizenship. Cities in Europe cultivate winning soccer teams in much the same way cities in the United States support their professional basketball or football teams. Feelings can run so high at soccer matches that violence among the fans erupts easily.

Soccer, known as football in most countries except the United States, is a global game, played regularly by more than 240 million people, as reported in 2000 by the world soccer fed-

eration, Fédération Internationale de Football Association (FIFA). During the 1999 women's World Cup matches played in the United States, it is possible that the entire population of the world saw some portion of the matches on worldwide television.

The oldest known precursors to soccer may have been kickball games played in China about 2500 years ago, but similar ball games were known in ancient times in the Americas (Figure 4.27). The Romans carried a version to England, where the game evolved. It emerged as the modern game of soccer in 1863, when the rules of the game were formalized. The first international match was played between teams from England and Scotland in 1872. British sailors then carried the game to mainland Europe, India, South America, and the South Pacific.

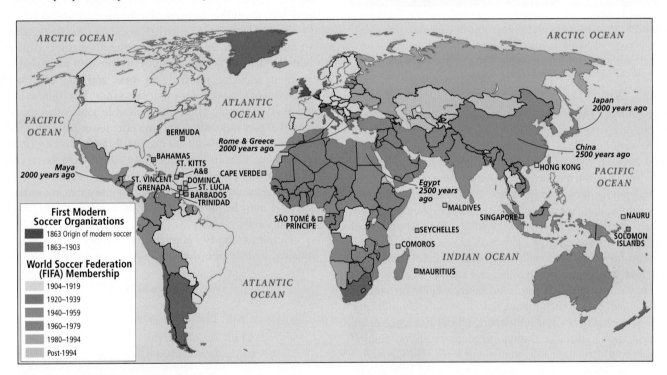

**FIGURE 4.27** **The origin and spread of soccer.** The map depicts the spread of modern soccer from England through the rest of the world. Kickball games originated in many parts of the world before modern soccer began in 1863; the locations of ancient versions of kickball are shown on the map.

the demands of assimilation. Since 1990, about a thousand Thai brides have arrived in Germany every year. Although they willingly take on the role of housewife as opposed to career woman, they usually live isolated in apartments and sorely miss their female relatives—who, in Thai culture, are a woman's best friends. Most feel obliged to drop all Thai ways of doing things because they sense that Germans view Thai culture as inferior. Few Thai wives know one another, and many confess to daydreaming about the Thai hospitality and family conviviality they once enjoyed.

The assimilation model is not always mildly acquiesced to (Box 4.3). In France, for example, wearing of the *hijab* by observant

Muslim schoolgirls became the center of a national debate about civil liberties, religious freedom, and national identity (Figure 4.28). Eventually, in order to justify the banning of the *hijab*, all symbols of religious affiliation were declared illegal in French schools. In 2006, movements to ban Muslim dress arose in the UK and the Netherlands. The degree of adjustment expected of immigrants in Europe differs from that expected of immigrants in North America, where the norm is **acculturation**; that is, enough adaptation to the host culture for members of the minority culture to function effectively and be self-supporting. Overall, in North America, immigrants are somewhat freer to keep their native cultures and customs

**FIGURE 4.28** Muslims in Europe. Muslims are a tiny minority in Europe, but many intend to make the region their permanent home. Just how, or if, assimilation of Muslims into largely secular Europe will proceed is a major topic of public debate. [AFP/Getty Images.]

than in Europe. Those who retain native dress and particularly exotic customs may encounter discrimination, but they will also find that American laws and attitudes actually encourage and reward a measure of cultural retention, so long as a serious effort is made to learn English and be self-supporting.

While public displays of resistance to outsiders are fairly common in Europe, there are also many signs that acceptance of a multicultural, multiracial Europe is growing. The relaxation of citizenship laws is one example. The establishment of ethnic Asian and African businesses, especially restaurants, in small rural towns and the local acceptance of the immigrant families who run them are others. Cross-cultural marriage is increasingly common, and the multiethnic offspring of such marriages will undoubtedly influence attitudes in their generation. Furthermore, official EU agencies are encouraging public dialogue through frequent forums on cultural integration held at the local and international levels. The Council of the European Union has drawn up guidelines for the legal acceptance of cultural diversity by new EU member states from Central Europe, and this project, which has set high standards, has forced western Europeans to reflect on and address their own record of ethnic prejudice as well.

## European Ideas About Gender

Gender roles in Europe have changed significantly from the days when most women married young and worked in the home, raising large families and tending to various agricultural duties. As we have noted, a large percentage of Europeans live in cities, marry late, and have only one child, and increasing numbers of European women are working outside the home, though the percentages vary considerably among different parts of the region (Figure 4.29).

As is the case nearly everywhere on earth, Europeans still generally accept the notion that men and women have different innate

abilities that determine the functions they should perform. European public opinion among both women and men largely holds that women are less able than men to perform the types of work typically done by men and that men are less skilled at domestic duties. Indications are that Europe is a world leader in changing these views—women increasingly work outside the home as factory laborers, service workers, professionals, government bureaucrats, and elected officials—but these ideas continue to influence European social institutions. In most places, men have greater social status, hold more managerial positions, earn higher pay, and have greater autonomy in daily life (more freedom of movement, for example) than do women. These male advantages have a stronger hold in Central and South Europe today than they do in West and North Europe.

Usually, women who work outside the home are expected to do most of the domestic work as well, a situation called the **double day.** United Nations research shows that in most of Europe, women's workdays, including time spent in housework and child care, are 3 to 5 hours longer than men's (Iceland and Sweden reported that women and men there share housework equally). Women burdened by the double day generally operate with somewhat less efficiency in a paying job than do men, and they tend to choose employment closer to home that offers more flexibility in the hours and skills required. These more flexible jobs (often erroneously classed as part time) almost always offer lower pay and less chance for advancement, though not necessarily fewer working hours, than typical male jobs. In Central Europe, the double day is especially taxing for women because labor-saving devices are not yet widely available: laundry is often done by hand, kitchens are rudimentary, and food must be purchased daily because small refrigerators will hold only a day's food.

Despite strong EU emphasis on policies encouraging gender equality, and despite the fact that well over half the university

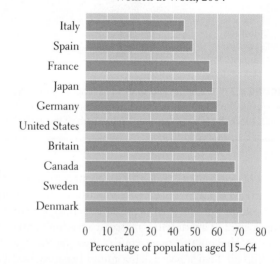

**FIGURE 4.29** Women at work in Europe (and selected other countries for comparison), 2004. A majority of women in Europe work outside the home, and their numbers are increasing. A related trend is declining birth rates in these countries. [Data from "Women and the world economy—A guide to womenomics," *The Economist,* April 12, 2006, p. 74.]

## BOX 4.3 | AT THE REGIONAL SCALE  Muslims in Europe

In the fall of 2005, riots erupted in and around Paris when young people of North African descent protested their lack of access to higher education, jobs, and decent apartments. Investigations by the media over the next few weeks revealed that these young people, most born in France, had a legitimate complaint. They were indeed living in poverty on the fringes of society and faced de facto exclusion from the routes out of poverty available to other French young people. And almost all were Muslim.

These rather violent protests came on the heels of several other incidents that revealed a deepening alienation among the children of immigrants, especially those who practiced Islam. A terrorist bombing in London was carried out by young British Muslims. A controversial media personality in the Netherlands who spoke out against the domestic abuse of Muslim women was killed by a man of Dutch and Moroccan descent. Then came the publication of cartoons in Denmark that insulted the Prophet Muhammad, which resulted in protests by Muslims around the world. All of these incidents highlighted the growing numbers of Muslim immigrants and their descendants in Europe (Figure 4.30) as well as growing worries that, for a variety of reasons, assimilation is not occurring. A public dialogue on the place of Islam in Europe is now in progress, with voices ranging from moderation to extremism on both sides. Even the highest officials are adjusting their attitudes. In 2006, Pope Benedict, during a goodwill visit to Turkey (a predominantly Muslim country), revoked an earlier statement critical of Islam and supported the admission of Turkey to the EU. The role of Islam in Europe is likely to be discussed for some years to come.

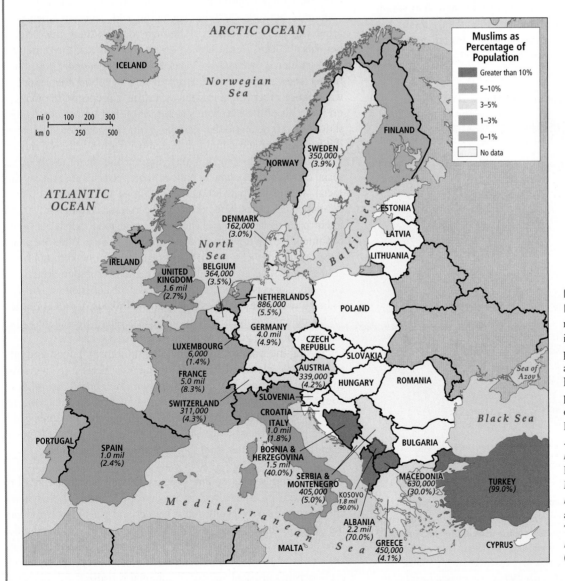

**FIGURE 4.30** Muslims in Europe. Muslims are a tiny minority in Europe, but many intend to make the region their permanent home. Just how, or if, assimilation of Muslims into largely secular Europe will proceed is a major topic of public debate. [Data from "Muslims in Europe," *The Times*, at http://images .thetimes.co.uk/TGD/picture /0,,216394,00.jpg; "Muslims in Europe: Country guide," *BBC News*, at http://news.bbc.co.uk/2 /hi/europe/4385768.stm#france; and "Muslims in Europe," *Financial Times*, at http://www.ft.com/cms/s /89b5eccc-f48c-11d9-9dd1- 00000e2511c8,ft_acl=s01=1.html.]

graduates in Europe are now women, the political influence and economic well-being of European women lag behind those of European men. As Figure 4.31 illustrates, in most European national parliaments, women make up less than a third of elected representatives; only in North Europe do women come anywhere close to filling 50 percent of the seats in the parliament or legislature. Although Germany recently elected Angele Merkel as its first woman prime minister, women generally serve in the lower ranks of government bureaucracies, in positions that give them little voice in the formation of national policies.

Because they are largely absent from policy-making positions, women have to rely on men to push for such legal measures as the right to work, equal job opportunities, and equal wages and fringe benefits. Progress has been slow. For example, in 2006, female unemployment was higher in every EU country than male unemployment, and 32 percent of women's jobs were part time, as opposed to only 7 percent of men's jobs. Throughout Europe, women are paid on average 15 percent less than men for equal work, and young women, who tend to be more highly educated than young men, still have higher unemployment rates than young men.

One might question why, given Europe's overall prosperity and generally progressive attitudes, women have not achieved

more equality with men. The explanation is that although EU policies have officially ended legal discrimination against women, old patriarchal attitudes persist. Furthermore, geographic separation, language differences, multiple ethnic and national loyalties, varied education levels, and long-standing social hierarchies keep women from collaborating to advance equality. This situation is changing as relaxed EU borders open up entrepreneurial options for women and encourage international exchanges and conferences as well as region-wide research on women's issues.

## Social Welfare Systems

In Europe, the term **social welfare** describes elaborate, tax-supported systems that serve all citizens in one way or another. The value of social welfare is more widely accepted in Europe than in the United States, where the term *welfare* has a narrower definition, referring to limited tax-supported services for the poor or disabled. In Europe, social welfare programs include many services for middle-class citizens, such as health care, free or low-cost higher education, tax-supported housing, disability payments, and generous unemployment and pension benefits. (Although some of these middle-class benefits also exist in the United States and Canada, there they are not considered "welfare" per se.) Europeans generally pay much higher taxes than North Americans, and in return they expect a wider range of services and safety nets. Still, Europeans do not agree on the goals of these welfare systems, or on just how generous they should be. Some argue that Europe can no longer afford high taxes if it is to remain competitive in the global market. Others maintain that Europe's high standards of living are the direct result of the social contract to take care of basic human needs for all; they say that social welfare is what distinguishes Europe in the community of developed countries. The debate has been resolved differently in different parts of Europe, and the resulting regional differences have become a source of worry in the European Union. With open borders, unequal benefits might encourage those in need to flock to a country with a generous welfare system and overburden the taxpayers there. Conversely, a country with fewer social benefits might find its best and brightest workers emigrating to more generous countries. Or the differences in tax systems might accentuate the differences in rates of economic growth in different parts of the EU.

Using the work of British sociologist Crescy Cannan and others, we have classified European welfare systems into five basic categories (Figure 4.32). Each category makes certain assumptions about gender roles. All five systems are based on the tacit assumption that the typical family has two parents and that the male is the sole or primary breadwinner, even though in most parts of Europe, as in the United States, such families are no longer typical.

1. **Social democratic welfare systems** are well developed in all of Scandinavia, but especially so in Sweden and Iceland. These systems attempt to achieve equality across gender and class lines by providing generous health care, education, housing, and child and elder care benefits to all citizens from cradle to grave. Citizens pay high taxes to provide an environment, from prenatal care on, in which every person's physical and social potential is realized. Child care is widely available, not just to help women enter the labor market, but

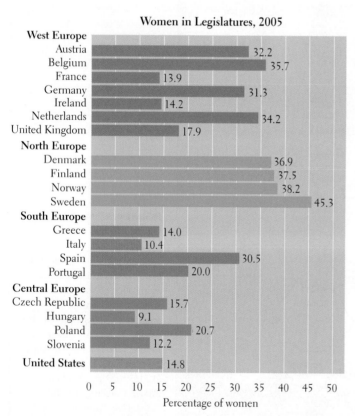

**Women in Legislatures, 2005**

**West Europe**
Austria — 32.2
Belgium — 35.7
France — 13.9
Germany — 31.3
Ireland — 14.2
Netherlands — 34.2
United Kingdom — 17.9

**North Europe**
Denmark — 36.9
Finland — 37.5
Norway — 38.2
Sweden — 45.3

**South Europe**
Greece — 14.0
Italy — 10.4
Spain — 30.5
Portugal — 20.0

**Central Europe**
Czech Republic — 15.7
Hungary — 9.1
Poland — 20.7
Slovenia — 12.2

**United States** — 14.8

Percentage of women (0 5 10 15 20 25 30 35 40 45 50)

**FIGURE 4.31** Women in the legislatures of selected countries, 2005. Women are about half the adult population of the EU, but they do not have anywhere near their fair share of representation in European legislatures; hence their influence on legislation is seriously restricted. Notice that some regions are closer to equity than others, and also notice how the United States compares. [Data from *United Nations Human Development Report 2005* (New York: United Nations Development Programme), Table 26.]

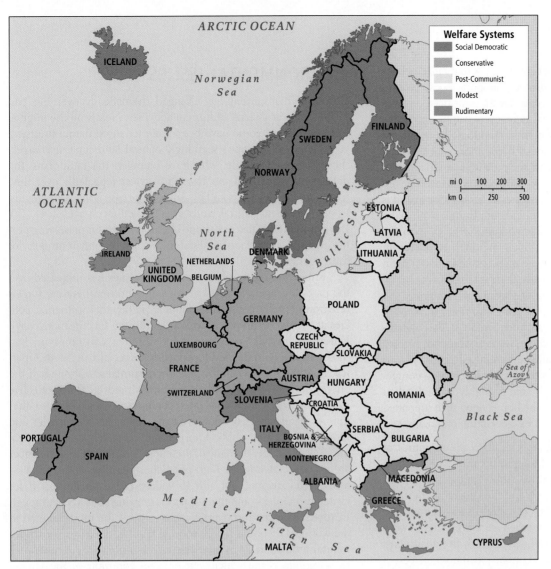

**FIGURE 4.32** European social welfare systems. Within Europe, there is clear regional variation in approaches to social welfare. The five basic categories of social welfare systems shown here and described in the text should be taken as only an informed approximation of the existing patterns.

also to provide the state with a mechanism for instilling in its citizens such values as polite behavior, good nutrition, the work ethic, and good study habits. Such early childhood training is a key feature of this system because it is meant to ensure that, in adulthood, every citizen will be able to contribute to his or her highest capabilities and that social problems will not develop. Although gender equality is a stated goal, traditional gender roles are officially emphasized throughout this lifelong support system, and women are not yet equal to men in the workplace and in public forums.

**2. Conservative welfare systems** are found in the countries of West Europe except for the United Kingdom and Ireland. They seek to provide a minimum standard of living for all citizens. Hence the state assists those in need, but it does not see its mission as assisting upward mobility. State programs may make a college education accessible to the children of poor families, but strict entrance requirements in some disciplines can be hard for the poor to meet. On the other hand, many state-supported services, such as health care and pension plans, are available to all economic classes. Like other public institutions in these countries, the welfare system reinforces the traditional "housewife contract" by assuming that women will stay home and take care of children and the elderly.

**3. Modest welfare systems,** presently found only in the United Kingdom, are designed to encourage individual responsibility and the work ethic. Welfare is thought to encourage dependency, and the ideal citizen is completely self-reliant. Welfare recipients are often stigmatized in jokes and conversation as being lazy and "on the dole." This type of welfare system has been evolving in the United Kingdom since conservative reforms in the 1980s and 1990s, and it is roughly similar to the welfare system in Canada (but more generous than the U.S. system). Benefits are modest, just enough for those who qualify to maintain a minimally adequate standard of living. The state claims no direct interest in the quality of its citizens' daily lives; how people live is thought to be a matter of individual choice in a free market, except that basic health care is provided free to all residents. Women are free to work outside the home or not, but the state supports neither option. Critics point out that in reality, women cannot enter the labor market on the same terms as men because of inequitable pay and the fact that women are assumed to be family caretakers, yet state-supported child care and elder care are minimal and their availability is unpredictable.

**4. Rudimentary welfare systems** are found primarily in South Europe—Portugal, Spain, Italy, Cyprus, Malta, and Greece—and in

Ireland. These countries do not accept the idea that citizens have inherent rights to government-sponsored support. Local governments provide some services or income for those in need, but the availability of such services varies widely, even within a country. Although fertility rates are now low in these countries, the traditional extended family and community are still common, and the state assumes that when people are in need, their relatives and friends will provide the needed support. The state also assumes that women are available to provide child care and other social services for free. It is true that less than 50 percent of women in South Europe work in the formal economy, but women in these countries actually form the bulk of the flexible, low-wage (and often informal) workforce in small family-owned industries or service agencies; hence few remain available to be family caretakers. Finally, there is the official assumption that the large informal (hidden, or gray-market) economy will provide some sort of employment for anyone who needs it. Some of the circumstances that have underlain rudimentary welfare systems are now changing. In Italy, for example, women who once worked part time for small family firms while performing caregiver tasks on the side are disappearing as these small firms close due to competition on the global market. Italy's young, highly trained urban women, who work full-time outside the home, need state-supported child care if they are to produce the next generation of Italians.

**5. Post-Communist welfare systems** prevail in the countries of eastern and southeastern Europe. During the Communist era, these systems were, in theory, comprehensive. Although the bureaucracy administering benefits was inefficient and the programs unevenly funded, the ideal resembled the cradle-to-grave social democratic system in Scandinavia, except that women were pressured to work outside the home. Benefits often extended to nearly free apartments and health care and to meals provided on the job at nominal cost to the worker. The big change in the post-Communist era is that state funding of these welfare systems has collapsed and people must cope with the loss of a wide range of benefits at the same time that jobs are being lost. In the transition to a market economy, state-provided benefits are continuing to shrink, and many people have gone without such basic necessities as food, heating fuel, and health care; there is thus a temptation to migrate to western Europe, where the basics are still provided.

As the European Union evolves and expands, Europe's social welfare systems will probably become more similar to one another, but it is still unclear just which models will prevail. Some observers suggest that the dominance of the German economy in Europe could mean that the conservative German model will spread. Others think that the entry of Scandinavian countries into the EU will inspire a shift to more activist welfare policies with more equal treatment of women. The latter position is supported by a 2006 report of the European Commission, which shows that there is increasing awareness across the EU of the interlinking of welfare provisions, women's economic participation, and general economic growth. Although the costs of welfare policies will be a factor, as will be pressures to limit benefits to immigrants, overall it is likely that the state provision of health care and a social safety net will remain a feature of life in Europe.

# ENVIRONMENTAL ISSUES

Europe's environment has changed dramatically over the past 10,000 years as a result of human activities. Nearly all the original forests are gone; some have been gone for more than a thousand years. Many of Europe's seemingly natural landscapes are largely the creation of people, who have changed the landforms, the drainage systems, and the vegetation cover repeatedly over time (compare Figures 4.33a and b). An extreme case is the Netherlands, where almost no natural landscapes are left (see pages 224–226). Environmental issues in Europe often focus less on preserving pristine natural areas than on establishing livable environments for future generations.

Public awareness of environmental issues has increased over the past 30 years, and in all European countries there are **Green** (environmentally conscious) political parties that influence policies at the national level as well as within the EU. In many ways, European lifestyles result in lower resource consumption than those elsewhere in the developed world, especially in North America. In 2006, the average European consumed about one-half the energy of the average North American. Moreover, Europeans live in smaller spaces, their yards often contain vegetable gardens rather than great expanses of mowed lawns, their cars are smaller and more fuel efficient, public transportation is widely used, and people walk or bicycle to many of their appointments. These practices are in part related to the high population densities and social customs of the region, but also to widespread explicit support for Green principles—and that support is growing (Box 4.4). Although Europe's air, seas, and rivers remain some of the most polluted in the world, 2005 opinion polls showed that EU citizens see environmental protection as an incentive for innovation, not a hindrance to economic performance—an opinion common in the United States. Successive EU environmental reports over the first few years of the 2000s show slow improvement in most environmental categories in nearly all 27 countries, but there is still a long way to go to meet the EU's various stated goals.

## Air Pollution

The whole of Europe produces about a quarter of the world's carbon dioxide emissions by burning fossil fuels (the United States produces another quarter). In hope of reducing their contributions to global greenhouse gas emissions, all European countries have signed the Kyoto Protocol, which sets incremental goals for emission control beginning in 2010. At present, there is significant air pollution across much of Europe, but it is particularly heavy over the North European Plain. This is a region of heavy industry, dense transport routes, high population densities, and affluent lifestyles, all of which lead to intense fossil fuel use and the production of $CO_2$ and other airborne pollutants. One consequence of this pollution is acid rain, which is a threat over most of Europe (see Figure 4.33b).

***Air Quality in Central Europe.*** According to geographer Brent Yarnal, the high level of air pollution in the former Communist states of Central Europe is due to the world's highest per capita

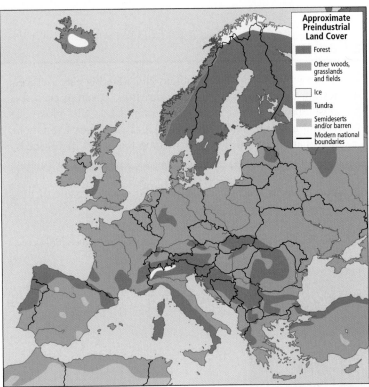

(a) Preindustrial

**FIGURE 4.33** Land cover (preindustrial) and human impact (2002). In the preindustrial era (**a**), humans had converted much forest to grassland and farmland, but there was as yet little impact from industrialization. By 300 years later (**b**), human impact on the land in Europe was much greater, with virtually no place devoid of human influence. [Adapted from *United Nations Environment Programme*, 2002, 2003, 2004, 2005, 2006 (New York: United Nations Development Programme), at http://maps.grida.no/go/graphic/human _impact_year_1700_approximately and http://maps.grida.no/go/graphic /human_impact_year_2002.]

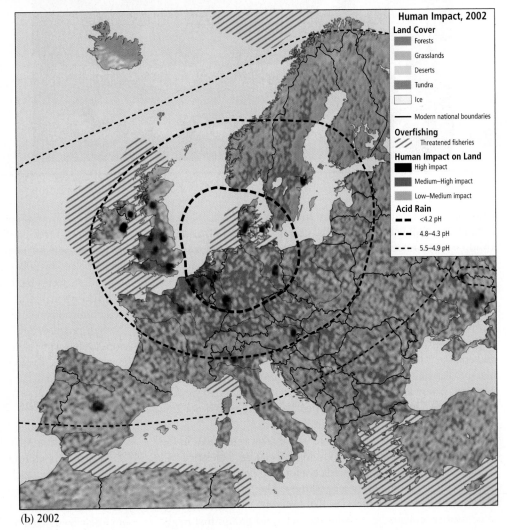

(b) 2002

## BOX 4.4 │ AT THE LOCAL SCALE The Guerrilla Gardeners of London

From the anarchic fringe of Europeans who favor innovation in environmentalism comes a report on the Guerrilla Gardeners of London. Will Pava, writing in the online London *Times* on April 8, 2006, tells of a movement that has quickly grown to embrace more than 500 activists and a Web site, http://www.guerrillagardening.org/. Under cover of darkness, these gardeners—bureaucrats, stock traders, and computer jockeys by day—sneak into Central London to plant colorful flowers and foliage in traffic islands and roundabouts (Figure 4.34). Bypassing the town councils (which tend to impose crippling rules), they make quick assaults late at night armed with trowel, spade, mulch, and watering can. Authorities are unable to stop the guerillas from covering even the most obscure bit of neglected urban land with blooming hyacinths, tulips, marigolds, shrubs, and even trees. Even more radical are the Seed Bombers, a group that packs flower seeds, soil, and water into compact parcels and tosses them into derelict patches of public land, where they shatter on impact, spewing forth seeds that will produce plants capable of outcompeting the weeds.

Source: National Public Radio, Rob Gifford, "Morning Edition," May 15, 2006. http://www.timesonline.co.uk/tol/news/uk/article703112.ece.

**FIGURE 4.34** Guerrilla Gardening in London. On this night in November 2006, a group of Guerrilla Gardeners went to Westminster Bridge Road to weed, clean up litter, and plant red tulips. Their work has the support of the local community. [Jonathan Warren/www.j-warren.co.uk.]

emissions from burning oil and gas. In Poland in the late 1980s, for instance, air quality in the industrial centers was so poor that young children were relocated to rural environments to preserve their health. Such severe environmental problems developed in part because the Marxist theories and policies promoted by the Soviet Union portrayed nature as existing only to serve human needs. Moreover, with only superficial democratic institutions in place during the Soviet era, there was little opportunity for public outrage at pollution to be channeled into constructive political activism and change. This is now changing, but unevenly. In Hungary, popular protest has resulted in air pollution abatement. In Poland, however, where air pollution remains a serious health risk, a 2004 study by Yale University indicated that Poles were not particularly interested in meeting EU air quality standards.

Reformers had thought that the new market economies in the former Soviet bloc countries would improve energy efficiency and reduce emissions, but Yarnal predicts a long lag between economic reforms and the reduction of air pollution. The reasons for this lag are several: government subsidies that lower the costs of fossil fuels to consumers encourage wasteful use, reformist groups that might push for environmental improvements are not well organized, and in Central Europe, antipollution activists find it hard to mobilize public support for regulations that might slow economic growth. Nevertheless, by 2005, emissions from power plants, factories, and agriculture were decreasing, but emissions from motorized transport had increased as economies grew. Those countries with the worst emissions records, such as Estonia and Latvia, were making the best progress. All but Slovenia, which had relatively low emissions to start with, were making progress toward the 2010 reduction levels set by the Kyoto Protocol.

## Pollution of the Seas

Europe is surrounded by seas: the Baltic Sea, the North Sea, the Arctic Ocean, the Atlantic Ocean, the Mediterranean Sea, and the Black Sea. Any pollutants that enter Europe's rivers, streams, and canals eventually reach those surrounding seas (Figure 4.35). The Atlantic Ocean, the Arctic Ocean, and the North Sea are able to disperse most pollutants dumped into them because they are part of, or closely connected to, the circulating flow of the world ocean. In

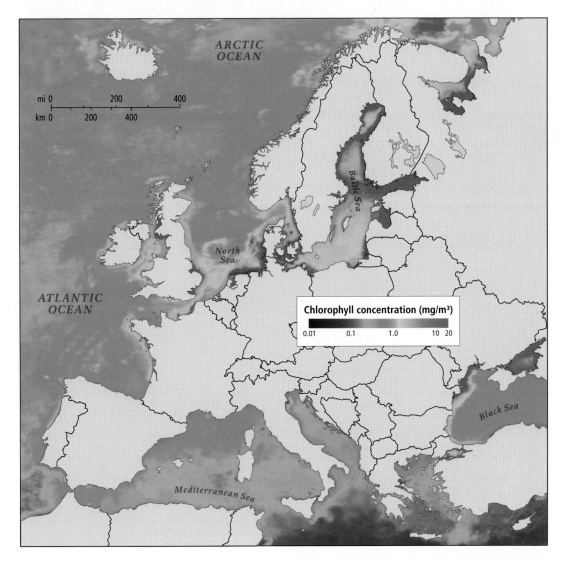

**FIGURE 4.35** Pollution of the seas. Europe is blessed with an exceptionally long and convoluted coast that affords easy access to the world's oceans. But pollution of the nearly landlocked Baltic, Black, and Mediterranean seas is causing increasing concern. Chlorophyll concentration is one measure of pollution levels, which is significant in the Atlantic and Arctic oceans as well. [Adapted from a NASA image created by Jesse Allen, Earth Observatory, using data provided courtesy of the SeaWiFS Project, NASA/Goddard Space Flight Center and ORBIMAGE, available at http://earthobservatory.nasa.gov /Newsroom/NewImages/images .php3?img_id=17332.]

contrast, the Baltic, Mediterranean, and Black seas are nearly landlocked bodies of water that do not have the capacity to flush themselves out quickly; all three are prone to accumulate pollution. The countries surrounding these seas are at different stages of development, which makes it difficult for them to cooperate on solutions.

The Mediterranean Sea and its **basin** (the land that drains into the sea) are the receptacle for multiple pollutants. Municipal and rural sewage (75 percent of it untreated), eroded sediments, agricultural chemicals, industrial wastes, nuclear contaminants, and oil spills are pouring into the water from adjacent lands. Solid waste floats on the sea surface, algal blooms resulting from chemical pollutants render the sea inhospitable to natural life-forms, fish catches are down, and beloved seaside resorts are unsafe for swimmers.

Pollution in the Mediterranean is exacerbated by the fact that it has just one tiny opening to the world ocean. Atlantic seawater flows in through the narrow Strait of Gibraltar and moves eastward, evaporating as it goes. Evaporation concentrates the natural salts in the water, and as a result, the water at the far eastern end of the Mediterranean is saltier and heavier than the water entering from the west. This heavier water sinks, flows back westward at a lower depth, and finally exits the Mediterranean at Gibraltar, many decades later. It takes about 80 years for a complete exchange of water to flush out pollutants. The ecology of the Mediterranean is attuned to this lengthy cycle, but the balance has been upset by the 320 million people now living in the countries surrounding the sea.

Since 1995, the relatively rich and industrialized European Union and relatively poor North Africa and Southwest Asia have been working to ameliorate pollution in the Mediterranean Basin. Although most of the pollution is generated by Europe, populations to the south and east of the sea are increasing rapidly, and soon their sheer numbers will also pose an environmental threat. A 2006 comprehensive report issued by the European Environment Agency and the UN documented the main environmental issues for each of the perimeter countries. The southern and eastern Mediterranean countries were cited for lacking adequate urban waste treatment and control of industrial chemical effluents, while the northern Mediterranean (European) countries, from which most pollutants now come, were cited for lack of political will to make changes, despite having the necessary prevention and corrective mechanisms and the legal framework to enforce regulations. The report noted that the first priority is to get regional antipollution agreements ratified by the 21 countries involved and then to enforce them—actions that are made particularly difficult by the conflicts in the eastern Mediterranean: in Iraq, Palestine, Israel, and Lebanon.

**TABLE 4.1** Human well-being rankings of countries in Europe and other selected countries[a]

| Country (I) | GDP per capita, adjusted for PPP[b] (GDP ranking among 177 countries), 2005 (II) | Human Development Index (HDI) ranking among 177 countries,[c] 2005 (III) | Gender Development Index (GDI) ranking among 140 countries, 2005 (IV) | Gender Empowerment Measure (GEM) ranking among 80 countries, 2005 (V) |
|---|---|---|---|---|
| **Selected countries for comparison** | | | | |
| Barbados | 15,270 (36) | 30 (high) | 29 | 30 |
| Japan | 27,967 (13) | 11 (high) | 14 | 43 |
| Kuwait | 18,047 (33) | 44 (high) | 39 | ND |
| United States | 37,562 (4) | 10 (high) | 8 | 12 |
| **North Europe** | | | | |
| Denmark* | 31,465 (5) | 14 (high) | 13 | 2 |
| Estonia* | 13,539 (39) | 38 (high) | 35 | 35 |
| Finland* | 27,619 (16) | 13 (high) | 10 | 5 |
| Iceland | 31,243 (6) | 2 (high) | 3 | 4 |
| Latvia* | 10,270 (52) | 48 (high) | 43 | 48 |
| Lithuania* | 11,702 (44) | 39 (high) | 36 | 39 |
| Norway | 37,670 (3) | 1 (high) | 1 | 1 |
| Sweden* | 26,750 (20) | 6 (high) | 4 | 3 |
| **West Europe** | | | | |
| Austria* | 30,094 (9) | 17 (high) | 19 | 13 |
| Belgium* | 28,335 (12) | 9 (high) | 9 | 6 |
| France* | 27,677 (15) | 16 (high) | 16 | ND |
| Germany* | 27,756 (14) | 20 (high) | 20 | 9 |
| Ireland* | 37,738 (2) | 8 (high) | 11 | 16 |
| Luxembourg* | 62,298 (1) | 4 (high) | 7 | ND |
| Netherlands* | 29,371 (11) | 12 (high) | 12 | 8 |
| Switzerland | 30,552 (8) | 7 (high) | 6 | 11 |
| United Kingdom* | 27,147 (18) | 15 (high) | 15 | 18 |

Belen Piniero: "All of us who have been cleaning these beaches, we know that even if we get everything clean, our land will be dead for many years. Maybe our children will be able to enjoy the beaches again. I don't know," she continued, but stopped short as tears welled in her eyes. "Sorry, but I don't want to talk any more."

It's December 2002. Belen is one of hundreds of volunteers who wade through the thick black oil that covers everything over 350 miles (563 kilometers) of Atlantic coastline in northwestern Spain. Millions of fish and birds are smothering in the gunk; dead and dying animals lie all about. Help is slow to come, so the volunteers are inventing their own methods for cleaning up the poisonous oil by hand. It is a hopeless and depressing task, and nerves are frayed.

A month earlier, a Greek-owned oil tanker, the *Prestige*, had broken apart in a heavy storm and sunk, spilling 20 million gallons of highly toxic crude oil (twice the amount dumped by the *Exxon Valdez* off Alaska in 1989). For the people of the United States, this oil spill was a blip on the nightly news,

## TABLE 4.1    (Continued)

| Country (I) | GDP per capita, adjusted for PPP[b] (GDP ranking among 177 countries), 2005 (II) | Human Development Index (HDI) ranking among 177 countries,[c] 2005 (III) | Gender Development Index (GDI) ranking among 140 countries, 2005 (IV) | Gender Empowerment Measure (GEM) ranking among 80 countries, 2005 (V) |
|---|---|---|---|---|
| **South Europe** | | | | |
| Cyprus* | 18,776 (28) | 29 (high) | 28 | 39 |
| Greece* | 19,954 (25) | 24 (high) | 24 | 36 |
| Italy* | 27,119 (19) | 18 (high) | 18 | 37 |
| Malta* | 17,633 (32) | 32 (high) | 32 | 58 |
| Portugal* | 18,126 (29) | 27 (high) | 26 | 21 |
| Spain* | 22,391 (23) | 21 (high) | 21 | 15 |
| **Central Europe** | | | | |
| Albania | 4,584 (94) | 72 (medium) | 56 | ND |
| Bosnia and Herzegovina | 5,967 (70) | 68 (medium) | ND | ND |
| Bulgaria* | 7,731 (61) | 55 (medium) | 45 | 29 |
| Croatia | 11,080 (47) | 45 (high) | 40 | 32 |
| Czech Republic* | 16,357 (35) | 31 (high) | 30 | 31 |
| Hungary* | 14,584 (37) | 35 (high) | 31 | 44 |
| Macedonia | 6,794 (70) | 59 (medium) | 49 | 41 |
| Montenegro[d] | 3,800 (106) | ND | ND | ND |
| Poland* | 11,379 (45) | 36 (high) | 33 | 27 |
| Romania* | 7,277 (63) | 64 (medium) | 51 | 56 |
| Serbia (including Kosovo)[d] | 4,400 (99) | ND | ND | ND |
| Slovakia* | 13,494 (40) | 42 (high) | 37 | 33 |
| Slovenia* | 19,150 (27) | 26 (high) | 25 | 30 |
| **European Union (average)** | 28,100 | ND | ND | ND |
| **World** | 8,229 | ND | ND | ND |

[a]Rankings are in descending order; i.e., low numbers indicate high rank.

[b]PPP = purchasing power parity, figured in 2003 U.S. dollars.

[c]The high, medium, and low designations indicate where the country ranks among the 177 countries classified into three categories by the United Nations.

[d]Data from *CIA World Factbook*, 2005; dates of CIA estimates vary.

*Designates a member of the European Union.

ND = No data available.

*Sources: United Nations Human Development Report 2005* (New York: United Nations Development Programme), Table 1 and Table 26, at http:/hdr .undp.org/reports/global/2005/pdf/HDR05_complete.pdf; *CIA World Factbook,* 2005, at https://www.cia.gov/cia/publications/factbook/index.html.

coming during the buildup for war in Iraq. For the people of Galicia, the oil spill was an enduring nightmare. Was the disaster unavoidable, just one of those unfortunate times when a ship gets caught in a storm?

*Watch the FRONTLINE/World video "The Lawless Sea" to learn the facts.* ■

## MEASURES OF HUMAN WELL-BEING

We have already observed that there is considerable disparity in wealth across Europe. Table 4.1 compares the well-being of people in Europe with that of people in Barbados, Japan, Kuwait, and the United States. The indices in the table are those introduced in Chapter 1 and used throughout this book.

In the category of gross domestic product per capita (Table 4.1, column II), you can see that most countries in North and West Europe compare well with Barbados, Japan, Kuwait, and the United States. The only exceptions are Estonia, Latvia, and Lithuania, which separated from the Soviet Union in the early 1990s; these three countries have notably lower per capita

GDPs than their neighbors in North Europe. South Europe is less affluent than either North or West Europe. The countries of Central Europe are, for the most part, way down the scale in GDP per capita, with the notable exception of Slovenia. In fact, on this measure, several of these countries fall well below several Middle and South American countries (compare Table 3.3 on pages 156–157). Yet despite these low GDP per capita figures, there was not great disparity in wealth between classes in Central Europe until recently, because the Communist system specifically sought to even out the distribution of wealth. As a result, GDP per capita figures for this subregion are a somewhat more meaningful measure of average well-being than they are in Middle and South America and elsewhere, where class disparities are huge. Even in the relatively wealthy country of Slovenia, the richest 20 percent of the population is just four times more wealthy than the poorest 20 percent. In Middle and South America, the richest 20 percent can be more than 20 times as wealthy as the poorest 20 percent.

In many parts of Europe, state subsidies keep the prices of necessities low and make salaries go further. All countries in North, West, and South Europe, except for Malta, Estonia, Latvia, and Lithuania, rank in the top 30 on the global United Nations Human Development Index, which is sensitive to more factors than just income (Table 4.1, column III). Iceland, Norway, and Sweden are at the top of the 177 countries ranked, undoubtedly because they have comprehensive social welfare systems. In contrast, the countries of Central Europe (except Slovenia) rank relatively low on the HDI (from 31 to 72) because, as the Communist system failed,

people lost not only their jobs but also their health care, housing, and food subsidies. Even those who retained their jobs saw their salaries become meaningless pittances as a result of high inflation rates. Air and water pollution increased, personal stress over uncertainties mounted, overall quality of life declined drastically, and life expectancies plummeted.

In making opportunities available to women (Gender Development Index or GDI, column IV) North and West Europe do the best, and the same is true for the Gender Empowerment Measure (GEM, column V). The exceptions are again Estonia, Latvia, and Lithuania, which rank comparatively low because they were long part of the Soviet Union and retain many customs and institutions that restrict women. The GDI and GEM rankings of South Europe are lower than those of North and West Europe: Greece, Cyprus, Italy, and Malta have some of the lowest GDI and GEM rankings in Europe, and Spain, Portugal, and Italy are not improving as rapidly as expected. In much of Central Europe, useful statistics have not been collected until recently, in part because of underfunded statistics departments and in part because, under Communism, the problem of gender inequity went unrecognized. Specific studies report that, compared with men, a disproportionate number of even well-educated young women in Central Europe are unemployed, while their older female relatives are being forced into early, inadequately funded retirement (see page 239). This inequity is one factor that may account for the not yet fully understood migration of young, educated Central European women into West and North Europe.

## III SUBREGIONS OF EUROPE

In this section we look at the subregions of Europe as separate units, examining in greater detail some of the issues that affect them. As we discussed earlier, the subregion in which a particular country is included is based on its location, on convention, and on our best judgment of the situation in Europe at this time. Good arguments for other groupings could be made.

Every subregion of Europe is undergoing changes. These changes result from the end of cold war politics, the struggles of Central European countries to establish national identities and viable economies, moves toward greater economic and political cooperation throughout the region, and global shifts in economic and political power.

## WEST EUROPE

Despite their economic success, the countries of West Europe (Figure 4.36)—the United Kingdom, the Republic of Ireland, France, Germany, Belgium, Luxembourg, the Netherlands, Austria, and Switzerland—are confronting several issues that will affect their development and overall well-being:

- The continuing trend toward economic and political union that is affecting all of Europe

- Social tensions, such as those generated by the influx of immigrants from outside the subregion

- The persistence of high unemployment despite general economic growth

- The need to increase the global competitiveness of the subregion's agricultural, industrial, and service sectors while at the same time maintaining costly, but highly valued, social welfare programs

### Benelux

Belgium, the Netherlands, and Luxembourg—often collectively called the Low Countries or Benelux—are densely populated countries that have achieved very high standards of living. These three countries are well located for trade: they lie close to North Europe and the British Isles and are adjacent to the commercially active Rhine Delta and the industrial heart of Europe. The coastal location of Benelux and its great port cities of Antwerp, Rotterdam, and Amsterdam give these countries easy access to the global

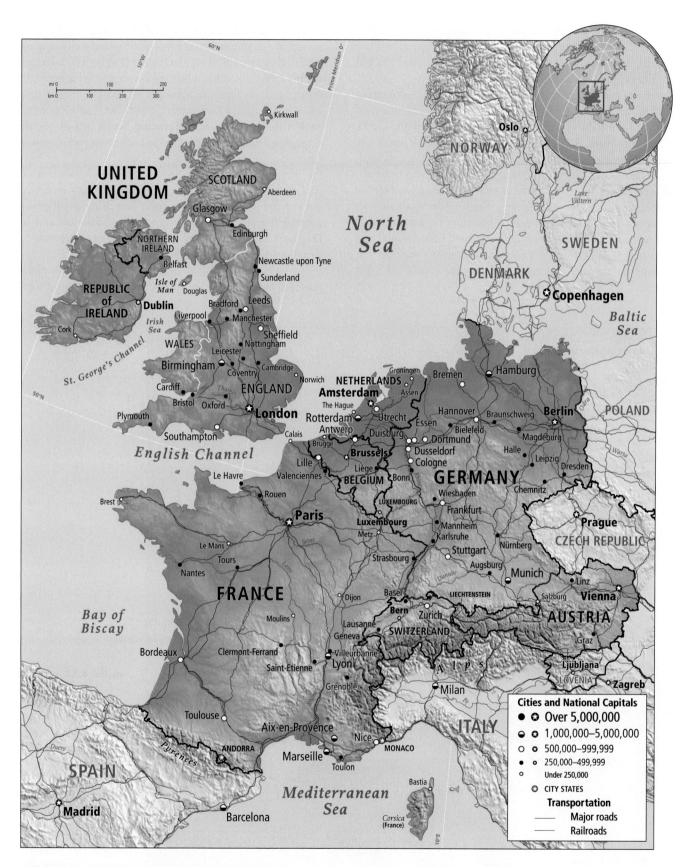

**FIGURE 4.36** The West Europe subregion.

marketplace. The Benelux countries have long played a central role in the European Union. Brussels, Belgium, is considered the EU capital, with most EU headquarters located there, and this EU activity draws other business to Brussels.

International trade has long been at the heart of the economies of Belgium and the Netherlands. Both were active colonizers of tropical zones: Belgium in Africa and the Netherlands in the Caribbean and Southeast Asia. Their economies benefited from the wealth extracted from their colonies and from global trade in tropical products such as spices, cacao beans, fruit, wood, and minerals. Private companies based in Benelux still maintain advantageous relationships with the former colonies, which supply raw materials for European industries. The Benelux countries occasionally find themselves embroiled in conflicts related to both their colonial past and pressures to integrate with their neighbors in the new EU era; the European "chocolate standards war" discussed in Box 4.5 is a case in point.

**The Netherlands: A Human-Made Place.** The largest Benelux nation, the Netherlands, is particularly noted for having reclaimed land that was previously under the sea (Figure 4.37). Today, its landscape is almost entirely a human construct. As its population grew during and after the medieval period, people created more living space by filling in a large natural coastal wetland (Figure 4.38a). To protect themselves from devastating North Sea surges, they built dikes, dug drainage canals, pumped water with windmills, and constructed artificial dunes along the ocean. Today, a train trip through the Netherlands between Amsterdam and Rotterdam takes one past raised rectangular fields crisply edged with narrow drainage ditches and wider transport canals. The fields are filled with commercial flower beds of crocuses, tulips, daffodils, and hyacinths (Figure 4.38b); there are also grazing cattle and vegetable gardens, which feed the primarily urban population.

The Netherlands has 16.3 million people, who enjoy a high standard of living. It is the most densely settled country in Europe

**FIGURE 4.37** Land reclamation areas in the Netherlands. The Netherlands is noted for having a landscape that is almost entirely human-made. As its populations grew, people created more living space by filling in a large natural coastal wetland. Filling in wetlands is now prohibited in many parts of the world because of the known negative environmental impacts. [Adapted from William H. Berentsen, *Contemporary Europe: A Geographic Analysis* (New York: Wiley, 1997), p. 317.]

(a)

(b)

**FIGURE 4.38 Coastal lands in the Netherlands.** These two photographs of the Netherlands, taken from the air, show the manipulation of natural land and water interactions in coastal zones and the intensive human uses of the reclaimed lands. (**a**) Part of the port of Rotterdam, looking northwest. Cranes transfer products between river vessels and oceangoing ships. The oil "tank farms" figure prominently in Dutch chemical industries. (**b**) Fields of flower bulbs on raised, drained lands close to the North Sea coast. Economically, the flowers, which are sold locally, are merely a by-product of the bulbs, which are marketed internationally. [Aeroview, Rotterdam.]

---

## BOX 4.5 | AT THE GLOBAL SCALE   The Chocolate Standards War

In Europe, chocolate is an important luxury product, and European high-end chocolates such as Godiva (a Belgian company) are sold around the world at very high prices. Chocolate first enters Europe in the form of dried cacao beans, which are grown mostly by small producers in Africa who barely eke out a living and have no power to negotiate the prices they are paid. Meanwhile, European cacao traders have grown wealthy from their earnings on the world cacao market, and European workers in chocolate factories also live well in comparison to cacao farmers. Some observers would argue that the European cacao traders and chocolate factory workers are rewarded with high incomes for their higher levels of skill and technology and for developing the chocolate market in the first place. Critics of global market systems might counter that the cacao farmers are also quite skilled, despite little access to schooling. And they might question whether it is right that education and technology should provide such overwhelming benefits to those who trade and process a product, in comparison with those who produce the raw materials.

The chocolate standards war isn't about equalizing profits, however, though profits will be affected by its outcome. The "war" itself is about how much non-cocoa oil can be used in European chocolates.

Belgium, France, and six other EU nations agree that real chocolate (so labeled) must be made only of cocoa butter derived from cacao beans, with no oil or fat additives. But the United Kingdom, Denmark, and several other countries routinely add cheaper oils to their mass-produced chocolate candy. The EU Commission, charged with defining common manufacturing standards for the entire EU, is trying to reconcile these different national chocolate traditions. Some contend that with open EU borders, the cheaper chocolate made with additives will have an unfair advantage over pure chocolate, both in the EU and abroad. Globally, six multinational firms, controlling well over half of all chocolate sales, argue that chocolate recipes should not be standardized. The purists, on the other hand, argue that European chocolate will be more competitive in global markets if it is a gourmet product that maintains the highest standards of purity.

This European chocolate debate has global implications. The West African countries that produce cacao beans—Ghana, Cameroon, and Côte d'Ivoire—are former European colonies that depend on cacao for at least a quarter of their export earnings. They estimate that the use of even as little as 5 percent non-cocoa oils in the EU will mean a 10 percent drop in world cacao consumption and will devastate their economies. These three African countries have petitioned their former colonizers to maintain the pure chocolate standards in accordance with agreements they made to help African economies rebound from colonial exploitation.

The chocolate story is taken up again in Chapter 7, where we will see how little control African producers have over the market in which they must operate.

(with the exception of Malta), and this density has consequences. There is no land left for the kind of high-quality suburban expansion preferred in the Netherlands without intruding on agricultural space. And there is not nearly enough space for recreation; bucolic as they appear, transport canals and carefully controlled raised fields are not a venue for picnics and soccer games. On one route, a train traveler might see huge cranes in the distance at the port of Amsterdam, high-tech office buildings to one side and a new satellite town to the other, and in the foreground an elderly woman biking along a canal. People now travel great distances to reach their jobs, and these long commutes add to air pollution and miserable traffic jams. For even weekend getaways, people usually leave the Netherlands for a neighboring country that has more natural areas. The choice to maintain agricultural space is largely a psychological and environmental one, as agriculture accounts for only 2 percent of both the GDP and employment.

## France

France has the shape of an irregular hexagon bounded by the Atlantic on the north and west, mountains on the southwest and southeast, Mediterranean beaches on the south, and the lowlands of Belgium on the northeast (see Figure 4.36). The capital city of Paris, in the north of the hexagon, is the heart of the cultural and economic life of France. Paris, as one of Europe's two world cities (London is the other), has an economy tied to the provision of highly specialized services—traditionally fine food, food processing, fashion, and tourism, but increasingly accounting, advertising, design, financial, scientific, and technical services. These service industries draw in a steady stream of workers from Europe and beyond.

Few would quarrel with the idea that French culture has a certain cachet and that France is an arbiter of taste. Its magnificent marble buildings, manicured parks, museums, and grand boulevards have made France the leading tourist destination on earth, with 11 percent of global tourism arrivals. In 2004, France attracted 75 million international tourists, more than the population of the entire country (60.7 million).

Paris, though elegant, is also a working city. It is the hub of a well-integrated water, rail, and road transport system, and its central location and transport links attract a disproportionate share of trade and businesses, so it qualifies as a primate city. Together with its suburbs, Paris has more than 11 million inhabitants—roughly a fifth of the country's total. In the 1970s, French planners decided that Paris was large enough, and they began diverting development to other parts of the country, especially to Toulouse and farther south to the Mediterranean.

To the west and southwest of Paris (toward the Atlantic) are less densely settled lowland basins that are used primarily for agriculture. France has the largest agricultural output in the EU, and globally, it is second only to the United States in agricultural exports. Its climate is mild and humid, though drier to the south. Throughout the country, farmers have found profitable specialties, such as wheat (France is ranked fifth in world production), grapes (Figure 4.39; France and Italy compete for first place in world wine production and wine exports), and cheese (France is second in world production). Despite France's leadership in agricultural production, agriculture accounts for only 2.5 percent of the French national GDP and employs only 4.1 percent of the population. While modernization was making ever higher agricultural production possible with

**FIGURE 4.39** The French countryside. The French wine country is a picturesque attraction for tourists. These vineyards are in the Champagne region of France, east of Paris. [David Barnes/Danita Delimont.]

ever lower labor inputs, the manufacturing sector quickly grew very large, and now the high-end service sector is growing by leaps and bounds as well. French industry now accounts for 21 percent of GDP, and services account for 76 percent.

The Mediterranean coast is the site of France's leading port, Marseille, and of the famous French Riviera. Here, development is booming. Tourism, in particular, has inspired the building of marinas, condominiums, and parks that threaten to occupy every inch of waterfront. The downside of this wealthy, densely populated region is that it produces large amounts of the pollutants that contribute to Mediterranean environmental problems (see the discussion on pages 218–219).

France derived considerable benefit and wealth from its large overseas empire, which it ruled until the mid-twentieth century. Many of the citizens of former French colonies in the Caribbean, North America, North Africa, sub-Saharan Africa, Southeast Asia, and the Pacific (and their descendants) now live in France and bring to it their skills and a multicultural flavor. Paradoxically, while the European French take pride in being cosmopolitan, they are also very protective of their distinctive culture and wish to guard what they consider to be its purity and uniqueness. This has led to a marginalization of immigrant minorities, who tend to live on the fringes of the big cities in high-rise apartment ghettos, where they are poorly served by public transport, have trouble accessing higher education or technical training, and suffer unemployment rates that are twice the national average.

France's strong cultural identity is both a blessing and a curse. Many people in France are deeply concerned about what they perceive to be a decrease in France's world prestige as its industrial competitiveness declines, unemployment rises, social welfare benefits decrease, and traditional French culture is diluted by an influx of migrants from former colonies. Such feelings have resulted in increasing popular support for the National Front, France's xenophobic right-wing party. But others in France are seeking to reinforce a multicultural identity in the potentially more egalitarian global community that France helped to create by being a model for democracy. From a purely practical point of view, because it has a low birth rate, France needs immigrants to keep its economy humming, its technology on the cutting edge, and its tax coffers full.

## Germany

The most famous image of Germany in recent times is that of a happy crowd dismantling the Berlin Wall in 1989. This symbolic end of Soviet influence in Central Europe had particular significance for Germany, which for 40 years had been divided into two unequal parts. At the end of World War II, Russian troops occupied the eastern third of Germany; by 1949, the Soviets and their German counterparts had created a Communist state, known as East Germany, in this area. Over the next decade or so, perhaps as many as 3 million East Germans fled west to the other part of the country, then called West Germany. To retain the remaining East German population, the Soviets literally walled off the border in the summer of 1961. East German skilled labor, mineral resources, and industrial capacities were used to buttress the socialist economies of the Soviet bloc and to support the military aims of the Soviet Union and its allies.

Because Germany was regarded as the perpetrator of two world wars, West Germany had to tread a careful path within Europe after 1945, seeing to its own economic and social reconstruction and rebuilding a prosperous industrial base, yet not seeming to become too powerful economically or politically. For the most part, West Germany played this complex role successfully. Since the early 1980s, it has been a leader in building the European Union, and, with the EU's largest economy, it has borne the greatest financial burden of the European unification process.

After the fall of the Berlin Wall, the two Germanys were reunited. But East Germany came home with a suitcase full of troubles that are proving expensive to fix. Its industries were outdated, inefficient, polluting, and in large part irredeemable. Much of its decaying or substandard infrastructure (bridges, dams, power plants, and housing) did not meet the standards of West Europe and had to be remodeled or dismantled and replaced. Its industrial products were not competitive in world markets, and East German workers, though considered highly competent in the Soviet sphere, were undereducated and underskilled by West European standards. Reunified Germany is Europe's most populous country (83 million people) and is a global leader in industry and trade, but the costs of absorbing the poor eastern zone have dragged Germany down in many rankings within Europe. Unemployment rose sharply (especially among women) in the 1990s and has remained above 10 percent into 2006. Not only have there been layoffs in the east (where unemployment is nearly double the national average), but workers throughout Germany, accustomed to high wages and generous benefits, are losing jobs as firms mechanize or move overseas, some to the United States.

The German multinational corporation DaimlerChrysler (Figure 4.40), headquartered in Stuttgart, is an example of a German firm that moved operations out of Germany. Starting with Daimler-Benz's purchase of the American auto company Chrysler in the late 1990s, DaimlerChrysler opened plants in the United States, Mexico, Poland, Hungary, India, Japan, Korea, China, Taiwan, and Malaysia, as well as in Germany. At the same time, it reduced its total workforce by 100,000. Its expansion into Asia came not only because factory costs are lower there, but also because the market for luxury cars such as Mercedes-Benz is likely to soar in Asia. As of 2007, the company was under reorganization and in May of that year sold 80.1 percent of Chrysler to a private equity firm.

## The British Isles

The British Isles, located off the northwestern coast of the main European peninsula, are occupied by two countries: (1) the United Kingdom of Great Britain (England, Scotland, and Wales) and Northern Ireland—often called simply Britain or the United Kingdom (UK), and (2) the Republic of Ireland (not to be confused with Northern Ireland).

***The Republic of Ireland and Northern Ireland.*** The only physical resources in the Republic of Ireland (usually called simply

**FIGURE 4.40** Mercedes-Welt (Mercedes World). The Mercedes-Welt Center, shown here, is part of the DaimlerChrysler installation in Berlin. With a showroom, service center, and restaurant, the Center is one of the city's most modern buildings. [AP Photo/Herbert Knosowski.]

*Ireland*) are its soil, abundant rain, and beautiful landscapes. This combination has contributed to a long dependence on agriculture and tourism, while industrialization lagged. As a result, the Irish people were for a long time the poorest in West Europe, and over the last two centuries some 7 million Irish emigrated to find a better life. In the 1990s, however, a remarkable turnaround began. Ireland attracted foreign manufacturing companies by offering cheap, well-educated labor, accessible air transport, access to EU markets, low taxes, and other financial incentives. Foreign firms such as Glaxo, Merck, Norsk Hydro, and Phillips Petroleum brought in small industries that specialized in software development, food and drink processing, and the manufacture of pharmaceuticals, chemicals, high-end giftware, building products, and textiles. The economy grew so quickly that soon Irish labor was in short supply.

To supply the workers required, Ireland invited its diaspora back, and other foreign workers were recruited in many locales (recall the story of Janis Neulans at the opening of this chapter). In 2000, 42,000 people immigrated to Ireland, 18,000 of whom were returning emigrants. Now Czech workers are packing meat in Ireland, and Filipino nurses are working in most Irish hospitals. The cost of living has risen sharply, however, and this rise could affect the country's ability to attract yet more foreign investment. Ireland did not favor the admission of Central European countries to the EU because it feared that those countries would attract investment away from Ireland by offering even cheaper labor pools. Nevertheless, Ireland has one of Europe's fastest-growing (though still small) economies, with industry playing nearly as important a role as services in the country's GDP. Ireland now has Europe's second highest per capita GDP (see Table 4.1 on pages 220–221).

Northern Ireland—six counties in the northeastern corner of the island—is distinct from the Republic of Ireland. Northern Ire-

land began to emerge as a political entity in the seventeenth century, when Protestant England conquered the whole of Catholic Ireland. England removed or killed many of the indigenous people and settled Lowland Scots and English farmers on the vacated land. The remaining Irish resisted English rule with guerrilla warfare for nearly 300 years, until the Republic of Ireland gained independence from the United Kingdom in 1921. Northern Ireland remained part of the United Kingdom. Here Protestant majorities held political control, and the minority Catholics experienced economic and social discrimination. Catholic nationalists unsuccessfully lobbied by constitutional (peaceful) means for a united Ireland. Other Catholic groups, the most radical being the Irish Republican Army (IRA), resorted to violence, including terrorist bombings, often against British peacekeeping forces, who were seen as supporting the Protestants, and civilians. The Protestants reciprocated with more violence. More than 3000 people had been killed in the Northern Ireland conflict by 1995.

In 1998, tired of seeing the development of the entire island—and especially of Northern Ireland—blighted by the persistent violence, the opposing groups finally reached a peace accord known as the Good Friday Agreement, which was overwhelmingly approved by voters in both Northern Ireland and the Republic of Ireland. The plan provides for more self-government shared between Catholics and Protestants, the creation of human rights commissions, the early release of convicted terrorist prisoners, the decommissioning of paramilitary forces (such as the IRA), and the reform of the criminal justice system. Since 1998, violence has periodically risen and then abated, and the society has slowly begun to accept peaceful coexistence. The city of Belfast remains largely divided along religious lines, but some neighborhoods are beginning to integrate.

***The United Kingdom.*** Like Ireland, the United Kingdom has a mild, wet climate and a robust agricultural sector—based, in its case, on grazing animals. Its usable land is extensive, and its mountains contain mineral resources, particularly coal and iron. In contrast to Ireland, however, the United Kingdom has been operating from a position of power for many hundreds of years. Beginning in the seventeenth century, Britain added resources from its colonies, first in Ireland and then in the Americas, Africa, and Asia, to its own resources. Together, these resources were sufficient to make Britain the leader of the Industrial Revolution in the early eighteenth century. By the nineteenth century, the British Empire covered nearly a quarter of the earth's surface. As a result, British culture was diffused far and wide, and English became the lingua franca of the world.

Britain's widespread international affiliations, set up during the colonial era, positioned it to become a center of international finance as the global economy evolved. Today, London is Europe's leading financial center, and it handles 31 percent of global foreign currency exchange, more than twice as much as New York City, its closest rival. The United Kingdom is no longer Europe's industrial leader, however. At least since World War II, and some say earlier, the United Kingdom has been sliding down from its high rank in the world economy toward the position of an average European nation. The discovery of oil and gas reserves in the North Sea gave the United Kingdom a cheaper and cleaner source of energy for industry than the coal on which it had long depended, but it did not stop the economic decline. Cities such as Liverpool and Manchester in the old industrial heartland and Belfast in Northern Ireland have experienced long depressions.

Beginning in the 1970s, a series of conservative governments tried to make the United Kingdom more competitive by instituting two decades of budget cutbacks, reducing social spending, selling off unprofitable government-run firms, and tightening education budgets. Although these measures were not so labeled, they were the same as the structural adjustment policies (SAPs) implemented in many other world regions (see Chapter 3). Unemployment rose. Eventually, foreign investment began to pour into the United Kingdom because, after years of cutbacks, skilled workers were willing to accept relatively low wages.

Britain has now successfully established technology industries in parts of the country that previously were not industrial—for example, near Cambridge and west of London in a region called Silicon Vale. But the service sector now dominates the economy (accounting for 73 percent of the GDP and 80 percent of the labor force). In the last decade, new, though not high-paying, jobs have been created in health care, food services, sales, financial management, insurance, communications, tourism, and entertainment. Even the movie industry has discovered that England is a cheap and pleasant place for film editing and production. Britain's past experience with a worldwide colonial empire has left it well positioned to provide superior financial and business advisory services (producer services) in a globalizing economy. Such service jobs, however, tend to go to young, educated, multilingual city dwellers, many of them from elsewhere in the European Union, not to the middle-aged, unemployed ironworkers and miners left in the old industrial UK heartland. Because of the EU's open borders, there are now more than 400,000 foreign nationals working in Greater London alone (Figure 4.41).

**FIGURE 4.41** A multiethnic scene on the main street in Southall, Greater London. [Christine Osborne Pictures/photographersdirect.com.]

Politically, the whole United Kingdom was governed until recently by a single parliament with delegates from the four countries. England, with the largest population by far, usually prevailed in national debates. In 1997, Scotland and Wales voted to institute their own elected governing bodies—a parliament for Scotland and a national assembly for Wales. In 1998, the Good Friday Agreement allowed for local Northern Irish government as well. This decision for home rule is seen as an example of **devolution,** the weakening of a formerly tightly unified state. It springs both from dislike of the structural adjustment policies of recent UK conservative governments—which brought high unemployment and losses of social programs to both Scotland and Wales—and from rising feelings that the wishes of the voters in both countries were consistently overwhelmed by the English majority in the British Parliament. It remains to be seen whether this change will result in the eventual independence of Scotland, Wales, and Northern Ireland from the United Kingdom or merely the delegation of some powers to what amount to subassemblies within the United Kingdom. Much will depend on decisions within the EU regarding allocation of political power and development funds.

The Jamaican poet and humorist Louise Bennett used to joke that Britain was being "colonized in reverse." She was referring to the many immigrants from the former colonies who now live in the United Kingdom. Indeed, it would be hard to overemphasize the international spirit of the United Kingdom. London, for example, has become an intellectual center for debate in the Muslim world. Salman Rushdie, the controversial Indian Muslim writer, lives there; many bookstores carry Muslim literature and political treatises in a variety of languages; Israelis and Palestinians talk privately about peace; and the Saudi Arabian government has a strong presence. On a leisurely stroll through London's Kensington Gardens, you will discover thousands of people from all over the Islamic world who now live, work, and raise their families in London. They are joined by many other people from the British Commonwealth: Indian, Malaysian, African, and West Indian families pushing baby carriages and playing games with older children. Impromptu soccer games may have team members from a dozen or more countries. A 2000 survey of ethnic diversity in Britain noted that Britons of African and West Indian heritage contribute $5 billion per year to the economy, and that over 200 languages are spoken in London daily.

## SOUTH EUROPE

The Mediterranean subregion of South Europe (Figure 4.42) was once unrivaled in wealth and power. First Athens and then Rome were imperial seats whose influence extended widely. In the medieval period, from the fifth to the fifteenth centuries, the Italian cities of Venice, Florence, and Genoa were major trading centers with contacts stretching across the Indian Ocean and by land through Central Asia to eastern China. During the sixteenth century, Spain and Portugal developed large colonial empires, primarily in the Americas but also in Africa and Southeast Asia. But times changed. As the rest of Europe's economy expanded to a global

scale with the development of colonies by England, France, and the Netherlands in distant world regions, the Mediterranean ceased to be a center of trade. South Europe declined, and the Industrial Revolution reached this subregion relatively late. Even today, much of South Europe remains poor in comparison to the rest of the continent: only the countries of Central Europe are poorer (see Table 4.1). In this section, we pay particular attention to Spain and Italy, contrasting the parts of each country that are now prospering with the parts that have remained poor.

In all six countries of South Europe—Portugal, Spain, Italy, Malta, Cyprus, and Greece—agriculture was the predominant occupation through most of the twentieth century. Farmers often lived in poverty, using simple tools to produce crops for local consumption. Since the 1970s, however, there has been an effort to modernize and mechanize agriculture across the subregion. The focus has been on irrigating large holdings to grow commercial crops. Irrigation has increased output and opened up new semiarid locations for cultivation, but not everyone welcomes the abundance of produce flowing from South Europe. Today, both Spain and Italy mass-produce vegetables and fruits in quantities that exceed demand in the European Union. The surplus has driven down prices, hurting smaller family farmers across Europe and creating ill feelings toward the European Union. Moreover, as in the western United States, irrigation schemes result in water shortages, depletion of reservoirs and natural aquifers, and increases in soil salinity over time. Farmers in northern Spain and Portugal complain because water originating there is transported by canal to irrigate more arid regions in the south.

Despite the extensive improvement projects, agriculture as a whole (measured as a percentage of GDP and as a percentage of the labor force) has declined in importance in South Europe as manufacturing, tourism, and other services have increased. For several decades, middle-class tourist money has brought new life to the economy of the region, but tourist money has done little to elevate the lives of poor rural people. Although some localities have grown prosperous as resort sites, many others have been bypassed. In Greece, for example, income from tourism has been confined to scenic coastal and island locations, and many interior rural areas remain poor and isolated. In much of South Europe, the very qualities that attract tourists are being threatened by the sheer numbers of visitors and the effects they have on coastal environments and on rural culture.

### Spain

In the last quarter of the twentieth century, Spain emerged from centuries of underdevelopment. Today, it is poised to take advantage of opportunities in a more integrated Europe, and its future is brighter. Just 40 years ago, however, Spain was a poor and underdeveloped country. By the beginning of the twentieth century, its once vast empire in the Americas and Asia was all but gone. This loss was followed by years of civil war and then a military dictatorship led by Francisco Franco. Even Spain's physical location was no longer an advantage, as commercial activity had shifted away

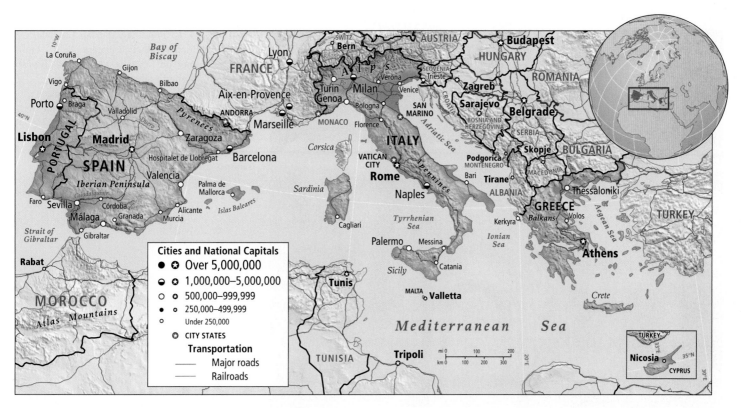

**FIGURE 4.42** The South Europe subregion.

from the Mediterranean to West Europe. The Pyrenees kept Spain isolated from the more prosperous parts of Europe, making travel and commerce difficult.

In 1975, Spain made a cautious transition to democracy, and in 1986 it joined the EU. Since then, its growth has accelerated with the help of EU funds aimed at bringing Spain into economic and social harmony with the rest of the EU through investment in infrastructure and in human resources. As a result of these improvements, foreign businesses are investing in industry and factories have been modernized. The cities of Barcelona and Madrid are now centers of population, wealth, and industry. Both are now linked to the rest of Europe by a network of roads, airports, and high-speed rail lines. Indeed, Spain's standing as a European cultural center has so improved that on August 10, 2003, a *New York Times* piece began with the headline "Spain Is the New France," meaning primarily that Spain is now a leader in stylish European lifeways and is an attractive tourist destination.

Nonetheless, some segments of Spanish society have not participated in this new creativity and prosperity. Many small farmers remain poor, particularly in the northwestern and southwestern provinces, which are arid and have few resources. Even in the relatively prosperous zones of Barcelona and Madrid, workers have been laid off as a result of factory modernizations. In 2005, unemployment hovered at 9.2 percent (at the high end for western Europe) and affected twice as many women as men. Underemployed Spanish workers were willing to work for significantly lower wages than were paid in North and West Europe.

Spain's low wage scales are one reason that it has attracted foreign investment from around the world. Other attractions include democratic institutions able to withstand crises (such as the terrorist bombing in Madrid in March 2004), an educated and skilled workforce (partly the result of EU funds for training), and the absence of stringent environmental and workplace regulations. Additionally, Spain has good ports on both the Atlantic and the Mediterranean. Spain now has an industrial base that includes food and beverage processing and the production of chemicals, metals, machine tools, textiles, and automobiles and auto parts (Ford, DaimlerChrysler, Citroën, Opel, Renault, MDI, and Volkswagen all have factories in Spain). An interesting indicator of Spain's growth is its increasing investment in its former colonies in Latin America. Between 1996 and 2000, Spanish corporations invested U.S.$14 billion in hotels, factories, technology, utilities, and banks in countries such as Argentina, Cuba, Mexico, and Venezuela. Spain's telecommunications giant, Telefonica, has invested more than U.S.$5 billion in Latin American telecommunications companies.

Regional disparities in wealth within Spain contribute to high and persistent levels of social discord in certain parts of the country. Two of Spain's most industrialized and now wealthiest regions—the Basque country in the central north (adjacent to France) and Catalonia in the east (also bordering France)—have especially strong ethnic identities. Their inhabitants often speak of secession as a response to long years of repression by the Spanish government; in addition, the Basques would like to avoid supporting Spain's poorer districts. A Basque separatist movement periodically resorts

**FIGURE 4.43** La Mezquita Cathedral, Córdoba, Spain. This Roman Catholic cathedral was first built as a Muslim mosque. Begun on the ruins of a Roman temple in 784, shortly after the Moors entered Europe, the mosque took more than 200 years to complete. It was the largest of more than a thousand mosques in Córdoba. It became a church in 1236, when Córdoba was taken from the Moors by the warriors of King Ferdinand III of Castile. [Toyohiro Yamada/Taxi/Getty.]

to violence. Meanwhile, Galicia, in the far northwestern corner of Spain, also has a strong cultural identity, and the preservation of the Galician language is a popular issue there. All three of these ethnic enclaves emphasize their distinctive languages as markers of identity and chafe against the use of Castilian Spanish as Spain's official language.

Migration is another important part of Spain's present situation. Spain's connections to North Africa date back more than 1300 years; Moorish Muslim conquerors came in 700 C.E. and stayed until the 1400s, deeply influencing Spanish culture, language, architecture, cuisine, and attitudes toward gender roles (Figure 4.43). Today, North Africans still have an influence, but now as low-wage, often illegal, immigrants who come to work in cleaning and maintenance or as factory workers; there are more than 600,000 North Africans living in Spain today. They are joined by several million people from sub-Saharan Africa and from Spain's former colonies in Middle and South America, many professionally trained people who despaired of making a decent living in Ecuador, Venezuela, or Argentina. But even more Spaniards are now leaving Spain than immigrants are entering; since the 1970s, several million Spanish workers have migrated to other European countries.

## Italy

Italy is by far the wealthiest country in South Europe and is, in fact, the seventh largest economy in the world, measured by GDP. Over

the millennia, Italy has contributed greatly to European culture and prosperity. By the thirteenth century, the northern trading center of Venice, on the Adriatic Sea, had commercial links reaching around the Mediterranean and into Africa and Arabia, to the Malabar Coast of India, and all the way to China. Marco Polo, from a wealthy trading family in Venice, took a trip through Central Asia to China during the late 1200s and returned after some 20 years to give medieval Europeans their first impressions of life and commerce in China. The wider view of the world that Polo brought home contributed to the European Renaissance in northern Italy. Wealthy families such as the Medicis patronized the arts and literature and invested in beautifying public and private spaces, and Italy's enlightened ideas about philosophy, art, and architecture spread throughout Europe. Italy's prominence faded, however, as Spain, England, Portugal, and the Netherlands grew rich on their colonies in the Americas and Asia. During the 1700s and 1800s, Italy got caught up in a long series of wars between its then richer and more powerful neighbors, France, Spain, and Austria.

Progress was rocky for Italy's ordinary people, who were buffeted by the political ambitions of politicians and religious leaders (the Pope, the leader of the Catholic Church, remains headquartered at the Vatican in Rome). Wide disparities in wealth and power left whole sections of the country lawless and in decline. This was especially true of the southern half of the peninsula and the islands of Sardinia and Sicily, which were collectively known as "the south." Absentee landlords and central government tax collectors from northern Italy hired *mafiosi* to collect fees from those who worked

**FIGURE 4.44** High fashion in Italy. At his Milan headquarters in September 2005, Giorgio Armani prepares for his Spring–Summer fashion show. [Alessandra Benedetti/CORBIS.]

the land but were given no chance for self-government. The south remains woefully behind the north economically.

Northern Italy, by comparison, has one of the most vibrant economies in the world, producing products renowned for their quality and design: Ferrari automobiles, Olivetti office equipment, and high-quality musical instruments. In Milan, one of the largest cities in Italy, fashion designers such as Giorgio Armani have surpassed even their rivals in Paris (Figure 4.44). However, the mainstay of the Italian economy has traditionally been the exporting power of thousands of small mom-and-pop factories making everything from machine valves to leather clothing—all of which can now be made more cheaply in China. Italy's share of global trade is dropping, and its budget shortfall has exceeded the limits that the EU places on its member countries (Germany and France also have budget shortfalls). Having adopted the euro as its currency, Italy cannot devalue the lira as it used to do to make its products more attractive on the world market. For all of Italy's stylish success as a current industrial and services leader in Europe and the world, its continued ability to attract investment has been damaged by corruption and arcane rules that in 2005 earned Italy a global competitiveness rank of 47, below that of many countries in Southeast Asia, Southwest Asia, and even Africa.

Italy has been a democracy since World War II, and it is a charter member of the EU. Yet because of its domestic politics, Italy is known as the "bad boy" of Europe. This designation, which is to be taken humorously, derives in large part from unfair stereotypes of Italians as clever operators on the fringes of legality. But it also derives from the fact that Italians vote governments in and out in rapid succession and are not above electing leaders who have been mixed up in shady dealings. In 2001, Silvio Berlusconi took office as prime minister while under investigation for money laun-

dering, perjury, and bribery. For 5 years, Berlusconi managed to use his position as leader of the parliamentary majority to have legislation passed that helped his cause, but in 2006, he was voted out of office. Italians have devised ways of living with such official indiscretions without descending into national crises. During the dozens of governmental emergencies since 1950, a quasi-government based on informal relationships—the *sottogoverno*, or "undergovernment"—has taken over whenever a predicament has developed, and daily life has proceeded apparently unimpaired.

Italy's strengths and its weaknesses alike arise from its strong faith in institutions based on the family or on personal relationships. Most businesses are family-run, including several of the largest corporations of northern Italy as well as thousands of small informal firms. And, as mentioned earlier in this chapter (see pages 215–216), the rudimentary welfare system found in Italy (and other countries of South Europe) assumes that strong families and personal networks will provide what the state cannot.

## NORTH EUROPE

North Europe (Figure 4.45) traditionally has been defined as the countries of Scandinavia—Iceland, Denmark, Norway, Sweden, and Finland—and their various dependencies. Greenland (see Figure 1.1 on page 1) and the Faroe Islands are territories of Denmark; the island of Svalbard in the Arctic Ocean is part of Norway. North Europeans themselves are now including the three Baltic states of Estonia, Latvia, and Lithuania in their region, and so we do so here. All three were part of the Soviet Union until September 1991, and these small Baltic states have many remaining links to Russia, Belarus, and other parts of the former Soviet Union.

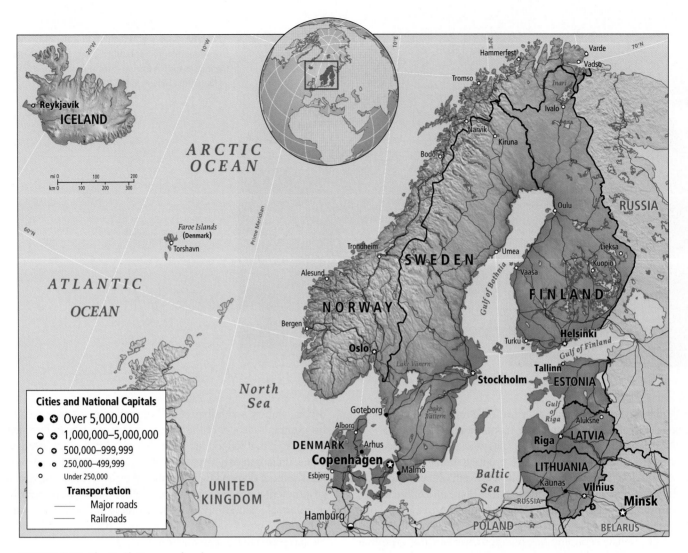

**FIGURE 4.45** The North Europe subregion.

Although their cultures are not Scandinavian, they are attempting to reorient their economies and societies in varying degrees to North Europe and the West.

The countries of North Europe are linked by their locations on the North Atlantic, the North Sea, and the Baltic Sea. Most citizens can drive to a seacoast within a few hours. The main cities of the region—Copenhagen, Oslo, Stockholm, and Helsinki—are vibrant ports that have long been centers of shipbuilding, fishing, and the warehousing of goods. They are also home to legal and financial institutions related to maritime trade.

## Scandinavia

The southern parts of Scandinavia are where most people in North Europe live and where economic activity is concentrated (see Figures 4.11 and Figure 4.20). These same warmer lowlands of southern Scandinavia are also dedicated to agriculture. Denmark, although a small country, produces most of North Europe's poultry, pork, dairy products, wheat, sugar beets, barley, and rye. Despite Denmark's agricultural successes, two-thirds of the Danish econ-

omy is nonagricultural, consisting of services (financial services and education, for example), high-tech manufacturing, construction and building trades, and fisheries. Sweden is the most industrialized nation in North Europe. The Swedes produce most of North Europe's transport equipment and two highly esteemed automobiles, the Saab and the Volvo (though some Volvo products are now built outside Europe), as well as furniture and housewares famed for their spare, elegant design (Figure 4.46). Helsinki, Finland, is home to some of the world's most successful information technology companies, such as TietoEnator, Nokia, and Novo Group.

The northern part of Sweden and almost all of Norway are covered by mountainous terrain, while Finland is a low-lying land dotted with glacial lakes, much like northern Canada. Sweden and Finland contain most of Europe's remaining forests. These well-managed forests produce timber and wood pulp. Norway, which has less usable forestland, had been considered poorly endowed with natural resources compared with the rest of Scandinavia, but the discovery of gas and oil under the North Sea in the 1960s and 1970s has been a windfall to that country. The exploitation of these resources dwarfs all other sectors of Norway's economy. Norway is

**FIGURE 4.46**  The IKEA store in Aelmhult, Sweden. Founded in Aelmhult, in 1943, IKEA now offers Swedish design in stores throughout Europe and North America and in over 40 countries worldwide. The company employs over 100,000 workers, 90 percent in Europe. [Sven Nackstrand/AFP/Getty Images.]

able to supply its own energy needs through hydropower, and it exports oil and gas to the rest of the EU. It is now one of Europe's wealthiest countries and, having used its wealth wisely, now ranks first globally in human well-being (see Table 4.1).

The fishing grounds of the North Sea, the North Atlantic, and the southern Arctic Ocean are also an important resource for the countries of North Europe (as well as for several other countries on the Atlantic). In recent decades, however, overfishing has severely reduced stocks. Rights of access to the fishing grounds remain a cause of dispute, but in 1994, the EU created a joint 200-mile (320-kilometer) coastal exclusive economic zone (EEZ) that ensures equal access and sets fishing quotas for member states. In January 2001, in order to give North Sea fish populations a chance to rebound, EU members plus Norway and Iceland agreed to an annual 3-month hiatus in cod fishing and in the taking of all juvenile fish to feed salmon in commercial farms. By 2006, however, it was clear that these short respites from fishing had not increased the cod population; a hiatus of as long as 12 years was called for over a greatly increased area of the North Sea. The thought now is that global warming may account for the drastic slump in cod catches.

Two important characteristics distinguish the Scandinavian countries from other European nations: their strong social welfare systems (see pages 214–215) and the extent to which they have achieved equality of participation and well-being for men and women. Iceland, Norway, Sweden, Denmark, and Finland hold the top five positions in the United Nations Gender Empowerment Measure (GEM) rankings (see Table 4.1).

### Social Welfare in Sweden.

Sweden's cradle-to-grave social welfare system provides stability and a safety net for every one of the country's 9 million people. This system is founded on three beliefs. First is the idea of security: that all people are entitled to a safe, secure, and predictable way of life, free from discomfort and unpleasant-

ness. This concern extends to interior design in subsidized housing, which is modern, elegant, and practical. Second, the appropriate life is ordered, self-sufficient, and quiet, not marred by efforts to stand out above others. Third, when the first two concepts are practiced as they should be, the ideal society, *folkhem* ("people's home"), is achieved. These three beliefs help to explain why Swedes are willing to pay for a social welfare system that, as of 2006, consumes more than half of their government's annual budget.

### Equal Rights in Norway.

Unlike most societies, Norway has directed much of its most innovative work on gender equality toward advancing men's rights in traditional women's arenas. For example, men now have the right to at least a month of paternity leave to stay home with a newborn or adopted child. This policy recognizes a father's responsibility in child rearing and provides a chance for father and child to bond during the early days of life. Even when addressing male violence against women, Norway recognizes in its public discussions that this global problem stems in part from cultural customs that suffocate sensitivity in individual men. Male children's games that include derision of boys who are not violent are now openly discouraged in Norway.

At the same time, equal rights for women are closer to being a reality in Norway than in any other country. In 1986, Gro Harlem Brundtland, a physician, became the country's first woman prime minister, and she appointed women to 44 percent of all cabinet posts. Norway became the first country in modern times to have such a high proportion of women in important government policy-making positions. By 2005, women made up 38.2 percent of Norway's parliament (second only to Sweden's 45.3 percent; Finland had 37.5 percent) (see Figure 4.31 on page 214). They also held nearly 33 percent of elected positions in municipal government and more than 40 percent of posts in county government. The growing policy-making power of women is perhaps best seen in the fact that, by the 1990s, women were already leaders of Norway's three major political parties.

It is not just in politics that gender equality is sought. Norway was the first country in the world to have an ombudsperson whose job it is to enforce the 1979 Equal Status Act. This act has a "60–40 rule" requiring a minimum of 40 percent representation by each sex on all public boards and committees. Even though the 60–40 rule has not yet been achieved in all cases, female representation averages over 35 percent for state and municipal agencies. In addition, employment ads are required to be gender neutral, and all advertising must be nondiscriminatory.

## The Baltic States

The three small Baltic states of Estonia, Latvia, and Lithuania are situated together on the east coast of the Baltic Sea. Culturally, these countries are distinct from one another, with different languages, myths, histories, and music. After World War II, all three shared the experience of being forced to become Soviet republics. When the Soviet Union collapsed in 1991, they regained their independence. Despite their recent connection to Russia, their cultural ties are traditionally with West and North Europe. Ethnic Estonians, who make up 60 percent of the people in Estonia, are nominally Protestant, with strong links to Finland; they view themselves as Scandinavians. Ethnic Latvians, who are just 50 percent of the population of Latvia, are mostly Lutheran; they see themselves as part of the old German maritime trade tradition. Lithuanians are primarily Roman Catholic and also see themselves as a part of West Europe.

Populations in all three countries are decreasing and aging because birth rates are generally very low and young people now have the option to migrate to North and West Europe. Ethnic Russians, who today make up about a third of the population in both Estonia and Latvia, are having more children than the indigenous populations. They could become the dominant culture group within a few decades—a crucial development, if these ethnic Russians continue to cultivate strong ties with Russia. In Lithuania, Poles, not Russians, are the dominant minority (somewhat under 20 percent), and ethnic Lithuanians will remain the dominant majority for some time to come. Ironically, although the Russian and Polish minorities were established in the Baltic states to secure the influence of the Soviet Union, under EU rules their rights as minorities had to be guaranteed before the Baltic countries could join the European Union.

Of all the countries of North Europe, the Baltic states have the most precarious economic outlook. Under Soviet rule, the Russians expropriated their agricultural and industrial facilities and used them primarily for Russia's benefit. Until 1991, 90 percent of the Baltic states' trade was with other Soviet republics. Since then, Estonia, with a light industrial base that is relatively easy to reorient, has undertaken the most radical economic change, moving toward a market economy and increasing its trade with the West. Estonia is the only one of the three Baltic states to experience real economic growth, and its accomplishments have attracted foreign aid and private investors. Heavily industrialized Latvia and Lithuania now trade more with the UK, Germany, and the West than with

**FIGURE 4.47 Industrial pollution in Estonia.** Gray snow floats down from the stacks of the Kunda cement factory. As in most other former Soviet republics, environmental degradation in Estonia resulted from the emphasis on industrial output as the key to economic growth and prosperity. [Larry C. Price.]

Russia. They have had to overcome low standards of quality, and many of their factories remain out of date and heavily polluting.

The ecological consequences of industrialization under the Soviets are widespread. One prominent example in Estonia is the oil shale production facility near Kohtla-Järve in the northeastern corner of the country. Smoldering hills of shale residue add to the noxious smoke from the plants. Oily acids seep into the groundwater and into the Baltic, causing skin ulcers on coastal fish. It is possible to ignite the chemicals floating on well water on nearby farms. Figure 4.47 shows the level of ground and air pollution near a cement factory in Estonia.

The Baltic states see national security as their major problem because their strategic position along the Baltic Sea has long been coveted by Russia, which has few easy outlets to the world's oceans. In this regard, the status of the Russian exclave of Kaliningrad is crucial. Kaliningrad is called an **exclave** because, although it is an actual part of Russia, it lies along the Baltic, far from the main part of that country, between two EU states: Lithuania and Poland (see Figure 4.1). A relatively ice-free port, Kaliningrad is headquarters for the Russian Baltic fleet. Russia's strategic interest in Lithuania is unlikely to diminish, and Lithuania fears that changing power relationships in Russia could result in a reinvasion of its territory. The countries of western Europe, unwilling to commit to military support of Lithuania, hope to resolve differences with the Russians without antagonizing them.

## CENTRAL EUROPE

No part of Europe is in a more profound period of change than Central Europe (Figure 4.48). In the euphoria at the end of the cold war, many people believed that democracy and market forces would quickly turn around the region's centrally planned economies, which were sluggish and highly polluting. Instead, some parts of the region have experienced violent political turmoil, and virtually all have experienced at least temporary drops in their standard of living. It appears that two tiers of countries are emerging. In the 1990s, those countries physically closest to western

Europe—Poland, the Czech Republic, Slovakia, Hungary, and Slovenia—elected new governments democratically, made rapid progress toward market economies, and began to attract foreign investment. Access to consumer goods improved markedly. In 2004, all of these countries joined the European Union, but conforming to the wide range of economic, environmental, and social requirements set forth by the EU has not been easy for them.

The formerly Communist countries that are experiencing the greatest economic and social difficulties are Albania, Bosnia and Herzegovina, Bulgaria, Croatia, Macedonia, Romania, Serbia (including the region of Kosovo), and Montenegro. For these countries, adjusting to democracy and privatization has proved difficult. Plagued by corruption and economic chaos, many governments have relaxed the pace of reforms as they struggle merely to maintain civil peace. In the 1990s, most of the countries that were once provinces of Yugoslavia (Bosnia and Herzegovina, Croatia,

**FIGURE 4.48** The Central Europe subregion.

Macedonia, Serbia, and Montenegro) became mired in a series of armed ethnic conflicts that eventually resulted in European (including NATO) and U.S. military intervention. After all this turmoil, not surprisingly, many people in these countries are looking back nostalgically to the Communist era, when jobs were stable, health care and education were free, and there was a strong, authoritative government. Still, many observers are optimistic about even the poorest and most politically unstable parts of Central Europe. The countries in this region have useful natural resources for both agriculture and industry, and all have large, skilled, yet inexpensive workforces that are beginning to attract foreign investment. Germany, France, the Netherlands, Italy, the United Kingdom, Russia, and the United States are their most important trade partners, but the combination changes from country to country (Serbia has only limited trade); venture capitalists from Hungary, the Czech Republic, and Slovenia are also beginning to invest in their neighbors to the south.

## The Difficulties of Economic Transition

The shift away from central planning toward a free market economy has been rocky in Central Europe. One problem has been that, because democratic institutions are still immature, old political bosses and high-level bureaucrats have been able to claim benefits for themselves at the expense of other citizens, as demonstrated in the following case studies of agricultural reform. To varying degrees, small farmers in the Czech Republic, Slovenia, Poland, and elsewhere have found it difficult to adjust to the requirements of a market economy.

*Case Study: Agricultural Reform in the Czech Republic.* In the Czech Republic in the 1990s, land that had been expropriated after World War II and turned into state farms and cooperatives was either reorganized into private owner-operated cooperatives or restored to its original owners and their descendants. The latter group, rather than farming the land themselves, usually leased their regained land, either in small parcels to those who wished to become family farmers or in large parcels to the new private cooperatives. The private cooperatives are owned by groups of workers who labor on the farms and who share their profits. Eventually, 1.25 million acres (500,000 hectares), representing 30 percent of the country's agricultural land, went to individual private farmers, large and small. Private owner-operated cooperatives now account for the majority of the agricultural land in the Czech Republic.

German economist Achim Schlüter, who studies Czech agriculture, discovered that land redistribution was only the beginning of the reform process. Successful privatized farms require assets in addition to land, such as animals, machinery, and buildings, and they need managers with marketing connections and agricultural expertise. Some former agricultural bureaucrats who had that expertise and understood the old system were able to make quick, informed decisions that allowed them to farm profitably in the new era, and they became executives of large privatized farm cooperatives. New small family farmers, on the other hand, soon gave up

because they had little understanding of market economies and were ill equipped to negotiate the best terms for themselves. Consequently, there has been a trend toward large quasi-corporate farms, which are preferred by the EU economic advisers who readied the Czech Republic for entry into the European Union. Larger, more efficient farms are preferred in part because overproduction, which could contribute to falling commodity prices across Europe, is easier to control on such farms. EU advisers have suggested that the smaller family farms find more lucrative uses for their land, such as farm tourism. They say that "dude farms," similar to dude ranches in the western United States, or living history demonstrations of traditional food cultivation and preparation techniques could attract affluent urban families from all over Europe, as Vera Kuzmic discovered (see the personal vignette below).

*Case Study: Agricultural Reform in Slovenia.* In Slovenia, unlike most of the rest of Central Europe, farms were not collectivized in the Communist era, so it has not been necessary to redistribute land. The problem instead is that farms are too small for efficient production. The average farm size is just 8.75 acres (3.5 hectares), and agriculture accounts for less than 3 percent of GDP. Although it has plenty of rich farmland, Slovenia is a net importer of food, mostly from EU countries such as Italy, Spain, and Austria. Nonetheless, the new emphasis on private entrepreneurship has encouraged some Slovene farmers to seek a niche in the domestic market. The case of Vera Kuzmic is illustrative.

**Personal Vignette**  Vera Kuzmic (a pseudonym) lives a 2-hour drive south of Ljubljana, Slovenia's capital. Her family has farmed 12.5 acres (5 hectares) of fruit trees near the Croatian border for generations. After Slovenia became independent in 1991, first her husband and then Vera lost their government jobs due to economic restructuring. The Kuzmic family decided to try earning its living in vegetable market gardening because vegetable farming could be more responsive to market changes than fruit tree cultivation. By 2000, the adult children and Mr. Kuzmic were working on the land, and Vera was in charge of marketing their produce and that of neighbors she had also convinced to grow vegetables.

Vera secured market space in a suburban shopping center in Ljubljana, where she and one employee maintained a small but orderly vegetable and fruit stall (Figure 4.49). Her produce had to compete with much less expensive Italian-grown produce sold elsewhere in the same shopping center—all of it produced on large corporate farms in northern Italy and trucked in daily. But Vera gained market share by bringing her customers special orders and by guaranteeing that no pesticides or herbicides were used on the fields, only animal manure. For a while, her special customer services and her organically grown produce kept her in business. But when Slovenia joined the EU in 2004, more had to be done to compete with produce growers and marketers across Europe.

Anticipating the challenges to come, the Kuzmics' daughter Lili completed a marketing degree at the University of Ljubljana. The family incorporated their business, and Lili is now its Ljubljana-based director, while Vera manages the farm. Lili's market research shows that it would be wisest to diversify. So, while continuing to focus on Ljubljana's expanding pro-

**FIGURE 4.49** Vera Kuzmic in her market stall in Ljubljana, Slovenia.
[Lydia Pulsipher.]

fessional population, whose food habits are changing and who are willing to pay extra for fine vegetables and fruits, the farm also produces gourmet preserves, marmalades, and spreads, and it sells as much as possible at the farm to save on transport costs. In a recently built banquet facility on the farm, Vera prepares special dinners for bus-excursion groups interested in traditional Slovene dishes made from homegrown crops.

*Source: Lydia Pulsipher's conversations with Vera Kuzmic and Dusan Kramberger, 1993 through 2006.* ■

**Challenges to Industrial Reform.** Because industrial production was a specialty of Central Europe when some of these countries were part of the Soviet bloc and others part of Yugoslavia, industry would seem to be their best hope for competing in the EU marketplace. Unfortunately, much of Central Europe's considerable industrial base is still burdened by legacies from the Communist era: a reliance on heavy industries that use energy and other resources inefficiently and hence are highly polluting, a dependence on imported raw materials, and an emphasis on coal and steel production rather than on consumer-oriented manufacturing and service industries. Upper Silesia, Poland's leading coal-producing area, provides an example of the extreme effects of industrial pollution: its forests have succumbed to acid rain, its soils yield contaminated crops, and water pollution is at deadly levels. Residents have experienced birth defects, high rates of cancer, and lowered life expectancies. Industrial pollution was one of Poland's greatest obstacles to entry into the European Union, and it barely achieved the EU standard.

Corruption has also inhibited post-Communist industrial reform. During the last years of the Soviet Union and during the post–cold war era, corruption mushroomed. In some countries, involvement by organized crime plagued the sale of formerly state-owned industries, especially in military sectors. One mobster based in Budapest reportedly took over virtually the entire Hungarian armaments industry and sold weapons globally. For a time, former Communist politicians had connections to international organized crime groups, and some foreign businesses eager to invest in Central Europe reluctantly paid bribes to those groups. But in recent years, the wish to join the EU has encouraged most countries to control organized crime. While industry in this subregion of Europe is still not particularly successful, better design has been encouraged, and regional trade shows are full of attractive, if as yet only locally produced, goods. High-end artisan-made products, such as handmade designer shoes, crystal glassware, luxury baby clothes, modern kitchens, stunningly attractive meticulously constructed prefabricated houses, and home furnishings appear to be finding wider markets.

**Continuing Hardship.** Throughout Central Europe, people suffered directly from the privatization or collapse of state-owned industries. Payrolls were unmet, and firms went bankrupt, resulting in widespread layoffs. Tax collection fell far behind schedule, and public funds for essential services were depleted. In reaction, highly skilled people moved to wealthier parts of the world.

The economic transition has been particularly hard on women. As state-run firms were sold off and made more efficient, women—who occupied the lowest ranks of the workforce—were the first to be laid off, despite the fact that in many cases they were a family's sole support. By 2004, most governments had official agencies that looked after the interests of women, who still suffered higher unemployment than men. Even young women, who by then were completing higher-education degrees in larger numbers than young men, continued to find employment more slowly than men and to be paid only 80 percent as much as similarly qualified men. Some older women were choosing to found small private businesses, while younger women were taking their skills and migrating to western Europe, New Zealand, Australia, or North America. Since all the countries in Central Europe have very low or negative population growth rates, losing these young future mothers is a serious blow.

## Progress in Economic Transition

For all the signs of trouble, the governments of Central Europe are taking "free-market-friendly" steps to help their economies grow more quickly. One strategy is to encourage foreign investment. Virtually all the countries of Central Europe actively recruit foreign investors and offer them a variety of incentives. Although there have been a number of false starts by investors unaware of the difficulties they would face, there have also been some successes. An example is Poland's joint venture with the French multinational glass and high-tech building materials company Saint-Gobain to construct one of the world's most modern glass factories. This factory, constructed in Silesia, near the Czech border, in Poland's most polluted region, is one of the most environmentally responsible plants of its kind in Europe, with emissions well below the standards set by Germany, the industry leader. All waste glass will be completely recycled. By 2006, Saint-Gobain had 19 factories and subsidiaries producing glass and high-tech materials in Poland.

Tourism is increasingly recognized as an important part of a well-rounded market-based economy. The Adriatic has long attracted tourists from western Europe seeking an economical sunny vacation. Medieval castles and wine festivals draw weekend visitors to rural areas. The cities of Central Europe, some of Europe's oldest and most distinguished municipalities, are becoming magnets for tourists from America, Japan, and western Europe. One example is Budapest, the capital of Hungary (Figure 4.50). Situated on the Danube River and built on the remnants of the third-century Roman city of Aquincum, Budapest was one of the seats of the Austro-Hungarian Empire, which ruled central Europe until World War I. By 2000, Budapest had reemerged as one of Europe's finest metropolitan centers, filled with architectural treasures, museums,

**FIGURE 4.50** The Fishermen's Bastion and its outdoor restaurant in Budapest. This structure was completed in 1905 on the site of the old fish market in memory of the fishermen who helped to defend the city during the medieval period. It is now one of Budapest's many tourist destinations. [Leo de Wys.]

concert halls, art galleries, discos, cabarets, theaters, and several universities of distinction. An added attraction is that Hungarians have revived their love affair with jazz, called "the music of the imperialists" during the Communist era. Since the change to a market economy, dozens of jazz clubs have opened. Hotels are booked solid for the popular May Jazz Festival.

## New Experiences with Democracy

The advent of democracy in Central Europe has reduced some of the tensions associated with the economic transition because people have been able to vote against governments whose reforms created high levels of unemployment and inflation. Most countries have now had several rounds of parliamentary and local elections, and voters are becoming accustomed to Western-style political campaigns. The voters' tendency to remove reformers before they can accomplish anything significant has dampened the enthusiasm of potential foreign investors who are waiting for real and sustained change before they offer their capital. On the other hand, in many cases, even crudely practiced democracy is leading to long-term economic growth and stability by encouraging governments to take a greater interest in the financial security of the population as a whole. For example, in Hungary and the Czech Republic, largely because of pressure from voters and from the EU, there are plans for new national pension systems to replace defunct Communist pension plans. Viable pension systems are important on two counts: they will give financial stability to the large cohort that labored under Communism and lost their safety net, and they will relieve younger workers who would have had to bear the burden of supporting impoverished aged parents.

## Southeastern Europe: Understanding an Armed Conflict

A conflict that erupted in southeastern Europe in 1991 took the lives of more than 250,000 people (see the discussion on pages 194–195). This mountainous region north of Greece, formerly known as the Balkans, is home to many ethnic groups of differing religious faiths (Figure 4.51). While many in the media have asserted that violence and ethnic hatred is an age-old way of life in this region, a deeper look reveals greater complexity.

Many of the so-called indigenous ethnic groups of southeastern Europe were themselves invaders over the last 1200 years or so, coming from central Europe, Russia, Central Asia, Greece, and beyond. These disparate groups occupied the Balkan Peninsula along with groups who had been there even longer, and none of them were particularly concerned with precise land rights. For the most part, the various ethnic groups managed to live together peacefully, intermarrying, blending, and realigning into new groups. Historian Charles Ingrao argues that coexistence was even easier in places such as Bosnia, where three or more groups living side by side prevented any one group from dominating the others. More recently, external wars exacerbated local ethnic tensions, as when the Nazis made an ally of Croatia in World War II, pitting it against ethnic Serbs and others. Nevertheless, over the last several hundred years, ethnic rivalries were no worse than those in the rest of Europe.

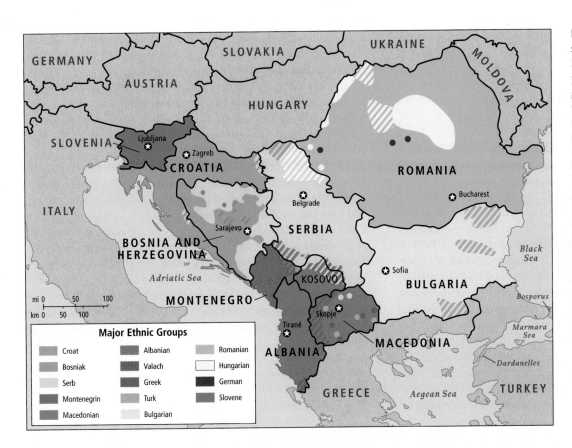

**FIGURE 4.51** Ethnic groups in southeastern Central Europe. Southeastern Central Europe is a patchwork of culture and religious groups who entered the region at various times over the last several thousand years. They had to contend with each other's differences and with conquerors from the north and the south. Normally relations were amicable and intermarriage was common, but occasionally hostilities arose, often as a result of outside pressures. [Adapted from Philippe Rekacewicz, UNEP/GRID-Arendal, *The Military Balance, 2002–2003* (London: The World Bank, 2003); and Catherine Samary and Jean-Arnault Dérens, *Les conflits yougoslaves de A á Z,* (Paris: Editions de l'Atelier, 2000), and available at http://maps.grida.no/go/graphic/ethnic_groups_in_the_south_eastern_europe.]

Marshal Josip Broz Tito, Yugoslavia's founder and leader until his death in 1980, tried through the power of his personal leadership to gloss over ethnic differences by encouraging pan-Yugoslav nationalism. Although there was ethnic peace during Tito's rule, there was no national dialogue about the benefits of ethnic diversity. Serbs were the most populous ethnic group in the military and in much of the federal government bureaucracy throughout the 1980s, and resentment of the Serb-dominated central government increased as the centrally planned economy of the country deteriorated. When Slovenia and Croatia, two of Yugoslavia's wealthiest provinces, voted to declare independence in 1991, they were responding to the deteriorating economy, exasperation with a government dominated by Serbs, and growing apprehension about burgeoning nationalism in Serbia. Partly in response to Serbian nationalism, Slovenes and Croats themselves became more nationalistic, and a spiral into competing nationalisms threatened.

Given the scale of the interethnic violence that occurred in the following years, it is not surprising that many people, especially outsiders, have seen ethnic hatred as the overriding cause of the conflict. However, other more likely primary causes include Serbian geopolitics, an earlier lack of guidance by Marshal Tito on multicultural affairs, invented myths about ethnically pure nation-states, and a lack of humanitarian leadership. The Serb-dominated government and military of Yugoslavia allowed the province of Slovenia to separate after just a short skirmish, in part because of its location close to the heart of Europe and far from Serbia and in part because there were no significant enclaves of ethnic Serbs in Slovenia. Nonetheless, Slovenia was a major loss to Yugoslavia

because it had been by far the most modernized and productive of the provinces. The Yugoslav government then feared that unless it made a strong show of force in dealing with the Croats and the Bosnians, the rest of the non-Serb populations of Yugoslavia would also move to secede and take valuable territory, resources, and skilled people with them. This fear became a reality when Croatia and Bosnia and Herzegovina attempted to secede from Yugoslavia, and the Serbs reacted brutally. Despite the undoubted culpability of the Serbs in starting the conflict and in perpetrating much of the brutality, they cannot be blamed exclusively for all the interethnic violence that has taken place. While suffering ethnic cleansing, Croats and Bosnians themselves used brutality and ethnic cleansing in retaliation, as did Albanians (also known as Kosovars) later.

Conflict sprang up again in the late 1990s in the province of Kosovo, the southern area of Serbia dominated by ethnic Albanians. In the late 1980s, in order to divert public attention away from the country's economic deterioration, Yugoslav President Slobodan Milosevic, himself a Serb, aggressively claimed that Serbs were being threatened by Kosovar Albanians. Throughout the 1990s, Milosevic's government in Serbia had been removing Albanians (Kosovars) from government positions and barring them from using public services. Increasingly, the Serbian government used arrests and violence to quell Kosovar activists. In response, the guerrilla Kosovo Liberation Army was formed, supported by expatriates of Albanian ethnicity. They also attempted to claim specific territory and to secede from Yugoslavia—perhaps to join the country of Albania, which lies just to the west. In spring 1999, Serbia forced most of the majority Albanian population out of the province of Kosovo. Several weeks later, in their first ever formal military action, NATO forces bombed Serbia in an effort to stop massacres and prolonged ethnic cleansing in Kosovo. The violence de-escalated, but hostilities between Kosovar Albanians and Serbs continue sporadically, fueled by covert aid to both sides from agents outside the region, some inspired by nationalism, some by Islamic fundamentalism.

The conflicts in and around Serbia have devastated the economies of southeastern Europe. They destroyed the infrastructure of commerce, interrupting trade and tourism for all countries in the region. Dozens of bridges were bombed, and the Danube River, a major transport conduit through Europe to the Black Sea, was blocked with wreckage. In 2000, an election in Serbia resoundingly removed Slobodan Milosevic from power, and in 2002 he was brought to trial for war crimes in the International Court of Justice (World Court) in The Hague (where he died in 2006). To forestall any future troubles, the European Union and the countries of southeastern Europe had earlier agreed that, once Milosevic was gone, they would enact the Balkan Stability Pact. Funded by the European Union, this pact is promoting the recovery of the region through public and private investment as well as through social training to enhance the public acceptance of multicultural societies and strengthen informal democratic institutions. In May 2006, the people of Montenegro voted to declare independence from Serbia, which thus far retains control over Kosovo.

## Reflections on Europe

Europe's position as a center of world economic and political power seems less formidable when the region is viewed up close. Europe is not a monolith. Its geographic patterns are highly variable, and its borders are fluid. The EU expanded to the southeast again in 2007 with the addition of Bulgaria and Romania, and other countries may join by 2010. The distribution of wealth, social welfare policies, approaches to environmental issues, treatment of outsiders, and participation of women in public and private affairs are just some of the aspects of life in Europe that reflect the region's geographic variation.

Many of the issues confronting people in Europe are similar to those encountered elsewhere around the world. For example, the effort at economic integration is exposing European countries to the same problems of structural adjustment now faced by so many, much poorer, developing countries. Europe's problems, although generally less severe and played out in more affluent circumstances, are no less potentially disruptive than those faced by Latin America, Asia, and Africa. Europeans must answer the same questions: How can a country create and retain satisfying jobs for its people when there are qualified workers willing to work for less within relatively easy reach? And if immigrants come, how can they be absorbed fairly and with the least disruption (a question very similar to that being debated in the United States). How can a society help the poor and unemployed to a better life while at the same time lessening the drain on public funds by welfare programs? How can poor parts of the region be stimulated to develop, and how can wealth be more equitably distributed? And how can development everywhere be made to conform to the requirements of an increasingly stressed environment? No less complex are the cultural questions of how countries with very different sets of traditions and mores can find sufficient common ground to cooperate economically and thus achieve greater prosperity for everyone. For all its current troubles, Europe's head start on development, its long-established world leadership, and the heightened awareness of global economic issues gained by its own efforts at integration all position Europe to retain its global influence.

Europe's questioning of U.S. foreign policy after September 11, 2001, was an important chapter in post–cold war geopolitics, and it will be interesting to see the extent to which this challenge to U.S. power actually affects global relationships over the long term. To people in the Americas at large, Europe's current preoccupation with economic integration should prove instructive. Many of the same issues must be confronted if the economic integration of the Americas is to proceed: the free movement of people and increasing immigration, the loss of jobs to low-wage areas, the need to address growing disparities in wealth, and the need to strengthen and coordinate social and environmental policies, to name just a few.

The future of the relationship between Europe and the world region we cover in Chapter 5—Russia and the newly independent states of eastern Europe (Belarus, Moldova, and Ukraine), Caucasia, and Central Asia—is not clear. Will these two regions eventually integrate their economies and societies and become one large Europe, perhaps stretching to the Pacific, with 800 million people? Will Russia's very serious economic and social problems make it too unattractive a partner for Europe, or will its energy resources make it attractive no matter what? The next chapter will discuss these and other geographic issues.

# Chapter Key Terms

acculturation 211
Age of Exploration 190
assimilation 210
basin 219
central planning 193
cold war 193
Council of the European Union 198

cultural homogenization 184
democratic institutions 193
devolution 230
double day 212
dumping 206
economic integration 194
economies of scale 200

ethnic cleansing 195
euro 200
European Commission 198
European Economic Community (EEC) 198
European Parliament 198
European Union (EU) 184

exclave  236

feudalism  189

Green  216

guest workers  208

Holocaust  193

humanism  190

humid continental climate  188

iron curtain  193

medieval period  189

Mediterranean climate  188

mercantilism  190

nationalism  193

North Atlantic Treaty Organization
  (NATO)  202

producer services  205

Protestant Reformation  190

Renaissance  190

resource base  196

Roma  193

Scandinavia  184

Schengen Accord  208

social welfare  214

subsidies  206

temperate midlatitude climate  187

welfare states  193

world cities  191

## Critical Thinking Questions

1. What are the push and pull factors that have led to migration *within* the European Union over the last 10 years?

2. What are some of the main factors that made it possible for Europe to wield such great influence globally starting in the 1500s?

3. Support or refute the statement that Europe is rich and powerful because of its rich natural resource base.

4. Discuss the reasons for Europe's low or, in some cases, negative population growth rates. What are some of the consequences of this pattern? Why might an understanding of a region's population age structure affect immigration policy?

5. What is the evidence that Europe is an economic rival to the United States? How might this position affect power relationships globally? What are some ways to ameliorate the rivalry?

6. Why and how might a game such as soccer serve as an international salve to ethnic and political rivalries?

7. Why might Europe's emphasis on human well-being through state-supported welfare systems place Europe in a position to establish strong global leadership on matters of civil society?

8. What are some indications of regional variations in gender equity in Europe? Where do women fare the best? Where do they fare the worst?

9. What is the evidence that European ways of life are less stressful on the environment than North American ways? What is the evidence that environmental issues are central to European political debate?

10. How do the issues of immigration in Europe compare with those in the United States? If you were a poor, undocumented immigrant searching for a way to support your family, would you choose the United States or Europe as a possible destination? Why?

Manchester
Birmingham
London
UNITED
KINGDOM
FRANCE
BELGIUM
Brussels
LUX
Luxembourg
Amsterdam
NETHERLANDS
GERMANY
Hamburg
Berlin
Rhine
Prague
CZECH
REP.
Vienna
POLAND
Bratislava
SLOVAKIA
Budapest
HUNGARY
ROMANIA
MOLDOVA
Chisinau
UKRAINE
Odesa

*North Sea*

*Norwegian Sea*

NORWAY
Oslo

DENMARK
Copenhagen

SWEDEN

*Baltic Sea*

Stockholm

Kaliningrad

Gulf of Bothnia

FINLAND
Helsinki
Tallinn
ESTONIA
Riga
LATVIA
LITHUANIA
Vilnius
Brest
Minsk
BELARUS
Homyel
Smolensk
Lviv
Chernobyl
Kiev
Mykolayiv
Kherson
Sevastopol

*Greenland*

North Pole
90°N

*ARCTIC OCEAN*

*Svalbard*

Franz Josef Land

*Barents
Sea*

Murmansk
Kirovsk
Kola Peninsula
White Sea

Arkhangelsk

North Land

Novaya Zemlya

*Kara
Sea*

Kara Strait

Dikson

*Laptev
Sea*

*Taymyr Peninsula*
Lena R.

Vorkuta

Salekhard
Yamal Peninsula
Pechora
Pechora
Basin
West Siberian Plain
Urengoy
Norilsk

*North Siberian Lowland*

Central
Siberian

S i b e

Olenek

Lake
Ladoga
Lake
Onega
St. Petersburg
Tver
Yaroslavl
Ivanovo
MOSCOW
Moskva
Bryansk
Kursk
Tula
Lipetsk
Ryazan
Nizhniy Novgorod
Voronezh
Penza
Kirov
Kama
Perm
Kazan
Volga
Saratov
Samara
Tolyatti
Ufa
Orenburg
Orsk
Magnitogorsk
Volgograd
Astrakhan

*North European Plain*

Divina

Pechora

*Ural Mountains*

Nizhniy Tagil
Yekaterinburg
Tyumen
Chelyabinsk

Surgut

Ob

*RUSSIAN FEDERATION*
*(RUSSIA)*

Plateau

Vilyuy
Res.

*Tunguska Basin*

Yenisey
Angara

Seversk
Tomsk

Omsk
Novosibirsk
Novokuznetsk

Krasnoyarsk

Gladkaya
Tayshet
Bratsk
Bratsk
Res.

*Irkutsk
Basin*

Lake
Baikal

*Caucasus Mts.*

GEORGIA
Batumi
Tbilisi
ARMENIA
Yerevan
AZER.
AZERBAIJAN
Baku
Groznyy

*Black Sea*

Rostov-na-Donu
Sea of
Azov
Donetsk
Dnipropetrovsk
Kharkiv
Don

*Caspian
Depression*

*Caspian Sea*

*Steppes*

Aktogay

*Aral
Sea*

Astana

KAZAKHSTAN

Qaraghandy

Leninsk

Syr Darya

Lake
Balkhash

W. Sayan Mts.
Kyzyl
E. Sayan Mts.
Angarsk
Irkutsk
Ulan-Ude
Yab

*Lake
Baikal*

TURKEY
Lake
Van
Van
Murat
IRAQ
Lake
Urmia

IRAN
Tehran
Esfahan
Mashhad

*Elbruz Mts.*

*Zagros Mts.*

TURKMENISTAN
Ashkhabad

Amu Darya

UZBEKISTAN
Samarkand
Tashkent
Dushanbe
TAJIKISTAN
Pamirs

Shymkent
Bishkek
Almaty
KYRGYZSTAN
Lake

*Tien Shan*
Tarim

Junggar Basin

Urumqi

*Altai Mts.*

MONGOLIA

Ulan Bator

Kulu

*Persian Gulf*

QATAR
Ad Dawhah
(Doha)
Abu Zaby
Dubayy
UAE

Muscat
OMAN

AFGHANISTAN
Kabul
Islamabad
Lahore
Faisalabad
PAKISTAN
Karachi

*Hindu Kush*

INDIA
New Delhi

*Tarim Basin*

*Taklimakan Desert*

C H I N A

Lanzhou

Chengdu

G

Bao

## Land Elevations

| meters | feet |
|---|---|
| 4877 | 16,000 |
| 3353 | 11,000 |
| 2134 | 7000 |
| 914 | 3000 |
| 305 | 1000 |
| 152 | 500 |
| 0 | 0 |

1:26,000,000
Azimuthal Equidistant Projection

# CHAPTER 5

# RUSSIA and the Newly Independent States

**Global Patterns, Local Lives** When a coffin arrived at the head offices of the Moscow catering firm Na Ilyinke, bearing the name of the still quite lively Alexei Alexeyevich Likhachev, director of the company, he sensed it was not a harmless prank. Soon the phones began ringing with condolence messages from shareholders who had received invitations to his memorial service. Likhachev and his partners knew immediately what the real message was: "Sell or else!"

Low-profile, medium-sized companies like Na Ilyinke have become the focus of Russian-style corporate raiders. Intimidation and bogus legal practices entailing corrupt lawyers, judges, bureaucrats, and police are used to frighten business owners into selling their successful firms at prices that are less than half their true value. Often it is the real estate that is the primary focus of the raiders. Na Ilyinke occupies prime property in the heart of Moscow. In the confusion during the Soviet collapse, Likhachev and his partners acquired the formerly state-owned catering firm in an attractive privatization deal, which may itself have been a bit shady.

Shortly after the coffin arrived, the partners were notified that their 58 percent share of the company had been sold, via a forged power of attorney, first to a woman in Ukraine and then to a man in New York. Likhachev, an elderly man, backed out at that point, selling his stock to his partners, but the other partners persevered in getting the government to investigate the fraud and received a favorable ruling. The raiders' assaults continued, however, in the form of three judgments returned against the company in court proceedings of which the partners were not notified and hence did not attend.

**FIGURE 5.1** Regional map of Russia and the Newly Independent States.

245

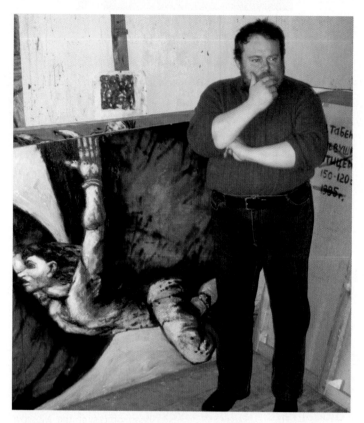

**FIGURE 5.2  Forced takeovers in Russia.** Lev Tabenkin, a Russian painter, stands in front of some of his canvases at a temporary studio. He lost possession of his old studio after the ownership was transferred without his knowledge. Tabenkin and the Union of Artists are battling through the courts to get back a number of studios at a Moscow artists' cooperative. [Peter Finn/*The Washington Post.*]

Unsure that official law enforcement agencies will help them, the Na Ilyinke partners now hope to fend off a physical takeover of their firm by turning it into an armed camp with armed guards, an alarm system, razor wire, and steel doors that can seal off sections of the building. As one partner puts it, "If you lose physical possession of your property, you are in serious trouble. So far we've kept them out." Other businesspeople, and even artists, have not been so lucky (Figure 5.2).

*Adapted from Peter Finn, "Hostile takeovers: Russian financial predators use legal tactics to seize prized real estate," The Washington Post National Weekly Edition (April 24-30, 2006): 18-19.* ■

Na Ilyinke operates in a region that has attempted to change its political and economic systems entirely in just a few short years. Less than two decades ago, the **Union of Soviet Socialist Republics (USSR,** more commonly known as the **Soviet Union)** was the largest political unit on earth, stretching from eastern Europe all the way across Central and East Asia to the Pacific Ocean, covering one-sixth of Earth's land surface. In December 1991, the Soviet Union ceased to exist. In its place stood fifteen independent countries joined in a loose economic alliance. The largest and most pow-

erful was Russia. But soon, three of the fifteen—Latvia, Lithuania, and Estonia—having chafed under Russian domination for many years, abandoned their close association with Russia. In this book, they are discussed in Chapter 4 along with the European Union, which they have now joined.

Thus the region we describe in this chapter presently consists of the Russian Federation (Russia and its internal ethnic republics, hereafter shortened to *Russia*) and 11 loosely allied countries: Belarus, Moldova, and Ukraine to the west of Russia; to the southeast, Georgia, Armenia, and Azerbaijan, referred to jointly as the Caucasian republics; and to the south, the five republics of Central Asia: Kazakhstan, Kyrgyzstan, Tajikistan, Uzbekistan, and Turkmenistan (see Figure 5.1). Russia and the Newly Independent States are in a part of the world where regional analysis is particularly difficult at this time, precisely because the major changes taking place may ultimately result in some new regional alignments. Some republics in the west may eventually join Europe, while the Central Asian republics may align themselves with neighbors in Southwest or South Asia with which they share cultural traditions and economic interests. The far eastern parts of Russia are already finding common trading ground with East Asia and Oceania. The map in Figure 5.3 shows these potential regional realignments.

The breakup of the Soviet Union triggered a transformation in the region's political and economic systems. Authoritarian government gave way to varyingly successful experiments with freedom and democracy, and an economy controlled by government bureaucrats was replaced by a system that encourages citizens to establish their own businesses, though high levels of corruption in government and business give mixed messages to those who do try entrepreneurship.

The demise of the Soviet Union also required a period of adjustment and reorganization around the globe because the cold war between the West (Western Europe and North America and their allies) and the East (the Soviet Union and its allies) had affected so many economic and political relationships worldwide. Although the tension between East and West diminished drastically in the 1990s, it is not yet clear just what roles the old members of the former Soviet Union will play in world affairs.

Russia's international sphere of influence has diminished because it no longer has the economic means to court alliances by giving loans and assistance, nor is it any longer the dominant trade partner among its allies. Yet Russia remains prominent in the region because of its size, world standing, oil and gas resources, and historical significance. The Newly Independent States maintain strong economic ties to Russia even as they forge new trading relationships with the outside world. Russia itself has had to make major adjustments to compete for resources and trading partners in the worldwide marketplace. Shoring up Russia's economic situation in recent years are dramatically higher prices for natural gas.

Today, internal economic, political, and social systems throughout the region are in flux. On the one hand, the changes have caused great hardship and anxiety for many, especially those over age 50, who are suffering dramatic drops in income and well-being. On the other hand, many people living in the former Soviet Union

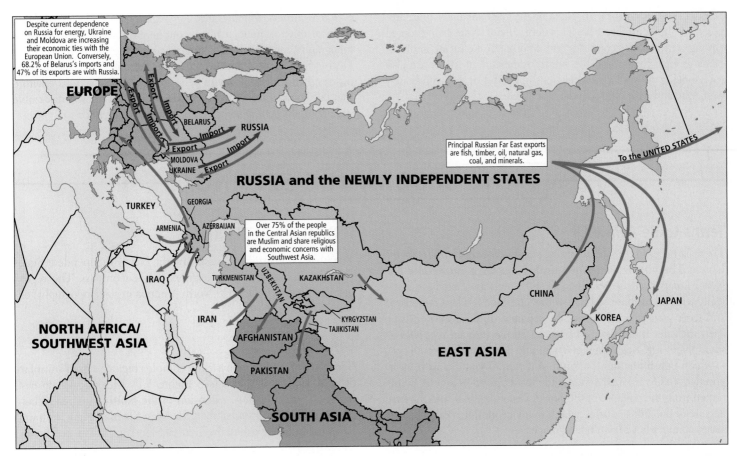

Despite current dependence on Russia for energy, Ukraine and Moldova are increasing their economic ties with the European Union. Conversely, 68.2% of Belarus's imports and 47% of its exports are with Russia.

Principal Russian Far East exports are fish, timber, oil, natural gas, coal, and minerals.

Over 75% of the people in the Central Asian republics are Muslim and share religious and economic concerns with Southwest Asia.

**FIGURE 5.3** Russia and the Newly Independent States: Contacts with other world regions. The region of Russia and the Newly Independent States is in flux as all of the component countries rethink their geopolitical positions vis-à-vis one another and adjacent regions. The Newly Independent States and Far Eastern Russia are developing new relationships with their neighbors on the periphery while still retaining some sort of connection to European Russia. The arrows indicate various types of contact, ranging from economic to religious and cultural.

are exhilarated by their new opportunities. The changes offer them greater freedom of expression and of movement, the satisfaction and challenge of entrepreneurship, the right to practice religion openly, and the availability of consumer products in far greater variety and quality than ever before.

## Themes to Explore in Russia and the Newly Independent States

Several themes appear throughout this chapter, many of which are associated with the changing internal and external relationships just mentioned.

1. **The move to a market economy.** The most remarkable change in the region is the effort to move rapidly from a centrally planned economy to a market economy based on private enterprise. The success of this change, under way since the late 1980s, is uneven across the region. Wider lifestyle choices and greater wealth for some are balanced against the loss of jobs and social services and the decline in overall quality of life for large segments of the population.

2. **New alignments of people and trade.** The breakup of the Soviet Union into smaller, more autonomous political units has resulted in population movements, new allocations of resources, and new trading patterns. These changes have left some parts of the old Soviet Union isolated and disconnected from former trading partners, needed resources, and the transportation infrastructure.

3. **The slow emergence of democracy.** Market liberalization did not lead automatically to more popular participation in government. The Newly Independent States have all had to forge governing institutions in unstable circumstances, and while regular and orderly elections are increasingly the norm at all levels, from local to federal, the democratization process is uneven.

4. **The reconfiguration of the institutions of a civil society.** The Soviet Union provided many social and civil safeguards for its citizens, but it did so in an authoritarian manner. During the transition to a market economy, crime and corruption have flourished. Progress in building trained bureaucracies and civic-minded volunteer and service organizations is slow.

5. Changing identities. In place of the regional unity once imposed by the centralized Soviet system, alliances are being forged among new, smaller countries and among ethnic, social, and religious groups. Russia and the Newly Independent States are also each looking outward to new relationships with Europe, with Southwest Asia, and with China, India, and Japan.

6. Environmental challenges. The region has long faced a difficult physical environment: a huge continental territory with difficult topography, a harsh climate, and few outlets to the world's oceans. During Soviet rule, efforts to make this difficult landscape prosperous resulted in many unwise decisions that led to serious environmental degradation. Now the environment is under considerable stress, but it is getting somewhat more public attention.

# I THE GEOGRAPHIC SETTING

## Terms to Be Aware Of

There is no entirely satisfactory new name for the former Soviet Union. The formal name used in this chapter is *Russia and the Newly Independent States*. Russia is still closely associated economically with all of these states, but they are independent countries, and their governments are legally separate from Russia's. Russia itself includes more than 30 (mostly) ethnic *internal republics*—such places as Chechnya, Tatarstan, and Tuva (also spelled Tyva) —which constitute about a tenth of its territory and a sixth of its population. Occasionally, we use the word *transition* to refer to the period from the collapse of the Soviet Union in 1991 into the first decade of the 2000s, during which many economic, political, and social changes have been taking place.

## PHYSICAL PATTERNS

The physical features of Russia and the Newly Independent States vary greatly over the huge territory they encompass. The region bears some resemblance to North America in size, topography, climate, and vegetation.

## Landforms

Because this is such a physically complex region, a brief summary of its landforms is useful (see Figure 5.1). Moving west to east, there is first the eastern extension of the North European Plain, then the Ural Mountains, then the West Siberian Plain, followed by an upland zone called the Central Siberian Plateau, and finally,

**FIGURE 5.4** Kamchatka. In autumn, the stone birch, alder, and elfin wood in Kamchatka's Nalychevo (Nalitchevo) Nature Park erupt in a spectacular display. The Nalychevo River valley is surrounded by snow-capped volcanoes, and there are hot and cold mineral springs in the upper reaches of the river. In 1996, the park was placed on UNESCO's World Heritage List. [© Christian Gluckman/http://www.edouardas.com.]

**FIGURE 5.5** Katun River in the Altai Mountains, Altai Republic, Russia. The Altai mountain range, where Siberia meets Kazakhstan, China, and Mongolia, is the highest mountain system in southern Siberia. Known for its diversity and beauty, the region exhibits taiga thickets, forest steppes, snowy peaks, deserts, and the Katun—the main river in Altai, known for its great rafting. [Jennifer Castner/Pacific Environment.]

in the far east, a mountainous zone bordering the Pacific. To the south of these territories, from west to east, there is an irregular border of mountains (the Caucasus), semiarid grasslands, or **steppes** (in western Central Asia), and barren uplands and high mountains (in eastern Central Asia). The adjacent regions of Southwest Asia, South Asia, and East Asia skirt this region to the south.

The eastern extension of the North European Plain rolls low and flat from the Carpathian Mountains in Ukraine and Romania 1200 miles (about 2000 kilometers) to the Ural Mountains. This part of the region is called *European Russia* because the Ural Mountains are traditionally considered part of the not very clear border between Europe and Asia. European Russia is the most densely settled part of the entire region and is its agricultural and industrial core. Its most important river is the Volga, which flows into the Caspian Sea. The Volga River is a major transport route; it is connected—via canals, lakes, and natural tributaries—to many parts of the North European Plain, to the Baltic and White seas in the north, and to the Black Sea in the southwest.

The Ural Mountains extend in a fairly straight line south from the Arctic Ocean into Kazakhstan. The Urals are not much of a barrier to humans or to nature: there are several easy passes across the mountains, and winds carry moisture all the way from the Atlantic and Baltic across the Urals and into Siberia. Much of the Urals' once-dense forest has been felled to build and fuel the new industrial cities, such as Yekaterinburg and Chelyabinsk, established in the Urals during the twentieth century.

The West Siberian Plain, lying east of the Urals, is the largest plain in the world. A vast, mostly marshy lowland about the size of the eastern United States, it is drained by the Ob River and its tributaries, which flow north into the Arctic Ocean. Long, bitter winters mean that in the northern half of this area a layer of permanently frozen soil (**permafrost**) lies just a few feet beneath the surface; in the far north, the permafrost comes to within a few inches of the surface. Because water doesn't sink through this layer, swamps and wetlands form in the summer months, providing habitats for many migratory birds. In the far north lies the **tundra,** where only mosses and lichens can grow. The West Siberian Plain has some of the world's largest reserves of oil and natural gas, although their extraction is made difficult by the harsh climate and the permafrost.

The Central Siberian Plateau and the Pacific mountain zone farther to the east together equal the size of the United States. Permafrost at varying depths prevails over almost this entire area. Notable exceptions are along the coast, where the Pacific Ocean moderates temperatures, and where additional heat is supplied by the many active volcanoes created as the Pacific Plate sinks under the Eurasian Plate. Places warmed by these forces are the Kamchatka Peninsula (Figure 5.4), Sakhalin Island, and Sikhote Alin; as yet only lightly populated, they are havens for wildlife.

To the south of the West Siberian Plain is an irregular band of grasslands (steppes) and deserts stretching from the Caspian Sea to the mountains bordering China. Farther to the south and east of these grasslands is a wide, curving band of mountains, including the Caucasus, Elburz, Hindu Kush, Tien Shan, and Altai (Figure 5.5). Their rugged terrain has not deterred cultural influences that have penetrated north and south for literally tens of thousands of years. People have persistently crisscrossed these mountains, exchanging plants (apples, onions, citrus fruits, rhubarb, wheat), animals (horses, sheep, cattle), technologies (cultivation, animal tending, portable shelter construction, rug and tapestry weaving), and religious belief systems (principally Islam and Buddhism, but also Christianity and Judaism).

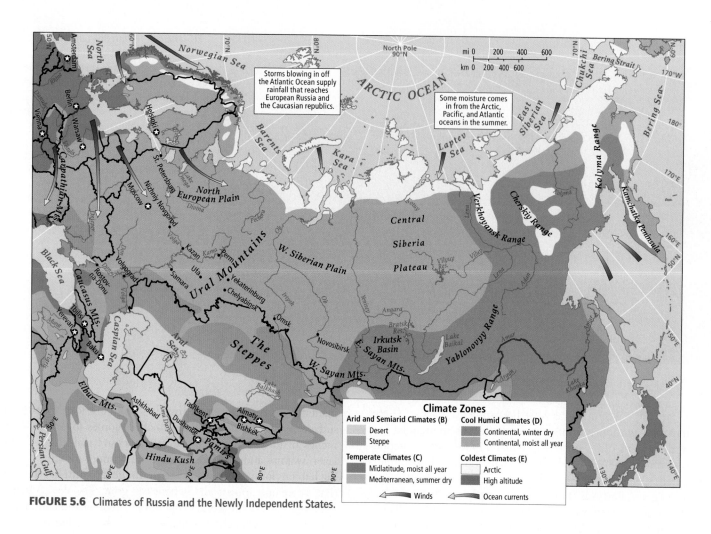

**FIGURE 5.6** Climates of Russia and the Newly Independent States.

## Climate and Vegetation

Other than Antarctica, no place on earth has as harsh a climate as the northern part of the Eurasian landmass occupied by Russia (Figure 5.6). Here is a prime example of a continental climate, in which the winters are long and cold, with only brief hours of daylight; the summers are short and cool to hot, with long days. Both the short summers and the northerly location with underlying permafrost restrict the crops that can be grown and curtail overall agricultural production. The seasonal continental "breathing" (described in Chapter 1 on page 23) is dominated in winter by a long, dry, frigid exhalation of air that flows out of the arctic zones. A short inhalation in summer brings in only modest amounts of moist, warm oceanic air from the distant Atlantic in the west and lesser amounts from the also distant Pacific in the east. The vast mountain ranges along the region's southern edge block access to warm, wet air from the Indian Ocean far to the south.

Most rainfall in the region comes from storms that blow in from the Atlantic Ocean far to the west. By the time these initially rain-bearing air masses arrive, most of their moisture has been squeezed out over Europe, but a fair amount of rain does reach Ukraine, European Russia, and the Caucasian republics. Across most of the region, agriculture is precarious at best, requiring expen-

sive inputs of labor, water, and fertilizer. The Caucasian mountain zones are some of the only areas in the region where rainfall adequate for agriculture coincides with a relatively warm climate and long growing seasons, and, together with Ukraine and European Russia, this area is the agricultural backbone of the region (Figure 5.7). The best soils are in a region stretching from Moscow south toward the Black and Caspian seas, including much of Ukraine and Moldova. Here the natural vegetation is open woodland and steppe. The soils in Belarus and the northern part of European Russia are generally not particularly fertile, and growing seasons are very short. The natural vegetation is mixed and boreal coniferous forest.

East of the Urals, the lands of Siberia receive moderate precipitation (primarily from the east) but experience long, cold winters. Huge expanses of Siberia are covered with **taiga** (northern coniferous forest), which stretches to the Pacific. Agriculture is generally not possible, though reindeer are tended in the tundra of the far north.

East of the Caucasus Mountains, the lands of Central Asia have semiarid to arid climates influenced by their location in the middle of a very large continent. The summers are scorching and short, the winters intense. The more southern areas support grasslands, which are used for herding and for agriculture where irrigation is possible.

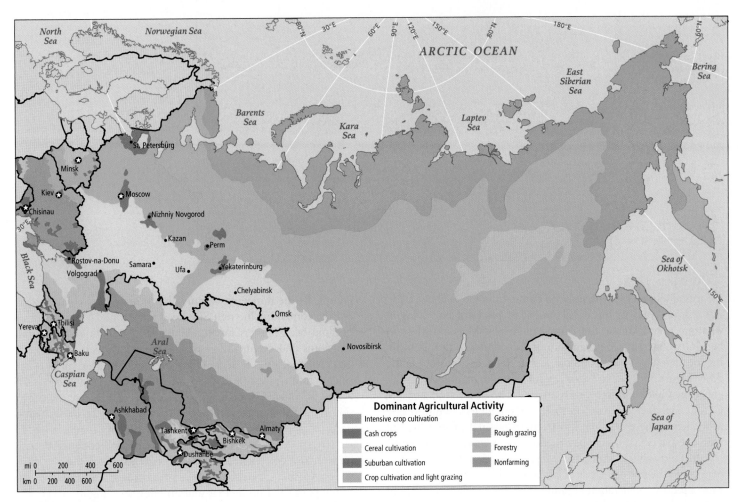

**FIGURE 5.7 Agriculture in Russia and the Newly Independent States.** Agriculture in this part of the world has always been a difficult proposition, partly because of the cold climate and short growing seasons, and partly because soil fertility or lack of rainfall are problems in all but a few places (Ukraine, Moldova, and Caucasia). [Adapted from Robin Milner-Gulland with Nikolai Dejevsky, *Cultural Atlas of Russia and the Former Soviet Union*, rev. ed. (New York: Checkmark Books, 1998), pp. 186–187, 198–199, 204–205, 216–217.]

## HUMAN PATTERNS OVER TIME

The core of the entire region has long been European Russia, the most densely populated area and the homeland of the ethnic Russians. Expanding gradually from this center, the Russians conquered a large area inhabited by a variety of other ethnic groups. These conquered territories remained under Russian control as part of the Soviet Union (1917–1991), which attempted to create an integrated social and economic unit out of the disparate territories. The breakup of the Soviet Union has reversed this gradual process of Russian expansion for the first time in centuries.

### The Rise of the Russian Empire

For thousands of years, the politically dominant people in the region were **nomadic pastoralists** who lived on the meat, milk, and fiber provided by their herds of sheep, horses, and other grazing animals. Their movements followed the changing seasons across the wide grasslands stretching from the Black Sea to the Central

Siberian Plateau. The nomads would often take advantage of their superior horsemanship and hunting skills to plunder settled communities. To defend themselves, permanently settled peoples gathered in fortified towns.

Towns arose in two main areas: the dry lands of Central Asia and the moister forests of Ukraine and Russia. As early as 5000 years ago, Central Asia was supporting settled communities, and sometimes even large empires, which were enriched by irrigated croplands and by trade along the famed Silk Road, the ancient trading route between China and the Mediterranean (Figure 5.8). About 1500 years ago, the **Slavs,** a group of farmers, including those known as the Rus (of possible Scandinavian origin), emerged in what is now Poland, Ukraine, and Belarus. They moved east, founding numerous settlements, including the towns of Kiev in about 480 and Moscow in 1100. By 600, Slavic trading towns were located along all of the rivers west of the Ural Mountains. The Slavs prospered from a lucrative trade route between Scandinavia, Constantinople (modern-day Istanbul), and Baghdad by way of the Volga River from 720 to 860. This trade route, combined with

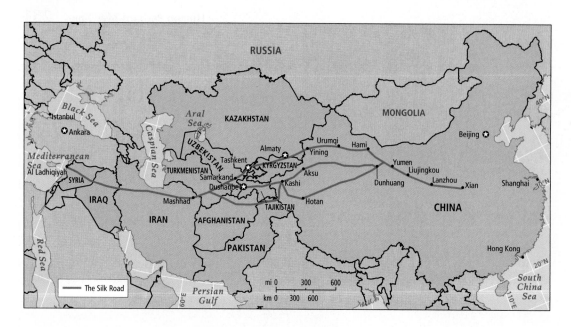

**FIGURE 5.8** The ancient Silk Road. Merchants who plied the Silk Road rarely traversed the entire distance. Instead, they moved back and forth along only part of the road, trading with other merchants to the east or west. [Adapted from *National Geographic* (March 1966): 14–15.]

the introduction of Christianity via Constantinople in about 1000, was instrumental in the development of powerful kingdoms in what are now Ukraine and European Russia. Kiev and Moscow became well-organized urban commercial centers, with Kiev forming the largest and most populous feudal city-state in Europe at the time (Figure 5.9). Slavic art and architectural traditions were influenced by Greek Christian missionaries, who also introduced the Cyrillic alphabet, still used in most of the region's countries. The influences of these missionaries on landscapes in the region are still visible, especially in religious structures and national monuments.

In the twelfth century, the Mongol armies of Genghis Khan conquered the forested lands of Ukraine and Russia. The **Mongols** were a loose confederation of nomadic pastoral people centered in eastern Central Asia and descended from much earlier nomadic

groups. By the thirteenth century, the Mongols had conquered an empire containing the rich Silk Road trade route. Moscow's rulers became tax gatherers for the Mongols, dominating neighboring kingdoms and eventually growing powerful enough to challenge local Mongol rule. Ivan the Terrible, the first Slavic (or Russian) ruler to assume the title of Czar, conquered the Mongols in 1552 at Kazan, in a fierce battle that marks a clear beginning of the expansion of the Russian empire; it is commemorated by St. Basil's Cathedral, a major landmark on Moscow's Red Square (Figure 5.10).

By 1600, Russians centered in Moscow had conquered many former Mongol territories, integrating them into their empire, which by then stretched eastward from the Baltic Sea over the thinly populated northern Eurasian landmass (Figure 5.11). The outlying parts of the empire, south and east of the Ural Mountains,

**FIGURE 5.9** Independence Square, or *Maidan Nezalezhnosti,* Kiev. According to legend, Kiev was founded by three brothers, Kyy, Shchek, and Khorev, and their sister Lybid in the fifth century. The bronze statue honoring them resides in Kiev's central square in front of the music academy, several fountains, and the Independence Column (partially visible behind the statue). [imagebroker/Alamy.]

**FIGURE 5.10** St. Basil's Cathedral, Moscow. Perhaps no other architectural symbol so vividly conjures up the image of Russia as St. Basil's. Originally dubbed "Cathedral of the Intercession of the Virgin," this remarkable structure was erected as a memorial to Ivan the Terrible between 1555 and 1561 to commemorate his defeat of the Mongols and the conquest of Kazan. The name was changed in the seventeenth century when a chapel was built over the grave of the ascetic St. Basil the Blessed. The central church, capped by a pyramidal tower, stands amid eight smaller churches with colorful onion domes. [Stock Connection Distribution/Alamy.]

oped by the Mongols, their expansion into Siberia in some ways resembled the spread of European colonial powers throughout Asia and the Americas. For example, much like Britain, France, and Spain, the Russian imperial colonists appropriated Siberian resources for their own use and invested little in local development. They upheld European notions of private property at the expense of indigenous patterns of communal land use. Perhaps most significantly, Russian expansion into Siberia in the eighteenth and nineteenth centuries made way for massive migrations of laborers from Russia, who soon far outnumbered indigenous Siberians. By the mid-nineteenth century, Russia had also conquered Central Asia to gain control of cotton, its major export crop.

The Russian empire was ruled by a **czar,** who, along with a tiny aristocracy, lived in splendor while the vast majority of the people lived in poverty. Many Russians were **serfs,** who were legally bound to live and farm on land owned by a lord. If the land was sold, the serfs were transferred with it. Even though serfdom ended legally in the mid-nineteenth century, the brutal inequities of Russian society persisted into the twentieth century, fueling opposition to the czar. By the early twentieth century, a number of violent uprisings were under way.

contained important deposits of mineral resources and were inhabited by many non-Russian peoples. The first major non-Russian area to be annexed was western Siberia (1598–1689). Although the Russians used many methods of conquest and administration devel-

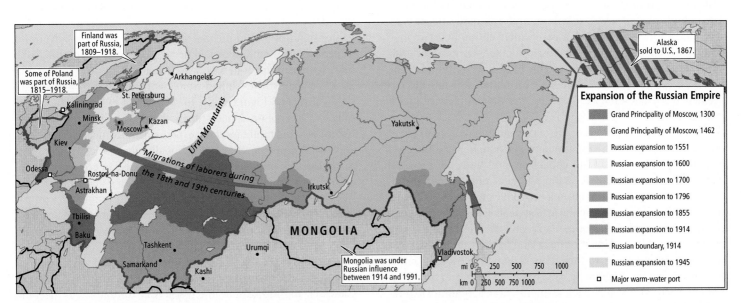

**FIGURE 5.11** Russian imperial expansion, 1300–1945. A long series of powerful entities expanded Russia's holdings across Eurasia to the west and east. Expansion was particularly vigorous after 1700, when Siberia, the largest area, was acquired. [Adapted from Robin Milner-Gulland with Nikolai Dejevsky, *Cultural Atlas of Russia and the Former Soviet Union,* rev. ed. (New York: Checkmark Books, 1998), pp. 56, 74, 128–129, 177.]

## The Communist Revolution and Its Aftermath

In 1917, at the height of Russian suffering during World War I, Czar Nicholas II was overthrown. Within a few months, a highly disciplined faction of revolutionaries called the **Bolsheviks** came to power. They were inspired by a common ideology, **Communism,** based largely on the writings of the German revolutionary philosopher Karl Marx. Marx criticized the societies of Europe as inherently flawed because the **capitalists**—a wealthy minority who owned the factories, farms, businesses, and other means of production—dominated the impoverished, propertyless majority, who worked for low wages that undervalued their labor. Communism called on workers to unite to overthrow the capitalists and establish a completely egalitarian society with no government and no money, in which people would work out of a commitment to the common good, sharing whatever they produced.

The Bolshevik leader Vladimir Lenin declared that the people of the former Russian empire needed a transition period in which to realize the ideals of Communism. Accordingly, Lenin's Bolsheviks formed the **Communist Party,** which set up a powerful government, centered in Moscow, that was to orchestrate the economy and bring about drastic redistributions of wealth while at the same time jump-starting industrialization. Ethnic Russians held most key government positions, even in non-Russian areas. However, government management of the economy did not work well, and soon production was not meeting demand.

After Lenin's death, Joseph Stalin began his 31-year rule of the Soviet Union as party chairman and premier (1922–1953), bringing a mixture of revolutionary change and despotic brutality that largely set the course for the rest of the Soviet Union's history. Stalin sought to cure production shortfalls by creating a **centrally planned,** or **command, economy** in which the state owned all real estate and means of production, while government bureaucrats in Moscow directed all economic activity: factory locations, production, management, distribution of products, and pricing. The idea was that under such a government-controlled economic system, which was termed **socialism,** the economy would grow quickly and hasten the transition to the idealized Communist state in which everyone shared equally. This notion was reflected in the new name chosen for the country: the Union of Soviet Socialist Republics.

Stalin used the powers of the command economy with fervor and cruelty. To increase agricultural production, he forced farmers to join large government-run collectives, and those who resisted were relocated or executed. Stalin saw increasing industrial production as the true key to achieving economic growth. Accordingly, he ordered massive government investments in gigantic development projects, such as factories, dams, and chemical plants, some of which are still the largest of their kind in the world. The labor was supplied largely by former farmers who had lost their farms to collectivization. Eventually, government-controlled companies monopolized every sector of the economy, from agriculture to mining to clothing design and production.

This strategy of government control resulted in both significant successes and massive failures. For those millions of farmers who peacefully went to work in urban industries, their wages brought higher standards of living. And the schools provided for their children made social mobility possible for future generations. During the Great Depression of the 1930s, the Soviet Union's industrial productivity grew steadily even while the economies of other countries stagnated. Production, however, was geared largely toward heavy industry (the manufacture of machines and transport equipment) and supplying the military with armaments, and little attention was paid to the demand for consumer goods and services that would have dramatically improved daily life for the Russian people and served as a source of employment and economic growth, as it did in Europe and North America. Meanwhile, Moscow gave scant consideration to the effects of heavy industrial development on either the natural environment or the health of Soviet citizens. During and after the Stalin regime, the Soviets perpetrated some of the world's worst industrial disasters (especially the unsafe disposal of hazardous materials) and created polluted and unlivable cities.

The most destructive aspect of Stalin's rule, however, was his ruthless use of the secret police, starvation, and mass executions to silence anyone who dared to oppose him. The lucky ones were those merely sentenced to labor camps in remote Siberia. Stalin's atrocities, which resulted in the deaths of at least 20 million people, created a climate of fear that squelched the possibility of political empowerment for the masses—a climate that was extended into Central Europe, and, as we shall see, lingers into the present in Russia.

## World War II and the Cold War

The Soviet Union did more than any other country to defeat Hitler's armies in World War II, but it bore the brunt of Nazi Germany's war machine, suffering an estimated 23 million casualties, more than all the other European combatants combined. After the war, in part because of these tremendous losses, the Soviet Union was determined to erect a buffer of allied Communist states that would be the battleground of any future war with Europe. This installation of undemocratic Communist regimes across eastern Europe was the first chapter of the cold war, which pitted the Soviet Union and its allies against the United States and its allies in global geopolitical rivalry (Figure 5.12). In an attempt to match the global military power of the United States, the Soviets diverted ever more resources to their military at the expense of much-needed economic and social development. Internationally, the Soviet Union spent scarce funds to promote the Communist model of economic development to countries near and far: China, Mongolia, North Korea, Afghanistan, Cuba, Vietnam, Nicaragua, and various African nations. The Western allies spent even more to promote the capitalist model. Closer to home, the Soviet Union maintained its pervasive influence through the use of political, economic, and military coercion. The economies of countries under its influence, such as Poland, East Germany, and Hungary in Central Europe, Georgia in Caucasia, and Kazakhstan in Central Asia, were tuned to meet the needs of the Soviet Union as a whole, and their political systems were tightly controlled from Moscow, with Moscow's will enforced by the presence of Soviet troops.

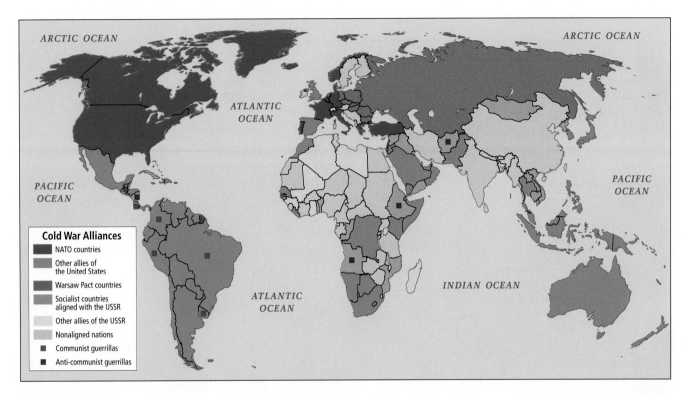

**FIGURE 5.12  The cold war in 1980.** The post–World War II contest between the Soviet Union and its allies and the United States and its allies eventually included many countries, as both protagonists sought to entice more allies through economic and military aid. Some countries managed to remain unaligned with either side. [Adapted from Clevelander, at http://en.wikipedia.org/wiki/image:cold_war_1980.png.]

Nonetheless, by the late 1960s, the economies and political systems of countries under Soviet influence were steadily drifting toward the free market, democratic model advocated by the United States and its allies. Then, in 1979, a war in Afghanistan, launched to prop up a Soviet-allied regime, severely drained resources and morale in the Soviet Union as a whole. The Soviets were badly beaten in Afghanistan by the mujahedeen, highly motivated Afghan freedom fighters who were aided by the United States and Pakistan. (We discuss the Soviet–Afghanistan conflict further in Chapter 8.) This defeat, along with social and political changes already under way in the region, made the 1980s a time of rapid and monumental change in the Soviet Union.

## The Post-Soviet Years

In 1985, a reform-minded president of the Soviet Union, Mikhail Gorbachev, responded to pressures for change by opening up public discussions of social and economic problems, an innovation known as **glasnost.** Gorbachev's efforts to revitalize the Soviet economy through **perestroika,** or restructuring, resulted in little real change. But when he also began to democratize decision making throughout the Soviet Union in the late 1980s, long-silenced resentment of the government in Moscow boiled over and long-suppressed political and interethnic tensions emerged. Independence movements surfaced, first in Poland and in the Baltic states of Estonia, Latvia, and Lithuania, and quickly spread. Gorbachev's liberalizing policies were strongly opposed by those who favored centralized control. After a failed coup d'état against Gorbachev in August 1991, the Soviet Union dissolved in an atmosphere of economic and political chaos.

In the years after 1991, one massive state was eventually replaced by 12 independent countries, each with its own agenda for reforming the failed systems inherited from the Soviet Union. The Russian Federation is the chief successor of the Soviet Union, and although it is much smaller now, it is still the largest country in the world, nearly twice the size of Canada (the second largest). Russia's efforts to maintain its global influence have hinged largely on its seat (with veto power) on the UN Security Council, which it inherited from the Soviet Union. Larger efforts at global leadership have foundered because the Russian leadership has been consistently unable to solve major internal problems. The transition to a free market economy has proceeded haphazardly, leaving once centrally planned economies drifting dangerously. In some parts of the region, the already sad record of environmental abuse has worsened due to lax supervision. Democratization advanced for a while, but in 2004, after a spate of deadly terrorist attacks, and with corruption and discontent over post-Soviet hardships growing, the public welcomed the strong-arm tactics of President Vladimir Putin. Putin, a former official in the Soviet KGB (roughly equivalent to the CIA and FBI), called for constitutional amendments that severely limited democratic participation and freedom of the press. His reelection in that year with a 72 percent majority ushered in a return to tightly centralized control and suppression of the news media.

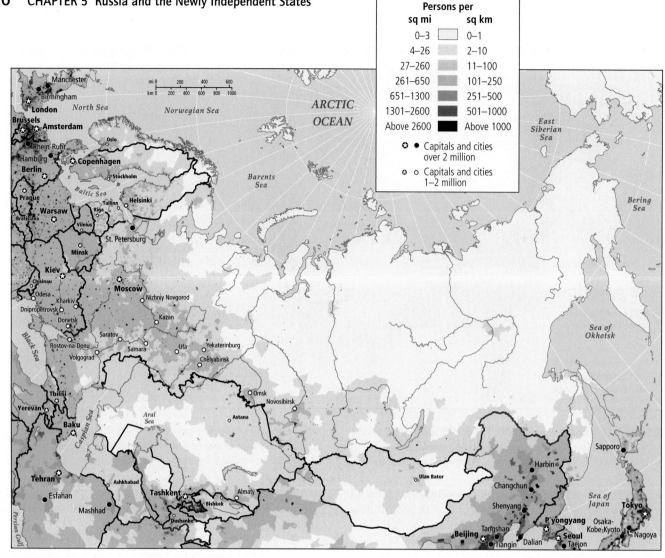

**FIGURE 5.13**  Population density in Russia and the Newly Independent States. [Data courtesy of Deborah Balk, Gregory Yetman, et al., Center for International Earth Science Information Network, Columbia University, at http://www.ciesin.columbia.edu.]

## POPULATION PATTERNS

European Russia is the most heavily settled zone within the region (Figure 5.13), with an average density of 22 people per square mile (8.5 per square kilometer). By comparison, the whole of the United States has an average density of 80 per square mile (30 per square kilometer). A broad area of moderately dense population forms a wedge stretching from Odesa in southern Ukraine on the Black Sea north to St. Petersburg on the Baltic Sea and east to Novosibirsk, the largest city in Siberia. This relatively dense area has most of Russia's usable agricultural soils, but the correlation of usable soils with population density does not hold everywhere: some industrial cities east of the Ural Mountains are located on particularly infertile land.

Beyond Novosibirsk on the Western Siberian Plain, settlement follows an irregular pattern of industrial and mining development across Siberia. These activities, in turn, are linked primarily to the course of the Trans-Siberian Railroad (see Figure 5.20 on page 264). The eastern industrial cities of Krasnoyarsk, Angarsk, and Irkutsk, all lying on the Trans-Siberian Railroad, each have several hundred thousand people. So Siberia, which is seen as a desolate, lonely land, is nonetheless a region where nearly 90 percent of the people are concentrated in a few large urban areas. The large cities don't dispel the impression of isolation, however. Beyond the cities, eastern Russia has an extremely low population density of just a few people per square mile.

In the west, a secondary spur of dense settlement extends south from the main wedge in European Russia into Caucasia, the region between the Black Sea and the Caspian Sea. Another patch of relatively dense settlement is centered on Tashkent and Almaty and along major rivers in the Central Asian republics. Here the development of irrigated agriculture (especially the cultivation of cotton) and mineral extraction has resulted in patches of high rural density, fueled partly by ethnic Russian immigration.

### Recent Population Changes

Before its breakup, the Soviet Union was considered a developed nation with fairly high standards of living and well-being. In the early 1990s, however, the well-being of the citizens of Russia and the Newly Independent States deteriorated significantly, and the region's population declined. Not only did infants and children die in large numbers, but adult life spans also fell. In Russia, for example, between 1990 and 2003 male life expectancy declined from 63.9 to 59 years, the shortest in any industrialized country, and female life expectancy dropped from 74.4 to 72 years. No other country has experienced a gap of 13 years between average male and female life expectancies during peacetime.

The apparent cause of declining life expectancies is the physical and mental distress caused by lost jobs and social disruption, particularly for males. By 1996, divorce rates—already high before the Soviet Union dissolved—had reached more than 6 out of every 10 marriages in Belarus, Russia, and Ukraine. Alcohol abuse shortens and saddens many lives in the region, especially those of men. The *Moscow Times* reported in 2005 that 7 million deaths per year in Russia are alcohol related and calculated that if alcohol were not available, the decrease in Russia's population would stop. In 2006, male suicide rates in Russia, Belarus, and Ukraine were among the highest on earth; studies show that suicides are related to alcohol abuse. By contrast, in Caucasia, despite armed conflict, severe economic reversals, and drastic reductions in living standards, both males and females have a significantly longer life span than they do in Russia or in Central Asia. This pattern was noticed well before the post-Soviet transition; it may be partly a result of longer agricultural growing seasons, a more nutritious diet, and much less consumption of alcohol (many obey the Muslim prohibition) in Caucasia, as well as a genetic predisposition to long life.

After 1991, many people in the region began to suffer nutritional deficiencies caused by sharply falling incomes and food scarcities, some of which were brought on by conflict in Russia's internal republics and in Caucasia. Overall, during the upheaval of the 1990s, people were eating fewer vegetables; less meat, eggs, and dairy products; and even less bread, long a staple of the diet in this region. The cost of food took up as much as half of family budgets. Although entrepreneurial farmers (especially from Ukraine and Caucasia) now supply attractive fresh food to the best city markets, ordinary people can afford these luxuries only rarely, and many grow their own produce (Box 5.1).

## BOX 5.1 | AT THE LOCAL SCALE   Urban Gardens

Urban gardens have long been an important source of nourishment across Europe and Asia. In Soviet times, city dwellers escaped the confinement of their small urban apartments by traveling to the urban fringe. Most Russian cities were surrounded by many small parcels of government-held land on which urbanites maintained small garden plots and sometimes second dwellings (called dachas), which provided them with a place to relax on weekends and to grow food. The vegetables and fruits they cultivated were cheaper than purchased food and less contaminated with pesticides and heavy metals.

Since the transition to a market economy, landholding and land use patterns have changed markedly. Middle-class and wealthy suburbs are beginning to ring the city. They are interspersed with the long-cultivated garden plots, some still maintained by inner-city residents who make the trip to cultivate them on weekends (Figure 5.14). Some inner-city high-rises now sport rooftop gardens. Given the instability of jobs in Moscow and other large cities, garden plots are likely to remain important supplements to family nutrition. But there is a limit to what urban and suburban people can produce, given the constraints of little time to cultivate, small available spaces, and short growing seasons. An interesting question is, as city life modernizes further, who will do the family gardening?

*Source:* Adapted from "St. Petersburg urban gardening club," http://spugc.echotech.org/index_en.html.

**FIGURE 5.14 Gardening by city dwellers.** Russians, like many Europeans, often journey on weekends to the urban fringe, where they cultivate small garden plots. Here, members of a Moscow cooperative gardening society are sowing potatoes. The small house in the background is where they store tools and where they may spend the night in order to extend their gardening hours. [TASS/Sovfoto.]

Environmental pollution also plays a large, if poorly understood, role in untimely illness and death, as we shall see later in this chapter. A third or more of the population may be affected by noxious airborne emissions, heavy metal pollution of the ground, air, or water, or radiation contamination.

A persistent decline in birth rates is another significant aspect of the region's population patterns. In the Soviet era, birth rates and fertility rates in Russia and Belarus were already low compared with those in the United States and Europe. Since the early 1990s, those rates have dropped still further, and the traditionally higher birth rates in the Caucasian and Central Asian republics are now dropping as well. Surveys showed that people were choosing to have fewer children mostly out of concern over the gloomy economic prospects for the near future: in 1992, 75 percent of Russian women who had decided against childbearing cited insufficient income as the reason. By 2006, many cited a desire to make money and have fun as the reason for postponing, if not rejecting, childbearing. The lack of interest in parenthood is complicated by the fact that contraceptives are in short supply and of poor quality; abortion remains a common method of birth control.

By 2005, for the many reasons just discussed, the populations of Russia, Belarus, Moldova, and Ukraine all were declining and aging. Russia alone lost 2.5 million people between 2003 and 2005. Population pyramids for several of the republics (Figure 5.15) show the

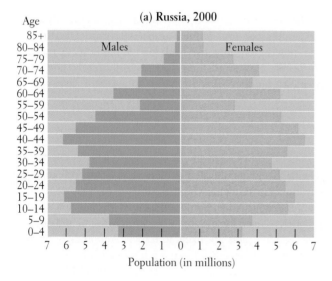

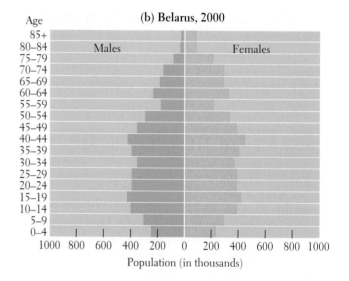

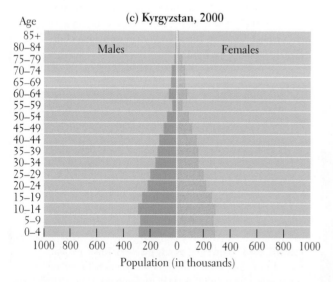

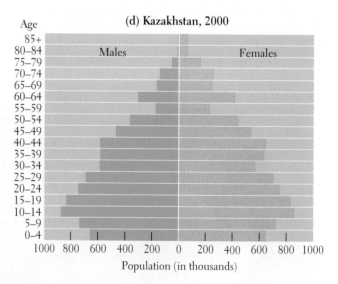

**FIGURE 5.15** Population pyramids for Russia, Belarus, Kyrgyzstan, and Kazakhstan. Note that the pyramid for Russia is at a different scale (millions) than the other three (thousands) because of Russia's larger total population. This difference, though, does not significantly affect the pyramid's shape. [Adapted from "Population pyramids of Russia," "Population pyramids of Belarus," "Population pyramids of Kyrgyzstan," and "Population pyramids of Kazakhstan," U.S. Census Bureau, International Data Base, at http://www.census.gov/ipc/www/idbpyr.html.]

overall population trends in the region and reflect geographic differences in patterns of family structure and fertility. The pyramids for Belarus and Russia resemble those of European countries (for example, Germany, as shown in Figure 4.13a on page 197). They are significantly narrower at the bottom, indicating that birth rates in the last several decades have declined sharply. The narrower point at the top for males in all four pyramids depicts their much shorter average life span. The bases of the pyramids for Kyrgyzstan and Kazakhstan, in Central Asia, are also narrowing as a drop in birth rates accompanies urbanization. But both pyramids still have substantial bases, reflecting the higher fertility rates common in a more rural, primarily Muslim society with a patriarchal family structure. All of the pyramids have a noticeable "waist" at the 55–59 age group, the result of high death rates and low birth rates during and just after World War II.

# II CURRENT GEOGRAPHIC ISSUES

The goals of the Soviet experiment begun in 1917 were unique in human history: to reform quickly and totally both a human society and its social, political, and economic systems. Now, in the aftermath of the Soviet Union's collapse, a second unique experiment is under way: the transformation of a Communist society and a centrally planned economy into its antithesis: a society based on democracy and a free market economy. But there is a lack of agreement on the wisdom of this experiment. At this point, the process has only begun, and for many, life is proceeding amid economic, political, and social uncertainty.

## ECONOMIC AND POLITICAL ISSUES

**Personal Vignette** The kindly, gray-haired teacher tapped the blackboard under the words *profit* and *inventory* as the first-graders struggled to pronounce the Russian words. They were reviewing the story of Misha, a bear who opens a honey, berry, and nut store in the forest. He soon outsmarts the overpriced government-run Golden Beehive Cooperative, becoming first a prosperous bear and eventually the finance minister of the forest.

*Source: Adapted from Sarah Koenig, "In Russia, teaching tiny capitalists to compete," New York Times (February 9, 1997).* ■

Teaching economics in the first grade may seem a bit surprising, and the practice may not be widespread, but this story illustrates the marked shift in focus that took place in the 1990s throughout Russia and the Newly Independent States. Schools now emphasize the teaching of free market economics instead of the ideas of Marx and Lenin. This shift in education is but one of many ambitious market reforms pursued in some former Soviet republics, especially Russia. To fully appreciate what free market reforms have meant to this region, it is important to know something of the Soviet institutions that were previously in place.

## The Former Command Economy

As we have seen, although the long-term goal of the command economy was to achieve Communism, its shorter-term goals were to end the severe deprivation suffered by so many under the czars and to distribute the benefits of economic development to all regions. To some extent, these goals were met: the Soviet economy grew rapidly until the 1960s, and abject poverty was largely conquered. Such basic necessities as housing, food, health care, and transportation were provided to all for free or at low cost—a remarkable accomplishment for any country.

**FIGURE 5.16** Political map of Russia and the Newly Independent States.

Nonetheless, the Soviet command economy was less efficient than market economies in allocating resources. With bureaucrats setting production goals for the whole Soviet Union, even small miscalculations of production types and amounts created nationwide shortages or gluts, and scarcities of food and raw materials occurred regularly throughout the Soviet era. At the same time, the lack of competition meant that producers had no incentive to use more efficient production methods or to supply higher-quality goods at prices lower than a competitor. Because each Soviet factory tended to have a monopoly on the production of a particular product, most consumer goods were of poor quality and available only at high cost to a privileged few. (Televisions and small refrigerators were an exception: they became widely available in the 1960s and 1970s.)

Another major problem was the Soviet Union's failure to keep up with the rest of the industrialized world in technological and managerial innovation. Granted, Soviet scientists and engineers developed advanced military and space exploration technologies, launching the world's first satellite in 1957 and the first manned spacecraft in 1961, and they made breakthroughs in physics, biology, metallurgy, computer technology, and a wide variety of other fields. In the civilian economy, however, the technology was often outdated, polluting, and inefficient. At the same time, Soviet leaders paid little attention to the need for rewarding hard work and innovation in the workforce; rewards went instead to those who had loyalty to and personal connections with the Communist Party.

### Soviet Regional Development Schemes.

A unique feature of Soviet central planning was the location of industrial projects in the farthest reaches of Soviet territory in order to serve political, rather than economic, purposes. Leaders wanted to bring the higher standards of living enjoyed in industrialized European Russia to all parts of the Soviet Union as well as to locate new developments close to necessary resources. Thus, for example, the metal-processing firm Norilsk Nickel was situated near mineral beds in far northern Siberia and a new city was built around it (Figure 5.17 and Box 5.2). The central planners and their military advisers also thought that dispersing industrial centers throughout the country would make them safer because an enemy would have to attack widely separated sites. Yet another factor in industrial location was the need to buttress Russia's claims to distant territories. Soviet leaders thought that economic development projects would placate remote ethnic minorities whose resources were being diverted to national uses, and that the migration of ethnic Russian managers into minority areas would promote conformity with the dominant socialist ethic and with modernized ways of life.

These ambitious regional development schemes never really succeeded, in large part because of transport problems, which we will discuss below. When the Soviet Union fell apart, this system of industrial location contributed to the drastic drop in national production because, once true costs were calculated, it was often too expensive to transport goods from remote factories to market.

### Transport Issues.

In locating industries in far-off corners of the Soviet Union, the central planners ignored the tremendous transport problems posed by this strategy, given the region's huge size and challenging physical geography. Water transport is the cheapest, but Russia's rivers run north-south, while its primary transport needs are east-west, and its ocean ports are few, remote, and often blocked with ice. Throughout the modern era, the development of land transport systems has been limited by long winters, permafrost, swampy forestlands, and complex upland landscapes, especially in Siberia. All of these conditions, plus a certain amount of bureaucratic inertia and the fact that few ordinary citizens could afford a car (by 2003 in Russia there were just 130 cars for every 1000 families), conspired against road building; hence the former Soviet Union, more than two and a half times the size of the United States, has less than one-sixth the number of hard-surface roads and virtually no multilane highways. Most of the roads that exist are in European Russia. Road traffic in this area is congested, in part because state firms have dissolved into smaller, more dispersed private firms that rely more on truck than on rail transport, and in part because, though car ownership is still low, it has recently tripled. Bottlenecks along the narrow two-lane roads, where drivers must cope with numerous railroad crossings, narrow bridges, and winding routes that take them through the middle of small settlements, are a major impediment to the expansion of commercial firms.

In eastern Russia and Central Asia, existing roads are still of such poor quality that cities are linked primarily by rail and air. Nearly every Russian city has an airport, but the Trans-Siberian Railroad is the chief connection between Moscow and Vladivostok, the main port city on the Russian Pacific coast (see Figure 5.20). The complete rail trip takes several days. Driving the distance via the mostly two-lane Trans-Siberian Highway, which parallels the railroad, is currently impossible because sections of the

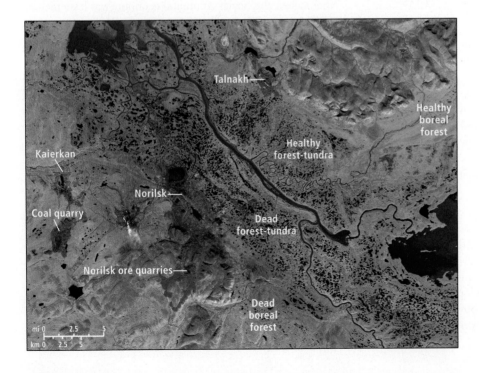

**FIGURE 5.17**  Satellite image of Norilsk, August 9, 2001. Lying 180 miles (300 kilometers) north of the Arctic Circle in northeastern Siberia, Norilsk (population 230,000 as of 2001) exists solely because of nickel, copper, and platinum mining operations. The shades of pink and purple in this false-color image indicate ground barren of vegetation, where there are cities, quarries, bare-rock formations, pollution damage, and in the case of deep purple, severely damaged ecosystems. [Jesse Allen/Earth Observatory/NASA.]

## BOX 5.2 | AT THE LOCAL SCALE   Norilsk: An Industrial City in Transition to a Global, Free Market Economy

Norilsk lies in Siberia, 200 miles (320 kilometers) north of the Arctic Circle (see Figure 5.1), where year-round average temperatures are below freezing. The city sits atop a rich deposit of minerals: 35 percent of the world's nickel supply, 10 percent of its copper, and 40 percent of its platinum. The city's only industry is Norilsk Nickel, Russia's largest metal mining company. In 1998, Norilsk Nickel employed 110,000 workers out of a total local population of about 270,000, and more than 18,000 of those workers were employed in growing food or providing social services (clinics, sports clubs, day-care centers, cafeterias) for the rest. Norilsk is important both to Russia and to the global economy. But Norilsk is also a site of major industrial pollution and natural habitat destruction; male life expectancy there is just 50 years, despite the availability of sports clubs and health care.

In 1994, a private firm acquired 38 percent of Norilsk Nickel and instituted cost-cutting measures in all its branches, especially the social services division. Nevertheless, in 1997, social services continued at a level that absorbed all the plant's profits, an esti-mated U.S.$260 million, or about U.S.$2000 per worker. By 2004, reports regarding Norilsk Nickel were mixed: it won first place in a nationwide competition for the most socially oriented company. However, the company had laid off 20,000 people, and others had left voluntarily. By 2006, Norilsk Nickel had attracted investment from U.S. and European banks, was producing 2 percent of the Russian GDP, had expanded into research on hydrogen energy, and was Russia's largest multinational, owning 51 percent of Stillwater Mining in the United States, and other investments around the world (Figure 5.18).

*Sources:* Adapted from *The Economist* (January 10, 1998): 59; "Russia's cooked books," *The Economist* (September 9, 2000); Marina Kamayeva, "Russia's Norilsk Nickel upgrades production facilities," *Business Information Service for the Newly Independent States* (Washington, D.C.: U.S. Department of Commerce, May 19, 2000), http://www.bisnis.doc.gov/bisnis/isa/000601nickel.htm; United Nations Conference on Trade and Development, World Investment Report, 2004, "The Shift Toward Services" (New York, 2004), p. 74; Norilsk Nickel Web page, http://www.nornik.ru/en/about/, 2006.

**FIGURE 5.18** Norilsk Nickel's worldwide operations. [Adapted from "About Norilsk Nickel," at http://www.nornik.ru/en/about.]

road in eastern Siberia have yet to be completed. In European Russia, rail service is regular, and the network is dense, connecting industrial cities and resource supplies. However, despite the low rate of car ownership, only six cities in Russia have light rail or underground commuter systems. During the 1990s, the Russian state transport agency, Rosavtodor, tried to assess the locations of the most likely markets for Russian products. As a result, Rosavtodor has recently increased road links to Europe by 10 percent, and it has laid plans to build roads to the Central Asian republics, China, Mongolia, and the Pacific coast. The rail, air, and highway systems are supplemented by oil and gas pipelines built in the late twentieth century.

## Reform in the Post-Soviet Era

Russia's recent economic reforms have been ambitious, but haphazard. Many key institutions of the command economy have been dismantled and sold to private interests, but so far this change has enriched only a small group of wealthy and politically connected individuals, many of whom are corrupt. The lives of the majority have become more difficult in many respects, and certainly more unpredictable.

Two key economic reforms have been privatization and the lifting of price controls. **Privatization** (sometimes called **marketization**) is the selling of industries formerly owned and operated by the government to private companies or individuals, who, it is hoped, will operate them more efficiently in response to the free market. Estimates vary, but it appears that by 2000, approximately 70 percent of Russia's economy was in private hands, a significant change from the 100 percent state-owned economy of 1991. Even the remaining state-owned industries have changed the way they operate to follow free market rules. As a result, prices are no longer kept artificially low to make goods affordable to all, but instead are determined by supply and demand.

The lifting of price controls has had unexpected consequences. When goods and services were first sold at their market value, prices skyrocketed because of scarcities, and managers and new private owners gained wealth. But the drastically higher prices forced many people to use their savings to pay for simple necessities. One consequence was that many individuals who would have liked to become small business owners lacked the capital to launch these enterprises. Eventually, as opportunities opened and competition developed (see the video vignette below), the supply of goods and services increased and prices fell, but in the interim people suffered.

Nor has privatization had the hoped-for outcome of increasing efficiency and improving life for all. Among the first enterprises privatized were the most lucrative ones: those producing raw materials for export, such as oil, timber, or metals. Officials in the administrative hierarchy allowed buildings, factories, and access to resources to be sold for a fraction of their real value in return for kickbacks or bribes. Those who managed to acquire ownership of resources such as oil or timber, or the rights to their development, profited immensely. These Russian versions of "robber barons" are called **oligarchs** because their money affords them political power. Now that so much industry has been privatized, the profits collect in the pockets of the oligarchs instead of funding government social services or pension programs. Social programs have become underfunded just as the need for them is increasing, as private companies reduce their workforces to save money and operate more efficiently and as increasing numbers of workers reach retirement age.

Vladislav Dudakov came to Moscow as a soldier assigned to guard Lenin's tomb. Now, in a new Russia where entrepreneurship is encouraged, he manages a shop for an investor who has opened up a chain of Starbucks-style coffee shops. One of the unique challenges of his job is managing employees who are used to working for the government. He has to teach Western business values, just as he was taught them during his first Western-style job: working for McDonald's. When McDonald's came to Russia, Vladislav traded the life of a soldier for that of a businessman, working his way from floor sweeper to store manager: "I was different from my peers. They would say, 'Why don't we extend our break a bit? Why do we have to scrub so hard?' My answer was: you must fulfill the task. At McDonald's we were always earning achievement awards. . . . The American style of management became my education." Vladislav now encourages the waitresses at his coffee shop to forget old Soviet habits and to "smile, smile, smile." His success has made Vladislav an example of a new class of Muscovite: young, entrepreneurial, and newly affluent.

*Hear more firsthand accounts of Russian capitalism as New York Times reporter Sabrina Tavernise interviews Russia's businesspeople in the FRONTLINE World video "Moscow: Rich in Russia: The Brave New World of Young Capitalists and Tycoons."* ∎

**Foreign Direct Investment (FDI).** The largest sources of capital for development in the whole region are foreign, primarily Euro-

pean, Asian, and North American. Foreign investors, however, are notoriously fickle when it comes to Russia, gaining and losing interest depending on changes within the country. Recently, high global energy prices have attracted foreign investment to Russia, Caucasia, and the Central Asian republics, despite the threat (primarily in Russia) of government takeovers of petroleum industries. Although many now express reservations about the region's dependence on its energy sector—prices are so high that potential buyers worldwide are actively pursuing alternatives to oil and gas—the region is likely to profit greatly from oil and gas over the next decade. One huge Soviet-era firm that successfully made the transition to private status and has now attracted foreign investors is Norilsk Nickel in Siberia (see Box 5.2).

***The Growing Informal Economy.*** To some extent, the new informal economy in the region is an extension of the old one that flourished under Communism. The black market of that time was based on currency exchange and the sale of hard-to-find luxuries. For example, in the 1970s, savvy Western tourists could enjoy a vacation on the Black Sea paid for by a pair or two of smuggled Levi's blue jeans and some Swiss chocolate bars. Today, however, the range of informal ventures involves more people and more occupations because many people who have lost stable jobs due to privatization now depend on the informal economy for their livelihoods. Often they operate out of their homes, selling cooked foods, vodka made in their bathtubs, electronics they have assembled, or pirated computer software, to name just a few products (Figure 5.19). Outright criminal activities, including sex trafficking, fraud, and blackmail, are also a part of the informal economy. Indeed, so much of the economy is in the informal sector that in many countries of the region, people may be better off financially than official gross domestic product (GDP) per capita figures suggest.

The informal economy is a major problem throughout the region because governments are unable to collect tax revenue from it, and the taxes generated by the formal economy are not sufficient to cover government obligations to the public. Furthermore, a highly active informal economy encourages a general scofflaw attitude on the part of the public. Tax dodging, which does not carry the stigma it does in other free market countries, has reduced tax revenues even from businesses in the formal economy. Only in the last several years have tax collections increased markedly as oil and gas revenues and incomes have risen.

***High Energy Prices End the Debt Crisis.*** Russia's foreign debt—a staggering 90 percent of its GDP in 1998—once dominated Russia's financial relationships with other countries. As a result of revenues from rising oil prices, Russia has been able to bring the debt down to levels that most lenders consider reasonable. While two-thirds of the debt had been accumulated by the Soviet Union, debt also mounted after 1991 when the new government had to borrow because tax collections waned at the same time exports lagged, industries closed, and unemployment rose. An expensive war in Chechnya also added to the debt. In early 2000, Russia defaulted on its debt payments and then requested debt reschedul-

**FIGURE 5.19** The informal economy in Russia. Imported (probably smuggled) VCRs and TVs are sold in informal non-taxpaying open markets like this one in Moscow. They are tended by men affecting, if not actually living, a "mafioso" lifestyle. Police are paid to look the other way. [Gerd Ludwig/National Geographic Image Collection.]

ing with all its creditors. Hoping to negotiate changes within the country, Western lenders tied rescheduling to a commitment to achieve a free market economy, reduce theft and corruption, and encourage Russian oligarchs to reinvest their profits in Russia, not simply keep them in foreign banks. In recent years, higher energy prices have greatly improved Russia's finances, increasing the GDP and tax revenues and thereby enabling an accelerated repayment of the debt, which the Russian finance ministry plans to have down to just 12 percent by 2008.

**New Trading Partners.** For decades, Russia had a colonial relationship with its allies in Central Europe, Caucasia, and Central Asia, dominating their militaries and foreign policies and appropriating their resources in uneven exchanges that left some republics poorer and more dependent on Russia than they would have been otherwise. Today, despite Russia's struggle to maintain what is left of these arrangements, Poland, the Czech Republic, Slovakia, Hungary, Romania, Bulgaria, and the former Soviet republics of Latvia, Lithuania, and Estonia are now part of the European Union. Although Ukraine and Moldova retain strong economic and political ties to Russia, both would like to be invited to join the EU, but an invitation is not likely to come soon. Belarus also retains close economic ties with Russia; the two countries trade freely without tariffs or customs duties, though in the evolving energy market the relationship is less smooth, with the two sparring over oil and gas prices and Russia's use of Belarus's pipelines to move fuel to Europe.

Other Newly Independent States remain economically connected with one another and with Russia by arrangements that may not last. For example, the cotton growers of the Central Asian republics, who used to sell raw cotton to textile mills located in Russia, which then exported cotton cloth back to Central Asia, are increasingly engaged in the global cotton trade. Central Asian cot-

ton producers are hoping to attract foreign investors to build cotton mills that will produce finished textiles to be exported directly from Central Asia to the global marketplace. Oil provides another example of changing relationships. Kazakhstan, Uzbekistan, and Turkmenistan, though in possession of smaller reserves of oil than Russia (see the personal vignette on LUKOIL on pages 264–265), all now export petroleum products and hence are competing with Russia. In 2006, Kazakhstan began delivering oil directly to Xinjiang Province, China, through a newly constructed pipeline (see Figure 5.21). India, another energy-hungry industrializing giant, is also lobbying for Central Asian oil and gas.

**Obstacles to International Trade.** As Russia and the Newly Independent States seek to gain wealth by increasing exports to the world market, they must overcome several roadblocks to new international trade relationships. First, they are accustomed to trading with one another under terms set by Soviet central planners who assigned each part of the region monopoly production of key products, such as oil, heavy equipment, cotton, electronics, and food crops, which were then redistributed to where they were needed. Now, with new national borders, each country must establish new trade agreements to ensure sufficient food and consumer goods for its populace and adequate resources for its industries. Second, all must improve their products and manage their production costs to compete globally with producers in Europe, the Americas, and Asia. Third, most former Soviet republics lack marketing expertise, and they must acquire these skills rapidly. Many of these countries now maintain Web sites on the Internet to familiarize the public with their products and to seek investors. Increasingly they are holding trade fairs and conferences to accomplish the same goals.

The underdeveloped transport system in the region has created further obstacles. Established rails and roads are not always in

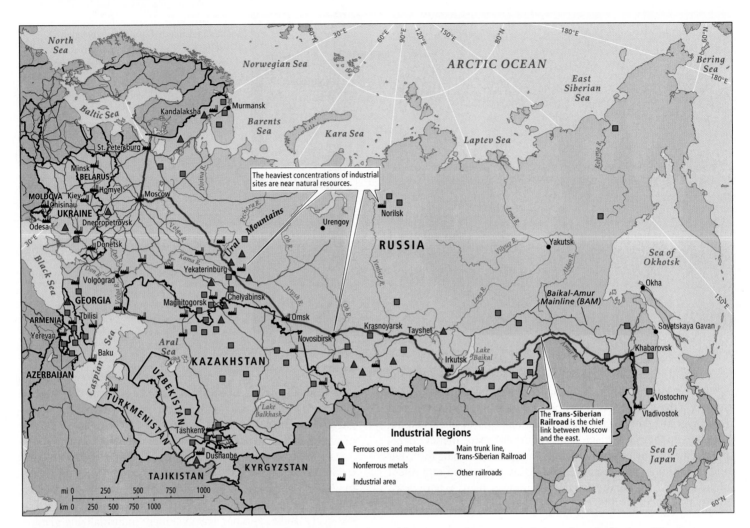

**FIGURE 5.20** Principal industrial areas and land transport routes of Russia and the Newly Independent States. The industrial, mining, and transport infrastructure is concentrated in European Russia and adjacent areas. The main trunk of the Trans-Siberian Railroad and its spurs link industrial and mining centers all the way to the Pacific, but the frequency of these centers decreases with distance from the borders of European Russia. [Adapted from Robin Milner-Gulland with Nikolai Dejevsky, *Cultural Atlas of Russia and the Former Soviet Union*, rev. ed. (New York: Checkmark Books, 1998), pp. 186–187, 198–199, 204–205, 216–217; and http://www .travelcenter.com.au/russia/images/trans-sib-map-v3.jpg.]

ideal locations for emerging patterns of foreign trade. For example, Kazakhstan and Uzbekistan are increasing their direct trade with the outside world, but they are landlocked and still have rail connections only with what are now Russian railroads (Figure 5.20). Their oil and gas pipelines flow primarily to Russia (Figure 5.21), and their air links, long focused on Moscow, have had to be refocused on Europe and Asia.

**Personal Vignette** In 2003, exciting changes were afoot at Paramgit Kumar's Manhattan gas station. He had managed the gas station for Getty Petroleum Marketing for several years, before Getty was purchased by Moscow-based LUKOIL. Kumar was one of the first to join LUKOIL's plan to rebrand the old Getty stations to make it obvious that Russian oil products are sold here. Russian president Vladimir Putin, in the United States for a meeting with President George W. Bush, stopped by to officially open Kumar's station, which sports a bright white and red logo.

The expansion of Russia's oil and gas industries into retail marketing in North America is one measure of their increasing global influence. In 2000, LUKOIL, Russia's largest oil company, purchased 72 percent of Getty Petroleum Marketing, a chain of 1300 gas stations in 13 eastern U.S. states (Figure 5.22). This purchase was the first public acquisition by a Russian corporation of a U.S. firm. Although the deal was relatively small ($71 million was paid), it was important psychologically and has already led to other Russian investments in the United States. LUKOIL plans to expand the chain to 3000 stations, more than doubling the number of its U.S. stations, and in May 2006, LUKOIL bought the assets of another U.S. firm, Marathon Oil.

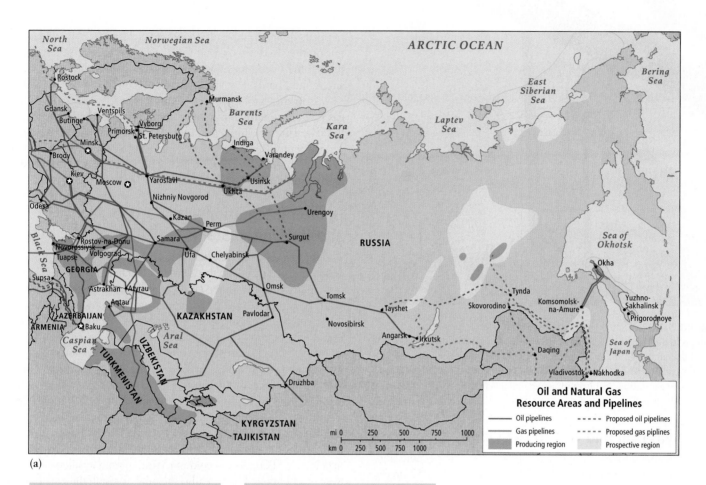

(a)

| Oil Production, 2003 | |
|---|---|
| **Region or Basin** | **1000 Barrels/day** |
| West Siberia | 5,882 |
| Volga-Urals | 1,887 |
| Precaspian | 679 |
| South Caspian | 454 |
| Timan-Pechora | 373 |
| Middle Caspian | 261 |
| South Turgay | 209 |
| Central Asia | 161 |
| North Caucasus | 72 |
| Far East | 65 |
| Azerbaijan Onshore | 32 |
| East Siberia | 32 |
| Baltic | — |
| Barents Sea | — |
| **Total Region** | **10,107** |
| **Total World** | **79,110** |

(b)

| Gas Production, 2003 | |
|---|---|
| **Region or Basin** | **Billion Cubic Meters** |
| West Siberia | 573.1 |
| Central Asia | 90.0 |
| Precaspian | 25.9 |
| Volga-Urals | 25.1 |
| South Caspian | 15.7 |
| East Siberia | 8.8 |
| Timan-Pechora | 3.6 |
| Far East | 1.9 |
| Azerbaijan Onshore | 0.4 |
| Barents Sea | — |
| **Total Region** | **744.5** |
| **Total World** | **2,618.5** |

(c)

**FIGURE 5.21**  Oil and natural gas: Resources and pipelines. [Adapted from "Russia Country Analysis Brief," May 2004, Energy Information Administration, U.S. Department of Energy, at http://www.eia.doe.gov/emeu /cabs/Russia/images/fsu_energymap.pdf.]

Originally, LUKOIL planned to retain the Getty brand name, but in the aftermath of September 11, 2001, Presidents Bush and Putin began discussing ways to decrease U.S. dependence on oil from Southwest Asia and to increase U.S. purchases of oil from Russia. Unexpectedly, a Russian identity for U.S. gas stations turned into a marketing advantage. To facilitate this business venture, LUKOIL made low-cost loans available to station managers like Kumar, who needed capital to make the costly upgrades necessary for the rebranding.

*Sources: Adapted from John Lofstock, "The Russians are coming," Convenience Store News, October 12, 2003, http://www.csnews.com/csnews/reports _analysis/oilwatch_display.jsp?vnu_content_id=1990275;http://energy.ihs .com/News/WW-News/Lukoil.htm, 2006.*  ■

**FIGURE 5.22** Russian oil in the United States. This is one of ten LUKOIL stations in the greater Philadelphia metro area. The Moscow-based conglomerate has continued its penetration into the northeastern United States by adding 800 new locations acquired from Mobil. LUKOIL now has some 5000 gas stations in 15 countries, about half of them in the United States. [AP Photo/Coke Whitworth.]

## Supplying Oil and Gas to the World

Crude oil and natural gas are the region's most lucrative exports and the keys to its economic future. Russia on its own is the world's largest exporter of natural gas, the second-largest oil exporter (Saudi Arabia is first), and the third-largest energy consumer. The state-owned energy company **Gazprom** is the tenth-largest oil and gas entity in the world.

Private oil companies and Gazprom already account for more than half of Russia's federal tax receipts. But much potential revenue has escaped because of the uncontrolled way in which Russia's oil and gas resources were privatized. In recent years, the central government has repurchased a controlling share of Gazprom, which had been partially privatized in the 1990s, and tightened its control over the entire oil and gas sector by confiscating some facilities and strengthening rules to ensure that the industry pays all its taxes regularly. As a consequence of the reestablishment of government control, the expansion of export facilities for Gazprom has proceeded at breakneck pace, but the development of similar facilities for private companies has been delayed by the government, thus giving the government-held firm the advantage.

A tug-of-war has developed between Russia and multinational energy companies over the rights to develop large oil reserves in Central Asia. Maneuverings to gain the rights to develop, transport, and sell or buy these resources have influenced international politics for many years. Countries with designs on Central Asian oil include Russia, Turkey, Iraq, China, India, Pakistan, Iran, the United Kingdom, and of course, the biggest energy consumer of all, the United States. Meanwhile, the holders of these considerable oil reserves, the new Central Asian republics, are not powerful or expe-

rienced enough to stave off the covetous advances of world powers, nor do they have the capital to develop the resources themselves, though their need for oil income is great. Hence they are between a rock and a hard place; suspicion abounds as leaders get bought off.

As of the early 2000s, contention over Central Asian oil and gas hinges primarily on pipeline routes; Russia, the United States, the EU, Turkey, China, and India have all developed or proposed various routes (see Figure 5.21), and as of 2006, the United States had military bases to protect its interests in Georgia, Uzbekistan, and Kyrgyzstan.

## Greater Integration with Europe and the United States

The international oil and gas trade has done much to improve the economies of Russia and the Newly Independent States and to connect them with the inner circle of industrialized nations. At the same time, the United States and Europe, worried that some of these countries might again become their geopolitical adversaries if their economic and political burdens are not eased, are seeking ways to bring them all into closer formal association with established trading institutions. Russia is scheduled to become part of the World Trade Organization in mid-2007 and is the newest member of the **Group of Eight (G8)**, an organization of the most highly industrialized nations. In part, such invitations are intended to reassure Russia, whose former allies have broken ties with it and are now members of the European Union or the North Atlantic Treaty Organization (NATO), the old cold war Western military alliance against the Soviet Union. They are also meant to forestall the possibility of a new cold war developing around issues such as oil, nuclear proliferation, the war on terror, and international migration, bringing possible global realignments that are as yet unclear.

Russia also craves closer association with the European Union, which accounts for over 50 percent of Russia's foreign trade. The EU continues to tempt the Newly Independent States with agreements on trade and the employment of workers in the EU in exchange for compliance with European trade-related standards on matters such as intellectual property rights, the personal safety of workers, product liability, and environmental issues. Europe sees upgrading conditions in the former Soviet Union as a way of developing a buffer zone between itself and unbridled illegal immigration from Central and East Asia and from such trouble zones as Afghanistan, Iraq, Iran, Ethiopia, Congo, and Somalia, as immigrants from all of these countries now use the area as a transit route. Underlying all these interactions is the fact that Europe is dependent on Russia for 20 percent of its natural gas and is interested in securing this supply and negotiating lower gas prices.

## Political Reforms in the Post-Soviet Era

Many observers in the West expected that democratization would go hand in hand with the introduction of a market economy in Russia and the Newly Independent States, but this has not happened. Although several countries have held elections in which there were competing candidates for office, many forms of authoritarian control remain. For example, in Belarus, Russia, and the Central Asian republics, elected representative assemblies often act as rubber stamps for very strong presidents and exercise only limited influence on policy. Across the region, even though officials are elected, they sometimes are overly influenced by oligarchs who wield power because of their enormous wealth. In Russia itself, President Putin exercises tight control and is thought to at least tacitly allow the repression and even elimination of rivals, dissidents, and critics of his policies.

To be fair, it is not clear whether a majority of the people in the region are prepared to participate more fully in the decision-making process: these people have never before had their own voice in governance, and the last 80 years have not encouraged civic responsibility. Even today, ordinary citizens tend to think of political leaders as patriarchs who can dispense favors and produce simple solutions to complex problems. The Russian electorate is increasingly differentiated geographically: older, rural, less educated Russians tend to vote in higher numbers and favor socialist parties and authoritarian leadership, while younger, urban, better educated, and white-collar Russians favor reform parties but are less likely to vote at all. Public participation in civil society in the form of social activism and public demonstrations for or against government policy is increasing, but is still low. In 2006 surveys, the Public Opinion Foundation found very low support in the Russian population for public protests, though in Moscow, small groups do protest fairly often.

***The Media and Political Reform.*** While the media developed considerable independence after the fall of the Soviet Union, recent years have seen a return to state control of the media. In the Soviet era, all communications media were under government control. There was no free press, and public criticism of the government was punishable. Yet it was journalists, risking punishment by engaging in vocal critiques of public officials and policies, who were instrumental in bringing an end to the Soviet Union. Between 1991 and the early 2000s, the communications industry was a center of privatization, and several media tycoons emerged. Privately owned newspapers and television stations regularly criticized the policies of various leaders of Russia and the other republics. It appeared that a free press was developing, and for a time, even President Vladimir Putin lauded a free press as vital to the growth of civil society.

A turning point was Putin's decision in 2000 to arrest the media tycoon Vladimir Gusinsky, whose newspapers and broadcast facilities relentlessly criticized the government. Gusinsky's media facilities were seized by the government. Since then, critical analysis of the government has become rare throughout Russia. Journalists who have openly criticized Putin's policies have been treated to various forms of censorship and violence. Since 2000, more than a dozen have been killed, including, in October 2006, Ana Politkovskaya, who frequently reported on Russia's oppressive policies in the internal Caucasian republic of Chechnya (see page 270). Within hours of her death, people began a continuous vigil outside her apartment building, where they posted signs accusing President Putin of having arranged for her death (Figure 5.23). Her killer has not yet been found.

**FIGURE 5.23** Silencing critics in Russia. The public outcry against the October 2006 assassination of journalist Ana Politkovskaya was immediate and strong. Signs accusing Vladimir Putin of having her silenced were posted on the wall of her apartment building. Flowers banked the sidewalk for weeks, and there was a constant vigil of citizens who mourned the demise of a free press. [© Jurji Fikfak.]

Closely related to the struggle to develop a free communications industry is the availability to the general public of communications technology (Table 5.1 and Figure 5.24). Television sets were widely available in the Soviet Union, but programming was tightly controlled by the government and was free of advertisements. Now there are advertisements, programming is less tightly controlled, and many people receive European and other stations via satellite dishes. Access to telephone landlines is still limited, but mobile phones have made instant personal communication possible in even the most remote cities of Siberia and Central Asia, where they are essential to many new enterprises. Mobile phone service, however, remains expensive and unreliable, and it is not available to most people.

Mobile phone companies are hoping to help the Newly Independent States leapfrog into the global economy by facilitating Internet commerce. Throughout the region, however, personal computers are still rare—Russia has one-tenth the number of computers per capita that the United States has. Internet users are far fewer in Central Asia and Caucasia than elsewhere in the region. Nonetheless, information technology is increasingly available, and demand is growing.

***The Military in the Post-Soviet Era.*** The future role of the military is one of the most important political issues facing the region today. The military was once the most privileged sector of society, and its officers had access to goods and services that ordinary citizens could

### TABLE 5.1 Increase in Internet use, 2000–2007

| Country | Internet users as percentage of population, 2007 | Percentage increase in Internet users since 2000 |
|---|---|---|
| Armenia | 5.1 | 400 |
| Azerbaijan | 8 | 5,556.7 |
| Belarus | 35 | 1,785.8 |
| Georgia | 4.0 | 778 |
| Kazakhstan | 2.7 | 471.4 |
| Kyrgyzstan | 5.2 | 442.6 |
| Moldova | 10.9 | 1,524 |
| Russian Federation | 16.5 | 664.5 |
| Tajikistan | 0.1 | 150 |
| Turkmenistan | 0.5 | 1,700 |
| Ukraine | 11.5 | 2,539.1 |
| Uzbekistan | 3.3 | 11,633.3 |
| United States | 69.4 | 114.7 |

*Source:* "Internet usage in Asia," at http://www.internetworldstats.com/stats3.htm; "Internet usage in Europe," at http://www.internetworldstats.com/stats4.htm; "Internet usage statistics for the Americas," at http://www.internetworldstats.com/stats2.htm#north.

only dream about. Now, after huge funding cuts, the Russian military has been reduced considerably. Military personnel returning home from far-flung posts all across Central Asia and Siberia actually pose a threat to civil society. With few skills, few job prospects, little decent housing for their families, and sometimes not even adequate clothing, these former soldiers form a huge reservoir of discontent. Nonetheless, aside from the botched coup d'état by top military commanders against President Gorbachev in August 1991, further military rebellions have not occurred.

A related concern is that the former Soviet nuclear arsenal could fall into the wrong hands if military cuts are poorly managed. There is some evidence that nuclear expertise and even some warheads may have been smuggled out of the region to such countries as North Korea and Iran, whose governments are distrusted by the international community. By 2000, Russia and other countries in the region were cooperating with the International Atomic Energy Agency in controlling nuclear material. In September 2006, the Central Asian republics all signed a treaty creating a nuclear-free zone. However, Russia remains a nuclear power (as does the United States), and in 2006 it began helping Iran build nuclear energy facilities.

***The Political Status of Women.*** Although they were granted equal rights in the Soviet Union's constitution, women never held much power in government and hence had little say in policy. In 1990, women accounted for 30 percent of Communist Party membership, but they made up just 6 percent of the governing Central Committee, and generally women did not have a strong political voice. Since the fall of the Soviet Union, now that they are no longer appointed by party officials, women have not fared well in the election process, perhaps because of lingering cultural biases against women exercising public authority. Women increased their representation in Belarus by 50 percent, and in Ukraine by 3 percent, but they lost political representation in the parliaments of Russia and the Caucasian republics, dropping from a 20 percent share in Soviet times to just 5.3 to 11.3 percent in 2005 (Figure 5.25). In Kyrgyzstan, women lost all representation when the country went from a bicameral to a unicameral legislature. In Tajikistan, women gained in representation, but in Turkmenistan, their representation decreased by 38 percent.

Nonetheless, the influence of women promises to increase. By the late 1990s, especially in Caucasia and the Central Asian republics, newly aware women were finding ways to influence policy by working through nongovernmental organizations and such agencies as the United Nations Development Programme (UNDP). Hence formal representation in parliament is no longer the only, or even the best, measure of women's political activity and influence.

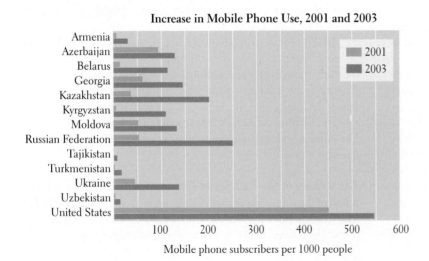

**Increase in Mobile Phone Use, 2001 and 2003**

Mobile phone subscribers per 1000 people

**FIGURE 5.24** Access to mobile phones in Russia and the Newly Independent States. Access to information throughout the region is low compared with that in more developed countries. In the past 5 years, however, people's access to mobile phones has increased significantly. [*United Nations Human Development Report 2003 and 2005* (New York: United Nations Development Programme), Tables 11 and 13.]

## Women in Legislatures, 2003 and 2005

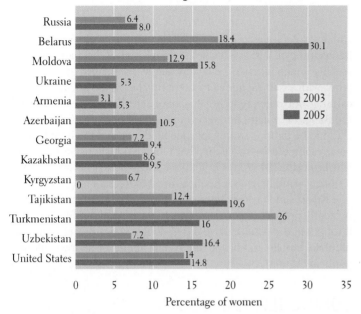

| Country | 2003 | 2005 |
|---|---|---|
| Russia | 6.4 | 8.0 |
| Belarus | 18.4 | 30.1 |
| Moldova | 12.9 | 15.8 |
| Ukraine | 5.3 | |
| Armenia | 3.1 | 5.3 |
| Azerbaijan | 10.5 | |
| Georgia | 7.2 | 9.4 |
| Kazakhstan | 8.6 | 9.5 |
| Kyrgyzstan | 0 | 6.7 |
| Tajikistan | 12.4 | 19.6 |
| Turkmenistan | 26 | 16 |
| Uzbekistan | 7.2 | 16.4 |
| United States | 14 | 14.8 |

Percentage of women

**FIGURE 5.25** Women legislators. This graph shows the percentage of legislators who are women in Russia and the Newly Independent States (with the United States for comparison). Belarus and Turkmenistan stand out as having the most women lawmakers, but both are authoritarian societies in which true democratic participation is rare. [Adapted from *United Nations Human Development Report 2005* (New York: United Nations Development Programme), Gender Empowerment Measure, Table 26, and http://www.ipu.org/wmn-e/classif.htm.]

However, despite their experiences of sex discrimination, support among women for women's political movements is not widespread. When interviewed, women expressed concern that working for women's rights and well-being would be seen as anti-male or as a rejection of traditional feminine roles, a fear that is itself a sign of their low political status.

## Russia's Internal Republics

Russia's centuries-long history of expansion into neighboring lands has left it with an exceptionally complex internal political geography. As the Russian czars and the Soviets pushed the borders of Russia eastward toward the Pacific Ocean over the past 500 years (see Figure 5.11 on page 253), they conquered a number of small non-Russian areas, incorporating them into the **Russian Federation.** These 30 republics and 10 autonomous regions now constitute 25 percent of Russia's land area (Figure 5.26). Many have significant

**FIGURE 5.26** Ethnic character of Russia and the Newly Independent States. Russia, with all its internal republics and autonomous regions, is formally called the Russian Federation. The ethnic character of many of the more than 30 internal republics was changed by the policy of central planning so that Russians now form significant minorities. As of 2002, the ethnic makeup of Russia was 79.8 percent Russian, 3.8 percent Tatar, 2 percent Ukrainian, 1.2 percent Bashkir, 1.1 percent Chuvash, and 12.1 percent other. [Adapted from James H. Bater, *Russia and the Post-Soviet Scene* (London: Arnold, 1996), pp. 280–281; Graham Smith, *The Post Soviet States* (London: Arnold, 2000), p. 75; and *The World Factbook 2007*, at https://www.cia.gov/cia/publications/factbook/.]

ethnic minority populations that trace their origins to peoples who spoke such languages as German, Turkish, Finnish, Ugric, or Persian. The peoples of some of the internal republics, particularly the more southerly ones, are followers of Islam, the dominant religion in Central Asia. The peoples of most other republics are Christian, but those of Kalmykia, between the Volga and Don rivers, are Buddhist, and some groups have animist beliefs that predate any of the organized religions.

Even under the czars, **Russification**—assimilation of all minorities to Russian (Slavic) ways—was the central policy. During the Soviet era, ethnic Russians and other Slavs were settled in the internal republics in order to "civilize" what were viewed as culturally marginal people. Slavic immigrants often outnumbered native people, and they tended to receive the best jobs and housing. During and after the Soviet period, minorities organized to resist Russification, and that resistance tended to enhance local ethnic identities. As a result, Russia has tried to avoid formally addressing the political status of the internal republics because, if these ethnic enclaves were to demand greater autonomy—or worse, independence—the territorial integrity of the vast Russian Federation would be threatened and its access to large, resource-rich areas diminished.

Shortly after the breakup of the Soviet Union, several internal republics did demand greater autonomy, and two of them, Tatarstan and Chechnya, declared outright independence. Whereas Tatarstan has since been placated by significant efforts to increase its economic and political viability and autonomy, Chechnya was at once more resistant and more stigmatized by its rebellion. Russia responded to the Chechen revolt with a major show of military force that has led to the worst bloodshed of the post-Soviet era. For the time being, most of the other ethnic enclaves are content to stay closely associated with Russia.

***The Conflict in Chechnya.*** Chechnya, a small internal republic on the northern flanks of the Caucasus Mountains, has a fertile, hilly landscape that is home to 800,000 people, about 130,000 of whom have been displaced by war since 1991. The Chechens converted to the Sunni branch of Islam in the 1700s, largely in response to Russian oppression. Ever since then, Islam has served as an important symbol of Chechen identity and of resistance against the Orthodox Christian Russians, who, following a long struggle, annexed Chechnya in the nineteenth century. After the Russian Revolution in 1917, the Chechens and other groups in the Caucasus formed the Republic of Mountain Peoples, hoping to separate themselves from Russia, but the Soviets abolished the republic in 1924. In the 1930s, Chechens were forced onto collective farms, which supplied produce to Soviet cities. During World War II, their reputation for resisting Russian policies resulted in accusations of collaborating with the Germans.

In 1991, as the Soviet Union was dissolving, Chechnya declared itself an independent state. To the Russians, this declaration represented a dangerous precedent that could spark similar demands by other cultural enclaves throughout Russia. In addition, Russia wished to retain the region's agricultural, oil, and gas wealth and was planning to cross Chechen territory with pipelines for transporting oil and gas to Europe. Since 1991, Chechen guerrillas seeking independence have repeatedly challenged the Russian army, which has responded by bombing the capital of Groznyy and carrying out other reprisals that have left the city destroyed and thousands of civilians homeless or dead. Although many Chechens claim to be fighting simply for independence and the right to their own resources, Russia claims that Chechen factions have links to Muslim extremist groups in Central Asia—perhaps even to global terrorists—and hence represent a threat to Russia's long-term security. Russia's ineffective attempts to address the troubles in Chechnya early on, before people resorted to terrorism, and its brutality throughout the conflict have raised doubts about Russia's commitment to human rights and its overall ability to address internal political dissent.

## SOCIOCULTURAL ISSUES

When the winds of change began to blow through the Soviet Union in the 1980s, political and economic repercussions were expected. But few anticipated that the transition away from central control of most aspects of life and toward marketization of the economy would happen so speedily and be so unstructured and so disruptive of social stability. On the one hand, the new freedoms have encouraged self-expression and individual initiative for some people, as well as more political participation and a cultural and religious revival. On the other hand, the collapse of the structure once provided by the socialist state has resulted in the loss not only of livelihoods, but also of housing, food, health care, and civil order. In the wake of uncontrolled change, widespread corruption took root.

### Cultural Dominance of Russians and Other Slavs

The Russians always regarded themselves as the heart of the Soviet Union, a point of view that originated with the czars when they extended the Russian empire over adjacent regions. Throughout the twentieth century, Russians felt closest culturally to the European Slavic republics of Ukraine, Moldova, and Belarus and to the Baltic states. The distinctive languages, religion (Islam), and cultural origins of Caucasia and Central Asia set those republics apart. The Soviets considered this distinctiveness to be a liability to the cohesiveness of the Soviet Union, yet the resources these republics contained, as well as their strategic locations, were indispensable. In order to ensure compliance with regional economic plans, to provide a skilled labor force, and to acculturate minorities to Slavic (especially Russian) customs and attitudes, the Soviets resettled Slavic people in all the internal republics as well as in the peripheral republics (now the Newly Independent States). Ethnic Slavic technicians, teachers, and professionals occupied the choice positions throughout the region. In most resettlement locations, Russians became significant and powerful minorities (see Figure 5.26).

Across the region, Russian culture dominated all aspects of public life. For example, although more than 40 legally recognized languages were spoken throughout the Soviet Union, Russian was the language of official business and was taught in all schools as the primary language. Users of local languages were forced to use the Cyrillic (Russian) alphabet. By the 1980s, the use of minority languages was in decline everywhere in the Soviet Union. The ancient local customs of non-Russian peoples—including religion, family organization, domestic architecture, manner of dress, farming, and diet—were considered outmoded and were suppressed. In the school curriculum, Russian culture was presented as the norm. The Russians were especially intolerant of Islam in the internal republics and in Caucasia and Central Asia.

Until the mid-1970s, migrants within the Soviet Union consisted primarily of ethnic Slavs engaged in Russification. Those from the periphery who did move to European Russia did so temporarily for education and training. Then, in the 1980s, migration between Russia and its neighbors slowed and shifted into reverse. In the 1990s, as many as a third of Russian and other Slavic emigrants returned home to Russia from the newly independent Caucasian and Central Asian republics. Conversely, ethnic minorities living in Russia went home to the Caucasian and Central Asian republics or to internal ethnic enclaves. Only in the newly independent European states (the Baltic states, Belarus, Ukraine, and Moldova) did Russians remain as a significant minority, apparently because they saw a brighter future there than in Russia.

## Cultural Revival in the Post-Soviet Era

Since the achievement of independence and the departure of many Russians, the Newly Independent States and the internal republics have, to varying extents, started to reassert their pre-Soviet cultural identities. Belarus has shown the least tendency to restore its culture and remains the most Russian of the Newly Independent States.

One aspect of this cultural revival is the resurgence of interest in religion. Under the Soviets, religious practice was discouraged because religious beliefs were thought to inhibit the commitment of the people to revolutionary change. In European Russia (as well as in Georgia and Armenia), most people have some ancestral connection to Orthodox Christianity, and those with Jewish heritage form a sizable minority. Religious observance by both groups increased markedly in the 1990s, and many sanctuaries were rebuilt and restored. A spectacular example of such restoration is Christ the Savior Cathedral in Moscow (Figure 5.27), which had been destroyed by the Soviets. A countertrend to the robust revival of Orthodox Christianity is the modest spread of evangelical Christian sects from the United States (Southern Baptists, Adventists, Pentecostals). This evangelical movement may be ephemeral, lasting only as long as the money the missionaries bring keeps flowing. However, as the following personal vignette illustrates, its teachings

(a)

(b)

**FIGURE 5.27** The Cathedral of Christ the Savior in Moscow. Geographer Dmitri Sidorov traced the history of this cathedral from the 1810s to the present. **(a)** The cathedral was destroyed by the Bolsheviks in 1931 to make way for a planned, but never built, Palace of the Soviets. **(b)** The restored cathedral. The restoration of this cathedral is often publicized as signaling the dismantling of the antireligious Soviet state and the beginning of a new era in Russian history. [(a) Dmitri Sidorov/E. I. Kirichenko/V. Mikosha; (b) AP Photo/Mikhail Metzel.]

that with faith comes economic success and that salvation of the soul helps the individual face the rigors of everyday life can be particularly comforting, not only during times of hardship but also as individuals adjust to new prosperity.

| Personal Vignette | Valerii, age 35, once a government research scientist, is now a new capitalist. He makes a comfortable living importing and exporting goods in the informal economy. Although he has to bribe officials and pay protection money to the *reketiry* (racketeers or mobsters), his income places his family of three—himself; his wife, Nina, age 30; and their son, Mikhail, age 10—at a much higher economic level than their longtime friends. Nina is the only woman among them who does not work outside the home. Their new wealth, the precariousness of their position in the informal economy, and the fact that their friends do not share their prosperity cause Nina and Valerii to be uneasy. In search of values that will guide them in these new circumstances, both have recently been baptized in an evangelical Christian sect. They say they chose this particular religious group because it promotes modesty, honesty, and commitment to hard work.

*Source: Adapted from Timo Piirainen, Towards a New Social Order in Russia: Transforming Structures and Everyday Life (Aldershot, UK, and Brookfield, Vermont: Dartmouth Press, 1997), pp. 171–179.* ■

In Russia's internal ethnic republics—such as Tatarstan, Chechnya, Dagestan, and Ingushetia—and in Azerbaijan and the Central Asian republics, many people are Muslims. Among them, observance of Islam is now more open and Muslim identity more politically important. In the Central Asian republics, however, the return to religious practices is often a subject of contention. Here, despite a predominant Muslim heritage, some local leaders, schooled in Soviet theory, view traditional Muslim religious practices as obstacles to social and economic reform. Other devout Muslim Central Asian leaders worry that the extremists among modern Islamic movements in the region may be agents of religious states such as Iran and Saudi Arabia and hence may endanger the security of Central Asian countries. Between 1992 and 1997, Tajikistan fought a civil war against Islamic insurgents with links to the Taliban in Afghanistan. In 2000, Uzbekistan and Kyrgyzstan joined to eliminate an extremist Islamic movement. But many religious leaders and human rights groups say that the fervor to eliminate radical insurgents has resulted in the persecution of ordinary devout Muslims, especially men. Human Rights Watch, which monitors human rights abuses worldwide, reports that at least 4000 Muslim men have been arrested and detained in Uzbekistan alone.

## The Social Effects of Unemployment and Loss of the Safety Net

Widespread unemployment and underemployment have been common throughout the region for more than a decade. Nearly all families are affected by job loss in some way. In fact, the number of people who no longer have paid employment is probably much higher than the official figures, as numerous state firms simply no longer have the money to pay employees who are still listed as

workers. It is estimated that in the mid-1990s, three-fifths of the Russian labor force was not being paid in full and on schedule. By 2006, official unemployment rates for the region ran from 1.9 percent in Belarus to 30 percent and higher in Caucasia and Central Asia. Benefits are nonexistent, and there is little job security because small private firms appear quickly and then fail. The rate of **underemployment,** which measures people who are working too few hours to make a decent living or who are highly trained yet working at menial jobs, runs as high as 25 percent in some parts of the region, so many people earn barely enough to survive. With no equal opportunity laws, job ads often contain wording such as "only attractive skilled people under 30 need apply."

It is especially devastating to lose a job in the former Soviet Union because for many decades the job was an individual's link not only to income, but also to necessities of life and general social services. Work was the center of community and social life. Thus, for many workers, when the job ended, there was no safety net and no social support group. With so many industries now retrenching, in some cases whole cities face not only widespread unemployment, but also the loss of many social benefits for their people (see Box 5.2).

The loss of jobs and status has been especially difficult for many men, and as we saw on page 257, some have turned to drinking and drug use. Death by alcohol poisoning in Russia rose 25 percent in the 1990s and is thought to have killed more Russians in that decade than the number that died during World War II.

## Gender and Opportunity in Free Market Russia

Soviet policy encouraged all women to work for wages outside the home, and most women in Russia and the Newly Independent States were, and are, laborers in factories and in the fields. By the 1970s, 90 percent of able-bodied women in Russia were working full time, giving it the highest rate of female paid employment in the world. Leninist theory had always argued that women should contribute fully to national development, although the traditional attitude that women are the keepers of the home persisted. The result was the "double day" for women. Unlike men, most women put in long days, working in a factory or office or on a farm for 8 hours, and then returned home to cook, care for children, and do the laundry and housework without the aid of household appliances. Because of shortages, they often had to stand in long lines to procure food and clothing for their families as well.

By the 1990s, the female labor force in Russia was, on average, better educated than the male labor force, and this was still true in 2006. The same pattern is emerging in Belarus, Ukraine, and Moldova and in parts of Muslim Central Asia, though in Central Asia educational attainment for both genders is lower. In Russia, the best-educated women commonly hold jobs as economists, accountants, scientists, and technicians, but they are unlikely to hold senior supervisory positions, and as recently as 2005, the wages of women workers averaged 36 percent less than those of men. Although three in four physicians and one in three engineers are women, these occupations do not have high status in Russia, and pay can be lower than for factory workers. On the other hand, executive physicians and supervisory engineers, who earn high salaries, are routinely male.

When market reforms reduced the number of jobs available to all citizens, President Gorbachev publicly stated that the primary role of Soviet women was domestic, so, despite their education and career ambitions, they could best serve their country by returning to their homes and leaving the increasingly scarce jobs to men. By the late 1990s, 70 percent of the registered unemployed were women. But many, if not most, of the women left jobless were—due to ill-

ness, death, or divorce—the sole support of their families (in the Russian Federation, two out of three marriages end in divorce, and the life expectancy for males is low). Moreover, women were often supporting three generations: themselves, their parents, and their children. The search for a new job is complicated by blatant gender bias in hiring and promotion practices. Job advertisements routinely specify gender, asking for an "attractive female receptionist" or a "male account executive." Women may be forced—sometimes literally—into marriage arrangements or sex work (Box 5.3).

---

## BOX 5.3 | AT THE REGIONAL SCALE    The Trade in Women

Among the less savory entrepreneurial activities in the new Russian market economy are those connected with the "marketing" of women. As observed at several other places in this book (especially Chapter 10), there is a growing demand for females as a commodity in global markets. Women from European Russia and Ukraine are deemed especially desirable. A relatively benign part of this market is the phenomenon of Internet-based mail-order bride services, which any viewer can encounter by going to Web sites related to Russia or other countries in the region. A woman in her late teens or early twenties pays about $20 to be included in an agency's catalog of pictures and descriptions (one Internet agency advertises 30,000 women listed). She is then interviewed by the prospective groom, who travels—usually to Russia or Ukraine—to choose from the women he has selected from the catalog.

A more unsavory practice is the kidnapping of unsuspecting females for sex work outside Russia and the Newly Independent States. In 2000, *The Economist* estimated that 300,000 such women were smuggled into the European Union yearly, many being held first at an intermediary stop in Central Europe. The sex business in Europe is thought to generate $9 billion a year, and the supply side seems to be dominated by Russian-speaking gangs. Usually, the women in this business, desperate for a job, sign up to work as domestic servants or waitresses in Europe, only to find upon arrival that they are being held by force and are expected to work as strippers, dancers, and prostitutes. Most often, the women kidnapped into sex work have no well-connected relatives or friends who will report them missing. Once ensnared, the women may be sold for a few thousand dollars to brothels in parts of the world where their European looks will be considered exotic. They are often unprotected from sexually transmitted diseases.

According to a 2005 *New York Times* report, one result of the tightening of EU borders, especially in Central Europe, is that Turkey, with its booming economy and lax borders, has become the "world's largest market for Slavic women." Natural or bottle blondes are preferred, with the most attractive being sent to Istanbul, the Black Sea, or the eastern Mediterranean coast or sold to agents in the United Arab Emirates. The flow of women is primarily from Russia, Ukraine, Belarus, Moldova, and Georgia

**FIGURE 5.28** The trade in women. Seventeen-year-old Lyubov looks out the window of the Neve Tirza women's prison as she awaits deportation from Israel as an illegal worker. Six months ago, Lyubov left her Russian coal mining city only to be sold unknowingly into prostitution for $9000. The importation of women is a big industry in Israel; women are recruited from the former Soviet Union and distributed to brothels throughout the country. [Reuters.]

(Figure 5.28). Turkey is now acting to curtail the trade and to provide hotlines and shelters for victimized women, but help reaches only a minority.

*Sources:* CBSNews.com, July 7, 2003, http://www.cbsnews.com/stories /2003/07/05/politics/main561828.shtml; "Trafficking in women: In the shadows," *The Economist* (August 26, 2000): 38–39; two studies by The Foundation of Women's Forum/Stiftelsen Kvinnoforum, Stockholm, August 1998, and The Angel Coalition Web site, 2004, http://www .angelcoalition.org/history.html; Craig S. Smith, "Turkey's growing sex trade snares many Slavic women," *New York Times*, June 26, 2005, section 4: 1; Islamica Community, http://www.islamicaweb.com/forums /showthread.php?t=36337.

## Corruption and Social Instability

Rampant corruption poses a serious threat to social stability in Russia and the Newly Independent States. After independence in 1991, gangsters took over portions of the economy in every republic, and literally every citizen had to deal daily with the effects of corruption. One example is the Caucasian republic of Georgia. During the presidency of Edvard Shevardnadze (1992–2003), the Georgian government bureaucracy was nearly nonfunctional until a bribe was paid. Additionally, because laws were poorly written and contradictory, the most capable and ambitious people simply ignored restrictions and regulations. Many adopted an attitude of noncompliance with all regulations—even those that clearly served the common good, such as speed limits, building codes, sanitary regulations, and limits on the transport and disposal of nuclear waste. It was a short step from such ingrained systemic corruption to protection rackets, outright robbery, and violence, especially after cheap weapons became readily available on the black market in 1991.

**Personal Vignette** In the mid-1990s, an American Fulbright scholar, John G. Stewart, was posted to teach public administration ethics in Georgia. He found himself learning from his students about the roots of corruption in daily life. They reminded him that, even before rampant inflation, civil servants routinely supplemented salaries too low to live on by accepting bribes, and the police essentially had a franchise from the government to collect their own salaries by shaking down the citizens. Stewart's students wondered how they, as future public administrators, could practice ethical values and survive in the real world. Since finishing the course, Stewart's 28 Georgian students have set up a support group to help one another implement ethical procedures and to find legitimate supplementary employment so that bribes need not become part of their income.

*Source: Adapted from Lydia Pulsipher's interviews with John Stewart, 1999, 2002, 2006.* ■

In 2004, Mikhail Saakashvili, a 35-year-old Georgian with a law degree from Columbia University in New York, was elected president on a platform stating that corruption was robbing Georgian children of a future. Once in office, Saakashvili tackled corruption on many levels. For instance, he fired more than 13,000 police and set up a system for replacing them with trained people who swore to uphold a code of ethics with zero tolerance for brutality. Police pay was increased nearly 20-fold to cover actual costs of living. Saakashvili himself admits that reported calls to the police are up, but he claims that this is not because of a crime wave, but because people are now interacting with the police in a healthy way, reporting crimes and asking for assistance whenever they need it without fear of having to pay a bribe.

## Impatience with the Present, Nostalgia for the Past

It is understandable that Russians would be impatient with the slow pace of political and economic reforms and with the job losses, declining services, and widening disparities of wealth they have

brought. Geographers Grigory Ioffe and Tatyana Nefedova have noted that Russia "is beginning to resemble an archipelago with islands of vibrant economic life [where the most modernized cities are] immersed in a sea of [rural] stagnation and decay." Declining rural populations have led to the loss of services in rural areas (such as road repair, postal services, and landline telephone service) and the abandonment of farmland and villages. The effect is stark rural-urban disparities of wealth and well-being that leave those on the periphery in the worst shape. Not surprisingly, nostalgia for the past, when at least the basics of life could be counted on, is common.

Nostalgia can play a political role, pushing elected lawmakers and officials to abandon reforms before they have had time to be effective. However, doing so may thwart the development of democratic institutions, a market-oriented economy, and the components of social infrastructure, such as housing, schools, and health care.

Ambivalence seems to be the best characterization of the way people in Russia and the Newly Independent States feel about the changes that have occurred since 1991. Although most people experience some nostalgia for the past, many of these same people also admit to liking a more open society with somewhat better access to information. Many are personally doing better, at least in some ways (income, career options), than during Soviet times. Still, the situation is in flux, and the constant threat of street crime, petty corruption, and sporadic anarchy make people nervous about the future. Many Russians have welcomed the return to more authoritarian government control under Vladimir Putin, even when that control restricts freedom of speech and democratic processes in general.

## ENVIRONMENTAL ISSUES

There is nothing particularly Communist about the idea that nature should be subservient to human economic desires. Capitalist economies operate largely under the same premise, although, as a result of public pressure, regulations often safeguard against the worst abuses. Nonetheless, a comment by Soviet leader Joseph Stalin provides some background for the environmental situation in the Soviet Union: "We cannot expect charity from Nature. We must tear it from her." Soviet ideology held that nature was the servant of industrial and agricultural progress, and the grander the evidence of human domination over nature, the better. Hence huge dams, factories, and other industrial facilities were built without regard to their effect on the environment or public health. Now Russia and the Newly Independent States have some of the worst environmental problems in the world, and governments, beset with myriad problems when the Soviet Union collapsed in 1991, have been reluctant to address them.

As one Russian environmentalist recently put it, "When people become more involved with their stomachs, they forget about ecology." Attracting potential investors from Europe, America, and Asia has been the highest priority for Russia and all the Newly Independent States because this money will create jobs. But such development will also increase industrial production and resource exploitation. In fact, investors may be attracted to the region precisely because legal protections for the environment are lax, saving them money in the short term. Although awareness that pollution

**FIGURE 5.29** Pollution in the former Soviet Union. "Everything rots, everything dies" are the sentiments expressed by this Azerbaijani woman, referring to her garden near the Baku oil fields. [Reza Deghati/National Geographic Image Collection.]

ultimately inhibits development is growing, leaders typically have not responded to complaints about pollution or other environmental problems because typically they themselves are being paid off by the polluters. By 2000, more than 35 million people in the former Soviet Union (15 percent of the population) were living in areas where the soil was poisoned (Figure 5.29) and the air dangerous to breathe, where birth defects such as missing limbs or hands were rampant, and where, by some estimates, only one-third of all schoolchildren enjoyed good health.

Pollution controls are further complicated by the fact that instead of one government in Moscow coordinating economic development, there are now 12 independent countries with widely differing policies and not enough money to correct even a few of the past environmental abuses. Figure 5.30 shows changes in human impacts on the environment over time; notice that by 2002, environmental degradation had crossed international borders, making the resolution of pollution issues especially difficult.

## Resource Extraction and Environmental Degradation

Russia and the Newly Independent States have considerable deposits of natural resources (see Figures 5.20 and 5.21). As we have mentioned, Russia alone has the world's largest natural gas reserves, major oil deposits, and forests that stretch across the northern reaches of the continent. Russia also has major deposits of coal and industrial minerals such as iron ore, lead, mercury, copper, nickel, platinum, and gold, and is the third-largest producer of hydropower in the world. The Central Asian republics share substantial deposits of oil and gas, which are centered on the Caspian Sea and extend east toward China.

For a while after the demise of the Soviet Union, general pollution levels fell and environmental degradation slowed, simply because the economy slowed markedly. For example, forest clearing in Siberia decreased by as much as half because both domestic demand and international prices for wood declined. This period of low demand gave some respite to Russian forest and water resources. But now, as the Russian economy rebounds, demand for all resources is growing, and air and water pollution levels are on the rise. In the case of Russia's timber resources, as global wood prices rise again, the economically strapped Russian government is issuing contracts to private foreign concessionaires for rapid and unsustainable clear-cutting, especially in Siberia, where environmental monitoring is difficult because of the size of the area.

In Siberia, some of the world's worst inland oil spills have contaminated lakes, rivers, and wildlife. Recently Lake Baikal, one of the world's largest and least damaged freshwater lakes, was threatened by the construction of a 2500-mile (4000-kilometer) pipeline to carry Russian oil to Asian Pacific markets. In April 2006, after local and international protests, President Putin unexpectedly agreed to divert the pipeline a safe distance around the lake. Still, this one-time reprieve does not mean that environmental policies are now generally more protective of the environment or that public environmental protests will enjoy future success.

## Urban and Industrial Pollution

Urban and industrial pollution was ignored during Soviet times as cities expanded quickly to accommodate new industries and workers flooded into them from the countryside. The dense concentrations of workers alone were enough to generate lethal levels of many pollutants. Even today, urban sewer systems are rare, and of those that exist, few actually process and purify the sewage that enters them. Moreover, because cities were often built with their residential areas located adjacent to industries producing harmful by-products, many people are exposed to high levels of industrial

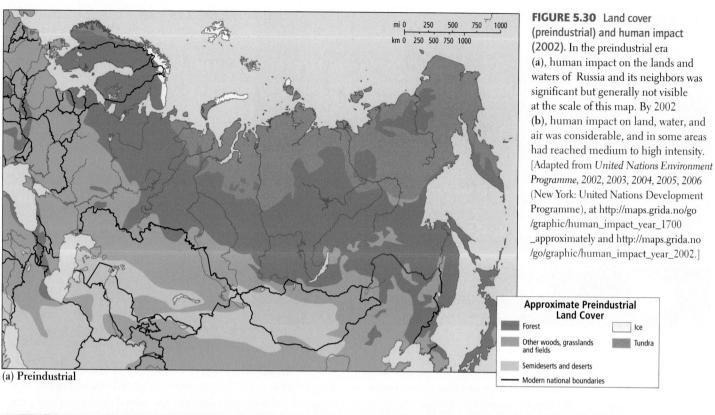

(a) Preindustrial

**Approximate Preindustrial Land Cover**

Forest     Ice

Other woods, grasslands and fields     Tundra

Semideserts and deserts

— Modern national boundaries

**FIGURE 5.30** Land cover (preindustrial) and human impact (2002). In the preindustrial era (**a**), human impact on the lands and waters of Russia and its neighbors was significant but generally not visible at the scale of this map. By 2002 (**b**), human impact on land, water, and air was considerable, and in some areas had reached medium to high intensity. [Adapted from *United Nations Environment Programme, 2002, 2003, 2004, 2005, 2006* (New York: United Nations Development Programme), at http://maps.grida.no/go /graphic/human_impact_year_1700 _approximately and http://maps.grida.no /go/graphic/human_impact_year_2002.]

(b) 2002

**Human Impact, 2002**

**Land Cover**

Forests

Grasslands

Deserts

Tundra

Ice

**Overfishing**

Threatened fisheries

**Human Impact on Land**

High Impact

Medium–High Impact

Low–Medium Impact

**Acid Rain**

- - <4.2 pH

- - - 4.8–4.3 pH

- - - 5.5–4.9 pH

— Modern national boundaries

pollution. One example comes from the city of Magnitogorsk in the Urals, whose economy is based on the largest steel mill in the world. One in three Magnitogorsk citizens has respiratory problems from breathing smoke and airborne chemicals. In all urban areas, air pollution resulting from the burning of fossil fuels is skyrocketing as more and more people purchase cars and as the industrial and transport sectors of the economy heat up.

It can be difficult to link urban pollution directly to specific health problems because the sources of contamination are diffuse. Often called **nonpoint sources of pollution,** they include untreated automobile exhaust, raw sewage, and agricultural chemicals that drain from fields into urban water supplies. Moscow, for example, is located at the center of a large industrial area, where infant mortality and birth defects are particularly high. Researchers are convinced that these effects result from a complex mixture of pollutants that are difficult to trace, but include all the nonpoint sources just mentioned.

Rural pollution can often be just as severe as urban pollution. People in the countryside are subjected both to drifting urban-generated airborne pollutants and to fertilizer and pesticide pollution from agriculture. East of the Urals, industrial cities send tons of chemicals and raw sewage into the rivers that traverse Siberia and provide habitats for fish and wildlife.

## Nuclear Pollution

Nuclear pollution in Russia and the Newly Independent States is the worst in the world, and its effects have spread well beyond the borders of the region. During the cold war, in the thinly populated deserts of Kazakhstan, the Soviet military set off almost 500 nuclear explosions. Local populations, mostly poor Kazakh herders, suffered radiation sickness and birth defects, but were not told the cause until 1989. The most famous nuclear disaster in the world occurred in north-central Ukraine in 1986, when one of four nuclear reactors at the Chernobyl power plant exploded. The explosion severely contaminated a vast area in northern Ukraine, southern Belarus, and Russia and spread a cloud of radiation over much of eastern Europe and Scandinavia. As a direct result of this incident, 5000 people died, 30,000 were disabled, and 100,000 were evacuated from their homes. Yet several Chernobyl-style reactors still operate in Ukraine, Russia, and Lithuania.

Even the pollution released at Chernobyl pales in comparison with the radiation leaking from former Soviet military sites such as Tomsk-7 (a closed city east of the Urals, now renamed Seversk), where the soil alone holds 20 times the amount of radiation released at Chernobyl. These facilities have been linked to radiation pollution recorded thousands of miles away in the Arctic, carried there by rivers and by migrating ducks. The Arctic Ocean and the Sea of Okhotsk in the northwestern Pacific (see Figure 5.1) are also polluted with nuclear waste dumped at sea. Although the Soviet government signed an international antidumping treaty, it sank 14 nuclear reactors and dumped thousands of barrels of radioactive waste in the world's oceans.

Russia and Kazakhstan have sought to earn money by taking in the nuclear waste of other countries eager to be rid of it (France

is a major client). According to experts in the field, no reliably safe system for storing nuclear waste has been found; yet the Russians and Kazakhs claim that they will safely and economically store the imported nuclear waste, using the earnings to clean up their own nuclear waste dumps. Reliable environmental impact studies, however, have not been done.

*Case Study: Irrigation and the Aral Sea.* Once the fourth-largest lake in the world, the Aral Sea is disappearing as a result of large-scale irrigation projects in Central Asia (Figure 5.31). For millions of years, this landlocked inland sea was fed by the Syr Darya and Amu Darya rivers, which brought snowmelt from the lofty Hindu Kush and Tien Shan to the northeast. In 1918, the Soviet leadership decreed that water diverted from the two rivers would irrigate millions of acres of land in Kazakhstan and Uzbekistan, which would be used to grow cotton. In 1956, millions more acres of cotton were planted in Turkmenistan, irrigated with Amu Darya water diverted into an open, 850-mile-long (1368-kilometer-long) desert canal. So much water was lost through evaporation that within 4 years, the Aral Sea had shrunk measurably. Yet planners, enticed by the income the cotton generated, irrigated still more land. By the early 1980s, no water at all was reaching the Aral Sea. By early 2001, the sea had lost 75 percent of its volume and had shrunk into three smaller lakes. The native fish died out due to increasing water salinity, and port cities were marooned far from the water (see inset map in Figure 5.31).

The shrinkage of the Aral Sea may have caused changes in climate and human health. The country around the sea has become drier, and there is some evidence that the summers are 2°F to 3°F (1.1°C to 1.6°C) hotter, while winters have become cooler and longer. The growing season is as much as 3 weeks shorter. Winds sweeping across the landscape pick up the newly exposed salt, chemical residues, and seabed sediment, creating poisonous dust storms. Drinking water is heavily polluted with agricultural chemicals. In rural northern Turkmenistan near the Amu Darya, a majority of the population suffers from chronic illnesses. At the southern end of the Aral Sea, in Uzbekistan, 69 of every 100 people report chronic illness, and in some villages, life expectancy is 38 years (the national average is 68 years).

What is being done to restore the Aral Sea? Efforts to stop irrigated agriculture have not succeeded because it generates so much income. Cotton is king: Uzbekistan is the world's fifth-largest cotton grower, cotton accounts for one-third of its total exports, and cotton production employs 40 percent of its national labor force. Uzbekistan draws from both rivers to irrigate 16 percent of its total cropland. In Turkmenistan and Kazakhstan, oil is now the leading export commodity, but cotton exports remain important. Kazakhstan has been most active in trying to reverse the degradation of the Aral Sea. In 2000, a dam was built to keep water in the northern Aral Sea (see inset map in Figure 5.31); by 2006, the water had risen 10 feet, and the fish catch was improving. Kazakh fishers note that there is now more open public debate about what to do next regarding the Aral Sea, whereas in Soviet times there was no questioning of the wisdom of grand-scale irrigation and no public awareness of its likely environmental effects.

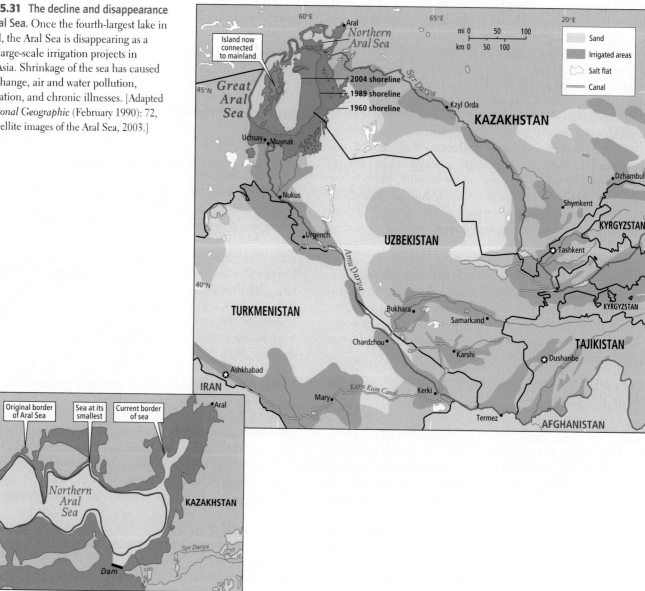

**FIGURE 5.31** The decline and disappearance of the Aral Sea. Once the fourth-largest lake in the world, the Aral Sea is disappearing as a result of large-scale irrigation projects in Central Asia. Shrinkage of the sea has caused climate change, air and water pollution, desertification, and chronic illnesses. [Adapted from *National Geographic* (February 1990): 72, 80–81; satellite images of the Aral Sea, 2003.]

## MEASURES OF HUMAN WELL-BEING

Because of the rapidly changing conditions in Russia and the Newly Independent States, in this chapter's human well-being table (Table 5.2) we are including statistics for the mid-1990s and the early 2000s to illustrate the variation in these measures over that period. As we have seen, average levels of well-being sank across the region after the breakup of the Soviet Union. At the same time, disparities in well-being increased as the transition to a market economy created opportunities for a few and troubles for many. But by 2003, living standards were improving in much of the region; in Russia, the per capita GDP was more than two times higher than in 1995 (see Table 5.2, column II). GDP per capita was up by smaller proportions in the rest of the Newly Independent States, except for Moldova, Uzbekistan, and Kyrgyzstan, where it continued to fall. However, with the exception of Russia,

all countries in the region have GDP per capita figures well below the world average of $7376 (2003).

The GDP per capita figures also show disparities among subregions. The explanation for economic decline or slow growth in Moldova, Caucasia, and Central Asia lies partly in the fact that Russia, Belarus, and Ukraine retained most of the former Soviet Union's industrial capacity and much of its transport and trade infrastructure (banks, company headquarters, government agencies that supply statistics, and customs services and that collect taxes), and these assets assisted their recovery after the Soviet collapse. The gaps between the poorest and richest countries in the region are increasing: Moldova, Uzbekistan, Tajikistan, and Kyrgyzstan all ranked lower on the United Nations Human Development Index (HDI) in 2005 than in 1998 (see Table 5.2, column III).

What these GDP per capita figures do not reveal is that there is also economic disparity within each republic. The former Com-

## TABLE 5.2 Human well-being rankings of Russia and the Newly Independent States and other selected countries[a]

| Country (I) | GDP per capita, adjusted for PPP[b] (GDP ranking among 177 countries) (II) | | Human Development Index (HDI) ranking among 177 countries[c] (III) | | Gender Empowerment Measure (GEM) and Gender Development Index (GDI) rankings, 2005 (IV) | |
|---|---|---|---|---|---|---|
| | 1995 | 2005 | 1998 | 2005 | GEM | GDI |
| **Selected countries for comparison** | | | | | | |
| Barbados | 6,560 | 15,270 (39) | 24 (high) | 30 (high) | 25 | 29 |
| Japan | 21,930 | 27,967 (13) | 8 (high) | 11 (high) | 43 | 14 |
| Kuwait | 17,390[d] | 18,047 (33) | ND | 44 (high) | 75 | 39 |
| United States | 26,966 | 37,562 (4) | 4 (high) | 10 (high) | 12 | 8 |
| World | 5,990 | 7,376 | | | | |
| **Russia and the Eurasian Republics** | | | | | | |
| Russian Federation | 4,531 | 9,230 (59) | 72 (medium) | 62 (medium) | 60 | ND |
| Belarus | 4,398 | 6,052 (82) | 68 (medium) | 67 (high) | ND | 53 |
| Moldova | 1,547 | 1,510 (143) | 113 (medium) | 115 (medium) | 53 | 91 |
| Ukraine | 2,361 | 5,491 (88) | 102 (medium) | 78 (medium) | 66 | 59 |
| **Caucasian Republics** | | | | | | |
| Georgia | 1,389 | 2,588 (117) | 108 (medium) | 100 (medium) | 67 | ND |
| Armenia | 2,208 | 3,671 (107) | 99 (medium) | 83 (medium) | ND | 62 |
| Azerbaijan | 1,463 | 3,617 (109) | 110 (medium) | 101 (medium) | ND | 77 |
| **Central Asian Republics** | | | | | | |
| Kazakhstan | 3,037 | 6,671 (77) | 93 (medium) | 80 (medium) | ND | 61 |
| Turkmenistan | 2,345 | 5,938 (84) | 103 (medium) | 97 (medium) | ND | ND |
| Uzbekistan | 2,376 | 1,744 (138) | 104 (medium) | 111 (medium) | ND | 86 |
| Tajikistan | 943 | 1,106 (152) | 118 (medium) | 122 (medium) | ND | 93 |
| Kyrgyzstan | 1,927 | 1,751 (137) | 109 (medium) | 109 (medium) | ND | ND |

[a]Rankings are in descending order, i.e. low numbers indicate high rank.

[b]PPP = purchasing power parity, figured in 2003 U.S. dollars.

[c]The high and medium designations indicate where the country ranks among the 177 countries classified into three categories (high, medium, low) by the United Nations.

[d]1998 data; 1995 data not available

ND = No data available.

Sources: *United Nations Human Development Report 1998* (New York: United Nations Development Programme), Tables 1–3, at http:/hdr.undp.org/reports /global/1998/en/. *United Nations Human Development Report 2005* (New York: United Nations Development Programme), Tables 1, 25, 26, at http://hdr.undp .org/reports/global/2005.

munist system's goal of equalizing the distribution of wealth is now being discarded as a few become very rich in the new market economy while the majority experience a decline in living standards, at least temporarily. Well-being remained as high as it did largely because citizens of this region helped one another to find shelter, food, and personal assistance. Another pattern is masked by the decade spread of GDP per capita and HDI statistics: progress has slowed across the region. Between 2003 and 2005, the HDI rank of every country but Russia and Armenia fell—those of Belarus, Georgia, Azerbaijan, Turkmenistan, and Uzbekistan fell by 10 points or more. The United Nations Gender Empowerment Measure (GEM) is available only for the Russian Federation, Moldova, Ukraine, and

t Index (GDI) figures are some-
.2, column IV). They are not
because they ignore actual party by females and measure only

the extent to which males and females have access to the same basic resources, such as food, health care, and income. As is the case in all parts of the world, in no part of this region are women approaching equality with men with respect to income.

# III SUBREGIONS OF RUSSIA THE NEWLY INDEPENDENT STATES

In this section, we discuss Russia first because it is the largest country in the region and because it has dominated the entire region for many hundreds of years in what most scholars now recognize as a quasi-colonial manner. The Newly Independent States of Belarus, Moldova, and Ukraine are discussed next because of their location and their close social, cultural, and economic associations with European Russia. They are followed by the Caucasian republics of Georgia, Armenia, and Azerbaijan. Finally we discuss the countries of Central Asia: Kazakhstan, Kyrgyzstan, Tajikistan, Turkmenistan, and Uzbekistan.

## RUSSIA

In the post-Soviet era, Russian influence remains strong, even as the Newly Independent States have begun to construct their own economies and political identities. Russia is still the largest country on earth in terms of area—nearly twice the size of Canada, the United States, or China. It is also a leader in population; with 143 million people, it ranks eighth in the world. It also ranks high in natural wealth. For convenience, we have divided Russia into three parts: European Russia; Siberian Russia, which comprises about half the landmass east of the Urals; and the Russian Far East.

### European Russia

European Russia is that area of Russia that shares the eastern part of the North European Plain with Latvia, Lithuania, Estonia, Belarus, Ukraine, and Moldova (Figure 5.32). It is usually considered the heart of Russia because it is here that early Slavic peoples established what became the Russian empire, with its center in Moscow. Although it occupies only about one-fifth of the total territory of present-day Russia, European Russia has most of the industry, the best agricultural land, and about 70 percent of the population of the Russian Federation (100 million). A tiny exclave of European Russia is Kaliningrad, located on the Baltic Sea between Lithuania and Poland. The Kaliningrad port is important to Russia because it is free of ice for more of the year than St. Petersburg. Kaliningrad is an important transshipment locale for trade, and Russia's Baltic fleet is stationed there.

The vast majority of European Russians live in cities, their parents and grandparents having left rural areas when Stalin collectivized agriculture and established heavy industry. Most of these people went to work in one of four major industrial regions chosen by central planners for accessibility and location near crucial mineral resources. One industrial region is centered on Moscow, another in the Ural Mountains and foothills (part of this zone lies in Siberian Russia), a third along the Volga River from Kazan to Volgograd, and the fourth just north of the Black Sea extending into Ukraine around Donetsk (see Figure 5.20 on page 264).

The most densely occupied part of European Russia—stretching south from St. Petersburg on the Baltic to Caucasia and east to the Urals (see Figure 5.13 on page 256)—coincides with that part of the continental interior that has the most favorable climate (see Figure 5.6 on page 250). Even so, because the region lies so far to the north (Moscow is at a latitude 100 miles [160 kilometers] north of Edmonton, Canada) and in a continental interior, the winters are rather long and harsh and the summers short and mild.

***Urban Life and Patterns in European Russia.*** Moscow remains at the heart of Russian life. The city has grown in the post-Soviet era; about 10.4 million people live within the city limits and another 3 million live in the metropolitan area, making Moscow the largest city in greater Europe. Always a center of Russian culture, Moscow has so changed since 1991 that a common saying is, "Moscow and Russia are two different countries." The availability of goods—food, electronics, furniture, appliances, clothing, automobiles, and especially luxury products—and entertainment of all types has exploded. Prices are higher here than in most Russian cities and rising at the rate of about 15 percent per year. The criminals are more concentrated, more innovative, and more violent in Moscow. Entrepreneurs are especially active. Moscow's booming market economy is privatizing rapidly: between 1992 and 1994, jobs in the private sector went from a quarter to a half of the city's jobs. After 1996, however, the growth of the private sector slowed, and by 2002, in an effort to quell social unrest among the jobless and underemployed, the government began to expand public employment, which had been shrinking as government offices closed. In Moscow, the retail sector and small-scale services to the public—such as street food vending and sales of inexpensive clothes, home furnishings, and small appliances—created the most private sector jobs, many in the informal economy. While the city is full of apparently affluent, educated young people, many are merely "dressed for success" and are living on credit. For those adjusting to the loss of "cradle-to-grave" security, life in Moscow can be painful, frightening, exhilarating, or all three, as the following account of the experiences of Natasha and her customers illustrates.

**FIGURE 5.32** The European Russia subregion.

**FIGURE 5.33** A stylish woman in Moscow, 2004.
[AP Photo/Sergey Ponomarev.]

**Personal Vignette** Natasha is an engineer in Moscow. She has managed to keep her job and the benefits it carries, but inflation has so diminished her buying power that in order to feed her family, she sells used household items and secondhand clothes in a street bazaar on the weekends. "Everyone is learning the ropes of this capitalism business," she laughs. "But it can get to be a heavy load. I've never worked so hard before!" Asked about her customers, Natasha says, "Many are former officials and high-level bureaucrats who just can't afford the basics for their families any longer." Pensioners are especially deprived; living on their $130 monthly pensions is impossible, so they resort to begging on Moscow's elegant shopping streets close to parked Bentleys and Rolls Royces. These newly poor elderly, often highly educated, buy warm used sweaters from Natasha and eat in nearby soup kitchens.

"But, you know," Natasha continues, brightening and changing the subject, "the bales of used clothes I buy now come from the United States. A guy drives a carload from a container ship in Amsterdam harbor every 2 weeks. And there are lots of sturdy clothes, especially for children, some quite new with the price tags still on; but [her voice registering disappointment in American styles] only occasionally is there a really fashionable item for a woman." (Moscow women of all economic strata are noted for making creative, even extreme, fashion statements, as Figure 5.33 demonstrates.)

*Source: This composite story is based on work by Alessandra Stanley, David Remnick, and David Lempert and Gregory Feifer.* ■

The privatization of real estate in Moscow gives an interesting insight into pre- and post-Soviet circumstances. During the Communist era, all housing was state owned, and there was a general housing shortage; thousands of families were always waiting for housing, some for more than 10 years. Most apartments were built after the mid-1950s in large, shoddily constructed gray concrete

**FIGURE 5.34**
St. Petersburg. The Hermitage Museum is the building on the right. [Iconotec/Alamy.]

blocks; the units included one to three rooms and housed one to four (or more) people. In 1991, the government allowed residents of Moscow to acquire, free, about 250 square feet (23 square meters) of usable living space per person—about the size of a typical American living room. Most families simply took ownership of the apartments in which they had been living; by 1995, a majority of Muscovite families owned their apartments. A lively real estate market quickly emerged, with a rising wealthy elite willing to pay as much as U.S.$100,000 to $250,000 for large remodeled apartments in the center city. Rents also skyrocketed. Those who could find alternative housing for themselves made tidy sums simply by renting out or selling their center-city apartments. The demand for commercial space for private businesses reduced the number of dwelling units, thus exacerbating the housing shortage. As the supply of space tightened, criminal efforts to acquire property through intimidation increased (as described at the opening of this chapter).

St. Petersburg, renamed Leningrad during most of the Communist period (1924–1991), is European Russia's second-largest city,

with about 5 million people. It is a shipbuilding and industrial city located at the eastern end of the Gulf of Finland, on the Baltic Sea. Czar Peter the Great ordered the city built in 1703 as part of his effort to Europeanize Russia; from 1713 to 1918, it served as the Russian capital and has been one of Russia's cultural centers. The city has a rich architectural heritage, including many palaces and public parks. Since 1991, it has been undergoing a renaissance: the transport system is being refurbished, urban malls are proliferating, and historic sites are being renovated. The czars' Winter Palace, now called the Hermitage Museum, houses one of the world's most important art collections (Figure 5.34). Urban planners working in St. Petersburg maintain an elaborate Web site with numerous pictures that can be reached at http://www.spb.ru/eng/.

## Siberian Russia

*Siberia* encompasses the West Siberian Plain beyond the Urals and about half the Central Siberian Plateau (Figure 5.35). It is so cold

**FIGURE 5.35** The Siberian Russia subregion.

**FIGURE 5.36** New and old in Siberian Russia. Elderly women wearing babushkas, representatives of the survival-oriented Communist days, mix with trendy young folks in Novosibirsk. [Richard Young, M.D.]

that 60 percent of the land is permafrost. This vast central portion of Russia is home to just 33 million people, many of whom moved into the region from European Russia. With the exception of Norilsk (see Box 5.2 on page 261), settlement is concentrated in cities in the somewhat warmer southern quarter. For millennia, Siberia was a land of wetlands and quiet, majestic forests. But during the Soviet era, it became dotted with bleak urban landscapes and industrial squalor—the legacy of the Soviet effort to establish industries to exploit Siberia's valuable mineral, fish, game, and timber resources. This policy is exemplified by Novosibirsk, Siberia's twentieth-century capital and financial center.

Novosibirsk, the largest Russian city east of the Urals (1.4 million people), is located more than 1600 miles (2500 kilometers) east of Moscow. This isolated commercial center has the highest concentration of industry between the Urals and the Pacific and has many young people, an opera and ballet, and an academic research center (Figure 5.36). During the cold war, it was a center for strategic industries such as optics and weaponry and contained more than 200 heavy industrial plants. In the post-Soviet era, these factories became infamous for obsolescence and high rates of pollution, but they also represent a potential for future profit if privatized and reorganized. Novosibirsk, with its banks and financial services, is drawing outside investors interested in acquiring factories in the vicinity. An example of a still mostly state-owned (though actively seeking private investors) industry that seems to be making a successful transition to a market economy is Novosibirsk Instruments (N.I.). N.I. once produced artillery for the Soviet army, then moved

into telescopes for civilian consumers in the 1980s, and most recently has focused on providing advanced amateur astronomers around the world with good-quality telescopes. But N.I. is still a center of high-precision weapons manufacture, especially night-vision small arms, raising the question of who around the world is buying the wide variety of weapons offered by N.I. on the Internet.

## The Russian Far East

The Russian Far East is an extensive territory of mountain plateaus with long coastlines on the Pacific and Arctic oceans (Figure 5.37). Here, almost 90 percent of the land is permafrost; the coastal volcanic mountains stop the warmer Pacific air from moderating the Arctic cold that spreads deep into the continent. This subregion makes up slightly over one-third of the entire country of Russia and is nearly two-thirds the size of the United States. If the population of 8.16 million were evenly distributed across the land, there would be just 2.3 persons per square mile (1 per square kilometer), but 75 percent of the people live in just a few cities along the Trans-Siberian Railroad and on the Pacific coast. The residents of this subregion are primarily immigrants and exiles, some sentenced to the region for perceived misdeeds in the Soviet era (some of the prison camps made famous in Aleksandr Solzhenitsyn's novel *The Gulag Archipelago* were along the Kolyma River). These Far Easterners worked in timber and mineral extraction enterprises and in isolated industrial enclaves. Since the 1980s, many immigrants have headed for cities along the

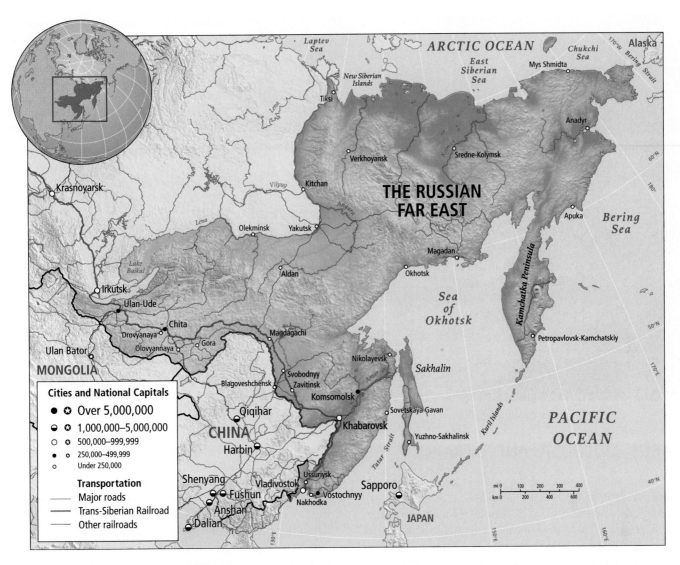

**FIGURE 5.37** The Russian Far East subregion.

Pacific coast, especially Nakhodka and Vladivostok, near North Korea. Since serfdom never intruded in the Far East, there is a spirit of independence that is rare elsewhere in Russia.

The interior of the Russian Far East has many resources, but it has not been developed because of its distance from European Russia and its difficult physical environments. Only 1 percent of the territory is suitable for agriculture, mostly along the Pacific coast and in the Amur River drainage basin. Relatively unexploited stands of timber cover 45 percent of the area. The wilderness supports many endangered species, such as the Siberian tiger, Far Eastern leopard, and Kamchatka snow sheep. Soviet central planners, and now private investors, have been attracted by the region's timber, coal, natural gas, oil, and minerals. Based on recent oil discoveries near Sakhalin Island, estimates are that 50 percent of Russia's oil reserves are in the Russian Far East; these reserves are strategically important, given the rising demand for oil in Asia.

The Russian Far East has great potential for linkages with other world regions. In recent decades, port facilities on the Pacific coast have made the region more accessible to such countries as Japan, Korea, and China (Figure 5.38). China and Japan, both of which need energy resources, are vying for the right to build an oil pipeline from Angarsk in Siberia to either Daqing in China or Nakhodka on the Russian Pacific coast, the closest port to Japan (see Figure 5.21). China's proposed line would be ready in just a few years, but Japan's would give Russia access to much wider petroleum markets in the western Pacific. Russia is also considering a rail link across China to North Korea and Vladivostok. Some geographers and economists forecast that, because of its rich resource reserves and its location so far from the Russian core, the Russian Far East is likely to integrate economically with other Pacific Rim nations, and perhaps eventually detach itself politically from European Russia.

**FIGURE 5.38** The Vostochny International Container Services (VICS) at Vostochny Port, in the Russian Far East. The port connects with the Trans-Siberian Railroad, offering the shortest route from East Asia to central China, Central Asia, Russia, and Europe at competitive rates and transit times. A shipment from Shanghai to Finland, for example, takes 23 days, roughly half as long as shipment totally by sea. [Courtesy of Vostochny International Container Services.]

## BELARUS, MOLDOVA, AND UKRAINE

Sandwiched between Russia and the eastern European countries that were once part of the Soviet Union's sphere of influence are the Newly Independent States of Belarus, Moldova, and Ukraine (Figure 5.39). Each of these countries is the home of a distinct Slavic people, but Russian residents form significant and influential minorities in all three. Given their location and history, these countries continue to maintain close ties with Russia, though Moldova and Ukraine would like closer association with Europe.

### Belarus

In size and terrain, Belarus resembles Minnesota: its flat, glaciated landscape is strewn with forests and dotted with thousands of small lakes, streams, and marshes that are replenished by abundant rainfall. Much of the land has been cleared and drained for collective farm agriculture. The stony soils are not particularly rich; nor, other than a little oil, are there many known useful resources or minerals beneath their surface. Belarus absorbed a large amount of radiation contamination after the Chernobyl nuclear accident in Ukraine in 1986; 20 percent of its agricultural land and 15 percent of its forestland were rendered unusable for the foreseeable future.

During the twentieth century, Belarus was rather thoroughly Russified by a relatively small but influential group of Russian workers and bureaucrats. Although 80 percent of the population of 9.8 million is ethnic Belarusian (a Slavic group) and only 13 percent is Russian, the Russian language predominates, and Belarusian culture survives primarily in museums and historical festivals. The drab urban concrete landscape looks and feels like Soviet Russia. The Belarusian economy remains dominated by state firms that sell to Russia; only a few retail shops are privatized. The country's important petrochemical industry uses primarily Russian petroleum for its raw material and energy. Belarus also earns some income by transshipping oil and gas from Russia to Europe.

Forced to accept independence by the collapse of the Soviet Union in 1991, Belarus remains tightly controlled by its Soviet-style leader, Alexander Lukashenko. He actively seeks to reintegrate Belarus with Russia, a desire welcomed by the remaining Communists in Russia and by those who wish to see the former Soviet empire reconstituted. They see Belarus as a useful buffer state against the growing influence of the EU and NATO. However, even Russian President Putin finds Belarus too slow to adopt a market economy, and within Belarus, opposition to the government is growing. A music and art underground is blossoming. "Belowood," a free theater-of-protest, and "Drum Ecstasy," a group of young musicians who use strong rhythmic beats but no words, have gained wide acclaim as effective underground voices against government policy, though they have also inspired violent reprisals by the police. In March 2006, large crowds of demonstrators in Minsk charged that the March 14 elections that retained Lukashenko in power were fraudulent. Many, including elderly citizens, were beaten by police (Figure 5.40).

**FIGURE 5.39** The Belarus, Moldova, and Ukraine subregion.

**Cities and National Capitals**

| | | |
|---|---|---|
| ● ✪ | Over 5,000,000 | |
| ◒ ✪ | 1,000,000–5,000,000 | |
| ○ ✪ | 500,000–999,999 | |
| ● ✪ | 250,000–499,999 | |
| ○ | Under 250,000 | |

**Transportation**

—— Major roads

—— Railroads

**FIGURE 5.40** Government repression in Belarus. Riot police beat protesters in the Belarusian capital, Minsk, in March 2006. Police blocked the path of about 1000 protesters heading to a jail where demonstrators arrested in previous protests were being held. When the crowd surged forward, the police attacked without regard to the age of the protesters. [AP Photo/Ivan Sekretarev.]

## Moldova and Ukraine

Moldova and Ukraine have maintained their Slavic cultures, including dance, cuisine, religious customs, and architectural forms. Their distinctive ethnic artistry (ornately painted Easter eggs, elaborate holiday breads and pastries, intricate embroidery and lace) is prominently displayed in homes and marketed at home and abroad. These two closely related countries seem to be tending toward closer association with the EU. They enjoy a warmer climate than Belarus and Russia, and, with good agricultural resources, they have the potential to increase their agricultural production significantly. Ukraine (47 million people), where agriculture accounts for one-fifth of the GDP and one-fourth of employment, is comparable to France in size, population, and agricultural output. Moldova, lying to the west and bordering Romania, is much smaller (4.2 million people) than Ukraine but produces many of the same products. It is now pursuing the possibility of emphasizing wine, nuts, and fruits to be marketed not only to Russia but also to the Baltic states and to Europe via Romania.

Much of Ukrainian agriculture remains collectivized, partly to ensure adequate food supplies. But increasingly families are growing half of their own food in small family gardens that have proliferated since independence, especially in urban areas. Many analysts believe that Ukraine—which supplied the Soviet Union with much of its food, especially after World War II, and which continues to trade heavily with Russia—could in the future enjoy a niche in the agricultural economy of Europe, giving France some serious competition in the production of vegetables, fruits, and wines. To do so, however, Ukraine would have to make many adjustments to meet EU standards and refurbish its collective farms, which for the most part are much less efficient than farms in western Europe. For the time being, Ukraine will probably continue to trade its agricultural products mostly with Russia, but as farms privatize and modernize, its participation in the global agricultural market is likely to increase.

The European Union has a strong interest in Ukraine and Moldova that is related less to their potential for future EU membership (which rises and falls with changes in EU relations with Russia) than to a number of other matters. First, undocumented immigrants on their way to western Europe from Sri Lanka, Bangladesh, Afghanistan, Chechnya, and even West Africa regularly travel through Ukraine and Moldova. Increasingly, the EU is providing financial aid to tighten border security in these two countries. Second, the Black Sea, which is ringed by countries (including Turkey to the south) that are in economic and social crisis, is increasingly a focus of concern in Europe. New EU members, such as Slovenia, Estonia, and Latvia, are being enlisted to offer counsel to Ukraine and Moldova (and the Caucasian republics) on how to reform their societies to meet EU standards. The assumption is that the desire to be favorably regarded in the EU will overcome long-standing undesirable habits and patterns. Finally, the EU was alerted to the inconvenience of its own reliance on Russian gas when, in the winter of 2006, Russia temporarily stopped the flow of gas to Ukraine in the midst of elections that brought a pro-Western government to power. Moldova and Ukraine are themselves potential routes for pipelines from Russia to EU countries.

# CAUCASIA: GEORGIA, ARMENIA, AND AZERBAIJAN

Caucasia is located around the rugged spine of the Caucasus Mountains, which stretch from the Black Sea to the Caspian Sea (the language map in Figure 5.41 serves as the map for this subregion). It includes a piece of the Russian Federation on the northern flank of the mountains as well as the three independent republics of Georgia, Armenia, and Azerbaijan, which occupy the southern flank of the mountains in what is known as Transcaucasia. Transcaucasia is a band of subtropical intermountain valleys and high volcanic plateaus that drop to low coastal plains near the Black and Caspian seas (see Figure 5.1). Whereas much of the land in the rest of Russia and the Newly Independent States is arid or cold, these treasured pieces of rugged mountain slope and narrow plain are blessed with warm temperatures and abundant

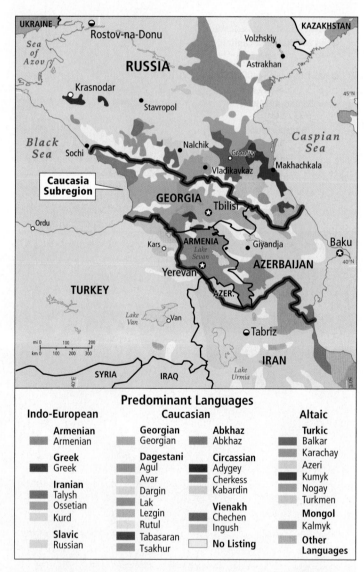

**Predominant Languages**

| Indo-European | Caucasian | | Altaic |
|---|---|---|---|
| **Armenian** | **Georgian** | **Abkhaz** | **Turkic** |
| Armenian | Georgian | Abkhaz | Balkar |
| **Greek** | **Dagestani** | **Circassian** | Karachay |
| Greek | Agul | Adygey | Azeri |
| **Iranian** | Avar | Cherkess | Kumyk |
| Talysh | Dargin | Kabardin | Nogay |
| Ossetian | Lak | | Turkmen |
| Kurd | Lezgin | **Vienakh** | **Mongol** |
| | Rutul | Chechen | Kalmyk |
| **Slavic** | Tabasaran | Ingush | **Other** |
| Russian | Tsakhur | No Listing | **Languages** |

**FIGURE 5.41** The Caucasia subregion. This map illustrates the diversity of ethnolinguistic groups in Caucasia and its adjacent culturally related neighbors. [Adapted from http://www.geocities.com/southbeach/marina/6150/ethno.jpg.]

**FIGURE 5.42** Aziza Mustafa Zadeh, an Azerbaijani singer and jazz pianist. Popular in Europe and the eastern Mediterranean, Zadeh's version of jazz is inspired by *mugam*, an ancient, mesmerizing modal system of traditional music from Caucasia. [Ms. Aziza Mustafa Zadeh, Jazziza Records/ www.azizamustafazadeh.de.]

moisture from the Black and Caspian seas. In these favorable conditions, farmers can grow crops, such as citrus fruits and even bananas, that can be produced in no other part of the vast region. Before 1991, most of the Soviet Union's citrus and tea came from Georgia, as did much of its grapes and wine. In May 2006, Russia unexpectedly banned the importation of Georgian wine, ostensibly because it was polluted with heavy metals, but trade experts suspect this was punishment for Georgia's resistance to Russian domination in seeking to market its wine and other products in the EU.

In this mountainous space the size of California live more than 50 ethnic groups—Armenians, Chechens, Ossetians, Karachays, Abkhazians, Georgians, and Tatars, to name but a few—most speaking separate languages (see Figure 5.41). The groups vary widely in size, from a few hundred (for example, the Ginukh) to over 6 million (the Turkic Azerbaijanis). Some are Orthodox Christians, many are Muslims, some are Jews, Roma pass through, and some retain ancient elements of local animistic religions. All these groups, including Chechens and others who live in Russian Caucasia (see Figure 5.26 on page 269), are remnants of ancient migrations. For thousands of years, Caucasia was a stopping point for nomadic peoples moving between the Central Asian steppes, the Mediterranean, and Europe. Other ethnic enclaves were created more recently by Soviet-instigated relocation and then return of minorities (the Tatars, for example). Today, many Caucasians maintain ties to Europe and North America, where hundreds of thousands of emigrants from the region live. Armenians may be the best known of these groups. Like native peoples in North America (the Nunavut) and Southeast Asia (the Acehese), some of the ethnic groups in Caucasia are using the Internet as a way to gain publicity for their unique geographic location and complex historical heritage. A good exam-

ple is the Web site maintained by the Republic of Abkhazia (http://www.abkhazia.org/) (Abkhazia is now part of Georgia).

In Caucasia, music plays a prominent role in daily life and in national identity. Azerbaijan serves as an example. One music center is Nagorno-Karabakh, where particular families maintain their own music, often blending the traditions of several culture groups. Elsewhere in Azerbaijan, families specialize in Western music and are so skilled that they win competitions in Europe and America. Aziza Mustafa Zadeh (Figure 5.42) is a classically trained jazz pianist, vocalist, and composer who uses Western and Central Asian musical themes in her compositions. She too comes from a long line of Azerbaijan musicians. To learn more about the musical families of Azerbaijan, look at this Web site: http://hajibeyov.com/bio/bio _life/aliyeva_farah/aliyeva_farah_eng/54_families.html.

As a result of external political maneuvering, members of a single ethnic group may now live in several different Caucasian republics, and because of their strong ethnic loyalties, this pattern can result in persistent conflict (Figure 5.43). After the Soviet collapse, the three culturally distinct republics of Georgia, Azerbaijan, and Armenia each obtained independence as nation-states (see page 274 on Georgia's rocky path to democracy). Then, very quickly, several ethnic groups within these already tiny states took up arms to obtain their own independent territories. The Abkhazians and the South Ossetians took up arms against Georgia. The Christian Armenians in Nagorno Karabakh (an exclave of Armenia located inside Azerbaijan) fought against the Muslim majority; 15,000 people died before a truce was signed in 1994. Although the ethnic strife across Caucasia has some local causes, it is often instigated by larger powers—Turkey, Russia, Iran—seeking their own national goals: access to strategic military installations or to agricultural and mineral resources.

**FIGURE 5.43** Areas of contention in Caucasia. [Adapted from European Center for Minority Issues, "Ethnopolitical map of the Caucasus," at http://info@ecmi.de/emap/m_caucasus.html.]

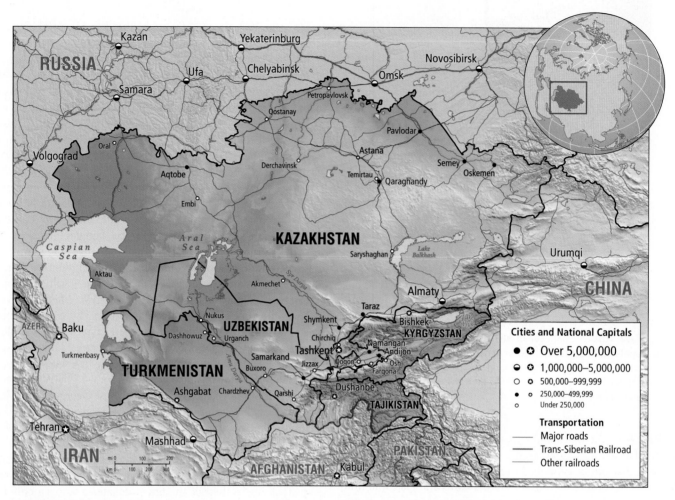

**FIGURE 5.44** The Central Asia subregion.

Access to oil and gas reserves has recently been another source of conflict in Caucasia. Estimates of Caspian Sea oil reserves vary widely, from 28 billion to 200 billion barrels, but a major issue is how to get the oil and gas out of Caucasia safely and into the world market. Pipelines, trucks, and ocean tankers would all have to pass through contested territory or across difficult terrain. A consortium of Western oil companies has considered several pipeline routes to the Black Sea. Transit through Chechnya is problematic because of its continuing guerrilla war with Russia. Pipelines through Georgia and Caucasian Russia exist now, and new ones are under construction. In May 2005, the Baku-Tbilisi-Ceyhan pipeline opened, carrying oil from Azerbaijan through Georgia to Turkey's Mediterranean port of Ceyhan. In June 2006, Kazakhstan joined the pipeline and has committed to shipping 7.5 million tons of oil annually. Turkey can use some of the energy supplied, but most will be sold to Europe.

## THE CENTRAL ASIAN REPUBLICS

Since the fall of 1991, following nearly 12 decades of Russian colonialism, five independent nations have emerged in Central Asia: Kazakhstan, Kyrgyzstan, Tajikistan, Turkmenistan, and Uzbekistan

(Figure 5.44). Consistent with the mosaic of cultures that is found here, each of these new nations has traditions that are recognizably different from those of the others. Yet all of them draw on the deep, common traditions of this ancient region. After decades of efforts to align their cultures and economies with those of the Soviet Union, the countries of Central Asia are poised for change brought on by political independence, by the rise of ethnic and religious (Islamic) sensitivities, and especially by the chance to profit from the exploitation of oil and gas reserves. The revival of Muslim traditions, the emergence of oil as a major resource, and the need for new economic partners have led these new Central Asian countries to consider renewed association with the countries of Southwest Asia and with Afghanistan, Pakistan, and western China. It is not unlikely that in the future, regional analysis will be applied to a set of countries stretching from the Mediterranean across Central Asia to western China.

## The Physical Setting

Central Asia is situated in the center of the Eurasian continent, in the rain shadow of the lofty mountains that lie to the south in Iran, Afghanistan, and Pakistan (see Figures 5.1 and 5.6). Its dry continental climate is a reflection of this location. What rain there is falls mainly in the north, on the Kazakh steppes, where wide grasslands support limited agriculture and large herds of sheep, goats, and horses. In the south are deserts crossed by rivers carrying glacial meltwater from the high peaks still farther to the south. These rivers are tapped to irrigate huge fields of cotton and smaller fields of wheat and other grains (see the discussion of irrigation and the Aral Sea on page 277 and Figure 5.31 on page 278). In many areas of the south, gardens are attached to houses and are walled to protect against drying winds. These gardens nurture tree crops (plums, pistachios, apricots, and apples, as well as onions and garlic, were first domesticated in Central Asia) and many types of melons.

Uzbekistan and Turkmenistan are largely low-lying plains. Kazakhstan has plains in the north and uplands in the south that grade into high mountains. Kyrgyzstan and Tajikistan lie high in the Hindu Kush, the Pamirs, and Tien Shan to the southeast of Kazakhstan. These two countries have exceedingly rugged landscapes: Tajikistan's elevations go from near sea level to more than 22,000 feet (6705 meters), and those in Kyrgyzstan are similar. Both offer little in economic potential; Tajikistan is the poorest country in the entire region and has undergone numerous changes in government and a civil war since 1991.

## Central Asia in World History

Civilization flourished in Central Asia long before it did in lands to the north. The ancient Silk Road, a continuously shifting ribbon of trade routes connecting China with the fringes of Europe, operated for thousands of years, diffusing ideas and technology from place to place (see Figure 5.8 on page 252). Modern versions of ancient trading cities still dot the land: two examples are Bukhara, renowned as a center of Islamic learning and culture, and Samarkand, which has been in existence for 5000 years. Commerce along the Silk Road diminished by the fifteenth century as traders shifted to shipping goods by sea. Central Asia then entered a long period of stagnation until, in the mid-nineteenth century, czarist Russia developed an interest in the region's major export crop, cotton.

The Russian imperial agents built modern mechanized textile mills, which quickly replaced small-scale textile firms that had employed people to do traditional cotton, wool, and silk weaving (Figure 5.45). As part of the Russification process, the new mills employed imported Russians and were often located in entirely modern Russian towns built alongside railroad lines. Consequently, these industries benefited few Central Asians. In rural areas, cotton fields often replaced the wheat fields that had fed local people. Occasional famines struck whenever people could not afford imported food or when supplies were interrupted.

## The Soviet and Post-Soviet Eras

Although Central Asia had been under Russian control since the mid-1800s, Russian domination intensified during the Soviet era. Only a few Central Asians actually joined the Communist Party; those who did were thoroughly Russified. The practice of Islam, though officially tolerated, was undermined by the Communist

**FIGURE 5.45** An Ersari rug from Turkmenistan. One of the most remarkable cultural traditions of Central Asia is carpet weaving, an art practiced primarily by women and children. The carpets probably originated as the highly portable and useful furnishings of nomadic yurts and *gers* (tents). Each region and culture group has distinctive patterns with particular combinations of colors. The practice of weaving rugs stretches from interior China to Caucasia, Turkey, and the Balkans. [Courtesy of Charles W. Jacobsen, Inc., Syracuse, N.Y.]

government's promotion of atheism and by restrictions on Islamic worship, pilgrimages to Makkah, and access to mosques.

An important episode in the story of Russian domination of Central Asia was the Soviets' 1979 decision to invade neighboring Afghanistan to prop up a pro-Soviet puppet government there that was encountering increasing popular resistance. The Soviet invasion inspired many disparate Afghan interests to join in opposition to it, and it served as a cause around which anti-Russian sentiment in Central Asia could solidify. The Soviets were eventually defeated in 1989 by a shaky coalition of Afghans and Afghan sympathizers, primarily the United States and Pakistan. (NATO reported 15,000 Soviet soldiers lost and over 1 million Afghans dead as a result of the 10-year war.) With Russia weakened by this war, the Central Asian republics quickly gained independence in the 1990s (Figure 5.46).

The transition since independence in 1991 has been rocky. Some conditions of daily life have deteriorated significantly, as reflected in lower HDI rankings (see Table 5.2 on page 279). Russia has sought to maintain a strong influence on Central Asia, but it is

**FIGURE 5.46 Kazakh nomads.** Once Kazakhstan gained independence in 1991, ethnic Kazakhs returned home from self-imposed exile in Mongolia. Kazakhs and Mongolians share a cultural history of nomadic herding. Their portable yurt homes are visible in the background. [Gerd Ludwig/National Geographic Image Collection.]

being challenged by growing international interest in the region's oil and gas resources. Transportation and services have deteriorated even from what they were in Soviet times. Poverty is now more widespread, while a few have prospered. Health standards are the lowest of any in the former Soviet Union; infant mortality rates are particularly high. Although elections are held and free markets are opening up opportunities for entrepreneurship, government bureaucracies remain authoritarian and patriarchal, and heads of state are either autocratic or dictatorial. Elections are routinely hijacked by fraudulent vote counts. Most Central Asian govern-

**FIGURE 5.47 Industrial pollution and waste hot spots in the Ferghana Valley.** In the remote Ferghana Valley, which touches both Kyrgyzstan and Tajikistan, mining and heavy industry have contaminated the soil with toxic heavy metals. Because these industries are essential to the local economies, however, government officials have taken a callous attitude toward controlling pollution. [Adapted from Viktor Novikov and Philippe Rekacewicz, UNEP/GRID-Arendal, April 2005, at http://maps.grida.no/go/graphic/industrial_pollution_and_waste_hotspots_in_the_ferghana_valley.]

ments fear Islamic fundamentalism, so even moderate Muslims endure repression. The revival of Islam, although increasingly influential, is thus far subdued, and believers are cautious.

A particularly troubling legacy of change in the modern era survives in the complex mountainous borderlands of Tajikistan and Kyrgyzstan, where Central Asia touches China, Pakistan, and Afghanistan. There, Soviet efforts to develop commercial agriculture and industry and to supply a labor force for this development resulted in high levels of pollution, including uranium-related contamination (Figure 5.47).

The original cultural mosaic of eastern Central Asia is ancient and can be accounted for partially by the complexity of the mountainous terrain, which kept groups cut off from one another. But modernization created new patterns of settlement. The forced movement of large portions of particular ethnic groups by Russian-directed industries broke up families and clans and changed the groups' economic arrangements and trading patterns. There are now, for example, new pockets of Tajiks in Uzbekistan and Afghanistan, Afghans in Tajikistan, and Kazakhs in Turkmenistan. The Soviets displaced people, purposely destroyed traditional economies, and did not prepare Central Asians to govern themselves. When Soviet authoritarian control abruptly diminished in the 1990s, armed conflict erupted. Instability along the southern borders of the Central Asian republics was fueled by the U.S.-NATO war in Afghanistan commencing in 2001 and stretching through the decade.

## The Free Market in Central Asia

Despite the overall slow pace of change and the difficulties of the transition to independence, the Central Asian republics appear to be finding their "capitalist legs" fairly quickly. As Communism fades and the Russians leave, Central Asians are taking up old market-based skills that have been part of their heritage since the ancient Silk Road days, when trade and long-distance travel were the heart of the economy. For example, a largely contraband trade in consumer goods (clothes, electronics, cars) and illegal substances (from drugs to guns) with Iran, Afghanistan, and western China is blossoming. New market methods, such as microcredit, are getting poor marginalized people involved in entrepreneurialism (Figure 5.48). There are also entrepreneurial opportunities to render services such as food, transport, and accommodations to tourists and to traveling businesspeople who hope to set up legitimate enterprises in the region (one such effort is described in the personal vignette below). Meanwhile, all five republics are negotiating with competing multinational oil companies to market Central Asian oil and gas to their best advantage. Iran, certain that shoppers will once again materialize from across the steppes, as they have for thousands of years, is establishing a free trade zone along its northern border with Turkmenistan.

**Personal Vignette**    Fahreeden is a taxi driver who operates out of Samarkand, Uzbekistan, in a dusty and battered old Fiat that he starts by hot-wiring the ignition. The tires are worn down to the steel fibers, and the brakes no longer work. His client today is an American lawyer pursuing an interest in religious history. The destination is the train station in Bukhara, Fahreeden's hometown. Bukhara is 4 hours away, a distance that will bring three flat tires and five halts by highway police, who expect bribes (baksheesh) at each stop. Fahreeden and the lawyer have agreed on a fee that includes the anticipated bribes, tire repairs, and a generous luncheon worthy of a Silk Road merchant.

The dusty ribbon of road through the semiarid landscape is conveniently punctuated with shops dedicated to tire repair. There are no signs advertising this service, just a bald tire sitting upright along the roadside. As you wait for your tire to be repaired, you may be treated to tea and conversation with the family of the repair person. Eventually, due to missed trains and changed plans, the lawyer will spend the night in the taxi driver's home, where, to the guest's surprise, the walls and floors are covered with beautiful and expensive handmade carpets. Dinner, prepared

**FIGURE 5.48** The free market in Central Asia. In Kyrgyzstan, Minavar Salijanova took advantage of a United Nations microcredit loan of about U.S.$100 to buy 60 chickens. Each day the chickens produced 60 eggs, which she sold in the market for about 6 cents each. Every 3 months she bought more chickens, and she now has 147. She has also bought some goats and is expanding her business. "The program has improved our vision for life," she says, and neighbors have begun to follow her example. [Staton R. Winter/New York Times Pictures.]

by Fahreeden's wife, Miriam, is shish kebab, fresh fruit, tomato salad, round loaves of bread, rice pilaf, tea, and lassi (a yogurt-based drink, taken from a common bowl). The guest will sleep in splendor on the same soft rugs on which his meal was graciously served.

*Source: From the travels of Ron Leadbetter, personal communication, Autumn 1996.* ■

Fahreeden's efforts to adapt to the post-Soviet era have led him to reinvent ancient ways of life in Central Asia. As was the case for many families in the days of the ancient Silk Road trade, his income is based on accommodating travelers and merchants and steering them safely through a large, arid, and potentially dangerous territory. Given a modicum of peace, he is likely to survive, and even thrive.

> ## Reflections on Russia and the Newly Independent States

The region comprising the Russian Federation, Belarus, Ukraine, Moldova, Caucasia, and Central Asia has gone through dramatic changes over the last decade. No other set of countries has ever attempted such a rapid and peaceful transformation to a free market economy. As the command economy disappears and localities direct their own economic affairs, regional and local differences and disparities are bound to increase. Although Russia retains its leadership position, its power and influence have weakened. Some of its closest allies in eastern Europe are no longer in its sphere of influence and are now closely connected with the increasingly powerful European Union. The Central Asian republics are likewise finding common cause with their neighbors to the west, south,

and east. As time passes, it may no longer be appropriate to combine these particular political units into one region, but just what the future holds is unclear.

It is likely that central governments will remain strong in most republics and that participatory democracies will evolve only slowly. Although some individuals are eager to learn from the West, there is widespread, and understandable, resistance to rapid cultural change and to excessive influence from abroad. Most republics in the region will probably have an ambivalent relationship with the West for some years to come, enthusiastically embracing many aspects of the new market economies and some aspects of democracy, yet mourning the loss of certain civilities and welfare guarantees of the old system. Rising crime, violence, and fraud, associated either with the breakdown of the old system or with the introduction of a market economy, have alienated many, especially in Russia, from Western models of development and have encouraged an acceptance of authoritarian government policies. In the Central Asian republics, the nuances of the market economy may be sorted out more rapidly because of the long heritage of entrepreneurial trade in the region. On the other hand, the religious values of a reviving Islamic tradition in Central Asia appear to be the nexus of anticapitalist and anti-Western feeling.

All parts of the region will be hampered for years to come by aging and inefficient industrial infrastructure and by the severe environmental degradation that has accompanied industrialization. As modernization and marketization proceed, the rapid private development of the region's rich resource base will probably lead to increased pollution. Yet, if societies continue to open up, there will be exhilarating opportunities for greater self-expression, for the possibility of entrepreneurial activities, and for the material well-being that market economies can provide.

## Chapter Key Terms

| | | |
|---|---|---|
| **Bolsheviks** 254 | **marketization** 261 | **serfs** 253 |
| **capitalists** 254 | **Mongols** 252 | **Slavs** 251 |
| **centrally planned economy** 254 | **nomadic pastoralists** 251 | **socialism** 254 |
| **command economy** 254 | **nonpoint sources of pollution** 277 | **Soviet Union** 246 |
| **Communism** 254 | **oligarchs** 262 | **steppes** 249 |
| **Communist Party** 254 | **perestroika** 255 | **taiga** 250 |
| **czar** 253 | **permafrost** 249 | **tundra** 249 |
| **Gazprom** 266 | **privatization** 261 | **underemployment** 272 |
| **glasnost** 255 | **Russian Federation** 269 | **Union of Soviet Socialist Republics (USSR)** 246 |
| **Group of Eight (G8)** 266 | **Russification** 270 | |

## Critical Thinking Questions

1. What were the social circumstances that gave rise to the Russian Revolution? Explain why the revolutionaries adopted a form of Communism grounded in rapid industrialization and central planning.

2. Describe the challenges posed by the physical environments (landforms and climate) of the various parts of this region: the North European Plain, the Caucasus, the Central Asian steppes, Siberia, and the various mountain ranges (Urals, Altai, Pamirs).

3. In general terms, discuss the ways in which the change from a centrally planned economy to a more market-based economy has affected career options and standards of living for the elderly, for young

professionals, for unskilled laborers, for members of the military, for women, and for former government bureaucrats.

4. Why do some scholars compare the historical record of the Russian empire to that of the colonizing empires of western Europe? In what ways was its behavior similar? How was it different?

5. Given Russia's crucial role in World War II and the defeat of Germany, discuss what led to the cold war between the Soviet Union and the West.

6. Discuss the reasons for the diminishing of Russia's sphere of influence after 1991 and briefly trace how Russia's role in global politics has changed through the 1990s into the present.

7. Discuss the evidence that its natural resources are becoming the foundation of a new role for Russia in the global economy.

8. The diffusion of Western business ideas and methods has affected Russia and the Newly Independent States. Give examples of how these ideas and methods are changing urban landscapes in the region.

9. Given the changing options for women in the region, what do you anticipate will be the trend in population growth in the future? Will birth rates remain low, or will they increase? Why?

10. Ethnic and national identities are affecting how the internal republics of Russia are thinking about their futures. How might the geography of Russia change if these feelings were to intensify? In what ways might the global economy be affected?

# CHAPTER 6

# NORTH AFRICA AND SOUTHWEST ASIA

**Global Patterns, Local Lives** It is April 2003, during the early stages of the Iraq war, and a group of men in a coffee shop in Alexandria, Egypt, are watching television. In the broadcast from a neighborhood in Baghdad, Iraq, a grief-stricken man whose brother's house has just been destroyed by a U.S. bomb is making a passionate statement: "I don't want this freedom, I don't want this democracy. My brother and his children are dead. Here I am drenched in blood to prove it."

The channel they are watching is Al Jazeera, the most controversial, criticized, and, with 50 million viewers, by far the most popular channel in Southwest Asia and North Africa (Figure 6.2). Al Jazeera has built a reputation as a critic of leaders and governments throughout the region since its founding in 1996 by journalists who had formerly worked for a TV station that was censored and then closed by the Saudi Arabian government. Now based in Qatar and funded by that country's emir, the network has, from time to time, been banned from operating inside Iran, Iraq, Jordan, and Bahrain. It has even offended many of its Arab viewers by interviewing officials of the Israeli government.

While Al Jazeera's broadcasts generally reflect popular sentiment among its viewers, such as the strong anti-U.S. sentiment that has swept through the region since the U.S. invasion of Iraq, some of the station's key personnel have political opinions with which most U.S. citizens would agree.

Senior Producer Samir Khader describes Al Jazeera's role as "to educate the Arab masses on something called democracy, respect of other opinions, free debate. . . . And to try by using all these things to shake up these rigid societies. To awaken them." Other reporters, such as Hassan Ibrahim, were educated in the United States and, while they are critical of

**FIGURE 6.1** Regional map of North Africa and Southwest Asia.

297

**FIGURE 6.2** Al Jazeera. News anchor Muntaha al-Rumahi presents a panel discussion on Osama bin Laden from Kandahar, Afghanistan, in October 2001. At the time, Al Jazeera was the only channel reporting live from Kabul and Kandahar. [AFP/Getty Images.]

the war in Iraq, are highly respectful of the U.S. Constitution and its acceptance of open political dissent.

Nevertheless, the U.S. government, which once praised Al Jazeera as a beacon of free speech for its unflinching coverage of corruption and human rights abuses by various governments in the region, is now the sta-

tion's most vocal critic. Former Secretary of Defense Donald Rumsfeld called its reports "vicious, inaccurate and inexcusable" and even accused Al Jazeera of using actors to stage broadcasts like the one described above to deliberately enrage Iraqis against the U.S. presence in Iraq.

Already hugely influential in the Arabic-speaking world for its willingness to criticize governments and expose corruption—a serious problem in the region—Al Jazeera is poised to become a global force with the launching of Al Jazeera International, an English-language news channel that will cover issues from an international perspective. There will be broadcast centers in Doha, Qatar; London; Kuala Lumpur, Malaysia; and Washington, D.C. An editor in each center will independently choose what to broadcast.

In 2004, Al Jazeera's success in the region inspired competition in the form of Al Arabiya (http://www.alarabiya.net), a satellite news network based in Dubai. Al Arabiya has correspondents in more than 40 cities and provides wide-ranging Arabic-language documentaries and uncensored debates on political and social issues to listeners in all parts of the world.

*Adapted from "Control Room" an award-winning documentary about Al Jazeera; http://english.aljazeera.net/NR/exeres/55ABE840-AC30-41D2-BDC9-06BBE2A36665.htm; "Rumsfeld's Al Jazeera Outburst," The Sunday Times-Online, November 27, 2005: http://www.timesonline.co.uk /article/0,,2089-1892464,00.html; Lydia Pulsipher's conversations with Josh Rushing, University of Tennessee, April, 2006; http://www.alarabiya .net/english.htm#003.* ■

Freedom of the press is a fragile concept in any part of the world, and it is not surprising that an independent press is only now developing in North Africa and Southwest Asia (Figure 6.3). The 21 countries in the region (see Figure 6.1) achieved political independence only in the twentieth century, many of them not until after World War II. (Western Sahara is now regarded as a dependency by Morocco, but it disputes this dependent status.) Although none were outright colonies of Europe, most were forced to submit to some type of formal control by European countries (and more recently by the United States) that resembled colonialism. These

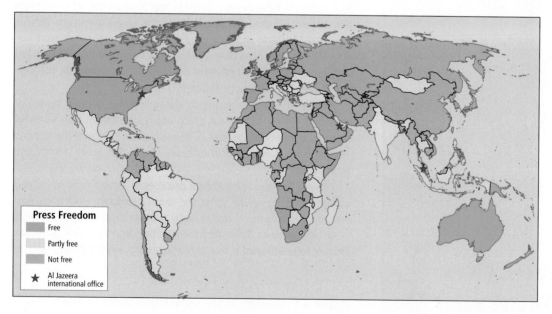

**Press Freedom**
- Free
- Partly free
- Not free
- ★ Al Jazeera international office

**FIGURE 6.3** Freedom of the press worldwide, 2006. The three categories of "freedom of the press" shown on the map are based on the legal environment in which the media operate, the degree of independence from government ownership and influence, economic pressures on news content, and violations of press freedom ranging from murder of journalists to extralegal abuse and harassment. [Adapted from "Map of Press Freedom 2006," Freedom House, at http://www .freedomhouse.org/uploads/maps /fop_current.pdf.]

foreign countries tried to appropriate the region's resources, especially oil and gas, at very low prices and to influence its economies, government officials, cultural institutions, and even religious practices. Throughout this era, democratic institutions, such as a free press, were not encouraged either by Western countries or by local elites. That history is part of the reason why many people in the region mistrust the West and see a need to protect their cultures from outside influences.

This region has long been subject to misrepresentation and unwarranted generalizations, both typical by-products of colonialism. Too often, outsiders see single issues as defining the entire region: the Arab–Israeli conflict, for example, or the isolated revival of particularly extreme versions of Islam, or the U.S. invasions of Iraq in 1991 and 2003. Some outsiders have such a sketchy mental image of this region that they may erroneously equate life in Morocco with that in Saudi Arabia, Sudan, or Iran. In fact, despite a common arid climate and a general religious preference for Islam, there is much geographic and cultural variety throughout this region. Oil is present in abundance in some countries but totally absent in others. In some parts of the region, nearly everyone lives in a city; in other parts, rural life is dominant. In some countries, civil laws are heavily influenced by religious law; others have secular legal systems. Women may be active in public life and prominent in government, or they may lead secluded domestic lives and have few educational opportunities. And ethnicity varies greatly within countries and across the region.

## Themes to Explore in North Africa and Southwest Asia

Several themes appear throughout this chapter, many of which are associated with the changing internal and external relationships just mentioned.

1. **The role of Islamic culture.** A major source of debate in the region is the degree of influence Islamic beliefs should have on government, law, gender roles, and social behavior. Although in many countries there is popular acceptance of a religious influence in government, there are strong voices in favor of modifying that stand. To some extent, the debate is between the two main branches of Islam, Sunni and Shi'ite, but there are moderates in both branches.

2. **The impact of oil.** The oil-rich countries of the Persian Gulf are major suppliers of fossil fuels to the rest of the world. Oil has provided wealth for a minority of citizens, but the economies of most countries in the region, especially the oil-rich countries, remain undiversified.

3. **Water scarcity.** A key challenge is to make this desert region habitable for an increasing population by obtaining more water for more people and expanding agriculture.

4. **Political conflict and violence.** In the aftermath of European de facto colonialism, rivalry for resources and territory and agitation for more direct democracy have resulted in conflict within and between countries, which frequently devolves into terrorism. Terrorists have struck in nearly every country in the region. Too often this conflict is simplistically attributed to religious rivalries alone.

5. **Global impacts of local issues.** The relationships between the countries in this region and countries around the world are difficult for a variety of reasons: the escalating Arab–Israeli conflict, undemocratic governments within the region, the mismanagement of oil wealth, and the presence of foreign military forces. Autocratic and unresponsive governments established under the influence of Europe and the United States have bred violence as a political strategy, and this violence has spread beyond the boundaries of the region to Europe, Africa, Southeast Asia, and the United States.

## I THE GEOGRAPHIC SETTING

## Terms to Be Aware Of

The common term *Middle East*, often used for the eastern Mediterranean countries of the region, is not used in this chapter because it describes the region only from a European perspective. To someone in Japan, the region lies to the far west, and to a Russian, it lies to the south. In addition, the term *Middle East*, as popularly used, does not usually include the western sections of North Africa, which we include in this region. Some people who live in the region, however, do use the term *Middle East* themselves, as do the press and some scholars. Although it is now common for the media to refer to the *Islamic world*, that term is not used in this book either because it carries a false implication that there is a cohesive Islamic community, the members of which could be expected to participate in common practices and have the same values. Furthermore, non-Muslims constitute significant minority communities throughout the region. The reader should be careful not to equate *Arab* with *Muslim*: the former term describes ethnicity; the latter, religion. There are non-Muslim Arabs and non-Arab Muslims across the region.

## PHYSICAL PATTERNS

Landforms and climate are particularly closely related in this region. The climate is dry and hot in the vast stretches of relatively low, flat land and somewhat moister where mountains are able to capture rainfall.

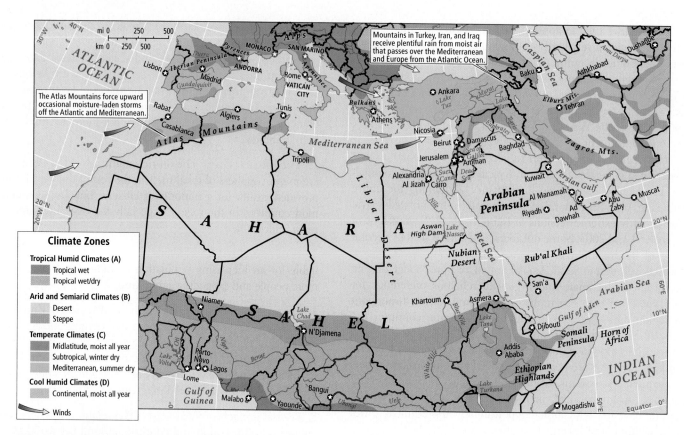

**FIGURE 6.4** Climates of North Africa and Southwest Asia.

## Climate

No other region in the world is as pervasively dry as North Africa and Southwest Asia (Figure 6.4). A global belt of dry air that circles the planet between roughly 20° and 30° N creates desert climates in the Sahara of North Africa, the Arabian Peninsula, Iraq, and Iran. The Sahara's size and southerly location make it a particularly hot desert region; in some places, temperatures can reach 130°F (54°C) in the shade at midday. In the nearly total absence of heat-retaining water or vegetation, nighttime temperatures can drop quickly to below freezing. Nevertheless, in even the driest zones, humans survive at scattered oases, where they maintain groves of drought-resistant date palms and some irrigated field crops. Traditionally, desert inhabitants have worn light-colored, loose, flowing robes that reflect the sunlight during the day and provide insulation against the cold at night. Many people continue this practice today.

On the uplands and margins of the deserts on both the African and Eurasian continents, enough rain falls to nurture grass, some trees, and limited agriculture. Such is the case in western Morocco, northern Algeria and Tunisia, Turkey, northern Iraq, and Iran. Generally too dry for cultivation, the rest of the area has long been the prime herding lands of the region's nomads, such as the Kurds of Southwest Asia, the Berbers in North Africa, and the Bedouin of the steppes and deserts on the Arabian Peninsula. Recently, the restrictions of a dry climate have been overcome with irrigation, but the general aridity of the region means that sources of irrigation water are scarce.

## Landforms and Vegetation

The undulating surfaces of desert and steppe lands cover most of North Africa and Saudi Arabia (compare Figures 6.1 and 6.4). Fringes of mountains, located here and there across the region, capture moisture and make it possible for plants, animals, and humans to survive and even flourish. In northwestern Africa, the Atlas Mountains stretch from Morocco on the Atlantic to Tunis on the Mediterranean. They block and lift moisture-laden winds from the Atlantic, creating conditions that lead to rainfall of more than 50 inches (127 centimeters) a year in some places (see Figure 1.23 on page 24). Snowfall is sufficient in some Atlas Mountain locations to support a skiing industry. Behind these mountains, to the south and east, spreads the great Sahara, where rain falls rarely and unpredictably.

Africa and Southwest Asia are separated by a rift formed between two tectonic plates—the African Plate and the Arabian Plate—that are moving away from each other (see Figure 1.20 on page 21). The rift, which began to form about 12 million years ago, is now occupied by the Red Sea. The Arabian Peninsula lies to the east of this rift. There, mountains bordering the rift in the southwestern corner rise to 12,000 feet (3658 meters), and the precipitation they capture from moisture-laden monsoon winds crossing Central Africa is the primary source of rain for the entire peninsula. Behind these mountains, to the east, lies the great desert region of the Rub'al Khali. Like the Sahara, it is virtually devoid of vegetation. Constantly moved by strong winds, the sand dunes of

**FIGURE 6.5** A wadi, or dry riverbed, in the Ahaggar Mountains of Algeria. [Victor Englebert/Photo Researchers.]

the Rub'al Khali are the world's largest, some reaching more than 2000 feet (610 meters) in height. To protect themselves from the persistent winds, blowing sands, and temperatures that vary radically from day to night, the few remaining desert dwellers still line their tents with animal skins, rugs, and tapestries.

The landforms of Southwest Asia are more complex than those of North Africa and Arabia. As the Arabian Plate, carrying the great desert peninsula of Arabia, rotates slowly to the northeast, away from the African Plate, it is colliding with the Eurasian Plate and pushing up the widely spaced mountains and plateaus of Turkey and the Zagros Mountains in Iran (see Figure 1.20 on page 21). Turkey, Iran, and adjacent lands are often the sites of earthquakes related to these tectonic plate movements. Because of its mountainous landforms, Turkey receives considerable rainfall from the moisture masses that pass over Europe from the Atlantic Ocean, but only a little moisture makes it over the mountains to the interior of Iran, which is very dry.

There are only three major rivers in the entire region, and all have attracted human settlement for thousands of years. The Nile flows north from the moist central East African highlands across mostly arid Sudan and desert Egypt to a large delta on the Mediterranean. The Euphrates and Tigris rivers both begin in the mountains of Turkey and flow southeast to the swamps of the Shatt al Arab estuary terminating in the Persian Gulf. A fourth and much smaller river, the Jordan, starts as snowmelt in the uplands of southern Lebanon and flows 223 miles (359 kilometers) through the Sea of Galilee to the Dead Sea. Most other streams are actually dry riverbeds, or wadis, most of the year, with water in them only after the infrequent and usually light rains that fall between November and April (Figure 6.5).

Although North Africa and Southwest Asia were home to some of the very earliest agricultural societies (Figure 6.6), overuse of soils and irrigation, as well as probable changes in climate, have made crop agriculture possible in only a few places today. In areas along the Mediterranean coast, where farmers can count on rain during the winter, they grow citrus fruits, grapes, olives, and many vegetables, often using supplemental irrigation. In the valleys of the major rivers, seasonal flooding and modern irrigation methods provide water for such crops as cotton, wheat, barley, and vegetables. In occasional oases, and in a few places where groundwater is pumped to the surface, such crops as dates, melons, apricots, and vegetables are grown. Across the region, small groups of nomadic pastoralists as well as some settled farmers keep grazing animals.

**FIGURE 6.6** Grazing animals once roamed the Sahara. These 8000-year-old paintings in a cave at Tassili-n-Ajjer, Algeria, reveal that animals once grazed in what is now the heart of the Sahara. Climate change and human activity have repeatedly affected the landscape over the millennia, and such animals would not survive there under present conditions. [Pierre Boulat/ Woodfin Camp and Associates.]

## HUMAN PATTERNS OVER TIME

Important advances in agriculture and the organization of human society, as well as urbanization, took place long ago in this part of the world. Later, three of the world's great religions were born here: Judaism, Christianity, and Islam. Starting in the seventh century C.E., Islam grew to become the region's dominant religion and the center of its private and public life. For many centuries, the region influenced Europe and the rest of the world through its advanced learning, effective trading strategies, and refined culture. As we shall see, Muslim empires waxed and waned, and then waxed again. In more recent times, especially since World War I, influences from Europe and the United States have been challenging the region's traditional mores regarding religion, gender roles, and the role of the state in society.

## Sites of Early Cultivation

Between 10,000 and 8000 years ago, nomadic peoples of this region founded some of the earliest known agricultural communities in the world. These communities were located in an arc formed by the uplands of the Tigris and Euphrates river systems (in modern Turkey and Iraq) and the Zagros Mountains of modern Iran. This zone is often called the **Fertile Crescent** (Figure 6.7) because of its plentiful fresh water, its open forests and grasslands, its abundant wild grains, and its fish, goats, sheep, wild cattle, and other large animals. As the early inhabitants adjusted to the region's environment, their emerging skills in domesticating its plants and animals allowed them to build more elaborate settlements. These settlements eventually grew into societies based on widespread irrigated agriculture in lowland locations along the base of the mountains and along the two rivers.

**FIGURE 6.7** The Fertile Crescent, one of the earliest known agricultural sites. About 10,000 years ago, people in the Fertile Crescent began domesticating cereal grains, legumes, and animals, especially sheep and goats. These uses of domesticated animals spread into Europe and Africa as agricultural peoples traded their surpluses for other goods or moved into other regions. Three major empires developed successively in the eastern extent of the Fertile Crescent: the Sumerian, the Babylonian, and the Assyrian. [Map adapted from Bruce Smith, *The Emergence of Agriculture* (New York: Scientific American Library, 1995), p. 50; photo, Patrick Syder/Lonely Planet Images/Getty.]

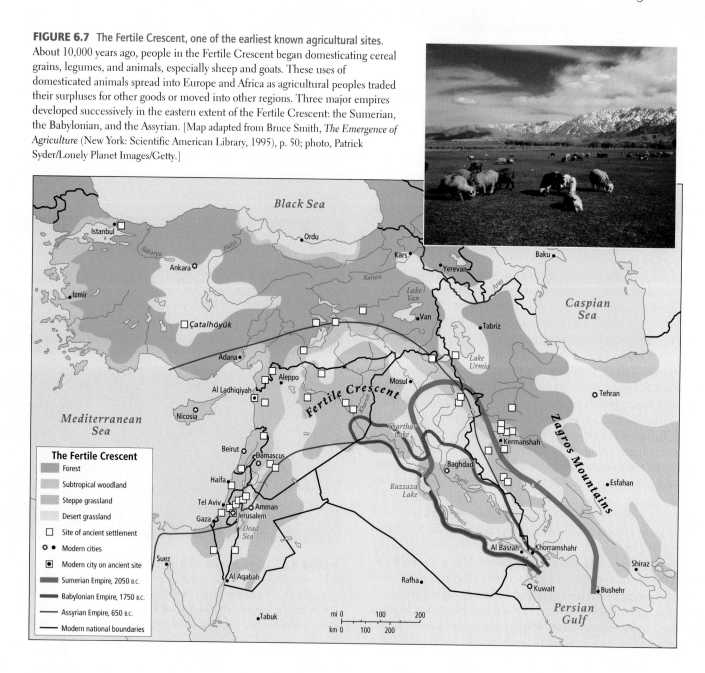

**The Fertile Crescent**
- Forest
- Subtropical woodland
- Steppe grassland
- Desert grassland
- □ Site of ancient settlement
- ✪ ● Modern cities
- ◉ Modern city on ancient site
- Sumerian Empire, 2050 B.C.
- Babylonian Empire, 1750 B.C.
- Assyrian Empire, 650 B.C.
- Modern national boundaries

Agriculture was also practiced in the Nile Valley and the Maghreb (northwestern Africa) about 6500 years ago, and in the eastern mountains of Persia (modern Iran). Societies of nomadic animal herders living in adjacent steppe zones traded animal products for the grain and manufactured goods produced in the settled areas. Eventually the agricultural settlements took on urban qualities: dense settlement, specialized occupations, concentrations of wealth, and centralized government and bureaucracies. The city of Sumer (in modern southern Iraq), for example, which existed 5000 years ago, gradually extended its influence over the surrounding territory. The Sumerians developed wheeled vehicles, oar-driven ships, and irrigation technology.

At times, nomadic tribes would band together, sweep over settlements with devastating cavalry raids, and set themselves up as a ruling class, though they tended to adopt the cultures of the sedentary peoples they conquered. Usually rulers tried to expand their domains through a combination of military aggression, administrative consolidation, and trading acumen. These strategies occasionally led to the creation of vast but unwieldy empires that became vulnerable to defeat by new waves of nomadic peoples. In this way, the civilization of Sumer was succeeded by the Babylonian and Assyrian empires, the latter reaching its zenith about 3000 years ago.

## The Emergence of Gender Roles

Some researchers think that the dawning of agriculture may mark the transition to markedly different roles for men and women. Archaeologist Ian Hodder reports that at the 9000-year-old site of Çatalhöyük, in south-central Turkey (see Figure 6.7), there is little evidence of gender differences. Families were small and men and women performed similar chores in daily life, had similar status and power, and both played key roles in social and religious life. Scholars think that after the development of agriculture, as wealth and property became more important in human society, a concern with family lines of descent and inheritance emerged, which led in turn to the idea that women's bodies needed to be controlled so as to prevent them becoming pregnant by more than one man and thus confusing lines of inheritance.

## The Coming of Judaism, Christianity, and Islam

The early religions of this region were based on a belief in many gods linked to natural phenomena, but several thousand years ago, **monotheistic** belief systems—those based on one god—emerged. The three major monotheistic religions—Judaism, Christianity, and Islam—all have their origins in the eastern Mediterranean (Figure 6.8).

(a) An Israeli bride and groom in the Gaza Strip participate in a Moroccan "Hena" ceremony prior to their wedding in 2005.

(b) A bride and groom leave the church following an Arab Christian wedding in Jerusalem.

(c) A Muslim bride and groom are serenaded during wedding ceremonies in Tehran.

**FIGURE 6.8** Weddings in the three major religions of the region. [(a) Ronen Zvulun/Reuters/Landov; (b) PonkaWonka; (c) Abbas/Magnum Photos.]

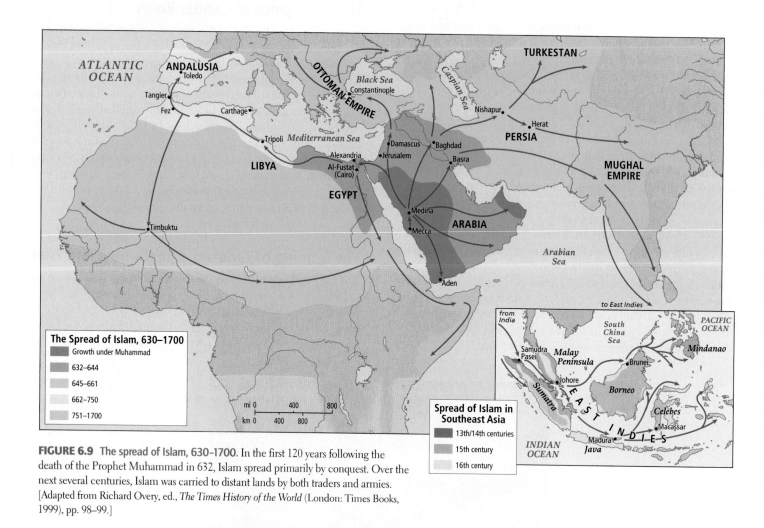

**FIGURE 6.9** The spread of Islam, 630–1700. In the first 120 years following the death of the Prophet Muhammad in 632, Islam spread primarily by conquest. Over the next several centuries, Islam was carried to distant lands by both traders and armies. [Adapted from Richard Overy, ed., *The Times History of the World* (London: Times Books, 1999), pp. 98–99.]

**Judaism** was founded approximately 4000 years ago. According to tradition, it was begun by the patriarch Abraham, who led his followers from Mesopotamia (modern Iraq) to the shores of the eastern Mediterranean (modern Israel and Palestine). Jewish religious history is recorded in the Torah (the Old Testament of the Bible) and is characterized by the belief in one God, Yaweh, a strong ethical code summarized in the Ten Commandments, and an enduring ethnic identity. After the Jews rebelled against the Roman Empire at the Masada fortress on the Dead Sea in 73 C.E., they were expelled from the eastern Mediterranean and migrated to other lands in a movement known as the **Diaspora** (the dispersion of an originally homogeneous people). Many Jews dispersed across North Africa and Europe, and others went to various parts of Asia. Jews were among the earliest European settlers in the Americas.

**Christianity** is based on the teachings of Jesus of Nazareth, a Jew, who gathered followers in the area of Palestine about 2000 years ago. Jesus, also known as Christ, taught that there is one God, whose relationship to humans is primarily one of love and support but who will judge those who do evil. This benevolent philosophy grew popular, and both Jewish (religious) and Roman (government) authorities of the time saw Jesus as a dangerous challenge to their power. After his execution in Jerusalem in about 32 C.E.,

Jesus' teachings (the Gospels) were written down by many followers. His ideas spread and became known as Christianity. By 400, Christianity was the official religion of the Roman Empire. In 1054, political tensions split Christianity into the Western tradition associated with Rome and the Eastern, or Orthodox, tradition headquartered in Constantinople (Istanbul). While Christianity maintained a strong foothold in Europe, the spread of Islam after 622 C.E. meant that only remnants of Christianity remained in the eastern Mediterranean, North Africa, and Central and South Asia. During the eleventh, twelfth, and thirteenth centuries, European Christians launched various military ventures, known as the Crusades, to retake the eastern Mediterranean from the Muslims. Christianity was later spread through European colonization and missionary activities to the Americas and parts of Africa and Asia.

**Islam** is now the overwhelmingly dominant religion in North Africa and Southwest Asia. Islam emerged in the seventh century C.E. when, according to the **Qur'an** (or **Koran**), the holy book of Islam, the archangel Gabriel revealed the principles of the religion to the Prophet Muhammad. Born in about 570, Muhammad was a merchant and caravan manager in the small trading town of Makkah (Mecca) on the Arabian Peninsula near the Red Sea. Islam was, and is, considered an outgrowth of Judaism and Christianity. Followers of

Islam, called **Muslims,** believe that Muhammad, who was called to be a prophet in 610, was the final and most important in a long series of revered prophets, which included Abraham, Moses, and Jesus. They also believe that the Qur'an contains the words God (**Allah**) revealed to Muhammad. (*Allah* is the Arabic word for "God"; thus Arabic-speaking Christians as well as Muslims use the word *Allah* in their prayers.) Because Jerusalem is the place from which Muhammad is believed to have ascended into heaven, that city is particularly holy to Muslims around the world, as it is to Jews and Christians. All Muslims are encouraged to undertake the **hajj,** the pilgrimage to the cities of Makkah and Al Madinah (Medina) (the sites of Muhammad's mosque and burial place) at least once in a lifetime. Although there have been several powerful Muslim empires, the religion of Islam, per se, has virtually no religious hierarchy or central administration. The world's one billion Muslims may communicate directly with God, not necessarily through a clerical intermediary, though there are numerous mullahs (clerical leaders) who help their followers interpret the Qur'an. As a result, the interpretation of Islam varies widely within and among countries and from individual to individual.

## The Spread of Islam

The Bedouin—nomads of the Arabian Peninsula—were among the first converts to Islam, and they were already spreading the faith and creating a vast Islamic sphere of influence by the time of Muhammad's death in 632. Over the next century, Muslim armies extended an Arab–Islamic empire over most of Southwest Asia, North Africa, and the Iberian Peninsula of Europe (Figure 6.9, Figure 6.10).

The new empire controlled both politics and trade, and it established bonds between government functions and Islam that continued after the empire's demise. Islamic rules of conduct guided such matters as taxation policy and particularly the system of law, which was administered by religious specialists. To legitimize their authority, the leaders of the empire often claimed a blood relationship with Muhammad. Arabs remained the dominant elite as Islam spread, but they set up local non-Arabs and non-Muslims as administrators of newly acquired territory. Thus they allowed local rule, but collected taxes that supported the empire. They were particularly tolerant of other monotheists, such as Jews and Christians.

While Europe was stagnating during the medieval period (450–1500), Muslim scholars traveled extensively throughout Asia and Africa. These scholars made many important contributions to the fields of history, mathematics, geography, medicine, and other academic disciplines, which flourished from Baghdad in Iraq through Damascus (Syria), Alexandria (Egypt), and Fez (Morocco) to Toledo in Spain. These centers of world learning arose in vibrant economies that benefited from the early Islamic development of financial institutions and practices connected to wide-ranging trade, such as banks, trusts, checks, and receipts. From India, Muslims

(a)    (b)    (c)

**FIGURE 6.10  The early Arab role in global trade.** The similarity between the ingredients for a Mexican *mole* (a) and the ingredients for an Indonesian curry dish, *sayur lodeh* (c), is in large part the result of Arab participation in the growing global trade networks in the medieval period. Spaniards brought new food crops, such as the chocolate in *mole*, from the Americas to the Mediterranean. Arab traders, using boats like the dhows shown in (b), linked many parts of Asia (for example, China, the Philippines, Indonesia, Burma, and India) with the Mediterranean. [(a) and (c) Ignacio Urquiza; (b) Eric Nathan/Alamy.]

imported the useful mathematical concept of the zero, further advancing trade, record keeping, and scientific knowledge in the region. By the eleventh century, Islamic medicine was already drawing on a wide variety of traditions that originated in places from China to West Africa and included such practices as medical record keeping and regular peer review of physicians—practices that remained far superior to those in Europe into the nineteenth century.

By the end of the tenth century, the Arab–Islamic empire began to break apart into smaller units as outlying provinces eluded the control of central authorities. From the eleventh to the fifteenth centuries, Mongols from eastern Central Asia conquered parts of the Arab-controlled territory, forming the Mughal Empire. Meanwhile, beginning in the 1200s, nomadic herders in western Anatolia (Turkey) had begun to forge the Ottoman Empire, the greatest Islamic empire the world has ever known. By the fifteenth century, the Ottoman Muslims had defeated the Christian Byzantine Empire (the successor to the Roman Empire) and taken over its capital, Constantinople, which they renamed Istanbul. Very soon the Ottomans controlled most of the eastern Mediterranean, Egypt, and Mesopotamia, and by the late 1400s they controlled much of southeastern and central Europe. At about the same time, the Arab Muslims lost their control of the Iberian Peninsula.

During these various empires, the Islamic faithful were found from Spain in the east, deep into North Africa, in Turkey and southeastern Europe (the Balkans), and across Central Asia to the Indus Valley and northern India (see Figure 6.9). Muslim traders and mystics carried the Islamic faith still farther, by land across Central Asia to western China and by sea to the East Indies, the southern Philippines, and even parts of coastal southeastern China. These areas, and those of the former Arab–Islamic empire, are still predominantly Islamic, with the exception of Spain, some parts of the Balkans, and southeastern China, where, nonetheless, remnants of Islamic culture persist.

The Ottoman Empire, like the Arab–Islamic empire before it, supported religious tolerance within its borders. Jews and Christians were allowed to practice their religion, although there were economic and social advantages to converting to Islam. The Ottomans prospered from the productivity of the empire's many different peoples. Drawing on trading networks that stretched from Europe across the Mediterranean to Morocco and Algeria and east to the Indian Ocean, Istanbul became a cosmopolitan capital, outshining most European cities until the nineteenth century.

The origins of cultural practices in Islam are often obscure and misunderstood. Many people believe, for instance, that Islam instituted restrictions on women, such as **seclusion** (the requirement that a woman stay out of public view) and **veiling** (the custom of covering the body with a loose dress and the head—and in some places, the face—with a scarf). There is ample evidence, however, that these practices predate Islam by thousands of years, and indeed, some Christians and Jews in the region adhere to similar customs. Nor are the antiforeign attitudes associated with some movements in Islam today inherent in that religion. On the contrary, some past Islamic empires, such as the Ottoman Empire, were known for their cosmopolitan culture, tolerance, and even appre-

ciation and adoption of foreign ways. Antiforeign sentiment is a more recent phenomenon that emerged after years of European and U.S. interference in the affairs of the region.

## Western Domination and State Formation

The Ottoman Empire ultimately withered in the face of a Europe made powerful by the Industrial Revolution. Throughout the nineteenth century, North Africa provided raw materials for Europe in a trading relationship dominated by European merchants, who sought cotton, phosphates, manganese, hides, and wool. In 1830, France became the first European country to exercise direct control over a North African territory when its military gained control of Algeria. France eventually administered that land almost as though it were a part of France, although, while the French had free access to Algeria, Algerians were not allowed the benefits of full citizenship in France. In the late nineteenth century, motivated in part by the fear that a rival European power would step in first, Britain gained control of Egypt (1882) and then Sudan, and France took over Tunisia (1881). France began control of Morocco in 1907 and established full control in 1912; Libya ceded control to Italy in that same year (Figure 6.11). In these countries, Europeans set up a form of dependence that they termed a **protectorate**: local rulers remained in place, but European officials made the important economic and trade-related decisions.

One result of European influence was that traditional, more egalitarian systems of landownership were undercut to the disadvantage of small cultivators. European officials privatized tribal collective lands and gave title only to the tribal leaders in return for their political support. Thus European domination disempowered many and encouraged the concentration of wealth and power in the hands of a relative few.

World War I (1914–1918) brought the fall of the already shrunken Ottoman Empire. The Ottomans had allied with Germany, and after the war ended in defeat for that country, the League of Nations (a precursor to the United Nations) allotted almost all the former Ottoman territories at the eastern end of the Mediterranean to France and Britain, which already controlled most of those areas before the war (Figure 6.12). Only Turkey was recognized as an independent country at that time. Syria and what is now Lebanon became *mandated territories* of France (dependencies thought incapable of self-rule). Palestine, much of which became Israel in 1948, and Iraq, out of which the kingdom of Jordan was carved, became mandated territories of Britain in 1920. On the Arabian Peninsula, Bedouin tribes were consolidated under the control of Sheikh ibn Saud in 1932, after which time Saudi Arabia began to emerge as an independent country. The smaller states bordering the Persian Gulf (Yemen, Oman, the United Arab Emirates, Qatar, Bahrain, and Kuwait) were British protectorates after World War I, and they eventually became independent countries.

The aftermath of World War II had further effects on the political development of North Africa and Southwest Asia. The United States supported those autocratic local leaders who were most sympathetic to U.S. cold war policies (as opposed to those of

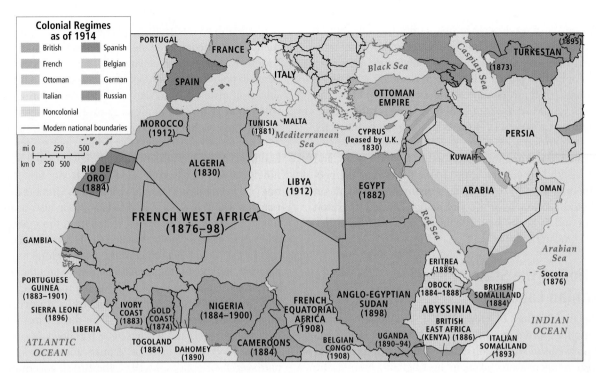

**FIGURE 6.11 Colonial regimes in 1914.** European powers began influencing the affairs of much of the region in the nineteenth century and expanded their control in the early twentieth century, even before the Ottoman Empire fell and its parts were allocated to various European and regional powers (the dates when they took control of each country are shown). [Adapted from *Hammond Times Concise Atlas of World History* (Maplewood, N.J.: Hammond, 1994), pp. 100–101.]

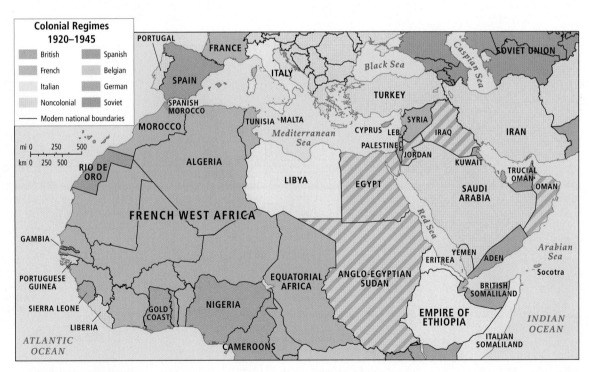

**FIGURE 6.12 Colonial regimes, 1920–1945.** Following World War I, what was left of the Ottoman Empire in the eastern Mediterranean was allotted as protectorates to the British and French. After World War II, virtually the entire region of North Africa and Southwest Asia was divided into independent countries. [Adapted from *Rand McNally Historical Atlas of the World* (Chicago: Rand McNally, 1965), pp. 36–37; and *Cultural Atlas of Africa* (New York: CheckMark Books, 1988), p. 59.]

the Soviet Union) and most likely to maintain a friendly attitude toward U.S. business interests. In Iran and Saudi Arabia, where vast oil deposits became especially lucrative by midcentury, European and U.S. oil companies played a key role in deciding who ruled. The governments of these countries undertaxed oil profits, and most of the revenues were gathered by a small ruling elite. Although some oil revenue went toward building roads, hospitals, and schools—facilities that had not previously existed—relatively little was invested in opening up opportunities for the masses, and disparities in wealth between the elite and the majority of the people increased dramatically. The region remained dependent on Europe and North America for the technology needed to exploit its oil reserves and to begin the mechanization of manufacturing, transportation, and agriculture.

## The Creation of the State of Israel on Palestinian Lands

In 1947, after the atrocities of World War II, the Western powers were searching for a place to settle the tens of thousands of European Jewish survivors of Nazi death camps. Although limited numbers of Jewish refugees were taken in by England, France, the United States, and countries in South America, no country stepped forward to offer a home to all Jews.

The idea of European Jews migrating to their ancestral homeland in Palestine, once known as Israel, was based on the Old Testament and on the history of the Jews' expulsion from the region beginning in 73 C.E. In the centuries of the Diaspora that followed, many Jews encountered persistent discrimination, especially in eastern Europe, where occasionally whole villages would be murdered in **pogroms** (episodes of ethnic cleansing). This deadly oppression culminated in the Nazi era before and during World War II, when more than 6 million Jews across Europe were imprisoned, enslaved, and eventually killed in gas chambers during the Holocaust. The notion of Palestine once again becoming a Jewish homeland had been gaining popularity among Jews since it was proposed in nineteenth-century Europe and Russia. By this time, however, Palestine had been home to non-Jewish peoples for more than 1500 years, many of whom had converted to Islam soon after its founding. The modern Palestinians were descended from those peoples and from other Arab groups who had joined them over the centuries.

In the late nineteenth century, a small group of Jews, known as **Zionists,** had begun to purchase land here and there from wealthy Palestinian landholders, establishing communal settlements called *kibbutzim.* Jewish and Arab populations lived intermingled at that time. Most Palestinians displaced by the Zionist land purchases were poor farmers and herders, who had been accustomed to using land held by village landlords. Zionist land acquisition increasingly narrowed their options. In 1917, the British government issued the Balfour Declaration, which committed it to supporting the establishment of "a national home for the Jewish people" in Palestine, but only if the civil and religious rights of non-Jewish communities in Palestine could be ensured. To safeguard their rights, Palestine was made a British mandate in 1923 (Figure 6.13a). It should be

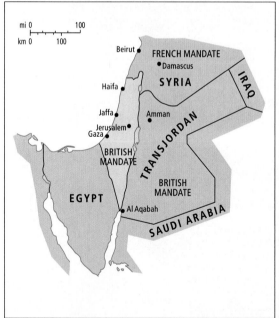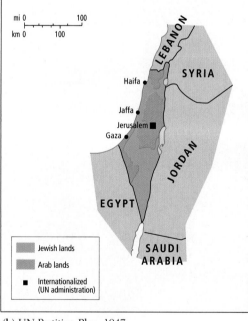

(a) Palestine, 1923    (b) UN Partition Plan, 1947    (c) Israel, 1949

**FIGURE 6.13** Israel and Palestine, 1923–1949. (a) In the 1920s, Britain controlled what is now Israel and Jordan (Transjordan was the precursor to Jordan). (b) After World War II, the United Nations developed a plan for separate Jewish and Palestinian (Arab) states. (c) The Jewish settlers did not agree to that plan; instead, they fought and won a war, creating the state of Israel. [Adapted from Colbert C. Held, *Middle East Patterns—Places, Peoples, and Politics* (Boulder, Colo.: Westview Press, 1994), p. 184.]

noted that the Balfour Declaration was itself a breach of previous agreements Britain had made with Arab leaders to guarantee them independent control over Arab lands. This breach is still cited as a cause for distrust of Europe and the United States.

Land purchases by European Jews continued. By 1946, strong sentiment had built among the world's Jews in favor of a Jewish homeland in Palestine, and many Jews already there took up arms to convince Britain to stand by the Balfour Declaration. The Palestinian Arabs and their primarily Muslim supporters in the region fiercely objected to the formation of the state of Israel, out of fear that Palestinians would lose their claim to the land they had long held under the Ottomans and British and would be denied a voice in their own governance. The conflict between Zionists and Palestinian Arabs was not over religion, but over land and water, and in large part, this remains the case today.

The Palestinians were not unified, and they lacked the cultural familiarity with Europe and the United States that many Jews had, so they were not as powerful in advancing their cause in the world community. By 1947, Europe and the United States had sided with the Jews. In that year, the United Nations recommended a partition plan that would have divided Palestine between Jews and Arabs, with the city of Jerusalem internationalized (Figure 6.13b). The Arabs rejected this plan because they would have lost land; the World Zionist Organization (Jews) accepted it, but only with reservations. When the British left in early May 1948, Jewish settlers fought a war and, in 1949, created a Jewish state from Palestinian land (Figure 6.13c). The resulting Israeli–Palestinian dispute has monopolized the politics of the region since that time; we will return to that ongoing conflict later in this chapter (see pages 327–329).

## POPULATION PATTERNS

The great size and aridity of this region have had important implications for its patterns of human settlement and interaction. Although the region as a whole is nearly twice as large as the United States, the areas that are useful for agriculture and settlement are tiny by comparison. The land supports 150 million more people than the United States, and population densities in livable spaces can be quite high—over 260,000 per square mile (100,000 per square kilometer), for example, in some of Egypt's urban neighborhoods. (The highest urban density in the United States—in New York City—is approximately one-tenth this figure.) Vast tracts of desert are virtually uninhabited. Most people live along coastal zones, in river valleys, and in upland zones that capture orographic rainfall (Figure 6.14). Efforts are being made to extend livable space into the desert, but doing so may require extremely costly and

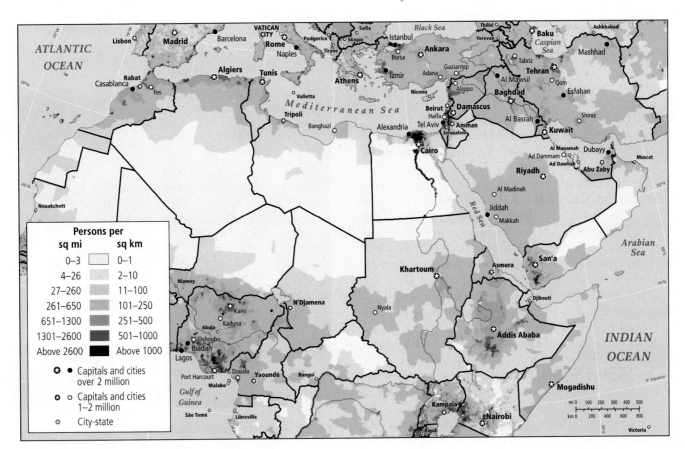

**FIGURE 6.14** Population density in North Africa and Southwest Asia. [Data courtesy of Deborah Balk, Gregory Yetman, et al., Center for International Earth Science Information Network, Columbia University, at http://www.ciesin.columbia.edu.]

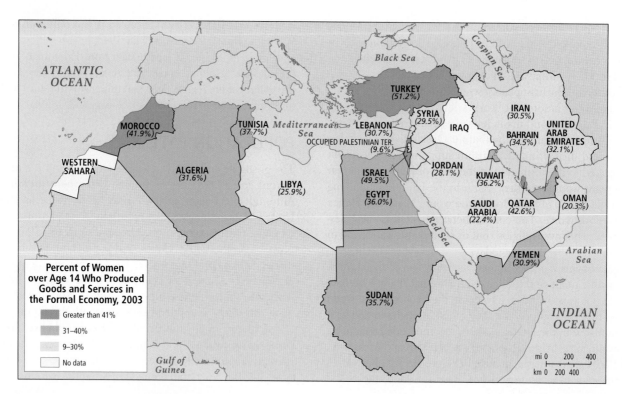

**FIGURE 6.15** Percentage of women who are wage-earning workers in the region's countries. [Adapted from Joni Seager, *Women in the World: An International Atlas* (New York: Viking Penguin, 1997), pp. 66–67; and *United Nations Human Development Report 2005* (New York: United Nations Development Programme), Table 25.]

environmentally questionable solutions to the problems of water scarcity and soil infertility.

In the 1960s, when most women married shortly after reaching puberty, some countries in the region had average fertility rates of close to 9 children per woman. By 2005, the average fertility rate for the region was reduced to around 3.5. Nonetheless, this rate is much higher than the world average of 2.7; the only region with a higher average is sub-Saharan Africa. At present rates of growth, the population of the region, now about 450 million, will reach 540 million by 2025. It is difficult to imagine how the area will support the additional people. Fresh water is already in extremely short supply, and most countries must import food at great cost. Furthermore, 32 million more jobs would have to be created within the next decade to employ the added population (compared with just 4.5 million needed in Europe, whose population is growing very slowly). Moreover, if the status of women changes more quickly than anticipated, the demand for jobs will increase even beyond the anticipated 32 million. As Figure 6.15 shows, by the year 2003, considerably less than 50 percent of women across the region (except in Israel and Turkey) worked outside the home at jobs other than farming (Figure 6.16).

Many inhabitants of the region believe that the industrialized world promotes population control in the developing world so that developed nations can continue to consume more than their fair share of the world's resources. On the other hand, many leaders

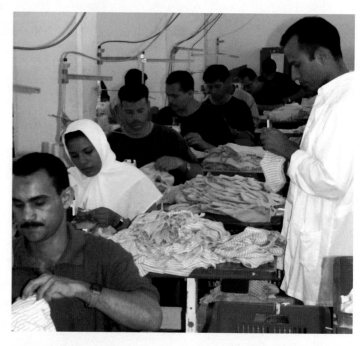

**FIGURE 6.16** Men and women work together in Conytex, a textile business in Egypt. Conytex, one of six Sekem Holding Co. businesses, designs and makes a variety of baby and household textiles from organically grown cotton provided by another of Sekem's businesses. [IFC.]

(especially in Iran, Jordan, Oman, Qatar, Turkey, Yemen, and all of North Africa except Libya) see the wisdom of population control and encourage family planning. Moreover, most specialists in Islamic law say that limiting family size is acceptable for a wide variety of reasons, so long as the motive is not to shirk parenthood altogether. Nevertheless, by 2005, in the region as a whole, only about 41 percent of women were using modern methods of contraception, and only 53 percent were using any methods of contraception at all.

## Gender Roles and Population Growth

In rural agricultural societies, fertility rates tend to be high because families depend on children as a labor supply. In urban contexts, especially where women have opportunities to work or study outside the home, they usually choose to have fewer children. In societies where fertility rates remain high despite a decline in agriculture, gender inequality may partially explain why many couples are still choosing to have so many children. Throughout North Africa and Southwest Asia, wherever men have a much higher education level than women, greater access to employment, and greater power to make family and community decisions, children are the most important source of family involvement and power for women. Where women are secluded in the home and undereducated, they have little chance to accumulate wealth or to enhance the prospects and reputation of the family in ways other than by having children (see pages 316–319).

Although women contribute their labor and goodwill to the reproduction, health, and well-being of families, one result of their lesser ability to contribute to the family financially and their lower social standing is a deeply entrenched cultural preference for sons. Families sometimes continue having children until they have a desired number of sons, and young females may not survive, as indicated by the missing females in the 0–4 age cohort of the population pyramids in Figure 6.17 (see also the discussion in Chapter 1, pages 38–39).

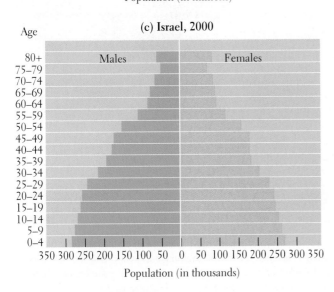

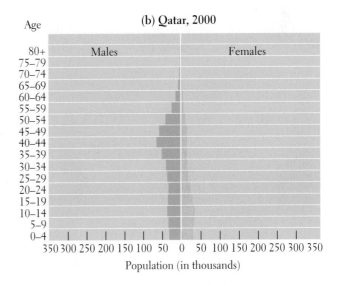

**FIGURE 6.17** Population pyramids for Iran, Qatar, and Israel. The population pyramid for Iran is at a different scale (millions) from those for Israel and Qatar (thousands). The imbalance of Qatar's pyramid in the 25–54 age groups is caused by the presence of numerous male guest workers. Note, too, that all three pyramids show missing females in the younger age groups. (This is most easily observed by drawing lines from the ends of the male and female age bars to the scale at the bottom of the pyramid and comparing the numbers.) [Adapted from "Population pyramids for Iran," "Population pyramids for Qatar," and "Population pyramids for Israel" (Washington, D.C.: U.S. Bureau of the Census, International Data Base, May 2000), at http://www.census.gov/ipc/www/idbpyr.html.]

## Migration and Urbanization

Emigration is increasingly common across North Africa and Southwest Asia. Because jobs are hard to find in the region, especially for workers with more than a basic education, many millions emigrate to other parts of the world in search of work. Most of those leaving the region are young men, because here women typically do not travel widely, or on their own. Tens of millions are guest workers in Europe, where, in 2004, Turkish guest workers alone numbered more than 3.3 million. Many emigrants intend to return home eventually, in part because it is difficult to gain citizenship in European countries. In the meantime, remittances sent home to their families significantly increase local standards of living.

In some parts of the region, immigration, rather than emigration, is the trend. Over the past 30 years, the oil-rich states of the Arabian Peninsula have recruited large numbers of guest workers to rapidly modernize the built environment and perform needed services. In Saudi Arabia, for example, the demand for skilled guest workers is large even though many young Saudis with education certificates are unemployed. Saudis lack the construction, technical, and service skills that are in high demand. By 2004, according to Saudi Arabia's own agency for labor relations, immigrants made up 88 percent of the labor force. Foreigners fill jobs at all skill levels, working in domestic service; teaching in schools and colleges; build-

ing roads, government buildings, housing, schools, water desalinizing plants, oil and gas production facilities, and modern irrigation systems; and running the completed systems. Similar patterns are found in other Persian Gulf countries. In the city of Dubayy (Dubai), for example, immigrants outnumber natives eight to one.

Guest workers come to the Arabian Peninsula from all over the world, but employers prefer Muslims. Over the last two decades, several hundred thousand workers—some technically skilled, many willing to do manual labor—have arrived from Palestinian refugee camps in Lebanon and Syria; others have come from Egypt and India. In the 1980s, 1.5 million came from Pakistan alone. Many, especially female domestic workers, came from Muslim countries in South and Southeast Asia—India, Sri Lanka, Bangladesh, Indonesia, and the Philippines—where chances for employment are limited; they were attracted to Saudi Arabia by the relatively high wages and the chance to make the pilgrimage to the holy cities of Makkah and Al Madinah.

Immigrant workers on the Arabian Peninsula are only temporary residents with no job security, often living in squalid conditions alongside the opulence of native lifestyles. During and after the Gulf War in 1991, millions of foreign workers fled or were summarily expelled from the Gulf countries. These shifts have had global effects. When migrant workers are forced back to refugee camps in Syria or to rural villages in Indonesia, Turkey, Mexico, or Egypt, the loss of

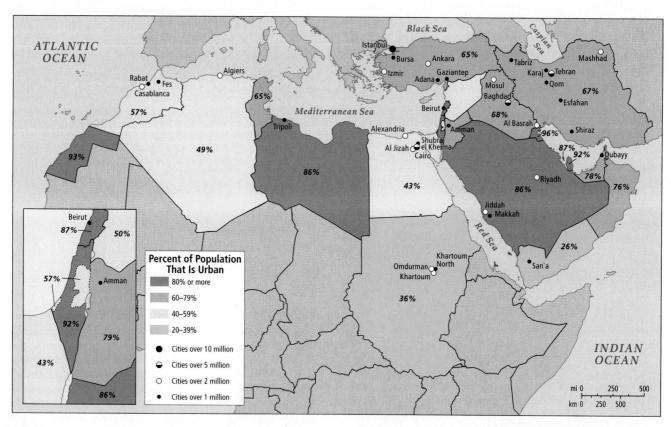

**FIGURE 6.18** Percentage of the population living in cities in North Africa and Southwest Asia. By 2005, more than 58 percent of the region's people lived in urban areas, although the pattern varied greatly across the region. In 2006, there were at least 35 cities with more than 1 million people. The inset map shows the Eastern Mediterranean countries. [Data from *2005 World Population Data Sheet* (Washington, D.C.: Population Reference Bureau) and *World Gazetteer*, at http://www.world-gazetteer.com/wg.php?x=&men =gcis&lng=en&dat=32&srt=npan&col=ahq&geo=-1.]

income often means a return to grinding poverty for their families. Although the numbers of migrant workers remain high today, some Arabian Peninsula governments are undertaking efforts to better prepare their own citizens to take over jobs, especially at managerial levels. As those trained in Europe and America return from their studies abroad, some gladly take on jobs that immigrants once performed, but others idly wait many years for a high-status job.

Israel also has many immigrants, but those immigrants are Jews settling there permanently, many of them fleeing persecution. Immigration surged in the 1990s, when more than 700,000 Jews were allowed to enter from the former Soviet Union. Another 25,000 fled the Ethiopian civil war. Despite already crowded conditions, Israel encourages Jewish immigrants, partly because its constitution grants sanctuary to Jews and partly because it believes that a large Jewish population is important for political weight in the region.

Internal migration from rural villages to urban areas has also been an important population pattern across the region. Until recently, most people lived in small settlements; by 2005, however, 58 percent of the region's people lived in urban areas, although the pattern varied greatly across the region (Figure 6.18). There are now more than 120 cities with populations of at least 100,000; 35 cities have more than one million people. In contrast to patterns in other regions, where the poor build rings of shantytowns on the perimeters of relatively modern cities, poor urban migrants in this region often occupy the medieval interiors of old cities. Here streets are narrow pedestrian pathways and there is little plumbing, sewage disposal, or clean water. The ancient dwellings, dating back 500 years or more, are worthy of historic preservation, but the inhabitants are far too poor to provide even routine maintenance.

Alongside these inner-city spaces used intensively by the poor are new modern urban spaces created by increasingly wealthy elites. Geographer Petra Kuppinger notes that Cairo, Egypt, like other cities in the region, is the site of new shopping malls, gated residential communities, tourist hotels, and private clubs that do not fit into the established urban landscapes, but rather are walled-off enclaves that only emphasize disparities in wealth and fears of political upheaval. These new urban spaces are financed by global capital, and they take little heed of local environmental conditions and only superficially use locally evolved architectural styles.

## Refugees

There are millions of refugees in North Africa, the eastern Mediterranean, Turkey, Iraq, and Iran. Usually they are escaping human conflict, but environmental disasters such as earthquakes or long-term drought also displace many people. When Israel was created, as many as 2 million Palestinians were displaced to refugee camps in Lebanon, Syria, Jordan, the West Bank, and the Gaza Strip. (The West Bank and Gaza Strip are remnants of once much larger Palestinian-controlled lands.) Palestinians still constitute the world's oldest and largest refugee population, numbering at least 5 million. The tiny country of Jordan bears a heavy refugee burden. By early 2007, Jordan was sheltering about 4 million Palestinian refugees plus 700,000 Iraqis, who fled there when the Iraq war and insurgency began to take enormous tolls. Iran is sheltering more than a million Afghans and Iraqis because of continuing violence and instability in their home countries. Across the region, even more people are refugees within their home countries: 1.7 million Iraqis are internal refugees, and in Sudan, between 5 million and 6 million are living in refugee camps.

Throughout the region, refugee camps often become semipermanent communities of stateless people in which whole generations are born, mature, and die. In Palestinian refugee camps, which are no better than shantytowns, conditions are extremely difficult (Figure 6.19). In the Gaza Strip, conditions are squalid, and 81 percent of the people live in poverty; 60 percent have incomes limited to $150 a month. Although residents of these camps may show enormous ingenuity in creating a community and an informal economy under adverse conditions, the cost in social disorder is high. Tension and crowding create health problems. Even birth

**FIGURE 6.19 Refugee life.** These children are on their way to school in Beach Camp, Gaza City, in May 2006. Beach Camp is one of the world's longest-standing refugee camps, in place since the late 1940s. Its residents live in extreme poverty. [© Zoriah/The Image Works.]

control is out of reach: women have an average of 5.78 children each. Children rarely receive enough schooling. Disillusionment is widespread. Years of hopelessness, extreme hardship, and lack of employment take their toll on youth and adults alike, leading some to turn to terrorism as revenge against those they see as responsible for their suffering. Moreover, even though international organiza-tions contribute some funds to support a basic level of services for refugees, these displaced people constitute a huge drain on the resources of their host countries. In Jordan, for example, the popu-lation of refugees and their children accounts for more than half the total population of the country, and they have changed life for all Jordanians.

# II CURRENT GEOGRAPHIC ISSUES

North Africa and Southwest Asia (Figure 6.20) are characterized by diversity in almost every regard, especially in physical landscapes, level of economic development, political and economic systems, social relations, and the role of religion.

## SOCIOCULTURAL ISSUES

In this section we'll examine a few examples that illustrate the broad social changes occurring in the region: in religion, family, gender roles and female seclusion, the lives of children, and cul-tural diversity as reflected in language.

### Religion in Daily Life

More than 93 percent of the people in the region are followers of Islam. The Five Pillars of Islamic Practice (Box 6.1) embody the central teachings of Islam. All but the first pillar are things to do in daily life, rather than articles of faith. (Pregnant women and others who are poor, young, or sick are exempted from the physically demanding pillars.) Islamic practice in the region is thus a consis-tent part of daily life.

Saudi Arabia occupies a prestigious position in Islam, as it is the site of two of Islam's three holy shrines, or sanctuaries: Makkah, the birthplace of the Prophet Muhammad and of Islam, and Al Madi-nah, site of the Prophet's mosque and his burial place. (The third holy shrine is in Jerusalem.) The fifth pillar of Islam, that all Mus-lims who are able should make a pilgrimage (hajj) to the two holy cities at least once in a lifetime, has placed Makkah and Al Madinah at the heart of Muslim religious geography. Each year, a large private sector service industry organizes and oversees the 5- to 7-day hajj for more than 2.5 million foreign visitors (Figure 6.21). Although oil now overshadows the hajj as a source of national income, the event is economically, as well as spiritually, important to Saudi Arabia.

Beyond the Five Pillars, Islamic religious law, called **shari'a**, "the correct path," guides daily life according to the principles of the Qur'an. Some Muslims believe that no other legal code is nec-essary in an Islamic society. There are, however, numerous inter-pretations of just what behavior meets and does not meet the requirements of shari'a. Insofar as the interpretation of shari'a is concerned, the Islamic community is split into two major groups: **Sunni** Muslims, who today account for 85 percent of the world community of Islam; and **Shi'ite** (or **Shi'a**) Muslims, who are found primarily in Iran, but also in southern Iraq and the south of Lebanon. The Sunni–Shi'ite split dates from just after the death of Muhammad, when there were already divisions of opinion over who should succeed the Prophet and have the right to interpret the Qur'an. In 661, the Shi'ite-favored successor, Caliph Ali, Muham-mad's son-in-law, was assassinated by the opposing faction, which came to be known as Sunni. This act gave the Shi'ite–Sunni split significance that has endured to the present, and the original ani-

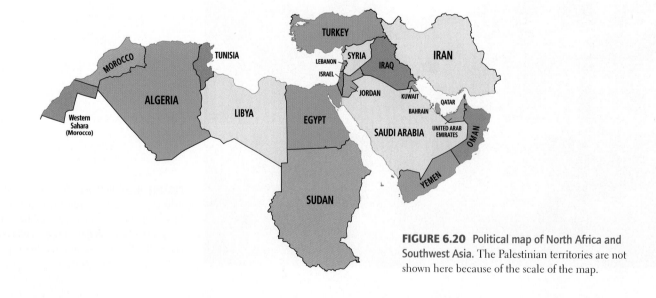

**FIGURE 6.20** Political map of North Africa and Southwest Asia. The Palestinian territories are not shown here because of the scale of the map.

**FIGURE 6.21** The Grand Mosque in Makkah, 2004. Devout Muslims pray as they circle the Kaaba (the black rectangular structure at the upper right) inside the Grand Mosque. [AP Photo.]

mosities have been exacerbated by countless local disputes over land, resources, and philosophies. In Iraq, for example, conflict between Sunni and Shi'ite Iraqis has been greatly intensified by U.S. intervention and the rivalry over political power and oil resources that resulted. Across the region, there are many different interpretations of shari'a within both Shi'ite and Sunni groups, and devoutly observant members of both groups can be found living a host of different lifestyles (Box 6.2).

## Role of Religion in Society

In Muslim communities across the world, a major debate is unfolding on the role of religion in society. Although the differences have led to violence when extremists have felt that governments do not pay proper attention to religion, for the most part the debate is taking place in civil society. Typically, Islam has not recognized a separation of religion and the state as that concept is understood in the West. Rather, there have been varying mixtures of government and religion. In several of the countries in this region—Saudi Arabia, Yemen, the United Arab Emirates (UAE), Oman, and Iran, for ex-

ample—the state is the defender and even the enforcer of the religious principles of Islam. In these **theocratic states,** Islam is the officially accepted religion, the leaders must be Muslim and are considered divinely guided, and the legal system is based on conservative interpretations of shari'a that claim to hark back to the time of Muhammad. Other countries—Algeria, Egypt, Morocco, Iraq, Turkey, and Tunisia—have declared themselves to be **secular states.** In these countries, theoretically, there is no state religion and no direct influence of religion on affairs of state. In practice, however, religion plays a public role even in the secular states. Most political leaders are at least nominally Muslim, and Islamic ideas play a role in government affairs.

Even in the non-Muslim countries of the region, religion is a significant factor in public life. Israel is not a theocratic state, but Judaism strongly influences government policy. Lebanon is a multireligious state with a Jewish minority, more than twelve versions of Christianity, and at least five versions of Islam. Throughout the region, however, as elsewhere, there is great variation in the stringency with which people practice their religion. Many people who consider themselves Muslims do not strictly observe the Five Pillars

---

**BOX 6.1 | CULTURAL INSIGHT**  The Five Pillars of Islamic Practice

1. A testimony of belief in Allah as the only God and in Muhammad as his Messenger (Prophet).

2. Daily prayer at one or more of five times during the day (daybreak, noon, midafternoon, sunset, and evening). Although prayer is an individual activity, Muslims are encouraged to pray in a group and in a mosque.

3. Obligatory fasting (no food, drink, or smoking) during the daylight hours of the month of Ramadan, followed by a light celebratory family meal after sundown.

4. Obligatory almsgiving (*zakat*) in the form of a progressive "tax" of at least 2.5 percent that increases as wealth increases. The

alms are given to Muslims in need. *Zakat* is based on the recognition of the injustice of economic inequity; although it is usually an individual act, the practice of government-enforced *zakat* is returning in certain Islamic republics.

5. Pilgrimage (hajj) at least once in a lifetime to the Islamic holy places, especially Makkah, during the twelfth month of the Islamic calendar. Rituals shared with the devout of all backgrounds, from around the world, reinforce the concept of *ummah*, the transcultural community of believers.

*Source*: Carolyn Fluehr-Lobban, *Islamic Society in Practice* (Gainesville: University of Florida Press, 1994).

## BOX 6.2 | AT THE GLOBAL SCALE The Diversity of Islam

When they envision Islam, most Westerners tend to think that the religion is uniform and extremist in its views. In fact, there is no "Islamic world" and no central Muslim authority; instead, there are many versions of Islam beyond the two main divisions of Shi'ite and Sunni. There are many interpretations of the Qur'an, and several sets of religious law—which is not surprising, given that there are now about 1.5 billion Muslims in the world. In fact, the Muslims in the region of Southwest Asia and North Africa make up less than a third of the world population of Muslims (see Figure 1.13 on page 12). Here are stories of how just three different Muslims live out their faith.

Ebrahim Moosa, a native of South Africa who was an activist in the anti-apartheid struggle there, was inspired to learn more about his Muslim faith when a fundamentalist Christian South African friend told him that Islam was false and Muhammad a fraud. In response, Moosa spent years studying in **madrasas** (Muslim schools) in Asia and learning Arabic so he could read and interpret the Qur'an himself. Later he traveled worldwide as a reporter for a Muslim magazine. His studies and experiences persuaded him that there are multiple versions of Islam. Now a professor of religious studies at Duke University in North Carolina, he is an activist for a progressive Islam—the idea that Muslims must find a modern path that eschews racism and sexism, one that works for human rights but does not succumb to Western secularism and materialist values.

Abdullah Shakr is a middle-aged Jordanian businessman who is fluent in English and favors well-made business suits. He favors the reshaping of the Muslim world through the reestablishment of the Caliphate, a pan-Muslim entity that would erase national borders and create a single Muslim state stretching from Morocco to Indonesia. The idea is modeled on the original Caliphate, a very loose and extremely decentralized geographic unit created after the death of Muhammad in 632, that expanded as a result of conquest, treaty, and trade to cover most of

Southwest and Central Asia and North Africa. Shakr, who feels that Islam has grown poor, weak, and corrupt under European colonialism and Western influence, hopes to spread his pan-Muslim philosophy peacefully through the spoken word. To this end, he and his fellow members of Hizb ut-Tahrir, the Party of Liberation, take time away from their business interests to stand on street corners from Pakistan to Jordan to Morocco, calmly passing out literature that explains their perspective.

In Istanbul, Turkey, Nasiye Wadud (a pseudonym) hopes to become a physicist, but she is presently blocked from continuing her studies by a government ban on head scarves. The government of Turkey, which is secular, banned the scarves and other displays of religiosity in all public institutions when an Islamist political party gained 20 percent of the vote in a national election. Nasiye chooses to wear a scarf and a long, lightweight coat in public because these garments signal that she is a devout and modest Muslim. Although Nasiye does not regard herself as a militant, she and her female friends are active in a student-led movement to eliminate the ban on scarves and to allow other practices of Islam, such as daily prayers, within the walls of the university. Nasiye finds herself in a quandary: by her choice of dress, she serves as a symbol of militant Islam—a movement that supports the idea that women should stay out of the public eye and play only familial roles—at the same time she is seeking a professional and public role as an educated woman.

*Sources: Washington Post National Weekly Edition* (January 19–25, 2004): 10–11; Yonat Shimron, *Raleigh News and Observer* (November 30, 2003), http://www.duke.edu/religion/home/moosa/moosa.html; James Brandon, "The Caliphate, one nation under Allah, with 1.5 billion Muslims," *The Christian Science Monitor*, May 10, 2006, http://www.csmonitor.com/2006/0510/p01s04-wome.html; FRONTLINE, "Portraits of ordinary Muslims," http://www.pbs.org/wgbh/pages/frontline/shows/muslims/portraits/turkey.html and http://www.pbs.org/wgbh/pages/frontline/shows/muslims/themes/.

or shari'a, and they prefer that Islam play a limited role in public life. The same can be said for many Christians and Jews in the region, many of whom trace their heritage back to pre-Islamic days.

## Family Values

In North Africa and Southwest Asia, the traditional multigenerational patriarchal family is still very much the norm. Nonetheless, patterns are changing: families are becoming smaller, and no longer do several generations of one family always share the same household.

In traditional Islamic culture, the family defined a physical space as well as a functional grouping. Physically, the family space was usually a walled compound focused inward on a courtyard, where food, shelter, and companionship were provided (Figure 6.22). The size and details of these compounds varied according to

social class and geographic location. The family group consisted of kin that spanned several generations, including elderly parents as well as adult siblings and cousins and their children. A system of interlocking duties, obligations, and benefits, often assigned by gender and age, provided a role for each individual and solidarity for the whole family. All accomplishments or misdeeds became part of a family's heritage. The responsibilities of family membership were enforced through informal social pressures that ensured no one became an obvious shirker. Whether one received a meal or something as grand as a university education from pooled family resources, the recipient knew that some measure of repayment would come due eventually. This informal contract between generations ensured the flow of financial remittances to elderly kinfolk; although these remittances were formally expected only from sons, daughters who were able sent them as well. These concepts of home and family and their attendant obligations are still very much

FIGURE 6.22 A view of El Oued, Algeria. Known as the "City of a Thousand Domes," El Oued is an oasis town over 1000 years old located in the Sahara. The large dome in the foreground tops a mosque. The small domes, covering rooms around central rectangular courtyards, help shield the more than 130,000 residents from the desert summer heat. As in many crowded cities in the region, houses seldom rise more than one or two stories, in part because this helps air circulation in a hot climate. [© ALTITUDE/Yann Arthus-Bertrand.]

part of value systems in the region, but everyday practice is changing as people move into apartments in cities and away from strict village social pressures.

## Gender Roles and Gender Spaces

Carefully specified gender roles are common in many cultures, and there is often a spatial component to these roles. In North Africa and Southwest Asia, the differences between male and female roles are reflected in the organization of space within the home and within the larger society, though practice varies widely from place to place.

In both rural and urban settings, men and boys go forth into *public spaces*—the town square, shops, the market. Here, men not only make a living, but also continually transact alliances with other men that will advance the interests of their families. It is through such networks, which span a range of social strata, that people get the best price for an appliance or a car, find a mate for a son or daughter, obtain a job or admission to a professional school, find some scarce item, or unsnarl a particularly nasty bureaucratic problem. For any favors they receive, they incur future obligations. If men are successful at making such arrangements, they also garner considerable respect and prestige within their own families.

Women used to inhabit primarily secluded *private spaces*. In the traditional family compounds described above, the courtyard was usually a private, female space within the home; the only men who could enter were relatives. For the urban upper classes, female space was an upstairs set of rooms with latticework or shutters at the windows from which it was possible to look out at street life without being seen (Figure 6.23). Today, the majority of people in the

FIGURE 6.23 A house in Jiddah, Saudi Arabia. The various types of louvered and latticed bays allow women to observe life on the streets from seclusion. Such architectural details are found wherever Islam has been an influence. [Hubertus Kanus/Photo Researchers.]

317

region live in urban apartments, yet even here there is a demarcation of public and private space, with one or two formal reception areas for nonfamily visitors and rooms deeper into the dwelling for family activities.

Now, women as well as men go out into public spaces, but just how women enter these spaces remains an issue, and customs vary not only from country to country but also from rural to urban settings and by social class. In some parts of the region, particularly in Saudi Arabia, seclusion remains strict. People believe that women should not be in public space except when on important business, and it is the custom that women be accompanied by a male relative in public. Elsewhere (in Egypt, for example), some classes may enforce seclusion more strictly than others. Some affluent urban women may observe seclusion more strictly than do rural women because, although rural women are traditional in their outlook, they have many tasks that they must perform outside the home: agricultural work, carrying water, gathering firewood, and marketing. In the more secular Islamic societies—Morocco, Tunisia, Libya, Egypt, Turkey, Lebanon, and Iraq—women regularly engage in activities that place them in public spaces. And, increasingly, female doctors, lawyers, teachers, and businesspeople are found in even the most conservative societies.

Figure 6.24 compares the various levels of restrictions on women across the region. Some of these restrictions are officially sanctioned, and some are the result of informal social pressures. The primary reason for veiling is that when a woman enters public space, her honor must be maintained; therefore, in much of the region, it is important that her street clothing be modest. What she wears in the privacy of her home among family and friends may be quite different. When going to a party, for example, a woman may wear a long coat and shawl to get there and then shed them on arrival, revealing a stylish outfit of either Western or Muslim design for her friends to admire (Box 6.3). As another means of maintaining their honor, it is common for women in public to travel in groups, or for a single woman to take a younger male sibling along on errands.

There is considerable controversy over the origin and validity of female seclusion and veiling as Muslim customs. Scholars of Islam say that these ideas predate Islam by thousands of years. They do not derive from the teachings of the Prophet Muhammad, who in fact may have been reacting against such customs when he advocated equal treatment of males and females and helped with domestic chores in his own home. His first wife, Khadija, was an independent businesswoman whose counsel he often sought. The Qur'an allows a woman, married or not, to manage her own property and keep her wealth or a monthly salary for herself. By contrast, the custom of transferring to her husband's control all the wealth a woman brought to a marriage was not legally ended in Britain until 1870 and in some parts of the United States until in the 1960s!

Many countries across the region (Tunisia and Morocco are examples) are enacting legislation to improve the status of women.

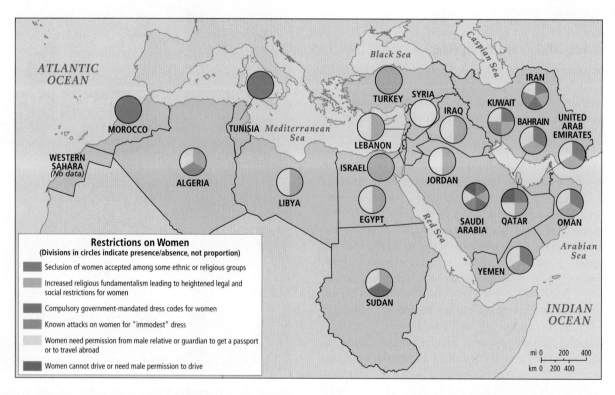

**FIGURE 6.24 Variation in restrictions on women.** There is considerable variation from country to country in the restrictions placed on women. Women's rights are perhaps most strongly protected in Turkey and Israel, where equality for women is constitutionally guaranteed, though religious fundamentalists are working to repeal these guarantees in both countries. [Adapted from Joni Seager, *The Penguin Atlas of Women in the World, Completely Revised and Updated* (New York: Viking Penguin, 2003), pp. 14–15.]

## BOX 6.3 | CULTURAL INSIGHT  The Veil

The veil allows a devout Muslim woman to preserve a measure of seclusion even when she enters a public space, and thus it actually increases the space she may occupy with her honor preserved. Some garments totally cover the body, including the face; some allow the eyes to be seen; and some are a mere head covering. Among women who have worn the all-encompassing *chador* (full body veil), there are those who find that it offers anonymity and satisfying privacy from the gaze of strangers. Women who wear some other version of religious dress, such as the *hijab*—which can mean a long, loose dress and a scarf that leaves the face exposed, or just the head scarf worn with any other dress, or even blue jeans and a T-shirt—speak of liking the signal it sends that "here is a devout Muslim woman." They add that when women don religious garb, they are reminding men to be religiously observant as well. Paradoxically, many women throughout the region wear high-fashion Western clothes under the *chador* or *hijab*, or at special gatherings that may include women only.

Veils may be bought in shopping malls across the region. But for women who think of the veil as providing a chance for a fashion statement, Wafeya Sadek runs an Islamic designer-veil studio out of her home in Cairo. Women come from across the city to buy a custom-made veil. One of her best-sellers is a leopard-print veil, favored by young women for daywear. Some of Sadek's designs for evening wear seem to call for the type of bold individualism that is frowned on by conservative Muslims: for example, a close-fitting black bonnet topped with a wildly colorful cockscomb of velvet. Operating in the informal economy, Madame Sadek counts on word-of-mouth advertising. As Figure 6.25 shows, fashionable veils are also available for Muslim women in other countries.

*Sources:* Amy Docker Marcus, *Wall Street Journal* (May 1, 1997); I-chun Che, "Clothing line offers modesty and beauty for Muslim women," *The Sun*, Sunnyvale, CA (October 15, 2003), http://www.svcn.com/archives/sunnyvalesun/20031015/sv-cover.shtml.

**FIGURE 6.25** Fashionable Islamic clothing in the United States. There are now about 8 million Muslims in the United States. Sadiya Shaikh, together with her mother and a friend, started Flippant, a design house in California that appeals to young Muslim women like these who wish to be both stylish and religiously observant. [Jim Gensheimer, *San Jose Mercury News.*]

In some places the need is great: contrary to the teachings of the Prophet Muhammad, women have been forced to marry and to undergo virginity tests, and they are unprotected in cases of domestic violence. Legislatures across the region are being petitioned by activist women to criminalize "honor killings" in which a woman or girl who is perceived to have dishonored her family in some way, or has been a victim of rape, is actually killed by her own relatives. (Such killing, thought to restore family honor, is strictly forbidden by Muslim doctrine.) In 1999, the new king of Morocco, Mohammed VI, began to implement a program that encourages participation by women in civil society and reforms the legal code to grant women legal rights. Tunisia recognizes the equal right of either spouse to seek divorce. It also requires mutual spousal obligation in household management and child care, and it no longer permits preteen girls to be married.

One source of contention in this region is the custom of **polygyny**: in many countries Muslim men are legally permitted to take more than one wife. Although polygyny is allowed under certain conditions by the Qur'an, it is not encouraged, nor is it a common modern practice. It is estimated, for example, that less than 4 percent of males in North Africa have more than one wife. The occurrence of polygyny is low partly because the Qur'an, which limits the number of wives to four, requires that all must be treated with scrupulous equality. Urbanization and modernization are also important factors. When agriculture was the main economic activity, multiple wives with several children each may have been more productive economically, but urban life, with its small living spaces and cash requirements, favors smaller families. And according to the Ibn Khaldun Center for Development Studies in Cairo, democratization in North Africa and Southwest Asia led to the successful bans against polygyny in Tunisia, Lebanon, and Palestine.

## The Lives of Children

Three sweeping statements can be made about the lives of children in the Islamic cultures of North Africa and Southwest Asia. First, children contribute to the welfare of the family starting at a very young age. In cities, they run errands, clean the family compound, and care for younger siblings. In rural areas, they do all these chores and also tend grazing animals, fetch water, and tend gardens. Second, their daily lives take place overwhelmingly within the family circle. Both girls and boys spend their time within the family compound and in nearby spaces with adult female relatives and with siblings and cousins of both sexes. Even teenage boys in most parts of the region identify more with family than with age peers. Only the poorest of children play in urban streets. In rural areas, prepubertal girls often have considerable *spatial freedom*—the ability to move about in public space—as they go about their chores in the village. (After puberty they may be restricted to the family compound.) The U.S. geographer Cindi Katz found that until puberty, rural Sudanese Muslim girls have considerably more spatial freedom than do girls of similar ages in the United States, who are rarely allowed to range alone through their own neighborhoods.

The third observation is that the lives of children in this region, like those of so many around the world, are increasingly circumscribed by school and television. Most children go to school; many boys go for a decade or more, and increasingly girls go for more than a few years. Even in rural areas, it is fairly common for even the poorest families to have access to a television, and it is often on all day, in part because it provides a window on the world for secluded women. TV may serve either to reinforce traditional cultural values or as a vehicle for secular values, depending on which channels are watched. Since September 2001, the exposure of children (and women) to broadcast news has taken a decided leap.

## Language and Diversity

The eastern Mediterranean is the ancient home of peoples who spoke Semitic languages, including the peoples of Assyria, Babylonia, and Phoenicia. Modern speakers of Semitic languages include the Arabs and Jews. Arabic is now the official language in all countries of North Africa and Southwest Asia except Iran, Turkey, and Israel (where Farsi, Turkish, and Hebrew, respectively, are spoken), but this uniformity of language masks considerable cultural diversity (Figure 6.26). There are numerous minorities within the region who have their own non-Arabic languages, such as Berbers, Tuaregs, Nubians, Kurds, and Turkomans. There are also many dialects of Arabic that indicate deep cultural variations across the region. Yet, in this era of modern communications, many dialects and minority languages are disappearing as standardized Arabic becomes the language of mass media culture. Nonetheless, the widespread use of Arabic facilitates region-wide communication and pan-Muslim solidarity. Meanwhile, the use of French and English as second languages, which is also very common, especially in urban areas, and the dominance of English on the Internet are contributing to a decrease in language diversity and to the intrusion of outside cultural influences.

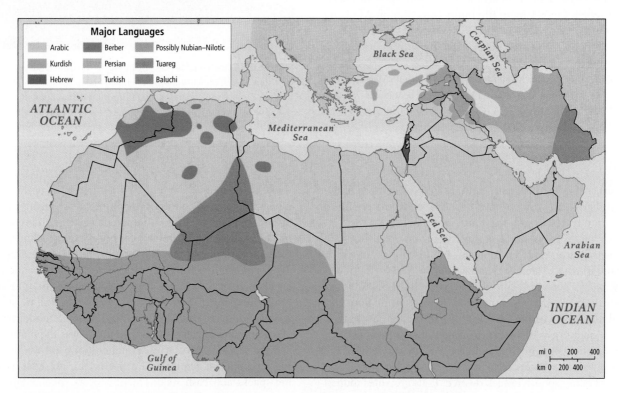

**FIGURE 6.26** Major languages of North Africa and Southwest Asia. [Adapted from Charles Lindholm, *This Islamic Middle East: An Historical Anthropology* (Oxford: Blackwell, 1996), p. 9.]

## Islam in a Globalizing World

Many contemporary Muslims see modern global culture, much of it originating in Europe and North America, as undermining important values, such as the duty of the individual to the family and community. Hence some Muslims object to the liberalization of women's roles, especially to their being active outside the home. Many Muslims also lament the global spread of Western culture because of its open sexuality and consumerism and what they see as its tendency toward hedonism. They worry that such ways will lead to family instability, alcoholism, drug addiction, and street crime by youth who have been poorly raised by inattentive parents. Consumerism is seen as leading to selfishness and to widening gaps between rich and poor (remember that one of the pillars of Islam is the duty to address the injustice of poverty). These concerns are similar to those of people of other religions around the world, such as Baptists in the southern United States and Hindu nationalists in India, who worry about the erosion of traditional values and ethical systems.

Particularly conservative Muslims, known as **Islamists** (also called **Islamic fundamentalists**), hark back to older, and hence what they feel are more accurate, interpretations of the Qur'an. They favor a simple, prayerful life focused on family and community, traditional gender roles, and respect for the elderly. Different Islamist factions vary greatly in the perspectives and fervor they bring to their causes. The majority simply seek to moderate Western influences; for example, most see the advantages of bringing their people into the computer age, but they are reluctant to accept some of what they see as negative Western influences that come from the Internet. The proper role of the press is also hotly debated, with some believing that a free press is essential and others that the press should not go beyond simple statements about events (Box 6.4). More moderate Muslims call for an open debate within Islam that includes a reevaluation of

---

## BOX 6.4 | AT THE REGIONAL SCALE    The Evolving Role of the Press

The press, including both print and broadcast media, did not play a major role in North Africa and Southwest Asia until very recently. The few respected newspapers in major cities rarely took strong stands in opposition to government policies. In many cases, the press was actually controlled by sitting governments and served primarily as a government information organ, informing the public about events and policies but exercising little critical analysis. That is still the case in Saudi Arabia, which we will discuss first; some exceptions to this general pattern follow.

In Saudi Arabia, there are more than a dozen newspapers on the newsstands every morning, and all are owned or controlled by the royal Saud family. The Saudi print press is said to be the most influential in the Arab world, yet all Saudi newspapers are constrained by the proviso that nothing critical may be said about Islam, the royal family, or the Saudi government, and investigative reporting is rare. When accidents happen or some malfeasance by a public official is revealed, the story is blandly reported with little or no effort to explore the causes of events or their possible effects, or to hold responsible officials accountable. Occasionally, critical public discourse will begin about an issue and then be preemptively shut down. When a fire in a girls' school killed 15 and injured 50, it was soon revealed that rescuers and firefighters were held back by men from the Commission for the Promotion of Virtue and Prevention of Vice—the country's government-paid religious police—because the girls had fled to the balconies without their *abiyas* (veils and robes). For a few days the Saudi press covered the public outcry and called for investigations of the religious police, but then the Minister of the Interior informed the editors that the stories were to stop, and they ceased immediately.

Saudis who seek a more objective approach to news from their country and elsewhere watch the Al Jazeera TV network, broadcast from neighboring Qatar (see pages 297–298), and Al Arabiya, broadcast from Dubayy, via illegal satellite TV hookups.

They also read the Saudi paper *Arab News*, which can be read online in English at http://www.arabnews.com/.

In 1997, Hisham Kassem founded the *Cairo Times*, an independent English-language weekly. Nonpartisan and meant to inform readers about current affairs in Egypt, the *Cairo Times* often carried articles critical of the Egyptian government, and soon Kassem lost his license to publish in Egypt. He then registered the *Cairo Times* in the U.S. state of Delaware, where he prepared each issue before flying the files to Cyprus for printing. He then carried the paper into Egypt as luggage, where it had to clear the censors, who often refused to let him sell a particular issue. Many times he was approached by the Egyptian government to become a pro-government "snitch," as he puts it, but he persisted with his critical stance, saying that a free press is essential to reforming what he calls the failed regimes of the Middle East. In 2004, Kassem closed the *Cairo Times* and, with the backing of several business leaders, founded *Al Masry Al Youm*, which specializes in reporting on domestic issues such as election fraud and judges striving for more independence from government pressure. The paper is popular with ordinary Egyptian citizens, and in less than 2 years it has broken even. Kassem says, "We work hard on the credibility of the content, making sure that it is well researched, that it serves the reader. We try to make the paper as reader-friendly as possible. At the same time, we run all the other departments—distribution, advertising, finance—professionally. If the editorial content is influenced by advertising, you deprive the country of its watchdog."

*Sources:* Lawrence Wright, "The Kingdom of Silence," *The New Yorker* (January 5, 2004): 48–73; "Fresh Air," interview by Terry Gross with Hisham Kassem, National Public Radio (February 4, 2004); Amira Salah-Ahmed, "Hisham Kassem," Business Today.com, May 2006, http://www.businesstodayegypt.com/article.aspx?ArticleID=6712; "Why Al Masri Al Youm matters," *The Arabist*, Feb. 24, 2006, http://arabist.net/archives/2006/02/24/why-al-masri-al-youm-matters/.

ancient shari'a interpretations of the Qur'an and a search for more progressive norms on human rights and gender equity.

## ECONOMIC AND POLITICAL ISSUES

There are major economic and political barriers to peace and prosperity within North Africa and Southwest Asia today. The oil wealth so prominent in Western minds is limited to a few elites; most people in the region are low-wage urban workers or relatively poor farmers or herders. The economic base of the region is unstable because oil and agricultural commodities, its main resources, are subject to wide price fluctuations on world markets. A more diverse range of industries is just beginning to emerge and is in need of investment. Meanwhile, hard times loom for many poorer non-oil-producing countries as large national debts are forcing governments to restructure their economies in the ways commonly associated with structural adjustment programs: the streamlining of production and the cutting of jobs and social services. Political barriers to peace and prosperity include the ongoing Israeli–Palestinian conflict that has affected life and attitudes throughout the region; the U.S.-instigated war in Iraq; and lack of substantial progress toward broad-based democratic participation in governments.

## The Oil Economy

The region contains some of the world's largest known reserves of fossil fuels (oil and natural gas). These deposits are located mainly around the Persian Gulf (Figure 6.27), and the oil wealth so often associated with the region as a whole is concentrated mostly in countries that border the Gulf: Saudi Arabia, Kuwait, Iran, Iraq, Oman, Qatar, and the UAE. Oil and gas are also found in the North African countries of Algeria, Tunisia, Libya, and Sudan.

Early in the twentieth century, European and North American companies were the first to exploit the region's oil reserves. These companies paid governments a small royalty for the right to explore and drill for oil. The oil was processed at on-site refineries owned by the foreign companies and sold at very low prices, primarily to Europe and the United States and eventually to other places, such as Japan.

The governments of the region did not assume control of their oil reserves until the 1970s, when they declared all oil resources and industries to be the property of the state. However, even before this, in the 1960s, the oil-producing countries organized a **cartel**, a group of producers strong enough to control production and set prices for its products. The oil cartel is called **OPEC**—the **Organization of Petroleum Exporting Countries.** OPEC now includes all

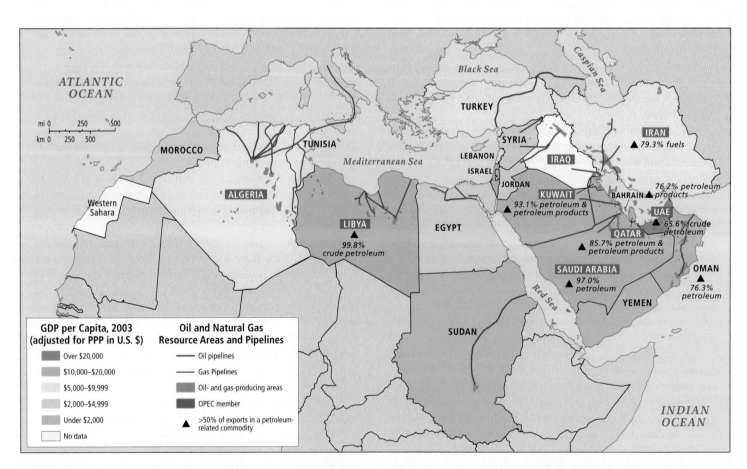

**FIGURE 6.27** Economic issues: Oil and gas resources in North Africa and Southwest Asia. [Adapted from Rafic Boustani and Philippe Farques, *The Atlas of the Arab World—Geopolitics and Society* (New York: Facts on File, 1990), pp. 85, 88, 89; Richard Overy, ed., *The Times History of the World* (London: Times Books, 1999), p. 304; *Hammond Atlas of the Middle East,* revised (Union, N.J.: Hammond, 2001), pp. 8–9; and U.S. Department of Energy, http://www.eia.doe.gov/emeu/cabs /Archives/africa/nafrica.pdf.]

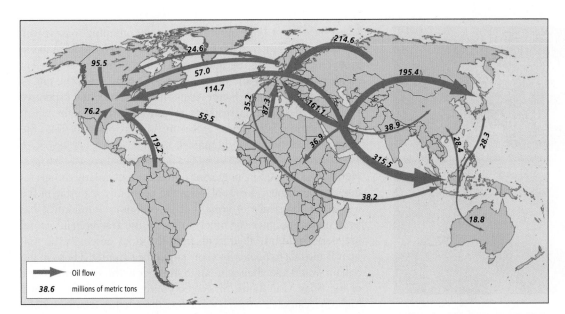

**FIGURE 6.28** World oil flow in millions of metric tons, 2003. Patterns of flow in the oil trade change constantly. In 2003, the United States got its supply from many sources. Europe received much of its oil from the Arab states, as did Japan and Southeast Asia, but the largest source of oil for Europe was Russia. [Adapted from "Flow of oil," *National Geographic Atlas of the World Eighth Edition* (Washington, D.C.: National Geographic Society, 2005), Plate 20.]

the oil-producing states marked with a blue box on Figure 6.27, plus Venezuela, Indonesia, and Nigeria. OPEC members cooperate to periodically restrict oil production, thereby significantly raising the price of oil on world markets. World petroleum prices result from a complicated set of factors, however. The OPEC countries often fail to reach agreement, and non-OPEC producers, such as Russia and Canada, can influence petroleum supplies and prices. Consumers, by reducing or expanding demand, can also affect prices. Furthermore, geopolitical events can have a major effect on oil prices, as demonstrated by the aftermath of the September 11, 2001, attacks, when oil prices rose and fell a number of times. By 2005, a major factor in oil prices was the increasing demand in rapidly industrializing China and India. Figure 6.28 shows the countries that presently buy oil from the region.

After the major oil-producing countries raised prices dramatically in 1973, oil income in Saudi Arabia alone shot up from U.S.$2.7 billion in 1971 to U.S.$110 billion in 1981. Yet because the OPEC countries did not invest much of their oil wealth at home in basic human resources, they have been unable to improve their economic bases. Like their poorer neighbors, they have remained dependent on the industrialized world for their technology, manufactured goods, and skilled labor. Nevertheless, oil has brought significant benefits to the region as a whole. Those countries that had the largest amounts of oil (Saudi Arabia, the UAE, and Kuwait) sharply increased their investment in long-neglected roads, airports, new cities, irrigated agriculture, and petrochemical industries, though they invested much less in education, social services, public housing, and health care. People from non-oil-producing countries in the region were also able to share in the wealth: many of the laborers on development projects in the oil-rich countries came from Egypt, Jordan, Syria, and Turkey.

Over the next several decades, oil prices will probably rise as demand for energy rises. At some point in the future, the depletion of oil resources and the development of new sources of energy will force OPEC countries to find other ways of earning an income or risk economic ruin.

## The Traditional Economy: Agriculture and Herding

Despite the abundance of oil and gas in some areas, much of the region remains poor and highly dependent on agriculture and herding. In several of the largest countries, 30 percent or more of the people are employed in agriculture and depend on their own efforts for most of their daily nutrition. Nonetheless, the actual market value of their production is very low. In Morocco, for example, agriculture accounts for only 22 percent of GDP, yet 40 percent of the labor force works in agriculture. Similar situations exist in Egypt, Iran, Turkey, and to a lesser extent, Saudi Arabia and Libya. When measured in dollar value, compared with the industry and service sectors, agriculture is not productive, but when measured in value to diets and family budgets, farming and herding are essential to national economies.

*Agriculture.* Crop agriculture in the region has, until recently, been confined to coastal zones, river valleys, and areas with a Mediterranean climate (see Figure 6.4 on page 300). Grains, cotton, rice, corn, peanuts, fruit, sesame, and sugarcane are raised in the Nile Valley of Sudan and Egypt. In the Tigris and Euphrates river valleys, Turkish farmers produce cotton, tobacco, sugar beets, and livestock, and Iraqi and Syrian farmers grow grains, nuts, tea, tobacco, and livestock. Iranians produce similar farm products, primarily in uplands. In the mountains of the Arabian Peninsula, farmers still produce coffee, which was first domesticated in this region (Figure 6.29). On the northern coast of Africa, grains, citrus fruits, olives, dates, and wine grapes are grown for home consumption and for export to Europe.

Ambitious irrigation schemes in Libya, Egypt, Saudi Arabia, Turkey, and Iraq are now expanding commercial, primarily export-oriented agriculture to neighboring areas that were previously too dry to support large-scale cropping. These attempts at expansion have not always been well conceived. Many state-sponsored irrigation projects have damaged soil fertility through **salinization,** which

**FIGURE 6.29** Buying coffee in the old city section of San'a, Yemen. Yemen is reportedly where coffee was first marketed and exported. Beginning in the seventeenth century, Mocca was the main port for Yemeni coffee exports and is the place where mocha coffee got its name (the port is no longer active). Most coffee in Yemen is grown in ten western provinces in populated narrow valleys or on high mountain terraces. [Danita Delimont/Alamy.]

occurs when large amounts of water evaporate, leaving behind salts and other minerals that accumulate over time and inhibit plant growth. Salinization is particularly likely to occur in arid lands because there is little rain to wash away the salt. Israel has developed more efficient techniques of drip irrigation that curtail salinization, but until very recently poorer states have been unable to afford the technology, and others are wary of depending on a technology developed by Israel, a country they deeply distrust.

The region's numerous political tensions have convinced many governments that they should try to be self-sufficient in food production regardless of the expense. To that end, Saudi Arabia and Libya have spent huge amounts of money pumping massive volumes of groundwater to the surface at highly unsustainable rates to grow crops for home use and export. Libya is mining an ancient aquifer for irrigation and other purposes at eight times the rate of natural replenishment. Saudi Arabia has even used the expensive process of desalinizing seawater to provide irrigation for wheat fields (Figure 6.30). Wheat grown in this way costs 10 times the price of wheat on the world market.

***Herding.*** The tending of grazing animals—particularly camels for transport and sheep for wool—was the economic mainstay of the region for thousands of years, but its economic importance has been in decline since the nineteenth century. Traditionally, camel breeders would take their herds from the sandy deserts of the Arabian Peninsula to cooler Syria or Iraq in the summer. Those in the Sahara would take their animals to summer in the southern fringes of the Atlas Mountains. Since the construction of railroads and highways, camels are used less and less, though they retain symbolic importance. The demand for sheep's wool continues. Sheep (and goats), however, cannot travel long distances, so the range of nomadic migrations of sheepherders was always smaller than that of camel herders. The settled herding of sheep is now relatively common even in urban fringe areas, and many families engaged in nonagricultural employment maintain a small herd of sheep that the boys of the family tend.

Nomadic herders, such as the Kurds who live at the juncture of Turkey, Iraq, and Iran, have lost financial and spatial independence as the region has modernized and national borders have become official dividing lines. In establishing large irrigated agriculture projects, many governments are seeking control of water sources that were previously available to the herders, and they may require the nomads to settle permanently in order to provide a labor pool for the irrigation projects. Herders are also forced to settle in one place because their tendency to cross national borders in search of grazing lands and water is now perceived as a national security threat. Finally, governments find nomads hard to tax or to

**FIGURE 6.30** An irrigated wheat field in the Saudi Arabian desert. [Ray Ellis/Photo Researchers.]

control in other ways; for example, it is difficult to enforce compulsory school attendance on their children. The requirement to settle permanently, however, drastically alters the economy and social structure of herding communities and can result in impoverishment and social strife.

## Economic Diversification and Growth

Greater **economic diversification**—expansion of an economy to include a wider array of activities—could bring economic growth and broader prosperity to the region and limit the damage that a drop in the price of oil or other commodities on the world market would bring. For the most part, however, diversification has failed to happen. By far the most diverse economy is that of Israel, which has a large knowledge-based service economy and a particularly solid manufacturing base. Israel's goods and services and the products of its modern agricultural sector are exported worldwide. Turkey is the next most diversified, with Egypt, Morocco, and Tunisia also starting to move into new economic activities. All depend on substantial government investment, and Israel and Egypt rely on significant aid from the United States.

The region's relative lack of economic diversification and its slow economic growth are largely a result of misguided economic development policies, a lack of private investment, political tensions, and war. Beginning in the 1950s, many governments, such as those of Turkey, Egypt, Iraq, Israel, Syria, Jordan, Tunisia, and Libya, established state-owned enterprises for the production of goods for local consumption (not export), such as machinery and metal items, textiles, toilet paper, cement, processed food, and paper and printing. They then protected these enterprises from foreign competition with tariffs and other trade barriers. With only small local markets to cater to, profitability was low and the goods were relatively expensive. Because the industries were protected from both foreign and domestic competition, their products were shoddy and unable to compete in the global marketplace. Furthermore, the extension of government control into so many parts of the economy nurtured corruption and bribery as personal connections with policy makers and regulators became an important factor in the success of any enterprise. Economic diversification and growth were also hindered by a lack of financing from private investors within the region, such as Saudi oil barons, who have generally preferred more profitable investments in Europe, North America, or Southeast Asia.

Finally, both international and domestic military conflicts and ensuing political tensions have resulted in some of the highest levels (proportionately) of military spending in the world. When military spending is looked at as a percentage of GDP, four of the world's top five spenders are in this region, and all but Tunisia and Libya are above the global average of 2 percent (Figure 6.31). High levels of military spending divert funds from other types of development.

All of these factors saddled the poorer governments of the region with crippling debt burdens that forced them to cut back their role in the economy. This trend has been under way since the 1970s, when international lending institutions imposed structural adjustment programs (SAPs) that shifted governments away from

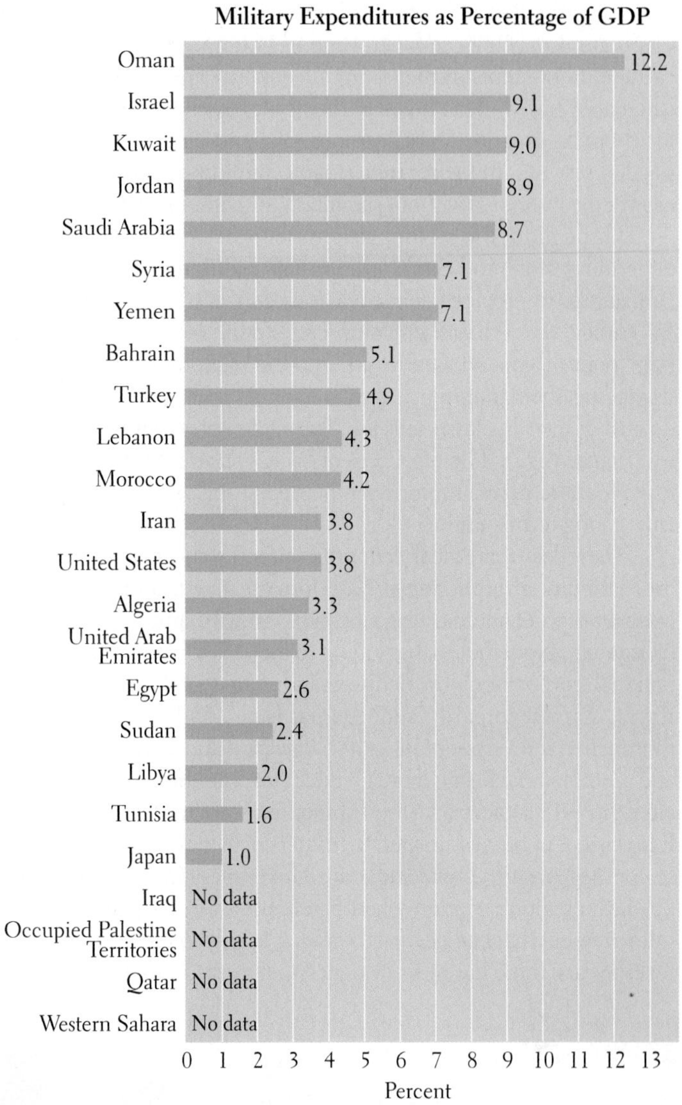

**Military Expenditures as Percentage of GDP**

| Country | Percent |
|---|---|
| Oman | 12.2 |
| Israel | 9.1 |
| Kuwait | 9.0 |
| Jordan | 8.9 |
| Saudi Arabia | 8.7 |
| Syria | 7.1 |
| Yemen | 7.1 |
| Bahrain | 5.1 |
| Turkey | 4.9 |
| Lebanon | 4.3 |
| Morocco | 4.2 |
| Iran | 3.8 |
| United States | 3.8 |
| Algeria | 3.3 |
| United Arab Emirates | 3.1 |
| Egypt | 2.6 |
| Sudan | 2.4 |
| Libya | 2.0 |
| Tunisia | 1.6 |
| Japan | 1.0 |
| Iraq | No data |
| Occupied Palestine Territories | No data |
| Qatar | No data |
| Western Sahara | No data |

**FIGURE 6.31** Military expenditures as a percentage of GDP, 2003. North Africa and Southwest Asia lead the world in military spending as a percentage of GDP. The world average is 2 percent of GDP; Tunisia is the only country in the region below that figure. The United States and Japan are included for comparison. [Adapted from *United Nations Human Development Report 2005* (New York: United Nations Development Programme), Table 20.]

state-led economic development and limited their ability to provide social services (see Chapter 3, page 139).

## Side Effects of Development Efforts

Overall, SAPs in this region have had significant negative effects. Their effect on the poor majority has amounted to a disinvestment in human capital, when in fact a healthier, more educated populace is what is most needed. In Egypt in the 1980s, for example, reductions in food and housing subsidies doubled poverty in rural areas and increased it by half in urban areas. To provide for their impoverished families, hundreds of thousands of men migrated to Jordan and Saudi Arabia to work under short-term construction labor contracts. Millions of other poor rural people were pushed into Egypt's cities and into the informal economy. The streets of Cairo are now so clogged with vendors that traffic cannot pass. Because entrepreneurs in the informal economy are already impoverished, they have little to invest in their enterprises, making it difficult for them to expand to create more jobs and pay taxes to support needed services. In Cairo, more than 6 million poor people live in crowded conditions, without sanitation and other services.

Throughout the region, governments in similar situations are depending more and more on international nongovernmental organizations to provide such services as education and health care. In Turkey, the Education Volunteers Foundation seeks to "leap-frog" poor urban and rural youth into the computer age by sharply upgrading their training in math, science, and language skills in dozens of mobile classrooms and learning centers across the country (Figure 6.32). The resources of NGOs, however, are too small to serve more than a minority of the needy and are unlikely to provide a long-term solution.

These less than ideal economic circumstances have added to the difficulty of attracting private foreign investment that would contribute to economic diversification, growth, and the transfer of technology. Nonetheless, beginning in the 1990s, a number of multinational corporations opened operations in Egypt, and Lebanon, Egypt, Turkey, Jordan, Saudi Arabia, Israel, and Kuwait all drew multinational investment in technology, food and beverage delivery, and auto manufacturing. Foreign direct investment in the region fell after the 9/11 attacks and the subsequent wars in Afghanistan and Iraq, except in services related to military activities. Tourism was the sector most negatively affected, but all investment interest cooled. By 2004, the region's economy had barely begun to rebound, and the improvement that had occurred was due largely to high oil prices. In recent years, there has been evidence that more oil wealth is staying

within the region as direct investments, as regional stock exchanges have expanded with the latest rises in oil prices.

## The Economic and Political Legacy of Outside Influence

Political and economic cooperation in the region has been thwarted by a complex tangle of hostilities between neighboring countries. Many of these hostilities are the legacy of the long history of outside interference in regional politics. Outside influences have continued to affect the region long after the maneuvering of Europeans described earlier in the chapter; the Iran–Iraq war of 1980–1988 and the Gulf War of 1990–1991 were, at least in part, instigated by pressures from outside the region. The same is true of the Iraq war that began in the spring of 2003, which was spearheaded not by Iraq's neighbors, but by the United States and a limited number of allies.

***The Iraq War (2003–Present).*** In searching for an explanation for the terrorist attacks of September 11, 2001, the U.S. administration of George W. Bush focused on Iraq. Iraq's President Saddam Hussein was thought to harbor a lingering hostility toward the United States since the Gulf War, in which the United States rolled back Iraq's invasion of Kuwait. Actual links between the 9/11 attacks and Iraq were never proved, and given that Saddam was a secular leader opposed to Islamic fundamentalism, it is unlikely that he and Al Qaeda leader Osama bin Laden could have found common ground. Nevertheless, the Bush administration was convinced that Iraq had an arsenal of nuclear and other weapons of mass destruction and posed an immediate threat to the United States. The Bush administration began war on March 20, 2003, with the goals of confiscating Iraq's weapons of mass destruction, removing Saddam from power, and turning Iraq into a democracy. Great Britain and a few other countries joined the United States as allies. After only a short time, on May 1, 2003, President Bush declared the war won, but soon terrorist bombs and insurgent attacks by Iraqis who resented the U.S. occupation were taking the lives of U.S. and allied soldiers at the rate of more than 2 per day, for a total of 3000 by January of 2007. The death toll for Iraqis (including civilians) has been much higher—at least 200,000, and perhaps as many as 650,000, according to an October 2006 report by the Johns Hopkins Bloomberg School of Public Health, with many deaths coming as a result of war conditions rather than directly at the hands of foreign troops or insurgents.

The Iraq war has had a variety of unintended consequences. It brought to the surface long-simmering tensions between Sunni in the northwest of Iraq, who had dominated the country under Saddam despite constituting only 32 percent of the population; Shi'ites in the south, who, at 60 percent of the population, have dominated politics since the war; and Kurds in the north, who were brutally suppressed under Saddam. (Saddam's treatment of the Kurds ultimately led to his conviction of mass murder and execution in 2006.) Since many Iraqi Shi'ites have ties to neighboring Iran, that country has had more influence on the postwar Iraqi government than it ever had in the past. Despite a steady increase in Iraqi civil-

**FIGURE 6.32** Education Volunteers Foundation of Turkey (TEGV). Here, TEGV students perform a skit in Charshamba Park, Istanbul. Not only do such skits serve as a vehicle for public education, but performing in them helps the students develop leadership skills. [Courtesy of Kevin Miller, Jr.]

ian deaths, however, recent research does not clearly point to an emerging civil war among Shi'ites, Sunni, and Kurds, but rather to a general pattern of chaos as a variety of groups jockey for power, some possibly hoping to exploit sectarian divisions for their own gain. Nevertheless, opinion surveys of Iraqis have been certain about one thing: the vast majority of Iraqis want all foreign military forces to leave the country as soon as possible.

*As it turned out, winning the war in Iraq was the easy part. Liberating the country from Saddam's brutal legacy of ethnic hatred is something else.*

Reporter Sam Kiley

It is February 2003. Kadijah, a blind, middle-aged Kurdish woman, is talking to reporter Sam Kiley. She explains she is blind as a result of the chemical warfare waged by Saddam Hussein against the Kurds in 1988. Saddam's goal was to eliminate the mostly agricultural Kurds who lived in the northern part of the country atop large oil reserves so that their houses and land could be used for other purposes. The day she was blinded by an attack on her village, in which chemicals were dropped from low-flying jets, four members of Kadijah's family were killed. In all, 182,000 Kurds went missing between 1988 and 2003. Saddam's troops actually videotaped the mass execution of many; others, like the children and brothers of Nabat, another middle-aged Kurdish woman, whose village was decimated, were killed in the presence of family members, or have simply never been seen again.

Once Kurdish homes and lands were cleared, Saddam had them reoccupied by various minorities who had been displaced elsewhere in Iraq. When American troops (just 50 Green Berets) at last arrived in Kirkuk, one of four large Kurdish cities, in April 2003, the Kurdish population was ecstatic. But soon their energies turned to venting their rage on the Arabs, Bedouins, and Turkomans who had been occupying Kurdish homes. For several anxious days, the 50 Green Berets were left with the task of controlling a city of 700,000 feuding people.

When asked if she could manage to live with the Arab and Bedouin settlers, now that Saddam Hussein is gone, Nabat said, "With Arab people? No, never. They might not all be responsible, but my heart would never allow it. It is better to live with our own people." An understandable attitude, but one that is making the building of a peaceful, self-governing Iraq especially difficult. This is why reporter Kiley says the hardest job will be resolving the hatred created by Saddam's regime and keeping the peace.

*To learn more about the history of the Kurdish people in Iraq, see the FRONTLINE/World video "The Road to Kirkuk."* ∎

### Understanding the Continuing Israeli–Palestinian Conflict.

The more than half-century-long Israeli–Palestinian conflict, which has spawned several major wars and innumerable skirmishes, is a persistent obstacle to political and economic cooperation in the region. Israel, the focus of major animosity in the region, has by far the most modern and diversified economy. Since the 1950s, its development has been facilitated by the immigration of relatively well-educated middle-class Jewish settlers from the United States,

**TABLE 6.1  Circumstances and state of human well-being among Palestinians and Israelis, 2005**

| | Population (in millions) | Infant mortality (per 1000 live births) | Unemployed (percent) | Percentage of population in poverty, 2003 |
|---|---|---|---|---|
| Palestinians | 3.8 | 21 | 50 | 75 |
| Israelis | 7.1 | 5.1 | 8.9 | 18 |

*Sources:* Population Reference Bureau, *2005 World Population Data Sheet*, 2005; *United Nations Human Development Report 2005* (New York: United Nations Development Programme); *Arab Human Development Report 2003.*

Russia, and South America; by large aid contributions from the United States and from private interests; and by the country's own excellent technical and educational infrastructure. The Palestinian people, on the other hand, are severely impoverished and undereducated after years of political, economic, and social repression and the loss of the bulk of their lands to Israel over the past 60 years (Table 6.1). Many Muslims throughout the world regard Israel as a threat to the entire region, given its past territorial expansion and strong backing by the United States.

Warfare between the Israelis and Palestinians began in 1948 on the very day that the last British soldier left. In the ensuing conflict between the Israelis and forces from neighboring Arab countries, Israel prevailed, and the Palestinians' land shrank yet further, with the remnants incorporated into Jordan and Egypt by 1949. In the repeated conflicts that followed—such as the Six Day War (1967) and the Yom Kippur War (1973)—Israel not only defeated alliances of its much larger neighbors, including Egypt, Syria, and Jordan, but also expanded Israeli military control into the territories of those neighbors. By 1973 (Figure 6.33a), the Israeli-occupied lands comprised the West Bank of the Jordan River (a part of the former Palestine that had become part of Jordan), the Gaza Strip and the Sinai Peninsula (which had become part of Egypt), and the Golan Heights (which had become part of Syria).

As a result of continuing hostilities, hundreds of thousands of Palestinians fled the war zones or were removed to refugee camps in nearby countries. Some Palestinians stayed inside Israel and became Israeli citizens, but they have not been treated as equals to Jewish Israelis by the state. The losses of land and political oppression led to uprisings (called **intifada**) among the Palestinians and to mounting terrorist incidents, which were then responded to by the Israeli military.

When Israel occupied Palestinian lands in 1967, the United Nations Security Council passed a resolution requiring Israel to return those lands, known as the *occupied territories*, in exchange for peaceful relations between it and neighboring Arab states. This resolution, later dubbed the *land-for-peace formula*, was only partially fulfilled when, in the 1993 Oslo Accords, the Palestine Liberation Organization (PLO), a Palestinian group formed with support from

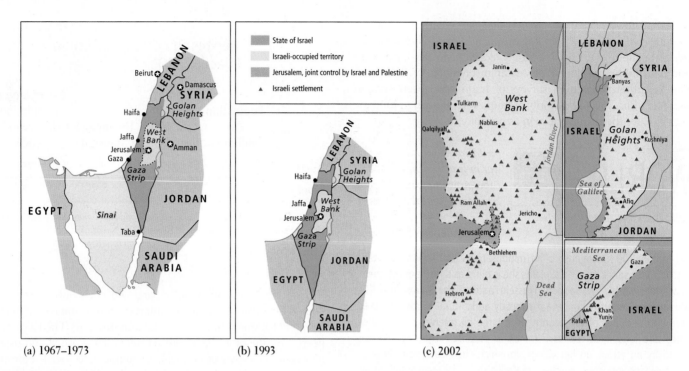

(a) 1967–1973          (b) 1993          (c) 2002

**FIGURE 6.33** Israel after 1949. When the state of Israel was created in 1949, many of its Arab neighbors were opposed to a Jewish state. (a) In 1967, Israel soundly defeated combined Arab forces and took control over the Sinai, the Gaza Strip, the Golan Heights, and the West Bank. (b) In subsequent peace accords, the Sinai was returned to Egypt, but Israel maintained control over the Golan Heights and the West Bank, claiming that they were essential to Israeli security. (c) Although the Palestinians were granted some autonomy in the Gaza Strip and the West Bank, during the 1990s the Israelis, contrary to verbal agreements in the Oslo Accords, began building Jewish settlements in the West Bank and Golan Heights. They also retained control of territory in the Gaza Strip, and in 2002 began building a wall around Palestinian territory on the West Bank. [Adapted from Colbert C. Held, *Middle East Patterns—Places, Peoples, and Politics* (Boulder, Colo.: Westview Press, 1994), p. 184; and "Israeli settlements in the occupied territories—2002," Foundation for Middle East Peace, at http://www.fmep.org/maps/map_data/settlements/occupied_territories2002.html.]

Israel's Arab neighbors with the goal of destroying Israel, acknowledged Israel's right to exist in return for gaining control over the Gaza Strip and the West Bank (Figure 6.33b). An important part of this agreement was Israel's commitment to stop settling Jews in the occupied territories. After seizing those areas in 1967, Israel had established hundreds of Jewish settlements, mainly on the West Bank and also in the Gaza Strip (Figure 6.33c), as part of its effort to secure Israeli control of the territories and the water in aquifers under the West Bank (see the discussion on pages 332–333, 335). Despite its agreements to desist, Israel continued building housing units on the West Bank into 2003 and began constructing a wall around the Palestinian enclave. Meanwhile, the Palestinian people, many of whom had been living in refugee camps for 40 years or more, mounted prolonged uprisings against Israel, known as the first intifada (1987–1993) and the second intifada (2000–present). Both periods were characterized by escalating violence. Palestinian suicide bombers targeted Israeli civilians, and the Israeli military used deadly force to quell demonstrations and to punish the families and communities of Palestinian activists.

A notable hiatus in the ongoing violence followed Israel's removal of some of its settlements and military personnel from the West Bank and Gaza Strip in 2005 (Figure 6.34). However, relations deteriorated and violence ensued after Palestinian elections in 2006 gave a parliamentary majority to Hamas, a political party and activist organization that had previously been associated with acts of violence against Israel. In response, the European Union, the United States, and Canada withdrew financial aid to Palestine. This financial pressure figured in the subsequent decision by Hamas to recognize the right of Israel to exist—a major concession—but other Palestinian groups derided Hamas's decision, and civil war threatened. In 2006, any progress was largely negated when Israel, in retaliation for the kidnapping of two Israeli soldiers by Hezbollah (an anti-Israeli militia in southern Lebanon), launched a brief war that destroyed most of Lebanon's infrastructure.

Two peace activists, Israeli Simona Sharoni and Palestinian Mohammed Abu-Nimer, in their 2004 book *Understanding the Contemporary Middle East*, write that, while most Israelis and Palestinians have concluded that violence is not the answer to their conflict and that diplomacy would serve them better, the U.S. mainstream media continue to misrepresent the conflict. Palestinians are presented as unaccountably prone to violence, and their acts are routinely referred to as "terrorist attacks." Meanwhile, the illegal extension of settlements by Israel into Palestinian lands, the building of the wall around Palestinian territory, and the use of deadly force by the state of Israel are depicted as normal actions taken in the name of "national security."

Little recognized in the press or among world leaders is the fact that ordinary citizens, both Israeli and Palestinian, have separately

**FIGURE 6.34** Israeli soldiers entering Neve Dekalim settlement in the Gaza Strip, 2005. In 2005, Israel began pulling out of Jewish settlements in the Gaza Strip. Jewish settlers (seen here in the background) often attempted to block the pullout; in this case, they set tires on fire in protest. [AP Photo/ Oded Balilty.]

and collaboratively designed ways to end the conflict, acknowledging the national aspirations and the right to land of both parties. Examples of these bottom-up peace initiatives include joint Israeli–Palestinian peace demonstrations and women's groups who have tried to end the Israeli occupation of the West Bank and Gaza Strip. Palestinian–Israeli Physicians for Human Rights have joined to address the medical problems of the overwhelmingly poor Palestinians. Groups from both sides hold youth camps so that Israeli and Palestinian children can, through personal friendship, break the cycle of hatred. Sharoni and Abu-Nimer write that, whereas "official representatives of the two [sides] viewed peace mostly as the absence of war and direct violence [or what might be called] *negative peace*, the grassroots activists within both communities envisioned peace as a transformative process grounded in the presence of justice (positive peace)." The advantage of positive peace strategies is that they proactively focus on justice and equality, rather than on mere stability and the absence of violence (Box 6.5).

## The Tension Between Religion and Democracy

Many fear that Islamic fundamentalism (Islamism) is the greatest threat to the region's political stability and economic development. Islamist movements are grassroots religious revivals that also seek political power. The leaders of these movements seek to take control of national governments currently dominated by secular parties. Islamism has been treated as a major threat since an Islamist revolution led by Ayatollah Ruhollah Khomeini overthrew the secular, though markedly authoritarian and corrupt, government of Iran in 1979. Although Islamist movements have not displaced any secular governments in the years since 1979, they may be gaining strength in some countries. Several secular governments in the region appear willing to sacrifice even small moves toward democracy in order to suppress Islamist movements. However, attempts to suppress these

assertive, yet largely peaceful, popular movements may only increase the likelihood of violent revolutionary reaction, as has occurred for some years in Algeria. The United States supports a number of governments (those of Egypt, Saudi Arabia, Algeria, Morocco, and Turkey) that from time to time actively suppress Islamist movements (such as the Muslim Brotherhood in Egypt). Many in the region see such foreign support as unwarranted antidemocratic meddling.

Understanding the popular base of Islamism can yield insights into its strength. Many of its recruits are young men from lower-class neighborhoods in the region's largest cities. Others are descendants of poor farmers and nomads who were forced off their land and into the cities by state-sponsored development programs and imported labor-saving agricultural technologies. Refugee camps in Palestine and elsewhere are also important sources of recruits. Given the crowded, polluted, and often chaotic living conditions in refugee camps and in the slums of cities such as Cairo and Algiers, it is not surprising that many of their inhabitants question the basic philosophy of the governments under which they live. Many Islamist activists, especially the leaders, come from the large pool of recent university graduates who are frustrated by their inability to find employment or to participate in the political process in their own countries and who are disenchanted with foreign interference in the region. Some Islamist leaders are respected religious men without much education who argue that all will be well if people return to fundamentalist versions of Islam and eschew secularism. Perhaps the greatest support for Islamism is gained by appealing to the widespread discontent of millions of increasingly poor people, whose plight has been further worsened by SAPs and economic upheavals brought on by the war in Iraq and other conflicts. So far, most Islamist movements are still relatively small and based primarily in particular countries. Al Qaeda, the organization blamed for the 9/11 attacks, is much more radical, violent, and internationalized than most Islamist movements.

## BOX 6.5 | AT THE GLOBAL SCALE Terrorism as an Economic and Political Strategy

**Terrorism** is the use, or the threat, of violence, intended to create a climate of fear in a given population. To this we might add that terrorism can be the ultimate effort to bring attention to a cause by those who feel powerless and invisible. State reactions against terrorism can take on some of the very qualities of terrorism itself when police or security personnel inflict violence on suspected perpetrators before due legal process has been served. There are also charges that some states surreptitiously sponsor terrorism by quietly encouraging and even funding radical elements, but these charges are difficult to prove because citizens of these states may act without actual government sanction.

In modern times in North Africa and Southwest Asia, the terrorist strategy was first used successfully by militant Jewish Zionists in 1946. They bombed the King David Hotel in Jerusalem in an effort to get Britain to fulfill its agreement to support the formation of the state of Israel on Palestinian lands; 91 people died. More recently, the word *terrorism* has been attached to the actions of Palestinians seeking redress of grievances against the state of Israel for taking their lands in 1948 and thereafter in a series of short wars (started by Arabs, but not without provocation). Groups in the federation known as the Palestinian Liberation Organization (PLO), acting on behalf of Palestinians, have launched suicide attacks on Israeli civilians since the late 1980s. Such groups say that they resort to violence only in response to the brutal Israeli military efforts to quell demonstrations and protect Jewish settlements constructed illegally in the contested West Bank and Gaza Strip territories.

Terrorism has also been used in North Africa—for example, by Islamists in Algeria protesting the failure of the military government to recognize the results of an election in 1991, and by the Algerian government against those suspected of being Islamists. In Egypt, protesters against the policies and Western alignment of the Sadat and Mubarak governments bombed innocent bystanders, and then were themselves squelched by brutal police tactics. In Saudi Arabia, terrorists launched car bomb attacks in May and November 2003 on upscale residential compounds in Riyadh, where foreign advisers were living. These events, and the implication that Saudi Arabia had at least tacitly encouraged the 9/11 attacks in the United States by allowing Saudi religious and school curricula to foster hatred for non-Muslims, led to a significant Saudi public education campaign exposing the horrors of terrorism, including pictures of the aftermath of terrorist attacks in different parts of the Saudi Arabia.

Terrorism, and its counterpart of violence against civilians by government police and officials, is a major destabilizer of civil society. The 9/11 attacks affected economies worldwide by inhibiting travel and commerce and by diverting massive public funds to security efforts. Aside from its immediate effects of death and destruction and economic recession, terrorism also increases hatred between neighboring groups, causing them to demonize one another—and so the losses proliferate. But there is an alternative. Scholars of terrorism point out that terrorist activity is much less likely when people see themselves as having access to open political participation and free elections and to timely justice and redress of grievances through legal means.

*Sources:* "Western team said to be in Libya on antiweapons mission," *New York Times* (January 1, 2004); "Talk of the Nation," National Public Radio, January 21, 2004; BBC News continuous special coverage, *In Depth: Israel and the Palestinians, 2000–2004,* http://news.bbc.co.uk/1/hi/world/middle_east/978626.stm; http://www.atimes.com/atimes/Middle_East/GB12Ak04.html.

---

Governments in the region have differing relationships with Islamist movements. For example, the Saud family in Saudi Arabia came to power, and remains there, by cooperating with the very conservative Wahhabi school of Islamist thinking. Iran, which is predominantly Shi'ite, is now governed by an Islamist dictatorship. Other governments, such as those of Egypt and Algeria, are using the threat of Islamism as an excuse to keep political power in the hands of a small, wealthy elite that is religiously conservative, but not radically so. On the other hand, non-Islamist groups dedicated to moderate reform in such places as Jordan, Bahrain, Oman, Qatar, Tunisia, and Morocco are attempting to avert political extremism by redistributing political power more broadly.

## Reform Efforts from Within the Arab Community

In 2002, an independent consortium of Arab governments and Arab scholars, bureaucrats, and activists began producing the *Arab Human Development Reports (AHDR),* a series of frank and revolutionary annual reports on human development within Arab countries. The consortium identified three deficits afflicting the larger Arab community (and other Muslim communities as well): in general human freedoms, in women's rights, and in access to knowledge. Enhancing democratic institutions was identified as the key to erasing these deficits, but it was recognized that domestic democratic reforms were dependent in large part on the development of a better-educated electorate. Hence the 2003 report focused on building a "knowledge society" and urged Arab leaders to introduce their people to the "global knowledge stream." The 2004 report sought a viable transition to freedom and good governance. The 2005 report focused on Arab women's economic, political, and social empowerment, noting how gender inequity affects Arab development.

Although critics say that Arab reformist leaders are doing little to truly increase democratic participation, there is some evidence to the contrary. The *AHDR 2003* promoted Morocco, Jordan, and Qatar as places that are effectively reforming autocratic institutions and advancing the knowledge society. King Mohammed VI of Morocco, for instance, is lauded for his efforts to liberalize the social climate for women (see page 319). In the tiny state of Qatar on the Persian Gulf, the emir, Sheikh Hamad bin Khalifa Al Thani, has instigated broad-based reform and ambitious education projects. Still, an extensive opinion poll of citizens of Algeria, Lebanon, Jordan, and Palestine, carried out as part of the *AHDR* for 2004, revealed a broad consensus that governments need to do more to lessen corruption, open up their political systems, and provide greater personal and political freedoms to all citizens.

# ENVIRONMENTAL ISSUES

*Salam*, the Arabic root of the word *Islam*, means peace and harmony. Islam therefore calls on its followers to live in peace and harmony with both human and natural systems. In Muslim societies, the numerous references in the Qur'an to the role of humans as stewards of Earth are typically interpreted to mean that humans have the right to use Earth's resources, but only within the general limitations that Islam places on greed and personal ambition. The Qur'an requires Muslims to avoid spoiling or degrading human and natural environments, to share resources such as water equally with all forms of life, and to conserve natural resources even if they are abundant.

In practice, Muslims are like people everywhere: they have not always followed their own religious teachings and cared for Earth. Urban crowding, mechanized agriculture and industry, and the pursuit of material goods have resulted in pollution, species extinctions, and degraded environments in North Africa and Southwest Asia (Figure 6.35). At present rates of use, the region has a very limited supply of arable land, water, forests, and even minerals. Yet people across the region expect to achieve higher living standards, which, along with population growth, will tax the region's resources mightily. Careful management of the environment will

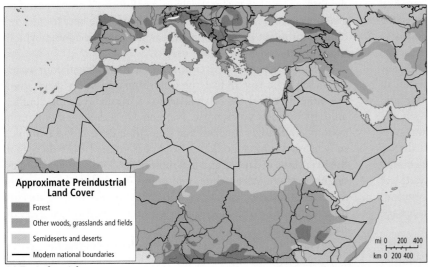

**Approximate Preindustrial Land Cover**

- Forest
- Other woods, grasslands and fields
- Semideserts and deserts
- — Modern national boundaries

mi 0  200  400
km 0  200 400

(a) **Preindustrial**

**FIGURE 6.35** Land cover (preindustrial) and human impact (2002). In the preindustrial era **(a)**, human impact on the environments of this region was localized and light and not visible at this scale. Population densities were very low except in patches along coastal zones. By 2002 **(b)**, even the most remote desert locations bore some human impact. The Nile Basin and many urban and coastal zones showed medium to high impact. [Adapted from *United Nations Environment Programme, 2002, 2003, 2004, 2005, 2006* (New York: United Nations Development Programme), at http://maps.grida.no/go/graphic/human_impact_year_1700_approximately and http://maps.grida.no/go/graphic/human_impact_year_2002.]

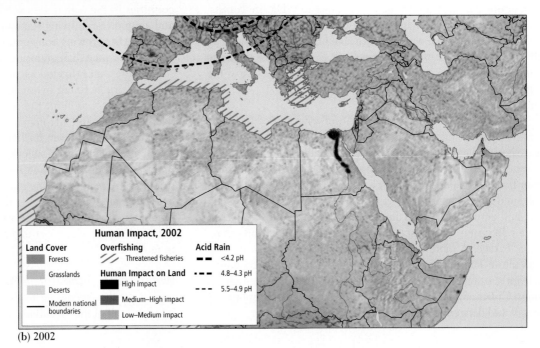

**Human Impact, 2002**

| Land Cover | Overfishing | Acid Rain |
|---|---|---|
| Forests | ⫽ Threatened fisheries | ▬ ▬ <4.2 pH |
| Grasslands | **Human Impact on Land** | ▬ ▪ ▪ 4.8–4.3 pH |
| Deserts | ■ High impact | ▪ ▪ ▪ 5.5–4.9 pH |
| — Modern national boundaries | Medium–High impact | |
| | Low–Medium impact | |

(b) **2002**

undoubtedly become an important issue over the next several decades. In this section, we look at water availability and use and desertification as two related environmental issues that are central to the region's future.

## Water Availability

Water has always been in short supply in this arid part of the world. Hence cultural attitudes toward water differ greatly from those of North Americans. Here, people expect to use very little water in daily life, and they have devised many strategies to cope with the limited supply, such as designing buildings to maximize shade and conserve moisture; capturing mountain snowmelt and moving it to dry fields and villages in underground water conduits (*qanats*); bathing in public baths; and practicing countless water recycling techniques. Water is crucial to Muslim religious practices. A few decades ago, a religious decree in Saudi Arabia approved the human use of wastewater if it had been completely and properly treated and impurities removed from it. This decree was important because it encouraged recycling of water and helped extend the life of existing freshwater aquifers.

Despite a long tradition of careful water use, several factors have exacerbated the water shortage in the region over the last few decades. Modernization has brought new ways to use water: in household plumbing, sewage treatment, industrial cooling and cleaning, and mechanized irrigated agriculture. Moreover, the number of people who must share the region's scarce water resources is growing rapidly. UN experts have established that to maintain basic human health and support development, a country must have no less than 1000 cubic meters (1300 cubic yards) of water available per capita per year. By 2025, virtually all countries in the region will have less water than the UN considers necessary. Figure 6.36 shows water stress globally.

The greatest use of water throughout the region is for irrigated agriculture, despite the fact that in most of the region's countries agriculture does not contribute significantly to GDP. In Tunisia, for instance, where agriculture accounts for just 14 percent of GDP yet employs 55 percent of the labor force, 88 percent of all the water used is used for agriculture, 9 percent for domestic purposes, and just 3 percent for industry. In addition, across the region, agricultural chemicals are polluting the water in rivers and streams.

Because the overwhelming majority of water is used for irrigation, even small increases in irrigation efficiency will yield large increases in the volume of fresh water available for domestic and other uses. Efficiency could be improved by applying drip, rather than flood, irrigation more widely. Other promising strategies involve reducing reliance on water-intensive crops such as rice, cotton, and citrus fruits; gradually removing government subsidies to large water users; and upgrading water distribution networks to reduce leakage. A more controversial strategy would be to give up agriculture entirely and concentrate instead on other economic activities, using the proceeds to import food from water-rich regions. Critics of this strategy argue that it could lead to a serious loss of jobs and dangerous dependency on imported food.

Efforts to increase water supplies have also involved the construction of dams and reservoirs on major river systems such as the Euphrates, the Tigris, the Jordan, and the Nile (Box 6.6). But these projects threaten the rights of both downstream and upstream water users. Damming a river removes water from huge tracts of land downstream and stops the deposition of fertility-enhancing silt during seasonal flooding. The artificial reservoir created upstream of a dam not only floods villages, fields, wildlife habitat, and historic sites, but the silt that used to serve as fertilizer downstream will fill up the reservoir behind the dam in just a few decades. In a warm climate, the standing water behind a dam becomes a breeding environment for disease-carrying mosquitoes and parasites. Furthermore, huge volumes of water are lost from the reservoir through evaporation or leakage. Meanwhile, irrigated soil downstream, repeatedly saturated and dried out, will eventually become salty and infertile. Other strategies for increasing water supply have problems as well. The production and use of energy for desalinization of seawater creates air pollution. The pumping of groundwater lowers the water table and causes the land to subside, often resulting in the intrusion of seawater into the aquifer, which then renders the groundwater useless.

Sharing the water of rivers that flow across national boundaries has never been easy. In fact, the words *river* and *rivalry* come from the same root, and the original sense of "rivalry" was the conflict among users of a watercourse. Naturally, in a disagreement, each country emphasizes the facts that strengthen its case. For example, in debates over who should get Euphrates water, Turkey notes that the river starts in Turkey and that most of the flow in the Euphrates originates there. Meanwhile, Iraq points out that the river travels the longest distance in Iraq, and that the Sumerians (the ancient inhabitants of what is now Iraq) were the earliest major users of Euphrates water.

A number of factors complicate the chances of reaching water allocation agreements along the Euphrates and the Nile. Most countries in the region are experiencing rapid population growth and a rapid rise in living standards and urbanization. One consequence is the growing consumption of meat, which requires more water to produce than grains or vegetables. Urbanites tend to consume more water (for indoor plumbing and flush toilets) and use more electricity (often hydroelectricity) than rural residents. Furthermore, the people in the region's river-basin states have long had territorial and ideological disputes with one another. To complicate matters even further, the international laws governing water rights along river systems evolved in parts of the world—mainly Europe—where water was plentiful, and these laws are ill suited to dry regions. In the case of Euphrates water, for the time being, the fact that Turkey has the upstream geographic location, much better political relations with Europe and the United States than its neighbors, and a strong military tilts the balance of power in its favor.

***Water Issues in the Israeli-Palestinian Conflict.*** When Palestinians say that according to the 1993 Oslo Accords, Israel must relinquish all of the West Bank and Gaza Strip it seized during the

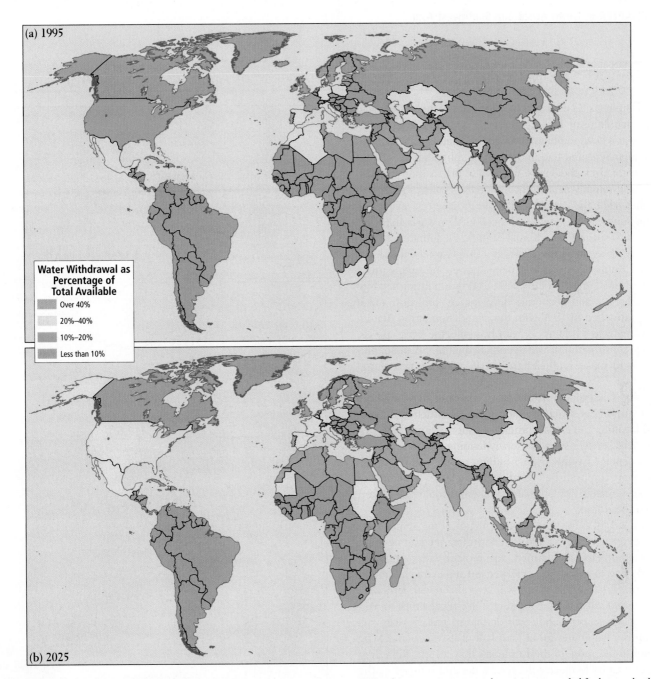

**Water Withdrawal as Percentage of Total Available**

Over 40%

20%–40%

10%–20%

Less than 10%

(b) 2025

**FIGURE 6.36** Global water availability, 1995 and 2025. Declining availability of potable water is a critical environmental issue for most countries in the region. An area experiences *water stress* when annual water supplies drop below 1700 cubic meters per person (approximately correlated with yellow on the map). *Water scarcity* occurs when the amount of water withdrawn from lakes, rivers, or groundwater and the annual rainfall supplies drop below 1000 cubic meters per person, the minimum needed for human health (orange on the map). The United Nations predicts that, by 2025, nearly all of the countries in the region will have reached the scarcity stage. Only Turkey will have sufficient water to maintain good health and meet development needs. [Adapted from "Freshwater Stress" map (New York: United Nations Environment Programme, 2002), at http://www.unep.org/vitalwater/21.htm.]

1967 war, what Israelis hear is that they must give up major water rights as well as settlements. The West Bank includes the west side of the Jordan Valley, which has better water resources than the east side. The Palestinians are willing to share some of the water in the aquifers of the West Bank, but they claim the lion's share of the resource because its **recharge area**—the area with the highest precipitation, which resupplies the aquifers with water—is located under the hills of this Palestinian territory. Israel says it should

retain control over water resources in the West Bank. It holds that desalinization projects should be undertaken to provide sufficient water for Israel, Jordan, and the Palestinians, instead of merely redistributing existing resources.

At present, Israel pumps the most water from the Jordan River and from the aquifers that underlie the occupied territories. It uses the water primarily in its technically advanced agricultural sector, but Israeli agriculture employs only 2.6 percent of the labor

**333**

## BOX 6.6 | AT THE REGIONAL SCALE Turkey's Southeastern Anatolia Project (GAP)

The Euphrates River begins life as trickles of melting snow in the mountains of Turkey, then gathers volume from several tributaries as it flows southeast through the Anatolian Plateau and then through Syria and Iraq. In Iraq, the Euphrates is joined by a second major river, the Tigris, which also originates in Turkey. The two run roughly parallel for hundreds of miles, joining just before they reach the Persian Gulf.

The Southeastern Anatolia Project (GAP in Turkish; Figure 6.37) was first suggested by Kemal Atatürk, the founder of modern Turkey. It began in the 1930s as a study on how to use the Tigris and Euphrates rivers to generate power and supply irrigation systems for Turkey. In the 1980s, GAP was transformed into a regional social and economic development program, and it was fully launched in 1990 with the completion of the Atatürk Dam. As many as 21 dams, 19 with hydroelectric power plants, are under construction or will eventually be built. A major aim of GAP is to make southeastern Turkey a major agricultural export center. This economic advance, seen as an important step in preparing Turkey for entry into the European Union, includes road and airport construction, reforestation, instruction in commercial farming, the establishment of fisheries, and the provision of social services such as literacy training, health care, and women's centers. Eventually, 3.5 million people will be served, many of them impoverished Kurds, part of a larger settlement of these once nomadic herders that straddles the borders of Turkey, Iraq, Syria, and Iran.

As important as the Southeastern Anatolia Project may be to Turkey, its effects on Syria and Iraq downstream could be disastrous. If completed as planned, the project will reduce the Euphrates flow to Syria by 30 to 50 percent and to Iraq by around 75 percent. Moreover, the water will be polluted with fertilizers, pesticides, and salts after having served to irrigate crops in Turkey. Syria, whose population will double in just 25 years, depends on the Euphrates for at least half of its water, and Syria also plans to use more of the Euphrates flow for irrigation. Iraq is the last country to receive the river water and is therefore the most vulnerable to negative effects from the project. Disastrous confrontation over water is forecast by many, but the WorldWatch Institute reports that, despite many water-based disputes in the region, the last war over water took place 4500 years ago. Since then, water conflicts have usually been resolved peacefully through cooperation.

*Source:* Stephen Kinzer, "Restoring the Fertile Crescent to its former glory," *New York Times* (May 29, 1997); Christine Drake, "Water resource conflicts in the Middle East," *Journal of Geography* (January/February 1997): 4–12; Hussein A. Amery, personal communication, August 2001; Southeast Anatolia Project (GAP) Web site, http://www.gap.gov.tr/gap_eng.php?sayfa=English/gapgb.html; WorldWatch Institute, *State of the World 2005 Trends and Facts—Water Conflict and Security Cooperation*, http://www.worldwatch.org/node/69.

**FIGURE 6.37** Dams of the Tigris and Euphrates drainage basins. Turkey's projects to manage the Tigris and Euphrates river basins through dam construction have international implications. Water that is retained in Turkey will not reach its neighbors. The main map shows Turkey's dams on the headwaters of the two rivers, as well as dams built in Syria and Iraq, all of which will have environmental effects, especially on the lower reaches of both rivers in Iraq. The smaller map shows the full extent of Turkey's Southeastern Anatolia Project. [Adapted from "Turning the Tides" map, *Vital Water Graphics: Problems Related to Freshwater Resources*, United Nations Environment Programme, at http://www.unep.org/vitalwater/22.htm.]

force and contributes just 2 percent to the GDP. By contrast, the Palestinian economy remains agriculturally based (33 percent of GDP, 13 percent of labor force) and will be damaged if the Palestinians are not allocated sufficient amounts of water. At present, Palestinians are banned from sinking wells or accessing the water they need.

## Desertification

The United Nations defines **desertification** as the ecological changes that convert nondesert lands into deserts. These changes include the loss of soil moisture, the loss of vegetation or the replacement of plant species by other species that are more adapted to dry conditions, the erosion of soil by wind, and the formation of sand dunes on formerly vegetated land. Desertification is occurring in many places; globally, 23,000 square miles (60,000 square kilometers, or an area about the size of West Virginia) becomes new desert every year. It is not yet known just how much of this loss is due to natural climate variations or changes and how much is due to human-induced climate change through greenhouse gas emissions, modification of forests and grasslands, and overuse of water resources, but clearly human activity plays a major role.

In North Africa and Southwest Asia, nomadic herders are often blamed for increasing desertification by overstocking rangelands and allowing their animals to overgraze those lands. With little grass remaining, the soil is unable to retain moisture. Studies are showing that the real story is more complicated, however. Geographer and veterinarian Diana Davis found that herders are usually careful to manage their herds so that the grasslands on which they depend are not destroyed. In her research among herders in Morocco, Davis observed that when times are good, herders are more likely to decrease rather than increase their herds, recognizing that it is during good times that rangelands should be allowed to regrow. Most likely, a wide array of economic changes has contributed to the general drying of grasslands bordering the Sahara. As groundwater levels are lowered through urban use and irrigated agriculture, plant roots can no longer reach sources of moisture. Furthermore, the encouragement of settled cattle ranching (rather than nomadic herding) by international development agencies leads to overgrazing of pastureland and to excessive demands on scarce water. When nomadic pastoralists settle permanently and take up modern ways of life, their per capita use of water increases.

## MEASURES OF HUMAN WELL-BEING

As we explained in Chapter 1, though the gross domestic product per capita is a less than adequate measure of the actual state of human well-being, it nonetheless gives some indication of the distribution of wealth across the region. In Table 6.2, you can see that in this region, as in others, the pattern of wealth as measured by GDP per capita (column II) is uneven. One-third of the region's countries, primarily the oil-producing nations (Saudi Arabia, Oman, the UAE, Qatar, Kuwait), as well as Bahrain and Israel, have relatively high GDP per capita figures. The other two-thirds

are well below the world average of about U.S.$8229. Even politically influential countries such as Turkey and Egypt have per capita incomes well below the world average. Because the GDP per capita figure is only an average, the unequal distribution of wealth in this region is masked. Many people live on much less than these figures, while a very few are extremely rich.

The United Nations Human Development Index (HDI) (Table 6.2, column III) combines three indicators to evaluate well-being in 177 countries and rank them. The indicators are life expectancy at birth, educational attainment, and income adjusted to purchasing power parity (PPP). You may remember that in other chapters we cited Kuwait's HDI figures in our comparisons of human well-being. Despite its relatively high GDP per capita, Kuwait does not rank very high on the HDI scale in comparison to other countries with similar per capita GDP figures, such as Barbados. Until the past decade, education was available to only a minority of females in the region. Thus, in Arab states across the region, there is a large population of older illiterate women. In Iraq, Yemen, Egypt, and Morocco, less than 50 percent of women can read basic texts.

Israel is the only country in this region with both a high GDP per capita and a high HDI rank (23). Israel's ranking reflects its ability to provide the basics of a decent life—adequate income, shelter, food, education, and health care—to all its citizens, regardless of ethnicity, sex, or religion. Israeli culture emphasizes education, and Israel has attracted highly educated Jewish immigrants from around the world. However, Israel has also regularly received development and military assistance from the United States—U.S.$360 per capita in development assistance in 2006. In that year, Egypt, the next largest regular recipient of U.S. development assistance after Israel, received just U.S.$27 per capita and ranks much lower on both the GDP per capita and HDI measures. The war in Iraq and the resulting insurgency make it difficult to calculate the actual size of U.S. nonmilitary aid to that country; an October 2006 U.S. State Department report listed $10.8 billion in aid for that year.

In this chapter, the Gender Empowerment Measure (GEM) and the Gender Development Index (GDI) are combined in Table 6.2 (column IV). For a variety of reasons, the statistics necessary to calculate GEM, which ranks countries on the extent to which women have opportunities to participate in economic and political life, are not available for two-thirds of the countries in this region (only 80 countries are ranked worldwide). The GDI measures only the access of males and females to health care, education, and income, not the extent to which males and females actually participate in policy formation and decision making within a country. As we have already observed for this region, women are important influences in the home, but only a very small percentage of women across the region participate in policy making or are leaders in the larger society. Israel's relatively high GEM rank of 24 is the result of its direct efforts to include females in all aspects of society, including the armed forces. Even in Israel, however, women held only 14.2 percent of parliamentary seats in 2006.

## TABLE 6.2 Human well-being rankings of countries in North Africa and Southwest Asia and other selected countries[a]

| Country (I) | GDP per capita, adjusted for PPP[b] (GDP ranking among 177 countries), 2005 (II) | Human Development Index (HDI) ranking among 177 countries,[c] 2005 (III) | Gender Empowerment Measure (GEM) and Gender Development Index (GDI) rankings, 2005 (IV) | Female literacy (percent), 2003 (V) | Male literacy (percent), 2003 (VI) |
|---|---|---|---|---|---|
| **Selected countries for comparison** | | | | | |
| Barbados | 15,270 (39) | 30 (high) | 25 (GDI = 29) | 99 | 99 |
| Japan | 22,967 (13) | 11 (high) | 43 (GDI = 14) | 99 | 99 |
| United States | 37,562 (4) | 4 (high) | 12 (GDI = 8) | 99 | 99 |
| World | 8,229 | | | | |
| **The Northeast** | | | | | |
| Turkey | 6,772 (75) | 94 (medium) | 76 (GDI = 70) | 81 | 96 |
| Iraq[d] | 3,197 | 126 (medium) | ND (GDI = 107) | 43 | 64 |
| Iran | 6,995 (69) | 99 (medium) | 75 (GDI = 78) | 70 | 84 |
| **The Eastern Mediterranean** | | | | | |
| Syria | 3,576 (110) | 106 (medium) | (GDI = 84) | 74 | 91 |
| Lebanon | 5,074 (92) | 81 (medium) | ND (GDI = 68) | 81 | 92 |
| Israel | 20,033 (25) | 23 (high) | 24 (GDI = 23) | 96 | 98 |
| Jordan | 4,320 (100) | 90 (medium) | ND (GDI = 73) | 85 | 95 |
| Occupied Territories | 1,026[e] | 102 (medium) | ND (GDI = ND) | 87 | 96 |
| **The Arabian Peninsula** | | | | | |
| Saudi Arabia | 13,226[f] (44) | 77 (medium) | 78 (GDI = 65) | 69 | 87 |
| Yemen | 889 (158) | 151 (low) | 80 (GDI = 121) | 29 | 70 |
| Oman | 13,584[g] (41) | 71 (medium) | ND (GDI = 60) | 65 | 82 |
| UAE | 22,420[g] (23) | 41 (high) | ND (GDI = ND) | 81 | 76 |
| Qatar | 19,844[g] (27) | 40 (high) | ND (GDI = ND) | 89[h] | 89[h] |
| Bahrain | 17,479[g] (36) | 43 (high) | 68 (GDI = 41) | 83 | 92 |
| Kuwait | 18,047[f] (33) | 44 (high) | (GDI = 39) | 81 | 85 |
| **The Nile** | | | | | |
| Egypt | 3,950 (105) | 119 (medium) | 77 (GDI = ND) | 44 | 67 |
| Sudan | 1,910 (130) | 141 (medium) | ND (GDI = 110) | 50 | 69 |
| **The Maghreb** | | | | | |
| Morocco | 4004 (104) | 124 (medium) | ND (GDI = 97) | 38 | 63 |
| Algeria | 6,107 (81) | 103 (medium) | ND (GDI = 82) | 60 | 80 |
| Tunisia | 7,161 (68) | 89 (medium) | ND (GDI = 69) | 65 | 83 |
| Libya | 11,400[h] (48) | 58 (medium) | ND (GDI = ND) | 71 | 92 |

[a]Rankings are in descending order; i.e., low numbers indicate high rank.

[b]PPP = purchasing power parity, figured in 2003 U.S. dollars.

[c]The high, medium, and low designations indicate where the country ranks among the 177 countries classified into three categories by the United Nations.

[d]All data for Iraq from *United Nations Human Development Report 2000*. Data were not available for 2005. All figures have undoubtedly declined.

[e]Estimate, not PPP.

[f]Estimate based on regression.

[g]Data refer to year other than 2003.

[h] Estimate, *CIA World Factbook*, 2007.

ND = No data available.

Source: *United Nations Human Development Report 2005* (New York: United Nations Development Programme), except as indicated above, at http:/hdr.undp.org/reports/global/2005/pdf/HDR05_complete.pdf.

# III SUBREGIONS OF NORTH AFRICA AND SOUTHWEST ASIA

The subregions of North Africa and Southwest Asia present a mosaic of the issues that have been discussed so far in this chapter. Although all countries in the region except Israel share a strong tradition of Islam, they vary in how Islam is interpreted in national life and the extent to which Westernization is accepted. They also vary in prosperity and in the extent to which wealth is evenly distributed in the general population. Conflict is not uncommon in this region, and the closer examination allowed by the subregional perspective may help you understand the factors that have led to discord and, in some cases, violence.

## THE MAGHREB

If you are interested in old movies, you have probably already formed intriguing images of North Africa: Berber or Tuareg camel caravans transporting exotic goods across the Sahara from Tombouctou to Tripoli or Tanger; the Barbary Coast pirates; Rommel, the World War II "Desert Fox"; the classic lovers played by Humphrey Bogart and Ingrid Bergman in *Casablanca*. These images—some real, some fantasy, some merely exaggerated—are of the western part of North Africa, what Arabs call the Maghreb ("the

place of the sunset"). The reality of the Maghreb, however, is much more complex than popular Western images of it.

The countries of the Maghreb stretch along the North African coast from Western Sahara (considered a dependency by Morocco, though Western Sahara contests this status) through Libya (Figure 6.38). A low-lying coastal zone is backed by the Atlas Mountains, except in Western Sahara and Libya. Despite the region's overall aridity, these mountains trigger sufficient rainfall in the coastal zone to support export-oriented agriculture, and in winter, a modest skiing industry. Algeria, Tunisia, and Libya have interiors that reach into the huge expanse of the Sahara.

The landscapes of the Maghreb reflect the long and changing relationships all the countries have had with Europe. European domination in this area lasted well into the twentieth century, during which time the people of North Africa took on many European ways: consumerism, mechanized market agriculture, and manners of dress, language, and popular culture. The cities, beaches, and numerous historic sites of the Maghreb continue to attract millions of European tourists every year, who come to buy North African products—fine leather goods, textiles, handmade rugs, sheepskins, brass and wood furnishings, and paintings—and to enjoy a culture that, despite Europeanization, seems exotic. The agricultural lands

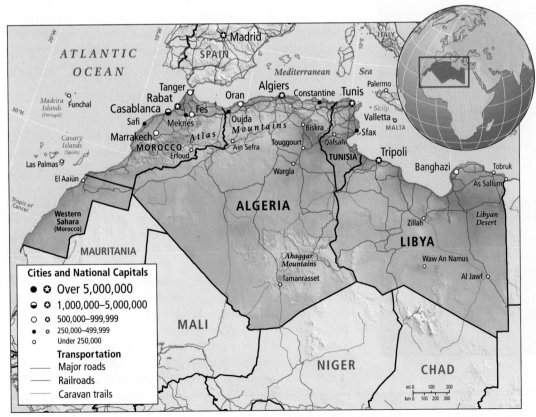

**FIGURE 6.38** The Maghreb subregion (Northwest Africa).

of the Maghreb (Figure 6.39) are strategically located close to Europe, where there is a strong demand for Mediterranean food crops: olives, olive oil, citrus fruits, melons, tomatoes, peppers, dates, grains, and fish. Oil, gas, and petroleum products from the Maghreb, which make up over 30 percent of GDP and 95 percent of total exports in both Algeria and Libya, supply about one-quarter of the oil and gas used by the European Union.

Europe is also a source of jobs. Local firms are supported by European investment and tourism, and millions of guest workers have migrated to Europe, many after losing jobs as a result of agricultural modernization. In 2006, during widespread debates in the EU over immigration, there were an estimated 6 million to 10 million North African migrants in Spain and France, many of them

**FIGURE 6.39** Dade's Gorge in Morocco. This is a narrow, dramatic valley in the Atlas Mountains, with lush villages and rich agricultural plots. [ALTITUDE/Yann Arthus-Bertrand.]

illegal. Between January and June of 2006, 7000 migrants arrived from North Africa in Spain alone (some had trekked from sub-Saharan Africa into North Africa first). For the migrants, the trip across the Mediterranean in leaky, flimsy craft is dangerous; every year scores of North Africans die trying to reach the Spanish coast. Thousands are apprehended and sent home. Their plights once in Europe are reminiscent of those of illegal Mexican migrants into the United States: low pay, poor housing, exploitation by unscrupulous employers, and a constant threat of deportation. And after the terrorist bombings in Spain in March 2004, sentiment against illegal North African immigrants increased.

In North Africa, settlement and economic activities are concentrated along the narrow coastal zone (see Figure 6.14 on page 309). Most people live in the cities that line the Atlantic and Mediterranean shores—Casablanca, Rabat, Tanger, Algiers, Tunis, and Tripoli—and in the towns and villages that link them. The architecture, spatial organization, and lifestyles of these cities mark them as cultural transition zones between African and Arab lands and Europe (Figure 6.40). The city centers retain an ancient ambience, with narrow walkways fronted by family compounds, but most people live in modern apartment complexes around the peripheries of these old cities. Multilane highways, shopping malls, restaurants serving international cuisine, and art galleries serve the citizens.

Many North Africans are seeking an identity that is less European and retains their own distinctive heritage. As historian John Ruedy puts it, for those people who were Westernized during the era of European domination and who now make up the educated middle class, independence from Europe meant the right to establish their own secular nations with constitutions influenced by European models. But by the 1990s, Islamist movements were seeking to challenge these independent secular states by linking them in a negative way to European domination. In the Maghreb, the rejection of Europeanization, including the reassertion of Islamic law, has been particularly hostile. Secular leaders and some citizens who are religious but who prefer a secular government are fearful of the outcome.

## Violence in Algeria

Algeria, with 33 million people, one-third under the age of 15, is arguably the Maghreb's most troubled country. The 1990s were a time of extreme civil violence between Islamists and the military-controlled government, and between 150,000 and 200,000 people died. The violence disrupted the society and ruined the already strained economy.

The roots of the troubles lie in the violent struggle with France for independence, won in 1962, after which Algeria was ruled by a single-party socialist military dictatorship that repressed all viable opposition. The new government took over the large European-owned farms that produced dates, wine, olives, fruit, and vegetables for export, and soon these farms ended up in the control of the Algerian elite, either as private farms or as large cooperatives. Most of the large agricultural population remained on small, infertile plots. By the early 1990s, nearly half the economy, including factories and utilities, was state owned and inefficiently operated. The

**FIGURE 6.40** Tripoli, Libya. The Al Fatah Tower business center and other modern structures dwarf the older buildings of Tripoli, many of which sport a satellite dish or two. [© Israel Images/Alamy.]

economy declined to the point that, by the late 1990s, unemployment among urban dwellers under the age of 30 was 80 percent or higher. The unequal division of wealth between those Algerians who replaced Europeans in positions requiring skills and education and those who remained poor, with only a few years of education, created social divisions. Those who are well-off and liberal in their attitudes toward economic reform, women's roles, and freedom of speech are often derided by the Islamists as too secular (though a significant number of Islamist leaders are themselves technically and professionally educated people).

During the early 1990s, the Algerian government shifted its policy regarding ownership of industries, and in an attempt to lure foreign investment and create jobs, industries were privatized. During the lag that often occurs between structural adjustment reforms and economic resurgence, average personal income fell by a third. In 1991, under these dismal economic conditions, an Islamist party won the country's first free parliamentary elections by an overwhelming margin. A peaceful transfer of power could have diffused the pent-up discontent of many Algerians. Instead, the government —motivated largely by fears that the newly elected government would pursue an extreme religious path—declared the elections to be null and void. Duly elected Islamist officials were jailed, and the country was plunged into a devastating civil war.

Although some Islamist advocates joined religiously based social service agencies and worked to improve life in Algeria's shantytowns, extremist factions of Islamists began brutal terrorist attacks. Particularly targeted were journalists, musicians and artists, and educated working women not observing seclusion or not wearing the veil. Several thousand such women were killed, often in drive-by shootings. In addition, perhaps as a scare tactic, the populations of whole villages were massacred, with apparently no attention paid to political or religious affiliation. Some studies and a few journalistic reports show that state-sponsored "militias" (troops in civilian clothing) also terrorized civilians to dissuade them from supporting the Islamists.

By mid-1997, public sentiment was turning against extremists in the Islamist movement. In a relatively fair election in June 1997, moderate Islamists, espousing nonviolence, won nearly one-third of the seats in the national assembly. Soon, violence had subsided, and more than 1000 militants surrendered under an amnesty program. In March 2006, an amnesty release of Islamist militant prisoners started the process of reconciliation, which includes the surrender of several remaining militants labeled as terrorists, as well as their weapons.

## THE NILE: SUDAN AND EGYPT

The Nile River begins its trip north to the Mediterranean in the hills of Uganda and Ethiopia in central East Africa. The countries of Sudan and Egypt share the main part of the Nile system (Figure 6.41), and it is their chief source of water. Although Sudan and Egypt have the Nile River, an arid climate, and Islam (Egypt is 94 percent Muslim, Sudan 70 percent) in common, they differ culturally and physically. Egypt, despite troubling social, economic, and environmental problems, is industrializing and plays an influential role in global affairs, whereas Sudan struggles with civil war and remains relatively untouched by development.

### Sudan

Sudan, slightly more than one-quarter the size of the United States, is the largest country in the region and, in fact, in all of Africa. Yet,

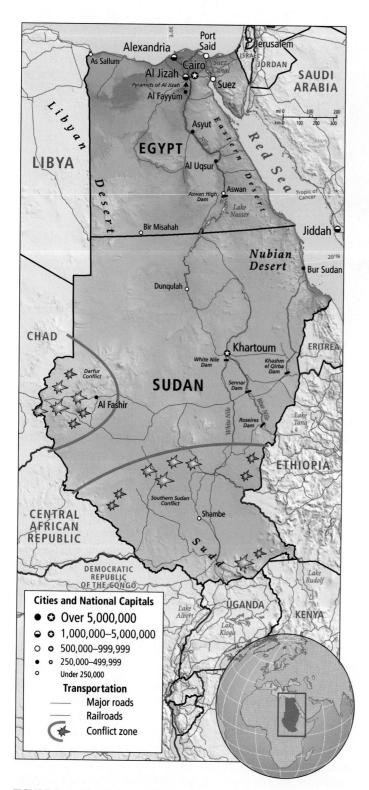

**FIGURE 6.41** The Sudan and Egypt subregion. [Conflict zones adapted from USAID, http://www.USAID.gov/locations/sub-saharan_africa/sudan /sudan_bombjuly.pdf; http://www.lib.uTEXAS.edu/maps/AFRICA/darfur _villages_0802_2004.Jpg.]

with 42 million people, it has about half the population of Egypt. The country has three distinct environmental zones that become drier as the altitude decreases. The upland south, called the Sudd (the first zone), consists of vast swamps fed by rivers bringing water from the rainy regions of equatorial Africa. North of, and somewhat lower than, the Sudd, the much drier steppes and hills of central Sudan (the second zone) are home to animal herders. Yet farther north and lower is the Sahara (the third zone), which stretches to the border with Egypt. Although most of Sudan is desert and dry steppe country, it does have the main stream of the Nile and its two chief tributaries, the White Nile and the Blue Nile. This river system brings water from the far upland south that is used to irrigate fields of cotton for export. Most Sudanese live in a narrow strip of rural villages along these rivers (see Figure 6.14 on page 309). Sudan's only cities are clustered around the famed capital of Khartoum, where the White and Blue Niles join.

There is animosity between southern and northern Sudan that has roots deep in the past and is fed by the politics of oil and religion. Islam is the religion of northern lowland Sudan. It did not spread to the upland south until the end of the nineteenth century, when Egyptians, backed by the British, subdued that part of the country despite fierce resistance. Sudan then became a joint protectorate of Britain and Egypt until its independence in 1955. The southern Sudanese people, for the most part, either are Christian or hold various indigenous beliefs, and most are black Africans of Dinka and Nuer ethnicity. In southwestern Sudan (Darfur) there are also Muslim, Arabic-speaking black Africans. Despite the religion and language they share with the north, these southwestern Sudanese, as well as the Christians of the south, are antagonistic toward the Muslim, Arabic-speaking, lighter-skinned Sudanese of the north, who for thousands of years raided the upland south for slaves. In 1983, the Muslim-dominated government decided to make Sudan a completely Islamic state and imposed shari'a on the millions of non-Muslim southerners. As a result, Sudan is now home to two civil wars and has been classified as a **failed state,** meaning that the government has lost control and is no longer able to defend its citizens from armed uprisings.

One civil war is between the Muslim Arab-dominated regime in Khartoum and black, non-Muslim rebels in the south who protest the imposition of shari'a but also want to stake a claim on Sudan's rich oil reserves (see Figure 6.27 on page 322). This civil war has claimed 2 million lives since 1983, forced 4 million people to abandon their homes, and orphaned many thousands of children, who were then forced to become soldiers for one side or the other. Since 2004, the two sides have come close to a peace pact that would allow for a sharing of power and oil wealth, but periodically conflict flares again. The effects of the conflict have spilled across the border into Uganda.

The second civil war began in Darfur, where the uniformly Muslim and Arabic-speaking population is nonetheless divided between those who identify with the Arab north and those who consider themselves black Africans. Both groups want a share of the oil, and both are afflicted by declining per capita supplies of water. When one faction attempted to establish a secular state, the

**FIGURE 6.42** Sudanese soldiers believed to be part of the *janjaweed* militia. This group, part of a military unit who call themselves, among other things, the Quick and the Horrible, comes every week to the animal market in Mistiria in north Darfur. [AP Photo/Ben Curtis.]

Khartoum government responded with force administered by an armed Arab militia called the *janjaweed*. Samantha Powers, a Harvard University specialist in genocide, reported after a trip to Darfur in 2004 that the government apparently gave the *janjaweed* free rein to rape, rob, and kill blacks and helped by bombing black villages just before the raiders arrived (Figure 6.42). Her findings are backed by similar stories from thousands of refugees and by Human Rights Watch. A desperately needed international peacekeeping presence has been slow to form, but aid for refugees has been sent by the United Nations, Egypt, Saudi Arabia, and numerous other countries, and a small peacekeeping force under the banner of the African Union (see Chapter 7, pages 378–379) has reduced some of the violence against ordinary people. In June 2006, a delegation from the UN Security Council met in Darfur to discuss deploying United Nations peacekeepers, but by that time 200,000 Darfurians were dead and more than 2 million had been displaced, many to the neighboring country of Chad, to which the conflict was spreading.

## Egypt

The Nile flows through Egypt in a somewhat meandering track more than 800 miles (about 1300 kilometers) long, from Egypt's southern border with Sudan north to its massive delta along the Mediterranean. Egypt is so dry that the Nile Valley and the Nile delta are virtually the only habitable parts of the country (see Figure 6.14). Ninety-six percent of Egypt is desert. Yet, for thousands of years, agriculture along the banks of the Nile has fed the country and provided high-quality cotton for textiles.

The Nile's flow is no longer unimpeded. At the border with Sudan, the river is captured by a 300-mile-long (483-kilometer-long) artificial reservoir, Lake Nasser. The lake stretches back from the Aswan High Dam (see Figure 6.41), which controls flooding along the lower Nile and produces hydroelectric power for Egypt's cities and industries. North of the dam, the downstream river environment is greatly modified by its effects. Because the floodplain is no longer replenished every year by floodwaters carrying a fresh load of sediment from upstream, irrigation and fertilizers are needed to maintain the productivity of the principal crops: cotton, grains, vegetables, and sugarcane, as well as animal feed (Figure 6.43). The

**FIGURE 6.43** Oasis of Bahariya, Egypt. A farmer works on an irrigation canal in this large oasis southwest of Cairo, at the edge of the Western (Libyan) Desert. [Sylvain Grandadam/The Image Bank/Getty.]

effects of the dam extend into the Mediterranean: the Nile delta, no longer replenished with sediment, is eroding, and eastern Mediterranean fisheries are being affected by the dearth of natural nutrients and the infusion of chemical fertilizers.

Egypt is the most populous of the Arab countries (74 million in 2005) and the most politically influential. Its geographic location bridging Africa and Asia gives it strategic importance, and the country plays an influential role in global issues—for example, in debates on world trade and in the peace process between Palestine and Israel. Egypt was part of the U.S.-led coalition in the 1991 Gulf War, but opposed the 2003 war in Iraq, saying it was likely to increase terrorism: "Iraq will produce 100 bin Ladens," ran a headline in Cairo. But Egypt is also a major recipient of U.S. aid, which limits its ability to take independent positions.

Egypt's limited resources and its high rate of population growth mean that poverty rates are high: 44 percent of the population lives on just $2 a day. Although it is the cities that are growing most rapidly (in 2006, Egypt was 43 percent urban), the number of people still living and working on the land has also increased. Despite this increase in available labor, the Nile fields are no longer sufficient to feed Egypt's people, and food must be imported. Because of the need for cash to buy this imported food, rural men often migrate to neighboring countries, where they work to supplement the family income, leaving the women in charge of farming and village life.

---

**Personal Vignette**  Mohammed Abdel Wahaab Wad has just spent a week in jail for protesting the loss of the land his family has leased for several generations. In the 1950s, land was redistributed from rich to poor farmers at very low rents so that they would be able to support their families. In the late 1990s, a new law reversed the process—the result of structural adjustment programs imposed by the World Bank and the International Monetary Fund. The purpose of the new law is to increase the land's productivity by cultivating it in large tracts managed with machinery and controlled by the original owners. This reorganization of the agricultural system is expected to boost Egypt's ability to feed itself and to grow water-intensive export crops such as cotton and rice. The change in control of the land is expected to affect 900,000 tenant farmers like Mr. Wad, who have continued to use labor-intensive techniques to cultivate crops of vegetables and wheat just sufficient to feed their families.

Mr. Wad, who has 11 children to help him cultivate, has never before even been asked for rent for the 8 acres (3.25 hectares) he occupies on the edge of the Sahara, northwest of the Aswan High Dam. Now the new law says that the owner can tell the family to leave. The landowner is Gamal Azzam, a member of the urban aristocracy, who is adamant that he will recover his land and that the Wad family must go. There are growing worries that the results of the new law will be, in the short run, an outbreak of armed conflict, and in the long run, another rush to the already crowded cities. Migration to the cities has resulted in a period of malnutrition for some while the agricultural system is reorganized. Undoubtedly modernized agriculture is producing more food per acre than traditional farmers have been able to produce, but that food will be too expensive for the dispossessed.

*Source: Douglas Jehl, "Egypt's farmers resist end of freeze on rents," New York Times (December 27, 1997): A5.* ∎

---

Egypt's two main cities, Cairo and Alexandria, are both in the Nile delta region. With 14 million people in its urban region, Cairo, at the head of the delta, is one of the most densely populated cities on earth. Alexandria, on the Mediterranean, has 4 million people. Egypt's crowded delta region faces a crisis of clean water availability and the threat of disease and pollution. In Chapter 4, we discussed the problem of pollution around the Mediterranean (see pages 218–219). To address Egypt's share of this problem, its recently created Ministry of the Environment is seeking international contractors to treat industrial, agricultural, and urban solid and liquid wastes, which for years have been dumped untreated into the Nile and the Mediterranean. But an example from Cairo illustrates a disturbing pollution linkage between Europe and North Africa. A major industrial complex in Cairo recycles used car batteries sent to Egypt for disposal from all over Europe. However, the recycling complex, where the Ministry of the Environment has yet to exert its influence, releases lead concentrations 30 times higher than allowed by world health standards into the air that blows across Cairo and into the eastern Mediterranean. Some children's playgrounds in Cairo are so polluted that they would be considered hazardous waste sites in the United States and Europe.

In the 1990s, Egypt's economy, long plagued with stagnation, inflation, and unemployment, appeared to be turning around. After years of structural adjustment programs, inflation had been brought under some control, and multinational companies, such as Microsoft, Owens-Corning, McDonald's, American Express, Löwenbräu of Germany, and three German automakers, had opened subsidiaries in Egypt. For the ordinary working people of Egypt, however, this flashy development did not translate quickly into prosperity; instead, they lost public services. By 2006, Egypt's economy was only marginally improved. Unemployment remained at around 20 percent and was expected to remain high due to the return of Egyptian workers from war-torn Iraq. Protests against the United States were widespread, especially among Egypt's underemployed educated young adults, who, partly because of the general economic and political malaise, but also because of their own personal values and their dismay over Western policies, are attracted to Islamism. Overall, conservative interpretations of Islam seem to be gaining popularity.

In the 2006 elections, the enduring government of Hosni Mubarak faced two opposition groups: a liberal party seeking economic liberalization and a yet more secular state, and the Muslim Brotherhood—both a political party and a service organization—seeking an Islamic state. The Mubarak government easily won reelection and then jailed the leader of the liberal opposition party, meanwhile loudly warning of the threat of the increasingly popular Muslim Brotherhood. Hisham Kassem, the editor of Egypt's one opposition newspaper, *Al Masry Al Youm* (see Box 6.4 on page 321), said in an interview on U.S. radio in June 2006 that the government uses the threat of Islamism to scare away democratic reforms and increase donations from the West.

## THE ARABIAN PENINSULA

The desert peninsula of Arabia has few natural attributes to encourage human settlement, and today large areas remain virtually unin-

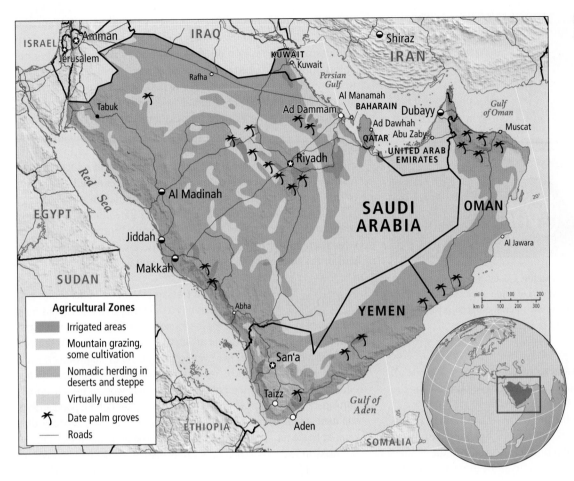

**FIGURE 6.44** The Arabian Peninsula subregion.

habited. The land is persistently dry and barren of vegetation over large areas (Figure 6.44). Streams flow only after sporadic rainstorms that may not come again for years. The peninsula as a whole has one significant resource—oil—which amounts to about 40 percent of the world's proven reserves. Saudi Arabia has by far the largest portion, with approximately 20 percent of the world's reserves.

Traditionally, control of the land was divided among several ancestral tribal groups led by patriarchal leaders called **sheikhs.** The sheikhs were based in desert oasis towns and in the uplands and mountains bordering the Red Sea. The tribespeople they ruled were either nomadic herders, who traveled over wide areas in search of pasture and water for their flocks, or poor farmers settled where rainfall or groundwater supported crops. Until the twentieth century, the sheikhs and those they governed earned additional income by trading with camel caravans that crossed the desert (once the main mode of transport) and by providing lodging and sustenance to those on religious pilgrimages to Makkah.

In the twentieth century, the sheikhs of the Saud family, in cooperation with conservative religious leaders of the Wahhabi sect, consolidated the tribal groups to form an absolutist monarchy called Saudi Arabia, which occupies the central part of the peninsula. The Saud family consolidated its power just as reserves of oil and gas were becoming exportable resources for Arabia. The resulting wealth has added greatly to the power and prestige of the Saud family and of their allies, the Wahhabi clerics. Their tight control over

Saudi Arabia has inhibited the development of opportunities for young people. Discontent, especially among young Saudi adults, is rising and is sometimes expressed by an embrace of fundamentalist Islam. A majority of the suicide hijackers involved in the 9/11 attacks in the United States were Saudi, as is the putative leader of the Al Qaeda movement, Osama bin Laden, who is a member of a wealthy Saudi family in the construction business. Violence such as the terrorist attacks in Riyadh in 2003 (see Box 6.5 on page 330) is likely to be repeated, in part because there is no civil forum for airing discontent. Increasingly, the regional press speculates that, sooner or later, political power will shift in this country.

Despite the overall conservatism of Arabian society here, oil money has changed landscapes, populations, material culture, and social relationships across the peninsula. Where there were once mud-brick towns and camel herds, there are now large modern cities served by airports and taxis. The population of the peninsula has burgeoned to 53 million people, half of whom live in Saudi Arabia. Pickup trucks not only have replaced camels, but are now used to transport them (Figure 6.45). Irrigated agriculture has made the peninsula nearly self-sufficient in food, though only in the short term, since the irrigation is unsustainable. Education is now promoted for young girls as well as boys (though classes are segregated by gender), and among the minority that attend high school and college, women now outnumber men, a fact that will surely influence future politics. Increasing numbers of women work outside

**FIGURE 6.45** Camels in modern Arabian life. Camels remain an important symbol of wealth, but as conveyors of cargo, they are being replaced by pickup trucks. Sometimes, in the interest of speed, a camel may get a ride. [Robert Azzi/Woodfin Camp.]

serves, are rapidly modernizing and are very affluent. Bahrain and Dubayy (Dubai) generate income not so much from oil as from services related to oil production and transport, and by providing entertainment, shopping, and manufactured goods for neighboring wealthy Saudis (Figure 6.46); the actual work is done by contract laborers from India and Southeast Asia. Dubayy is also the destination for some young women trafficked out of Russia and Ukraine for prostitution. The modest oil and gas reserves of Oman and Qatar are sufficient to generate relatively high standards of living. The emir of Qatar and his wife are generous patrons of education and openly encourage social reform. Yemen occupies about one-fourth as much space as Saudi Arabia, but has a population nearly as large (21 million compared with Saudi Arabia's 25 million); it has only small and as yet undeveloped oil reserves. Yemen's standards of education and of living remain by far the lowest on the peninsula and in the entire region (see Table 6.2 on page 336).

**Personal Vignette**   Raufa Hassan al-Sharki is in her forties and holds a Ph.D. in social communications from the University of Paris. In 1996, she founded the Empirical Research and Women's Studies (ERWS) Center at San'a University in Yemen. As Yemen's most outspoken feminist, her overarching goal is to help women learn how to vote independently; Yemen's Islamist party, Islah, supports her efforts. Despite having the right to vote, Yemeni women usually do not participate in the political process. Nearly 70 percent of the population is still rural, and in rural areas only 1 in 10 women can read. Typically, girls work at home rather than going to school. Women perform many essential tasks in the rural economy, such as herding cattle, grinding wheat, and carrying water. After marriage, the average woman bears 6 children.

their homes, usually in work spaces secluded from men. Despite all this apparent progress, economic and social development for the great majority of peninsula citizens has been slow.

The six smaller nations on the perimeter of the Arabian Peninsula, now ruled by ancestral clans that resisted the Saudi family expansion in the 1930s, have varying profiles. Kuwait and the United Arab Emirates, each with close to 10 percent of the world's oil re-

**FIGURE 6.46** Ski Hall in the Mall of the Emirates, Dubayy. Malls in Dubayy (Dubai) are noted for their opulence. In a specially insulated section of this mall, families can actually ski on 6000 tons of artificial snow made by controlling the climate inside the ski facility, even though temperatures outside may be above 100°F (38°C). Users can rent clothing and equipment and enjoy a 400-meter downslope run. A new facility with a mountain range is planned for 2008. [Dorothea Schmid/Bilderberg/Aurora Photos.]

Dr. al-Sharki has found that husbands generally keep their wives' and daughters' voter registration certificates because both men and women believe the women would be likely to lose them. As a result, men often control whether or not, and how, a woman votes.

Sheikh Ahmed Abdulrahman Jahaf, whose daughter is running for Parliament, worries that what he calls the "backward villagers" will think ill of her. He believes that an educated woman such as his daughter is more attractive as a bride than as a public official. According to Sheikh Jahaf, "the nature of women leads us to the conclusion that a woman's right place is home." In these waters, Dr. al-Sharki wades carefully, always getting the sheikhs' permission before she talks to the village women. But lately the sheikhs, too, seem to be changing their views. They see that they must begin to respond to the demands of their people, both men and women, if they wish to stay in power. They even recognize that if they support women's right to vote and encourage them to do so, the sheikhs' own sons may profit in future elections when female voters are more numerous.

For all her caution, Dr. al-Sharki eventually moved too fast for Yemen. In 1999, after she organized a successful international conference at ERWS, the government reorganized the center with an all-male staff and board of directors and removed al-Sharki as its director. She continues her work by explaining gender issues on the Arabian Peninsula to audiences in American universities.

*Sources: Daniel Pearl, "Yemen steers a path toward democracy with some surprises," Wall Street Journal (March 28, 1997): A1, A11; Beloit College Web site: "Noted human rights activist named 2006 Weissberg Distinguished Professor in International Studies," http://www.beloit.edu/~pubaff/releases/05-06/0506weissberg_profile.htm.* ■

# THE EASTERN MEDITERRANEAN

Jordan, Lebanon, Syria, Israel, and the Palestinian people (the Palestinians have no actual political state, but live within all these countries) have all been preoccupied over the last 60 years with political and armed conflict, which arose out of the establishment of the state of Israel but encompasses wider issues of access to land and resources and religious rivalry (Figure 6.47). In the 1970s, the Arab–Israeli conflict spilled over Lebanon's borders and exacerbated discord between Christian and Muslim Lebanese, resulting in a long civil war. Syria has conflicts with most of its neighbors. It has interfered in local politics in Lebanon for years, is accused of sponsoring assassinations there, and had troops stationed there until 2005. Syria also has conflicts with Iraq and Turkey over water rights and with Israel over possession of the Golan Heights. The core of the subregional conflict, between Israel and the Palestinians, was described on pages 327–329.

If these tensions were resolved, the Eastern Mediterranean would have the potential for economic leadership in the region. The countries of this subregion are strategically located adjacent to the rich markets of Europe and the potentially lucrative markets of Central Asia and the Persian Gulf states. Their climate makes them suitable for tourism and subtropical agriculture. Jordan already exports vegetables, citrus fruits, bananas, and olive products; if irrigation water could be found (not an easy matter; see Box 6.6

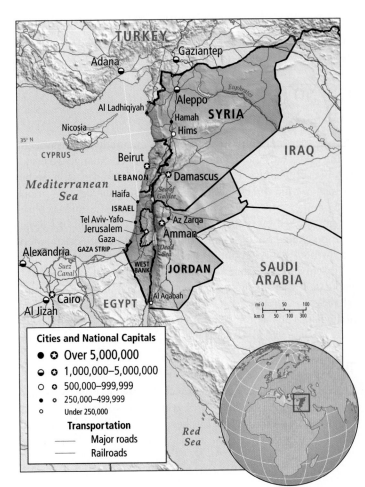

**FIGURE 6.47** The Eastern Mediterranean subregion.

on page 334), Syria could do likewise. Syria lost an important trading partner after the breakup of the Soviet Union; since then (1991), private investors, especially from the Gulf states, have been helping Syria expand its industrial base to include pharmaceuticals, food processing, and textiles, in addition to gas and oil production. After the Lebanese civil war ended in 1990, that country was on an active path to rebuilding its economy and infrastructure until an armed conflict between Hezbollah (an anti-Israeli militia) and Israel in 2006 destroyed much of southern Lebanon.

Israel has the most educated, prosperous, and healthy population in the region, a pool of unusually devoted immigrants, and financial resources contributed by the worldwide Jewish community and a number of Western governments, particularly that of the United States. A range of sophisticated industries are located along the coast between the large port cities of Tel Aviv and Haifa. Israeli engineering is world renowned, and Israeli innovations in cultivating arid land have spread to the Americas, Africa, and Asia. Some people argue that Israel could be a model of development for neighboring countries because it has managed to develop despite its rather meager resources, but this possibility has been hindered by Israel's continual conflict with its neighbors. The years of politically fueled violence have turned Israel into an armed fortress and have discouraged Israelis to the point that emigration now exceeds immigration. In the first two months of 2004 alone, there was a net loss of 13,000 people.

## THE NORTHEAST: TURKEY, IRAN, AND IRAQ

Turkey, Iran, and Iraq (Figure 6.48) are culturally and historically distinct from one another. For example, a different language is spoken in each country: Farsi in Iran, Turkish in Turkey, and Arabic in Iraq. Yet the three countries have some similarities. At various times in the past, each was the seat of a great empire, and each was deeply affected by Islam. Each occupies the attention of Europe and the United States because of its location, resources, or potential threats. All three countries share some common concerns, such as how to allocate scarce water and how to treat the large Kurdish population that occupies a zone overlapping all three countries. Moreover, each country has experienced a radical transformation at the hands of idealistic reformist governments, although the paths pursued have varied dramatically.

### Turkey

Turkey is more closely affiliated with Europe and the West than any other country in the region, with the possible exceptions of Israel and Lebanon. Turkey has been a strong ally of the West for most of the twentieth century, and a strong faction in Turkey seeks to join the European Union. Turkey's affinity with Europe is increased by the fact that many Turks have spent years in Europe as guest workers. The official reasons given for why Turkey has not yet been accepted into the EU are that it has not sufficiently marketized its economy or installed constitutional human rights guarantees, including freedom of religion and protections for minorities, such as the Kurds. Less publicly stated are European worries about absorbing a predominantly Muslim state into an at least nominally Christian Europe and the security risk Islamic fundamentalists would pose were Turkey legally a part of Europe.

Turkey was once the core of the Ottoman Empire, which, as we saw earlier, was dismantled after World War I. At the end of a civil war, in 1923, Turkey undertook a path of radical Europeanization, led by a military officer, Mustafa Kemal Atatürk, who is revered as the father of modern Turkey. Kemal Atatürk and his followers declared Turkey a secular state, modernized the bureaucracy, encouraged women to discard the veil and the custom of seclusion, promoted state-sponsored industrialization, and actively sought to establish connections with Europe. However, backlashes and simple inertia have been obstacles to change. Although the state has sponsored industrialization and modernization in western Turkey, much of eastern Turkey remains agricultural and relatively poor. Islam, long deemphasized as a matter of state policy, is nevertheless an overriding influence on daily life, and fundamentalist versions of the religion are experiencing a resurgence.

Turkey straddles the Bosporus Straits, a narrow passage from the Black Sea to the Mediterranean that is described (with exagger-

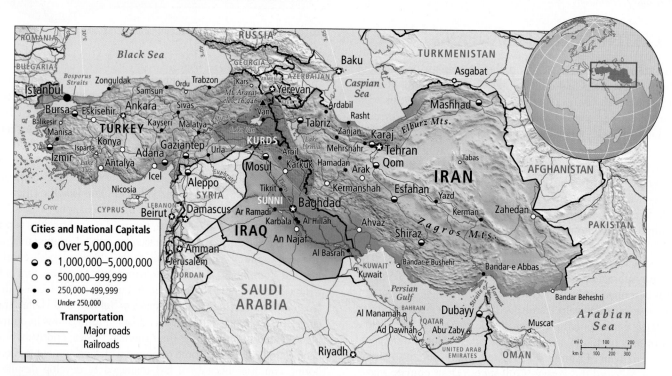

**FIGURE 6.48** The Northeast subregion and the Kurds. The area of Kurdish concentration overlaps three countries. The brown triangle shows the concentration of Sunni Muslims in Iraq, now the center of intense resistance to the U.S. occupation. [Adapted from Edgar O'Ballance, *The Kurdish Struggle 1920–1994* (New York: St. Martin's Press, 1996), p. 235.]

**FIGURE 6.49** Kurds demonstrating in southeastern Turkey. In April 2006, Kurdish demonstrators flashed victory signs behind a barricade separating them from Turkish security forces. [AP Photo/Murad Sezer.]

ation) as the separation between Europe and Asia. During the cold war, Turkey's location made it a strategic member of the North Atlantic Treaty Organization (NATO). Today, this location gives Turkey potential advantages if it can establish economic links with Europe, Caucasia, and Central Asia. Istanbul, already a booming city of more than 11 million, is now the regional headquarters for hundreds of international companies. There are more than 250 U.S.-based companies alone, virtually all of them joint ventures of one kind or another: Coca-Cola, IBM, Eastman Kodak, Kraft Foods, Levi Strauss, and Citibank, to name just a few. Of Turkey's 71 million people, 59 percent are city dwellers. Many of them once lived in rural areas of Turkey, then traveled to Europe to work in factories and on farms; they have returned eager to pursue a standard of living similar to what they observed in Europe. Remittances from Turkish guest workers in Europe add substantially to the country's GDP; the landscape of western Turkey is now dotted with large new houses built with remittances.

Turkey's future depends on its ability to manage its relatively abundant water resources. With its mountainous topography, Turkey receives the most rainfall of any country in the region, and the headwaters of the economically and politically important Tigris and Euphrates rivers are in the mountains of southeastern Turkey. Agriculture employs 40 percent of Turkey's workforce and accounts for 12 percent of its GDP (mostly through exports of cotton and luxury food items). As new irrigation and power generation systems on the Euphrates come on line (see Box 6.6 on page 334), agricultural production will increase dramatically. Cheap hydropower will also expand industry, which already accounts for 80 percent of Turkey's exports. Despite a lack of electrical power, the country's diversified economy manages to produce a wide range of goods, including apparel, foodstuffs, textiles, metal items, transport equipment, leather goods, cement, and tires, all of which are exported to Europe and Central Asia.

Perhaps the greatest obstacle to Turkey's stability is its ongoing conflict with its Kurdish minority (Figure 6.49). The Kurds are tribal peoples (some of them nomadic) who have lived in the mountain borderlands of Iran, Iraq, and Turkey (see Figure 6.48) for at least 3000 years. The division of the Ottoman Empire after World War I by France and Britain dispersed Kurdish lands among Turkey, Iran, Iraq, and Syria, leaving the Kurds without a state of their own. All four countries have had hostile relations with their Kurdish minorities. The conflict springs from complex disputes over control of territory and resources, but is exacerbated by cultural and language differences and the resistance of the Kurds to control by the state. In Turkey, the conflict has escalated to the point of nearly continuous armed strife and repression of those Kurds who are noncombatant. The war in Iraq has added to the conflict, in part because the loyalties of the transnational Kurdish community are in question.

## Iran

Iran occupies a transitional geographic position in this subregion: its western parts are occupied by people of Arab, Turkish, Kurdish, and Caucasian heritage, its eastern parts by people with roots in South Asia, especially Afghanistan and Pakistan. Because its northern flank borders the Caspian Sea, Iran shares in the debate over the future of this inland sea and its resources, especially water and oil. Iran is also close to Central Asia, and its northern border regions have language and ethnic features in common with Turkmenistan. During the cold war, Iran was a zone of contention between the United States and the Soviet Union.

Like Turkey, Iran has abundant natural resources, a strategic location, and a large and growing population (70 million) that could make it a regional economic power. Iran's large petroleum reserves give it influence in the global debate on energy use and pricing. Yet social and economic turmoil have long held the country back.

Iran's present situation in the region and prominence in global politics results from its geographic location, its recent history, and its status as a theocratic state based largely on Islamism. The theocracy was formed under the leadership of the Islamic fundamentalist Ayatollah Ruhollah Khomeini, a Shi'ite spiritual leader who, in 1979, led a revolution against Shah Reza Pahlavi. Pahlavi's father had seized control of the country in the 1920s and introduced secular and economic reforms patterned after those instituted by Kemal Atatürk in Turkey; he abdicated in favor of his son in 1941. In 1953, intervention by the United States helped Shah Reza Pahlavi retain power during a reformist coup attempt. Through U.S. support, Pahlavi continued his father's efforts at Europeanization. But his emphasis on military might and on royal grandeur paid for with oil money overshadowed any genuine efforts at agricultural reform

and industrialization. Disparities in wealth and well-being in Iran grew ever wider.

With the revolution of 1979, political conditions in Iran changed rapidly and radically. Those who had hoped for a more democratic and secular Iran were crushed by the new theocratic state. Resisters were imprisoned, and many were executed. Among those most affected were women. Many upper-class women had lived emancipated lives under the shah's reforms, studying abroad and returning to serve in important government posts. Now all women past puberty had to wear long black *chadors*, and they could no longer travel in public alone, drive, or work at most jobs. Yet even some highly educated women supported the return to seclusion, seeing it as a way to counter the unwelcome effects of Western influences. The turmoil of the revolution was characterized by several episodes in which Westerners were taken hostage and by general resentment against the West, which in turn led to decades of isolation and conflict. During the 1980s, Iran engaged in a devastating war with Iraq (discussed below).

As of 2001, Iran began to change both internally and in its relationships with the outside world. Iran sought to rejoin the community of nations by liberalizing its internal policies and announcing an end to its active nuclear armament program, and punitive economic sanctions that had been imposed by the United States and Europe in response to Iran's suspected funding of terrorism in the West were eased for a while. Elections in May 1997 and again in 2001 brought the victory of a moderate leader, Mohammad Khatami, favored by women and young voters for his liberalizing reforms. Because of strong conservative resistance, however, the reforms have come

**FIGURE 6.50** A store window display of saffron in Mashhad, Iran. Iran's second largest city, and one of Shi'ite Islam's most holy cities, Mashhad is located in the rich agricultural Khorasan province in northeastern Iran. Iran produces 80 percent of the world's saffron, and most of it is grown in Khorasan. Reputedly the world's finest, Iran's saffron accounts for 13.5 percent of the country's non–oil export income. [© Hans Rossel.]

slowly and sporadically. In January 2004, the conservative clerics who made up the Council of Guardians and had final say over who can stand for election rejected more than 8000 would-be candidates. Protests followed, but liberal and moderate candidates were not restored to the ballot, and Mahmud Ahmadi-Nejad, a political Islamist, was elected president. Under his direction, Iran has sought increased stature in the international com-munity by developing a nuclear program that at times is depicted as merely aimed at power generation, at times as an effort to develop nuclear weapons. Currently, the West regards Iran as a dangerous emerging nuclear power deserving of sanctions by the United Nations.

Although many essential products must still be imported, the Iranian government has begun to work toward economic diversification and industrialization. The country's out-of-date transport system is being updated so that Iran can begin again to participate in trade with its Arab and Central Asian neighbors. Agriculture, except for saffron (Figure 6.50), remains Iran's weakest economic sector; even though it accounts for 25 percent of the GDP and employs one-third of the workforce, Iran must import a large part of its food.

Partly because of Iran's location at the borders of several regions undergoing social and economic transitions (Russia and the newly independent states, Southwest Asia, and South Asia), it has the unhappy distinction of being host to more than 1 million international refugees, especially Iraqis and Afghans.

## Iraq

Iraq, home to one of the earliest farming societies on earth and to the Babylonian Empire of biblical times, was carved out of the dying Ottoman Empire after World War I. Most of Iraq's 28 million people live in the area of productive farmland in the country's eastern half, on the floodplains of the Tigris and Euphrates rivers. While the chaos of the present war makes accurate figures on population numbers and distributions difficult to obtain, generally speaking, Sunni Muslims (at least 7 million) reside in the northern two-thirds of the plains, around the capital of Baghdad. About 6 million ethnic Kurds, who are also Sunni, live mostly in the northern mountains in the regions bordering Turkey, Syria, and Iran. The southern third of the country is occupied by at least 11 million Shi'ite Muslims concentrated around Al Basrah at the head of the Persian Gulf. Iraq's main oil fields are also in the south. These distributions are not exclusive; minority pockets of all groups live throughout the country.

Iraq remained a British mandate until 1932, when it became an independent monarchy. Although about half of the Iraqi people are Shi'ite Muslims, since 1932 the government and most of the wealth have been controlled by the less numerous non-Kurdish Sunni Muslims. Following independence, the Iraqi monarchy, like many governments in the region, maintained strong alliances with Britain and the United States and gradually lost touch with its people. A tiny minority monopolized the wealth generated by increasing oil production, but invested little in development. The Shi'ites living in the south, where the oil is located, benefited little from oil earnings.

In 1958, a group of young military officers, including Saddam Hussein, overthrew the monarchy and created a secular socialist republic that prospered over the next 20 years despite occasional political disruptions. Saddam founded Iraq's secret police and then assumed leadership of the Ba'ath Party and eventually the presidency of Iraq. Large proven oil reserves (second in size only to those of Saudi Arabia) became the basis of a very profitable state-owned oil industry, and the profits from oil financed a growing industrial base. Agriculture on the ancient farmlands of the Tigris and Euphrates floodplain also prospered, though environmental problems mounted. The proceeds from oil and agriculture produced a decent standard of living for most Iraqis, who also benefited from excellent government-sponsored education and health-care systems, but corruption and political repression increased. Friction between Sunni and Shi'ites was particularly strong under Saddam Hussein, and since his ousting, strife over political power and access to water and petroleum resources has drastically increased Sunni–Shi'ite hostilities.

Like several other countries in the region, Iraq has been in a state of war and crisis for several decades. The two major events that we will describe below—the Iran–Iraq war (1980–1988) and the Gulf War (1990–1991)—and the 10 years of severe economic sanctions that followed crippled the country. By 2000, European countries were lifting their sanctions, and Iraq was tacitly allowed to sell some oil to buy necessities, but U.S. sanctions remained in place. Then, in 2003, the United States launched a war against Iraq that was still ongoing at the time of publication (see pages 326–327).

***The Iran–Iraq War (1980–1988).*** In 1980, Iraq invaded Iran with the intent of acquiring territory along the Shatt al Arab estuary and in the fertile lowlands between the Zagros Mountains and the Tigris River. President Saddam Hussein calculated correctly that other countries would not come to Iran's aid. Historian Bernard Lewis points out that this war had an ethnic dimension (Arabs against Persians), a sectarian dimension (Sunni Muslims versus Shi'ite Muslims), and an economic dimension (control over oil extraction and sale). It was also a war between the newly revived Islamic traditionalism of Iran and the secularism of Iraq. And it was a geopolitical war, a battle over territory and regional domination. A further geopolitical dimension was that the United States and Arab countries, thinking of Iraq as a buffer zone against the spread of Islamism from Iran, had financed Saddam Hussein's regime and equipped Iraq with sophisticated weapons. But Saddam, expecting a relatively quick victory, had not realized the extent to which Iranian leaders were willing to use the bodies of their religiously inspired young men against the war machine of Iraq. Iran lost hundreds of thousands of people, many of them young boys barely past puberty. Eventually, with heavy financing from wealthy Arab neighbors and with U.S. military and intelligence support, Saddam forced Iran to the peace table. Saddam himself emerged in a marginally better position, but the infrastructures and oil industries of both countries were left in serious disrepair.

***The Gulf War (1990–1991).*** Just 2 years after the end of the war with Iran, Iraq, using the weapons and knowledge it had garnered from the West, invaded Kuwait and initiated the Gulf War. A major impetus for the war was probably Kuwait's unwillingness to restrain its oil production in accordance with OPEC guidelines, thereby driving down the global price of oil at a time when Iraq was badly in need of oil revenues to fund its recovery from the war with Iran. But Iraq was also interested in acquiring more territory along the Persian Gulf, of which it has only a small strip (see Figure 6.48). Iraq's decision to invade Kuwait ultimately proved disastrous because it brought retaliation from a coalition of European and Arab countries led by the United States in an operation known as Desert Storm. The result was a resounding defeat for Iraq and its withdrawal from Kuwait; the United States suffered relatively few deaths (148) and injuries (458), while Iraq lost more than 100,000 civilian and military personnel. Despite its victory, the United States was criticized throughout the region for having rejected several opportunities to solve the conflict peacefully.

In the years following the Gulf War, Iraq's refusal to give up its lethal chemical weapons, thought by some to be capable of killing on a massive scale, led the UN to impose a full trade embargo—no imports, no exports—the most crippling economic sanctions ever leveled against a country. These measures stopped nearly all sales of the country's oil, and the ensuing economic depression hit the poor, the ill, the elderly, and children the hardest. Although basic foods and medicines were supposedly not covered by the embargo, shortages of food, medicine, and other necessities developed in Iraq, causing massive deaths of civilians, especially children. (UNICEF estimates that about 500,000 children died as a result of the sanctions.) Even doctors and lawyers, unemployed because of empty state coffers, worked as street vendors. Meanwhile, smugglers, many of them desert people with a knowledge of ancient transport routes into Jordan and Syria, formed a new economic elite. Another group of the newly rich bought up state-owned industries at bargain prices. Money from such sales, plus political control through extreme repression of the general populace and imprisonment and execution of its enemies, enabled the regime of Saddam Hussein to survive until 2003, when the United States invaded again.

## Reflections on North Africa and Southwest Asia

The region of North Africa and Southwest Asia looms large in global politics. It is the site of two major ongoing violent confrontations: the Israeli–Palestinian conflict and the U.S. war and the insurgency in Iraq. It is also the source of much of the world's oil and gas in an era when fossil fuel energy is essential for emerging and established economies alike. In addition, the region is the site of a major debate over the role of religion in civil society. Because Islam often takes strong positions on gender roles, the outcome of this debate has great implications for the daily lives and public roles of women in the region.

Most of the region has endured some form of foreign domination over the last several hundred years and is presently in transition from outside control to regional or local systems of governance and economic development. Although not united since the Ottoman Empire, the 21 countries described in this chapter share long histories that have intersected repeatedly and often violently. The success of OPEC has led to other efforts to set up associations that work for common goals. Although the nations of North Africa and Southwest Asia are far from achieving economic or political union, the possibilities are occasionally discussed within the region, and there is unanimity around the concept of resisting any future domination by the West.

The trends to look for in the news about this region are efforts to resolve the two ongoing violent conflicts; a growing pan-Arab movement loosely defined by anti-Western, anti-globalization concerns; increased attention to environmental issues; political maneuverings over water rights; efforts to develop an educated and informed citizenry capable of participatory democracy; and a decline in authoritarian governments juxtaposed with contravening efforts to establish Islamist governments. Birth rates can be expected to continue declining as gender roles are slowly but relentlessly redefined. As always, the geographic patterns of these trends will be uneven, but will be related to earlier social and physical patterns.

## Chapter Key Terms

Allah 305

cartel 322

Christianity 304

desertification 335

Diaspora 304

economic diversification 325

failed state 340

Fertile Crescent 302

hajj 305

intifada 327

Islam 304

Islamic fundamentalism (Islamism) 321

Islamists 321

Judaism 304

madrasa 316

monotheistic 303

Muslims 305

OPEC (Organization of Petroleum Exporting Countries) 322

pogroms 308

polygyny 319

protectorate 306

Qur'an (or Koran) 304

recharge area 333

salinization 323

seclusion 306

secular states 315

shari'a 314

sheikhs 343

Shi'ite (or Shi'a) 314

Sunni 314

terrorism 330

theocratic states 315

veiling 306

Zionists 308

# Critical Thinking Questions

1. What are the social forces in North Africa and Southwest Asia that modify the power of religion?

2. Discuss how people in this region have affected world diets (including especially the Americas), first through domestication of plants and animals and then through trade.

3. To what extent is the present-day map of North Africa and Southwest Asia related to the dismantling of the Ottoman Empire after World War I?

4. Consider the various factors that encourage relatively high fertility in this region and design themes for a public education program that would effectively encourage lower birth rates. Which population groups would you target? How would you incorporate cultural sensitivity into your project?

5. Consider the new forces that are affecting urban landscapes: immigration and globalization. What might be some of the expected effects on ordinary people of these abrupt changes in traditional living spaces?

6. Compare and contrast the public debate over the proper role of religion in public life in your country and in one country in this region (for example, Turkey, Morocco, Egypt, or Saudi Arabia). Contrast the roles of religious fundamentalists in the debates in your country and in the country chosen.

7. Gender is a complex subject in this region. Choose a rural location and an urban location and make a list of the forces in each that would affect the future of a 20-year-old woman. Describe those hypothetical futures objectively; that is, without using any judgmental terminology.

8. Why is it important to know that in some countries of this region agriculture may produce only a small amount of the GDP, yet employ 40 percent or more of the people? What are some of the things such a relationship would indicate about the state of development in that country? What public policies would be appropriate in these circumstances—for example, should agriculture be deemphasized?

9. Describe the circumstances that led to support in Europe and the United States for the formation of the state of Israel. Why did the West overlook the Palestinian people in this political undertaking?

10. Discuss the possibilities that scarcity of water is or will become a cause of violence in the region. What is the evidence against this happening?

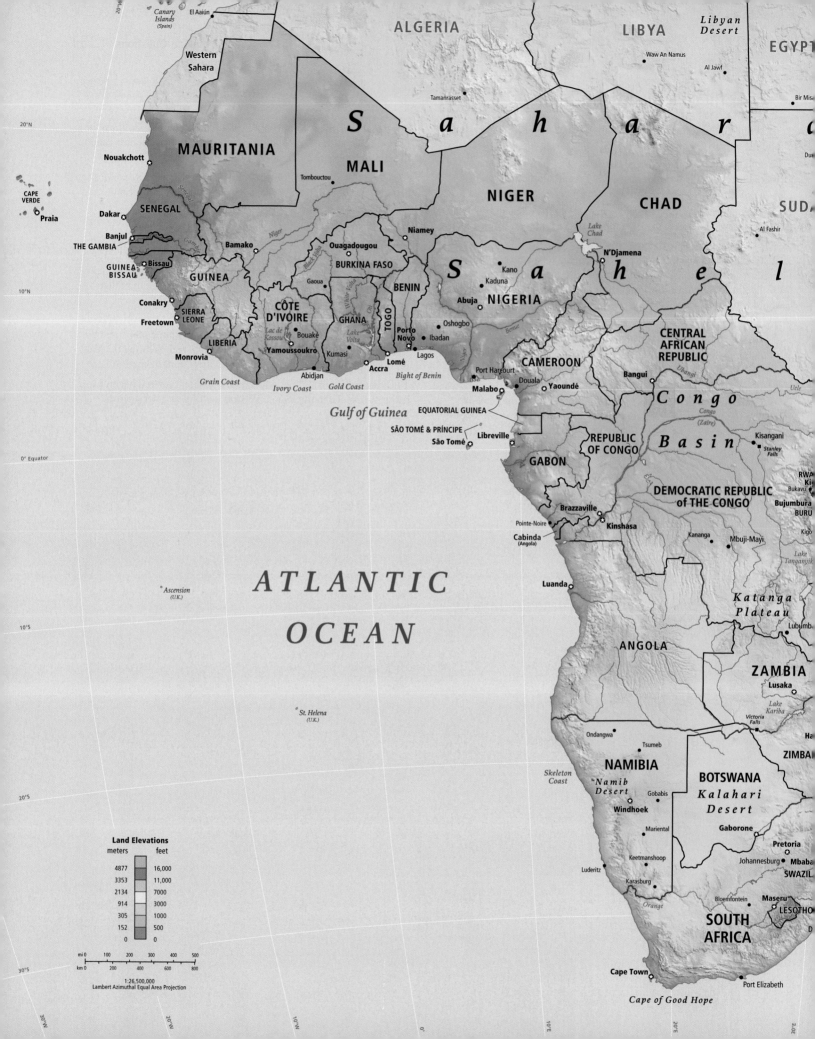

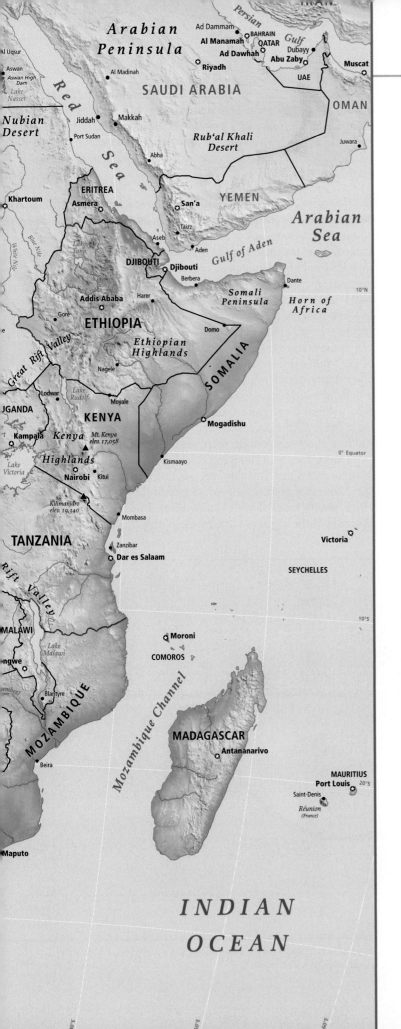

# CHAPTER 7

# SUB-SAHARAN AFRICA

**Global Patterns, Local Lives** Liberian environmental activist Silas Siakor is an affable and unassuming fellow. But his casual style conceals a fierce dedication to his homeland and a remarkable ability at sleuthing.

At great personal risk, Siakor uncovered evidence that seventeen international logging companies were paying Liberia's president, Charles Taylor, so that they could illegally log and sell Liberia's tropical hardwoods. Taylor used the money and weapons he received to fund a 14-year civil war that took the lives of 150,000 civilians. His armies were made up largely of armed cadres of kidnapped children. The logging companies—based in Europe, China, and the Middle East—paid Taylor large sums and supplied him with weapons; in return, they reaped huge fortunes from the tropical forests. The largest firms had their own private militias, which were responsible for rapes, murders, and kidnappings in Liberia.

Liberia was Africa's first republic, founded by freed slaves from the American South in 1822. They hoped Liberia would be a shining example of democracy in an Africa then mostly controlled by outsiders, but instead, civil wars have filled the nearly two centuries since the country's founding. Liberia's forests cover nearly 12 million acres (an area the size of Vermont); they are among the last old-growth rain forests in West Africa and harbor a large population of forest elephants and other endangered species. But despite Liberia's rich timber and mineral (diamond) resources, 80 percent of its people are unemployed; those that work earn only a few hundred dollars a year, and adult literacy rates are below 56 percent. In 2005, the average life expectancy at birth was 42 years.

Siakor secretly prepared a clear, well-documented report substantiating the massive logging fraud (Figure 7.2) and showing the connections between the timber industry, illegal weapons trade, and the funds that gave Taylor the power to wage civil war. In response to Siakor's report, the United Nations Security Council voted to impose sanctions to stop the timber trade (Figure 7.3) and prosecute some of the people involved. Charles Taylor fled to Nigeria, but eventually was turned over to face a war crimes tribunal in The Hague, Netherlands, where he awaits trial. Democratic elections followed in Liberia, and Africa's first elected woman president,

**FIGURE 7.1** Regional map of sub-Saharan Africa.

**FIGURE 7.2** Silas Siakor at work. Siakor is documenting illegally taken logs that are being salvaged by the Ana Woods Liberia Corporation in Liberia's Rivercess County. [Goldman Environmental Foundation.]

Ellen Johnson-Sirleaf, took office in January 2006. In a move that was bold given the poverty and political instability of her country, she cancelled all contracts with timber companies pending a revision of Liberian forestry law. Silas Siakor, who won the international Goldman Environmental Prize in April 2006, says the most rewarding outcome of his work is seeing that power for change can still lie with the little fellow.

*Adapted from "Silas Kapanan 'Ayoung Siakor: A voice for the forest and its people," Goldman Environmental Prize Web site, http://www.goldmanprize.org/node/442; Scott Simon, "Reflections of a Liberian environmental activist," "Weekend Edition Saturday," National Public Radio, April 29, 2006, http://www.npr.org/templates/story/story.php?storyId=5370987. ■*

The story of Silas Siakor illustrates some of sub-Saharan Africa's challenges in the aftermath of colonialism and extreme racial discrimination: poverty, lack of education, lack of infrastructure needed for economic progress, and the hijacking of rich resources by corrupt leaders and international corporations. But it also illustrates that Africans themselves hold the promise of solutions to these challenges.

Sub-Saharan Africa, the region discussed in this chapter (see Figure 7.1), is home to about 750 million people. It contains several of the fastest-growing economies in the world, as well as some of the world's richest deposits of oil, gold, platinum, copper, and other strategic minerals. Yet the news that comes out of Africa is usually about disease, environmental devastation, war and genocide, political corruption, and poverty. All too often, these reports fail to convey both how outsiders are implicated in Africa's problems and how effectively Africans themselves are devising solutions.

During the era of European colonialism (1850s–1950s), wealth flowed out of Africa, and although colonialism officially ended with the granting of political independence in the 1950s, 1960s, and 1970s, wealth is still flowing out of Africa. Investors from the rich countries of the world continue to reap Africa's wealth by extract-

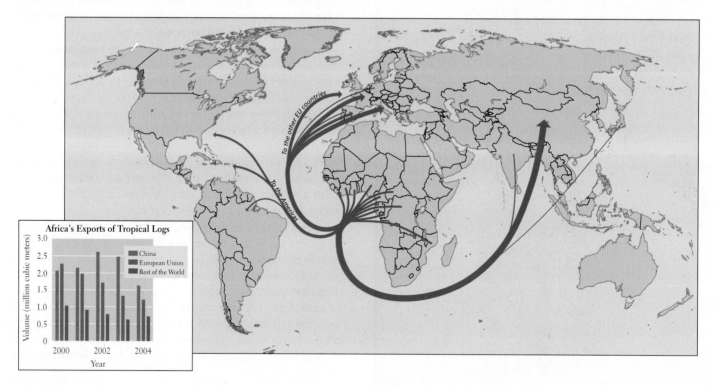

**FIGURE 7.3** Destinations of Africa's exported tropical logs, 2000–2004. Most end-point consumers of Africa's tropical woods are in North America and Europe; much of the wood sold to China also ends up in furniture sold by

firms like Ikea in the United States and Europe as well as in China. [Data from "Exports by, and imports from, Africa" (Global Timber.Org.UK), at http://www.globaltimber.org.uk/africa.htm.]

ing minerals and other natural resources under unfair terms of trade—a practice called **neocolonialism.** An example is provided by the fishing industry of three countries: Senegal, the Gambia, and Guinea-Bissau, all lying on the northwestern coast of the continent. The fish are plentiful here, fed by nutrients pumped in by the cool Canary Current. In the 1990s, high-tech fishing fleets from around the world discovered this resource, and by 1996, 97 percent of the fish caught in these waters were taken and sold abroad by fleets from Asia, Russia, and Europe. In the case of Guinea-Bissau, the foreign fleet owners paid just U.S.$11 million a year for licenses to take a catch worth U.S.$130 million a year on the global market. Senegal and the Gambia struck similarly lopsided deals. Local fishers and workers in related industries were soon impoverished. By June of 2006, some out-of-work fishers were using their boats to carry an estimated 10,000 undocumented West African migrants to the Canary Islands (possessions of Spain), from where they hoped to reach mainland Europe and find work. Fish-eating Europeans, unaware of what was happening to West African fishers, were puzzled as to why so many Africans were risking so much to find work in their part of the world.

It is not surprising, then, that Africa is impoverished and often at war with itself. The average per capita income in sub-Saharan Africa is the lowest in the world. From time to time, violent conflicts, often arising over access to resources or ill-advised government policies, have turned the generally low standard of living into outright destitution. But today, amid the turmoil, there are many hopeful signs that the conflicts can be resolved and the well-being of the majority enhanced. In recent years, a number of Africans have been recognized as among the best leaders and champions of democracy the world has to offer; two of the best known in the West are South Africa's former president, Nelson Mandela, and former UN Secretary-General Kofi Annan, from Ghana. But there are countless others: just beginning to gain international recognition, for example, are the women leaders who have emerged in Liberia, Rwanda, Kenya, and elsewhere across the continent. Since independence, African countries and societies have struggled to determine appropriate pathways to economic development and progress. Although many have not yet found a sustainable path, some are now experiencing significant growth: Angola, Botswana, Chad, the Gambia, Kenya, Nigeria, Senegal, and South Africa all posted GDP growth rates of 5 percent or better in 2005.

## Themes to Explore in Sub-Saharan Africa

Several themes are developed in this chapter:

1. **The roots of African poverty.** Modern global trade patterns, based in part on past exploitative colonial relationships with Europe, keep Africa a provider of cheap raw materials and low-cost labor, a place where resources are used unsustainably and disparities in wealth are on the increase.

2. **A disadvantageous position in the global economy.** As primarily a supplier of cheap resources and unskilled labor, Africa has little power in the global economy. Pervasive poverty has kept its citizens from developing marketable skills. Funds needed for education and social services have gone instead to large debt payments to governments and banks in the developed world. Corruption, ethnic clashes, and religious conflict also inhibit economic development.

3. **Population dynamics and disease patterns.** Generally, Africa is not densely populated, but birth rates remain high, and in some places population densities are too high to be supported by local resources. Diseases have long been a factor limiting growth, but now the HIV-AIDS epidemic is particularly severe—reducing growth, shortening life expectancies, affecting social conditions, and slowing improvements in standards of living.

4. **Distinct gender roles.** Across Africa, it is common for men and women to have distinctly different roles. At least since colonial times, women have performed most of the labor connected with growing and preparing food, most local transport duties, and most marketing of surpluses. Men have prepared fields for cultivation, produced most cash crops, and often migrated far from their families to work in mines or urban factories.

5. **Encouraging developments.** Although many hurdles remain, African countries are joining together to define the region's problems and design solutions for them. The private sector, both informal and formal, is emerging as a vital force in the economy as direct government involvement in agriculture and industry decreases. Political reform is building as women assume leadership roles and democratic elections become a reality in more than half of sub-Saharan African countries.

## ⊢ **I THE GEOGRAPHIC SETTING** ⊣

## Terms to Be Aware Of

The naming of African countries can often be confusing. There are two neighboring countries called Congo: the Democratic Republic of the Congo and the Republic of Congo. Because these designations are both lengthy and easily confused, they will be abbreviated in this text. The Democratic Republic of the Congo (formerly Zaire) will carry the name of its capital in parentheses: Congo (Kinshasa), as will the Republic of Congo: Congo (Brazzaville). Check the regional map (Figure 7.1) to note the locations of these countries and capitals. Another potential confusion is between the country of South Africa and the subregion of Southern Africa, which contains ten countries.

## PHYSICAL PATTERNS

The African continent is big—the second largest after Asia. At its widest point, it stretches 4000 miles (6400 kilometers) from east to west, and its length from the Mediterranean to its southern tip is nearly 5000 miles (8000 kilometers). But Africa's great size is not matched by its surface complexity. More than one-fourth of the continent is covered by the Sahara, which comprises many thousands of square miles of what, to the unpracticed eye, appears to be a homogeneous desert landscape. (Most of this area was discussed in Chapter 6; this chapter concentrates on Africa south of the Sahara.) Africa has no major mountain ranges, but it does have several high peaks, including Mount Kilimanjaro (19,324 feet [5890 meters]) and Mount Kenya (17,057 feet [5199 meters]), both near the equator. Both have permanent snow and ice at their peaks (now melting due to global warming) and temperature-altitude zones similar to those on mountains in the tropics of South America (see Figure 3.7 on page 121).

## Landforms

Geologists usually place Africa at the center of the ancient supercontinent Pangaea (see Figure 1.20 on page 21). As several landmasses broke off from Africa and moved away—North America to the northwest, South America to the west, and India to the northeast—Africa readjusted its position only slightly, drifting gently northeast into Southwest Asia. Although Africa remains in approximately the same place on the globe's surface that it has occupied for more than 200 million years, it is still breaking up, especially along its eastern flank. The Arabian Plate has split away from Africa and drifted to the northeast, leaving the Red Sea, which separates Africa and Asia. The Great Rift Valley, another series of developing rifts, curves inland from the Red Sea and extends more than 2000 miles (3200 kilometers) south to Mozambique, near the east coast (see Figure 7.1). At some future time, Africa is expected to break apart along these rifts (Figure 7.4).

The surface of the continent of Africa can be envisioned as a raised platform, or plateau, bordered by fairly narrow and uniform coastal lowlands and covered by an ancient mantle of rock in various stages of weathering. The platform slopes downward to the north. Thus the southeastern third of the continent is an upland region with several high peaks, whereas the northwestern two-thirds of the continent is a lower-lying landscape, interrupted only here and there by uplands and mountains. Despite their lack of complexity, the landforms of Africa have obstructed transport and hindered connections to the outside world. Routes from the plateau to the coast must negotiate steep **escarpments** (long cliffs) around the rim of the continent, and the long, uniform coastlines have few natural harbors.

## Climate

Most of sub-Saharan Africa has a tropical climate (Figure 7.5) because 70 percent of the continent lies between the Tropic of Cancer and the Tropic of Capricorn. Average temperatures generally stay above 64°F (18°C) year-round everywhere except at the more temperate southern tip of the continent and in cooler upland zones (hills and plateaus). Seasonal climates in Africa differ more by the amount of rainfall than by temperature.

Most rainfall comes to Africa by way of the **intertropical convergence zone** (**ITCZ**), a band of atmospheric currents that circles the globe roughly around the equator (see inset, Figure 7.5). At the ITCZ, warm winds converge from both north and south and push against each other, causing the air to rise, cool, and release moisture in the form of rain. The rainfall produced by the ITCZ is most abundant in central and western Africa near the equator. There, in places such as the Congo Basin, the frequent rainfall nurtures the dense vegetation of tropical rain forests.

**FIGURE 7.4** Geologic history in the making. Ethiopian geologist Dereje Ayalew and his colleagues from Addis Ababa University examine crevices that opened during an earthquake in the Afar Triangle, Ethiopia, near the northern end of Africa's Great Rift Valley, in 2006. The earthquake occurred just as their helicopter was landing at the site; crevices began racing toward the researchers like a zipper opening up. The Arabian Plate, the Nubian African Plate, and the Somalian African Plate are pulling apart, and eventually ocean water will fill this place. [Dereje Ayalew/Addis Ababa University.]

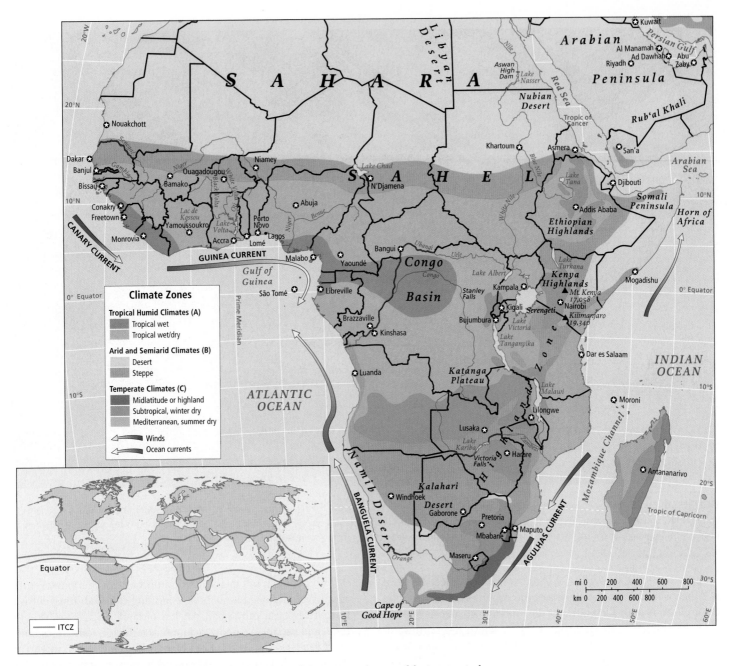

**FIGURE 7.5**  Climates of sub-Saharan Africa. The inset shows the position and range of the intertropical convergence zone (ITCZ). [Adapted from http://www.astrosurf.com/luxorion/meteo-tropicale.htm.]

The ITCZ shifts north and south seasonally, generally following the area of Earth's surface that has the highest average temperature at any given time. Thus, during the height of the Southern Hemisphere summer in January, the ITCZ might bring rain far enough south to water the dry grasslands, or steppes, of Botswana. During the height of the Northern Hemisphere summer in August, the ITCZ brings rain as far north as the southern fringes of the Sahara—an area called the **Sahel,** where steppe and savanna grasses grow. Poleward of both of these extremes, a belt of descending dry air blocks the effects of the ITCZ and creates the deserts found in Africa (and on other continents) at roughly 30°N latitude (the Sahara) and 30°S latitude (the Namib and Kalahari deserts).

Across far western Africa and central Africa, the tropical wet climates that support equatorial rain forests are bordered on the north, east, and south by seasonally wet/dry tropical woodlands and beyond that by moist tropical savanna or steppe, where tall grasses and trees intermingle in a semiarid environment. Both tropical wet and wet/dry climates have provided suitable land for agriculture for thousands of years (Figure 7.6). Farther to the north and south lie the true desert zones of the Sahara and the Namib and Kalahari. This banded pattern of African ecosystems is modified in many areas by elevation and wind patterns. For example, winds blowing north along the east coast keep ITCZ-related rainfall away from the **Horn of Africa,** the triangular peninsula that juts out from

**FIGURE 7.6  Market gardens along the Sénégal River in Mali.** The river is an important source of moisture in this Sahel region, providing water for nearly 10 million people. Women transport water in containers up the steps from the river to their gardens, where they water individual plants so as to conserve moisture. This high-altitude photo shows the tiny walled garden plots. [Yann Bertrand-Artaud/ALTITUDE.]

northeastern Africa below the Red Sea. As a consequence, the Horn of Africa is one of the driest parts of the continent. And along the coast of the Namib Desert, moist air from the Atlantic is blocked from moving over the desert by cold air above the northward-flowing Benguela Current. Like the Peru Current off South America, the Benguela is chilled by its passage past Antarctica. Rich in nutrients, it supports a major fishery along the west coast of Africa.

The climate of Africa presents a number of challenges to human habitation. Parasites and insects that breed prolifically in warm, wet climates cause and spread debilitating diseases such as river blindness, schistosomiasis, and malaria. In drier tropical climates, water for drinking, farming, and raising animals is often in short supply, and soils are not particularly fertile, even if irrigated. Wherever both temperature and moisture are high, organic matter in the soil decays rapidly. In a standing forest, the organic matter shed every day provides a continual source of nutrients for plant growth. But if the forest is removed, the source of nutrients is also gone, and the soil quickly deteriorates. The minerals are **leached** (washed out) into groundwater and runoff, and fine soil particles collect downslope, where the sun bakes them into a permanently hard surface called **laterite.**

To maintain soil quality, cultivators of the wet tropics over the ages have developed a method of farming called **shifting cultivation.** They clear only small patches of land, an acre or two at a time, and use the cleared vegetation as fertilizer, sometimes burning it to release nutrients. They plant their gardens with many varied species that cover the soil quickly to prevent it baking hard in the hot sun. Because the soil loses fertility quickly, the small plots produce well for only 2 or 3 years, and they are then allowed to revert to forest for several decades. Typically, commercial agriculture that depends on large cleared fields and long-term production has not succeeded for extended periods in tropical zones because of soil infertility, tropical diseases, and pests.

There is evidence that Africa has experienced repeated climate changes over the last 20,000 years. Many scientists think that the changes occurring today are at least partly the result of global warming caused by the rising levels of greenhouse gases produced in the industrialized parts of the world. Scientists expect the present climate patterns in Africa to intensify: hot, wet places may become hotter and wetter; hot, dry places may become hotter and drier. Grasslands on the desert margins may become deserts; low-lying coastal areas may be inundated by seawater as melting glaciers at the poles raise sea levels. Because so many Africans live in arid or low-lying zones, as many as one-fourth of them could eventually find their homelands uninhabitable.

## HUMAN PATTERNS OVER TIME

Africa's rich past has often been misunderstood and dismissed by people from outside the region—a strategy that has helped some outsiders to defend their exploitative ventures there. European slave traders and colonizers called Africa the "Dark Continent" and assumed it was a place where little of significance in human history had occurred. The substantial and elegantly planned city of Benin in western Africa, which European explorers encountered in the 1500s, never became part of Europe's image of Africa, nor did the city of Loango in the Congo Basin. Because of the pervasive influence of Europe, even today, most people are unaware of Africa's internal history or its contributions to world civilization.

### The Peopling of Africa and Beyond

Africa is the original home of the human species. It was in eastern Africa that the first human species evolved more than 2 million years ago, although they differed anatomically from human beings today. These early tool-making humans are known to have ventured as far as Dmanisi, Georgia (between the Caspian and Black seas), as early as 1.8 million years ago. Recently analyzed archaeological evidence recovered in Ethiopia shows that anatomically modern humans (*Homo sapiens*) evolved from a separate species in Africa about 200,000 years ago, and fossil skulls show that by at least 90,000 years ago, modern humans had reached the eastern Mediterranean. Evidence is mounting that modern humans, like earlier humans, radiated out of Africa, spreading across mainland and island Asia and eventually into Europe, probably intermingling with other human populations they encountered rather than replacing them.

### Early Agriculture, Industry, and Trade in Africa

In Africa, people began to cultivate plants as early as 7000 years ago in the southern belt of the Sahara, which was less arid at that time. The highlands of present-day Sudan and Ethiopia in eastern Africa were another early center of plant and animal domestication and food production. The Bantu peoples seem to have brought plant cultivation to equatorial Africa when they migrated from western Africa to the Congo Basin sometime within the last 2500 years. About 1500 years ago, descendants of farmers from both coastal West Africa and the highlands of eastern Africa were converging and migrating into southern Africa, displacing local hunter-gatherers with their agricultural systems.

Political economist Samir Amin makes the point that Africa had complex and varied social and economic systems well before the modern era. Trade routes spanned the African continent and reached around the Mediterranean to Rome and east to India and China. For example, the Kush people, who lived in the upper Nile River region 4000 years ago, served as trading middlemen between Africans in the interior of the continent and Egyptians. Gold, ele-

phant tusks, and timber from tropical Africa were exchanged with Mediterranean people for a variety of goods: brass, copper, iron (not yet available in Africa), olive oil, animals, dyes, food, and manufactured goods. In East Africa, in about 300 B.C.E., the kingdom of Aksum, in what is present-day Ethiopia, took over from the Kush and expanded the trade with Egypt and Assyria (in modern Iraq) to include both Rome and India, and even China.

About 2500 years ago, people in northeastern Africa learned how to smelt iron. Thereafter, iron smelting, and eventually steel production, spread slowly from the highlands of eastern central Africa to West Africa (steel was not invented in Europe until the 1700s). The slow spread of ironworking is explained in part by the fact that the continent was only lightly populated. By 1300 years ago, when Europe was in the Dark Ages and Islam was just awakening, a remarkable iron-producing and trading civilization with advanced agricultural and gold-mining technology had developed in the highlands of southeastern Africa in what is now Zimbabwe (Figure 7.7). Known now as Great Zimbabwe, this empire traded with merchants from Arabia, India, Southeast Asia, and China, exchanging the products of its mines and foundries for silk, fine porcelain, and exotic jewelry. Then, around 1500, for reasons as yet little understood, the Great Zimbabwe empire collapsed. Nineteenth-century European archaeologists who discovered the ruins of Great Zimbabwe did not believe that indigenous African civilizations could have existed, let alone traded so widely, so they mistakenly credited the Great Zimbabwe culture to outsiders, not Africans, and then looted much of the material evidence that might have revealed the true history of the site.

**FIGURE 7.7** Aerial view of the Great Zimbabwe National Monument. The Great Zimbabwe complex covered some 1800 acres (728 hectares) at its height between 1200 and 1300 and may have been home to as many as 40,000 people. [David Wall/Lonely Planet Images.]

In West Africa, several influential centers made up of dozens of linked communities developed in the forest and the savanna. One of several powerful kingdoms was the empire of Ghana (700–1000 C.E.), which was located not in the modern country of Ghana, but in present-day Mali north of the headwaters of the Niger River. The Ghana empire was succeeded by other kingdoms that successively enlarged the empire's territorial jurisdiction. The kingdoms' wealthy and prominent converts to Islam periodically sent large entourages on pilgrimages to Makkah (Mecca), where their opulence was a source of wonder.

In both eastern and western Africa, Africans traded not only their ivory, iron, animals, and gold, but also fellow Africans. There was a long-standing custom in Africa of enslaving people captured as a result of hostilities between two or more ethnic groups. The treatment of slaves within Africa, governed by local custom, was usually reasonably humane, though by no means egalitarian. At some unknown time in the past, but certainly long before the beginning of Islam, a slave trade with Arab and Asian lands to the east developed (Figure 7.8). Scholars generally agree that, after the spread of Islam began around 700, close to 9 million African slaves were exported to parts of Asia (Persia, India, Southeast Asia) and to Islamic areas around the Mediterranean. Protections afforded slaves in Africa were lacking when slaves were traded to non-Africans. Male slaves were often castrated before they were offered for sale to Muslim traders. Despite this practice, numerous sources report that African slaves who worked in the homes and fields of wealthy Mediterranean Muslims were often treated like distant and lesser-status relatives and generally not brutalized.

## Europeans and the Slave Trade

The course of African history shifted dramatically in the mid-1400s, when Portuguese sailing ships began to appear off Africa's west coast. The names given to stretches of this coast by the Portuguese and other early European maritime powers reflected their interest in acquiring Africa's resources: the Gold Coast, the Ivory Coast, the Pepper Coast, the Slave Coast.

By the 1530s, the Portuguese had organized the slave trade with the Americas. The trading of slaves by the Portuguese, British, Dutch, and French was more widespread and brutal than either the small-scale taking of slaves that had occurred within Africa or the slave trade conducted with Muslims and Asians. Its most distinctive feature was the commercial motivation of the Europeans, who needed cheap labor for their American plantations, which in turn supplied raw materials and money for the Industrial Revolution in Europe. Slaves were often treated strictly as a commodity.

To acquire slaves, the Europeans established forts on Africa's west coast and paid nearby African kingdoms with weapons, trade goods, and money to make slave raids into the interior. As in the pre-European slave trade, some of the captives were taken from enemy kingdoms in battle. Many more, however, were kidnapped from their homes and villages in the forests and savannas. Most slaves traded in the international market were male because the raiding kingdoms preferred to keep captured women for their reproductive capacities. Between 1600 and 1865, about 12 million captives were packed aboard cramped and filthy ships and sent to the Americas. Up to a fourth of them died at sea. Those who arrived in

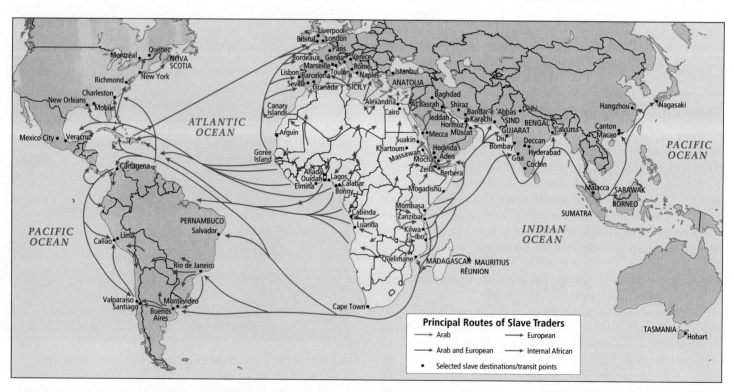

**FIGURE 7.8** African slave trade. [Adapted from the work of Joseph E. Harris, in Monica Blackmun Visona et al., eds., *A History of Art in Africa* (New York: Harry N. Abrams, 2001), pp. 502–503.]

the Americas went primarily to plantations in South America and the Caribbean; about one-fourth were sent to the southeastern United States (see Figure 7.8).

The European slave trade severely drained the African interior of human resources and set in motion a host of damaging social responses within Africa that are not well understood even today. It made Africans dependent on European trade goods and technologies, especially the guns used by the raiding kingdoms. Enslavement of Africans by Africans persists into the present, especially in West and Central Africa, where UNICEF estimates that 200,000 children are sold into slavery each year as agricultural laborers, domestic servants, and prostitutes.

## The Scramble to Colonize Africa

The British transatlantic slave trade officially ended in 1807, but other European nations continued trading in slaves until 1865. By that time, Europeans found it more profitable to use African labor *in Africa* to extract raw materials for Europe's growing industries, and they had established formal colonies by the late nineteenth century. European interests extended inland to include the exploitation of fertile agricultural zones, areas of mineral wealth, and places with large populations that could serve as sources of labor.

Colonial powers competed avidly for territory and resources. The result was the virtually complete seizure and partition of the continent by the time of World War I (Figure 7.9). Africans were not consulted about the partitioning of their continent, and only two African countries retained independence: Liberia on the west coast, because it was populated by former slaves from the United States, and Ethiopia (then called Abyssinia) in East Africa, because its strong monarchy defeated early Italian attempts to colonize it. Otto von Bismarck, the German chancellor who convened the 1884 Berlin Conference at which the competing powers first partitioned Africa, declared: "My map of Africa lies in Europe." With some notable exceptions, the boundaries of most African countries today derive from the colonial boundaries set up between 1884 and 1916 by European treaties. These territorial divisions lie at the root of many of Africa's current problems.

A few basic geographic patterns in the European domination of Africa during the colonial period can help us to visualize how the geography of Africa was changed by colonization. The geographer Robert Stock, a specialist in sub-Saharan Africa, identifies three such geographic patterns:

1. European settlers occupied land at relatively high densities in only a few places. These were mainly areas with especially attractive resources or places where Europeans considered the climate comfortable, such as the relatively cool highlands of Kenya in East Africa and the upland plateau of South Africa (see the green patches in Figure 7.5). In these places, Europeans forced indigenous farmers

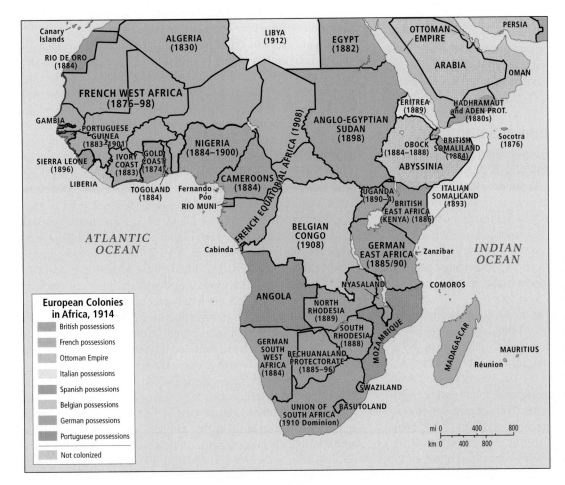

**FIGURE 7.9** The European colonies in Africa in 1914. The dates on the map indicate the beginning of officially recognized control by the European colonizing powers, which took various legal forms. Countries without dates were informally occupied by colonial powers for a few centuries. Abyssinia (modern Ethiopia) and Liberia were the only territories on the African continent not colonized. [Adapted from Alan Thomas, *Third World Atlas* (Washington, D.C.: Taylor & Francis, 1994), p. 43.]

and herders into reservations on marginal land or made them labor on European-owned farms and plantations. Areas taken over by Europeans were privileged: taxes were kept low, and roads were constructed to provide ready access to local and foreign markets.

2. Africans continued to farm small plots in agricultural regions and places the Europeans considered disagreeable—though they still controlled those regions. African farmers were often directed to switch from their traditional food crops to export crops. Africans suffered malnutrition as more and more land was shifted from food production to cash crop or livestock production. These indigenous farming areas received little assistance from Europeans, transportation was only modestly upgraded, and Africans were often required to pay higher taxes than the Europeans who had taken over their land.

3. Remote areas that were difficult to exploit economically were treated as labor reserves. Many young men were taken from these areas to work on government projects such as railroad construction. Separated from their families, subjected to dangerous working conditions, and paid very little, many died of stress, disease, and malnutrition. One of these areas was the Belgian Congo (now Congo [Kinshasa]), where, during the reign of King Leopold of Belgium (1865–1909), as many as 10 million people starved to death.

In all areas colonized by Europe, the main objectives of colonial administrations were to extract as many raw materials as possible, create markets in Africa for European manufactured goods, and keep the costs and the commitments of European-administered governments to an absolute minimum. It is useful to examine the case of South Africa to appreciate how the European incursion into Africa led to the expropriation of land, the subjugation of African peoples, and in this case, the infamous system of racial segregation known as **apartheid**.

## Case Study: The Colonization of South Africa

In the 1650s, the Dutch took possession of the Cape of Good Hope, at the southern tip of Africa, from the Portuguese. By 1673, Dutch settlers were expanding into the lands of the native Khoi-Khoi people. The Dutch immigrant farmers, called Boers, practiced herding and farming techniques that used large tracts of land. As they spread into the interior, the Boers pushed indigenous people off their land, despite strong resistance (Figure 7.10). After a smallpox epidemic in 1713 left insufficient KhoiKhoi labor for their agricultural ventures, the Boers enslaved laborers elsewhere in Africa and brought them to work on their farms.

The British were also interested in the wealth of South Africa, and in 1795 they gained control of areas around the Cape of Good Hope. Slavery was outlawed throughout the British Empire in 1834, prompting large numbers of slave-owning Boers to elude British control by migrating to the northeast in what was called the Great Trek (1835). They often came into intense and violent conflict with the Africans in these interior areas. In the 1860s, extremely rich diamond deposits were unearthed in these areas, and in the 1880s, gold secured the economic future of this land, which became known as the Orange Free State and the Transvaal. Africans were forced to

**FIGURE 7.10** **Boers returning from hunting.** This painting by Samuel Daniell (London, 1804) shows some of the details of daily life in Boer-dominated South Africa. Notice the status differences of the people, the various roles of animals, and the architecture of the substantial Boer house. [The British Library/Art Resource, NY.]

work in the diamond and gold mines under disastrously unsafe conditions and for minimal wages. They lived in unsanitary compounds that travelers of the time compared to large cages.

Britain, eager to claim the wealth of the mines, invaded the Orange Free State and the Transvaal in 1899, waging the bloody Boer War. The war gave them control of the mines briefly, until resistance by Boer nationalists forced the British to grant independence to South Africa in 1910. This independence, however, applied to only a small minority of whites: the Boers and some British who chose to remain. Black South Africans, more than 80 percent of the population, lacked legal political rights until 1994.

In 1948, apartheid laws were enacted to reinforce the longstanding segregation of Boer society. These laws required everyone except whites to carry passbooks and live in racially segregated rural townships, specific sections of cities, or workers' dormitories attached to mines and industries. Eighty percent of the land was reserved for the use of Europeans, who at that time made up just 15 percent of the population. Blacks were assigned to ethnicity-based "homelands" that were considered independent enclaves within the borders of, but not legally part of, South Africa, though the government exerted strong influence in them. South Africans who were racially mixed (designated "Coloreds") and expatriate (mostly Indian) merchants held an intermediate status that was never clearly defined, but they were clearly subservient to whites.

The fight to end racial discrimination in South Africa began even before the apartheid laws were formally introduced in 1948. The African National Congress (ANC), the first and most important organization participating in this struggle, was formed in 1912 to work nonviolently for civil rights for black Africans. After the apartheid laws were passed, the ANC began to recruit more outspoken

**FIGURE 7.11** Slegs vir Blankes (Reserved for Whites). This sign at a Port Elizabeth, South Africa, beach, photographed around 1988, clearly demonstrates how apartheid was enforced in several local languages. [David Turnley/CORBIS.]

**FIGURE 7.12** Nelson Mandela votes in the first post-apartheid election. On April 27, 1994, the first election open to all adults was held in South Africa. Nelson Mandela, only recently released from prison, was elected president. It was also the first time he was able to vote. [David Turnley/CORBIS.]

leaders. Overt resistance to the apartheid system intensified in the 1960s and grew steadily into the 1990s, despite heavy-handed repression and segregation (Figure 7.11). Among the key leaders who fought for racial justice in South Africa were Steven Biko, an attorney who died while in police custody; Anglican Archbishop Desmond Tutu; and Nelson Mandela, an ANC leader who was jailed by the South African government for 27 years. The resistance was carried out and supported by millions of ordinary people, including schoolchildren. Finally, in the early 1990s, the white-dominated government realized that it could no longer resist majority rule. It released Nelson Mandela and began negotiating with the ANC and other political organizations for an end to apartheid. Although officially independent from Britain since 1910, South Africa gained its independence from white minority rule only in 1994, when, at age 75, Mandela was elected the country's president (Figure 7.12).

## The Aftermath of Independence

In Africa, the era of formal European colonialism was relatively short. In most places it lasted for about 80 years, from roughly the 1880s to the 1960s. In 1957, Ghana, in West Africa, became the first African colonial state to achieve its independence. The last sub-Saharan African country to gain independence was Eritrea, in 1993, although Eritrea won its independence not from a European power, but from its neighbor Ethiopia, after a 3-year civil war.

The road to nation building has been a rocky one for the African countries, but most of them have been on this road for less than 40 years. Like their colonial predecessors, most independent African governments became authoritarian, antidemocratic, and dominated by privileged and Europeanized elites. In the last several years, however, pro-democracy movements have begun to spread across sub-Saharan Africa. When strict criteria are applied—universal adult suffrage; regularly scheduled, competitive, multiparty, secret-ballot elections; and access of all parties to the public via open media channels—23 of 47 countries qualify as democracies (see Figure 7.35 on page 384).

Africa enters the twenty-first century with a complex mixture of enduring legacies from the past and looming challenges for the future. Although Africa has been liberated from colonial domination, it is still strongly influenced by the residue of oppressive colonial policies and by neocolonialism exercised by the world's wealthy countries and multinational firms. Because of its reliance on exports and imports, it remains inextricably linked to the global economy, in which it has long operated at a disadvantage. Many countries remain economically dependent on their former European colonizers. Poverty has expanded rapidly, while solutions to Africa's problems have been slow in coming. In the recent past, Africa has faced declining economic productivity and rising debt; severe periodic drought and famine; major health problems, including the world's worst HIV-AIDS epidemic; and the challenge of having a very young and fast-growing population. The brightest spots in Africa's future emerging from the recent decades of rapid and often wrenching change are the willingness of many Africans to consider innovative alternatives to conventional solutions and the new socially minded leaders, both men and women, who are arising among educated Africans.

## POPULATION PATTERNS

A look at the population density map of sub-Saharan Africa (Figure 7.13) will surprise many readers, who may have the erroneous impression that Africa is densely populated. In fact, the population is distributed very unevenly, but generally sparsely, over the continent. Only a few places exhibit the densities that are widespread in Europe, India, and China. Nonetheless, there are serious population problems in Africa. Some countries (for example, Rwanda, Burundi, and Nigeria) have pockets of very high density, but have not developed existing resources to support their people. A number of countries (for example, Chad, Liberia, Mali, Niger, and Madagascar) have annual population growth rates above 3 percent, and their populations could double by 2030.

### Africa's Carrying Capacity

The standard of living experienced by a population usually depends in large part on a region's **carrying capacity**: the maximum

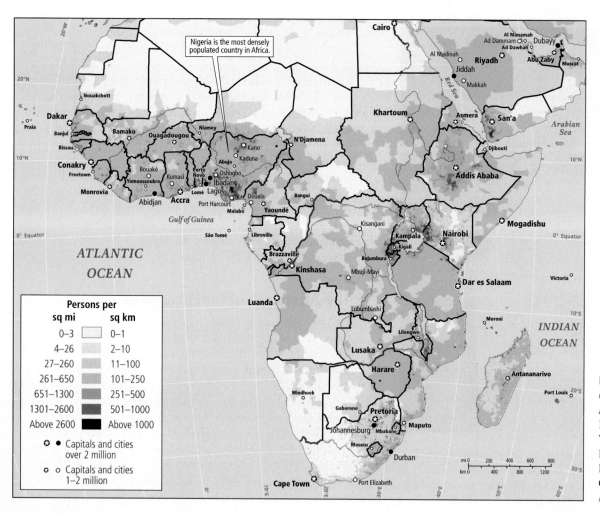

**FIGURE 7.13** Population density in sub-Saharan Africa. [Data courtesy of Deborah Balk, Gregory Yetman, et al., Center for International Earth Science Information Network, Columbia University, at http://www.ciesin.columbia.edu.]

**FIGURE 7.14** Refugees in Africa. These Bantu people from Somalia are waiting in Kenya to participate in a resettlement program that will bring them to the United States. [AP/Wide World Photos.]

number of people it can support sustainably with food, water, and other essential resources. Carrying capacities in Africa vary widely across the continent, and the factors that limit carrying capacity also vary from place to place. In some persistently dry places, such as areas bordering the Sahara, the lack of water limits cultivation and grazing, which in turn limits human settlement. In some persistently wet places, such as the Congo Basin, the leached soils cannot sustain long-term cultivation. Countries with alternating wet and dry seasons—such as South Africa, Kenya, and other countries in eastern Africa—can support fairly dense populations as long as people use the land and water sustainably. Some otherwise habitable African ecological zones, such as the high plains of the Serengeti southeast of Lake Victoria, are sparsely populated because they are breeding grounds for disease-carrying organisms, especially tsetse flies and mosquitoes.

Ultimately, the carrying capacity of any place is affected by cultural, social, economic, and political factors as well as by physical features. For example, in most of Africa, people must depend largely on the local agricultural carrying capacity of the land for subsistence because there is insufficient wealth to import items such as food, building materials, and industrial raw materials that would allow many more people to be supported. To understand why this is so, compare an American suburban family of four living on a half-acre lot, supporting themselves on the adults' annual salaries of $50,000 to $100,000, with a rural African family of seven or eight subsisting almost entirely on 4 acres (1.6 hectares) of land, often with little education and no reliable outside source of cash. While the Africans must literally live off their land, the American suburbanites, like the inhabitants of other wealthy societies, can tap into resources around the globe (including Africa); thus their standard of living is only partially linked to the carrying capacity of their country's environment.

Political unrest places additional burdens on a region's carrying capacity. War and oppression often force people to leave their

homes as refugees. Over the last decade of the twentieth century, for example, people in Chad, Sudan (see Chapter 6), Somalia, Ethiopia, Uganda, Liberia, Sierra Leone, Congo (Kinshasa), Congo (Brazzaville), Rwanda, and Mozambique poured back and forth across borders to escape genocide, a tactic usually instigated by a corrupt state and carried out by one ethnic or political faction against another. According to the United Nations High Commission for Refugees, in 2003, Africa hosted about 33 percent of the world's 10.3 million international refugees (Figure 7.14). The United Nations estimates that, if people who are displaced within their home countries are also counted, Africa has half the world's refugee population, and that about three-fourths of Africa's refugees are women and children. As difficult as life is for these refugees, the burden on the countries that host them is also severe. Even with help from international agencies, the host countries find their own development plans complicated by the arrival of so many distressed people, who must be fed, sheltered, and given health care. Large portions of economic aid to Africa have been diverted to deal with the emergency needs of refugees.

## Population Growth

African populations are growing faster than any others on earth. In less than 50 years, the population of sub-Saharan Africa more than tripled, reaching nearly 657 million in 2000. Between 2000 and 2006, the population increased by nearly 100 million to 752 million. This rapid growth is the main threat to human well-being in places where the carrying capacity has already been reached or exceeded. The addition of increasing numbers of individuals is outstripping even the best efforts to improve nutrition, education, housing, health care, and employment possibilities. In some places, Africa's already low standards of living are declining further. The UN projects that the number of primary school students (aged 6–11) in sub-Saharan Africa will increase from 76.9 million in 1985 to 195.3 million by 2025. This increase is significant because schools are already in extremely short supply and underequipped across the continent.

The geographer Ezekiel Kalipeni has found that many Africans are not yet choosing to have smaller families because they view children as both an economic advantage and a spiritual link between the past and the future. Childlessness is considered a tragedy. Not only do children ensure a family's genetic and spiritual survival, they still do much of the work on family farms. In regions with high infant mortality, parents have many children in the hope of raising a few to maturity.

In a handful of countries, however—those that are most highly developed—fertility rates have declined significantly. In South Africa, Botswana, Seychelles, Réunion, and Mauritius (the last three are tiny island countries off East Africa), circumstances have changed sufficiently to make smaller families desirable. First, in all five countries, the education of women has improved and gender role restrictions have been relaxed (as measured by the United Nations Gender Empowerment Measure). Hence women are choosing to use contraception because they have life options beyond motherhood. Second, where infant mortality is now relatively low (only in Seychelles, Réunion, and Mauritius), parents can expect their children to live to adulthood. However, the rate of contraception use in sub-Saharan Africa as a whole is still less than half that in all other world regions. In short, in most places in Africa, the demographic transition—the sharp decline in births and deaths that usually accompanies economic development—is only beginning and is developing slowly (see Figure 1.35 on page 40).

---

**Personal Vignette** Mary, a Kenyan farmer, has just had her third son. The father of Mary's children works in a distant city and visits the family only several weeks a year. He supports himself with his earnings and buys occasional nonessentials for the family. Mary owns only one cow and a small piece of land that can't be further divided, so all she can provide for her children is an education. Mary says that she can afford to educate only three children. She tells an interviewer that three children are enough for happiness, and so today, at age 29, she is having surgery that will prevent conception.

Such attitudes are spreading in Kenya, where food, health care, and jobs are in short supply. Mary plans to augment her farm income by starting a sanitary pit toilet construction business. She has applied for a small loan (U.S.$150) for this purpose. The success of her business could mean that her children will become well educated and that she herself will gain prestige. Studies of the effects of microcredit opportunities have shown that female participants tend to increase their use of birth control markedly. In addition, when women like Mary accomplish their goals, they become role models for other women, who then limit their families so that they can also become self-sufficient owners of small businesses.

*Sources: Jeffrey Goldberg, New York Times Magazine (March 2, 1997): 39; World Resources, 1996–1997 (New York: Oxford University Press, 1996), p. 5. Information on the role of microcredit in fertility patterns is from Fiona Steele, Sajeda Amin, and Ruchira T. Naved, "The impact of an integrated micro-credit program on women's empowerment and fertility behavior in rural Bangladesh," Policy Research Division Working Paper no. 115 (New York: Population Council), 1998.* ■

Recent research has shown that, contrary to the popular perception, African men are not resistant to limiting family size; when they are included in public health education, some men will actually suggest birth control to their wives. The inclusion of men is especially important in cultures in which men are assumed to be the major decision makers. For example, among some groups in Africa, the custom of abstaining from sex for several years after a birth is an effective strategy for limiting family size—obviously a method that men must consent to. But if this strategy results in men going to sex workers in the interim, then HIV infection is a peril for all in the family.

Men in polygynous relationships, which exist in parts of West Africa and elsewhere (see page 390), tend to produce more children than monogamous males. Polygynous men tend to be less well educated and to see multiple wives and many children as status symbols, because traditionally wives and children produced wealth. Educating polygynous men about the realities of equipping children for success in modern times can lead them to choose fewer children and perhaps even fewer, better educated wives who have only one or two children and then work outside the home.

The population pyramids in Figure 7.15 demonstrate the contrast between countries experiencing rapid growth (such as Nigeria, Africa's most populous country) and countries where growth has slowed markedly (such as South Africa, its most developed country). South Africa's pyramid has contracted at the bottom because its birth rate has dropped from 35 per 1000 total population to 23 per 1000 just since 1990. This decrease is largely the consequence of economic and educational improvements and social changes that have come about since the end of apartheid in the early 1990s. But this pattern in the South Africa pyramid may also be a consequence of the spread of HIV among young adults and the babies they bear. HIV-AIDS, the consequences of which are discussed further in the next section, is now the main cause of the

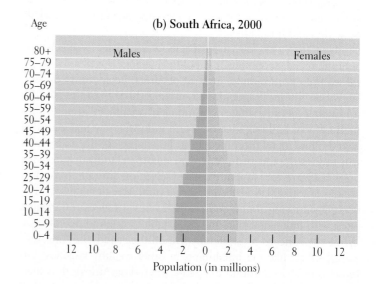

**FIGURE 7.15** Population pyramids for Nigeria and South Africa. Nigeria, with a total population of 134 million, has a population growth rate of 2.8 percent. South Africa, with a population of 44 million, has a growth rate of 0.9 percent. [Adapted from "Population pyramids for Nigeria" and "Population pyramids for South Africa" (Washington, D.C.: U.S. Census Bureau, International Data Base), at http://www.census.gov/ipc/www/idbpyr.html.]

slowing of population growth rates in Africa, especially in the Southern Africa subregion.

## Population and Public Health

Sub-Saharan Africa has long been troubled by infectious diseases that have been particularly harmful and difficult to control; they include schistosomiasis, sleeping sickness, malaria, river blindness, and cholera. Observers report that while in developed countries it is noncommunicable diseases, such as heart disease and cancer, that are the leading causes of death, in Africa, infectious diseases (including HIV-AIDS) are by far the largest killers, responsible for about 50 percent of all deaths (Figure 7.16). Some infectious diseases are linked to particular ecological zones. For example, people living between the 15th parallels north and south of the equator are most likely to be exposed to sleeping sickness (trypanosomiasis), which is spread among people and cattle by the bites of tsetse flies that inhabit the vegetation near rivers and lakes and in woodlands and grasslands. The disease attacks the central nervous system and, if untreated, results in death. Several hundred thousand Africans suffer from sleeping sickness, and most of them are not treated because they cannot afford the expensive drug therapy.

Schistosomiasis (bilharziasis), Africa's most common chronic tropical disease, and malaria, its second most common, are linked to standing fresh water, and their incidence has increased with the construction of dams and rice paddies. Schistosomiasis is a debilitating, though rarely fatal, disease affecting about 170 million sub-Saharan Africans. It develops when a flatworm carried by a particular freshwater snail enters the skin of a person standing in water.

Malaria, which is spread by the anopheles mosquito (which lays its eggs in standing water) is even more costly. The disease kills at least one million sub-Saharan Africans annually, most of them children under the age of 5. But malaria has serious social and economic effects beyond the loss of these beloved children. Millions of adult Africans are feverish, lethargic, and unable to work efficiently because of malaria. The disease also inhibits trade and scares off investors and travelers, and the treatment, which can include blood transfusions, has increased the incidence of HIV-AIDS.

Until recently, little research in Africa was focused on how best to control the most common chronic tropical diseases. That is now changing; for example, more than 60 research groups are working on a vaccine that will prevent malaria in most people. In the meantime, however, an African child dies every 30 seconds for lack of malaria medication, while the world's short supply of the medication goes mostly to Western people traveling and working in tropical zones. On the brighter side, river blindness (onchocerciasis), which has caused approximately 18 million sub-Saharan Africans to go blind, has been successfully eradicated from 11 African countries through the use of pesticides against the blackfly that spreads it. The effort has allowed the reoccupation of about 250,000 square kilometers of agriculturally productive river valley, previously abandoned due to blackfly infestations.

## HIV-AIDS in Africa

The epidemic of acquired immunodeficiency syndrome (AIDS), caused by the human immunodeficiency virus (HIV), is the most severe public health problem in sub-Saharan Africa. In 2006, sub-Saharan Africa had an estimated 24.5 million HIV-infected people, 63 percent of the estimated worldwide total of 38.6 million (Figure 7.17). Infection rates for adults in Africa are the world's highest at 6.1 percent. In Southern Africa, where, until the HIV-AIDS epidemic, development prospects were brightest, rates are even higher: 18 percent in South Africa, 40 percent in Botswana. Young adults, especially women, have the highest rate of infection. The rate of infection for women ages 15 to 24 is three times that for men of this age.

The disease threatens to change sub-Saharan Africa's population growth and life expectancy patterns drastically. In Botswana, Africa's second most well-off country, HIV-AIDS reduced the average life expectancy from 59 years in 1990 to just 35 years in 2005. Zimbabwe's life expectancy went from 58 years in 1990 to 41 years in 2005, the same as it was 200 years ago. Zimbabwe's population growth rate was 3.2 percent in 1993, but by 2005 it was only 1.1 percent. If the epidemic is not stopped, Zimbabwe could lose 20.5 percent of its population by 2010. The demographic effects of HIV-AIDS are comparably dramatic in many of the other countries of sub-Saharan Africa.

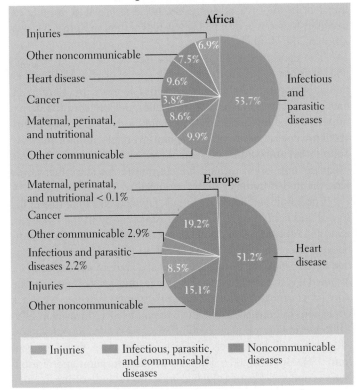

**FIGURE 7.16** Leading causes of death in Africa and Europe, 2002. [Adapted from *Global Burden of Disease Estimates, 2002* (Geneva: World Health Organization).]

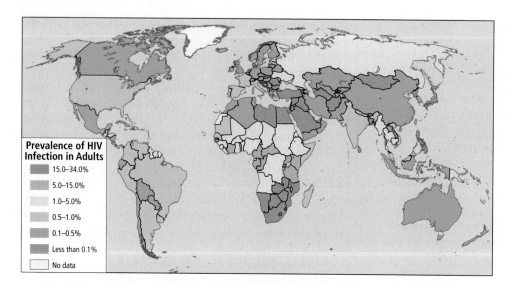

**FIGURE 7.17** Global prevalence of HIV-AIDS. [Adapted from *UNAIDS 2006 Report*, Chapter 2, at http://www.unaids.org/en/HIV_data/2006GlobalReport/default.asp.]

In Africa, HIV-AIDS affects women even more than men: 55 percent of HIV-infected adults in sub-Saharan Africa are women, and four-fifths of the world's women infected with HIV are in Africa. (Figure 7.18 shows the dramatic effect of HIV-AIDS on female life expectancy in selected countries of East and Southern Africa.) The reasons for this pattern are related to the social status of women. Young mothers are often infected by their husbands, who may visit sex workers when they travel for work or business. Men often have multiple sexual partners; nevertheless, they may coerce their wives to have sex. The wives have little power to insist

that their mates use condoms. In Uganda, one-fourth of married urban women, who are normally considered at low risk, are HIV-positive. Where HIV-AIDS education is lacking, some men think that only sex with a mature woman can cause the disease, so very young girls are increasingly sought as sex partners (this is sometimes referred to as "the virgin cure"). Finally, for many poor women, especially recent urban migrants removed from their village support systems, occasional sex work is part of what they do to survive economically. In some cities, virtually all sex workers are infected. In the 5 years after 2000, because of AIDS, the life expectancy for women dropped below that of men in Kenya, Malawi, Zambia, and Zimbabwe.

The rapid urbanization of Africa (see the discussion on pages 386–387) has contributed to the swift spread of HIV-AIDS, which the latest statistics (2005) show is definitely more prevalent in urban areas. Transportation between cities and the countryside has improved, and bus and truck drivers are thought to be major carriers of the disease, as are urban migrants who return home for visits. Young men and women also encounter each other more easily in urban settings, often without the community pressure that would have precluded intimacy in the countryside.

Education has played a contradictory and changing role in the spread and control of HIV-AIDS in Africa. Early in the epidemic, educated Africans were actually more susceptible to infection, partly because they were likely to live in urban areas. Now that the epidemic has grown, those who can read and understand explanations of how HIV is spread have an advantage, and rates of infection are now lower among educated people than among people who lack education. In Senegal, where the decision to invest heavily in HIV-AIDS prevention was made in the 1980s (when infection rates across Africa were still low), levels of infection held at less than 1 percent from 1990 through 2005. Through public education and distribution of condoms, Uganda also had great success in lowering the incidence of new HIV infections from 15 percent in the early 1990s to just 4.1 percent in 2004. Elsewhere, notably in South Africa in the late 1990s, top officials denied that HIV-AIDS was a problem, and infection rates soared to 28 percent of all pregnant women by 2003.

**Change in Female Life Expectancy, 1990–2015**

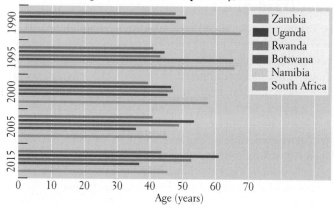

**FIGURE 7.18** Current and projected effects of HIV-AIDS on female life expectancy in selected countries. As the effects of HIV-AIDS began to be felt in sub-Saharan Africa, life expectancy for women dropped dramatically in several countries (after 1990 in Zambia, Uganda, and Rwanda and by 2000 in Botswana, Namibia, and South Africa; there are no 1990 data for Botswana). It is hoped that as sex education programs expand, medical research continues, and drugs become more affordable and available, female life expectancy will rise again. Recovery is likely to be slow, however, with female life expectancy in Uganda and Rwanda recovering the fastest (notice the lengths of these bars in 2005 and 2015). [Adapted from U.S. Census Bureau, International Programs Center, unpublished tables, Slide 23, at http://www.census.gov/ipc/www/slideshows/hiv-aids/TextOnly/Slide23.html.]

While prevention through education is the first bulwark against HIV-AIDS, given the number of people that are now infected, treatment is the most effective strategy for reducing AIDS-related illness and death. Unfortunately, the overwhelming majority of Africans cannot afford the costly combination of drugs that keeps victims alive in such places as North America, Europe, and Japan. The antiretroviral "drug cocktail" produced by American and European drug companies costs about U.S.$10,000 per year per patient. In 2000, in an attempt to lower this cost, a global movement began to challenge the patents of the drug companies. By the end of that year, the hope for cheaper drug treatments was given a boost when drug firms in Cuba and India began to offer a regimen that cost just $1.00 a day—but even that price is still too high for people in most countries where infection rates are high. By 2003, several private foundations and governments were seeking ways to provide affordable drug treatment for HIV-AIDS. In Africa, access to antiretroviral drugs varies by country. For example, in 2004, the Central African Republic made the drugs available at affordable rates with the help of $25 million donated by developed nations, including the United States. Nevertheless, in the region as a whole, as of 2005, barely 11 percent of AIDS patients had access to anti-retroviral drugs.

Across the continent, the consequences of the HIV-AIDS epidemic are enormous. Nelson Mandela has said that he now views HIV-AIDS as a challenge greater than apartheid. As of 2005, more than 20 million sub-Saharan Africans had already died, and as many as 80 million more AIDS-related deaths are expected by 2025. Young adults, parents, teachers, skilled farmers, craftspeople, and trained professionals have been lost, and over 12 million orphans have been created, many without any family left to care for them or to pass on vital knowledge and skills (Figure 7.19). Elderly grandparents or neighbors must provide for orphaned dependent children (who may themselves have been infected with HIV at birth). Rural households may sell land and animals to pay for care or funeral expenses and thus lose the wherewithal for future sustenance. Time, money, and health-care professionals are being chan-

**FIGURE 7.19** AIDS orphans in Kenya. The majority of these children in a makeshift community school in Nairobi are HIV-positive. Kenyan officials estimate that in 2005 there were over a million AIDS orphans in the country. [Reuters/Corinne Dufka.]

neled to HIV-AIDS prevention and treatment and away from education and the control of the other diseases, such as malaria, that actually threaten an even larger percentage of sub-Saharan Africans.

## II CURRENT GEOGRAPHIC ISSUES

Most countries of sub-Saharan Africa (Figure 7.20) have been independent of colonial rule for about 50 years. Because the effects of the colonial era in Africa run so deep, however, it is too soon to tell just how some of the African countries will find ways to become politically, economically, and socially viable. Still, some countries are achieving higher standards of living for their citizens, and many researchers are optimistic about the future.

### ECONOMIC AND POLITICAL ISSUES

Today, Africa is the poorest region of the world and exhibits widespread political turbulence. British development specialist Eliza-

beth Francis gives five reasons for Africa's difficulties over the past 50 years:

1. The long-term effects of colonialism positioned Africa in the global market as a supplier of cheap resources and low-cost labor. That role persists today, leaving African countries in competition with other suppliers of cheap labor and raw materials and with little influence on market prices.

2. Political leaders and citizens of the new states had little or no experience in allocating resources, power, and opportunity fairly. Consequently, corruption became widespread, and people were unable to assert their will through democratic channels.

**FIGURE 7.20** Political map of sub-Saharan Africa.

**3.** Civil unrest and wars have retarded both social and economic progress. During the cold war, interventions by the United States, France, and the Soviet Union and its allies intensified wars in Africa as these world powers attempted to draw the new African nations into alliances with them.

**4.** Steep oil price increases in the 1970s forced African governments into debt to pay for fuel.

**5.** The World Bank and the International Monetary Fund promoted ill-advised economic reorganization and structural adjustment programs in the 1980s and 1990s. As a result, government financial support for education and health was cut off just when the improvement of human skills and well-being was crucial for further development.

Today, however, there are a number of positive developments: fresh leaders, especially women, are emerging; there are promising new sources of investment funds; the region's oil resources are producing needed national income; economies are diversifying; locally based environmental and social activism is increasing (Box 7.1), and educated migrants are returning to help their homelands.

## Agricultural and Natural Resources

*Subsistence Agriculture Remains a Major Source of Livelihood.* Most Africans, 66 percent of whom live in rural areas, produce their own food by farming small plots, raising livestock, or both. Many also fish, hunt, and gather some of their food. Small-scale cash cropping, adopted during the colonial era to make money to pay taxes, is often part of the system of production. Today people may grow cash crops of peanuts, cacao beans, rice, or coffee to obtain money to pay school fees or purchase a bicycle or other useful items.

Subsistence farms usually occupy 2 to 10 acres (1 to 4 hectares). Fields given over to food production are often intercropped (planted with many different species) in a planned arrangement that is rotated over the course of the field's use. After a few years of cultivation, a field may be allowed to lie fallow (remain uncultivated) so that the soil can be replenished by the growth of natural vegetation. In some places, however, the intensity of land use has increased to the point that sufficient fallow times are no longer possible, and soil fertility is decreasing. Permanent cultivation requires a high degree of specialized knowledge and care to sustain soil fertility. Sometimes commercial fertilizer is used.

Most African farmers practice **mixed agriculture** (Figure 7.21); that is, in addition to crops, they also raise some livestock, but they usually limit their holdings to a cow or two, some poultry, and a few goats and sheep. Herding, or **pastoralism,** is practiced primarily in savannas, on desert margins, or in the mixture of grass and shrubs called open bush. Herders use the milk, meat, and hides of their animals, and they typically trade for grain, vegetables, and other necessities with settled farmers. In recent decades, drought and the encroachment of irrigated field agriculture has forced many herders to reduce or divide their herds.

Urban farming is an increasingly significant part of the lives of city dwellers in Africa. Tiny vegetable gardens and poultry pens can be seen even in the hearts of central business districts. A study has shown that in Mombasa, Kenya, 15 percent of households grow food on urban land. These urban gardens provide nutrition to city dwellers; generate employment, especially for women; and put derelict urban land to sustainable, productive use.

*Exports of Raw Materials to the Global Economy.* Even after independence, the economies of African countries were still centered around the export of one or two raw materials, and this pattern

**FIGURE 7.21** Small farm plots surrounding a Kikuyu village near Mount Kenya. The foothills around Mt. Kenya north of Nairobi, settled by the Kikuyu about 400 years ago, are today one of the most intensively farmed areas in the country. Most of the plots are small, but the farmers have moved beyond bare subsistence and are producing saleable surpluses. [Russell Middleton/University of Wisconsin–Madison, African Studies Program.]

## BOX 7.1 | AT THE LOCAL SCALE   Big Changes Come to Myeka High School

An example of how investment in education can lead to multiple strands of development is provided by the case of Myeka High School, which lies deep in the hinterland of South Africa and far from sources of electricity. For years, the school was destitute; few students graduated because of the lack of educational materials as well as general demoralization. Then the U.S.-based SELF foundation invested in solar power equipment for the school, and Dell contributed computers, while Infosat Telecommunications set up Internet service. Graduation rates soared, and students left with employable skills

Intrigued by the idea of alternative power sources, a science teacher and several students devised a biogas system to utilize waste from the school's toilets to create methane gas, which now powers the school's electrical system (Figure 7.22). The spin-offs are impressive: sanitation problems are solved, and the manure fertilizes the school's gardens, which feed the many AIDS orphans in attendance. Teachers and students have mastered science, math, and technical skills that will serve them for years to come.

*Sources:* David Lipshultz, "Solar power is reaching where wires can't," *New York Times*, Business Section, September 9, 2001; Myeka High School Web page, http://www.solarengineering.co.za/myeka_html1.htm, 2004.

**FIGURE 7.22** A biogas digester being installed at Myeka High School. After earlier experiments with solar power, students and teachers at Myeka High School built a biogas plant as an alternative energy source. The students were amused by the idea of cooking and powering their computers with gas generated from cattle manure and human waste, but they have been pleased to find that it works very well. [Courtesy of Will Caywood and Richman Simelane.]

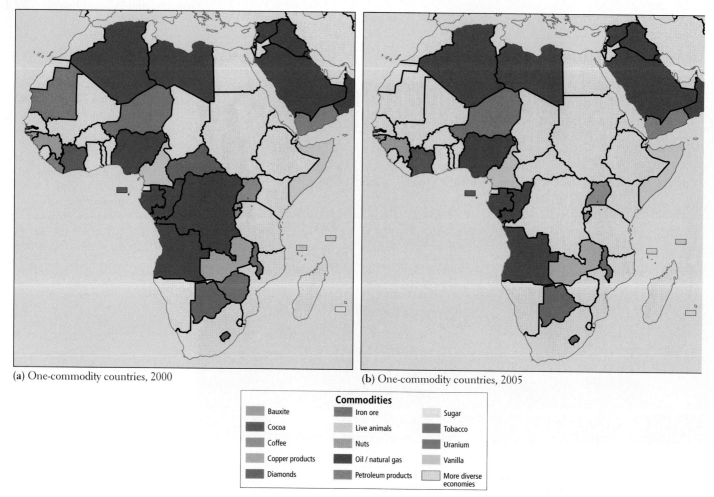

(a) One-commodity countries, 2000

(b) One-commodity countries, 2005

**Commodities**

| | | |
|---|---|---|
| Bauxite | Iron ore | Sugar |
| Cocoa | Live animals | Tobacco |
| Coffee | Nuts | Uranium |
| Copper products | Oil / natural gas | Vanilla |
| Diamonds | Petroleum products | More diverse economies |

**FIGURE 7.23** One-commodity countries. (a) African countries that depended on just one commodity for more than 50 percent of their export earnings in 2000. (b) Five years later, ten more African countries had diversified their economies sufficiently to no longer depend on just one main commodity. [Adapted from George Kurian, ed., *Atlas of the Third World* (New York: Facts on File, 1992), p. 76; data from *United Nations Human Development Indicators 2000* (New York: United Nations Development Programme); 2005 data from the *CIA World Factbook, 2006*, at https: //www.cia.gov/cia/publications/factbook/index.html; *People Daily Online*, at http://english.people.com.cn/200509/20/eng20050920_209525.html.]

is only now beginning to change (Figure 7.23). Although these countries are beginning to diversify their economies, for the most part they remain yoked to raw materials exports. Southern Africa exports a wide range of minerals; it produces 27 percent of the world's gold and 50 percent of its platinum. West Africa exports oil (primarily from Nigeria and neighboring countries), and Côte d'Ivoire produces cacao beans, nearly all of which supply the European chocolate industry. (The Africa–Europe link in cacao production and the chocolate trade is discussed in Chapter 4.) East Africa exports small percentages of the world production of copra (dried coconut, used to make oil), palm oil, and soybeans; Central Africa exports some tropical hardwoods.

Development of commercial agriculture and industries that process agricultural products are sorely needed to meet the demands of growing populations and to supply more profitable products for export, yet there are impediments to these goals. Large-scale cultivation of tree crops (such as palm oil), rice, and various dessert fruits is increasingly common as African countries search for crops that will sell well in the global marketplace (Figure 7.24). However, getting access to consumers in developed countries can be dif-ficult because, despite free trade initiatives, the United States and the European Union often place high tariffs on African products to protect their own or neighboring producers. Africa's marginally fertile soils and underdeveloped infrastructure are additional drags on commercial agriculture. Still, in such places as Botswana and South Africa, production has increased thanks to improved crop varieties, more effective environmental management techniques, and greater investment in the infrastructure needed for agriculture. Commercial farms now produce cashews, cotton, coffee, tea, corn (maize), groundnuts (peanuts), poultry, meat, and dairy products. But commercial farming requires substantial investment in land, as well as expensive mechanized equipment and chemical fertilizers and pesticides, leaving most small farmers out of the game.

Governments may try to help small farmers acquire these costly necessities, but there is much debate over how involved governments should be. In Zimbabwe, for example, the promotion of *smallholder horticulture* for export is linked with a highly controversial land reform program instituted by the government of President Robert Mugabe. In this program, wealthy white farmers, many from South Africa, had their highly productive farms confiscated and

**FIGURE 7.24** Processing vanilla beans in Madagascar. Vanilla beans are Madagascar's most lucrative commodity export, producing four and a half times more export income than its next largest commodity, cloves. Notice that many school-aged youngsters are sorting the beans. [Julie Larsen Maher © WCS.]

redistributed to poor Zimbabweans, some of whom were fighters in the war of independence in the 1970s but lacked agricultural skills. Many Zimbabweans saw the white owners as usurpers of African lands because their holdings were originally taken from African people during the colonial period. Critics of Mugabe pointed out that while land reform programs can do much to improve production by smallholders, the best and largest of the confiscated farms did not in fact go to smallholders, but to friends and associates of Mugabe, who then used the land for their own enrichment. The results of this program have been disastrous. As of 2006, food production had dropped so low that a majority of Zimbabweans were on food aid, around 80 percent were unemployed and living below the poverty line, and public protests were counteracted with violence.

*Botswana: A Case Study.* Since its independence in 1966, Botswana has been hailed as the fastest-rising economic star in sub-Saharan Africa. In 1960, the GDP per capita was U.S.$535, and in 1966 Botswana was one of the 25 poorest countries in the world. By 2005, the GDP per capita was U.S.$5516 ($8714 when adjusted for PPP)—second only to that of South Africa. How was an arid, landlocked country able to prosper so well and so rapidly? In 1966, 40 percent of Botswana's GDP derived from agriculture. Now agriculture accounts for only 2.4 percent of GDP, while mineral resources dominate the economy. Diamond deposits were found shortly after independence, and by 1995 Botswana was the world's third largest diamond producer (Figure 7.25). Unlike the profits of diamond industries in other parts of Africa, which have financed

**FIGURE 7.25** Jwaneng diamond mine, Botswana. The Jwaneng diamond mine is one of the world's richest. The diamond-laden earth is loaded into trucks and taken to a treatment plant for crushing and sifting. The diamonds are then separated into industrial-quality and gem-quality stones. [World Diamond Council, www.diamondfacts.org.]

civil wars and terrorism, much of Botswana's diamond wealth has gone toward infrastructure improvements. The largest diamond mining company is Debswana, an equal joint venture between the DeBeers Group and the government. Diamonds account for over 63 percent of Botswana's government revenues and 70–80 percent of its export dollars. When this wealth is statistically apportioned among just 1.6 million people, the per capita GDP is at the medium rank by world standards.

The actual allocation of wealth in Botswana, however, is highly uneven. A typical diamond worker earns only $85 a month, or one-seventh of the average per capita GDP. Meanwhile, income for the richest 10 percent in Botswana is 78 times that for the poorest 10 percent, one of the highest wealth disparities on earth. Perhaps as many as 20 percent of the people have reasonably well-paid positions in government or the mining industry; the remaining people work in agriculture (16 percent) or in primarily low-paid, menial urban jobs (64 percent); 30 percent suffer serious poverty. This disparity is reflected in Botswana's rather low ranking of 131 on the Human Development Index—which, nonetheless, is relatively high for Africa. Still, there is a trend toward investment in human capital, exemplified by the government's commitment to educating children at least through the tenth grade, which could lead ultimately to a more prosperous society.

The burgeoning HIV-AIDS crisis, however, threatens Botswana's prosperity. Close to 40 percent of the adult population is afflicted. While HIV-AIDS is perhaps more widespread in urban areas, the disease also has dire consequences in rural areas, where the poor are concentrated. At least 100,000 agricultural workers and extension agents have been disabled, and, as elsewhere, the youngest generation is left without the necessary life skills or support. To address the issue, Botswana is considering lighter plows and tools that can be used by children and frail adults, seed varieties that require less labor, and garden species that mature at different times of the year so as to spread out the work. And, with the deaths of so many male farmers, customs restricting equal rights to land for women are also being reformed. In the diamond industry, so important to Botswana's export earnings, 35 percent of the employees are infected. Debswana has formed a trust fund to pay for antiretroviral drugs for employees and their spouses, with the result that deaths due to AIDS have declined. By the end of 2005, the government was providing antiretroviral drugs to an additional 55,000 citizens.

***South Africa: An Economic Leader.*** South Africa has managed to create a large and diversified economy with a total output in 2005—U.S.$533.2 billion, PPP—equivalent to more than a third of the output of all other sub-Saharan African countries combined. The country's stock exchange is the tenth largest in the world. A major reason for South Africa's present state of development is that for centuries it had a well-off minority European population (about 16 percent of the total population) with the skills and external connections to foster economic development. After independence in 1910, the Dutch (Boers) and British who remained in the country established many lucrative industries, such as the refining of minerals, gemstones, and metals from the country's rich mines. These

enterprises kept profits in the country, allowing the construction of a large industrial economy; however, only a small minority reaped the benefits of this effort.

Although their labor helped build an exceptional economy, 84 percent of black South Africans lived at a bare subsistence level under the apartheid system, suffering low wages and poor education, housing, and health care. With the end of apartheid, this pattern has begun to change, but only slowly. In 2005, fully 25 percent of the people remained unemployed, and 50 percent were below the poverty line.

## The Current Economic Crisis

As a result of the low prices paid for their products on the global market and the high prices of the industrial products they import, most African countries are quite poor (Figure 7.26). Their efforts to get out of poverty through the development of mechanized agriculture, small industries, hydroelectric plants, roads, and so forth have faltered, largely because of corruption and mismanagement. Worse yet, most governments have borrowed heavily from international banks to finance these projects, and now the task of repaying the loans is reducing the revenue available for education and other employment-related improvements.

***Structural Adjustment Programs in Africa.*** In the early 1980s, the major international banks and financial institutions that had made loans to African governments developed a plan to obtain repayment. Working through the World Bank and the IMF, these institutions threatened to stop all further lending to the debtor countries unless they enacted structural adjustment programs (SAPs). As explained in more detail in Chapter 3 (see pages 139–140), SAPs attempt to reduce government economic involvement by selling off government-owned utilities and other enterprises and by slashing government payrolls through cuts in social and agricultural programs—all with the aim of boosting tax revenues, which are then used for loan repayment.

In Africa, the SAPs did accomplish some good. They tightened bookkeeping procedures and thereby curtailed corruption and waste in bureaucracies. They reduced the power of the elites to commandeer resources for their own profit. They eliminated some inefficient and corrupt state agencies. And they stopped the practice of capping food prices to appease urban dwellers, with the result that local food producers received fairer prices for their crops and were encouraged to produce more. Where they were well managed, SAPs closed corrupt state-owned monopolies in industries and services, and they opened some sectors of the economy to medium- and small-scale business entrepreneurs. They also made tax collection more efficient.

SAPs have also created problems. Most researchers on Africa agree that SAPs have made it harder for the poor majority to make a decent living and stay healthy. They have also failed to reduce the debt burden. On the contrary, debt has continued to grow, despite the fact that the region as a whole is now spending more on debt payments than on health care and education combined—more

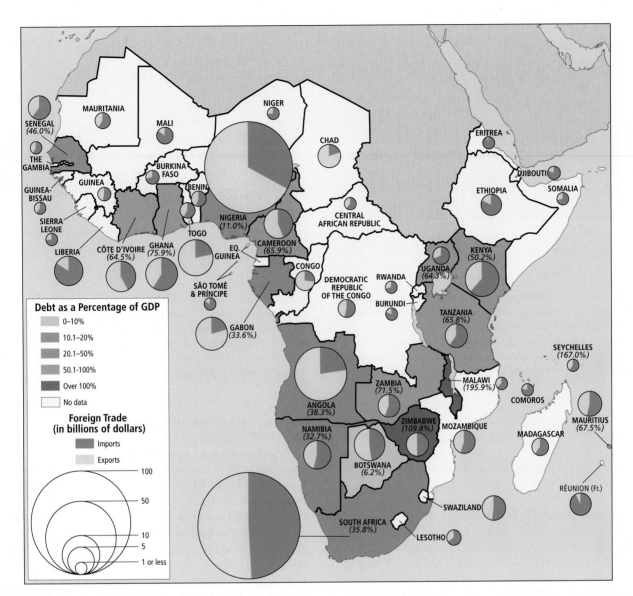

**FIGURE 7.26** Economic issues: Public debt, imports, and exports. In most sub-Saharan African countries, imports exceed exports—even in South Africa, which has a large, diverse economy. As these countries borrow money for development, their public debt increases. In ten countries rich in natural resources (principally petroleum or mineral deposits), exports are greater than imports (see pie diagrams). However, even in these countries (except for Nigeria), public debt is at least one-third of the GDP. [Debt and trade data from the *CIA World Factbook, 2006,* at https://www.cia.gov/cia/publications/factbook/index.html.]

than U.S.$13 billion a year. One reason debt persists is that SAPs have generally eliminated more jobs than they have created as large state bureaucracies have been significantly reduced, state-owned businesses sold to the private sector, and protections for local industries removed.

Moreover, contrary to SAP theory, Africa did not attract foreign direct investment (FDI) once the cuts had been made. As of 2005, foreign direct investment in the whole of Africa was less than 1.5 percent of the world total (Figure 7.27). Oil and minerals attracted the most investment, with most FDI funds going to South Africa, Nigeria, and Angola. Prospective investors in sub-Saharan Africa have been discouraged by problems that SAPs either ignored

or made worse. Loss of public funds for schools perpetuated an underskilled workforce, which in turn led to unemployment, disaffection, and political instability. Deteriorating infrastructure provided poor-quality water, transport, health care, banking, and utility services. The result has been economic decline and rising unemployment, which has left more than two-thirds of the population in poverty and prone to social unrest.

There are indications that conditions for investment in Africa may be improving as African expatriates themselves are beginning to invest in Africa. Between 2000 and 2003, for the first time, remittances by Africans working abroad amounted to more than total foreign direct investment in the region. Africans are sending money

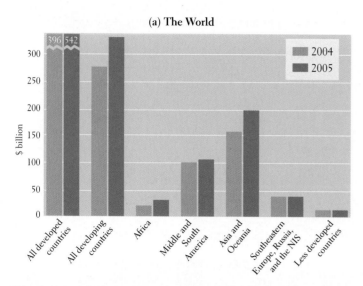

**(a) The World**

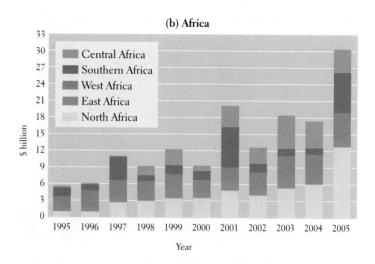

**(b) Africa**

**FIGURE 7.27** Foreign direct investment (FDI) inflows: The world and Africa. (a) Both developed and developing countries around the world had increases in FDI in 2005. FDI inflows to the 50 least developed countries rose to U.S.$10 billion, but were still marginal compared with the growth of FDI in the developing countries of Asia, Oceania, Latin America, and the Caribbean. (b) Distribution of FDI inflows among Africa's subregions for an 11-year period, 1995–2005. According to the United Nations, FDI inflows to Africa in 2005 represented nearly 20 percent of the region's capital, indicating that Africa was not generating much capital internally. Nevertheless, Africa still received only 3 percent of global FDI. [Adapted from *World Investment Report 2006* (New York: UN Conference on Trade and Development), pp. 39–40, at http://www.unctad.org/en/docs/wir2006ch2_en.pdf.]

home to build houses for their families, to start small businesses, to fund education for children, to help the needy in their communities, and even to found universities. These remittances are a more stable source of investment than foreign direct investment, and are much more likely to reach poorer communities.

***Agriculture and Economic Restructuring.*** Economic restructuring has had dramatic effects on the agricultural sector, which employs 70 percent or more of Africans in one way or another. Modern agricultural development programs provide incentives for investment in cash crops for export. These incentives may include lower taxes on profits, decreased government regulation, and low-cost loans for equipment and land.

At the same time that export cropping has been encouraged, however, structural adjustment programs have inadvertently reduced the availability of food for Africa's own consumption. Between 1961 and 2005, per capita food production in Africa decreased by 14 percent, making it the only region on earth where people are eating less well than they used to (see Figure 1.38 on page 44). For example, the often recommended policy of **currency devaluation** lowers the value of African currencies relative to the currencies issued by other countries. Currency devaluation promotes the sale of African export crops by making them cheaper on the world market, but it also makes all imports more expensive. Therefore, farmers who grow food for the African market must spend much more on seeds, fertilizers, pesticides, and farm equipment, few of which are produced within the region. Furthermore, with the shift to export crops, many farmers no longer have the time or space to grow their own food or food for local markets, so increasingly, Africans must pay for expensive imported food. Meanwhile, because SAPs have also mandated export crop production increases in many countries

in other world regions, world market prices for agricultural export crops have declined, eroding farm incomes.

Another issue in agricultural restructuring relates to gender roles. Agricultural scientists, with their European and North American backgrounds, typically neglected to consider that in Africa, women grow most of the food for family consumption. Hence they did not consult women or include them in development projects, except possibly as field laborers. The displacement of women from the land and food production, as well as the focus on export crops, contributed to the economic decline in sub-Saharan African countries after 1960 and to the loss of sustainable farming practices. In Nigeria, Dr. Bede Okigbo, a botanist at the International Institute of Tropical Agriculture (IITA) in Ibadan, had long been aware that traditional farmers (many of them women) who grow complex tropical gardens are highly skilled systems analysts. They can successfully cultivate 50 or more species of plants at one time, sustainably producing more than enough food for their families on a continuous basis. Okigbo observed that his own institute was less capable of growing productive gardens than were the local farmers. By the late 1990s, Okigbo and other scientists were beginning to persuade international aid agencies to design farm aid projects along lines suggested by experienced local farmers.

***Industry and Economic Restructuring.*** Africa's debt crisis was triggered by miscalculations on the part of both African and foreign development "experts." One example is the World Bank's attempt in the 1980s to develop an export-oriented shoe factory in Tanzania. The country's large supply of animal hides was to be used to manufacture high-fashion shoes for the European market, using expensive imported machinery. Due to EU protective tariffs and miscalculations by both Tanzanians and the World Bank, however, the factory never managed to export shoes to Europe. Unfortunately, even though there is a large market in Africa for good shoes, the machines could not produce the practical shoes Africans want to wear. The unused factory deteriorated, but Tanzania was still expected to repay the cost to the World Bank.

Whether export-oriented manufacturing industries in Africa can succeed is uncertain. The rich industrialized countries place tariffs on African manufactured goods such as footwear and clothing, so Africa lacks access to prosperous markets in Europe and North America. But if Africa sells its products only to other poor regions, the demand for its products may be highly unpredictable because poor people's access to cash fluctuates. Some aspects of SAPs have actually worked against the African export-oriented manufacturing industries they were intended to help. For example, the removal of tariffs on textiles from Asia resulted in a flood of imported cloth from China, which in turn led to factory closings and crucial job losses in the textile industries of Tanzania, Zimbabwe, and Uganda. Meanwhile, throughout Africa, currency devaluations have made imported equipment and spare parts for locally focused industries prohibitively expensive.

Economic and industrial recovery began in Africa around 1995, and some observers are now describing African countries as "the last emerging markets." A sign of increased confidence in the potential for business growth is the financial stock markets that are opening across Africa. But recovery has yet to gain the momentum necessary to raise standards of living and improve the well-being of the majority of Africa's people. Investment from both domestic and foreign sources will be required if African countries are to broaden and diversify their industrial bases. This is beginning to happen: India and China, with their rapidly expanding economies, are showing an interest in Africa. For example, the TATA Group, a large industrial conglomerate in India, is investing throughout the region in telecommunications, power generation, real estate, tourism facilities, automobile production, metal processing, and consumer products. If they succeed, these are the types of investments that could greatly increase African standards of living because they foster local job creation.

So far, China has invested in projects that foster raw materials exports, especially minerals and timber, but oil and gas are also a target as China seeks resources that will keep its own growth on course. Such investment could help Africa enormously—for example, to facilitate the extraction of resources, China paved 80 percent of the main roads in Rwanda—but cautions are in order. Chinese companies have shown few reservations about funding brutal local leaders in order to get access to resources, as the opening of this chapter suggests. Moreover, Chinese trade agreements have been known to undercut African-made products, and when China invests in extractive industries such as mining, it imports Chinese mine workers rather than employing Africans.

***The Informal Economy and Economic Restructuring.*** For many of Africa's working people, the escape hatch from the hardships resulting from SAPs has been the informal economy. Most people who work in the informal economy perform useful and productive tasks, such as selling prepared food, vegetables grown in one of Africa's prolific urban gardens, or craft items (Figure 7.28). Others make a living by smuggling scarce or illegal items, such as drugs, weapons, endangered animals, or ivory.

In most African cities, informal trade once supplied perhaps one-third to one-half of all employment; now it often provides more than two-thirds. This surge of economic activity outside of formal systems is evidence that Africans are working hard to better their own lot during hard times. Informal employment has offered some relief from abject poverty, but it cannot resolve the overarching problems of African economies. Because the informal economy is very difficult to tax, it does not contribute revenue to pay for government services or to repay debt. An informal entrepreneur has trouble building the collateral to expand and create more jobs or to qualify for credit. Some entrepreneurs would gladly register their businesses in order to gain the advantages of being aboveboard, but licensing fees alone are, on average, twice the average annual per capita income, and the bureaucratic snarl can make registering a business take as long as 3 years.

The profits of informal businesses are unreliable, often declining over time as more and more people compete to sell goods and services to people with less and less disposable income. One result is a decline in living standards that is disproportionately severe for women and children. When large numbers of men lose their jobs in factories or the civil service, they crowd into the streets and bazaars as vendors, displacing the women and young people who formerly dominated there. Women and children may then turn to contract work, manufacturing clothing, shoes, or other items in their homes. Often children miss school, working long hours for

**FIGURE 7.28** An informal economy that works. Eunice Manyika, a "Doily Mama," is one of hundreds of Zimbabwean women who make doilies at home and as they ride on the train to Johannesburg, where they sell their crocheted handiwork in the market. There, they buy other goods to take home and resell in Zimbabwe for a profit. [AP Photo/Cobus Bodenstein.]

little pay and using hazardous chemicals or techniques. With families disintegrating under the pressure of economic hardship, some women turn to sex work—a growing sector in the informal economy—putting themselves at high risk of contracting HIV-AIDS. More and more children from these disintegrating families must fend for themselves on the streets. Cities such as Nairobi, Kenya, which had very few street children as recently as 1985, now have thousands.

## Alternative Pathways to Economic Development

For all the reasons we have described, Africans are seeking alternatives to past development strategies. Regional economic integration along the lines of the European Union, South America's Mercosur, or Southeast Asia's ASEAN is one such strategy. Another is locally designed and locally based grassroots development.

***Regional Economic Integration.*** Many African countries, especially in West Africa, are simply too small to function efficiently in the world economy. Their national markets and their resources cannot nurture a significant industrial base. These countries have remained heavily dependent on Europe; the bulk of their trade is often with former colonial powers. A measure of this dependency is that only 11 percent of the total trade of sub-Saharan Africa was conducted between African countries in 2002. Furthermore, it is difficult to establish regional trade links because the necessary transport and communication networks are not in place. Air travel, and even long-distance telephone calls, from one African country to another must often be routed through Europe, and roads are poor, even between major cities. Each of the 47 countries of sub-Saharan Africa has its own tangle of bureaucratic regulations for business, trade, and work permits, and only a few (Ghana and Tanzania, for example) have attempted to streamline court systems to handle disputes quickly. These impediments make doing business in Africa about 50 percent more expensive than doing business in Asia, so even many Africans choose to go elsewhere with their investment money.

Although governments in Africa are reluctant to surrender their economic policy authority to outside entities, there are a number of subregional organizations working toward economic integration (Figure 7.29). These organizations share a number of goals: reducing tariffs between members, forming common currencies, reestablishing peace in war-torn areas, upgrading transportation and communication infrastructure, and building regional industrial capacity. Full-scale economic union along the lines of the European Union is an eventual target.

- **ECOWAS (Economic Community of West African States)** is dominated by Nigeria and focuses on trade, currency, infrastructure, and political stability.
- **CEEAC (Economic Community of Central African States)** is a regional economic cooperation zone that has been hampered by wars and violence in its member states.

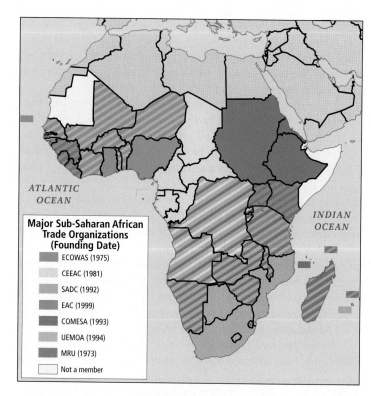

**FIGURE 7.29** Principal trade organizations in sub-Saharan Africa. Observe that some countries belong to more than one organization.

- **SADC (Southern African Development Community)** works on trade, currency, infrastructure, and political stability in Southern and East Africa.
- **EAC (East African Community)** works toward a EU-style economic and political union between Kenya, Tanzania, and Uganda.
- **COMESA (Common Market for Eastern and Southern Africa)** aspires to become a free trade area.
- **UEMOA (West African Economic and Monetary Union)** promotes economic integration among countries that share a common currency.
- **MRU (Mano River Union)** establishes economic integration and a customs union in the subregion comprising Guinea, Liberia, and Sierra Leone.

All of these efforts at subregional integration owe a debt to the Organization of African Unity (OAU), a pan-African movement founded by President Kwame Nkrumah of Ghana in the 1960s. The **African Union (AU)**, which grew out of the OAU in 2002, includes all the countries on the African continent, plus nearby islands. Broadly conceived as an antidote to European dominance, it is now an increasingly active, but still loose, union that promotes economic cooperation and general social welfare. Its activities continue to evolve: at the annual AU summit meeting in 2003, a Pan-African Parliament was formed, with five members from each of the countries on the continent. The main goals of the AU and the as yet

**FIGURE 7.30** A common method of transport in Africa. These women are returning from the market at Kisangani, Congo (Kinshasa). Women throughout Africa are responsible for the transport of a large percentage of African trade goods. [Tom Friedmann/Photo Researchers.]

only advisory Pan-African Parliament are to help member states and the continent as a whole by advancing economic integration, civil society institutions, women's rights, and environmental conservation. In 2004, the AU formed a small peacekeeping force to aid the UN in calming the situation in southern Sudan and Darfur.

***Grassroots Development.*** Another alternative, yet complementary, development strategy for Africa is **grassroots economic development,** aimed at providing sustainable livelihoods in the countryside as an alternative to urban migration as well as coping strategies for poor urban residents. One of the most promising ideas is **self-reliant development,** which consists of small-scale self-help projects, primarily in rural areas. These projects use local skills, create local jobs, produce products or services for local consumption, and maintain local control so that participants retain a sense of ownership. One district in Kenya has more than 500 such self-help groups. Most members are women who terrace land, build water tanks, and plant trees. They also form farm cooperatives for small-scale production; build houses, schools, and barns; run bookshops, nurseries, and restaurants; and form credit societies. Nonetheless, there are limitations to such efforts. For one thing, they require numerous facilitators with extraordinary skills in managing people. And, ironically, their success has tempted governments not to give rural communities their fair share of financial support from tax revenues.

The issue of rural transport illustrates how an Africa-centered perspective and a focus on local needs might generate improvements that differ from those advocated by large external development agencies. When non-Africans learn that transport facilities in Africa are in need of development, they usually imagine building and repairing roads, railroads, airports, buses, and the like. But a recent study that analyzed transport on a local level—the time spent, the distances traveled, the loads carried—discovered that women provide a major part of village transport and that most of the goods moved *are carried on their heads!* Women "head up" firewood from

the forests, crops from the fields, and water from wells, carrying these loads to their homes, as well as carrying goods bound for rural markets (Figure 7.30). Eighty-seven percent of this domestic load bearing is on foot; on average, an adult woman moves the equivalent of 44 pounds (20 kilograms) more than 1.25 miles (2 kilometers) each day and spends about 1.5 hours doing so. As forests are depleted and water becomes scarcer, the time spent, loads carried, and distances traveled by women increase. Mothers usually recruit their daughters to help them carry their loads, thus cutting into the girls' time in school. Men tend to carry significantly fewer of these local burdens and use various conveyances to move them. In Kasama, Zambia, for example, an adult woman transports the equivalent of 35.7 tons (32.4 metric tons) a year, compared with just 7.1 tons (6.4 metric tons) transported by the typical man, no matter what his conveyance. Yet even when development agencies have examined transport at the household and village level, they have focused exclusively on transport by men, suggesting bicycles, wheelbarrows, donkey carts, and pickup trucks to help men out.

Analysis of grassroots development shows that tiny changes can make improvements that are less disruptive and more sustainable than the changes wrought by large-scale development. Giving women access to low-tech transport such as donkeys and bicycles would free some of their time for education, short training courses, and perhaps small-scale businesses. An added benefit would be that these improvements might forestall migration to cities.

***Technological Development.*** A generation of modern technological entrepreneurs is emerging in Africa. Well educated and familiar with global communication, they are buoyed by Africa's potential in human and natural resources rather than intimidated by its past problems. These entrepreneurs, who are in their twenties, thirties, and forties, were often educated in Europe or North America and first experienced success as young professionals in those wealthy economies. Wishing to avoid becoming part of Africa's

---

**BOX 7.2** | **AT THE LOCAL SCALE** Cell Phones Change Lives and Societies in Sub-Saharan Africa

In Kinshasa, Iyombe, a young man who had hoped for a career as an electrician, is using his technical skills as a cell phone airtime transferrer in the informal economy. He mans a tiny stand on a busy street where, for cash, he moves minutes from his phone to the phones of friends, relatives, or business partners of his customers.

Cell phones are opening up Congolese society in important ways by giving people access to services that previously were unavailable. Landline phones are virtually nonexistent in Congo, but an estimated 3.2 million Congolese, like people throughout Africa, use their cell phones daily for such activities as sending money to merchants, talking with relatives in remote regions, reaching a medical doctor, or confirming the deposit of funds in a bank (Figure 7.31). By July 2006, 75,000 former soldiers in Congo's long civil war were receiving monthly payments, authorized by cell phone contact, for turning in their AK-47 rifles and attending job training. The experimental program is funded by the World Bank and EU donors.

*Source:* Kevin Sullivan, "Bridging the digital divide," *Washington Post National Weekly Edition* (July 17–23, 2006): 10–11.

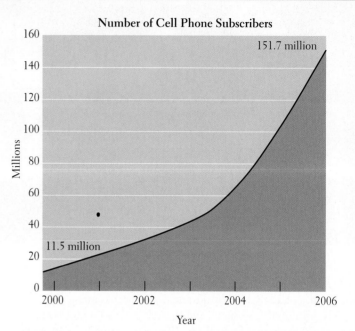

**FIGURE 7.31** Increase in cell phone use in Africa, 2000–2006. Africa has few landlines, but cell phone use is growing rapidly. Between 2000 and 2006, cell phone subscribers increased by more than tenfold to nearly 152 million. (Data from the second quarter of each year are listed.) [Adapted from *Washington Post National Weekly Edition* (July 17–23, 2006): 11.]

---

brain drain, they desire to find similar opportunities in their own countries and have established businesses in a variety of areas, from telecommunications to cut-flower exports (Box 7.2). For example, in 1994, Africa Online, now the continent's largest Internet service provider (ISP), was founded by three young Kenyans who returned from the United States after graduating from MIT, Harvard, and Princeton, respectively. Africa Online has become the leading pan-African Internet service, with operations in Côte d'Ivoire, Swaziland, Tanzania, Kenya, Uganda, Ghana, Namibia, and Zimbabwe. Although Internet service is still largely confined to the main cities, by the end of 2000, all 54 countries on the African continent had achieved some permanent connectivity and had installed some local full-service dial-up ISPs.

## Political Issues: Colonial Legacies and African Adaptations

Few places in the world today are as politically turbulent as sub-Saharan Africa. In 2002, 28 percent of the population lived in countries considered to be at high risk of armed conflict. The news media bring us frequent reports of civil wars, genocide, and military despots. We also see the lingering effects of corrupt presidents such as Mobutu Sese Seko, who plundered the national treasury of

Congo (Kinshasa) for decades while millions of Congolese went without food, shelter, health care, and education.

Africa's conflicts are partly a legacy of European colonialism and cold war geopolitics and partly the result of Africa's own ethnic and regional issues, which have been manipulated by unscrupulous elites in efforts to accrue personal power and wealth. Fortunately, political and economic changes are reducing the power of entrenched elites and forcing them to allow democratic freedoms that, in the future, may bring more responsible government and better economic potential. For example, in July 2006, presidential and parliamentary elections were held for the first time in 46 years in Congo (Kinshasa). UN and Carter Center observers reported a largely peaceful vote with 49 percent turnout, mostly free of fraud; the incumbent, Joseph Kabila, won by 58 percent.

***Origins of Conflict.*** In many ways, the prevalence of conflict in Africa is related to its experience with European colonial rule. The colonizers dismantled the continent's traditional systems of governance and diverted its resources to other world regions. National borders, which were established by Europeans during the colonial era, split some ethnic groups; in other cases, very different and sometimes hostile groups ended up sharing the same country (Figure 7.32). In too many cases, years of carnage followed independence as civil wars

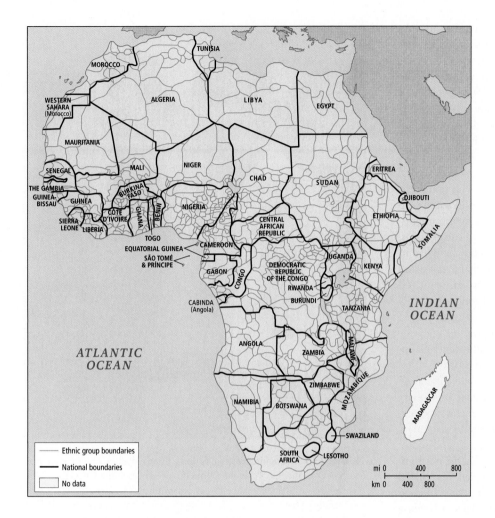

**FIGURE 7.32** Ethnic groups in sub-Saharan Africa. This map shows the large number of ethnic groups spread across the continent of Africa. Superimposed on this pattern are the present national boundaries, which were, for the most part, imposed by European colonizers; very rarely do ethnic group and national boundaries match. [Adapted from James M. Rubenstein, *An Introduction to Human Geography* (Upper Saddle River, N.J.: Prentice Hall, 1999), p. 246.]

arose from attempts to crush ethnic or regional separatist movements and from general repression of disadvantaged minorities. The state encouraged interethnic violence bordering on genocide in Congo (Kinshasa), Uganda, Rwanda, and Burundi. In places such as Congo (Kinshasa), Nigeria, Liberia, and Sierra Leone, conflicts were worsened by military establishments that were only loosely under the control of civilian governments, sometimes seizing control through coups d'état. (We examine the conflict in Nigeria below.)

The European colonial strategy of using local elites to control local populations often created or exacerbated ethnic tensions within a colony and led to minority rule after independence. During the colonial era in Belgian Rwanda and Burundi, for example, the elite Tutsi were empowered to take advantage of the more numerous Hutu. This practice destabilized long-standing relations between the two groups and was partly responsible for the ethnic bloodshed in these two countries in the 1990s.

**Postcolonial Elite Rule.** For a time after independence, African elites in a number of countries continued colonial traditions of authoritarian and corrupt rule. Often a single political party monopolized power, and its leaders repressed any viable political opposition. The elites argued that the open political debates of Western democracies were incompatible with African traditions of commu-

nal identity and collective decision making. Similarly, they argued that democracy tended to warp African traditions of patronage (wherein the leader is expected to look after the needs of the people) and to create huge, inefficient government bureaucracies, with corruption penetrating to the highest levels. But again, some optimism is in order: in the first decade of the 2000s, several African countries (South Africa, Botswana, Mozambique, and even Nigeria, to some extent) enjoy responsible leadership that has steered them through difficult times and has still managed to achieve economic growth and social stability.

**Case Study: Conflict in Nigeria.** Nigeria is the economically dominant country in West Africa because of its large population, its oil reserves, and its relatively diversified economy. It is the most populous country in Africa, and its 123 million people outnumber the total of those who live in the rest of West Africa (see Figure 7.13 on page 364). Like the other countries of West Africa, Nigeria is a culturally complex nation. It presents an interesting case study of how colonial and postcolonial rule have led to increasing tensions, which are too often attributed to conflict among the various ethnic groups that have long occupied the area.

As geographer Robert Stock observes, "Nigeria was, and still remains, a creation of British imperialism." The British deliberately split up some ethnic groups and forced other disparate groups into one unusually diverse country. Today Nigeria contains people who speak 395 indigenous languages in 11 language groups.

In present-day Nigeria, there are four main ethnic groups. The Hausa (21 percent of the population) and Fulani (9 percent), both predominantly Muslim, have, until recently, lived mostly as herders in the north, where the moist grassland grades into dry savanna and then semidesert. Today the Fulani and Hausa are not only herders but also traders and dryland farmers. Since independence in 1960, Hausa and Fulani elites have dominated both the government of Nigeria and its military. The Yoruba (20 percent), who practice a complex animist religion as well as Christianity, live in southwestern Nigeria in the moister woodland and grassland environments. Once nearly all Yoruba were settled farmers, growing cacao, yams, peanuts, and other crops on a shifting basis, but many now live in urban areas and are laborers, tradespeople, professionals, artists, and craftspeople. The Igbo (17 percent) are primarily farmers, centered in the southeast in and near the tropical rain forest zone. Originally animists, many are now also Christian. Hundreds of other ethnic groups live among and between these larger groups.

The colonial administration reinforced a north-south dichotomy, especially in education. Among the Hausa, the British ruled via local Muslim leaders who did not encourage public education. In the south, among the Yoruba and Igbo, Christian missionary schools open to the public were common. At independence, the south had more than ten times as many primary and secondary school students as the north and was more prosperous, and southerners held most government civil service positions. Yet the northern Hausa, who tended to be poorer and less well educated, dominated the top political posts. Over the years, bitter disputes have erupted between the southern Yoruba-Igbo and northern Hausa regarding the distribution of development funds and jobs, the validity of census data, and the fairness of elections (the video vignette below illustrates such a clash).

In 1966, after an Igbo-led military coup d'état and a Hausa-led countercoup, the slaughter of 30,000 Igbos ignited a 3-year civil war in which the Igbo portion of southeastern Nigeria tried unsuccessfully to secede. More than 200,000 people died as a result of the war and ensuing food shortages.

An even more important factor than ethnicity in Nigerian politics is oil, a major resource whose exploitation has brought what geographer Michael Watts calls "rapacious frontier capitalism" and environmental violence. Oil is Nigeria's primary resource: 90 percent of its foreign trade earnings come from the sale of oil. Nigeria's oil reserves have the potential to make it the richest country in Africa, but this resource has been mismanaged to enrich local military leaders and their associates as well as foreign oil companies, especially Royal Dutch Shell. A Nigerian commission reported that between 1990 and 1994, the military regime in power stole $12 billion in oil money, and a subsequent regime is thought to have taken another $12 billion. Officially, Nigeria earns about $10 billion a year from oil, so corrupt officials robbed the Nigerian people of more than half of the country's earnings.

Much of Nigeria's oil is located on lands occupied by the Ogoni people, which lie along the northeastern edges of the Niger

**FIGURE 7.33** **The politics of oil in Nigeria.** Sunday Badon, a member of the Movement for the Survival of the Ogoni People, stands in front of a lake of oil seeping from an abandoned Shell wellhead in June 2006. The Ogoni tribe successfully lobbied to remove Shell from this part of the Niger delta in 1993, but the consequences of Shell's oil extraction activities remain. The Ogoni protested that they had received nothing in return for four decades of oil production. Today, the same grievances are fueling a revolt across the entire oil heartland of Nigeria, Africa's largest oil producer. [Reuters/Tom Ashby.]

River delta. However, virtually none of the profits from its production, and very little of the oil itself, go to the state in which this land lies, much less to the Ogoni's homeland, Ogoniland. Although Ogoniland has received few benefits from oil extraction, it has suffered from the resulting pollution (Figure 7.33). Oil pipelines crisscross Ogoniland, and spills and blowouts are frequent; between 1985 and 1994 there were 111 spills. Oil spills many times the size of the *Exxon Valdez* spill in Alaska occur repeatedly. Natural gas, a by-product of oil drilling, is burned off, even though it could be used to generate electricity—something many Ogoni go without.

In the 1990s, a prominent Ogoni writer and businessman, Ken Saro-Wiwa, organized a protest movement proposing major changes in Nigerian federal organization, expansion of democracy, and direct action against Royal Dutch Shell and Chevron Corporation installations. By its own admission, Shell netted $200 million in profits yearly from Nigeria, but in 40 years it paid only $2 million to the Ogoni community whose oil it had appropriated. In 1995, the world community was outraged when Ken Saro-Wiwa was summarily executed with nine other Ogoni by the Nigerian government.

*It is a vicious circle—you cannot get investment while there is violence and killing and disturbance and then if you don't get investment which [would create] employment, then the killing, the violence will go on. So it is a very, very bad vicious circle.*

President Olusegun Obasanjo of Nigeria, interview with the BBC, 2002

In 2002, Amina Lawal, then pregnant, was sentenced to death by stoning for adultery. She lived in a Muslim village in northern Nigeria, where shari'a had replaced the corrupt secular legal system 3 years before, when Nigeria's military dictatorship ended. The democratically elected government in Abuja, in the Christian south, declared she would never be executed. Her case exacerbated tensions between the different ethnic groups in the north and south. This simmering conflict erupted during preparations for the Miss World pageant, which was to be held in Abuja that year. When a Christian fashion writer suggested that the Prophet Muhammad would surely have chosen one of the beauty contestants for his wife, the northern Muslims were offended. Riots between Muslims and Christians broke out in the northern city of Kaduna, and the pageant, which Christian Nigerian officials had hoped would help to burnish Nigeria's global image, packed up and left.

Amina Ladan-Baki, a banker and world rights activist, believes that the discord between ethnic and religious groups in Nigeria is manipulated to suit the agendas of politicians. "The politicians mislead people. They use religion. They use diverse cultures to unite or disunite. 'Vote for me because I am of your stock.'" President Olusegun Obasanjo, quoted above, believes the violence is due to poverty.

In February 2004, a shari'a court of appeals overturned Amina Lawal's conviction. However, other women have been charged with adultery in the 12 Nigerian states that follow shari'a, and they face the possible sentence of death by stoning. Meanwhile, continuing political and economic tensions between the north and the south make it likely that, as a matter of pride and identity, the north will retain its conservative approach to Islam.

*To learn more about Amina Lawal and the Miss World pageant, watch the FRONTLINE/World video "The Road North: What the Miss World Riots Reveal about a Divided Country."* ■

Can geographic strategies resolve the tensions in Nigeria? While enduring these divisive and violent civil disputes, some Nigerians have continued to pursue solutions to unwieldy ethnic and political spatial patterns. One strategy has been to create more political states (Nigeria now has 30) and thereby reallocate power to smaller local units, something Ken Saro-Wiwa had advocated. Recently, large, wealthy states have been subdivided to ensure more equitable distribution of development money and oil profits and to prevent any one state from seizing control of the central government. Although dividing the country into more states has increased administrative costs, it also seems to have eased regional ethnic and religious hostilities. Another strategy to defuse ethnic conflict has been the relocation of the capital city from Lagos on the Atlantic coast to Abuja, an interior village north of the confluence of the Niger and Benue rivers, in neutral territory away from the heart-

lands of the Hausa, Yoruba, and Igbo. Abuja is a forward capital: a capital city built to draw migrants and investment to a previously underdeveloped area. However, the cost of establishing Abuja has increased Nigeria's debt by $4.5 billion, and it turns out that most civil servants are unwilling to move there from the vibrant cosmopolitan city of Lagos.

***Effects of the Cold War in Africa.*** The cold war between the United States and the former Soviet Union deepened and prolonged many conflicts in sub-Saharan Africa. After independence, some African governments sought to address the problems caused by economic underdevelopment and exploitation during the colonial period by turning to socialism, often receiving economic and military aid from the Soviet Union. Still gripped by bitter anti-Communism, the United States (with its allies) tried to undermine these governments by arming and financing rebel groups—as it did in Angola, Congo (Kinshasa), and Namibia—or by pressuring other governments in the region to intervene. In the 1970s and 1980s, the United States encouraged South Africa's apartheid government to carry out military interventions against socialist governments in Namibia (considered at the time a sort of colony of South Africa), Angola, and Mozambique (Figure 7.34). Another major area of cold war tension was the Horn of Africa, where Ethiopia received aid from the United

**FIGURE 7.34** One legacy of colonialism in Africa. During Mozambique's civil war, the world's arms makers supplied enough weapons to arm nearly every person in the country with a rifle. More than 6 million AK-47s alone entered Mozambique, a country with only 18 million people. In a move to reduce the number of guns among the population, the Anglican Church began a Guns for Plowshares program, in which guns can be exchanged for tools and agricultural implements, sewing machines, bicycles, schoolbooks, and other useful items. [João Silva/CORBIS SYGMA.]

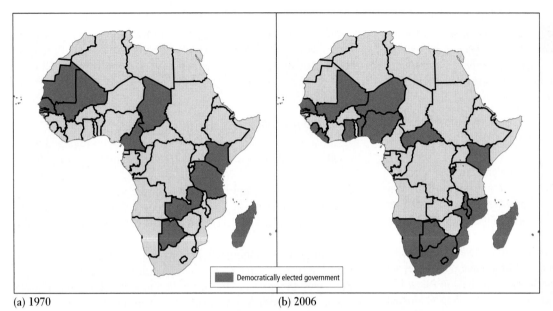

(a) 1970                                        (b) 2006

**FIGURE 7.35** Democratically elected governments in sub-Saharan Africa. Between 1970 and 2006, the number of democratically elected governments in sub-Saharan Africa rose and fell, but by 2006, the number had reached 23. It is important to note, however, that the criteria for democratic elections have been strengthened significantly since 1970. [Adapted from Barbara McDade, "Freedom in Africa today," in *Freedom in the World 2006*, at http://www.freedomhouse.org/uploads/special_report/36.pdf; and from http://en.wikipedia.org/wiki/Image:Freedom_House_electoral_democracies_2006.png.]

States and Somalia received aid from the Soviet Union. Much of the aid went to fund fighting between those two African nations.

## Shifts in African Geopolitics

Signs of progress toward democratic systems are visible in patches across West and Southern Africa. In Southern Africa, the white-dominated government of South Africa tried to prolong white rule throughout the subregion for years. Now that white political control has ended in South Africa, that country's leaders are pursuing friendlier and more cooperative relations with neighboring countries, bringing greater stability to Southern Africa as a whole. In West Africa, after years of backsliding and turmoil, elected governments are once again in charge in a number of countries (Figure 7.35).

Across Africa, legal opposition parties have been able to express dissent more openly and freely. There is hope for a "second inde-

---

**BOX 7.3**  **AT THE REGIONAL SCALE**  African Women Become Active in Politics and Civil Society

Charity Kaluki Ngilu will never forget the day she became a professional politician. She was washing dishes in her kitchen in Kitui, east of Nairobi, Kenya, when she saw a group of women who had worked with her on community health projects approaching her back door. Mrs. Ngilu answered the knock on the door, drying her hands on an apron. The women said they wanted her to run for parliament in Kenya's first multiparty elections. She assumed they were joking.

That was in 1992. Ngilu beat the governing party's incumbent, then became a major advocate for women's issues, and in 1997, became the first woman to run for president in Kenya. She didn't win, but men were among her strongest supporters because they believed she was capable of making bigger changes than a man could. By 2004, she was minister of health, in charge of Kenya's greatest challenge: responding to the HIV-AIDS epidemic.

Charity Ngilu started an avalanche. By 2005, women across Africa were assuming positions that gave them the policy-making powers that could bring big changes. The opening of this chapter describes the changes that President Ellen Johnson-Sirleaf of Liberia has been making in that country. Rwanda, so recently devastated by genocide and mass rapes of women and girls, was suddenly inundated with women leaders: at the local level, women now make up 40 percent of the mayors and council members and serve as important education and health activists. In the parliament, women hold 45 percent of the seats, and in the judiciary, Anne Gahongayire serves as secretary general of the supreme court. In Somalia, where conditions for women are extremely dire (women have been repeatedly subjected to rape and mutilation on the street and in the market by gangs of men), women legislators and activists are struggling to change the laws, such as one that requires a woman who charges a man with rape to supply his food while he is in jail awaiting trial. Increasingly across the continent, women who run businesses are joining with women who run education and health NGOs and those elected to political office to challenge all types of gender discrimination.

*Sources:* James C. McKinley, Jr., "A woman to run in Kenya? One says, 'Why not?'" *New York Times* (August 3, 1997): 3; Kennedy Graham, ed., *The Planetary Interest: A New Concept for the Global Age* (London: Rutgers University Press, 1999); "Women taking control of power in Africa," "Talk of the Nation," National Public Radio, April 20, 2006.

pendence" as autocratic, corrupt ruling elites are replaced or moderated by political parties that enjoy the electoral support of the majority of the people. Botswana, Kenya, and South Africa are examples of this trend. At least one encouraging statistic supports hopes for more democratic governments across Africa: in 1970, 11 states had held some sort of elections, but by 1990, only 4 countries in sub-Saharan Africa were true pluralist democracies. By 2006, however, according to Freedom House, an organization that monitors elections globally, 23 sub-Saharan African states had held open, multiparty, secret-ballot elections with universal suffrage (Box 7.3). The trend toward democracy promises greater prosperity for the region because democracy allows for more transparency in government proceedings, public criticism of leaders, chances for the opposition to speak, and opportunities to replace leaders through elections.

Despite this trend, there remain several states—including Ethiopia, Uganda, and Congo (Kinshasa)—where violence or fraud make fair elections difficult or impossible. Two countries, Côte d'Ivoire and Zimbabwe, have experienced **devolution** from democracies with growing economies into autocratic governments sponsoring economic decline and degenerating well-being. Côte d'Ivoire suffered a military coup d'état in 1999. As of April 2007, Zimbabwe continues to struggle against a long downward slide—the result of an authoritarian, repressive government, inappropriate economic strategies, spreading poverty, and increasing social unrest—all the result of rigged elections in 2002 that left Robert Mugabe in power.

## SOCIOCULTURAL ISSUES

To the casual observer, it may appear that a majority of sub-Saharan Africans live traditional lives in rural villages. A closer look, however, reveals the imprint of the Western colonizers and the modern world on such aspects of traditional culture as religion, gender roles, health, and community life. Furthermore, a growing number of people are moving to cities, where they find their lives transformed as they exchange isolated villages for crowded shantytowns. Yet even in the cities, traditional African culture persists and is visible in religious patterns, domestic life, the informal economy, and especially in the plentiful urban subsistence gardens.

### Settlement Patterns

This section discusses the two main settlement patterns found in sub-Saharan Africa: small rural villages and dense low-rise cities. At the same time, there is an overall trend toward more urban settlement and the high-rise model of European, Asian, and American cities.

**Rural Settlements.** More than 65 percent of sub-Saharan Africa's people still live in villages, making this region a very rural area. There are thousands of versions of the African village, with many different types of houses and village arrangements, all depending on the cultural heritage of the inhabitants (Figure 7.36). But there are some constants in African villages, too. People usually live in extended-family compounds, consisting of several houses arranged around a common space in which most activities take place. These

(a) One of the very colorful Ndebele villages in Zimbabwe.

(b) Cattle are returned to a Masai village compound in East Africa after a day's grazing.

(c) Granaries with removable thatch roofs sit alongside houses in this Dogon village in Mali.

**FIGURE 7.36** Three of the many types of villages found in sub-Saharan Africa. [(a) Walter Bibikow/Index Stock Imagery/PictureQuest; (b) Frans Lanting/Minden Pictures; (c) Henning Christof/Das Fotoarchiv.]

compounds may stand alone, dispersed across an agricultural landscape, or they may be grouped with other compounds and surrounded by fields designated for each family. Villages, and the compounds that make them up, are economic units. Subsistence agriculture—production for family sustenance—is almost always an important component of village life, with labor carefully apportioned by custom. Increasingly, rural people also work seasonally for pay on commercial farms or have several other ways of earning small amounts of cash in the formal and informal economies.

***Urbanization.*** Although the majority of Africans live in rural areas, cities are not a recent settlement pattern in Africa. Many cities have long histories, extending back to the age of the great empires of West Africa and the kingdoms of East and Southern Africa. In the past, the cultural and social institutions of these cities focused on maintaining social and economic continuity with the surrounding rural areas. In recent decades, however, African cities have grown very quickly, and their links with rural areas are now more tenuous. The average annual growth rate of African cities—more than 5 percent —exceeds even that of cities in Asia (3 percent) and Latin America (2.5 percent). In the 1960s, only 15 percent of Africans south of the Sahara lived in cities; now about 34 percent do, but the percentage

of the population that is urban varies considerably among countries (Figure 7.37).

African cities have attracted massive migration for at least two reasons. First, life in rural villages and towns is perceived as offering few jobs and little opportunity for upward mobility. Second, there is a widely held misconception that life in the cities offers quick access to money and prestige. In reality, a majority of Africa's city dwellers work at very low wages in the informal economy. On the other hand, urbanization can add a new dynamism to economic development by attracting talented young people to educational institutions and then to jobs in industry and services. African cities are also attempting to recruit back to Africa those who have migrated to Europe and North America by advertising the high-tech or managerial jobs and urban lifestyles that are now available.

As is often the case in developing countries, most African countries have one very large primate city, usually the capital, which attracts virtually all migration. (The primate city phenomenon is discussed in Chapter 3 on pages 131–133.) Kampala, Uganda, for example, with 1.4 million people, is almost ten times the size of Uganda's next largest city, Gulu. The population of another typical primate city, Kinshasa in the Democratic Republic of the Congo, soared from 450,000 in 1960 to an estimated 4.6 million in 2000.

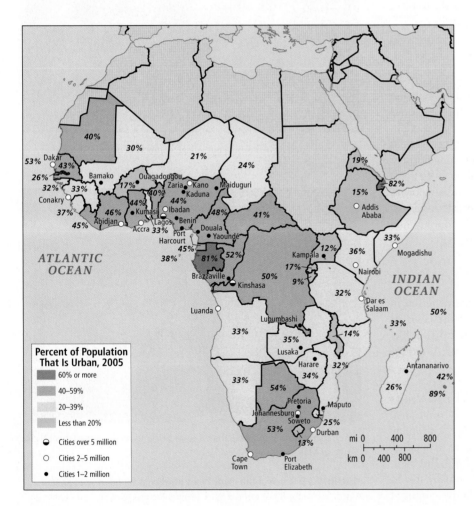

**FIGURE 7.37** Percentage of the population living in cities in sub-Saharan Africa, 2005. Slightly over 36 percent of the region's population is urban, but there is great variation from place to place. Thirty-six sub-Saharan African cities now have populations over 1 million, and seven of them have over 3 million people. [Data from 2007 *World Population Data Sheet* (Washington, D.C.: Population Reference Bureau); and *World Gazetteer*, at http://www .worldgazetteer.com/wg.php?x=&men=gcis&lng =en&dat=32&srt=npan&col=ahq&geo=-1.]

**(a)** Closely packed houses in a shantytown near downtown Lagos, Nigeria, leave little room for ventilation in the tropical heat. The dwellings surround a parking lot for the cars of office workers in nearby buildings.

**FIGURE 7.38** Urban Africa. African urban landscapes vary radically, ranging from oppressive shantytowns to modern mid-rise office buildings, to single-family homes and apartment towers, to elegant villas set in landscaped grounds. [(a) L. Gilbert/CORBIS SYGMA; (b) © Brian J. McMorrow /www.pbase.com/bmcmorrow.]

**(b)** A local ferry passes in front of the twin towers of the Bank of Tanzania in Dar es Salaam.

The population of Lagos, Nigeria, was 48 times larger in 2000 (11 million) than it was in 1950 (230,000). Such rapid and concentrated urbanization usually overwhelms a country's urban infrastructure.

Typically, governments have paid little attention to the housing needs of the throngs of poor urban migrants. Most migrants have had to construct their own shanties on illegally occupied land, using materials that would be discarded as trash in prosperous developed countries (Figure 7.38). They live in vast squatter settlements surrounding the older urban centers. Transport in these huge and shapeless settlements is a jumble of government buses and private vehicles. People often have to travel long hours through extremely congested traffic to reach distant jobs, getting most of their sleep while sitting on a crowded bus. Africa's mid-level officials are aware of the transport problem because they deal with it themselves.

For some members of the small but growing middle class, new homes and jobs may be located in a single development. In South Africa and Botswana, a few industries are providing housing or locating close to housing developments, thus resolving the home-to-job transport problem for at least some of their workers. The housing NGO Habitat for Humanity, active in many parts of Africa, aims to build not just houses, but self-sustaining communities.

## Religion

Religion is particularly important in African daily life. The region's rich and complex religious traditions derive from three main sources: indigenous African belief systems, Islam, and Christianity. Hinduism is also present, having been introduced, like Islam and Christianity, from outside sub-Saharan Africa. Its influence is found primarily in specific locations along Africa's east coast and on the eastern coastal islands, where there are significant Indian or mixed African-Indian populations.

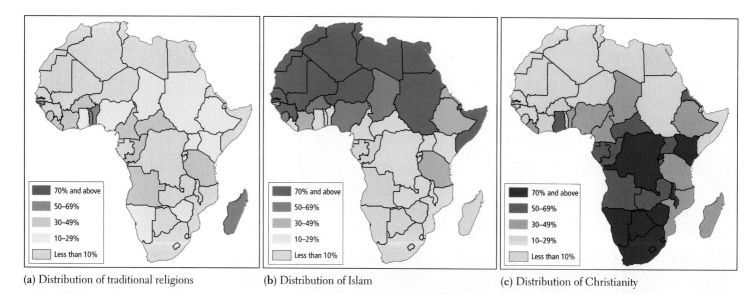

(a) Distribution of traditional religions

(b) Distribution of Islam

(c) Distribution of Christianity

**FIGURE 7.39** Religions in Africa. Notice that the various religions in Africa overlap in distribution; in many countries, one is dominant, but others are present. [Adapted from Matthew White, "Religions in Africa," in *Historical Atlas* *of the Twentieth Century* (October 1998), at http://users.erols.com/mwhite28 /afrorelg.htm; revised with new data from the *CIA World Factbook, 2006*, at https://www.cia.gov/cia/publications/factbook/index.html.]

***Indigenous Belief Systems.*** Traditional African religions, sometimes classed as **animism,** probably have the most ancient heritage of any religion on earth and are found in every part of Africa. Figure 7.39a shows the countries in which traditional beliefs remain particularly strong. According to these beliefs, all entities—plants, animals, and specific landscape features—are part of a whole that includes the entire environment. Traditional beliefs and rituals seek to bring the vast host of departed ancestors into contact with people now alive—who, in turn, are the connecting links in a timeless spiritual community that stretches into the future. The future is reached only if present family members procreate and perpetuate the family heritage in other ways, such as by storytelling, coming-of-age rituals, and making offerings to the ancestral spirits. The people in the living community are usually led by a powerful man, or occasionally a powerful woman, who combines the roles of politician, patriarch or matriarch, and spiritual leader.

According to traditional African beliefs, the spirits of the deceased are all around—in trees, streams, hills, and art objects, for example (Figure 7.40). In return for respect (expressed through ritual), these spirits offer protection from life's vicissitudes and from the ill will of others. Rosalind Hackett, a scholar of African religions and art, writes that "African religions are far more pragmatically oriented than Western religions, being concerned with explaining, predicting and controlling misfortune, sickness and accidents." African religions remain fluid and adaptable to changing circumstances. For example, Osun, the god of water, traditionally credited with healing powers, is now also invoked for those suffering economic woes.

Religious beliefs in Africa, as elsewhere, are not static, but evolve continually as new influences are encountered. If Africans who practice traditional beliefs convert to Islam or Christianity, they commonly retain parts of their indigenous religious heritage, such as reverence for ancestors, and blend them with aspects of their new faith, creating a fresh entity—a process sometimes referred to as **syncretism** (or **fusion**). The three maps in Figure 7.39 show a spatial overlap of belief systems, but they do not convey the philosophical

**FIGURE 7.40** Coffin art in Ghana. Ghana's coffin carvers accomplish an apparently seamless interweaving of religion, art, and modern life. They are experts at interpreting the life of the deceased artistically: a bus for a bus driver, a Mercedes-Benz for a chauffeur, a plane for a pilot. Both rich and poor prize the coffin carver's art; people may commission a coffin long before death comes, often spending their life savings to buy a niche in the cultural memory of the community. [Carol Beckwith and Angela Fisher/ Robert Estall Photo Library, UK.]

blending of two or more faiths, which is widespread. In the Americas, the African diaspora has participated in the creation of new belief systems developed from the fusion of Roman Catholicism and African beliefs: Voodoo in Haiti; Obeah in Jamaica; Condomble in Brazil; and, more recently, Santeria in Cuba, Puerto Rico, and North America. In the 1990s, African versions of evangelical Christianity began to influence American evangelicals by sending African missionaries to places such as Atlanta and Birmingham.

***Islam and Christianity.*** Islam began to extend south of the Sahara, especially into East Africa, soon after Muhammad's death in 632 (see the discussion of Islam and Christianity in Chapter 6). As Figure 7.39b shows, Islam is now the predominant religion throughout North Africa and is important in much of West and East Africa, including the central coastal zone on the Indian Ocean, where Islamic traders were especially active. When the British began their formal colonization of West Africa in the late 1800s, they obtained the help of Islamic leaders in governing the countryside and towns, especially in the drier northern areas. For this reason, the British did not encourage Christian missionaries in the interior away from the coast; hence, today, the elite descendants of these Islamic administrators are still politically powerful in Nigeria and some other parts of West Africa.

Christianity came to Ethiopia in the fourth century, well before it spread through Europe, and the Ethiopian Coptic Church still maintains practices derived from this long heritage. Christianity began to spread through the rest of the region in the nineteenth century, when Methodist missionaries from Europe and America became active along the west coast of Africa; Roman Catholic missionaries followed. Today Christianity is prevalent along the coast of West Africa and in Central and Southern Africa (Figure 7.39c). The missionaries often refused to allow Africans to incorporate their folk beliefs into their new faith, a rejection that helped to spawn independent, charismatic Christian sects that made more room for folk traditions (Box 7.4).

Christian missionaries in Central and Southern Africa found a niche as providers of the education and health services that colonial administrators had neglected. In East Africa today, Christians are either the majority or an important minority. There, as in West Africa, colonial administrators in the British-held territories discouraged Christian missionaries from going north into Muslim territories (such as northern Sudan) because Islamic leaders were already aiding the colonization process.

In the 1980s, old-line established churches began to gain adherents in East and West Africa. For example, the Anglican Church (Church of England) grew so rapidly in Kenya, Uganda, and

---

## BOX 7.4 | AT THE REGIONAL SCALE    The Gospel of Success

We first encountered the "gospel of success" in Chapter 3. In Africa, this adaptable version of evangelical Christianity combines certain interpretations of Christianity and capitalism with central and western African beliefs in the importance of sacrificial gifts to ancestors and the power of miracles (Figure 7.41). Its message is simple: according to preachers such as those at the Miracle Center in Kinshasa, "The Bible says that God will materially aid those who give to Him.... We are not only a church, we are an enterprise. In our traditional culture you have to make a sacrifice to powerful forces if you want to get results. It is the same here." The supplicants accept the message that generous gifts to such churches will bring divine intervention to alleviate their miseries, whether physical or spiritual. They donate food, television sets, clothing, and money—one woman gave 3 months' salary in the hope that God would find her a new husband. The collection plates are large, sturdy plastic bags. By combining spiritual ministration with a touch of hucksterism, the leaders of this largely urban movement receive the respect and adulation that Africans used to reserve for rural village leaders.

Like all religious belief systems, the gospel of success is best understood within its cultural context. The movement isn't as blatantly materialistic as it first appears. Many of the believers, new to the city, feel isolated and are seeking a supportive community to replace the one they left behind. People view their material contributions to the church as similar to the labor and

goods they previously donated to maintain their standing in their home village. In return for these dues and volunteer services, members receive social acceptance and community assistance in times of need.

**FIGURE 7.41** Worshipers at the Miracle Center in Kinshasa, Democratic Republic of the Congo. [Robert Grossman/CORBIS SYGMA.]

Nigeria that by 2000 there were more Anglicans in these countries than in the United Kingdom. African Anglicans are conservative on social issues—they are against female priests and the acceptance of homosexuals—but often liberal on economic issues. For example, in 1998, African Anglican bishops persuaded the worldwide Anglican Communion to oppose holding the world's poor countries to crippling debt payments. The Anglican Church in Africa attracts the educated urban middle class, whereas modern evangelical versions of Christianity appeal to the less educated, more recent urban migrants—the most rapidly growing cohort of African populations.

## Gender Relationships

Long-standing African traditions dictate a fairly strict division of labor and responsibilities between men and women. The exact allocation of work, power, and family wealth varies from place to place. In general, however, women are responsible for domestic activities, including rearing the children, tending the sick and elderly, and maintaining the house. Women carry about 90 percent of the water, collect 80 percent of the firewood, and produce and prepare nearly all the food. Men are usually responsible for preparing land for cultivation. In the fields intended to produce food for family use, women sow, weed, and tend the crops as well as process them. In the fields where cash crops are grown, men perform most of the work. Women may help to weed and harvest these fields, but it is usually understood that any earnings belong to the men.

When husbands in search of cash income migrate to work in the mines or in urban jobs, women take over nearly all agricultural work. Studies have shown that the majority of agricultural laborers in Africa are women, and that they contribute about 70 percent of the total time spent on African agriculture. Rural women are the primary rice cultivators in countries where rice is a staple food, and they handle about 50 percent of the care of livestock. They tend to work with hand tools in the fields and in the home, and during their reproductive years, they often do field labor with a child strapped on their backs. When there are small agricultural surpluses or handcrafted items to trade, it is women who transport them to and sell them in the market. Throughout Africa, married couples tend to keep separate accounts and manage their earnings as individuals, so when a wife sells her husband's produce at the market, she usually gives the proceeds to him.

In rural areas, African men, on average, do not have as many responsibilities as women, nor do they work as hard or for as many hours. The reasons for this situation are complicated, related in part to the fact that so many men migrate to the cities or the mines to work, at least seasonally. The retreat of men from virtually all tasks directly related to supporting domestic life seems to have started with European colonialism. European Christians colonized Africa during the Victorian era, when the prevailing attitude in Europe was that women were lesser creatures who should remain in the home and let their husbands deal with the outside world. This attitude influenced the colonial policy in Africa of recruiting men for cultivating cash crops and doing work for wages. At first, men took on such activities to meet the colonial administration's requirement that taxes be paid in cash; now, men work for cash to pay for chil-

dren's school fees, basic electricity, certain consumer goods, and even food, when families lose access to land due to agricultural modernization. Women are left to shoulder all the domestic work as well as what was formerly the shared work of subsistence agriculture.

In the precolonial past, there were social controls that tempered gender relationships. Most marriages were social alliances between families; therefore husbands and wives spent most of their time doing their tasks with family members of their own sex, rather than with each other. For most of the day, women were influenced primarily by other women. Traditional gender relationships were modified wherever more patriarchal Muslim or Christian religious and cultural practices were introduced. In most cases, it seems that as men gained power, women lost freedom. Muslim women, restricted to domestic spaces, could no longer move about at will, trade in the markets, or engage in public activities. In Islamic Africa, the strong patriarchy left many women in a sort of servitude to male wishes, with ambiguous legal rights. While Christian women operate in the public sphere more than Muslim women do, they too are socialized to restrict their activities to the home. It is important to note that having multiple wives—the practice of **polygyny** (Figure 7.42)—is more common in sub-Saharan Africa, where it has ancient pre-Muslim roots, than in Muslim North Africa. Nevertheless, only a small minority of Africans are in polygynous marriages today. Those countries with the highest rates of polygyny—Senegal, Burkina Faso, Mali, and Cameroon—have relatively small populations.

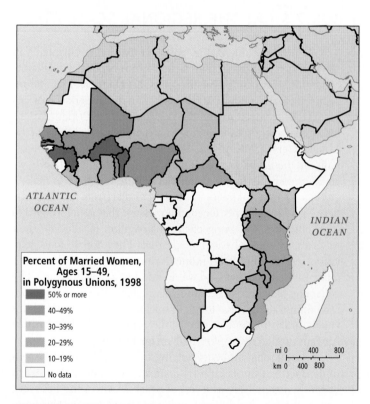

**FIGURE 7.42**  Polygyny in sub-Saharan Africa. Polygyny is less common in East Africa than in West Africa, and it is uncommon in Southern Africa. [Adapted from *The World's Women, 2000: Trends and Statistics* (New York: United Nations, Department of Economic and Social Affairs, 2000), pp. 27–28.]

## Female Circumcision

A practice known as **female circumcision,** popularly called *cutting,* is widespread in at least 27 countries throughout the central portion of the African continent (Figure 7.43). Often mistakenly thought to be a Muslim practice, this custom was reported by Herodotus in 500 B.C.E., predating both Christianity and Islam. It is now practiced by some African Muslims, Christians, and followers of ancient traditional African religions. The practice is probably intended to ensure that a female is a virgin at marriage (and hence not pregnant by another male) and that she thereafter has a low interest in intercourse (other than with her husband), but it also has many complex symbolic meanings. In the procedure, which is a much more radical operation than male circumcision and usually performed without anesthesia, the labia minora and the clitoris are removed, and in the most extreme cases (called *infibulation*), the vulva is stitched nearly shut. Female circumcision eliminates any possibility of sexual stimulation for the woman. Because the practice results in the permanent removal of a healthy organ, many physicians refer to it as *female genital mutilation* or *female genital excision.* For those who have undergone female circumcision, urination and menstruation are difficult, intercourse is painful, and childbirth is particularly devastating because the scarred flesh is inelastic. A 2006 medical study conducted with the help of 28,000 women in six African countries showed that women who had undergone circumcision even years earlier were 50 percent more likely to die during childbirth, and their babies were at similarly high risk. Prior research had shown that the practice also leaves women exceptionally susceptible to HIV infection.

In some culture groups, such as the Kikuyu of Kenya, nearly all females would have been circumcised 40 years ago; today only about 40 percent of Kikuyu schoolgirls have been. But as many as 130 million girls and women presently living have undergone the procedure (2 million a year, 6000 per day). The practice has spread beyond Africa, also taking place surreptitiously among African emigrants in Europe and North America. In Boston, Dr. Nawal Nour heads a center at Brigham and Women's Hospital that handles the special needs of (primarily immigrant) women who have previously undergone the procedure.

Defenders of the custom cite its symbolic importance as ritual purification; often girls who have not undergone circumcision are considered unclean and unmarriageable. The *New York Times* reporter Celia W. Dugger interviewed a 12-year-old in Côte d'Ivoire, who said that she wants more than anything to be "cut down there," gesturing to her lap. All her friends have had it done, and afterward they were showered with gifts and money and there was a huge celebration for relatives and friends. The father of this girl said that if he did not have it done to her, he would not be allowed to speak in village meetings, and no man would marry her. Nonetheless, the girl's mother, who had undergone circumcision, said she hated the custom because it deprives a woman of sexual sensitivity.

Many African and world leaders, both men and women, have concluded that the custom constitutes an extreme human rights abuse, and it is now criminalized in 16 countries, but because it is so deeply ingrained in some value systems, the most successful

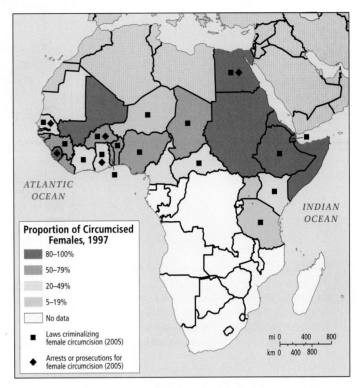

**FIGURE 7.43** Female circumcision in Africa, 1997–2005: Prevalence and legality. This practice, mapped here using 1997 data, occurs all across the center of the African continent in spite of government policies against it. The practice has been declared illegal in the countries indicated (the dates when the laws were enacted are shown in parentheses), but enforcement is lax in many. [Adapted from Joni Seager, *The State of Women in the World: An International Atlas* (London: Penguin, 1997), p. 53; additional data from Amnesty International, at http://www.amnesty.org/ailib/intcam/femgen/fgm9.htm; http://www.crlp.org/pub_fac_fgmicpd.html; *The World's Women, 2000: Trends and Statistics* (New York: United Nations, Department of Economic and Social Affairs, 2000), pp. 159–161.]

eradication campaigns are those that emphasize the threat circumcision poses to a woman's health. Thus the 2006 study is expected to be particularly helpful in changing people's views of the practice and in creating acceptable alternative practices. A growing number of Kenyan families, for example, are turning to a new rite known as "circumcision through words." This new ritual involves a weeklong program of counseling, ending with a community celebration and affirmation of a girl's passage to adulthood. In Uganda, a health education program seeks to show the public that some cultural practices can change without compromising the society's values. It is reported that in just 2 years (1994 to 1996), female circumcision in Uganda declined by more than 30 percent, largely as a result of this program; in 2003, according to Amnesty International, only 5 percent of Ugandan girls were cut.

## Ethnicity and Language

*Ethnicity,* as we have seen, refers to the shared language, cultural traditions, and political and economic institutions of a group. The

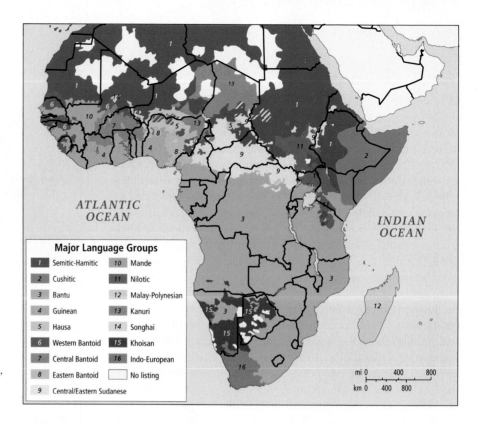

**FIGURE 7.44** Major language groups of Africa.
[Adapted from Edward F. Bergman and William H. Renwick, *Introduction to Geography—People, Places, and Environment* (Englewood Cliffs, N.J.: Prentice Hall, 1999), p. 256; and Titus Didactica, Frankfurt, Germany, http://titus.uni-frankfurt.de/didact/karten/afr/afrikam.htm.]

map of Africa presented in Figure 7.44 shows a rich and complex mosaic of ethnic languages. Yet despite its complexity, this map does not adequately depict Africa's cultural diversity (compare it with Figure 7.32 on page 381). Most ethnic groups have a core territory in which they have traditionally lived, but very rarely do groups occupy discrete and exclusive spaces. Often several groups share a space, practicing different but complementary ways of life and using different resources. For example, one ethnic group might be subsistence cultivators, another might herd animals on adjacent grasslands, and a third might be craft specialists working as weavers or smiths. On the other hand, people can share an ethnic identity, yet have little in common culturally or in the spaces they occupy. In Kenya, for example, Kikuyu villagers in the hinterland live very different daily lives from Kikuyu urbanites who reside in Nairobi.

People may also be very similar culturally and occupy overlapping spaces, but identify themselves as being from different ethnic groups. Hutu and Tutsi cattle farmers in Rwanda share an occupation, similar languages, and similar ways of life. However, as a result of Belgian, German, and eventually British colonial policy, which exaggerated ethnic differences, Hutu and Tutsi now think of themselves as having very different ethnicities. In the 1990s and again in 2004, the Hutu and Tutsi people engaged in several episodes of mutual genocide. Now those seeking reconciliation note that cultural training must play a central role in changing the way Rwandans think about each other. Not surprisingly, deciding which language shall be used in school instruction is a highly politicized and crucial debate.

Some countries in Africa have only one or two ethnic groups; others—such as Nigeria, Tanzania, and Cameroon—have many. Cameroon, sometimes referred to as a microcosm of Africa because of its ethnic complexity, officially has six main ethnic groups, but they are subdivided into 250 different smaller ethnic groups. Anthropological research has shown that the values and practices of the dif-

ferent groups concerning such things as sexual ethics may differ in the extreme, making the development of a cohesive national consciousness difficult. Nonetheless, the vast majority of African ethnic groups have peaceful and supportive relationships with one another.

To a large extent, language correlates with ethnicity. More than a thousand languages, falling into more than a hundred language groups, are spoken in Africa, and most Africans speak at least two: their native tongue and a **lingua franca** (language of trade). Some languages are spoken by only a few dozen people; others by many, such as Hausa, which is spoken by millions of people from Côte d'Ivoire to Cameroon. Language use everywhere is always in flux, so some African languages are now dying out and being replaced by languages that better suit people's needs or have become politically dominant. Increasingly, a few lingua francas are taking over, such as Hausa in West Africa and Arabic and Swahili in East Africa. (Swahili, also called Kiswahili, is an Arabic-influenced Bantu language.) Former colonial languages such as English, French, and Portuguese (all classed as Indo-European languages in Figure 7.44) are also widely used in commerce, in education, on the Internet, by pan-African congresses, and in the exchange of scientific knowledge. The new Pan-African Parliament now organizing under the auspices of the African Union is debating whether to use Swahili or English as its official language.

## ENVIRONMENTAL ISSUES

Africa is home to many unique animal and plant species that warrant protection and preservation. But attempts to ensure the well-being of Africa's human citizens often threaten the habitats that support endangered animals and plants (Figure 7.45). This is especially true where increasing population densities are placing pressure

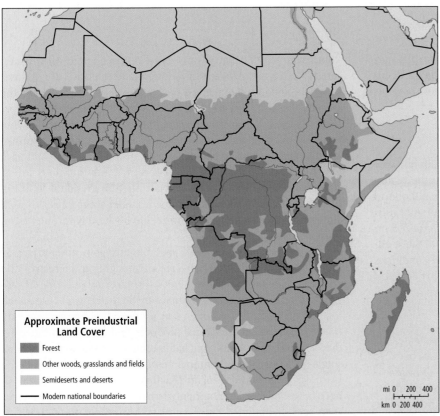

(a) Preindustrial

**FIGURE 7.45** Land cover (preindustrial) and human impact (2002). Although human impact was significant in the preindustrial era (**a**), its effects were not visible at this scale. By 2002, (**b**), however, human impact was pervasive; notice especially the impact on the forests of Central Africa, on coastal fisheries, and on the distribution of acid rain. [Adapted from *United Nations Environment Programme, 2002, 2003, 2004, 2005, 2006* (New York: United Nations Development Programme), at http://maps.grida.no/go/graphic/human_impact_year_1700 _approximately and http://maps.grida.no/go/graphic/human _impact_year_2002.]

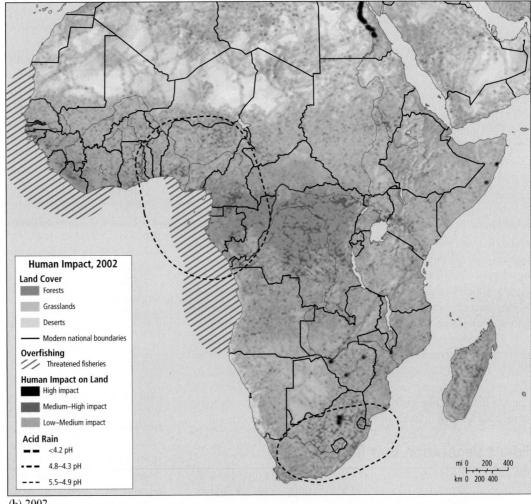

(b) 2002

on local water and fuelwood resources. The wisest African planners are designing strategies that both provide for basic human needs and address Africa's environmental issues. This section presents some of the major environmental issues facing Africa today.

## Desertification

Climatologists think that parts of Africa are now in a natural cycle of increased aridity. The effects are most dramatic in the region called the Sahel (Arabic for "shore" of the desert), a band of arid grassland 200 to 400 miles (320 to 640 kilometers) wide that runs east-west along the southern edge of the Sahara (see Figure 7.5 on page 357). Over the last century, the Sahel has shifted to the south in the process known as **desertification,** by which arid conditions spread to areas that were previously moist. For example, the *World Geographic Atlas* in 1953 showed Lake Chad situated in a tropical rain forest well south of the southern edge of the Sahel. By 1998, the Sahara itself was encroaching on Lake Chad.

The Sahel, other dry grasslands, and some dry forested areas in Africa are what geographers call **fragile environments** because they exist in zones that are barely sufficient for the needs of native plants and animals. Rainfall is already low in these grasslands, at 10–20 inches (25–50 centimeters) per year, and there are only low levels of organic matter in the soil to provide nutrients. Consequently, any further stress, such as fire, plowing, or intensive grazing, may cause drought-adapted native grasses to give way to introduced species, such as tough bunchgrass and thorny shrubs, that are more tolerant of stress. But these alien species do not cover the soil as well as the native plants, so they allow rain to evaporate more quickly, and they are not useful for grazing. Soon, the dry, denuded soil is blown away by the wind, and the remaining sand piles up into dunes as the grassland becomes more like a desert. The Ethiopian highlands, for example, are home to 85 percent of Ethiopia's population and 75 percent of its livestock. Erosion in this fragile environment resulting from deforestation and desertification has damaged 80 percent of the highlands, some areas so severely that food can no longer be produced there.

Indigenous animal herders are often blamed for desertification. However, long-term natural cycles are often the underlying cause, with human activity speeding up the cycle or triggering the process and broadening the area affected. Moreover, economic development activities can advance desertification. For example, agencies such as the World Bank have sometimes advised pastoralists to raise cattle instead of their traditional animals, such as camels and goats, because cattle can be sold as meat in distant urban markets. But raising cattle places greater stress on native grasslands than more traditional herding practices. Irrigated agriculture can also lead to desertification when minerals from the water build up in the soil over time, an effect called salinization (see Chapter 6, pages 323–324). The overharvesting of trees and shrubs for fuelwood also leads to drier soil and advances desertlike conditions.

## Forest Vegetation as a Resource

Like forests all over the world, the forests of Africa are disappearing. By 2000, an estimated 1.15 million square miles (3 million square kilometers)—more than 60 percent—of the original African rain forests were gone. Africans have long made extensive use of forest products, and much of the loss of African forests is attributed to the demand for farmland and fuelwood by growing populations, especially in West Africa. But in recent years, timber companies that formerly operated in Asia, where they were castigated for unsustainable tree harvesting practices, have moved into Africa and have lent financial support to some of the most unscrupulous rulers. Some environmentally aware officials, such as President Johnson-Sirleaf of Liberia, have banned foreign timber companies, but such strong measures are difficult to implement where governments are weak or corrupt.

In recognition of the rapid loss of forests, governments in East and West Africa and in the Sahel are encouraging **agroforestry**—the raising of economically useful trees—to take the pressure off old-growth forests. By practicing agroforestry on the fringes of the Sahel, a family in Mali, for example, can produce building and fencing materials, fuelwood, medicinal products, and food all on the same piece of land, thus doubling what its annual income would be from single-crop agriculture or animal herding. In 2004, the Nobel Peace Prize was awarded to Wangari Maathai of Kenya for her contributions to sustainable development, democracy, and peace. Nearly 30 years ago, she founded the Green Belt Movement, through which poor women planted more than 30 million trees, thus protecting their environment, promoting their own environmental awareness, and improving their self-esteem. Maathai is known for saying that a healthy environment is essential for democracy to flourish.

So far, logging for export is only a minor part of African economies, but it is responsible for the severe depletion of forests in easily accessible countries such as Liberia (see the story of Silas Siakor at the opening of this chapter) and Côte d'Ivoire in West Africa. The harvested logs are sold as raw wood, even though the financial return would be much greater if the trees were harvested sustainably and processed in Africa into refined wood products such as paneling or furniture. The most extensive remaining tracts of African rain forest are in the Congo Basin, most of which lies in the country of Congo (Kinshasa), where sustainable forest management is generally not practiced. As the armed conflict that has engulfed this country and its neighbors since the mid-1990s subsides, commercial logging is expected to expand and the loss of forested land to increase.

**Dry forests**—those forests that lose their leaves during extended dry seasons—once covered nearly twice as much of Africa as rain forests. Now dry forests are being destroyed even faster than rain forests because they lie closer to settlements and their products are in great demand by growing populations. Their greatest use is as fuelwood. Africans still use wood or charcoal to supply nearly all their domestic and industrial energy. Making and selling charcoal—a prepared, high-heat wood product—is an important industry in the informal economies of most African countries. Much of the wood used is harvested free because Africans traditionally have considered forests to be a resource held in common. In urban areas, wood and charcoal remain the cheapest fuels available. Even in Nigeria, which is a major oil producer, most people use fuelwood because they cannot afford petroleum products.

For a decade or more, many Africans have recognized the need to use fuelwood more sustainably. Fast-growing trees such as leucana are being farmed for making charcoal. Economic development planners have tried to promote fuel-efficient stoves, but users find that the stoves lack the convenience and nostalgic associations of old-fashioned cooking fires. Some Africans are turning to alternative energy sources—hydroelectric power, bottled gas, solar power, and wind power—to meet their growing energy needs, but the trend is not yet strong.

## Wildlife and National Parks

Africa's wildlife has long played a role in the global and continental economy. African animal products such as skins, taxidermy specimens, and ivory have been important exports to Europe, China, and the Americas for several hundred years, and live animals from Africa continue to supply the world's zoos. Furthermore, wildlife has always played an important role in the lives of Africa's human populations. Many Africans still depend on wild game (often called "bushmeat") and fish as their main sources of protein, and as populations increase faster than opportunities to earn cash for purchasing food, pressure on the continent's wildlife has become extreme. Some species are being reduced to critical levels, such as mountain gorillas in Congo (Kinshasa), where the ongoing war has nearly eliminated this large primate. Reports documenting the diminishing numbers of gorillas, elephants, lions, zebras, and giraffes and the decline of their rain forest and grassland habitats have raised public concern about Africa's diminishing wildlife heritage (Figure 7.46). Although increasing international interest has been helpful, it may not be enough to save many animals. The

**FIGURE 7.46 African wildlife in a not so wild moment.** Some lions in the Queen Elizabeth National Park in Uganda have adapted to sleeping in trees in order to evade poachers. [Rhett A. Butler/mongabay.com.]

cheetah, for example, is very near extinction, with only 12,000 individuals left in the wild. About 7000 have been killed in the last 10 years, mostly by farmers on the fringes of nature reserves who kill cheetahs to protect their livestock. While the use of dogs to protect livestock, an old African strategy, has peacefully resolved the conflict between cheetahs and some farmers, the loss of habitat to agriculture will further reduce the number of cheetahs.

Several strategies are emerging to address this many-sided crisis. Interest in game farming as a means of improving domestic food supplies and reducing expensive imports is growing. Wild animals are more resistant to African diseases and better adapted to the African climate than domesticated animals. Another strategy is to use ecotourism as both a way to conserve wildlife and a promising source of income. Africa now has one-third of the world's preserved national park land, covering about 10 percent of Earth's surface. Africa is unique in that its national parks are home to the only remaining concentrations of migratory plains animals in the world, and the potential to develop ecotourism is large in much of East and Southern Africa. But the idea that African national parks can both conserve wildlife and earn income will need careful nurturing. The parks are often islands in a sea of rural poverty, and because they are underfunded, it is easy for poachers to intrude. Because there has been too little effort to explain how preservation of wildlife could benefit local people or contribute to their income, hostility toward the parks has produced a number of violent attacks on foreign tourists. News of these attacks eclipses, at least temporarily, the potential of the parks to attract more tourists. In addition, Africa's economic crisis has further reduced the revenues available to maintain the parks. Many Africans believe that the parks are unlikely to be self-sustaining—much less profitable—in the near future, but that they are nonetheless worth the investment because of the valuable assets they protect. One African initiative seeks to establish a $250 million fund to be supported by African governments, the business community, local people, and international donors to ensure that protected areas provide clear benefits to African communities.

| Personal Vignette | In the early 1970s, new college graduates John and Terese Hart visited what would eventually (in 1992) become the Okapi Wildlife Reserve in northeastern Congo (Kinshasa) (Figure 7.47a). They returned in the 1980s to do their doctoral research. By 1996, they and Congolese graduates of the University of Kisangani and local wildlife experts had set up a research center in the Ituri Forest that trained 20 Congolese scientists each year and employed scores of local people. Then the war in Congo reached their doorstep, and they and some of the staff fled just before the research center was looted of all its equipment by marauding militias. But they soon heard reports that the local students and employees had remained behind at great peril and were reassembling the center bit by bit and continuing the conservation work for which they had been trained.

Corneille Ewango, a Congolese botanist (Figure 7.47b), remained at the center, connected to email via a satellite hookup and solar panel. As he reported on the bloody battles, he also noted who was poaching elephants, how many were killed, and who was opening new illegal mines in the reserve. His work made it possible to focus on what could be saved and

(a) The okapi, known as the "forest giraffe," is a rare and threatened species that lives mainly in the Okapi Wildlife Reserve, in the Ituri Forest, in northeastern Congo (Kinshasa).

(b) Corneille Ewango helped lead the effort to protect and preserve the Okapi Wildlife Reserve through 10 years of civil war. He is now a graduate student at the University of Missouri, St. Louis.

**FIGURE 7.47** Okapi Wildlife Reserve. [(a) Reuters; (b) Goldman Environmental Foundation.]

to put local pressure on the rebel leaders to control their militias and respect the Okapi Wildlife Reserve.

The Harts belatedly realized that as expatriate researchers, they had failed their colleagues by fleeing. Conservation could proceed under war conditions as long as workers remained on site, as the Congolese had done. These local conservationists had been fishers, farmers, traders, and miners, and one was even a former poacher. But all had been converted through education to a lifelong dedication to the study and protection of precious natural areas, no matter what the cost to themselves.

*Source: John and Terese Hart, "Rules of engagement for conservation: Lessons from the Democratic Republic of Congo," Conservation in Practice 4(1) (Winter 2003), http://www.conbio.org/cip/article41RUL.cfm.* ■

## Water

In a growing number of African countries, the amount of available water per capita is declining rapidly. In arid Africa, water is being pumped from aquifers faster than it is being replaced by nature; water is also being diverted from natural wetlands. Large economic development projects may redirect water to industrial, agricultural, or urban uses, leaving rural inhabitants of arid lands with insuf-

ficient water and causing the loss of complex ecosystems. Kenya, for example, has only 830 cubic yards (635 cubic meters) of water available per capita per year, compared with the minimal acceptable global standard of 1300 cubic yards (1000 cubic meters); its water supply is expected to drop to just 250 cubic yards (190 cubic meters) per capita per year by the year 2050.

Safe water is scarce even in the moist areas of the continent, where it is being polluted by salinization, human wastes, and chemical poisons. In rural areas, most households must carry all their water from springs, wells, pools, or streams; in urban areas, water is often drawn from standpipes connected to municipal water supplies (Figure 7.48). As sources become depleted or polluted, women (the traditional procurers of water) must walk farther and farther to collect safe water. The difficulty of carrying water limits the amounts available for use, and illness is spread when insufficient water is used to wash dishes, diapers, clothing, and other materials. Because plumbing and sewage treatment are usually not available, even in cities, human wastes often find their way into water sources.

Saving what is now being wasted is perhaps the greatest source of "new" water for most places on earth. Agriculture accounts for

**FIGURE 7.48** Carrying drinking water home. A child leads her younger sister while carrying drinking water in Lagos, Nigeria, in June 2005. [AP Photo/George Osodi.]

about 70 percent of global water use. In Africa, many large agricultural projects are irrigated, and most modern irrigation systems lose a great deal of water through leakage and evaporation. In addition, standing pools of water are sources of insect- or parasite-borne diseases such as malaria and schistosomiasis. Culturally sensitive development planners say that the solutions to Africa's water problems lie less in high-tech applications than in rediscovering local methods of water conservation. For example, a decade or two ago, West African women who delivered irrigation water directly to the roots of their plants by using human water brigades (see Figure 7.6 on page 358) were seen as inefficient by foreign development planners. Today, these water-saving practices (and the camaraderie they foster) are better appreciated. The installation of "old-fashioned" sanitary roof catchments and cisterns in place of complex piped delivery systems can save construction costs as well as water, as can the use of sanitary hand-flushed pit toilets in place of public sewer systems.

## MEASURES OF HUMAN WELL-BEING

The very low gross domestic product (GDP) per capita figures in Africa (Table 7.1, column II) indicate a desperately poor conti-

nent. However, GDP per capita does not include economic activities that take place outside the formal market system: sharing, bartering, participation in the informal economy, raising food in family plots, and conserving resources. Moreover, within the low range of GDP per capita figures, there is still considerable variation. Botswana, Cape Verde, Equatorial Guinea, Gabon, Mauritius, Namibia, South Africa, Swaziland, and Seychelles all have per capita GDPs above U.S.$3000 per year, but together they account for only 55 million people, less than 7 percent of Africa's population. Moreover, within each of these countries, there is a wide range of incomes. Many people live on much less than the official per capita GDP.

Most African countries rank in the lowest third of world nations on the Human Development Index (HDI; see Table 7.1, column III). Because their GDP figures are so low, there is little tax money to invest in education, basic health care, and sanitation. The literacy figures (columns V and VI) reflect both a generally low level of education and discrimination against women. During the 1980s and 1990s, per capita government spending fell, often in response to cutbacks mandated by SAPs, and the introduction of fees limited the number of people who could afford services such as education and health care. One result has been that more mothers and infants have died during and after childbirth. Spending on education fell from U.S.$11 billion in 1994 to U.S.$7 billion in 1998. When school fees were introduced, enrollment in elementary schools fell from 77.1 percent to 66.7 percent. School fees are increasingly recognized as counterproductive because when they are in place, governments are incapable of enforcing school attendance. The decision to educate is left up to often illiterate parents who may lack the funds or the understanding of just how education can help their children out of poverty.

For various reasons, such as the low status of women in African countries and official hesitancy to reveal the data, the statistics necessary to calculate the Gender Empowerment Measure (GEM) are available for only five countries, so this measure is not included in Table 7.1. Generally, women in Africa have not been able to participate in public policy formulation and decision making. Nonetheless, progress for women is building. In addition to the presidency of Liberia, women hold ministerial posts in Rwanda and Uganda. In South Africa, Namibia, Uganda, Rwanda, Tanzania, Eritrea, and Mozambique, women hold more than 20 percent of the seats in parliament (in the United States, women hold just 15 percent of the seats in Congress). Women occupy 50 percent or more of professional and technical positions in Botswana (53 percent), Namibia (55), and Swaziland (61). And women have an enormous economic role as entrepreneurs everywhere in Africa, especially in the informal economy.

On the Gender Development Index (GDI), all of the 35 sub-Saharan African countries ranked were near the bottom (a total of 140 countries are ranked; see Table 7.1, column IV). This index shows that in sub-Saharan Africa, women have starkly lower access than men to health care, education, and income, but both genders are seriously deprived in these categories.

**TABLE 7.1** Human well-being rankings of countries in sub-Saharan Africa and other selected countries[a]

| Country (I) | GDP per capita, adjusted for PPP[b] (GDP ranking among 177 countries), 2005 (II) | Human Development Index (HDI) ranking among 177 countries,[c] 2005 (III) | Gender Development Index (GDI) ranking among 140 countries, 2005 (IV) | Female literacy (percent), 2003 (V) | Male literacy (percent), 2003 (VI) |
|---|---|---|---|---|---|
| **Selected countries for comparison** | | | | | |
| Barbados | 15,270 | 30 (high) | 29 | 99 | 99 |
| Japan | 22,967 | 11 (high) | 14 | 99 | 99 |
| United States | 37,562 | 4 (high) | 8 | 99 | 99 |
| World | 8,229 | | | | |
| **West Africa** | | | | | |
| Benin | 1,115 | 162 (low) | 126 | 22.6 | 46.4 |
| Burkina Faso | 1,174 | 175 (low) | 138 | 8 | 18.5 |
| Cape Verde Islands | 5,214[e] | 105 (medium) | 81 | 68 | 85 |
| Côte d'Ivoire | 1,476 | 163 (low) | 128 | 38.2 | 60 |
| The Gambia | 1,859[e] | 155 (low) | 119 | 30.9 | 45 |
| Ghana | 2,238[e] | 138 (medium) | 104 | 45.7 | 63 |
| Guinea | 2,097 | 156 (low) | ND | 19[d] | 36[d] |
| Guinea-Bissau | 711[e] | 172 (low) | 135 | 24.7 | 55 |
| Liberia | 1,100[d] | ND | ND | 22[d] | 54[d] |
| Mali | 994 | 174 (low) | 136 | 12 | 27 |
| Mauritania | 1,766[e] | 152 (low) | 118 | 43.4 | 59.5 |
| Niger | 835 | 177 (low) | 140 | 9.4 | 19.6 |
| Nigeria | 1,050 | 158 (low) | 123 | 59.4 | 74.4 |
| Senegal | 1,648 | 157 (low) | 120 | 29.2 | 51 |
| Sierra Leone | 548 | 176 (low) | ND | 20.5 | 39.8 |
| Togo | 1,696[e] | 143 (medium) | 112 | 38.3 | 68.5 |
| **Central Africa** | | | | | |
| Cameroon | 2,118 | 148 (low) | 113 | 64.1 | 84.7 |
| Central African Republic | 1,089[e] | 171 (low) | ND | 33.5 | 64.8 |
| Chad | 1,210[e] | 173 (low) | 137 | 12.7 | 40.6 |
| Congo (Brazzaville) | 965 | 142 (medium) | 108 | 77.1 | 89 |
| Congo (Kinshasa) | 697 | 167 (low) | 131 | 52 | 80 |
| Djibouti | 2,086[e] | 150 (low) | ND | ND | ND |
| Equatorial Guinea | 19,780[e] | 121 (medium) | 95 | 76.4 | 92 |
| Eritrea | 849[e] | 161 (low) | 125 | 45.6 | 68 |
| Ethiopia | 711[e] | 170 (low) | 134 | 33.8 | 49 |

TABLE 7.1 (*Continued*)

| Country (I) | GDP per capita, adjusted for PPP[b] (GDP ranking among 177 countries), 2005 (II) | Human Development Index (HDI) ranking among 177 countries,[c] 2005 (III) | Gender Development Index (GDI) ranking among 140 countries, 2005 (IV) | Female literacy (percent), 2003 (V) | Male literacy (percent), 2003 (VI) |
|---|---|---|---|---|---|
| **Central Africa** (*continued*) | | | | | |
| Gabon | 6,397 | 123 (medium) | ND | ND | ND |
| São Tomé and Príncipe | 1,231[d] | 126 (medium) | ND | 62[d] | 85[d] |
| Somalia | 600[d] | ND | ND | 14[d] | 36[d] |
| **East Africa** | | | | | |
| Burundi | 648[e] | 169 (low) | 132 | 52 | 66.8 |
| Kenya | 1,037 | 154 (low) | 117 | 70.2 | 78 |
| Rwanda | 1,268[e] | 159 (low) | 122 | 59 | 70.5 |
| Tanzania | 621 | 164 (low) | 127 | 62.2 | 77.5 |
| Uganda | 1,490 | 147 (low) | 117 | 58 | 78 |
| **East African Islands** | | | | | |
| Comoros | 1,714[e] | 132 (medium) | 101 | 49 | 63.5 |
| Madagascar | 809 | 146 (low) | 116 | 65.2 | 76.4 |
| Mauritius | 11,287 | 65 (medium) | 54 | 80.5 | 88 |
| Seychelles | 10,232[d] | 51 (high) | ND | 60[d] | 56[d] |
| **Southern Africa** | | | | | |
| Angola | 2,344[e] | 160 (low) | 124 | 53.8 | 82 |
| Botswana | 8,714 | 131 (medium) | 100 | 81.5 | 76.1 |
| Lesotho | 2,561[e] | 149 (medium) | 114 | 90.3 | 73.7 |
| Malawi | 605 | 165 (low) | 116 | 54 | 75 |
| Mozambique | 1,170 | 1768 (low) | 133 | 31 | 62.3 |
| Namibia | 6,180 | 125 (medium) | 96 | 83.5 | 86.8 |
| South Africa | 10,386[e] | 120 (medium) | 92 | 81 | 84 |
| Swaziland | 4,726 | 147 (medium) | 115 | 78 | 80 |
| Zambia | 877 | 166 (low) | 130 | 59.7 | 76 |
| Zimbabwe | 2,443[d] | 145 (low) | 111 | 86.3 | 93.8 |

[a]Rankings are in descending order; i.e., low numbers indicate high rank.

[b]PPP = purchasing power parity, figured in 2003 U.S. dollars.

[c]The high, medium, and low designations indicate where the country ranks among the 177 countries classified into three categories by the United Nations.

[d]Data are for a date other than 2003 because data for 2003 were not available.

[e]Estimate based on history and multiple factors.

ND = No data available.

*Source: United Nations Human Development Report 2005* (New York: United Nations Development Programme), except as indicated above, at http://hdr.undp.org/reports/global/2005/pdf/HDR05_complete.pdf.

# III SUBREGIONS OF SUB-SAHARAN AFRICA

We divide sub-Saharan Africa into four subregions that reflect primarily geographic location: West Africa, Central Africa, East Africa, and Southern Africa. Such factors as traditions and the current geopolitical situation have also guided our placement of a particular country in a particular subregion. For each subregion, we provide a short discussion of the physical setting (see Figure 7.1) and then concentrate on specific circumstances that have led to current conditions.

## WEST AFRICA

West Africa, which occupies the bulge on the west side of the African continent, includes 15 countries (Figure 7.49). West Africa is physically framed on the north by the Sahara, on the west and south by the Atlantic Ocean, and on the east by Lake Chad and the mountains of Cameroon. The region can be thought of as a series of horizontal physical zones grading from dry in the north to moist in the south. The north-south division also applies to economic activities—herding in the north, farming in the south—and, to

some extent, to religions and cultures—Muslim in the north, Christian in the south. Most of these cultural and physical features do not have distinct boundaries, but rather zones of transition and exchange.

Stretching across West Africa from west to east are horizontal vegetation zones that reflect the horizontal climate zones shown in Figure 7.5 (page 357). Remnants of tropical rain forests still exist along the Atlantic coasts of Côte d'Ivoire, Ghana, and Nigeria, although much forest has been lost as a result of intensifying settlement and agriculture. To the north, this moist environment grades into drier woodland mixed with savanna. Farther north still, as the environment becomes drier, the trees thin out and the savanna dominates. Only where annual flood cycles create wetlands is moisture sufficient to foster fishing and cultivation as well as grazing (Box 7.5). The savanna blends into the yet drier Sahel region of arid grasslands. In the northern parts of Mali, Niger, and Mauritania, the arid land becomes actual desert, part of the Sahara.

Along the coast, and inland where rainfall is abundant, crops such as coffee, cacao, yams, palm oil, corn, bananas, sugarcane,

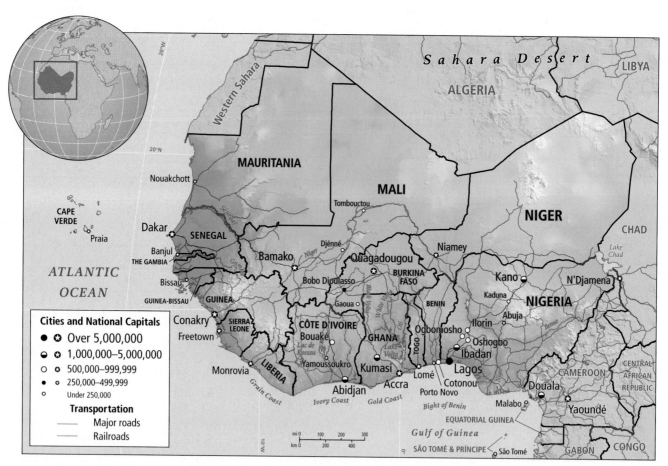

**FIGURE 7.49** The West Africa subregion.

## BOX 7.5 | AT THE REGIONAL SCALE  Water Management in the Niger River Basin

The Niger is the most important river of West Africa. It carries summer floodwaters northeast into the normally arid, clay-lined lowlands of Mali and Niger (Figure 7.50). There the waters spread out into lakes and streams that nourish wetlands. For a few months of the year (June through September), these wetlands ensure the livelihoods of millions of fishers, farmers, and pastoralists, who share the territory in carefully synchronized patterns of land use that have survived for millennia. These desert wetlands along the Niger produce eight times more plant matter per acre than the average wheat field and provide seasonal pasture for millions of cattle, goats, and sheep—they are home to the highest density of herds in all of Africa.

Because of mounting population pressure, international experts are advising the governments of Mali and Niger to dam the river and channel its water into irrigated agriculture projects that will help to feed the more than 20 million overwhelmingly poor people of Mali and Niger. Even though 80 percent of the people are farmers, the two countries already must rely on imported food. But the dams will forever change the seasonal rise and fall of the river that has supported an intricate mix of wildlife and human uses for thousands of years. Their effects will be felt in Europe as well, because the Niger is a seasonal migration stopover for birds that spend the summer in Europe, north of the Mediterranean.

**FIGURE 7.50** Threatened seasonal wetlands. Every rainy season, the Bani River, a major tributary of the Niger, floods and turns the city of Djénné, Mali, into a series of islands. Djénné, with a population of 20,000, is the oldest known urban settlement in sub-Saharan Africa (about 2200 years old). The annual flood revitalizes the surrounding soil and the river's marine life and permits the population to feed itself. A proposal to build a dam upstream threatens the city's self-sufficiency. [Sarah Leen/Matrix.]

and cassava are common. Farther north and inland, millet, peanuts, cotton, and sesame grow in the savanna environments. Cattle are also tended in the savanna and north into the Sahel; farther south, however, the threat of sleeping sickness spread by tsetse flies prevents cattle raising (see page 367). People in the Sahel and Sahara are generally not able to cultivate the land, except along the Niger River during the summer rains. Food, other than that provided by animals and the fields along the riverbanks, must come from trade with the coastal zones.

West African environments have dried out as more and more people clear the forests and woodlands for cultivation, grazing, and fuelwood. When the forest cover is gone, sunlight evaporates moisture, and desiccating desert winds blow south all the way to the coast. The drying out of the land is exacerbated by the recurrence of a natural cycle of drought. The ribbons of differing vegetation running west to east are not as distinct as they once were, and even coastal zones occasionally experience dry, dusty conditions reminiscent of the Sahel.

The geographies of religion and culture in West Africa reflect the north-south physical patterns to some extent. Generally, the north is Muslim, and Arab and North African influences are apparent. Southern West Africa is populated by people who practice a mix of Christian and traditional religions, and whose background and culture are central African. Some large ethnic groups once occupied distinct zones, but as a result of migration linked to colonial and postcolonial influences, members of specific groups live today in many different areas and in a number of West African countries. There are hundreds of smaller ethnic groups, each with its own language; the map depicting ethnic languages in Africa (see Figure 7.44) shows the cultural complexity of the entire West Africa subregion. Nigeria illustrates the potential difficulties faced by a nation attempting to unite a multitude of ethnic groups (see pages 381–383).

## Conflict in Coastal West Africa

The 1990s and early 2000s have seen continuing bloody conflict in the far West African countries of Guinea, Sierra Leone, Liberia, and Côte d'Ivoire. The wars in these four countries are intertwined, but each conflict is fed by local disputes that ultimately can be traced back to poverty and horrendously bad leadership. In the 1980s, Liberia was a country with a reasonably bright future. Then, in 1989, Liberian Charles Taylor, an American-educated, Libyan-trained-and-armed, would-be revolutionary, tried to stage a coup d'état. He failed, but then joined with a Sierra Leone rebel who had forced children to form an army. At least 200,000 people were murdered in Sierra Leone and Liberia as the child army captured diamond mines, which, along with the region's tropical timber, became the source of funding for further aggression. Taylor returned to Liberia with his AK-47-armed child soldiers, and in 1997, terrorized Liberians elected him president. He then invaded Guinea under the pretext that Liberian dissidents were ensconced there. With American military aid, Guinea pushed Taylor back, but only after many more deaths.

In 2002, the conflict spilled into Côte d'Ivoire, formerly one of the most stable West African countries, but one nonetheless troubled by regional disparities. Abidjan, the former capital on the southern coast, has fine roads, skyscrapers, French restaurants, and a modern seaport, but the north is much poorer. The north is full of cacao plantations that employ immigrant agricultural workers at very low wages and, in some cases, enslaved children from nearby Burkina Faso and Mali. Investigators say that slavery and abusive wages resulted when unscrupulous Côte d'Ivoire cacao planters started looking for a way to cut costs in the midst of a depressed world cacao market (see the discussion in Box 4.5 on page 225), but the basis of the civil conflict in Côte d'Ivoire is more far-reaching. The extreme poverty of the north and the large immigrant labor population led to a rebellion against the south, with support coming from Burkina Faso, the home of many workers in Côte d'Ivoire. Support for the southern faction came from Liberia, already a scene of conflict. Again, child soldiers were armed and told to support themselves by looting, which in relatively wealthy Côte d'Ivoire was a tempting proposition for poor, illiterate, hungry boys and young men. In 2003, UN peacekeepers were belatedly sent to Liberia.

By 2006, there were many hopeful signs. As we saw in the opening of this chapter, a highly respected, reform-minded woman, Ellen Johnson-Sirleaf, was elected president of Liberia. Under Johnson-Sirleaf's leadership, Liberia is beginning to rebuild, but the cost of that rebuilding is expected to total $500 billion. Meanwhile, human rights groups and chocolate eaters in Europe and elsewhere (Africans consume very little chocolate or cacao products) successfully pressured international chocolate producers to address a wide range of issues, including child and slave labor on cacao plantations and unsustainable cacao production practices. In May 2006, the first African international cacao summit in Abuja, Nigeria, resulted in a general consensus that African cacao interests must stop overproduction and improve the product so that African growers—who produced 80 percent of the cacao for the $75 billion annual global chocolate market in 2005—can live decent lives.

## CENTRAL AFRICA

If you were to look at a map of Africa without any prior knowledge, you might reasonably expect the countries of Central Africa—Cameroon, the Central African Republic, Chad, Congo (Brazzaville), Equatorial Guinea, Gabon, São Tomé and Príncipe, and Congo (Kinshasa)—to be the hub of continental activity, the true center around which life on the continent revolves (Figure 7.51). Paradoxically, Central Africa plays only a peripheral role on the continent. Its dense tropical forests, tropical diseases, and difficult terrain make it the least accessible part of the continent, and recent armed conflict over resources and power has left it in a deteriorating state that discourages visitors and potential investors.

Central Africa consists of a core of wet tropical forestlands surrounded on the north, south, and east by bands of drier forest and then savanna (see Figure 7.5 on page 357). The core is the drainage basin of the Congo River, which flows through this area and is fed by a number of tributaries. Although rich in resources, the forests form a barrier that has isolated Central Africa from the rest of the continent for centuries and has impeded the development of transportation, commercial agriculture, and even mining. Moreover, the soils of the forests are not suitable for long-term cash-crop agriculture. Most Central Africans are rural subsistence farmers who live along the rivers and the occasional railroad line. The forests themselves are home to nomadic hunter-gatherers, such as the Mbuti and Pygmy people, but these people and their way of life are highly endangered. Cities are only beginning to grow; there is little urban employment, and too often, the newcomers are people escaping civil violence in the countryside.

The region's colonial heritage has contributed to these difficulties. In what is now known as the Democratic Republic of the Congo (Kinshasa), for example, the Belgian colonists left a terrible legacy of violence. In 1885, Belgium's king, Leopold II, with a personal fortune to invest, obtained international approval for his own personal colony, the Congo Free State, with the promise that he would end slavery, protect the native people, and guarantee free trade. He did none of these things. The Congo Free State consisted of 1 million square miles (2.5 million square kilometers) of tropical forestland rich in ivory, rubber, timber, and copper, with a population of 10 million people, whose labor King Leopold would soon appropriate to extract Congo's resources for his own profit. The cruel behavior of Belgian colonial officials was witnessed by the young writer Joseph Conrad when he obtained a job on a steamer headed up the river in 1890. Conrad's novella, *Heart of Darkness* (published in 1902), and a pamphlet by Mark Twain, "King Leopold's Soliloquy" (published in 1905), awakened Europeans and Americans to Leopold's perfidy, and efforts began to end his regime. However, French colonists in what is now Congo (Brazzaville), Gabon, and the Central African Republic; German colonists in Cameroon; and Portuguese colonists in neighboring Angola used tactics similar to Leopold's. For example, a chronicler of the 1890s recorded that after a group of Africans rebelled against the French, their heads were used as a decoration around a flower bed in front of a French official's home.

The indigenous people of Central Africa form an overlapping ethnic mix, and they speak some 700 languages. The artificial political borders imposed by the European policy of deliberately pitting Africans against one another exacerbated tensions between ethnic groups. When Europeans removed local chiefs and installed new leaders who would do their bidding, indigenous governance systems were disrupted and long-standing traditional restraints on unwise leadership were lost, opening the door for the abuses of power that developed in the twentieth century.

When the countries of Central Africa were granted their independence in the 1960s, the Europeans stayed on to continue their economic domination. Their tax-free profits were sent home to France, Belgium, or Portugal; rarely were profits reinvested in Central Africa. With no money to develop an infrastructure, no experi-

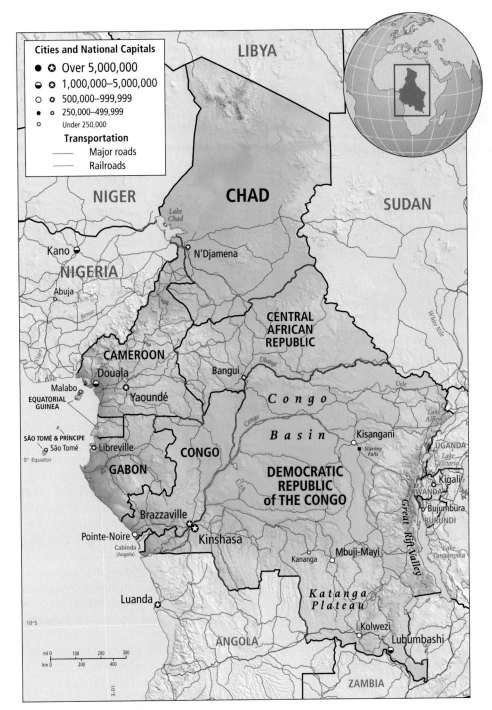

**FIGURE 7.51** The Central Africa subregion.

ence with multiparty democracy, and no educated constituency, many Central African countries, such as Congo (Kinshasa), fell into the hands of corrupt leaders. The political upheavals that resulted when some groups were left out of representation in government were accompanied by downward-spiraling economies. As in other parts of sub-Saharan Africa, the average person's purchasing power has decreased over the last few decades. Tax concessions to foreigners as well as government corruption have reduced tax revenues; hence maintenance of infrastructure is negligible. Few

industries, either government or private, can survive due to the lack of electricity and raw materials. People are so poor that there is little market for any type of consumer product.

Yet the subregion has an enormous wealth of untouched natural resources: copper, gold, diamonds, oil, and rare tropical plants and animals. Congo (Kinshasa) is sub-Saharan Africa's fourth largest oil producer, and oil is its major foreign trade commodity. It also has enough undeveloped hydroelectric potential to supply power to every household in the country and to those of some neighboring

**FIGURE 7.52** Rain forest along the Mpivie River in Loango National Park, Gabon. [Rhett A. Butler/mongaby.com.]

countries as well. Perhaps of greater importance are the natural wonders of the tropical rain forests (Figure 7.52), in which live rare species such as white rhinos, Congo peacocks, various types of gorillas, elephants, and okapi (see the personal vignette on the Okapi Wildlife Reserve on pages 395–396).

## Congo (Kinshasa): A Case Study of Devolution

When independence came to Congo (Kinshasa) in 1960, political and economic collapse was imminent as Belgian colonial officials withdrew, leaving untrained Congolese in charge. The new government of Patrice Lumumba tried to correct the inequalities of the colonial era by nationalizing foreign-owned companies in order to keep their profits in the country. Lumumba's strategies, adopted during the height of the cold war, and the support he received from the Soviet Union raised the red flag of Communism for the West. The Western allies arranged for the killing of Lumumba and supported and armed politicians who opposed him, such as the extremely corrupt Mobutu Sese Seko.

After a violent political struggle, Mobutu Sese Seko seized power in 1965. Mobutu also nationalized the country's institutions, businesses, and industries, but rather than using the profits for the good of the Congolese people, he and his supporters began to extract personal wealth from the newly reorganized ventures. Foreign experts, trained technicians, and capital left the country. Congo's economy, physical infrastructure, educational system, food supply, and social structure declined rapidly, and by the 1990s, the vast majority of Congolese people were impoverished. Mobutu was finally overthrown in 1997 by Laurent Kabila, a new dictator supported by Uganda and Rwanda, neighbors to the east who coveted Congo's mineral wealth. Kabila's inept regime quickly led to widening strife in Central Africa. By 1999, Angola, Namibia, Zimbabwe, Chad, Rwanda, and Uganda all had troops in Congo and

were aggressively competing for parts of its territory and resources. More than 4 million Congolese died as a result of the conflict, and 1.6 million remain displaced.

Despite the presence of UN peacekeeping forces, the violence ebbed and flowed into the first decade of the twenty-first century, with few meaningful changes in social or economic policies. But in 2006, the UN Security Council officially recognized that the ongoing conflicts in the Great Lakes region of Africa (which includes the point where Congo, Sudan, Uganda, Rwanda, and Burundi come together) were all connected and linked to the illegal exploitation of natural resources, the illicit trade in those resources, and the proliferation and trafficking of arms. It urged all countries to disarm and demobilize their militias and armed groups, especially northern Uganda's Lord's Resistance Army, and to promote the lawful and sustainable use of natural resources. In July 2006, Congo held an election, which was won by Laurent Kabila's son Joseph.

## EAST AFRICA

The subregion of East Africa (Figure 7.53) occupies the Horn of Africa, which reaches along the southern shore of the Red Sea and the Gulf of Aden, and includes the countries of Djibouti, Eritrea, Ethiopia, and Somalia. To the south of the Horn, it includes the coastal countries of Kenya and Tanzania; the interior highland countries of Uganda, Rwanda, and Burundi in the Great Rift Valley; and the islands of Madagascar, Comoros, Seychelles, Mauritius, and Réunion in the Indian Ocean across the Mozambique Channel.

Across the subregion, the relatively dry coastal zones contrast with moister interior uplands. A vital and productive farming economy has long existed in the uplands, and today across the region nearly 80 percent of the population makes a living from farming. A service economy is beginning to grow in urban areas and tourism zones along the coast, in Madagascar, and around the national parks of Kenya, Uganda, and Tanzania. Industry is only beginning to develop, primarily in Kenya and Tanzania. On the coast, there is a lively trading economy in the port cities, which have a long history of trade-related services and links across the Arabian Sea and Indian Ocean.

Africa's ability to support its people is being tested in East Africa, where growing consumption is straining the region's carrying capacity. The entire Horn of Africa has suffered periodic famines since the mid-1980s, caused by naturally occurring droughts but made worse by ineffective governments and political bosses who manipulated the situation for personal gain. The result in Somalia and Ethiopia has been an environment of chaos and lawlessness for ordinary citizens. Militant Islamic organizations have attempted to impose order and conformity, but their rigid rules, especially those affecting women, have led to further oppression. Furthermore, their rise led to worry on the part of other nations that a center of Islamic terrorism was evolving. In 1993, after long years of intermittent civil violence, the United States, in an effort to quell the conflict, stepped in on the side of unscrupulous warlords. This action resulted in the debacle in and around Mogadishu, Somalia, immortalized in the 2001 film *Black Hawk Down*. The United States withdrew, but conflict flared again in 2006, when militant Islamists

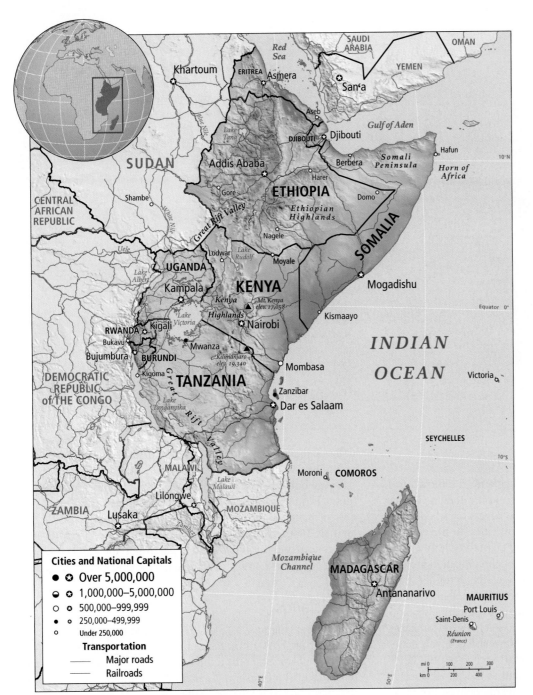

**FIGURE 7.53** The East Africa subregion.

seized control of Somalia, instigating an invasion by Ethiopia. Once again, large numbers of traumatized refugees needed international protection.

## The Interior Valleys and Uplands

East Africa's interior is dominated by the Great Rift Valley, an irregular series of tectonic rifts curving inland from the Red Sea and extending southwest through Ethiopia and into the highland Great Lakes region, then bending around the western edges of the subregion and extending into Mozambique to the south of the subregion. The Great Rift Valley, which contains numerous archaeological sites that have shed light on the early evolution of humans and other large mammals, is today covered primarily with temperate and tropical savanna, although patches of forest remain on the high slopes and in southwestern Tanzania. The semiarid highlands are drained by deep river valleys, including the valley where the headwaters of the Blue Nile lie in Ethiopia. Rain falls during the summer, and conditions vary with elevation. Most rural people make a living by cattle herding (Figure 7.54), but in the moister central uplands of Ethiopia and Kenya and in the woodlands of southern Tanzania, herding and cultivation are combined in mixed farming: animal manure is used to fertilize a wide variety of crops that are closely adapted to particularly fragile environments. Pastures and fields are rotated systematically. Agronomists have only recently begun to appreciate the precision of these ancient and specialized mixed agriculture systems, which are now threatened by civil disorder, development schemes, and population pressure.

**FIGURE 7.54** Tsemai and Ari herders bring cattle to drink from the Weita River in the Omo Valley, Ethiopia. The Omo Valley in southwestern Ethiopia, which runs along the Great Rift Valley near the border with Kenya, is where some of the earliest remains of human species have been found. [© Ariadne Van Zandbergen/www. photographersdirect.com.]

## The Coastal Lowlands

To the east and south of the Ethiopian highlands is a broad, arid apron that descends to the Gulf of Aden and the Indian Ocean. Along the coasts of the Horn of Africa are large stretches of sand and salt deserts, a few rivers, and a few port cities. In the country of Somalia, which occupies the low, arid coastal territory in the Horn, people eke out a meager living through cultivation around oases and herding in the dry grasslands. Somalia's 8 million people, virtually all Muslim and ethnically Somali, are aligned in six principal clans. It is clan warlords, Islamic fundamentalists, and political adventurers who have engaged in a violent rivalry for power over the last few decades. The resulting turmoil has driven many cultivators from their lands and has blocked access to some lands that are crucial to the food production system, so the land that is still available is being overused. Thus it is primarily politics, not rapid population growth, that has increased pressure on arable land in all countries of the Horn of Africa. Nonetheless, at present rates of growth, populations in the Horn are expected to double and even triple by 2050, and conditions will make it difficult to accommodate them all.

East Africa's shores from southern Somalia southward have long attracted traders from around the Indian Ocean—from China, Indonesia, Malaysia, India, Persia (Iran), Arabia, and elsewhere in Africa. The result has been a grand blending of peoples, cultures, languages, plants, and animals. In the sixth century, Arab traders brought Islam to East Africa, where it remains an important influence in the north and along the coast. Traders, speaking the lingua franca of Swahili, established networks that linked the coast with the interior. These networks penetrated deeply into the farming and herding areas of the savanna and into the semifeudal cultures of the highland lake country. Today, trade across the Indian Ocean continues: Arab countries are East Africa's most important trading partners, but Asia, China, India, and Japan are increasingly important to East African trade, and trade with Indonesia and Malaysia is also increasing. East Africa exports mostly agricultural products (live animals, hides, bananas, coffee, cashews, processed food, tobacco, and cotton) and imports machinery, transportation and communication equipment, industrial raw materials, and consumer goods.

## The Islands

Madagascar (Figure 7.55), which lies off the coast of East Africa, is the fourth largest island in the world; only Greenland, New Guinea, and Borneo are larger. Its unique plant and animal life—the result of *diffusion* (transmission by natural or cultural processes) from mainland Africa and Asia and subsequent evolution in isolation on the island—is highly prized by biologists, increasingly attractive to tourists, and greatly threatened by environmental degradation. The other East African island nations—the Comoros, the Seychelles, and Mauritius—are much smaller and less physically complex than Madagascar. All the islands have a cosmopolitan ethnic makeup, the result of thousands of years of trade across the Indian Ocean. The people and languages of Madagascar, for example, have ancient origins in Southeast Asia. Further cultural mixing in the islands took place during European (primarily French and British) colonization. During the colonial era, these islands supported large European-owned plantations, worked by laborers brought in from Asia and the African mainland. More recently, islanders have added tourism to the economic mix. Most visitors come from the African mainland, from Asian countries surrounding the Indian Ocean, and from Europe.

**FIGURE 7.55** A traditional village north of Antananarivo, Madagascar. [Yann Bertrand-Artoud/ALTITUDE.]

## Kenya and Tanzania: Case Studies in Contested Space

Kenya and Tanzania are modestly industrializing countries in East Africa that remain largely dependent on agriculture for export; tea, coffee, cacao, beer, cement, petroleum products, horticultural products, and processed food are their major exports. Both countries are regional hubs of trade, yet both remain among the poorest countries in the world (see Table 7.1 on pages 398–399).

In ancient times, traders in this part of Africa exchanged commodities—slaves, ivory, rare animals—that can no longer be legally traded. By the early nineteenth century, the depletion of elephant herds (killed for their ivory tusks) and changing attitudes toward slavery had pushed the region into a period of stagnation that made it easier for the British to gain control. The British colonists relocated native highland farmers in Kenya, such as the Kikuyu, to make way for coffee plantations, and they displaced pastoralists in Kenya and Tanzania, such as the Masai, from the savanna. Some of the lands of the Masai became game reserves designed to protect the savanna wildlife for the use of European hunters and tourists.

Today, the former colonial game reserves of Kenya and Tanzania are among the finest national parks in Africa, offering protection to wild elephants, giraffes, zebras, and lions as well as many less spectacular but no less significant species. Tourism, mostly in the form of visits to these parks, accounts for 20 percent of foreign exchange earnings, bringing Kenya about $600 million a year and Tanzania a similar amount (2004 figures). The future of the parks is precarious, however, because of competing demands for the land by herders and cultivators and the indigenous tendency to view wild animals as a useful consumable resource. In Kenya, a living

elephant is worth close to $15,000 a year in income from tourists who come to see it. Nonetheless, some still see hunting elephants for their ivory as a profitable activity, even though the profit to a hunter may be no more than $1000 per dead animal. The Masai complain that elephants destroy their grazing lands and gardens, and hence they see no reason to protect elephants, which can quickly overpopulate an area and stress its environment. Modern park policy in Kenya now advocates public education that clearly points out the long-term value to local communities of saving endangered species.

After Kenya's independence from Britain in the early 1960s, government policies resettled African farmers on highland plantations formerly owned by Europeans, giving the country a strong base in export agriculture through the production of coffee, tea, sugarcane, corn, and sisal. Under postcolonial policies, nomadic pastoralists have fared less well than farmers. Government strategies for managing the national parks have inadvertently resulted in the spread of tsetse fly infestations to the local pastoralists' herds. Disputes over scarce water supplies have broken out between cultivators who wish to irrigate their crops and herders who need regular sources of water for their animals. So far, the government has sided with the farmers, even to the point of policing the nomadic herders from armed helicopters.

Agriculture, especially export agriculture, is important to Kenya and Tanzania, but their dependence on this sector has put both people's livelihoods and the famous national parks at risk. Part of the problem is population growth. In the mid-twentieth century, populations throughout East Africa began to grow extremely rapidly; in 1946, Kenya had just 5 million people, but by 2006, it had 34 million. In the past, the primarily arid land could adequately support relatively small populations when managed according to traditional herding and subsistence cultivation practices. Today, the land's carrying capacity is severely stressed. Competition for arable land has pushed thousands of small farmers out of the more fertile uplands into the drier, more fragile, marginally productive savannas, which had served both as communal rangelands and as habitats for wild animals. Even as farmers in Kenya and Tanzania try desperately to increase agricultural production, volatile world market prices for some export crops can send the national economies plunging, as happened when world coffee prices fell sharply in 2001. These problems are threatening both Kenyan and Tanzanian game reserves and parks. Population pressure has forced farmers to plow and overgraze wildland buffer zones surrounding the parks. Hungry people have begun to poach park animals for food, and there is now a large underground trade in bushmeat.

## SOUTHERN AFRICA

At the beginning of the twenty-first century, both optimism and worry are in the voices heard across Southern Africa (Figure 7.56). In South Africa, optimism is based on the fact that it is now free of the scourge of apartheid, racial conflict has abated, and civil peace has brought substantial economic growth. South Africa is the wealthiest and perhaps the best-run country in the whole of sub-Saharan Africa. Other countries in Southern Africa—Mozambique, Namibia, Botswana, Lesotho, Swaziland, and Angola—are enjoying new stability and economic growth as well. Only Zimbabwe, once thought of as having great promise, has become a worrisome failed state because of ill-conceived development strategies and bad governance (see pages 372–373). But the main cause for worry in the region is the HIV-AIDS epidemic, which by 2005 threatened 20–40 percent or more of the adult population in every country in Southern Africa (see the discussion on pages 367–369).

Southern Africa is a plateau ringed on three sides by mountains and a narrow lowland coastal strip. At its center is the Kalahari Desert, but for the most part, Southern Africa is a land of savannas and open woodlands. Its population density is low; the highest densities are found in the urban north-central and southeastern parts of South Africa (see Figure 7.13 on page 364). Africa's mineral wealth is concentrated in Southern Africa, where there are rich deposits of diamonds, gold, chrome, copper, uranium, and coal.

**FIGURE 7.56** The Southern Africa subregion.

**FIGURE 7.57** Fruit for sale along a country road in Malawi. [© Jim Whitmer/ www. photographersdirect.com.]

Angola and Mozambique are both former Portuguese colonies. Their wars of independence evolved into civil wars, with Marxist, Soviet-supported governments on one side and rebels supported by South Africa and the United States on the other. Mozambique ended its civil war in 1994, and by 1996, its once-devastated economy was experiencing one of the highest rates of growth on the continent (7 percent in 2005). In Angola, where sporadic fighting was ongoing in 2006 (especially in Cabinda, a coastal exclave between the two Congos and a main oil-producing area), stability has been elusive. Liberation movements fighting to free Angola from Portuguese rule split into factions in the early 1970s. A Marxist faction held the coastal regions, where most Angolans live and where the country's capital and oil and fishing resources are located. Another faction, the Union for the Total Independence of Angola (UNITA), drew its strength and financial support from the mineral-rich south-central plateau, but also from links with the notoriously corrupt Mobutu regime in Congo (Kinshasa) and from the United States, which feared a Communist takeover. UNITA held the upper hand for years, controlling 70 percent of the country until 1998, when a peace treaty among most Angolan forces was signed. Now the country is experiencing extraordinarily rapid growth in its GDP (17 percent in 2005) because of the development of its oil resources.

The three interior countries of Malawi, Zimbabwe, and Zambia are still overwhelmingly rural. Agriculture and related food-processing industries form an important part of their economies and their exports. Over the last two decades, however, agricultural productivity has fallen. The case of Malawi is illustrative. Malawi had enough food for its people and surpluses for export until the 1980s, when land productivity began to decline as a result of overuse. After independence from Britain in 1964, the government encouraged large plantations that grew cash crops (tobacco, tea, sugar) for export. Fragile tropical soils that need years of rest between plantings were expected to produce continuously. After a few years, the de-

pleted soil caused plantation production to decline. Smallholders, who lost land to these commercial producers, found themselves with insufficient land to allow for the necessary fallow periods, and their productivity began to fall as well. By the 1990s, 86 percent of Malawi's rural households had less than 5 acres (2 hectares) of land, yet they produced nearly 70 percent of the food for rural and urban consumption in Malawi (Figure 7.57). Now HIV-AIDS is keeping skilled farmers out of the fields because they are sick themselves or must tend the sick, so harvests are further depleted.

The economies of Zambia and Zimbabwe remain focused on extractive industries, such as commercial agriculture and mining. In Zimbabwe, chromium, gold, coal, nickel, platinum, and silver mining accounts for 40 percent of the country's exports; however, the economy has failed because of misplaced advice from international lending agencies, the decline of commercial agriculture, and the inept and now illegitimate government of Robert Mugabe. Zambia's rich copper resources have drawn investment and management by the Chinese, whose need to bring electricity to their rapidly industrializing cities has created a large demand for copper. The Chinese-run mines are efficiently run, and production has soared, but very little of the profit reaches Zambians: local miners earn just U.S.$50 a month, while many miners are imported Chinese workers. Working conditions can be deadly; no health-care services are provided; and, unlike the state-run mines, these mines provide no education for the miners' children. Zambian activists note that, as in colonial times, the extraction of Zambia's resources is primarily benefiting foreign investors. The thwarting of Botswana's promising economic development by a burgeoning HIV-AIDS epidemic was discussed at some length earlier in this chapter.

## South Africa: A Model for Southern Africa

South Africa is a country of beautiful and striking vistas. Just inland from its long coastline lie uplands and mountains that

**FIGURE 7.58** The Mzimkulu River valley, a dairy and farming community in the foothills of the Drakensberg Mountains, South Africa. [© Bernie Olbrich/AfriPics.com. All Rights Reserved.]

afford dramatic views of the coast and the ocean. The interior is a high rolling plateau reminiscent of the Great Plains in North America (Figure 7.58). South Africa is not tropical; it lies entirely within the midlatitudes and has consistently cool temperatures that range between 60°F and 70°F (16°C and 20°C). The extreme southern coastal area around Cape Town has a Mediterranean climate: cool and wet in the winter, dry in the summer.

About twice the size of Texas (whose population is 22 million), South Africa has a population of 44 million people: 75 percent are of African descent, 14 percent of European descent, and 2 percent of Indian descent; the rest have mixed backgrounds. When Nelson Mandela was elected president in 1994, after more than 300 years of European colonization and 50 years of apartheid, he insisted that in order to heal the racial divide, the country must face the reality of what had happened under apartheid. Truth and Reconciliation Commissions asked those who had been brutal on both sides to step forward and admit what happened in detail; then they were formally forgiven for their misdeeds.

South Africa is now considered to be one of the world's top ten emerging markets. It has a balanced economy that accounts for one-third of sub-Saharan Africa's production. It has built competitive industries in communications, energy, transport, and finance (it has a world-class stock market) and supplies goods and services to neighboring countries; the other Southern African countries now get the majority of their imports from South Africa (though China's share is growing). In return, products from all over Africa are sold in South Africa's markets, and many people throughout Southern Africa are dependent on jobs in South Africa.

Economic improvements in Angola, Mozambique, Namibia, and Botswana; peace in Angola; and racial reconciliation and political and economic progress in South Africa, as well as in the tiny countries of Swaziland and Lesotho, have prompted forecasters to speculate that Southern Africa, more than any other subregion, holds the potential to lead sub-Saharan Africa into a more prosperous age. Certainly, poverty is still widespread, resources are overstressed, and the severe HIV-AIDS epidemic is bringing terrible losses. But with creative—indeed, courageous—leadership and outside help, South Africa and a few of its neighbors may be able to move directly from agrarian economies into the information age of the twenty-first century, bypassing the industrial phase experienced by Europe, North America, and Japan.

## Reflections on Sub-Saharan Africa

Late one evening in a restaurant in Central Europe, after a lengthy conversation that touched on some of the world's perplexing problems, my colleague leaned across the elegant white linen tablecloth and asked, "But don't you think that Africa is, after all, better off for having been colonized by Europeans?"

How does one reply to such a question? Africa, the ancient home of the human race, is of all regions on earth the one in greatest need of attention. Africa is poor and, by some measures, getting poorer. Yet the reasons for Africa's poverty are not immediately apparent to the casual observer. Africa is blessed with many kinds of resources—agricultural, mineral, and forest—but too often the income from them goes to outsiders. And although most of Africa is not densely occupied, and birth rates are dropping, population growth, especially in urban areas, is thwarting efforts to improve standards of living. Africa's people work hard, but their productivity, at least as measured by the standards of the developed world, is low.

An understanding of the reasons for Africa's present poverty and its social and political instability begins to emerge only through an exploration of its history over the last several centuries of European colonialism and a careful analysis of how that history is still affecting the organization of African economies and societies. Colonialism methodically removed Africans from control of their own societies and lives, and it turned Africa's people and resources to the service of distant countries. Even today, with colonialism officially dead for more than three decades, outsiders in the global marketplace continue to view African resources as available for the taking at less than a fair price. In view of all this, it is hard to believe that anyone would think Africa is better off for having been colonized.

Africa is changing, however. Africans are beginning to devise new economic development strategies and political institutions to replace the ones imposed by outsiders over the last 500 years. Although Africa remains a continent of countries created out of foreign perceptions of how Africans should be organized politically, African leaders are now articulating wider visions. Pan-African movements for economic integration and regional cooperation are emerging.

It is tempting to suggest that the rest of the world should leave Africa to the Africans for the next era. But that view may be too simplistic. Despite the fact that Africa had plenty of help getting into its current predicament, and although it is primarily outsiders who have prospered from its wealth, it has received very little foreign aid or foreign direct investment from the developed countries to help it move on. Many free market economists, such as Jeffrey Sachs, director of the Center for International Development at Harvard University, think wealthy nations should provide financial support for Africa's comeback so that Africans can themselves become a market for imported products. Experts suggest several strategies. First, the cancellation of African debt to foreign governments, international lending agencies, and some private lenders should go well beyond the modest levels agreed to by the G8 countries (the world's most developed economies) in 2005. Then tax revenues could once again go to support schools, health care, and social services. American Baptists, Presbyterians, and Methodists and the international Anglican and Roman Catholic leadership are publicly on record as supporting this idea. A second suggestion is that the developed world lower tariffs against African manufactured products to foster development of African industries. This proposal is opposed by people who worry about losses of jobs in the developed world. A third suggestion is that future aid to Africa be designed to take advantage of indigenous skills and knowledge and that development planning and aid money address local needs as defined and managed by local experts. This fundamentally conservative perspective acknowledges the need to help, but puts its faith in the ultimate ability of the African people to determine their own future.

## Chapter Key Terms

African Union (AU) 378

agroforestry 394

animism 388

apartheid 362

carrying capacity 364

Common Market for Eastern and Southern Africa (COMESA) 378

currency devaluation 376

desertification 394

devolution 385

dry forests 394

East African Community (EAC) 378

Economic Community of Central African States (CEEAC) 378

Economic Community of West African States (ECOWAS) 378

escarpment 356

female circumcision 391

fragile environment 394

fusion 388

grassroots economic development 379

Horn of Africa 357

intertropical convergence zone (ITCZ) 356

laterite 358

leaching 358

lingua franca 392

Mano River Union (MRU) 378

mixed agriculture 370

neocolonialism 355

pastoralism 370

polygyny 390

Sahel 357

self-reliant development 379

shifting cultivation 358

Southern African Development Community (SADC) 378

syncretism 388

West African Economic and Monetary Union (UEMOA) 378

## Critical Thinking Questions

1. If you were to investigate the origins of lumber and lumber products sold where you live, where would you start? What would make you think that you are or are not participating in the African timber trade? How is this trade related to the concept of neocolonialism?

2. Why does being an exporter of raw materials place a country at a disadvantage in the global economy? What would be some ways to amend the situation?

3. What are the environmental links to the spread of the chronic communicable diseases of Africa? Why might some scientists argue that malaria is a worse threat to Africa's development than HIV-AIDS?

4. Why is rapid population growth still an issue in Africa? What are the factors that contribute to this growth? How might the developed countries help the situation?

5. After reflecting on the many passages throughout this chapter regarding gender roles in Africa, develop a statement of what you think are the most salient points to remember.

6. What do you judge to be the most important ways that women are exerting an influence on African political issues?

7. Technology is changing life in Africa. Describe what you think are the most crucial ways in which technology will modify the power of ordinary people to better their lives and to participate in civil society.

8. What are the various circumstances that make young heterosexual women so susceptible to HIV-AIDS in sub-Saharan Africa? How does the social and economic status of women (versus men) influence the spread of HIV in women?

9. Looking at the various African locations where food is in short supply, explain the main factors that account for this situation. How would you respond to those who say that drought and climate are the principal causes of famine?

10. If you were designated to make a speech to the Rotary Club in your town on hopeful signs out of Africa, what would you include in your talk? Which pictures would you show?

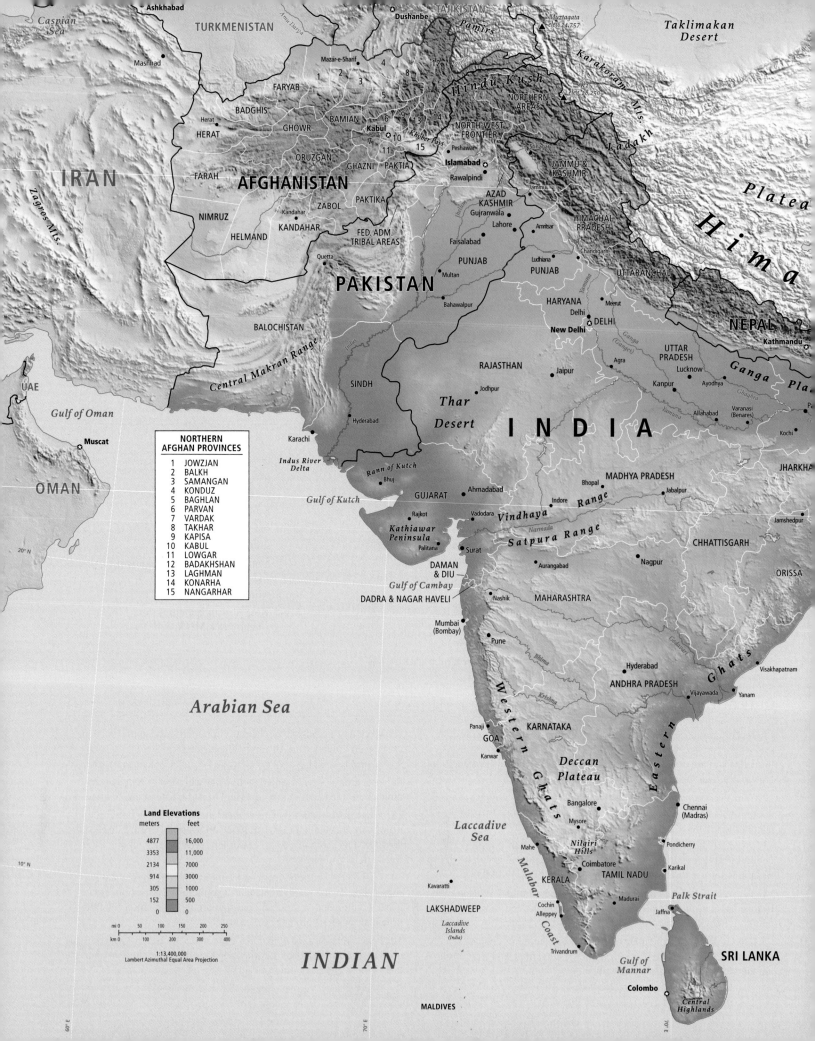

## CHAPTER 8

# SOUTH ASIA

**Global Patterns, Local Lives** It's April 16, 2006. Narendra Modi, the governor of the state of Gujarat in India, draped in garlands by well-wishers, is embarking on a hunger strike. He is protesting a decision by the national government in New Delhi to limit the height of the Sardar Sarovar Dam on the Narmada River in the neighboring state of Madhya Pradesh. Damming the river has become the centerpiece of Gujarat's efforts to deal with the periodic droughts that affect as many as 50 million of its citizens.

Far away, in New Delhi, Medha Patkar, the leader of the "Save the Narmada" movement, is in day 18 of her hunger strike in protest against the same dam, which threatens the livelihoods of 320,000 farmers and fishers who are to be relocated to make way for the rising waters. Although Indian law requires resettling the hundreds of thousands of people displaced by the Sardar Sarovar Dam, so far only a fraction of them have received usable land. As an act of protest, some of the farmers refused to move, even as the rising waters of the reservoir consumed their homes. Indian police forcibly removed the farmers and relocated them to refugee camps. Many have since relocated to crowded urban slums.

Environmental problems have also been at the heart of the Sardar Sarovar controversy, which has been going on since the project began in 1961 (Figure 8.2). The Narmada River is already in ecological trouble. Once a placid, slow-moving, easily crossed river—and one of India's most sacred—its natural cycles have already been disturbed by previous dams, causing massive die-offs of aquatic life and high unemployment among fishers. Furthermore, despite huge investments of Gujarati public funds, nearly 80 percent of the areas most vulnerable to drought in that state will get no water from the Sardar Sarovar project. The World Bank withdrew its funding of the dam in the mid-1990s due to concerns that the economic benefits would be negated by the environmental costs. Ecologists say that far less costly water management strategies, such as rainwater harvesting, groundwater recharge, and watershed management, would be better options for the drought-stricken farmers of Gujarat.

Less than a day after beginning his hunger strike, Gujarat's governor ended it, because the Indian Supreme Court ruled that the Sardar Sarovar dam could be raised higher. The following day, Medha Patkar ended her fast as well, because in the same decision the Supreme Court ruled that all people displaced by the dam must be adequately relocated. Furthermore, the decision confirmed that human impact studies are required for dam

**FIGURE 8.1** Regional map of South Asia.

413

**FIGURE 8.2** The Sardar Sarovar Dam on the Narmada River, south of the Indian city of Ahmadabad. This dam is held up as a model for the development of modern India, but is also reviled as a symbol of how the rights of the poor are trampled as they are forced to relocate with little or no compensation. [Dave Amit/Landov/Reuters.]

projects (there are currently over 700 such projects in India); no such study was provided for the Sardar Sarovar Dam.

*Adapted from India eNews at its Best, April 16, 2006, http://www .indiaenews.com/politics/20060416/4531.htm; "Dam protester's health gets worse," BBC News Online, April 4, 2006, http://news.bbc.co .uk/2/hi/south_asia/4876110.stm; Rahul Kumar, "Medha Patkar ends fast after court order on rehabilitation," One World South Asia, http: //southasia.oneworld.net/article/view/131077/1/2220; "Narmada's revenge," Frontline, India's National Magazine 22(9) (2005), http: //www.flonnet.com/fl2209/stories/20050506002913300.htm; "Water harvesting, addressing the problem of drinking water," http: //www.narmada.org/ALTERNATIVES/water.harvesting.html.* ■

The recent history of water management in the Narmada River valley highlights some key issues now facing South Asia and other regions with developing economies. Across the world, large, often poor, populations are depending on increasingly overtaxed environments, and their expectations for even a marginally better standard of living nearly always require more water and energy. The strategies that emerge to meet these urgent needs often make neither economic nor environmental sense, but are driven to completion by political and social pressures. In this Indian case, the wealthier, more numerous, and more politically influential farmers of Gujarat have tipped the scales in favor of a dam project that is almost certain to produce more problems than it solves—certainly for those who are being displaced. But the same can be said for dam projects discussed elsewhere in this book (Figure 8.3). In many developing countries, dam projects have been protested by

those who stand to lose their homes and livelihoods to the schemes, as well as by ecologists and economists who say the costs outweigh the benefits. Yet even as international lending agencies have backed away from such projects, national, state, and local governments, sometimes aided by private investors (as in Gujarat), have stepped in to fund them. Rarely have the developers asked, "Development for whom?"

South Asia is a relatively easy region to define. It is surrounded by the Indian Ocean on the east, south, and west and by the world's highest mountains to the north. Physically, South Asia is far smaller than Africa: it would fit into the African continent five times. Yet, with nearly 1.64 billion people, it has nearly twice Africa's population. The countries that make up the region are Afghanistan and Pakistan in the northwest; the Himalayan states of Nepal and Bhutan; Bangladesh in the northeast; India; the island country of Sri Lanka; and several groups of islands in the Arabian Sea and the Bay of Bengal (see Figure 8.1). Because its clear physical boundaries set it apart from the rest of the Asian continent, the term **subcontinent** is often used to refer to the entire Indian peninsula, including eastern Pakistan and Bangladesh.

Although it is complex culturally and politically, South Asia has a number of unifying features: the village is one; the common experience of British colonization is another. The hunger strikes highlighted by the story of the Sardar Sarovar Dam reveal yet another commonality: nonviolence as a strategy for gaining political ends. Nonviolence has ancient roots in South Asia, and during India's struggle for independence from Britain, one of the country's most revered leaders, Mohandas Gandhi, developed it into a powerful and now widely used political tactic.

## Themes to Explore in South Asia

Here are some themes to follow as you read this chapter:

1. **The ancient and layered pattern of cultural influences.** South Asia has experienced multiple waves of cultural and religious influences since prehistoric times. The most recent is globalization.

2. **The importance of village life.** South Asia has many large cities, but 70 percent of the region's people live in hundreds of thousands of villages, and rural modes of spatial organization and interaction persist even in the cities.

3. **The lingering influence of British colonization.** Although British colonial rule ended more than 50 years ago, it has left its mark on the landscape and culture of the region. South Asians are struggling to define national identities that can accommodate lingering British and other outside influences while retaining strong South Asian traditions and moral values.

4. **Extremes between rich and poor.** Disparities in wealth and startling contrasts between traditional and technically advanced ways of life are commonplace across the region and pose dangers to the continuation of democracy in these pluralistic societies.

5. **Continuing population growth.** South Asia now has a population higher than that of China. High population growth puts stress

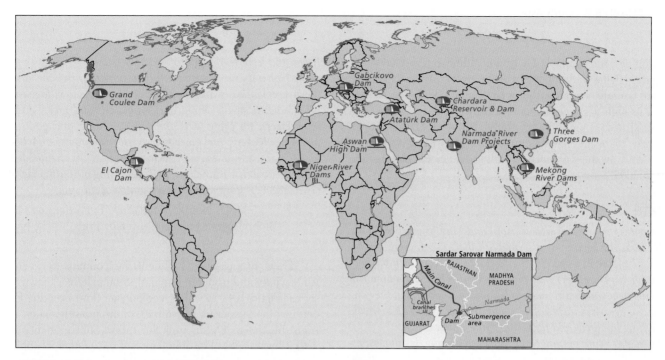

**FIGURE 8.3** Major dams around the world. This map shows the location of dams featured or mentioned in this book. All were built with the intention of improving human circumstances, yet all have also resulted in considerable environmental damage and dislocation and hardship for people.

on resources, and the region faces a number of challenges in providing its people with the basics of life.

6. Environmental concerns and conflicts. As in other densely populated regions experiencing economic growth and rising consumption, environmental issues are foremost on people's minds. In addition to the increasing problems of land degradation, shortages of water, and pollution, a number of conflicts have emerged over how resources are to be used and the effects of large-scale development on local people.

## I THE GEOGRAPHIC SETTING

## Terms to Be Aware Of

South Asians are reclaiming the original names of many of their important places, which had been changed under British colonial rule. In the last decade, Indians, especially, have begun to use the old names. The city of Bombay, for example, is now officially *Mumbai*, Madras is *Chennai*, Calcutta is *Kolkata*, Benares is *Varanasi*, and the Ganges River is the *Ganga River.*

## PHYSICAL PATTERNS

Many of the landforms, and even the climates, of South Asia are the result of the huge tectonic forces that positioned the Indian subcontinent along the southern edge of the Eurasian continent, where the warm Indian Ocean surrounds it and the massive mountains of the Himalayas shield it from cold air flows from the north.

## Landforms

The Indian subcontinent and the territory surrounding it contain some of the most spectacular landforms on earth and illustrate

dramatically what can happen when two tectonic plates collide. About 180 million years ago, the Indian-Australian Plate, which carries India, broke free from the eastern edge of the African continent and drifted to the northeast. As it began to collide with the Eurasian Plate about 60 million years ago, India became a giant peninsula jutting into the Indian Ocean. As the relentless pushing from the south continued, both the leading (northern) edge of India and the southern edge of Eurasia crumpled and buckled, forming the world's highest mountains—the Himalayas, which rise more than 28,000 feet (8500 meters)—as well as other very high mountain ranges to the east and west. The continuous compression also lifted up the Plateau of Tibet, which rose up behind the Himalayas to an elevation of more than 15,000 feet (4500 meters) in some places.

South and southwest of the Himalayas are the Indus and Ganga river basins, also called the Indo-Gangetic Plain. Still farther south is the Deccan Plateau, an area of modest uplands 1000–2000 feet (300–600 meters) in elevation, interspersed with river valleys. This upland region is bounded on the east and west by two moderately high mountain ranges, the Eastern and Western Ghats. These mountains descend to long but narrow coastlines interrupted by extensive river deltas and floodplains. The river valleys and coastal zones are densely occupied; the uplands only slightly less so. Because of its high degree of tectonic activity and deep crustal fractures, South Asia is prone to devastating earthquakes, such as the magnitude 7.7 quake that shook the state of Gujarat in western India in 2001, or the 7.6 quake that hit the India–Pakistan border region in 2005.

## Climate

*The end of the dry [winter] season [April and May] is cruel in South Asia. It marks the beginning of a brief lull that is soon overtaken by the annual monsoon rains. In the lowlands of eastern India and Bangladesh, temperatures in the shade are routinely above a hundred degrees; the heat causes dirt roads to become so parched that they are soon covered in several inches of loose dirt and sand. Tornadoes wreak havoc, killing hundreds and flattening entire villages....*

*This is also a time of hunger, as with each passing day thousands of rural families consume the last of their household stock of grain from the previous harvest and join the millions of others who must buy their food. Each new entrant into the market nudges the price of grain up a little more, pushing millions from two meals a day to one.*

Alex Counts, *Give Us Credit*
(New York: Times Books, 1996), p. 69

*From mid-June to the end of October [summer] is the time of the river. Not only are the rivers full to bursting, but the rains pour down so relentlessly and the clouds are so close to village roofs that all the earth smells damp and mildewed, and green and yellow moss creeps up every wall and tree....Cattle and goats become aquatic, chickens are placed in baskets on roofs, and boats are loaded with valuables and tied to houses.... As the floods rise villages become tiny islands, ... self-sustaining outpost[s] cut off from civilization ... for most of three months of the year.*

James Novak, *Bangladesh: Reflections on the Water*
(Bloomington: Indiana University Press, 1993), pp. 24–25

These two passages highlight the contrasts between South Asia's winter and summer dominant wind patterns, known as **monsoons** (Figure 8.4). In winter, cool, dry air flows from the Eurasian continent to the ocean; in summer, warm, moisture-laden air flows from the Indian Ocean over the Indian subcontinent, bringing with it heavy rains. The abundance of this rainfall is amplified by the intertropical convergence zone (ITCZ), where air masses moving south from the Northern Hemisphere converge near the equator with those moving north from the Southern Hemisphere and produce copious precipitation as they rise and cool. As described in Chapter 7 (see pages 356–357), the ITCZ shifts north and south seasonally. Climatologists now think that the intense rains of South

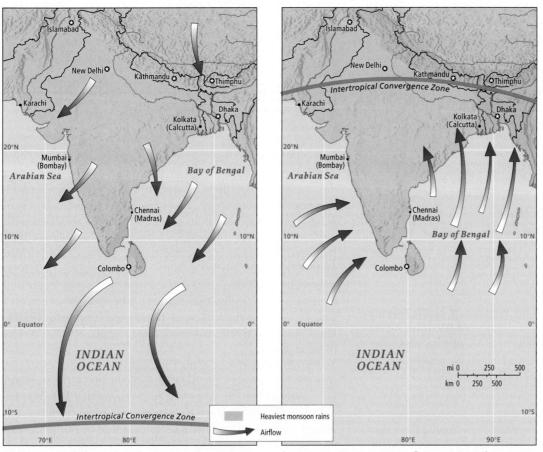

**FIGURE 8.4** Winter and summer monsoons in South Asia. **(a)** In the winter, cool, dry air blows from the Eurasian continent south across India toward the ITCZ, which in winter lies far to the south. **(b)** In the summer, the ITCZ moves north across India, picking up huge amounts of moisture from the ocean, which are then deposited over India.

**(a) Winter.** Cool, dry air flows from Asian subcontinent.

**(b) Summer.** Warm, moist air flows to Asian subcontinent.

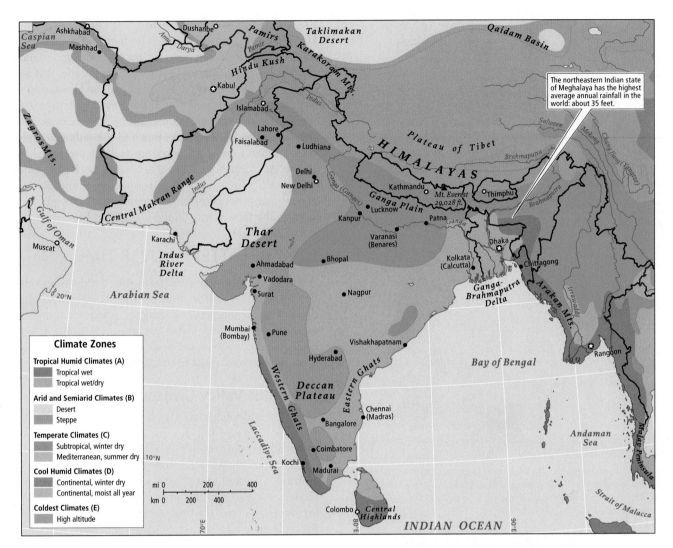

The northeastern Indian state of Meghalaya has the highest average annual rainfall in the world: about 35 feet.

**Climate Zones**

**Tropical Humid Climates (A)**
Tropical wet
Tropical wet/dry

**Arid and Semiarid Climates (B)**
Desert
Steppe

**Temperate Climates (C)**
Subtropical, winter dry
Mediterranean, summer dry

**Cool Humid Climates (D)**
Continental, winter dry
Continental, moist all year

**Coldest Climates (E)**
High altitude

**FIGURE 8.5** Climates of South Asia.

Asia's summer monsoon are due to the ITCZ being sucked onto the land by a vacuum created when huge volumes of air over the Eurasian landmass heat up in the summer and rise into the upper atmosphere.

The monsoons are major influences on South Asia's climate. In early June, the warm, moist ITCZ air of the summer monsoon first reaches the mountainous Western Ghats. The rising air mass cools as it moves over the mountains, releasing rain that nurtures dense tropical rain forests and tropical crops in the central uplands. Once on the other side of India, the monsoon gathers additional moisture and power in its northward sweep up the Bay of Bengal, sometimes turning into tropical cyclones. As the monsoon system reaches the hot plains of West Bengal and Bangladesh in late June, columns of warm rising air create massive, thunderous cumulonimbus clouds that drench the parched countryside. Precipitation is especially intense in the foothills of the Himalayas, which hold the world record for annual rainfall—about 35 feet, even though there is no rain at all for half the year. Monsoon rains run in a band parallel to the Himalayas that reaches across northern

India all the way to northern Pakistan by July. These patterns of rainfall are reflected in the varying climate zones (Figure 8.5) and agricultural zones (see Figure 8.32 on page 439) of South Asia.

By November, the cooling Eurasian landmass sends cooler, drier air over South Asia. This heavier air from the north pushes the warm, wet air back south to the Indian Ocean. Although very little rain falls in most of the region during this winter monsoon, parts of southeastern India and Sri Lanka receive winter rains as the ITCZ drops moisture picked up on its now southward pass over the Bay of Bengal.

The monsoon rains deposit large amounts of moisture over the Himalayas, much of it in the form of snow and ice. Meltwater feeds the headwaters of the three river systems that figure prominently in the region: the Indus, the Ganga, and the Brahmaputra. All three rivers begin within 100 miles (160 kilometers) of one another in the Himalayan highlands near the Tibet–Nepal–India borders (see Figure 8.1). The Ganga flows generally south through the mountains and then east across the Ganga Plain to the Bay of Bengal. The Brahmaputra flows first east, then south, and then west to join

(a) **Premonsoon Stage:** The river flows in multiple channels across the flat plain.

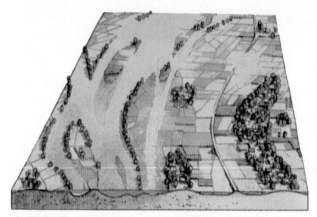

(b) **Peak Flood Stage:** During peak flood stage, the great volume of water overflows the banks and spreads across fields, towns, and roads. It carves new channels, leaving some places cut off from the mainland.

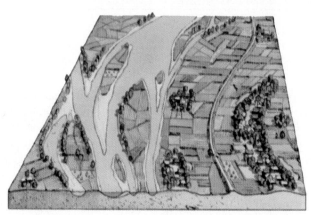

(c) **Postmonsoon Stage:** The river returns to its banks, but some of the new channels persist, changing the lay of the land. As the river recedes, it leaves behind silt and algae that nourish the soil. New ponds and lakes form and fill with fish.

**FIGURE 8.6** The Brahmaputra River in Bangladesh at various seasonal stages. People who live along the river have learned to adapt their farms to a changing landscape, and along much of the river farmers are able to produce rice and vegetables nearly year-round. [Adapted from *National Geographic* (June 1993): 125.]

the Ganga in forming the giant Ganga-Brahmaputra delta, also known as the Delta of Bengal. The Indus, in the far west, takes a southwesterly course across arid Pakistan and empties into the Arabian Sea. These rivers, and many of the tributaries that feed them, are actively wearing down the surface of the Himalayas; thus they carry an enormous load of sediment, especially during the rainy season, when their volume increases. Their velocity slows when they reach the lowlands, and much of the sediment settles out as silt. It is then repeatedly picked up and redeposited by successive floods. As illustrated in the diagram of the Brahmaputra River (Figure 8.6), the seasonally replenished silt nourishes much of the agricultural production in the densely occupied plains of Bangladesh. The same is true on the Ganga and Indus plains.

## HUMAN PATTERNS OVER TIME

A variety of groups have migrated into South Asia over the millennia, many of them as invaders who conquered peoples already there. The merging and blending of different cultural, religious, social, and political elements has given South Asia a level of cultural diversity that few other world regions can equal, and has made it one of the most contentious places in the world.

### The Indus Valley Civilization

There are indications of early humans in South Asia as long as 200,000 years ago, but the first evidence of modern humans is about 38,000 years old. The first substantial settled agricultural communities, known as the **Indus Valley civilization** (or **Harappa culture**), appeared about 4500 years ago along the Indus River in modern-day Pakistan and northwest India. The architecture and urban design of this very early civilization were quite advanced for the time: there were multistory homes with piped water and sewage disposal; planned towns with wide, tree-lined boulevards laid out in a grid; and a high degree of consistency in building materials over a region that covered more than 1000 square miles (2500 square kilometers). Fine beadwork and jewelry, evidence of a trade network that extended to Mesopotamia and eastern Africa, continue to fascinate students of this region, as does a writing system that has yet to be deciphered. Much of the Indus Valley civilization's agricultural system survives to this day, including techniques for storing monsoon rainfall to be used for irrigation in dry times; methods for cultivating wheat, barley, oilseeds, cotton, and other crops adapted to arid conditions; and the use of wooden plows drawn by oxen. Vestiges of its language, and possibly biological traits as well, survive today among the Dravidian peoples of southern India, who originally migrated from the Indus region.

The reasons for the decline of the Indus Valley civilization after about 800 years (3700 years ago) are debated. Some scholars believe that complex geologic and ecological changes brought about a gradual demise; others argue that foreign invaders brought a swift collapse. In any case, many aspects of Harappa culture blended with subsequent foreign influences to form the foundation of modern South Asian religious beliefs, social organization, linguistic diversity, and cultural traditions.

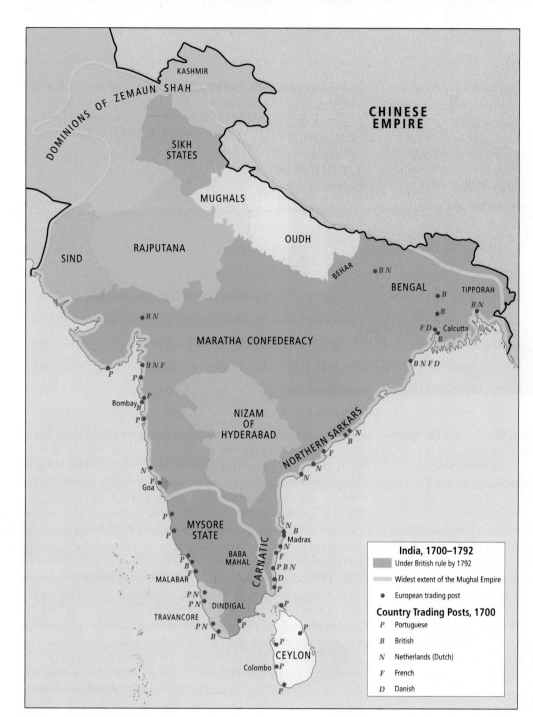

**FIGURE 8.7** Precolonial South Asia. By 1700, several European nations had established trading posts along the coast of India and Ceylon. After the death of the Mughal ruler Aurangzeb in 1707, the ability of the Mughals to assert strong central rule throughout South Asia declined. A number of emergent regional states competed with one another for territory and power. Among the strongest was the Maratha Confederacy, composed of a number of small states dominated by the Maratha peoples (known for their martial skills). Weakness at the center and constant rivalry paved the way for British conquest by the end of the eighteenth century. [Adapted from William R. Shepherd, *The Historical Atlas* (New York: Henry Holt, 1923–1926), p. 137; and Gordon Johnson, *Cultural Atlas of India* (New York: Facts on File, 1996), p. 111.]

Within the map legend:

**India, 1700–1792**

Under British rule by 1792

Widest extent of the Mughal Empire

● European trading post

**Country Trading Posts, 1700**

P   Portuguese

B   British

N   Netherlands (Dutch)

F   French

D   Danish

## A Series of Invasions

The first recorded invaders of South Asia, the **Aryans,** moved into the rich Indus Valley and Punjab from Central and Southwest Asia 3500 years ago. Many scholars believe that the Aryans, in conjunction with the indigenous Harappa culture and other indigenous cultures, instituted some of the early elements of classical Hinduism, the major religion of India. One of those elements was the remarkably influential caste system, which divides society into hereditary hierarchical categories.

The ability to accommodate immense diversity as new influences blended with those already present is a distinctive feature of the cultural tradition of South Asia. In addition to the Aryans, other invaders from greater Central Asia included the Persians, the armies of the Greek general Alexander the Great, and numerous Turkic and Mongolian peoples. Starting about 1000 years ago, Arab traders and religious mystics came by sea to the coasts of southwestern India and Sri Lanka, introducing Islam into the region.

The invasion in 1526 by the **Mughals,** a group of Turkic Persian people from Central Asia, intensified the spread of Islam. The Mughals reached the height of their power and influence in the seventeenth century, controlling the north-central plains of South Asia. After the last great Mughal ruler (Aurangzeb) died in 1707, a number of regional states and kingdoms rose in power and competed with one another (Figure 8.7), creating an opening for yet another invasion. Several European trading companies competed to gain a

**FIGURE 8.8** The great fortress at Agra. Built by the Mughals in 1565–1571, the fortress has walls 72 feet (22 meters) high enclosing an area about 1.5 miles (2.4 kilometers) in circumference. [J. H. C. Wilson/Robert Harding Picture Library.]

foothold in the region. Of these companies, Britain's was the most successful. By 1857, the British East India Company, acting as an extension of the British government, had subdued most of the regional and competing European powers and supplanted the last of the Mughals through manipulation of political rivalries, strategic trading alliances, and military conquest.

One legacy of the Mughals is the more than 420 million Muslims now living in South Asia. The Mughals also left a unique heritage of architecture, art, and literature that includes the Taj Mahal, the fortress at Agra (Figure 8.8), miniature painting, and the tradition of lyric poetry. The Mughals helped to produce the **Hindi** language, which became the language of trade of the northern subcontinent and is still used by more than 400 million people. In the northern reaches of South Asia, the Islamic ideals of the Mughals made a lasting mark on aesthetics, social interactions, gender roles, and religious architecture.

## The Legacies of Colonial Rule

Britain, the most recent influential invader of South Asia, was already a strong trading and military presence in the 1750s, when the British East India Company established its control first over a number of coastal trading centers and then over interior regions. The British controlled most of South Asia from the 1830s through 1947 (Figure 8.9), profoundly influencing the region politically, socially, and economically. Even areas not directly ruled by the British felt the influence of colonialism. Afghanistan repelled British attempts at military conquest, but the British continued to meddle there, wishing to establish Afghanistan as a "buffer state" against expanding Russian interests. Nepal remained only nominally independent during the colonial period, and Bhutan became a protectorate of the British Indian government.

***Economic Influence.*** British colonial policy in South Asia, as elsewhere, was to use the region's resources primarily for the benefit of

Britain. This policy often resulted in detrimental effects on South Asia. A typical example was the fate of the textile industry in Bengal (modern-day Bangladesh and the Indian state of West Bengal). Bengali weavers, long known for their high-quality muslin cotton cloth, initially benefited from the greater access British traders gave them to overseas markets in Asia, in the Americas, and on the European continent, as well as in Britain itself. The high quality of Bengali cloth and the great demand for it reflected the advanced manufacturing economy of South Asia, which in 1750 produced 12 to 14 times more cotton cloth than Britain alone and more than all of Europe combined. However, as Britain's own highly mechanized textile industry developed during the second half of the eighteenth century, cheaper British cloth replaced Bengali muslin, first in world markets and eventually throughout South Asia. The British East India Company further supplanted indigenous textile manufacturing by severely punishing those who continued to run their own looms. Thus, as one British colonial official put it, while the mills of Yorkshire prospered, "the bones of Bengali weavers bleached the plains of India."

The British-induced reversal of India's fortunes in the eighteenth century profoundly affected its people's lives in the nineteenth century. Many people who were pushed out of their traditional livelihood in textile manufacturing were compelled to find work as landless laborers in an economy that already had an abundance of agricultural labor. Others migrated to emerging urban centers. Increasingly, bandits roamed the countryside, while small landowners lost their land to large landowners because they could not pay or bribe their way out of the land taxes instituted by the British East India Company in some parts of India. In the 1830s, a drought exacerbated the inequities of colonial rule, and more than 10 million people starved to death. It was during these trying times that South Asian workers were pressed into joining the stream of indentured laborers emigrating to other British colonies in the Americas, Africa, Asia, and the Pacific (Box 8.1).

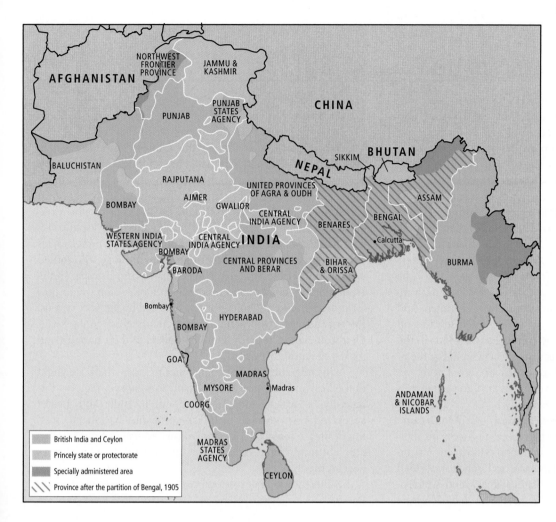

**FIGURE 8.9** British Indian Empire, 1860–1920. After winning control of much of South Asia, Britain controlled lands from Baluchistan to Burma, including Ceylon and the islands between India and Burma. [Adapted from Gordon Johnson, *Cultural Atlas of India* (New York: Facts on File, 1996), p. 158.]

Map legend:
- British India and Ceylon
- Princely state or protectorate
- Specially administered area
- Province after the partition of Bengal, 1905

What little economic development did take place in South Asia was allowed only if it benefited Britain. The production of tropical agricultural raw materials, such as cotton, jute (a fiber used in making gunnysacks and rope), tea, sugar, and indigo (a blue dye) was encouraged so that it could supply Britain's other colonies and Britain itself. Industrial development, which might have competed with Britain's own industries, was discouraged.

The economic historian Dietmar Rothermund argues that British colonial economic policy encouraged population growth. Had farmers been able to become industrial workers and business owners, they might have favored smaller families, like their industrial counterparts elsewhere, rather than producing as many children as possible to supply agricultural labor. Nonetheless, the British Empire did bring some benefits. Trade with the rest of the empire brought prosperity to a few areas, especially the large British-built cities on the coast, such as Bombay (now Mumbai), Calcutta (Kolkata), and Madras (Chennai). It built a railroad system that boosted trade within South Asia and greatly eased the burden of personal transport. In addition, English became a common language for South Asians of widely differing backgrounds, assisting both trade and cross-cultural understanding.

South Asian governments retain institutions put in place by the British to administer their vast empire, and these governments have also inherited many of the shortcomings of their colonial forebears, such as highly bureaucratic procedures, a resistance to change, and a tendency to remain aloof from the people they govern. Nonetheless, these governments have proved functional in most cases. In particular, democratic government, though it was not instituted on a large scale until the final days of the empire, has given people an outlet for voicing their concerns and has enabled many peaceful transitions of elected governments since 1947.

***Independence and Partition.*** The early years of the British Indian empire were characterized by great social unrest; nationalist movements sprang up across much of India, but they were always checked. Agitation against British rule continued through the nineteenth century, and political protests were common by the 1920s. Mohandas Gandhi, who as a young merchant working in South Africa had protested the treatment of Indians and Africans by the white South African government, emerged as a leader when he brought his tactics of **civil disobedience** back to India. Gathering a large group of quietly peaceful protesters, he would notify the government that the group was about to break a discriminatory colonial law. If the authorities ignored the act, the demonstrators would have made their point and the law would be rendered moot. On the other hand, if the government used force against the peaceful demonstrators, they would lose the respect of the masses. Throughout the 1930s and 1940s, this technique was used to slowly but surely undermine British authority across South Asia.

# BOX 8.1 | AT THE GLOBAL SCALE   The Indian Diaspora

From the island of Bali in Indonesia to Queens in New York City, one can hear South Asian temple bells ringing and catch the pungent fragrance of spices being ground for curry and kabob dishes. The people maintaining these customs belong to what is called the **Indian diaspora,** the set of all people of South Asian origin living (and often born) abroad (Figure 8.10). (The term *Indian diaspora* is used because for many centuries India was the popular name for the whole region of South Asia.)

Archaeological evidence shows that trade between the people of the Indus Valley (in Pakistan) and eastern Africa and Mesopotamia occurred as early as 4000 years ago. Indian traders brought Hindu culture to Bali (in Southeast Asia) over 2000 years ago. Under British colonial rule, from about 1850 to 1920, hundreds of thousands of South Asian people were recruited, and sometimes even kidnapped, to labor as indentured servants on British plantations in such places as Trinidad and Guyana in the Caribbean, Fiji in the Pacific Ocean, and Malaysia in Southeast Asia. Others built railroads in East Africa or served with the British army in other parts of the world. Many of these migrants remained in their new homes permanently.

After World War I, migrants from the west coast of India came to the coast of East Africa, where they became successful merchants and exporters. In the 1970s, in response to violent African nationalist movements, they left places such as Uganda and settled in the United Kingdom, the United States, Canada, and Australia. The films *Mississippi Masala* (1992) and *Miss India Georgia*

(1998) illustrate South Asians' struggles to adapt to life in the United States; *East Is East* (1999) and *Bend It Like Beckham* (2000) portray South Asian immigrant life in England.

When the oil-rich nations of the Arabian Peninsula began to flourish in the 1970s, they attracted tens of thousands of Muslim workers from India, Pakistan, Bangladesh, and Sri Lanka. Others found jobs in Southeast Asian factories and plantations and in England's factory towns, or as taxi drivers and construction workers in Canadian and U.S. cities. Probably the most socially visible members of the South Asian diaspora are those professionals in medicine, science, and information technology who settled in North America, Australia, New Zealand, and Europe. In the United States, citizens and residents of South Asian origin now form the wealthiest ethnic group, as defined by the U.S. census. Recent migrants, who maintain particularly close contact with their homelands, remit millions of dollars to their families and make regular return visits.

This migration stream contributes to brain drain: the flight of the best and brightest South Asians to wealthier regions. But recently, increasing foreign investment in India and greater access to foreign markets by Indian companies have been creating desirable jobs in India, and young, educated South Asians are returning home, where they are contributing to India's development.

*Source:* Adapted from material contributed by geographer Carolyn Prorok.

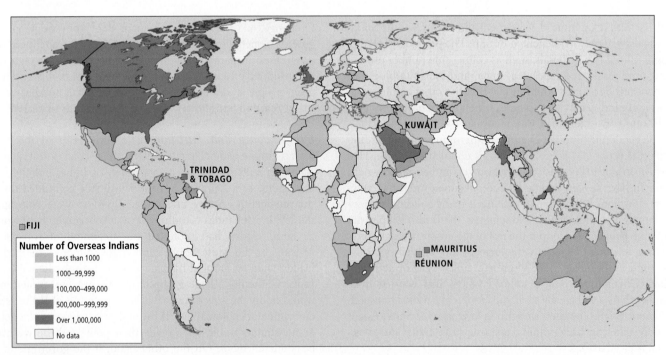

**FIGURE 8.10** Indian diaspora communities around the world. This map makes clear that people of South Asian origin now live virtually everywhere on earth. They engage in many occupations, but are found particularly in the professions, technology development, commerce, and agriculture. [Data from *Report of the High Level Committee on The Indian Diaspora,* at http://indiandiaspora.nic.in/contents.htm, specifically "Estimated size of overseas community: Countrywise," at http://indiandiaspora.nic.in /diasporapdf/part1-est.pdf.]

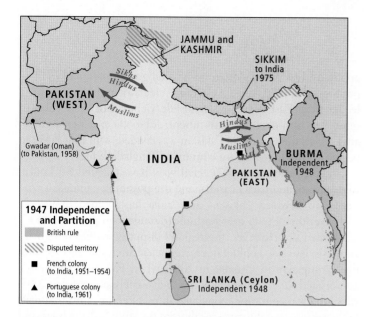

**FIGURE 8.11** Independence and Partition. India became independent of Britain in 1947, and by 1948, the old territory of British India was partitioned into the independent states of India and East and West Pakistan. The Jammu and Kashmir region was contested space, and remains so today. Sikkim went to India, and both Burma and Sri Lanka became independent. Following additional civil strife, East Pakistan became the independent country of Bangladesh in 1971. [Adapted from *National Geographic* (May 1997): 18.]

In 1947, independence was granted to British India, which was divided into two independent countries: predominantly Hindu India and Muslim Pakistan (Figure 8.11) (Afghanistan, Bhutan, and Nepal were never officially British colonies; Ceylon [now Sri Lanka] became independent in 1948). The **Partition,** which Gandhi greatly lamented, was perhaps the most enduring and damaging outcome of colonial rule. The idea of two nations was first suggested by some Muslim political leaders who were con-

cerned about the fate of a minority Muslim population in a united India with a Hindu majority. Although it was more a political strategy to ensure the interests of Muslims than a real political demand, the idea took hold and became part of the independence agreement between the British and the Indian National Congress (India's principal nationalist party). It was decided that northwestern and northeastern India, where the population was predominantly Muslim, would become a single country consisting of two parts, known as West and East Pakistan, separated by India. Although both India and Pakistan maintained secular constitutions, with no official religious affiliation, the general understanding was that Pakistan would have a Muslim majority and India a Hindu majority. Fearing that they would be persecuted if they did not move, some 4 million Hindus and Sikhs migrated from their ancestral homes in Pakistan to India; similarly, another 4 million Muslims left their homes in India for Pakistan (Figure 8.12). In the process, families and communities were divided, looting and rape were widespread, and more than a million people were killed in innumerable skirmishes.

Many historians argue that the Partition could have been avoided had it not been for the "divide-and-rule" tactics the British used throughout the colonial era to heighten tensions between South Asian Muslims and Hindus, thus creating a role for themselves as indispensable and benevolent mediators. For example, British local administrators commonly favored the interests of minority communities in order to weaken the power of majorities that could have threatened British authority. The legacy of these "divide-and-rule" tactics includes not only the Partition, but also the repeated wars and skirmishes, strained relations, and ongoing arms race between India and Pakistan.

***Since Independence.*** In the more than 60 years since the departure of the British, South Asians have experienced both progress

**FIGURE 8.12** A great migration. As a result of the Partition, a great two-way migration took place—from India to Pakistan and from Pakistan to India—as both Hindus and Muslims panicked, fearing persecution. Violence did indeed occur, largely because of a vacuum of leadership, as the British withdrew. Over a million people died in skirmishes between Hindus, Muslims, and Sikhs in the aftermath of the Partition. [© 1947 Margaret Bourke-White/Time & Life Pictures/ Getty.]

and setbacks. Democracy has expanded steadily, albeit somewhat slowly. India has maintained its status as the world's most populous democracy and is gradually dismantling age-old traditions that hold back poor, low-caste Hindus, women, and other disadvantaged groups. Agricultural advances have brought relative prosperity to some rural areas. Industry and (especially) services now constitute a far larger share of the GDP than agriculture, and the information technology (IT) sector is growing fast. Nonetheless, as of 2005, the South Asian countries had a collective annual GDP per capita of only about U.S.$2000, adjusted for PPP (only sub-Saharan Africa's was lower), and poverty persists for the majority in both rural and urban areas throughout the region.

Although threats from outside the region have diminished, the danger of violence originating from within has steadily increased. After 1947, East and West Pakistan, despite their common religion, struggled to overcome their stark cultural differences, economic disparity, and geographic separation. East Pakistan was the more populous, but West Pakistan dominated government and military positions. In 1970, the first general election in 12 years gave a leader from East Pakistan a landslide victory, but West Pakistan refused to turn over the government. A bloody civil war resulted, after which the country of Pakistan was divided, with East Pakistan becoming the independent country of Bangladesh in 1971.

In the years since 1971, civil wars have plagued Sri Lanka, Afghanistan, and parts of India, and the possibility of nuclear confrontation between Pakistan and India has loomed ominously whenever the dispute over Kashmir (discussed on pages 446–447) has heated up. However, the region's disputing parties have been known to lay aside conflict in emergencies. Such was the case after the earthquakes in Gujarat in 2001, when Pakistan sent much-appreciated aid to India, and on the India–Pakistan border in 2005, when the two countries cooperated in relief efforts. Future threats to the region's stability may result from the stresses that a huge and still growing population is placing on the region's already precari-

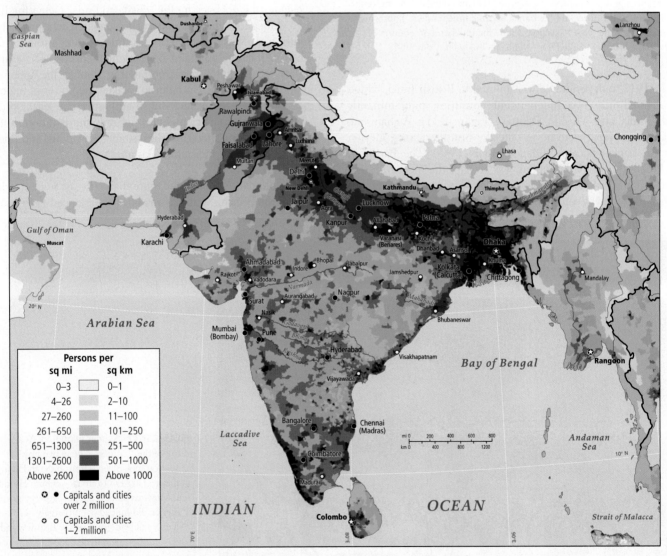

**FIGURE 8.13** Population density in South Asia. [Data courtesy of Deborah Balk, Gregory Yetman, et al., Center for International Earth Science Information Network, Columbia University; at http://www.ciesin.columbia.edu.]

**FIGURE 8.14** A bicycle rickshaw and driver in Delhi. [Lindsay Hebberd/ Woodfin Camp & Associates.]

ous natural environment and from the uneven distribution of resources and economic prosperity.

## POPULATION PATTERNS

South Asia is one of the most densely populated regions in the world (Figure 8.13). In cities such as Peshawar, New Delhi, and Mumbai, commuters cling to the outsides of packed buses as they travel lopsidedly through the streets, and dense throngs pack urban sidewalks. Although at most 30 percent of the population is urban, South Asia has several of the world's largest cities: Mumbai with 19 million, Kolkata with 15 million, Delhi with 16 million, Dhaka with 13 million. People come to South Asian cities for all the reasons that cities attract people everywhere: business opportunities, better jobs, education, or training. Some seek anonymity and individualism—perhaps even a rise in caste status—that are not possible in close-knit rural communities. Many are pushed into urban migration by agricultural modernization that reduces rural incomes and employment opportunities or by population growth that makes access to land and resources ever more difficult. Some are refugees who have left drought-stricken or flooded countrysides.

One consequence of South Asia's rapid urban growth is that employment and availability of affordable housing have not kept pace. Thus many people simply live on the streets. In a National Public Radio interview several years ago, an Indian journalist asked a bicycle rickshaw driver (Figure 8.14) about himself as he peddled her through Delhi. He replied that his belongings—a second set of clothes, a bowl, and a sleeping mat—were under the seat where she was sitting. He had come to Delhi from the countryside 14 years before, and he had never found a home. He knew virtually no one, he had few friends and no family, and no one had ever inquired about him before. He worked virtually around the clock and slept here and there for 2 hours at a time.

*Source: Gagan Gill, "Weekend Edition," National Public Radio (August 16, 1997).* ∎

## Population Growth Factors in South Asia

South Asia's rapid population growth will continue to strain efforts to improve quality of life in the region. South Asia already has more people (1.44 billion) than China (1.33 billion). By 2020, India alone is expected to overtake China, whose rate of natural increase (0.6 percent) is less than half of India's (1.7 percent). Each year, India adds nearly 18 million people to its ranks. To accommodate these new Indians adequately, each year the country would need to build 127,000 new village schools, hire 373,000 new schoolteachers, build 2.5 million new homes (at 7 people per home), create 4 million new jobs, and produce 180 million new bushels of grain and vegetables. Thus far, food production in India has kept pace with population growth, but its ability to continue doing so is by no means secure.

South Asia has been trying to reduce birth rates since 1952. India spends over a billion dollars a year on population control programs. Unlike many other countries, which rely on financial aid from developed countries to fund such programs, India pays for nearly all of this effort on its own. Fertility rates have indeed declined significantly in India and Bangladesh, and especially in Sri Lanka, but much less in Pakistan and Nepal (Figure 8.15). What accounts for these patterns? The answers are similar to those we have given in earlier chapters.

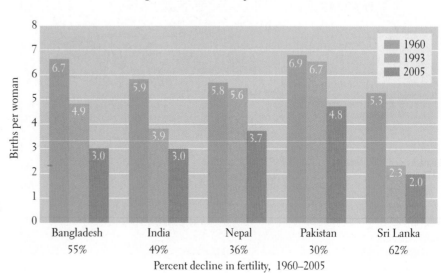

**FIGURE 8.15** Declines in total fertility rates in South Asia. All South Asian countries have experienced substantial declines in fertility since 1960. Many factors have contributed to this decline. [Adapted from *A Demographic Portrait of South and Southeast Asia* (Washington, D.C.: Population Reference Bureau, 1994); *2005 World Population Data Sheet* (Washington, D.C.: Population Reference Bureau).]

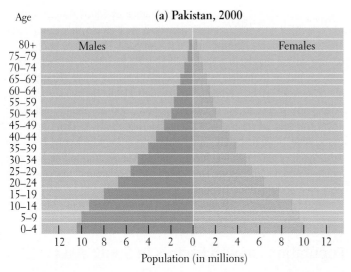

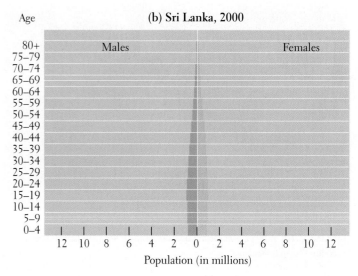

**FIGURE 8.16** Population pyramids for Pakistan and Sri Lanka, 2000. Both Pakistan and Sri Lanka have experienced decreasing fertility rates since 1960 (see Figure 8.15). However, the great size and youth of Pakistan's population ensure that it will continue to grow for years, while Sri Lanka's population is more or less stable. [Adapted from "Population pyramids for Pakistan" and "Population pyramids for Sri Lanka" (Washington, D.C.: U.S. Census Bureau, International Data Base), at http://www.census.gov/ipc/www/idbpyr.html.]

One factor is the age of the population. More than a third of South Asia's people are under age 15. The population pyramid for Pakistan (Figure 8.16a), for example, shows that a significant portion of that country's population is in the early reproductive years, so even a one-child-per-family policy would result in population growth for years to come. A factor of equal importance for poor, rural, uneducated people is the contribution of children to the family economy. Babies quickly become productive family members (Box 8.2), and grown children are the only retirement plan that most South Asians will ever have. Because access to health care is limited, infant mortality rates in the region average 67 per 1000 live births (only sub-Saharan Africa's are higher); hence couples often choose to have more than two or three children to ensure that at least some will reach maturity.

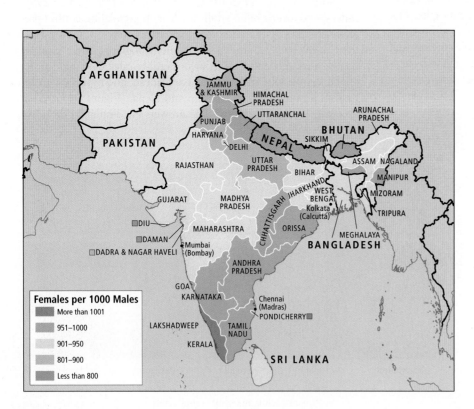

**FIGURE 8.17** Ratio of females to males in the South Asia region, 2000–2001. Maps are particularly helpful in showing the spatial results of gender discrimination. Here the green, khaki, yellow, and light orange colors show where females are unusually underrepresented in the population. Only Kerala and Pondicherry have somewhat larger numbers of females than males. The text offers some explanations for these patterns. [Adapted from "Sex ratio map of India—2001," at http://www.mapsofindia.com/census2001/sexratio/sexratio-india.htm.]

Sri Lanka has a much lower infant mortality rate (11 per 1000 live births) than Pakistan (85 per 1000). Sri Lanka's low rate is related to high rates of education for women (discussed below) and development policies aimed at more egalitarian wealth distribution, which have fostered the growth of the middle class. As we have seen, the incentive to have a lot of children diminishes with economic development, and that pattern is reflected in Sri Lanka's population pyramid (Figure 8.16b).

Another reason many South Asian families favor a large number of offspring is that they—like people in many regions of the world—view sons as more likely to contribute to a family's wealth than daughters. A popular toast to a new bride is "May you be the mother of a hundred sons." Many middle-class couples who wish to have sons hire high-tech laboratories that specialize in identifying the sex of an unborn fetus, then abort the fetus if it is female. Several Indian states have banned this use of technology, but enforcement is difficult. The practices of selective abortion, neglect of girl children, and female infanticide have resulted in an odd circumstance throughout India: among adults, men outnumber women. (As a result of their longer natural life span, adult women outnumber adult men in most places.) The 2001 Indian census showed 933 females for every 1000 males. In India, females outnumber males only in the state of Kerala, where there are 1058 females for every 1000 males (Figure 8.17).

Indian social scientists explain the exception of Kerala by noting that its entire population is more educated. The adult literacy rate in Kerala is 90 percent, versus 61 percent in India as a whole, and the elected Communist state government has for many years

---

## BOX 8.2 | AT THE REGIONAL SCALE   Should Children Work?

Beginning in the mid-1990s, a number of reports questioned the exploitation of child labor in Asia, especially in the handwoven carpet industry, which has been the subject of reports of outright enslavement of kidnapped children. Carpet weaving is an ancient artistic and economic enterprise in Central and East Asia as well as in South Asia. Traditionally, it has been a family-run enterprise, with women and children as the weavers and men as the merchants. International trade has increased the demand for fine handwoven carpets, but it is South Asian middlemen and foreign traders from Europe and America, not the weavers, who profit from that demand, and some unscrupulous carpet merchants have apparently resorted to forced labor to produce carpets.

Children have always participated in the home-based carpet industry (Figure 8.18). For thousands of years, young children have learned weaving skills from their parents and have become proud members of the family's production unit. Yet in today's society, children must balance their role as family workers with the need to attend school, where they will learn the skills that will enable them to survive in a modern economy.

Should children be allowed to work? It is clear that kidnapping and enslavement must be stopped, but is there room for different cultural views of what childhood should be like? When consumers buy goods made by children, are they supporting family values or greedy factory owners and middlemen? The United Nations and South Asian governments are now addressing these questions. They have instituted an active program to curb child labor abuses while remaining open to the positive experience that learning a skill and being part of a family production unit can be for a child. Until the modern era, such work was considered an important part of a child's training for adulthood, even in the United States and Europe. In India, there is now a national system to certify that exported carpets are made in shops where the children go to school, have an adequate midday meal, and receive basic health care. Such carpets will bear the label "Kaleen." If abuses can be eliminated, the custom of child labor is likely to continue because of its significant psychological and economic benefits to children and their families.

*Sources:* "Weekend Edition," National Public Radio (March 17, 2001); Janet Hilowitz, "Social labelling to combat child labour: Some considerations," *International Labour Review* 136 (1997): 215–232.

**FIGURE 8.18 Children at work.** Two young boys tie knots at a carpet loom in India. [Cary Wolinsky/Stock, Boston.]

provided education and broad-based health care for all. Education is credited with the fact that 63 percent of women in Kerala use contraception; in India as a whole, the rate is only 48 percent.

It is perhaps useful to note that, despite concern about overpopulation in South Asia, there are those who speculate that as workforces in Europe, Japan, and North America age, South Asia's youthful population could turn out to be a dividend—a young, creative workforce ready to seize as yet undreamed of opportunities in the technology economy. Of course, for the population dividend to pay off, young South Asians must be healthy and well educated.

## HIV-AIDS in South Asia

The HIV-AIDS epidemic has reached South Asia, but for a variety of reasons, the infection rate among adults is hovering at 0.7 per-

cent, which is about one-ninth the rate for sub-Saharan Africa. The risk factors that contribute to the spread of HIV are the same in South Asia as elsewhere: 84 percent of infection is by sexual transmission, which results from unprotected sex and low condom use, migration and mobility that brings strangers into contact, men who have both homosexual and heterosexual relationships, thus spreading infection to their wives, and the low status of women that keeps them from demanding safe sex. In addition, the general stigma of having HIV-AIDS discourages testing and encourages secrecy when infection is discovered. The rate is as low as it is because of several encouraging developments, including growing government involvement in HIV control, the participation of private companies in educating their employees, and the development, manufacture, and distribution in South Asia of a variety of anti-HIV drugs.

## ‖ CURRENT GEOGRAPHIC ISSUES

For the Westerner first learning about South Asia (Figure 8.19), the size of the region and the immense diversity of its people and landscapes may seem overwhelming. Here we try to catch a glimpse of the whole region, first by looking at the texture of daily life in both village and city, and then by examining the cultural characteristics that touch all lives, yet vary greatly in practice across the region: gender roles, ethnicity, religious beliefs, and social stratification.

Finally, we discuss the region's complex and interrelated economic, political, and environmental issues.

## SOCIOCULTURAL ISSUES

Many travelers to South Asia say that life there is best understood if it is observed in the intimate setting of the village or city neighborhood, where relationships among individuals, and within and between groups, are easier to discern (Figure 8.20). We begin our coverage of sociocultural issues by visiting two villages and an urban neighborhood to absorb something of the rhythm of life.

### Village Life

The writer Richard Critchfield, who has studied village life in more than a dozen countries, writes that the village of Joypur (Bangladesh) in the Ganga-Brahmaputra delta is set in "an unexpectedly beautiful land, with a soft languor and gentle rhythm of its own." In the heat of the day, the village is sleepy: naked children play in the dust, women meet to talk softly in the seclusion of courtyards, and chickens peck for seeds. Here and there, under the trees, people ply their various trades: a tinker mends pots; the village tailor sews school uniforms on his hand-cranked sewing machine. But in the early evening, mist rises above the rice paddies and hangs there, "like steam over a vat." It is then that the village comes to life, at least for the men. The men and boys return from the fields, and after a meal in their home courtyards, the men come "to settle in groups before one of the open pavilions in the village center and talk—rich, warm Bengali talk, argumentative and humorous, fervent and excited in gossip, protest and indignation" as they discuss their crops, an upcoming marriage, or national politics.

The anthropologist Faith D'Aluisio and her colleague Peter Menzel give us another peek into village life as night falls. Ahraura is a village in the state of Uttar Pradesh in north-central India. In

**FIGURE 8.19** Political map of South Asia.

**FIGURE 8.20** Several generations of a northern Indian family from the village of Dharamkot in Himachal Pradesh province. [David Morgan.]

the enclosed women's quarters of a walled compound, Mishri is finishing up her day by the dying cooking fire as her year-old son tunnels his way into her sari to nurse himself to sleep. Mishri, aged

27, lives in a tiny world bounded by the walls of the courtyard she shares with her husband and five children and several of her husband's kin. Like most villages of northern India, her village observes the practice of purdah, in which women keep themselves apart from male gazes. That Mishri can observe purdah because she need not help her husband in the fields is a mark of status. Within the compound, she works from sunup to sundown, only chatting for moments with two women who cover their faces and scurry from their own courtyards to hers for the short visit. Mishri is devoted to her husband, who was chosen for her by her family when she was 10; out of respect, she never says his name aloud.

The vast majority—about 70 percent—of South Asians live in hundreds of thousands of villages like Joypur and Ahraura. Even many of those now living in South Asia's giant cities were born in a village or visit an ancestral rural community on occasion.

## City Life

South Asian cities are often depicted as chaotic, crowded, and violent, with overstressed infrastructures and dilapidated housing. They are also full of creative and spiritual people who are modern, liberal in their thinking, literary, and multicultural. But were one to observe life for the masses of urban citizens in places as widely separated and culturally different as Chennai, Mumbai, Kathmandu, and Peshawar, one would discover that beyond the main avenues, these cities are actually thousands of tightly compacted, reconstituted villages, where daily life is intimate and familiar, not anonymous as in Western cities. Such a place is Koli, an ancient fishing village that predates the city of Mumbai and is now squeezed between Mumbai's elegant high-rises and the Bay of Mumbai (Figure 8.21). Koli is a labyrinth of low-slung, tightly packed homes ringed by fishing boats. The screeching of taxis and buses is soon lost in quiet calm as one ducks into a narrow covered walkway. At

**FIGURE 8.21** The village of Koli. Now surrounded by the vibrant city of Mumbai, Koli was once a small fishing settlement on a beautiful bay. There are still some fishermen in Koli, but most residents are educated and work in the city. [Courtesy of Alex Pulsipher.]

first, Koli appears impoverished, but inside, the homes are well appointed, some with marble floors, TVs, and computers. This is no warren of shanties, but rather a community of educated bureaucrats, tradespeople, and artisans who work throughout Mumbai.

## Language and Ethnicity

Within the life of one South Asian village or urban neighborhood, there is often room for considerable cultural variety. Differences based on caste, economic class, religion, and even language are usually accommodated peacefully by long-standing customs that guide cross-cultural interaction. One equalizing factor is that everyone in South Asia is, in one way or another, a minority. As the Indian writer and diplomat Shashi Tharoor writes:

*A Hindi-speaking male from … Uttar Pradesh might cherish the illusion that he represents the "majority community," … but he does not. As a Hindu he belongs to the faith adhered to by some 82 percent of the population, but a majority of the country does not speak Hindi; a majority does not hail from Uttar Pradesh; and if he were visiting, say, Kerala, he would discover that a majority is not even male.… Our archetypal Hindu has only to step off a train and mingle with the polyglot polychrome crowds thronging any of India's five major metropolises to realize how much of a minority he really is. Even his Hinduism is no guarantee of majorityhood,*

*because his caste automatically places him in a minority as well: if he is a Brahmin, 90 percent of his fellow Indians are not; if he is [of the] Yadav [caste], 85 percent of Indians are not, and so on.*

Shashi Tharoor,
*From Midnight to Millennium,* 1997, p. 112

There are many distinct ethnic groups in South Asia, each with its own language, dialect, and subdialect. In India alone, 18 languages are officially recognized, but there are actually hundreds of separate languages and literally thousands of dialects. The complexity of the distribution of languages results from the region's history of multiple invasions from outside, the relentless movements and rearrangements of people, and the long periods of isolation experienced by particular groups. In Figure 8.22, number 21 indicates some of the most ancient culture groups in the region. Remnants of Austro-Asiatic languages in these groups that were once more widely distributed were left as isolated pockets when sweeping cultural changes were brought by invaders. These languages are distantly akin to others found farther east in Southeast Asia. The languages represented by numbers 1–12 are linked to various groups of Aryan people who entered South Asia from Central Asia during prehistory.

The Dravidian language-culture group, represented by numbers 14–19, is another ancient group that predates the Aryan inva-

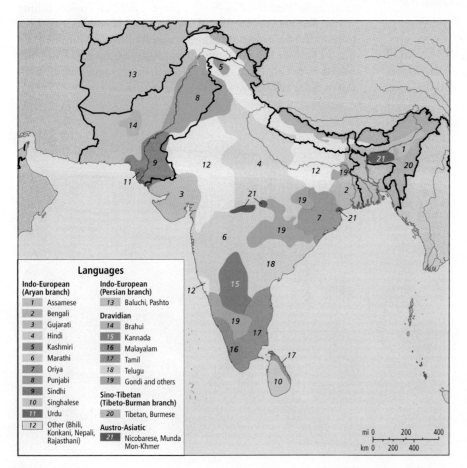

**FIGURE 8.22** Major language groups of South Asia. The modern pattern of language distribution in South Asia is a testimony to the fact that this region has long been a cultural crossroads. [Adapted from Alisdair Rogers, ed., *Peoples and Cultures* (New York: Oxford University Press, 1992), p. 204.]

sions by a thousand years or more. Today, Dravidian languages are found mostly in southern India, but a small remnant of the extensive Dravidian past can still be found in the Indus Valley in south-central Pakistan.

By the time of British colonization, Hindustani—an amalgam of Persian and Sanskrit-based northern Indian languages—was the lingua franca of all of northern India and what is today Pakistan. The Muslims wrote Hindustani in a form of Arabic script and called it Urdu, whereas the Hindus and other groups wrote it in a script derived from Sanskrit and called it Hindi. Today, Urdu is the national language of Pakistan (though it is the first, or native, language of only a minority) and Hindi the national language of India, but people who speak the common forms of these languages can understand one another in the same way that speakers of American and British English do. Hindi, because of its origins in Hindustani and the popularity of Hindi-language films, is understood by most Pakistanis and by about 50 percent of India's population. Only Chinese and English are spoken by more people worldwide. But, as Shashi Tharoor noted, Hindi is not the first language of the majority; no Indian language can make that claim.

English is a common second language throughout the region. For years, it was the language of the colonial bureaucracy, and it remains a language used at work by professional people. From 10 to 15 percent of South Asians speak, read, and write English.

## Religion

The main religious traditions of South Asia are Hinduism, Buddhism, Sikhism, Jainism, Islam, and Christianity (Figure 8.23). Islam, Christianity, and Judaism are discussed in detail in Chapter 6, so we will not cover those belief systems here.

***Hinduism.*** **Hinduism** is a major world religion practiced by approximately 900 million people, 800 million of whom live in India. It is a complex belief system, full of seeming contradictions that often make it difficult for outsiders to understand. These contradictions result largely from the fact that Hinduism includes a broad range of beliefs and practices that have their roots in highly localized folk traditions (known as the Little Tradition) as well as in a classical Hindu system based on literary texts (known as the Great Tradition). Both traditions are based on the amalgam of Harappan and Aryan ritual beliefs recorded in ancient Sanskrit texts. For example, most Hindus worship a number of gods and goddesses, but many of these are found only in one region, in one village, or even in one family. Hence, although local areas have retained their own deities, over time some local gods have become incorporated into the classical Hindu pantheon.

A major tenet of classical Hindu philosophy, as described in the 4000-year-old Hindu scriptures called the *Vedas*, is that all gods are merely illusory manifestations of the ultimate divinity, which is

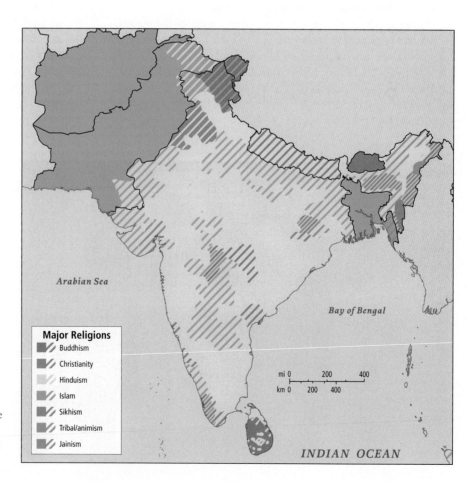

**FIGURE 8.23** Major religions in South Asia. Notice the overlapping patterns in many parts of the region. [Adapted from Gordon Johnson, *Cultural Atlas of India* (New York: Facts on File, 1996), p. 56.]

**Major Religions**
Buddhism
Christianity
Hinduism
Islam
Sikhism
Tribal/animism
Jainism

*Arabian Sea*

*Bay of Bengal*

mi 0    200    400
km 0    200    400

*INDIAN OCEAN*

formless and infinite. Many devout Hindus worship no gods at all, but may engage in meditation, yoga, and other spiritual practices designed to liberate them from illusions and bring them closer to the ultimate reality, described as infinite consciousness. The average person, however, is thought to need the help of personified divinities in the form of gods.

Hindus usually observe caste identities, which are deeply interwoven with classical Hindu ritual and distinguished by dietary rules. Nevertheless, there are some things that almost all Hindus have in common, such as the belief in reincarnation: the idea that any living thing that desires the illusory pleasures (and pains) of life will be reborn after it dies. A reverence for cows, which are seen as only slightly less spiritually advanced than humans, also binds all Hindus together. This, along with the Hindu prohibition on eating beef, may stem from the fact that cattle have been tremendously valuable in rural economies as the primary source of transport, field labor, dairy products, fertilizer, and fuel (people in the region burn animal dung).

***Geographic Patterns in Religious Beliefs.*** There is an uneven and overlapping distribution pattern of religions in South Asia, as Figure 8.23 shows. Hindus, as we have observed, are the most numerous, and they are found mostly in India. The Ganga Plain is considered the **hearth** (place of origin) of Hinduism, and every 12 years, during India's largest religious festival, millions of Hindus converge on the city of Allahabad to bathe at the confluence of the Ganga and Yamuna rivers as an act of devotion.

Other religions are important in various parts of the region. **Buddhism** began about 2600 years ago as a reform and reinterpretation of Hinduism. Its origins are in northern India, where it flourished early in its history before spreading eastward to East and Southeast Asia. Only 1 percent of South Asia's population—though this is about 10 million people—are Buddhists; they are a majority in Bhutan and Sri Lanka.

The 420 million Muslims (followers of Islam) in the region form the majority in Afghanistan, Pakistan, Bangladesh, and the Maldives. Muslims are also a large and important minority in India, numbering about 120 million, mostly in the northwestern and central Ganga Plain but also scattered throughout the country.

**Sikhism** was founded in the fifteenth century by Guru Nanak as a challenge to contemporary socioreligious systems, but the new religious philosophy was inspired by both Hindu and Islamic ideals. Sikhs espouse belief in one God, high ethical standards, and meditation. Philosophically, Sikhism rejects the idea of caste, but accepts the Hindu idea of reincarnation. (In everyday life, however, caste continues to play a role in Sikh identity.) The 18 million Sikhs in the region live mainly in Punjab, in northwestern India, but they are also found elsewhere. Their influence in India is greater than their numbers because many Sikhs hold positions in the military and police. More Sikhs live in diaspora communities than live in India itself, and some have financed a Sikh separatist movement (see the discussion of the conflict in Punjab on pages 445–446).

**Jainism,** like Buddhism, originated as a reformist movement within Hinduism more than 2000 years ago. Jains (about 6 million people, or 0.6 percent of the region's population) are found mainly in western India and in large urban centers throughout the region (Figure 8.24). They have more influence than their numbers would suggest, especially within the Hindu community. They are known for their educational achievements, nonviolence, and strict vegetarianism.

**Parsis,** though few in number, are a highly visible minority in India's western cities, where they have distinguished themselves in business and finance, politics, and the arts. Parsis are descended from Persian Zoroastrian migrants who left Iran over 1000 years ago to escape persecution after the Islamic conquest of Persia. Zoroastrian and Parsi ideas on such things as human rights and dedicating one's life to generous selfless acts have influenced Christianity, Judaism, Islam, and modern international law.

**FIGURE 8.24** The Vimala Vasahi temple, one of the Jain temples in Dilwara at Mount Abu, Rajasthan. The temple was built entirely of marble in 1021; the carvings are elaborate, and each pillar is different. [Robert Holmes/CORBIS.]

The first Christians in the region are thought to have arrived in the far southern Indian state of Kerala with St. Thomas, the Apostle of Christ, in the first century C.E. Today, Christians are an important minority along the west coast of India, on the Deccan Plateau, and in northeastern India, where in some places more than half the descendants of the ancient aboriginal inhabitants are Christian, the result of British colonial mission efforts in the late nineteenth and twentieth centuries. Christians are also a small minority elsewhere in India, as well as in Sri Lanka, Pakistan, and Bangladesh.

Small communities of Jews are found along the Malabar Coast (Cochin) and in such major cities as Mumbai, Kolkata, and Ahmadabad. Some Jews in South Asia are thought to be the descendants of ancient migrants who arrived perhaps 2000 years ago or even earlier. Others came via Europe in the modern period.

Animism is practiced throughout South Asia, especially in central and northeastern India, where there are indigenous people whose occupation of the area is so ancient that they are considered aboriginal inhabitants. (Animism is discussed in Chapter 7 on page 388.) In South Asia, animist beliefs often incorporate aspects of Hinduism, Islam, Buddhism, or Christianity.

*The Hindu-Muslim Relationship.* The different religious traditions of South Asia have influenced one another, and this is especially true of the two largest faiths. Where Hindus have lived in close association with Muslims—often within the same villages—they have absorbed Muslim customs, such as honoring Muslim saints. For their part, Muslims have adopted some of the Hindu ideas of caste. Although in India, Muslims make up only 12 percent of the population, the political and social relationships between Hindus and Muslims are enormously complex. The great independence leaders Mohandas Gandhi and Jawaharlal Nehru both emphasized the common cause that once united Muslim and Hindu Indians: throwing off British colonial rule. Since independence, members of the Muslim upper class have been prominent in Indian national government and the military. Muslim generals have served India willingly, even in its wars with Pakistan after the Partition. Hindus and Muslims often interact amicably, and they occasionally marry each other.

But there is a darker side to the Hindu-Muslim relationship. Especially in Indian villages, some upper- and middle-caste Hindus regard Muslims as members of low-status castes. Religious rules about food are often the source of discord because dietary habits are a primary means of distinguishing caste. Hindus regard the cow as sacred, and although they do not kill cows for food or hides, some at the lowest rungs of Hindu society process and consume individual animals that die by other means. Muslims, on the other hand, run slaughterhouses and tanneries (though discreetly), eat beef, and use cowhide to make shoes and other items. To some Hindus, therefore, some Muslims appear to have offensive customs. Also fueling this perception is the occasional conversion to Islam of entire low-caste or tribal Hindu villages seeking to escape the hardships of being members of a disadvantaged social category.

The Hindu-Muslim relationship is no less complex in Bangladesh. After the separation of Bangladesh from Pakistan, many upper-class Muslims moved back to Pakistan, where they had been bureaucrats and administrators. The exodus left Bangladesh with a preponderance of poor Muslim farmers. Meanwhile, some Hindu landowners remained, and some lower-caste Hindus converted to Islam. Thus, in Bangladeshi villages, Muslims are usually a majority, but the Hindus are often somewhat wealthier. Although the two groups may coexist amicably for many years, they view themselves differently, and conflict resulting from religious differences —euphemistically called **communal conflict**—can erupt over seemingly trivial events.

| **Personal Vignette** | The sociologist Beth Roy, a specialist in conflict resolution who studies communal conflict in South Asia, recounts an incident in the village of Panipur (a pseudonym), Bangladesh, in her book *Some Trouble with Cows* (Figure 8.25).

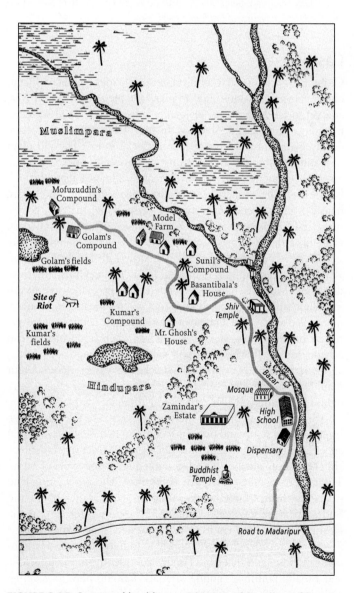

**FIGURE 8.25** Some trouble with cows. This map of the village of Panipur (a pseudonym) illustrates how intimately the separate Muslim and Hindu communities were connected. The map shows Muslim and Hindu areas, the area where the riot took place, and numerous other features of village life. [Courtesy of the University of California Press.]

The incident started when a Muslim farmer either carelessly or provocatively allowed one of his cows to graze in the lentil field of a Hindu. The Hindu complained, and when the Muslim reacted complacently, the Hindu seized the offending cow. Although this was not an unusual incident, in this case, by nightfall, Hindus had allied themselves with the lentil farmer and Muslims with the owner of the cow. More Muslims and Hindus converged from the surrounding area, and soon there were thousands of potential combatants lined up facing each other. Fights broke out. The police were called. In the end, a few people died when the police fired into the crowd of rioters. Relationships in the village were deeply affected by the incident. In the words of Roy, the dispute "delineat[ed] distinctions of caste, class, and [religious] culture so complex they intertwine[d] like columbines climbing on an ancient wall."

*Source: Beth Roy, Some Trouble with Cows—Making Sense of Social Conflict (Berkeley: University of California Press, 1994), pp. 18–19.* ■

## Caste

**Caste,** the ancient system of dividing society into hereditary hierarchical categories, seems alien to many people outside South Asia, and yet social division and inequality are common in all societies, including Europe and America. In fact, all human groups have deeply ingrained concepts of relative social status. The old often have more authority than the young; conversely, as in U.S. pop culture, the young sometimes have greater influence than the old. Nearly everywhere on earth, social difference is indicated by clothes, hairstyle, body decorations, manner of speaking, material possessions, residential location, gender, and religion. In some quarters, race still carries overtones closely akin to those of caste in South Asia.

Caste is a custom associated primarily with Hindu India, but all religious communities in the region (including Christians, Buddhists, and Muslims) have incorporated elements of the caste system into their cultures. One is born into a given subcaste, or community (called a *jati*), and that happenstance largely defines one's experience for a lifetime—where one will live, where and what one can eat and drink, with whom one will associate, one's marriage partner, and sometimes one's livelihood. The classical caste system has four main divisions or tiers, called *varna*, within which are many hundreds of *jatis* and sub-*jatis*, which vary from place to place.

**Brahmins,** members of the priestly caste, are the most privileged in ritual status and thus must conform to those behaviors that are considered most ritually pure (for example, strict vegetarianism, abstention from alcohol, and abstention from certain types of work). Then, in descending rank, are **Kshatriyas,** who are warriors and rulers; **Vaishyas,** who are landowning farmers and merchants; and **Sudras,** who are low-status laborers and artisans. A fifth group, the **Harijans** (also called Dalits—"the oppressed"—or untouchables), is actually considered to be so lowly as to have no caste. Harijans perform those tasks that caste Hindus consider the most despicable and ritually polluting: killing animals, tanning hides, sweeping, and cleaning. A sixth group, also outside the caste system, is the **Adivasis,** who are thought to be descendants of the region's ancient original inhabitants.

Although *jatis* are associated with specific subcategories of occupations, in modern economies this aspect of caste is more symbolic than real. Members of a particular *jati* do, however, follow the same social and cultural customs, dress in a similar manner, speak the same dialect, and tend to live in particular neighborhoods or villages. This spatial separation arises from the higher-caste communities' fears of ritual pollution through physical contact or sharing of water or food with lower castes. When one stays in the familiar space of one's own *jati*, one is enclosed in a comfortable circle of families and friends that becomes a mutual aid society in times of trouble. This cohesion within *jatis* and their attachment to place help to explain the persistence of a system that seems to put such a burden of shame and poverty on the lower ranks.

Although it seems rigid, the caste system is quite dynamic. The particular hierarchy of *jatis* in a given locale is often disputed by the various groups themselves and may change over time. *Jatis* are constantly jockeying with one another for position and status. Caste identity has also asserted itself in different ways at different times throughout history. Many scholars argue that British policies entrenched and politicized caste (and religious) identity in ways that are still influential today. Most recently, caste has become almost "ethnicized" and has taken on a renewed importance in both national and local politics.

It is important to note that caste and class are not the same thing. Class refers to economic status, and there are class differences within caste groups because of differences in wealth. Historically, upper-caste groups (Brahmins and Kshatriyas) owned or controlled most of the land and lower-caste groups (Sudras) were the laborers, so caste and class tended to coincide, but there were many exceptions. Today, as a result of expanding educational and economic opportunities, caste and class status are less connected. Some Vaishyas and Sudras have become large landowners and extraordinarily wealthy businesspeople, while some Brahmin families struggle to achieve a middle-class standard of living. By and large, however, Harijans remain very poor.

In the twentieth century, Mohandas Gandhi began an official effort to eliminate discrimination against "untouchables." As a result, India's constitution bans caste discrimination. In the late 1940s, India began an affirmative action program that reserves a portion of government jobs, places in higher education, and parliamentary seats for Harijans (referred to as "Scheduled Castes") and Adivasis ("Scheduled Tribes"). Together, Scheduled Castes and Scheduled Tribes now constitute approximately 23 percent of the Indian population and are guaranteed 22.5 percent of government jobs. In 1990, this program was extended to include other socially and educationally "Backward Castes" (such as very disadvantaged *jatis* of the Sudras caste), reserving an additional 27 percent of government jobs for these groups. This program has not been without controversy, however (Figure 8.26).

Among educated people in urban areas, the campaign to eradicate discrimination on the basis of caste has been remarkably successful. Some Harijans throughout the country are now powerful officials. Members of high and low castes now ride city buses side by side, eat together in restaurants, use the same restrooms, drink from the same water fountains, and attend the same schools and universities. More remarkably, for some urban Indians—especially educated professionals who meet in the workplace—caste is disap-

**FIGURE 8.26 Caste still matters.** In May 2006, medical students and many younger doctors across India protested government plans to increase enrollment quotas in elite institutions for lower-status castes and tribal groups. The government's aim was to increase opportunity and raise living standards for traditionally excluded people. Students argued that admission should be based only on merit. In August 2006, the India Supreme Court ruled against the government (in favor of the students' position), and the issue may be resolved for now. [AP Photo/Bikas Das.]

pearing as the crucial factor in finding a marriage partner. Nonetheless, it would be incorrect to conclude that caste is now irrelevant in India. Less than 5 percent of registered marriages cross even *jati* lines. Nearly everyone notices social clues that reveal an individual's caste, and in rural areas, where the majority of Indians still reside, the divisions of caste remain prevalent.

## Geographic and Social Patterns in the Status of Women

While South Asia has a number of women in very high positions of power, the overall status of women in the region is notably lower than the status of men. Women's status and welfare are lowest in the belt that stretches from the northwest in Afghanistan across Pakistan, western India, and the Ganga Plain into Bangladesh (see Table 8.1). Women fare better in eastern, central, and southern India and in Sri Lanka, where different marriage and inheritance practices—possibly originating in ancient Dravidian matrilineal social organization—have given women greater access to education and resources.

Urban women generally enjoy greater individual freedom than rural women, and in rural areas, middle- and upper-caste women are more restricted in their movements than are lower-caste women. Lower-caste women, however, must contend with sexual harassment and exploitation from upper-caste men. Young women today have significantly greater educational and employment opportunities than those of a generation ago, and they are entering the skilled workforce in large numbers. The socioeconomic status of Muslim women is notably lower than that of their Hindu and Christian counterparts. A recent national survey in India reported that Muslims on the whole have an average standard of living well below that of most Hindus. This disparity translates into educational levels for Muslim women that are significantly below the national average. In India, Muslim women's workforce participation rates also tend to be the lowest. Low rates of education and workforce participation for women also prevail in Muslim-dominated countries such as Pakistan and Afghanistan, but the arrival of microcredit practices (discussed on page 443) and export processing industries in Bangladesh (similar to maquiladoras in Mexico; see Chapter 3, page 139) may be expanding opportunities for low-income women there.

***Culturally Enforced Gender Inequity.*** Women in Afghanistan have arguably had the most difficult lives in the region since an archconservative Islamist movement, the **Taliban,** gained control of the government there in the mid-1990s (see the discussion of the Taliban on pages 447–448). The Taliban (meaning "God's students") supported strict and distorted interpretations of Islamic law. Both men and women had to follow strict dress codes (men had to wear beards and robes), and females, even urban professional women, had to live in seclusion: girls and women were not allowed to work outside the home or attend school, and they had to wear a completely concealing heavy garment, called a *burqa*, whenever they were out of the house. The Taliban even decreed that women must whisper and not make noise as they walked, because the sound of their footsteps was distracting and potentially erotic to men. Although the Taliban were driven from power in November of 2001, cultural and religious conservatism continues to adversely affect Afghan women. Concerted efforts to change their lives for the better are under way.

| Personal Vignette | From behind her microphone at Radio Sahar (Dawn), Nurbegum Sa'idi speaks to a female audience on a wide range of topics. Radio Sahar, located in the city of Herat, is the latest in a network of independent women's community radio stations to spring up in Afghanistan since early 2003. Radio Sahar has a broadcast radius of 30–45 miles (50–70 kilometers) and provides

**FIGURE 8.27** Women's community radio in Afghanistan. Radio journalist Parwin of Radio Quyaash, a sister station to Radio Sahar, interviews the Afghan head of Women's Affairs on Children's Day. [© Leslie Knott.]

**FIGURE 8.28** Material culture of purdah. Lattice screens known as *jalee* are an architectural feature in South Asian areas where women are secluded. Like the louvers and latticed bay windows of Saudi Arabia (see Figure 6.23 on page 317), *jalee* allow ventilation and let in light, but shield women from the view of strangers. [Lindsay Hebberd/Woodfin Camp & Associates.]

2 hours of daily programming consisting of educational items that address cultural, social, and humanitarian matters as well as music and entertainment. Radio Sahar is supported by a community radio advisory board, and the women members—none of whom are radio professionals—run the radio station themselves.

According to local activists, initiatives such as Radio Sahar and others (Figure 8.27) are vital to improving the condition of women in Afghanistan. Given the high percentage of illiterate women with little or no access to education, radio provides one of the most powerful ways to reach and educate women, allowing them to connect with one another in this conservative, male-dominated society. As Sa'idi attests, "It's great when you feel you can bring about change. The feedback we have been getting from listeners tells us that Sahar is providing new hope for the women in Herat."

*Source: United Nations Office for the Coordination of Humanitarian Affairs Integrated Regional Information Network, "Afghanistan: New radio station to tackle women's problems," November 20, 2003.* ∎

**Purdah,** the practice of concealing women from the eyes of nonfamily men, especially during their reproductive years, is observed in various ways across the region (Figure 8.28). The practice is strongest in the northwest and across the Indo-Gangetic Plain, where it takes the form of seclusion of women and of veiling or head covering in both Muslim and Hindu communities. Purdah is weaker in central and southern India, but even here, separation between unrelated men and women is maintained in public spaces. The custom is generally not observed by Adivasis or by low-caste Hindus, but that is changing. In recent decades, as low-status households increase their economic standing, their ability to seclude women signals surplus wealth and increased ritual purity. Unfortunately for women, the effect of this trend is to limit their economic independence and autonomy in the long run.

Throughout South Asia, most women are partners in marriages arranged for them, often without their wishes being con-

sulted. In most cases, the bride's family pays the groom's family a sum of money called a **dowry** at the time of the marriage. Usually a bride (who may be as young as 12) goes to live in her husband's family compound, where she becomes a source of domestic labor for her mother-in-law. Most brides work at domestic tasks for many years until they have produced enough children to have their own crew of small helpers, at which point they gain some prestige and a measure of autonomy. Motherhood in South Asia is actually more significant than in the West. A woman's power and mobility increase when she has grown children and becomes a mother-in-law herself. But in some communities, the death of a husband, no matter what the cause, is a disgrace to a woman and can completely deprive her of all support and even of her home, children, and reputation. Widows may be ritually scorned and blamed for their husbands' deaths. Widows of higher caste rarely remarry, and in some areas, they become bound to their in-laws as household labor or may be asked to leave the family home.

***Bride Burning and Female Infanticide.*** For many years, the phenomenon of bride burning, or dowry killing, in which a husband and his relatives stage an "accident," such as a kitchen fire, that kills his wife, has been reported, especially in India. The wife's death enables the widower to marry again and collect a new dowry. The National Crime Records Bureau of India reported that 7026 such deaths occurred in 2005. This practice is an extreme example of the domestic violence that is pervasive throughout the region. In some cases, the threat of an accident is used to extort further dowry from a wife's family. More commonly, a young bride may be made to feel undeserving of her status in the groom's family if the family feels cheated in the marriage negotiations.

Changing customs regarding dowry appear to be a cause of the growing incidence of both bride burning and female infanticide—in which girl babies, deemed unaffordable because of the dowry investment they will require, are killed—in India. Until the last several decades, it was the custom among the lower castes to pay a **bride price,** not a dowry: the groom paid the family of the bride a relatively small sum that symbolized the loss of their daughter's work to her family's economy and the gain of her labor by his. Dowry, by contrast, originated as an exchange of wealth between landowning, high-caste families. With her ability to work reduced by purdah, an upper-caste female was considered a liability. The dowry that went with a bride to her new husband gave her dignity as a wife and leverage in her husband's household, because the dowry had to be returned if the marriage was dissolved. Moreover, the payment of a substantial dowry could increase the status of a bride's family by attracting a groom of higher status and could set the tone for the future marriage settlements of siblings. Eventually, caste and class became conflated as dowry essentially became a nonrefundable endowment paid by the bride's family to a higher-status groom and his family.

Oddly, increasing education for males and increasing family affluence reinforced the custom of dowry. Young men, even those of low caste, came to feel that their diplomas increased their worth as husbands and put them in the category of those who deserved a bride with a substantial dowry. This upgrade in status through education gave them the power to demand larger and larger dowries. Soon the practice spread through lower-caste families wanting to upgrade their status, and now the poorest of families are crippled by the dowries they must pay to get their daughters married, which in turn influences the matches they make for their sons.

Despite the fact that the Indian government bans the practice of dowry, the social duty to provide a dowry for a daughter is taken extremely seriously because the stigma of having an unmarried daughter is huge. Some ambitious families are giving their daughters a graduate school education, with its promise of earning power, in lieu of a dowry. But for an increasing minority of poor families, the birth of more than one daughter threatens the family with hopeless impoverishment, whereas the birth of sons promises future daughters-in-law who will bring dowry wealth with marriage. A village proverb captures this inequitable relationship: "When you raise a daughter you are watering another man's plant." Some families view the birth of a daughter as such a calamity that they are led to the desperate act of poisoning second and third daughters soon after birth.

***Education, Earning Power, and the Status of Women.*** Oxfam, India Literacy Project, and South Asian Women's Network are a few of the agencies seeking to diminish the influence of purdah and improve the status and earning opportunities of South Asian women. These organizations believe that freeing women from purdah and other ancient strictures will encourage lower fertility and greater economic growth because women will channel less of their energy into reproduction and more into such activities as business, innovative agriculture, recycling of resources, entrepreneurship, teaching, and other service occupations. In turn, these changes will increase the educational attainments of all of South Asia's children and improve the health and nutrition of families. United Nations researcher Martha Nussbaum found that women who can read often seek a way to earn some income and generally invest their earnings in food, medicine, and schooling for their children (Figure 8.29).

In Mazār-e-Sharif, Afghanistan, the state governor has supported the recent establishment of a women's bazaar, where all the shops are owned and operated by women but the customers are of

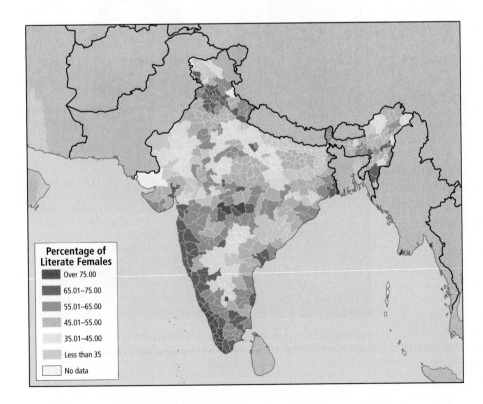

**Percentage of Literate Females**

- Over 75.00
- 65.01–75.00
- 55.01–65.00
- 45.01–55.00
- 35.01–45.00
- Less than 35
- No data

**FIGURE 8.29** Female literacy in India, 2005. Female literacy lags behind male literacy in nearly all parts of the country. Female literacy is crucial to improving the lives of children: women who can read often seize opportunities to earn some income and nearly always use this income to help their children. The white lines indicate district boundaries. [Adapted from "India female literacy districtwise," *Maps of India,* at http://www.mapsofindia.com/census2001/femaleliteracydistrictwise.htm.]

**FIGURE 8.30** Internet access for women. Across India, women are increasingly using Internet cafés to access information and communicate with family and friends. At "Cybercafe" in Bangalore, these two young women send e-mail to family members traveling abroad. [Chris Stowers/Panos.]

both sexes. So far, only four women have been brave enough to withstand the harassment of men who resent the bazaar. Fariba Majid, the Women's Ministry representative who founded the bazaar, hopes for at least 20 shopkeepers in time. She remarks that, for the bazaar to survive, the women must quickly turn a profit; otherwise, their husbands will put a stop to it. But for now, one such entrepreneur, Ms. Barmaki, who runs her handicraft shop out of a cargo container, has won the consent of her husband to employ their 12-year-old daughter.

The Internet has become a forum for discussion within South Asian communities about issues related to the low status of women. Web sites such as India Family Net Talk (http://indiafamily.net/), and access to global Internet connections in general, lend strength to social change efforts (Figure 8.30).

***Gender Equality at the Village Level and Beyond.*** Although women in India face a number of constraints and disadvantages, a strong activist movement in India has led to fairly enthusiastic enforcement of constitutional protections, at least after the fact. In the 1980s, Prime Minister Rajiv Gandhi introduced *panchayati raj* (village governing councils) to encourage gender equality in village life. Thirty percent of the seats on these local councils must be reserved for women during a given election cycle. Furthermore, since 1993, there has been training for women *panchayati raj* members to improve their effectiveness on the councils. At the parliamentary level, however, Indian women remain very poorly represented; just 9 percent of the Indian Parliament was female in 2006, when a confederation of Muslim and Hindu women's groups met with the prime minister to ask for legislation that would reserve one-third of the seats in the lower house of Parliament and in state assemblies for women for a 15-year trial period. Such quotas are already in place in Pakistan (22 percent), Nepal (33 percent), and Bangladesh

(14 percent). It is hoped that these temporary quotas will afford women sufficient political experience to win elections without the aid of quotas once they have expired.

A minority of upper- and middle-class urban professional women in India, Pakistan, Sri Lanka, and Bangladesh may have more in common with their counterparts in Europe and America than they do with village women in their own countries. In the region's major cities, there are growing numbers of highly successful businesswomen, female directors of companies, highly qualified female technicians, high-ranking female academics, and women who serve prominently in government (Figure 8.31). India, Bangladesh, Sri Lanka, and Pakistan have all had women heads of state.

## ECONOMIC ISSUES

South Asia is a region of startling economic contrasts, where a country (India) that is home to hundreds of millions of desperately poor people can also foster a growing computer software industry and a space program. Although the British colonial system deepened the extent of South Asia's poverty and widened the gap between rich and poor, the current wealth disparities in the region have resulted mostly from economic policies favored by postindependence leaders. Despite India's celebrated democratic traditions, its poor have often been left out of the political process and bypassed or hurt by economic reforms.

**FIGURE 8.31** Naina Lal Kidwai, CEO of HSBC Bank in India. In 2005, *Fortune* listed Kidwai as one of the 50 most powerful women in business worldwide. [Santosh Verma/Bloomberg News/Landov.]

Agriculture remains the basis of South Asian economies, but rapid industrialization and self-sufficiency have been the dream since independence. Recent strategies adopted to encourage economic development include an emphasis on information technology in India and innovative financing strategies pioneered in Bangladesh. The service sector has expanded more rapidly than either agriculture or industry, however, in almost all South Asian countries over the past decade.

## Agriculture and the Green Revolution

While 60 percent of the region's population is engaged in agricultural labor, the contribution of agriculture to most national economies hovers below 25 percent of the GDP. Although production per unit of land has increased dramatically over the past 50 years, agriculture is the least efficient economic sector, meaning that it gets the lowest return on investments of land, labor, and cash.

Figure 8.32 shows the distribution of agricultural zones in South Asia. Except in the far northwest and in highland areas, double-cropping is common throughout the region: crops adapted to dry conditions are planted in winter, and those adapted to wet conditions are planted in summer, when monsoon rains are plentiful. Rice is the main crop wherever rainfall is plentiful and occupies about one-third of the total land planted in grain. Wheat, grown in the west-central and northwestern parts of the region, is

the second most important crop, and its cultivation is spreading with the use of irrigation systems. Animal grazing is common in the driest areas of Pakistan, Afghanistan, and northwestern India.

Until the 1960s, agriculture across South Asia was based largely on traditional small-scale systems that managed to feed families in good years, but often left them hungry in years of drought or flooding. Moreover, these systems did not produce sufficient surpluses for the region's growing cities, which even now import food from outside the region. Much of South Asia's agricultural land is still cultivated by hand, and overall, agricultural development has been neglected in favor of industrial development, especially in India. Nonetheless, by the 1970s, important gains in agricultural production had begun.

Beginning in the late 1960s, a so-called **green revolution** boosted grain harvests dramatically through the use of new agricultural tools and techniques—seeds selected for high yield and for resistance to disease and wind damage, fertilizers, mechanized equipment, irrigation, pesticides, herbicides, and double-cropping —and to a lesser extent, an increase in the amount of land under cultivation. Where the new techniques were used, yield per unit of farmland improved by more than 30 percent between 1947 and 1979, and both India and Pakistan became food exporters. International loans financed the building of dams to store monsoon water for irrigation and to create hydroelectric power. These projects, in turn, boosted industrial growth and created jobs. India was able to

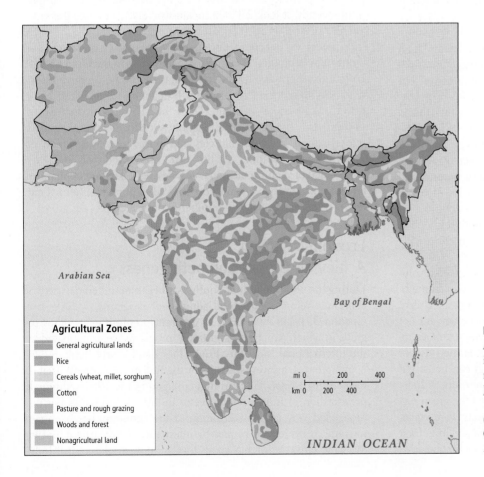

**Agricultural Zones**

- General agricultural lands
- Rice
- Cereals (wheat, millet, sorghum)
- Cotton
- Pasture and rough grazing
- Woods and forest
- Nonagricultural land

*Arabian Sea*

*Bay of Bengal*

mi 0    200    400

km 0    200    400

*INDIAN OCEAN*

**FIGURE 8.32** Agricultural zones in South Asia. Agricultural zones in this region form a complex pattern influenced by landforms, climate, cultural customs, and recent development theories. [Adapted from Gordon Johnson, *Cultural Atlas of India* (New York: Facts on File, 1996), p. 34; "Afghan Economy (map)," SESRTCIC (Statistical, Economic and Social Research and Training Centre for Islamic Countries) InfoBase, at http://www .sesrtcic.org/members/afg/afgmapec.shtml.]

repay its green revolution–related loans, which boosted its reputation for creditworthiness.

The green revolution, however, has not made India reliably self-sufficient in food. Although food production on a per capita basis has increased since 1960, scholars and policy makers are concerned about India's long-term ability to maintain this trend, especially in the face of climate change and decreasing soil fertility. In addition, the benefits of the green revolution have been very uneven. Some states, such as Punjab and Haryana, which have extensive irrigation networks, have gained tremendously while other states have lagged behind. Many poor farmers who were unable to afford the special seeds, fertilizers, pesticides, and new equipment were forced off rented or borrowed land and into low-wage farm labor in the early decades of the green revolution (1960s and 1970s). Although some rural incomes did expand, those farmers who lost access to land have migrated to the cities, where they have difficulty finding jobs of any sort.

Moreover, increased food supplies have not eliminated hunger and malnutrition. Between 1970 and 2001, the amount of food produced per capita in South Asia increased 18 percent, and the proportion of undernourished people dropped from 33 percent of the population to 22 percent. However, many undernourished people have not benefited from increased supplies of food because much of the food is exported or sold to those who can pay for it. Poor rural people also lose out during food scarcities because South Asian governments place a high priority on ensuring sufficient supplies of food to the cities, where urban unrest poses a threat to the interests of government and the middle and upper classes.

Meanwhile, increasing soil salinity and other kinds of environmental damage created by chemical fertilizers, pesticides, and high levels of irrigation are reducing yields in many areas. One such area is the Pakistani Punjab, that country's most productive, but highly irrigated, agricultural zone.

Green revolution technologies have also inadvertently reduced the utility of many crops for the rural poor, especially women. The new varieties of rice and wheat, for example, yield more grain, but less of the other components of the plant previously used by women, such as the wheat straw used to thatch roofs, make brooms and mats, and feed livestock. Moreover, women's already low status in agricultural communities often erodes further as their contribution to household production is supplanted by new technologies, such as small tractors and mechanized grain threshers, which are usually controlled by male members of the family.

A potential remedy for some of the failings of the green revolution style of agriculture is **agroecology**: the use of traditional methods to fertilize crops and natural predators to control pests. Unlike green revolution techniques, the methods of agroecology are not disadvantageous to poor farmers because the necessary resources are readily available in most rural areas. To participate, farmers do not need access to cash, but only to knowledge, which can be taught orally to small groups and over the radio. Studies in southern India that compared agroecology techniques with those of the green revolution found that their productivity and profitability were equal, but that agroecology techniques reduced soil erosion and loss of soil fertility, two undesirable effects of green revolution techniques.

## Industry over Agriculture: A Vision of Self-Sufficiency

After independence from Britain in 1947, South Asia's new leaders favored industrial development over agriculture. Influenced by socialist ideas, they believed that government involvement in industrialization was necessary to ensure the levels of job creation that would cure poverty. Another of their goals was to reduce the need to import manufactured goods from the industrialized world. The new South Asian leaders engineered government takeovers of the industries they believed to be the linchpins of a strong economy: steel, coal, transport, communications, and a wide range of manufacturing and processing industries.

For the most part, South Asian industrial policies in the decades after independence failed to meet their goals. The emphasis on self-sufficiency in industry was ill suited to countries that had been primarily agricultural for years. In India, for example, governments invested huge amounts of money in a relatively small industrial sector that even today employs only 12 percent of the population (compared with the 60 percent employed in agriculture). Since only a small portion of the population directly benefited from this investment, industrialization failed to significantly increase South Asia's overall prosperity.

Another problem was that the measures intended to boost employment often contributed to inefficiency and ignored market incentives. One policy encouraged industries to employ as many people as possible, even if they were not needed. So, for example, until recently, it took 250,000 Indian workers to produce the same amount of steel as 8000 Japanese workers; consequently, for years, Indian steel was not competitive in the world market, a situation that is only now beginning to change. In addition, as in the former Soviet Union, decisions about which products should be produced were made by ill-informed government bureaucrats rather than being driven by consumer demand. Until the 1980s, items that would improve daily life for the poor majority, such as cheap cooking pots, buckets, cheap yet sturdy bicycles, and simple tools, were produced in only small quantities. At the same time, there was a relative abundance of vacuum cleaners, watches, TVs, kitchen appliances, and cars, but only a very few could afford to purchase them. That, too, is now changing as the number of service employees grows.

## Economic Reform: Achieving Global Competitiveness

During the 1990s, much of South Asia began to undergo a wave of economic reforms (structural adjustment programs, or SAPs; see Chapter 3, pages 139–140). In contrast to countries in other regions in which structural adjustment programs were mandated by the International Monetary Fund (IMF) and World Bank, India's economic reforms were initiated by the Indian government itself in response to a financial crisis that emerged in the 1980s. Although privatization of India's public sector industries and banks has not proceeded very far, marketization of the economy has been significant, as evidenced in part by international investment and by the wide range of foreign goods that are now available. But India

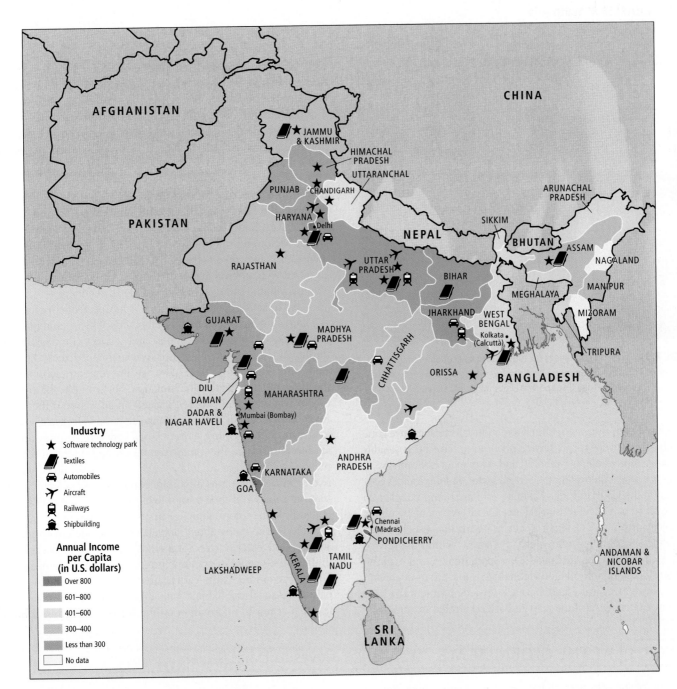

**FIGURE 8.33** Per capita annual income and industrial and information technology centers in India. The presence of industry is often associated with higher incomes, and planners may seek to bring industry to low-income places for this reason. In some areas, poverty may be so great that even fairly intensive industry is able to raise average incomes above base levels only slowly. [Data from Directorate of Economics & Statistics of the respective state governments, 2006.]

itself is now developing world-class companies that are investing not only in India (Figure 8.33), but also in Europe, the United States, Africa, and Southeast Asia. Such companies include TATA (engineering, oil, finance, manufacturing, tourism), Mittal (the world's largest steel producer by volume), and United Breweries of Bangalore.

In this more competitive environment, productivity has increased in the export and industrial sectors of the South Asian economy. Unlike China, whose growth was fueled by cheap labor, India, in particular, has achieved spectacular successes in high technology. Recent years have seen strong economic growth (7 percent and higher) driven by cutting-edge activity in the cities of Mumbai, Ahmadabad, and Bangalore. Pakistan and Bangladesh, with growth rates in the 6–7 percent range, are also showing overall economic improvement, though much less industrialization.

Like SAPs in Middle and South America and Africa, however, the new economic policies are producing wider disparities in income. The urban and industrial elites have received most of the gains, while little regard has been paid to the effects of reform on the poor, many of whom have lost jobs and access to social services.

**FIGURE 8.34** Training employees for India's service economy. An instructor in a Cochin outsourcing firm teaches other employees the skills necessary to provide technical support to U.S. clients. [© Stephen Voss.]

While South Asia's agricultural and industrial sectors have gone through ups and downs, its service economies have been steadily expanding. As a whole, between 20 and 40 percent of South Asian workers are now employed in the service sector, but this sector's contribution to the GDP in India, Pakistan, Sri Lanka, and Bangladesh is well over 50 percent and rising. Hence the diverse and increasingly technology-driven service sector is seen as having the best chance of competing successfully in the global economy. Within the service sector, facilities that engage in trade, transport, storage, and communication (including information technology) show the most growth; finance, insurance, real estate, business services, and tourism have also grown quickly. All these activities are connected in some way with international commerce and benefit from India's success at developing information technology (Figure 8.34). Tourism also has considerable potential as South Asia's own middle class expands and outsiders are attracted to the region. Tourism sites abound in both coastal and interior mountain locations, especially in India, and South Asian companies such as the TATA group are also investing in tourism sites abroad—in South Africa, for example.

***Differing Views of Globalization.*** Through trade, British colonization, and labor migration, South Asia has been connected to the global economy for thousands of years. Today, the growth of the service sector and information technology in South Asia foretells increasing connections to the global marketplace. This development is viewed with ambivalence in the region, particularly in India, the country with the strongest links to the global economy.

Globalization is desired because it is expected to increase the number of high-paying jobs, raise standards of living, and fuel local production of consumer products and services (Box 8.3). Cities throughout the region have already experienced these benefits. And some small rural places also see information technology as the answer to their geographic isolation. For example, the community of Dhab (40,000 people), located on an island in the Ganga River downstream of Varanasi (Benares), has no electricity and few jobs. The shifting course of the Ganga between the rainy and dry seasons has isolated the island community and inhibited development. Dhab's citizens are lobbying for dikes that would control the river's course and, more important, would support electricity and Internet cables and thus bring Dhab from the era of candles and oil lamps to the information age in just a few months.

Those who worry about globalization include environmental activists and conservative community leaders, who occasionally find common cause in opposing projects like the one proposed for Dhab. Environmentalists warn against unduly modifying the river's natural seasonal oscillations with dikes, and community leaders fear that access to information on the Internet will cause a break-

---

## BOX 8.3 | AT THE GLOBAL SCALE   Wall Street Moves to Dalal Street

One of the effects of globalization has been a migration of jobs from industrialized countries in the West to countries in Asia, Africa, and Latin America, as a way for corporations to lower costs. Manufacturing jobs were the first to go in the 1980s. Since the early 1990s, an increasing number of call center, back office, and even high-tech jobs have found their way to countries such as India that have large, college-educated, low-cost workforces. India's pharmaceutical and high-tech industries have attracted a number of American jobs in recent years. Now Wall Street firms such as J. P. Morgan, Goldman Sachs, and Lehman Brothers are beginning to seek out India's highly skilled workers, especially those of Mumbai's "Wall Street," Dalal Street, to provide finance and accounting services, including taxes, payroll, and internal auditing.

Stiff global competition makes cost cutting imperative for Wall Street firms. India's relatively low real estate costs and sala-ries seem to provide a solution. For example, a junior analyst from an Ivy League school costs $150,000 a year in the United States, but a graduate of a top Indian business school costs only $35,000 a year in India. Yet, for that Indian employee, this salary translates into over 1 million rupees, which buys a much higher standard of living than the U.S. employee would enjoy.

Total outsourcing of business service jobs by American companies is expected to reach $136 billion over the next decade. Of the 3.3 million jobs that will be created by American companies, 1 million will move abroad. India, along with China, Russia, and the Philippines, is expected to gain the most.

*Sources:* Saritha Rai, "As it tries to cut costs, Wall Street looks to India," *New York Times* (October 8, 2003); "Business sectors involved in job outsourcing," *Maps of India*, 2005, http://www.mapsofindia.com/outsourcing-to-india/index.html.

down of traditional social and economic relationships and bring a flood of Westernization. Television is another agent of globalization that is having a dramatic cultural impact, as the following vignette illustrates.

The tiny kingdom of Bhutan, couched in the Himalayas between China and India, had for centuries been more or less secluded from the rest of the world. That all changed in June 1999, when a royal decree legalized television, making Bhutan the last country in the world to "plug in." Within a short time, several entrepreneurs, such as Rinzy Dorji, whom the Bhutanese call "The Cable Guy," were in business. For $5 a month, the price of a bag of chilis, Rinzy provides Bhutanese households with 45 cable TV channels—everything from the BBC to *Baywatch.*

Although Rinzy's business is booming, not everyone welcomes the new technology. As Kinley Dorji, editor of Bhutan's only newspaper, describes, "Soon after television started, we started getting letters . . . from children, children who seemed very [anxious]. The letters actually specifically asked about this World Wrestling Federation program, 'Why are these big men standing there and hitting each other? What is the purpose of it?' They didn't understand. . . . Now, a few months later my son jumps on me one morning and says, 'I am Triple H, and you can be Rock.' And, suddenly we are fighting. Suddenly these [TV wrestlers] are new heroes for our children."

Foreign Minister Lyonpo Jigma Thinley thoughtfully observes, "People have suddenly realized that there are so many things they desire, which they were not even aware of before. The truth is that most of these television channels are commercially driven. And some of the Bhutanese people are driven towards consumerism. And, that is inevitable. It's unfortunate, but inevitable." On the other hand, he has also heard people saying, "My God! We didn't know that we are living in a peaceful country. There seems to be violence and crime everywhere in the world." He concludes, "So, in a way, the positive thing is that people realize how good a life they are living in this country."

*To learn more about the impact of satellite and cable TV in Bhutan, see the FRONTLINE/World Video "Bhutan: The Last Place." ■*

### Economic Development and Poverty Rates.
Poverty continues to decline in South Asia, but much more slowly since 1991 than it did from about 1968 to 1990. High population growth over the last 50 years has meant that, despite reductions in the poverty rate from nearly 45 percent of the population in 1952 to less than 25 percent by 2000, there are now nearly half again as many people who qualify as poor—simply because there are now so many more people overall. In 2000, there were 256 million people who could not afford a diet with the minimal caloric intake needed to sustain life (the regional definition of poverty), whereas 50 years ago there were only 180 million people in such straits.

### Innovative Help for the Poor.
In recent years, South Asians have developed some promising strategies for helping poor people. One of these is **microcredit,** a strategy that makes very small loans available to poor would-be business owners in both rural and urban areas. Throughout South Asia, and indeed, in much of the world,

**FIGURE 8.35** Microcredit for poor entrepreneurs. In this village in Bangladesh, women borrowers and participants in the local Grameen Bank gather weekly, usually at someone's house, to discuss business and pay their loan installments. For many, this is an unusual and treasured social outing. [IFAD Photo by Anwar Hossain.]

poor people have a difficult time obtaining loans from banks, so they must rely on small-scale moneylenders, who often charge interest rates of 30 percent or more *per month.* In the late 1970s, Mohammed Yunnis, an economics professor in Bangladesh, started the Grameen Bank, or "People's Bank," which makes small loans, mostly to people in rural villages who wish to start businesses. The loans often pay for the start-up costs of small enterprises such as cell phone–based services, chicken raising, small-scale egg production, or construction of pit toilets. The problem of collateral is resolved by having potential borrowers organize themselves into small groups that are collectively responsible for paying back the loans. If one member fails to repay a loan, then everyone in the group will be denied loans until the loan is repaid. This system, reinforced with weekly meetings, creates incentives to repay the loans. The weekly meetings are often the only time that women in purdah leave the confines of their homes and may be their first contacts with women (and sometimes men) who are not their kin (Figure 8.35). The repayment rate on the loans is extremely high, averaging around 98 percent—much higher than most banks achieve.

So far, the Grameen Bank has been an enormous success in Bangladesh, where it has loaned over U.S.$5.6 billion to more than 6 million borrowers. Similar microcredit projects have been established in India and Pakistan and throughout Africa, Middle and South America, North America, and Europe. In 2006, the Nobel Peace Prize was awarded to Dr. Yunnis for his work in microcredit.

**Personal Vignette** In a small hamlet in Bangladesh, not too far from the Indian border, is the house of Mosamad Shonabhan, a 32-year-old married woman whose life has been changed by her 11-year participation in the Grameen Bank. Everyone agreed that she had been the smartest of her family's children, but because her father earned only 50 cents a day as a farm laborer, she could not go to school, and was instead married at the age of 14 to a young barber. For a year she lived in her father-in-law's house, but then financial problems forced her to move back into her father's house. There she faced increasingly dire circumstances as her father's health deteriorated. After a few years, a local political leader suggested that she join the Grameen Bank's lending program. She was afraid to go, as she had never handled money and had heard a local rumor that the bank's real purpose was to convert people to Christianity. Nevertheless, she went, and eventually took out a loan for $40.00 that would allow her to set up a small rice-husking operation in her father's backyard. Eleven years and eleven loans later, Mosamad earns about $1.50 every day—three times what her father had made—and is a pillar of the local community. Her main source of income is a small shop inside her father's old house, which she bought from her siblings after his death. She also leases an acre of land, which produces enough rice to feed her family and the numerous guests and friends who now come by to see her. She plans to open another shop that her husband will run. She is being taught to read by her 15-year-old daughter, whom she plans to send to university.

*Source: Adapted from Alex Pulsipher's field notes, 2000.* ■

## POLITICAL ISSUES

Since independence in 1947, South Asian countries have peacefully resolved many conflicts, smoothed numerous potentially bloody transfers of power, and nurtured vibrant public debate over the issues of the day. India is often cited as a bastion of democracy that serves as an example of political enlightenment to the rest of the developing world. In recent years, however, there have been increasing signs that corruption, demagogic leadership, and violence are eroding democracy in the region. Shifting patterns of authority have increased tensions between upper- and lower-caste groups. The rise of religious nationalism also threatens peaceful relations both within and among the region's nations.

### Caste and Democracy

Since the beginning of democracy in India in the 1930s, caste has been a defining yet contradictory factor in both local and national politics. At the local level, most political parties design their vote-getting strategies to appeal to subcaste (*jati*) loyalties. They often secure the votes of entire *jati* communities with such political favors as new roads, schools, or development projects. These arrangements fly in the face of the official ideologies of the major political parties, which deny any caste loyalties, and of Indian government policies, which actively work to undermine discrimination on the basis of caste. Currently, the role of caste in politics seems to be increasing, as several new political parties that explicitly support the interests of low castes have emerged. This assertion of political rights by low castes has been met with a backlash from upper-caste groups, resulting in a number of violent clashes in recent years. Hence caste has been woven into the political system in ways that create and maintain tension and conflict.

### Religious Nationalism

Increasingly, people frustrated by government inefficiency, recent scandals surrounding corruption, and the failure of governments to deliver on their promises of broad-based economic development and prosperity are joining religious nationalist movements (Figure 8.36). **Religious nationalism** is the belief that a particular religion is

**FIGURE 8.36** Religious nationalism. Indian holy men participate in the eleventh religious parliament in the Hindu temple town of Vadtal, near Ahmadabad, India, in 2006. The parliament was organized by the Hindu nationalist group VHP, which takes extreme positions toward Muslims. The painting behind the group shows, at the left, the Hindu goddess Durga killing Pakistani president Pervez Musharraf, and at the right, the Hindu god Lord Shiva dancing on the body of Osama bin Laden. [AP Photo/Ajit Solanki.]

## BOX 8.4 | AT THE REGIONAL SCALE  Babar's Mosque or Ram's Temple: The Geography of Religious Nationalism

Proponents of religious nationalism often try to gain mass support through political campaigns and acts of terrorism that interweave with South Asian history, mythology, and landscapes. In late 1992 and early 1993, a series of riots occurred throughout South Asia that were the culmination of a long campaign waged by India's leading proponents of Hindu nationalism, the VHP (World Hindu Council), and have affected Indian society up to the present. The riots were triggered by the destruction of a Muslim mosque by some 300,000 Hindus brought to the town of Ayodhya on the Ganga Plain by the VHP. The Mughal emperor Babar had built the mosque in 1528–1529, purportedly on the ruins of a Hindu temple believed to mark the birthplace of the Hindu god Ram. After the mosque's demolition, mobs commanded by urban Hindu nationalist political parties and social organizations (many of them funded by Indian Hindus living abroad) burned and looted selected Muslim businesses and homes throughout India, often with the complicity of the police. Nearly 5000 people died.

Violence erupted again in 2002, when a train full of Hindus returning to Gujarat from the disputed site at Ayodhya was attacked and burned by a Muslim mob. Ensuing riots killed between 2000 and 2500 people, most of whom were Muslims. Then, in 2005, five Muslim terrorists from Pakistan-administered Kashmir attacked the site of the mosque, now occupied by a makeshift Hindu temple. All of the terrorists were killed in the attack, but only a few people died in the ensuing riots.

The VHP's selection of the Ayodhya mosque-temple site as a locus of protest was calculated to elicit violent reactions from both Muslims and Hindus. The locations and tactics chosen reflected the urban base of Hindu nationalist parties and the violent gangster-style tactics still common in local South Asian politics.

*Sources:* Alex Pulsipher's field notes, India, 1993; Stuart Corbridge and John Harriss, *Reinventing India* (Malden, Mass.: Blackwell-Polity Press, 2000); *Amnesty International Report 2003*, AI index: POL 10/003/2003, http://www.amnesty.org; "Unrest in India following attack in Ram Mandir," *India Daily* (July 7, 2005); "Ayodhya: Jail custody for militants arrested in JK," *India Info PTI* (Thursday, August 4, 2005).

strongly connected to a particular territory, perhaps even to the exclusion of other religions, and that those who share that religion should have control over their own political unit—be it a neighborhood, part of a country, or a separate country. Although both India and Pakistan were formally created as secular states, religious nationalism has long been a reality, shaping relations between people and their governments in those countries. India is increasingly thought of as a Hindu state and Pakistan and Bangladesh as Muslim states. Many people in the dominant religious group strongly associate their religion with their national identity.

Hindu nationalism in India is predominantly supported by urban men from middle- and upper-caste groups who fear the erosion of their castes' political influence and who particularly resent the extension of the quota system for government jobs and seats in universities to lower-caste groups (see pages 434–435). This sentiment has been fueled by alliances among politically mobilized lower castes who are no longer willing to follow the dictates of the dominant castes.

Political parties based on religious nationalism have gained popularity throughout South Asia, often fueling conflicts between religious majorities and minorities within a particular state (Box 8.4). Although their members think of these parties as forces that will purge their country of corruption and violence, they are, in fact, usually only slightly less corrupt, and certainly no less violent, than other parties.

## Regional Political Conflicts

The most intense armed conflicts in South Asia today are **regional conflicts,** in which nations dispute territorial boundaries or a minor-

ity actively resists the authority of a national or state government (Figure 8.37). Two regional conflicts in the neighboring Indian states of Punjab and Kashmir can serve as examples. (Other regional conflicts, in Assam and in Sri Lanka, are discussed on page 463 and pages 468–469, respectively.) Both Punjab and Kashmir are located in far northwestern India along the border with Pakistan, and both conflicts have religious as well as political components. Although the Punjab crisis has faded in recent years, Kashmir continues to be a source of unrest that has the potential to destabilize the region.

***Conflict in Punjab.*** Punjab is the ancestral home of the Sikh community. When India and Pakistan were partitioned in 1947, the state of Punjab was divided between the two countries. Caught between majority populations of Muslims and Hindus during the violence that followed the Partition, large numbers of Sikhs chose to live in the Indian part of Punjab or elsewhere in India because they thought the secular Indian constitution would better allow them to preserve their unique identity. Since that time, however, Sikhs have felt alienated from the rest of India, despite the fact that they are a wealthy and influential minority in many cities outside of Punjab. In 1973, the Akali Dal, the moderate Sikh political party, issued demands for religious concessions, equitable water rights, and greater political autonomy, which were ignored by the national government of India. In an attempt to keep the Akali Dal fractured, the ruling Congress Party cultivated relations with Sikh extremists.

This strategy backfired in the early 1980s as the extremists gained strength and barricaded themselves in the holiest Sikh shrine, the Golden Temple in Amritsar, which they then used as a base of operations to agitate for an independent Sikh nation. In

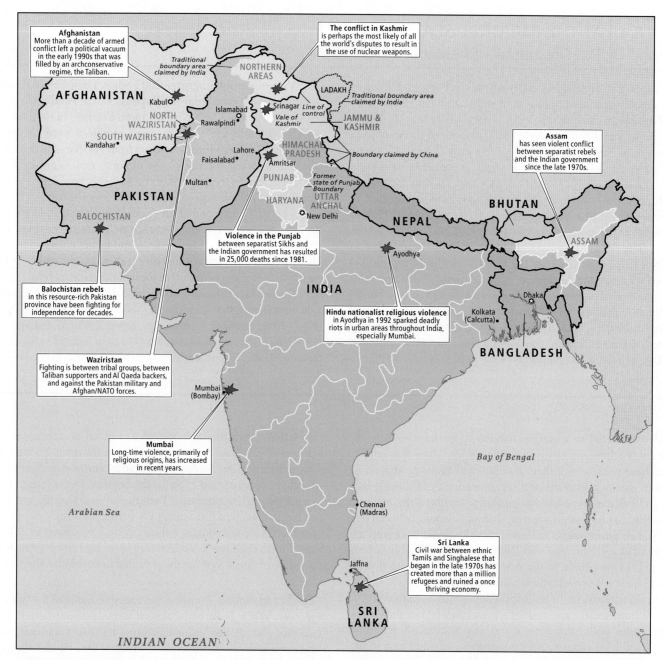

**Afghanistan**
More than a decade of armed conflict left a political vacuum in the early 1990s that was filled by an archconservative regime, the Taliban.

**The conflict in Kashmir**
is perhaps the most likely of all the world's disputes to result in the use of nuclear weapons.

*Traditional boundary area claimed by India*

NORTHERN AREAS

LADAKH

*Traditional boundary area claimed by India*

AFGHANISTAN

Kabul ⊕

NORTH WAZIRISTAN

SOUTH WAZIRISTAN

Kandahar •

Islamabad ⊕

Rawalpindi •

Srinagar ✦
*Vale of Kashmir*

*Line of control*

JAMMU & KASHMIR

HIMACHAL PRADESH

*Boundary claimed by China*

**Assam**
has seen violent conflict between separatist rebels and the Indian government since the late 1970s.

Lahore •

Faisalabad •

Amritsar •

PUNJAB

PAKISTAN

Multan •

*Former state of Punjab Boundary*

UTTAR ANCHAL

BHUTAN

ASSAM

BALOCHISTAN

HARYANA

⊕ New Delhi

NEPAL

**Violence in the Punjab**
between separatist Sikhs and the Indian government has resulted in 25,000 deaths since 1981.

**Balochistan rebels**
in this resource-rich Pakistan province have been fighting for independence for decades.

INDIA

Ayodhya ✦

Dhaka ⊕

Kolkata (Calcutta) •

**Hindu nationalist religious violence**
in Ayodhya in 1992 sparked deadly riots in urban areas throughout India, especially Mumbai.

**Waziristan**
Fighting is between tribal groups, between Taliban supporters and Al Qaeda backers, and against the Pakistan military and Afghan/NATO forces.

BANGLADESH

Mumbai (Bombay) ✦

**Mumbai**
Long-time violence, primarily of religious origins, has increased in recent years.

*Bay of Bengal*

*Arabian Sea*

Chennai (Madras) •

**Sri Lanka**
Civil war between ethnic Tamils and Singhalese that began in the late 1970s has created more than a million refugees and ruined a once thriving economy.

Jaffna •

SRI LANKA

*INDIAN OCEAN*

**FIGURE 8.37** Conflict in South Asia. Some of the most well known violent conflicts in South Asia, such as those over Kashmir and Punjab, are rooted in disputes that have remained unsettled since the Partition of India and Pakistan. Some conflicts, such as those in Afghanistan and Pakistan, are linked to international disputes and some to religious differences; nearly all are in some way related to disputes over land and resources.

1984, government forces attacked the shrine, damaged the temple, and killed the militants as well as numerous innocent pilgrims caught in the crossfire. This incident deeply alienated and further radicalized the Sikh community. Shortly thereafter, Prime Minister Indira Gandhi, who had called for the attack, was assassinated by two of her Sikh bodyguards. Riots and organized mob violence spread throughout India over the next few days, resulting in the deaths of more than 2700 Sikhs.

Since then, agreements acceding to many Sikh demands for water rights and control of religious sites have been signed, but not implemented, which alienated Sikhs yet further and led to more political violence and the deaths of 25,000 people. Further violence has been avoided for the time being, but the Sikh situation remains sensitive.

**Conflict in Kashmir.** Of all the armed disputes in South Asia, many experts fear that the situation in Kashmir most threatens global peace. Kashmir has long been a Muslim-dominated area, and in 1947 Pakistan's leaders believed that it should be turned over to Pakistan based on the rules for the Partition established by the British. However, the maharaja (king) of Kashmir at the time, a Hindu, wanted Kashmir to remain independent, and its most popular political leader and much of the populace favored joining India. When Pakistan-sponsored raiders invaded western Kashmir in 1947, the maharaja quickly agreed to join India. A brief war between Pakistan and India resulted in a cease-fire line (line of control) that became a tenuous boundary.

A popular vote for or against joining India was never held due to India's resistance. Pakistan attempted another invasion of Kashmir in 1965, but was defeated. The two countries are technically still waiting for a UN decision on where the final border will be, but Pakistan effectively controls the thinly populated mountain areas north and west of the densely populated Vale of Kashmir. India holds nearly all the rest, where it maintains an ominous presence with more than 500,000 troops. A more limited border dispute between India and China involves the Ladakh region of Kashmir (see Figure 8.37).

Since 1947, between 60,000 and 100,000 people have been killed in the conflict over Kashmir (Figure 8.38). Civil war has erupted repeatedly because, after years of military occupation, many Kashmiris support independence from both India and Pakistan. As in Punjab, much of the conflict has centered around the right of Kashmiris to run their own affairs, as the national government in New Delhi has often appointed its own favorites in an attempt to maintain strong central control. Anti-Indian Kashmiri guerrilla groups, equipped with weapons and training from Pakistan, have carried out many bombings and assassinations. Blunt counterattacks launched by the Indian government have killed large numbers of civilians and alienated the local police force, which is now seen as sympathetic to the militants. Sporadic fighting between India and Pakistan continues along the boundary line, where at one location the two countries intermittently clash in the world's highest battle zone, at an altitude of 20,000 feet (6000 meters).

Another complication in the Kashmir dispute is the increased involvement throughout the 1990s of Islamist militants known as mujahedeen, soldiers who played a major role in Afghanistan's war of resistance against the Soviet Union in the 1980s. Many mujahedeen receive weapons and training from Pakistan (as do the Kashmiri militants), but since 2001, some have also been supported by terrorist networks such as Al Qaeda.

The conflict is complicated still further by the fact that both India and Pakistan—which came close to war in 1999 and again in 2002—have nuclear weapons. Although some international security analysts believe that the conflict in Kashmir is the one most likely in the world to result in the use of nuclear weapons because of the nationalistic fervor of the protagonists, others argue that the existence of nuclear weapons on both sides of the border has actually deterred an all-out war in much the same way the United States and the Soviet Union were deterred from engaging in armed conflict during the cold war period.

## War and Reconstruction in Afghanistan

In the 1970s, political debate in Afghanistan became polarized between urban elites, who favored industrialization and democratic reforms, and rural conservative religious leaders, whose positions as landholders and ethnic leaders were threatened by the proposed reforms. Some urban elites allied themselves with the Soviets, who, fearing that a civil war in Afghanistan would destabilize the Central Asian states of the Soviet Union, invaded Afghanistan in 1979. The anti-Soviet mujahedeen were formed by the rural conservative leaders (often erroneously labeled "warlords") and their followers, who, as resistance to Soviet domination increased, became ever more strongly influenced by militant Islamist thought. The United States and its allies Pakistan and Iran overlooked this development because it was still the cold war era, and they gave considerable support to any anti-Soviet movement. The mujahedeen were tenacious fighters, and in 1989, after heavy losses, the Soviets gave up and left the country. Anarchy prevailed for a time as the Afghan factions fought one another, but the rural conservatives eventually defeated the reformist urban elites.

In the early 1990s, the radical religious-political-military movement called the Taliban emerged from among the mujahedeen. For the most part, the Taliban are illiterate young men from remote villages, led by students from the *talibs* (Islamist schools of philosophy and law). The Taliban saw their role as controlling corruption and crime, minimizing extreme Western ways introduced or reinforced by the Russian occupation and seen as licentious by rural Afghans, and bringing stability and peace by strictly enforcing shari'a, the Islamic social and penal code (see Chapter 6, page 315). In addition to the restrictions they placed on women (see page 435), efforts by the Taliban to purge their society of non-Muslim influences included restricting education for everyone, especially girls and young women; destroying 1500-year-old Buddhist sculptures in the Bamiyan Valley; and banning the production of opium, to which many Afghan men had become addicted.

**FIGURE 8.38 Conflict over Kashmir.** In Jammu, India, policeman Randhir Singh's widow, left, mourns as she touches her husband's body after its arrival from Srinagar. Singh was one of seven policeman killed during a gunfight in October 2006, in which Indian security forces battled Islamist militants holed up in a hotel in the heart of Srinagar, Indian Kashmir's main city. [AP Photo/Channi Anand.]

**FIGURE 8.39** Reconstruction in Afghanistan. By May 2003, reconstruction was in full swing in Kabul, which was heavily damaged after three decades of war. Here, against a backdrop of brick kilns, Afghan men move building materials for a structure that will be a restaurant. [AP Images.]

By 2001, the Taliban had been in power for just 6 years but controlled 95 percent of the country, including the capital, Kabul. They were moving north, where an alliance of various ethnic groups (the Northern Alliance) was mounting a counteroffensive. The events of September 11, 2001, focused the United States and its allies on removing the Taliban, who were at the time giving shelter to Osama bin Laden and his Al Qaeda network, the apparent perpetrators of the terrorist attacks. By late 2001, the Taliban were overpowered, and the United Nations stepped in to help establish an interim coalition government. A national assembly was convened to designate a new government and appoint a head of state in 2002, and to ratify a new constitution in 2004.

Postwar reconstruction and the establishment of democratic government in Afghanistan face a number of challenges, including the country's ethnic and linguistic diversity, poverty, lack of infrastructure, and a landscape devastated by war (Figure 8.39). Furthermore, the ability of the new government to ensure security and to meet the needs of people outside Kabul has been thwarted by the continuing influence of the Taliban, by ethnic leaders (some of whom are funded by opium production), and by widespread distrust of the international military forces and private security personnel stationed in Afghanistan since 2001. A 2003 report by ACTIONAID (an antipoverty NGO based in the United Kingdom) indicated that the majority of Afghanistan's rural poor favor democratic government based on Islam, but feel that their concerns have not been adequately addressed.

## The Future of Democracy

Although there are many political hot spots in South Asia, some of its countries have elected democratic governments regularly since their independence. Yet the status of democracy varies throughout the region. Signs of democratic expansion in India include the emergence of a more competitive multiparty system with freer and more

peaceful elections, and a clearer focus on reducing corruption and violence and providing opportunities for women and other disadvantaged groups. Despite Sri Lanka's democratic institutions and relatively high standard of living, that country continues to be wracked by violence between the Singhalese and Tamil factions. Although Pakistan has had elections, it functions as a military dictatorship. In Bangladesh, after years of military dictatorship, democratic elections have occurred with some regularity, though government corruption is a recurring cause for public protest. The tiny Himalayan kingdom of Bhutan granted its people the right to elect local representatives in 2002, and elections were held in 2005. The king still retains considerable power, though he favors reducing his influence in politics.

In Nepal, an elected legislature and multiparty democracy were introduced in 1990. Intended to reduce the king's power and give ordinary citizens a greater political voice, these reforms did not result in either improved economic conditions for the Nepalese people or real political change. Revolutionaries inspired by the ideals of the late Chinese leader Mao Zedong (but with no support from China) waged a "people's war" against the Nepalese monarchy and controlled much of the countryside. In 2006, the king dissolved the elected parliament, thereby losing the support of the middle class, which then mounted a major peaceful demonstration. The king relented, agreed to reinstate the parliament, and in April 2007, several Maoists joined the interim government, controlling 5 of the 21 ministries.

## ENVIRONMENTAL ISSUES

South Asia has been occupied by people for millennia, but in 1700 (just prior to British colonization), population density and human environmental impacts were low (Figure 8.40a). By the beginning of the twenty-first century, population density and human impacts were vastly greater (Figure 8.40b). Today, South Asia has a range of

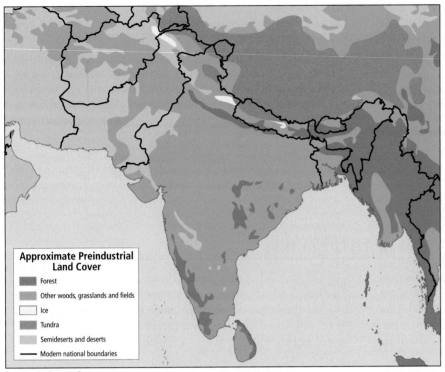

(a) Preindustrial

**FIGURE 8.40** Land cover (preindustrial) and human impact (2002). South Asia has been occupied by humans for thousands of years; part (**a**) shows land cover in the preindustrial era, when much of the forest had already been removed. By 2002 (**b**), the human impact on land, water, and air was extensive throughout the region, but variable in intensity. Remote and sparsely occupied parts of Afghanistan and Pakistan were impacted by war. [Adapted from *United Nations Environment Programme, 2002, 2003, 2004, 2005, 2006* (New York: United Nations Development Programme), athttp://maps.grida.no/go/graphic/human _impact_year_1700_approximately and http://maps.grida .no/go/graphic/human_impact_year_2002.]

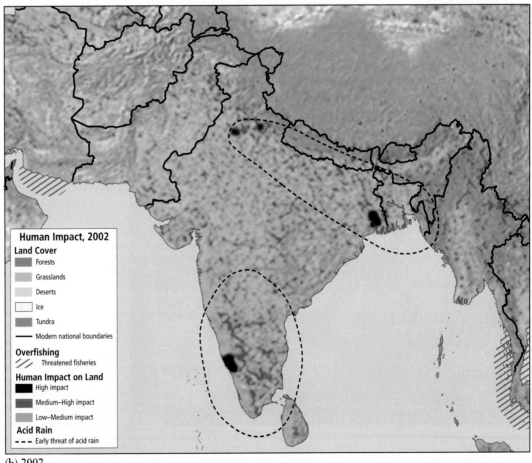

(b) 2002

serious environmental problems, including deforestation, water shortages, water pollution, and industrial pollution. However, the region is also implementing a broad array of solutions, some of which mix modern environmental research and technologies with the deep cultural roots of South Asia's people.

## Deforestation

Deforestation is not new to the Indian subcontinent. The spread of agriculture at the time of the Indus Valley civilization, as well as the expansion of developing Hindu kingdoms after the Aryan migrations, resulted in the clearing of forested lands beginning more than 3000 years ago. Ecological historians Madhav Gadgil and Ramachandra Guha have shown that the western regions of the subcontinent (from India to Afghanistan) became drier and drier as the forests vanished. Starting in the mid-nineteenth century, perhaps a million trees a year were felled for use in building railroads alone.

In the twenty-first century, the subcontinent's forests are still shrinking due to commercial logging and village populations that are expanding into forestlands for living and cultivation space. Forests are also being cleared to make trails for tourist trekkers in mountainous zones, such as those in Nepal, Kashmir, and the Nilgiri Hills of southern India, and the trekkers consume wood for cooking and heat. Among the results of forest clearing are mas-

---

### BOX 8.5 | AT THE LOCAL SCALE   A Visit to the Nilgiri Hills

The Mudumalai Wildlife Sanctuary in the Nilgiri Hills (part of the Western Ghats), in Tamil Nadu, harbors some of the last scraps of forest in southern India. Here live a few of the last wild Bengal tigers (Figure 8.41) and a dozen or more other rare species, such as leopards, panthers, sloth bears, pythons, flying squirrels, and barking deer.

Longwood Shola, another Nilgiri reserve, is a tiny remnant of ancient tropical evergreen forest where two local naturalists, members of ethnic minorities native to the Nilgiri Hills, manage the 287-acre (116-hectare) park. This small space harbors 13 mammal species, 52 bird species, and 118 plant species, many of them found only in the Nilgiris. One of those bird species, the grey jungle fowl, is ancestor to the domestic chicken. Tiny as it is, the forest of Longwood Shola protects three perennial streams that supply fresh water to 16 downstream villages. Among the projects the naturalists have initiated is a reforestation effort for which they themselves are growing the seedlings from local native species.

Phillip Mulley, a naturalist, Christian minister, and leader of the Badaga ethnic group, helps visitors to understand the issues that face the people native to the Nilgiri Hills. The Badaga, Toda, Kota, and Kurumba indigenous peoples must now compete for space not only with a growing tourist industry (1.7 million visitors in 2005), but also with huge tea plantations that were recently cut out of forestlands by the state government to provide employment for Tamil refugees from the ongoing conflict in Sri Lanka. So while one branch of the state government (the forestry department) and citizen naturalists are trying to preserve forestlands, another branch (the social welfare department), faced with a huge refugee population, is cutting them down.

*Sources:* Lydia and Alex Pulsipher's field notes, Nilgiri Hills, June 2000; *Tamil Nadu Human Development Report*, 2003, at http://www.undp.org.in/hdrc/shdr/TN/.

**FIGURE 8.41 Endangered wildlife.** This photo was taken during an unexpected encounter with a Bengal tiger in the Mudumalai Wildlife Sanctuary in the Nilgiri Hills in southwestern India. The Bengal tiger is a rare and endangered species. [Courtesy of Alex Pulsipher.]

sive landslides that often close railroad lines and roads during the rainy season. These landslides increase erosion, and the eroded mountain environments are less able to absorb rainfall during future summer monsoons, leading in turn to increased flooding and silt deposition downstream in lowland places such as Bangladesh.

In 1973, in the Himalayan district of Uttar Pradesh, India, a sporting-goods manufacturer planned to cut down a grove of ash trees so that his factory, in the distant city of Allahabad, could use the wood to make tennis racquets. The trees were sacred to nearby villagers, however, and when their protests were ignored, a group of local women took dramatic action. When the loggers came, they found the women hugging the trees and refusing to let go until the threat to their grove ended. Soon the manufacturer located another grove. The women's action grew into the **Chipko** (literally, "hugging") or **social forestry movement,** which has spread to other forest areas, slowing deforestation and increasing ecological awareness.

Unlike China and many other nations facing similar problems, South Asia has a healthy and vibrant climate of environmental activism that brings the consequences of deforestation to the attention of the public. Activism focused on saving forests is a reaction to a pattern found throughout South Asia, in which the resources of rural areas are channeled to urban industries without consideration of the needs of local rural people. The proponents of the social forestry movement argue that management of forest resources should be turned over to local communities. They say that people living at the edges of forests possess complex local knowledge of those ecosystems gained over generations—knowledge about which plants are useful for building materials, for food, for medicinal uses, and for fuel. These people have the incentive to manage forests carefully because they want their progeny to benefit from forests for generations to come.

The fact is, though, that it is difficult to convince impoverished people to conserve forest resources. Burgeoning local populations themselves contribute to deforestation as they try to obtain ever more firewood for themselves and fodder for their animals, and their need for income predisposes them to collaborate with poachers of rare forest products. Moreover, the powerful industrial and government interests that monopolize forest reserves are not likely to yield control to local people easily. Despite these problems, several natural reserves have been established in India, and local residents are gaining increasing influence in decisions about control of these areas (Box 8.5).

## Water Issues

One of the most controversial environmental issues in South Asia today is the use of water. South Asia has more than 20 percent of the world's population, but only 4 percent of its fresh water. It is not surprising, then, that India and Bangladesh have disputes over access to the waters of the Ganga River.

***Conflicts over Ganga River Water.*** In recent years, during the dry season, India has diverted 60 percent of the Ganga's flow to Kolkata to flush out channels where silt is accumulating and hampering river traffic. India's policies, however, deprive Bangladesh of normal freshwater flow. Less water in the delta allows salt water from the Bay of Bengal to penetrate inland, ruining agricultural fields. The diversion has also caused major alterations in Bangladesh's coastline, damaging its small-scale fishing industry. Thus, to serve the needs of Kolkata's 15 million people, the livelihoods of 40 million rural Bangladeshis have been put at risk, triggering protests in Bangladesh. In the late 1990s, India signed a treaty promising a fairer distribution of water, but as of 2006, Bangladesh was still receiving a considerably reduced flow of water. As with other major environmental problems, a solution has been hard to achieve because the population adversely affected is not only poor and rural, but is located in a different region—in this case, in a different country—from the politicians and bureaucrats who are in a position to respond to protests.

Similar water use conflicts occur between states within India and between the wealthier and poorer sectors of the population. For example, just 17 five-star hotels in Delhi use about 210,000 gallons (800,000 liters) of water daily, enough to serve the needs of 1.3 million slum dwellers. At the state level, Haryana's diversion of water from the Ganga deprives farmers downstream in Uttar Pradesh of the means to irrigate their crops.

***Water Purity of the Ganga River.*** Water purity is an issue in historic religious pilgrimage towns such as Varanasi, where each year millions of Hindus come to die, be cremated, and have their ashes scattered over the Ganga River. As the number of such final pilgrimages has increased, wood for cremation fires has become scarce, and incompletely cremated bodies are being dumped into the river, where they pollute water used for drinking, cooking, and bathing. In an attempt to deal with this problem, the government recently installed an electric crematorium on the riverbank. It is attracting quite a bit of business, as a cremation here costs 30 times less than a traditional funeral pyre.

Of greater concern now is the large amount of industrial waste and sewage dumped into the river (Figure 8.42). Most sewage enters the river in raw form because Varanasi's sewage system (built by the British early in the twentieth century) long ago exceeded its capacity. Pumps have been installed to move the sewage up to a new and expensive sewage processing plant, but the plant is so overwhelmed by the volume of water during the rainy season that it can process only a small fraction of the city's sewage.

Veer Bhadra Mishra, a Brahmin priest and professor of hydraulic engineering at Banaras Hindu University in Varanasi, is on a mission to clean up the Ganga using unconventional methods. He is working with engineers from the United States to build a series of processing ponds that will use India's heat and monsoon rains to clean the river at half the cost of other methods. In addition, he preaches a contemporary religious message to the thousands who visit his temple on a bank of the sacred Ganga. The belief that the Ganga is a goddess that purifies all she touches leads many Hindus to think that it is impossible to damage this magnificent river. Mishra reminds them that because the Ganga is

**FIGURE 8.42** Pollution of the Ganga. The city of Varanasi dumps waste into the Ganga River through many sewage pipes like this one. The Ganga is India's most sacred and, at the same time, most polluted river. [AP Photo/ John McConnico.]

their symbolic mother, it would be a travesty to smear her with sewage and industrial waste.

## Industrial Pollution

In many parts of South Asia, the air as well as the water may be endangered by industrial activity. Emissions from vehicles and coal-burning power plants are so bad that breathing Delhi's air is equivalent to smoking 20 cigarettes a day. The acid rain caused by industries up and down the Yamuna and Ganga rivers is destroying good farmland and such great monuments as the Taj Mahal. M. C. Mehta, a Delhi-based lawyer, became an environmental activist partly in response to the condition of the Taj Mahal. For more than 20 years, he has successfully promoted environmental legislation that has removed hundreds of the most polluting factories from the river valleys. His efforts are also a response to a horrible event that took place in central India in 1984, when an explosion at a pesticide plant in Bhopal produced a gas cloud that killed at least 3000 people and severely damaged the lungs of 50,000 more. The explosion was largely the result of negligence on the part of the U.S.-based Union Carbide Corporation, which owned the plant, and the local Indian employees who ran it. In response to the tragedy, the Indian government launched an ambitious campaign to clean up poorly regulated factories.

## MEASURES OF HUMAN WELL-BEING

Gross domestic product (GDP) per capita in South Asia (Table 8.1) is nearly as low as it is in Africa—and in some cases lower—and a large proportion of people in the region are extremely poor. Most people in this region, however, are frugal and resourceful. They recycle nearly everything, they are entrepreneurs in the informal economy, they grow their own food whenever possible, and they strictly limit cash expenditures through reciprocal exchange agreements with one another. Through such efforts, the people of South Asia manage to give themselves a somewhat higher standard of living than the GDP per capita figures would indicate.

South Asia does not exhibit the wide variations of GDP per capita among countries seen in some world regions (East Asia, Southeast Asia, and Oceania, for example). The Maldives, which has the highest GDP per capita, has only a tiny population and depends on tourism for much of its income. Sri Lanka's high GDP per capita results from a physical environment that is favorable to agriculture (tea is a chief product) and from the presence of exportable minerals. More important, a history of investing in the education and health of its citizens has helped Sri Lanka to equalize wealth distribution among its 19 million people. The country would undoubtedly be even more prosperous if conflict between the Tamil and Singhalese ethnic groups had not hindered its development for several decades.

On the United Nations Human Development Index (see Table 8.1, column III), all countries in the region except Afghanistan had advanced to the medium ranking as of 2005—a notable accomplishment, given the huge populations involved (more than one-fifth of Earth's population). However, the rankings are based on averages, so they do not reflect the staggering numbers of people who remain very poor. With the exceptions of the Maldives and Sri Lanka, both literacy (columns V and VI) and life expectancy (col-

**TABLE 8.1**  Human well-being rankings of countries in South Asia and other selected countries[a]

| Country (I) | GDP per capita, adjusted for PPP[b] (GDP ranking among 177 countries), 2005 (II) | Human Development Index (HDI) ranking among 177 countries,[c] 2005 (III) | Gender Development Index (GDI) ranking among 140 countries, 2005 (IV) | Female literacy (percent), 2003 (V) | Male literacy (percent), 2003 (VI) | Life expectancy, 2005 (VII) |
|---|---|---|---|---|---|---|
| **Selected countries for comparison** | | | | | | |
| Barbados | 15,270  (36) | 30  (high) | 29 | 99 | 99 | 75 |
| Japan | 22,967  (13) | 11  (high) | 14 | 99 | 99 | 82 |
| Kuwait | 18,047  (30) | 44  (high) | 39 | 81 | 84.7 | 77 |
| United States | 37,562  (4) | 4  (high) | 8 | 99 | 99 | 77 |
| World | 8,229 | | | | | |
| **South Asia** | | | | | | |
| Afghanistan[d] | 800 | Not ranked | ND | 21 | 51 | 47 |
| Bangladesh | 1,770  (133) | 139  (medium) | 105 | 31.4 | 50.3 | 63 |
| Bhutan | 1,969  (129) | 134  (medium) | ND | 34[d] | 60[d] | 55[d] |
| India | 2,892  (114) | 127  (medium) | 98 | 47.8 | 73.4 | 63 |
| Maldives | 4,798[e]  (95) | 96  (medium) | ND | 97 | 97 | 67 |
| Nepal | 1,420  (146) | 136  (medium) | 106 | 35 | 63 | 62 |
| Pakistan | 2,097  (125) | 135  (medium) | 107 | 35 | 62 | 63 |
| Sri Lanka | 3,778  (106) | 93  (medium) | 66 | 88.6 | 92.2 | 74 |

[a]Rankings are in descending order; i.e., low numbers indicate high rank.

[b]PPP = purchasing power parity, figured in 2003 U.S. dollars.

[c]The high, medium, and low designations indicate where the country ranks among the 177 countries classified into three categories by the United Nations.

[d]Data from *CIA World Factbook*, 2006; not available in UNHDI report, 2005.

[e]World Bank preliminary estimate; no UN GDP data available.

ND = No data available.

*Sources: United Nations Human Development Report 2005* (New York: United Nations Development Programme), except as indicated above, at http:/hdr.undp.org/reports/global/2005/pdf/HDR05_complete.pdf. *CIA World Factbook*, 2006, at https://www.cia.gov/cia/publications/factbook/index.html.

umn VII) are very low in this region, especially for women. And, as in Africa, governments have not provided even the most basic services to large portions of the population. For example, in all countries except Sri Lanka, a secondary education is provided for only 50 percent or less of the eligible children. In Afghanistan under the Taliban in 1995, just 12 percent of those eligible were in school, none of them girls. That figure improved markedly by 2005, when millions of Afghan children were back in school, 1.5 million of them girls. But that progress was hard to maintain. After a resurgence of the Taliban in 2006, hundreds of thousands of children again left

school, again, many of them girls. Sanitation is another basic service that is crucial to well-being. In the entire region, less than half of the people have access to sanitary toilets, including outhouses and pit toilets. The lack of these facilities contributes to the generally low life expectancy.

The region's generally low rankings on the United Nations Gender Development Index (see Table 8.1, column IV) (India ranked 98 out of 140 countries reporting; Bangladesh, 105; Pakistan, 107) indicate a serious and pervasive lack of basic opportunities in health care, education, and earning power for women.

# III SUBREGIONS OF SOUTH ASIA

The subregions of South Asia are grouped roughly according to their physical and cultural similarities, so in several cases, parts of India are grouped with adjacent countries.

## AFGHANISTAN AND PAKISTAN

Afghanistan and Pakistan (Figure 8.43) share location, landforms, and history, and they have both been involved in recent political disputes over global terrorism. Many cultural influences have passed through these mountainous countries into the rest of South Asia: Aryan migrations, Alexander the Great and his soldiers, the continual infusions of Turkic and Persian peoples, and the Turkic-Mongol influences that culminated in the Mughal invasion of the subcontinent at the beginning of the sixteenth century. Today, both countries are primarily Muslim and rural; 78 percent of Afghan-

istan's people and 66 percent of Pakistan's live in villages and hamlets. Nonetheless, there are large cities in both countries, including Kabul, Herat, Karachi, Islamabad, and Peshawar. Both countries must cope with arid environments, scarce resources, and the need to find ways to provide rapidly growing urban and rural populations with higher standards of living. Both countries are home to conservative Islamist movements. Afghanistan was ruled until late 2001 by a fundamentalist religious group, the Taliban; Pakistan is ruled by a military dictatorship that needs the support of fundamentalists to stay in power. Afghanistan's more extreme poverty is made worse by the armed strife from which it has suffered for nearly 30 years (see pages 447–448).

The landscape of this subregion is best considered in the context of the ongoing tectonic collision between the landmasses of India and Eurasia. At the western end of the Himalayas, the colli-

**FIGURE 8.43** The Afghanistan and Pakistan subregion.

**FIGURE 8.44** The Pamir Mountains. The Pamirs, in the north of Afghanistan, rise abruptly from narrow valleys to such a height that a person must bend backward to see the summits. Seasonal rivers often course through the dry and windswept landscapes past tiny villages huddled next to near-vertical slopes. [© Luke Powell.]

sion uplifted the lofty Hindu Kush and Pamir and Karakoram mountains of Afghanistan and Pakistan. This system of high mountains and intervening valleys swoops away from the Himalayas and bends down to the southeast toward the Arabian Sea. Landlocked Afghanistan, bounded by Pakistan on the east and south, Iran on the west, and Central Asia on the north, is entirely within this mountain system. Pakistan has two contrasting landscapes: the north, west, and southwest are in the mountain and upland zone just described; the central and southeastern sections are arid lowlands watered by the Indus River and its tributaries. In both countries earthquakes are the common result of continued tectonic pressure and slippage.

## Afghanistan

The Hindu Kush and the Pamir Mountains are rugged and steep-sided, in the north of Afghanistan (Figures 8.43 and 8.44). The mountains extend west, then fan out into lower mountains and hills, and eventually into plains to the north, west, and south. In these gentler but still arid landscapes, characterized by steep, sparsely vegetated meadows and pasturelands, most of the country's people struggle to earn a subsistence living from cultivation and the keeping of grazing animals. The main food crops are wheat, fruit, and nuts. Opium poppies are native to this region, but they were not an important cash crop until the Soviet invasion in 1979; the ensuing wars created the environment for opium production.

As of mid-2005, there were about 30 million Afghans, 45 percent of whom were 14 years of age or younger. Life expectancy is only 42 years (down from 47 in 2003), yet the population is growing by 2.6 percent per year (women bear more than six children on average). Literacy rates in Afghanistan are among the lowest on earth and are increasing only very slowly; just 21 percent of women and 51 percent of men over age 15 can read.

The less mountainous regions in the north, west, and south are associated with Afghanistan's main ethnic divisions (Figure 8.45). Very few of these groups could properly be called tribes. In the north along the Turkmenistan, Uzbekistan, and Tajikistan

**Ethnolinguistic Groups**
- Baluch
- Pashtun
- Haraza
- Nuristani
- Ismaelien
- Turkmen
- Uzbek
- Tajik
- Kyrgyz
- Other

**FIGURE 8.45** Language and ethnicity in Afghanistan. The overlapping of ethnolinguistic groups illustrates how, over the millennia, particular groups have been divided and moved to different places. [Phillippe Rekacewicz and Cecile Marin, "A tangle of nations," *Le Monde diplomatique*, January 2000, at http://www.monde.diplomatique.fr.]

borders are the Afghan Turkmen, Uzbek, and Tajik ethnic groups, who share culture and language traditions with the people of these three countries and are intermingled with them. The Hazara, who are concentrated in the middle of the country, trace their ancestry to the Mongolian invasions of the past, yet are closely aligned with Iranian culture and languages. In the south, the Pashto-speaking Pashtuns and Baluchis are culturally akin to groups farther south across the Pakistan border. Within each group there is significant variation, especially regarding views on religion, education, modernization, and gender roles. For example, the Taliban are primarily illiterate Pashtuns, but educated Pashtuns, long important in Afghan governance, played an important role in the resistance against the Taliban and in creating the new Afghan government.

The ethnic diversity of Afghanistan and its neighbors has thwarted many efforts to unite the country under one government. Although the various ethnic groups have remained separate and competitive, contrary to Western media reports, they have not been at continuous war with one another. However, the events of the last several decades in the aftermath of the Russian invasion of 1979 have disturbed long-standing relationships and feelings of trust. More important, the sheer devastation of almost three decades of war has left the country crippled. Although the traditional rural subsistence economy of Afghanistan proved remarkably resilient and self-sufficient in the face of ongoing civil strife, the sufficiency of that system has been compromised by an ongoing drought that threatens over a million people with starvation.

The cities of Afghanistan form a contrast with the countryside and with one another. By the 1970s, Kabul was a modernizing city with utilities and urban services, tree-lined avenues and parks, and lifestyles similar to those in cities elsewhere in South Asia. Most women were not veiled; girls attended schools and wore the latest fashions. Then the Russians invaded and brought to Kabul an extreme amoral version of Western culture that quickly alienated the more conservative countryside. After the mujahedeen defeated the Russians and the Taliban came to power, modernization became anathema. Kabul suffered destruction first as the Russians were driven out and then as the Taliban were themselves driven out in 2001. Today, Kabul has more than 3 million people, but the built environment is much diminished by war. The cities of Herat in the west and Kandahar in the south were always more traditional and culturally conservative—more in tune with the hinterlands. Both suffered destruction during the late twentieth-century conflicts, and little rebuilding has taken place in recent years.

## Pakistan

Although not much larger in area than Afghanistan, Pakistan has more than five times as many people (162 million). Pakistanis, too, live primarily in villages, though the country contains nine cities with more than 1 million people each. Some villages sprinkled throughout the arid mountain districts are associated with herding and subsistence agriculture, but it is the lowlands that have attracted the most settlement. Here, the ebb and flow of the Indus River and its tributaries during the wet and dry seasons form the rhythm of agricultural life. The river brings fertilizing silt during floods and provides water to irrigate millions of cultivated acres during the dry season.

**FIGURE 8.46** Cottage industry in rural Pakistan. Young rural Sindhi women embroidering pillowcases in the traditional colors and patterns of the Sindh Province of Pakistan. [© Stephanie O'Connor.]

Pakistan's favorable annual economic growth rate of 6 percent or better from the 1980s to the present was fueled in part by agriculture in the Indus River basin, especially in the Pakistani Punjab, where cash crops such as cotton, wheat, rice, and sugarcane are grown in large irrigated fields. Pakistan has been irrigated for many thousands of years, but recent irrigation projects have overstressed the system, resulting in waterlogged and salinized soil. As a result, growth in agricultural production has slowed in recent years.

Other problems in Pakistan are related to its military spending. Part of the reason that Pakistan's GDP per capita ($2097) is lower than India's ($2892) is its huge investment in the military rather than in education, health care, and economic modernization, which has meant that economic growth has not benefited the poor majority. Moreover, a nuclear weapons program, thought necessary primarily because of Pakistan's disputes with India over Kashmir (see pages 446–447), resulted in international sanctions that limited job creation. After 9/11, however, the United States, which had cut off military aid, repaired relations with Pakistan, offering significant military and social aid in return for Pakistan's support against the Taliban in Afghanistan and against terrorism generally.

The fact that two-thirds of the Pakistani population lives on less than U.S.$2 a day demonstrates that the benefits of the country's overall economic growth have bypassed the majority. The productive farmland of Punjab is owned by a small elite, and although textile, yarn-making, and embroidery industries are growing around the cities of Lahore and Karachi, providing jobs for some rural people (Figure 8.46), the wealth generated has not resulted in higher wages. Often the profits are invested outside of Pakistan. In addition, international drug dealing and corruption are tolerated by officials. Arif Nizami, editor of the Lahore newspaper *The Nation*, is quoted by *New York Times* correspondent John Burns as saying, "Pakistan is a country where millions cannot get two square meals a day, yet the prime minister has a fleet of planes, flies to his place in the country in a personal helicopter, and lives in a palace [that would] shame the White House."

## HIMALAYAN COUNTRY

The Himalayas form the northern border of South Asia (Figure 8.47). The regional map (see Figure 8.1) shows this mountainous zone running from west to east through the northern borderlands of India and continuing through Nepal and Bhutan. Notice that the country of India actually extends east in a narrow corridor running between Nepal and Bhutan to the north and Bangladesh to the south, then balloons out in a far eastern lobe bordering Burma. Physically, this mostly mountainous subregion grades from relatively wet in the east to dry in the west because the main monsoons move up the Bay of Bengal and strike the east first, and hardest (see Figure 8.4 on page 416).

This strip of Himalayan territory can be viewed as having three zones: (1) the high Himalayas, (2) the foothills and lower mountains to the south, and (3) a narrow strip of southern lowlands running along the base of the mountains. This band of lowlands forms the northern fringe of the Indo-Gangetic Plain in the west and the narrow Assam Valley of the Brahmaputra River in the east. Although some people manage to live in the high Himalayas, most people live in the foothills, where they cultivate strips and terraces of land in the valleys or herd sheep and cattle on the

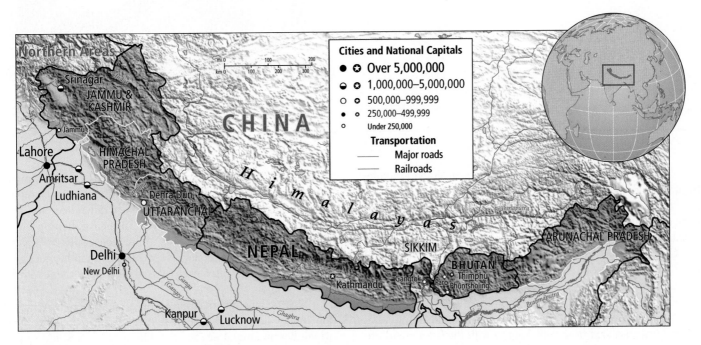

**FIGURE 8.47** The Himalayan Country subregion.

## BOX 8.6 | AT THE LOCAL SCALE  Salt for Grain and Beans: A Geography of Trade

The Dolpo-pa people (Figure 8.48) are yak herders and caravan traders who live in the high, arid part of Nepal. In that difficult environment, they can produce only enough barley and corn to feed themselves for half the year. Through trade, the Dolpo-pa parlay a half-year's supply of grain into enough food to feed them for a whole year. Here is how it works: At the end of the summer harvest, they load a portion of their grain onto yaks and head north to Tibet. There, they trade grain to Tibetan nomads in return for salt, a commodity in short supply in Nepal. They leave some of the salt in their village for their own use; the rest they carry on to the villages of the Rong-pa people in the central foothills of Nepal. There they trade this salt for enough grain to last them through the winter. Bargaining is fierce, and it can take days before a price is agreed upon; prices average 3 to 5 measures of corn for 1 measure of salt. The trading season ends in November, so the Dolpo-pa usually winter over in a convenient field, paying for the privilege.

In addition to farming, the Rong-pa people herd sheep and goats, and salt is a necessary nutrient for both them and their animals. In recent years, because of salt shortages in Tibet, the Dolpo-pa have not brought enough salt to meet all the Rong-pa's needs. The Rong-pa, therefore, load goats and sheep with bags of red beans and set out for Bhotechaur, in western Nepal, where they meet Indian traders at a large bazaar and trade their beans for iodized Indian salt; a good price is 1 measure of beans for 3 measures of salt. Often they will also sell a sheep or two to buy cloth, or perhaps a copper pan, to take home.

*Source:* Adapted from Eric Valli, "Himalayan caravans," *National Geographic* (December 1993): 5–35.

**FIGURE 8.48** Grain for salt. The Dolpo-pa people (left) trade with Tibetan nomads (right), exchanging their grain for salt. [Eric Valli.]

hills. Some of this area is extremely rural in character: for example, Bhutan's capital city, Thimphu, with just 53,000 inhabitants, is the largest town in that country. Nepal, however, is rapidly urbanizing: Kathmandu, the capital city, and its suburbs now have about 1.5 million people.

Culturally, Himalayan Country is Muslim in the west and Hindu and Buddhist in the middle; animist beliefs are important throughout, but are especially strong in the far eastern portion. Throughout the subregion, but especially in valleys in the high mountains and foothills, indigenous people continue to live in traditional ways, isolated from daily contact with the broader culture (see the video vignette on page 443). The Indian state of Arunachal Pradesh, at the eastern end of the subregion, is a good example of this pattern: there, a population of less than 1 million speaks more than 50 languages. Despite these language differences, the Himalayan people have learned to survive in their difficult mountain habitat by relying on one another. An example of this cross-cultural reciprocity comes from central Nepal, where two indigenous groups engage in a complicated cycle of trade that links them with both Tibet and India (Box 8.6).

Most people in Himalayan Country are very poor. Statistics for the Indian states in this subregion hover near those for Nepal, which has an annual per capita GDP of just U.S.$1420, adjusted for PPP. In Nepal, the average life expectancy is 62, literacy is

**FIGURE 8.49** Arunachal Pradesh, in far northeastern India. This landscape illustrates the thick and colorful vegetation that results from warm temperatures, consistent abundant rainfall, and relatively little human impact in this lightly settled, mountainous state. The lake was created in 1950 by an earthquake. [Courtesy of Dipankar Mitra.]

about 47 percent, and nearly half of the children under 5 years of age are malnourished. At present, only 18 percent of the land is cultivated, and the country's high altitude and convoluted topography make agricultural expansion unlikely. The government's strategy, therefore, is to improve crop productivity and lower the population growth rate.

The Indian state of Arunachal Pradesh is one of this subregion's more lightly populated and more prosperous areas. Moist air flowing north from the Bay of Bengal brings plentiful rain as it lifts over the mountains. This is one of the most pristine regions in India. Forest cover is abundant, and a dazzling sequence of flora and fauna occupies habitats at descending elevations: glacial terrain, alpine meadows, subtropical mountain forests, and fertile floodplains (Figure 8.49). Conditions at different altitudes are so good for cultivation that lemons, oranges, cherries, peaches, and a variety of crops native to South America—pineapples, papayas, guavas, beans, corn (maize), and potatoes—are now grown commercially for shipment to upscale specialty stores.

Over the past 20 years, Himalayan Country has experienced numerous changes brought about by the increasing numbers of tourists trekking and climbing the mountains and seeking spiritual enlightenment at its numerous holy sites. Tourism in Nepal, in particular, has been a mixed blessing, creating economic opportunities for some people, but having a devastating effect on the culture and the landscape.

## NORTHWEST INDIA

Northwest India stretches almost a thousand miles from the states of Punjab and Rajasthan, at the border with Pakistan, eastward to encompass the famous Hindu holy city of Varanasi (Figure 8.50). It is dry country, yet contains some of the wealthiest and most fertile areas in India.

In the western part of this subregion, there is so little rainfall that houses can safely be made of mud with flat roofs. Widely spaced cedars and oaks are the only trees, and the landscape has a dusty khaki color. Yet filling the landscape between the trees are fields of barley and wheat, potatoes, and sugarcane, plowed by villagers using oxen and a humpbacked breed of cattle. Along the northern reaches of the subregion, the rivers descending from the Himalayas compensate for the lack of rainfall. The most important is the Ganga, which, with its many tributaries, flows east and waters the state of Uttar Pradesh, bringing not only moisture but also fresh silt from the mountains.

The western half of the subregion contains one of India's poorest states—Rajasthan—as well as one of its wealthiest—the Indian part of Punjab. Rajasthan, with only a few fertile valleys, is dominated by the Thar (Great Indian) Desert in the west, which covers more than a third of the state. Historically, small kingdoms were established wherever water could be contained (Rajasthan means "land of kings"). Seminomadic herders of goats and camels still cross the desert with their animals, selling animal dung or trading it for grazing rights as they travel and thus keeping farmers' fields fertile and their fires burning. Perhaps the best known of these people are the Rabari, about 250,000 strong today (Figure 8.51). By the 1990s, however, there were more and more small farmers occupying former pasturelands. Although these farmers still want dung, they cannot afford to lose one bit of greenery to passing herds, so the increasing population pressure means that the benefits of reciprocity between herders and farmers are being lost.

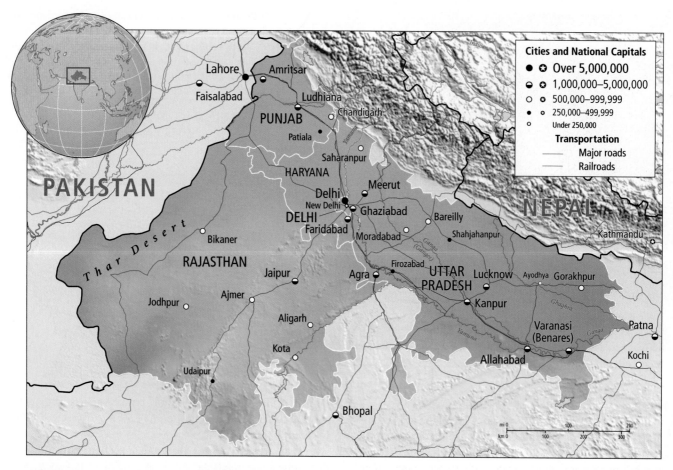

**FIGURE 8.50** The Northwest India subregion.

In arid Rajasthan, less than 1 percent of the land is arable, yet agriculture, poor as it is, produces 50 percent of the state's GDP. It is not surprising that the annual per capita GDP is just U.S.$500. The crops include rice, barley, wheat, oilseeds, peas and beans, cotton, and tobacco. A thriving tourist industry, focused on the palaces and fortresses constructed by warrior princes of the past who fought off invaders from Central Asia, accounts for much of the other half of the GDP.

Punjab, one of India's most productive agricultural states, has an annual per capita GDP of U.S.$1105, more than double that of Rajasthan. These differences in wealth result partly from differences in physical geography and from the introduction of green revolution technology. Unlike Rajasthan, Punjab receives water and fresh silt carried down from the mountains by rivers. Nearly 85 percent of Punjab's land is cultivated, and in some years Punjab alone provides nearly two-thirds of India's food reserves. The typical crops are corn, potatoes, sugarcane, peas and beans, onions, and mustard.

Although agriculture is the most important economic activity throughout the subregion and employs about 75 percent of the people, industry is also important in the union territory (city-state) of Delhi and on the Ganga Plain. Here, as elsewhere in the subregion, typical industries are sugar refining and the manufacture of

cotton cloth and yarn, cement, and glass. Residents also make craft items and hand-knotted wool carpets. Urban areas have more modern forms of industry. In the city of Chandigarh, for example, there are 15 medium- to large-scale industrial facilities that produce electronic and biomedical equipment, household appliances, tractor parts, and cement tiles and pipes.

The city of New Delhi, India's capital, is located approximately in the center of Northwest India. The city center was built by the British in 1931 just south of the old city of Delhi—an important Mughal city—amid the remains of seven ancient cities. It has all the monumental hallmarks of an imperial capital, and all the problems one might expect of a big city where so many are poor. The Delhi metropolitan area (including the old and new cities and outlying suburbs) has more than 15 million people and attracts a continuing stream of migrants. Most have left nearby states to escape conflict or poverty, or both, such as those who have fled Tibet to escape Chinese oppression and others who have fled the conflicts with Pakistan since the 1980s. Annual per capita GDP in Delhi (U.S.$1335, adjusted for PPP) is below the average for India (U.S.$2892), but actual incomes for most people are lower still; Delhi's tiny minority of the extremely wealthy pulls up the average. The literacy rate for the Delhi metropolitan area—just 82 percent—is held down by the

**FIGURE 8.51** Camels taken out to graze in the Rabari village of Bhopavand, near the border of Rajasthan and Gujarat. Some Rabari still lead a seminomadic life, traveling northward through Rajasthan and across the Thar Desert in the dry season. [Dilip Mehta/Contact Press Images.]

continual arrival of migrants from poor rural areas, but Delhi also has far fewer schools than it needs for its population. In fact, the city has difficulty providing even the most basic services: there are insufficient water, power, and sewer facilities, and 75 percent of the city structures violate local building standards. Many people have no buildings to inhabit at all; they live in shanties constructed from found materials.

Delhi has been designated the world's fourth most polluted metropolitan area, with an annual pollution-related death toll of 7500. Most of the pollution comes from the more than 3 million unregulated motor vehicles: taxis, trucks, buses, motorized rickshaws, and scooters—most without pollution control devices and all competing for space, cargo, and passengers.

Despite this grim picture, life in Delhi is vibrant and upbeat, in part because the middle class is growing and many of the poor sense that there is a chance for upward mobility. The global market economy, evident across the nation, is especially visible in cities like New Delhi, where swanky clubs, restaurants, department stores, boutiques, cinemas, and video parlors share space with traditional businesses, and Western products are easily available.

## NORTHEASTERN SOUTH ASIA

Northeastern South Asia, in strong contrast to Northwest India, has a wet tropical climate. This subregion bridges national and state boundaries, encompassing the states of Bihar, Jharkhand, and West Bengal in India; the country of Bangladesh; and the far eastern states of India—all clustered at the north end of the Bay of Bengal (Figure

8.52). Its dominant features are the Ganga and Brahmaputra rivers and the giant drainage basin and delta region created by those rivers. The wet climate and fertile land have nourished a population that is now among the densest on earth (see Figure 8.13 on page 424), often struggling to support itself in overcrowded conditions. Here, we discuss several parts of the Indian portion of this subregion before moving on to Bangladesh.

## The Ganga-Brahmaputra Delta

The Ganga-Brahmaputra delta is the largest delta on earth. Every year, the two rivers deposit enormous quantities of silt, building up the delta so that it extends farther and farther out into the Bay of Bengal. The rivulets of the delta change course repeatedly until the bay is periodically flushed out by a huge tropical cyclone. The people of the delta have learned never to regard their land as permanent, and they have adapted their dwellings, means of transport, and livelihoods to drastic seasonal changes in water level and shifting deposits of silt. Villages sit on river terraces or, in the lowlands, are raised on stilts above the high-water line. Moving about in small boats, people fish during the wet season; when the land emerges from the floods, they turn to farming. Called *nodi bhanga lok* ("people of the broken river"), those who occupy the constantly shifting silt of the delta region are looked down upon by more permanent settlers on slightly higher ground. Because they must often flee rising floodwaters, they are less secure financially and are thought by their neighbors to lack the qualities of thrift and good citizenship that come from living in one place for a lifetime. Even beyond the delta, however, most people have to cope

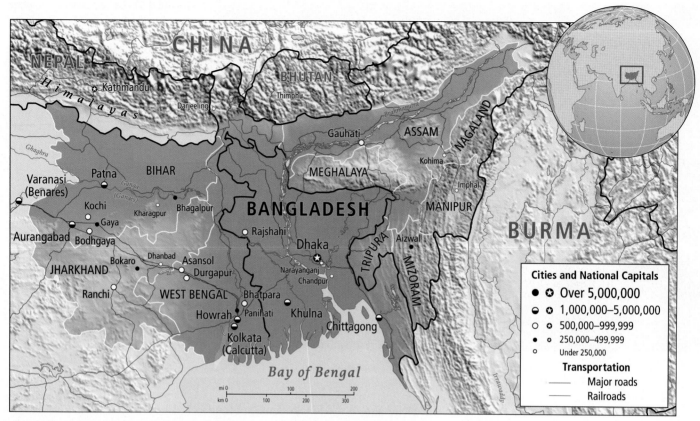

**FIGURE 8.52** The Northeastern South Asia subregion.

with flooding at some time of the year, either when the summer monsoon rains come or, in the foothills of the Himalayas, when the spring snowmelt in the mountains swells rivers beyond their banks.

## West Bengal

With a population of 80 million packed into an area slightly larger than Maine, West Bengal is India's most densely occupied state. Its population was swelled by refugees from eastern Bengal (now part of Bangladesh) after the Partition in 1947, and even more immigrants have come in since the Chinese takeover of Tibet and the Pakistani civil war that gave Bangladesh its independence. Today, better employment and farming opportunities draw a continual flow of Bangladeshi migrants (most of them illegal). The result is a mix of cultures that are frequently at odds: there are often Hindu and Muslim religious demonstrations and disputes with indigenous people over land occupied by new migrants.

Nearly 75 percent of the people in this crowded state earn a living in agriculture, yet agriculture accounts for only 35 percent of West Bengal's GDP. In addition to growing food for their own consumption, many people work as rice and jute cultivators or as tea pickers—all labor-intensive but low-paying jobs. Twenty-five percent of India's tea comes from West Bengal; the plant is grown in the far north of the state around Darjeeling, a name well known to

tea drinkers. One consequence of the dense population's dependence on agriculture is that continuous intensive cultivation often overstresses the land severely and soil fertility declines over time. In addition, the surrounding woodlands are depleted by impoverished farm laborers who must gather firewood because they cannot afford kerosene.

Kolkata (Calcutta), a giant, vibrant city of 13 million people, is famous in the West for Mother Teresa's ministrations to its poor at Nirmal Hriday (Home for Dying Destitutes). Bengalis, however, are proud of their two Nobel laureates (Rabindranath Tagore and Amartya Sen) and Academy Award–winning filmmaker Satyajit Ray. Kolkata was built on a swampy riverbank in 1690 and served as the first capital of British India. But its sumptuous colonial environment became lost in the squatters' settlements that surrounded the city and have invaded its parks and boulevards. The city's decline began in the late 1940s, when the Partition and agricultural collapse sent millions of people pouring into Kolkata and other cities. Soon the crowds overwhelmed the city's housing and transport facilities. Outmoded regulations limited incentives to start businesses and to improve private property. By the late 1990s, the physical decline, economic inertia, and lack of opportunity in Kolkata sent young professionals fleeing to other parts of the world. By 2005, however, those who had fled were returning, and new graduates were staying—but, unfortunately, not because old Kolkata was being refurbished, as Box 8.7 explains.

## BOX 8.7 | AT THE LOCAL SCALE Kolkata's Building Boom

By 2005, developers from all across Asia were building massive high-rise apartments and offices for Kolkata's elites, surrounded by wide boulevards, parks, swimming pools, and private clubs, on wetlands to the south and west of the city. There are at least 50 such large projects under way, and it is estimated that in a few years the developments will attract enough people to bring the total population of the metropolitan area to 21 million (Figure 8.53).

Kolkata's real estate boom has resulted in a scarcity of skilled workers, especially civil engineers, plumbers, and electricians, and as the new buildings are readied for occupancy, the demand for service workers of all types continues to soar. So far, those worrying about the fate of old Kolkata, or about the environmental ramifications of such massive building on wetlands within the Ganga-Brahmaputra delta, are not getting much coverage in the press.

**FIGURE 8.53 New suburbs, Kolkata.** On the outskirts of the old colonial city of Kolkata, new suburban developments are rapidly rising. This picture from promotional literature shows a complex of three independent towers and a retail/office center built around a large landscaped garden. The complex is meant to attract first-class technology companies and Kolkata's expanding technology-savvy middle class. [Courtesy of DLF Ltd.]

## Far Eastern India

Far Eastern India has two physically distinct regions: the river valley of the Brahmaputra where it descends from the Himalayas, and the mountainous uplands stretching south of the river between Burma on the east and Bangladesh on the west. Although migrants have recently arrived here from across South Asia, this subregion has traditionally been occupied by ancient indigenous groups that are related to the hill people of Burma, Tibet, and China.

The Indian state of Assam encompasses the Brahmaputra River valley, eastern India's most populated and most productive area. Hindu Assamese make up two-thirds of the population of 29 million, and indigenous Tibeto-Burmese ethnic groups make up another 16 percent; the rest are recent migrants. The Indian government has attempted to reduce the proportion and influence of the Assamese people, who have continually objected to being under Indian control, by making large tracts of land available to outsiders, such as Bengali Muslim refugees, Nepalese dairy herders, and Sikh merchants. Since the late 1970s, there have been violent disputes between the Assamese and the new settlers and between Assam and India's national government. In the fall of 2004, outbreaks of violence resulted in more than 50 deaths.

More than half the people in Assam work in agriculture; another 10 percent are employed in the tea industry or in forestry (forests cover about 25 percent of the land area). Two-thirds of the cultivated land is planted in rice, but tea is the main cash crop—Assam produces half of India's tea. Assam also has oil: by the 1990s, Assam's oil and natural gas accounted for more than half that produced in all of India. Given the country's shortage of energy, this alone could explain India's efforts to dominate the Assamese politically.

Colorful names such as "Land of Jewels" and "Abode of the Clouds" convey the exotic beauty of the emerald valleys, blue lakes, dense forests, carpets of flowers, and undulating azure hills in the mountainous sections of eastern India that surround Assam. In these uplands, occupied largely by indigenous ethnic groups, people produce primarily for their own consumption (80 percent or more of the inhabitants make a living from the land). Over the

**FIGURE 8.54** Living root bridge, Meghalaya, India. In the state of Meghalaya, the land is highly convoluted, with deep, narrow valleys. For hundreds of years, the people have built bridges across these chasms using the living roots of the ancient Indian rubber tree (*Ficus elastica*). The tree roots are trained through hollowed-out logs laid across the chasm. Once the roots are securely rooted on the other side—a process that may take 15 years—some are bound together to form a footbridge; other roots form the handrails. A mature bridge can hold 50 people at once and can last for more than 500 years. The bridge in the picture has two spans and is known as "the Double Decker Root Bridge." [© Kazi N. Fattah.]

millennia, these people have devised ingenious ways to make use of local natural resources (Figure 8.54).

Although literacy rates are not high in most of Far Eastern India, the state of Mizoram ranks second in India, at 88 percent, because of the influence of Christian missionary schools.

## Bangladesh

Bangladesh is one of South Asia's poorest countries and the world's most densely populated agricultural nation. More than 144 million people live in an area slightly smaller than Alabama, with 77 percent of them in rural areas, trying to manage as farmers on severely overcrowded land. The caloric intake for 43 million Bangladeshis is insufficient to meet the minimum energy requirements for daily life, and half the children are underweight for their age.

Nevertheless, Bangladesh is making improvements, bit by bit. The percentage of rural people living in poverty is dropping slowly, according to the U.S. Agency for International Development. Adult literacy went from 34 percent in 1990 to 41 percent in 2003, and enrollment of eligible children in secondary school increased by 50 percent in the same period (45 percent are enrolled). Mean-while, contraceptives became more available (58 percent of women now use some form of birth control), which brought a significant reduction in fertility rates, from seven children per woman in 1974 to three in 2005. Infant mortality has also dropped, from 128 per 1000 in 1986 to 65 per 1000 in 2005 (still double the rate of Vietnam or El Salvador).

Economically, there are some signs that the textile industry is reviving; for example, Bangladesh consistently increases its shipments to the U.S. market each year. Nevertheless, its total share of the U.S. textile market is shrinking as India and China grab greater shares; to compete against these two giants, the quality of its textiles will have to improve. The real competitive edge for Bangladesh, however, may lie in further expanding opportunities for microcredit, an innovation of its Grameen Bank that is now contributing to progress among the poor worldwide (see pages 443–444).

## CENTRAL INDIA

The Central India subregion stretches across the widest part of India, from Gujarat in the west to Orissa in the east (Figure 8.55).

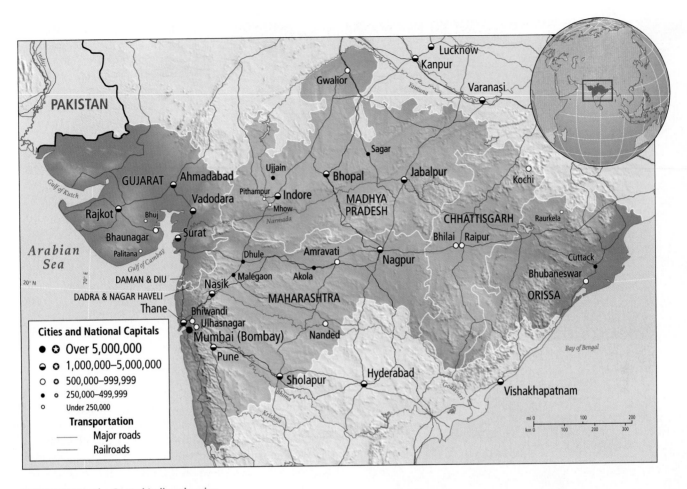

**FIGURE 8.55** The Central India subregion.

It contains India's last untouched natural areas as well as much of its industry. The Narmada River, site of many hydroelectric dams, flows across the subregion and empties into the Gulf of Cambay.

Some of India's most significant environmental battles are being fought in the central highlands of this subregion; the protests over the damming of the Narmada River, described at the opening of this chapter, are only one example. Central India has most of India's remaining forest cover and is home to a concentration of national parks and sanctuaries (most notably tiger reserves). However, because the forest cover is patchy, not continuous, many wild plants and animals are isolated in such small populations that their extinction becomes inevitable, even if it is somewhat delayed by protection of the parks. In recognition of this problem, there are now plans to reconstitute forest corridors between parks. Estimates of future human population growth, however, do not bode well for the future of wildlife anywhere in this subregion.

Central India is notable for its several industrial areas. In the state of Gujarat, on the Arabian Sea, the service and industrial sectors account for 75 percent of the GDP; the rate is even higher, at 83 percent, in the neighboring state of Maharashtra. Even the central plateau and eastern areas of Central India, although more rural in character, have pockets of industrial activity. An example is the city of Indore, a commercial and industrial hub whose residents think of it as a mini-Mumbai. Nearby, the town of Pithampur, known as India's Detroit, houses several automobile plants, a steel plant, a container plant, and small appliance factories (Figure 8.56). Industry and urbanization are clearly connected, and in fact, the western part of the subregion is exceptionally urbanized for India. In Gujarat, about 37 percent of the people live in urban areas; in Maharashtra, the location of the megacity Mumbai, the figure is about 43 percent.

Even though Central India might be considered newly industrializing, it could still lose industries if it fails to compete effectively in the mobile global economy. For example, the city of Ahmadabad in Gujarat, with 5 million people, used to be known as the Manchester of India because of its large textile mills. Most of the old mills are closed today, however, having lost out to newer mills and cheaper labor elsewhere in Asia. But another aspect of globalization may help Gujarat keep its economic edge. Many well-educated Gujaratis live abroad in a worldwide diaspora of merchants and businesspeople who maintain ties with their homeland. This diaspora could prove to be a major source of foreign investment for Gujarat.

**FIGURE 8.56** An automobile plant in Pithampur. The photo shows the final stages of vehicle assembly in the Eicher Motors plant. [© 2004 Eicher Goodearth Ltd.]

Bombay is the name by which most Westerners know Maharashtra's capital, but since 1995 its official name has been Mumbai, after the Hindu goddess Mumba. In the sixteenth century, when a local sultan gave the bay, with its seven small islands, to the Portuguese, it became known as Bom Bahia in Portuguese, or "Beautiful Bay." Eventually the British joined the islands with bridges and landfill and built the largest deepwater harbor on India's west coast. Mumbai's metropolitan area, with close to 20 million people, is now India's most prosperous city. It hosts India's largest stock exchange and the nation's central bank. It pays about a third of the taxes collected in the entire country and brings in nearly 40 percent of India's trade revenue; its annual per capita GDP is three times that of Delhi.

Mumbai's wealth is most evident when one looks up at the elegant high-rise condominiums built for the city's rapidly growing middle class. But at street level, the urban landscape is dominated by the large numbers who are living on the sidewalks, in narrow spaces between buildings, or in large, rambling shantytowns. The largest of these communities, Dharavi, houses more than 600,000 people on less than 1 square mile, and most of those people have no plumbing (Figure 8.57). People work at three or four jobs, and the community is known for its inventive entrepreneurs. One young man, for example, collects and sells aluminum cans that once held ghee, India's form of butterfat. He says he makes about 15,000 rupees a month (U.S.$480), nearly twice the salary of the average

**FIGURE 8.57** Mumbai's Dharavi section, Asia's largest slum. Women who must wash clothes in the watercourse flowing through Dharavi have a difficult time because garbage and raw sewage pollute the water daily. [Steve McCurry/National Geographic Image Collection.]

college professor in India and much more than he made as a truck driver. Hence, despite widespread poverty, Mumbai has more than a few success stories.

Mumbai is known popularly in India as "Bollywood" because it produces popular Hindi movies portraying love, betrayal, and family conflicts. The stories, played out on lavish sets and accompanied by popular music and dance, serve to distract their huge audiences from the physical difficulties of daily life, at least temporarily. Mumbai produces many more films than Hollywood, on much smaller budgets. The stars make six or more films a year and are so popular that movie posters are everywhere, in public and private spaces alike.

## SOUTHERN SOUTH ASIA

Southern South Asia encompasses the southernmost part of India and the country of Sri Lanka (Figure 8.58). It resembles the rest of the region in that the majority of people—ranging from about 70 percent in the west to somewhat more than 50 percent in the east—work in agriculture. But this subregion is set apart by its relatively high proportion of well-educated people, its advanced technology sectors, its strong tradition of elected Communist state governments (primarily in Kerala), its focus on environmental rehabilitation and preservation, and the higher status of women. The cultural mix here also sets it apart: it is the center of ancient Dravidian cultures and languages that predate Aryan influences.

This part of India receives consistent rainfall and is well suited for growing rice, peanuts, chilis, limes, cotton, cinnamon and cloves, and castor oil plants (used in medicines). The southwestern coast of India (known as the Malabar Coast) has a narrow coastal plain backed by the Western Ghats. The sea-facing slopes of these mountains are some of the wettest in India; they support forests containing teak, rosewood, and sandalwood, all highly valued furniture woods. Small parts of the Deccan Plateau, a series of uplands to the east, are also forested. Here, dry deciduous forests yield teak, eucalyptus, cashews, and bamboo. Several large rivers and numerous tributaries flow eastward across this plateau and form rich deltas along the lengthy and fertile coastal plain facing the Bay of Bengal.

## Bangalore and Chennai

Bangalore, long known for its vibrant information technology activity, which has drawn thousands of well-trained technical employees from all over India, continues to evolve. It now has many IT rivals throughout South Asia, but it remains competitive (Figure 8.59). Meanwhile, manufacturing has gained markedly; the products produced in and around the city range from appliances and home furnishings to solar-powered lighting and pharmaceuticals. Chennai, 200 miles (320 kilometers) to the east on the Bay of Bengal, is known for its innovative engineering and research institutes and is cultivating a role as a prime destination for jobs outsourced from Europe. It now also emphasizes manufacturing and is becoming a center for Italian-designed high-fashion shoe

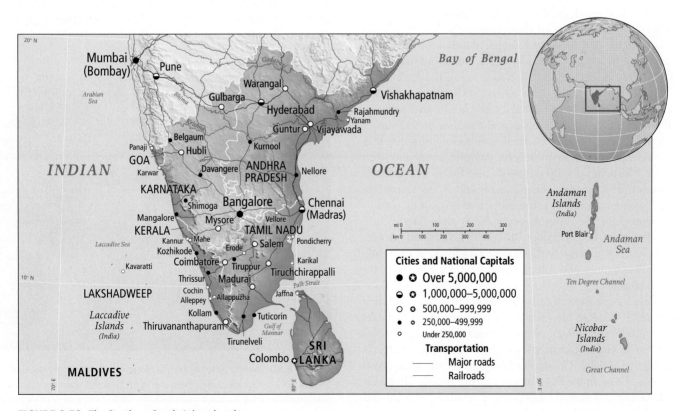

**FIGURE 8.58** The Southern South Asia subregion.

**FIGURE 8.59** The newly built Microsoft technology center in Bangalore's IT campus, December 2005. Bangalore claims to be the information technology capital of the supercharged and ever-expanding economy of India, but its competitors are becoming more numerous. [© Benjamin Lowy/CORBIS.]

manufacturing. Because styles for these high-end shoes change so rapidly, the Chennai factories are assured of the continuous production of about 2 million pairs per year.

## Kerala

Kerala is a well-watered, primarily coastal state in far southwestern India. It is often cited as an aberration on the Indian subcontinent because its people enjoy a higher standard of well-being than the rest of the country. Just why Kerala stands out in this way is not entirely understood, but the state has a unique history. Since the 1950s, it has had a series of elected Communist governments that have strongly supported broad-based social services, especially education for both males and females. In addition, it has long had a variation on the traditional family structure that gives considerable power to women; for example, a husband often resides with his wife's family (a pattern also found in Southeast Asia) instead of the reverse. Female seclusion is less stringently practiced; women are often on the street, and it is not unusual to see a woman going about her shopping duties alone. Agriculture employs less than half the population of 31.8 million people. Remittances from migrants to the Middle East figure prominently in the state's economy, and fishing is an important economic sector as well (Box 8.8).

Another feature that may have added to the openness of society and greater freedom for women is Kerala's significant contact with the outside world. Beginning thousands of years ago, traders from around the Indian Ocean, and especially Southeast Asia, made calls along the Malabar Coast, possibly bringing new ideas about gender roles. Then, in the first century C.E., the Apostle Thomas is said to have come to spread Christianity. Eventually, Muslim and Jewish traders brought their faiths by sea as well. This history of cul-

tural and religious pluralism seems to have led to somewhat more tolerance than exists in neighboring states.

## Sri Lanka

Sri Lanka, known as Ceylon until 1972, is a large island country off India's southeastern coast known for its beauty. From the coastal plains, covered with rice paddies and with nut and spice trees, the land rises to hills where tea and coconut plantations are common. At the center is a mountain massif that reaches to nearly 8200 feet (2500 meters) in elevation. The summer monsoons bring abundant moisture to the mountains, giving rise to lush forests and several unnavigable rivers that produce hydroelectric power. Close to 30 percent of the land is cultivated, and just over 25 percent remains forested.

The original hunter-gatherers and rice cultivators of Sri Lanka, today known as Veddas, now number fewer than 5000. They were joined several thousand years ago by settlers from northern India who built numerous city-kingdoms. Today known as Singhalese, these descendants of northern Indians make up about 74 percent of the population of 19.7 million. The Singhalese brought Buddhism to Sri Lanka, and today 70 percent of Sri Lankans (most of them Singhalese) are Buddhists.

About a thousand years ago, Dravidian people from southern India, known as Tamils, began migrating to Sri Lanka; by the thirteenth century they had established a Hindu kingdom in the northern part of the island (see Figure 8.22 on page 430). Later, in the nineteenth century, the British imported large numbers of poor Tamils from the state of Tamil Nadu in southeastern India to work on British-managed tea, coffee, and rubber plantations. Known as "Indian" Tamils, the plantation-based Tamils share linguistic and

**FIGURE 8.60** Jewelry stores along Sea Street in Colombo, the capital of Sri Lanka. Gemstones have long been one of Sri Lanka's resources, and polishing and setting them is a major industry. Other businesses include textiles and clothing. [Steve McCurry/National Geographic Image Collection.]

religious traditions with the "Sri Lankan" Tamils of the north and east, but each community considers itself distinct because of its divergent historical, political, economic, and social experiences. Together, the two Tamil populations make up about 18 percent of the total population of Sri Lanka. The Sri Lankan Tamils have done well, dominating the commercial sectors of the economy, whereas the Indian Tamils have remained isolated and largely poverty-stricken laborers on interior plantations.

Sri Lanka once had a thriving economy, led by a vibrant agricultural sector, and a government that made significant investments in health care and education. It was thought to be poised to become one of Asia's most developed economies (Figure 8.60). But conditions in rural areas took a drastic turn for the worse in the 1960s as declining prices for the chief agricultural exports prompted the government to shift investment away from rural development and toward urban manufacturing and textile industries, dominated by Singhalese. This shift exacerbated a political conflict that already existed because the legal status of the plantation laborers

imported by the British had not been guaranteed when the British withdrew as a colonial government. The Singhalese-dominated government alienated all of Sri Lanka's Tamils not only by threatening to bar Indian Tamils from participation in elections, but also by making only Singhala, not Tamil or English, an official language. Moreover, Buddhist Singhalese were privileged in obtaining university admissions, job opportunities, and other social services. Reaction to this favoritism took the form of protests by Tamil plantation workers as well as brutal acts of terrorism by guerrilla forces, known as the Tamil Tigers, operating mostly in Sri Lanka's far northeast; news reports often confused the two. In 1987, Indian troops intervened at the request of the Sri Lankan government, but that action failed to end the violence.

Meanwhile, resentment grew in southern Sri Lanka in response to the government's lack of attention to growing poverty and corruption. In 1989, Sri Lanka appeared to be near collapse as the result of an insurrection by southern Singhalese guerrillas and the withdrawal of the Indian forces in the north. Armed conflict flared

---

## BOX 8.8 | AT THE LOCAL SCALE   Kerala's Fishing Industry

Geographer Holly Hapke has studied how recent structural changes in the fishing industry have altered family economics among fishers in and around Trivandrum, in the far south of Kerala. Before independence in 1947, fishing in Kerala was carried out almost entirely on a small scale. Thousands of fishers worked from boats that held crews of 1 to 40 men. The catches were small and consisted of multiple species of fish. Once the boats had returned, the wives of the fishers sorted and cleaned the fish, then sold them on the beach to wholesalers or in town at the market. Unlike most women elsewhere in South Asia, the Kerala fishmongers regularly dealt with the public as businesswomen; at the same time, they performed multiple household and child-rearing tasks

In the 1950s, the Indian government, hoping to improve the incomes of Kerala's fishers by increasing the catch for the local market, introduced large mechanized vessels. Very soon, the project shifted to harvesting prawns for the world market, but mechanization also spread to other fisheries along the coast. By the mid-1970s, the mechanized fleet was encroaching on traditional fishers, damaging their gear, and competing for their catch. The intensification of fishing has led to declining fish stocks and pronounced decreases in income for Trivandrum's fishing villages, which continue to use traditional technology. Men no longer harvest sufficient fish, so the women no longer have fish to clean and sell. Although the men occasionally find wage labor in the mechanized fishing industry, fishers are no longer bringing in sufficient income, so families have become increasingly dependent on the women's earnings. The women, therefore, have started buying fish from male wholesalers and then reselling them in the market or to regular customers (Figure 8.61). To compete as fish vendors against the larger oper-

ators, the women must themselves deal in larger volumes; this requires investment cash, which they must borrow at the high rates charged by moneylenders. The exorbitant interest depletes the women's earnings, leaving them operating on the margins, with their family economies at risk. Despite their considerable entrepreneurial spirit, their situation is likely to worsen as the traditional sector disappears.

*Source:* Adapted from Holly M. Hapke, *Fish Mongers, Markets, and Mechanization: Gender and the Economic Transformation of an Indian Fishery* (Syracuse, N.Y.: Syracuse University, 1996).

**FIGURE 8.61** Women selling fish in a Trivandrum neighborhood market. [Courtesy of Holly Hapke.]

repeatedly throughout the 1990s. The civil war has produced more than a million refugees and severely impeded Sri Lanka's economic development. Urban industrial development has lagged despite efforts to create a free trade zone to attract industries. Repeated terrorist bombings have frightened off international investors as well as potential tourists. As the conflict continues, security remains tight on major highways and at the airport.

## Reflections on South Asia

A common expression in South Asia is that any statement one may make about such a complex region can be matched by an opposite statement that is equally true. For example, South Asia has some of the largest and most crowded cities on earth, yet more than 70 percent of the population lives in rural villages. Poverty is endemic, and the gaps between rich and poor are widening, but South Asia is also the home of one of the most potentially empowering and far-reaching development strategies ever conceived: microcredit. Although telephones are missing from most homes, information technology is flourishing. India is the world's largest democracy, yet it is also a place where religious intolerance has led to thousands of deaths in the last few years. And while religious conflict between Muslims and Hindus threatens to precipitate nuclear war between Pakistan and India, the region is a mecca for those seeking spiritual enlightenment. It is also the home of the highly effective strategy of nonviolent social resistance, begun by Mohandas Gandhi and adopted by advocates for human rights all over the world. In South Asian democracies, a person's sex significantly determines whether that person has access to sufficient food, education, income, opportunity, and indeed, to life itself. Yet in no other region have more women been elected heads of state, and South Asian women are among the most articulate supporters of women's rights in the global forum.

Perhaps one of the most provocative characteristics of South Asia is that it has spawned eloquent and prolific writers who frequently publish their work in English. Many Indian and Bangladeshi writers make the best-seller lists in Europe and North America (see the bibliography on this textbook's Web site at http://www.whfreeman.com/pulsipher). Most have something to add to the global conversation about poverty, human rights, and development. This literary tradition is exemplified by Bengali Nobel laureate Rabindranath Tagore, who wrote prophetically before his death in 1941 about the need for South Asia to carefully assess its Western-led development paths:

*We have for over a century been dragged by the prosperous West behind its chariot, choked by the dust, deafened by the noise, humbled by our own helplessness and overwhelmed by the speed. We agreed to acknowledge that this chariot-drive was progress, and the progress was civilization. If we ever ventured to ask, "progress towards what, and progress for whom," it was considered to be peculiarly and ridiculously Oriental to entertain such [reservations] about the absoluteness of progress. Of late, a voice has come to us to take count not only of the scientific perfection of the chariot but of the depth of the ditches lying in its path.*

Rabindranath Tagore, "Crisis of Civilization,"
in *Collected Works of Rabindranath Tagore*, vol. 18

It took more than three decades for Tagore's critique of development policy to gain wider acceptance. The "progress for whom" question is only now being asked in the highest halls of policy formulation. Strategies for aiming development at the poorest, rather than at those who are already reasonably well off, are finally being invented; as we have seen, South Asians are some of the most innovative creators of these strategies. Readers can discover in their own towns and cities small groups of cooperative borrowers emulating the million-plus clients of the Grameen Bank of Bangladesh. At the close of a chapter about a region that is arguably one of the poorest on earth, it is illuminating to note that this region is also a leader in inventive ideas for development as well as in provocative thought about the present trajectory of human society.

## Chapter Key Terms

| | | |
|---|---|---|
| Adivasis 434 | Harijans 434 | Partition 423 |
| agroecology 440 | hearth 432 | purdah 436 |
| Aryans 419 | Hindi 420 | regional conflict 445 |
| Brahmins 434 | Hinduism 431 | religious nationalism 444 |
| bride price 437 | Indian diaspora 422 | Sikhism 432 |
| Buddhism 432 | Indus Valley civilization 418 | social forestry movement 451 |
| caste 434 | Jainism 432 | subcontinent 414 |
| Chipko movement 451 | *jati* 434 | Sudras 434 |
| civil disobedience 421 | Kshatriyas 434 | Taliban 435 |
| communal conflict 433 | microcredit 443 | Vaishyas 434 |
| dowry 436 | monsoon 416 | *varna* 434 |
| green revolution 439 | Mughals 419 | |
| Harappa culture 418 | Parsis 432 | |

## Critical Thinking Questions

1. Dams and the reservoirs they create are sources of irrigation water and generators of electricity. What are the factors that have rendered so many places in South Asia, as well as elsewhere around the world, in need of markedly more water and electricity?

2. Explain why some would say that India has been part of globalization for thousands of years. Tie this history to what is happening in the present.

3. What are some of the lingering features of the British colonial era in South Asia? To what extent is globalization reinforcing or erasing these features?

4. Development projects are supposed to improve the lives of the people where they take place. To what extent has the development of India's IT industries lived up to this ideal?

5. Name some possible ways in which India's large population of young people could be an advantage in the global economy.

6. Describe some trends in South Asia that are increasing environmental degradation and some other trends that are increasing awareness among the citizenry of the need to protect the environment.

7. The monsoon cycle creates starkly different landscapes over the course of a year. What are some of the ways in which the monsoons of South Asia affect daily life, agricultural patterns, and the material culture of the region?

8. Industrialization in South Asia was stifled by the British and remained dormant for a long time. Describe industrialization in the pre-British era and its postindependence fate as well as how it is reviving in the 2000s.

9. Why do you think so many agencies place such great emphasis on teaching poor women to read? What are the effects they seek?

10. What are the factors attracting the Indian diaspora back to the homeland?

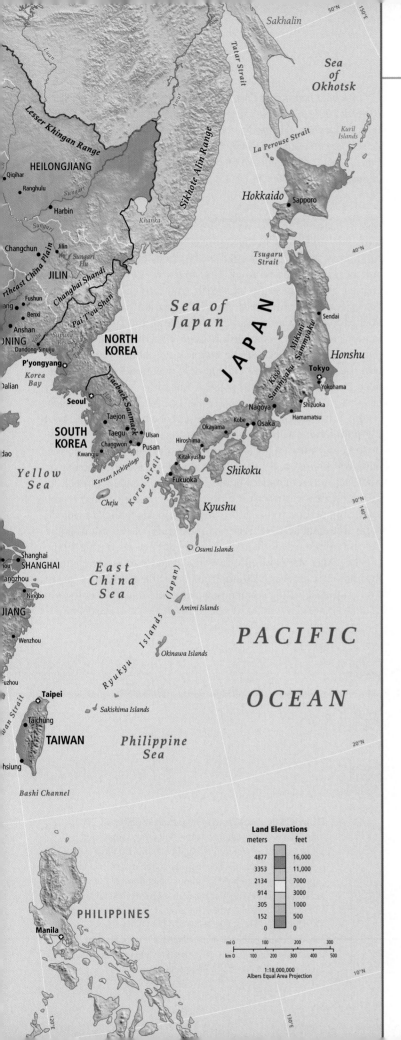

**Land Elevations**

| meters | feet |
|--------|------|
| 4877 | 16,000 |
| 3353 | 11,000 |
| 2134 | 7000 |
| 914 | 3000 |
| 305 | 1000 |
| 152 | 500 |
| 0 | 0 |

mi 0   100   200   300
km 0   100   200   300   400   500

1:18,000,000
Albers Equal Area Projection

# CHAPTER 9

# EAST ASIA

**Global Patterns, Local Lives** In 2000, at age 18, Lee Xia (Lee is her family name) left her farming village in China's Sichuan Province for the city of Dongguan in Guangdong Province on the Pearl River delta. She was accompanied by two friends. A few months earlier, the government had taken their families' farmland for an urban real estate project, paying compensation of just U.S.$2000 per family. The three young women accepted an offer to work in a Dongguan toy factory so they could send money back to their families, who now had to buy food and housing. Like so many migrants, they did not have the necessary official permission to leave their home territory. But they were excited. On the 4-day bus trip, they entertained dreams of one day shopping in the Wal-Mart supercenter located in Dongguan.

When the young women arrived in the city, they joined 17 million other migrants (Figure 9.2), at least 60 percent of whom were "floating," meaning they were illegal migrants with no residency rights, dependent on their employers for housing. They soon found that the labor recruiters had lied about their wages: they would be paid U.S.$30, not U.S.$45, a month. In addition, they would work 12-hour shifts in 100°F heat, get no overtime pay, and have just one day off a month. But there was no point in protesting. Their families in Sichuan needed the money they would send home, and the recruiters purposely brought in thousands of surplus workers, so complainers could be easily replaced.

Xia felt better when she saw that the toy factory was a clean, modern, white building known locally as "Palace of Girls" (nearly all 3500 employees were female). Within a day, she had completed her training, signed a 3-year contract, and mastered her task of putting eyes on stuffed animals. But her enthusiasm faded when she learned that she and her friends would be spending much of their money on canteen food and would be sharing one small concrete room and a tiny bath with eight other women and a rat or two. Her hopes of one day becoming a factory supervisor were dashed when she discovered that to obtain such a position she might have to grant sexual favors to the plant manager.

In less than 6 months, Xia broke her contract and returned home to her village in Sichuan, where—using ideas and assertiveness she had gained from her time in the south—she opened a snack stand. In just a month, she made ten times her investment of U.S.$12. But Xia yearned to return to the southern coast to try again. So, a few months later, leaving the stand to her sister's husband (who also cared for the couple's baby), she returned to Dongguan with her sister.

**FIGURE 9.1** Regional map of East Asia.

473

**FIGURE 9.2 Workers in the new Chinese economy.** Chinese women produce Santa Claus dolls at the Fly Ocean Toy factory on the outskirts of Guangzhou (Canton) in Guangdong Province. This factory is similar to the one where Lee Xia worked. [Guang Niu/Getty Images.]

Since her first arrival, the city had grown by 20 percent and now had 1400 foreign companies trying to hire thousands of workers. Through connections, Xia and her sister easily found jobs in a Taiwanese-owned factory at twice the wages Xia had originally earned. Just a year after her first trip to Dongguan, Xia was making about U.S.$100 a month, enough to live relatively comfortably with just three roommates. She was sending money home and once again saving to start a business. This time she dreamed of opening a bar back in her village.

*Adapted from Kathy Chen, "Boom-town bound," Wall Street Journal (October 29, 1996): 1, A6; "Life lessons," Wall Street Journal (July 9, 1997); Peter S. Goodman, "A turn from factories, China's labor pool is shifting," Washington Post National Weekly Edition (October 11–17, 2004): 19; Louisa Lim, "The end of agriculture in China," Reporter's Notebook, National Public Radio, May 19, 2006.* ■

Lee Xia's first trips to Dongguan on China's booming southeastern coast illustrate some features of the radical process of transformation under way in the vast region of East Asia. The needs of rural areas are being subverted to the needs of China's industries and burgeoning cities. Rural land grabs by developers have increased 15-fold since the mid-1990s, and farmers rarely get a fair deal. Meanwhile, hundreds of millions of rural young adults have flocked to East Asia's coastal cities—some of the largest and most rapidly growing cities on earth (Figure 9.3). In China, most were migrating illegally because the **hukou** (household registration) **system** (begun in the Maoist era and only now under revision) effectively tied rural people to the place of their birth. In desperate search of work, over 100 million people, like Lee Xia, ignored the *hukou* system, but by doing so, they became part of what is called the **floating population,** with no rights to housing, schools, or health care. So these migrants are not only working in menial low-wage jobs, but have left children and spouses behind and will make agonizing sacrifices to send remittances home. While the remittances help, the migration deprives poor rural places of the energy, initiative, and leadership of their young adults and contributes to the growing disparity in wealth between East Asia's poor rural places and its booming and overcrowded urban centers (see discussion on page 493).

Lee Xia's story also illustrates how gender roles are changing throughout East Asia. Although many of the old restrictions on women persist, young women now constitute one-third of the esti-

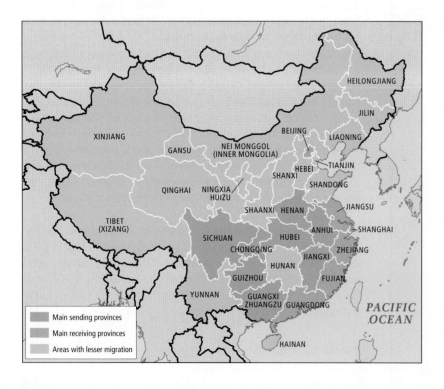

Main sending provinces
Main receiving provinces
Areas with lesser migration

**FIGURE 9.3 Rural-to-urban migration in China.** By 2003, some 114 million migrant workers had moved from rural areas to urban jobs in China. About one-third of these migrants were women. The map shows the underdeveloped central and western provinces from which most of the rural workers came and the metropolitan, primarily coastal areas and nearby townships where they found work. [Adapted from Zhan Shaohua, "Rural labour migration in China: Changes for policies," *Policy Papers/10* (United Nations Economic, Social and Cultural Organization, 2005), p. 14, at http://unesdoc.unesco .org/images/0014/001402/140242e.pdf.]

mated 250 million workers in the factories of China's coastal provinces and special economic zones across the country. The pattern of young women leaving home to find work in distant factories and offices is now global. Some inadvertently get caught up in human trafficking and become enslaved in the sex trade. Only a few ever achieve the financial freedom to truly seek their own fortunes.

Lee Xia's story also hints at a surprising change in the labor market that is under way in China: the labor surplus is disappearing! Turnover of low-paid, ill-treated workers is very high, and firms are beginning to offer better wages and other amenities to retain workers. The next time we encounter Lee Xia, in a personal vignette in the Economic and Political Issues section (see page 495), the continuation of her story will reveal how improving circumstances for workers are rippling into other aspects of life in China and will probably bring higher prices to malls worldwide.

China is part of the region of East Asia, home to nearly one-fourth of humanity. This vast territory stretches from the Taklimakan Desert in far western China to Japan's rainy Pacific coastline, and from the frigid mountains of Mongolia in the north to the subtropical forests of China's southeastern coastal provinces (see Figure 9.1). This world region comprises the countries of China, Mongolia, North Korea, South Korea, Japan, and Taiwan (the last has been independent since World War II, but never recognized by China), which are home to several of the world's most rapidly growing economies. East Asia's natural environments are among the most threatened worldwide, and the region's burgeoning rates of consumption will add to local and global environmental stresses. For all these reasons, East Asia will figure prominently in global relationships for years to come. Because of China's great size, enormous population, and huge economy, we give it particular emphasis in this chapter. Although much smaller in every way, Japan,

as another major global economic power—third after the United States and China in the total value of its goods and services—also gets particular emphasis.

| Themes to Explore in East Asia |

Current trends in the economy, migration patterns, population, and environment of East Asia are reflected in the following themes, which appear repeatedly throughout this chapter:

1. **Rapid economic growth.** Since China implemented economic reforms, its economy has become one of the fastest growing in the world. Other better-established market economies in the region (those of Japan, Taiwan, and South Korea) are much more productive per capita, but are growing at less than half the rate of China's.

2. **Influence of ideas from traditional Chinese thought.** Ancient Chinese forms of government and philosophy still influence economies, bureaucracies, and gender roles throughout East Asia, though their influence is manifested differently across the region.

3. **Population concentration on coasts and in lowlands.** The population is unevenly distributed, with the greatest concentrations in coastal areas and other lowlands, where overcrowding can be extreme.

4. **Urban-rural disparities.** Disparities between rural and urban areas in wealth and well-being are pronounced, especially in China, where growing cities gain special economic concessions from the state and attract millions of rural migrants.

5. **Environmental stress.** The demands of large, densely settled populations pursuing modern consumerist lifestyles are resulting in dangerous levels of water and air pollution and losses of soil fertility and natural habitats.

---

# I THE GEOGRAPHIC SETTING

## Terms to Be Aware Of

East Asian place-names can be very confusing to English-speaking readers. Whenever possible, we will give place-names in English transliterations of the appropriate Asian language. We will also avoid redundancies. For example, *he* and *jiang* are both Chinese words for *river*. Hence the Yellow River (so called because it is yellowed by its heavy sediment load) is the Huang He, and the Long River is the Chang Jiang (also called the Yangtze); it is redundant to add the term *river* to either name. The same is true of the word *hai* for "sea." The word *shan* appears in many place-names and usually means "mountain." Pinyin (a spelling system based on Chinese sounds) versions of Chinese place-names are now commonplace. For example, the city once called Peking in English is now Beijing, and Canton is Guangzhou. The region popularly known as Manchuria is here referred to simply as China's Far Northeast. Inner Mongolia, a province within China, is also known as Nei Mongol. Although China refers to Tibet as Xizang, people around the world who support the idea of Tibetan self-government

avoid using that name. This text uses Tibet for the region (with Xizang in parentheses), and Tibetans for the people who live there.

## PHYSICAL PATTERNS

A quick look at the regional map of East Asia (see Figure 9.1) reveals that its topography is perhaps the most rugged in the world. Powerful tectonic forces have produced a wide range of complex landforms. East Asia's varied climates result from a dynamic interaction between huge warm and cool air masses and the land and oceans. Its rapidly expanding human populations have affected the variety of ecosystems that have evolved there over the millennia and that still contain many important and unique habitats.

## Landforms

There are few flat surfaces to be found in the landscapes of East Asia. A simple way to visualize the region's varied landforms is to

**FIGURE 9.4** The Taklimakan Desert. This desert, the warmest and driest in China, sits in the Tarim Basin between the Kunlun Mountains and the Plateau of Tibet to the south and the Tien Shan (Celestial Mountains) to the north. [AP Photo/Eugene Hoshiko.]

think of them as analogous to the shapes that would be formed in a huge carpet if a grand piano (representing the Indian subcontinent) were shoved deeply into one side of it. The mountain ranges, plateaus, depressions, fissures, and bulges across Eurasia have resulted from the slow-motion collision of the Indian-Australian Plate (which carries the Indian subcontinent) with the southern edge of Eurasia (the Eurasian Plate) that began roughly 60 million years ago and continues today. The Himalayas are the most dramatic result of this collision. The area on the north side of the Himalayas absorbed some of the pressure by bulging upward and spreading outward to the northeast, forming the Plateau of Tibet (also known as the Xizang–Qinghai Plateau or Northern Plateau). To the north of the plateau, in the Xinjiang Uygur Autonomous Region and Qinghai Province, the land responded to the tectonic forces by sinking to form basins, including the Qaidam Basin and the sea-level Tarim Basin (Figure 9.4). Farther north, it buckled again to form the mountains south of the Junggar Basin and those of western Mongolia and southern Siberia. To the east and west of the Himalayan impact zone, wrinkled mountain and valley formations curve away to the southwest and southeast from the Plateau of Tibet.

The landforms of East Asia form four descending steps, moving roughly west to east. The top step is the Plateau of Tibet (depicted in gray with gold in Figure 9.1). Many of the rivers of China and the Southeast Asian mainland have their headwaters along the eastern rim of this plateau.

The second step down is a broad arc of basins, plateaus, and low mountain ranges (depicted in yellowish tan). These landforms include the deep, dry basins and deserts of Xinjiang and Qinghai, to the north of the Plateau of Tibet, and the broad, rolling highland grasslands and deserts of the Mongolian Plateau northeast of

Xinjiang. East of Xinjiang, this step also includes the upper portions of China's two great river basins, through which flow the Huang He and, farther east and south, the Chang Jiang. Far to the south is the rugged Yunnan–Guizhou Plateau, which is dominated by a system of deeply folded mountains and valleys that bends south through the Southeast Asian peninsula. The middle portions of the Nu (Salween), Mekong, and Red rivers are found here.

The third step, directly east of this upland zone, consists mainly of broad coastal plains and the deltas of China's great rivers (shown in shades of green in Figure 9.1), with intervening low mountains and hills (shown as light brown) toward the south. Starting from the south is a series of three large lowland river basins: the Zhu Jiang (Pearl River) basin, the massive Chang Jiang basin, and the Huang He lowland basin on the North China Plain. China's Far Northeast, the Korean Peninsula, and the westernmost parts of southern Siberia are also part of this third step.

The fourth step consists of the continental shelf, covered by the waters of the Yellow Sea, the East China Sea, and the South China Sea. Numerous islands—including Hong Kong, Hainan, and Taiwan—are anchored on this continental shelf; all are part of the Asian landmass.

The islands of Japan have a different geological origin: they are volcanic, rather than being part of the continental shelf. They rise out of the waters of the northwestern Pacific in the highly unstable zone where the Pacific, Philippine, and Eurasian plates grind against one another. Lying along a portion of the Pacific Ring of Fire (see Figure 1.21 on page 22), the entire Japanese island chain is particularly vulnerable to disastrous eruptions, earthquakes, and **tsunamis** (seismic sea waves). The volcanic Mount Fuji, perhaps Japan's most recognizable symbol (Figure 9.5), last erupted in 1707. However, in 2001, there were deep internal rumblings, suggesting that the mountain's period of dormancy may be ending.

**FIGURE 9.5** Mount Fuji. Japan's highest peak, at 12,388 feet (3776 meters), provides a stately distant backdrop to a woman working in tea fields. At one time considered a sacred mountain, today Mount Fuji attracts some 200,000 climbers annually, 30 percent of them foreigners. [Chad Ehlers/Alamy.]

The East Asian landmass has so few flat portions, and so many of them are so dry or so cold, that the large numbers of people who occupy the region have had to be particularly inventive in creating spaces for agriculture. They have cleared and terraced entire mountain ranges using only simple hand tools, irrigated drylands with water from melted snow, drained wetlands using elaborate levees and dams, and applied their complex knowledge of horticulture and animal husbandry to help plants and animals flourish in difficult conditions.

## Climate

East Asia has two contrasting climate zones (Figure 9.6): the dry interior west and the wet (monsoon) east. Recall from Chapter 8 that the term *monsoon* refers to the seasonal reversal of surface winds that flow from the Eurasian continent to the surrounding oceans during winter and from the oceans inland during summer.

***The Dry Interior.*** The dry western zone lies in the interior of the East Asian continental landmass. Because land heats up and cools off more rapidly than water, locations in the middle of large landmasses in the midlatitudes tend to experience intense cold in winter and intense heat in summer. The western part of China, which includes the Mongolian Plateau, the basins and mountains of Xinjiang and Qinghai, and the Plateau of Tibet, is an extreme example of such a midlatitude continental climate. Because there is too little vegetation or cloud cover to retain the warmth of the sun after nightfall, the difference between summer daytime and nighttime temperatures may be as much as 100°F (55°C).

Grasslands and deserts of several varieties cover most of the land in this dry region. Only scattered forests grow on the few relatively well-watered mountain slopes and in protected valleys supplied with water by snowmelt. In all of East Asia, humans and their effects are least conspicuous in the large, uninhabited portions of the deserts of Tibet (Xizang), the Tarim Basin in Xinjiang, and the

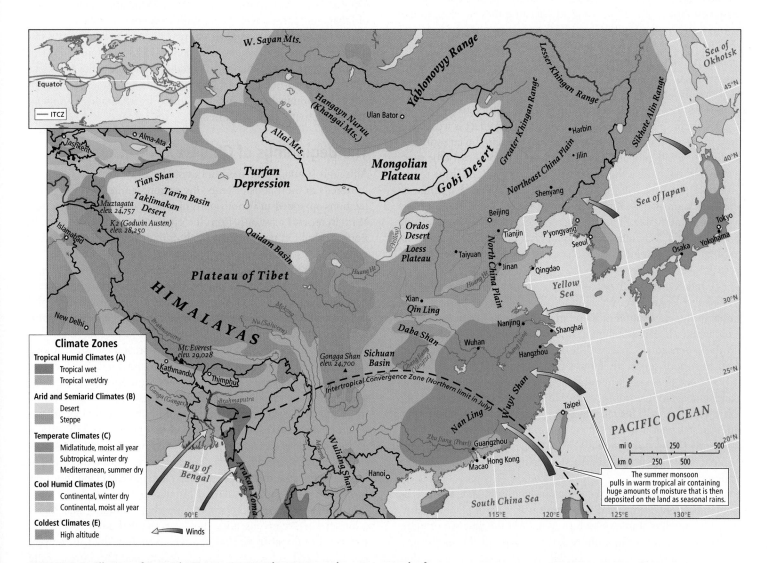

**FIGURE 9.6** Climates of East Asia. Two contrasting climate zones characterize much of East Asia: the dry continental west and the moist (monsoon) east.

Mongolian Plateau. The grasslands of Mongolia and northwestern China traditionally supported only scattered groups of nomadic pastoralists. However, they are increasingly being put to more intensive uses, such as irrigated agriculture in the Tarim and Junggar basins. In Mongolia, many former nomads now live on large-scale, stationary livestock production cooperatives, also supported with irrigation.

***The Monsoon East.*** The monsoon climates of the east are influenced by the extremely cold conditions of the huge Eurasian landmass in the winter and the warm temperatures of the surrounding seas and oceans in the summer. In what is commonly called the winter monsoon, dry, descending frigid arctic air sweeps south and east through East Asia, producing long, bitter winters on the Mongolian Plateau, on the North China Plain, and in China's Far Northeast. The winter monsoon also causes occasional freezes in southern China. Central and southern China have shorter, less severe winters because they are protected from the advancing arctic air by the east-west ranges of the Qin Ling and because they lie close to the warm waters of the South China Sea.

In the summer, as the continent warms, the air above it ascends, pulling in wet tropical air from the Pacific Ocean and its adjacent seas (see Figure 9.6). The warm, wet air from the ocean deposits moisture on the land as seasonal rains. As the summer monsoon moves northwest, it must cross numerous mountain ranges and displace cooler air. Consequently, its effect is weakened toward the northwest. Thus the Zhu Jiang basin in the far southeast is drenched with rain and enjoys warm weather for most of the year, whereas the Chang Jiang basin, which lies in central China to the north of the Nan Ling range, receives only about 5 months of summer monsoon weather. The North China Plain, north of the Qin Ling and Dabie Shan ranges, receives only about 3 months of monsoon rain. Very little monsoon rain reaches Inner Mongolia (in China) or the southern Mongolian Plateau. Northwestern China also gets very little rain.

China's Far Northeast is wet in summer, and neighboring Korea and Japan have wet climates all year because of their proximity to the sea. But all of these areas still have hot summers and cold winters because of their northerly location and their exposure to the continental effects of the huge Eurasian landmass. Japan and the more southerly Taiwan actually receive monsoon rains twice: once in spring, when the main monsoon moves toward the land, and again in autumn, as the winter monsoon forces warm air off the continent. This retreating warm air picks up moisture over the coastal seas, which is then deposited on the islands. More northerly Japan, however, has a much longer and more severe winter than subtropical Taiwan. Much of Japan's autumn precipitation falls as snow.

The eastern areas of the Eurasian landmass, watered by seasonal rains, once supported rich forests with diverse plant and animal life, much of it unique to East Asia. Over the past two or three millennia, however, agriculture has transformed the landscape. Many lowland ecosystems were wiped out as farmers expanded into any well-watered or irrigable land, be it flat or hilly. Hills and low mountainsides were rarely left covered with undisturbed forest; they were logged continuously, planted with orchards, or completely cleared and terraced for agriculture. Today, the few undisturbed natural areas in the humid zone are remote, deeply convoluted topographically, and increasingly threatened by development.

## HUMAN PATTERNS OVER TIME

East Asia is home to some of the most ancient civilizations on earth. Settled agricultural societies have flourished in China for over 7000 years. A little more than 2000 years ago, the basic institutions of government that still exist across the region today were established in eastern China. Until the twentieth century, China was the main source of wealth, technology, and culture for people across East Asia.

Chinese civilization evolved from several hearths, including the North China Plain, the Sichuan Basin, and the lands of interior Asia that were inhabited by Mongolian nomadic pastoralists. On East Asia's eastern fringe, the Korean Peninsula and the islands of Japan and Taiwan were profoundly influenced by the culture of China, but they were isolated enough that each developed a distinctive culture and maintained political independence most of the time. These characteristics proved crucial during the twentieth century, when these areas (except for North Korea) leapt ahead of China economically and militarily, partly by integrating European influences that China disdained.

### The Beginnings of Chinese Civilization

Although humans and their ancestors have lived in East Asia for hundreds of thousands of years, the region's earliest complex civilizations appeared in various parts of China about 4000 years ago. Written records exist only from the civilization that was located in north-central China. There, a small, militarized feudal aristocracy controlled vast estates on which the mass of the population worked and lived as impoverished farmers and laborers. The landowners usually owed allegiance to one of the petty kingdoms that dotted northern China. These kingdoms were relatively self-sufficient and well defended with private armies, and they often proved insubordinate to any aspiring central authority.

Between 400 and 221 B.C.E., a long period of war ensued among these petty kingdoms; finally, a single dominant kingdom emerged. What gave this kingdom, known as the Qin empire, its advantage was the use of a trained and salaried bureaucracy to aid the military in extending the monarch's authority into the countryside (Figure 9.7). The old feudal allegiance system was scrapped, and the estates of the aristocracy were divided into small units and sold to the previously semi-enslaved farmers. The kingdom's agricultural output increased because the people worked harder to farm land they now owned. In addition, the salaried bureaucrats who replaced their former masters were more responsible about building and maintaining levees, reservoirs, and other tax-supported public works that reduced the threat of natural disasters. Although the Qin empire was short-lived, subsequent empires maintained Qin bureaucratic ruling methods, which have proved essential to the governing of China as a whole right up to the present.

**FIGURE 9.7** An emperor's army. After unifying China in 221 B.C.E., Qin Shi Huang became its emperor. One of his many public works was the creation of a life-sized terra-cotta army of thousands of lifelike soldiers, each with a distinctive face and body, in a vast underground chamber that became his mausoleum. Since 1974, archaeologists have been excavating, restoring, and preserving this site, near modern-day Xian. It attracts more than 2 million visitors per year. [Courtesy of Mark Samols.]

## Confucianism

Closely related to China's bureaucratic ruling tradition is the philosophy of **Confucianism.** Confucius, who lived about 2500 years ago (prior to the Qin empire), was an idealist interested in reforming government and eliminating violence from society. He espoused the view that human relationships should involve a set of defined roles and mutual obligations, and that if each person understood his or her proper role and acted accordingly, a stable, uniform, and enduring society would result. Confucian values include respect for parents (filial piety) and government officials, courtesy, loyalty (to family and government), knowledge (moral wisdom), and integrity. Saving face and giving face (allowing others to maintain their personal dignity) is an important cultural norm in East Asia that stems from Confucian values.

The model for Confucian philosophy was the patriarchal extended family, in which the oldest male held the seat of authority as well as the responsibility to provide for the well-being of everyone in the family. All other family members were aligned under the patriarch according to age and sex and owed their allegiance and obedience to him. Extending these conventions, Confucianism held that commoners must obey imperial officials, and that everyone must obey the emperor, who, in turn, was obliged to ensure the welfare of society. As the supreme human being, the emperor was seen as the source of all order and civilization—in a sense, the grand patriarch of all China.

Over the centuries, Confucian philosophy penetrated all aspects of Chinese society, altering its social, economic, and political geography, deeply influencing gender roles, and circumscribing the lives of the masses at the base of the hierarchical pyramid. Concerning the ideal woman, for example, a student of Confucius wrote, "A woman's duties are to cook the five grains, heat the wine, look after her parents-in-law, make clothes, and that is all! When she is young, she must submit to her parents. After her marriage, she must submit to her husband. When she is widowed, she must submit to her son." The system also rested on the labor of large numbers of low-status men and children, who worked in agriculture and as servants.

Confucian ideology had a lasting effect on China's human geography because of the way it was utilized to maintain power and position. Elite groups emphasized the Confucian ideal of obedience and loyalty to maintain the status quo when it suited their interests, even at the expense of others in society. To block the emergence of competing powerful segments of society, elites attempted to limit merchants' wealth wherever possible. In parable and folklore, merchants were characterized as a necessary evil, greedy and disruptive of the social order. At certain times over the past 2000 years, the expropriation of farmers' and artisans' wealth through high taxes and the curtailment of merchants' prosperity left these groups little incentive or investment capital for agricultural improvements, industrialization, or entrepreneurship. At other times, however, the anti-merchant aspect of Confucianism was less influential, and trade and entrepreneurship flourished. Confucianism was never as strong in southern coastal China, which had a vibrant maritime society in which trade was important. Reflect on these historical points as you consider the modern geography of China.

Although the Confucian bureaucracy allowed Chinese empires to expand (Figure 9.8), its rigidities often led to periods of imperial decline. The heavy tax burden on farmers led to periodic revolts that weakened imperial control. In addition, invasions by nomadic peoples from what is today Mongolia and western China dealt the final blow to several successive Chinese empires despite defenses such as the Great Wall, built along China's northern border. So

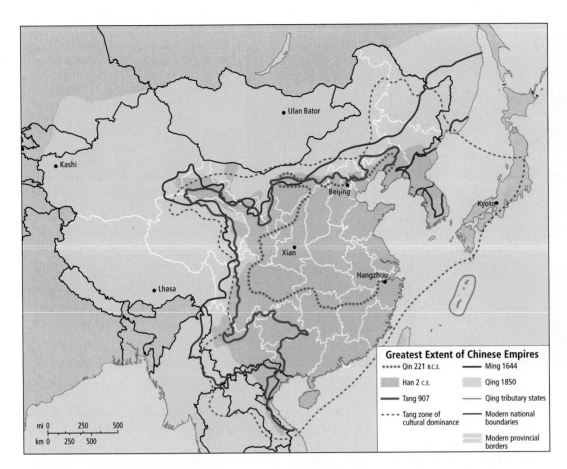

**FIGURE 9.8** The extent of Chinese empires, 221 B.C.E.–1850 C.E. The Chinese state has expanded and contracted throughout its history. [Adapted from *Hammond Times Concise Atlas of World History*, 1994.]

weakened were the Chinese in the 1100s, 1200s, and early 1300s that the Mongolian military leader Genghis Khan and his descendants were able to conquer all of China and then push west as far as Hungary and Poland (see Chapter 5, page 252). It was during this empire (the Yuan, or Mongol) that traders such as the Venetian Marco Polo made the first direct contacts between China and Europe.

## China's Preeminence

By the tenth century, China was the world's most developed region: it had the world's wealthiest economy, the largest cities, and the highest living standards. Improved strains of rice allowed dense farming populations to expand throughout southern China and supported large urban industrial populations. Metallurgy flourished. As early as 1078, people in northern China were producing twice as much iron as people in England did 700 years later. Nor was innovation lacking: Chinese inventions included papermaking, printing, paper currency, gunpowder, and improved shipbuilding techniques. All these advances contributed to the growth of remarkable urban areas. When Marco Polo wrote about his travels in China between 1271 and 1295 (during the Mongol empire), Europeans were stunned to learn that the imperial city of Hangzhou (near modern Shanghai) was 100 miles in circumference; had a population of 900,000 (many housed in apartment buildings); was dotted with parks, criss-crossed with boulevards, and served by large canals, sewer systems, and a fire department; and that the city's principal markets attracted 40,000 to 50,000 people daily.

During the Ming empire (1368–1644), Zheng He, a Chinese Muslim admiral from Yunnan Province, ventured far into the outside world, sailing in ships that were several times larger than those of Columbus. He ventured throughout Southeast Asia, across the Indian Ocean, and along the coasts of India, continuing into the Persian Gulf and down the coast of Africa. His voyages, which lasted from 1405 to 1433, were benign, apparently aimed at diplomacy and genial exchange, not conquest. For unknown reasons, the Ming rulers did little to build on Zheng's overtures to the outside world, though Chinese ships continued to ply Southeast Asian waters and the Indian Ocean. The Chinese empire on land experienced unprecedented economic expansion under the Ming rulers, but their domestic policies favored elite interests at the expense of ingenuity and change and ultimately left China ill prepared to respond to the challenge posed by Europe after 1600.

In 1644, invaders from Manchuria in far northeastern China overran Beijing and established the Qing (Manchu) empire. The Manchus unified the country, expanded its borders, and remained in power until challenged by European and Japanese incursions in the 1800s and internal revolts in the early 1900s.

## European Imperialism in East Asia

By the mid-1500s, Spanish and Portuguese traders interested in acquiring silks, spices, and ceramics had found their way to East Asian ports. In exchange, they brought a number of new food crops from the Americas (corn, peppers, peanuts, potatoes) that initiated a spurt of economic expansion and population growth. By the mid-1800s, China's population stood at over 400 million, and many Chinese were migrating to the frontiers of the empire—Yunnan in the central southwest and Manchuria in the northeast.

By the nineteenth century, European influence had increased markedly as European merchants gained access to Chinese markets. British merchants exchanged opium from India for Chinese wares, such as silks and fine ceramic dinnerware, that were much prized in Britain. Between 1839 and 1860, the Opium Wars broke out between Britain and China over China's unsuccessful attempt to crack down on the illegal importation of opium. China lost these wars and paid dearly. Hong Kong became a British possession and Macao a Portuguese possession; British merchants were granted unfettered access to trade in key coastal cities such as Shanghai; and foreigners (Europeans and Americans) were granted extensive rights to establish communities and conduct business on Chinese soil.

The final blow to China's preeminence in East Asia came in 1895, when a rapidly modernizing Japan won a spectacular naval victory over China in the Sino-Japanese War. After its defeat by the Japanese, the Qing empire made only halfhearted attempts at modernization, and it collapsed in 1912 after a coup d'état. During the decline of the Qing empire and until China's Communists took control in 1949, much of the country was governed by provincial warlords in rural areas and by a mixture of Chinese, Japanese, and European administrative agencies in the major cities (Figure 9.9). During this era, radical ideologies gained popularity in new Western-style universities as intellectuals searched for a new basis of political authority to replace Confucianism. Of particular interest were various forms of socialism and Communism.

## China's Turbulent Twentieth Century

Two rival reformist groups arose in China in the early twentieth century. One was the Nationalist party, known as the Kuomintang (KMT), led by Chiang Kai-shek. The KMT united the country in 1924 and at first reorganized it politically and economically according to the pro-worker, pro-farmer socialist views of Vladimir Lenin. After 1927, however, the KMT increasingly served the interests of the urban upper and middle classes. The rival group, the Chinese Communist Party (CCP), appealed most to the far more numerous rural laborers. Japan took advantage of these internal struggles to seize control of China's Far Northeast (Manchuria) in 1931. Then, in 1937, while the Western powers were distracted by European Fascism, Japan invaded China, conquering most of its major cities and coastal provinces and killing approximately 10 million Chinese in the process. For a while, the KMT and CCP tried to unite against this common enemy, but once Japan surrendered to the Allied forces at the end of World War II (in 1945), the two Chinese reformist parties resumed their conflict. Although the KMT was backed by the United States, it was quickly pushed out of the country by the then very popular CCP, led by Mao Zedong. The KMT and many of its urban supporters fled to the Chinese island of Formosa (now called Taiwan). There they formed a government-in-exile opposed to the mainland Communists. For half a century Taiwan was recognized as a separate country (and even deemed to stand in for all of China) by the West, but was considered merely a rebellious province by China (see the discussion on pages 484–485).

On the Chinese mainland, Mao Zedong's revolutionary mobilization of the rural majority resulted in the historic proclamation of the People's Republic of China on October 1, 1949, in Beijing's Tiananmen Square. Mao's government became the most powerful and expansionist government China had ever had, dominating the outlying areas of China's Far Northeast, Inner Mongolia, and Xinjiang Uygur and launching a brutal occupation of Tibet (Xizang). The People's Republic of China was in many ways similar to a traditional Chinese empire. The Confucian bureaucracy was replaced by the Chinese Communist Party, and Mao Zedong became a sort of emperor with unquestioned authority.

The chief early beneficiaries of the Communist revolution were the masses of Chinese farmers and landless laborers. On the eve of the revolution in 1949, huge numbers lived in abject poverty. Famines were frequent, infant mortality was high, and life expectancy was low. The vast majority of women and girls held low social status and spent their lives in unrelenting servitude.

The revolution drastically changed this picture. All aspects of economic and social life became subject to central planning by the Communist Party. Land and wealth were reallocated, often resulting in an improved standard of living for those who needed it most. Heroic efforts were made to improve agricultural production and to ameliorate floods and droughts. The masses, regardless of age, class, or sex, were mobilized to construct huge public works projects (roads, dams, canals, terraced mountains) almost entirely by hand. Industries were founded in many regions with an eye to providing

**FIGURE 9.9** "Twelfth-Night Cake for Kings and Emperors." This French cartoon lithograph, made in 1898, shows Britain, Germany, Russia, and Japan imperialistically carving up China, as France watches. A stereotypical Chinese official throws up his hands to show he is powerless to stop them. [The Granger Collection, New York.]

jobs to a broad segment of society. "Barefoot doctors" with rudimentary medical training dispensed basic medical care, midwife services, and nutritional advice to the remotest locations. Schools were built in the smallest of villages. Opportunities for women opened up, and some of the worst abuses against them—such as the crippling binding of women's feet to make them small and childlike—stopped. Until recent decades, famine had occurred somewhere in the country every few years. Since the mid-1970s, however, China has achieved a remarkable ability to feed its people. The vast majority of Chinese who are old enough to have witnessed these changes say that, materially, life is now better by far than before the revolution.

Nonetheless, this progress came at enormous human and environmental costs. During the **Great Leap Forward** (a government-sponsored program of massive economic reform initiated in the 1950s), 30 million died from famine brought on by the rush to fulfill poorly planned development objectives. Meanwhile, deforestation, soil degradation, and agricultural mismanagement became widespread. In the aftermath, some Communist Party leaders tried to correct the inefficiencies of the centrally planned economy, only to be demoted or jailed as Mao Zedong attempted to stay in power. In 1966, the **Cultural Revolution,** a series of highly politicized and destructive mass campaigns, enforced support for the Communist government and punished dissenters. Hundreds of millions were required to study the "Little Red Book" of Mao's sayings (Figure 9.10). Educated people and intellectuals were a main target of the Cultural Revolution because they were thought to instigate

**FIGURE 9.10  China's Cultural Revolution.** Taken in the late 1960s, this photo shows a group of Chinese children in uniform in front of a picture of Chairman Mao Zedong. The children each hold a copy of Mao's "Little Red Book." [Hulton Archive/Getty Images.]

dangerously critical evaluations of Communist Party central planning. Many Chinese scientists and scholars were sent to labor in mines and industries or to jail, where many of them died. Children were encouraged to turn in their parents. Petty traders were accused of being capitalists and severely punished, as were those who adhered to any type of organized religion. The Cultural Revolution so disrupted Chinese society that by Mao's death in 1976, the Communists had been thoroughly discredited.

Two years later, a new leadership formed around Deng Xiaoping. In the early 1980s, he initiated a series of reforms to give China's economy some of the characteristics of an open market while at the same time maintaining Communist Party political control and intolerance of minor dissidence. However, the pace and direction of reform did not keep up with popular expectations. In April and May of 1989, more than 100,000 students and workers agitating for greater freedom of expression and government accountability marched on Tiananmen Square in Beijing. The government suppressed the demonstrations by firing into the crowd, killing hundreds and injuring thousands. Arrests, hasty trials, and executions followed. Today, although remarkable levels of economic growth have been achieved, making China's economy the second largest in the world, after that of the United States, many observers are concerned that civil and human rights are often not respected and that political activity is too tightly controlled.

## Japan Becomes a World Leader

Although China is the behemoth of East Asia, Japan, with only one-tenth the population and 5 percent of the land area of China, has developed into one of the world's most economically powerful countries, with its third largest economy. Modern Japanese populations are descended from migrants from the Asian mainland, the Korean Peninsula, and the Pacific islands. By 300 C.E., the society was divided into military clans that had established their rule over most of the islands that are now part of Japan. Settlement was concentrated in central Honshu and on the southern islands (see Figure 9.1). Ideas and material culture imported from China and the Korean Peninsula strongly influenced the everyday lives of Japan's people and their use of the land. These imports included Buddhism, Confucian bureaucratic organization, architecture, Chinese writing, the arts, and agricultural technology.

Between 800 and 1300, Japanese society turned inward, becoming more feudal and rigidly structured, and influences from the Asian mainland waned. Active trade with Portugal, beginning in 1543, brought new ideas and technology that strengthened the wealthier feudal lords, who then unified Japan under a military bureaucracy. Between 1600 and 1868, a hereditary line of elite military rulers (known as **shoguns**) imposed a strict four-tier social class system. Rejecting all European influences, the shoguns expelled foreigners and imposed isolation once again.

In 1853, a small fleet of U.S. naval vessels commanded by Commodore Matthew C. Perry arrived near the mouth of Tokyo Bay. The foreigners, carrying military technology far in advance of Japan's, forced its government to open the economy to international trade and political relations. In response to this foreign military threat as well as domestic pressure for political and economic change, a

**FIGURE 9.11** Infrastructure development in early twentieth-century Japan. This street scene in Tokyo in about 1905 shows that modernization was in full swing: electric streetcars, electric utility poles, several different kinds of vehicles, and crowds of people, mostly men, can be seen. [Hulton Archive/Getty Images.]

group of reformers (the Meiji) seized control of the Japanese state, setting the country on a crash course of modernization and industrial development (Figure 9.11). They sent Japanese students abroad and recruited experts from around the world, especially from Western nations, to teach everything from foreign languages to modern technology.

Meanwhile, Japanese settlement expanded from central Honshu and the southern islands to northern Honshu and Hokkaido, where new farmlands and cities were developed and strategic resources, such as coal, were exploited for heavy industry. From 1895 to 1945, to fend off European imperialism, to expand its resource base further, and to gain a labor force for its mines and factories, Japan colonized Korea, Taiwan (then known as Formosa), and Manchuria, then pushed farther into China and into Southeast Asia (Figure 9.12). Its policies were colonial in nature and quite brutal.

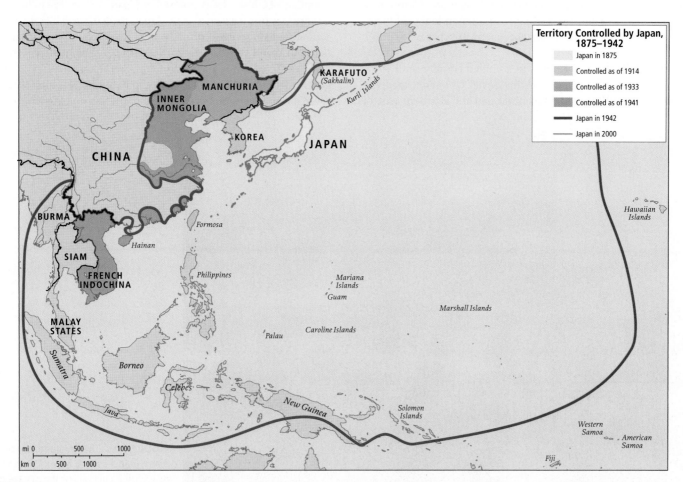

**FIGURE 9.12** Japan's expansions, 1875–1945. Japan colonized Korea, Taiwan (Formosa), Manchuria, China, parts of Southeast Asia, and several Pacific islands to further its program of economic modernization and to fend off European imperialism in the early twentieth century. [Adapted from *Hammond Times Concise Atlas of World History*, 1994.]

Japan's imperial ambitions ended with its defeat in World War II, and the country was occupied by U.S. military forces from 1945 until 1952. The U.S. government imposed many social and economic reforms and required Japan to create a democratic constitution and reduce the emperor to symbolic status. Japan's postwar constitution allows only a very limited military, and for more than 50 years the country has relied primarily on U.S. forces based in Japan, South Korea, and the western Pacific to protect it from attack. Japan rebuilt rapidly after World War II with U.S. support, and it eventually became a giant in industry and global business, exporting automobiles, electronic goods, and many other products to the developed world and investing in developing economies worldwide. By 2001, Japan's economy was the world's most technologically advanced after that of the United States, though China was soon to take over as the second largest producer of goods and services.

## Conflict and Transfers of Power in East Asia

The history of the entire region is to some extent grounded in what transpired in China and Japan. All East Asian countries, for example, have been affected by ancient Confucian philosophy, the relics of which can be found in present-day political and social policy. North Korea, and for a while Mongolia, joined China in the Communist experiment. Taiwan and South Korea have chosen to model themselves after Japan's state-aided market economy (see pages 487–489).

**The Korean Peninsula.** Korea was a unified country until 1945, when, at the end of World War II, the Soviet Union declared war against Japan and invaded Manchuria and the northern part of Korea. To prevent the entire country coming under Soviet control, the United States proposed dividing Korea at the 38th parallel. The United States took control of the southern half of the peninsula, and the Soviet Union took the northern half. The United States withdrew its troops in the late 1940s, after elections in South Korea, but in the north, Korean and Soviet Communists established a military government. In June 1950, North Korea attacked South Korea, and the United States came to the south's defense, leading a United Nations force that fought a 3-year war against North Korea and its allies, the Soviet Union and China. The Korean War ended in a truce and the establishment of a demilitarized zone (DMZ) at the 38th parallel in 1953 (Figure 9.13), after great loss of life on both sides and devastation to the peninsula's infrastructure. North Korea closed itself off from the rest of the world and to this day remains impoverished and defensive, occasionally gaining international attention by rattling its nuclear arsenal. South Korea has developed a prosperous market economy on the Japanese model.

**Taiwan.** Taiwan is a modern, crowded, and highly industrialized society that has played a leading role, through investment and technology diffusion, in the rapid transformation of East Asian and Southeast Asian economies. Long a poor agricultural island on the periphery of China, it was annexed by Japan in 1895, when the Qing empire was in a weakened state. Japan treated Taiwan as a colony, exploiting its resources for more than half a century until Japan's defeat in World War II. Then, in 1949, the Chinese nationalists (the KMT), pushed out of mainland China, set up an anti-Communist government on Taiwan, naming it the Republic of China (ROC). Over the next 50 years, with U.S. aid and encouragement, it be-

**FIGURE 9.13** The demilitarized zone. Taken from the South Korean perspective in 2005, this photo shows the gateway to the DMZ, which looks something like a theme park. [© 2005 Albert H. Teich.]

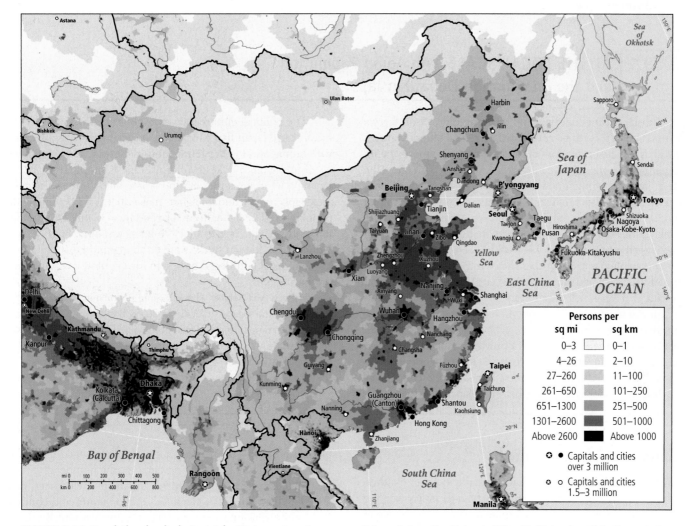

**FIGURE 9.14** Population density in East Asia. [Data courtesy of Deborah Balk, Gregory Yetman et al., Center for International Earth Science Information Network, Columbia University, at http://www.ciesin.columbia.edu.]

came a powerhouse industrialized economy and operated as an independent country, but China never relinquished claim to Taiwan and still considers it a province of China.

Taiwan today is a geopolitical hot spot. Its status is ambiguous: the United Nations does not recognize it as a country because of China's opposition, yet Taiwan operates as an independent country in nearly every way. The United States tacitly supports Taiwan and sells it arms, but its strong interest in trade with mainland China and its desire to avoid military entanglements keep it from taking an overt stand on behalf of Taiwan.

*Mongolia.* In the distant past, Mongolia's nomadic horsemen periodically posed a threat to China, and the Mongolian ruler Genghis Khan (ca. 1162–1227) and his successors eventually conquered most of China. As a consequence, controlling its northern neighbor became a Chinese obsession. China did control Mongolia from 1691 until the 1920s. A Communist revolution began soon thereafter, and Mongolia continued as an independent Communist country under Soviet guidance until the breakup of the Soviet Union in 1989 launched it on a difficult road to a free market economy.

## POPULATION PATTERNS

East Asia is the most populous world region. As you can see in Figure 9.14, however, people are not evenly distributed on the land. China, with 1.3 billion people, has more than one-fifth of the world's population, but many areas are very lightly settled because of their difficult terrain or climate. Ninety percent of China's people are clustered on only one-sixth of the total land area. They extract a very high level of agricultural production from this land, though at considerable environmental cost. People are concentrated especially densely in the eastern third of China: in the North China Plain, the Sichuan Basin, the middle and lower Chang Jiang (Yangtze) basin, and the delta of the Zhu Jiang (Pearl River) in southeastern China. The west and south of the Korean Peninsula are also densely settled, as are northern and western Taiwan. In Japan, settlement is concentrated in a band stretching from the cities of Tokyo and Yokohama on Honshu south through the coastal zones of the Inland Sea to the islands of Shikoku and Kyushu. This urbanized region is one of the most extensive and heavily populated metropolitan zones in the world, accommodating well over half of Japan's total population. The rest of Japan is mountainous and more lightly settled.

**485**

## BOX 9.1 | AT THE LOCAL SCALE An Aging Population and the Immigration Debate in Japan

With increasing longevity and falling fertility, Japan faces a situation soon to be confronted by several other countries. Japan has the largest proportion of elderly people among the industrialized nations (20 percent over 65), while its birth rate is the lowest. Japanese women now have an average of only 1.25 children. If this trend continues, Japan's population will plummet from the current 127.7 million to just over 100 million by 2050, shrinking the country's labor pool by more than a third and dragging down the country's national wealth.

One solution is to throw open the country to immigrants from Asia, Africa, and Latin America. A recent United Nations report estimates that Japan will have to import over 640,000 immigrants per year just to maintain its present workforce and avoid a 6.7 percent annual drop in its GDP. "It's been obvious for a while now that Japan desperately needs an influx of workers from outside," says Tony Lazlo, director of an organization that researches multicultural issues in Japan. "People in business have been forced to recognize this, but it's not really widely known in Japanese society."

The political culture, however, clings to a vision of an ethnically pure Japan that can avoid the social problems associated with immigration in Western countries. Resistance to higher levels of immigration takes two forms. One is simply resistance to having a lot of foreigners around—prejudices against foreigners are deeply rooted, and foreigners make up just 1.2 percent of the population. The other form of resistance is milder, claiming Japan just needs time to adjust to being a multicultural society. Nonetheless, foreign workers are dribbling into Japan, in a multitude of legal and illegal guises, to fill the most dangerous and low-paying jobs.

*Sources:* Asia Society, "Japan's aging population: A challenge for its economy and society" (October 7, 2003), http://www.asiasource.org; Howard W. French, "Insular Japan needs, but resists, immigration," *New York Times* (July 24, 2003); "Aging Japan," http://www.Economist.com, June 5, 2006.

## Declining Population Growth

Throughout East Asia, people are having markedly fewer children than in the past. China has a one-child-per-family policy (see the discussion on page 498), and the Chinese are now reproducing at a rate considerably lower than the world average: 12 births per 1000 people yearly, as opposed to the world average of 21 per 1000. If the one-child family pattern continues, China's growth will eventually slow and possibly even stop—but not until perhaps 2050, simply because so many Chinese are just entering their reproductive years. Elsewhere in East Asia, rates of natural increase hover around China's rate of 0.6 percent (the same as the U.S. rate). Japan has the lowest rate in the region, at 0.1 percent. In fact, Japan's population is growing so slowly that it will not double for 700 years. Thus Japan's population, like China's, is rapidly aging. By 2025, Japan will have a ratio of one pensioner for every two workers—comparable to that of several European countries. Only in Mongolia (with 2.6 million people) and North Korea (with 22.9 million) are women still averaging 2 or more children each. But even in those two countries, increasing opportunities for women outside the home mean that family size is shrinking, family life and gender roles are changing, and the numbers of dependent elderly are growing (Box 9.1).

## HIV-AIDS in East Asia

According to the Joint United Nations Program on HIV-AIDS, East Asia is experiencing a concentrated and potentially explosive HIV-AIDS epidemic. Although rates of infection are relatively low by global standards—0.1 percent of the population aged 15 to 49 is infected—this figure masks a large and growing number of infections (140,000 per year, according to the Population Reference Bureau). By 2005, between 870,000 and 1.3 million adults and children in the region were infected with HIV. Doctors without Borders predicts that, if present conditions and attitudes continue, by 2010 there will be at least 10 million people living with the disease in East Asia. The stigma against HIV-AIDS is very strong in this region, preventing people from getting tested and treated and thus increasing the likelihood of the infection spreading.

The majority of HIV infections are in China, where concentrated epidemics have been under way for many years in certain regions (Yunnan, Xinjiang, Guangxi, Sichuan, Henan, and Guangdong) and are poised to spread rapidly in several others. The number of reported HIV-AIDS cases has increased significantly in recent years, particularly among migrants, injecting drug users, and sex workers. Japan is also witnessing a steady increase in the rate of new HIV infections; the number of new HIV-AIDS cases reported annually has doubled since the 1990s to more than 900 per year in 2005. This rise has been accompanied by an increase in other sexually transmitted infections over the same period and may be related to evidence of more widespread sexual activity among Japanese youth. Estimates are that there are 17,000 total cases of HIV infection in the country (less than 0.05 percent of the population aged 15 to 49). The incidence of HIV-AIDS in South Korea is somewhat lower than in Japan. In Mongolia, where 50 percent of the population is under 23 and drug use and sex work are increasing with greater global interaction, it is surprising that only about 500 adult cases have been reported thus far.

# II  CURRENT GEOGRAPHIC ISSUES

Although the countries of East Asia adopted new economic systems only after World War II, most of them are making progress in creating a better life for their citizens. In fact, Japan, Taiwan, and China's Hong Kong Special Administrative Unit have among the highest standards of living in the world. The great challenge in this part of the world will be to achieve and maintain improved living standards for large numbers of people in a manner that does not place too much stress on the environment. Balancing economic prosperity with environmental integrity is an issue every region faces, but East Asia's large population makes finding solutions to this problem especially imperative.

## ECONOMIC AND POLITICAL ISSUES

After World War II, the countries of East Asia (Figure 9.15) established two types of economic systems. The Communist regimes of China, Mongolia, and North Korea relied on central planning to set production quotas and to allocate goods among their citizens. In contrast, Japan, Taiwan, and South Korea established **state-aided market economies** with the assistance and support of the United States and Europe. This type of economic system is similar to the free market system of the West in that market forces, such as supply and demand and competition for customers, determine many economic decisions. However, the government maintains a distinct interventionist role in a state-aided market economy.

More recently, the differences among East Asian countries have diminished as China and Mongolia have set aside strict central planning and adopted reforms that allow market forces, rather than government quotas, to set production quantities and prices. This transition has resulted in rapid economic growth, but has caused hardship for many people because that growth has been un-

even, bringing income inequality, and because government belt-tightening has resulted in the loss of social services to millions. At the same time, the older state-aided market economies of Japan, South Korea, and Taiwan have experienced much slower economic growth since the late 1990s.

## The Japanese Model

Throughout the nineteenth century, the economies of Japan, Korea, and Taiwan were minuscule compared with China's. Then, during the twentieth century, all three grew tremendously. By the 1980s, Japan's economy was the second largest in the world, behind only that of the United States. Following economic setbacks in the 1990s and China's rapid growth, Japan's economy (measured as GDP, adjusted for PPP) is now the third largest in the world. South Korea has the world's sixteenth largest economy and Taiwan the nineteenth largest. Credit for the economic success of Taiwan and South Korea belongs mostly to the Japanese model of a state-aided market economy.

***Japan's Economic Rise.*** Japan has been renowned for its successful export industries, which turn out such products as automobiles, heavy equipment, cameras, computers, and video and audio equipment. Japan's rise to world prominence in trade and technology since the middle of the nineteenth century (when the country's doors were closed tightly to all trade and international travel and its industrial technology was relatively primitive) is one of the most remarkable tales in modern history—all the more so because of its limited base of natural resources and its crippling defeat at the end of World War II.

It would be hard to overstate the extent of devastation in Japan at the end of the war. Except for Kyoto, which was spared because of its historical and architectural significance, all of its major cities had been mercilessly bombed and destroyed by the United States —most notably Hiroshima and Nagasaki with nuclear weapons and Tokyo with incendiary bombs. Key to Japan's rapid recovery was a strong state bureaucracy that helped wealthy private capitalists to start the industrial enterprises that became the foundation of its early economic development—a strategy later replicated in Taiwan and South Korea. The government provided advice, financial assistance, and protection from internal and external competition. Also important were the opening of markets abroad for Japanese products; a hard-working, educated, and skilled citizenry; and the security that came from guaranteed lifetime employment for most urban workers.

Innovation in manufacturing was another factor in Japan's success. Japanese manufacturers organized firms in highly efficient spatial arrangements. Firms that supply automobile parts, for example, are often clustered around a single company where the final assembly of automobiles takes place. This proximity allows suppliers

**FIGURE 9.15**  Political map of East Asia.

**FIGURE 9.16** Modern Tokyo at twilight. The Shinjuku business and shopping district in the west of Tokyo is known for its over-the-top sight and sound experiences. Every conceivable product or service is available for money. [Peter Adams/CORBIS.]

to deliver their products "just in time," literally minutes before they are needed. Fewer defective parts are produced because production lines are constantly surveyed for errors. Called the continuous improvement system (or the *kaizen* system), this method results in ever more efficient production and in less space being taken up by warehouses. The *kaizen* system has influenced economic geography globally as a growing number of firms now pursue a strategy of locating the assembly of both components and final products in a single area, either abroad or at home. Toyota Motor Corporation, for example, implemented this model in Kentucky (see Chapter 2, pages 77–78).

For much of the post–World War II period, Japan's economy grew roughly 10 percent a year. Heavy industry and electronics manufacturing were the leading sectors. Japanese brand names such as Sony, Panasonic, Nikon, and Toyota became household words around the world. However, since 1992, employment has shifted significantly from manufacturing to services, such that the service sector now employs 68 percent of the workforce and constitutes 72 percent of Japan's GDP. Banking and financial services are especially important industries that have contributed to Tokyo's status as a world city (Figure 9.16), yet recently they have undergone changes via mergers and consolidations.

Japan's "economic miracle" continues to have an immense worldwide effect. Resources from all parts of the world are shipped to Japan, and Japanese purchases are mainstays of local economies in the supplying countries, effectively making Japan something of a global employer. Japanese know-how has led to innovations that turn imported raw materials into high-quality manufactured goods in demand throughout the world. Toyota, for example, has added marine products, agricultural biotechnology, and the design and building of high-quality, energy-efficient prefabricated homes to its automotive production (Figure 9.17). Japanese profits from these

industries are often invested abroad in factories, hotels, and resorts, thus securing multiple markets for the country.

***Recession and Subsequent Low Economic Performance.*** In the early 1990s, the Japanese economy showed signs of strain, as did East Asia's other state-aided market economies. The situation was worsened by a financial crisis in the late 1990s that swept through Southeast Asia, an important market and site of invest-

**FIGURE 9.17** Innovation beyond automobiles. Toyota Motor Corporation, best known for automotive production, has been building houses for the Japanese market since the mid-1970s. In 2000, it introduced a house that has an earthquake-resistant steel frame, excellent thermal insulation, and solar-powered roofing panels. The emphasis is on superior construction and environmental considerations. This photo shows a Toyota housing gallery in Tokyo. [AP Photo/Shuji Kajiyama.]

ment for Japan, South Korea, and Taiwan. However, many critics of the Japanese model of economic development argue that the long-standing close relationship between government and industry was at fault because it constrained the economy with overly ambitious real estate investments at home and abroad and overexpansion of industrial productive capacity. Equally damaging was the unwillingness on the part of bureaucrats and politicians to let unprofitable companies go bankrupt. Corruption also contributed to the Japanese economic crisis: the close relationship between government and industry had nurtured favoritism that resulted in inefficiency. Yet another problem was that, between 1988 and 1992, Japan increased its productive capacity by an amount equal to the entire economy of France—just when the demand for its products began to diminish at home and abroad.

Some experts on the Japanese economic model link economic strains and lack of growth to overprotected workers who have little incentive to be inventive. In the United States and Europe, even activities that are eccentric, if they lead to innovation, are encouraged with higher pay; in Japan, they can result in censure or marginalization. Conformity, rather than cutting-edge creativity, leads to job security, though not to significant salary raises. Other critics say that it is the protection of certain outdated sectors of the economy, such as farming and retail, from global competition that is slowing growth in Japan and burdening consumers with unnecessarily high prices. One consequence of Japan's relatively low wages is that middle- and lower-class consumers cannot buy many of the high-quality items they themselves produce, so high-end goods are sold instead to consumers in North America and Europe. It is argued that workers should be paid more, with raises contingent on performance, so that there will be incentives for those workers to generate more of their own technological innovations. Furthermore, more prosperous workers could afford to buy Japan's products, thus strengthening the economy from within. Another drag on the economy is the high savings rate of people over 60. This group is thought to be hoarding enormous surplus funds that could fuel economic growth if the money were spent on consumer goods.

***Economic Challenges.*** Even though the Japanese economic model has changed somewhat in the aftermath of recession, there is debate over whether that model can continue to be successful. The Japanese people, long accustomed to tolerating extended workdays, overcrowding, pollution, high prices, and only modest buying power in the interest of rapid economic growth, are agitating for changes that will improve their living standards. The greater participation of women in the workforce and the need for employees to be more creative and innovative are leading to changes in the workplace. Moreover, increased life expectancy is producing an aging population that will soon require more social spending, further burdening the working-age population (see Box 9.1).

Internationally, China's rising economic power promises opportunities for investment by Japan, South Korea, and Taiwan. All three have significant projects in China, but China also poses a competitive threat to these countries in the global marketplace. Competition is also emerging from the newly industrializing countries of Southeast Asia—all countries with strong export economies

that have received infusions of Japanese investment funds. Meanwhile, the United States has been pushing Japan to remove its barriers against foreign imports. Since 1950, Japan has consistently blocked many incoming trade goods.

## The Communist Command Economy

The Communist economic systems of China, Mongolia, and North Korea emerged after World War II as a way to salvage societies that had been badly damaged by a grossly self-serving elite, by Japanese and European colonial domination, and by civil war. The idea was that centralizing the management of all facets of the economy would achieve maximum efficiency. All three countries abolished private property, and the state took control of agricultural and industrial production; construction; service industries such as transportation and distribution, utilities, social services, and education; and sales of food and consumer goods. Economies directed by central government planning agencies in this way are called command, or centrally planned, economies (see the discussion of the Soviet command economy in Chapter 5 on pages 259–260). These sweeping changes ultimately proved less successful than was hoped.

By design, most people in the Communist economies were not allowed to consume more than the bare necessities. On the other hand, the "iron rice bowl" policy guaranteed nearly everyone a job for life and the means to obtain the essentials. The emphasis on job security and hard work, rather than on creativity, led to overwhelming conformity and a lack of innovation; hence productivity remained low.

***The Commune System.*** When the Communist Party first came to power in China in 1949, its top priority was to make great improvements in both agricultural and industrial production. Similar early goals were held by the Communist regimes in North Korea and Mongolia, though they had much smaller populations and resource bases to work with. In China, the first strategy was land reform, and the initial effort was to take large, unproductive tracts out of the hands of landlords and put them into the hands of the millions of landless farmers. By the early 1950s, much of China's agricultural land was divided into tiny plots. But it soon became clear that these small plots were not going to produce enough to feed the people who were leaving agriculture to work in the many expanding industries in the cities. Communist leaders decided to band small landholders together into cooperatives so that they could pool their labor and resources to increase agricultural production. In time, these cooperatives became full-scale communes, with an average of 1600 households each. The communes, at least in theory, took care of all aspects of life. They were the basis of Communist Party political organization, provided health care and education, and built rural industries to supply themselves with such items as fertilizers, gunnysacks, pottery, and small machinery. The rural communes also had to fulfill the ambitious expectations of the leaders in Beijing for better flood control, expanded irrigation systems, and the generation of surplus funds for investment elsewhere.

The Chinese commune system had several difficulties. Farmers were required to spend so much time building roads, levees, and drainage ditches or working in the new rural industries that they had too little time to farm. Local Communist Party administrators often compounded the problem by overstating harvests to impress their superiors in Beijing. The leaders in Beijing responded by requiring larger food shipments to the cities, which created food shortages in the countryside.

***Focus on Heavy Industry.*** The Communist leadership believed that before China produced consumer goods, it needed to improve its infrastructure—to build roads, railways, and dams—all of which required heavy equipment. So the government emphasized mining for coal and other minerals, producing iron and steel, and building heavy machinery. (Similar policies were followed in North Korea and Mongolia.) Funds for this development were to come from the already strained agricultural sector: farmers were required to sell their products to the state at artificially low prices, and the state directed the profits gained in reselling those products toward industrialization. Other funds for industry came from profits in mining and forestry.

***Regional Disparity in China.*** For centuries, China's interior west has been poorer than its coastal east. The interior west has been locked into herding and agricultural economies while, even in the most restricted times, the economies of the east have benefited from trade and industry. This spatially uneven development has continued to plague the Chinese in modern times. Right after the revolution, economic policy was focused on **regional self-sufficiency,** with each region encouraged to develop independently, building both agricultural and industrial sectors, in the hope of creating jobs and evening out the national distribution of income. During the cold war, Communist leaders also believed that dispersed industrial development and regional self-sufficiency would foil an enemy's effort to destroy China's productive capacity.

Government funds were used to set up industries in nearly every province, regardless of practicality. For example, a steel industry was established in Inner Mongolia at great expense, although other wealthier provinces already had steel industries that could have been improved with a far greater payoff. Similarly, huge mechanized grain farms were developed on cleared forestlands in far northern Heilongjiang, although the same effort elsewhere would have yielded better results and allowed the forests to be saved for other purposes. This policy of regional self-sufficiency also encouraged individual communities to develop backyard smelters, kilns, tool and die factories, or their own tiny tractor factories. Many outdated, highly polluting small industries across China date from this era. More recently, in yet another effort to keep people in the interior, the government has encouraged a wide range of often foreign-financed industries to open in interior towns and cities (see page 494).

## Globalization and Market Reforms in China

By the late 1970s, it was clear that China's command economy was not going to make dramatic improvements in living standards. In the 1980s, China's leaders enacted market reforms that changed the country's economy in four ways. First, economic decision making was decentralized. Second, farmers and small businesses were permitted to sell their produce and goods in competitive markets, thus improving the efficiency with which food and goods were produced and distributed. Third, regional specialization, rather than regional self-sufficiency, was encouraged. Finally, the government allowed foreign investment in Chinese export-oriented enterprises and the sale of foreign products in China. This four-part shift to a market-based economy transformed the region, and indeed the whole world, as China became a participant in the global economy. In the 1990s, China emerged as a significant producer of manufactured goods for the world market. Equally important, it now represents a market of more than 1.3 billion potential consumers; nearly every major company in the world is eager to sell its goods to customers in China. Meanwhile, Mongolia has participated in this revolution only modestly and North Korea, not at all.

***Regional Specialization.*** Decentralizing decision making encouraged regional specialization. Managers of many state-owned enterprises now had the right and responsibility to improve the efficiency of their operations by setting production levels and prices for

**FIGURE 9.18** Regional specialization in agriculture. Nearly half of China's crop production takes place along the fertile banks of the Chang Jiang (Yangtze River). Among the crops grown are rice, wheat, barley, corn (maize), beans, cotton, and hemp. Note the terrace-style fields throughout the landscape. [Michael S. Yamashita/CORBIS.]

goods and services according to the demands of consumers in the open market. And to some extent, anyone could create a new enterprise. In response, managers and entrepreneurs (sometimes with assistance from the state) took advantage of the different resources and opportunities offered by different areas of the country. The old colonial city of Shanghai, for example, once again became a center of trade and finance, and the Zhu Jiang (Pearl River) delta quickly evolved into a massive industrial center for the production of export goods.

***Trends in Agriculture.*** The reforms that began in the late 1970s and China's subsequent integration into the global economy have brought new opportunities to some agricultural sectors and disadvantages to others. The **responsibility system** returned agricultural decision making to the household unit, subject to the approval of local authorities. Farmers are now encouraged to organize family-sized or larger operations that can meet at least local food demand, ideally producing some excess for urban centers and in some cases for export. They can choose crops and marketing strategies, but must accept responsibility for success or failure. Because agricultural potential varies greatly across China, regional specialization in agriculture has increased markedly (Figure 9.18). Many farmers have shown entrepreneurial skills by assessing the market for lucrative niches they can fill. In places where the growing season is long

enough, such as the far southern provinces of Yunnan and Guangdong, farmers have prospered by growing fresh produce for distant northern cities as well as for local cities, often with labor seasonally imported from yet poorer areas. Other farmers have failed to modernize sufficiently or have lost their land to urban sprawl. Nonetheless, it is remarkable that China, once plagued by famine, now produces prolifically for domestic consumption and has become a major exporter of high-value, labor-intensive agricultural products to other countries in Asia, particularly Japan and South Korea. Exports of fresh and preserved fruits, vegetables, fish, animal products, and manufactured foods and the production of organic foods are likely to increase as China expands its comparative advantage in such products.

At the same time, changing patterns of food production and consumption, urban expansion, and trade are leading to shortages in land and water that raise new challenges for China's **food security**—its ability to supply sufficient basic food to all its people consistently over the long term. Only a portion of China's vast territory can support agriculture because much of the land is too cold, too dry, or too steep to cultivate (Figure 9.19). China's huge population has already stretched the productive capacity of many fertile zones well beyond sustainability, and demands are only increasing. As cities grow, they consume prime agricultural land, and as urban populations grow more affluent, their taste for meat and other

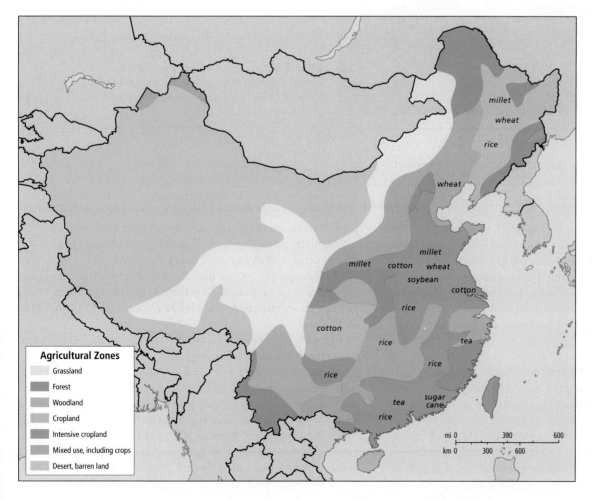

**FIGURE 9.19** China's agricultural zones. As part of the economic reforms instituted in recent years in China, greater regional specialization in agricultural products is taking place. [Adapted from "World agriculture," *National Geographic Atlas of the World, Eighth Edition* (Washington, D.C.: National Geographic Society, 2005), p. 19; and "China—Economic, minerals," *Goode's World Atlas 21st Edition* (New York: Rand McNally, 2005) p. 207.]

Agricultural Zones
- Grassland
- Forest
- Woodland
- Cropland
- Intensive cropland
- Mixed use, including crops
- Desert, barren land

## BOX 9.2 | AT THE GLOBAL SCALE China's "Button Town" Fights Itself for Global Market Share

Wang Chunqiao is a button millionaire. His factories are in Qiaotou, a small Chinese town southeast of Shanghai that has about 200 factories and 20,000 migrant workers. Buttons used to be made in small factories all over the world from seashells, wood, leather, brass, and copper; Italy, Turkey, and France were important producers. But like the garment industry, button production has largely relocated to Asia. And Qiaotou has cornered 60 percent of the world production of buttons (Figure 9.20)!

Now Qiaotou's button manufacturers have to compete with themselves to gain more customers, cannibalizing one another through buyouts and drastic price cuts. Price cutting is painful because operating costs are rising. So many foreign firms have rushed to China recently that good workers can demand higher wages—workers in the button factories now earn at least U.S.$120 per month. To attract and keep them, bosses must also provide food, housing, and cultural and sports activities. Furthermore, energy costs have increased, as have costs of supplies; copper prices have more than doubled in a year. Still savoring his rags-to-riches rise with a big smile, Wang Chunqiao tells a reporter that he had not anticipated that his profit margin could get so narrow so quickly.

*Source:* Louisa Lim, "Chinese 'button town' struggles with success," "Morning Edition," National Public Radio, August 22, 2006, http://www.npr.org/templates/story/story.php?storyId=5686805.

**FIGURE 9.20** China's "button town." Workers finish baskets of buttons at a factory in the small town of Qiaotou, China, which has cornered the world's button market. [© Louisa Lim.]

---

animal products requires more land and resources than the plant-based national diet of the past. In addition, world market prices can now have a negative effect on food production. Since farmers can choose their crops, many shy away from growing grain when world market prices are down; overall, grain production has decreased since about 1999. On the other hand, like Japan, China now has the economic capacity to buy the grain and other food it needs from other countries. While some researchers and policy makers are concerned about China's increasing dependence on foreign suppliers, others view the availability of food from multiple sources as a positive development in China's ability to maintain its food security.

***A Market Focus for Rural Enterprises.*** One of the most remarkable recent developments resulting from decentralization of economic control in China is the rapid growth of new rural enterprises aimed at providing a wide variety of goods and services—from operating mines to making cooking pots to assembling electronic equipment. These enterprises have become the mainstay of many rural economies, often to the detriment of agriculture, especially in the eastern and southern coastal provinces of Jiangsu, Fujian, and Guangdong. Such enterprises may still be township or village collectives, but increasingly they are privately owned. All of them leave major decisions to managers (rather than to the collectives) and price their products according to market demand. Rural enterprises now constitute a quarter of the Chinese economy, and they may actually produce over half of the country's industrial output and 40 percent of its exports. These enterprises employ more than 128 million people—more than the Chinese government itself—and account for more than 30 percent of farm household income, as some family members now work in rural enterprises rather than on the land.

Despite the importance of rural enterprises in the economy, their growth has been accompanied by some problems. They are significant contributors to environmental stress. In the mountains west of Beijing, for example, industrial pollution exceeds that in the city, and clouds of exhaust from the trucks that link the rural industries with their markets contaminate the air. Paper mills pollute waterways, farmers' fields, and aquifers. Corruption is also a problem: some managers use their wealth to get elected to village councils, which they then control; others steal funds from their own enterprises, evade taxes, and pay officials to look the other way when they breach environmental regulations. Although the government is moving to enact legislation to alleviate these problems, it is wary of driving rural enterprises out of business. In fact, their continued competitive viability in the global market is in question because they are beginning to experience the previously unheard-of problem of a labor shortage and have little flexibility to withstand commodity price hikes (Box 9.2).

***The Persistence of Spatial Disparities.*** Since the reforms of the last 25 years, China's per capita GDP has doubled several times, from U.S.$300 in the mid-1980s to U.S.$5033 by 2003 (both figures adjusted for PPP). However, long-standing regional and rural-urban disparities in wealth have increased as a result of the reforms. Only

a few rural provinces have increased their per capita GDP significantly (Figure 9.21). Remote provinces, such as Tibet (Xizang) and Yunnan, and those with little mineral or agricultural wealth, such as Inner Mongolia and Qinghai, have had difficulty generating rural enterprises and have suffered a labor drain to the cities.

Public recognition of regional disparities has led to widespread social unrest. In 2003 alone, there were 58,000 public protests. In an effort to calm the unrest, the government relieved taxes on farms and farm products, improved farm subsidies, and raised market prices. In 2006, rural incomes overall grew 7 percent faster than urban incomes, but because rural incomes started so much lower and growth for rural areas in coastal zones far exceeded that in interior zones, the overall regional disparities remained. Still, the possibility

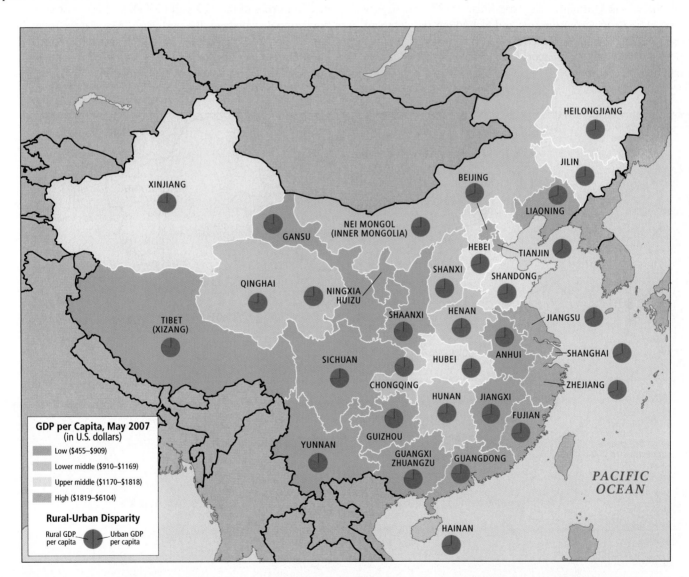

**FIGURE 9.21** China's regional and rural-urban GDP disparities, 2004. The colors show that the average range of GDP per capita at the provincial level generally decreases from high (orange), along the coast, to lower (green), in the interior. Within the provinces there is also a disparity between urban (higher) and rural (lower) GDP per capita, as shown by the pie diagrams. Notice that on the coast the rural-urban disparities are less than in the interior, where the dark green (rural) wedges are a smaller portion of the pie. [Adapted from "Exhibit 4. Classification of provinces by per-capita GDP," *China Human Development Report 2005* (Beijing: United Nations Development Programme China, 2006), p. 148, at http://www.undp.org.cn/modules.php?op =modload&name=News&file=article&catid=18&topic=40&sid=242&mode =thread&order=0&thold=0; province data from http://www.china.org.cn/english /features/ProvinceView/148894.htm.]

of making a living on the land has drawn some urban workers back to the farm.

### International Trade and Special Economic Zones.

In the early 1980s, China's government, wary of the disruption that could result from a rapid opening of the economy to international trade, selected just five coastal cities to function as free trade zones, or **special economic zones (SEZs):** Zhuhai, Shantou, and Shenzhen in Guangdong (near Hong Kong); Xiamen in Fujian; and Haikou on Hainan. Industries in these cities were allowed to recruit foreign investors and use capitalist management methods that had not yet been permitted in the rest of the country. In the late 1990s, the program was expanded to 32 provincial locations in the interior of China, stretching all the way to Urumqi in Xinjiang (Figure 9.22). These new locations were designated **economic and technology development zones (ETDZs).** Like SEZs, the ETDZs provide footholds for international investors and multinational companies eager to establish operations in the country.

Today the SEZs and ETDZs plus old urban centers such as Beijing are major **growth poles** in the country, meaning that their success is drawing yet more investment and migration (Figure 9.23). In just 25 years, many coastal cities, such as Shenzhen, Qingdao, Wenzhou, and Dongguan, have grown from medium-sized towns or even villages into some of the largest urban areas in the country. By 2005, the SEZs and ETDZs were so successful that there were growing shortages of managers, skilled technical and construction workers, and even factory laborers.

### Shifts in China's Employment Scene.

In the opening of this chapter, we met Lee Xia, who is typical of the millions of young migrants who leave their sheltered village lives to work in the highly competitive industries of the SEZs and ETDZs. (Xia's story continues in the personal vignette on the next page.) Many arrive intending not only to succeed as individuals, but also to send money home to their impoverished families and possibly even to return home one day to improve their communities. As Xia learned, city life has many short-

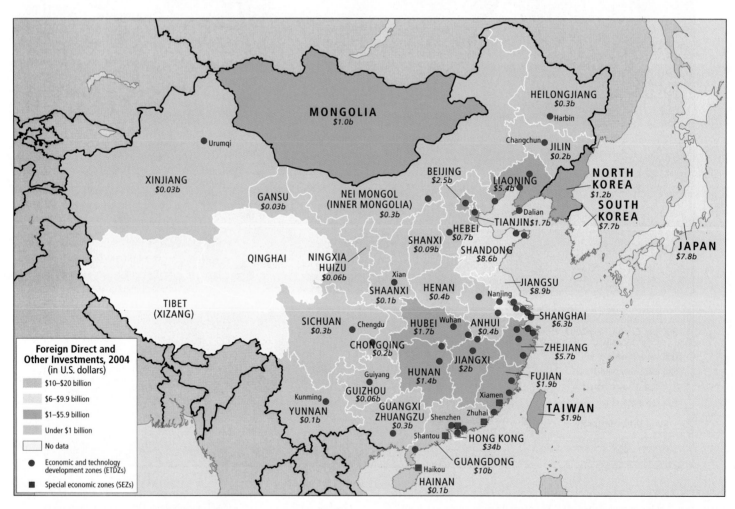

**FIGURE 9.22** Foreign investment in East Asia. The map shows China's original and recently designated special economic zones (SEZs) and economic and technology development zones (ETDZs). The colors on the map reflect levels of foreign investment (direct and otherwise) in each of the countries of the region and in each of China's provinces. [FDI data from FDI Invest in China, at http://www.fdi.gov.cn/common/info.jsp?id=ABC00000000000022787. Specific Web site no longer available without registering at http://www.fdi.gov.cn/pub/FDI_EN/Statistics/default.htm.]

**FIGURE 9.23** DaimlerChrysler in Beijing. DaimlerChrysler AG formally opened its first factory in China in September 2006. The factory, in suburban Beijing, makes Mercedes-Benz and Chrysler sedans for the upscale market. DaimlerChrysler is joining a rush of foreign automakers for a share of the booming Chinese car market with its U.S.$1.9 billion investment in China. [AP Photo/Elizabeth Dalziel.]

**Personal Vignette** When Lee Xia and her sister went home to rural Sichuan the second time, Xia first tried to open a bar in the front room of her parents' house. She hoped to introduce the popular custom of karaoke singing she had enjoyed in Dongguan, but people couldn't stand the noise, and family tensions rose. Her sister, delighted to be home with her husband and baby, quickly found a job in a new small fruit processing factory. Her husband began farming again to fill the new demand for organic vegetables among the middle class in the Sichuan city of Chongqing. The government offered him a subsidy to learn new cultivation techniques if he would guarantee regular delivery of salad greens to a central purchasing depot.

Amid all this success and her bar's failure, news that training was now available in Dongguan for skilled electronics assembly convinced Xia to try again. Word was that the plant would be air-conditioned and that this time she could eventually earn U.S.$400 a month, enough to afford some serious shopping and perhaps one of the stylish new apartments built by the electronics firm for skilled workers.

*Sources: Paul Wiseman, "Chinese factories struggle to hire," USA Today, April 11, 2005, file:///Users/lydia/Desktop/%204th%20edition%201st %20drafts/East%20Asia/USATODAY.com%20-%20Chinese%20factories %20struggle%20to%20hire.webarchive; Louisa Lim, "The end of agriculture," "Reporter's Notebook," National Public Radio, May 19, 2006, http://www .npr.org/templates/story/story.php?storyId=5411325; Mei Fong, "A Chinese puzzle: Surprising shortage of workers," Wall Street Journal (August 16, 2004): B1; David Barboza, "Labor shortage in China may lead to a trade shift," New York Times (April 3, 2006), business section, 1.* ■

comings: pay is often too low to accommodate a decent standard of living, and workdays are often 12 hours long, with only an occasional day off. Employers compensate by offering their workers free housing, but conditions are rough, with little or no privacy and often strict codes of conduct (some employers dock their workers a day's pay for failing to make their beds).

Such circumstances are changing rapidly, however. The spectacular success of China's SEZs and ETDZs, which was based on cheap, flexible labor—flexible in that workers would put up with almost any abuse to earn a little cash—has, despite China's huge population, produced a shortage of labor. So many factories, businesses, housing projects, hotels, and shopping malls (Wal-Mart is completing 55 supercenters in China) have been built or are under construction that experienced and skilled workers of all types are increasingly in short supply. Those with management skills are especially scarce.

To attract employees, some factory owners are now paying U.S.$100 per month, air-conditioning their plants, building health clubs, adding libraries, shortening the workday, and increasing time off. Some offer training. Those that provide such amenities are attracting workers, but the extra costs will eventually be attached to the prices of products, perhaps rendering China far less competitive in the world market. Factories are already moving to Vietnam, the Philippines, and countries in Africa with yet cheaper and still more flexible labor. But China's share of the world market for manufactured goods is so large that prices in Wal-Marts around the world are likely to rise eventually.

***China in the Global Economy.*** China's dramatic economic changes over just 25 years, its strong growth rate, and its ability to attract investment while maintaining a low ratio of external debt to GDP led to its admission to the World Trade Organization in 2001. (Taiwan was admitted the same year.) By 2004, China had one of the fastest-growing economies in the world, partly because it received more foreign direct investment than any other developing country (Figure 9.24).

China's activity in the global economy has had far-reaching effects. For example, 80 percent of toys sold in the United States are now manufactured in China. The sheer size of its population and demand for raw materials gives China tremendous clout in global financial and commodities markets. When Chinese premier Wen Jiabao announced in April 2004 that the Chinese government would take strong action to slow the excessive growth of its economy, the value of shares of U.S. mineral companies immediately plummeted due to fears that this Chinese policy would decrease demand for items such as copper, aluminum, and iron ore. The following day, U.S. tanker company shares also dropped. China is a huge importer of oil—it accounts for 30 percent of the growth in global demand for oil—and its growing demand for gasoline has already translated into higher prices at the pump. This influence is likely to increase as the number of automobiles on its roads reaches a projected 100 million by 2014 (one car for every 15 people).

Many governments and large multinational corporations see China's participation in the WTO as essential to the goal of global

**FIGURE 9.24** Top origins of investment in China in 2005. Total foreign direct investment in China in 2005 was U.S.$60.3 billion. The graph depicts the origin of the investors by country or locale and the amounts invested. Although definitely high, the figures may be inflated by "round-tripping," which refers to capital that actually originates in China, is sent to a foreign tax haven, or to Macao or Hong Kong, and then returns as "foreign" investment. [FDI data from FDI Invest in China, at http://www.fdi.gov.cn/common /info.jsp?id=ABC0000000000003075. Specific Web site no longer available without registering at http://www.fdi.gov.cn/pub /FDI_EN/Statistics/default.htm.]

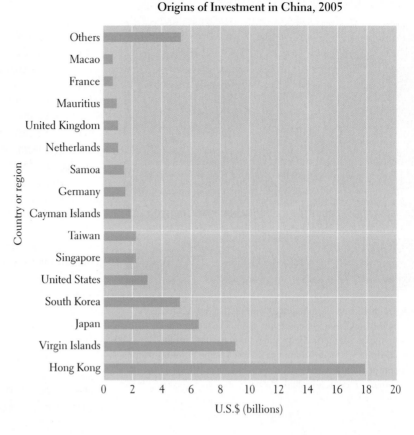

Origins of Investment in China, 2005

economic integration. Its inclusion in the organization remains controversial, however, because so much of its growth is based on environmentally destructive activities and on abuses of workers, both of which effectively lower production costs so that its goods can sell at a lower price. Moreover, China's brutal suppression of separatist movements in Tibet (Xizang) and Xinjiang, its human rights abuses against citizen groups agitating for greater political freedom, and its clumsy attempts to censor the Internet just when millions have the chance to get connected to the wider world (Box 9.3) have led some consumers and environmentalists to argue that China's entry into the WTO should have been conditional upon improvements in human rights, worker safety, freer access to information, and environmental protection. Protests against the WTO grew after China's admission as workers worldwide lost their jobs, partly as a result of competition from China. High-income countries and some business leaders were willing to ignore these problems to avoid angering the Chinese government and jeopardizing trade. But developing (low-income) countries are now the majority of the WTO, and they can determine whether new trade rules are adopted or not. The perception that Chinese goods are unfairly competing with the industries of developing countries could result in constricted markets for China. Just how China will align itself in the WTO is uncertain.

China's precarious position in relation to other newly industrializing countries has brought out a charming and benign side of China not previously seen. Some analysts call this "**soft power**": China is doing everything possible to change its image, soft-pedaling its competitive side and emphasizing its helpful, accommodating, investor side. Glamorous cultural exhibits and free language lessons are offered to lure visitors and students to China. In order to encourage future trade with Southeast Asia, Africa, South Asia, and Latin America, jovial emissaries are sent to assure governments that China will be a trading partner and investor that will not meddle in their internal affairs or make difficult demands that impinge on their sovereignty. Human rights and democratization are represented as purely internal affairs. In 2006, China signed 20-year deals to buy oil from Brazil, Venezuela, and Sudan and metals from Chile, and it invested billions in telecommunications in Argentina and Venezuela. In Africa, China is selling its textiles and household consumer products (to the detriment of local producers) and is particularly interested in extracting energy and raw materials. To get access to those resources, China is extending development aid to places such as Mozambique, Ethiopia, Zambia, and Zimbabwe. China is also promoting a new East Asian trade bloc (to include all the East Asian plus ASEAN countries), meant to counter NAFTA and the European Union.

## SOCIOCULTURAL ISSUES

East Asia's economic progress has led to social and cultural change throughout the region. Successful attempts to control population growth are creating new problems—an aging society, an imbalance in the numbers of males and females, and controversies over immigration. Modernized economies are changing work patterns and

Ever since 1949, the central government in China has controlled the news media there. By the late 1990s, however, the expanding use of electronic communication devices was loosening central control over information. By 2001, 23 million people were connected to the Internet in China, and by 2006, 123 million were connected (Figure 9.25). Just a few years ago, telephones were very rare, but now anyone can buy temporary cell phone access without showing identification. As a result, millions of Chinese have anonymous access to an international network of information. Greater public access to information—which most agree is essential in a market economy—has had a number of important social effects.

One such effect is that it is difficult for the government to promulgate inaccurate explanations for disasters or problems caused by inefficiency and corruption. Reporters can now easily check the accuracy of government explanations by calling witnesses or the principal actors directly on cell phones and then placing the explanations on the Internet. Analysts inside and outside China see the availability of the Internet to ordinary Chinese citizens as a watershed event that will surely affect public awareness of political issues.

However, the Internet in China is not the open forum that it is in most Western countries. Internet researcher Rebecca MacKinnon wrote in *Yale Global Online* that early in June 2005, when people writing blogs on a Microsoft-hosted service used the words "democracy," "freedom," or "human rights," they got a rude reminder: "The title must not contain prohibited language, such as profanity. Please type a different title." MacKinnon observed that this warning equating the words "democracy" and "freedom" with profanity "marks a new milestone in the continuing battle for free expression in China." Later in 2005, it was revealed that the U.S. technology firms Google and Microsoft had allowed the Chinese government to use software to block access to certain Web sites for users of their services in China. U.S. monitors determined that, rather than pornography sites, political sites were the most common target. As McKinnon observed, Chinese censorship has "undermined earlier optimism about the Internet [in China] ushering in an era of free expression."

*Sources:* Craig S. Smith, "China arrives at a moment of truth," *New York Times* (April 1, 2001): 5; Rebecca MacKinnon, *Yale Global Online*, June 28, 2005, front page, at http://yaleglobal.yale.edu/display.article?id=5928.

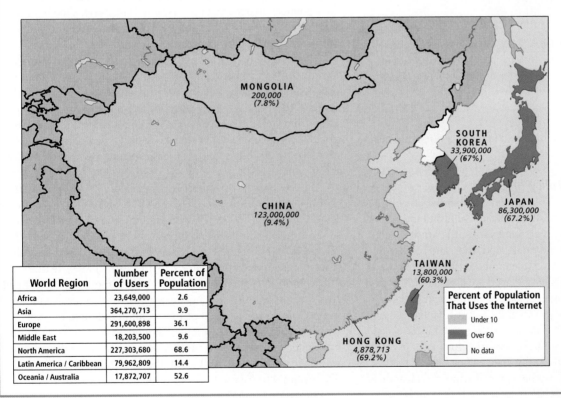

| World Region | Number of Users | Percent of Population |
|---|---|---|
| Africa | 23,649,000 | 2.6 |
| Asia | 364,270,713 | 9.9 |
| Europe | 291,600,898 | 36.1 |
| Middle East | 18,203,500 | 9.6 |
| North America | 227,303,680 | 68.6 |
| Latin America / Caribbean | 79,962,809 | 14.4 |
| Oceania / Australia | 17,872,707 | 52.6 |

**FIGURE 9.25** Internet use in East Asia, 2006. Some parts of East Asia have very high Internet use (rust), while China and Mongolia have relatively low rates. Nonetheless, in these two countries, diffusion of information coming from the Internet has significant influence on public awareness, something the government tries to control, not always successfully. [Data from http://www.internetworldstats.com/stats3.htm.]

family structures. In this section, we will look at how countries in the region are responding to these changes.

## Population Policies and the East Asian Family

Population patterns in East Asia began to change after World War II. By the 1970s, government policies, aided by rapid urbanization, had reduced fertility in Japan and Taiwan, but less so in South Korea. By 2000, women in all three countries were bearing fewer than 2 children on average. In China and North Korea, fertility rates remained very high into the 1970s at nearly 5 children per woman; in Mongolia, 7 children per woman was the average. Drastic action was taken in China, where leaders realized that the rapidly rising population was sapping the country's ability to make

**FIGURE 9.26** China's one-child-per family policy. A man walks past a sculpture in Beijing promoting China's one-child-per-family policy. Enforcement of the policy, introduced in the 1970s to control China's population growth, has been relaxed a bit in recent years. [AP Photo/Greg Baker.]

economic progress. By 1980, China had adopted a one-child-per-family policy (Figure 9.26).

### The One-Child-per-Family Policy in China.
Chinese demographers and the Chinese people at large realized that the one-child-per-family policy would dramatically affect family life and the fabric of society. For example, within two generations, the kinship categories of sibling, cousin, aunt and uncle, and sister- and brother-in-law disappeared from all families that complied with the policy—a major loss in a society that has long placed great value on the extended family. But it is the prospect of the one child being a daughter, with no possibility of having a son in the future, that has caused families the most despair. For years, the makers of Chinese social policy have sought to eliminate the preference for males by empowering women economically and socially. They believe that female children will be just as desirable as males when it is clear that well-trained, powerful daughters can bring honor to the family name and earn sufficient income for their families. However, the preference for sons remains strong.

Low population growth creates another problem: a very large elderly population that must be supported by a relatively small group of people of working age (see Box 9.1 on page 486). In China, the elderly are especially dependent on the younger generation. There was never a nationwide pension system, and pensions are even less common now, since the closing of many communes during the economic reforms of the 1980s and 1990s. The offspring in one-child families, whether male or female, have few kinfolk with whom to share family responsibilities when they reach adulthood. Hence, as the one-child family spreads through the generations, one adult may have to shoulder alone the care of elderly parents and any children.

Although the one-child family is now the most common family form in China, the one-child-per-family policy has never been very popular, both for the reasons just discussed and because the policy is enforced unevenly. Rural families are exempted from the policy (though most stop at two children), and from time to time the government has offered incentives to limit family size in some parts of the country but not in others. In some urban areas, for instance, couples who have only one child receive a monthly subsidy and a housing allowance. In other areas, complying couples receive special chances for promotion. In some cases, those who have additional children may lose these benefits, receive demotions, and have to pay a fine; elsewhere, a mother pregnant with her second child may be forced to abort. In other cases, the infractions may be ignored. Some Chinese, such as those newly affluent in the market economy and some private-enterprise farmers, may have additional children and simply pay the substantial fines. Ethnic minorities are officially exempted from family size restrictions, presumably to answer political charges that the majority Han group is too dominant, but there is evidence that family size limits have been imposed on ethnic Tibetan women, sometimes through forced abortions.

There is no doubt that population control has been successful in reducing births: between 1950 and 1990 the Chinese birth rate dropped dramatically, and by 2005 it was 0.6 per 1000. However, some of the drop is simply the result of modernization of life in China and would have happened without the one-child-per-family policy. Most people worldwide who learn to read, move to the city, and take up a lifestyle based on cash income will choose to have fewer children.

The population pyramid in Figure 9.27a reflects two sharp past declines in birth rates. The first took place during the Great Leap Forward (1959–1961), when famine and malnutrition led to lower fertility. After that era, China's birth rate grew rapidly. The second decline began 30 years ago and continued as the result of strict population control efforts in the early 1980s. In the mid-1980s, birth rates increased when market reforms were instituted. Increasing prosperity and a general relaxing of controls prompted some Chinese to have two or more children despite official sanctions. By the mid-1990s, another significant decline in birth rates was under way, perhaps inspired by the realities of raising children in urban situations as well as by the one-child-per-family policy. The pyramid in Figure 9.27b projects the long-term effect of current trends on China's population in the year 2050. At that time, a smaller proportion of young people will be supporting a relatively larger elderly population, as is currently the situation in wealthy countries in the West.

### Missing Females in Cultural Context.
The pyramid in Figure 9.27a shows another interesting phenomenon. If you look carefully, you can see that up to age 30 or more, the numbers of males exceed the numbers of females by small but significant amounts. In 2000, there were roughly 44 million girls but over 50 million boys

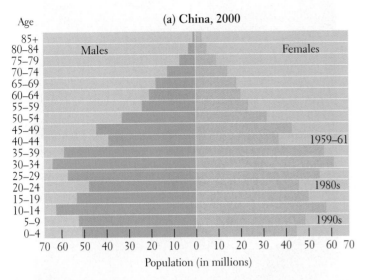

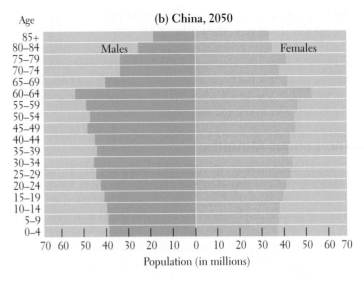

**FIGURE 9.27**  Population pyramids for China, 2000 and 2050. [Adapted from U.S. Census Bureau, International Data Base (Washington, D.C.), at http://www.census.gov/cgi-bin/ipc/idbpyry.pl.]

aged 0–4, indicating a deficit of about 6 million girls, or a ratio of 114 boys to every 100 girls. (The normal sex ratio at birth is 105 boys to 100 girls.) What happened to the missing girls?

There are several possible answers. Given the preference for male children, the births of these girls may simply have gone unreported as families hoped to conceal their daughter and try again for a son. There are many anecdotes of girls being raised secretly or even disguised as boys. Adoption records indicate that girls are given up for adoption much more often than boys. Or the girls may have died in early infancy, either through neglect or actual infanticide. Finally, some parents have access to medical tests that can identify the sex of a fetus, and there is evidence that in China, as elsewhere around the world, some of these parents choose to abort a female fetus.

A side effect of the preference for sons is that there is now a growing shortage of women of marriageable age throughout East Asia. In China alone, there was an estimated deficit of 10 million women aged 20-35 in 2000. Females are also effectively "missing" from the marriage rolls because many educated young women are too busy with career success to meet eligible young men. Some are even too busy to attend the singles parties arranged to bring them together with males of compatible education and income, so they send their parents! The mother of one such daughter away on a business trip recently attended a singles gathering in Shanghai where she patiently recited to interested males the vital statistics of her daughter, including her age, height, weight, hobbies, salary, and career ambitions. No mention was made of future children.

Elsewhere in East Asia, the cultural preference for sons persists, with a deficit of girls appearing on the 2000 population pyramids for Japan, the Koreas, Mongolia, and Taiwan. Nonetheless, evidence shows that attitudes may be changing. In Japan, South Korea, and Mongolia, the percentage of women receiving secondary education equals or exceeds that of men. Comparable data are not available for North Korea, China, and Taiwan. In the latter two countries, however, educated young women are increasingly

recruited for middle management positions, suggesting that soon they could have incomes equal to those of young men.

## Family and Work in Industrialized East Asia

Life in a small city apartment is very different from life among a host of relatives in a farming village. Nonetheless, throughout East Asia, urban wives still perform most domestic duties and primarily look after husbands and children. Often they must also care for elderly parents or parents-in-law. Surveys show that, although there is a small but growing group of trained young women who seek careers and are as ambitious as men, even many university-educated married women say they wish to earn only supplementary income for the household.

One reason that many women have not been able to take on work outside the home as well as the duties of family life is that jobs in East Asian industrial economies are particularly demanding. Workdays are long, and commuting often adds more than 3 hours to the time away from home. Even more important is the "culture of work," which in Japan, Taiwan, and South Korea (less so in China) is based on male camaraderie and demands an especially high level of loyalty to the firm, often at the expense of family. In Japan, for example, after-work leisure time often must be spent with business colleagues if one is pursuing a promotion. It is considered disloyal to refuse overtime, and employees are discouraged from taking time off to spend with a spouse or to help care for children or elderly relatives. Women, when hired, are usually there to support men as secretaries and assistants, not to participate as full members of corporate teams.

By 2000, the East Asian urban work ethic and family structure were being publicly challenged, not least by the male workers themselves. In Tokyo, for example, a group calling itself Men Concerned About Child Care meets regularly to discuss ways to participate more fully in home and community life. Another group has formed to lobby for 4-hour workdays for both men and women, so that both can spend time with the family. This group has filed lawsuits against employers who routinely require employees to accept job transfers to distant locations where they cannot take their families. It is particularly significant that urban men are the ones challenging the extremes of the East Asian work ethic. They are the group that has been most regimented and most deprived of personal time and family life, but they are also the group that has the most power to change the system.

## Indigenous Minorities

Cultural diversity exists throughout East Asia, even though most countries have one dominant ethnic group. In China, for example, 93 percent of Chinese citizens call themselves "people of the Han." The name harks back about 2000 years to the Han empire, but it gained currency only in the early twentieth century, when nationalist leaders were trying to create a mass Chinese identity. The term *Han* does not denote an actual ethnic group, but rather connotes people who share a general way of life, pride in Chinese culture, and a sense of superiority to ethnic minorities and outsiders. The main language spoken by the Han is Mandarin, although it is just one of many Chinese dialects. (All Chinese dialects use the same writing system.)

Even though the non-Han minorities make up only 7 percent of China's population, they number about 117 million people. There are more than 55 different minority groups scattered across the vast expanse of China; as Figure 9.28 shows, most of the major groups

**FIGURE 9.29** A Silk Road market. At the market in Kashi, you can still buy a camel, a fine Oriental rug, the original bagel, or the skin of an endangered animal. You can also get a haircut or get your teeth fixed, without a painkiller. [Reza/National Geographic Image Collection.]

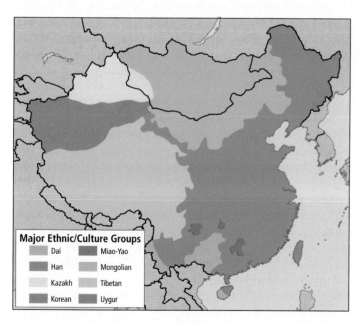

**FIGURE 9.28** Major ethnic groups of China. Although the most populated areas in the east and southeast are dominated by the Han, China is home to a number of distinct ethnic groups. This map shows the spaces traditionally occupied by the Han and the ethnic minorities. It does not show the recent resettling of Han in Xinjiang and Tibet, nor does it show the Hui, people from many ethnic groups whose ancestors converted to Islam and who are found in disparate locations along the old Silk Road and in coastal southeastern China. [Adapted from Chiao-min Hsieh and Jean Kan Hsieh, *China: A Provincial Atlas* (New York: Macmillan, 1995), p. 12. The Web site http://www.index-china.com/minority /minority-english.htm supplies a comprehensive survey of minorities in China.]

**Major Ethnic/Culture Groups**
- Dai
- Han
- Kazakh
- Korean
- Miao-Yao
- Mongolian
- Tibetan
- Uygur

live in areas outside the Han heartland of eastern China. Some of these areas have been designated *autonomous regions*, where these groups theoretically manage their own affairs. In practice, however, the Han-dominated Communist Party has not allowed them self-government. The central government in Beijing controls the fate of the minorities, especially those who live in zones that are believed to pose security risks or that have resources of economic value.

***Turkic-Speaking Peoples.*** One autonomous region is Xinjiang Uygur, where Turkic-speaking peoples, such as the Uygurs and Kazakhs, live. Many of these peoples remain nomadic, and they have revived trade across the borders of Central Asia. The old Silk Road market of Kashgar (now Kashi) is back in business (Figure 9.29). Because this part of northwestern China has oil and other mineral resources the Beijing government wishes to claim for national development, it has sent troops and hundreds of thousands of Han settlers to Xinjiang. The Han settlers fill most managerial jobs in mineral extraction, the military, nuclear testing, and power generation. An important secondary role of the Han is to dilute the power of Uygurs and Kazakhs within their own lands. The minorities are urged to rid themselves of "unacceptable" cultural practices and distinctive national identities. For most, assimilation to Han ways would include giving up Islam. The people of Xinjiang have responded with resistance. Assimilation may be the long-term outcome, but for now there has been a rebirth of Islamic culture, inspired in part by the resurgence of Islam in Central Asia in the aftermath of the breakup of the Soviet Union. Islamic prayers are again heard publicly, Muslim women are again wearing Islamic

**FIGURE 9.30** The Qinghai-Tibet Railway. Opened in July 2006, the world's highest railway—including nearly 596 miles (1000 kilometers) built at an altitude above 13,123 feet (4000 meters)—is a controversial engineering feat. It now connects Beijing to Lhasa, a trip that takes 48 hours at a cost of U.S.$50–U.S.$150. In addition to transporting goods and people to and from Tibet, the railway is expected to add to Tibet's main industry, tourism. Tibetans worry that it will become a conduit for more Han migration to Tibet. [Photo © Wu Hong/epa/CORBIS.]

*On map:*
XINJIANG · Qaidam Basin · Delingha · Xining · **Golmud-Lhasa is the newest section of the railway** · Golmud · Qinghai Hu · Lanzhou · QINGHAI · GANSU · Tanggulashan · C H I N A · Anduo · Plateau of Tibet · Nagqu · Nu (Salween) · TIBET (XIZANG) · SICHUAN · Damxung · Yangbajain · Lhasa · H i m a l a y a s · Gongga Shan elev. 24,700 ▲

dress, and Islamic architectural traditions are being revived. The Beijing government keeps the upper hand by harshly punishing those who are suspected of Islamic fundamentalism.

***The Hui.*** The original Hui people were descended from ancient Turkic Muslim traders who traveled the Silk Road across Central Asia from Europe to Kashgar to Xian (see Figure 5.8 on page 252). Most Hui today are descended from non-Turkic peoples who converted to Islam. They still live as distinct groups, some numbering a million or more, on the western Loess Plateau in Ningxia Huizu Autonomous Region and Gansu Province; in small pockets of the North China Plain; and in Sichuan, Yunnan, Hunan, Qinghai, and Xinjiang. The Hui, who have a long tradition of commercial activity, have been particularly successful in China's new free market economy. They are using their money not only for consumption of luxury goods, but also to revive religious instruction and to fund their mosques, which are now more obvious in the landscape.

***The Tibetans.*** In contrast to the prosperous Hui are the Tibetans, an impoverished ethnic minority of nearly 5 million individuals scattered thinly over a huge and high mountainous region in western China. Their homeland is widely known as Tibet, but since Chinese troops invaded in 1950, the Chinese government has referred to it as the Xizang Autonomous Region. The Chinese government suppressed the Tibetan Buddhist religion, long the mainstay of most Tibetans' daily lives, by destroying thousands of temples and monasteries and massacring many thousands of monks and nuns. The spiritual leader of Tibet, the Dalai Lama, was forced into exile in India, along with thousands of his followers. To dismantle traditional Tibetan society further, hundreds of thousands of Han Chinese were resettled in Tibet, where they control the economy and the major cities, exploit mineral and forest resources, and force native Tibetans to adopt Chinese ways. A new railway link connecting Tibet more conveniently to the rest of China was completed in 2006 (Figure 9.30). The Han in Tibet see the railway as a public service that will promote Tibetan development, but Tibetan activists see it as a conveyor belt for more Han dominance.

***Indigenous Groups in Southern China.*** In Yunnan Province in southern China, there are more than 20 groups of ancient native peoples living in remote areas of the deeply folded mountains that stretch into Southeast Asia. These groups speak many different languages, and many have cultural and language connections to the indigenous people of Tibet, Burma, Thailand, or Cambodia. Interestingly, women and men are treated more equally here than among the Han. A crucial difference may be that among several groups, most notably the Dai, the husband moves in with the wife's family at marriage and provides her family with labor or income. A husband inherits from his wife's family rather than from his birth family, and female children are valued just as highly as males. Similar patterns of family structure and gender roles are noted among indigenous peoples in Southeast Asia, as we will see in Chapter 10.

**FIGURE 9.31** Ethnic minorities in Japan. This Ainu chief lives and works in a traditional Ainu living history village in the Akan National Park in eastern Hokkaido, Japan. He is working with wood that he carves to sell to visitors. [© F. Staud/www.phototravels.net.]

*Aboriginal Peoples in Taiwan.* In Taiwan and the adjacent islands, the Han account for 95 percent of the population, but this region is also home to 60 indigenous minorities. Some have cultural characteristics—languages, crafts, and agricultural and hunting customs—that indicate a strong connection to ancient cultures in far Southeast Asia and the Pacific. The mountain dwellers among these groups have resisted assimilation better than the plains peoples; both are now protected and may live in mountain reserves if they choose. But increasingly, these Taiwanese aboriginal peoples are attending schools and being absorbed into mainstream urbanized Taiwanese life.

*The Ainu in Japan.* There are several indigenous minorities in Japan. Not surprisingly in a country with a particularly strong sense of cultural solidarity and a tendency to suppress difference, they have suffered considerable discrimination, although the topic is sensitive and rarely discussed. A small and distinctive minority group is the **Ainu.** Now numbering only about 16,000, the Ainu are a racially and culturally distinct group who are thought to have migrated many thousands of years ago from the northern Asian steppes. They once occupied Hokkaido and northern Honshu, living by hunting, fishing, and some cultivation, but they are being displaced by forestry and other development activities. Few full-blooded Ainu remain because, despite prejudice, they have been steadily assimilated into the mainstream Japanese population. Some now make a living by demonstrating traditional crafts to tourists visiting Ainu living history villages (Figure 9.31). Some Ainu attempt to revive the "Ainu spirit" by teaching children the Ainu language and traditional ways.

## ENVIRONMENTAL ISSUES

Some of East Asia's environmental problems result from high population density and rapid economic development, others from long-term patterns of poor resource management or more recent ineffective planning. Japan, Taiwan, and South Korea are affected by problems that result primarily from their high degree of industrialization, their efforts to overcome limited resource bases, and their high rates of urbanization.

It is now apparent that China has some of the most severe environmental problems on the planet (Figure 9.32). The very accomplishments that have improved China's social and economic situation, such as industrialization, improved transport, increased agricultural production, better housing, and widespread home heating, are producing air pollution, water pollution, water scarcity, resource depletion, and erosion. Adding to the problem is the continuing population shift to dense urban areas. Of the many environmental problems that exist in China, we focus here on just two: air pollution and water availability and quality.

### Air Pollution in China

Since the late 1970s, industrial output in China has been growing at an average annual rate of 18 percent, and as a consequence, China's use of fossil fuel energy has been rising steadily. Coal provides 65 percent of China's energy needs, and its coal consumption now accounts for 40 percent of the world total, making China the world's largest consumer of coal. It is also the world's largest coal producer, and has large coal reserves for future growth. Between 1975 and 2005, China's coal consumption more than quadrupled. By 2006, it was bringing a new coal-fired power plant on line every week. Its use of hydroelectricity, such as that generated by the new Three Gorges Dam (Box 9.4), and increases in its use of oil, gas, and other sources of energy are unlikely to decrease the demand for coal, simply because China has such an energy deficit. Yet coal burning is the primary cause of China's poor air quality and is blamed for a critical toll in respiratory ailments. The combustion of coal releases high levels of two pollutants: suspended particulates and sulfur dioxide ($SO_2$). In Chinese cities, these emissions

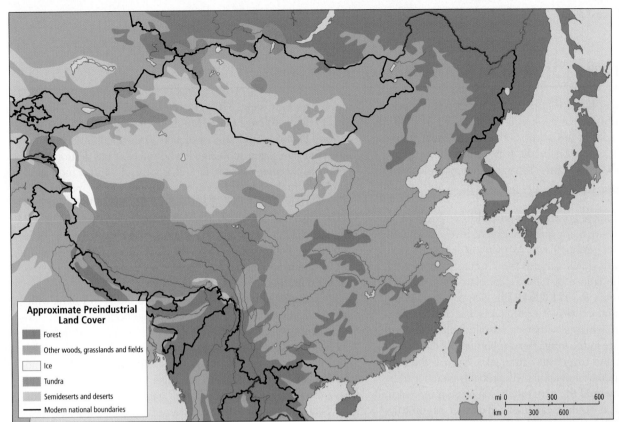

**Approximate Preindustrial Land Cover**

- Forest
- Other woods, grasslands and fields
- Ice
- Tundra
- Semideserts and deserts
- Modern national boundaries

mi 0   300   600
km 0   300   600

(a) Preindustrial

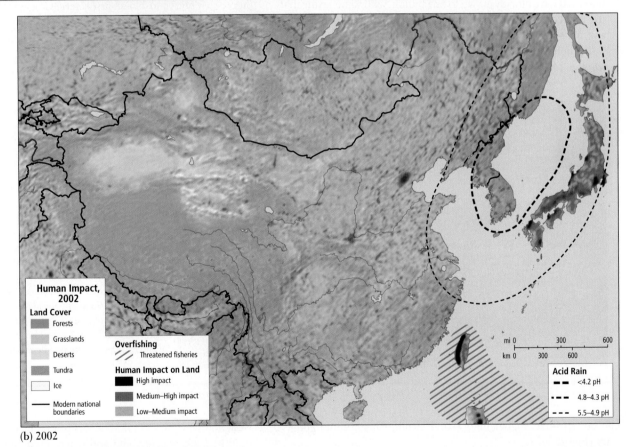

**Human Impact, 2002**

**Land Cover**
- Forests
- Grasslands
- Deserts
- Tundra
- Ice
- Modern national boundaries

**Overfishing**
- Threatened fisheries

**Human Impact on Land**
- High impact
- Medium–High impact
- Low–Medium impact

mi 0   300   600
km 0   300   600

**Acid Rain**
- <4.2 pH
- 4.8–4.3 pH
- 5.5–4.9 pH

(b) 2002

**FIGURE 9.32** Land cover (preindustrial) and human impact (2002). East Asians had already changed land cover extensively by 1000 years ago (a), and by 2002 (b), human impact had significantly intensified. Nearly every location was affected in one way or another, and in some places the impact was extreme. [Adapted from *United Nations Environment Programme, 2002, 2003, 2004, 2005, 2006* (New York: United Nations Development Programme), at http://maps.grida.no/go/graphic/human_impact_year_1700_approximately and http://maps.grida.no/go/graphic/human_impact_year_2002.]

## BOX 9.4 | AT THE LOCAL SCALE The Three Gorges Dam

The Three Gorges Dam (Figure 9.33) is the largest engineering project in history, and when it is completed in 2009 it will be the largest dam in the world: 600 feet (183 meters) high and 1.4 miles (2.3 kilometers) wide. It is designed to improve navigation on the Chang Jiang (Yangtze), generate electricity equivalent to about 3 percent of the national total, and control flooding.

Despite the Chinese government's unfailing enthusiasm for the Three Gorges Dam, many qualified experts involved with the project in China see serious flaws. As many as 80 cracks have appeared in the dam, which suggests shoddy construction of the type that led to the failure of China's Banqiao Dam in 1975, which was responsible for 171,000 deaths due to flooding and ensuing famine. Another worry is earthquakes that could seriously weaken the dam, which sits above a seismic fault. The enormous weight of water in the long reservoir behind the dam could lubricate the fault and trigger an earthquake. Even if the dam holds, its power generation potential will probably be reduced by the buildup of silt behind the dam due to soil erosion upstream. Any failure of the dam would mean financial disaster, as construction costs have already grown to $25 billion, with some saying that the real costs may be three times this figure, due in part to massive theft from the project by corrupt officials.

Then there are the incalculable costs associated with relocating the 1.9 million people who once lived where the dam now forms a reservoir some 370 miles (600 kilometers) long. Thirteen major cities, 140 large towns, hundreds of small villages, as many as 1600 factories, and at least 62,000 acres (25,000 hectares) of farmland have been submerged. The reservoir has destroyed important archaeological sites, as well as some of China's most spectacular natural scenery. There are significant environmental costs as well. Only one of myriad examples is the fate of the giant sturgeon, a fish that can weigh as much as three-quarters of a ton and is as rare as China's giant panda. Sturgeon used to swim more than 1000 miles (1600 kilometers) up the Chang Jiang past the location of the dam to spawn, but now the sturgeon's reproductive process has been irretrievably interrupted.

International funding sources such as the World Bank withdrew their support of the dam decades ago because of these and other shortcomings of the project. However, the construction of *Da Ba* (the Big Dam) is supported by eastern industrialists who need the energy, the construction industry that wanted to build the dam, and government officials eager to impress the world and leave their mark on China with this mega-project.

*Sources:* Egram from British Embassy, Beijing, December 21, 2005, http://www.bbc.co.uk/blogs/opensecrets/3Gorges3.pdf; Human Rights Watch, http://www.hrw.org/reports/1995/China1.htm; http://news.bbc.co.uk/1/hi/world/asia-pacific/844786.stm; "'Cracks' in China's Three Gorges Dam," BBC News, http://news.bbc.co.uk/1/hi/world/asia-pacific/1925172.stm

**FIGURE 9.33** The Three Gorges Dam. This 2006 photo shows a distant view of the world's largest hydropower dam project under construction, on the Chang Jiang (Yangtze River). At the time it was taken, there were less than 3000 cubic meters of concrete left to be placed before the dam would be completed—9 months ahead of schedule, according to official Chinese sources. The dam is situated near Xiling Gorge, the easternmost of the Three Gorges on the middle reaches of the Chang Jiang. [AP Photo/Xinhua, Du Huaju.]

can be ten times higher than World Health Organization guidelines (Figure 9.34).

The geographic pattern of air pollution in China correlates with patterns of population density, urbanization, and industrialization. In the lightly populated plateaus and mountains of the north, north-

west, and west, air pollution occurs primarily in urban areas, where industries are concentrated and where there are large numbers of homes to be heated. In the Sichuan Basin and in most of eastern China, air pollution is prevalent in both cities and rural areas. Air pollution is the worst in northeastern China because homes and

## Ambient Concentrations of Air Pollutants

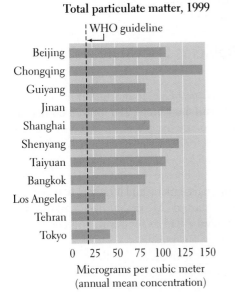

Total particulate matter, 1999

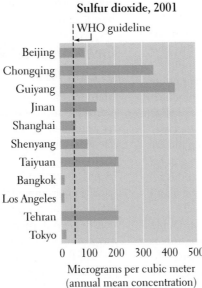

Sulfur dioxide, 2001

**FIGURE 9.34** Air pollutant concentrations in selected Chinese and other large cities, 1999 and 2001. The vertical dashed lines indicate World Health Organization guidelines for safe concentration levels. [Adapted from http://devdata .worldbank.org/wdi2005/Table3_13.htm#about.]

industries are in close proximity and both depend heavily on coal for fuel. Less home heating is necessary in the warmer climate of southeastern China, but the air is still laden with sulfur dioxide and suspended particles because there are many industries in this region and they burn a softer, more polluting type of coal. Sulfur dioxide from coal burning contributes to acid rain, which is displaced to the east by prevailing winds, reaching Korea, Japan, Taiwan, and beyond. Particulates from China's coal burning are transported globally by high-altitude west-to-east-flowing jet streams, affecting air quality on the west coast of North America and even reaching as far as the Mediterranean Sea.

Throughout eastern China, large numbers of people are exposed to high levels of gaseous emissions from vehicles. The use of personal cars in China has hardly begun; Beijing has just one-tenth the vehicles that Los Angeles has. Nevertheless, the pollution attributable to vehicles in Beijing already equals that of Los Angeles. China's vehicles have traditionally had very high rates of lead and carbon dioxide emissions, but the government is increasingly "green" in its energy policies. By 2008, new Chinese cars will have to conform to higher mileage and lower emissions standards than U.S. cars.

## Water in China:
## Too Much, Too Little, and Polluted

*Flooding.* During the summer monsoon, huge amounts of rain are deposited on eastern China, often causing catastrophic floods along the major rivers and their tributaries, particularly along the Chang Jiang and Huang He. Engineers have constructed elaborate systems of dikes, dams, reservoirs, and artificial lakes to help control flooding as well as to provide water for irrigation and hydroelectric power for industries and urban populations. China has built 22,000 to 24,000

large dams—half of all the large dams in the world, including the Three Gorges Dam, which is now the world's largest—most of them since 1949. But dams bring their own problems, as Box 9.4 explains.

Despite these efforts, catastrophic flooding occurs somewhere in China almost every year. Along the Chang Jiang, heavy rains in July and August of 1998 and 2004 caused some of the worst flooding in two centuries. Two hundred and forty million people were affected, with 3656 drowned, 14 million left homeless, and 25 million hectares (about 62 million acres) of farmland flooded. Farther north, along the Huang He, the rains are not as heavy, but levels in water channels are still reaching the highest marks ever recorded because silt carried by the water is building up in the riverbed. The situation is so precarious that the silt must be removed by hand to prevent levees from bursting—a disaster that could kill more than a million people.

Although flooding is a natural phenomenon, it has been made worse by human activities. Many environmental scientists attribute the sharp rise in catastrophic floods in the last three decades to deforestation (25 percent of China's forests were clear-cut during that time), cultivation, urbanization, mining, and overgrazing—all human activities that remove water-absorbing vegetation and soil. Another suggested factor is a change in rainfall patterns caused by global warming.

*Drought.* At the other end of the water availability scale, droughts occur somewhere in China every year and often cause more suffering and damage than any other natural hazard. Droughts are triggered by periods of abnormally low rainfall and abnormally high temperatures, but their effects are made worse by many of the same human factors that contribute to flooding. When people begin to live or farm in dry environments, as many millions have done in China during the twentieth century, they clear natural vegetation and tap underground aquifers for water. The removal of vegetation increases rates of water evaporation from the soil, and the growing demand for water depletes the aquifers, setting the stage for water shortages in the future. However, droughts are occurring with regularity even in the wetter parts of southern China. Some scientists attribute this phenomenon to shifting weather patterns related to global warming.

An example of what most of China may face in the near future is occurring in the North China Plain. In this area, which produces half of China's wheat and a third of its corn, the water table is falling more than 10 feet a year. Almost every year since 1972, the lower sections of the Huang He have completely dried up during the dry season, due to increased withdrawals of water for irrigation and for urban uses. Meanwhile, the rapidly growing urban areas in this region, such as China's capital, Beijing, are facing severe water shortages that constrain industrial expansion and economic growth.

China's urban water demand is projected to almost double over the first decade of this century, as a result of growing urban populations and industries. Industrial water needs are expected to expand by 62 percent. The United Nations reports that in China's cities, water shortages are already curtailing industrial production by more than U.S.$10 billion a year.

With its aquifers being depleted, China has limited options: find new sources of water or find ways to conserve. Conservation is the less expensive option by far and is much likelier to reap results. But can it save enough water to make a difference?

***Water Pollution.*** The United Nations reports that pollution of China's water, as well as its overall scarcity of water, is causing health problems costing U.S.$4 billion a year. Fully one-third of the population does not have safe drinking water. Raw sewage and chemical pollution have contaminated lakes, reservoirs, and 29,000 miles (47,000 kilometers) of China's waterways. Much of the pollution comes from synthetic nitrogen fertilizers used in agriculture, but some of the worst water polluters are the village enterprises that, since 1978, have been so successful in creating jobs and alleviating rural poverty. To add to the problem, only 24 percent of the Chinese population has access to adequate sanitation. In Sichuan Province, for instance, only 37 percent of the 9 million people who live in the city of Chengdu are connected to any sort of sewer system. In rural areas, 90 percent of the people have only latrines or even more primitive facilities. In the east, salt water is intruding into aquifers because coastal cities have withdrawn so much water from them. Some Chinese scientists estimate that half of the groundwater supplying Chinese cities is contaminated.

***Efforts to Improve Environmental Health.*** The desire for improved environmental health is fueling efforts at both national and local levels. The 1998 and 2004 floods along the Chang Jiang prompted a major public protest against inept planning and inadequate enforcement of flood control regulations, both acknowledged by the government. One result was the beginning of a Green movement. Higher literacy rates and decreased media censorship are creating a more informed public that is pressing for environmental cleanup. Although only 30 percent of urban water is now recycled, many of China's cities are taking steps to improve recycling procedures that will process wastewater and make it safe to use elsewhere.

A system of permits, incentives, and penalties aimed at removing pollutants from wastewater discharges is being imposed on Chinese industry and farming at all levels.

## Environmental Problems Elsewhere in East Asia

Many of the environmental problems found in China also exist elsewhere in East Asia. Japan, the Koreas, and Taiwan do not suffer water deficits because they are located in the wet northwestern Pacific. However, in the late 1990s, deforestation in North Korea led to immediate flooding and long-term soil depletion. That country has also recently suffered a series of devastating crop failures related to environmental mismanagement, and it is thought that many thousands of people have starved. Mongolia, like northern and northwestern China, has always had to cope with arid conditions; much of the country is desert and grassland.

All these countries have experienced the ills of water and air pollution connected with modern agriculture, industrialization, and urbanized living, but to varying degrees. Mongolia and North Korea have lower overall levels of pollution only because they are not yet heavily industrialized. Individual factories and vehicles are highly polluting. Air pollution in the largest cities of Japan (Tokyo and Osaka), Taiwan (Taipei), and South Korea (Seoul) and in adjacent industrial zones is severe enough to pose public health risks. Antipollution legislation is being passed, and enforcement is increasing, but high population densities and rising expectations for

**FIGURE 9.35** Pollution in Taiwan. A family masked against air pollution goes for a drive in a park in Kaohsiung, in southwestern Taiwan. [Jodi Cobb/ National Geographic Image Collection.]

better standards of living make it difficult to improve environmental quality. Taiwan is a case in point.

Taiwan's extreme population density of 1600 people per square mile (615 per square kilometer) and its high rate of industrialization have exacerbated pollution and related environmental problems. In the far north, around the metropolitan area of Taipei, seven cities house a total of 3.26 million people, and densities can rise to 5000 people or more per square mile (1930 per square kilometer). As standards of living have increased, so has per capita consumption of water, sewage facilities, energy, and material goods of all sorts. Automobile ownership has been rising rapidly in Taiwan, to the point that there are now four motor vehicles (cars or motorcycles) for every five residents, which adds up to more than 16.5 million exhaust-producing vehicles on this small island. In addition, there are nearly eight registered factories for every square mile (three per square kilometer), all emitting waste gases. As a result, Taiwan has some of the dirtiest air on earth—publicly acknowledged by the government as six times dirtier than that of the United States or Europe (Figure 9.35).

In the category of natural hazards, the entire coastal zone of East Asia is intermittently subjected to **typhoons** (tropical cyclones). Japan's location along the northwestern edge of the Pacific Ring of Fire means that it has many volcanoes, and it also experiences earthquakes and tsunamis. These natural hazards are a constant threat in Japan; the heavily populated zone from Tokyo southwest through the Inland Sea is particularly endangered. Earthquakes are also a critical natural hazard in Taiwan.

## MEASURES OF HUMAN WELL-BEING

In East Asia there is extreme variation from country to country, as well as within countries, on all indices of human well-being (Table 9.1). (Hong Kong and Macao, until recently European colonies and now both part of China, are included here merely for comparison purposes.)

The gross domestic product (GDP) per capita per year, adjusted for purchasing power parity (PPP), ranges from U.S.$1800 in North Korea to U.S.$27,967 in Japan (see Table 9.1, column II). Yet this apparently enormous regional disparity in wealth is somewhat misleading. Communist governments in North Korea, Mongolia, and China have attempted for half a century to keep poverty at bay by providing basic necessities for their citizens. In fact, mass abject poverty such as that found in India and Bangladesh or Africa

**TABLE 9.1** Human well-being rankings of countries in East Asia and other selected countries[a]

| Country (I) | GDP per capita, adjusted for PPP[b] (GDP ranking among 177 countries), 2005 (II) | Human Development Index (HDI) ranking among 177 countries,[c] 2005 (III) | Gender Empowerment Measure (GEM) and Gender Development Index (GDI) rankings, 2005 (IV) | Female literacy (percent), 2003 (V) | Male literacy (percent), 2003 (VI) | Life expectancy, 2005 (VII) |
|---|---|---|---|---|---|---|
| **Selected countries for comparison** | | | | | | |
| United States | 37,562 (4) | 10 (high) | 8 (GEM = 12) | 99 | 99 | 77 |
| Mexico | 9,168 (56) | 53 (medium) | 46 (GEM = 38) | 89 | 92 | 75 |
| **East Asia** | | | | | | |
| China | 5,003 (93) | 85 (medium) | 64 | 86.5 | 95 | 72 |
| Hong Kong | 27,179 (17) | 22 (high) | 22 | 90 | 97 | 82 |
| Japan | 27,967 (13) | 11 (high) | 14 (GEM = 43) | 99 | 99 | 82 |
| North Korea[d] | 1,800 | ND | ND | 99 | 99 | 71.6 |
| South Korea | 17,971 (34) | 28 (high) | 27 (GEM = 59) | 99 | 99 | 77 |
| Macao | 22,000 | ND | ND | 92 | 97 | 82 |
| Mongolia | 1,850 (132) | 114 (medium) | 90 | 98 | 98 | 64 |
| Taiwan[d] | 29,000 | ND | ND | 94 | 98 | 77 |

[a]Rankings are in descending order; i.e., low numbers indicate high rank.

[b]PPP = purchasing power parity, figured in 2003 U.S. dollars.

[c]The high, medium, and low designations indicate where the country ranks among the 177 countries classified into three categories by the United Nations.

[d]Data from *CIA World Factbook*, 2006; not available in UNHDI report, 2005.

ND = No data available.

*Sources: United Nations Human Development Report 2005* (New York: United Nations Development Programme), Table 1, Table 25, and Table 26, at http:/hdr.undp.org/reports/global/2005/pdf/HDR05_complete.pdf. *2005 World Population Data Sheet* (Washington, D.C.: Population Reference Bureau).

has not existed in China or Mongolia for several decades, and those two countries now fall into the medium ranking on the Human Development Index. North Korea and Taiwan are not ranked by the UN. Although North Korea is known to be extraordinarily poor, the adult life expectancy of almost 72 years indicates that, at least until recently, some basic needs were being met (North Korea experienced a devastating famine from 1995 to 2001). Nonetheless, North Korea's infant mortality rate (21 per 1000 live births in 2005) is four times higher than South Korea's (5 per 1000) though lower than the rates for China (27) and Mongolia (58). Complete well-being figures for Taiwan are not available because of its unresolved political status vis-à-vis China. Nonetheless, Taiwan has a relatively high GDP per capita and provides many social services for its citizens. Were Taiwan ranked on the HDI, it would rank high, close to Hong Kong.

Because gender empowerment and gender development figures for the region are quite incomplete, it is difficult to assess progress toward gender equity; the patterns appear to be irregular at best. China did not report statistics on primary and secondary school enrollment in 2005, and just 14 percent of women obtained college or technical training, compared with 16 percent of men. Chinese female workers earn about two-thirds of what male workers

earn. In Mongolia, women get more education than men—almost twice as many women as men go beyond high school. Yet, despite being more qualified, Mongolian women also earn about two-thirds of what men earn. In the industrialized societies of Japan and South Korea, nearly all men attend high school, as do all Japanese women, but only 88 percent of South Korean women do so. Yet more South Korean women (61 percent) go to college or technical school than in Japan (47 percent). Japanese and South Korean women earn only about half of what men earn.

China has made the most spectacular improvements in human well-being. Since the 1970s, China has reduced infant mortality by 50 percent, and life expectancy has risen from 63 to 72 years. These advances are the result of efforts to improve basic health-care delivery, including massive immunization and family planning programs, programs to control infectious and parasitic diseases, and improved nutrition. Also contributing to these improvements in well-being are better housing, improved water quality, and literacy training. These advances were mirrored for a time in Mongolia and North Korea, but both countries lost ground in the 1990s. In Japan, Taiwan, and South Korea, basic health indicators such as life expectancy and infant mortality are comparable to, if not better than, those for Europe and the United States.

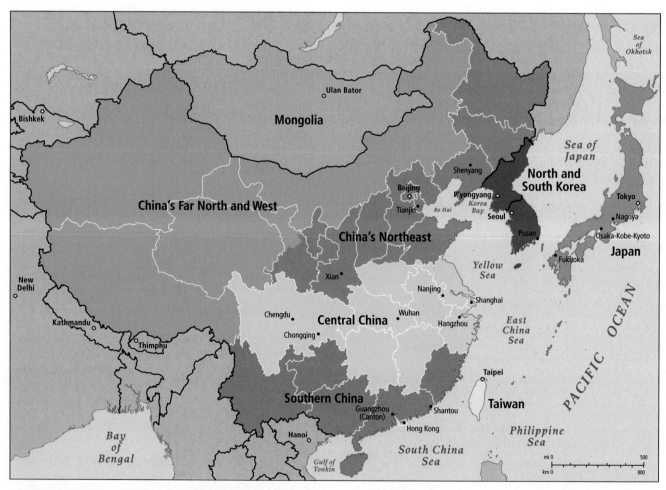

**FIGURE 9.36** Subregions of East Asia. This map shows the four subregions of China as well as the other four subregions of East Asia: Taiwan, Japan, North and South Korea, and Mongolia.

# III SUBREGIONS OF EAST ASIA

This section discusses Japan, Taiwan, North and South Korea, and Mongolia as four distinct subregions of East Asia. The remaining country, China, is so large and diverse that it is divided into four subregions (Figure 9.36).

## CHINA'S NORTHEAST

China's Northeast consists of the Loess Plateau, the North China Plain, and the Far Northeast (Figure 9.37). The Loess Plateau and the North China Plain are the ancient heartland of China. By the eighth century, the city of Chang'an (now encompassed by the city of Xian) had 2 million inhabitants and may have been the largest city on earth. After 900, the center of Chinese civilization shifted to the North China Plain, but the Loess Plateau remained a crucial part of China. Xian served as the eastern terminus of the Silk Road that connected China with Central Asia and Europe (see Figure 5.8 on page 252).

The Loess Plateau and the North China Plain are linked physically as well as culturally. Both are covered by fine yellowish **loess,**

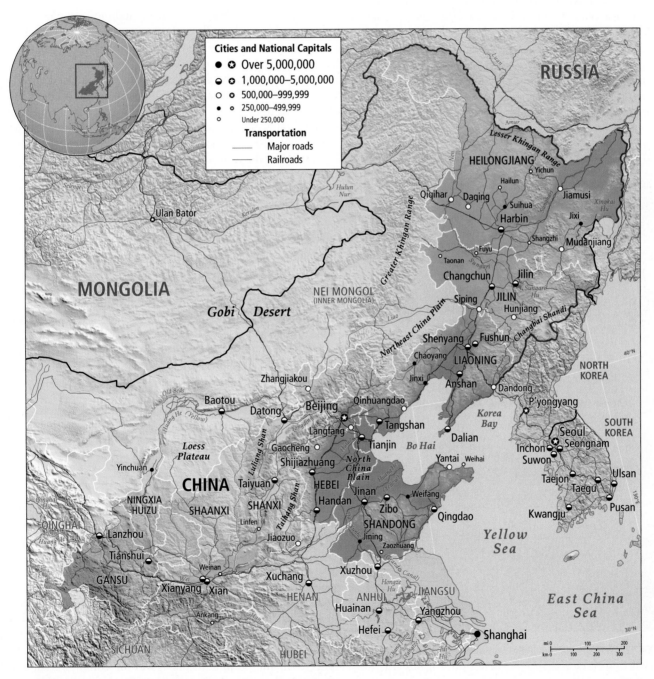

**FIGURE 9.37** China's Northeast subregion.

or windblown silt. For millennia, dust storms have picked up loess from the surface of the Gobi and other deserts to the north and west and carried it east. The loess drifted into, and over time filled up, deep mountain valleys in Shanxi and Shaanxi provinces, creating an undulating plateau. The Huang He, in a second massive earth-moving process, yearly transports millions of tons of loess sediment from the Loess Plateau to the coastal plain. Over the millennia, this river has created the North China Plain by depositing its heavy load of loess sediment in what was once a much larger Bo Hai Sea.

## The Loess Plateau

An unusually diverse mixture of peoples from all over Central Asia and East Asia have found the Loess Plateau a fertile, though challenging, place to farm and herd. Among the many culture groups that share this populous and productive plateau are the Hui and the Han of China and, particularly in the western reaches, Mongols, Tibetans, and Kazakhs. Many of the inhabitants farm cotton and millet in irrigated valley bottoms or raise sheep on the drier grassy uplands. China's largest coal reserves are also found here.

**FIGURE 9.38** Loess. In some places in Shanxi Province, the loess is so thick and firm that people excavate energy-efficient, cavelike houses in it. [Wolfgang Kaehler.]

Soil erosion is a particularly severe problem in the Loess Plateau. Thousands of years of human occupation have stripped the land of its ancient forests, leaving only grasses to anchor the loess. Because the loess is so thick—hundreds of feet deep in many areas (Figure 9.38)—deep gullies form after torrential rains. Such gullies now cover much of the landscape. Hillsides terraced for agriculture are prone to landslides, so many terraces have been abandoned. For decades, the government has maintained a reforestation campaign to stabilize slopes, but cropland is too badly needed to restore all the land to forests.

## The North China Plain

The North China Plain is the largest and most populous expanse of flat, arable land in China. Since the sixth century it has been home to most of the imperial families. From this region, the Han Chinese have dominated the country.

Today the North China Plain is one of the most densely populated spaces in the country: 370 million people—more than in the whole of North America—live in a space about the size of France. The majority are farmers who work the relatively small fields that cover the plain. Most of them produce wheat, which grows well in the relatively dry climate. The forests that once blanketed the plain were cut down millennia ago, and now the only trees that survive are those surrounding temples and planted as windbreaks.

The Huang He, which created the North China Plain, remains its most important physical feature, serving as a major transportation artery and a source of irrigation water, but also producing disastrous floods. After the river descends from the Loess Plateau, its speed slows dramatically on the flat surface of the plain, and its load of sediment begins to settle out. This sediment raises the level of the riverbed year after year until, in some places, it lies higher than the plain. Occasionally, during the spring surge, the river breaks out of the levees that have been built to control it and rushes across the surrounding densely occupied plain. The spring surge has helped the Huang He cut many new channels over time (Figure 9.39), and it also endows the plain with a new layer of fertile soil as much as a yard (a meter) thick each year. Because it both enriches and destroys, the river is referred to as both the Mother of China and China's Sorrow.

## China's Far Northeast

The northeastern corner of China, once known as Manchuria, was long considered a peripheral region, partly because of its location and partly because of its harsh climate. The winters are long and bitterly cold, the summers short and hot. The growing season is less than 120 days. Nonetheless, China's Far Northeast found an important niche in the country's economy in the 1950s, when its rich mineral resources—oil, coal, gas, gold, copper, lead, and zinc—made major industrialization possible and its fertile state farms began to produce wheat, corn, soybeans, sunflowers (for oil), and beets. The Far Northeast was known for its steel (20 percent of national output), coal, and oil production (it was the country's leading producer of oil) and for its "iron man" workers who labored heroically

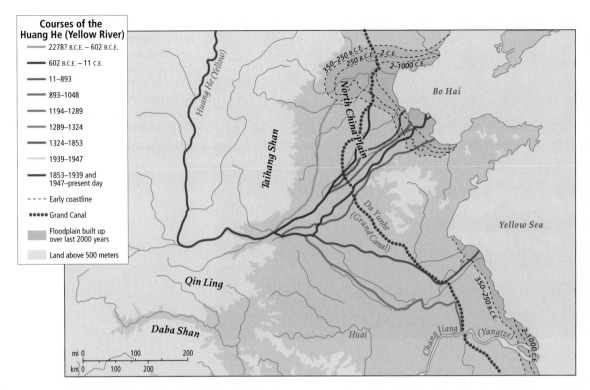

**FIGURE 9.39 The Huang He's changing course.** The lower course of the Huang He has changed its direction of flow many times over the last several thousand years. In 2000 B.C.E., it flowed north and entered the Bo Hai, south of Beijing. Then it repeatedly shifted to the east, then southeast, like the hand of a clock, until it joined the Huai, and finally cut south all the way to the Chang Jiang delta at Shanghai. Now it once again flows into the Bo Hai. The Da Yunhe (Grand Canal), considered the oldest and longest canal in the world, was completed about 610 C.E. and originally was built to carry troops and supplies north. Later, in the eighteenth century, it was used to carry taxes (in the form of grains) to the capital, Beijing. [Adapted from Caroline Blunden and Mark Elvin, *Cultural Atlas of China*, rev. ed. (New York: Checkmark Books, 1998), p. 16.]

for the revolution in return for meager wages and barracks-like housing. The Heilong Jiang (Amur River), which runs along more than 1000 miles (1600 kilometers) of the border between Russia and China, became a conduit for increasing trade with Russia, other parts of China, and Japan. The Far Northeast was slated to make the entire country industrially self-sufficient.

Beginning in the 1980s, however, when Communist Party policy began to favor marketization, the geographic focus of development shifted to Guangdong Province in Southern China, where light industries aimed at foreign markets began to spring up in cities close to the international banking center of Hong Kong (then a British colony). Next, it was the Chang Jiang (Yangtze) delta and Shanghai that began to attract government and foreign investment. Then development in far western China became a focus. By the 1990s, China's Far Northeast was labeled a rust belt. Its out-of-date industrial facilities are being dismantled to make way for new factories that can compete in the global economy. Its millions of laid-off workers are being evicted from slum dwellings that will be replaced with upscale apartments for the skilled high-tech workers being enticed from elsewhere in the country. The intraregional gap in income and well-being is thus being widened further.

Much of the renovation in cities such as Dalian on the coast is being financed by Korean and Japanese firms that train young people to do telephone sales or other information technology (IT)

or manufacturing work. The Japanese occupied China's Far Northeast (then known as Manchuria) between 1932 and 1945. Their regime is remembered as brutal, but they were the first to introduce industrial development to the region. Now, because they are bringing investment cash and a chance for economic renewal, they are welcomed.

## Beijing and Tianjin

Beijing, China's capital city of 11.5 million people, lies between the North China Plain and the Far Northeast. It is the administrative headquarters of the People's Republic of China, has several of the nation's most prestigious universities, and, with the nearby port city of Tianjin, is an industrial and transport center. The choice of Beijing as host of the 2008 Summer Olympics reflects the city's—and China's—new position as a global leader.

Beijing was once a grand imperial city full of architectural masterpieces—notably the Forbidden City, the former imperial headquarters—but much of its character was lost under the Communists, who tore down many monuments. Nowadays, in the context of China's acceptance of capitalism, much of Beijing is being reconstructed, this time as a center for international commerce, with new neighborhoods of high-rise office towers and hotels replacing older neighborhoods of tile-roofed single homes and Communist-era

**FIGURE 9.40** Beijing's new Olympic Stadium, under construction. The 2008 Summer Olympics in Beijing inspired a building frenzy costing $40 billion. In the rush to prepare facilities that would bring recognition to China's modernizing landscapes, thousands of historic structures were destroyed and corruption flourished. Eventually, the organizers building China's Olympic facilities had to sign anticorruption pledges. [AP Photo/ Elizabeth Dalziel.]

apartment blocks (Figure 9.40). Construction for the 2008 Summer Olympics is competing for labor, materials, and space with this already extremely ambitious building program.

## CENTRAL CHINA

Central China consists of the upper, middle, and lower portions of the Chang Jiang (Yangtze River) basin (Figure 9.41). Like many of Eurasia's rivers, the Chang Jiang starts in the Plateau of Tibet. It descends to skirt the south of Sichuan Province, then flows through the Witch Mountains (Wu Shan) and over the Three Gorges Dam, leaving the upper basin for the middle basin in China's densely occupied central plain. It then winds through the coastal plain (the lower basin) and enters the Pacific Ocean in a huge delta, the site of the famous trading city of Shanghai.

### Sichuan Province

Sichuan is China's most populous province, with 107.2 million people, and one of its richest in resources. It has fertile soil, a hospitable climate, inventive cultivators, and sufficient natural resources to support diversified industries. For years Sichuan embodied the ideal of complementary agricultural and industrial sectors, but it has not kept pace with the coastal industries that have attracted its young people for years. Now old industrial cities like Chongqing are recruiting foreign investors to enhance their competitive edge.

The heart of the province is the Sichuan Basin, also called the Red Basin because of the underlying red sandstone. It is a region of hills and plains crossed by many rivers that drain south toward the Chang Jiang. Because the basin is on a south-facing slope, it receives comparatively direct sunlight during much of the year. Furthermore, the basin is surrounded by mountains that are highest in the north and west, forming a barrier against the arctic blasts of winter and trapping the moist, warm air moving up from the southeast year-round. For these reasons, the Sichuan climate is generally mild and humid, and the basin is so often cloaked in fog or low cloud cover that it is said, "A Sichuan dog will bark at the sun."

Over thousands of years, Sichuan's relatively affluent farmers have cleared the native forests and manicured the landscape, so that only a few patches of old-growth forest are left in the uplands and mountains. The rivers have been channeled into an intricate system of irrigation streams. Irrigation in such a wet climate may seem surprising, but by diverting the river water into many small sluiceways, the farmers control flooding and make water available during the winter dry season, when the climate is still warm enough for some crops. Moreover, this is a region of **wet rice cultivation,** in which the roots of the plants are submerged in water early in the growing season. The geographer Chiao-Min Hsieh vividly describes how the wet "rice fields in [the lowlands] are shaped like squares on a chessboard. Everywhere one can hear water gurgling like music as it brings life and growth to the farms." The rice yields are abundant.

The population density of Sichuan is more than 800 people per square mile (300 per square kilometer). Most of them are farmers, growing rice, wheat, and corn and raising animals such as silkworms, pigs, and poultry as a sideline. In the new market economy, it is often these sidelines that earn them the most cash income (Box 9.5). In the mountain pastures to the west of the basin, Tibetan herders raise cattle, yaks, sheep, and horses. (Sichuan has the largest population of Tibetans outside Tibet—close to 1 million.)

The two main cities of the Sichuan Basin are Chengdu, the provincial capital (population 4.6 million), and Chongqing (population 7.6 million). Chengdu is a transport hub and is home to light industries, especially food processing and the manufacture of textiles and precision instruments. Some of its higher-quality products

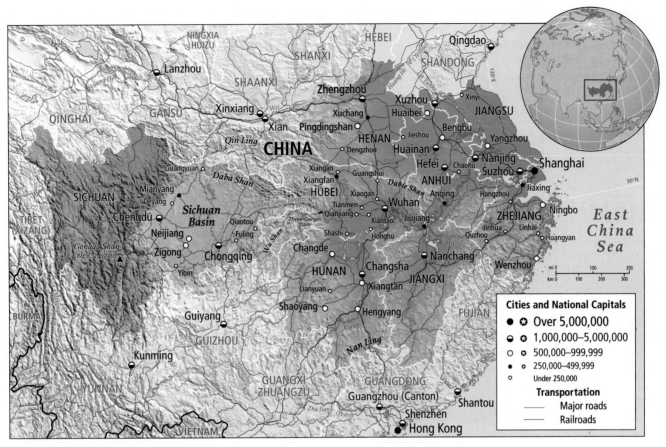

**FIGURE 9.41** The Central China subregion.

---

# BOX 9.5 | AT THE LOCAL SCALE   Silk and Sericulture

Silk cloth has long figured prominently in China's interactions with the outside world—indeed, it gave its name to the Silk Road, the trade route between China, Central Asia, and Europe. Long an important crop in central and southern China and in Japan, silk is now a major industry in many parts of South and Southeast Asia as well. Although the economic importance of silk has been reduced, both because it has been replaced by cotton and synthetic fibers and because China now has many other industries, silk is still an important commodity in Chinese commerce.

Silk fiber is made by caterpillars of the *Bombyx* moth family, also known as silkworms, which produce a fine, strong, continuous fiber from which they spin a cocoon. In a process called sericulture, the caterpillars, raised from eggs, are fed on the leaves of the mulberry tree. The caterpillars that hatch from just one ounce of eggs will consume the leaves of 25 to 30 trees! To make silk, the completed cocoons, with the caterpillars inside, are plunged into hot water to kill the insects. The cocoons are then carefully dried and unwound so that the fiber can be used to weave cloth. Sericulture is primarily a sideline cottage industry run by women—one of the ways that Chinese women, and now women throughout East Asia, have gained access to their own cash (Figure 9.42).

**FIGURE 9.42** Silkworm cultivation. This woman, working in Shajiao Commune, is skilled in the ancient art of sericulture. [Gina Corrigan/Robert Harding.]

are sold in the global market. Chongqing, with its thousands of iron and steel manufacturers and machine-building industries, is an old industrial city now in decline, but targeted for renewal as Sichuan attempts to attract workers back to the interior from the coastal provinces. As a result of the construction of the Three Gorges Dam (see Box 9.4 on page 504), Chongqing, which lies at the western end of the dam reservoir, is also in line to become a hub for shipping and other new economic activities.

Sichuan's productivity has enabled it to support its dense population, but growing development is resulting in degradation of its environment. The basin's tendency to retain warm, moist air allows industrial and vehicle emissions to build up, leading to intolerable pollution levels. For example, in 2001, sulfur dioxide emissions in Chongqing exceeded World Health Organization standards by nearly seven times (see Figure 9.34 on page 505).

## The Central and Coastal Plains

After the Chang Jiang emerges from its concrete shackles behind the Three Gorges Dam, it traverses an ancient, undulating lake

---

## BOX 9.6 | AT THE LOCAL SCALE  Shanghai's Urban Environment

Shanghai has a long history as a trendsetter. The opening of its gates to Western trade in the early nineteenth century spawned a period of phenomenal economic growth and cultural development that led it to be called "the Paris of the East." As a result of China's recent reentry into the global economy, Shanghai is undergoing a vigorous rejuvenation. In less than a decade, the city's urban landscape has been remade by the construction of over a thousand business and residential skyscrapers, subway lines and stations, highway overpasses, bridges, and tunnels. The changes, however, have occurred too fast for planners to keep up with them. For hundreds of miles into the countryside, suburban development linked to Shaghai's economic boom is gobbling up farmland, and displaced farmers have rioted. The excessive opulence of Shanghai is causing discontent even among those who live there. Mortgage payments on new apartments leave many middle-class workers with no discretionary funds.

Pudong, the new financial center for the city, sits across the Huangpu River from Shanghai's famous Bund—the elegant row of big brownstone buildings that served as the financial capital of China until half a century ago. Previously a maze of dirt paths and sprawling neighborhoods of simple tile-roofed houses, Pudong has been transformed by soaring high-rises (Figure 9.43). The grandest is Jin Mao Tower, the fifth tallest building in the world. Atop the tower, Hyatt Corporation has opened its first grand hotel in China, and some 45 banks have settled into the building's Lujiazui Finance and Trade Zone.

With a population now exceeding 14 million, Shanghai is increasing the area of green space in the city to improve the local environment for its residents. All of this activity is part of the local government's plan to turn Shanghai into an international metropolis that will attract further investment and economic development.

*Sources:* Asko Ahokas, "Lost in modern Shanghai," *Design Forum* (December 19, 2002); Louisa Lim, "Hot real estate market has China worried," "Morning Edition," National Public Radio, August 24, 2006, http://www.npr.org/templates/story/story.php?storyId=5699783; Louisa Lim, "Opulent offices trigger debate, anger in rural China," "All Things Considered," National Public Radio, June 12, 2006.

(a)

(b)

**FIGURE 9.43** Shanghai, many cities in one. Shanghai is China's largest and most modern urban area, yet it still retains many characteristics of the old city. **(a)** In one of the oldest parts of Shanghai, hundreds of traditional shops and stalls nonetheless exhibit many modern symbols. **(b)** Nearby stands the Jin Mao Tower, currently China's tallest building, ranking fifth highest in the world. [(a) © M. Lines; (b) © Khai Y. Chan.]

bed that is interrupted in many places by low hills. This lake bed forms the middle basin of the Chang Jiang. It and the river's lower basin, leading to the coast, are rich agricultural regions dotted with industrial cities.

The middle basin and the lower basin (a coastal plain) have been filled with **alluvium** (river-borne sediments) carried down from the Plateau of Tibet and the Sichuan Basin. Other rivers entering the Chang Jiang Basin from the north and south also bring in loads of silt, which are added to the main river channel. The Chang Jiang carries a huge amount of sediment—as much as 186 million cubic yards (142 million cubic meters) per year—past the large industrial city of Wuhan. This sediment, which under natural conditions was deposited on the floor of the basins during annual floods, formerly enriched agricultural production. But because the floods often destroyed people, animals, and crops, the Chang Jiang, like the Huang He, now flows between levees. Still, it occasionally breaches the levees and floods both rural and urban areas, as it did most recently in 2004. As the river approaches the Pacific, it deposits the last of its sediment load in a giant delta, at the outer limits of which lies the old and rapidly reviving trading city of Shanghai.

The climate of the middle and lower basins, though not as pleasant as Sichuan's, is milder than that of the North China Plain. The Qin Ling range and other, lower hills that extend eastward across the northern limits of the basin block some of the cold northern winter winds. The mountains also trap warm, wet southern breezes, so the basins retain significant moisture during most of the year. The growing seasons are long: 9 months in the north and 10 in the south. The natural forest cover has long since been removed to provide agricultural land for the ever-increasing rural population. Summer crops are rice, cotton, corn, and soybeans; winter crops include barley, wheat, rapeseed (canola), sesame seeds, and broad beans. There are more than 400 million people in the middle and lower basins of the Chang Jiang. In the past, many of them were farmers, but every year more people migrate to the many industrialized cities of the region. Shanghai overshadows them all in importance.

For many centuries, Shanghai's location facing the East China Sea, with the whole of Central China at its back, positioned the city well to participate in whatever international trade was allowed. When the British forced trade on China in the 1800s, Shanghai became their base of operations. Before the Communist revolution, Shanghai was among the most cosmopolitan cities on earth, home to well-educated literati, wealthy traders, and a goodly number of underworld figures as well. Then, after the revolution, the Communists designated the city as the nexus of capitalist corruption, and it fell into disgrace.

When China began to open up to world trade in the 1980s, Shanghai entrepreneurs were well situated to take advantage of government incentives (such as reduced taxes, help in preparing sites, and permission to take profits out of the country) to build factories, workers' apartment blocks, international banks, luxury apartment houses, shopping malls, and elegant postmodern-style villas (Box 9.6). Today, some of the Overseas Chinese (Box 9.7; see also Chapter 10, pages 557–558)—particularly those who fled the repression of the Communist Party to places such as Taiwan, Singapore, and Malaysia—have returned to invest in everything from television stations and discotheques to factories, shopping centers, and entertainment parks. Europeans and both North and South Americans are attracted to Shanghai too, hoping to find joint

## BOX 9.7 | AT THE GLOBAL SCALE  The Overseas Chinese

China has had an impact on the rest of the world not only through its global trade, but also through the migration of its people to nearly all corners of the world. The first recorded emigration by the Chinese took place over 2200 years ago. Ever since then, China's contacts with the outside world have been continuous, if occasionally aloof. These contacts spread eastward to Korea and Japan, westward into Central and Southwest Asia via the Silk Road, and by the fifteenth century, to the south and southwest along Admiral Zheng He's maritime route, which led from northern Vietnam through Southeast Asia to coastal India, Arabia, and Africa (see page 480).

Trade was probably the first impetus for Chinese emigration. Eventually, across the whole of Southeast Asia and into the Indian Ocean, Chinese merchants, artisans, sailors, and workers played such a significant economic role that the region came to be known in Chinese as *Nanyang*, the South Seas. Most of the travelers to the South Seas hailed from the southeastern coastal provinces of Fujian, Guangdong, and Zhejiang. Taking

their families with them, some settled permanently on the peninsulas and islands of what are now Indonesia, Thailand, Malaysia, and the Philippines, and came to be known as the "Overseas Chinese."

In the nineteenth century, economic hardship in China and a growing international demand for labor spawned the migration of as many as 10 million Chinese to countries on all the inhabited continents. After the Communist revolution, they were joined by others fleeing the repression of the Communist Party. As a result, Chinatown communities are present in places as widely scattered as Singapore, London, Central Europe, São Paulo, the Caribbean, San Francisco, and Toronto. The term "Overseas Chinese" has been extended to apply to Chinese emigrants and their descendants in all of those locations.

*Sources:* Wei Djao, *Being Chinese: Voices from the Diaspora* (Tucson: University of Arizona Press, 2003); Josh Kurlantzick, "China's charm offensive in Southeast Asia," *Current History* (Carnegie Endowment for International Peace, September 2006).

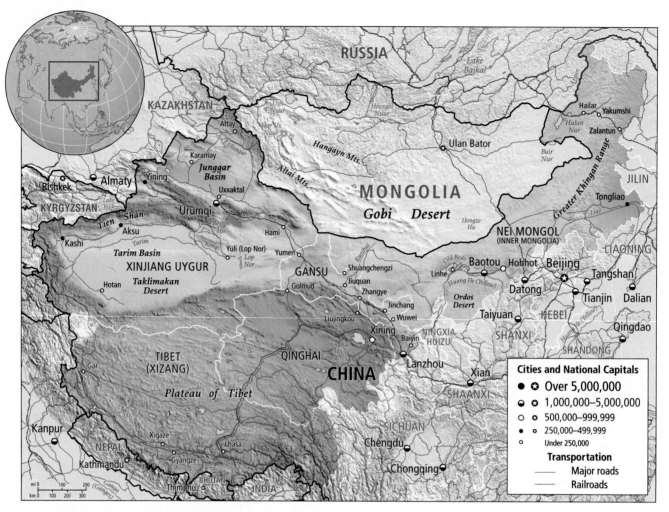

**FIGURE 9.44** China's Far North and West subregion.

ventures with Chinese partners so that they can tap into the huge pool of consumers emerging in China.

## CHINA'S FAR NORTH AND WEST

The Far North and West of China (Figure 9.44) once occupied a central place in the global economic system. Traders carried Chinese and Central Asian products such as silks, rugs, spices and herbs, and ceramics over the Silk Road to Europe, where they were exchanged for gold and silver. Today, the subregion is recovering some of its ancient economic vitality, but just who will profit is not yet clear.

Despite its trading past, this large interior zone has long been considered backward by the rulers of eastern China because of its dry, cold climate, vast grasslands, long history of nomadic herding, and persistent adherence to Islam. Settlements are widely dispersed, and most agriculture requires irrigation. Three of the subregion's four political divisions have been designated autonomous regions (not provinces) because of the high percentage of ethnic minority populations there (Nei Mongol, also called Inner Mongolia, is per-

haps the best known of these autonomous regions), but their people do not enjoy real autonomy. The central government in Beijing retains control over political and economic policies because it sees this subregion as crucial to China's future: it has energy and other resources for industrial development, it is close to the emerging oil-rich economies of Central Asia and Russia, and it affords a place to resettle some of China's surplus population.

### Xinjiang

The Xinjiang Uygur Autonomous Region, in northwestern China, is the largest of China's political divisions, accounting for one-sixth of China's territory. Xinjiang has only 17.5 million inhabitants; their roots lie mainly in Central Asia, and most are Muslims. The most numerous, at 8 million, are the Turkic-speaking Uygurs; there are also Mongols, Persian-speaking Tajiks, Kazakhs (Figure 9.45), Kirghiz, Manchu-speaking Xibe, and Hui. New migrants of Han Chinese origin number about 6 million. The peoples native to the subregion once made their living as nomadic herders and animal traders, moving with their herds and living in **yurts** (or **gers**)—round, heavy felt tents stretched over collapsible willow frames (see Figure 1.18b on page 17). These cozy houses can be folded and carried on horseback or in horse-drawn carts. In the last few decades, many nomadic people have taken jobs in the emerging oil industry and now live in apartments provided for laborers. The Han migrants live primarily in the cities. They work as bureaucrats and managers; in the oil, gas, and nuclear power industries; in state-owned agricultural colonies; and on highway and railroad projects. The Han tend to be much better paid, so there is a noticeable disparity in wealth and well-being between them and the indigenous people, especially farmers (farmers here earn only one-third the income of farmers in the eastern regions).

Xinjiang consists of two dry basins: the Tarim Basin, occupied by the Taklimakan Desert, and the smaller Junggar Basin to the northeast. Both are virtually surrounded by 13,000-foot-high (4000-meter-high) mountains topped with snow and glaciers. In this distant corner of China, far from the world's oceans, rainfall is exceedingly sparse. Snow and glacial meltwater from the high mountain peaks are important sources of moisture. Much of the meltwater makes its way to underground rivers, where it is protected from the high rates of evaporation on the surface. Long ago, people built conduits called *qanats* deep below the surface to carry groundwater dozens of miles to areas where it was needed. *Qanats* have made some of the hottest and driest places on earth productive. The Turfan Depression, situated between the Junggar and the Tarim basins, is a case in point. Temperatures often reach 104°F (40°C), and evaporation rates are extremely high. But because the *qanats* bring irrigation water, this area produces some of China's best luxury foods: melons, grapes, apples, and pears. The produce is sold to urban populations in eastern China.

In Xinjiang today, the local and the global, the very traditional and the very modern, confront each other daily. Herdspeople still living in yurts and *gers* dwell under high-tension electric wires that supply new oil rigs. Tajik women weave traditional rugs that are sold to merchants who fly from far-away developed countries to the ancient trading city of Kashi for the Sunday market (see Figure 9.29 on page 500). With the breakup of the Soviet Union, citizens of the new republics of Central Asia are eager to revive their trading heritage, and China is welcoming them and attracting outside investors from Europe and the Americas by establishing ETDZs in cities such as Kashi and Urumqi.

The Uygur and other ethnic leaders of Xinjiang are wary of Beijing, fearing that the central government's true intent is not to improve life for the local people, but rather to exploit Xinjiang's oil and gas resources and to appropriate land for settling eastern China's excess population. They are also concerned about the thinly disguised eastern prejudices against Uygur ethnicity and religion. Uygur leaders have been issuing warnings about the situation on the Internet. For example, one young Uygur man in Urumqi, Xinjiang's capital, wrote of his discontent: "I am a strong man, and well-educated. But [Han] Chinese firms won't give me a job. Yet go down to the railroad station and you can see all the [Han] Chinese who've just arrived. They'll get jobs. It's a policy to swamp us." Also on the Internet are stories of harsh treatment of dissidents, who have been arrested and charged with being Islamic fundamentalists. Resistance continues to grow.

**FIGURE 9.45** People of the Xinjiang Uygur Autonomous Region. In Yining, in northwestern Xinjiang, a Kazakh girl plays "kiss the maiden." In the game, a suitor gallops after the girl and tries to steal a kiss. If he succeeds, she chases him and tries to beat him with a whip. The contest tests the riding skills of both future spouses. [Jay Dickman.]

## The Plateau of Tibet

Situated in far western China, the Plateau of Tibet is the traditional home of the Tibetan people. Officially, it includes the Xizang Autonomous Region and Qinghai Province. Tibet (Xizang) and Qinghai lie an average of 13,000 feet (4000 meters) and 10,000 feet (3000 meters) above sea level, respectively. They are surrounded by mountains that soar thousands of feet higher. They have cold, dry climates (late June can feel like March does on the American Great Plains) because of their high elevation and because the Himalayas to the south block warm, wet air from moving in. Across the plateau, but especially along the northern foothills of the Himalayas, snowmelt and rainfall are sufficient to support a short growing season for barley and such vegetables as peas and broad beans. Snowmelt also forms the headwaters of some major rivers: the Indus, Ganga (Ganges), and Brahmaputra begin in the western Himalayas, and the Nu (Salween), Irrawaddy, Mekong, Chang Jiang, and Huang He all begin along the eastern reaches of the Plateau of Tibet.

Traditionally, the economies of Tibet and Qinghai have been based on the raising of grazing animals. The yak is the main draft animal, and it also provides meat, milk, butter, cheese, hides, and hair, as well as dung and butterfat for fuel and light. Other animals of economic importance are sheep, horses, donkeys, cattle, and dogs. Animal husbandry on the sparse grasses of the plateau has required a mobile way of life so that the animals can be taken to the best available grasses at different times of the year. Yet for several

**517**

decades, the Chinese government has pressured Tibetan herds-people to settle in permanent locations so that their wealth can be taxed, their children schooled, their sick cared for, and their dissidents curtailed. Still, throughout the plateau, many native (non-Han) peoples continue to live mobile yet solitary lifestyles, as they have for centuries, occasionally adopting some of the accoutrements of modern life and adapting to its restrictions (Figure 9.46).

The history of Tibet's political status vis-à-vis China is long and complex, characterized by both cordial relations and conflict. During China's imperial era (prior to the twentieth century), Tibet maintained its own government and sent representatives to the Chinese imperial court, but faced the constant threat of invasion and Chinese meddling in its affairs. In the early 1900s, Tibet declared itself separate and free from China and conducted its affairs as an independent country. It was able to maintain this status until 1949–1950, when the Chinese Communist army "liberated" Tibet, promising a "one-country two-systems structure." A Tibetan uprising in 1959 led China to abolish the Tibetan government and violently reorder Tibetan society, as described on page 501.

By the 1990s, the Beijing government's strategy was to overwhelm the Tibetans with secular social and economic modernization and Han Chinese settlers rather than outright military force (though a Chinese military presence is maintained). To attract trade, China is spending hundreds of millions of dollars on housing and on roads, railroads, and a tourism infrastructure that capitalizes on European and American interest in Tibetan culture. China sees its

**FIGURE 9.46** Life in western Tibet. Young men in the town of Gyanze, Tibet, spend much of their leisure time playing billiards. Often the billiard tables sit outdoors in the cool, crisp, sunny air of the Tibetan Plateau. [Michael D. Gan, Netherlands.]

actions in Tibet as part of its overall strategy to integrate the entire country economically and socially. Schools are being built and jobs opened up to young Tibetans. Increasingly, Tibetans are accepting the Chinese presence and are channeling their Tibetan cultural pride into efforts to preserve the Tibetan language and religion.

In contrast to China, women in Tibet have always had a relatively high position in society. Among the nomadic herders, they were free to have more than one husband, just as men were free to have more than one wife. In addition, the custom of the husband joining the wife's family allowed women to attain a higher status

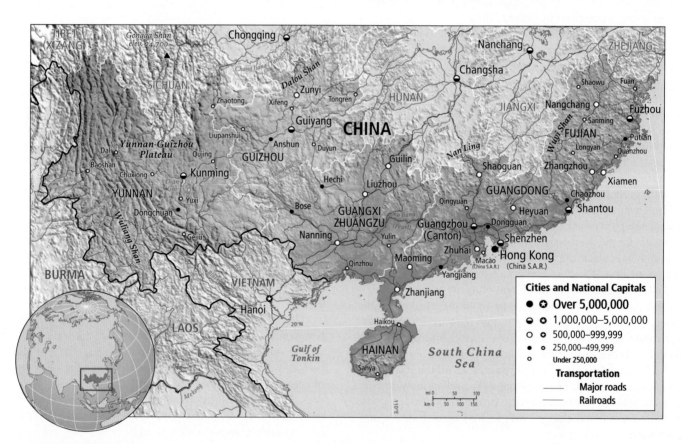

**FIGURE 9.47** The Southern China subregion.

**FIGURE 9.48** The natural riches of Yunnan. Rainy and warm Yunnan Province is increasingly important nationally for its market gardening and the raising of small animals. Here a family transports its wares in the produce baskets typical of the subregion. [Eastcott/Momatiuk/Woodfin Camp & Associates.]

than they had among the Han. Although Buddhism introduced patriarchal attitudes from outside Tibet, it encouraged female independence. At any given time, up to one-third of the male population was living a short-term monastic life, so Tibetan Buddhist women have become particularly self-sufficient. Chinese culture has typically regarded the women of the western minorities as barbarian, precisely because their roles were not circumscribed: they were not secluded, and they rode horses, worked alongside the men in herding and agriculture, and were assertive.

# SOUTHERN CHINA

Southern China has two distinct sections (Figure 9.47). The first is made up of the mountainous and mostly rural provinces of Yunnan and Guizhou to the southwest. The second includes the Guangxi Zhuangzu Autonomous Region and Guangdong and Fujian provinces on the southeastern coast, where the booming cities of China's evolving economic revolution are located.

## The Yunnan–Guizhou Plateau

The provinces of Yunnan and Guizhou share a plateau noted for its natural beauty and mild climate and for being the home of numerous indigenous groups that are culturally distinct from the Han Chinese. The plateau is a rough land of deeply folded mountains that trend north-south, through which the Nu (Salween) and Mekong rivers flow. The heavily forested valleys may be as deep as 5000 feet (1500 meters), yet only 1300 feet (400 meters) across; although people can call to one another across the valleys, it may take more than a day's difficult travel to reach the other side. In some places, rope and bamboo bridges have been slung across the chasms. The landforms here are unstable, and earthquakes cause heavy damage to the stairstep rice paddy terraces.

The valuable natural resources of the Yunnan–Guizhou Plateau are primarily biotic. Yunnan Province, with its fertile soil, is an important source of produce and meat for the bustling cities on the southeastern coast (Figure 9.48). The province is also required to supply wood for China's industrialization, but nonetheless it is still one of the most heavily forested regions in the country. Yunnan is called "the national botanical garden" because many of China's plant species and one-third of its more than 400 bird species are native to the area's exotic landscapes (Figure 9.49). Both flora and fauna are now threatened by recent human disturbance of the environment. A British team of biologists traveling in Yunnan in the mid-1980s reported that by then "birds were absent even in the reserves." Until the 1970s, the tropical forests of far southern Yunnan (close

**FIGURE 9.49** The fantastic landscape of the Yunnan–Guizhou Plateau. The lower-lying, eastern parts of the plateau are composed of karst (limestone) deposits that are eroded by water to form jagged peaks jutting out from surrounding flat valleys. This phenomenon is found in its most extreme form in Guilin, Guizhou, where a karst "forest" of 90-foot-high (27-meter-high) jagged limestone columns is interspersed with lakes. [Keren Su/China Span.]

to Burma, Thailand, and Laos) harbored elephants, bears, porcupines, gibbons, and boa constrictors.

Kunming, capital of Yunnan Province, lies at the heart of a booming heroin trade that flows from major producers in the mountains of Burma and Laos to China's northern and eastern coastal cities, where the heroin is shipped to the global market. The cheap local price has left many local youths addicted, and Kunming is now also a center for the treatment of drug addiction, where herbal medicine is the chief therapy.

## The Southeastern Coast

The southeastern coastal zone of China has long been a window to the outside world. Its ports began launching ships that journeyed as far as the Persian Gulf and the northern and eastern coasts of Africa during the Tang empire (618–907). Arab traders began to visit this part of China around the same time. By the fifteenth century, some of the first Europeans in the region described a string of flourishing trading towns all along the coast. The overwhelming majority of Overseas Chinese (see Box 9.7 on page 515) have their roots along China's southeastern coast. In the 1980s, the central government decided to take advantage of this long tradition of outside contact by designating several of the old coastal fishing towns as SEZs, with special rights to conduct business with the outside world and to attract foreign investors. The quick result was a chain of cities attracting millions of migrants to work in light industries that supply the world with all manner of consumer products.

The principal river of Southern China, the Zhu Jiang (Pearl River), is joined by several tributaries to form one large delta below the city of Guangzhou (Canton) in Guangdong Province. The lowlands along the rivers and delta have a subtropical climate and a perpetual growing season. The rich delta sediment and the interior hinterlands are used to cultivate sugarcane, tea, fruit, vegetables, herbs, mulberry trees for sericulture (the raising of silkworms), and timber, all of which are sold in the various SEZs and exported to global markets.

## Hong Kong

Hong Kong, one of the most densely populated cities on earth, has packed most of its 6.9 million people into just 23 square miles (60 square kilometers) of the city's total of 380 square miles (985 square kilometers). This very small place has the world's eighth largest trading economy and the world's largest container port. It is also the world's largest producer of timepieces. Hong Kong residents have China's highest per capita income; its annual GDP per capita (adjusted for PPP) was more than U.S.$27,000 in 2003.

Hong Kong was a British crown colony until July 1997, when Britain's 99-year lease ran out and Hong Kong became a special administrative region (SAR) of China. Many wealthy citizens fled, worried that China would absorb Hong Kong and no longer allow it economic and political freedom. Oppression has been curtailed, however, because of Hong Kong's long-important role as China's unofficial link to the outside world of global trade: before 1997, some 60 percent of foreign investment in China was funneled through Hong Kong. Since 1997, Hong Kong has had less autonomy, but

**FIGURE 9.50** The Wynn Macau hotel and casino. This sleek Las Vegas–style casino, which opened in 2006, is part of Macao's bid to transform itself from a tacky target for mainland day-tripping gamblers into a major global tourism destination. The casino, owned by American gaming mogul Stephen Wynn, reputedly cost U.S.$1.2 billion to build. Macao is the only place in China where casino gambling is permitted. [AP Photo/Kin Cheung.]

dramatic changes in its fortunes are unlikely. Given the very obvious success of many cities along China's southeastern coast, from Macao to Shanghai, Hong Kong will probably continue in its role as a financial hub in the development of this very rapidly growing region. Its new state-of-the-art airport is a symbol of its importance to China.

## Macao

Macao (also spelled Macau), built on a series of islands across the Zhu Jiang estuary from Hong Kong, is the oldest permanent European settlement in East Asia. Portuguese traders arrived in about 1516, and by 1557 had established a raucous colonial trading and gambling center. Macao remained a Portuguese colony until the Chinese government regained control in 1999. Macao grew rapidly after 1949 with the influx of refugees from the Communist revolution, but today, its population has shrunk to just 500,000 —primarily Chinese from nearby southern provinces who earn a living mainly from the manufacture of clothing, textiles, toys, and plastic products and from the tourism industry. Macao remains China's only gambling center, but since marketization its casinos have been managed by two Las Vegas firms (Figure 9.50). Although Macao is not as prosperous as Hong Kong, its GDP per capita is still four times that of the mainland (see Table 9.1 on page 507).

## JAPAN

Japan consists of a chain of four main islands and hundreds of smaller ones (Figure 9.51). Prone to severe earthquakes, tsunamis,

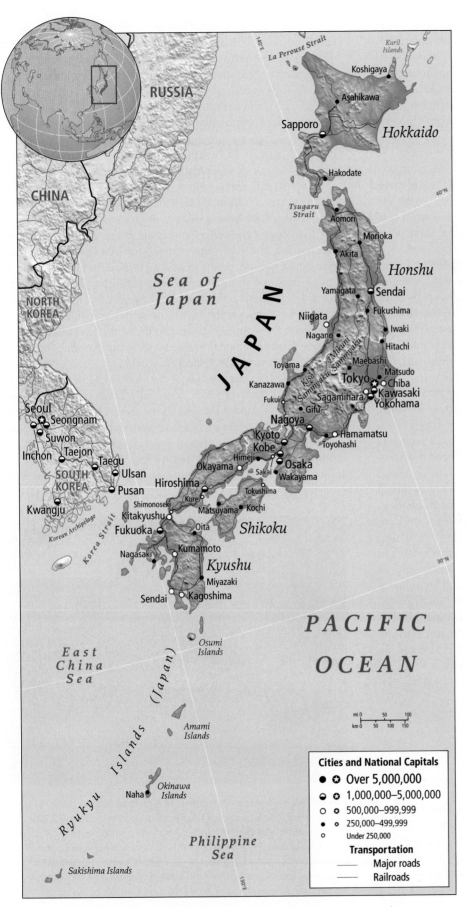

**FIGURE 9.51** The Japan subregion.

**Cities and National Capitals**

| | | |
|---|---|---|
| ● | ✪ | Over 5,000,000 |
| ◒ | ✪ | 1,000,000–5,000,000 |
| ○ | ✪ | 500,000–999,999 |
| • | ✪ | 250,000–499,999 |
| ○ | | Under 250,000 |

**Transportation**

—— Major roads

—— Railroads

and typhoons, and so mountainous that only 18 percent of its land can be cultivated, the Japanese archipelago might seem an unlikely place to find 128 million people living in affluent comfort. Yet its citizens have learned to cope with these limitations and have made the most of crowded conditions in cities and countryside alike.

Counting its tiny southernmost islands, Japan stretches over a range of latitudes roughly comparable to that between Canada's province of Nova Scotia and the Florida Keys (46°N–24°N), and, although half a world away, it has a similar range of climates. The climate is cool on the far northern island of Hokkaido; temperate on the islands of Honshu, Kyushu, and Shikoku; and tropical in the southern Ryukyu Islands. Despite the moderating effect of the surrounding oceans, the great seasonal climate shifts of nearby continental East Asia give Japan a more extreme seasonal variation in temperature than would be the case if it lay farther out to sea.

Honshu, the largest and most densely populated of Japan's islands, has the most mountains and forests. Because it has gained access to forest products from Southeast Asia and other parts of the world, Japan has been able to keep much of its own forestland in reserves. Today, the interior mountains of Honshu (and Hokkaido) remain largely forested, although many of the slopes are planted with a monoculture of *sugi* (Japanese cedar) to be harvested commercially. The few flat lowlands and coastal areas of northern and southern Honshu were once intensively cultivated and supported a dense rural population. Today, industrial cities fill many of these lowlands, especially in the south. Tokyo–Yokohama is one of the largest urban agglomerations in the world, with 36.7 million people (28 percent of Japan's total). It lies on the east-central coast of Honshu on the Kanto Plain, Japan's largest flatland. Kobe–Osaka–Kyoto, Japan's second largest metropolitan area with 17.5 million people, is one of several major urban areas located on the flatlands that ring the Inland Sea.

Japan's other three main islands are Hokkaido, the least populated island, referred to as Japan's northern frontier, and Kyushu and Shikoku, two smaller southern islands. The northern slopes of Kyushu and Shikoku, facing the Inland Sea, have narrow strips of land put to agricultural, industrial, and urban uses. The southern slopes are more rural and agricultural. Off Kyushu's southern tip, the very small, mountainous Ryukyu Islands (which include Okinawa) stretch out in a 650-mile (1000-kilometer) chain, reaching almost to Taiwan. Although many areas are densely settled and are important for tourism and specialty agriculture, such as the growing of pineapples, the Ryukyu Islands are considered a poor rural backwater of the four big islands.

## Food Production and the Challenge of Limited Resources

Japan has nearly the largest ratio of people to farmland in the world: more than 7000 people depend on each square mile (more than 2700 per square kilometer) of cultivated land. Thus Japanese agriculture depends heavily on high-yield varieties of rice and other crops as well as on irrigation, fertilization, and mechanization. Because flat land is scarce, Japanese farmers have had to make efficient use of hill slopes by planting tea bushes and orchards and by terracing for rice paddies.

Although the agricultural way of life plays a large part in Japanese national identity, today only 4 percent of the population farms for a living. Japanese farms are highly productive in terms of output per acre; however, their produce is some of the most expensive in the world. A single cantaloupe can cost as much as U.S.$10, a head of lettuce U.S.$5, and a pound of fine beef more than U.S.$50. These high prices are partly a reflection of the very superior quality of the produce, but they are also a consequence of government efforts to encourage Japanese farmers through tariffs that reduce competition from imports and subsidies that result in scarce production.

The seas surrounding Japan, where warm and cold ocean currents mix, are an especially important source of food. There are some 4000 coastal fishing villages and tens of thousands of small craft that work these waters and bring home great quantities of tuna, halibut, mackerel, salmon, and other fish. In addition, a large aquaculture industry produces oysters, seaweed, and other foods in shallow bays and freshwater fish in artificial ponds. Tokyo's early morning fish market is the largest in the world. Japan also sends out large fishing ships, complete with canneries and other processing equipment, to harvest oceans around the world. Japan has received severe criticism from environmentalists for its consumption of endangered marine species, especially whales.

Nevertheless, Japan is one of the world's largest consumers of foreign agricultural products, especially grains and frozen and processed foods, and it has one of the lowest levels of **food self-sufficiency** (the percentage of food consumed daily that is supplied by domestic production) and some of the highest food prices in the world (Figure 9.52). In 2002, Japan's food imports totaled U.S.$42 billion. The Japanese government has expressed concern over in-

**FIGURE 9.52** Street food, an answer to high food costs in Japan. A vendor in Tokyo sells inexpensive food from a small vehicle designed to maneuver through traffic efficiently. Urban office workers form neat queues at the curb and wait eagerly to buy the wide variety of foods and hot and cold drinks the vendor has prepared. Agricultural policies make food especially expensive in Japan; this method of vending saves the consumer the costs added on by restaurants. [© Hans van Oort, Netherlands.]

creasing food imports and their implications for food security. Hoping to limit Japan's dependence on food imports, the government has set a goal of raising its food self-sufficiency from the current level of 41 percent to 45 percent by the year 2010.

## Living in Japan

Perhaps more than any other country on earth, Japan is renowned for its ability to confound foreigners. Here, in an attempt to penetrate some of Japan's cultural complexity, we look at its unique cultural patterns, its urban life, its home life, and its aging population.

***Distinctive Culture.*** A substantial foundation of long-standing customs and traditions is topped with a plentiful mix of borrowings from around the world to create the unique blend of contrasts and apparent contradictions that is Japanese culture. We see the mix in many ways: in the kinds of clothes people wear, in the foods they eat, in the music they listen to, in the sports they enjoy, and even in how they practice religion. Thus, among other contrasts, Japan is a land of kimonos *and* blue jeans, sushi *and* hamburgers, the koto (a traditional stringed instrument) *and* rap music, sumo wrestling *and* baseball. Japanese couples who marry in traditional wedding kimonos often change their clothing after the ceremony to a Western-style tuxedo and a white bridal dress for the reception.

Some experts trace these contrasts to the forced opening of Japan by the United States in 1853. In the years that followed, many Japanese, particularly in Tokyo, took on exaggerated international trappings to accommodate foreign visitors, at the same time protecting what was truly Japanese. There is often a distinction in Japanese life between what is displayed on the outside, *omote*, and what is private on the inside, *ura*. The former often protects the latter, as in the contrasts between *omote-ji*, the outer layer of cloth in a kimono, and *ura-ji*, the layer closest to the skin, and between *Omote-Nippon*, the urban-industrial eastern (Pacific Ocean) side of Japan that trades with the world, and *Ura-Nippon*, the more traditional and secluded western (Sea of Japan) side of the country.

***Urban and Suburban Japan.*** Nearly all of Japan's major cities are located along the coastal perimeter. The cities are where Japan's rapid industrialization has taken place, and their coastal locations facilitate the import of raw materials and the export of finished products. Ideas from the outside world can also penetrate easily, aiding technological advancement and cultural synthesis. Tokyo, for example, has the world's second largest stock exchange, numerous centers for research and development, and some of the world's most beautiful modern architecture, and it is also a major international cultural center.

At the same time, Japanese cities suffer from overcrowding and pollution. In Tokyo, it is not uncommon for a middle-class family of four, plus a grandparent or two, to live in a one-bedroom apartment. Japanese cities cannot easily expand to relieve crowding because they are limited by the surrounding mountains and ocean, by building codes minimizing potential damage from earthquakes, and by regulations protecting Japan's scarce agricultural lands. Every bit of land for suburban development is hard won. Also contributing to over-crowding are corruption and a lack of competition in the construction industry, which keep housing in short supply. Although there is growing pressure for change, most Japanese stoically endure minuscule, expensive apartments and long commutes to work. In the Tokyo–Yokohama metropolitan area, it is common to travel 2 or even 3 hours one way in trains that are so overcrowded that stations employ "shovers," who physically push as many passengers as possible into a single car.

The greatest source of discomfort for urban dwellers is pollution. Several cities have endured major episodes of poisoning from mercury and polychlorinated biphenyls (PCBs), and all large cities suffer chronic air pollution from automobiles and factories, as well as noise pollution. Local environmental movements have led to the first serious government efforts to limit pollution and to establish moderately effective environmental oversight. During Japan's boom years in the 1980s and early 1990s, many cities, including badly congested Tokyo, were able to expand public park space and open other recreational amenities such as artificial beaches, sports fields, and new aquariums.

***The Home.*** Japanese homes are known for their simple, elegant aesthetics and their highly functional use of space. Traditional homes are made of wood and are usually roofed with heavy tiles. Even in larger rural homes, interior spaces are much smaller than most Westerners are accustomed to. The three or four rooms of a typical middle-class home are used for many purposes during the course of the day, transformed as needed by sliding doors of paper and wood or by folding decorative room dividers. Furniture is simple, with much of family life centered on a low table and floor cushions. For sleeping, a firm futon is often rolled out on whatever floor space is available. While these home features are still the norm, some newer houses and apartments, though small, have Western-style arrangements of rooms and furniture. Most Japanese kitchens are equipped with a full range of electrical appliances, including the all-important automatic rice cooker. Almost every Japanese family has a color TV, video or DVD player, and sound system. Many also have personal computers, cell phones, and other consumer items, but urban families often eschew cars because parking spaces are prohibitively expensive.

***An Aging Society.*** Because of Japan's healthy low-fat, high-protein diet and good medical care, among other factors, Japanese life expectancy is the longest in the world: for men, 78 years; for women, 85 years. This longevity, combined with a low birth rate, has given Japan the world's largest proportion of elderly people (see Box 9.1 on page 486). According to estimates for 2005, 20 percent of the Japanese population is 65 years old or older, compared with 17 percent for the United Kingdom, 12 percent for the United States, and 8 percent for China. Thus Japan has a disproportionately large and growing social security obligation to its elders that is becoming harder to fund over time. The country also faces a challenge in providing a full range of services and facilities for the aged.

It has long been the practice in Japan that most older people who cannot fully care for themselves are cared for by their adult children. This pattern is being eroded, however, as younger generations find that they lack space in their dwellings for their parents or grandparents or prefer privacy to traditional family obligations. Thus a relatively new industry of nursing homes and other elder care facilities is growing in Japan.

## Working in Japan

Japan's distinctive culture is reflected in the working lives of its people as well as in their attitudes toward their own work and toward the outsiders who perform many of the most menial jobs.

*Traditional Work Patterns.* The Japanese are renowned for their hard work, high-quality output, and devotion to their jobs. People's lives are shaped by the companies that employ them. The norm has been that an employer provides lifetime employment, regular paid vacations, a pension upon retirement, and often, subsidized housing and perhaps even a graveyard for employees. Job-hopping is considered reprehensible, and loyalty to the employer is reinforced by personnel management that instills camaraderie and conformity. The prototypical Japanese corporate employee is a "salaryman"—a male, usually with a family he rarely sees, who lives to work. Individualism is regarded as selfish, and making innovative suggestions is discouraged. Employees who have good ideas may not present them for several years, or may give credit to someone else. The Japanese language does not even have a word for "entrepreneur"; only recently has the English term become a buzzword in Japanese.

This norm is now changing in the face of larger economic and social changes. During the recession of the 1990s, many companies had to reduce their workforces and cut back employee benefits. Moreover, some younger male and female graduates in business and technological fields have turned away from the old pattern in favor of the freedoms of temporary employment. Although they may settle for a lower standard of living and give up job-related benefits, women enjoy less gender discrimination in temporary jobs, and temporary workers do not feel compelled to acquiesce to their superiors or work overtime. This freelance approach to the world of work allows them time to start their own businesses and perhaps spend more time with their families. However, whether by choice or necessity, accepting a temporary or part-time job is risky. The old loyalty system makes it difficult to find a new job, and firms worried about industrial security are often unwilling to hire someone who has worked for a competing company.

*Foreign Workers.* Foreign workers from poor countries come to Japan as guest workers to take some of the hard, dirty, dangerous, and low-paying jobs that today's Japanese workers avoid. They come from places such as China, the Philippines, Pakistan, Bangladesh, Sri Lanka, Iran, Peru, Bolivia, Brazil, and Africa. Some Latin American immigrants are descendants of Japanese who went abroad in search of work some generations ago, when Japan was a poor country. Foreign men often work in construction, in factories, and as dishwashers in restaurants; women work as hotel maids, cleaners, and in other low-status jobs. Many foreign women are also employed as hostesses in bars and as sex workers. The recruitment of

women for sex work has become a controversial issue because unscrupulous "agents" force women into sex-industry jobs after promising them other kinds of work.

## TAIWAN

Taiwan is located a little over 100 miles (160 kilometers) off China's southeastern coast (Figure 9.53). With an area of 14,000 square miles (36,000 square kilometers) and almost 23 million people, Taiwan is a crowded place. A mountainous spine runs from the northeastern corner to the southern tip of the island, with a rather steep escarpment facing east and a long, gentler slope facing west. Most of the population lives on the western side, especially at lower elevations along the coastal plain. Increasingly, as farming declines, people are becoming concentrated in a few urban centers. The greatest concentration is in the far north, in the area surrounding Taipei, the capital.

Despite its small size and small population, Taiwan is one of the most prosperous countries of the **Asia–Pacific region,** a huge trading area that includes all of Asia and the countries around the Pacific Rim. It is fourteenth globally in the size of its foreign trade, and its GDP per capita in 2003 was U.S.$29,000. After the Communist revolution in China, Taiwan took advantage of its ardent anti-Communist stance, its burgeoning refugee population, and its geographic location close to the mainland to draw aid and investment from Europe and America and eventually Japan. The island's economy quickly changed from overwhelmingly rural and agricultural to mostly urban and industrial. Many industries made products for Taiwan's impressive export markets. Then, slowly, the economic

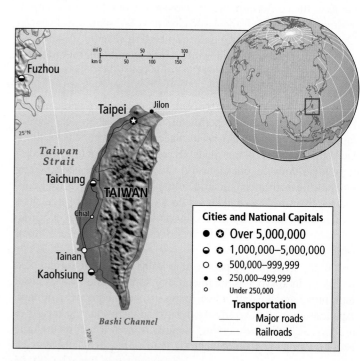

**FIGURE 9.53** The Taiwan subregion.

emphasis changed again, from labor-intensive industries to high-tech and service industries requiring education and technical skills. By 1999, only 8 percent of the people were still working in agriculture.

By the mid-1990s, the domestic economy was expanding rapidly, and local buyers were absorbing a major proportion of Taiwan's own production: home appliances, electronics, automobiles, motorcycles, and synthetic textiles. The Taiwanese performed a neat pirouette to make this high rate of local consumption possible. As Taiwan lost its labor-intensive industries to cheaper labor markets throughout Asia—including mainland China—Taiwanese entrepreneurs built factories and made other investments in the very places that were giving them such stiff competition: Thailand, Indonesia, the Philippines, Malaysia, Vietnam, and the new SEZs in Southern China. Hence Taiwan remained competitive as a rapidly growing economy with strong export markets and also profited from dealing with the less advanced, but emerging, economies in the region.

Taiwan's likely future role in the wider world is a hotly debated question. As mainland China emerges as a huge world power, tiny Taiwan can no longer promote the idea that it speaks for (or is) China. On the other hand, Taiwan is perfectly situated to partici-

pate in and enhance the development of mainland Chinese markets and related efforts to integrate the Asia–Pacific economy.

## KOREA, NORTH AND SOUTH

Some of the most enduring international tensions in East Asia have focused on the Korean Peninsula. As we have seen, after centuries of unity under one government, the peninsula was divided in 1953. The two resulting countries are dramatically different. Communist, inward-looking, poor North Korea does not participate in the global economy, while more cosmopolitan, newly democratic, affluent South Korea has pursued a model of state-aided capitalist development.

Physically, Korea juts out from the Asian continent like a downturned thumb (Figure 9.54). Two rivers, the Yalu (Amnok in Korean) and the Tumen, separate the peninsula from the Chinese mainland and a small area of Russian Siberia. Low-lying mountains cover much of North Korea and stretch along the eastern side of the peninsula into South Korea, covering nearly 70 percent of the peninsula. There is little level land for settlement in this mountainous

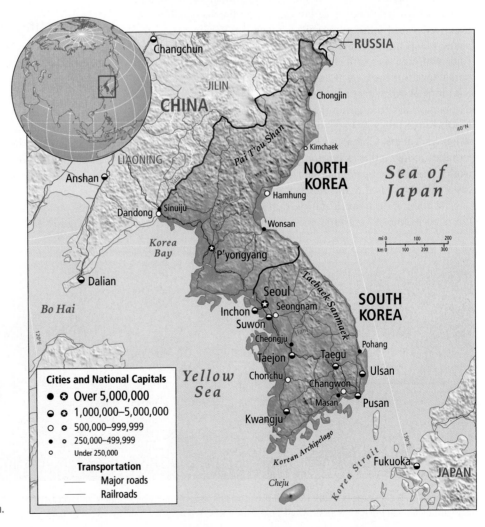

**FIGURE 9.54** The Korea subregion.

zone. The rugged terrain disrupts ground communications from valley to valley. Along the western side of Korea, floodplains slope toward the Yellow Sea, and most people live on these western slopes and plains. Although the peninsula is surrounded by water on three sides, its climate is essentially continental because it lies so close to the huge Asian landmass. The same cyclical monsoons that "inhale" and "exhale" over the Asian continent (see pages 477–478) bring hot, wet summers and cold, dry winters.

The Korean Peninsula was a unified country as early as 668 C.E. Scholars believe that present-day Koreans are descended primarily from people who migrated from the Altai Mountains in western Mongolia because the Korean language appears to be most closely related to languages from that region. Other groups—Chinese, Manchurians, Japanese, and Mongols—invaded the peninsula, sometimes as settlers, other times as conquerors. The Buddhist and Chinese Confucian values brought by some of these groups influenced Korea's educational, political, and legal systems. Korea is noted for early advances in mathematics, medicine, and printing.

## Contrasting Political Systems

Although an armistice brought the Korean War to an end in 1953, North Korea and South Korea each assumed a strong nationalist stance, which has resulted in 50 years of hostile competition between the two. Both governments adopted the Korean concept of *juche*, which means self-reliance or "the right to govern yourself in your own way." In North Korea, *juche* was interpreted as unquestioned loyalty to the Great Leader, Kim Il Sung, and later to his son, Kim Jong Il, who succeeded him in 1997. There is no pretense of democracy, and the government remains a military dictatorship. North Korea limits its trade and other involvement with its continental neighbors, China and Russia. Its restrictive internal and external economic policies make it one of the poorest nations in the world. Despite its apparent desire for isolation, North Korea has occasionally threatened South Korea and its allies, especially Japan.

In 2002, North Korea provoked global concern when it announced its intent to resume production of enriched plutonium, forced UN monitors to leave the country, and then, in early 2003, withdrew from the Treaty on the Non-Proliferation of Nuclear Weapons. Most intelligence assessments confirm that North Korea tested its first nuclear device on October 9, 2006, but the extent of its capacity and the number of weapons it has remain uncertain. Regardless of whether Kim Jong Il is using the nuclear issue as a ploy to negotiate aid from the West (substantial aid was granted in 2007) or is preparing for an attack he believes the United States plans to launch, it is generally agreed that the presence of nuclear arms in North Korea is cause for concern worldwide.

In South Korea, *juche* has come to mean vigorous individualism, coupled with pride in and loyalty to one's own people and nation. Social criticism and even aggressive protests by labor unions and other social movements are allowed, with the understanding that one's ultimate loyalty is still to South Korea. Politically, South Korea allied itself with the United States, Japan, and Europe shortly after World War II and sought economic revival through foreign aid and capitalist development. But despite its economic success, South Korea was governed by a series of military dictatorships until 1997, when democratic elections were held. Kim Dae Jung, an outspoken critic of the military who had endured years of harassment and imprisonment for his convictions, was elected president.

## Contrasting Economies

The two Koreas differ dramatically in their economic resources. North Korea (with a population of 23 million) has better physical resources for industrial development, including forests and deposits of coal and iron ore. The many rivers descending from its mountains have considerable potential to generate hydroelectric power. South Korea (with a population of 48 million) has better resources for agriculture because its flatter terrain and slightly warmer climate make it possible to grow two crops of rice and millet per year. Nevertheless, in recent decades, South Korea has surpassed North Korea in both agricultural and industrial development. Its electronic, automotive, chemical, and shipbuilding industries successfully compete with those in the Americas, Europe, and Japan.

**FIGURE 9.55** A cityscape in South Korea. Taken from an observation tower (shadow at lower left), this late afternoon panoramic photo of the harbor and city of Pusan, South Korea, reveals the modern city at its best. [© Tan Yilmaz.]

**FIGURE 9.56** Life in North Korea. The government of North Korea seeks to divert people's attention from the harsh realities of life with extravagant ceremonies and festivals. The annual Arirang Festival, running 6 nights a week for 2 months, celebrates the history of North Korea and the administration of its late president, Kim Il Sung, who died in 1994 but retains the title of Eternal President. [© David Astley.]

A major reason for South Korea's economic success has been the formation of huge corporate conglomerates known as *chaebol*. These conglomerates include such internationally known companies as Samsung and Hyundai. South Korea's government has assisted the *chaebol* by making credit easily available to them and by helping them purchase foreign patents so that Koreans could focus on product quality and marketing rather than on inventing. Nevertheless, the *chaebol* have come under increasing criticism in recent years because their close connections with the government have led to corruption. Unwise loans and investments have disrupted the South Korean economy. Laws restricting *chaebol* finance deals were enacted in 2000. One of the largest *chaebol*, Daewoo, was dismantled because it was deeply in debt and had too many layers of management.

Despite these problems, the South Korean economic system has worked well enough that South Koreans, who were extremely poor at the end of the Korean War, can now afford the products they formerly only exported. For example, even in rural areas, where incomes are comparatively low, 90 percent of households have a refrigerator, an electric rice cooker, and a propane gas range, and many have telephones and color television sets. Meanwhile, emerging urban landscapes, built in recent years, testify to South Korea's success in the global economy (Figure 9.55).

North Koreans have benefited from some Communist social policies, particularly access to basic health care and education. In general, however, as the economy has deteriorated, these services have also declined. North Korea's industries are inefficient and its workers poorly motivated. Furthermore, in curtailing trade with the Soviet Union, the country lost markets and access to raw materials, fuel, and technical training. Poor harvests on collective farms and recurrent cycles of floods and droughts brought extensive famines from 1995 through 2001. Although the government has been unwilling to release information, as many as 2 million people may have died of hunger over that 6-year period. Life expectancy has declined, and the infant mortality rate increased between 2001 and 2003. The government has responded to these dire situations by diverting the public's attention with grand synchronized dance and music extravaganzas featuring 100,000 performers (Figure 9.56). Attendance is required.

The future of the Koreas remains uncertain. Closer cooperation and even reunification are episodically discussed in between crises over nuclear armaments. Reunification would bring economic benefits to both sides. For example, both countries need more energy—North Korea to begin development, South Korea to grow its industrialized economy so as to better compete with Japan, Taiwan, and China. South Korean investment would allow for the building of hydroelectric dams on sites in North Korea, to the benefit of both countries. However, rejoining with North Korea would require many sacrifices from South Koreans, just as it has from West Germans as they attempt to merge their economy with that of much poorer, formerly Communist East Germany. Because of the stark difference in wealth between the two Koreas, money that might have been used to develop South Korea further would have to be used to pay for basic improvements in North Korea. Poor refugees might flood into South Korea, lowering standards of living there even more.

## MONGOLIA

**Personal Vignette**  As their truck descended onto a broad plain within the high Altai Mountains of southern Mongolia, the visitors saw two horsemen herding hundreds of sheep. One of the horsemen, a sun-tanned man in his sixties, rode over, dismounted, and greeted the two Americans (and their interpreter) with a smile and a query about their state of health. He wore a traditional knee-length coat, fastened at the side and the shoulder with buttons and belted with a silk sash, and Western-style black leather boots. Anthropologists Cynthia Beall and Melvyn Goldstein were to discover that his easy, confident demeanor was typical of Mongolians. He seemed amused to have found these strangers popping up among the sheep on this normally lonely sweep of grazing land.

They had spoken for only a few moments when the Mongolian herder said cheerfully, "You know, I heard on the radio that your Foreign

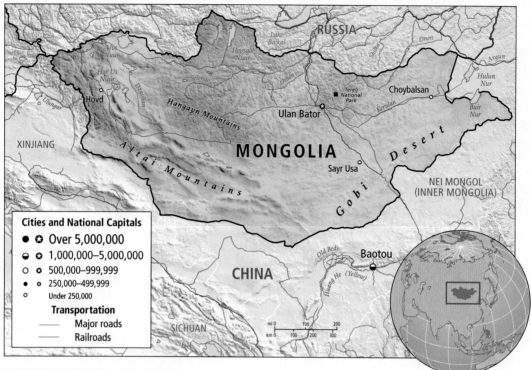

**FIGURE 9.57** The Mongolia subregion.

Minister has visited our capital and that our two countries are now friends. That is good. Please come to visit my camp later. It's not far from [where you are headed]. We can talk more then. I have many questions to ask you about America, and I have a lot to say about [how things are changing in Mongolia]." Goldstein and Beall admit that this encounter shattered their assumptions that Mongolians would be wary of outsiders, hesitant to express themselves, and unaware of the outside world.

*Source: Melvyn C. Goldstein and Cynthia M. Beall, The Changing World of Mongolia's Nomads (Berkeley: University of California Press, 1994).* ■

With just 2.6 million people, Mongolia occupies a territory so large (Figure 9.57) that its average population density is nearly the lowest on earth, at 4 people per square mile (1.5 per square kilometer). For thousands of years, the economy has been based on the nomadic herding of sheep, goats, camels, horses, and yaks. Today, nearly half the people are still engaged in rural activities in some way related to this traditional lifestyle. The other half now live in cities, primarily the capital, Ulan Bator. Urban workers are employed in a wide range of services and in industries related to the processing of minerals from Mongolia's mines and such animal products as hides, fur, and wool. In the 1990s, Mongolia's transition from Communist central planning to a market economy was complicated by an economic recession that forced some people working in urban jobs to return to the countryside.

The Mongolian Plateau lies in the heart of Central Asia, directly north of China and south of Siberia. It is high, cold, and dry, with an extreme continental climate. The physical geography of Mongolia can be broken down into four major zones. In the far south and extending across the border into China is the Gobi Desert—actually a very dry grassland that grades into true desert in especially dry years or where it is overgrazed. To the west and northwest of the Gobi is a huge, rolling, somewhat moister grassland. The remaining two zones are Mongolia's two primary mountain ranges: the forested Hangayn Mountains, in north-central Mongolia, and the grass- and shrub-covered Altai Mountains, which sweep around to the west and south and into the Gobi Desert (Figure 9.58).

## History

The Mongolians are apparently descended from groups of people who have occupied the massive mountains and plains of Central Asia for more than 40,000 years. Nomadic pastoralism is one of the most complex agricultural traditions in the world and dates at least as far back as the domestication of plants (8000 to 10,000 years). Nomadic pastoralists must understand the intricate biological requirements of the animals they breed, and they must also know the ecological and vegetation cycles in the landscapes they traverse with their herds.

The present country of Mongolia is the north-central part of what was once a much larger culture area in eastern Central Asia known by the same name. Between 1206 and 1370, the Mongols, under Genghis Khan and his son Kublai Khan, created the largest ever land-based empire, stretching from the eastern coast of China to central Europe, including parts of modern Russia and Iran. While in control of China, the Mongols improved the status of farmers, merchants, scientists, and engineers; fostered international trade;

**FIGURE 9.58** Terelj National Park. This national park is located about 50 miles (80 kilometers) north of Mongolia's capital, Ulan Bator, in the Hentiyn Mountain range, which extends into Siberia. The park has a triple purpose: one part is closed to the public in order to preserve natural landscapes for wildlife; a second part is open for enthusiasts of "extreme" ecotourism, especially hiking, rock climbing, rafting, and skiing; and a third part is open to modern tourism, complete with car parks, *gers* (traditional felt tents used by nomads), a meat-based diet, TV, discos, and overpriced goods. [Andrew McConnell/Alamy.]

and broke the control of traditional elites by abolishing their automatic access to privilege. They refined the manufacture of textiles, jewelry, and blue and white porcelain, and trade in these wares along the Silk Road flourished. Although the Mongols brought many important reforms to China, they also practiced authoritarian rule and discrimination against ethnic Chinese. Deposed by the Chinese in 1366, the Mongols retreated to the north; over the next 300 years, their control of territory in Central Asia and Europe dwindled. Eventually, the southern part of Mongolia became part of China (Inner Mongolia).

By 1900, after centuries of Chinese rule, Mongolians were among some of the poorest people in Asia. Profits from herding went to the Chinese feudal elite, and trade was monopolized by Chinese merchants who sent their profits home rather than investing in Mongolian development. Influenced by the Russian Revolution of 1917, and after considerable turmoil, Mongolia first declared independence from China in 1921 and 3 years later became a Communist republic under the Soviet sphere of influence.

## The Communist Era in Mongolia

From 1924 to 1989, Mongolia sought guidance from Soviet advisers and technicians. They helped set up a system of central planning that reorganized the nomadic pastoral economy according to socialist policies, but without drastic disruption of traditional lifeways. Rural households of extended families became economic collectives. Some households continued to herd and breed animals, others engaged in sedentary livestock production, and still others farmed crops. During this time, Mongolia also began to industrial-

ize and urbanize. By 1985, nomadic herding and forestry were declining, accounting for less than 18 percent of the Mongolian GDP and less than 30 percent of the labor force, whereas employment in mining, industry, and services accounted for about half of the GDP and a third of the labor force.

The Mongolian economy was supported by the Soviets, who subsidized a wide array of social services. By 1989, there was nearly universal education through middle school, and adult literacy had reached 93 percent. Health care was available to all, and life expectancy had risen dramatically. The previously high infant mortality rates dropped, and then fertility decreased. That erudite middle-aged nomadic herder who impressed the American anthropologists in the early 1990 s was undoubtedly a beneficiary of these educational and social reforms.

## Recent Economic Issues

The collapse of the Soviet Union in 1991, coupled with several natural disasters and lower world prices for the minerals that Mongolia produces, led to immediate budget problems and declines in economic growth, human well-being, and social order. The Soviet advisers left without training Mongolian replacements to run the economy. Layoffs and factory closings caused a severe decline in living standards. Between 1990 and 1992, education, for example, was cut by 69 percent. Kindergartens and schools closed, and many older students left school to help their families. Throughout the 1990s, young adults remained unemployed and rootless, some turning to substance abuse for solace. Street children, the outcasts of disintegrating families, proliferated in Ulan Bator. By 2003, about

**FIGURE 9.59** Tradition and modernity in Ulan Bator. In July 2006, poor families newly arrived in Ulan Bator, having given up a life of nomadic pastoralism on the lightly populated grasslands of Mongolia, pitched their *gers* in the shadow of the modern industrial "wilderness" of the capital city. [Barry Lewis/CORBIS.]

one-third of the population lived in poverty (Figure 9.59). Now there are some signs of a better future. Mongolia's new membership in the World Trade Organization is bringing increased international attention and development assistance, and private foreign investors are beginning to take an interest in the country, especially in its minerals. China's booming economy just to the south will be a ready market.

## Gender Roles

Traditionally, Mongolian women enjoyed a status approximately equal to that of men, but outside forces, such as Lamaist Buddhism in the sixteenth century and Chinese rule in the seventeenth and eighteenth centuries, eroded their status. The Communist era restored, or even enhanced, egalitarian gender attitudes. The daily work of women herders was valued by the government, and, like their husbands, they were eligible for old-age pensions. A woman could leave an unsuccessful marriage because there was support for her and her children. A few women became teachers, judges, and party officials, and soon women were represented in institutions of higher education.

In today's modern nomadic pastoral economy, as in the past, women usually choose which stock to breed, tend the mother animals and their young, preserve meat and milk, and produce many essentials from animal hides and hair. Women provide the materials for the construction of *gers* and also dye and weave beautiful furnishings for the interiors. Both boys and girls are taught to ride horses at an early age, and both help with the herding. As they mature, women become more responsible than men for the routines of daily life in the home. Men engage primarily in pasturing the herds and arranging the marketing of the animals and animal products. Men also perform other tasks—such as the occasional planting of barley and wheat—that are carried out beyond the home compound.

As a result of the jolting changes when the Soviets left and the economy collapsed, women lost some of the status they had enjoyed, but they are now regaining legal protections and access to education, jobs, health care, housing, and credit. And women entrepreneurs are emerging. As in so many other countries that are moving from Communism to a market economy, the informal economy—in which women are particularly active—has been an essential component of the Mongolian economic transition.

## Reflections on East Asia

The opening of China to the global economy and to outside influences presents challenges for all of the world's economies, but especially for those of China's East Asian neighbors—Japan, Taiwan, and South Korea. Already invested in China's interior and coastal SEZs and ETDZs, their expectation of profiting from China's debut into capitalism must be tempered with caution because it is not yet clear how the rapid social changes occurring in China will affect either its own economy or the global economy. Economic liberalization in China thus far has increased wealth disparities drastically, bringing real hardship to millions. The shortages of skilled labor just becoming apparent could benefit workers because competition for scarce labor is likely to produce better wages and working conditions. This change will also hasten a shift toward more technologically sophisticated and higher-quality goods as China tries to maintain an edge in the global economy relative to the cheaper labor markets of Southeast Asia and Africa.

As Asians grow older, some of the standard popular images of population issues in Asia are now largely obsolete, even though China clearly has more people than it can support sustainably if its standards of living increase. The one-child-per-family policy has reduced growth drastically and has apparently produced an urbanizing, upwardly mobile generation that seriously considers having *no* children at all. China's population is likely to begin a downturn in the foreseeable future; Japan's is already shrinking (down just a bit in 2006). To cope with labor shortages, shrinking family sizes, an increasing proportion of elderly people, and the fact that working women are no longer available to care for children and elders, governments may have to spend more on social services and be content with slower economic growth.

Security across the region is in some question. The nuclear rumblings of North Korea pose the only explicit threat. After threatening Taiwan's sovereignty in 2005, China has backed off,

and Taiwan is unlikely to initiate trouble with a country in which its citizens have invested heavily. Japan, legally banned from having an army since World War II, is cautiously building a modest peacekeeping force so that it can participate in UN collective security efforts.

Finally, East Asia, along with the rest of the industrialized world, will be considering how to balance the desire for a clean, safe environment with the desire for consumerist lifestyles. In East Asia, at once the place where so many of the world's consumer goods are produced and where so much pollution is generated, something will have to give.

## Chapter Key Terms

Ainu 502

alluvium 515

Asia–Pacific region 524

*chaebol* 527

Confucianism 479

Cultural Revolution 482

economic and technology development zones (ETDZs) 494

floating population 474

food security 491

food self-sufficiency 522

*ger* 516

Great Leap Forward 482

growth poles 494

*hukou* system 474

*kaizen* system 488

loess 509

*qanats* 517

regional self-sufficiency 490

responsibility system 491

shogun 482

soft power 496

special economic zones (SEZs) 494

state-aided market economy 487

tsunami 476

typhoon 507

wet rice cultivation 512

yurt 516

## Critical Thinking Questions

1. Why is the interior west of continental East Asia (western China and Mongolia) so dry and subject to extremes in temperature?

2. Why do the island and peninsula countries of Japan, the Koreas, and Taiwan still experience some continental climate conditions?

3. Discuss how an understanding of the ancient history of East Asia can be useful in analyzing its present circumstances. Think about issues of political hierarchies, land use, attitudes toward the environment, gender relationships, and the power of the people.

4. What were the motives of European countries in making overtures to trade with East Asian countries in the seventeenth and eighteenth centuries?

5. Discuss some of the early strategies used by the Chinese Communist Party to reorganize the Chinese economy in the areas of agricultural policy, industrial policy, labor management, and urbanization. Assess the success of these efforts and compare them with current policies in these four categories.

6. Contrast Japan's pre–World War II policies in East Asia with its present role in the region. What are the principal similarities or differences?

7. Judging from what you have read about the Koreas, explain why there are now two countries on the Korean Peninsula. Then compare and contrast the two countries.

8. Contrast Taiwan's economic interests in mid-twentieth-century China with its present relationship to the mainland.

9. China used to be afflicted with recurring famines. What is the situation now? What are the present worries about food security? China now exports food, so shouldn't that mean that the country has sufficient food?

10. China continues to use a lot of coal despite the fact that its air is already polluted by coal burning. What are its future prospects for stopping the use of coal?

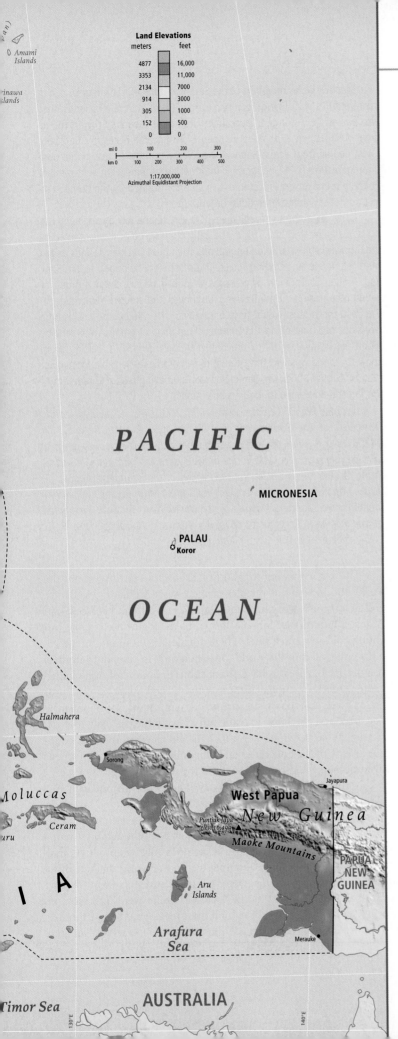

**Land Elevations**

| meters | feet |
|---|---|
| 4877 | 16,000 |
| 3353 | 11,000 |
| 2134 | 7000 |
| 914 | 3000 |
| 305 | 1000 |
| 152 | 500 |
| 0 | 0 |

1:17,000,000
Azimuthal Equidistant Projection

PACIFIC

MICRONESIA

PALAU
Koror

OCEAN

Halmahera

Sorong

Moluccas

Ceram

uru

West Papua

New Guinea

Puntjak Jaya
elev. 16,499

Maoke Mountains

Jayapura

I A

Aru
Islands

PAPUA
NEW
GUINEA

Arafura
Sea

Merauke

AUSTRALIA

Timor Sea

# CHAPTER 10

# SOUTHEAST ASIA

**Global Patterns, Local Lives** In December 2005, a group of indigenous forest people in the state of Sarawak, on the island of Borneo, in Malaysia, attended a public meeting wearing orangutan masks. They carried signs informing on-lookers that, although the government protects orangutans, it ignores the basic right of indigenous people to be safe in their own ancestral lands.

Over the years, Sarawak forest dwellers have tried many tactics to save their lands from destruction by logging companies and oil palm and pulpwood plantations. The state began licensing logging companies and plantations in the mid-twentieth century, and by the 1980s, the forest people were experiencing landslides, polluted water, meager hunts, and decreased fish catches. To stop the destruction, they blocked logging roads with felled trees and their own bodies. But with government support, the companies removed the blockades and had the protesters arrested. Although news stories were widely circulated, the outside world paid little attention to the fracas, and the forest removal continued until 90 percent of Sarawak's lowland forests had been degraded and 30 percent had been clear-cut.

In the 1990s, a group of citizens in Berkeley, California, organized the Borneo Project. They offered to become a "sister city" to the Uma Bawang longhouse (a longhouse is a Sarawak forest group that lives in a traditional linear residential structure), giving help wherever it was needed. It soon became evident that an inventory of how the Uma Bawang used the forest could be crucial in validating their claim to their land and perhaps could keep timber extractors out.

The forest dwellers explained in detail to the people from Berkeley how they used thousands of forest resources. They plotted where the wild boar feed and where deer gather for salt licks. Photographs documented the oral information, but most useful was a community-based mapping project. In 1995, using rudimentary compass and tape techniques, the Uma Bawang, with the help of the Berkeley group, began mapping both the extent and content of their forest home (Figure 10.2). The practice spread, and in 2001, similar maps helped the Iban of central Sarawak win a precedent-setting court case. (The ruling was overturned in 2005, but a final decision is pending at the Malaysian federal level.)

**FIGURE 10.1** Regional map of Southeast Asia.

(a)

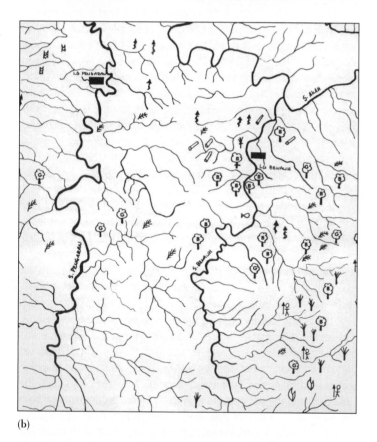

(b)

**FIGURE 10.2** Villagers mapping their home territory. (a) These villagers first gathered to discuss and draw the boundaries of their lands (in Penan, Sarawak). (b) Then they added important features such as huts (some of which are abandoned), sago palms, hunting areas, and graveyards, as well as physical features. [Bruno Manser Fonds.]

Meanwhile, as members of the Sarawak state legislature grew rich in joint ventures with foreign companies logging on indigenous lands, they also passed new stringent land survey standards meant to thwart the indigenous mappers. Upgraded mapping skills became necessary. By 2005, forest people from across Sarawak were learning how to use global positioning systems (GPS), geographic information systems (GIS), and satellite imagery to make sophisticated maps, an effort funded by the Borneo Project and other nongovernmental organizations.

Early on a spring morning in 2006, it happened again, as it has across Sarawak since the 1980s: the early morning quiet was shattered by the unmistakable roar of an engine and then loud cracking sounds as tall trees fell in the surrounding forest. The machinery belonged to the Samling Timber Corporation, which sells to British buyers. Soon a group of young men from the Uma Bawang longhouse had erected a barricade. In the past, the protest would have ended with their quick arrest. This time word got out quickly via the Internet. Hundreds of emails were sent by global environmentalists to the British timber-buying firm, letting them know the world was watching. Could the emails be responsible for the fact that, in August 2006, the barricade was still making a sizable remnant of the forest inaccessible to logging operations?

*Adapted from The Borneo Wire, Spring 2006, Jessica Lawrence, editor, Newsletter of The Borneo Project, 1771 Alcatraz Avenue, Berkeley, CA 94703, http://Borneoproject.org; "Rainforest dwellers successfully maintain logging road blockade in one of Malaysia's last virgin jungle areas," Bruno Manser Fonds, Society for the Peoples of the Rainforest. August 15, 2006, http://www.bmf.ch/en/en_index.html; Mark Bujang, "A community initiative: Mapping Dayak's customary lands in Sarawak," presented at the Regional Community Mapping Network Workshop, Nov. 8–10, 2004, Diliman, Quezon City, Philippines.* ◼

The Uma Bawang longhouse has distant cousins around the world: the indigenous peoples of the forests, wildlands, wetlands, and deserts of North and South America, North and sub-Saharan Africa, Central Asia, Siberia, the Himalayas, and the remote hill country of India and China. There are literally thousands of indigenous groups left in the world, but often only in remnant numbers (Figure 10.3). All of us have connections to Southeast Asia's native people, who, in ancient times, domesticated some of our most prized foods and spices: sugarcane, nutmeg, cloves, yams, bananas, rice, pigs, and chickens, to name but a few. In the past half-century, the region has become a key link in the global economy, and it supplies many of the consumer and electronic goods sold at inexpensive prices in the shops and malls of the world.

In the 1970s, after long decades of European colonialism and then war and political violence, Southeast Asia entered an era of rapid modernization. Governments and local investors began to aggressively market the minerals and forests of the region's mainland and islands, and they built factories making products that eventually became highly competitive in the world market. Businesses and governments were courting investors from Japan, Europe, Australia, and North America. Some were interested in profiting from the region's natural resources, some from its industrial and tourism potential. By the 1990s, the forests of Thailand and the Philippines were depleted, and logging was in full swing

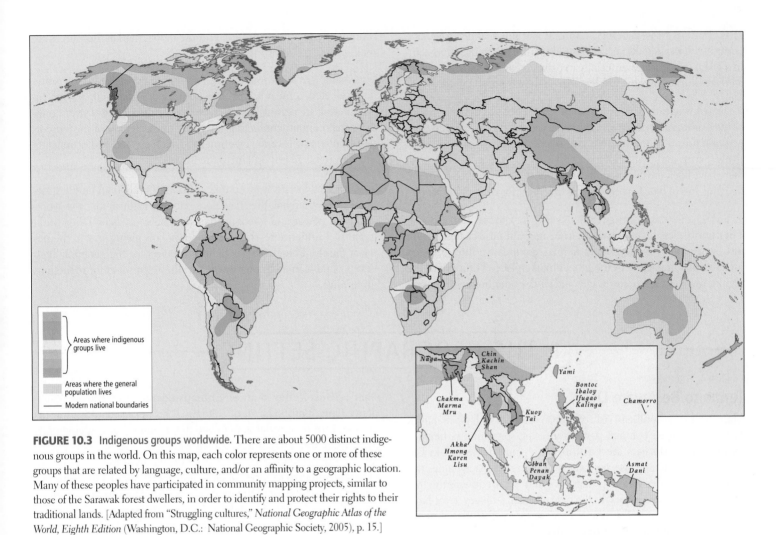

**FIGURE 10.3** Indigenous groups worldwide. There are about 5000 distinct indigenous groups in the world. On this map, each color represents one or more of these groups that are related by language, culture, and/or an affinity to a geographic location. Many of these peoples have participated in community mapping projects, similar to those of the Sarawak forest dwellers, in order to identify and protect their rights to their traditional lands. [Adapted from "Struggling cultures," *National Geographic Atlas of the World, Eighth Edition* (Washington, D.C.: National Geographic Society, 2005), p. 15.]

in Malaysia and Indonesia. Timber flowed out of the region's forests toward the port of Singapore and on to building sites in Europe, Japan, and the Americas. Millions of farmers and even some reclusive forest people came streaming into towns, quickly turning them into crowded cities festooned with shantytowns. Many found work in factories, and although their wages of a few dollars a day were very low by Western standards, they were enough to raise standards of living substantially. With increasing government revenues and private profits, a building boom ensued. Magnificent banks, hotels, and office buildings began to rise in capital cities across the region.

By the early 1990s, Southeast Asia was regarded as a model of rapid development that other regions might emulate, but by the late 1990s, that model was tarnished. First, Western consumers became aware of the sweatshop conditions under which the region's products were manufactured. Protests followed on U.S. college campuses and at meetings of the World Bank and World Trade Organization, which were seen as encouraging the growth of such conditions. Second, the environmental costs of rapid industrialization were revealed: deforestation to make room for rubber trees and oil-producing palms; soil erosion caused by loss of forest cover;

and heavy air pollution from fossil fuels and the burning of brush and peat on cutover forestlands. Finally, in 1997, a widespread economic recession hit. Thousands of factories closed, at least temporarily, and families that depended on factory wages had to turn to the informal economy and migration to pay for necessities. Logging, however, expanded throughout the recession as foreign demand for fine tropical hardwoods remained strong. By 2005, Southeast Asian factories were once again increasing production, and logging was expanding still further.

## Themes to Explore in Southeast Asia

This chapter discusses the characteristics of Southeast Asia and the successes and problems the region has faced in recent years. Our discussion will include the following themes:

1. **Distinct culture groups.** Many distinct culture groups in Southeast Asia have lived side by side for centuries, absorbing a spectrum of cultural influences from the outside world, yet retaining their uniqueness.

2. Disparities in well-being. Cultural and physical diversity have contributed to wide disparities in wealth and well-being, often within a given country and especially in urban areas. Nonetheless, overall disparities in income are less than in other world regions.

3. Transformation of the tropical environment. Both the rich soils of river valleys and deltas and the much poorer soils of cleared tropical rain forests in the uplands and islands are being farmed intensively. Farms and plantations support large numbers of people, but there have been environmental consequences, including soil erosion, flooding, loss of biodiversity, and loss of habitat for indigenous people.

4. Labor and resources geared toward the global marketplace. The countries of the region range from agricultural societies that are just beginning to modernize to fiercely competitive industrialized societies that are active in the global marketplace. Throughout the region, workers receive low wages and work under unhealthy condi-tions while producing products for the global market. Resources are being extracted at unsustainable rates.

5. Challenges to national unity. Authoritarian governments—some fairly elected, some not, and often buttressed by the military—are encountering increasing difficulty in maintaining control over countries that contain conflicting interest groups. Separatist movements are challenging several governments, which are accused of violating human rights in responding to the threats.

6. Changes in family structure. Although traditional family structures are patriarchal, they have some unique features that give women more power and freedom than they have in many other world regions. Furthermore, the extended family is giving way to the nuclear family. These changes help to account for the generally high levels of education among women in the region and for reductions in fertility.

---

## I THE GEOGRAPHIC SETTING

### Terms to Be Aware Of

Many people in Southeast Asia choose to dispense with place-names that connote their colonial past. As a result, some names that are familiar to Westerners are not the names now preferred in the region. We generally use the officially preferred place-name; the older, more familiar Western name is given in parentheses the first time a place-name appears. In the case of Burma, however, we have made an exception to this practice: Myanmar is the official name used by the military government that seized control in a coup d'état in 1990, but Burma, the country's traditional name, is preferred by the citizenry; hence we use *Burma* in this text. Borneo is a large island that is shared by three countries: Kalimantan is part of Indonesia, Sarawak and Sabah are part of Malaysia, and Brunei is a very small, independent, oil-rich country.

### PHYSICAL PATTERNS

The physical patterns of Southeast Asia have a continuity that is not immediately obvious on a map of the region. On a map, one sees a unified mainland region that is part of the Eurasian continent and a vast and complex series of islands arranged in chains and groups. The apparent contrast between mainland and islands obscures the fact that these landforms are related in origin. Moreover, despite the large territory it covers, most of the region shares a tropical or subtropical climate.

### Landforms

Southeast Asia is a region of peninsulas and islands (see Figure 10.1). Although the region stretches over an area larger than the continental United States, most of that space is ocean; the area of all the region's land amounts to less than half that of the contiguous United States. The large mainland peninsula, sometimes called Indochina, that extends to the south of China is occupied by Burma, Thailand, Laos, Cambodia, and Vietnam. This peninsula itself sprouts a long, thin peninsular appendage that is shared by outlying parts of Burma and Thailand, a part of Malaysia, and the city-state of Singapore, which is built on a series of islands at the southern tip. The **archipelago** (a series of large and small islands) that fans out to the south and east of the mainland is grouped into the countries of Malaysia, Indonesia, and the Philippines. Indonesia alone has some 17,000 islands, and the Philippines has 7000. The independent country of Brunei shares the large island of Borneo with Malaysia and Indonesia. Timor-Leste (East Timor), which gained independence in 2002, shares the island of Timor with Indonesia.

The irregular shapes and landforms of the Southeast Asian mainland and archipelago are the result of the same tectonic forces that were unleashed when India split off from the African Plate and crashed into Eurasia (see pages 21–22). As a result of this collision, which is still under way, the mountainous folds of the Plateau of Tibet, which reach heights of almost 20,000 feet (6100 meters), bend out of the high plateau and turn south, descending rapidly and then fanning out to become the Indochina peninsula. The gorges widen into valleys that stretch toward the sea, each containing hills of 2000 to 3000 feet (600 to 900 meters) and a river or two flowing from the mountains of China to the north. The major rivers of the peninsula are the Irrawaddy and the Salween in Burma; the Chao Phraya in Thailand; the Mekong, which flows through Laos, Cambodia, and Vietnam and forms the border of Thailand and Laos for more than 620 miles (1000 kilometers); and the Black and Red rivers of northern Vietnam.

The curve formed by Sumatra, Java, the Lesser Sunda Islands (from Bali to Timor), and New Guinea conforms approximately to the shape of the Eurasian Plate's leading edge (see Figure 1.20 on page 21). As the Indian-Australian Plate plunges beneath the Eurasian Plate along this curve, hundreds of earthquakes and volcanoes

(a) Banda Aceh shoreline before June 2004.

(b) After the tsunami, the original shoreline is gone.

**FIGURE 10.4** Banda Aceh, Indonesia, before and after the December 2004 tsunami. When the tsunami hit, this community, at the western end of Sumatra, was totally destroyed. [Digital Globe.]

occur, especially on the islands of Sumatra and Java. Volcanoes and earthquakes also occur in the Philippines, where the Philippine Plate is pushing against the eastern edge of the Eurasian Plate. The volcanoes of the Philippines are part of the Pacific Ring of Fire (see Figure 1.21 on page 22). Volcanic eruptions, and the mudflows and landslides that occur in their aftermath, endanger and complicate the lives of many Southeast Asians. In the long run, though, the volcanic material creates new land and provides minerals that can enrich the soil for farmers. Earthquakes are especially problematic

because of the tsunamis they can set off. The tsunami of December 2004, triggered by a giant earthquake just north of Sumatra, swept east and west across the Indian Ocean, taking the lives of 230,000 people and injuring as many more. It is thought to be one of the deadliest natural disasters in recorded history (Figure 10.4).

The now-submerged shelf of the Eurasian continent that extends under the Southeast Asian peninsulas and islands was above sea level during the recurring ice ages of the Pleistocene epoch, during which much of the world's water was frozen in glaciers (Figure 10.5).

**FIGURE 10.5** Sundaland 18,000 years ago. The now-submerged shelf of the Eurasian continent that extends under Southeast Asia's peninsulas and islands was exposed during the last ice age and remained above sea level until about 16,000 years ago. [Adapted from Victor T. King, *The Peoples of Borneo* (Oxford: Blackwell, 1993), p. 63.]

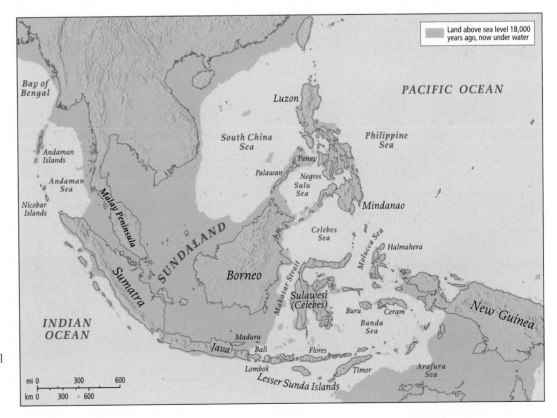

The exposed shelf, known as Sundaland, allowed ancient people and Asian land animals (such as elephants, tigers, rhinoceroses, proboscis monkeys, and orangutans) to travel south to what became the islands of Southeast Asia.

## Climate

The tropical climate of Southeast Asia is distinguished by continuous warm temperatures in the lowlands—consistently above 65°F (18°C)—and heavy rain (Figure 10.6). The rainfall is the result of two major processes: the monsoons (seasonally shifting winds) and the intertropical convergence zone (ITCZ), the band of rising warm air that circles Earth roughly around the equator (see pages 356–357). The wet summer season extends from May to October, when the warming of the Eurasian landmass sucks in moist air from the surrounding seas. Between November and April, there is a long dry season on the mainland, when the seasonal cooling of Eurasia causes dry air from the interior continent to flow out toward the sea. On the many islands, however, the winter can also be wet because the air that flows from the continent picks up moisture as it passes south and east over the seas. The air releases its moisture as rain after ascending high enough to cool. With rains coming from both the monsoon and the ITCZ, the island part of Southeast Asia is one of the wettest areas of the world.

Irregularly every 2 to 7 years, the normal patterns of rainfall are interrupted, especially in the islands, by the El Niño phenomenon

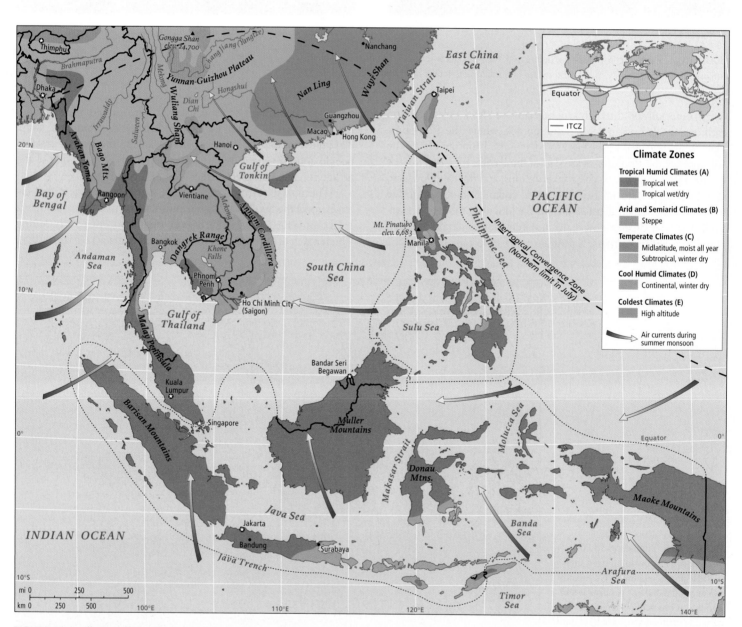

**FIGURE 10.6** Climates of Southeast Asia. The inset shows the intertropical convergence zone (ITCZ), which is partly responsible for Southeast Asia's being one of the wettest regions of the world.

(see Figure 11.9 on page 589). In an El Niño event, the usual patterns of air and water circulation in the Pacific are reversed. Ocean temperatures are cooler than usual in the western Pacific near Southeast Asia. Instead of warm, wet air rising and condensing as rainfall, cool, dry air sits at the ocean surface. The result is severe drought, with often catastrophic results for farmers.

The soils in Southeast Asia are typical of the tropics. Although not particularly fertile, they will support dense and prolific vegetation when left undisturbed for long periods. The warm temperatures and damp conditions promote the rapid decay of **detritus** (dead organic material) and the quick release of useful minerals. These minerals are taken up directly by the living forest rather than enriching the soil. Some of the world's most impressive rain forests thrived in this region until very recently. Today, these forests are being cleared at record rates, especially by multinational logging companies like those described in the opening of this chapter, but also by local people seeking farmland. Often the refuse is burned, creating smoke that contributes to severe air pollution under certain conditions.

## HUMAN PATTERNS OVER TIME

First settled in prehistory by migrants from the Eurasian continent, Southeast Asia later was influenced by Chinese, Indian, and Arab traders. Later still, it was colonized by Europe (1500s to early 1900s) and Japan (World War II). By the late twentieth century, colonial domination had abated, and the region sought to profit from selling manufactured goods to its former colonizers. Although poverty is still a problem, signs of the region's economic success include declining birth rates and longer life expectancies (the demographic transition), rapid industrialization, growing urbanization, and rising rates of literacy.

## The Peopling of Southeast Asia

The modern indigenous populations of Southeast Asia arose from two migrations widely separated in time. In the first, **Australo-Melanesians,** a group of hunters and gatherers from the present northern Indian and Burman parts of southern Eurasia, moved into the exposed landmass of Sundaland about 40,000 to 60,000 years ago. They were the ancestors of the indigenous peoples of New Guinea, Australia, and Indonesia's easternmost islands. Very small numbers of their descendants still live in small pockets in upland areas elsewhere, notably in the Philippines, Borneo, Java, Sumatra, the Malay Peninsula, and the Andaman Islands.

In the second migration, people from southern China began moving into Southeast Asia about 10,000 years ago, at the end of the last ice age (Figure 10.7). Their migration gained momentum about 5000 years ago, when a culture of skilled farmers and seafarers from southern China, named **Austronesians,** migrated first to Taiwan, then to the Philippines, and then into island Southeast Asia and the Malay Peninsula. Some of these sea travelers eventually moved westward to southern India and to Madagascar off the east coast of Africa, and eastward to the far reaches of the Pacific islands (see Chapter 11).

## Cultural Influences

Southeast Asia has been and continues to be shaped by a steady stream of cultural influences, both internal and external. The surrounding seas and the spring and summer monsoon winds from the

**FIGURE 10.7 Hmong women returning to their village.** One of the last groups to move from China to Southeast Asia was the Hmong, who probably left China during the nineteenth century. Although many Hmong were displaced by the Vietnam War, some still live in Sapa, in northwestern Vietnam, a remote tourist destination, among other Southeast Asian locales. [QT Luong/terragalleria.com.]

west brought seaborne traders, religious teachers, and occasionally even invading armies from the coasts of India and Arabia. These newcomers brought religions (Hinduism, Buddhism, Islam, and Christianity), trade goods such as cotton textiles, and food plants such as mangoes and tamarinds deep into the Indonesian archipelago. They penetrated as far as Cambodia, Laos, Vietnam, and China on the mainland as well as reaching the islands farther south. The merchant ships sailed home on the autumn and winter monsoon winds from the northeast, carrying Southeast Asia's people, as well as spices, bananas, sugarcane, root crops, silks, domesticated pigs and chickens, and other goods, to the wider world.

Islam came to the region mainly through South Asia, after India fell to Muslim (Mughal) conquerors in the fifteenth century. Islam is a major religion in Southeast Asia today, and Indonesia has the world's largest Muslim population. China was also influential, not through imperial expansion, but through Chinese traders and laborers who brought cultural influences to coastal zones. Those influences remain obvious today in such places as Singapore and parts of Malaysia, where Buddhist, Taoist, and Confucian temples exist side by side with Islamic mosques and Hindu temples. Other areas are more homogenous, such as Burma, Thailand, and Laos, which are largely Buddhist, and Indonesia, which is mostly Muslim.

As important as external influences have been, indigenous cultural characteristics have also shaped the character of the region. Several powerful urban empires emerged in Southeast Asia, including Angkor (800–1400), Pagan (800–1100), and Srivijaya (600–1000).

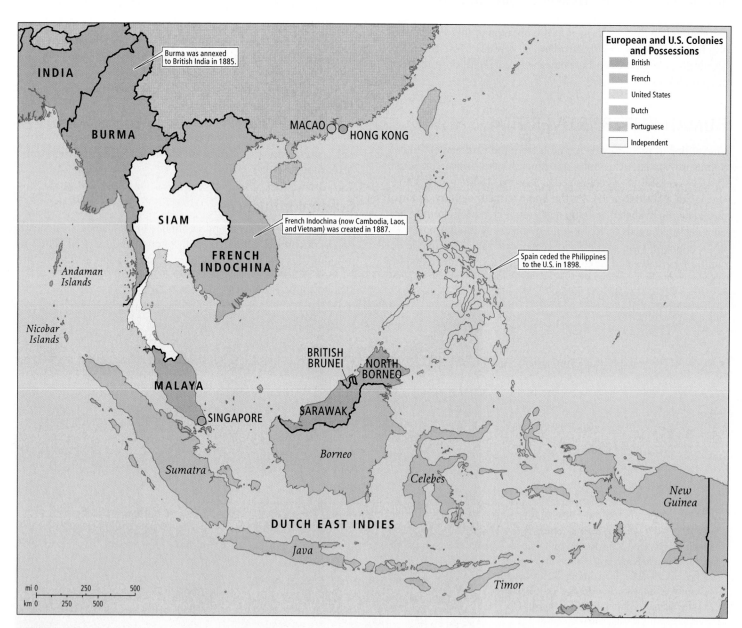

**FIGURE 10.8** European and U.S. colonies in Southeast Asia, 1914. Of the present-day countries in Southeast Asia, only Thailand (Siam) was never colonized. [Adapted from *Hammond Times Concise Atlas of World History* (Maplewood, N.J.: Hammond, 1994), p. 101.]

The cities of these empires were impressive for their size and architecture. At its zenith in the 1100s, the city of Angkor, in present-day Cambodia, was among the largest in the world, and its ruins are a World Heritage Site today (see Figure 10.20 on page 550).

## Colonization

Over the last five centuries, several European countries and the United States established colonies or quasi-colonies in Southeast Asia (Figure 10.8). Drawn by the region's fabled spice trade, the Portuguese sailed around Africa, and by 1511, they had established the first permanent European settlement at the port of Malacca on the southwestern coast of the Malay Peninsula. Although better ships and weapons gave the Portuguese an advantage, their anti-Islamic and pro-Catholic policies provoked strong resistance in Southeast Asia. Only in Timor-Leste (East Timor) did the Portuguese establish Catholicism as the dominant religion.

By 1540, the Spanish had established trade links across the Pacific between the Philippines and their colonies in the Americas, especially Mexico. Like the Portuguese, they practiced a style of colonial domination grounded in Catholicism, but they met less resistance because of their greater tolerance of non-Christians. The Spanish ruled the Philippines for more than 350 years, and as a result, the Philippines is the most deeply Westernized and certainly the most Catholic part of Southeast Asia.

The Dutch were the most economically successful of the European colonial powers in Southeast Asia. From the sixteenth to the nineteenth centuries, under the auspices of the Dutch East India Company, they extended their control of trade over most of what is today called Indonesia. Eventually, however, the Dutch became interested in growing cash crops for export. Between 1830 and 1870, the Dutch diverted farmers from producing their own food to working part time without pay in Dutch enterprises, especially coffee, sugar, and indigo plantations. The resulting disruption of local food production systems caused periodic severe famines, provoking resistance that often took the form of Islamic religious movements, thereby hastening the spread of Islam throughout Indonesia. As in South Africa, the Dutch settlers made little effort to spread Christianity.

Like their Dutch rivals, the British colonizers were commercially motivated. In the early nineteenth century, the British East India Company, operating with government and military powers sanctioned by the British Crown, controlled a few key ports on the Malay Peninsula. The British held these ports both for their trade value and to protect the Strait of Malacca, the passage for sea trade between China and Britain's empire in India. In the nineteenth century, Britain extended its rule over the rest of modern Malaysia to benefit from its tin mines and plantations and added Burma to its empire. By controlling Burma, Britain controlled access to forest resources and overland trade routes to China.

The French entered the region first as Catholic missionaries in the early seventeenth century. They worked mostly in the eastern mainland area of Southeast Asia—the modern states of Vietnam, Cambodia, and Laos. Then, in the late nineteenth century, spurred by rivalry with Britain and other European powers for greater access to the markets of nearby China, the French colonized the area, which became known as French Indochina.

In all of Southeast Asia, the only country not to be colonized was Thailand (then known as Siam). Like Japan, it protected its sovereignty through both diplomacy and a massive drive toward European-style modernization.

## Struggles for Independence

Agitation against colonial rule began in the late nineteenth century, when Filipinos fought first against Spain and then against the United States, after that country took control of the Philippines in 1898. However, independence for the Philippines and the rest of Southeast Asia had to wait until the end of World War II. By then, Europe's ability to administer its colonies had been weakened, partly because its attention was diverted by the devastation of the war. Moreover, Japan had exploded the myth of European superiority by conquering most of European-held Southeast Asia and holding it for several years, until its defeat by the United States. By the mid-1950s, the colonial powers had granted self-government to most of the region.

The most bitter battle for independence took place in the French-controlled territories of Vietnam, Laos, and Cambodia. Although all three became nominally independent in 1949, France retained political and economic power. Various nationalist leaders led resistance movements against continued French domination. Although they did not begin as Communists, and although they shared ancient antipathies toward China for its previous efforts to dominate trade, the resistance leaders accepted military assistance from Communist China and the Soviet Union. Thus the cold war was brought to mainland Southeast Asia.

In 1954, the French were defeated by Ho Chi Minh at Dien Bien Phu, in northern Vietnam (Figure 10.9). The United States, increasingly worried about the spread of international Communism, stepped in. The Vietnamese resistance, which controlled the northern half of the country, attempted to wrest control of the southern

**FIGURE 10.9** Remembering Dien Bien Phu. On May 7, 2004, Vietnamese performers in the Dien Bien Phu town stadium commemorated the fiftieth anniversary of the defeat of the French in Indochina. [AP Photo/Richard Vogel.]

half from the United States and a U.S.-supported South Vietnamese government. The pace of the war accelerated in the mid-1960s. After many years of brutal conflict, public opinion in the United States forced U.S. withdrawal from the conflict in 1973. The civil war continued in Vietnam, finally ending in 1975, when the North defeated the South and established a new national government. More than 4.5 million people died during the Vietnam War, including more than 58,000 U.S. soldiers. Another 4.5 million on both sides were wounded, and bombs, napalm, and defoliants ruined much of the Vietnamese environment; land mines continue to be a hazard to this day. The U.S. withdrawal from Vietnam

in 1973 ranks as one of its most profound defeats. After the war, the United States crippled Vietnam's recovery with severe economic sanctions that lasted until 1993. Since then the United States and Vietnam have become significant trading partners.

In Cambodia, where the Vietnam War had spilled over the border, a particularly violent revolutionary faction called the Khmer Rouge seized control of the government in the mid-1970s. Inspired by the vision of a rural Communist society, they attempted to destroy virtually all traces of European influence. They targeted Western-educated urbanites in particular, forcing them into labor camps, where over 2 million Cambodians—one-quarter of the population—

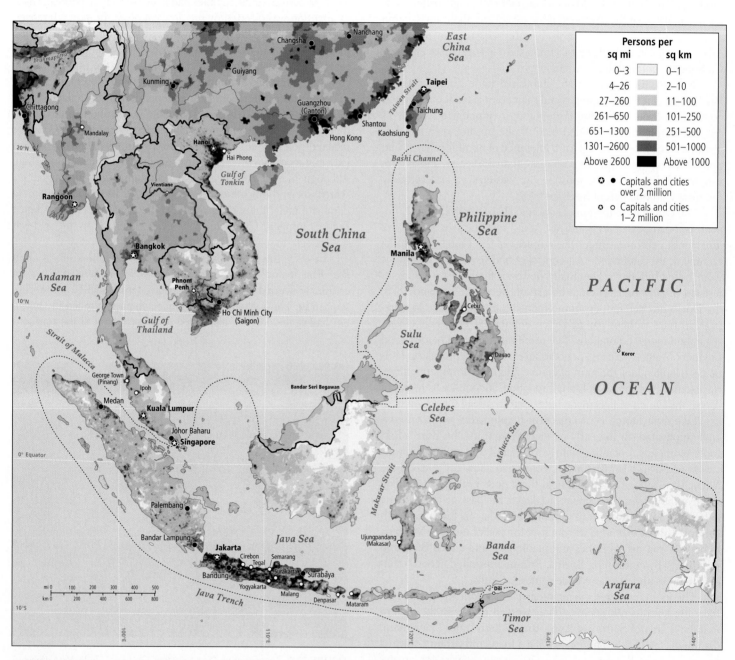

**FIGURE 10.10**  Population density in Southeast Asia. [Data courtesy of Deborah Balk, Gregory Yetman et al., Center for International Earth Science Information Network, Columbia University, at http://www.ciesin.columbia.edu.]

starved or were executed. In 1978, Vietnam deposed the Khmer Rouge and ruled Cambodia through a puppet government until 1989. A 2-year civil war then ensued. Despite a massive United Nations effort to establish multiparty democracy in Cambodia throughout the 1990s, the country remains plagued by political tensions between rival factions and by government corruption, and none of the Khmer Rouge leaders have ever been tried or held accountable for their actions.

 Former Khmer Rouge village chief, Choch: "It's not true. If I had done those things, how could I live here now?"

In Cambodia, victims of the Khmer Rouge reign of terror now often live as close neighbors to those who tortured and killed their loved ones. The perpetrators have never had to stand trial or face punishment because little forensic work has been done on the murders and massacres. Those responsible can easily deny having been involved.

One night in 1977, soldiers came to the home of Samrith Phum and took her husband away. She thought he was just going to a meeting, but he never came home. Samrith was then only 20 years old and had three young children, one a newborn. With her infant in her arms, she went to talk to the Khmer Rouge village chief, a man named Choch, and asked him, "Brother, do you know where my husband is?" But the village chief told her not to worry about other people's business. Says Samrith, "I didn't ask him any more after that. I was hopeless. I knew my husband was dead." A short while later, Choch appeared at Samrith's door and said he would take her to see her husband. Instead, he drove to the nearby prison and locked her up with her baby. She was released a year later, only after the Vietnamese drove the Khmer Rouge from power. Today, Samrith still lives just down the street from Choch. For his part, Choch denies any involvement in the killings or even ever being at the prison.

*To learn more about the aftermath of the Khmer Rouge, watch the FRONT-LINE/World Video "Cambodia: Pol Pot's shadow" and read Amanda Pike's "Reporter's diary: In search of justice," at http://www.pbs.org/frontlineworld/stories/cambodia/diary03.html.* ■

Although independence has brought violence to some areas, it has also brought relative peace and economic development to much of Southeast Asia. Since the 1960s, the economies of Thailand, Malaysia, Singapore, Indonesia, and the Philippines have grown considerably, aided largely by industries that export manufactured products to the rich countries of the world. Some critics regard this situation as a form of neocolonialism because the countries of the region remain dependent on markets in distant wealthy nations and compete with one another to sell their products to their former colonizers. This stiff competition often results in wages that are too low to provide a decent standard of living and in weak pollution controls and worker safety regulations.

## POPULATION PATTERNS

Today more than half a billion people occupy the peninsulas and islands of Southeast Asia. Twice as many people as live in the United States are packed into a land area half its size. The population map in Figure 10.10 reveals, however, that relatively few people live in the upland reaches of Burma, Thailand, and northern Laos, where the land is particularly rugged and difficult to traverse, or in much of Cambodia. Parts of Malaysia and Indonesia and Mindanao in the Philippines are also lightly settled because of wetlands, dense forests, mountains, and geographic remoteness. Small groups of indigenous people have lived in forested uplands for thousands of years, supported by shifting cultivation and by hunting, gathering, and small-plot permanent agriculture. Over the last several decades, however, in previously lightly settled places such as Sumatra, Kalimantan (on Borneo), Sulawesi (Celebes), the Moluccas, and West Papua (until recently called Irian Jaya), thousands of new settlers have been lured by promises of farmland and jobs in commercial agriculture and in extraction of forest products and minerals.

About 60 percent of the people of Southeast Asia live in patches of particularly dense rural settlement along coastlines, on the floodplains of major rivers, and in the river deltas of the mainland. In the islands, settlement is most concentrated on Luzon, in the northern Philippines, and on Java. These places are attractive because the rich and well-watered soils allow intensive agriculture. About 38 percent of the region's people live in cities; Jakarta, Manila, Bangkok, and many medium-sized cities of the region are among the most rapidly growing metropolitan areas on earth (Table 10.1; reasons for this growth are discussed on pages 561–562). Rural migrants stream into the increasingly crowded and dense slum and squatter areas that exist in all of the region's cities. Only Singapore, an unusually wealthy country composed almost entirely of one city

**TABLE 10.1 Recent growth in Southeast Asia's metropolitan areas with 5 million or more inhabitants**

| Rank, 2007 | City/Country | Population 2000* | Population 2007 (est.) |
|---|---|---|---|
| 1 | Manila, Philippines | 9,906,000 | 18,491,000 |
| 2 | Jakarta, Indonesia | 15,961,000 | 18,267,000 |
| 3 | Bangkok, Thailand | 6,320,200 | 10,230,000 |
| 4 | Kuala Lumpur, Malaysia | 4,428,800 | 6,933,000 |
| 5 | Bandung, Indonesia | 3,416,000 (1995) | 5,980,000 |
| 6 | Singapore–Johor Baharu (Malaysia) | 4,500,000 (2001) | 5,490,000 |
| 7 | Ho Chi Minh City, Vietnam | 3,316,500 | 5,183,000 |

* Unless otherwise specified.
*Source: World Gazetteer,* http://www.world-gazetteer.com.

**FIGURE 10.11** Singapore. This view from atop the Raffles Hotel captures the active and prosperous spirit of Singapore. In the foreground, you can see the old colonial town center and, across the river, the modern banking district. The pleasure boat harbor is to the left, port facilities at upper left. Singapore is the world's largest port. [D. Saunders/TRIP.]

(Figure 10.11), is an exception. Indeed, in some cities, as much as one-third of the population resides in these temporary, usually self-built, settlements.

## Population Dynamics

Overall, Southeast Asians have smaller families today than they did in the past and are experiencing small drops in fertility annually, but the patterns vary geographically. In some places, fertility rates have declined sharply since the 1960s (Figure 10.12). In 1960, Singapore had a birth rate of 4.9 children per adult woman; by 2003, that rate was just 1.3, which is below replacement level (2.1 children per adult woman). The Singapore government is so concerned about the low fertility rate that it offers young couples various incentives for marrying and procreating. Despite these low birth rates, Singapore's population continues to grow a bit because the city attracts a steady stream of highly skilled immigrants.

Thailand, with one of the region's most successful family planning programs, has the next lowest fertility rate in the region: 1.7 children per adult woman. Compared with Singapore, Thailand is very poor, with just one-third the income per capita. Nevertheless, rapid economic change has made couples feel that smaller families are best. Although wages are low, Thai citizens have many employment opportunities and are allowed to move freely throughout the country to find work. High literacy rates for both men and women and Buddhist attitudes that accept the use of contraception have also been credited for the decline in Thailand's fertility rate.

Two extremely poor countries in the region show the usual correlation between poverty and high fertility. In Cambodia and Laos, women average between 4 and 5 children each, and infant mortality rates are 95 per 1000 births for Cambodia and 88 per 1000 for Laos. Equally poor Vietnam, however, has a relatively low fertility rate (2.2 children per adult woman) and infant mortality rate (18 per 1000 births). Vietnam's lower rates are explained by the fact that this

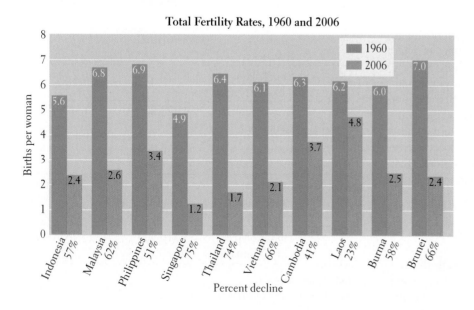

**FIGURE 10.12** Total fertility rates, 1960 and 2006. Total fertility rates declined during this period for nearly all Southeast Asian countries. Timor-Leste is the exception, but current data for that country are not available. [Data from *A Demographic Portrait of South and Southeast Asia* (Washington, D.C.: Population Reference Bureau, 1994), p. 9; Globalis, at http://globalis.gvu.unu.edu/; *2006 World Population Data Sheet* (Washington, D.C.: Population Reference Bureau).]

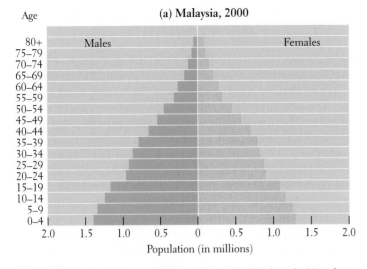

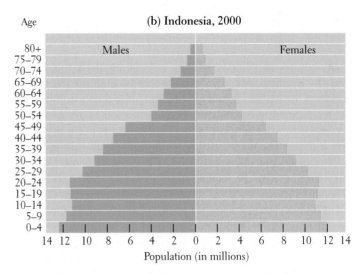

**FIGURE 10.13** Population pyramids for Malaysia and Indonesia. Note that the scale for the population axis is roughly eight times greater for Indonesia than for Malaysia. In 2000, Indonesia had 206.1 million people, while Malaysia had just 22.7 million people. [Adapted from "Population pyramids for Malaysia" and "Population pyramids for Indonesia" (Washington, D.C.: U.S. Census Bureau, International Data Base, at http://www.census.gov/ipc/www/idbpyr.html).]

socialist state provides basic education and health care to all of its people, regardless of income. In Vietnam, literacy rates are more than 94 percent for men and 87 percent for women, whereas only 60 percent of women in Cambodia and only 61 percent in Laos can read. In addition, Vietnam's rapidly growing economy is attracting foreign investment; hence employment for women is expanding, replacing child rearing as the central role in women's lives.

Southeast Asia's population is young, as the wide bases of the population pyramids for Malaysia and Indonesia in Figure 10.13 show. Analysts anticipate that fertility rates in the region will drop markedly in the next 25 years, but because so much of the population is now young—overall, those 15 and younger make up 30 percent of the population—the number of people in Southeast Asia will reach 766 million by 2050. Another trend is that more people will live longer lives—possibly a mixed blessing for younger workers who must provide care and support to the elderly. Given Southeast Asia's relatively small land area, the rising level of consumer demand, and the fragility of its tropical environments, population pressure on land, soil, fresh water, and forest resources will remain a major public issue for many years.

## Southeast Asia's HIV-AIDS Tragedy

As in sub-Saharan Africa (see Chapter 7), HIV-AIDS constitutes a significant public health issue in Southeast Asia (Figure 10.14). Infection rates are growing across the region: at present, Cambodia,

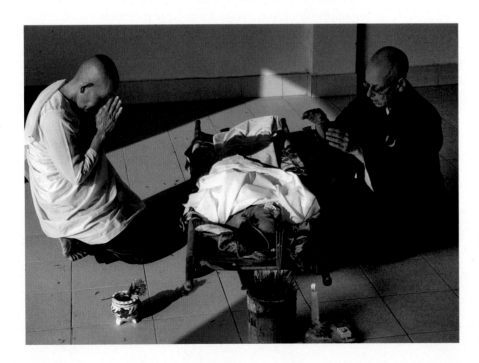

**FIGURE 10.14** Caring for AIDS patients in Cambodia. Two Buddhist nuns, one Cambodian (Lok Yay, left) and one American (Beth Goldring, right), care for AIDS patients at the Maryknoll hospice in Phnom Penh. Beth Goldring founded Brahmavihara, a small Buddhist organization that provides care and spiritual sustenance to AIDS sufferers, coordinating their work with the Catholic Maryknoll AIDS program. [Bennett Stevens.]

**545**

Thailand, Burma, and West Papua in Indonesia have the highest rates. In Thailand, AIDS is the leading cause of death, overtaking accidents, heart disease, and cancer, and more than a million people are thought to be infected. Men between the ages of 20 and 40 are the most common victims, but the number of women victims is rising. The disease may soon increase rapidly in rural areas and in secondary cities just coming into closer economic and social contact with the region's big cities, where infection is most prevalent.

There are several reasons for this gloomy forecast. Conservative religious leaders and faith-based international agencies restrict public sex education and AIDS prevention programs, such as the promotion of condom use, because these activities are viewed as promoting promiscuity. At the same time, popular customs that support sexual experimentation (at least among men), the high mobility of young adults, the reluctance of women to insist that their husbands and boyfriends use condoms, and intravenous drug use (primarily by men) make aggressive prevention programs all the more essential. In Thailand, for example, a visit to a brothel has long been a rite of passage for young men, and brothels are found in most neighborhoods. The development of sex tourism and associated human trafficking has also contributed to the spread of HIV among young women (this topic is discussed further on page 552). Studies in the 1990s showed that more than 60 percent of female sex workers in urban locations were infected with HIV; during the same year, 80 percent of young male soldiers visited brothels. In addition, just as in South Asia and Africa, truck drivers in Southeast Asia spread HIV because they may have sexual partners in several different locales. In Thailand, a promising government education campaign to promote condom use lost momentum due to cuts in international funding, and now HIV is spreading to more diverse populations. The greatest increase is among young women. Buddhist nuns have led the movement to take care of AIDS victims, and when antiretroviral drugs were developed, Buddhists were among the first to make them available to destitute patients).

# II CURRENT GEOGRAPHIC ISSUES

Not surprisingly, several issues identified in other world regions are also found in Southeast Asia (Figure 10.15). Like the Americas, Africa, and South Asia, Southeast Asia has a history of colonial rule; has experienced uneven development resulting in disparities in wealth; is expanding its links to the global economy through manufacturing and migration; and is home to a wide array of culture groups and religions, with social conflict in the face of rapid change an increasing reality. Environmental concerns—especially those related to urbanization and to unsustainable extraction of forest and soil resources—are also present. Nonetheless, there are many issues unique to the region.

## ECONOMIC AND POLITICAL ISSUES

The economic and political situation in the region has changed dramatically in recent years. From the mid-1980s to the late 1990s, Southeast Asian countries had some of the highest economic growth rates in the world. Several countries, such as Malaysia, Singapore, and Thailand, earned a reputation as economic "tigers" by aggressively embracing capitalism and achieving remarkable levels of economic growth. But for a few years beginning in 1997, economic growth stagnated and political instability increased as the foreign investors who had supported much of the region's growth withdrew. By 2003, economic recovery had helped reduce the number of people living in poverty, but only in mainland Malaysia and in Thailand. By 2005, poverty levels in Indonesia, for example, still remained higher than in the 1990s.

### Agriculture

Compared with services and industry, agriculture's role in the region's economy has been declining, and it currently accounts for less than one-sixth of total GDP. However, more than 60 percent of Southeast Asians still live in rural settlements and depend on agriculture for part of their support, though many of them also fish, make and sell crafts, perform services, or work part time for wages.

**FIGURE 10.15** Political map of Southeast Asia.

Commercial agriculture also yields important products for export (especially rice, rubber, sugar, coconut products, and palm oil) and provides important components of urban diets (rice, fruits and vegetables, fish, meat, and dairy products). Figure 10.16 shows patterns of agriculture throughout the region.

Several forms of agriculture are practiced in Southeast Asia. Shifting cultivation (described in Chapter 3 on page 123) is practiced by subsistence farmers in the hills and uplands of mainland Southeast Asia and in coastal Sumatra, Borneo, Sulawesi, the southern Philippines, and West Papua, where population densities are

relatively low. Farmers move their fields every 3 years or so, as soil fertility is depleted by that time and can take decades to replenish. Hence larger areas are needed to support human populations than with other agricultural systems.

The most productive form of agriculture in the region is **wet** (or **paddy**) **rice cultivation.** Wet rice cultivation is permanent, as opposed to shifting. It entails the planting (usually by women) of rice seedlings by hand in flooded terraced fields that are first cultivated (usually by men) with hand-guided plows pulled by water buffalo. Wet rice cultivation has transformed landscapes throughout

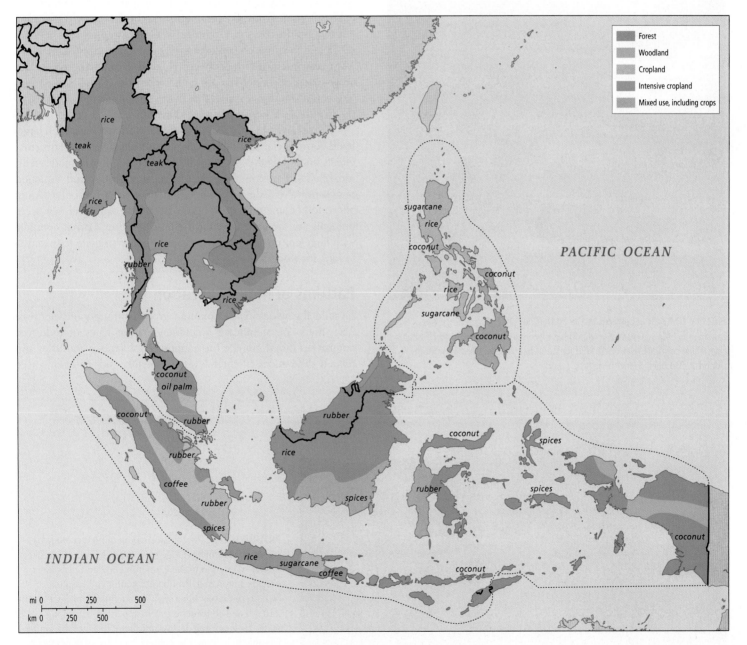

**FIGURE 10.16** Agricultural patterns in Southeast Asia. Tropical forests and crops, rice production, and shifting cultivation dominate the agricultural patterns of Southeast Asia. [Adapted from *Hammond Citation World Atlas* (Maplewood, N.J.: Hammond, 1996,) pp. 74, 83, 84.]

**FIGURE 10.17** Rice terraces in the Philippines. The Ifugao rice terraces on Luzon are at least 2000 years old and have been designated a World Heritage Site. They continue to provide food for the inhabitants of these mountains. [John Elk III/Lonely Planet Images.]

Southeast Asia (Figure 10.17). It is practiced where streams bring a yearly supply of silt and on rich volcanic soils on islands such as Java and Sumatra in Indonesia and Luzon in the Philippines.

During the recent decades of relative prosperity, small farms that were once operated by families have been combined into large commercial farms owned by local or multinational corporations. These farms produce cash crops for export, such as rubber, palm oil, bananas, pineapples, tea, and rice (Figure 10.18). This style of agriculture entails combining huge tracts of land into one system; clear-cutting patches of forest that once occupied the fringes of farms; deep plowing of the soil; usually planting one species of crop plant over many square miles; bolstering soil fertility with chemicals; using mechanized equipment; and, in the case of rice, using large quantities of water. Commercial farming reduces the need for labor, and the objective is quick profit, not long-term sustainability. Many commercial farmers have achieved dramatic boosts in harvests (especially of rice) by using high-yield crop varieties, the result of green revolution research that has been applied in many parts of the world (see Chapter 8, pages 439–440).

As we have noted in relation to other world regions, such large-scale commercial farming has significant negative environmental effects, including the loss of wildlife habitat and hence biodiversity, increased soil erosion, flooding, chemical pollution, and depletion of groundwater resources. In addition, poor farmers usually cannot afford to become green revolution farmers because the new technologies are too expensive for them. Often they cannot compete with large commercial farms and are forced to migrate to the cities to look for work.

## Patterns of Industrialization

Despite the widespread economic growth throughout Southeast Asia, there are significant differences in the types of industrial enterprises that dominate the economies of various countries and related differences in their levels of wealth and well-being (see

**FIGURE 10.18** The transformation of rural ecology by commercial farming. Large-scale agricultural systems like this tea plantation in the Cameron Highlands, Malaysia, provide a profitable livelihood for farm and corporate owners, but reveal signs of environmental damage. Erosion in fields and construction sites (upper left) and along roadsides is causing deadly landslides. Because the land is less able to hold moisture, there is increased flooding during the rainy season and drought during the dry season. [Alex Pulsipher.]

TABLE 10.2 Disparities of wealth for selected countries in Southeast Asia and Middle and South America, as shown through income spread ratios[a]

| Country | Income spread ratio, 2003 | Income spread ratio, 2005 |
|---|---|---|
| Indonesia | 5.2 | 5.2 |
| Malaysia | 12.4 | 12.4 |
| Singapore | 9.7 | 9.7 |
| Thailand | 8.3 | 8.3 |
| Vietnam | 5.6 | 5.5 |
| Brazil | 29.7 | 26.4 |
| Chile | 19.3 | 18.7 |
| Costa Rica | 11.5 | 12.3 |
| Mexico | 17.0 | 19.3 |
| Peru | 11.7 | 18.4 |

[a]Ratio of wealthiest 20 percent to poorest 20 percent. The higher the number, the greater the discrepancy between wealthy and poor.

*Sources: United Nations Human Development Report 2003, Table 13; United Nations Human Development Report 2005, Table 15 (New York: United Nations Human Development Programme).*

Table 10.4 on page 568). In Vietnam, Laos, Burma, and Cambodia and in the rural parts of the region's other countries, many people still depend on agriculture, fishing, and forestry. However, even in some rural areas, construction and maintenance jobs on such projects as new hydroelectric dams are raising incomes. In cities and towns, labor-intensive pursuits such as garment and shoe making, food processing, and many types of light manufacturing are expanding. In the urban and suburban areas of the wealthier countries—such as Singapore, Malaysia, and Thailand—and in a few areas in Indonesia and the Philippines, more technologically based manufacturing, including automobile assembly, chemical and petroleum refining, and assembly of computers and other electronic equipment, is important.

From the 1960s to the 1990s, national governments in Southeast Asia restricted and controlled foreign investment in the region. Using their own tax revenues, along with a modest amount of regulated foreign investment, governments tried to nurture economic development that would steadily create new jobs and be shielded from the ups and downs of global economic cycles. They favored *import substitution industries*—local industries that produce consumer goods for local use (see the discussion in Chapter 3 on pages 137, 139)—and used tariffs to exclude competing goods made outside the region. Export agriculture was also favored. For the most part, these policies resulted in strong and sustained economic growth. Standards of living increased markedly, especially in Malay-

sia, Indonesia, and Thailand. Unlike Middle and South America, where similar efforts have been less successful, these countries were aided by a higher proportion of local (versus foreign) capital and the strong support of government and economic elites, whose interests were tied to successful industrial development. The focus on cheap consumer goods rather than expensive luxury goods created a broad-based domestic market that could fuel further economic expansion. One important result was relatively less extreme disparities in wealth (Table 10.2).

***EPZs.*** In the 1970s, governments in the region adopted an additional strategy for encouraging economic development, this one aimed at manufacturing products for export, primarily to developed countries. They attracted foreign multinational corporations through the establishment of **Export Processing Zones (EPZs)**, specially designated free trade areas in which foreign companies can set up export-oriented industries using inexpensive labor and, sometimes, inexpensive local raw materials to produce items only for export. Subcontractors, often locally owned, also may set up factories to supply needed parts. Taxes are eliminated or greatly reduced. Since the 1970s, EPZs have expanded economic development in Malaysia, Indonesia, Vietnam, and the Philippines for the production of electronics, garments, textiles, and shoes.

Between 80 and 90 percent of the workers in the EPZs are women, not only in Southeast Asia but in other regions as well (Figure 10.19). So widespread is this phenomenon around the world that scholars and activists view the **feminization of labor** as a distinct characteristic of globalization over the past three decades

**FIGURE 10.19** Patterns of industrialization. Employees inspect printed circuit boards at Integrated Microelectronic, Inc. (IMI), in Santa Rosa, a suburban city south of Manila. The company has operations in the Philippines, California, and Singapore and serves clients in Japan, Europe, and the United States. [AFP Photo/Joel Nito/Newscom.]

(see Chapter 3, pages 164–165, and Chapter 9, pages 473–475). Bosses prefer to hire young, single women because they are perceived as the least troublesome employees. Statistics do show that, at least for now, women will work for lower wages than men, will not complain about poor and unsafe working conditions, will accept being restricted to certain jobs on the basis of sex, and are not as likely as men to agitate for promotions.

**The Southern Growth Triangle.** Large transnational economic zones, called **growth triangles,** are a trend in the region. The Singapore–Johor–Riau (or Southern) Growth Triangle was created in about 1989, when low-tech manufacturing began to be relocated from Singapore to Johor, just across the border in southern Malaysia, and eventually to the nearby Riau Archipelago in Indonesia. The moves were prompted by Singapore's relatively high labor costs, its land shortages, and its expensive fresh water. A single multinational firm, for example, can locate its core high-technology telecommunications, managerial, and shipment activities in Singapore to take advantage of its highly skilled labor force and high-tech business and port infrastructure. Its less technical activities (for example, textiles and electronics manufacturing) can be located in Johor, where workers earn about a third of what Singaporeans earn and where land and water are plentiful. Assembly operations are usually located in Riau, where labor is half again as cheap. This strategy of allocating different activities to different places based on the quality of the infrastructure and the cost of labor is now used throughout the global economy. While it improves corporate profit margins, it often produces wide disparities in income and well-being for workers.

**Working Conditions.** In general, the benefits of Southeast Asia's "economic miracle" have been unequally apportioned. In the region's new factories and other enterprises, it is not unusual for assembly-line employees to work 10 to 12 hours a day, 7 days a week, for less than the legal minimum wage and with no benefits. Labor unions that would address working conditions and wage grievances are frequently repressed by governments, and international consumer pressure to improve working conditions has been only partially effective. For example, Asian employees of the U.S. shoe manufacturing firm Nike were frequently exposed to hazardous chemicals, as well as to physical abuse and psychological cruelty on the job. Nike's labor tactics became widely known to U.S. university students when student activists revealed that many campus stores were selling Nike equipment and garments made under these substandard conditions in Southeast Asian factories. Although the students' exposure of abuses resulted in some improvements in working conditions, those improvements were relatively minor.

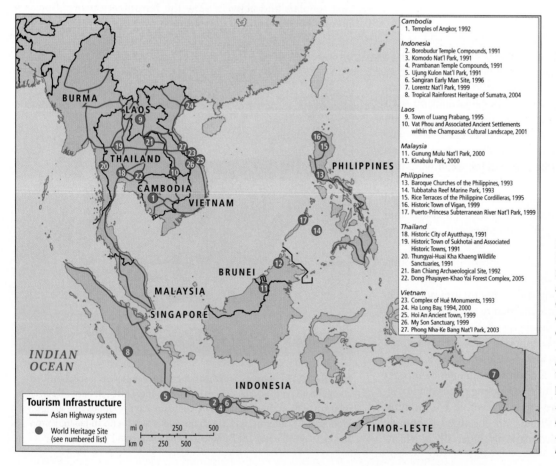

**FIGURE 10.20** Transportation infrastructure for tourism. These Southeast Asian World Heritage Sites are located along the partially completed Asian Highway, which will connect Europe to Indonesia. A map of the entire system is at http://www .unescap.org/ttdw/common/TIS/AH /maps/AHMapApr04.gif. [Adapted from UN World Heritage Convention, at http://whc.unesco.org/en/list/; http://www .unescap.org/ttdw/common/tis/ah /tourism%20attractions.asp; http://www .atimes.com/atimes/Asian_Economy /images/highways.html.]

A much more powerful force driving up pay and working conditions is the service sector, which is growing throughout the region. In all but Burma, Laos, Cambodia, Vietnam, and Timor-Leste, the service sector is rapidly catching up to the manufacturing sector as a contributor to GDP. Business-to-business services, financial services, data processing, customer services, and especially tourism are important components of this sector. These jobs require at least a high school education, and competition for workers means that wages and working conditions are somewhat better. In Brunei, Burma, Malaysia, the Philippines, and Thailand, women are acquiring tertiary education faster than men (see Table 10.3 on page 561). This should eventually increase the participation of women in the service sector, their earning ability, and their role and status in society.

**Tourism Development.** The fastest-growing industry in Southeast Asia, as in the world overall, is tourism. Since the 1970s, many Southeast Asian countries have viewed tourism as an opportunity to broaden the base of their economies and to spread industrial development more evenly throughout their territories. Between 1991 and 2001, the number of international visitors to the region doubled to over 40 million, which accounted for nearly 6 percent of the world's total. Earnings from tourism contribute tens of billions of U.S. dollars to Southeast Asian economies, and tourism has inspired new levels of regional cooperation to improve the transport infrastructure. One such project is the Asian Highway, a web of standardized roads looping through the mainland and connecting it with Malaysia, Singapore, and Indonesia (the latter via ferry) (Figure 10.20). Eventually the Asian Highway will facilitate ground travel through 32 Eurasian countries from Moscow to Indonesia and from Turkey to Japan.

The surge of tourism in Southeast Asia has also raised concerns. Too great a dependency on tourism leaves economies vulnerable to events that precipitously stop the flow of visitors or damage the tourism infrastructure. Examples are natural disasters such as the tsunami of December 2004 and human-made disasters such as the terrorist bombings in Bali (2002, 2005) and in southern Thailand (June and September 2006). In addition, mass tourism threatens the long-term survival of cultural heritage sites and local cultures, as discussed in Box 10.1.

## BOX 10.1 | AT THE REGIONAL SCALE   Tourism Development in the Greater Mekong Basin: Economic Boom or Cultural Bust?

By 5:30 A.M., the saffron-clad Buddhist monks were headed down Sisavangvong Road toward Luang Prabang, the ancient capital of a kingdom that once covered present-day Laos, southern China, and northeastern Thailand. In the light of dawn, they were a strip of moving orange silk, going through their morning ritual of collecting alms from the community. White-haired old men respectfully knelt and bowed as they offered glutinous rice to the monks. The ethereal moment was broken when a young girl in a too-loud voice invited tourists to buy some rice wrapped in a banana leaf to offer to the monks, commercializing a Buddhist religious ritual.

Luang Prabang, rich with monasteries and monuments, is a UNESCO World Heritage Site—meaning that it is of outstanding cultural or natural importance to the common heritage of humankind. Luang Prabang, one of several such sites along a newly developed heritage route for tourists in the Greater Mekong Basin is one of the new "hot" spots in Southeast Asian tourism. Airline routes, boat tours from Thailand, and now the Asian Highway (see Figure 10.20) bring more than 500 tourists a day to the town, making it a key contributor to the tourism dollars that Laos earns.

Tourism destination development is part of a large international initiative to develop the Greater Mekong Basin and enhance economic relations among the six Mekong countries (Burma, Cambodia, China, Laos, Vietnam, and Thailand) and get money into the hands of their people. The scheme is financed by the Asian Development Bank with support from tourism promoters such as the World Tourism Organization and the Pacific Asia Travel Association. But critics asking the "development for whom?" question argue that the people themselves will have little choice in deciding the pace or direction of this development. It is unclear how a small geographic area, isolated for centuries, can deal with a sudden influx of outsiders who, by their sheer numbers, will strain the infrastructure and services. The presence of these tourists could also damage cultural assets and make local people feel unwelcome in their own home spaces.

In Luang Prabang, the effects are already visible. Every second building along Phothisarat Street, in the center of Luang Prabang's old town, is a restaurant, shop, bakery, guest house, or Internet café. Plastic bottles and garbage litter the area, and shreds of plastic cling to the tall weeds lining the bank of the Mekong River that leads out of town. Tensions are rising between those who make tourism bucks and those, such as farmers on the other side of the river, who do not. Before, young people in the town aspired to be teachers or doctors; now they want to be tour guides.

*Sources:* Teena Amrit Gill, "Locals lose out as tourism booms," *Asia Times* Online, March 12, 2002, http://www.atimes.com/se-asia /DC12Ae02.html; Bui Nguyen Cam Ly, "As hordes of tourists come, heritage goes," *Inter Press Service News*, 2003, http://www.ipsnews.net /mekong/stories/heritage.html; Gulfer Cezayirli, "Fast-growing Asian tourism should enlist help of the urban poor," *Asian Development Bank*, 2003, http://www.adb.org/Media/Articles/2003/2009_Regional_Asian _Tourism_Should_Enlist_Help_of_the_Urban_Poor/default.asp.; "Asian Highway network gathers speed," *Asia Times* Online, June 14, 2006, http://www.atimes.com/atimes/Asian_Economy/HF14Dk01.html.

*The Sex Industry.*   **Sex tourism,** geared toward visitors who pay for sex (Figure 10.21), is in some ways an extension of the sexual entertainment industry that served foreign military troops stationed in Asia after World War II and during and after the conflicts in Korea and Vietnam. Now mostly civilian men arrive from around the globe to live out their fantasies during a few weeks of vacation in Thailand, Cambodia, Indonesia, Vietnam, or the Philippines. The industry is found throughout the region, but is most prominent in Thailand. In 2000, 9 million tourists visited Thailand alone, up from 250,000 in 1965, and some observers estimate that as many as 70 percent were looking for sex. In recent years, Thai government officials have encouraged sex tourism to create jobs, even though it is illegal, and they praised the sector's role in helping the country weather the economic crisis of 1997. Corrupt public officials also support sex tourism because it provides them with a source of un-taxed income from bribes.

One result of the "success" of sex tourism, as well as high local demand for sex workers, is that organized crime has entered the field, and girls and women are being coerced into sex work. Research indicates that some girls may have been sold by their families to pay off family debt. Others have been kidnapped, often at a very young age. Demographers estimate that 20,000 to 30,000 Burmese girls taken against their will (some as young as 12) are working in Thai brothels; their wages are too low to make buying their own freedom possible. In the course of their work, they must service more than 10 clients a day and are routinely exposed to physical abuse and sexually transmitted diseases, especially HIV.

**FIGURE 10.21**  Thailand's sex industry. Young Thai prostitutes try to attract customers in one of Bangkok's downtown red light districts. Thailand's sex workers send an estimated U.S.$300 million home annually to support their families in rural areas of the country. [AFP Photo/Stephen Shaver/Newscom.]

**Personal Vignette**   Twenty-five-year-old Watsanah K. (not her real name) awakens at 11 every morning, attends afternoon classes in English and secretarial skills, and then goes to work at 4:00 P.M. in a bar in Patpong, Bangkok's red light district. There she will meet men from Europe, North America, Japan, Taiwan, Australia, Saudi Arabia, and elsewhere who will pay to have sex with her. She leaves work at about 2:00 A.M., studies for a while, and then goes to sleep.

Watsanah was born in northern Thailand to an ethnic minority group whose members, like many others in the area, are poor subsistence farmers who have recently become involved in the global drug trade by growing opium poppies. Watsanah married at 15 and had two children shortly thereafter. Several years later, her husband developed an opium addiction. She divorced him and left for Bangkok with her children. There she found work at a factory that produced seat belts for a nearby automobile plant. In 1997, Watsanah lost her job as the result of the financial crisis that ripped through Southeast Asia. She became a sex worker to feed her children.

Although the pay, between $400 and $800 a month, is much better than the $100 a month she earned in the factory, the work is dangerous and demeaning. Sex work, though widely practiced and generally accepted in Thailand, is illegal, and the women who do it are looked down on, so Watsanah must live in constant fear of going to jail and losing her children. Moreover, she cannot always make her clients use condoms, which puts her at high risk of contracting AIDS or other sexually transmitted diseases. "I don't want my children to grow up and learn that their mother is a prostitute," says Watsanah, "that's why I am studying. Maybe by the time they are old enough to know, I will have a respectable job."

*Sources: Adapted from the field notes of Alex Pulsipher and Debbi Hempel, 2000; "Sex industry assuming massive proportions in Southeast Asia," International Labor Organization News (August 19, 1998); coverage of the HIV-AIDS conference in Thailand, July 11–16, 2004, by the Kaiser Family Foundation, http://www.kaisernetwork.org/aids2004/kffsyndication.asp?show=guide.html.* ∎

## Economic Crisis and Recovery: The Perils of Globalization

The financial crisis that swept Southeast Asia in the late 1990s forced millions of people back into poverty and created tensions that disrupted political order. A major cause of the crisis was the rapid shift of most Southeast Asian economies away from government regulation and toward the free market. Another factor was widespread corruption.

By the late 1980s, after a long period of government protection of regional business interests, most governments were opening up their national economies to foreign products and foreign-owned factories, as in the EPZs. Government trade officials from Japan, North America, Europe, and elsewhere negotiated with Southeast Asian governments to gain local markets for their countries' products. Multinational corporations saw opportunities to save on labor, land, transportation, and resource costs by locating their facilities and selling their products within the region. Thus factories manufacturing products for export in the EPZs were joined by foreign-owned factories manufacturing products they hoped to sell in Southeast Asia.

In opening national economies to the free market, governments focused especially on the crucial financial sector, which they hoped would provide needed investment capital for Southeast Asian firms eager to expand. Southeast Asian banks were given greater freedom to determine the kinds of investments they would make and were allowed to accept more money from foreign investors. Soon these banks were flooded with money from investors in the rich countries of the world, who hoped to profit as the banks invested in the region's growing economies. As a result, the stock prices of many public companies were inflated beyond their value. In addition, to make quick profits, the bankers made risky loans to real estate developers, often for high-rise office building construction. As a result, by the mid-1990s, many Southeast Asian cities had a glut of office space; in Bangkok alone, there was U.S.$20 billion worth of unsold office space.

Lifting the controls on investments was also made problematic by the extent of a kind of corruption known as **crony capitalism.** In most Southeast Asian countries, as elsewhere, corruption is encouraged by the close personal and family relationships between high-level politicians, bankers, and wealthy business owners. For example, in Indonesia, the most lucrative government contracts and business opportunities were for decades reserved for the children of former president Suharto, who ruled the country from 1967 to 1997. His children became some of the wealthiest people in Southeast Asia. This kind of corruption expanded considerably with the new foreign investment money, which was easily diverted to bribery or unnecessary projects that brought prestige to political leaders.

The cumulative effect of crony capitalism and the pursuit of quick returns on risky investments was that many ventures failed to produce any profits at all. In response, foreign investors withdrew their money on a massive scale. Before the crisis, in 1996, there was a net inflow of U.S.$94 billion to Southeast Asia's leading economies. In 1997, there was a net outflow of U.S.$12 billion. The panicked withdrawal of investment money was the immediate cause of the economic crisis.

The International Monetary Fund (IMF) made a major effort to keep the region from sliding deeper into recession. In 1998, the IMF implemented bailout packages worth some U.S.$65 billion, much of it used to rescue Southeast Asian banks. The IMF considered these banks essential to the region's economic stability, but critics argued that the bailouts gave the banks little incentive to improve their disastrous record. Moreover, the IMF's bailout packages were designed to protect foreign investors more than local economies and often came at the expense of local residents and laborers. The packages included structural adjustment programs, which, like those discussed in earlier chapters (see pages 139–140 and 374–376), required countries to cut government spending (especially on social services), reduce tariffs, and abandon other policies intended to protect domestic industries. Between 2001 and 2005, Southeast Asian economies registered growth rates between 4 and 5 percent, which marked an improvement over the negative growth rates of 1997–1999 but still fell short of pre-crisis levels. Local currencies regained their value, and trade expanded in all nations. Southeast Asian countries successfully expanded their older strategies, such as the establishment of EPZs, to attract additional multinational corporations to Southeast Asia. By 2006, the crisis, while still serving as an ominous reminder of the risks of globalization, had been more or less overcome.

While many effects of the Southeast Asian economic crisis still linger, its main long-term effect was to reduce the region's ability to compete with China in attracting new industries and investors. China now attracts more than twice as much foreign direct investment as Southeast Asia. Southeast Asia's wealthier countries, such as Singapore and Malaysia, lost investment to China and now must position themselves as locations offering more highly skilled labor and more high-tech infrastructure than China. On the other hand, China's rise in economic power is benefiting some of Southeast Asia's poorer countries, such as Vietnam, where wages are now considerably lower than China's. Vietnam has thus been able to attract investment that had been going to China.

The greatest challenges may be faced by Indonesia, where labor is not significantly cheaper than in China and where corruption and bureaucratic red tape have significantly reduced foreign investment in recent years. Nonetheless, Indonesia was actually helped out of the economic crisis by China's burgeoning demand for timber, cement, fuel, copper, and other resources, many of which come from Indonesia (and other Southeast Asian countries).

## The Association of Southeast Asian Nations

Southeast Asian countries trade more with the rich countries of the world than they do with one another (Figure 10.22). They export food, timber products, minerals, processed commodities, and manufactured finished goods and components to Japan, Australia, New Zealand, Europe, the Arab states, and the United States. They import manufactured products that they do not make themselves, including high-fashion consumer products, industrial materials, machinery, parts, and fuel (the region is a net importer of oil). Trade among countries within the region has been inhibited by the fact that they all export similar goods and traditionally have imposed tariffs against one another.

It is in this context that a region-wide free trade zone was created by the **Association of Southeast Asian Nations (ASEAN),** an increasingly important bulwark of both economic growth and political cooperation in the region. Started in 1967 as an anti-Communist, anti-China association, ASEAN now includes all the states in the region except Timor-Leste. It focuses on nonconfrontational accords that strengthen regional cooperation. An example is the Southeast Asian Nuclear Weapons–Free Zone Treaty signed in December 1995 by all ten Southeast Asian nations.

In 1992, ASEAN launched the **ASEAN Free Trade Association (AFTA),** patterned after the North American Free Trade Agreement and the European Union. Most tariffs will be reduced to 5 percent or less by 2010 for the older ASEAN members and by 2015 for Cambodia, Laos, Burma, and Vietnam. It is hoped that this reduction will lower production costs and make ASEAN's manufacturing industries more efficient, thus allowing ASEAN products to be priced lower and be more competitive in the global market.

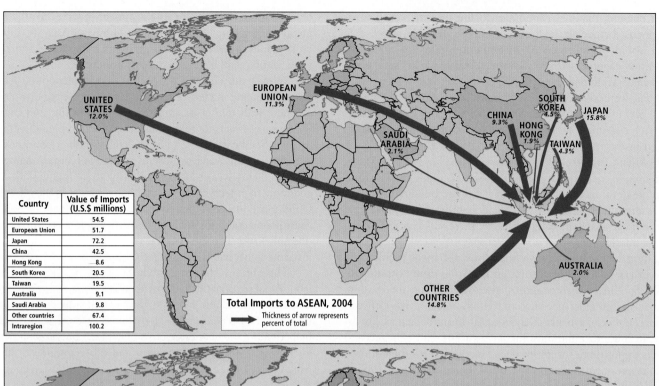

| Country | Value of Imports (U.S.$ millions) |
|---|---|
| United States | 54.5 |
| European Union | 51.7 |
| Japan | 72.2 |
| China | 42.5 |
| Hong Kong | 8.6 |
| South Korea | 20.5 |
| Taiwan | 19.5 |
| Australia | 9.1 |
| Saudi Arabia | 9.8 |
| Other countries | 67.4 |
| Intraregion | 100.2 |

**Total Imports to ASEAN, 2004**
Thickness of arrow represents percent of total

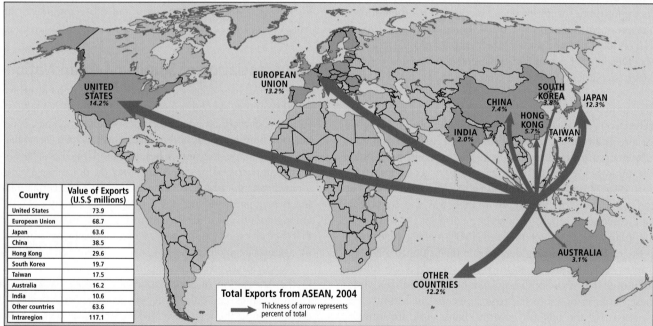

| Country | Value of Exports (U.S.$ millions) |
|---|---|
| United States | 73.9 |
| European Union | 68.7 |
| Japan | 63.6 |
| China | 38.5 |
| Hong Kong | 29.6 |
| South Korea | 19.7 |
| Taiwan | 17.5 |
| Australia | 16.2 |
| India | 10.6 |
| Other countries | 63.6 |
| Intraregion | 117.1 |

**Total Exports from ASEAN, 2004**
Thickness of arrow represents percent of total

**FIGURE 10.22** ASEAN imports and exports. The ASEAN countries are very active in world trade as both importers and exporters, but trade within the region (intraregional) is also very important. [Data from ASEAN *Statistical Yearbook* 2005, Chapter 5, at http://www.aseansec .org/SYB2005/chapter-5.pdf.]

An even greater shift in Southeast Asia's economies will occur if a free trade agreement is reached between ASEAN and China, which is expected to occur by 2010. Tariffs have already been reduced on a range of goods since talks on the potential free trade agreement began in 2002, and trade between ASEAN and China has grown faster than trade between ASEAN nations, so this relationship is actually hindering regional integration. China is now the region's fourth largest trade partner (well behind the EU, the United States, and Japan), buying primarily natural resources, and has increased its involvement in the region's economic development by providing loans and technical assistance on a number of infrastructure projects.

## Pressures For and Against Democracy

Although there are strong demands for a greater public voice in the political process in Southeast Asia, significant barriers to democratic participation exist (Figure 10.23). Undemocratic socialist regimes

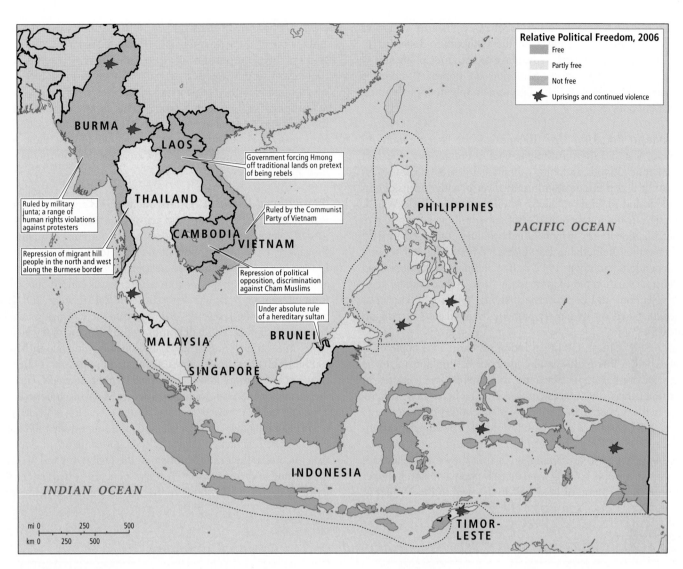

**FIGURE 10.23** **Relative political freedom in the region.** The state of political freedom in Southeast Asia is debatable, with most countries experiencing political violence or election fraud of some sort over the last decade. The categories here come from the nongovernmental organization Freedom House. [Adapted from "Map of Freedom 2006," at http://www.freedomhouse.org/template.cfm?page=363&year=2006.]

control Laos, Cambodia, and Vietnam, a military dictatorship runs Burma, Brunei is a sultanate, and in 2006 Thailand devolved (at least temporarily) from a parliamentary to a military government. There are few truly democratic institutions even in the slightly wealthier countries, such as Malaysia and the Philippines, or even in the rich city of Singapore. Some Southeast Asian leaders, such as Singapore's former prime minister Lee Kuan Yew, have argued that Asian values are not compatible with Western ideas of democracy. Yew and other leaders assert that Asian values are grounded on the Confucian view that individuals should be submissive to authority; hence Asian countries should avoid the highly contentious public discourse of electoral politics. But this position conveniently overlooks the Confucian expectation that governments should rule with justice and the needs of all society in mind and that citizens should act against corruption and misuse of power to restore the ideal social order. In fact, many people living under undemocratic, authoritarian regimes in the region have rebelled.

Neither economic development nor economic crisis seems reliably linked to progress toward democracy, which seems to move in fits and starts. In the late 1990s, during the economic crisis, Thais successfully agitated for constitutional provisions that reduced corruption and provided for regular elections. However, in 2006, when the Thai economy was doing modestly well, a military coup d'état, favored by the middle and upper classes, overthrew a democratically elected, but corrupt, populist prime minister, dissolved parliament, and changed the constitution to increase the role of the military in government. In Indonesia, after three decades of semidictatorial rule by President Suharto, the economic crisis spurred massive demonstrations that led to Suharto's resignation and a series of democratic parliamentary and presidential elections. People in Burma have been futilely protesting the rule of a military regime for more than a decade. The regime refused to step aside when the people elected Aung San Suu Kyi to lead a civilian reformist government in 1990. Suu Kyi has been under house arrest for most of the last

16 years, despite having won the Nobel Peace Prize in 1991, and the military remains in control. In the Philippines, disparities of wealth and privilege give the advantage to extremist politicians—alternately elitists or populists. Minorities are oppressed, unrest is common, and the military plays a prominent role in implementing government policies.

***Is Indonesia Breaking Up?*** Indonesia is the largest country in Southeast Asia and the most fragmented: it comprises more than 3000 inhabited islands stretching over 3000 miles (8000 kilometers) of ocean. It is also the most culturally diverse, with dozens of ethnic groups and multiple religions. Until the end of World War II, Indonesia was not a nation at all, but rather a loose assemblage of distinct island cultures, which Dutch colonists managed to hold together as the "Netherlands East Indies." When Indonesia became an independent country in 1945, its first president, Sukarno, hoped to forge a new nation out of these many parts. To that end, he articulated a famous and controversial national philosophy known as ***Pancasila,*** based on tolerance, particularly in matters of religion (Box 10.2).

In recent years, instability in Indonesia has many people wondering whether this multi-island country of 220 million might be headed for disintegration. Separatist movements have sprouted in four distinct areas. One of these movements has been successful, resulting in the formation of the independent country of Timor-Leste in 2002, but only after a long, bloody conflict. The large far eastern province of West Papua has a growing separatist movement that formed in response to discrimination against the local Melanesian population by people coming from densely populated Java in hopes of finding cheap, plentiful land. Just to the west in the Moluccas,

clashes between Christians and Muslims in 2002 killed thousands and encouraged the revival of a Christian-led separatist movement. In the far western province of Aceh in Sumatra, separatists battled Indonesian security forces for decades, largely because most of the wealth yielded by their province's resources, especially oil, was going to the central government in Jakarta. This dispute was defused after the disastrous tsunami of December 2004 focused international attention on Aceh. Jakarta tries sporadically to appease these movements by giving people in separatist areas greater control over their own affairs.

***International Terrorism.*** During the late 1990s, a series of bombs exploded in the Philippines and across Indonesia, and small terrorist cells were discovered in Malaysia, far southern Thailand, and Singapore. Until the bombing of the Sari Hotel in Bali (Indonesia) in October 2002, which killed nearly 200 foreign tourists, terrorist activity in the region was local in nature. It was carried out by local militant groups pursuing domestic political agendas and grievances—most often revenge for government campaigns against Muslim separatists. The Bali bombing, the bombing of the Marriott Hotel in Jakarta in August 2003, and ongoing violence in Muslim southern Thailand have drawn attention to apparent connections between local groups and international terrorist networks. However, the election of secular political parties in Malaysia, Indonesia, and Thailand in 2004 showed that most Southeast Asians, who are accustomed to a tolerant version of Islam, do not support the militants.

Despite heightened efforts by state governments in Southeast Asia to combat the rise of terrorism in the region, the task they face is daunting and complicated. A major concern is that, rather than trying to address local conditions that fuel militancy, states are resort-

---

## BOX 10.2 | CULTURAL INSIGHT *Pancasila:* Indonesia's National Ideology

The five precepts of *Pancasila,* the ideology at the heart of Indonesian life, are belief in God and the observance of *conformity, corporatism, consensus,* and *harmony.* These last four precepts could be interpreted as discouraging dissent or even loyal opposition, and they seem to require a perpetual stance of boosterism. For some people, the strength of *Pancasila* is the emphasis on harmony and consensus that seems to ensure there will never be either an Islamic or a Communist state. Others note that conformity and corporatism counteract the extreme ethnic diversity and geographic dispersion of the country. But conformity and corporatism also have a chilling effect on participatory democracy and on criticism of the president and the army. Indonesia's two political parties are state controlled and, in fact, there was no orderly democratic change of government from Indonesia's founding after World War II until the first presidential election in 2004.

An example of how *Pancasila* becomes part of the daily lives of people and reinforces traditional relationships is provided by the state-sponsored women's organization Dharma Wanita (Women's

Duty). The organization is modeled on an organization of U.S. military wives, and membership is obligatory for all female government employees and the wives of male government employees. Its purpose is to teach women to serve and support the careers of their husbands and male colleagues, regardless of the women's actual marital or career status. There is a chapter in every government office, and a woman holds office in Dharma Wanita according to her husband's position in an agency or firm. The ideals of Dharma Wanita have spread with immigrant Indonesian women to other parts of the region and to North America.

*Sources:* Adapted from Saskia Wieringa, "The politicization of gender relations in Indonesia," *The Colonial Widow* (The Hague: The Digital City Project, 1995); http://www.indonesiachicago.org/dharmawanita /events.htm; Dr. Kathryn Robinson, Senior Fellow, Department of Anthropology, Australian National University, "Gender equity and the transition to democracy in Indonesia," December 4, 2001, Washington, D.C., at http://www.usindo.org/Briefs/Kathryn%20Robinson%20Dec%204 .htm; *The Economist* (October 7, 2004).

ing to repressive measures such as illegal deportations and indefinite detention of dissidents without trial. Governments in the region have also used the international campaign against terrorism as a cover to crack down on opposition groups that merely challenge the government's authority and policies.

## SOCIOCULTURAL ISSUES

The cultural complexity of Southeast Asia supplies an array of strategies for responding to rapid and pervasive change. But culturally complex societies can often be in precarious balance if national mores fail to help citizens relate well across ethnic and religious lines. Furthermore, attitudes about family size and gender roles are changing in response to new economic circumstances.

### Cultural Pluralism

Southeast Asia is a place of **cultural pluralism** (or **cultural complexity**) in that it is inhabited by groups of people from many different backgrounds who have lived together for a long time yet have remained distinct, partly because they lived in isolated pockets separated by rugged topography and spans of ocean. Small groups of descendants of the Australo-Melanesians who came into the region some 40,000 to 60,000 years ago still live primarily by hunting, gathering, and shifting cultivation in the interior forests of several larger islands, primarily Sumatra, Borneo, Sulawesi, New Guinea, and the Philippines, and on the Malay Peninsula. Far more numerous today are descendants of the Austronesians who migrated from southern China about 5000 years ago. The descendants of these two major divisions of the earliest inhabitants of the region are usually referred to collectively as indigenous peoples. They are distributed in small patches from northern Burma and Thailand to southern Indonesia, living in both coastal zones and mountainous interiors. Their past isolation has contributed to distinctive cultural differences, with language one of the clearest markers of ethnicity among groups of both Australo-Melanesian and Austronesian descent.

Over the last 2000 years, newer immigrants have arrived, first from Arabia, Central Asia, India, China, Japan, and Korea. Then, after 1500, Europeans began colonizing the region. Today, Southeast Asia is one of the most ethnically diverse regions in the world.

Most recently, Southeast Asia's cultural mix has been broadened by ideas from around the globe: mercantilism, free market capitalism, Communism, nationalism, consumerism, and environmentalism. These ideas and the material culture associated with them—from advertising to music to spandex to green politics—have also modified the life and landscapes of Southeast Asia (Figure 10.24). Given the introduction of such new ideas, one might expect that by now Southeast Asia would be a homogenized melting pot. To some extent, notably in the cities, it is showing some signs of moving in that direction. In some of the region's countries, however, dozens of different languages are still spoken. (Indeed, it is estimated that 1000 of the world's 6000 or so still active languages are spoken here.) People practice many different religions, and neighboring families may trace their roots to remarkably different

**FIGURE 10.24** Too Phat, Malaysia's hip-hop duo. These young musicians, performing here in Singapore in 2004, are popular in Southeast Asia and are gaining global appeal. They have made several albums, some in English, which can be purchased online or in music stores in the United States. [Reuters/Bazuki Muhammad/Newscom.]

racial and ethnic origins. Today, the barriers separating groups are falling as modernization and industrialization draw large numbers of people into the cities and resettlement projects move several ethnic groups together onto newly opened lands. Migration and cross-cultural marriage are creating ethnically mixed groups who are unsure of their heritage and may never have seen their ancestral landscapes.

***The Overseas Chinese.*** One group that is prominent beyond its numbers in Southeast Asia is the Overseas (or ethnic) Chinese (Figure 10.25) (see Box 9.7 on page 515). Small groups of traders from southern and coastal China have been active in Southeast Asia for thousands of years, and over the centuries, there has been a constant trickle of immigrants from China. The ancestors of most of today's Overseas Chinese, however, began to arrive in large numbers during the nineteenth century, when the European colonizers needed labor for their plantations and mines. Later, many of those who fled China's Communist revolution after 1949 sought permanent homes in Southeast Asia's trading centers. Today, over 26 million Overseas Chinese live and work in Southeast Asia.

Chinese commercial activity throughout the region has reinforced the perception that the Chinese are diligent, clever, and civic-minded businesspeople, often working very long hours. At the same

**FIGURE 10.25** Overseas Chinese in Malaysia. Malay women sell breakfast from a portable stall in front of a store operated by a Malaysian of Chinese ethnicity in Lenggeng. Most businesses in Lenggeng and across Malaysia are run by ethnic Chinese (who are Malaysian citizens), and most of their customers are ethnic Malays. [Reuters/Bazuki Muhammad/Newscom.]

time, despite the fact that most have modest incomes, the Chinese in Southeast Asia have the reputation of being rich and influential in government and commerce, yet strongly loyal to their own group, often importing kin as employees rather than hiring local people. Many low- and middle-income Southeast Asian workers who were hurt by the financial crisis of the late 1990s blame their sorrows on the Overseas Chinese. The Chinese are indeed industrious, and with their region-wide connections and access to start-up money, they have been well positioned to take advantage of the new growth sectors of modernizing economies. Sometimes new Chinese-owned enterprises have put out of business older, more traditional establishments that many local people depended on. For example, in the town of Klaten in central Java, an open-air bazaar where local people once sold an array of goods has been shut down to make way for a government-sponsored, air-conditioned shopping mall. Only the Chinese can afford the expensive store rents in the new mall.

During the economic crisis, resentment against the Chinese brought a wave of violence. Chinese people were assaulted, their temples desecrated, their homes and businesses destroyed. Indeed, conflicts involving the Overseas Chinese have occurred in Vietnam, Malaysia, and many places in Indonesia (Sumatra, Java, Kalimantan, and Sulawesi). Some Overseas Chinese, increasingly aware that the global economy presents dangers as well as opportunities to their group, are attempting to alleviate their situation through public education. There are Overseas Chinese study centers in nearly all the countries of the region, and the Overseas Chinese now maintain a higher profile as civic participants within their local communities. At the same time, they are establishing more formal international connections and services, such as investment banking companies, that will give them financial flexibility and spatial mobility should life in one particular place become too inhospitable.

Complicating the situation of the Overseas Chinese are ventures by China to curry favor with people in many parts of Southeast Asia through what are called "soft power" initiatives. Chinese trade emissaries, museum exhibits on Chinese culture, language courses, study trips to China, and development loans are just some of China's efforts to build profitable future relationships with Southeast Asian countries.

## Religious Pluralism

The major religious traditions of Southeast Asia include Hinduism, Buddhism, Confucianism and Taoism, Islam, Christianity, and animism (Figure 10.26). All originated outside the region, with the exception of the animist belief systems of the indigenous peoples. Animism takes many different forms, but in general such natural features as trees, rivers, crop plants, and the rains all carry spiritual meaning and are the focus of festivals and rituals to give thanks for bounty and to mark the passing of the seasons.

The other religions were brought primarily by traders, priests (Brahmin and Christian), and colonists, and their patterns of distribution reveal an island-mainland division. Buddhism is dominant on the mainland, especially in Burma, Thailand, and Cambodia. In Vietnam, people practice a mix of Buddhist, Confucian, and Taoist beliefs that originated in China. Islam is dominant in Indonesia (the world's largest Muslim country), on the southern Malay Peninsula, and in Malaysia, and it is growing in the southern Philippines. Roman Catholicism is the predominant religion in Timor-Leste and in the Philippines, where it was introduced by Portuguese and Spanish colonists, respectively. Hinduism first arrived with Indian traders thousands of years ago and was once much more widespread; now it is found only in small patches, chiefly on the islands of Bali and

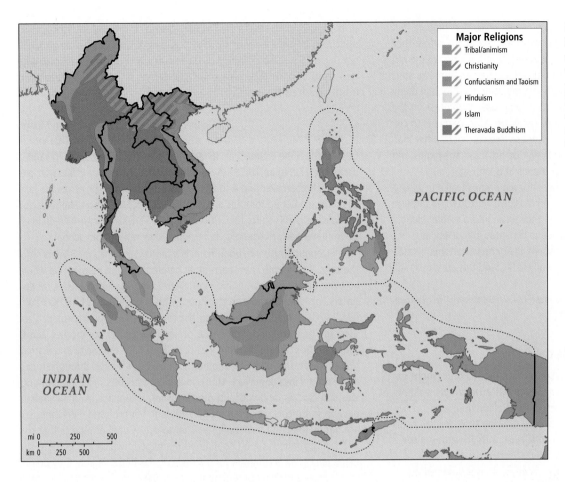

FIGURE 10.26 Religions of
Southeast Asia. Southeast Asia is
religiously very diverse: five of the
world's six major religions are
practiced there. Animism, the oldest
belief system, is found in both island
and mainland places. [Adapted from
*Oxford Atlas of the World* (New York:
Oxford University Press, 1996), p. 27.]

Lombok, east of Java. Recent Indian immigrants who came as laborers in the twentieth century during the later part of the European colonial period have reintroduced Hinduism to Burma, Malaysia, and Singapore, but only as minority communities.

All of Southeast Asia's religions have undergone change as a result of exposure to one another. Many Muslims and Christians believe in spirits and practice rituals that have their roots in animism. Hindus and Christians in Indonesia, surrounded as they are by Muslims, have absorbed ideas from Islam, such as the seclusion of women. On the other hand, Muslims have absorbed ideas and customs from indigenous belief systems, especially ideas about kinship and marriage, as illustrated in the following vignette.

**Personal Vignette**  Marta is a *Pemaes*, a woman who prepares a bride for a traditional Javanese wedding (Figure 10.27). Nearly all Javanese are Muslims, but Islam does not have elaborate marriage ceremonies, so colorful animist rituals have survived. Usually, the bride and groom have never met, their families having worked out the match. Hence their first meeting at the wedding ceremony is surrounded by a great deal of mystery. Marta's job is to reinforce this mystery and to prepare the couple for a life together. She bathes and perfumes the bride, puts on her makeup, and dresses her, all the while making offerings to the spirits of the bride's ancestors and counseling her about how to behave as a wife and how to avoid being dominated by her husband.

FIGURE 10.27 The Dahar Klimah (Dahar Kembul). In this phase of a traditional Javanese wedding, the *Pemaes* gives the bride a bundle of food (yellow rice, fried eggs, soybeans, and meat). Then the bridegroom makes three small balls of the food and feeds it to the bride; she then does the same for him. The ritual reminds them that they should share their belongings joyfully together. [Courtesy of Dirk Vranken, http:users.skynet.be/sky86158.]

The groom, meanwhile, is undergoing ceremonies that prepare him for marriage as well. Each will sign the marriage certificate before they meet. In the traditional Javanese view, it is best that a young couple not be in love—that way, they will not fall out of love. Rather, from the start they will hold each other at arm's length, not investing too much of themselves in a relationship that is bound to change over the course of the decades as they mature and their family grows up. Marta's role is to create a magically reinforced bond between these two strangers. After marriage, as before, a man's closest friends will be his male age-mates and kin, and a woman's will be her female friends and kin.

Despite these elaborate preparations for marriage, divorce in Malaysia and Indonesia is fairly common among Muslims, who often go through one or two marriages early in life before they settle into a stable relationship. Although the prevalence of divorce is lamented by society, it is not considered outrageous. Apparently, ancient indigenous customs, predating Islam, allowed for mating flexibility early in life, and this attitude is still tacitly accepted in Java.

*Adapted from Walter Williams, Javanese Lives (Piscataway, N.J.: Rutgers University Press, 1991), pp. 128–134.* ■

## Family, Work, and Gender

Family organization in Southeast Asia is variable, as one might expect in a region of so many cultures. Some patterns are remarkably different from the patriarchal patterns found elsewhere around the world. Nonetheless, there is a patriarchal overlay that derives from many sources—Islam, Hinduism, and Christianity, to name a few.

***Family Patterns.*** Throughout the region, it is common for a newly married couple to reside with the wife's parents. Along with this custom is a range of behavioral rules that empower the woman in a marriage, despite some basic patriarchal attitudes. For example, a family is headed by the oldest living male, usually the wife's father. When he dies, he passes on his wealth and power to the husband of his oldest daughter, not to his own son. (A son goes to live with his wife's parents and inherits from them.) Hence a husband may live for many years as a subordinate in his father-in-law's home. Instead of the wife being the outsider under the hegemony of her mother-in-law, as, for example, in Southwest Asia and India, it is the husband who must kowtow. The inevitable tension between the wife's father and the son-in-law is resolved by the custom of ritual avoidance. The wife manages communication between the two men by passing messages and even money back and forth. Consequently, she has access to a wealth of information crucial to the family and has the opportunity to influence each of the two men.

In urban families influenced by modernization, the young couple may choose to live apart from the extended family. This arrangement takes the pressure off the husband in daily life. Because this nuclear family unit is often dependent entirely on itself for support, wives usually work for wages outside the home. The drawback of this compact family structure, as many young families have discovered in Europe and the United States, is that there is no pool of extended kin available to help working parents with child care and housework. And since no one is left to help elderly par-

ents maintain the old rural family home, increasingly, tiny urban apartments must accommodate a young family and one or more aged grandparents in need of care. On the brighter side, working grandmothers in their vigorous middle years can often help their adult children financially.

**Personal Vignette** Buaphet Khuenkaew, 35, lives in Ban Muang Wa, a village near the northern Thai city of Chiang Mai. She married at 18 and has two children: a son, 10, and a daughter, 17 (Figure 10.28). A Buddhist with a sixth-grade education, she is both a homemaker and a seamstress. Six days a week she drives the family motor scooter 30 minutes to her job in Chiang Mai, where she sews buttonholes in men's shirts for 2800 baht (U.S.$118) a month. The children perform weekday household chores when they return from school.

Buaphet's husband, Boontham, is a farmer who is about 5 years older than she. The couple knew each other before they were married. He would visit her at her parents' home, and eventually they fell in love. One day he and his parents came to the house with a bride price of 10,000 baht (U.S.$420) in gold, and he asked her to marry him. She accepted. They used the gold to build their house on land owned by her mother, across the street from where Buaphet was born. They have electricity and a small television set. Drinking water comes from a well, is filtered through stones, and is stored in ceramic jars. Many household activities, including bathing, washing clothes, and washing dinner dishes, take place in small shelters outside, but meals are eaten inside. Although her husband feels that men are rightly regarded as superior in Thai society, Buaphet reports that she and her husband have an egalitarian marriage in which all decisions are made jointly. As is common for married couples in the region, they do not spend much of their leisure time together. She regularly spends time with her female friends and relatives, he with his male friends and family. Buaphet says she is happy with her life, but also regularly complains about not having appliances and more up-to-date furnishings as her friends have.

*Adapted from Faith D'Alusio and Peter Menzel, Women in the Material World (San Francisco: Sierra Club Books, 1996), pp. 228–239.* ■

**FIGURE 10.28** Buaphet sharing the morning meal with her family before setting off to work. [Joanna Pinneo/Material World.]

**TABLE 10.3** Gender comparisons for Southeast Asia and the United States: Income, education level, and labor force participation

| Country (I) | Estimated earned income (in U.S.$, adjusted for PPP), 2003 (II) | | Eligible students enrolled in post-secondary education (percent), 2002/2003 (III) | | Labor force participation outside the home (percent), 2003 (IV) | |
|---|---|---|---|---|---|---|
| | Female | Male | Female | Male | Female | Male |
| Brunei (HDI 33[a]) | 11,716 | 26,122 | 17 | 10 | 51 | 80 |
| Burma (HDI 129) | 1,011[b] | 1,389 | 15 | 9 | 66 | 88 |
| Cambodia (HDI 130) | 1,807 | 2,368 | 2 | 5 | 80 | 83 |
| Indonesia (HDI 110) | 2,289 | 4,434 | 15 | 19 | 56 | 82 |
| Laos (HDI 133) | 1,391 | 2,129 | 4 | 9 | 75 | 88 |
| Malaysia (HDI 61) | 6,075 | 12,869 | 33 | 26 | 49 | 79 |
| Philippines (HDI 84) | 3,213 | 5,409 | 34 | 27 | 50 | 81 |
| Singapore (HDI 25) | 16,489 | 32,089 | ND | ND | 50 | 78 |
| Thailand (HDI 73) | 5,784 | 9,452 | 42 | 36 | 73 | 86 |
| Vietnam (HDI 108) | 2,026 | 2,964 | 9 | 11 | 73 | 81 |
| United States (HDI 10) | 29,017 | 46,456 | 96 | 70 | 60 | 72 |

[a]HDI refers to the Human Development Index (see Table 10.4).

[b]Data from *CIA World Factbook*, 2005.

*Source: United Nations Human Development Report 2005* (New York: United Nations Development Programme), Table 25, Table 27, and Table 28.

**Work Patterns.** As the economies of Southeast Asia are transformed, so are the work lives of its people. The changing pattern of work in Southeast Asia is more complex than a simple shift from agricultural work to factory work or from rural to urban occupations. Agriculture is changing from small-plot family farming to large-scale commercial operations. The larger farms need fewer, but more highly skilled, workers. Rural communities, which used to be inward-looking and self-sufficient, now produce crops primarily for export.

Men work primarily in agriculture and manufacturing, but are increasingly finding jobs in the service industry. They have a higher rate of paid employment than women throughout the region (Table 10.3, column IV), but these data miss all the work that women do in the home, in the informal economy, and in traditional agriculture. When they do work in the formal economy, women's wages average only about half those of men, except in the very low-wage countries of Burma, Cambodia, and Vietnam (Table 10.3, column II), where women's wages are about two-thirds of men's. Changes may be on the way: in Brunei, Burma, Malaysia, the Philippines, and Thailand, significantly more women than men are completing training beyond secondary school (Table 10.3, column III). Hence, if training qualifications were the sole consideration, women would appear to have an advantage over men for future employment in those Southeast Asian industries, such as information technology, in which education is an advantage.

## Migration

There is a long-standing culture of migration in Southeast Asia. Well before the era of European colonization, seafarers and traders from Arabia, India, and China contributed to the region's cultural mix. Today, many people leave the region in search of work or move to new areas within the region as part of vast **resettlement schemes.** Others come into the region from elsewhere in Asia to work in low-wage agricultural jobs or in high-wage technology-related careers. Recently, as in so many places worldwide, rural-to-urban migration within the region has been particularly significant.

*Rural-to-Urban Migration.* Southeast Asia as a whole is just 38 percent urban, but the rural-urban balance is changing quickly in response to global market forces. Malaysia is already 62 percent urban, the Philippines 48 percent, and Singapore 100 percent. Often the focus of migration is the capital of a country, which may become a primate city—one that is at least two or three times the size of the second largest city and overwhelmingly dominates the economic and political life of the country. Bangkok is nearly 10 times larger than Thailand's next largest metropolitan area, Chiang Mai, and Manila is 9 times larger than Cebu, the Philippines' second city.

Individuals and families leave rural areas to escape poverty in the countryside. Commercial agriculture often displaces subsistence cultivators, who must then obtain cash to buy what they previously provided for themselves. These are the **push factors** in migration. **Pull factors,** in contrast, are those that attract people to the city, such as higher-paying and more reliable employment. Rural-to-urban migrants often leave behind young and aged relatives, whom they must then support with remittances. Rarely, however, can primate cities provide sufficient jobs, housing, and services for all the new arrivals. Of all the primate cities in Southeast Asia, only Singapore provides well for nearly all of its citizens.

| Personal Vignette | Mak (age 33) and Lin (age 27), husband and wife, left their two sons in the care of her parents in a village north of Bangkok. They could not afford the school fees for even one of their sons on their rural wages; yet they hoped to educate both boys, even though educating just the eldest is the custom. So Mak and Lin traveled by bus for 10 hours to reach Bangkok, where—after several anxious days—they both found grueling work unloading bags of flour from ships in the harbor.

Each day when they finished their work, they walked miles to their quarters in one part of a tiny houseboat anchored, with thousands of others, on the Chao Phraya River. That river is Bangkok's low-income housing site, its source of water, its primary transport artery, and its sewer. Mak and Lin had to step gingerly across dozens of boats to get to theirs, intruding repeatedly on the privacy of their fellow river dwellers. They bathed in the dangerously polluted river, washed their sweaty, flour-covered work clothes by hand, and cooked their dinner of rice over a Coleman stove.

Their only entertainment was provided by the private lives of their too numerous and too near-at-hand neighbors, who were often drunk on cheap liquor. For 2 years, Mak and Lin sent remittances to their family and managed to save enough to buy a bicycle, to pay the school fees, and to tide them over for several months at home. Although they were glad to be out of Bangkok, both nevertheless agreed that they would eventually go back if they could not find work near their village.

*Source: Alex Pulsipher's field notes, 2000.* ∎

**Resettlement.** Governments in Southeast Asia have developed resettlement schemes to move large numbers of rural people from one part of a country to another for a variety of reasons. Beginning in 1904, Dutch colonists in Indonesia began to move people from the islands of Java, Bali, and Madura to land only lightly occupied by indigenous people on the outer islands, on Sumatra and Sulawesi, in Kalimantan, and in West Papua. These immigrants supplied labor on plantations growing such export crops as rubber, coconuts, and palm oil. Much larger resettlement efforts on these same islands began after 1950. More than 5 million people have been relocated, making the Indonesian effort one of the largest land resettlement schemes ever.

The reasons for resettlement in Indonesia have changed over time, from promoting food production, especially national self-sufficiency in rice, to rural development, to population redistribution. Resettlement also dovetails with government efforts to assimilate Indonesia's outlying indigenous culture groups into mainstream society. Such policies are carried out by relocating people, especially Javanese, from Indonesia's densely occupied cultural heartland in western Java to outlying areas occupied by indigenous people (Figure 10.29). It is also Indonesian policy to disperse people from the crowded heartland to avoid political unrest there. Many thousands of people who are not part of the formal resettle-

**FIGURE 10.29** Results of resettlement. Designated SP6, this new settlement, carved out of the West Papua rain forest, is part of the Indonesian government's resettlement program. People who elect to move here receive a one-way air ticket, a house, 5 acres (2 hectares) of land, and a year's supply of rice. Estimates are that 5 million to 6.5 million people have participated in transmigration in Indonesia since 1950. [George Steinmetz.]

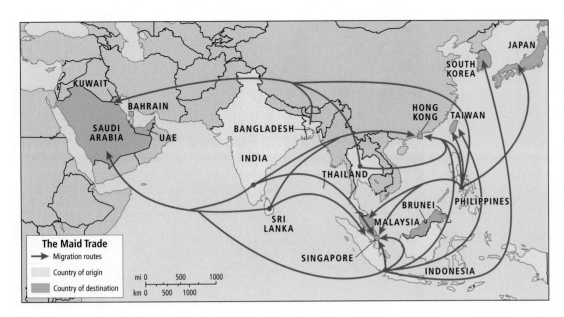

**FIGURE 10.30** Interregional linkages: The "maid trade." In the 1990s, between 1 million and 1.5 million Southeast Asian women were working elsewhere in Asia (including the Arab states) as domestic servants. It is estimated that, by 2005, the number had more than doubled, with the majority coming from the Philippines, Indonesia, and Sri Lanka. [Adapted from Joni Seager, *The Penguin Atlas of Women in the World* (New York: Penguin Books, 2003), p. 73; with updated information from the Migration Policy Institute, at http://www.migrationinformation.org/Profiles/display.cfm?ID=364.]

ment schemes have joined the stream of people to the outer islands. The results have been unfortunate. In some cases, thousands of informal migrants moving onto the lands of indigenous peoples have built shantytowns and destroyed native habitats for plants and animals. Many thousands of acres of forest have been cleared for agricultural resettlement in areas where the tropical soils are too fragile to sustain cultivation for more than a year or two. Severe environmental damage, ethnic discord, and breaches of human rights have been common.

In 2000, the indigenous people began to fight back, sometimes violently, in the Moluccas, on Sulawesi, and in Kalimantan. For example, the indigenous Dayak people of Kalimantan, many of whom continue to live in traditional ways, launched an attack on settlers from Madura. The Dayaks complain that their lands were taken without their consent and for far too little payment. Dayaks have killed hundreds of Madurese settlers.

***Extraregional Migration.*** **Extraregional migration,** or migration to countries outside the region, is especially important in Southeast Asia because of its effects on families, income, and employment. Increasingly, remittances from migrants around the world are a major contributor of **foreign exchange**—foreign currency, such as U.S. dollars, that Southeast Asian countries need to purchase imports. Filipinos working abroad are that country's largest source of foreign exchange, sending home over U.S.$6 billion, increasing household income by an average of 40 percent.

Extraregional migration may be either short-term or permanent. Recently, women have constituted well over 50 percent of the more than 8 million migrants. Many skilled nurses and technicians from the Philippines work in European, North American, and Southwest Asian cities. Southeast Asian Muslim women (chiefly from Indonesia) have become part of the "maid trade" of 1 million or more who work in the homes of wealthy citizens in the Arab states of Saudi Arabia, Kuwait, and the United Arab Emirates (Figure 10.30). Whereas

about 10,000 Indonesian women migrated to Saudi Arabia each year in the 1980s, by 1998 this number had risen to over 380,000 per year. This exponential increase was fueled in part by the Southeast Asian economic crisis of the late 1990s, when many women lost their factory jobs and had to seek alternative employment opportunities. Skilled male workers from Southeast Asia are especially well known in the world merchant marine. They typically serve in the lower echelons of shipboard occupations, working for 6 months or more at a stretch as seamen, cooks, or engine mechanics. In recent years, some have advanced to officer status. Their earnings are usually sufficient to provide a middle-class lifestyle and education for their families.

***Refugees from Conflict and Natural Disaster.*** In this region, forced migration still plays a role in the movement of people; violence and natural disasters are the most prevalent causes. During the last half of the twentieth century, millions of mainland Southeast Asians fled into neighboring states to escape protracted conflict. Thailand, Laos, and Cambodia, in particular, received many refugees during and after the Vietnam War. Currently, repression by Burma's military government has resulted in at least 680,000 refugees, with 500,000 displaced within Burma, and 180,000 having fled to Thailand (Figure 10.31a), Bangladesh, India, and Malaysia. Discord on the various islands of Indonesia has also sent people fleeing, usually to neighboring islands. Although it may appear that this discord is between ethnic or religious groups, in fact disputes usually arise not over cultural differences, but over issues such as resettlement (in the Moluccas, Sulawesi, and Kalimantan), political autonomy (in Timor-Leste), and unsustainable extraction of resources from peripheral provinces by the central government (in West Papua and Aceh). The tsunami of December 2004 complicated the refugee picture when it displaced well over 130,000 people in Sumatra alone (in addition to killing an equal number of people), plus several thousand in Thailand and Burma. Most went to Malaysia or to another part of their home country, and many have

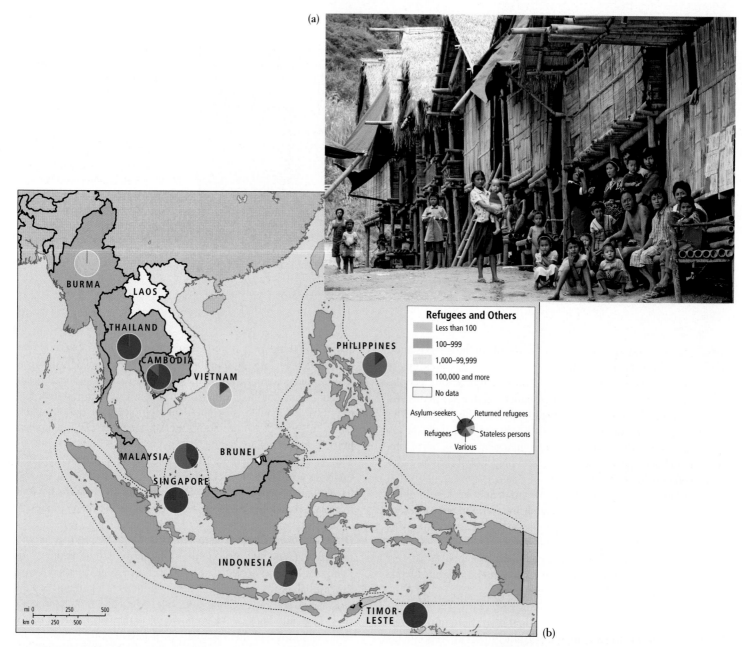

**FIGURE 10.31** Refugees and other "persons of concern" at the end of 2005. (a) Karen refugees from Burma at Tham Hin camp near Bangkok. (b) Nearly every country in the region has refugees. The data on these persons are generally provided by each government based on its own definitions and collection methods and are therefore provisional. [(a) Sukree Sukplang/ Reuters/Landov. (b) Data from "2005 global refugee trends: Statistical overview of populations of refugees, asylum-seekers, internally displaced persons, stateless persons, and other persons of concern to UNHDR," Table 1, June 2006, at http://www.unhcr.org/cgi-bin/texis/vtx/events/opendoc.pdf?tbl=STATISTICS&id =4486ceb12.]

now returned home. Figure 10.31b shows the numbers of refugees reported by each country.

## ENVIRONMENTAL ISSUES

Virtually everywhere in Southeast Asia, environments are severely affected by population pressure and by such commercial activities as mining, logging, and export agriculture (Figure 10.32). Most Southeast Asian countries were once rich in natural resources, including oil and natural gas, timber and other forest products, minerals (gold, gemstones, tin, and copper), and fertile soils in river valleys and volcanic areas. These resources have been rapidly depleted—often extracted by foreign-owned companies and sold abroad, often at unsustainably low prices, with only a fraction of the profits reinvested in the host countries. In this region, as elsewhere, it is useful to ask the question, "development for whom?"

People in Europe, North America, and Japan are on the demand side of the Southeast Asian resource equation. We may, for

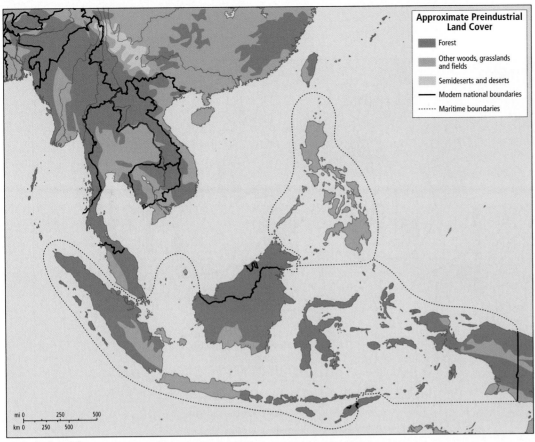

**Approximate Preindustrial Land Cover**

- Forest
- Other woods, grasslands and fields
- Semideserts and deserts
- Modern national boundaries
- Maritime boundaries

(a) Preindustrial

**FIGURE 10.32** Land cover (preindustrial) and human impact (2002). Southeast Asia has been occupied by humans for hundreds of thousands of years. Though the land cover was modified by forest clearing and agricultural terracing in the preindustrial era (**a**), much of the human impact was not apparent at this scale. By 2002 (**b**), human impact was intensive and extensive, affecting land, water, and air. [Adapted from *United Nations Environment Programme, 2002, 2003, 2004, 2005, 2006* (New York: United Nations Development Programme), at http://maps.grida.no /graphic/human_impact_year_1700 _approximately and http://maps.grida.no /graphic/human_impact_year_2002.]

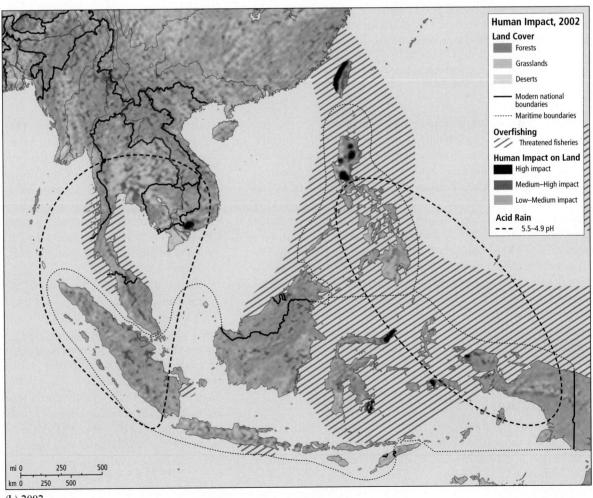

**Human Impact, 2002**

**Land Cover**
- Forests
- Grasslands
- Deserts
- Modern national boundaries
- Maritime boundaries

**Overfishing**
- Threatened fisheries

**Human Impact on Land**
- High impact
- Medium–High impact
- Low–Medium impact

**Acid Rain**
- 5.5–4.9 pH

(b) 2002

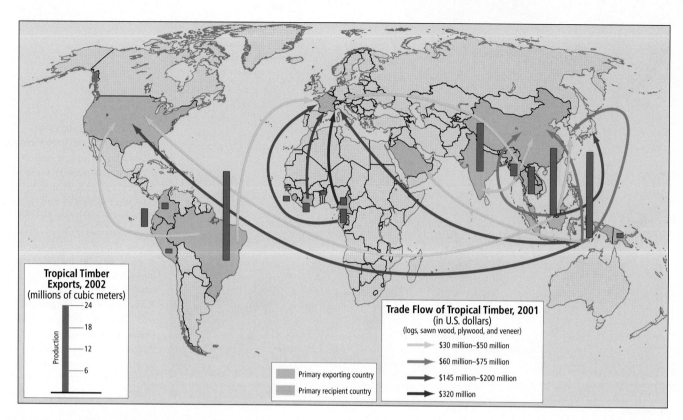

**FIGURE 10.33** Trade in tropical timber. Most of the consumer demand for tropical timber is in North America, Europe, and Japan. Large quantities of tropical timber sold to China are made into furniture, plywood, and flooring, which are then sold to consumers in the developed world. The trade shown here is mostly legal, but investigative reporters say that much of the wood fueling China's wood-processing industries is harvested illegally in Southeast Asia, North Korea, Russia, and Africa. [Adapted from United Nations Conference on Trade and Development (New York: United Nations, 2004), pp. 44–45, at http://www.unctad.org/en/docs/ditccom20041ch22_en.pdf; and Peter S. Goodman and Peter Finn, "A corrupt timber trade," *Washington Post National Weekly Edition*, April 9–15, 2007, pp. 6–9.]

example, have beautiful tropical hardwood paneling in our homes, purchased at an attractively low price at the local branch of a multinational home improvement warehouse, or we may use computer paper by the case. Although we live far from Southeast Asia, we often support our lifestyles with its resources. In the United States, for example, we consume 690 pounds (313 kilograms) of paper per capita, the highest rate in the world. Japan's per capita paper use is 500 pounds (225 kilograms); and most of Europe consumes 220 to 440 pounds (100 to 200 kilograms) per person. By contrast, most Southeast Asian countries consume less than 67 pounds (30 kilograms) per capita. Much of the wood fiber in paper products comes from Brazil, the Philippines, Indonesia, and other tropical pulpwood producers. The map in Figure 10.33 shows the destinations for Southeast Asia's timber products. Although we have paid for our paneling and paper, we must ask whether the price was fair, especially in light of the environmental damage that results from logging.

## Deforestation

Southeast Asia has the second highest rate of deforestation after sub-Saharan Africa. Environmentalists estimate that 13 to 19 square miles (34 to 50 square kilometers) of Southeast Asia's rain forests are destroyed each day, a rate 50 percent faster than in the Amazon. Figure 10.34 shows the activities that result in deforestation. Companies that have legal rights to log the land over a period of 25 years sometimes choose to "cut and run" in just a few months and thus avoid fulfilling already woefully ineffective conservation agreements. Moreover, illegal logging, such as that described at the opening of this chapter, often outpaces legal logging, and its impact on the environment is even worse because it is done secretly and in a matter of hours, using heavy equipment, and with no attention to preserving natural habitat.

Indirectly, much of the deforestation in Southeast Asia is linked to population growth and poverty. Traditional, nonintensive shifting cultivation on forested lands was especially effective and sustainable when the population density was low enough to allow the use of small plots and long fallow periods, during which the forest could fill in the clearings. But because of Southeast Asia's recent rapid population growth—to 557 million in 2005 (more than 50 times that in 1800)—cultivation is expanding even to the steepest of slopes. The resulting erosion is drastically changing the shape of the denuded land. Without a forest cover, water moves across the landscape too rapidly to be absorbed, carrying away precious topsoil, which is very thin in tropical environments. Watercourses cut

**Sources of Legal Deforestation**

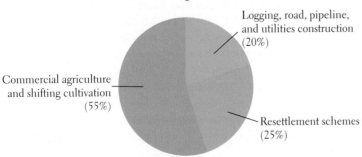

Logging, road, pipeline, and utilities construction (20%)

Commercial agriculture and shifting cultivation (55%)

Resettlement schemes (25%)

**FIGURE 10.34** Legal deforestation in Southeast Asia, 2004. This pie diagram is just an estimate of the proportion of responsibility borne by various legal activities for deforestation in Southeast Asia. Much of the deforestation, however, is the result of illegal logging. The World Wildlife Fund for Nature estimates (2004) that 83 percent of timber production in Indonesia stems from illegal logging. [Data from "Forestry issues—deforestation: Tropical forests in decline," at http://www.fao.org/forestry/site/33592/en; http://www.panda.org/about_wwf/what_we_do/forests/problems/forest_illegal_logging/index.cfm.]

new gullies and ravines in the uplands and deposit tons of silt in the lowlands and surrounding marine environments.

The Philippines serves as an example of widespread deforestation. There, logging for timber is the primary culprit, though clearing for commercial agriculture is next. In 1900, 75 percent of the Philippines was covered with **old-growth forest,** dominated by large trees in the mature stages of their life cycle. By 1990, only 2.3 percent of the country was covered with old-growth forest, with most of it confined to small patches in remote mountainous areas. On parts of Sumatra and Borneo, even secondary forest has disappeared, and in many places former forestlands are so degraded that they are useless for agriculture and are covered by scrub brush or just bare earth.

## Mining

Mechanical strip-mining is probably the extractive activity that most disrupts the land. This technique is increasingly used in Southeast Asia to extract such minerals as copper, silver, and gold. The land is cleared of forest cover, and heavy equipment then peels away the layers of soil and rocks until the desired mineral is exposed. In the United States, mining companies were required until 2004 to rehabilitate (at least partially) the landscapes they bulldozed and to dispose of mining and processing waste without polluting watercourses. One reason that international mining companies are interested in Southeast Asia is that environmental regulations do not exist or are not enforced there.

The most publicized mine in the region is the 13,000-foot-high (4000-meter-high) "Mine in the Clouds," an open-pit multi-mineral mine on Grasberg Mountain in West Papua (Figure 10.35). The mountain contains the world's largest known gold reserve. Every day in late 2005, New Orleans–based Freeport-McMoRan Copper and Gold Inc. extracted U.S.$11.5 million worth of copper, gold, and silver from this mine. Freeport-McMoRan, Indonesia's largest foreign taxpayer, now has contracts to mine over 9 million acres (3.6 million hectares) in the mountain range and plans to continue mining there for at least another century. Local villagers, however, complain that they have been driven off their land without compensation. And the Komoro people, who live on the coast about 50 miles away, have demonstrated that tailings from the mine have polluted the Aikwa River, causing flooding and ruining their stands of sago palm. Mining by Freeport-McMoRan has also threatened the nearby Lorenz National Park, in direct violation of Indonesian law.

In Indonesia (in Kalimantan and West Papua) and the Philippines (on Mindanao), national armed forces have been called out to quell protests against indigenous peoples' losses of lands to mining operations. Such clashes reflect the conflict between the interests of local, often remote, politically weak forest dwellers and traditional cultivators on the one hand and of national governments focused on rapid economic development on the other.

## Air Pollution

Usually, we think of places surrounded by the sea as pristine and their air as invigorating. Since the 1990s, however, the islands of Southeast Asia have been regularly covered with a putrid and persistent cloud of pollution. Although large cities such as Kuala Lumpur and Jakarta experience the worst pollution, air pollutants surge beyond safe levels even in remote parts of the islands. On the worst days, people are urged to wear masks, and airplanes cannot land.

One of the main causes of the poisonous cloud is smoke from fires set on logged forestland to prepare it for the planting of oil palm trees and luceana, a fast-growing tree used for paper pulp

**FIGURE 10.35** The open-pit mine on Grasberg Mountain in West Papua. [George Steinmetz.]

production. Unfavorable winds and drought can combine to spread the fires. At times, a high-pressure air mass (possibly linked to El Niño) can keep the smoky haze, as well as the region's normal industrial air pollution, from diffusing into the upper atmosphere. In October 2006, smoke from fires in Sumatra descended around the skyscrapers of Singapore, and the pilot of a 737 jet missed the runway in Borneo because of the smoke.

Some people suggest that the repeated occurrence of preventable environmental disasters, such as intense air pollution, highlights the weaknesses of Southeast Asian social institutions. Environmental regulations are not strongly enforced, and those who pay bribes may avoid them entirely. Meanwhile, the political systems do not encourage public debate about the problem. Throughout the region, great value is placed on gentle, consensual persuasion rather than on legal sanctions. This is especially true in the case of

environmental issues, which—as elsewhere in the world—consistently take a backseat to economic development. Ultimately, success in addressing Southeast Asia's environmental problems will depend both on consumers outside the region giving serious thought to the environmental consequences of their purchases and on the Southeast Asian public becoming more aware of how they can change unsustainable development policies.

## MEASURES OF HUMAN WELL-BEING

Human well-being is a complicated measure that is determined by sociocultural and political factors as much as by economic ones. The countries of Southeast Asia, unlike those of sub-Saharan Africa and South Asia, have greatly varying GDP per capita figures (Table 10.4,

**TABLE 10.4   Human well-being rankings of countries in Southeast Asia and other selected countries[a]**

| Country (I) | GDP per capita, adjusted for PPP[b] (GDP ranking among 177 countries), 2005 (II) | Human Development Index (HDI) ranking among 177 countries,[c] 2005 (III) | Gender Empowerment Measure (GEM) ranking among 80 countries, 2005 (IV) | Female literacy (percent), 2003 (V) | Male literacy (percent), 2003 (VI) |
|---|---|---|---|---|---|
| **Selected countries for comparison** | | | | | |
| Japan | 27,967  (13) | 11 (high) | 43 | 99 | 99 |
| Kuwait | 18,047  (30) | 44 (high) | ND | 81 | 85 |
| Mexico | 9,168  (60) | 53 (medium) | 38 | 89 | 92 |
| United States | 37,562  (4) | 10 (high) | 12 | 99 | 99 |
| **Southeast Asia** | | | | | |
| Brunei | 19,210  (29) | 33 (high) | ND | 90 | 95 |
| Burma | 1,700[d]  (140) | 129 (medium) | ND | 86 | 94 |
| Cambodia | 2,078  (64) | 130 (medium) | 73 | 60 | 77 |
| Indonesia | 3,361  (111) | 110 (medium) | ND | 83 | 93 |
| Laos | 1,759  (135) | 133 (medium) | ND | 61 | 77 |
| Malaysia | 9,512  (58) | 61 (medium) | 51 | 85 | 92 |
| Philippines | 4,321  (99) | 84 (medium) | 46 | 93 | 93 |
| Singapore | 24,481  (21) | 25 (high) | 22 | 87 | 97 |
| Thailand | 7,595  (66) | 73 (medium) | 63 | 98 | 99 |
| Timor-Leste | 400[d] | 140 (medium) | ND | ND | ND |
| Vietnam | 2,490  (120) | 108 (medium) | ND | 87 | 94 |

[a]Rankings are in descending order; i.e., low numbers indicate high rank.

[b]PPP = purchasing power parity, figured in 2003 U.S. dollars.

[c]The high, medium, and low designations indicate where the country ranks among the 177 countries classified into three categories by the United Nations.

[d]Data from CIA World Factbook, 2005.

ND = No data available.

*Sources: United Nations Human Development Report 2005* (New York: United Nations Development Programme), Table 1, Table 25, and Table 26, at http:/hdr.undp.org/reports/global/2005/pdf/HDR05_complete.pdf; *2005 World Population Data Sheet* (Washington, D.C.: Population Reference Bureau).

column II). The GDP per capita figures for some countries (Singapore and Brunei) are very high; for others (Malaysia, Thailand, and the Philippines) they are only moderate; and for still others (Burma, Cambodia, Laos, and Vietnam) they are very low. Singapore and Brunei enjoy widespread prosperity, but elsewhere in the region GDP per capita figures mask wide variations in well-being. To survive, those with the lowest incomes depend on subsistence cultivation, reciprocal exchange of labor and services, migration, and the informal economy. But often these very poor people live in close proximity to a vastly better-off minority who live in fine houses, travel widely, and consume at high levels. Protracted war in Vietnam, Laos, and Cambodia has hindered development in those countries, though their economies, especially Vietnam's, are expected to improve as their low wages attract foreign investment. In Burma, a military dictatorship and widespread corruption have been major factors in the continuing poverty of that country, which is among the richest in the region in natural resources.

As with GDP per capita, the countries of Southeast Asia vary in their rankings on the United Nations Human Development Index (HDI) (see Table 10.4, column III). Singapore's HDI ranking (25) is very high indeed, whereas Cambodia (130), Laos (133), and Burma (129) all rank far lower. In those countries with the region's lowest HDI rankings (Laos and Cambodia), literacy figures (columns V and VI) are low, especially for women. Governments in low-ranking countries have not provided sufficient education or health-care services. United Nations Gender Empowerment Measure (GEM) figures (see Table 10.4, column IV) are available for only half the Southeast Asian countries: Cambodia (73), Malaysia (51), the Philippines (46), Singapore (22), and Thailand (63). Because only 80 of the possible 177 countries are ranked on the GEM, these scores place all but Singapore in the bottom half of the rankings. (But notice that even very rich countries can have relatively low GEM rankings—as does Japan, with a ranking of 43.) Women in Singapore have many opportunities for employment, though at lower wage scales than men (see Table 10.3 on page 561). In Malaysia, a few well-connected women have access to high-paying jobs as administrators and as professional and technical workers. As shown in Table 10.3, women lag well behind men in education level (column III), wages (II), and access to jobs (IV) in most of the region, but even where women are more highly educated, their greater preparation has not yet led to equity in employment or wages.

---

# III  SUBREGIONS OF SOUTHEAST ASIA

The subregions of Southeast Asia share many similarities: tropical environments; ethnically diverse populations, with Overseas Chinese minorities that are especially active in commerce; vestiges of European colonialism (except for Thailand); difficult political trade-offs between democracy and dictatorship; and the importance of religion. Some obvious differences among the subregions are in their pace of modernization and in their varying responses to cultural and ethnic diversity—sometimes tolerated and occasionally celebrated, sometimes the cause of strife. Countries also differ in their approaches to achieving national unity, a notable concern throughout Southeast Asia.

## MAINLAND SOUTHEAST ASIA: BURMA AND THAILAND

Burma and Thailand occupy the major portion of the Southeast Asian mainland and share the long, slender peninsula that reaches south to Malaysia and Singapore (Figure 10.36). Although Burma and Thailand are adjacent and share similar physical environments, Burma is poor, depends on agriculture, and has a repressive military government, whereas Thailand is rapidly industrializing and has had a more open, less repressive society, despite occasional military rule. Both countries trade in the global economy, but in different ways: Burma supplies raw materials (including illegal drug components), while Thailand provides low- to medium-wage labor, hosts many multinational manufacturing firms, and has a thriving tourism industry with a worldwide clientele.

The landforms of Burma and northeastern Thailand consist of a series of ridges and gorges that bend out of the Plateau of Tibet and descend to the southeast, spreading out across the Indochina peninsula. The Irrawaddy and Salween rivers originate far to the north on the Plateau of Tibet and flow south through narrow valleys. The Irrawaddy forms a huge delta at the southern tip of Burma. The Chao Phraya flows from Thailand's northern mountains to the large plain in central Thailand. Most agriculture takes place in the Burmese interior lowlands around Mandalay and in Thailand's central plain.

Ancient migrants from southern China, Tibet, and eastern India settled in the mountainous northern reaches of Burma and Thailand. The rugged topography has protected these indigenous peoples—the Shan, Karen, Mon, Chin, and Kachin are among the largest such groups—from outside influences. Hence they follow traditional ways of life and practice animism. By contrast, urban development, Buddhism, and modernization are found in the southern Burmese valleys and lowlands and in Thailand's central plain. Burma is named for the Burmans, who constitute about 70 percent of the population and live primarily in the lowlands. Thailand is named for the Thais, a diverse group of indigenous people who also originated in southern China.

### Burma

Burma is rich in natural resources. From 70 to 80 percent of the world's remaining teakwood still grows in its interior uplands; a single tree can be worth U.S.$200,000. Other resources include oil,

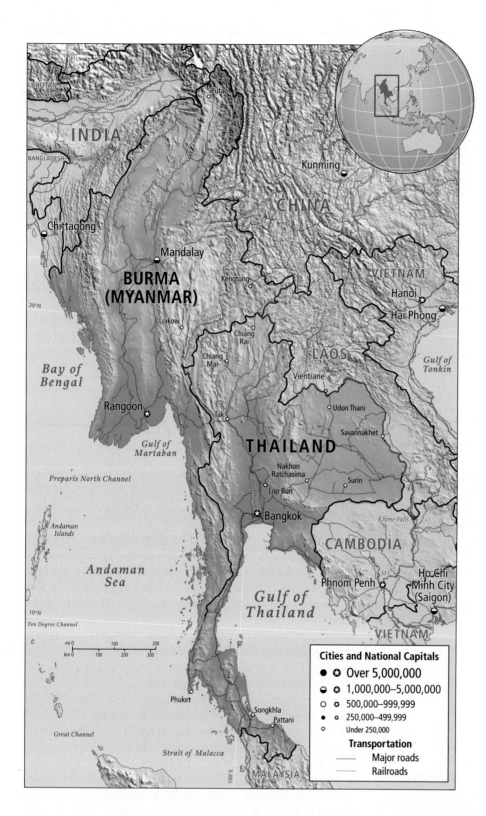

**FIGURE 10.36** Mainland Southeast Asia: The Burma and Thailand subregion.

natural gas, tin, antimony, zinc, copper, tungsten, limestone and marble, and gemstones. However, in part because of corruption and repression by a military junta, Burma ranks as one of the region's poorest countries, with an estimated per capita annual income of U.S.$1700 and an HDI rank of 129. About 70 percent of the people still live in rural villages, with livelihoods based primarily on wet rice cultivation and the growing of corn, oilseeds, sugarcane, and legumes and the logging of teak and other tropical hardwoods. They travel to nearby markets to sell or trade their produce for necessities (Figure 10.37).

**FIGURE 10.37** A market at Inle Lake, Burma. Several indigenous groups live in villages around the lake, growing vegetables, flowers, and rice. Men produce items made of silver, brass, and pottery, and women weavers have made the area Burma's second largest producer of silk products. [Toby Williams/Alamy.]

Burma is estimated to supply opium for over 60 percent of the heroin market in the United States, and it alternates with Afghanistan as the world's largest producer. Opium poppy cultivation, along with methamphetamine production, is the major source of income for a number of indigenous ethnic groups. The Wa, for example, who number about 700,000, depend on illegal drug production for 90 percent of their income and are said to protect their territory in northern Burma with surface-to-air missiles. The military government of Burma manipulates and profits from the drug traffic, especially in the lightly settled uplands and mountains near the Thai border. When indigenous inhabitants have protested, they have been silenced by assassination, plundering, resettlement, and the abduction and sale of their young women into the sex industries in neighboring countries. This situation has resulted in more than 100,000 Burmese refugees, who are now living in camps in Thailand and Bangladesh. Since 1997, the international community has applied economic sanctions against the military government, including a 2003 U.S. embargo on trade and investment in Burma.

## Is Thailand an Economic Tiger?

During the boom years of the mid-1990s, Thailand was known as one of the Asian "tiger" economies, meaning that it was rapidly approaching widespread modernization and prosperity. Thailand takes pride in having avoided colonization by European nations and in having transformed itself from a traditional agricultural society into a modern nation. According to the United Nations, between 1975 and 1998, Thailand had the world's fastest-growing economy, averaging an annual growth in GDP per capita of 4.9 percent (U.S. annual growth then averaged 1.9 percent). At the same time, the country kept unemployment relatively low for a developing country—at 6 percent—and inflation in check at 5 percent or lower. The soaring economic growth resulted from the rapid industrialization that took place in Thailand when the government provided attractive conditions for large multinational corporations, which located in Thailand to take advantage of its literate yet low-wage workforce and its lenient laws on environmental impacts, manufacturing, and trade.

The Bangkok metropolitan area, which lies at the head of the Gulf of Thailand, contributes 50 percent of the country's wealth and contains 16 percent of its population. Still, cities elsewhere in Thailand are also growing and attracting investment: Chiang Mai in the north, Khon Kaen in the east, and Surat Thani on the Malay Peninsula. And despite a significant economic downturn during the regional economic recession of the late 1990s, the country rebounded faster than its neighbors, and its middle class continued to expand, consume, and invest.

Nevertheless, about 50 percent of Thai working people are still farmers, who, according to official statistics, produce less than 10 percent of the nation's wealth. (Remember, though, that much of what farmers, especially women, produce is not counted in the statistics.) Thirty-one percent of the population, or 20 million people, live in cities, and many of them endure crowded, polluted, and often impoverished conditions. The life circumstances of Buaphet Khuenkaew, described in the personal vignette on page 560, illustrate a standard of living common among Thai people who live on the urban fringe. Many of them retain some agricultural ways of life and at the same time take up opportunities for employment in the city.

Industrialization has had unanticipated detrimental side effects in Thailand. Thousands of rural people have been drawn to the cities seeking jobs, status, and an improved quality of life. Many of them, like Mak and Lin, described in the personal vignette on page 562, arrive only to find a difficult existence and meager earnings. A group called the Assembly of the Poor began to convene on the streets of Bangkok in 1995 to protest poor living and working conditions as well as large development programs, such as the Yadana pipeline (Box 10.3), that have forced people off the land and into the city. So far the group has been effective: half the problems on its list have been resolved to its satisfaction. Thailand has generally protected the right to public protest (in contrast to many other countries in the region). Yet, after the military coup d'état of September 2006,

---

## BOX 10.3 | AT THE LOCAL SCALE   The Yadana Project: A Joint Venture with Mixed Results

Bangkok, the capital of Thailand, needs power. Neighboring Burma has natural gas reserves. It would seem that the Yadana natural gas field, located off Burma's west coast in the Andaman Sea, with reserves of some 5 trillion cubic feet (142 billion cubic meters), could be a huge boon to the people of Burma, who are among the world's poorest. But in a joint venture to tap the gas, Total of France holds a 31 percent interest, UNOCAL of California 28 percent, and the Burmese government oil company just 15 percent. The Petroleum Authority of Thailand holds the remaining 26 percent. The liquefied petroleum gas (LPG) passes through 155 miles (250 kilometers) of buried pipeline to a power plant outside Bangkok. For the next 25 years, the gas from the Yadana field is expected to generate up to 2800 megawatts of electricity per day, which is enough to power a large to medium-sized city in the United States.

From the Yadana project, Burma gets a natural gas facility near Rangoon providing 125 million cubic feet (3.5 million cubic meters) a day for domestic consumption, a gas-fired electricity generating plant, and a fertilizer plant (fertilizer is a by-product of LPG production). The project has additional benefits for Burma: some 2000 jobs, improved medical care and renovated hospitals, new schools, rebuilt roads, improved village water systems, increased electricity, and some village-scale development projects.

And Burmese villagers displaced by the pipeline are to be compensated. However, Burma's 15 percent of the deal, amounting to an estimated $450 million annually, goes to the repressive military government.

Although Burma has much to gain from the Yadana project, the military government acquired land for the pipeline forcibly by removing whole villages and then used forced labor on the project. Meanwhile, clearing and bulldozing for the pipeline have destroyed fields and forests with no compensation to the inhabitants. At shareholder meetings, critics in the United States protested UNOCAL's participation. They filed court cases stating that the company indirectly participated in human rights abuses by the Burmese government against its own citizens in connection with the pipeline project. UNOCAL's position is that it is simply an oil company and is not involved in, nor should it be held accountable for, Burma's political problems. The company argues that its expenditures to bring jobs and improvements in quality of life to local villagers should outweigh the project's other, negative effects.

*Sources:* Adapted from "Country report: Myanmar," *The Economist* (November 11, 1999): 16; http://www.earthrights.org/burma.shtml; Christopher Hopson, Burmanet News, Feb. 25, 2005, http://www.burmanet .org/news/2005/02/25/upstream-gunning-for-french-giant-over-burma -christopher-hopson/.

---

public protest was squelched—a development that shocked Thais, who are used to speaking their minds fairly freely.

What is the evidence for Thailand as an Asian tiger? Although its infrastructure is not as developed as those of other middle-income countries, it is linked in several ways to the emerging Asian Highway network (see Figure 10.20 on page 550); it now has three international airports; and more than one-third of the population, or 26 million people, now have cell phones. On the other hand, Thailand has only recently begun to address inequalities in education, basic health care, and housing. Even though literacy rates are above 98 percent, only 33 percent of Thai children attend secondary school, and only a tiny minority go on to higher education. As a result, too many of Thailand's workers are stuck at the low end of the global wage market, and there is a deficit of highly trained scientists and engineers. Part of the reason that Thailand lags behind countries such as South Korea and Taiwan may be that it received much less foreign aid during the cold war era to fund education, infrastructure improvements, and land reform.

## MAINLAND SOUTHEAST ASIA: VIETNAM, LAOS, AND CAMBODIA

On the map in Figure 10.38, the countries of Vietnam, Laos, and Cambodia might appear to be ideally suited for peaceful cooperation, sharing as they do the Mekong River and its delta on the

Southeast Asian mainland. Nonetheless, these three countries have suffered a disruptive half-century of war that has pitted them against one another. Until the end of World War II, all three were colonies, known collectively as French Indochina, and all three have fought a long struggle to transform themselves into independent nations (see the discussion on pages 541–543). In the late 2000s, they remain essentially Communist states; yet free market capitalism has been allowed in measured doses. Visitors and investors began arriving in the 1990s, and by 2000, once-somber city streets were abuzz with people and enterprises, especially in Vietnam.

A long, curved spine of mountains runs through Laos and into Vietnam. In the southern part of the subregion, these mountains are flanked on the west by the broad floodplain of the Mekong that is occupied by Thailand and Cambodia and on the east by the 1000-mile-long coastline and fertile river deltas of Vietnam. Vietnam is by far the most populous of these three countries, with a population of 83 million. In contrast, Cambodia has 13 million and Laos, almost entirely mountainous, has only 6 million. The rugged mountainous territory of northern Vietnam and Laos is the least densely occupied area in mainland Southeast Asia. Most of the subregion's population lives along the coastal zones of Vietnam and in the Mekong delta, where people accommodate the seasonal floods by building their homes on stilts (Figure 10.39), as in the Ganga-Brahmaputra delta in India. Farmers take advantage of the wet tropical climate and flat terrain to cultivate rice. The Red River delta in northern Vietnam is also an important rice-growing area. Throughout the subregion,

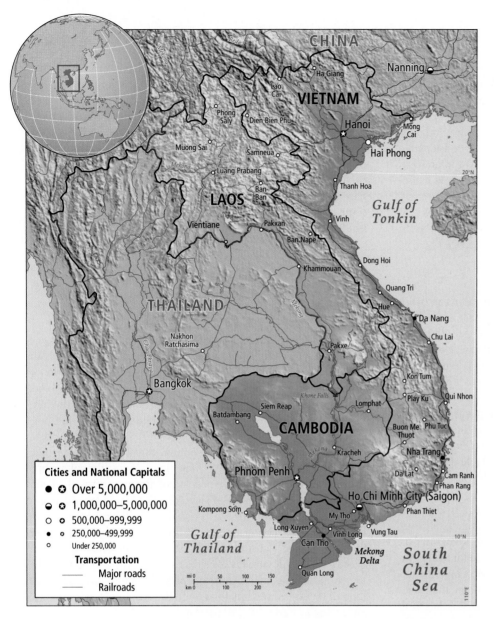

**FIGURE 10.38** Mainland Southeast Asia: The Vietnam, Laos, and Cambodia subregion.

(a)

(b)

**FIGURE 10.39** Living in the wetlands. (**a**) People who live along the Mekong River in Cambodia and southern Vietnam usually build their houses on stilts. (**b**) During the rainy season, the river can rise 10 feet (3 meters) or more, broadening into a lazy flow 2 to 3 miles (3 to 5 kilometers) wide. [Michael Yamashita/Woodfin Camp & Associates.]

at least 75 percent of the people support themselves as subsistence farmers; in Cambodia, only 15 percent of the people live in cities.

The main trading partners for Laos and Cambodia are Thailand and Vietnam (as sources of imports) and the United States and Thailand (as destinations for exports of timber, vegetables, and inexpensive manufactured goods). Thai investors wield considerable power in all three countries because they have broad interests in such resources as timber, gypsum, tin, gold, gemstones, and hydroelectric power and in the production of rice, vegetables, coffee, sugarcane, and cotton for export. Some investors are hoping to promote the Greater Mekong Basin as a secondary destination for European tourists who visit Thailand, as discussed in Box 10.1 (page 551).

## Vietnam

Vietnam is the most developed of the eastern Southeast Asian mainland countries. After the United States withdrew its military forces in 1973 (see page 542), the Communist government, assisted by the Soviet Union, began to invest aggressively in health care, basic nutrition, and basic education. By 2005, 87 percent of children reached the fifth grade, and adult literacy was above 90 percent. Fifty-six percent of the Vietnamese people are cultivators, and they have extensive knowledge of such practical matters as useful plants (including medicinal herbs) and animals, home building and maintenance, and fishing. The Vietnamese have also developed a national flair for excellent cuisine; when times are good, the dinner table is filled with artfully prepared fish, vegetables, herbs, rice, and fruits—a spread that is now popular with tourists.

Vietnam has mineral resources (phosphates, coal, manganese, offshore oil) and at one time had a lush forest cover, much of which was destroyed by defoliants in the later years of the Vietnam War. Despite its resources, Vietnam's economy languished for nearly two decades after the war, partly because of economic sanctions imposed by the United States and partly because of the inefficiencies of Communism. In the mid-1980s, the Hanoi leadership began to introduce elements of a market economy in a program of economic and bureaucratic restructuring called *doi moi* (very similar to perestroika in the Soviet Union or the structural adjustment programs imposed on many countries by international lending agencies). With the lifting of the U.S. embargo in 1994, firms from all over the world began to open branches in Vietnam. By the late 1990s, about 90 percent of the country's industrial labor force worked in the private sector, producing such exportable products as textiles and clothing, cement, fertilizers, and processed food.

Wages remain extremely low, partly because of an oversupply of workers (30 percent of Vietnam's population is under the age of 15). Some multinational firms are moving from China to Vietnam to take advantage of the million or more new workers entering the labor force every year and the several hundred thousand agricultural workers displaced annually. Multinationals that are exploring new or expanded investment in Vietnam include Bassett Furniture, Boeing, Citigroup, Cisco Systems, and Procter & Gamble.

Although economic growth in Vietnam ranged between 7 and 8 percent per year throughout the 1990s and early 2000s, the factors that brought growth also caused considerable dislocation in some parts of society. One focus of *doi moi* has been the privatization of land and other publicly held assets. The farmers who own newly privatized land have greater security and hence pay more attention to conservation and efficient production. But the process often leaves out the poorest people, who do not get land and so end up in the informal economy. New enterprises that promise to bring in direct foreign investment have sometimes taken precedence over local interests. For example, in the late 1990s, small-scale Vietnamese farmers felt the squeeze when their taxes were raised and some of their lands were confiscated to be used in such foreign-funded enterprises as tourist hotels, golf courses, and oil refineries.

## ISLAND AND PENINSULAR SOUTHEAST ASIA: MALAYSIA, SINGAPORE, AND BRUNEI

Malaysia and neighboring Singapore and Brunei are the most economically successful countries in Southeast Asia. Malaysia was created in 1963 when the previously independent Federation of Malaya (the lower portion of the Malay Peninsula) was combined with Singapore (at the peninsula's southernmost tip) and the territories of Sarawak and Sabah (on the northern coast of the large island of Borneo to the east) (Figure 10.40). All had been British colonies since the nineteenth century. Singapore became independent from Malaysia in 1965. The tiny and wealthy sultanate of Brunei, also on the northern coast of Borneo, refused to join Malaysia and remained a British colony until its independence in 1984. A look at Table 10.4 will confirm that Singapore is one of the richest countries on earth, with Brunei not far behind. (Brunei does not release statistics, but its estimated per capita income is U.S.$19,210, and virtually all citizens have a high standard of living supported by wealth from its oil and natural gas.) Malaysia, though much less wealthy than these two neighbors, still ranks relatively high (at 61) in GDP per capita.

## Malaysia

Malaysia is home to 25 million ethnically diverse people. Nearly 60 percent of Malaysia's people are Malays, and nearly all Malays are Muslims. Ethnic Chinese make up 24 percent of the population, and they are mostly Buddhist. Eight percent are Tamil- and English-speaking Indian Hindus, and 2 percent are forest-dwelling indigenous peoples, who live primarily in Sarawak and Sabah. Until the 1970s, conflicts among these groups divided the country socially and economically. Malaysians have worked hard to improve relationships among these groups, and they have had noteworthy success.

Most Malaysians (86 percent) live on the Malay Peninsula, which has 40 percent of the country's land area. Throughout most of its history, peninsular Malaysia was inhabited by Malays and a small percentage of Tamil-speaking Indians. The Indians were traders who plied the Strait of Malacca, the ancient route from India to the South China Sea. Even before the eastward spread of Islam in the thirteenth century, Arab traders began using this route in the ninth and tenth centuries, stopping at small fishing villages along the way to

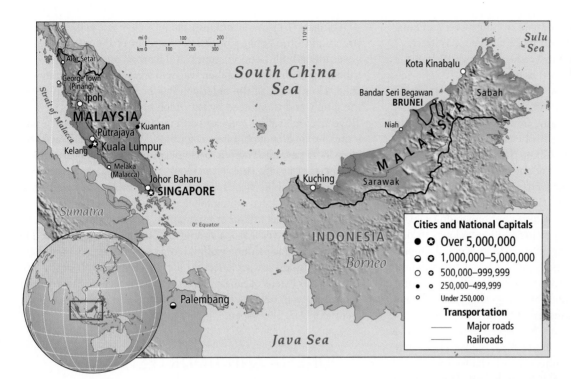

**FIGURE 10.40** Island and peninsular Southeast Asia: The Malaysia, Singapore, and Brunei subregion.

replenish their ships. By 1400, virtually all ethnic Malay inhabitants of peninsular Malaysia had converted to Islam.

During the colonial era, the British brought in Chinese Buddhists to work as laborers on peninsular Malaysian plantations. These Overseas Chinese eventually became merchants and financiers, while Indians achieved success in the professions and in small businesses. The far more numerous Malays remained poor village farmers and plantation laborers. As economic disparities widened, antagonism between the groups increased. After independence, animosities exploded with the onset of widespread rioting by the poor in 1969. Political rights were suspended, and it took 2 years for the situation to calm down. The violence so shocked and frightened Malaysians that they agreed to address some of its fundamental social and economic causes.

After a decade of discussions among all groups, Malaysia launched a long-term affirmative action program, called *Bumiputra*, in the early 1980s. Its core was a new economic policy designed to help the Malays and indigenous peoples gain economic advancement. The policy required Chinese business owners to have Malay partners. It set quotas that increased Malay access to schools and universities and to government jobs. As the program evolved in the 1980s, the goal became to bring Malaysia to the status of a fully developed nation by the year 2020. The program has succeeded in narrowing some social inequalities fairly rapidly; *Asiaweek*, a regional newsmagazine, reported that in late 1997, the real standard of living in Malaysia—adjusted for per capita purchasing power—was more than double what it had been at the beginning of the 1990s.

Nonetheless, disparities between rich and poor remain wider in Malaysia than in any other Southeast Asian country (Figure 10.41; see also Table 10.2 on page 549). The capital, Kuala Lumpur, boasts

**FIGURE 10.41** Wealth disparity in Malaysia. A woman collects her laundry at a simple wooden house in the Kampung Chendana district of Kuala Lumpur. In the background are the luxurious Petronas Towers, less than a mile away. [Mark Fallander/OnAsia.com.]

beautifully designed skyscrapers like those in Singapore, including the two tallest buildings in the world in 2000—Petronas Towers I and II. Yet, at least one-fifth of the city's residents are squatters, some on land, some on raft houses on rivers or bays.

The economy of Malaysia has traditionally relied on export products such as palm oil, rubber, tin, and iron ore. In contrast, much of the growth of the 1990s came from Japanese and U.S. investment in manufacturing, especially of electronics, and from offshore oil production in the South China Sea. Timber exports are also important (especially in Sarawak and Sabah, as discussed at the opening of this chapter), and the land cleared by logging is developed into additional oil palm and rubber tree plantations. Now Malaysia is hoping to benefit from a new multibillion-dollar high-tech manufacturing corridor 10 miles wide by 30 miles long (16 kilometers wide by 48 kilometers long) just south of Kuala Lumpur. Called the Multimedia Super Corridor (MSC), it includes two new cities: the new federal administrative center, called Putrajaya, and another city named Cyberjaya. The MSC is expected to help the nation spring fully prepared into a prosperous twenty-first century.

Like many Asian economic tigers, Malaysia is subject to volatile economic patterns. Currency values and stock market prices may fluctuate widely. Lavish building projects may be halted for years, as they were in Kuala Lumpur from 1998 through 2001. Overall, however, it does seem that Malaysia is likely to achieve increasing prosperity and continue to foster cultural diversity while discouraging extremist factions, religious or ethnic.

## Singapore

Singapore occupies one large and many small hot, humid, flat islands just off the southern tip of the Malay Peninsula. Its 4.3 million inhabitants, all of them urban, live at a density of over 17,946 people per square mile (11,155 per square kilometer), yet it is one of the wealthiest countries on earth, and its landscapes are elegant (see Figure 10.11 on page 544). Singapore's wealth is derived from the manufacture of pharmaceutical, biomedical, and electronic products; financial services; oil refining and petrochemical manufacturing; and oceanic transshipment services. Unemployment is low and poverty unusual, except among temporary and often illegal immigrants, primarily from Indonesia. Singaporeans seem to share the notion that financial prosperity is a worthy priority, and while the free market economy reigns, government plays a strong role.

Ethnically, Singapore is overwhelmingly Chinese (76.7 percent). Malays account for 14 percent of the population, Indians for 7.9 percent, and mixed or "other" peoples for 1.4 percent. Singapore's people are religiously diverse: 42 percent are Buddhist or Taoist, 18 percent Christian, 16 percent Muslim, and 5 percent Hindu; the rest include Sikhs, Jews, and Zoroastrians. Like that of Malaysia, the government encourages unity among this diversity. Singapore officially subscribes to a national ethic not unlike *Pancasila* in Indonesia. The nation commands ultimate allegiance. Loyalty is to be given next to the community and then to the family, which is recognized as the basic unit of society. Individual rights are respected, but not championed. The emphasis is on shared values, community consensus rather than conflict, and racial and religious harmony.

Singapore has a meticulously planned cityscape with safe, clean streets and little congestion. Eighty percent of the people live in government-built housing estates, and workers are required to contribute up to 25 percent of their wages to a government-run pension fund. Home care of the elderly, however, is the responsibility of the family and is enforced by law. Each child, on average, receives 10 years of education and can continue further if grades and exam scores are high enough. Literacy for those younger than 70 averages close to 95 percent. There is strict government control over virtually all aspects of society. Permits are required for almost any activity that could have a public effect: having a car radio, owning a copier, working as a journalist, having a satellite dish, being a sex worker, or performing as a street artisan or entertainer. Law and order are strictly enforced. A few years ago, a visiting U.S. teenager was sentenced to a caning for spray-painting graffiti. Drug users are severely punished, and drug dealers are sentenced to life imprisonment or death. But virtually all citizens of Singapore seem to have accepted this control and strictness in return for a safe city and the next highest per capita income in Asia after Japan.

## INDONESIA AND TIMOR-LESTE

Indonesia is a recent amalgamation of island groups, inhabited by people who, before the colonial era, never thought of themselves as a national unit (Figure 10.42). Rivalry between island peoples remains strong, and many resent the dominance of the Javanese in government and business. This resentment and other burgeoning issues of identity and allegiance are serious threats to Indonesia's continued unity (see the section "Is Indonesia Breaking Up?" on page 556). Emblematic of the fragile state of this subregion is the fact that Timor-Leste, the eastern half of the small island of Timor, is now independent. In 1999, after more than 20 years of armed conflict with Indonesia, in which the infrastructure was destroyed and as many as 250,000 people lost their lives, the people of eastern Timor voted in a UN plebiscite to split from Indonesia. For 2 more years, the Indonesian military tried to enforce cohesion, in part because of Timor-Leste's oil and gas deposits, but in 2002, after 1500 more deaths, Timor-Leste became an independent country.

The archipelago of Indonesia consists of the large islands of Sumatra, Java, and Sulawesi; the Lesser Sunda Islands east of Java; Kalimantan, which shares the island of Borneo with Malaysia and Brunei; West Papua on the western half of New Guinea; and the Moluccas. In all, it contains some 17,000 islands, but many of them are small, uninhabited bits of coral reef. The term *Indonesia* was coined in 1850 by James Logan, an Englishman residing in Singapore, from two Greek words: indos (Indian) and nesoi (islands). Indonesians themselves now refer to the archipelago as *Tanah Air Kita*, meaning "Our Land and Water." This name conveys a sense of the archipelago environment, but falsely indicates a universal feeling of togetherness and environmental concern.

Java and Sumatra are the two most economically productive islands and are home to 80 percent of Indonesia's population. Java alone has 60 percent of the nation's population (222 million people in 2005), but only 7 percent of the country's available land. Over

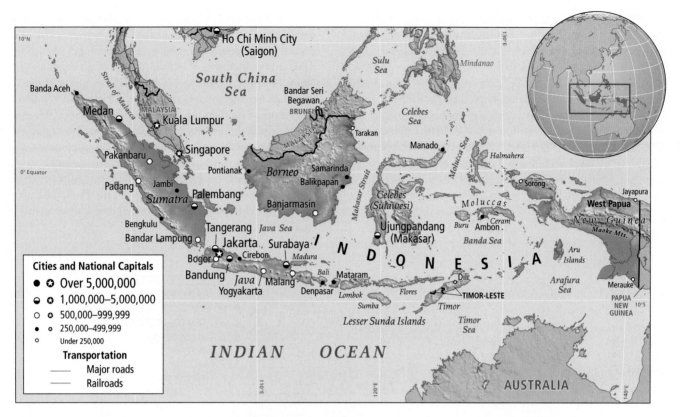

**FIGURE 10.42** The Indonesia and Timor-Leste subregion.

**FIGURE 10.43** People driven off their land by logging. Two members of the Dani, an indigenous group in West Papua, try to adjust to urban life in the highland trading town of Wamena. The Dani have little in common with recent Muslim migrants from Java, who were promised acreage carved out of the Dani's forested lands. [George Steinmetz.]

thousands of years, the ash from Java's 17 volcanoes has made the soil rich and productive, capable of supporting large numbers of people. Another 21 percent of the people live on the adjacent volcanic island of Sumatra. Recently, nearly 3 million people were resettled from Java to Sumatra, most under government sponsorship (see the discussion on pages 561–563).

The resettlement of Javanese to other islands in Indonesia is an increasingly contentious issue because indigenous people resent being inundated with Javanese culture and concepts of economic development (Figure 10.43). They see the resettlement as a Javanese effort to gain access to and profit from indigenously held natural resources, such as oil, precious metals, and forestlands. The government's argument that resettlement provides jobs is true, strictly speaking. But often the habitat is degraded beyond repair, and settlers are left in impoverished circumstances, in tiny shacks on muddy tracts of land, with little sense of community (see Figure 10.29 on page 562). The central government ignores these hardships and instead invokes the principles of *Pancasila*, arguing that resettlement unifies the people as Indonesians by spreading modernization and eliminating ways of life that differ from the government's vision of the norm.

## THE PHILIPPINES

The Philippines, lying at the northeastern reach of the Southeast Asian archipelago, comprises more than 7000 islands spread over about 500,000 square miles (1.3 million square kilometers) of ocean

(Figure 10.44). The two largest islands are Luzon in the north and Mindanao in the south. Together, they make up about two-thirds of the country's total land area, which is about the size of Arizona. The Philippine Islands, part of the Pacific Ring of Fire, are volcanic. The violent eruption of Mount Pinatubo in June 1991 devastated 154 square miles (400 square kilometers) and blanketed most of Southeast Asia with ash. Volcanologists had predicted the eruption, and precautions were taken, so although more than a million people were threatened, only 250 died. Over time, volcanic eruptions have given the Philippines fertile soil and rich deposits of minerals (gold, copper, iron, chromate, and several other elements).

The Philippines had 7 million people when it became a U.S. protectorate in 1898. Just over 100 years later, it has about 86 million.

Close to 50 percent of Filipinos live in cities, a high proportion compared with Southeast Asia as a whole (38 percent). Urban densities exceed 50,000 people per square mile (31,250 per square kilometer). The metropolitan area of Manila, the country's capital, with a population of nearly 18 million, is one of the largest and most densely settled urban agglomerations on earth (Figure 10.45). Some of the newest urban dwellers are people displaced by the 1991 eruption of Mount Pinatubo; others have been displaced by rapid deforestation, dam projects, and the mechanization of commercial agriculture. Many people in the cities are unemployed squatters living in shelters built out of scraps. The masses of urban poor (estimated to make up between 28 and 40 percent of the Philippine population) represent a particular political and civil threat because, even though

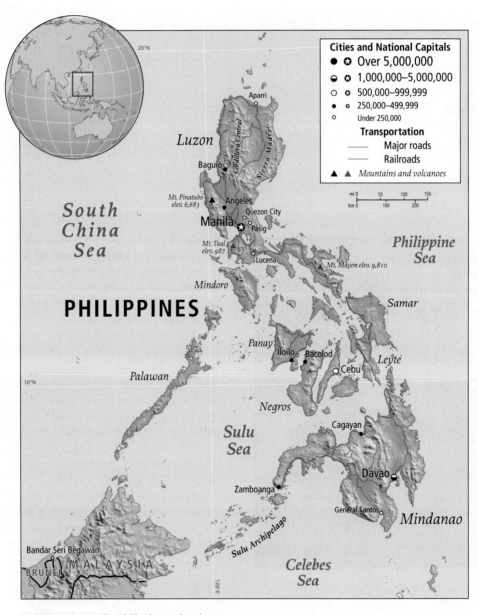

**FIGURE 10.44** The Philippines subregion.

**FIGURE 10.45** Manila's crowded streets.
[David Paul Morris.]

the excesses of the dictator Ferdinand Marcos have ended, the people's faith in their government has not yet been restored. Marcos maintained a flamboyant and brutal regime from 1965 until he was deposed in 1986. Since that time, there has been little economic progress, and violent social unrest has been very much a part of life in the Philippines.

The cultural complexity of Philippine society is illustrated by its ethnic profile. Within the indigenous majority—almost 96 percent of the population—there are at least 60 distinct ethnic groups. The Chinese make up about 1.5 percent of the population, and the remaining 3 percent includes Europeans, Americans, other Asians, and indigenous peoples from other Southeast Asian islands. Most Filipinos are Roman Catholic (83 percent), but the southern and central portions of the island of Mindanao and the Sulu Archipelago have been predominantly Muslim for a long time. Over the last few decades, the central government has been resettling many thousands of Catholics in Mindanao, apparently to dilute the Muslim population. Possibly in retaliation, in the 1990s, Muslim extremists, perhaps linked to international terrorist movements, began operating in the southern Philippines and carrying out bombings that have killed a number of people.

In the Philippines, as elsewhere in Southeast Asia, wealth and poverty are often associated with certain ethnic groups. The vast majority of the wealthy are descendants of Spanish and Spanish-Filipino plantation landowners or of Chinese financiers and businesspeople. It is estimated that 30 percent of the top 500 corporations in the islands are controlled today by ethnic Chinese Filipinos, who make up less than 1 percent of the population. Meanwhile, the poor are overwhelmingly indigenous people. By 1972, family conglomerates controlled 78 percent of all corporate wealth in the Philippines.

In contrast to Malaysia, where ethnic disparities in wealth became an openly discussed social issue, the Philippine government under Marcos did not embark on a new economic policy to help disadvantaged ethnic groups or improve the country's infrastructure. Instead, after violent protests erupted in the early 1970s, President Marcos declared martial law in 1972 and managed to hang onto power for another 14 years. During that time and since, there has been little progress in infrastructure development, job creation, or wealth redistribution.

Filipinos are keenly aware that a better social and political system is possible. The country's long association with the United States gave the Philippine people experience with U.S. democratic institutions and culture. But some of the worst aspects of U.S. culture were also imported (such as drug use, sex work, environmental abuse, and wasteful consumerism), especially around the six U.S. military bases that were maintained in the country until 1992. Nonetheless, the Subic Bay Naval Base alone employed 32,000 local people and indirectly created 200,000 jobs. Many of those workers, especially women office workers, were exposed to information, education, and opportunities to travel and migrate that other Southeast Asians did not have.

In their efforts to settle social unrest and attract investment, Marcos's successors have been hampered by the triple economic disasters of the Mount Pinatubo eruption, the closing of the U.S. military bases, and a drastic cut in U.S. foreign aid in 1993. One of the few bright spots in the Philippine economy is the success of microcredit programs that encourage the founding of small businesses by the poor, as described in the following personal vignette.

**Personal Vignette**   When Jesusa Ocampo made her first batch of *macapuno* candy for sale in 1992, she had no idea that the venture would one day grow into a full-fledged business. Neighbors snapped up all 20 packets of her coconut-based sweet. It wasn't long before Ocampo was making over 100 packets of candy daily. There was just one problem: she was too poor to pay for the expansion needed to meet the rising demand. Then a friend told her about a microcredit program financed by the Philippine government. Beginning with a loan of just U.S.$145 obtained through a local cooperative, Ocampo gradually built up her tiny operation. She flourished, and so did her credit rating. In the spring of 1997, she obtained a loan of U.S.$3000, her nineteenth loan. She and her family are now building their own home on land they have purchased outside Manila. ■

The case of the Philippines under Marcos shows how poor leadership and corrupt government can hold a country back. Despite the rampant sale of its once-rich timber resources, the country's economy has been slow to grow. The agricultural sector—which produces rice, coconuts, corn, sugarcane, bananas, pineapples, mangoes, fish, and meat—has dropped in productivity over the past 20 years, sinking from 25 percent of the total economy in the 1980s to just 14 percent in 2005, yet agriculture employs at least 36 percent of the labor force. Meanwhile, in 2005, industry accounted for 33 percent of the economy, but employed just 16 percent of the labor force. Perhaps the most meaningful figure, however, is the rate of unemployment. Although officially only 9 percent of the population is unemployed or underemployed, real unemployment figures may be as high as 30 percent, and 40 percent of the population lives below the poverty line.

Paradoxically, because the Philippines remained underdeveloped longer than its neighbors, the country was less affected than other countries across the region by the financial crisis and turmoil of the late 1990s. Foreign investors continued to build facilities, construction in Manila grew, and an industrial park rose on the former U.S. naval base at Subic Bay. Although the country is less developed than most of its neighbors, between 2001 and 2005 the Philippines' economy grew at nearly 5 percent, a rate comparable to Malaysia's. Importantly, it was remittances from workers abroad that alleviated poverty and helped to keep the economy viable during this period.

## Reflections on Southeast Asia

Southeast Asia is often held up as a model for other developing regions in terms of its rate of growth, distribution of wealth, and rate of personal savings. Table 10.2 shows, for instance, that the distribution of wealth in Southeast Asia has been far more equitable than in Middle and South America. It is important to note, however, that what is often called the miracle of Southeast Asian development is almost always calculated in economic terms. Those who do not think only in those terms (ordinary people, religious leaders, social and environmental activists) have advocated a more thorough analysis of the "miracle." They would also look at how rapid economic progress is affecting the environment, cross-cultural relations, human rights and political participation, investment in education, families, and the status of women.

The major economic downturn of the late 1990s, along with periodic pollution episodes, turned the region's attention to bad bank loans based on widespread corruption and rampant environmental exploitation. Soon it seemed as if the region's competitive edge had been founded on wildly unsustainable practices. None of that information was really new; it was just that the supposed economic prowess of the region had always overridden indications that all was not well.

Now that the economy has begun to rebound, will the various economic, environmental, political, religious crises in Southeast Asia accompanying that rebound lead to a reevaluation of just how development should proceed? Will Southeast Asians remember to ask the "development for whom" question and seek an answer that is inclusive, or will elitism continue? Recent elections and political events, such as the coup d'état in Thailand in 2006, indicate that support for democracy is fragile. Continuing deforestation and violations of the rights of indigenous forest people would seem to indicate that here, too, the goal of socially and physically sustainable development is receding.

## Chapter Key Terms

archipelago 536

ASEAN Free Trade Association (AFTA) 553

Association of Southeast Asian Nations (ASEAN) 553

Australo-Melanesians 539

Austronesians 539

crony capitalism 553

cultural complexity 557

cultural pluralism 557

detritus 539

*doi moi* 574

Export Processing Zones (EPZs) 549

extraregional migration 563

feminization of labor 549

foreign exchange 563

growth triangles 550

old-growth forest 567

*Pancasila* 556

pull factors 562

push factors 562

resettlement schemes 561

sex tourism 552

wet (paddy) rice cultivation 547

# Critical Thinking Questions

1. What role have national governments played in the destruction of forests in Southeast Asia?

2. Indigenous people in Southeast Asia face the same issues that indigenous people face all around the world. What factors might account for these commonalities?

3. Relate the landform features of Southeast Asia to those of South Asia and East Asia by describing the tectonic processes by which they were formed.

4. Account for the mild, wet climate of Southeast Asia and explain where the most extreme annual ranges in temperature are likely to be found.

5. Discuss the various factors that account for the fact that Southeast Asian soils tend not to be very fertile once they have been cleared of forest cover. Explain the role that temperature plays.

6. How do the cultural characteristics of Southeast Asia influence the spread of HIV-AIDS there?

7. Several Southeast Asian countries used to be commonly described as economic "tigers." What did this mean, and to what extent is this still a useful way to refer to these countries?

8. What factors have contributed to the fact that today agriculture contributes just one-sixth of the GDP in Southeast Asia, yet employs 60 percent of the people? How are these contradictions reflected in rural landscapes?

9. Gender roles and mating relationships in Southeast Asian families are not necessarily what an outsider might expect. What are some of the characteristics that interested you most, and why? Discuss some of the ways in which gender roles vary from those encountered in South Asia or East Asia.

10. The Overseas Chinese are a distinct culture group in Southeast Asia. What are their origins, and why do they stand out as a special case?

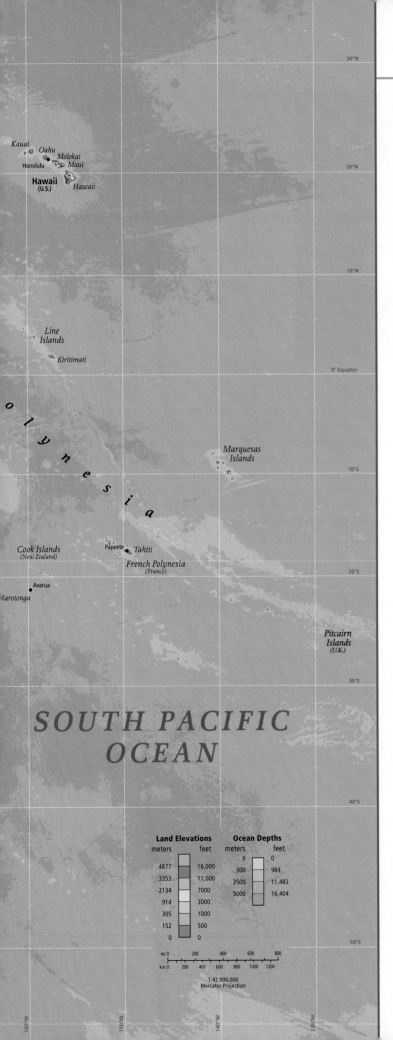

**Land Elevations**

| meters | feet |
|--------|------|
| 4877 | 16,000 |
| 3353 | 11,000 |
| 2134 | 7000 |
| 914 | 3000 |
| 305 | 1000 |
| 152 | 500 |
| 0 | 0 |

**Ocean Depths**

| meters | feet |
|--------|------|
| 0 | 0 |
| 300 | 984 |
| 3500 | 11,483 |
| 5000 | 16,404 |

mi 0    200    400    600    800

km 0    200  400  600  800  1000  1200

1:42,000,000
Mercator Projection

# CHAPTER 11

# OCEANIA: Australia, New Zealand, and the Pacific

**Global Patterns, Local Lives**   October 2003, Brisbane, Australia: The crowd of 47,000 waits eagerly as New Zealand's national rugby team takes the field opposite the team from Tonga, a Polynesian archipelago in the South Pacific. The famous Haka is about to begin.

A highly emotional and physical dance traditionally performed by the Maori of New Zealand, and today by other indigenous peoples of the islands of Oceania, to motivate fellow warriors and intimidate opponents before entering battle, the Haka has become an integral part of rugby in New Zealand. Before almost every international match for the past century, the New Zealand team, The All Blacks, has performed the Haka, chanting, screaming, jumping, stomping their feet, poking out their tongues, widening their eyes to show the whites, and beating their thighs, arms, and chests in unison. When a whole team performs a Haka it is a powerful sight (Figure 11.2). Until now, only the New Zealand team performed the Haka, but tonight is a different story. Halfway through New Zealand's Haka, the Tongan team responds with its own Haka. The crowd roars its approval of this scene, a revival of the traditional prelude to battle throughout Oceania's long history.

That this is all happening on a rugby field is emblematic of the colonial legacy in Oceania. Rugby, a full contact sport out of which American- and Australian-rules football developed, was brought to Oceania by the British, who colonized Australia, New Zealand, and many of the islands of the Pacific where rugby is now the most popular sport (Figure 11.3). After Tonga took on the New Zealanders in 2003, the Haka spread quickly to other sports and places far beyond the South Pacific. The University of Hawaii's football team now regularly performs a Haka before its games, as does Brigham Young University in Utah and a number of college football teams in Mexico. Even a high school football team in Trinity, Texas, which has many players of Tongan descent, now regularly performs a

**FIGURE 11.1** Regional map of Oceania.

583

**FIGURE 11.2** The Haka, a Maori tradition. The fierceness of the Haka challenge is visible on the faces of the New Zealand All Blacks rugby team as they face the Australian Wallabies in Christchurch, New Zealand, on July 8, 2006. [AP Photo/NZPA, Pool.]

Haka before its games. While the Haka is a powerful and often intimidating thing to witness, it may or may not lead to victory. Back in Brisbane, Tonga's Haka appears to have had little effect on the New Zealand team, who beat them 91-7.

To see videos of a Haka, go to http://www.Youtube.com and type in "haka."

*Adapted from an article by Phil Wilkins, "Tonga can only match the Kiwis in the Haka," Brisbane, Australia, October 25, 2003, at http://www.rugbyheaven.smh.com.au/articles/2003/10/24/1066974323766.html.* ∎

Oceania, which comprises Australia, New Zealand, Papua New Guinea, and the myriad Pacific islands (see Figure 11.1), has been dominated politically and economically by people of European descent for more than 400 years. Reviving customs such as the Haka is increasingly important in Oceania because dramatic economic and social shifts are reorienting the region not only away from old colonial powers and allies in Europe and North America and toward Asia and the global economy, but also toward concerns specific to Oceania. Although a minority of the Pacific islands still remain under the jurisdiction of a Western power,

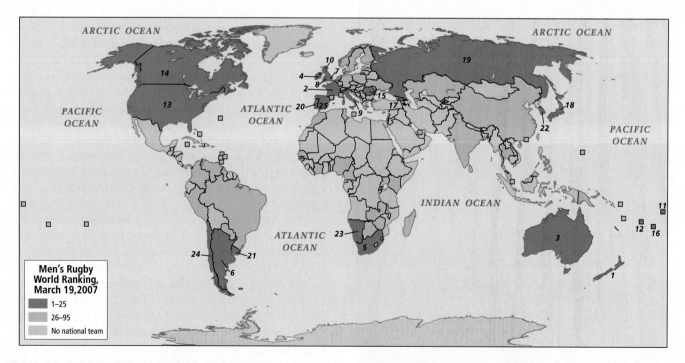

Men's Rugby World Ranking, March 19, 2007
- 1–25
- 26–95
- No national team

**FIGURE 11.3** Rugby around the world. Rugby is the world's third most popular sport. It is played by women, men, boys, and girls in over 100 countries, 95 of which are members of the International Rugby Board, the sport's governing body. Men's World Cup rugby competition began in 1987, when it was won by New Zealand. Rankings are based on a system of points for match results, relative team strength, and margin of victory, and an allowance for home field advantage; rankings can change weekly. [For additional information, see http://www.irb.com/EN/IRB+Organisation/.]

flowing ocean currents in the mid–South Pacific circulation pattern, shunting warm water to the south, where it warms the southeastern coast of Australia. In and around the reef, there is a great profusion of aquatic environments, some protected and warm, some cool and wildly active with pounding surf. Today, the Great Barrier Reef is dying in a number of places. One cause is thought to be the degradation since European settlement of the fresh water feeding the reef, which now contains chemicals, organisms, and sediments that are damaging the coral polyps. Warming of the ocean due to human-caused climate change is another cause of the reef's deterioration.

## Island Formation

The islands of the Pacific were created (and are being created still) by a variety of processes related to the movement of tectonic plates. The islands found in the western reaches of Oceania—including New Guinea, New Caledonia, and the main islands of Fiji—are remnants of the Gondwana landmass; they are large, mountainous, and geologically complex. Other islands in the region are volcanic in origin and form and are part of the Ring of Fire (see Figure 1.21 on page 22). Many of this latter group are situated in boundary zones where tectonic plates are either colliding or pulling apart. For example, the Mariana Islands east of the Philippines are volcanoes that were formed when the Pacific Plate plunged beneath the Philippine Plate. The two much larger islands of New Zealand were created when the eastern edge of the Indian-Australian Plate was thrust upward by its convergence with the Pacific Plate. The Hawaiian Islands were produced through another form of volcanic activity associated with **hot spots**: places where particularly hot magma moving upward from Earth's core breaches the crust in tall plumes. Over the past 80 million years, the Pacific Plate has moved across one of these hot spots, creating the string of volcanic islands known as Hawaii.

Volcanic islands exist in three forms: volcanic high islands, low coral atolls, and raised or uplifted coral platforms known as *makatea*. **High islands** are usually volcanoes that rise above the sea into mountainous, rocky formations that contain a rich variety of environments (Figure 11.6). New Zealand, the Hawaiian Islands, Tahiti, and Easter Island are among the many examples of high islands. An **atoll** is a low-lying island formed of coral reefs that have built up on the circular or oval rim of a submerged volcano (Figure 11.7). These reefs are arranged around a central lagoon that was once the volcano's crater. As a consequence of their low elevation, atoll islands tend to have only a small range of environments and very limited supplies of fresh water.

## Climate

Although the Pacific Ocean stretches nearly from pole to pole, most of Oceania is situated within the tropical and subtropical latitudes of that ocean. The tepid water temperatures of the central Pacific bring year-round mild climates to nearly all the inhabited parts of the region (Figure 11.8). The seasonal variation in temperature is greatest in the southernmost reaches of Australia and New Zealand.

***Moisture and Rainfall.*** With the exception of the arid interior of Australia, much of Oceania is warm and humid nearly all the time. New Zealand and the high islands of the Pacific receive copious rainfall and once supported dense forest vegetation, although much of that forest is gone after 1000 years of human impact (see Figure 11.29b on page 608). Travelers approaching New Zealand, either by air or by sea, sometimes notice a distinctive long, white cloud that stretches above the two islands. A thousand years ago, the

**FIGURE 11.6** Molokai, Hawaii, a high island. Some of the world's highest sea cliffs rise almost vertically more than 2000 feet above Molokai's Kalaupapa Peninsula (from which this photo was taken). The cliffs, in the Kalaupapa National Historical Park, have been designated a national natural landmark. [Mac Goodwin.]

**FIGURE 11.7** An atoll in the Tuamotu Archipelago of French Polynesia. As is the case here, the land area of an atoll is usually not continuous, but instead forms a sort of necklace of flat islets, known as *motu*, around a central lagoon. Often the necklace surrounds one or more islands that are the remnants of the old volcanic core. [David Doubilet.]

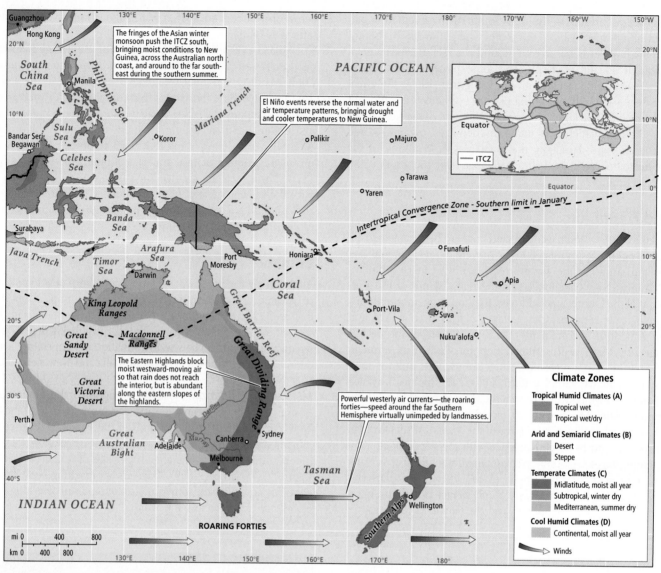

**FIGURE 11.8** Climates of Oceania.

Maori settlers also noticed this phenomenon, and they named the place *Aotearoa*, "land of the long white cloud." The distinctive mass of moisture is brought in by the legendary **roaring forties** (named for the 40th parallel south), powerful air and ocean currents that speed around the far Southern Hemisphere virtually unimpeded by landmasses. These westerly winds (blowing west to east) deposit 130 inches (330 centimeters) of rain a year in the New Zealand highlands and more than 30 inches (76 centimeters) a year on the coastal lowlands. At the southern tip of New Zealand's North Island, the wind averages more than 40 miles per hour (64 kilometers per hour) about 118 days a year. Cabbages have to be staked to the ground or they will blow away.

By contrast, two-thirds of the continent of Australia is overwhelmingly dry. The Great Dividing Range blocks the movement of moist easterly winds (blowing east to west), so rain does not reach the interior. As a result, a large portion of Australia receives less than 20 inches (50 centimeters) of rain a year, and humans have found rather limited uses for this territory. But the eastern (windward) slopes of the highlands receive more abundant moisture. This relatively moist eastern rim of Australia was favored as a habitat by the indigenous people, as well as by the Europeans who displaced them after 1800. During the southern summer, the fringes of the monsoon that passes over Southeast Asia and Eurasia (see Figure 10.6 on page 538) bring moisture across Australia's northern coast. There, rainfall in different years varies from 20 to 80 inches (50 to 200 centimeters).

Overall, Australia is so arid that it has only one major river system—in the temperate southeast, where most Australians live. There, the Darling and Murray rivers drain one-seventh of the continent and flow into the ocean at Adelaide. A measure of the over-

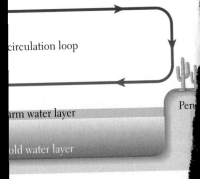

circulation loop

arm water layer

old water layer

Peru

**ns.** Water in the equatorial western Pacif…
…) is warmer than water in the eastern Pac…
…g wind patterns, the warm water piles u…
…his warm-water bulge in the western Pa…
…g air cools and, once in the higher atm…
…ction. In the east, the dry cool air desc…

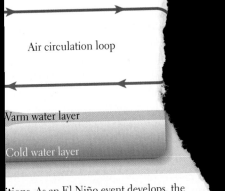

Air circulation loop

Warm water layer

Cold water layer

**itions.** As an El Niño event develops, the
…ge (orange) begins to move east. The air rising
…ons, one circulating east to west in the upper
…st.

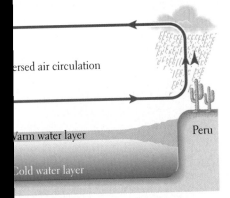

…rsed air circulation

Warm water layer

Peru

Cold water layer

Slowly, as the bulge of warm water at the
…st, it forces the whole system into the fully
…the water surface and in the upper atmos-
…rmal (**a**). Instead of warm, wet air rising over the
…d condensing as rainfall, cool, dry, cloudless
…'s surface. Meanwhile, in the east, the normally
…ences clouds and rainfall.

…have
…e 11.9).
…cts of this
…ea had received
…ed, springs and streams
…er-dry forests. The cloudless
…p and away from elevations above
…eters), so temperatures at high elevations dipped
…t night for stretches of a week or more. Tropical
…people unaccustomed to chilly weather sickened.
…ng the Pacific coasts of North, Central, and South
America, the warmer than usual weather brought unusually strong
storms, high ocean surges, and damaging wind and rainfall. Recently, an opposite pattern, in which normal conditions become
unusually strong, has been identified and named La Niña, though
scientists have barely begun to study it.

## Flora and Fauna

The fact that Oceania comprises an isolated continent and numerous islands has had a special effect on its animal life (*fauna*) and
plant life (*flora*). Many of its species are **endemic:** they exist in a particular place and nowhere else on earth. This is especially true of
Australia, but many Pacific islands also have endemic species.

***Plant and Animal Life in Australia.*** The uniqueness of Australia's
plant and animal life is the result of the continent's long physical isolation, its large size, its relatively homogeneous landforms, and its
strikingly arid climate. Since Australia broke away from Gondwana
more than 65 million years ago, its plant and animal species have
evolved in isolation. One spectacular result of this long isolation is
the presence of more than 144 living species of endemic marsupial

birds, storms, or ocean currents. On
their new home, they may evolve ove
unique to one island. High, wet islan
ied species because their more cor
provide niches for a wider range of
range of circumstances for evolution.

The flora and fauna of islands a
abitants once they arrive. In prehis
ceangoing sailing canoes brought
d breadfruit and animals such as p
man activities from tourism to mi
ntinue to change the flora and fau

Generally, the diversity of land a
western Pacific, near the larger la
, where the islands are smaller ar
ral rain forest flora of New Zealar
ds of the Pacific is rich and a
nd and the Pacific islands inclu
almost no indigenous reptiles, a
frogs. The reason for the absen
ealand and the islands were ne
sia by a land bridge that anin
ross. On the other hand, indig
l varied. Two examples are Nev
bird that grew up to 12 feet (3.
of food for the Maori people
fore Europeans arrived. Today
untry with the most introduced
, nearly all brought in by Europe

animals.
a very immature
with nipples. The b
species include wombat
marsupials fill ecological ni
are occupied by rats, badgers, moles,
ers), and bears. The **monotremes,** egg-laying mam
the duck-billed platypus (Figure 11.10) and the s
endemic to Australia and New Guinea. Some of th
birds known in Australia migrate in and out, but more than 325 species are endemic.

Most of Australia's endemic plant species are adapted to dry
conditions. Many of the plants have deep taproots to draw moisture from groundwater and small, hard, shiny leaves to reflect heat
and to hold moisture. Much of the continent is grassland and
scrubland, with bits of open woodland; there are only a few true
forests, found in pockets along the Eastern Highlands, the southwestern tip, and in Tasmania (Figure 11.11). Two plant genera
account for nearly all the forest and woodland plants: *Eucalyptus*
(450 species, often called "gum trees") and *Acacia* (900 species,
often called "wattles").

### Plant and Animal Life in New Zealand and the Pacific Islands.

The prehuman biogeography of the Pacific islands has long interested those who study evolution and the diffusion of plants and animals, including Charles Darwin, who formulated many of his ideas
about evolution after visiting the Galápagos Islands of the eastern
Pacific (see Figure 3.1 on pages 114–115). Islands have to snare
their plant and animal populations from the sea and air around
them as organisms are carried from larger islands and continents by

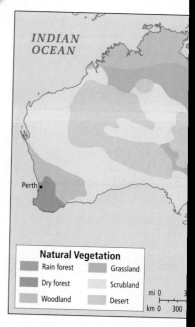

**FIGURE 11.11** Australia's natural vegetatio
grassland and scrubland; a few forests can be f
in the far southwest, and in Tasmania. [Adapte
*Oceania* (Englewood Cliffs, N.J.: Prentice Hall,

# HUMAN PATTERNS OVER TIME

| Personal Vignette | *With courage, you can travel anywhere in the world and never be lost. Because I have faith* |

*in the words of my ancestors, I'm a navigator.*

Mau Piailug

In 1976, Mau Piailug made history by sailing a traditional Pacific island voyaging canoe across the 2400 miles (3860 kilometers) of deep ocean between Hawaii and Tahiti (149°E, 17°S). He did so without a compass, charts, or other modern instruments, using methods passed down through his family. He relied mainly on observations of the stars, the sun, and the moon to find his way. When clouds covered the sky, he used the patterns of ocean waves and swells, as well as the presence of seabirds, to tell him of distant islands over the horizon.

Piailug reached Tahiti 33 days after leaving Hawaii and made the return trip in 22 days. His voyage settled a major scholarly debate over how people settled the many remote islands of the Pacific without navigational instruments, thousands of years before the arrival of Europeans. Some thought that navigation without instruments was impossible and argued that would-be settlers simply drifted about on their canoes at the mercy of the winds, most of them starving to death on the seas, with a few happening on new islands by chance. It was hard to refute this argument because local navigational methods had died out almost everywhere. However, in isolated Micronesia, where Piailug is from, indigenous navigational traditions had survived.

Piailug learned his methods from his grandfather in secret because the Germans who first colonized Micronesia, and later the Japanese, banned long-distance navigation to prevent their Micronesian forced

**FIGURE 11.12** A traditional Pacific island voyaging canoe. In 1999, the *Hokule'a* sailed from Hawaii to Easter Island and back—a round trip of about 14,000 miles (22,530 kilometers)—guided only by traditional navigational techniques, such as the reading of wave patterns. [Cary Wolinsky.]

laborers from sailing away. After World War II, however, Piailug was free to sail, and he began making small voyages of several hundred miles. Eventually he attracted the attention of some Hawaiians who were building a traditional voyaging canoe, *Hokule'a* (Figure 11.12), with the aim of proving that traditional technologies were adequate for long-distance Pacific travel. Since the successful 1976 voyage, Piailug has trained several students in traditional navigational techniques, which have become a symbol of cultural rebirth and a source of pride throughout the Pacific.

*Source: Richard Nile and Christian Clerk, Cultural Atlas of Australia, New Zealand, and the South Pacific (New York: Facts on File, 1996), pp. 63–65.*

## The Peopling of Oceania

The longest-surviving inhabitants of Oceania are Australia's **Aborigines,** who migrated from Southeast Asia 50,000–70,000 years ago (Figure 11.13). Amazingly, some memory of this ancient journey may be preserved in Aboriginal oral traditions, which recall mountains and other geographic features that are now submerged under water. At about the same time that the Aborigines were settling Australia, related groups were settling nearby areas. These were the **Melanesians,** so named for their relatively dark skin tones, a result of high levels of the protective pigment melanin. The Melanesians spread throughout New Guinea and other nearby islands, giving this area its name, Melanesia. They lived in isolated pockets. One indication of the great age and isolation of the Melanesian settlement of New Guinea is the existence of hundreds of distinct yet related languages on that island. Like the Aborigines, the Melanesians survived mostly by hunting, gathering, and fishing, although some groups—especially those inhabiting the New Guinea highlands—also practiced agriculture. There, agricultural features excavated in the terrain of a swampy basin have been dated at 9000 years ago, making this region one of the earliest places on earth where plant domestication has been firmly documented (Figure 11.14).

Much later, about 5000 to 6000 years ago, groups of linguistically related **Austronesians** migrated out of Southeast Asia and continued the settlement of the Pacific. By about a thousand years ago, some Austronesians had reached most of the remaining far-flung islands of the Pacific, sometimes mixing with the Melanesian peoples they encountered. These Austronesians were renowned for their ability to navigate over vast distances. They were fishers, hunter-gatherers, and cultivators who developed complex cultures and maintained trading relationships among their widely spaced islands.

In the millennia that have passed since first settlement, humans have continued to circulate throughout Oceania. Some apparently set out because their own space was too full of people and conflict, food reserves were declining, or they wanted a life of greater freedom. It is also likely that Pacific peoples were enticed to new locales by the same lures that later attracted some of the more romantic explorers from Europe and elsewhere: sparkling beaches, magnificent blue skies, beautiful people, scented breezes, and lovely landscapes.

The vast area settled by these people can be divided into three distinct cultural regions (see Figure 11.13). **Micronesia** refers to the

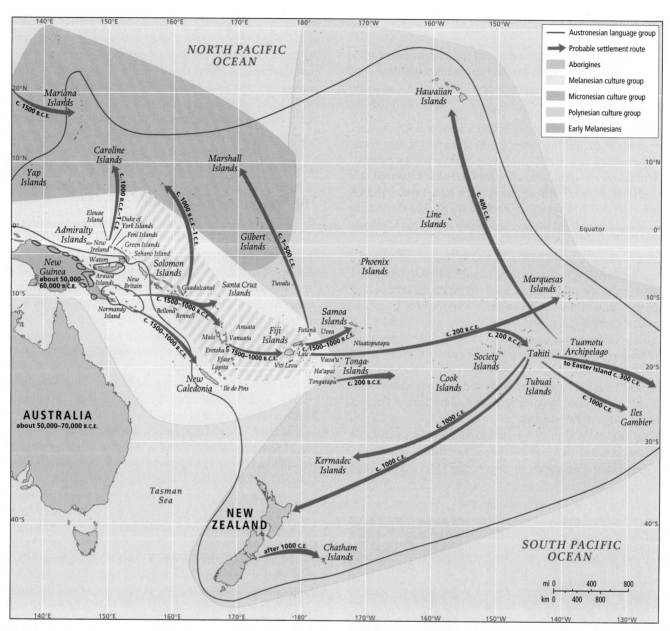

**FIGURE 11.13** Primary culture groups in the island Pacific. By about 25,000 years ago, people were spread across a large part of New Guinea and had even begun moving across the ocean to nearby Pacific islands. Archaeologists have dated a site on Buka in the northern Solomon Islands to about 26,000 years ago. Movement into the more distant Pacific islands apparently began with the arrival of the Austronesians, who went on to inhabit the farthest reaches of Oceania. [Adapted from Richard Nile and Christian Clerk, *Cultural Atlas of Australia, New Zealand, and the South Pacific* (New York: Facts on File, 1996), pp. 58–59.]

small islands lying east of the Philippines and north of the equator. **Melanesia** includes New Guinea and the islands south of the equator and west of Tonga (the Solomon Islands, New Caledonia, Fiji, and Vanuatu). **Polynesia** refers to the numerous islands situated inside a large irregular triangle formed by New Zealand, Hawaii, and Easter Island (a tiny speck of land in the far eastern Pacific, at 109°W, 27°S, not shown in the figures in this chapter). Polynesia is the most recently settled part of the Pacific; some Polynesian influence remains in Melanesia in such places as Fiji and the Solomon Islands.

## Arrival of the Europeans

The earliest recorded contact between Pacific peoples and Europeans took place in 1521, when the first Europeans to cross the Pacific, led by the Portuguese explorer Ferdinand Magellan (working for Spain), landed on the island of Guam in Micronesia. That encounter ended badly. The islanders, intrigued by European vessels, tried to take a small skiff. For this crime, Magellan had his men kill the offenders and burn their village to the ground. A few

months later, Magellan was hims[...] [...]ea and the other larger
became the Philippines, which he [...] [...] plentiful resource base made it pos-
theless, by the 1560s, the Spanish had set up a lucrative trade route [...] [...] people to live in small, simple societies, less subject to the
between Manila in the Philippines and Acapulco in Mexico. stratification and class tensions seen in so much of the world. On
Explorers from other European states followed, first taking an inter- the smaller islands of Micronesia and Polynesia, land and resources
est mainly in the region's valuable spices. The British and French were scarcer. On these islands, many societies were hierarchical,
explored extensively in the eighteenth century. with layers of ruling elites at the top and undifferentiated common-

The Pacific was not formally divided among the colonial pow- ers at the bottom. Moreover, many of the peoples of Oceania coex-
ers until the nineteenth century, and by that time, the United States, isted in a state of moderate antagonism. Although warfare occurred,
Germany, and even Japan had joined France and Britain in taking hostilities were often settled ritualistically and by means of annual
control of various island groups. European colonization of Oceania tribute-paying ceremonies, rather than by resorting to mortal vio-
proceeded according to the models developed in Latin America, lence. Individual rulers rarely amassed large territories or controlled
Africa, South Asia, and Southeast Asia, with the major emphasis be- them for long.
ing on extractive agriculture and mining. Native people were often
displaced from their lands or exposed to exotic diseases to which
they had no immunity, and their populations declined.

Many enduring notions about the Pacific arose from the European explorations of the eighteenth and nineteenth centuries. During this time, European thinkers were debating whether or not
civilization actually improves the quality of life for human beings.
Some argued that civilization corrupts and debases people, and they
glorified what they termed "primitive" people living in distant
places supposedly untouched by corrupting influences; they coined
the term **noble savage** to describe such people. Explorers of the
Pacific who were influenced by such ideas were caught off guard
when, from time to time, the islanders rebelled, armed themselves,
and attacked those who were taking their lands and resources. Usually the surprised Europeans quickly revised their opinions and rela-
beled the "noble savages" as brutish and debased.

The realities of life in Oceania were much more balanced and
complex than the positive and negative extremes that Europeans

## The Colonization of Australia and New Zealand

Although all of Oceania has experienced European or American
rule at some point, the most Westernized parts of the region are
Australia and New Zealand. The colonization of these two countries
by the British has resulted in many parallels with North America. In
fact, the American Revolution was the major impetus for "settling"
Australia because, once the North American colonies were independent, the British needed somewhere else to send their convicts.
In early nineteenth-century Britain, a relatively minor theft—for
example, of food or a piglet—might be punished with a term of
7 years' hard labor in Australia; a death sentence imposed on a poor
12-year-old girl for a violent attack on another child in London
might be commuted if she agreed to be shipped to Australia. After
their sentences were served, most former convicts chose to stay in the
colony, some founding prominent families. A steady flow of English

and Irish ... ar II, however, the European
they are given ... ...nce to Japan's invasion of much
spirit. They were joined by a ... ... northern Australia.
grants from the British Isles who were ac... ...ed States became the dominant power
inexpensive farmland. Waves of these immigra... ... ...U.S. investment became increasingly important
World War II. New Zealand was settled somewhat later, in the mid- to the economies of Oceania. Australia and New Zealand joined
1800s. Although its population also derives primarily from British the United States in a cold war military alliance, and both fought
immigrants, New Zealand was never a penal colony. alongside the United States in Korea and Vietnam, suffering con-

Another similarity among Australia, New Zealand, and North siderable casualties and experiencing significant antiwar activity at
America was the treatment of indigenous peoples by European home. U.S. cultural influences were strong, too, as North Ameri-
settlers. In both Australia and New Zealand, native peoples were can products, technologies, movies, and pop music penetrated
killed outright, annihilated by infectious diseases, or shifted to the much of Oceania.
margins of society. The few who lived on territory deemed unde- By the 1970s, another shift was taking place as many of the
sirable by Europeans were able to maintain their traditional way of island groups were granted self-rule by their European colonizers
life, but the vast majority who survived lived and worked in grind- and Oceania became steadily drawn into the growing economies of
ing poverty, either in urban slums or on cattle and sheep ranches. Asia. Since the 1960s, Australia's thriving mineral export sector has
Today, native peoples still suffer from pervasive discrimination and become increasingly geared toward supplying Japan's burgeoning
maladies such as alcoholism and malnutrition. Even so, some pro- manufacturing industries. Similarly, since the 1970s, New Zea-
gress is being made toward improving their lives, as discussed on land's wool and dairy exports have gone mostly to Asian markets. As
pages 599–601. we shall see, these transformations have been accompanied by con-
siderable cultural and economic strain. Nonetheless, despite occasional backlashes against "Asianization," Australia, New Zealand,

## Oceania's Shifting Ties with Other Countries

and the rest of Oceania are becoming more open to Asian influences. Many Pacific islands have significant Chinese, Japanese, Fil-
During the twentieth century, Oceania's relationship with the rest of ipino, and Indian minorities; even the small Asian minorities of
the world changed at least three times: from a predominantly Euro- Australia and New Zealand are increasing. On some Pacific islands,
pean focus to identification with the United States and Canada to such as Hawaii, Asians are now a majority of the population.
the currently emerging linkage with Asia. Up until roughly World
War II, the colonial system gave the region a European orientation.
In most places, the economy depended largely on the export of raw

## POPULATION PATTERNS

materials to Europe. Thus, even when a colony gained independ-
ence from Britain, as Australia did in 1901 and New Zealand did in Oceania occupies a huge portion of the planet, yet it has just 34 mil-
1907, people remained strongly tied to their mother countries— lion people—fewer than the state of California (Figure 11.16). They
even today the Queen of England remains the titular head of state

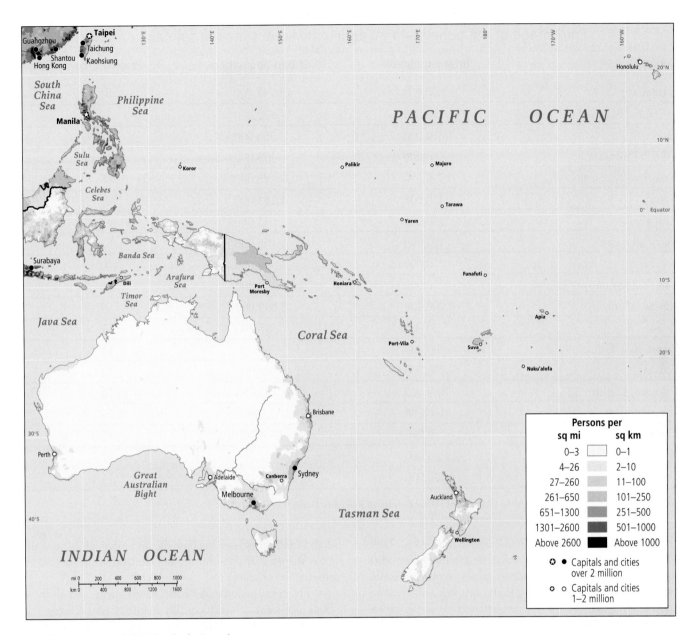

**FIGURE 11.16** Population density in Oceania.

live on a total land area slightly larger than the contiguous United States, but spread out in bits and pieces across an ocean larger than the Eurasian landmass. The Pacific islands, including Hawaii, have somewhat over 4 million people; Australia has 20.4 million, Papua New Guinea 5.9 million, and New Zealand 4 million.

## Population Growth and Distribution in Oceania

Two different population patterns exist in Oceania. The Pacific islands, with their younger, more rapidly growing populations and life expectancies in the sixties or low seventies, are similar to many developing countries. In contrast, Australia and New Zealand are more like the wealthier countries of the world, with their relatively older, more slowly growing populations and life expectancies around 80. The overall trend throughout the region, however, is toward smaller families and aging populations.

Figure 11.16 indicates the uneven distribution of people in this region. Most Australians live in a string of cities along their country's well-watered and relatively fertile eastern and southeastern coasts—Brisbane, Newcastle, Sydney, Canberra, Melbourne, Adelaide—and on the southwestern tip of the continent in and around the city of Perth. New Zealand has one large city, Auckland, on North Island, and several medium-sized cities on both North Island and South Island. Australia and New Zealand have among the highest percentages of city dwellers outside Europe: in 2005, 91 percent of

TABLE 11.1 Percentage of the population living in cities and towns in selected places in Oceania

| Country | Total population | Urban population | Percent urban |
|---|---|---|---|
| Australia | 20,600,000 | 18,746,000 | 91 |
| Federated States of Micronesia | 100,000 | 22,000 | 22 |
| Fiji | 800,000 | 368,000 | 46 |
| French Polynesia | 300,000 | 159,000 | 53 |
| Guam | 200,000 | 186,000 | 93 |
| Hawaii | 1,300,000 | 1,183,000 | 91 |
| Kiribati | 100,000 | 43,000 | 43 |
| Marshall Islands | 100,000 | 68,000 | 68 |
| Nauru | 10,000 | 10,000 | 100 |
| New Caledonia | 200,000 | 142,000 | 71 |
| New Zealand | 4,100,000 | 3,649,000 | 89 |
| Palau | 20,000 | 15,400 | 77 |
| Papua New Guinea | 6,000,000 | 780,000 | 13 |
| Samoa | 200,000 | 44,000 | 22 |
| Solomon Islands | 500,000 | 80,000 | 16 |
| Tonga | 100,000 | 23,000 | 23 |
| Tuvalu | 10,000 | 4,700 | 47 |
| Vanuatu | 200,000 | 42,000 | 21 |
| Totals | 34,840,000 | 25,565,100 | 73.4 |

Source: 2006 World Population Data Sheet (Washington, D.C.: Population Reference Bureau).

Australians and 86 percent of New Zealanders lived in cities (Table 11.1). The vast majority in these two countries live in modern affluence and work in a range of occupations typical of highly industrialized societies. Nonetheless, overall densities in both countries are low: Australia averages just 7 people per square mile (3 per square kilometer) and New Zealand 39 per square mile (15 per square kilometer).

The population densities of the smaller Pacific islands vary widely. Some are sparsely settled or uninhabited; others—including some of the smallest, such as the Marshall Islands and Tuvalu —have 800 to 1000 people per square mile (307 to 386 per square kilometer). The highest density in Oceania is found on the tiny island of Nauru, with 1529 people per square mile (570 per square kilometer). The causes of this density and the resulting environmental disaster are discussed on page 611.

## Urbanization in the Pacific Islands

The global trend of migration from the countryside to towns and cities is highly visible in the Pacific islands, where a steadily growing proportion of the population lives in urban areas (see Table 11.1). Throughout the Pacific, urban centers have transformed natural landscapes, and in some small states, such as Palau and the Mar-

shall Islands, they have become the dominant landscape. Although cities are places of opportunity, they can also be sites of both cultural change and conflict.

The great majority of Pacific island towns, and all the capital cities, are located in ecologically fragile coastal settings. Many of these towns were established during the colonial era for access to shipping and were situated in places suitable for only limited numbers of people. Consequently, little land is available for development, and access to housing is limited. Squatter settlements have been a visible feature of the region's urban areas for several decades. The discharge of untreated sewage and other wastes into coastal waters and lagoons has damaged marine environments, reduced the productivity of subsistence fisheries, and periodically led to the outbreak of diseases such as cholera. Air pollution is a new phenomenon in the Pacific islands, as is noise pollution.

Cultural changes, too, have resulted from urbanization. Although many urban residents were born outside the towns and maintain close connections to their rural home communities, some urban islanders have disavowed rural life, ethnic identity, and cultural commitments. Increasingly, people are marrying in town and across language divisions, creating new patterns of social alliances and networks. These changes, along with the adoption of urban lifestyles, are creating new social tensions and changing the very nature of social life in the island Pacific. Urban unemployment and crime are on the rise, and low economic growth restricts the revenue available to governments to manage urban development.

The urban poor of Oceania, whether they live in the developed economies of Australia and New Zealand—where they tend to have Aboriginal or Maori origins—or in cities such as Port Moresby in Papua New Guinea, Suva in Fiji, Papeete in Tahiti, or Honolulu in Hawaii, may have higher cash incomes than rural Pacific Islanders, but they have much lower standards of living. In the cities, there is little they can do to supplement their incomes with self-produced food or other necessities.

# II CURRENT GEOGRAPHIC ISSUES

Many current geographic issues in Oceania (Figure 11.17) are related to the transition now under way from European to Asian and inter-Pacific cultural influences. This change in cultural orientation is related to the larger transition to a global economy now under way everywhere. Oceania's old relationships were built on historical factors, such as the settlement of Australia and New Zealand by Europeans. Its new relationships, in contrast, are influenced by economic and geographic considerations, such as physical proximity to Asia and its recently opened markets—especially in China, but also in South and Southeast Asia.

Emerging technologies such as e-mail and the Internet and rapid air travel are having a special effect on the far-flung places of Oceania. These new ways of interacting encourage cultural sensitivity and the modification of old prejudices because they give people greater access to information about unfamiliar ways of life and the chance to experience distant places personally. One result of

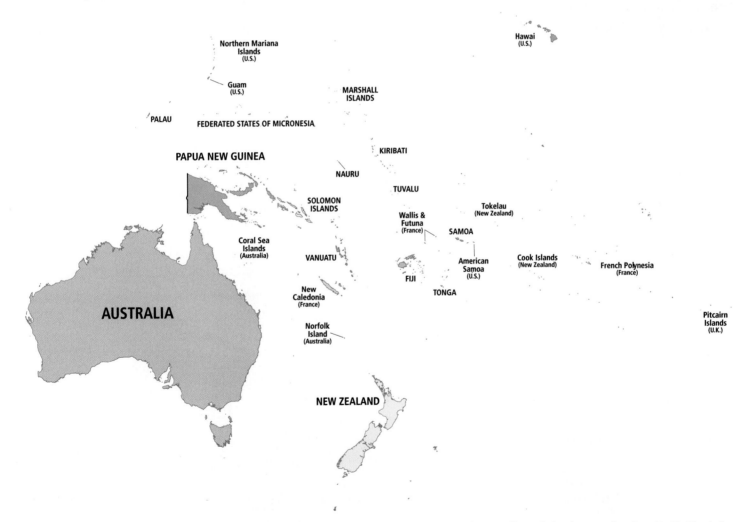

**FIGURE 11.17** Political map of Oceania. Thirteen of these entities are independent countries; the others are territories of or otherwise affiliated with other nations. Not depicted on the map are the Pacific Island Wildlife Refuges, a widely scattered, essentially uninhabited, group of northern Pacific islands that constitute a U.S. territory.

this greater interaction is that New Zealand and the Pacific islands are finding philosophical grounds for a closer mutual identity, including the fostering of greater public awareness of environmental issues and the renewal and acceptance of their common Polynesian as well as European cultural heritage. Australia participates in this new closeness too, but some regional policing and administrative duties that it has inherited from the British occasionally place it in a dominant role in Oceania that causes resentment in other parts of the region.

## SOCIOCULTURAL ISSUES

The cultural sea change away from Europe and toward Asia and the Pacific has been accompanied by new respect for indigenous peoples: the Aborigines of Australia; the Maori of New Zealand; and the Melanesians, Micronesians, and Polynesians of the Pacific islands. In addition, a growing sense of common economic ground with Asia has heightened awareness of the attractions of Asian culture.

### Ethnic Roots Reexamined

***Weakening of the European Connection.*** Until very recently, most people of European descent in Australia and New Zealand thought of themselves as Europeans in exile. Many considered their lives incomplete until they had made a pilgrimage to the British Isles or the European continent. In her book *An Australian Girl in London* (1902), Louise Mack wrote: "[We] Australians [are]

packed away there at the other end of the world, shut off from all that is great in art and music, but born with a passionate craving to see, and hear and come close to these [European] great things and their home[land]s."

These longings for Europe were accompanied by racist attitudes toward both indigenous peoples and Asians. Most histories of Australia written in the early twentieth century failed to even mention the Aborigines, and later writings described them as amoral. The prevailing idea was that both Australia and New Zealand should preserve European culture in this nether region of the Southern Hemisphere; thus, in the 1920s, immigrants from Asia, Africa, and the Pacific were legally barred in both countries. Trading patterns further reinforced connections to Europe: until World War II, the United Kingdom was the primary trading partner of both New Zealand and Australia.

When migration from the British Isles slowed after World War II, both Australia and New Zealand began to lure immigrants from southern and eastern Europe, many of whom had been displaced by the war. Hundreds of thousands came from Greece, parts of the former Yugoslavia, and Italy. The arrival of these non-English-speaking people began a shift toward a more multicultural society. The election of more liberal governments led to a loosening of restrictions against Asian migrants in the late 1950s, and in the 1960s the whites-only immigration policies were set aside entirely. There was a large increase in Asian immigrants in the early 1970s after the United States withdrew from Vietnam, when many Vietnamese refugees sought a safe haven. The more recent development of

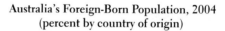

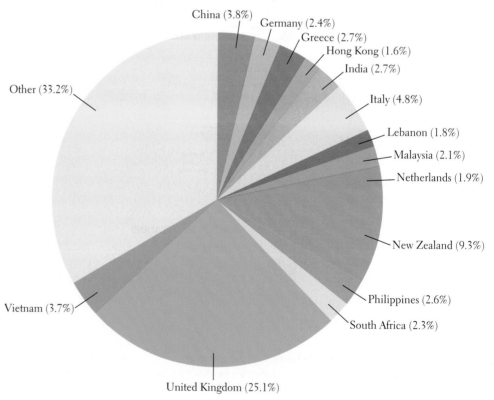

**Australia's Foreign-Born Population, 2004**
**(percent by country of origin)**

China (3.8%)
Germany (2.4%)
Greece (2.7%)
Hong Kong (1.6%)
India (2.7%)
Italy (4.8%)
Lebanon (1.8%)
Malaysia (2.1%)
Netherlands (1.9%)
New Zealand (9.3%)
Philippines (2.6%)
South Africa (2.3%)
United Kingdom (25.1%)
Vietnam (3.7%)
Other (33.2%)

**FIGURE 11.18** Australia's cultural diversity in 2004. About 24 percent (4.75 million) of Australia's people were born in other places, making Australia one of the world's most ethnically diverse nations. [Data from Year Book Australia, 2006, at http://www.abs.gov.au/Ausstats/abs@.nsf/bb8db737e2af84b8ca2571780015701e/0D98319CE458B364CA2570DE0006A39B?opendocument.]

high-tech industries has brought the recruitment of skilled workers from India (8135 in 2003–2004 alone).

While Asians remain a small percentage of the total population in both Australia and New Zealand, new immigration policies are increasing the numbers of new Asian immigrants, especially from China, Vietnam, and India (Figure 11.18). People of European descent are declining as a proportion of the population, but are expected to remain the most numerous segment throughout the twenty-first century.

### The Social Repositioning of Indigenous Peoples in Australia and New Zealand.
The makeup of the populations of Australia and New Zealand is also changing in another respect. In Australia, for the first time in recent memory, the number of people who claim indigenous origins is increasing. Between 1991 and 1996, the number of Australians claiming Aboriginal origins rose by 33 percent. In New Zealand, the number claiming a Maori background rose by 20 percent. These increases are probably linked to more positive attitudes toward indigenous peoples in the region. First, more people now understand that colonial attitudes are largely responsible for the low social standing and impoverished state of indigenous peoples; hence those who have some Aborigine or Maori ancestors are more willing to claim them. Another factor is that relationships between European and indigenous peoples are more common and more open, and so the number of people with mixed heritage is increasing.

A third factor may be increased respect for Aboriginal culture. Aborigines base their way of life on the idea that the spiritual and physical worlds are intricately related. The dead are everywhere present in spirit, and they guide the living in how to relate to the physical environment. *Dreamtime* refers to the time of creation when the human spiritual connections to rocks, rivers, deserts, plants, and animals were made clear (Figure 11.19) Aboriginal people who have remained close to their heritage still read the landscape as a complex sign system conveying spiritual meaning. Particular tribal groups are associated with specific animals or landscape features, from which they gather solace and inspiration. When people die, they are said to "go into the country." Unfortunately, very few Aboriginal people know their own culture, and many live in impoverished conditions.

In 1988, during a bicentennial celebration of the founding of white Australia, a contingent of some 15,000 Aborigines protested that they had little reason to celebrate. During the same 200 years, they had been excluded from their ancestral lands, experienced a suppression of their ancient cultural practices and beliefs, lost basic civil rights, and had effectively been erased from Australian national consciousness. Into the 1960s, Aborigines had only limited rights of citizenship, and it was even illegal for them to have a drink of alcohol. Until 1993, Aborigines were assumed to have no prior claim to any land in Australia. British documents indicate that during colonial settlement, all Australian lands were deemed to be available for British use because the Aborigines were thought to be too primitive to have concepts of land tenure, since their nomadic cultures had "no fixed abodes, fields or flocks, nor any internal hierarchical differentiation." After the Australian High Court de-

**FIGURE 11.19** Aboriginal art in Kakadu National Park, Northern Territory, Australia. For the Aborigines, the spiritual and physical worlds are intimately connected, as illustrated here by this representation of Djawok, a creator who left his image as a cuckoo on the rock. [Belinda Wright/National Geographic Image Collection.]

clared this position void in 1993, Aboriginal groups began to win some land claims, mostly for land in the arid interior previously claimed only by the Australian government. Court cases to restore Aboriginal rights and lands continue to be heard. Figure 11.20 shows the Aboriginal Embassy, a permanent installation in Canberra that advocates for Aborigines.

In New Zealand, relations between the majority European-derived population and the indigenous Maori have proceeded only somewhat more amicably. When the Maori signed the Waitangi Treaty with the British in 1840, they thought they were granting only rights of land usage, not ownership. The Maori did not regard land as a tradable commodity, but rather as an asset of the people as a whole, used by families and larger kin groups to fulfill their needs. Geographer Eric Pawson writes: "To the Maori the land was sacred … [and] the features of land and water bodies were woven through with spiritual meaning and the Maori creation myth." The British, on the other hand, assumed that the treaty had transferred

**FIGURE 11.20** The Australian Aboriginal Embassy. The Embassy, a collection of colorful wooden structures, was erected in front of Old Parliament House in Canberra as a protest against the seizure of Aboriginal lands by European settlers at the end of the eighteenth century and the continuing refusal of the Australian government to acknowledge that injustice. The protest began in 1972 and was continuing in 2001, when the authors visited. Nearby is a permanent encampment of Aboriginal people who staff the facility and welcome the many visitors. [Mac Goodwin.]

Maori lands to them, giving them *exclusive* rights to settle the land with British migrants and to extract wealth through farming, mining, and forestry.

By 1950, the Maori had lost all but 6.6 percent of their former lands; European settlers and the government owned and occupied the rest. Maori numbers shrank from a probable 120,000 in the early 1800s to 42,000 in 1900, and they came to occupy the lowest and most impoverished rung of New Zealand society. In the 1990s, however, the Maori began to reclaim their culture (Figure 11.21), and they established a tribunal that forcefully advances Maori interests through the courts. New Zealand formalized its efforts to right past wrongs and bring greater equality and social participation to the Maori and other minority groups in the country in 1996, when the government agreed to settle several long-standing Maori claims to land and fishery rights.

Maori numbers rebounded during the twentieth century, reaching an estimated 650,000 by 2006. This number includes the many New Zealanders who previously hid their Maori origins but are now proud to claim them. Overall, New Zealand may lead the world in addressing past mistreatment of indigenous peoples. Nonetheless, the Maori still have notably higher unemployment, lower educational attainments, and poorer health than the New Zealand population as a whole.

### Balancing Rights in the Pacific Islands.

The Republic of the Fiji Islands in the southern Pacific exemplifies an issue that is common in the Pacific islands: how to balance the rights of indigenous people with those of "outsiders" who may have lived there for several generations. Fiji, one of the first colonial domains in the Pacific to achieve independence from Britain (in 1970), has a population that is about evenly divided between indigenous Fijians and the descendants of indentured sugar plantation workers brought in by the British from India more than a century ago. Fijians of Indian origins hold significant economic and political power, especially in the western urban centers and in areas of tourism and sugar cultivation.

On the other hand, the indigenous Fijians, whose community affairs are governed by traditional chiefs, tend to live on rural islands in the east of the island group and to be less prosperous. Complicating matters is a codified system established by the British whereby

**FIGURE 11.21** Maori warrior issuing a traditional challenge in a sacred open meeting area. The challenge was a VIP welcome for Britain's Prince Charles during his 1994 visit to Waitakere City, New Zealand. [Tim Graham/Alamy.]

land rights are held by indigenous clans and land cannot be taken or sold. Indian Fijians who grow sugarcane or run businesses do so on leased land.

In 1987 and again in 2000, indigenous Fijians attempted coups d'état against legally elected governments dominated by Indian Fijians. Violence, looting, and rioting were curtailed by Australian peacekeepers, but the conflict retarded tourism and set back economic and social development for some years. In 2006, indigenous Fijians in the military staged a successful coup d'état that put them in charge into 2007. Many Indian Fijians have left the islands, fearing that the rise of indigenous rights would weaken egalitarian policies. Political leaders around the region, from Australia to Hawaii, have reminded indigenous Fijians that indigenous rights cannot be held superior to fundamental human rights.

## Forging Unity in Oceania

In embracing its Maori roots, New Zealand has in effect also begun to embrace its connection to the wider Pacific, especially Polynesia. Although the majority of immigrants to New Zealand continue to come from the United Kingdom, the second largest influx has come from Polynesia. Auckland now has the largest Polynesian population of any city in the world.

A sense of unity with Oceania as a whole is developing throughout the region as people begin to appreciate the region's cultural complexity and cooperate in activities ranging from schooling to sports to environmental activism. One way in which this unity is manifested is through interisland travel. Today, people travel in small planes from the outlying islands to hubs such as Fiji, where jumbo jets can be boarded for Auckland, Melbourne, or Honolulu. Cook Islanders call these little planes "the canoes of the modern age." New Zealanders can migrate to Australia to teach or train, a businessman from Kiribati in Micronesia can fly to Fiji to take a short course at the University of the South Pacific, or a Cook Islands teacher can take graduate training in Hawaii.

***Languages in Oceania.*** The Pacific islands—most notably in Melanesia—have a rich variety of languages. In some cases, the islands in a single chain have several different languages. A case in point is Vanuatu, a chain of 80 mostly high volcanic islands to the east of northern Australia. At least 108 languages are spoken by a population of just 180,000—an average of one language for every 1600 people! Another example is New Guinea, the largest, most populous, and most ethnically diverse island in the Pacific. No fewer than 800 languages are spoken on New Guinea by a populace of 5.5 million.

Languages are both an important part of a community's cultural identity and a hindrance to cross-cultural understanding. In Melanesia, and elsewhere in the Pacific, the need for communication with the wider world is served by several **pidgin** languages that are sufficiently similar to be mutually intelligible (Figure 11.22). Pidgins are made up of words borrowed from the several languages of people involved in trading relationships. Over time, they can grow into fairly complete languages, capable of fine nuances of expression. When a particular pidgin is in such common use that mothers talk to their children in it, then it can literally be called a

**FIGURE 11.22** A poster written in pidgin English. *Mi No Puret Mi Gat Banis* is pidgin for "I'm not afraid, I have protection." The poster is part of Papua New Guinea's program to combat an alarming increase in HIV-AIDS. [Torsten Blackwood/AFP Photo.]

"mother tongue." In Papua New Guinea, a version of pidgin English is the official language.

***The Pacific Way.*** The **Pacific Way** is a term used since 1970 to convey the idea that Pacific Islanders and their governments have a regional identity growing out of their own particular social experience. The concept first emerged when Pacific Islanders were grappling with the problems presented by school curricula imposed by the colonial powers, which suppressed the use of native languages and often depicted the region as peripheral and backward. They developed the Pacific Way as a philosophy to guide them in writing their own school texts. It embodied the idea that Pacific island children should first learn about their own cultures and places before they studied Britain, France, or the United States.

The Pacific Way includes the idea that Pacific Islanders have the ability to control their own development and solve their own problems, and increasingly it has grown into an integrated approach to economic development and environmental issues, with an emphasis on consensus as a traditional approach to problem solving. In some ways, the concept resembles *Pancasila* in Southeast Asia (see Chapter 10, page 556), and it has some of the same potential for abuse. But constructive steps have been taken to implement the Pacific Way. The South Pacific Regional Environmental Program (SPREP), in existence since 1980, emphasizes regional cooperation and grassroots environmental education. SPREP programs routinely incorporate traditional environmental knowledge into efforts to promote sustainable livelihoods, often based on traditional crafts.

***Sports as a Unifying Force.*** Sports and games are a major feature of daily life throughout Oceania, and the region has both shared them with and borrowed them from cultures around the world. Long-distance sailing, now a world-class sport, was an early skill in this region. Surfing evolved in Hawaii from ancient navigational

**FIGURE 11.23** Sports as a unifying force. In the Trobriand Islands, near New Guinea, the British game of cricket has been reformulated to include local traditions. Village teams of as many as 60 men dress in traditional garments and decorate their bodies in ways that are reminiscent of British cricket uniforms, such as the painting-on of white shin pads, yet also include elements of magical decoration. Chants and dances are part of Trobriand cricket matches. The matches can go on for weeks, and they attract crowds of young onlookers, as seen in this picture. [Robert Harding Picture Library, London.]

customs that matched human wits against the power of the ocean. On hundreds of Pacific islands and in Australia and New Zealand, the rugby field, the volleyball court, the soccer field, and the cricket pitch (Figure 11.23) are important centers of community activity. Baseball is a favorite in the parts of Micronesia that were U.S. trust territories. Women compete in the popular sport of netball, which is a little like basketball but without a backboard.

Sports competitions, including native dances such as the Haka described at the opening of this chapter, are the single most common and resilient link among the countries of Oceania. Such competitions encourage regional identity, and they provide opportunities for ordinary citizens to travel extensively around the region and to sports venues in other parts of the world. The South Pacific Games—featuring soccer, boxing, tennis, golf, and netball, among other sports—are held every 4 years. The Micronesians hold periodic games that incorporate tests of many traditional skills, such as spearfishing, climbing coconut trees, and racing outrigger canoes.

## Women's Roles in Oceania

Perhaps the most enduring myth Europeans created regarding Oceania was their characterization of the women of the Pacific islands as gentle, simple, compliant love objects. (Tourist brochures still promote this notion.) Although there is ample evidence to suggest that Pacific Islanders did have more sexual partners in a lifetime than Europeans did, the reports of unrestrained sexuality

**FIGURE 11.24** *Arearea* (Amusement) by Paul Gauguin. In this 1892 painting, Tahitian women are rendered in a European Romantic pastoral style that emphasizes their gentle, compliant demeanor. [Musée d'Orsay, Paris. Photograph by Erich Lessing, Art Resource, New York.]

related by European sailors were no doubt influenced by the exaggerated fantasies one might expect from all-male crews living at sea for months at a time. The notes of Captain James Cook are typical: "No women I ever met were less reserved. Indeed, it appeared to me, that they visited us with no other view, than to make a surrender of their persons." Over the years, such notions about Pacific island women have been encouraged by the paintings and prints of Paul Gauguin (Figure 11.24), the writings of novelist Herman Melville (*Typee*), and the studies of anthropologist Margaret Mead (*Coming of Age in Samoa*), as well as by movies and musicals such as *Mutiny on the Bounty* and *South Pacific*.

In reality, gender roles in the Pacific islands varied considerably from those in Europe, but not in the ways European explorers imagined. Women often exercised a good bit of power in family and clan, and their power increased with motherhood and advancing age. In Polynesia, a woman could achieve the rank of ruling chief in her own right, not just as the consort of a male chief. In everyday Polynesian life, men were the primary cultivators of food as well as the usual cooks. Women were primarily craftspeople, but they also contributed to subsistence by gathering fruits and nuts and by fishing. And in some places—Micronesia, for example—lineage was established through women, not men.

Today there is great variation in gender roles in Oceania, and they are changing significantly throughout the region. In Australia and New Zealand, women's access to jobs and policy-making positions has improved over the last few decades. Increasingly, young women are choosing careers and postponing marriage until their thirties. Many of these women have mates, but choose not to marry them. Nonetheless, Australian and New Zealand societies continue to reinforce the housewife role for women in a variety of ways. For example, the expectation is that women, not men, will interrupt their careers to stay home to care for young or elderly family members.

In the Pacific, gender roles and relationships vary greatly from island to island, and they change over the course of a lifetime. Today, many young women fulfill traditional roles as mates and mothers and practice a wide range of domestic crafts, such as weaving and basketry. Then, in middle age, they may return to school and take up careers. Some Pacific women, with the aid of government scholarships, pursue higher education or job training that takes them far from the villages where they raised their children. Accumulating age and experience may boost Pacific women into community positions of considerable power.

Throughout their lives, Pacific women contribute significantly to family assets through the formal and informal economies. Most traders in marketplaces are women, and the items they sell are usually made and transported by women. Yet, like women everywhere, they have trouble obtaining credit to expand their businesses. For example, of the 2039 loans approved by the Agricultural Bank of Papua New Guinea in January 1991, only 4 percent went to women.

## Being a Man: Persistence and Change

Because of its cultural diversity, Oceania encompasses many roles for men. In the Pacific islands, men traditionally were cultivators, deepwater fishers, and masters of seafaring. In Polynesia, they also were responsible for many aspects of food preparation, including cooking. Men fill many positions in the modern world, but idealized male images continue to be associated with vigorous activities.

In Australia and New Zealand, the supermasculine white working-class settler has long had prominence in the national mythologies. In New Zealand, he was a farmer and herdsman. In Australia, he was more often a many-skilled laborer—a stockman, sheep shearer, cane cutter, or digger (miner)—who possessed a laconic, laid-back sense of humor. He went from station (large farm) to station or from mine to mine, working hard but sporadically, gambling, and then working again, until he had enough money or experience to make it in the city (Figure 11.25). There, he often felt ill at ease and chafed to return to the wilds. Now immortalized in songs, novels, films such as the *Crocodile Dundee* series, and U.S. TV advertisements, these men are portrayed as a rough and nomadic tribe whose social life was dominated by male camaraderie and frequent brawls. No small part of this characterization derived from the fact that many of Australia's first immigrants were convicts.

Today, as part of larger efforts to recognize the diversity of Australian society, new ways of life for men are emerging and are breaking down the national image of the tough male loner. Nonetheless,

**FIGURE 11.25** Stereotypical Australian male: The last swagman. Drew Kettle (1920–2004) was an ex-marathon runner who walked the length and breadth of Australia raising money for the Flying Doctor Service. (Isolated Australian communities depend on the service for medical emergencies.) He always traveled with his dog Laddie and collected money in his "billy can" (see photo). Kettle was in his late 70s at the time this photo was taken, in 1997. [Claire Liembach/Robert Harding.]

the old model persists and remains prominent in the public images of Australian businessmen, politicians, and movie stars.

## ECONOMIC AND POLITICAL ISSUES

Although Oceania's reorientation away from Europe and toward Asia is having many cultural ramifications in the region, the process is driven largely by global economic forces such as trade, tourism, and migration.

### From Export to Service Economies

For decades, natural resources and agricultural products have supported the national economies of Australia and New Zealand. Australia, for example, is the world's largest exporter of coal, bauxite, and a number of other minerals and metals and also supplies about 50 percent of the world's wool used for clothing. New Zealand specializes in dairy products, meat, fish, wool, and timber products. Neither country has been a major supplier of more profitable manufactured goods to the world market. Today, both economies are dominated by diverse and growing service sectors.

While the shift to service economies is happening all over the world, it is especially dramatic in Australia and New Zealand because of their historical dependence on raw material exports. Although their manufacturing and materials processing capacity has grown over time, unprocessed raw materials still compose the bulk of exports from the two countries. The shift toward trading more with Asia than with Europe has had little effect on this pattern because most Asian economies have a much greater need to import raw materials than manufactured goods.

Despite this lack of manufacturing, Australia and New Zealand have managed to prosper. One reason is that extraction of minerals,

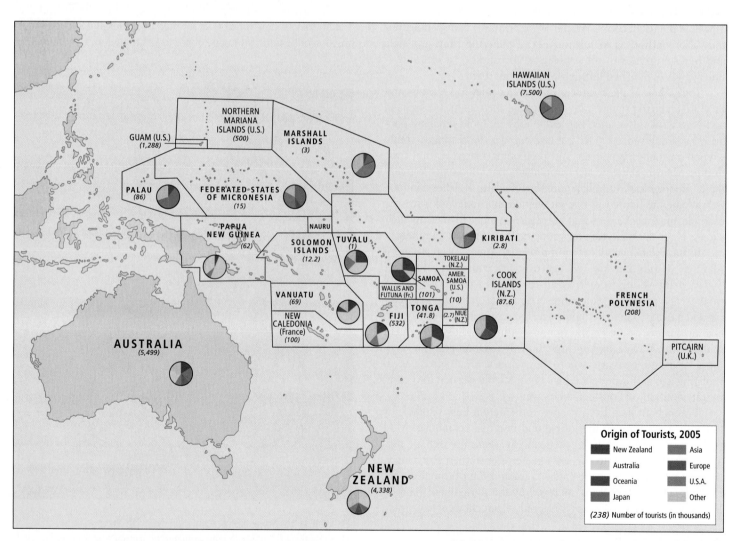

**FIGURE 11.26** Global issues: Tourism in Oceania. Tourism plays a major role in the economies of all countries in Oceania. The 10.5 million visitors to the region in 2005 contributed some U.S.$24.5 billion to local economies. The origins of the tourists reflect changing trade patterns in the region, with most, by far, coming from Asia. [Adapted from World Tourism Organization, at http://www.world-tourism.org/facts/tmt.html; *Financial Times World Desk Reference* (New York: Dorling Kindersley), 2004; Cook Islands, at http://166.122.164.43 /archive/2004/March/tcp-ck.htm; Guam, at http://www.travelweekeast.com/articles /standard.asp?isArticle=1924&pCat=6&rmenu=articles.]

management of herds, and agriculture have become technologically sophisticated enterprises that depend on a dynamic service economy and an educated workforce. Australia is now a world leader in providing technical and other services to mining companies, sheep farms, and winemakers. Meanwhile, New Zealand's well-educated workforce and well-developed marketing infrastructure has helped it break into luxury markets for dairy products, meats, and fruits. At the same time, the service sector in both countries has become increasingly independent of extractive industries with the emergence of globally competitive investment, finance, and insurance sectors. For example, the Australian Macquarie Bank is gaining prominence on Wall Street by buying the rights to operate large infrastructure projects such as ports, tunnels, airports, and toll roads in the United States, Canada, Britain, China, and 15 other countries.

In general, the Pacific islands also depend on extractive industries, such as mining and fishing, for their exports. On many of the islands, however, the economic scene is similar to that in the indigenous communities of Australia and New Zealand, in which some people still engage in subsistence lifestyles. On the islands of Fiji, for example, part-time subsistence agriculture engages more than 60 percent of the population, though it accounts for just under 17 percent of the economy.

## Tourism in Oceania

Tourism is a growing part of the economy throughout Oceania (Figure 11.26). The tourists come increasingly from Japan, Korea, Taiwan, Southeast Asia, the Americas, and Europe. In many Pacific island groups, the number of tourists far exceeds the island population. On Guam, for example, on average, there are over 650 tourists for every 100 residents. The ratio for the Cook Islands is 351 to 100; for Palau, 304 to 100; and for the Northern Mariana Islands, 603 to 100. These visitors create problems for island ecology, place extra burdens on water and sewer systems, and require expensive accommodations and services that are out of reach for local people. Perhaps nowhere in the region are the issues raised by tourism clearer than in Hawaii.

***The Hawaiian Case.*** Since the 1950s, travel and tourism has been the largest industry in Hawaii, producing 22 percent of the gross state product in 2005. (By comparison, travel and tourism accounts for 10.4 percent of GDP worldwide.) In 2005, tourism employed one out of every five Hawaiians and accounted for 23 percent of state tax revenues.

The point of origin of Hawaii's visitors has been shifting from North America to Asia, and by 1995, 40.3 percent of all visitors to Hawaii came from Asia, mostly from Japan. Thus the dramatic slump in Asian economies in the late 1990s, which led fewer people to travel for pleasure, had a major effect on the economy of Hawaii. The decline in tourism hurt not only the tourist industry itself, but also the construction industry, which had been thriving on the building of condominiums, hotels, and resort and retirement facilities. The terrorist attacks of September 11, 2001, also hurt Hawaii's economy by discouraging air travel. Although the industry had recovered by June 2002, these slumps illustrate the vulnerability of tourism to economic downturns and political events.

Sometimes the tourism industry can seem like an invading force to ordinary citizens. For example, an important segment of the Honolulu tourist infrastructure—hotels, golf courses, specialty shopping centers, import shops, nightclubs—is geared to visitors from Japan, and many such facilities are owned by Japanese investors. Hawaiian citizens and other non-Japanese shoppers can be made to feel out of place. Across Hawaii, the demand for golf courses by Asian tourists has brought about what Native (indigenous) Hawaiians view as desecration of sacred sites. Land that was once communally owned and cultivated, and then confiscated by the colonial government, has been sold to Asian golf course developers. Now the only people with access to the sacred sites are fee-paying tourist golfers.

**Personal Vignette** Despite tourism's important economic role in the Pacific, it can threaten local ways of life. An example is provided by an effort to build a retirement home near Honolulu, Hawaii. In the 1990s, the officials of a major U.S. mainland Protestant Christian denomination voted to build a retirement home for their church members on the island of Oahu. The idea was to acquire land and build a multilevel care facility to which church members from the mainland could retire, living independently until they needed nursing home care. This type of retirement relocation to sunny climes is often called "residential tourism." The church officials proceeded to look for affordable land close to Honolulu, yet with landscapes of rural tropical beauty. They found a suitable tract in Pauoa Valley, one of the last valleys near Honolulu where rural people of Native Hawaiian origins still live in extended family compounds and grow their traditional gardens (Figure 11.27).

If the Native Hawaiians were the only inhabitants of Pauoa Valley, their solidarity and Hawaiian laws would have protected the land from

**FIGURE 11.27** Upper Pauoa Valley, Honolulu, Hawaii. [Mac Goodwin.]

being sold. However, new immigrants from North America and Asia now share the valley, and some were eager to sell. If the retirement home were built, life would change irreversibly, even for those residents who chose not to sell. The ecology and ambience of the valley would be transformed by the introduction of a large complex with 141 apartments and a 106-bed nursing home, large concrete parking lots, and manicured grounds. Once the church officials understood the issues raised by the Native Hawaiians, they decided to build the retirement home in the mainland United States rather than in Pauoa Valley.

*Source: Conrad "Mac" Goodwin's research notes, Pauoa Valley, Oahu, Hawaii.*

*Sustainable Tourism.* Some islands have attempted to deal with the pressures of tourism by adopting the principle of sustainable tourism, which aims to decrease the imprint of tourism and minimize disparities between hosts and visitors. Samoa, for example, has created the Samoa Tourism Authority in conjunction with the South Pacific Regional Environment Programme. (The name *Samoa* refers to the independent country that was formerly known as Western Samoa; that country is politically distinct from American Samoa, a U.S. territory.) With financial aid from New Zealand, the Authority develops and monitors sustainable tourism components: beaches, wetland and forest island environments, and knowledge-based tourism experiences for visitors (information-rich explanations of island political, social, and environmental issues).

## New Asian Orientations

The increasing influence of Asia in Oceania is most apparent in the realm of economics. As we have just seen, Asians dominate the tourist trade, both as tourists and as investors, making a vital contribution to Pacific economies and affecting land uses significantly. In addition, increasing numbers of Asians are taking up residence in the region. In Hawaii, for example, the strongest force favoring orientation toward Asia in the future may be the simple fact that Asians, many of whose families have lived there for generations, now make up 57 percent of the population and have widespread economic influence.

The rise in Asian influence is also seen in Oceania's increasing exports of raw materials to Asia (Figure 11.28). Coconut, forest, and fish products from the Pacific islands are sold mostly to Asian markets, and increasingly, Asian companies control these industries. Fishing fleets from Asia regularly ply the offshore waters of Pacific island nations. Asia buys 71 percent of Australia's exports and 44 percent of New Zealand's—mainly wool, dairy products, meat, and hides, as well as coal and iron.

At the same time, because there is little manufacturing in Oceania, most manufactured goods are imported from Japan, South Korea, Taiwan, Singapore, Malaysia, Thailand, and Indonesia. The few manufacturers that exist have to mount aggressive "buy Pacific" appeals to maintain local markets for their wares, which almost always carry higher prices than imports. Although Asia's rising influence in Oceania is associated with growing prosperity, this growing trade relationship makes Oceania increasingly vulnerable to economic downturns in Asia.

*The Stresses of Reorientation.* The shift toward greater trade with Asia poses challenges for Oceania. Throughout the region, local industries that used to enjoy protected or "preferential" trade with Europe have lost that advantage because new European Union regulations stemming from the EU's membership in the World Trade Organization prohibit such arrangements. Now these industries face competition from much larger firms in Asia.

Australia and New Zealand provide good examples of the economic stress that results from reorientation away from protected markets in Europe and toward more open markets in Asia. Their extractive industries have maintained a competitive edge largely through mechanization and thus employ an ever smaller proportion of the population. Most people now work in other sectors of the economy, such as services or manufacturing. However, as trade barriers fall worldwide and as Asian manufacturing industries continue to become more competitive, the economies of Australia and New Zealand are under increasing pressure to reduce their numbers of employees and streamline their operations even further, while their governments are under pressure to cut social services and eliminate subsidies and tariffs. These two countries, long known for their egalitarian and protective policies, are changing into more competitive places.

Economic reform measures have been especially traumatic for the Australian and New Zealand labor movements, which historically have been among the world's strongest. Australian coal miners' unions successfully agitated for the world's first 35-hour work week. Other labor unions won a minimum wage, pensions, and aid to families with children long before such programs were enacted in many other industrialized countries. For decades, these arrangements were highly successful: both Australia and New Zealand enjoyed living standards comparable to those in North America, but with a more egalitarian distribution of income. Since the 1970s, however, competition from Asia has meant that increasing numbers of workers have lost jobs and many hard-won benefits. In Australia, unemployment rose from 2 percent in the 1960s to around 5 percent by 2006. Meanwhile, previously high rates of social spending were cut across the board to address mounting government deficits. The loss of social support, especially for those who have lost jobs, has contributed to rising income disparity and poverty in recent years. Australia now has the second highest poverty rate in the industrialized world, after that of the United States.

New Zealand's loss of its protected UK market when the United Kingdom joined the European Common Market in 1973 sent the economy into decline. By the mid-1980s, the government was forced to cut agricultural subsidies, funding for the welfare state, and the government payroll. It was hoped that the earlier dropping of New Zealand's whites-only immigration policies would attract Asian immigrants, especially wealthy professionals who would reinvigorate the economy with their savings and investments. This has happened, but some longtime residents of New Zealand resent the presence of wealthy newcomers at a time when they themselves are jobless or strapped for cash, and the resulting social tensions are pronounced.

Many Pacific islands are cushioned from the stresses of reorientation to Asia and to the global economy by customs that contribute to self-sufficiency. Official income figures (see Table 11.2 on page 614) do not reflect the fact that many households still rely on fish-

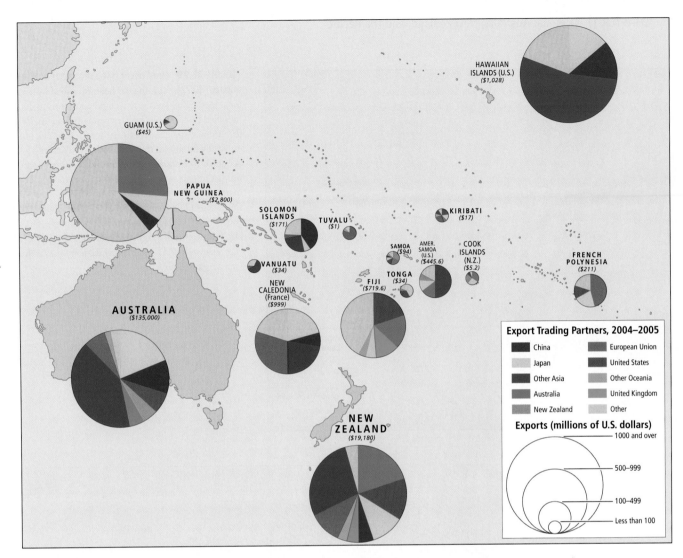

**FIGURE 11.28** Exports from Oceania. The colors of each pie chart indicate a country's export trading partners. The "Other" sections can include trade with Canada, Mexico, the Caribbean, non-EU Europe, sub-Saharan Africa, and other locales, some of them new trade partners. The volume of trade is indicated in millions of U.S. dollars. (Figures for Hawaii do not include exports to other parts of the United States.) [Major source: https://www.cia .gov/cia/publications/factbook/geos/Australia: http://www.dfat.gov.au /publications/stats.html New Zealand: http://www.stats.govt.nz/externaltrade Hawaii: http://www.ita.doc.gov/td/industry/otea/state_reports/hawaii .html#Markets.]

ing and subsistence cultivation for much of their food supply, nor do they include income from thriving informal economies and from remittances. These resources mean that Pacific Islanders can have a safe and healthy life with relatively little formal income. Islanders who can be self-sufficient while saving extra cash for travel and occasional purchases of manufactured goods are sometimes said to have achieved **subsistence affluence.** If there is poverty, it is related to geographic isolation, which often means lack of access to information and opportunity. But computers and the new global communication networks (satellite media, cell phones, and the Internet), although not yet widely available in the Pacific islands, have the potential to alleviate this isolation for some island groups.

At the same time, some places—Papua New Guinea, the Solomon Islands, and Tuvalu, for example—find that they are no longer cushioned by local customs. Once they too had vibrant local cultures, but in comparison to other parts of Oceania, they are now poor and undereducated (see Table 11.2), in large part because of the ways in which colonialism affected their societies. They have

high birth rates, low literacy rates, and low life expectancies. Conditions on many of the smaller Pacific islands typify what has been termed a **MIRAB economy**—one based on migration, remittance, aid, and bureaucracy. Foreign aid from former or present colonial powers supports bureaucracies that supply employment for the educated and semiskilled.

***The Future: A Mixed Asian and European Orientation?*** The Asian economic recession of the late 1990s emphasized the need for Oceania to maintain broad contacts with economies outside Asia, especially in Europe and the United States. Another factor supporting Western connections is the lingering fear of Chinese aggression, justified to some extent by expansionist moves that China has made toward Taiwan. Despite a recent move by China to expand diplomatic and cultural relations with Australia, and despite increasing trade links between the two countries, it is likely that both Australia and New Zealand will retain a cultural affinity with the West for generations to come.

**607**

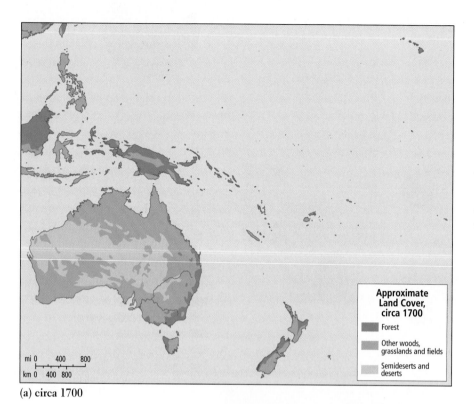

(a) circa 1700

**FIGURE 11.29** Land cover (pre-European) and human impact (2002). Australia has been occupied by humans for at least 60,000 years, so its land cover was no doubt modified by 1700, before Europeans arrived (**a**). By 2002 (**b**), human impact was evident across Australia, but was especially intense in the southeast. New Zealand was occupied by humans only a few thousand years ago, but human impact is now intense and the same can be said for many of the Pacific islands. [Adapted from *United Nations Environment Programme*, 2002, 2003, 2004, 2005, 2006 (New York: United Nations Development Programme), at http://maps.grida.no/go/graphic/human_impact_year_1700 _approximately and http://maps.grida.no/go/graphic/human _impact_year_2002.]

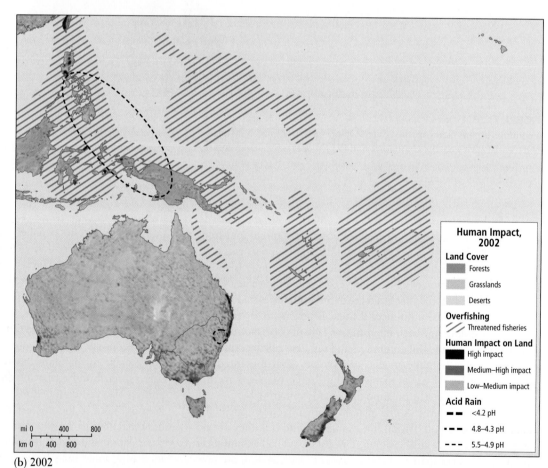

(b) 2002

Tens of thousands of indigenous subsistence cultivators were forced into new mining market towns, where their skills were of little use and where they needed cash to buy food and pay rent (Figure 11.32). Eventually, 30,000 of these people sued the Australian parent mining company, BHP, for U.S.$4 billion. Two villagers, Rex Dagi and Alex Maun, traveled to Europe and the United States to explain their cause and meet with international environmental groups. They and their supporters convinced U.S. and German partners in the Ok Tedi mine to divest their shares. In 1996, the parties reached an out-of-court settlement that set several important precedents: (1) the plaintiffs won U.S.$125 million in trust funds for damage mitigation; (2) the plaintiffs won a 10 percent interest in the mine; and (3) the settlement provided that any further disputes would be heard in Australian rather than Papua New Guinea courts, making it less likely that the mining company could manipulate the justice system.

The most extreme case of environmental disaster due to mining took place on the once densely forested island of Nauru (one-third the size of Manhattan, located north and a bit east of the Solomon Islands). During most of the last half of the twentieth century, the people of Nauru lived in prosperity; in fact, for a few years, the country had the highest per capita income on earth, thus attracting many immigrants. Nauru's wealth was based on the proceeds from the strip-mining of high-grade phosphates used in the manufacture of fertilizer, derived from eons of bird droppings (guano). The mining companies were owned first by Germany, then Japan, and finally Australia. But the phosphate reserves are now depleted, the proceeds ill-spent, and the environment destroyed—junked mining equipment sits on miles of bleached white coral, where the rain forest once stood. To avoid financial disaster, the government has been exploring shady ways of making money, including money laundering for Russian oligarchs; no-record off-shore banking services for illegal rain forest loggers in Africa, South America, and Southeast Asia; and fake passports.

***Nuclear Pollution.*** Another major environmental issue for the Pacific islands is radioactive pollution from nuclear weapons tests and reactor waste. During World War II, U.S. nuclear bomb experiments destroyed Bikini in the Marshall Islands and caused cancer among nearby islanders. Similarly, Mururoa in French Polynesia has been the site of 180 nuclear weapons tests and the recipient of numerous shipments of nuclear waste from France. The result has been widespread cancer, infertility, birth defects, and miscarriages among the populations of nearby islands. One response to this pollution was the 1985 Treaty of Rarotonga, which established the South Pacific Nuclear Free Zone. Most independent countries in Oceania have signed this treaty, which bans nuclear weapons testing and nuclear waste dumping on their lands. As a result of political pressure from France and the United States, however, French Polynesia and U.S. territories such as the Marshall Islands have not signed the treaty. Waste dumping and weapons testing continue; Japan, North Korea, South Korea, France, and the United States have all explored the possibility of depositing their radioactive waste in the Marshall Islands. Since the 1990s, Pacific island leaders have coordinated their efforts against nuclear testing and waste dumping through the United Nations Working Group on Indigenous Populations (UNWGIP).

***The United Nations Convention on the Law of the Sea.*** The 1994 UN Convention on the Law of the Sea (Law of the Sea Treaty) establishes rules governing all uses of the world's oceans and seas. The treaty has been ratified by 157 countries; the United States is not one of them. The treaty is based on the idea that all problems of the world's oceans are interrelated and need to be

**FIGURE 11.32** A side effect of mining in Oceania. Forest dwellers displaced by mining operations are now patronizing supermarkets, such as this one in Tari in the Papua New Guinea highlands. The stores are doing away with long-established personal trading networks and are rapidly bringing the impersonal global economy into one of the last frontiers on the planet. Pacific people are happy to have access to the goods, though many worry about the social and environmental changes that result. [Sassoon/ Robert Harding Picture Library, London.]

addressed as a whole. But it also reveals how the globalization of Pacific island economies has thwarted environmental protection for the islands. The treaty allows islands to claim rights to ocean resources 200 miles (320 kilometers) out from the shore, and island countries can now make money by licensing privately owned fleets from Japan, South Korea, Russia, and the United States to fish within these offshore limits. However, there is no overarching enforcement agency, and protecting the fisheries from overfishing by these rich and powerful licensees has turned out to be an enforcement nightmare for tiny island governments with few resources. Similarly, it has proved difficult to monitor and control the exploitation of seafloor mineral deposits by foreign companies. Even mining operations conducted legally outside the 200-mile limit have the potential to pollute fisheries and other ocean resources.

### Environmental Effects of Tourism.
Even tourism, which until recently was considered a "clean" industry, has been shown to create environmental problems. Foreign-owned tourism enterprises have often accelerated the loss of wetlands and worsened beach erosion. Tourism increases the use of scarce water resources, the production of sewage, and the consumption of environmentally polluting products such as gasoline, kerosene, fertilizers, plastics, and paper. The widely promoted concepts of ecotourism and cultural tourism, discussed in other chapters, are now common elements of development in the Pacific, but they have quickly become commercialized by mass tourism. All tourist ventures, no matter how ecologically sensitive, have some effect on the local environment, and the invasive presence of visitors invariably undermines local cultures.

### Global Warming and Ozone Depletion.
Global warming, which may raise sea levels by melting glaciers and ice caps (see Chapter 1, pages 45–47), is of obvious concern to residents of islands that already barely rise above the waves (Figure 11.33). If sea levels rise the

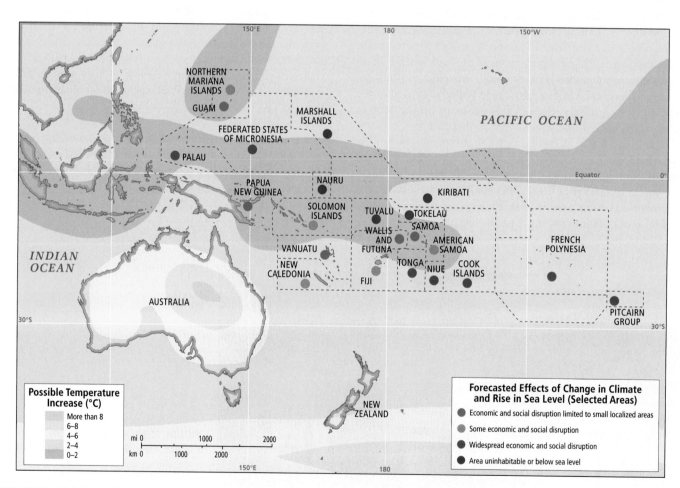

**FIGURE 11.33** Global warming as it may affect Oceania. According to the Intergovernmental Panel on Climate Change, "Sea-level rise … has contributed to erosion of … beaches and barriers; loss of coastal dunes and wetlands; and drainage problems in many low-lying, mid-latitude coastal areas. Highly diverse and productive coastal ecosystems, coastal settlements, and island states will continue to be exposed to pressures whose impacts are expected to be largely negative and potentially disastrous in some instances." [Map adapted from Richard Nile and Christian Clerk, *Cultural Atlas of Australia, New Zealand, and the South Pacific* (New York: Facts on File, 1996), p. 223; quote from Robert T. Watson and the Core Writing Team, *Climate Change Report: Synthesis Report* (Cambridge: Cambridge University Press, 2001), at http://www.grida.no/climate/ipcc_tar/vol4/english/index.htm.]

4 inches (10 centimeters) per decade predicted by the International Panel on Climate Change, many of the lowest-lying atolls, such as Tuvalu, will simply disappear under water. Other islands, some with already very crowded coastal zones, will be severely reduced in area and will become more vulnerable to storm surges and cyclones. (The highest point on the densely occupied island of Nauru, for example, is only 200 feet, or 61 meters, above the current sea level.) The people of the Pacific islands have little direct control over the forces that are creating these precarious circumstances because they themselves produce only a very small percentage of Earth's greenhouse gases. Some islanders are quietly migrating to Australia and New Zealand; others are lobbying to resettle their entire populations in Australia or New Zealand should their islands be submerged; and some, such as Tuvalu, have threatened legal action against both Australia and the United States for their huge greenhouse gas emissions and their failure to sign the Kyoto Protocol (see Chapter 1, page 48).

People in the whole of Oceania are also at high risk from depletion of the **ozone layer**. Ozone is a type of oxygen molecule that is normally heavily concentrated in a layer of the upper atmosphere, where it filters out biologically harmful ultraviolet radiation emitted by the sun. Since the mid-1980s, the ozone layer has been thinning, and a hole in the layer has appeared periodically over Antarctica. The cause is a buildup of a class of manufactured chemicals called chlorofluorocarbons (CFCs), which float up to the ozone layer and destroy it. The increasing amounts of ultraviolet radiation reaching Earth's surface through the damaged ozone layer are likely to increase the prevalence of skin cancer. Even though the rate of ozone depletion has slowed as the result of an international ban on CFCs, the hazard remains serious. Australia already has the highest incidence of skin cancer in the world, which is explained by the fact that its largely white, transplanted population has little protective skin pigment (melanin), yet lives under intense sunlight.

## MEASURES OF HUMAN WELL-BEING

The overall status of human well-being in Oceania is probably higher than the statistics indicate, especially for the Pacific islands, because subsistence agriculture and the informal economy remain important in everyday life. Statistics for Australia, New Zealand, and several of the larger islands of the Pacific are easy to obtain, but data for many of the smaller island groups are missing, or only partial data are available. Keep in mind that the sizes and populations of many of the island states are very small.

As discussed throughout this text, GDP per capita (Table 11.2, column II) is at best a crude indicator of well-being. Places in Oceania, like those in most other regions, have a wide variation in GDP per capita. Australia and New Zealand provide their citizens with a high standard of living that is further enhanced by publicly financed social programs such as health care and subsidized housing. The high standard of living in these societies is demonstrated by their long life expectancies and high literacy rates. Generally, the Pacific islands have relatively low GDP per capita figures. Still, even the lowest figures of U.S.$800–$1000 probably support an adequate standard of living for most people. The GDP per capita statistics do not reflect income from the informal economy, remittances from family members overseas, or the fact that many households practice subsistence fishing and agriculture.

The United Nations Human Development Index (HDI) (see Table 11.2, column III) is less useful for Oceania because statistics are missing for many places. However, it is noteworthy that Australia and New Zealand rank high—third and nineteenth, respectively. Hawaii is not ranked because it is part of the United States, but if it were, Hawaii would have a higher score than the United States as a whole (which is ranked seventh) because it ranks near the top of the U.S. states in life expectancy, educational attainment, and income. Fiji's rank dropped from 66 in 2000 to 92 in 2005, partly because of the continuing civil disturbances discussed on page 601. All other places in Oceania ranked by the United Nations are in the lower range of the medium category on the HDI index. These rankings are partly the result of the difficulty of providing services such as medical care and advanced training to very small, remote populations. Changing diets (more sugar and fat) and lifestyles (more sedentary) have provoked an increase in health problems such as diabetes and heart disease, resulting in significantly lowered life expectancies (column VII). Specialized, high-tech health-care facilities are generally not to be found outside New Zealand, Australia, and Hawaii.

The United Nations Gender Development Index (GDI) rankings (column IV) show that Australia nearly tops the world (it ranks second behind Norway) in leveling the playing field for males and females by supporting gender equality in health care and access to education. Australian women, however, earn on average nearly $10,000 less per year than men. New Zealand ranks fairly high on the GDI (17); this rank results partly from the high percentage of women in Parliament in 2005 (women held 29 percent of the seats) and from the fact that more than one-third of its administrators and managers, and more than half of its professional and technical workers, are women. New Zealand has had two women prime ministers since 1997. Nonetheless, like Australia's, New Zealand's income disparity between men and women is close to $10,000. The GDI figures are missing for most of the Pacific islands, but as we have observed, women participate actively in island life and enjoy increasing prestige as they age. Literacy rates are high throughout Oceania, with the exception of Papua New Guinea and the Solomon Islands in Melanesia. High literacy rates indicate that most of the citizenry is equipped to participate in democratic processes and in public debate on regional issues.

### Reflections on Oceania

One way to assess Oceania and its prospects is to think of it in relation to its nearest neighbors. In other chapters, we have noted the economic vigor of East and Southeast Asia's economies, the energy and potential they seem to have for future expansion, and the possibility that they will either subvert or pull along other economies around the globe. We have also considered the many challenges these regions face, not the least of which are crony capitalism,

**TABLE 11.2** Human well-being rankings of selected places in Oceania and other selected countries[a]

| Country (I) | GDP per capita, adjusted for PPP[b] (GDP ranking among 177 countries), 2005 (II) | Human Development Index (HDI) ranking among 177 countries,[c] 2005 (III) | Gender Development Index (GDI) ranking among 140 countries, 2005 (IV) | Female literacy (percent), 2003 (V) | Male literacy (percent), 2003 (VI) | Life expectancy (years), 2005 (VII) |
|---|---|---|---|---|---|---|
| **Selected countries for comparison** | | | | | | |
| Japan | 27,967 (11) | 11 (high) | 14 | 99 | 99 | 82 |
| United States | 37,562 (4) | 10 (high) | 8 | 99 | 99 | 78 |
| Kuwait | 18,047 (33) | 44 (high) | 39 | 81 | 84 | 78 |
| **Australia** | 29,632 (10) | 3 (high) | 2 | 99 | 99 | 80 |
| **Melanesia** | | | | | | |
| Fiji | 5,880 (85) | 92 (medium) | 71 | 91 | 94 | 68 |
| Papua New Guinea | 2,619 (116) | 137 (medium) | 103 | 51 | 63 | 55 |
| Solomon Islands | 1,753 (136) | 128 (medium) | ND | 62 | 62 | 62 |
| Vanuatu | 2,944 (113) | 118 (medium) | ND | ND | ND | 67 |
| New Caledonia (French overseas territory) | 15,000 (39) | ND | ND | 90 | 90 | 73 |
| **Micronesia** | | | | | | |
| Guam (U.S. territory) | 15,000 (39) | ND | ND | 99 | 99 | 78 |
| Federated States of Micronesia | 2,300 (122) | ND | ND | 88 | 91 | 67 |
| Kiribati | 1,900 (130) | ND | ND | ND | ND | 63 |
| Marshall Islands (U.S. territory) | 2,900 (113) | ND | ND | 94 | 94 | 68 |
| Northern Mariana Islands (U.S. territory) | 12,500 (45) | ND | ND | 96 | 97 | 76 |
| Nauru | 5,000 (93) | ND | ND | ND | ND | 61 |
| Palau | 7,600 (66) | ND | ND | 90 | 93 | 70 |
| **Polynesia** | | | | | | |
| French Polynesia (French Overseas Lands) | 17,500 (36) | Shares French rank of 17 (high) | ND (Shared French rank of 31 in 1998) | 98 | 98 | 68 |
| Hawaii (U.S.) | 37,425 (4) | Shares U.S. rank of 10 (high) | Shares U.S. rank of 8 | 99 | 99 | 77 |
| Tonga | 6,992 (72) | ND | ND | 99 | 99 | 71 |
| Tuvalu | 1,100 (153) | ND | ND | ND | ND | 64 |
| Samoa | 5,854 (85) | 74 (medium) | ND | 99 | 99 | 73 |
| American Samoa (U.S. territory) | 5,800 (86) | ND | ND | 97 | 98 | 76 |
| **New Zealand** | 22,582 (22) | 19 (high) | 17 | 99 | 99 | 78 |

[a]Rankings are in descending order; i.e., low numbers indicate high rank.

[b]PPP = purchasing power parity, figured in 2003 U.S. dollars.

[c]The high, medium, and low designations indicate where the country ranks among the 177 countries classified into three categories by the United Nations.

ND = No data available.

*Sources: United Nations Human Development Report 2005* (New York: United Nations Development Programme); at http:/hdr.undp.org/reports/global/2005 /pdf/HDR05_complete.pdf, *2005 World Population Data Sheet* (Washington, D.C.: Population Reference Bureau); *CIA World Factbook*, 2006, at https://www .cia.gov/cia/publications/factbook/index.html.

corruption, inattention to environmental issues, and lack of commitment to enhancing democratic processes. Will Oceania's economic relationships with Asia leave it wealthier and more secure—or more precarious economically, environmentally, and socially?

Some people have suggested that Oceania might implement the Pacific Way effectively and, perhaps, teach it to others. These observers are encouraged by Oceania's somewhat positive environmental record, its leadership in the movement to ban nuclear weapons, and the strength of Pacific indigenous cultures. The Pacific Way is based on ancient techniques for conflict resolution through cooperation and consensus. It is, however, more of an ideal and a legend than a reality, because traditional Pacific cultures were not particularly democratic. Many of the region's multi-island nations are not accustomed to collective decision making for the nation as a whole, and the disruptive forces of political factionalism and ethnic conflict have yet to be overcome. But it appears that most parts of Oceania are at a stage where cooperation and consensus are at least considered desirable.

Meanwhile, Australia and New Zealand are newly aware of their "Pacificness." Why not, then, become a region with a splendid mission—a place where value is placed on sustainable development, where affluence is defined in nonmaterial ways that focus on the quality of essentials rather than the quantity of nonessentials, where the primary goal is to enhance the quality of life, and where collaboration and common consent are honored? Such a leadership role for the smallest region on earth is not inconceivable. But it can hardly succeed if present consumption levels are maintained in urban Oceania and if the larger and more dominant developed regions—North America, Europe, East Asia, and Southeast Asia—literally swamp Oceania with pollution and rising seawater.

## Chapter Key Terms

Aborigines  591

atoll  587

Austronesians  591

El Niño  589

endemic  589

Gondwana  585

Great Barrier Reef  586

high islands  587

hot spots  587

marsupials  590

Melanesia  592

Melanesians  591

Micronesia  591

MIRAB economy  607

monotremes  590

noble savage  593

ozone layer  613

Pacific Way  601

pidgin  601

Polynesia  592

roaring forties  588

subsistence affluence  607

## Critical Thinking Questions

1. As Australia and New Zealand move away from intense cultural and economic involvement with Europe, new policies and attitudes have had to evolve to facilitate increased involvement with Asia. If you were a college student in Australia or New Zealand, how might you be experiencing these changes? Think about fellow students, career choices, language learning, and travel choices.

2. Discuss the emerging Pacific identity of the Pacific islands and take note of the extent to which Australia and New Zealand share or do not share in this identity. What factors are helping to forge a sense of unity across Polynesia and beyond? (First review the spatial extent of Polynesia.)

3. Discuss the many ways in which Asia has historic, and now increasingly economic, ties to Oceania. Include in your discussion patterns of population distribution, mineral exports and imports, technological interactions, and tourism.

4. Describe the main concern in Oceania related to global warming. Which parts of the region are likely to be most affected? To what extent can the countries of Oceania exercise control over their likely future as the climate warms?

5. Australia and New Zealand differ from each other physically. Compare and contrast the two countries in relation to water, vegetation, and prehistoric and modern animal populations.

6. Indigenous peoples worldwide are beginning to speak out on their own behalf. Discuss how the indigenous peoples of Australia, New Zealand, and the Pacific islands are serving as leaders in this movement and what measures they are taking to reconstitute a sense of cultural heritage.

7. Oceania is one of the most urbanized regions on earth. Discuss how and why this fact varies from popular impressions of the region, taking note of how urban life varies across the region from Hawaii, to Fiji, to Nauru, to Papua New Guinea, to Australia and New Zealand.

8. Sports competitions, ideas of subsistence affluence, and the concept of "the Pacific Way" are all helping to knit together the far-flung parts of Oceania. Discuss how these phenomena have given Oceania a distinctive identity and contemplate how emerging connections to Asia may interact with these unifying forces.

9. Gender roles have some interesting features in Oceania, and to some extent, women of the Pacific islands have stronger roles than those in Australia and New Zealand. Explain how colonization affected gender roles in various places across the region.

10. Most countries that depend significantly on the export of raw materials rather than processed goods are not particularly prosperous, but that is not the case in Australia and New Zealand, which both rank high on the HDI. Explain why they do so well and discuss whether or not this pattern is likely to persist.

# Glossary

**Aborigines** (p. 591) the longest surviving inhabitants of Oceania, whose ancestors, the Australoids, migrated from Southeast Asia possibly as early as 50,000 years ago, over the Sundaland landmass that was exposed during the ice ages

**acid rain** (p. 93) acidic precipitation that has formed through the interaction of rainwater or moisture in the air with sulfur dioxide and nitrogen oxides emitted during the burning of fossil fuels

**acculturation** (pp. 149, 211) adaptation of a minority culture to the host culture enough to function effectively and be self-supporting; cultural borrowing

*adivasis* (p. 434) a social group outside the caste system thought to be descendants of the ancient original inhabitants of South Asia

**African Union (AU)** (p. 378) a political organization consisting of all the countries on the African continent and some nearby islands promoting economic cooperation and social welfare

**age distribution or age structure** (p. 38) the proportion of the total population in each age group

**Age of Exploration** (p. 190) a period of accelerated global commerce and cultural exchange facilitated by improvements in European navigation, shipbuilding, and commerce in the fifteenth and sixteenth centuries

**agribusiness** (p. 71) the business of farming conducted by large-scale operations that produce, package, and distribute agricultural products

**agroecology** (p. 440) the practice of traditional, nonchemical methods of crop fertilization and the use of natural predators to control pests

**agroforestry** (p. 394) the raising of economically useful trees

**Ainu** (p. 502) an indigenous cultural minority group in Japan characterized by their light skin, heavy beards, and thick, wavy hair who are thought to have migrated thousands of years ago from the northern Asian steppes

**air pressure** (p. 23) the force exerted by a column of air on a square foot of surface

**Allah** (p. 305) the Arabic word for "God," used by Arabic-speaking Christians and Muslims in their prayers

**alluvium** (p. 515) river-borne sediment

**altiplano** (p. 173) an area of high plains in the central Andes of South America

**animism** (pp. 388, 409) a belief system in which natural features carry spiritual meaning

**apartheid** (p. 362) a system of laws mandating racial segregation in South Africa, in effect from 1948 until 1994

**aquifer** (p. 95) a natural underground water reservoir

**archipelago** (p. 536) a group, often a chain, of islands

**Aryans** (p. 419) an ancient people of Southwest Asia; the first recorded invaders of South Asia

**ASEAN Free Trade Association (AFTA)** (p. 553) a free trade association of Southeast Asian countries launched in 1992 by ASEAN and patterned after the North American Free Trade Agreement and the European Union

**Asia-Pacific region** (p. 524) a huge trading area that includes all of Asia and the countries around the Pacific Rim

**assimilation** (pp. 149, 210) the loss of old ways of life and the adoption of the lifestyle of another culture

**Association of Southeast Asian Nations (ASEAN)** (p. 553) an organization of Southeast Asian governments established to further economic growth and political cooperation

**atoll** (p. 587) a low-lying island, formed of coral reefs that have built up on the circular or oval rims of submerged volcanoes

**Australo-Melanesians** (p. 539) a group of hunters and gatherers who moved from the present northern Indian and Burman parts of southern Eurasia into the exposed landmass of Sundaland about 60,000 to 40,000 years ago

**Austronesians** (pp. 539, 591) a Mongoloid group of skilled farmers and seafarers from southern China who migrated south to various parts of Southeast Asia between 10,000 and 5000 years ago

**autonomy** (p. 14) the capacity of a people or a country to control their own affairs

**average population density** (p. 37) the average number of people per unit area (for example, per square mile or square kilometer)

**Aztecs** (p. 123) native people of central Mexico noted for advanced civilization before the Spanish conquest

**baby boomer** (p. 91) a member of the largest age group in North America, the generation born in the years after World War II, from 1947 to 1964, in which a marked jump in the birth rate occurred

**barriadas** (p. 132) see **favelas**

**barrios** (p. 132) see **favelas**

**basin** (p. 219) a bowl-shaped depression in the earth's surface that drains toward a central point

**birth rate** (p. 38) the number of births per 1000 people in a given population, per unit of time, usually per year

**Bolsheviks** (p. 254) a faction of communists who came to power during the Russian Revolution

**boreal forest** (p. 108) northern coniferous forests

**Brahmins** (p. 434) members of the Hindu priestly caste, which is the most privileged caste in ritual status

**brain drain** (p. 133) the migration of educated and ambitious young adults to cities or foreign countries, depriving the sending communities of talented youth in whom they have invested years of nurturing and education

**bride price** (p. 437) a price paid by a groom to the family of the bride to compensate them for the loss of her labor and companionship

**brownfields** (p. 80) old industrial sites whose degraded conditions pose obstacles to redevelopment

**Buddhism** (p. 432) a religion of Asia that originated in northern India in the sixth century B.C.E. as a reinterpretation of Hinduism; it emphasizes modest living and peaceful self-reflection leading to enlightenment

**Canadian Shield** (p. 108) the vast glaciated territory in Canada lying north of the Great Plains; characterized by thin soils, innumerable lakes, and large rivers

**capital** (p. 32) wealth in the form of money or property used to produce more wealth

**capitalists** (p. 254) usually a wealthy minority that owns the majority of factories, farms, businesses, and other means of production

**carrying capacity** (pp. 41, 364) the maximum number of people that a given territory can support sustainably with food, water, and other essential resources

**cartel** (p. 322) a group that is able to control production and set prices for its products

**cash economy** (p. 40) an economic system in which the necessities of life are purchased with monetary currency

**caste** (p. 434) an ancient Hindu system for dividing society into hereditary hierarchical classes

**central planning** (p. 193) a central bureaucracy dictating prices and output with the stated aim of allocating goods equitably across society according to need

**centrally planned economy** (p. 254) see **command economy**

*chaebol* (p. 527) huge corporate conglomerates in South Korea that receive government support and protection

**chain migration** (p. 83) a pattern in which immigrants to a new country encourage their family and friends to join them, thus creating a community of culturally similar immigrants in a particular place

**Chipko movement** (p. 451) a grassroots Indian environmental movement that attempts to slow down deforestation, reforest cleared land, and increase ecological awareness

**Christianity** (p. 304) a monotheistic religion based on belief in the teachings of Jesus of Nazareth, a Jew, who described God's relationship to humans as primarily one of love and support, as exemplified by the Ten Commandments

**civil disobedience** (p. 421) the breaking of discriminatory laws by peaceful protesters

**clear-cutting** (p. 62) the cutting down of all trees on a given plot of land, regardless of age, health, or species

**climate** (p. 22) the long-term balance of temperature and precipitation that characteristically prevails in a particular region

**cold war** (p. 193) the contest that pitted the United States and Western Europe, espousing free market capitalism and democracy, against the USSR and its allies, promoting a centrally planned economy and a socialist state

**cold war era** (p. 48) the period from 1946 to the early 1990s, when the United States and its allies in Western Europe faced off against the Union of Soviet Socialist Republics and its allies in Eastern Europe

**colonias** (p. 132) see **favelas**

**command economy** (pp. 254) an economy in which government bureaucrats plan, allocate, and manage all production and distribution

**Common Market for Eastern and Southern Africa (COMESA)** (p. 378) a subregional organization that promotes economic cooperation among 20 African states

**communal conflict** (p. 433) a euphemism for religion-based violence in South Asia

**Communism** (p. 254) an ideology, based largely on the writings of the German revolutionary Karl Marx, that calls on workers to unite to overthrow capitalist and establish an egalitarian society where workers share what they produce

**Communist Party** (p. 254) the political organization that ruled the USSR from 1917 to 1991. Other communist countries, such as China, Mongolia, North Korea and Cuba, also have communist parties.

**Confucianism** (p. 479) a Chinese philosophy that teaches that the best organizational model for the state and society is a hierarchy based on the patriarchal family

**contested space** (p. 142) any area that several groups claim or want to use in different often conflicting ways, such as the Amazon or Palestine

**contiguous regions** (p. 9) regions that lie next to each other

**conurbation** (p. 131) an interconnected group of urban areas and suburban developments; a metropolitan area; see **megalopolis**

**Council of the European Union** (p. 198) an administrative component of the EU that represents the individual member states and consists of ministers from the national governments of those states, who attend meetings according to topics under discussion: economic, finance, agriculture, foreign affairs, environment, and so forth

**country** (p. 49) a common unit of geographic analysis

**Creole** (p. 127) a European person, usually of Spanish descent, born in the Americas

**Creole cultures** (p. 118) distinctly American cultures that emerged from blending immigrant cultures such as those of Spain, Africa, Holland, Germany, Britain, China, and Japan.

**crony capitalism** (p. 553) a type of corruption in which politicians, bankers, and entrepreneurs, sometimes members of the same family, have close personal as well as business relationships

**cultural complexity** (p. 557) see **cultural pluralism**

**cultural diversity** (p. 14) differences in ideas, values, technologies, and institutions among culture groups that cohabit an area

**cultural homogeneity** (p. 13) uniformity of ideas, values, technologies, and institutions among culture groups in a particular area

**cultural homogenization** (p. 184) the tendency toward uniformity of ideas, values, technologies, and institutions among associated culture groups

**cultural identity** (p. 13) a sense of personal affinity with a particular culture group

**cultural marker** (p. 14) a characteristic that helps to define a certain culture group

**cultural pluralism** (p. 557) the cultural identity characteristic of a region where groups of people from many different backgrounds have lived together for a long time, yet have remained distinct

**Cultural Revolution** (p. 482) a series of highly politicized and destructive mass campaigns launched in 1966 to force the entire population of China to support the continuing revolution

**culture** (p. 12) all the ideas, materials, and institutions that people have invented to use to live on earth that are not directly part of our biological inheritance

**culture group** (p. 12) a group of people who share a particular set of beliefs, a way of life, a technology, and usually a place

**currency devaluation** (p. 376) the lowering of a currency's value relative to the U.S. dollar, the Japanese yen, the European euro, or other currency of global trade

**czar** (p. 253) title of the ruler of the Russian empire; derived from the word "caesar," the title of the Roman emperors

**death rate** (p. 38) the ratio of total deaths to total population in a specified community, usually expressed in numbers per 1000 or in percentages

**delta** (p. 22) the triangle-shaped plain of sediment that forms where a river meets the sea

**democratic institutions** (p. 193) institutions—such as constitutions, elected parliaments, and impartial courts—through which the common people gained a formal role in the political life of the nation

**demographic transition** (p. 40) the change from high birth and death rates to low birth and death rates that usually accompanies a cluster of other changes, such as change from a subsistence to a cash economy, increasing education rates, and urbanization

demography (p. 35) the study of population patterns and changes

deposition (p. 22) the settling out of sand and soil particles carried by wind or moving water when the speed of their flow slows

desertification (pp. 335, 394) a set of ecological changes that converts nondesert lands into deserts

detritus (p. 539) dead organic material (such as plants and insects) that collects on the ground

development (p. 33) usually used to describe economic changes like greater productivity of agriculture and industry that lead to better standards of living or simply increased mass consumption

"Development for whom?" (p. 33) a question that asks whether a particular development strategy will truly improve average standards of living for most local people, or if it will raise production levels and increase profits for only a few, those who are already well off

devolution (pp. 230, 385) the weakening of a formerly tightly unified state

Diaspora (p. 304) the dispersion of Jews around the globe after they were expelled from the eastern Mediterranean by the Roman Empire beginning in 73 C.E.; can now refer to other dispersed culture groups

digital divide (pp. 11, 74) the discrepancy in access to information technology between small, rural, and poor areas and large, wealthy cities that contain major government research laboratories; and universities

doi moi (p. 574) economic and bureaucratic restructuring in Vietnam

domestication (p. 25) the process of developing plants and animals through selective breeding to live with and be of use to humans

double day (p. 212) the longer workday of women with jobs outside the home who also work as caretakers, housekeepers, and cooks for their families

dowry (p. 436) a price paid by the family of a bride to the groom (opposite of bride price), formerly a custom practiced only by the rich

dry forests (p. 394) forests that lose their leaves during the dry season

dumping (p. 206) the cheap sale on the world market of overproduced commodities, lowering global prices and hurting producers of these same commodities elsewhere in the world

early extractive phase (p. 134) a phase in Central and South American history, beginning with the Spanish conquest and lasting until the early twentieth century, characterized by a dependence on trade in raw materials

earthquake (p. 22) a catastrophic shaking of the land surface, often caused by the shifting and friction of tectonic plates

East African Community (EAC) (p. 378) an organization formed by Kenya, Tanzania, and Uganda to promote economic links among the countries of East Africa

Economic Community of Central African States (CEEAC) (p. 378) a subregional organization that promotes economic cooperation among 11 Central African states

Economic Community of West African States (ECOWAS) (p. 378) an organization of West African states working toward forming an economic union

economic core (p. 60) the dominant economic region within a larger region; in nineteenth-century North America the core included southern Ontario and the north-central part of the United States (chiefly Illinois, Indiana, Ohio, New York, New Jersey, and Pennsylvania)

economic diversification (p. 325) the expansion of an economy to include a wider array of economic activities

economic integration (p. 194) the free movement of people, goods, money, and ideas among countries

economic restructuring (p. 117) reorganization of an economy to encourage economic growth in markets free of government controls

economic and technology development zones (ETDZs) (p. 494) zones in China with fewer restrictions on foreign business, established to encourage foreign investment and economic growth

economies of scale (p. 200) reductions in the unit costs of production that occur when goods or services are efficiently mass produced, resulting in a rise in profits per unit

economy (p. 30) the forum in which people make their living, including the spatial, social, and political aspects of how resources are recognized, extracted, exchanged, transformed, and reallocated

ecotourism (pp. 154, 373) nature-oriented vacations often taken in endangered and remote landscapes, usually by travelers from industrialized nations

El Niño (pp. 122, 589) periodic climate-altering changes, especially in the circulation of the Pacific Ocean, now understood to operate on a global scale

endemic (p. 589) belonging or restricted to a particular place

erosion (p. 22) the process by which fragmented rock and soil are moved over a distance, primarily by wind and water

escarpment (p. 356) long linear cliffs

ethnic cleansing (p. 195) the systematic removal of an ethnic group or people from a region or country by deportation or genocide

ethnic group (pp. 12, 336) a group of people who share a set of beliefs, a way of life, a technology, and usually a geographic location

ethnicity (pp. 86, 336) the quality of belonging to a particular culture group

euro (p. 200) the official (but not required) currency of the European Union as of January 1, 1999

European colonialism (p. 9) the practice of taking over the human and natural resources of often distant places to produce wealth for Europe

European Commission (p. 198) an administrative component of the EU that represents the interests of the EU as a whole and is independent of national governments

European Economic Community (EEC) (p. 198) an economic community created in 1958, when Belgium, Luxembourg, the Netherlands, France, Italy, and West Germany agreed to eliminate certain tariffs and promote mutual trade and cooperation in the interest of achieving stronger economic ties within Europe; a precursor to the EU.

European Parliament (p. 198) an administrative component of the EU that is directly elected and represents EU citizens; the number of seats each country has is based on its population, much as with the U.S. House of Representatives

European Union (EU) (p. 184) a supranational institution including most of West, South, North, and Central Europe, established to bring economic integration to member countries

evangelical Protestantism (p. 151) a Christian movement that focuses on personal salvation and empowerment of the individual through miraculous healing and transformation; some practitioners preach the "gospel of success" to the poor—that a life dedicated to Christ will result in prosperity for the believer

exchange or service sector (p. 30) barter or trade for money, goods, or services

exclave (p. 236) a portion of a political unit separated from the main part

Export Processing Zones (EPZs) or free trade zones (pp. 139, 549) specially created legal spaces or industrial parks within a country where, to attract foreign-owned factories, duties and taxes are not charged

extended family (p. 149) a family consisting of related individuals beyond the nuclear family of parents and children

external debts (p. 139) the debts a country owes to other countries or international financial institutions

external processes (geophysical) (p. 20) landform-shaping processes that originate at the surface of the earth, such as weathering, mass wasting, and erosion

extraction (p. 30) the acquisition of a material resource through mining, logging, agriculture, or other means

extractive resource (p. 30) a resource such as mineral ores, timber, or plants that must be mined from the earth's surface or grown from its soil

extraregional migration (p. 563) short-term or permanent migration to countries outside a home region

failed state (p. 340) a state whose government has lost control and is no longer able to defend its citizens from armed uprisings or other harm

fair trade (p. 33) trade that values equity throughout the international trade system; now proposed as an alternative to free trade

favelas (p. 132) Brazilian urban slums and shantytowns built by the poor; called colonias, barrios, or barriadas in other countries

female circumcision (p. 391) the removal of the labia and the clitoris and sometimes the stitching nearly shut of the vulva

feminization of labor (p. 549) the increasing representation of women in both the formal and informal labor force

Fertile Crescent (p. 302) an arc of lush, fertile land formed by the uplands of the Tigris and Euphrates river systems and the Zagros Mountains, where nomadic peoples began the earliest known agricultural communities

feudalism (p. 189) a social system once prevalent in Europe and Asia and elsewhere in which a class of professional fighting men, or knights, defended the monarch and the peasants or serfs, who cultivated the lands of their protectors.

floating population (p. 474) jobless or underemployed people who have left economically depressed rural areas for the cities and move from place to place looking for work

floodplain (p. 22) the flat land around a river where sediment is deposited during flooding

food security (p. 491) the ability of a state to supply a sufficient amount of basic food to the entire population consistently

forced migration (p. 530) the movement of people against their wishes

foreign direct investment (FDI) (p. 134) the amount of money invested in a country's businesses by citizens, corporations, or governments of other countries

foreign exchange (p. 563) foreign currency that countries need to purchase imports

formal economy (p. 30) all aspects of the economy that take place in official channels

formal institutions (p. 12) associations such as official religious organizations; local, state, and national governments; nongovernmental organizations; and specific businesses and corporations

forward capital (pp. 179, 361) a capital city built in the hinterland to draw migrants and investment for economic development

fragile environment (p. 394) an area that contains barely enough water, soil nutrients, or other resources essential to meet the needs of plants and animals; human pressure in such environments may result in long-term or irreversible damage to plant and animal life

free trade (p. 32) the movement of goods and capital without government restrictions

frontal precipitation (p. 25) rainfall caused by the confrontation and interaction of large air masses of different temperatures and densities

fusion (p. 388) see **syncretism**

Gazprom (p. 266) in Russia, the state-owned energy company; it is the tenth-largest oil and gas entity in the world

gender structure (p. 38) the proportion of males and females in each age group of a population

genetic engineering (p. 72) the genetic manipulation of crops to produce strains with certain features or with resistance to pests or disease

Geneva Conventions (p. 68) treaties concerning the conduct of war, some of which protect the rights of prisoners of war

genocide (p. 49) the deliberate destruction of an ethnic, racial, or political group

gentrification (p. 81) the renovation of old urban districts by middle-class investment, a process that often displaces poorer residents

geopolitics (p. 48) the use of strategies by countries to ensure that their best interests are served

*ger* (p. 516) see **yurt**

glasnost (p. 255) literally, "openness"; the opening up of public discussion of social and economic problems that occurred in the Soviet Union under Mikhail Gorbachev in the late 1980s

global economy (p. 29) the worldwide system in which goods, services, and labor are exchanged

global scale (p. 8) the level of geography that encompasses the entire world as a single unified area

global warming (p. 45) the predicted warming of the earth's climate as atmospheric levels of greenhouse gases increase

globalization (p. 10) the growth of interregional and worldwide linkages and the changes they are bringing about

Gondwana (p. 585) the great landmass that formed the southern part of the ancient supercontinent Pangaea

government subsidies (p. 75) amounts paid by the government to cover part of the production costs of some products in order to help domestic producers to sell their goods for less than foreign competitors

grassroots economic development (p. 379) economic development projects designed to help individuals and their families achieve sustainable livelihoods

Great Barrier Reef (p. 586) the longest coral reef in the world, located off the northeastern coast of Australia

Great Basin (p. 108) the dry, upland region in North America between the Rocky Mountains and Pacific Coastal zone

Great Leap Forward (p. 482) an economic reform program under Mao Zedong intended to quickly raise China to the industrial level of Britain and the United States

Green (political parties) (p. 216) in Europe, environmentally conscious political parties that influence policies at the national level as well as within the EU

green revolution (p. 439) increases in food production brought about through the use of new seeds, fertilizers, mechanized equipment, irrigation, pesticides, and herbicides

gross domestic product (GDP) (p. 31) the market value of all goods and services produced by workers and capital within a particular country's borders and within a given year

gross domestic product (GDP) per capita (p. 33) the market value of all goods and services produced by workers and capital within a particular country's borders and within a given year divided by the number of people in the country

**Group of Eight (G8)** (p. 266) an organization of highly industrialized countries: France, the United States, Britain, Germany, Japan, Italy, Canada, and Russia

**growth poles** (p. 494) zones of development whose success draws more investment and migration to a region

**growth rate** (p. 38) see **rate of natural increase**

**growth triangles** (p. 550) large transnational economic regions in which firms attempt to find the best skills and resources for the best prices

**guest workers** (p. 208) immigrants from outside Europe, often from former colonies, who come to Europe (often temporarily) to fill labor shortages; the expectation is that they will return home when no longer needed

**hacienda** (p. 135) a large agricultural estate in Middle or South America, more common in the past; usually not specialized by crop and not focused on market production

**hajj** (p. 305) the pilgrimage to the city of Makkah (Mecca) that all Muslims are encouraged to undertake at least once in a lifetime

**Harappa Culture** (p. 418) see **Indus Valley civilization**

**Harijans** (p. 434) members of a social group in Hindu India considered so lowly as to have no caste; also known as Dalits or untouchables

**hazardous waste** (p. 92) nuclear, chemical, or industrial wastes that can have damaging environmental consequences

**hearth** (p. 432) metaphorically, place of origin

**high islands** (p. 587) volcanoes that rise above the sea into mountainous, rocky formations that contain a rich variety of environments

**Hindi** (p. 420) the language of the Hindu people, which became the language of trade of the northern subcontinent of India and is still used by more than 400 million people

**Hinduism** (p. 431) a major world religion practiced by approximately 900 million people, 800 million of whom live in India

**Hispanic** (p. 56) a loose ethnic term that refers to all Spanish-speaking people from Latin America and Spain; equivalent to Latino

**Holocaust** (p. 193) a massive ethnic cleansing of 6 million Jews (and several million others: Roma, Slavs, the infirm, and political dissidents) perpetrated primarily by the Nazi government in Germany and the Fascist government in Italy

**Horn of Africa** (p. 357) the part of Africa that juts out from East Africa and wraps around the southern tip of the Arabian Peninsula

**hot spots** (p. 587) individual sites of upwelling material (magma) originating deep in the mantle of the earth and surfacing in a tall plume; hot spots tend to remain fixed relative to migrating tectonic plates

**hub-and-spoke network** (p. 72) the organization of air service in North America around hubs, strategically located airports used as collection and transfer points for passengers and cargo traveling from one place to another

*hukou* **system** (p. 474) the system in China by which citizens' permanent residence is registered

**human geography** (p. 3) the study of various aspects of human life that create the distinctive landscapes and regions of the world

**human well-being** (p. 33) various measures of the extent to which people are able to obtain a healthy life in a community of their choosing

**humanism** (p. 190) a philosophy and value system that emphasizes the dignity and worth of the individual

**humid continental climate** (p. 188) a mid-latitude climate pattern in which summers are fairly hot and moist, and winters are longer and colder the deeper into the interior of the continent one goes

**import quota** (p. 32) a limit on the amount of a given item that may be imported into a country over a given period of time

**import substitution industrialization (ISI)** (p. 137) a form of industrialization involving the use of public funds to set up factories to produce goods that previously had been imported

**Incas** (p. 123) Native American people who ruled the largest pre-Columbian state in the Americas, with a domain stretching from southern Colombia to northern Chile and Argentina

**income disparity** (p. 134) the gap in wealth and resources between the richest 10 (or 20) percent and the poorest 10 (or 20) percent of a country's population

**Indian diaspora** (p. 422) the set of all people of South Asian heritage living (and often born) outside South Asia

**indigenous** (p. 116) native to a particular place or region

**Indus Valley civilization** (p. 418) the first substantial settled agricultural communities in South Asia, which appeared about 4500 years ago along the Indus River in modern-day Pakistan and along the Saraswati River in modern-day India

**industrial production** (p. 30) the manufacture of goods for sale

**Industrial Revolution** (p. 31) a series of inventions, innovations and ideas that allowed manufacturing to be mechanized

**informal economy** (pp. 31, 139) all aspects of the economy that take place outside official channels

**informal institutions** (p. 14) ordinary or casual associations, such as the family or a community

**information technology (IT)** (p. 73) the part of the service sector that relies on the use of computers and the Internet to process and transport information; includes banks, software companies, medical technology companies, and publishing houses

**institutions** (p. 12) all the associations, formal and informal, that help people get along together

**internal (geophysical) processes** (p. 20) processes that, like plate tectonics, originate deep beneath the surface of the earth

**International Monetary Fund (IMF)** (p. 32) a financial institution funded by the developed nations to help developing countries reorganize, formalize, and develop their economies

**Internet** (p. 73) a computer network that allows for the electronic transfer of all kinds of information in seconds

**interregional linkages** (p. 9) economic, political, or social connections between regions, whether contiguous or widely separated

**Interstate Highway System** (p. 72) the federally subsidized network of highways in the United States

**intertropical convergence zone (ITCZ)** (p. 356) a band of atmospheric currents circling the globe roughly at the equator; warm winds from both north and south converge at the ITCZ, pushing air upward and causing copious rainfall

**intifada** (p. 327) a prolonged Palestinian uprising against Israel

**iron curtain** (p. 193) a fortified border zone that separated Western Europe from Eastern Europe during the cold war

**Islam** (pp. 304, 407) a monotheistic religion that emerged in the seventh century C.E. when, according to tradition, the archangel Gabriel revealed the tenets of the religion to the Prophet Muhammad

**Islamic fundamentalism (Islamism)** (p. 321) a grassroots movement to replace secular governments and civil laws with governments and laws guided by Islamic principles

**Islamists** (p. 321) fundamentalist Muslims who favor a religion-based state that incorporates conservative interpretations of the Qur'an and strictly enforced Islamic principles into the legal system

**isthmus** (p. 167) a narrow strip of land joining two large land areas

**Jainism** (p. 432) originally a reformist movement within Hinduism, Jainism is a faith tradition that is more than 2000 years old; found mainly in western India and in large urban centers throughout the region, Jains are known for their educational achievements, nonviolence, and strict vegetarianism

*jati* (p. 434) in Hindu India, the subcaste into which a person is born, which largely defines the individual's experience for a lifetime

**Judaism** (p. 304) a monotheistic religion characterized by the belief in one God, Yaweh, a strong ethical code summarized in the Ten Commandments, and an enduring ethnic identity

*kaizen* (continuous improvement) system (p. 488) a Japanese manufacturing system in which suppliers are clustered around assembly plants to permit just-in-time inventory and in which production lines are constantly monitored; this method results in more efficient production and less space taken up by warehouses

**knowledge economy** (p. 73) markets based on the management of information, such as finance, media, and research and development

**Kshatriyas** (p. 434) members of the Hindu warrior and ruler caste

**land reform** (p. 137) a policy that breaks up large landholdings for redistribution among landless farmers

**landforms** (p. 20) physical features such as mountain ranges, river valleys, basins, and cliffs

**ladino** (p. 167) a local term for mestizo used in Central America

**laterite** (p. 358) a permanently hard surface left when minerals in tropical soils are leached away

**Latino** (p. 56) a loose ethnic term that refers to all Spanish-speaking people from Middle and South America and Spain; equivalent to Hispanic

**latitude** (p. 5) the distance in degrees north or south of the equator; lines of latitude run parallel to the equator, and are also called parallels

**leaching** (p. 358) the washing out into groundwater of soil minerals and nutrients released into soil by decaying organic matter

**liberation theology** (p. 151) a movement within the Roman Catholic Church that uses the teachings of Christ to encourage the poor to organize to change their own lives and the rich to promote social and economic equity

**lingua franca** (pp. 16, 392) a common language used to communicate by people who do not speak one another's native languages; often a language of trade.

**living wages** (p. 32) minimum wages high enough to support a healthy life

**local scale** (p. 9) the level of geography that describes the space where an individual lives or works; a city, town, or rural area

**loess** (pp. 57, 509) windblown dust that forms deep soils in North America, central Europe, and China

**long-lot system** (p. 99) a settlement system of long, narrow plots of land stretching back from the edge of the St. Lawrence River, giving French Canadian settlers access to resources from both river and land

**longitude** (p. 5) the distance in degrees east and west of Greenwich, England; lines of longitude, also called meridians, run from pole to pole (the line of longitude at Greenwich is 0° and is known as the prime meridian)

**machismo** (p. 150) a set of values that defines manliness in Middle and South America

**Madrasa** (p. 316) Muslim religious school

**Mano River Union (MRU)** (p. 378) an organization for economic integration that administers a customs union in the subregion comprising Guinea, Liberia, and Sierra Leone

**maquiladoras** (pp. 112, 139) foreign-owned tax exempt factories, often located in Mexican towns just across the U.S. border from U.S. towns, that hire workers at low wages to assemble manufactured goods that are then exported for sale

*marianismo* (p. 150) a set of values based on the life of the Virgin Mary that defines the proper social roles for women in Middle and South America

**marketization** (p. 261) the process of developing a free market economy

**marsupials** (p. 590) mammals that give birth to their young at a very immature stage and nurture them in a pouch equipped with nipples

**material culture** (p. 12) all the things, living or not, that humans use

**medieval period** (p. 189) the period in Europe circa 450–1300 C.E., during which civil society declined and commerce ceased as the Roman Empire collapsed; by 1250, town life, trade, and commerce began to revive and diffusion from outside influences and European innovation encouraged the flourishing of the arts, philosophy, and architecture

**Mediterranean climate** (p. 188) a climate pattern of warm, dry summers and mild, rainy winters

**megalopolis** (p. 81) an area formed when several cities expand so that their edges meet and coalesce

**Melanesia** (p. 592) New Guinea and the islands south of the equator and west of Tonga (the Solomon Islands, New Caledonia, Fiji, Vanuatu)

**Melanesians** (p. 591) a group of Australoids named for their relatively dark skin tones, a result of high levels of the protective pigment melanin; they settled throughout New Guinea and other nearby islands

**mercantilism** (p. 127, 190) the policy by which the rulers of Spain and Portugal, and later of England and Holland, sought to increase the power and wealth of their realms by managing all aspects of production, transport, and trade in their colonies

**Mercosur** (p. 141) a free trade zone created in 1991 that links the economies of Brazil, Argentina, Uruguay, and Paraguay to create a common market

**mestizo** (p. 127) a person of mixed European and Native American descent

**metropolitan areas** (p. 80) cities with population of 50,000 or more and their surrounding suburbs

**microcredit** (p. 443) a program based on peer support that makes very small loans available to very low income entrepreneurs

**Micronesia** (p. 591) the small islands lying east of the Philippines and north of the equator

**Middle America** (p. 118) in this book, a region including Mexico, Central America, and the islands of the Caribbean

**MIRAB economy** (p. 607) an economy based on migration, remittance, aid, and bureaucracy

**mixed agriculture** (p. 370) the raising of a variety of crops and animals on a single farm, often to take advantage of several environmental niches

**Mongols** (p. 252) a loose confederation of nomadic pastoral people centered in eastern Central Asia, who by the thirteenth century had established by conquest an empire stretching from Europe to the Pacific

**monotheistic** (p. 303) pertaining to the belief that there is only one god

**monotremes** (p. 590) egg-laying mammals, such as the duck-billed platypus and the spiny anteater

**monsoon** (pp. 25, 416) a wind pattern in which in summer months warm, wet air coming from the ocean brings copious rainfall, and in winter, cool, dry air moves from the continental interior toward the ocean

**mother country** (p. 127) usually a European colonizing country

**Mughals** (p. 419) a dynasty of Central Asian origin that ruled India from the sixteenth to the nineteenth century

**multiculturalism** (p. 14) the state of relating to, reflecting, or being adapted to diverse cultures

**multinational corporation** (p. 32) a business organization that operates extraction, production, and/or distribution facilities in multiple countries

**Muslims** (p. 305) followers of Islam

**nation** (p. 49) a group of people who share a language, culture, political philosophy, and usually a territory

**nation-state** (p. 49) a political unit, or country, formed by people who share a language, a culture, and a political philosophy

**national identity** (p. 49) the feeling of strong allegiance to a particular country; often expressed as nationalism

**nationalism** (p. 193) devotion to the interests or culture of a particular country, nation, or cultural group; the idea that a group of people living in a specific territory and sharing cultural traits should be united in a single country to which they are loyal and obedient

**neocolonialism** (pp. 137, 355) modern efforts by dominant countries to control economic and political affairs in other countries to further their own aims

**Neolithic Revolution** (p. 25) a period 20,000 to 8000 years ago characterized by the expansion of agriculture and the making of polished stone tools

**New Urbanism** (p. 81) the growing popularity of urban areas

**noble savage** (p. 593) a term coined by European Romanticists to describe what they termed the "primitive" peoples of the Pacific, who lived in distant places supposedly untouched by corrupting influences

**nomadic pastoralists** (p. 251) peoples whose way of life and economy are centered on the tending of grazing animals who are moved seasonally to gain access to the best grasses

**nongovernmental organization (NGO)** (p. 49) an association outside the formal institutions of government, in which individuals, often from widely differing backgrounds and locations, share views and activism on political, economic, social, or environmental issues

**nonmaterial resources** (p. 30) skills and knowledge of economic value

**nonpoint sources of pollution** (p. 277) diffuse sources of environmental contamination, such as untreated automobile exhaust, raw sewage, or agricultural chemicals that drain from fields into water supplies

**North American Free Trade Agreement (NAFTA)** (pp. 75, 141, 202) a free trade agreement made in 1994 that added Mexico to the 1989 economic arrangement between the United States and Canada

**nuclear family** (p. 89) a family consisting of a father and mother and their children

**Ogallala aquifer** (p. 95) the largest North American natural aquifer, which underlies the Great Plains

**old-growth forests** (p. 567) forests that have never been logged and therefore contain diverse ecosystems

**oligarchs** (p. 262) in Russia, those who acquired great wealth during the privatization of Russia's resources and who use that wealth to exercise power

**organic matter** (p. 338) the remains of any living thing

**Organization for Economic Cooperation and Development (OECD)** (p. 47) the highly industrialized countries of North America, Europe, East Asia and Oceania (Australia, Austria, Belgium, Canada, Czech Republic, Denmark, Finland, France, Germany, Greece, Hungary, Iceland, Ireland, Italy, Japan, Korea, Luxembourg, Mexico, Netherlands, New Zealand, Norway, Poland, Portugal, Slovak Republic, Spain, Sweden, Switzerland, Turkey, United Kingdom, United States)

**OPEC (Organization of Petroleum Exporting Countries)** (p. 322) a cartel of oil-producing countries—including Algeria, Angola, Indonesia, Iran, Iraq, Kuwait, Libya, Nigeria, Qatar, Saudi Arabia, the United Arab Emirates, and Venezuela—that was established to regulate the production, and hence the price, of oil and natural gas

**orographic rainfall** (p. 24) rainfall produced when a moving moist air mass encounters a mountain range, rises, cools, and releases condensed moisture that falls as rain

**ozone layer** (p. 613) a layer of the upper atmosphere that filters out biologically harmful ultraviolet radiation emitted by the sun

**Pacific Rim (Basin)** (p. 65) the countries that border the Pacific Ocean on the west and east

**Pacific Way** (p. 601) the idea that Pacific Islanders have a regional identity and a way of handling conflicts peacefully, which grows out of their particular social experience

*Pancasila* (p. 556) the Indonesian national philosophy based on tolerance, particularly in matters of religion; its precepts include belief in God and the observance of conformity, corporatism, consensus, and harmony

**Pangaea hypothesis** (p. 20) the proposal based on scientific evidence that about 200 million years ago all continents were joined in a single vast continent, called by geologists Pangaea

**Parsis** (p. 432) a highly visible religious minority in India's western cities; Parsis are descendants of Persian migrants who did not give up their traditional religion of Zoroastrianism when Iran became Muslim

**Partition** (p. 423) in this context, the breakup following Indian independence that established Hindu India and Muslim Pakistan

**pastoralism** (p. 370) a way of life based on herding; practiced primarily in savannas, on desert margins, or in the mixture of grass and shrubs called open bush

**perestroika** (p. 255) literally, "restructuring"; the restructuring of the Soviet economic system in the late 1980s in an attempt to revitalize the economy

**permafrost** (p. 249) permanently frozen soil a few inches or feet beneath the surface

**permeable national borders** (p. 112) borders subject to easy flow of people and goods

**physical geography** (p. 3) the study of the earth's physical processes: how they work, how they affect humans, and how they are affected by humans

**pidgin** (p. 601) a language used for trading; made up of words borrowed from the several languages of people involved in trading relationships

**plantation** (p. 136) a large estate or farm on which a large group of resident laborers grow (and partially process) a single cash crop

**plate tectonics** (p. 21) the scientific theory that the earth's surface is composed of large plates that float on top of an underlying layer of molten rock; the movement and interaction of the plates create many of the large features of the earth's surface, particularly mountains

**pluralistic state** (p. 49) a country in which political power is shared among groups, each defined by common language, ethnicity, culture, or other characteristics

**pogroms** (p. 308) episodes of persecution, ethnic cleansing, and sometimes massacre, especially conducted against European Jews

**political ecologist** (p. 41) a geographer who studies power allocation in the interactions among development, human well-being, and the environment

**polygyny** (pp. 319, 390) the taking by a man of more than one wife at a time

**Polynesia** (p. 592) the numerous islands situated inside an irregular triangle formed by New Zealand, Hawaii, and Easter Island

**population pyramid** (p. 38) a graph that depicts the age and gender structures of a country

**populist movements** (p. 151) popularly based efforts, often seeking relief for the poor

**price supports** (p. 200) legal minimum prices set as a means of maintaining higher prices for a commodity despite an overabundance of supply

**primate city** (p. 131) a city that is vastly larger than all others in a country and in which economic and political activity is centered

**privatization** (p. 261) the sale of industries formerly owned and operated by the government to private companies or individuals

**producer services** (p. 205) a section of the service economy tailored to the needs of governments and businesses to provide advice, information, testing, licensing, and strategic planning

**protectorate** (p. 306) a relationship of partial control assumed over a dependent country, while maintaining the trappings of local government

**Protestant Reformation** (p. 190) a European reform (or "protest") movement that challenged Roman Catholic practices in the sixteenth century and led to the establishment of Protestant churches

**pull factors** (p. 562) positive features of a place that attract people to move there

**purchasing power parity (PPP)** (p. 34) the amount that the local currency equivalent of U.S.$1 will purchase in a given country

**purdah** (p. 436) the practice in South Asia of concealing women, especially during their reproductive years, from the eyes of nonfamily men

**push factors** (p. 562) negative features of the place where people are living that impel them to migrate

**pyroclastic flow** (p. 119) a type of volcanic eruption characterized by blasts of superheated rocks, ash, and gas that move with great speed and force

**qanats** (p. 517) underground conduits, built by ancient cultures and still used today, that carry groundwater for irrigation in dry regions

**Québecois** (p. 56) the French Canadian ethnic group or members of that group; also, all citizens of Québec, regardless of ethnicity

**Qur'an (or Koran)** (p. 304) the holy book of Islam, believed by Muslims to contain the words Allah revealed to Muhammad through the archangel Gabriel

**race** (p. 19) a social or political construct based on apparent characteristics such as skin color, hair texture, and face and body shape; but of no biological signficance

**racism** (p. 20) the negative assessment of people, often those who look different, primarily on the basis of skin color and other physical features

**rain shadow** (p. 24) the dry side of a mountain range, facing away from the prevailing winds

**rate of natural increase (growth rate)** (p. 37) the rate of population growth measured as the excess of births over deaths per 1000 individuals per year without regard for the effects of migration

**recharge area** (p. 333) the area receiving precipitation that refills an aquifer

**region** (p. 6) a unit of the earth's surface that contains distinct patterns of physical features and/or of human development

**regional conflict** (p. 445) especially in South Asia, a conflict created by the resistance of a regional ethnic or religious minority to the authority of a national or state government

**regional geography** (p. 4) the analysis of the geographic characteristics of particular places

**regional self-sufficiency** (p. 490) an economic policy in communist China that encouraged each region to develop independently in the hope of evening out the national distribution of production and income

**regional trade bloc** (p. 32) an association of neighboring countries that have agreed to lower trade barriers for one another

**religion** (p. 15) formal and informal institutions that embody value systems. Most have deep roots in history, and many include a spiritual belief in a higher power (God, Yaweh, Allah) as the underpinning for their value system.

**religious nationalism** (p. 444) the belief that a certain religion is strongly connected to a particular territory and that adherents should have political power in that territory

**remittances** (pp. 10, 527) earnings sent home by immigrant workers

**Renaissance** (p. 190) a broad European cultural movement in the fourteenth through sixteenth centuries that drew inspiration from the Greek, Roman, and Islamic civilizations; it marked the transition from medieval to modern times

**resettlement schemes** (p. 561) government plans to move large numbers of people from one part of a country to another to relieve urban congestion, disperse political dissidents, or accomplish other social purposes

**resource** (p. 30) anything that is recognized as useful, such as mineral ores, forest products, skills, or brainpower

**resource base** (p. 196) the selection of raw materials and human skills available in a region for domestic use and industrial development

**responsibility systems** (p. 491) economic reforms that gave the managers of Chinese state-owned enterprises the right and the responsibility to make their operations work efficiently

**Ring of Fire** (p. 22) the tectonic plate junctures around the edges of the Pacific Ocean; characterized by volcanoes and earthquakes

**roaring forties** (p. 588) powerful westerly air and ocean currents at about 40° south latitude that speed around the far Southern Hemisphere virtually unimpeded by landmasses

**Roma** (p. 193) the now preferred term in Europe for Gypsy

**Russian Federation** (p. 269) Russia and its political subunits, which include 30 internal republics and more than 10 so-called autonomous regions

**Russification** (p. 270) assimilation of all minorities to Russian (Slavic) ways

**Rust Belt** (p. 101) the economic core region of North America that is characterized by obsolete industries and abandoned factories

**Sahel** (p. 357) a band of arid grassland that runs east–west along the southern edge of the Sahara

**salinization** (p. 323) the impregnation of the soil by salts and other minerals left by the evaporation of water, damaging soil fertility

**scale** (pp. 2, 9) the proportion that relates the dimensions of a map to the dimensions of the area it represents; also variable-sized units of geographical analysis from the local scale to the regional scale to the global scale

**Scandinavia** (p. 184) Iceland, Denmark (including Greenland and the Faroe Islands), Sweden, Norway, and Finland

**Schengen Accord** (p. 208) an agreement signed in the 1990s by the European Union and many of its neighbors that called for free movement across common borders

**seclusion** (p. 306) the requirement (as in Purdah) that a woman stay out of public view, a regional cultural practice that predates Islam

**secular states** (p. 315) countries that have no state religion and in which religion has no direct influence on affairs of state or civil law

**secularism** (p. 15) a way of life informed by ethics and values that are not necessarily derived from a religious tradition

**self-reliant development** (p. 379) small-scale development schemes in rural areas that focus on developing local skills, creating local jobs, producing products or services for local consumption, and maintaining local control so that participants retain a sense of ownership

**serfs** (p. 253) persons legally bound to live on and farm land owned by a lord

**service sector** (p. 73) economic activity that involves the sale of services

**sex tourism** (p. 552) the sexual entertainment industry that services primarily men who travel for the purpose of living out their fantasies during a few weeks of vacation

**shari'a** (pp. 314, 421) literally, "the correct path"; Islamic religious law that guides daily life according to the interpretations of the Qur'an

**sheiks** (p. 343) patriarchal leaders of tribal groups on the Arabian Peninsula

**shifting cultivation** (pp. 123, 358) a productive system of agriculture in which small plots are cleared in forestlands, the dried brush is burned to release nutrients, and the clearings are planted with multiple species; each plot is used for only two or three years and then abandoned for many years of regrowth

**Shi'ite (or Shi'a)** (p. 314) the smaller of two major groups of Muslims with different interpretations of shari'a; Shi'ites are found primarily in Iran and southern Iraq

**shogun** (p. 482) a member of the military elite of feudal Japan

**Sikhism** (p. 432) a religion of South Asia that combines beliefs of Islam and Hinduism

**silt** (p. 119) fine soil particles

**Slavs** (p. 251) a group of farmers who originated between the Dnieper and Vistula rivers in modern-day Poland, Ukraine, and Belarus

**smog** (p. 92) a combination of industrial air pollution and car exhaust (smoke + fog)

**social forestry movement** (p. 451) see **Chipko movement**

**social welfare** (p. 214) in Europe, elaborate tax-supported systems that serve all citizens in one way or another

**socialism** (p. 254) a social system in which the production and distribution of goods are owned collectively and ideally political power is exercised by the whole community

**soft power** (p. 496) as described by some analysts, the attempt by China to soften its image in places it would like to invest

**Southern African Development Community (SADC)** (p. 378) an organization of 14 countries in Southern and East Africa working together for regional development and freer trade

**sovereignty** (p. 49) the capability of a country to manage its own affairs

**Soviet Union** (p. 246) see **Union of Soviet Socialist Republics**

**spatial analysis** (p. 2) the study of how and to what extent people, objects, or ideas are related to one another across an area

**special economic zones (SEZs)** (p. 494) free trade zones within China

**state-aided market economy** (p. 487) an economic system based on market principles such as private enterprise, profit incentives, and supply and demand, but with strong government guidance; in contrast to the free market (limited government) economic system of the United States and Europe

**steppes** (p. 249) semiarid, grass-covered plains

**structural adjustment policies (SAPs)** (p. 32) policies that require economic reorganization toward less government involvement in industry, agriculture, and social services; sometimes imposed by the World Bank and the International Monetary Fund as conditions for receiving loans

**subcontinent** (p. 414) term used to refer to the entire Indian peninsula, including Pakistan and Bangladesh

**subduction** (p. 22) the sliding of one lithospheric (tectonic) plate under another

**subduction zone** (p. 119) the zone where one tectonic plate slides under another

**subregions** (p. 10) smaller divisions of the world regions delineated to facilitate study of patterns particular to the area

**subsidence** (p. 130) sinking (of land)

**subsidies** (p. 206) monetary assistance granted by a government to an individual or group in support of an activity, such as farming or housing construction, that is viewed as being in the public interest

**subsistence affluence** (p. 607) the ability to maintain a safe and healthy lifestyle on relatively little formal income

**subsistence economy** (p. 40) circumstances in which a family produces most of its own food, clothing, and shelter

**Sudras** (p. 434) members of the Hindu caste of low-status laborers and artisans

**Sunni** (p. 314) the larger of two major groups of Muslims with different interpretations of shari'a

**sustainable agriculture** (p. 41) farming that meets human needs without poisoning the environment or using up water and soil resources

**sustainable development** (p. 41) improvement of standards of living in ways that will not jeopardize those of future generations

**syncretism (or fusion)** (p. 388) the blending of elements of a new or newly introduced faith with elements of an indigenous religious heritage

**taiga** (pp. 108, 250) subarctic forests

**Taliban** (p. 435) an archconservative Islamist movement that gained control of the government of Afghanistan in the mid-1990s

**tariff** (p. 32) a tax imposed by a country on imported goods, usually intended to protect industries within that country

**technology** (p. 17) an integrated system of knowledge, skills, tools, and methods upon which a culture group's way of life is based

**temperate midlatitude climate** (p. 187) as in south central North America and China and in Europe, that is moist all year with relatively mild winters and long hot summers

**temperature-altitude zones** (p. 121) regions of the same latitude that vary in climate according to altitude

**terrorism** (p. 330) the use, or the threat, of violence, intended to create a climate of fear in a population

**theocratic states** (p. 315) countries that require all government leaders to subscribe to a state religion and all citizens to follow rules decreed by that religion

**thermal inversion** (p. 92) a warm mass of stagnant air that is temporarily trapped beneath heavy cooler air

*tierra caliente* (p. 122) low-lying "hot lands" in Middle and South America

*tierra fria* (p. 122) "cool lands" at relatively high elevations in Middle and South America

*tierra helada* (p. 122) very high elevation "frozen lands" in Middle and South America

*tierra templada* (p. 122) "temperate lands" with year-round springlike climates at moderate elevations in Middle and South America

**total fertility rate (TFR)** (p. 38) the average number of children women in a country are likely to have at the present rate of natural increase

**trade deficit** (p. 77) the extent to which the money earned by exports is exceeded by the money spent on imports

**trade winds** (p. 122) winds that blow in a generally westerly direction across the Atlantic

**tsunami** (p. 476) a large sea wave caused by an earthquake

**tundra** (pp. 108, 249) a treeless area, between the ice cap and the tree line of arctic regions, where the subsoil is permanently frozen

**typhoon** (p. 507) tropical cyclone or hurricane

**underemployment** (p. 272) the condition in which people are working too few hours to make a decent living or are working at menial jobs even though highly trained

**Union of Soviet Socialist Republics (USSR)** (p. 246) the nation formed from the Russian empire in 1922 and dissolved in 1991

**United Nations (UN)** (p. 49) an assembly of 185 member states that sponsors programs and agencies that focus on scientific research, humanitarian aid, planning for development, fostering general health, and peacekeeping assistance

**urban growth poles** (p. 178) cities that are attractive to investment, innovative immigrants, and trade and thus stimulate further development

**urban sprawl** (p. 82) the encroachment of suburbs on agricultural land

**Vaishyas** (p. 434) members of the Hindu landowning farmer and merchant caste

*varna* (p. 434) the four hierarchically ordered divisions of society in Hindu India underlying the caste system: Brahmins (priests), Kshatriyas (warriors/kings), Vaishyas (merchants/landowners), and Sudras (laborers/artisans)

**veiling** (p. 306) the custom of covering the body with a loose dress and the head—and in some places, the face—with a scarf

**volcano** (p. 22) an area between plates or a weak point in the middle of a plate where gases and molten rock, called magma, can come to the earth's surfaces through fissures and holes in the plate

**weather** (p. 22) the short-term (day-to-day) expression of climate

**weathering** (p. 22) the physical or chemical decomposition of rocks by sun, rain, snow, wind, ice, and the effects of life-forms

**welfare state** (p. 193) a social system in which the state accepts responsibility for the well-being of its citizens

**West African Economic and Monetary Union (UEMOA)** (p. 378) an organization that works to promote economic integration among countries in West Africa that share a common currency

**wet rice (or paddy) cultivation** (pp. 480, 547) a prolific type of rice production that requires the submersion of the plant roots in water for part of the growing season

**World Bank** (p. 32) a global lending institution that makes loans to countries that need money to pay for development projects

**world cities** (p. 191) cities of worldwide economic and/or cultural influence

**world region** (p. 8) a part of the globe delineated according to criteria selected to facilitate the study of patterns particular to the area

**World Trade Organization (WTO)** (pp. 32, 468) a global institution made up of member countries whose stated mission is the lowering of trade barriers and the establishment of ground rules for international trade

**yurt (or ger)** (p. 516) a round, heavy felt tent stretched over a collapsible willow lattice frame used by nomadic herders in northwestern China and Mongolia

**Zionists** (p. 293) European Jews who worked to create a Jewish homeland (Zion) on lands once occupied by their ancestors in Palestine

# Index

Note: **Boldface** indicates a glossary term; *italics* indicate a subregion, foreign word, or book title; CAPS AND SMALL CAPS indicate a region; and *i* and *t* indicate illustrations and tables, respectively.

**Aborigines, 591**–592, 592*i*, 594, 598–601
Abortion, selective, 427
Abuja (Nigeria), 383
Abu-Nimer, Mohammed, 329–330
**Acculturation,** 149, **211**–212
Aceh (Indonesia), 13–14
**Acid rain, 93,** 93*i*
  in North America, 93
Acquired immunodeficiency syndrome.
    *See* HIV-AIDS
*Adivasis,* **434**
*Afghanistan* [SOUTH ASIA], 454–456, 454*i*,
    455*i*
  ethnic groups in, 455–456
  gender inequality in, 435–436
  history of, 454
  human well-being in, 452–453, 453*t*,
      455–456
  landforms of, 455–456
  population patterns in, 455–456
  postwar reconstruction in, 448
  Soviet Union's war with, 255, 291
  Taliban in, 435–436, 447–448, 454, 456
  U.S. incursion into, 448
Africa. *See* NORTH AFRICA AND SOUTHWEST
    ASIA; SUB-SAHARAN AFRICA
Africa Online, 380
African National Congress (ANC), 362–363
**African Union (AU), 378**–379
**Age distribution, 38**
**Age of Exploration, 190**
Age of revolutions, in Europe, 192–193
**Age structure, 38**
Aging population. *See also* **Population
    pyramids**
  in East Asia, 498, 523–524
  in Europe, 197, 197*i*
  in Middle and South America, 91
  in North Africa and Southwest Asia, 311,
      311*i*
  in North America, 91, 91*i*
  in Russia and the Newly Independent
      States, 258*i*, 259
Agnew, John, 3*i*
Agra fortress, 420, 420*i*
**Agribusiness, 71**
Agriculture
  agroecology and, 440
  animal husbandry in. *See* Animal
      husbandry

collectives in, 207
commune system in, 490–491
contested space and. *See* **Contested space**
corporate farming in, 71–72, 206–208
deforestation and. *See* Deforestation
development of, 25–28, 189
  gender roles and, 303
domesticated plants in, 25, 126*t*
dumping in, 206
in East Asia, 489, 491–492, 491*i*, 512, 515,
    522–523
  irrigation in, 506, 517
in Europe, 189, 206, 207–208, 226, 228,
    229, 238–239
fair trade agreements and, 142
genetically modified crops and, 206
green revolution in, 439–440
herding in, 25, 370. *See also* Herding
indigenous methods in, 123, 123*f*, 173
irrigation in. *See also* Water resources
  in East Asia, 506, 517
  in North Africa and Southwest Asia,
      323–324, 324*i*, 332, 341–342, 341*i*
  in North America, 96, 103, 108*i*
  in Russia and the Newly Independent
      States, 277, 278*i*
  salinization and, 323–324
  in sub-Saharan Africa, 396–397, 401
in Middle and South America, 123,
    124–125, 126*t*, 166–167, 166*f*, 168,
    173, 178
mixed, 370
in North Africa and Southwest Asia, 299,
    299*i*, 302–303, 302*i*, 323–325,
    341–342, 341*i*
in North America, 59–62, 70–72, 71*i*, 83,
    96, 99, 104–106, 108, 108*i*, 109, 111
in Oceania, 604, 609–610
plantations and, 59–60, 136
in Russia and the Newly Independent
    States, 250, 251*i*, 254, 256, 257, 263
  irrigation in, 277, 278*i*
shifting cultivation in, 123, 358
in South Asia, 439–440
in Southeast Asia, 546–548, 547*i*, 548*i*
in Soviet Union, 207
in sub-Saharan Africa, 358–359, 365,
    370–374, 371*i*–373*i*
  colonization and, 361–362
  debt restructuring and, 376

desertification and, 394
development of, 359, 360
irrigation in, 396–397, 401
subsidies in, 206
sustainable, 41–43
urban gardens and, 257, 370
**Agroecology, 440**
**Agroforestry, 394**
*Agua para Todos,* 155
Ahraura (India), 428–429
AIDS. *See* HIV-AIDS
Ainu, 502, 502*i*
Air pollution. *See also* Environmental issues
  in East Asia, 502–505, 505*i*
  in Europe, 217–218, 236, 236*i*, 239–240
  in North America, 92–93, 93*i*
  in Russia and the Newly Independent
      States, 275–277
  smog in, 92
  in Southeast Asia, 567–568
**Air pressure, 23**
Air travel, 72–73
Al Jazeera, 297–298, 321
Al Qaeda, 66–67, 327, 329, 330, 448
Albania, 221*t*, 237–242, 237*i*
Alcohol abuse, 63
Alexander the Great, 454
Alexandria (Egypt), 343
Algeria, 336*t*, 337–339, 337*i*
  economy of, 338–339
  violence in, 338–339
**Allah, 305**
Allende, Salvador, 147
**Alluvium, 515**
Alps, 186, 186*i*
Altai Mountains, 249, 527–528
Al-Thani, Hamad bin Khalifa, 331
**Altiplano, 173**
Altruism, 20
Amazon Basin, 115*f*, 119–121, 119*f*, 177, 177*f*
  environmental problems in, 151–155,
      152*f*–155*f*
Amazon Highlands, 115*f*–116*f*, 119
Amazon Lowlands, 115*f*–116*f*, 119–120, 119*f*
Amazon River, 119–120
America. *See* MIDDLE AND SOUTH AMERICA;
    NORTH AMERICA
American, definition of, 56
*American South (Southeast)* [NORTH
    AMERICA], 102–104, 103*i*, 104*i*

Amin, Samir, 359
Amnok River (Yalu River), 525, 525*i*
Amur River (Heilong Jiang), 511
Andean Community (CAN), 141
Anderson, E. N., 28
Andes, 119
Angkor empire, 540
Anglican Church, in sub-Saharan Africa, 389–390
Angola, 399*t*, 408–410, 408*i*
Animal(s)
  genetically modified, 72
  of Oceania, 589–590
Animal husbandry, 25. *See also* Agriculture
  in Australia, 608–609
  in East Asia, 517–518
  in North Africa and Southwest Asia, 324–325, 335
  in North America, 104
  in sub-Saharan Africa, 370
**Animism, 388**
  in South Asia, 431*i*, 433
  in Southeast Asia, 558, 559–560
  in sub-Saharan Africa, 388
Antigua and Barbuda, 158–160, 159*f*
  human well-being in, 256*t*
  tourism in, 160
*Aotearoa*, 588
**Apartheid, 362**–363, 383, 410
Appalachian Mountains, 56
**Aquifers, 95**–96, 95*i*
*Arab Human Development Reports*, 330–331
Arab reformists, 330–331
*Arabian Peninsula* [NORTH AFRICA AND SOUTHWEST ASIA], 300, 336*t*, 343–345, 343*i*
Arabian Plate, 300, 356
Arab-Israeli conflict, 308–309, 308*i*, 327–329
Arabs, vs. Muslims, 299
Aral Sea disaster, 277, 278*i*
**Archipelago, 535**
Arctic Ocean, pollution of, 218–219, 219*i*
Argentina
  capital of, 131, 175*f*, 176, 176*f*
  cultural diversity in, 149
  debt crisis in, 139
  economy of, 157*t*, 174–176, 174*f*
  globalization and, 137, 141, 146
  information technology in, 147
  landforms of, 119
  politics and government in, 146
  urbanization in, 131, 175*f*, 176, 176*f*
  wars and civil unrest in, 175–176
Armenia, 246, 247*i*, 279*t*
Arunachal Pradesh, 458–459, 459*i*
**Aryans, 419**–420, 430–431
**ASEAN (Association of Southeast Asian Nations), 553**–554, 554*i*

ASEAN Free Trade Association (AFTA), 553–554
Asian "tiger" economies, 571
**Asia-Pacific region, 524**
Assam (India), 463
**Assimilation, 149, 210**–212
**Association of Southeast Asian Nations (ASEAN), 553**–554, 554*i*
Atacama, 122
Atatürk, Kemal, 334
Atlanta, GA (USA), 66
Atlantic Ocean, pollution of, 218, 219*i*, 220–221
Atlantic Plate, 119
Atlas Mountains, 301
**Atolls, 587,** 587*f*
Australia, 582*i*, 583–615, 596*t*, 614*t*. *See also* OCEANIA
  colonization of, 593–594
  environmental issues in, 608*i*, 609–610
  independence for, 594
*Australian Girl in London, An* (Mack), 598
**Australo-Melanesians, 539**
Austria, 220*t*
**Austronesians, 539, 591**
Automobile industry, 77–78, 77*i*
  in China, 495*i*
  in Germany, 227
  in Japan, 77, 487–488
  in Korea, 527
**Autonomy, 14**
**Average population density, 37**
Azerbaijan, 246, 247*i*, 272, 279*t*
**Aztecs, 123**

Babar's Mosque, 445
**Baby boomers, 91,** 91*i*
Bachelet, Michelle, 176
Bahamas, 156*t*, 158–160, 159*f*
Bahrain, 343*i*, 344
Balfour Declaration, 308–309
Bali, terrorism in, 557
Balkans, 186, 240–242, 241*i*
Baltic Sea, 236
  pollution of, 218, 219*i*
Baltic states, 233–234, 234*i*, 236–237
  human well-being in, 220*t*, 221–222
  Russia and, 233, 236–237
Baltimore, MD (USA), 61, 81, 82
Banda Aceh (Indonesia), 537, 537*i*
Bangalore (India), 467–468, 467*i*
Bangladesh, 462*i*, 464
  creation of, 424
  human well-being in, 452–453, 453*t*
  microcredit in, 443–444, 443*i*
  water resources of, 451
Banqiao Dam, 504
Barbados, 131, 156*t*, 159*f*, 162, 162*f*
  human well-being in, 156*t*

information technology in, 147*i*, 160, 162
  population patterns in, 131
  tourism in, 160
  vs. Haiti, 161–162, 162*i*
Barcelona (Spain), 231
**Barriadas, 132**
**Barrios, 132**
**Basin, 219**
Batuque, 179
Beall, Cynthia, 527–528
Beijing (China), 475, 511–512, 512*i*
Belarus, 279*t*, 286, 287*i*
*Belarus* [RUSSIA AND THE NEWLY INDEPENDENT STATES], 246, 247*i*, 279*t*, 286, 287*i*
Belgium, 222–223, 223*i*, 224
  in European Union, 224
  human well-being in, 220*t*
Belize, 157*t*, 166–168, 166*i*, 167*t*
Benelux, 222–226, 223*i*–225*i*
Benin, 398*t*, 400–402, 400*i*
Bennett, Louise, 230
Bering land bridge, 57–58
Bhutan, 457–459, 457*i*
  human well-being in, 452–453, 453*t*
  television in, 443
Biaguiaje, Colon, 115–116
Biko, Steven, 363
Bin Laden, Osama, 66, 67, 326, 343, 448
Biological diffusion, 406
Birmingham, AL (USA), 66
Birth control. *See* Population control
**Birth rate, 38.** *See also* Population patterns
Bjibouti, 398*t*
Black Death, 35, 36*i*
Black River, 534*i*, 536
Black Sea, pollution of, 218–219, 219*i*
Blue Nile River, 405
Boers, 362
Bolivia, 171–174, 172*i*
  agriculture in, 144
  environmental issues in, 155
  globalization and, 137, 141
  human well-being in, 157*t*
  politics and government in, 145
Bollywood, 467
**Bolsheviks, 193, 254**
Bombay (Mumbai, India), 429–430, 429*i*, 466–467
Borders, 49
  permeable, 112
**Boreal forests, 108**
Borneo, 534*i*, 536
  deforestation in, 533–535
Bosnia and Herzegovina, 221*t*, 237–242, 237*i*
  ethnic cleansing in, 195, 240–242
Boston, MA (USA), 60, 99
Botswana, 399*t*, 408–410, 408*i*
  diamond mining in, 373–374, 373*i*

HIV-AIDS in, 374
Boundaries, regional, 6–7
Boys. *See* Males
Brahmaputra River, 417, 418, 418*i*, 461–462, 462*i*, 463
**Brahmins, 434**
**Brain drain, 133–**134
Brasília (Brazil), 179–180
*Brazil* (MIDDLE AND SOUTH AMERICA), 126, 176–180
   agriculture in, 144–145
   capital of, 179–180
   climate of, 121
   colonization of, 124, 125*i*
   cultural diversity in, 149
   debt crisis in, 139
   drug trade in, 147
   economy of, 137, 139, 141
   forestry in, 153–155, 155*i*
   globalization and, 139, 141
   information technology in, 147
   liberation theology in, 151
   politics and government in, 146
   poverty in, 141
   urbanization in, 178–180
Brazilian Highlands, 119
Bride burning, 436–437
**Bride price, 437**
Britain. *See* United Kingdom
British East India Company, 419
**Brownfields, 80–**81
*Brunei* [SOUTHEAST ASIA], 555, 555*i*, 568*t*, 574, 575*i*
Budapest, 240, 240*i*
**Buddhism, 432**
   in East Asia, 519
   in South Asia, 431*i*, 432
   in Southeast Asia, 559–560, 559*i*
Buenos Aries (Argentina), 131, 175*f*, 176, 176*f*
Bulgaria, 221*t*, 237–242, 237*i*
Bullard, Robert D., 92
*Bumiputra,* 575
Burkina Faso, 398*t*, 400–402, 400*i*
*Burma* [SOUTHEAST ASIA], 534*i*, 536, 569–571, 570*i*
   human well-being in, 568*t*, 569
   politics and government in, 555–556, 555*i*
*Burqa,* 435
Burundi, 399*t*, 404–408, 405*i*
Bush, George W., 67, 326
Busto Rosalino, Aguilar, 142–143
Button industry, 492

CAFTA, 168, 169
Cahokia, IL (USA), 59, 59*i*
Cairo (Egypt), 343
*Cairo Times,* 321
Calcutta (India), 462–463, 463*i*

Calderon, Felipe, 145
Calgary, AB (Canada), 108
Caliphate, 316
*Cambodia* [SOUTHEAST ASIA], 534*i*, 536, 572–574, 573*i*
   colonization of, 541
   economy of, 572–574
   human well-being in, 568*t*, 569
   independence movements in, 542–543
   Khmer Rouge in, 542–543
Cambridge, MA (USA), 99
Cameroon, 398*t*, 402–404, 403*i*
Canada. *See also* NORTH AMERICA
   federal system in, 79
   gross domestic product of, 96–97, 96*t*
   health care in, 78–79, 78*t*
   human well-being in, 96–97, 96*t*
   immigrants in, 53–55, 54*i*, 83–84, 85
   indigenous peoples of, 63
   population patterns in, 64, 64*i*, 65*i*
   Québecois in, 56
   relationship with U.S., 69, 70*i*
   trading partners of, 69, 70*i*, 74–77, 74*i*
**Canadian Shield,** 52*i*–53*i*, 57, 100, 107–**108,** 107*i*
Cancún (Mexico), 141, 141*i*, 155–156, 158
Cannan, Crescy, 214
Cape Verde Islands, 398*t*, 400–402, 400*i*
**Capital, 32**
**Capitalism, 254**
   crony, 553
Captions, 5
Caracas (Venezuela), 170
Carbon dioxide emissions, 45–48, 47*i*, 48*i*. *See also* Global warming
*Caribbean* (MIDDLE AND SOUTH AMERICA), 131, 158–162, 159*i*, 161*i*, 162*i*
   countries of, 159*i*
   Cuba vs. Puerto Rico, 160–161
   economy of, 159–160
   Haiti vs. Barbados, 161–162
   tourism in, 154, 158–160
Caribbean Plate, 119
Caroline Islands, 582*i*, 584
Carpet weaving, child labor in, 427
**Carrying capacity, 41,** 364
**Cartels, 322**
**Cash economy, 40**
Caste, 432, **434–**435
   democracy and, 444
   religion and, 433
Castro, Fidel, 160
Cattle, in Hinduism, 432, 433–434
*Caucasia* [RUSSIA AND THE NEWLY INDEPENDENT STATES], 246, 247*i*, 278–280, 279*t*, 284–285, 288–290, 288*i*, 289*i*
Caucasus Mountains, 249
Censorship

   in North Africa and Southwest Asia, 297–298, 298*i*, 321
   in Russia and the Newly Independent States, 267, 267*i*, 268*t*
*Central Africa* [SUB-SAHARAN AFRICA], 402–404, 403*i*
   armed conflicts in, 404
   colonial legacy in, 402–403
   devolution in, 404
   ethnic groups in, 402–403
   forests of, 402
   history of, 402
   human well-being in, 398*t*–399*t*
   natural resources of, 403–404
   physical geography of, 402, 403*i*
Central African Republic, 398*t*, 402–404, 403*i*
*Central America* [MIDDLE AND SOUTH AMERICA], 166*i*, 167*i*, 167–171, 167*t*
   economy of, 167–168
   environmental issues in, 168
   landforms of, 167
   population patterns in, 167*t*
   poverty in, 167
   wars and civil unrest in, 168–169
Central American Free Trade Association (CAFTA), 168, 169
*Central Andes* [MIDDLE AND SOUTH AMERICA], 171–174, 172*i*
*Central Asian Republics* [RUSSIA AND THE NEWLY INDEPENDENT STATES], 246, 247*i*, 270, 272, 290–294, 290*i*
   environmental issues in, 293, 293*i*
   free market in, 293–294, 293*i*
   history of, 291–293
   human well-being in, 278–280, 279*t*
   landforms of, 290–291
   Russification and, 270, 271
   in Soviet and post-Soviet eras, 291–293
*Central China* [EAST ASIA], 512–516, 513*i*
   Central and Coastal Plains, 514–516
   Sichuan Province, 512–514
Central China Plain, 513*i*, 514–516
*Central Europe* [EUROPE], 185, 185*i*, 237–242, 237*i*
   agriculture in, 207–208, 238–239
   corruption in, 239
   economic transition in, 237–240
   environmental issues in, 218, 239–240
   ethnic diversity in, 240, 241*i*
   European Union and, 202–203, 203*f*, 237
   gender issues in, 239
   human well-being in, 221–222, 221*t*
   industrialization in, 239
   migration and, 239
   politics and government in, 193, 237–240
   wars and civil unrest in, 240–242
   welfare systems in, 215*i*, 216
*Central India* [SOUTH ASIA], 465–467, 465*i*
Central Siberian Plain, 248–249, 250*i*

**Centrally planned (command) economy,**
   193, 253, **254**, 259–261, 489–490
Chad, 402–404, 403*i*
*Chador*, 319, 319*i*
**Chaebol, 527**
**Chain migration, 83**
Chang Jiang (Yangtze River), 475, 476, 512,
   513*i*, 515
   damming of, 502, 504*i*, 514
Chao Phraya River, 534*i*, 536, 562
Chavez, Hugo, 139–140, 145–146, 170
Chechnya, 248, 270, 272
Chengdu (China), 512–514, 513*i*
Chennai (India), 467–468, 467*i*
Chernobyl nuclear accident, 277
Chevron Texaco Corp., 115–116
Chiang Kai-shek, 481
Chiao-Min Hsieh, 512
Chicago, IL (USA), 61, 101–102
Childhood landscape maps, 2, 2*i*
Children
   HIV-AIDS and, 369
   in labor force, 427
   in Middle and South America, 150–151
   in North Africa and Southwest Asia, 320,
      329
   in South Asia, 427
   in sub-Saharan Africa, exploitation of,
      401–402
Chile, 122, 147, 157*t*, 174–175, 174*i*
   climate of, 122
   cultural diversity in, 149
   economy of, 139, 141, 174–176
   globalization in, 139, 141
   information technology of, 147
   politics and government in, 146
   wars and civil unrest in, 175–176
China
   advanced civilizations of, 480–481
   agriculture in, 489–492
   ASEAN and, 553–554, 554*i*
   biodiversity in, 519–520
   Central, 512–516, 513*i*
      Central and Coastal Plains, 514–516
      Sichuan Province, 512–514
   climate of, 477–478, 477*i*
   colonies in, 480–481
   Communist, 481–482
   Cultural Revolution in, 482
   early civilizations of, 478
   economic development in, 489–496
   energy resources in, 503–505
   environmental issues in, 502–506
   ethnic groups in, 500–502, 500*i*, 516–519
   Far North and West, 516–519, 516*i*
   Plateau of Tibet, 415, 516*i*, 517–519
   Xinjiang in, 516–517
   food security in, 491–492
   forests in, 519–520

   globalization and, 490–496
   Great Leap Forward in, 482
   hearths in, 478
   history of, 478–482
   HIV-AIDS in, 486
   Hong Kong and, 520
   human well-being in, 507–508, 507*t*
   immigration and, 515
   industrialization in, 481–482
   Internet in, 497
   landforms of, 476, 476*i*
   migration in, 473–474, 474*i*, 517
   Northeast
      Beijing, 511–512
      Far Northeast, 509*i*, 510–511
      Loess Plateau, 509–510, 509*i*
      North China Plain, 509–510, 509*i*
      Tianjin, 511–512
   opening to trade, 515
   population patterns in, 485–486, 485*i*,
      497–499
   preeminence of, 480, 480*i*
   relations with Japan, 481
   sociocultural issues in, 497–502
   soil erosion in, 510
   Southern, 518*i*, 519–520
      Hong Kong, 519*i*, 520
      Macao, 507*t*, 519*i*, 520
      Southeastern Coast, 519*i*, 520
      Yunnan-Guizhou Plateau, 519–520,
         519*i*
   subregions of, 509–520
   Taiwan and, 524
   Tibet and, 501, 517–519
   in 20th century, 481–482
Chinese, overseas, 515, 557–558
**Chipko movement, 451**
Chlorofluorocarbons, ozone layer and, 613
Chocolate industry, European Union and,
   225
Chongqing (China), 512, 513*i*, 514
**Christianity, 304**
   in North America, 89
   in South Asia, 431*i*, 433
   in Southeast Asia, 559–560, 559*i*
   in sub-Saharan Africa, 389
**Cities.** *See also* Urbanization
   evolution of, 191
   primate. *See* **Primate cities**
   world, 191, 226
**Civil disobedience, 421**
Civil unrest. *See* Wars and civil unrest
**Clear-cutting, 62**, 109, 110*i*. *See also*
   Deforestation; Logging
Cleveland, OH (USA), 61
**Climate, 22–25**
   of East Asia, 477–478, 477*i*
   El Niño and, 538–539
   of Europe, 187–188, 187*i*

   humid continental, 187*i*, 188
   Köppen classification of, 25, 26*i*–27*i*
   Mediterranean, 187*i*, 188
   of Middle and South America, 120*f*, 120*i*,
      121–122, 121*i*
   of North Africa and Southwest Asia, 300,
      300*i*
   of North America, 56–57, 57, 58*i*
   of Oceania, 587–589, 588*i*, 589*i*
   of Russia and the Newly Independent
      States, 250, 250*i*
   of South Asia, 416–418, 416*i*–418*i*
   of Southeast Asia, 538–539, 538*i*
   of sub-Saharan Africa, 356–359, 357*i*
   temperate midlatitude, 187–188, 187*i*
Climate change. *See* Global warming
Clinton, Bill, 79
Coal-burning plants. *See also* Energy
      production
   in China, 502–504
Coastal China Plain, 514–516
Coca/cocaine, 144, 146–147, 146*i*, 171.
      *See also* Drug trade
Cocos Plate, 119
Coffee, fair trade, 142
Cohen, Jeffrey, 166
**Cold war, 48, 193**–194, 254–255, 255*i*
   Eastern Europe and, 193–194, 254–255,
      255*i*
   North Africa and Southwest Asia and,
      306–308
   sub-Saharan Africa and, 383–384
Collective farms, 207
Colombia, 146, 157*t*, 168, 169*i*, 170–171
   agriculture in, 144
   drug trade in, 147, 171
   forestry in, 153
   globalization and, 141
   landforms of, 119
   politics and government in, 145
   wars and civil unrest in, 170
**Colonialism, 9**–10
   language and, 16
   mandated territories and, 306
   in Middle and South America, 124–129,
      124*i*
   in North Africa and Southwest Asia,
      306–308, 307*i*
   in North America, 59–60
   in Oceania, 593
   protectorates and, 306
   religion and, 15
   in South Asia, 420
   in Southeast Asia, 540*i*, 541
   in sub-Saharan Africa, 354–355, 361–364,
      361*i*
      legacy of, 369, 380–383, 390
   trade and, 31–32
**Colonias, 132**

Colorado River, 96
Columbus, Christopher, 116, 124
*Coming of Age in Samoa* (Mead), 603
**Command economy,** 193, **254,** 259–261, 489–490
**Common Market for Eastern and Southern Africa (COMESA), 378**
**Communal conflict, 433**
Commune system, 489–490
Communications industry. *See also* **Internet/information technology**
  in East Asia, 497
  in Russia and the Newly Independent States, 267, 268*i*, 268*t*
  in sub-Saharan Africa, 380
**Communism,** 254–255
  anti-Soviet uprisings in, 194
  in China, 481–482
  command economy in, 183, 254, 259–261, 489–490
  in East Asia, 481–482
  in Europe, 193
  in Korea, 526–527
  in Middle and South America, 160–161
  in Mongolia, 529
  origins of, 254
  rise and fall of, 254–255
  in Southeast Asia, 541–543
  in Soviet Union, 254–255
**Communist Party, 254**
Comoros, 399*t*, 404–408, 405*i*
Computers. *See* **Internet/information technology**
Condomble, 179
**Confucianism,** 479–480, 555
  in Southeast Asia, 559–560, 559*i*
Congo (Brazzaville), 355, 398*t*, 402–404, 402*i*
Congo (Kinshasa), 355, 386, 398*t*, 402–404, 402*i*
  devolution in, 404
Congo Free State, 402
Conrad, Joseph, 402
**Contested space, 142**
  deforestation and, 155
  in Middle and South America, 142–145, 154
**Contiguous regions, 9**
Continental drift, 21–22, 21*i*
*Continental Interior* [NORTH AMERICA], 107–108
Continental plates, 21–22, 21*i*
Contraception. *See* Population control
**Conurbation, 131**
Convention on the Law of the Sea, 611–612
Conway, Dennis, 166
Corporate farming, 71–72, 206–208
Corruption
  in Middle and South America, 146
  in Russia and the Newly Independent

States, 274
  in Southeast Asia, 553
  in sub-Saharan Africa, 274, 382–383, 385
Costa Rica, 157*t*, 166–168, 166*i*, 167*t*, 169
  agriculture in, 168
  environmental issues in, 168
  haciendas in, 142–143
  information technology in, 147
  population patterns in, 167*t*
Côte d'Ivoire, 398*t*, 400–402, 400*i*
  devolution in, 385
  wars in, 401–402
**Council of the European Union, 198**
**Countries, 49**
Cows, in Hinduism, 432, 433–434
Cree people, 100
**Creole cultures, 118**
**Creoles, 127**
Critchfield, Richard, 428
Croatia, 221*t*, 237–242, 237*i*
*Crocodile Dundee*, 603
**Crony capitalism, 553**
Crops. *See* Agriculture
Crowley, William, 133*i*
Cuba, 156*t*, 159*i*, 160–161, 161*i*
  cultural diversity in, 149
  health care in, 145–146
  human well-being in, 156*i*
  politics and government in, 145, 147
  population patterns in, 131
  tourism in, 160, 161*i*
  vs. Puerto Rico, 160–161
**Cultural complexity, 557–558**
**Cultural diversity, 14**
  in East Asia, 500–502, 500*i*, 516–519
  in Europe, 193
  in Middle and South America, 149
  in North Africa and Southwest Asia, 320, 320*i*
  in North America, 86–88, 87*i*, 99
  in Oceania, 598–604, 598*i*
  in Russia and the Newly Independent States, 269–270, 269*i*
  in South Asia, 420, 430–431, 455–456, 455*i*
  in Southeast Asia, 539–541, 557–558
  in sub-Saharan Africa, 380–381, 381*i*
Cultural geography. *See* Sociocultural issues
**Cultural hearths, 432**
**Cultural homogeneity, 13**
**Cultural homogenization, 184,** 203
**Cultural identity, 13**
**Cultural Insight Boxes**
  The Five Pillars of Islamic Practice, 315
  The Hungarian Sausage Experience and the European Union, 203
  Pancasilia: Indonesia's National Ideology, 556
  The Veil, 319
**Cultural markers, 14–20**

**Cultural pluralism, 557–558**
**Cultural Revolution, 482**
**Culture, 12**
  ethnicity and, 12–13. *See also* Racial/ethnic groups
  gender issues and, 18–19, 18*t*
  globalization and, 13–14
  language and, 16–17
  material, 12
  race and, 19–20
  religion and, 15
  values and, 14–15
**Culture groups, 12–13**
Curitiba (Brazil), 132, 132*i*, 155
**Currency devaluation, 376**
Cyprus, 221*t*, 230, 231*i*
  welfare system in, 215–216
**Czars, 253**
Czechoslovakia/Czech Republic, 221*t*, 237–242, 237*i*
  agriculture in, 238
  anti-Soviet uprisings in, 194
  economic transition in, 238

Da Silva, Luiz Inácio Lula, 144, 146
Dagestan, 272
Dai, 501
DaimlerChrysler, 227
Dal, Akali, 445
Dalits, 434
D'Aluisio, Faith, 428–429
Dams
  in East Asia, 502, 504*i*, 505, 514
  global locations of, 415*i*
  on Nile, 341–342
  in South Asia, 413–414, 414*i*, 415*i*
  on Tigris and Euphrates, 332, 334
  on Yangtze River, 502, 504*i*, 514
Danube River, 185*i*, 187
Darfur (Sudan), 340–341
Dark Ages, 189–190
Darling River, 589
Davis, Diana, 335
**Death rate, 38.** *See also* Population patterns
Debt crises. *See also* Structural Adjustment Policies (SAPs)
  in Middle and South America, 138*i*, 139, 178
  in North Africa and Southwest Asia, 325–326
  in Russia and the Newly Independent States, 262–263
  in Southeast Asia, 552–553
  in sub-Saharan Africa, 374–377
Deccan Plateau, 415–416
Deforestation
  clear-cutting in, 62
  in East Asia, 510, 515
  in Europe, 187, 216, 217*i*

Deforestation(*continued*)
   global warming and, 46, 47, 154
   in Middle and South America, 152*i*,
      153–155, 155*f*
   in New Zealand, 610
   in North America, 94*i*, 98, 109, 110*i*
   social forestry movement and, 451
   in South Asia, 450–451
   in Southeast Asia, 533–535, 566–567, 566*i*,
      567*i*
   in sub-Saharan Africa, 353–354, 394–395,
      401
**Delta, 22**
**Democratic institutions, 193**
   in Europe, 192–194, 240
   Islam and, 329–330
   in Middle and South America, 145–147
   in North America, 79–80
   in Russia and the Newly Independent
      States, 255
   in South Asia, 423, 444, 448
   in Southeast Asia, 554–557, 555*i*
   in sub-Saharan Africa, 384–385, 384*i*
Democratic Republic of the Congo, 386, 398*t*,
   402–404
**Demographic transition, 40, 40*i***
   in Middle and South America, 131
**Demography, 35**
Deng Xiaoping, 482
Denmark, 233–235, 234*i*
   currency of, 200
   human well-being in, 220*t*
   Muslims in, 213
**Dense nodes, 81**
Denver, CO (USA), 105
**Deposition, 22**
**Desertification, 335, 394**
   in North Africa and Southwest Asia, 335
   in sub-Saharan Africa, 394, 401
Deserts, 26*i*
   in East Asia, 476, 476*i*, 510, 517, 528, 528*i*
   in North Africa and Southwest Asia,
     298–299, 298*i*
   in sub-Saharan Africa, 300, 357, 358, 394
**Detritus, 539**
Detroit, MI (USA), 61, 100
**Development, 32–35**
   economic, 33–34, 134–140. *See also*
     Economic development
   environmental impact of, 33. *See also*
     Environmental issues
   human well-being and, 33, 34–35. *See also*
     **Human well-being**
   self-reliant, 379
   sustainable, 41–45
   tourism in, 606
**Development for whom?, 33, 41**
**Devolution, 230, 385**
   in sub-Saharan Africa, 385, 404
Dhab (India), 442

Dharma Wanita, 556
**Dialects, 16.** *See also* Language(s)
Diamond mining, 373–374, 373*i*
**Diaspora, 304**
   Indian, 422
   Jewish, 304
Diffusion, biological, 406
**Digital divide, 11, 74.** *See also*
     **Internet/information technology**
Dingo fence, 609–610, 609*i*
Diseases. *See* Health and disease
Divorce. *See also* Marriage
   in Southeast Asia, 560
Djénné (Mali), 401
Djibouti, 402–404, 403*i*
**Doi moi, 574**
Dolpo-pa, 458
**Domestication, 25.** *See also* Agriculture
Dominica, 156*t*, 158–160, 159*i*
Dominican Republic, 158–160, 159*i*
   human well-being in, 156*t*
   politics and government in, 146–147
   poverty in, 159
Dongguan (China), 473–474
Dorji, Kinley, 443
Dorji, Rinzy, 443
**Double day, 212**
**Dowry, 436–437**
Dravidians, 418, 430–431, 468
*Dreamtime*, 599
Drought. *See* Climate; Water resources
Drug trade
   in East Asia, 520
   in Middle and South America, 144,
     146–147, 146*i*, 150–151, 150*i*, 171
**Dry forests, 394**
Dudakov, Vladislav, 262
Dugger, Celia W., 391
**Dumping, 206**
Dust Bowl, 62, 62*i*
Duvalier, "Papa Doc," 162

**Early extractive phase, 134–137**
**Earthquakes, 22**
   in Ring of Fire, 22, 22*i*
   in Southeast Asia, 537
   tsunamis and, 476, 537, 537*i*, 563–564
*East Africa* [SUB-SAHARAN AFRICA], 404–408,
   405*i*
   coastal lowlands of, 406
   contested spaces in, 407–408
   human well-being in, 399*t*
   interior valleys and uplands of, 405,
     405*i*
   islands of, 399*t*, 405*i*, 406
   physical geography of, 404–406, 405*i*
**East African Community, 378**
EAST ASIA, 473–531
   agriculture in, 489, 491–492, 491*i*, 512,
     515, 522–523

   irrigation in, 506, 517
   automobile industry in, 77, 487–488
   biodiversity in, 519–520
   climate of, 477–478
   conflicts and power transfers in, 484–485
   Confucianism in, 479–480
   dams in, 502, 504*i*, 505, 514
   deserts in, 517, 528, 528*i*
   economic issues in, 77–78
   economy of, 487–496
     in China, 489–496
   Communist command, 489–490
   globalization and, 490–496
     Japanese model for, 487–489
   environmental issues in, 502–507
   ethnic groups in, 500–502, 500*i*, 516–519
   family and work in, 499–500
   fishing in, 522–523
   forests in, 519
   gender issues in, 499–500, 507*i*, 508,
     518–519, 530
   globalization and, 490–496
   growth poles in, 494
   herding in, 517–518, 527–528
   history of, 478–485
   HIV-AIDS in, 486
   housing in, 17*i*, 510*i*, 516–517, 523
   human impact on, 503, 503*i*
   human well-being in, 507–508, 507*t*
   income disparity in, 493–494
   industrialization in, 481–482
   information technology in, 78
   landforms of, 475–477, 476*i*
   migration in, 473–474, 474*i*, 524
   natural hazards in, 507
   natural resources of, 510–511, 519
   physical geography of, 475–478, 476*i*
   politics and government in, 481–482, 527
   population patterns in, 485–486, 485*i*
   regional map of, 472*i*–473*i*
   religion in, 479–480, 500–501, 519
   sociocultural issues in, 497–502
   subregions of, 509–530, 509*i*
     Central China, 512–516, 513*i*
     Far North and West China, 516–519,
       516*i*
     Japan, 520–524, 521*i*
     Korea, 525–527, 525*i*
     Mongolia, 527–530, 528*i*
     Northeast China, 509–510, 509*i*
     Southern China, 518*i*, 519–520
     Taiwan, 524–525, 524*i*
   terminology for, 475
   transportation in, 501, 501*i*
   wars and civil unrest in, 481, 482
East Germany, 227
East Pakistan, 423, 424. *See also* Bangladesh
East Timor. *See* Timor-Leste
Eastern Europe, 186
Eastern Ghats, 415–416

*Eastern Mediterranean* [NORTH AFRICA AND
SOUTHWEST ASIA], 336*t*, 345–346, 345*i*
**Economic and technology development
zones (ETDZs), 494**
**Economic Community of Central African
States (CEEAC), 378**
**Economic Community of West African
States (ECOWAS), 378**
**Economic core, 60–61, 100–102, 101***i***, 102***i**
Economic development, 33–34, 134–140
  colonialism and. *See* **Colonialism**
  devolution and, 385
  early extractive phase of, 134–137
  in East Asia, 487–496
  in Europe, 194, 197–205. *See also*
    **European Union (EU)**
  Export Processing Zones in, 549–550
  growth triangles in, 550
  import substitution industrialization phase
    of, 137–139
  microcredit in, 443–444, 443*i*
  in Middle and South America, 73, 75–77,
    116–117, 134–145, 138*i*, 141–142,
    164–165
  in North Africa and Southwest Asia,
    322–326
  in North America, 63, 73, 75–77, 138*i*,
    141–142, 164–165
  in Oceania, 604–608
  in Pakistan, 457
  on reservations, 63
  resource consumption and, 45
  in Russia and the Newly Independent
    States, 259–266
  in South Asia, 438–444
  in Southeast Asia, 546–554
  structural adjustment phase of, 32, 139–140
  in sub-Saharan Africa, 369–380
  sustainable, 41–45
**Economic diversification, 325**
**Economic integration, 194**
**Economic restructuring, 117**
**Economies of scale, 200**
**Economy, 30–31**
  cash, 40
  centrally planned (command), 193, 253,
    254, 259–261, 489–490
  components of, 30–31
  formal, 30–31
  global, 29–30, 31–32
  informal. *See* **Informal economy**
  knowledge, 73
  MIRAB, 607
  resource base and, 196
  service, 73
  state-aided market, 487
  substinence, 40
**Ecotourism, 154, 155**
ECOWAS
Ecuador, 115–116, 157*t*, 171–174, 172*i*, 173*i*

agriculture in, 144
cuisine of, 173*i*
cultural diversity in, 149
ecotourism in, 154, 154*i*
globalization and, 139, 141
land reform in, 174
migration from, 129
oil industry in, 115–116, 116*i*
politics and government in, 145
Education. *See also* Literacy
  human well-being and, 33–34
  in North Africa and Southwest Asia, 320,
    344–345
  in Oceania, 613, 614*t*
  in population control, 427–428
  in South Asia, 427–428, 437–438, 437*i*,
    447, 453, 453*t*
  in Southeast Asia, 545, 561*t*, 562, 568*t*
  in sub-Saharan Africa, 371, 397
  of women, 335, 344–345
Egypt, 339, 340*i*, 341–342, 341*i*
  economy of, 342
  human well-being in, 335, 336*t*, 342
  income disparity in, 342
  politics and government in, 342
  urbanization in, 342
**El Niño, 122**, 538–539, **589**, 589*i*
El Salvador, 157*t*, 166–168, 166*i*, 167*t*,
  169
  liberation theology in, 151, 168
  population patterns in, 167*t*
  wars and civil unrest in, 168
Elbaje, Miguel, 155–156
Elburz Mountains, 249
Elizabeth II, Queen of England, 79
Energy production. *See also* Oil and gas
    industry
  in East Asia, 510–511
  in Europe, 205, 229, 234–235
  geopolitical aspects of, 69
  in Middle and South America, 115–116,
    164, 170, 178
  in North Africa and Southwest Asia, 308,
    322–323, 322*i*, 323*i*, 343, 344,
    349–350
  in North America, 108, 110, 111
  in Russia and the Newly Independent
    States, 262, 263, 264–266, 265*i*, 275
  in Southeast Asia, 572
English language, in South Asia, 431
Environmental issues
  agribusiness, 70, 71–72, 206–208
  air pollution, 93*i*. *See also* Air pollution
  brownfields, 80–81
  carrying capacity, 364–365
  deforestation, 47–48, 152*i*, 153–155, 155*i*.
    *See also* Deforestation
  in East Asia, 502–507
  ecotourism, 154, 155
  in Europe, 216–221, 236, 236*i*, 239–240

fragile environments, 394
global warming, 45–48, 45–48, 122,
  154
  deforestation and, 47, 48, 154
  in Europe, 188
  flooding and, 188, 188*i*
  floods and, 188, 188*i*
  in Oceania, 612–613
  tropical storms and, 122
globalization and, 442–443
  in Middle and South America, 116, 116*i*,
    151–155, 152*i*–155*i*, 168
  in North Africa and Southwest Asia,
    331–335, 331*i*, 333*i*
  in North America, 92–96, 93*i*–95*i*
  in Oceania, 608*i*, 609–613, 612*i*
  population growth, 35–40
  in Russia and the Newly Independent
    States, 258, 274–277
  soil erosion, 104, 510, 515
  in South Asia, 413–414, 448–452
  in Southeast Asia, 533–535, 548, 551,
    564–568
  in sub-Saharan Africa, 353–355, 358, 371,
    392–397, 401
  tourism, 154, 551, 606, 612
  urbanization, 43–45. *See also* Urbanization
  water pollution, 92–93, 93*i*. *See also* Water
    pollution
Equator
  rain belt of, 25
  temperature at, 23
Equatorial Guinea, 398*t*, 402–404, 403*i*
Eritrea, 398*t*, 402–404, 403*i*
**Erosion, 22,** 104. *See also* Deforestation
  in East Asia, 510, 515
**Escarpments, 356**
Estonia, 233–234, 234*i*, 236–237
  human well-being in, 220*t*, 221–222
  Russia and, 236
Ethiopia, 398*t*, 402–404, 403*i*
**Ethnic cleansing, 195**
  in Eastern Europe, 308
  in sub-Saharan Africa, 392
  in World War II, 193, 308
  in Yugoslavia, 195, 240–242
**Ethnic groups, 12**–13. *See also* **Culture;**
  Racial/ethnic groups
**Ethnicity, 86,** 391–392
Eugene, OR (USA), 82
Euphrates River, 301, 302*i*, 332
  Southeastern Anatolia Project and, 334
**Euro, 200**
EUROPE, 183–242
  age of exploration in, 190–191, 191*f*
  age of revolutions in, 192–193
  agriculture in, 189, 206, 207–208, 226, 229,
    238–239
  anti-Soviet uprisings in, 194
  cities in, evolution of, 191, 192*i*

EUROPE *(continued)*
  climate of, 187–188
  in cold war, 193–194
  colonies of, 9–10, 190–191, 191*f*, 193–194.
     *See also* **Colonialism**
  countries of, 184–186, 185*i*, 199*f*
  cultural diversity in, 193, 203, 209–210
  cultural sources for, 189
  currency of, 200
  deforestation in, 187, 216, 217*i*
  deindustrialization in, 204*i*, 205, 229
  eastern, 186
  economy of, 194, 197–205. *See also*
     **Economic Union (EU)**
    service jobs in, 205
  energy resources in, 205, 229, 235
  environmental issues in, 216–221, 236,
     239–240
  ethnic cleansing in, 195
  gender issues in, 212–214, 214*i*, 216,
     220*t*–221*t*, 222, 235
  Greens in, 216
  history of, 189–195
  human impact in, 217*i*
  human well-being in, 220*t*–221*t*, 221–222
  information technology in, 205, 207*i*, 229
  landforms in, 186–187, 186*i*
  migration in, 193, 208–212, 209*i*, 230
  nation-states of, 193
  open borders and, 208
  physical geography of, 186–188, 187*i*
  political unification in, 200
  politics and government in, 193, 230,
     240–242
  population patterns in, 195–198, 195*i*–197*i*
  regional map of, 182*i*–183*i*
  religion in, 190
  rise of towns in, 190
  soccer in, 211
  sociocultural issues in, 208–216
  southeastern, 186, 240–242, 241*i*
  subregions of, 184–186, 199*f*, 222–242
     Central Europe, 185, 185*i*, 237–242,
      237*i*
     North Europe, 185–186, 185*i*, 233–237,
      234*i*
     South Europe, 185, 185*i*, 230–233, 231*i*
     West Europe, 185, 185*i*, 222–230, 223*i*
  tourism in, 197–198, 226, 240
  transportation in, 205, 206*i*
  urbanization in, 191, 192*i*, 196–197
  wars and civil unrest in, 193–194, 228
  welfare systems in, 193, 214–216, 215*f*, 235
  western, 186
**European colonialism,** 9–10. *See also*
  **Colonialism**
**European Commission,** 198–199
**European Common Agricultural Policy,** 206
**European Economic Community,** 198

**European Parliament,** 198
European Regional Development Fund, 200
*European Russia* [RUSSIA AND THE NEWLY
    INDEPENDENT STATES], 249, 280–283,
    281*i*
  history of, 251–255
**European Union (EU),** 183, **184,** 194
  capital of, 224
  Central Europe and, 202–203, 203*i*
  cultural homogenization and, 184, 203
  economic and social integrative role of,
    199–200
  food standards and, 203, 225
  future paths for, 203
  globalization and, 200–202
  goals of, 198–202
  governing institutions of, 198–199
  history of, 198
  membership in, 198, 201*i*, 237
    standards for, 203
  political integration and, 200
  trade share of, 202*i*
  United States and, 202
**Evangelical Protestantism,** 151, 389
Ewango, Corneille, 395
**Exchange sector,** 30
**Exclaves,** 236
**Export processing zones (EPZs),** 139,
    549–550, 552
**Extended families,** 149
**External debts,** 139. *See also* Debt crises
**External processes,** 20
**Extraction,** 30
**Extractive resources,** 30. *See also* Resources,
    extraction/consumption of
**Extraregional migration,** 563
*Exxon Valdez* oil spill, 108, 116, 220–221

**Failed states,** 340
**Fair trade,** 32
Fair trade coffee, 142
Faith, Mike, 106
Families. *See also* Children; Gender issues;
    Marriage
  in East Asia, 499–500
  in Europe, 212
  extended, 149
  in Middle and South America, 149–150
  in North Africa and Southwest Asia,
    316–317
  in North America, 89–91, 89*i*
  nuclear, 89–90, 89*i*
  in Russia and the Newly Independent
    States, 272
  single-parent, 89, 90
  in South Asia, 436
  in Southeast Asia, 559–560
  in sub-Saharan Africa, 390
Family planning. *See* Population control

*Far North and West China* [EAST ASIA],
    516–519, 516*i*
  Plateau of Tibet, 516*i*, 517–519
  Xinjiang, 516–517, 516*i*
Farming. *See* Agriculture
**Favelas,** 132–133, 132*i*
Female(s). *See also* Gender
  in agricultural development, 25
  average income of, in North America, 96*t*
  education of. *See also* Education
    in North Africa and Southwest Asia, 335,
     336*t*, 344–345. *See also* Literacy
    in South Asia, 437
  gender hypothesis and, 18
  in government, 214*i*
  income disparity for, 18, 18*t*, 96*t*
  in labor force, 549–550
    in East Asia, 473–475, 499–500, 508
    in Europe, 212–214, 216
    globalization and, 549–550
    in Middle and South America,
     164–165
    in North Africa and Southwest Asia,
     310, 310*i*
    in Southeast Asia, 549–550, 551, 552,
     560, 561*t*
  loadbearing by, in sub-Saharan Africa, 379,
    379*i*
  as mail-order brides, 273
  microcredit banks and, 443–444, 580
  as migrants, 133, 164. *See also* Migration
  mistreatment of, 19, 273
  mortality rates for, 39, 39*i*
  one-child policy and, 498–499
  in politics, in sub-Saharan Africa, 384
  seclusion of
    in North Africa and Southwest Asia,
     306, 317–318, 318*i*
    in South Asia, 429
  in sex industry, 273, 524, 552
  United Nations Gender Empowerment
    Measure for, 156*t*–157*t*, 158
  veiling of, 306, 318–319, 319*i*
**Female circumcision,** 391
Female infanticide
  in East Asia, 499
  in South Asia, 426*i*, 427, 436–437
**Fertile Crescent,** 302, 302*i*
**Feudalism,** 189
Fiji, 596*t*, 600–601, 614*t*
Finland, 233–235, 234*i*
  human well-being in, 220*t*
Fishing
  in East Asia, 522–523
  in North America, 93–95, 94*i*, 98
  in North Europe, 235
  in Oceania, 611–612
  in South Asia, 469
**Floating population,** 474

Floodplain, 22
Floods, 22
   in East Asia, 505
   in Europe, 188, 188*i*
   in sub-Saharan Africa, 401
Fonseca, James, 81
**Food security, 491**
**Food self-sufficiency, 522**
**Foreign direct investment (FDI), 134**
   in Russia and the Newly Independent
      States, 262
**Foreign exchange, 563**
Forests. *See also* Deforestation
   dry, 394
   in East Asia, 519
   in Europe, 187, 216, 217*i*
   in Middle and South America, 152*i*,
      153–155, 155*i*
   old-growth, 567
   in Russia and the Newly Independent
      States, 250
   in South Asia, 450–451
   in Southeast Asia, 533–535
   in sub-Saharan Africa, 394–395, 402
**Formal economy, 30–31**
**Formal institutions, 12**
**Forward capital, 179–180**
   Abuja as, 383
Fox, Vincente, 145, 166
**Fragile environments, 394**
France, 223*i*, 226
   agriculture in, 226
   Algeria and, 338–339
   colonies of, 226, 538–539, 541
   cultural identity of, 226
   economy of, 226
   French Revolution in, 192–193
   human well-being in, 220*t*
   immigrants in, 210, 226
   tourism in, 226
**Free trade, 32.** *See also* Globalization
Free Trade Agreement of the Americas
      (FTAA), 77, 141, 168
Free trade zones, 139
French Guiana, 157*t*, 169–170, 169*i*
French Polynesia, 596*t*, 614*t*
French Revolution, 192–193
**Frontal precipitation, 25**
Frontline World Video Vignettes
   Bhutan "The Last Place," 443
   Cambodia "Pol Pot's Shadow," 543
   Guatemala/Mexico "Coffee Country,"
      142
   India, "Hole in the Wall," 11
   Iraq, "The Road to Kirkuk," 327
   Mexico, "A Death in the Desert," 86
   Moscow, "Rich in Russia: The Brave New
      World of Young Capitalists and
      Tycoons," 262

   Nigeria "The Road North: What the Miss
      World Riots Reveal about a Divided
      Country," 383
   Spain, "The Lawless Sea," 220
FTAA, 141, 168
Fulani, 382
Fundamentalism, Islamic, 321–322
**Fusion, 388**

G8, 141
G22, 141
Gabon, 399*t*, 402–404, 403*i*
Gahongayire, Anne, 384
Gambia, 398*t*, 400–402, 400*i*
Ganga River, 415, 417, 461–462, 462*i*
   pollution of, 451–452
   water conflicts over, 451
Ganga-Brahmaputra delta, 461–462, 462*i*
Garcia, Matias, 86
Gaza Strip, 328, 328*i*, 332–335
**Gazprom, 266**
**Gender, 18.** *See also* Female(s); Males
   U.N. indices for, 34–35, 335
Gender Development Index (GDI), 35, 335.
      *See also* **Human well-being**
Gender Empowerment Measure (GEM), 35,
      335. *See also* **Human well-being**
Gender issues, 18–19
   agricultural development and, 303
   in East Asia, 499–500, 507*i*, 508, 518–519,
      530
   education and, 18–19, 18*t*
   in Europe, 212–214, 214*i*, 216, 220*t*–221*t*,
      222, 235, 239
   HIV-AIDS and, 368, 368*i*
   income disparity and, 18–19, 18*t*
   Islam and, 306, 317–320, 390
   in Middle and South America, 131,
      149–150, 164
   in North Africa and Southwest Asia, 303,
      311, 317–320, 330, 335, 336*t*,
      343–345
   in North America, 79–80, 89–91, 90*i*, 96*t*
   in Norway, 235
   in Oceania, 602–604, 613, 614*t*
   polygyny and, 319–320, 366
   in Russia and the Newly Independent
      States, 268–269, 269*i*, 272–273,
      279–280, 279*t*
   in South Asia, 426*i*, 427, 435–438, 447,
      453, 453*t*
   in Southeast Asia, 552, 556, 560–561
   in sub-Saharan Africa, 365–366, 379, 383,
      384, 390–391, 397, 398*t*–399*t*
**Gender structure, 38**
General Motors, 78
**Genetic engineering, 72**
**Geneva Conventions, 68**
Genital mutilation, 391

Genocide, 49
**Gentrification, 81**
Geography
   definition of, 2–3
   human, 3
   interdisciplinary nature of, 3*i*
   overview of, 2–4
   physical. *See* Physical geography
   regional, 4
   subdisciplines of, 3–4
Geomorphology, 20
**Geopolitics, 48–50**
Georgetown, KY (USA), 77–78
Georgia, 246, 247*i*, 279*t*
Germany, 223*i*, 226–227
   division of, 226–227
   economy of, 227
   in European Union, 227
   gender issues in, 214
   human well-being in, 220*t*
   reunification of, 194, 227
*Gers*, 17*i*, 515–516
Ghana, 398*t*, 400–402, 400*i*
   empire of, 360
   independence for, 363
Ghandi, Mohandas, 421, 433, 434
Ghandi, Rajiv, 438
Ghat Mountains, 415–416
Ghengis Kahn, 480, 528
Girls. *See* Female(s); Gender
Glaciers, in North America, 57
**Glasnost, 255**
**Global economy, 29–32.** *See also under*
      Economic; Globalization
   colonies in, 31–32. *See also* **Colonialism**
   free trade and, 32–33
**Global scale, 8,** 9*i*
**Global warming, 45–48**
   deforestation and, 47, 48, 154
   in Europe, 188
   flooding and, 188, 188*i*
   Oceania and, 612–613
   sea level changes and, 46, 612–613
   tropical storms and, 122
**Globalization, 10,** 10*i*, 32–33, 74–78
   cultural change and, 13–14
   East Asia and, 490–496
   environmental issues in, 442–443
   Europe and, 200–202. *See also* **European**
      **Union (EU)**
   European Union and, 200–202
   feminization of labor and, 549–550
   immigration and, 106
   in meatpacking industry, 106
   Middle and South America and, 134
   multinational corporations and, 32
   North America and, 74–77
   outsourcing and, 78, 442
   protectionism and, 201

**Globalization** (*continued*)
   resource consumption and, 45
   Russia and the Newly Independent States
     and, 285
   South Asia and, 465–467
   Southeast Asia and, 552–554
   trends in, 10–11
   Wal-Mart and, 75, 76
   World Trade Organization and, 201
Gobi Desert, 510, 516*i*, 528, 528*i*
Gold mining, 567
Golden Temple attack, 445–446
Goldman, Mara, 3*i*
Goldring, Beth, 545*i*
Goldstein, Melvyn, 527–528
**Gondwana, 585**
Goode's interrupted homolosine projection,
   5, 6*i*
Gorbachev, Mikhail, 255, 273
Gosar, Anton, 3*i*
Gospel of success, 151, 389
Gottman, Jean, 81
Government. *See* Politics and government
**Government subsidies, 75**
Grameen Bank, 443–444, 443*i*
Grandmother hypothesis, 18
**Grassroots economic development, 379**
Grazing, in Australia, 609–610
Great Australian Bight, 585
Great Barrier Reef, 586–587, 586*i*
**Great Basin,** 107*i*, 108
Great Britain. *See* United Kingdom
Great Dividing Range, 582*i*, 585, 588
Great Indian Desert, 459
Great Lakes, 57
**Great Leap Forward, 482**
*Great Plains Breadbasket* [NORTH AMERICA],
   104–106, 105*i*
   dust storms in, 62, 62*i*
   settlement of, 61–62
Great Rift Valley, 356, 405, 405*i*
Great Zimbabwe, 359
Greater Mekong Basin, tourism in, 551
Greece, 230, 231*i*
   ancient, 189
   economy of, 202
   human well-being in, 221*t*
   tourism in, 231
   welfare system in, 215–216
**Green revolution, 439–440**
Greene, Mark, 82
Greenhouse gases, 45–48, 47*i*, 48*i*. *See also*
   Global warming
   in North America, 92
**Greens, 216**
   in East Asia, 506
   in Europe, 216
Grenada, 156*t*, 158–160, 159*i*
Grenadines, 156*t*, 158–160, 159*i*

**Gross domestic product (GDP),** 31. *See also*
   Economic development
   in East Asia, 507–508, 507*i*
   in Europe, 220*t*–221*t*, 221–222
   in Middle and South America, 155–167,
     156*t*–157*t*
   in North Africa and Southwest Asia, 325,
     325*i*, 335, 336*t*
   in North America, 96–97, 96*t*
   in Oceania, 613, 614*t*
   population growth and, 39–40, 40*i*
   in Russia and the Newly Independent
     States, 257–258, 278–280, 279*t*
   in South Asia, 423, 453*t*
   in Southeast Asia, 568*t*
   in sub-Saharan Africa, 397, 398*t*–399*t*
**Gross domestic product (GDP) per capita,**
   33–34
**Group of Eight, 264**
**Growth poles, 494**
**Growth rate, 38**
**Growth triangles, 550**
Guadeloupe, 156*t*, 158–160, 159*i*
Guam, 596*t*, 614*t*
Guangzhou (China), 475
Guantánamo (Cuba), 160–161
Guatemala, 142, 142*i*, 157*t*, 166*i*, 167*t*, 168,
   169
   cultural diversity in, 168
   fair trade coffee in, 142, 142*i*
   liberation theology in, 151
   population dynamics in, 167*t*
   wars and civil unrest in, 168
Guerrilla Gardeners, 218
**Guest workers,** 208. *See also* Migration
   in Europe, 208
   in Japan, 524
   from North Africa and Southwest Asia,
     312
Guiana Highlands, 119
Guianas, 169–170, 169*i*
Guinea, 398*t*, 400–402, 400*i*
   wars in, 401–402
Guinea-Bissau, 398*t*, 400–402, 400*i*
Gulf of Carpentaria, 585
Gulf of Mexico, 57
Gulf Stream, 57, 58*i*
Gulf War, 326, 349–350
Gulu (Uganda), 386
Guyana, 157*t*, 169–170, 169*i*
Gypsies, 203

Habitat loss, in North America, 93–95, 94*i*
**Haciendas, 135**
Hackensack, NJ (USA), 81
Hackett, Rosalind, 388
Haiti, 156*i*, 158–162, 159*i*, 161–162, 162*i*
   HIV-AIDS in, 158
   poverty in, 158–159

   vs. Barbados, 161–162, 161*i*
**Hajj, 305**
Haka, 583–584, 584*i*
Hamas, 328
Han empire, 500
Hapke, Holly, 469
**Harappa culture, 418**
Harden, Carol, 3*i*
**Harijans, 434**
Hart, John, 395–396
Hart, Terese, 395–396
Hausa, 382, 383
Hawaii, 596*t*, 614*t*
   tourism in, 605–606
Hays-Mitchell, Maureen, 140
**Hazardous waste, 92**
Health and disease. *See also* HIV-AIDS
   in Middle and South America, 124, 131,
     156*t*–157*t*, 158
   in North America, 78–79
   smallpox, 124
   in sub-Saharan Africa, 367–369, 367*i*
*Heart of Darkness* (Conrad), 402
**Hearths, 432**
   in East Asia, 479
   in South Asia, 432
Heilong Jiang (Amur River), 511
Helms-Burton Act, 160
Herding, 370. *See also* Agriculture; Animal
   husbandry
   desertification and, 394
   in East Asia, 517–518, 527–528
   in North Africa and Southwest Asia,
     324–325, 335
   in sub-Saharan Africa, 370, 394
Hermack, Tim, 109–110
Hernandez, Orbalin, 165
Herzegovina. *See* Bosnia and Herzegovina
**High islands,** 587, 587*f*
*Hijab*, 319, 319*i*
*Himalayan Country* [SOUTH ASIA], 457–459,
   457*i*–459*i*
Himalayas, 415, 417, 457–459, 457*i*, 476
   monsoons and, 417
**Hindi,** 420, 430
Hindu Kush, 249, 455
**Hinduism,** 431–434, 431*i*
   nationalism and, 444–445
Hindustani, 431
Hispanic culture, in Southwest U.S., 112
**Hispanics,** 56
   in meatpacking industry, 106
Hispaniola, 124
HIV-AIDS. *See also* Health and disease
   in East Asia, 486
   global distribution of, 368*i*
   in Middle and South America, 158
   in South Asia, 428
   in Southeast Asia, 545–546

in sub-Saharan Africa, 367–369, 368*i*, 374, 408, 409
Ho Chi Minh, 541
Hodder, Ian, 303
*Hokule'a*, 591, 591*i*
**Holocaust, 193**, 308
Honduras, 157*t*, 166–168, 166*i*, 167*t*
    agriculture in, 168
    environmental issues in, 168
    population dynamics in, 167*t*
    wars and civil unrest in, 168
Hong Kong, 507–508, 507*t*, 519*i*, 520
    colonization of, 481, 520
Honor killings, 319
**Horn of Africa, 357**–358, 404–405, 405*i*
**Hot spots, 587**
Housing
    culture and, 17, 17*i*
    in East Asia, 17*i*, 510*i*, 516–517, 523
    in Middle and South America, 131–133, 131*i*
    in North Africa and Southwest Asia, 317–318, 317–320, 317*i*
    in North America, 98, 98*i*
    in sub-Saharan Africa, 385–386, 385*i*, 387
Huang He (Yellow River), 475, 476, 510, 511*i*
**Hub-and-spoke network, 72**
Hui, 501, 516, 517*i*
*Hukou*, 474
Human Development Index, 34–35, 335. *See also* **Human well-being**
**Human geography, 3**
Human immunodeficiency virus infection. *See* HIV-AIDS
Human impact, environmental. *See* Environmental issues
Human trafficking, 273. *See also* Sex industry
**Human well-being, 33**
    in East Asia, 507–508, 507*t*
    in Europe, 220*t*–221*t*, 221–222
    measures of, 34–35, 34*t*, 156*t*–157*t*, 335
    in Middle and South America, 155–167, 156*t*–157*t*
    in North Africa and Southwest Asia, 335, 336*t*
    in North America, 96–97, 96*t*
    in Oceania, 613, 614*t*
    in Russia and the Newly Independent States, 257–258, 278–280, 279*t*
    in South Asia, 452–453, 453*t*
    in Southeast Asia, 568–569, 568*t*
    in sub-Saharan Africa, 397, 398*t*–399*t*
    U.N. Gender Development Index and, 35, 335
    U.N. Gender Empowerment Measure and, 35, 335
    U.N. Human Development Index for, 34–35, 157–158, 335
**Humanism, 190**

Humbeck, Eva, 210
**Humid continental climate, 187*i*, 188**
Hungary, 221*t*, 237–242, 237*i*
    anti-Soviet uprisings in, 194
    in European Union, 203
    tourism in, 240
Hunting and gathering, 25
Hurricane Katrina, 53–54
Hurricanes, 122
Hussein, Saddam, 67, 326–327, 349, 350
Hutu, 392
Hydroelectric power, in North America, 110. *See also* Dams; Energy production

Ibraham, Hassan, 297
Iceland, 220*t*, 233
    human well-being in, 220*t*, 222
    welfare system in, 215–216, 216*i*
Igbo, 382, 383
Illness. *See* Health and disease; HIV-AIDS
Immigration. *See* Migration
**Import quotas, 32**
**Import substitution industrialization (ISI), 137**
*In the Land of God and Man: Confronting our Sexual Culture* (Paternostro), 158
**Incas, 123**, 123*i*, 171, 173
**Income disparity**
    definition of, 134
    in East Asia, 493–494, 493*i*
    gender-based, 18, 18*t*, 96*t*
    health status and, 131
    Internet access and, 11, 74
    in Middle and South America, 125–129, 127*i*, 134, 141, 150–151, 167, 174–175, 178
    in North Africa and Southwest Asia, 322, 325–326, 342
    in North America, 80–81, 87–88, 87*i*, 103
    race and ethnicity and, 87*i*
    for single-parent families, 89, 90
    in South Asia, 425, 440–443, 441*i*, 457, 466
    in Southeast Asia, 549, 549*t*, 575–576, 575*i*
    in sub-Saharan Africa, 374
India
    central, 465–467, 465*i*
    far eastern, 462*i*, 463–464
    human well-being in, 452–453, 453*t*
    immigration and, 420, 422
    independence movement in, 421–423
    languages of, 430–431
    northwest, 459–461, 460*i*, 461*i*
    politics and government in, 438
    relations with Pakistan, 423, 424, 446–447, 446*i*
    water resources of, 413–414
**Indian diaspora, 422**
Indian Parliament, women in, 438

**Indigenous peoples, 118.** *See also* **Cultural diversity; Racial/ethnic groups**
    of East Asia, 500–502
    global distribution of, 535*i*
    mapping of, 145
    of Middle and South America, 115–117, 120–121, 145, 149, 173
    as noble savages, 593
    of North America. *See* Native Americans
    of Oceania, 591–592, 592*i*, 594, 598–601
    of South Asia, 420, 455–456, 455*i*
    of Southeast Asia, 539, 569
Indochina, 536
Indo-Gangetic Plain, 415
*Indonesia* [SOUTHEAST ASIA], 534*i*, 536, 576–577, 577*i*
    human well-being in, 568*t*, 569
    national ideology of, 556
    politics and government in, 555*i*, 556
    resettlement programs in, 562–563, 562*i*
    terrorism in, 557
    tsunami in, 537, 537*i*
    wars and civil unrest in, 555–557
Indus River, 415, 417, 418
**Indus Valley civilization, 418**
**Industrial production, 30**
**Industrial Revolution, 31**
    in Britain, 192–193
    in Europe, 192
Industrialization. *See also* Economic development
    in East Asia, 481–482, 487–490
    in Europe, 204*i*, 205
    import substitution, 137
    in Middle and South America, 137–138
    in North America, 60–61, 73, 100–102, 101*i*, 102*i*
    population growth and, 35–36, 36*i*
    in Russia and the Newly Independent States, 254, 259–260
    in South Asia, 440–443, 441*i*, 465
    pollution from, 452
    in Southeast Asia, 548–552, 571–572
    in sub-Saharan Africa, 359
    debt restructuring and, 376–377
Infant mortality. *See also* **Human well-being**
    in East Asia, 508
Infanticide, female, 39
    in East Asia, 499
    in South Asia, 426*i*, 427, 436–437
Infibulation, 391
**Informal economy, 31**, 140
    microcredit in, 443–444, 580
    in Middle and South America, 140
    in Russia and the Newly Independent States, 140
    in South Asia, 443–444
    in Southeast Asia, 580
    in sub-Saharan Africa, 377–378, 377*i*

**Informal institutions, 12**
**Information technology.** *See*
    **Internet/information technology**
Ingushetia, 272
Inner Mongolia, 516
Institutional Revolutionary Party (PRI), 145
**Institutions, 12**
    formal, 12
    informal, 12
**Internal processes, 20**
International cooperation, 49–50
International Court of Justice, 242
International date line, 5
**International Monetary Fund (IMF), 32,**
    74, 176, 553. *See also* **Structural**
    **adjustment policies (SAPs)**
**Internet/information technology, 73–74**
    access to, 11, 74
    in East Asia, 438, 497
    in Europe, 205, 207i, 229
    in Middle and South America, 147, 148f,
        166
    in North Africa and Southwest Asia, 321
    in North America, 78, 110, 112
    outsourcing of, 78
    in Russia and the Newly Independent
        States, 267, 268i, 268t
    in South Asia, 438
    in Southeast Asia, 533–534
    in sub-Saharan Africa, 379–380, 380
**Interregional linkages, 9, 10i**
**Interstate Highway System, 72, 103**
**Intertropical convergence zone (ITCZ),**
    **356**–358, 538
    monsoons and, 416–417
**Intifada, 327**
Ioffe, Grigory, 274
Iran, 346i, 347–349
    borders of, 347–348
    economy of, 349
    Islamist revolution in, 329, 348
    natural resources of, 348
    population pyramid for, 311i
    as theocracy, 348
    wars and civil unrest in, 326, 348, 349
Iran-Iraq war, 326, 349
Iraq, 346i, 349–350
    Kurds in, 326–327
    wars and civil unrest in, 326, 349–350
        Gulf War, 326, 349–350
        Iran-Iraq war, 326, 349
        U.S. invasion, 67–68, 326–327
Ireland, 227–228
    economy of, 202, 228
    immigration in, 183–184
    wars and civil unrest in, 228
    welfare system in, 216
Irish Republican Army, 228
**Iron Curtain, 193**

Iron production, development of, 359
Irrigation. *See also* Agriculture; Water
    resources
    in East Asia, 506, 517
    in North Africa and Southwest Asia,
        323–324, 324i, 332, 341–342, 341i
    in North America, 96, 103, 108i
    in Russia and the Newly Independent
        States, 277, 278i
    salinization and, 323–324
    in sub-Saharan Africa, 396–397, 401
**Islam, 304**–306. *See also* **Muslims**
    in daily life, 314–315
    democracy and, 329–330
    diversity in, 316
    environmental issues and, 331–332
    family values and, 316
    Five Pillars of, 314, 315
    gender issues and, 306, 317–320, 390
    globalization and, 321–322
    in public life, 315–316, 329–330
    shari'a and, 314, 383
    spread of, 305–306
    slavery and, 360
    Taliban and, 435–436, 447–448
**Islamic fundamentalists (Islamists), 321**–322,
    329–330
    in Algeria, 338–339
    in Saudi Arabia, 343–344
Islamic world, 299
Island(s)
    East African, 399t, 405i, 406
    formation of, 587, 587f
    Pacific. *See* Oceania
    Southeast Asian, 572–574, 573i
*Island and Peninsular Southeast Asia*
    *(Malaysia, Singapore, Brunei)*
    [SOUTHEAST ASIA], 572–574, 573i
Island archipelagos, 535
Israel, 345–346, 345i
    conflict with Palestine, 308–309, 308i,
        327–329, 332–335, 345
    creation of, 308–309, 308i, 327, 328i
    economy of, 325
    human well-being in, 327t, 335, 336t,
        345–346
    population pyramid for, 311i
    water resources of, 332–335
    Zionist and, 308
**Isthmus, 167**
Italy, 230, 231i, 232–233
    economy of, 201–202
    human well-being in, 221t
    welfare system in, 215–216
Ivan the Terrible, 252

J curve, 36, 36i
**Jainism,** 431i, **432**
*Jalee*, 436

Jamaica, 124, 156i, 156t, 158–160, 159f, 159i
    cultural diversity in, 149
    natural resources in, 160
    tourism in, 160
Jamestown, VA (USA), 59
*Janjaweed*, 341, 341i
*Japan* [EAST ASIA], 520–524, 521i
    aging population in, 498, 523–524
    automobile industry in, 77, 487–488
    climate of, 477i, 478
    economic development in, 487–489
    environmental issues in, 506
    ethnic groups in, 502, 502i
    food production in, 522–523
    guest workers in, 524
    history of, 482–484
    homes in, 523
    human well-being in, 256t, 507–508, 507t
    immigration in, 486
    imperial, 483–484, 483i
    landforms of, 476, 476i
    physical geography of, 521–522
    population patterns in, 485–486, 485i,
        523–524
    relations with China, 481
    sociocultural issues in, 497–502,
        523–524
    urbanization in, 523
    work patterns in, 524
*Jati*, **434, 444**
Jazz, 240
Johnson-Sirleaf, Ellen, 354, 384, 402
Jordan, 345–346, 345i
    politics and government in, 330
Jordan River, 333–335
    West Bank of, 328, 328i, 332–335
*Joypur* (Bangladesh), 428
*Juche*, 526
**Judaism, 304.** *See also* Israel
    origins of, 304
    Zionism and, 308
Junggar Basin, 516i, 517

Kabila, Joseph, 380
Kabila, Laurent, 404
Kabul (Afghanistan), 448, 456
*Kaizen*, **488**
Kalinigrad (Russia), 236
Kalipeni, Ezekiel, 365
Kamchatka, 248i
Kampala (Uganda), 386
Kansas City, MO (USA), 105
Kashmir, armed conflicts in, 446–447, 447i
Kassem, Hisham, 321
Katun River, 249, 249i
Katz, Cindi, 320
Kazakhs, 500, 516, 517i
Kazakhstan, 246, 247i, 277, 279t, 290–294,
    290i

Kendall, Ann, 173
Kenya, 404–408, 405i
   female politicians in, 384
   gender issues in, 384
   human well-being in, 399t
   politics and government in, 384
   Tanzania and, 407–408
Kerala, 467i, 468, 469
Khader, Samir, 297
Khmer Rouge, 542–543
Khomeini, Ayatollah Rudollah, 329, 348
*Kibbutzim*, 308
Kiev (Russia), 252, 252i
Kiley, Sam, 327
Kim Dai Jung, 526
Kim II Sung, 526
Kim Jong II, 526
Kinshasa (Democratic Republic of the Congo), 386, 402
Kirchner, Néstor, 176
Kirghiz, 516, 517i
Kiribati, 596t, 614t
**Knowledge economy, 73**
Koli (India), 429–430, 429i
Kolkata (India), 462–463, 463i
Köppen classification, of climate regions, 25, 26i–27i
**Koran, 304**
*Korea* [EAST ASIA], 525–527, 525i
   automobile industry in, 77
   climate of, 477i, 478
   economies of, 527
   environmental issues in, 506
   human well-being in, 507–508, 507t
   partitioning of, 484
   physical geography of, 525–526
   politics and government in, 527, 529
   relations with Japan, 483
   reunification of, 527
Kosovo, 195, 241
**Kshatriyas, 434**
Kublai Kahn, 528
Kuhlken, Robert, 3i
Kumming (China), 520
Kuomintang (KMT), 481
Kurds, 13, 13i, 349
   repression of, 327
   Turkey and, 347
Kuwait, 343i, 344
   in Gulf War, 326, 349–350
   human well-being in, 256t, 335, 336t
Kyoto Protocol, 48
Kyrgyzstan, 246, 247i, 272, 279t

Ladan-Baki, Amina, 383
**Ladino, 167**
Lagos (Nigeria), 387, 387i

Lake Baikal, 275
Lake Nasser, 341
**Land reform, 137.** *See also* Contested space
**Landforms, 20–22**
   of East Asia, 475–477, 476i
   of Europe, 186–187, 186i
   of Middle and South America, 115i–116i, 117i, 119–121, 119i
   of North Africa and Southwest Asia, 300–301
   of North America, 56–57, 56i
   of Oceania, 584–587
   of Russia and the Newly Independent States, 248–249, 248i, 249i
   of South Asia, 415–416
   of Southeast Asia, 532i, 536–538, 537i
   of sub-Saharan Africa, 356, 357i
Language(s), 16–17, 16i. *See also* **Cultural diversity; Racial/ethnic groups**
   culture and, 16–17
   dialects and, 16
   distribution of, 16i
   of North Africa and Southwest Asia, 320, 320i
   of Oceania, 601
   of South Asia, 420, 430–431, 430i, 455, 455i
   of sub-Saharan Africa, 391–392, 392i
   of trade, 16
*Laos* [SOUTHEAST ASIA], 534i, 536, 572–574, 573i
   colonization of, 541
   economy of, 572–574
   human well-being in, 568t, 569
   tourism in, 551
Laramie, WY (USA), 108
**Laterite, 358**
Latin America, 118. *See also* MIDDLE AND SOUTH AMERICA
**Latinos, 56**
   Hispanic culture in Southwest U.S. and, 112
   in meatpacking industry, 106
**Latitude, 5,** 5i
Latvia, 233–234, 234i, 236–237
   human well-being in, 220t, 221–222
Lawal, Amina, 383
**Leaching, 358**
Lebanon, 345–346, 345i
Lee Xia, 473–474
Legends, 5
Lenin, Vladimir, 254
Leningrad (Russia), 282i, 283
Leopold, King of Belgium, 402
Lesotho, 399t, 408–410, 408i
**Liberation theology, 151,** 168
Liberia, 400–402, 400i
   environmental issues in, 353–354
   female politicians in, 354, 384, 402

human well-being in, 398t
   wars in, 401–402
Libya, 336t, 337–339, 337i
Likhachev, Alexi Alexeyevich, 245–246
Lima (Peru), 131
**Lingua franca, 16, 392**
Literacy
   birth rates and, 427–428
   in East Asia, 507t
   in Middle and South America, 158t–159t
   in North Africa and Southwest Asia, 335, 336t
   in Oceania, 613, 614t
   in South Asia, 427, 437–438, 437i, 453, 453t, 455
   in Southeast Asia, 545, 568t
   in sub-Saharan Africa, 397, 398t–399t
Lithuania, 233–234, 234i, 236–237
   human well-being in, 220t, 221–222
   Russia and, 236–237
**Living wage, 32**
**Local scale, 9,** 9i
**Loess, 57, 509–510**
Loess Plateau, 509–510, 509i
Logging. *See also* Deforestation
   clear-cutting in, 62
   in East Asia, 510, 515
   in North America, 62, 98, 109, 110i
   in Southeast Asia, 566–567, 566i, 567i
   in Sub-Saharan Africa, 353–354
London (U.K.), 229
Long River (Chang Jiang), 475, 476
**Longitude, 5,** 5i
**Long-lot system, 99**
Lopez Obrador, Andres, 145
Los Angeles, CA (USA), 80, 111
Lowell, Francis, 60
Lowell, MA (USA), 60
Luang Prabang (Laos), 551
LUKOIL, 264–265
Lumumba, Patrice, 404
Luxembourg, 220t, 222–224, 223i

Macao, 507t, 519i, 520
Macedonia, 195, 221t, 237–242, 237i
*Machismo,* 131, **150**
Machu Picchu (Peru), 171
MacNeish, Richard, 25
Madagascar, 399t, 404–408, 405i
**Madrasas, 316**
Madrid (Spain), 231
Magellan, Ferdinand, 593
*Maghreb* [NORTH AFRICA AND SOUTHWEST ASIA], 336t, 337–339, 337i
Magna Carta, 190, 193
Mail-order brides, 273
*Mainland Southeast Asia (Burma and Thailand)* [SOUTHEAST ASIA], 569–572, 570i

*Mainland Southeast Asia (Vietnam, Laos, Cambodia)* [SOUTHEAST ASIA], 572–574, 573*i*

Majid, Fariba, 438

Malaria, 367

Malawi, 399*t*, 408–410, 408*i*

*Malaysia* [SOUTHEAST ASIA], 534*i*, 536, 572–574, 573*i*
    deforestation in, 533–535
    economy of, 574–576
    human well-being in, 568*t*, 569, 573*i*, 574–575
    income disparity in, 575–576

Maldives, 452–453, 453*t*

Males. *See also* Gender issues
    preference for, 19, 39, 426*i*, 427, 436–437, 499
    stereotypical Australian, 603–604, 603*i*

Mali, 398*t*, 400–402, 400*i*
    wetlands in, 401

Malta, 221*t*, 230, 231*i*
    welfare system in, 215–216

Managua (Nicaragua), 131

Manchu empire, 480

Manchuria, 483–484, 509*i*, 510–511

Mandated territories, 306

Mandela, Nelson, 363, 363*i*, 369

**Mano River Union (MRU), 378**

Manufacturing. *See also* Economic development; Industrialization
    brownfields and, 80–81
    continuous improvement system for, 488
    in North America, 73, 100–102, 101*i*, 102*i*

Mao Zedong, 481, 482

Maori, 588, 591–592, 592*i*
    Haka dance of, 583–584
    indigenous peoples of, 594, 598–601

Map projections, 5–6, 6*i*

Mapping, indigenous, 145

Maps, 4–6. *See also specific regions*
    captions for, 5
    childhood landscape, 2, 2*i*
    distortion in, 5–6
    legends for, 5
    scale of, 2, 4–5, 4*i*
    title of, 5

**Maquiladoras, 112,** 138*i*, 139, 142, 164–165, 165*i*

Marcos, Ferdinand, 579, 580

Mariana Islands, 614*t*

***Marianismo,* 131, 150**

**Marketization, 261**

Marriage
    arranged, 436
    caste and, 435
    divorce and, 89, 560
    dowry in, 436–437
    in East Asia, 499

    polygynous, 319–320, 366, 390, 390*i*
    in South Asia, 436
    in Southeast Asia, 559–560

Marshall Islands, 596*t*, 614*t*

**Marsupials, 590**

Martinique, 156*t*, 158–160, 159*i*

Marx, Karl, 193, 254

**Material culture, 12,** 17

Mato Grosso, 177

Mauritania, 398*t*, 400–402, 400*i*

Mauritius, 399*t*, 404–408, 405*i*

Mazār-e-Sharif (Afghanistan), 437–438

Mead, Margaret, 603

Meatpacking industry, 106

**Medieval period, 189**

**Mediterranean climate, 187***i*, **188**

**Megalopolis, 81–82,** 81*i*

Mekong River, 534*i*, 536, 572, 573*i*, 574

**Melanesia, 592**

**Melanesians, 591**

Melville, Herman, 603

Men. *See* Males

Menchu, Rigoberta, 168

Menzel, Peter, 428–429

**Mercantilism, 127, 190**

Mercator, Gerhardus, 5

Mercator projection, 5, 6*i*

Merchant marine, 127

**Mercosur, 138,** 141–142

Meridians, 5, 5*i*

Merkel, Angela, 214

**Mestizos, 127,** 149

Metallurgy, development of, 359, 480

**Metropolitan areas, 80–81.** *See also* Urbanization

Mexico City, 130, 130*i*, 131

*Mexico* [MIDDLE AND SOUTH AMERICA], 123, 131, 143–145, 162–166, 163*i*–165*i*, 167*t*
    Aztecs in, 123
    economic development in, 138*i*, 139, 142, 163–164, 164–165, 165*i*
    human well-being in, 157*t*
    income disparity in, 131
    land reform in, 143–144
    maquiladoras in, 112, 138*i*, 139, 164–165, 165*i*
    migration and urbanization in, 129, 131–134, 150–151, 164–165, 165*i*
    NAFTA and, 73, 75–77, 138*i*, 141–142, 164–165
    political systems in, 145
    population dynamics in, 167*t*

**Microcredit, 443–444,** 580

**Micronesia, 591–592,** 596*t*, 614*t*

**Middle America, 118**

MIDDLE AND SOUTH AMERICA, 113–178
    agriculture in, 123, 124–125, 126*t*, 166–167, 166*i*, 168, 173, 178
    children in, 150–151

    climate of, 120*f*, 121–122, 121*f*
    colonialism in, 124–129, 127*i*
    contested spaces in, 142–145, 154
    cultural diversity in, 149
    drug trade in, 144, 146–147, 146*i*, 150–151, 150*i*
    early inhabitants in, 122–123, 123*i*
    economy of, 73, 75–77, 116–117, 134–145, 138*i*, 141–142, 164–165
    environmental issues in, 116, 116*i*, 151–155, 152*i*–155*i*, 168
    European conquest of, 124
    family structure in, 149–150
    gender roles in, 131, 149–150, 151, 164
    health and disease in, 124, 131, 156*i*–157*i*, 158
    human well-being in, 155–167, 156*i*–157*i*
    income disparity in, 125–129, 131, 134, 150–151, 174–175, 178
    indigenous peoples of, 115–117, 120–121, 149, 173
    information technology in, 147, 148*i*
    landforms of, 115*i*–117*i*, 119–121, 119*i*
    map of, 112*i*–115*i*
    migration and urbanization in, 131–134, 150–151
    oil industry in, 115, 178
    overview of, 115–117
    physical geography of, 119–122
    politics and government in, 145–147
    population patterns in, 128*i*, 129–131, 129*i*
    regional commonalities in, 117
    regional map of, 112*i*–113*i*
    religion in, 151, 179
    resource extraction in, 115–116, 117, 124, 173
    sociocultural issues in, 148–151
    subregions of, 158–180
        Brazil, 126, 176–180
        Caribbean, 158–162, 159*i*, 161*i*, 162*i*
        Central America, 166–171, 166*i*, 167*i*, 167*t*
        Central Andes, 171–174, 172*i*
        Mexico, 123, 131, 143–145, 162–166, 163*i*–165*i*, 167*t*
        Northern Andes and Caribbean Coast, 169–171, 169*i*
        Southern Cone, 174–180, 175*i*–177*i*
    terminology for, 118
    tourism in, 154, 158–160, 161*i*
    underdevelopment in, 125–129, 127*i*
    U.S. intervention in, 147, 160–161, 168, 169
    wars and civil unrest in, 143–145, 168–169, 170–171, 172, 174, 176

Middle East, 299

Migration. *See also* Urbanization
    acculturation and, 211–212
    assimilation and, 210–212

brain drain and, 133–134
chain, 83
cultural homogenization and, 184
in East Asia, 473–474, 474*i*, 486, 515, 524
environmental impact of, 43–45
in Europe, 183–184, 193, 208–212, 209*i*, 228, 229
extraregional, 563
globalization and, 106
guest workers and, 208
in Middle and South America, 129, 131–134, 150–151, 165–166
in North Africa and Southwest Asia, 312–313, 312*i*, 338
in North America, 53–55, 54*i*, 57, 61–62, 63–64, 65–66, 81, 82–86, 83*i*, 84*i*, 99, 103, 106, 110, 112, 515
open borders and, 208
permeable national borders and, 112
pull factors in, 562
push factors in, 562
remittances and, 10
in Russia and the Newly Independent States, 285–286
in South Asia, 420, 422
in Southeast Asia, 561–564
in sub-Saharan Africa, 386–387
urban growth poles in, 178
Military. *See also* Wars and civil unrest
in Russia and the Newly Independent States, 267–268
Military spending, in Pakistan, 457
Milosevic, Slobodan, 241
Minerals. *See* Resources
Ming empire, 480
Minifundios, 167*i*
Mining. *See also* **Resources,** extraction/consumption of
diamond, 373–374, 373*i*
in Oceania, 610–611
in Southeast Asia, 567, 567*i*
Minneapolis, MN (USA), 105
Minorities. *See* **Cultural diversity;** Racial/ethnic groups
**MIRAB economy, 607**
Misahaulli, Puerto, 154
Mishra, Veer Bhadra, 451–452
Missing female phenomenon, 39. *See also* Female infanticide
Mississippi delta, 56*i*, 57
Mitra, Sugata, 11
**Mixed agriculture, 370**
*Moldova* [RUSSIA AND THE NEWLY INDEPENDENT STATES], 246, 247*i*, 279*t*, 286–288, 287*i*
*Mongolia* [EAST ASIA], 527–530, 528*i*
economy of, 528, 529–530
environmental issues in, 506

gender issues in, 530
herding in, 527–528
history of, 485, 528–529
human well-being in, 507–508, 507*t*
physical geography of, 528, 528*i*, 529*i*
Mongolian Plateau, 476, 528, 528*i*
**Mongols, 252**
**Monotheistic religions, 303–305**
**Monotremes, 590,** 590*i*
Monroe Doctrine, 147
**Monsoons, 25,** 416–418, 416*i*, 418*i*, 477, 478, 526, 538
Monte Verde, 123
Montenegro, 221*t*, 237–242, 237*i*
Montserrat, 118*i*, 119
Moosa, Ebrahim, 316
Morales, Evo, 144, 144*i*
Moran, Emilio, 151
Mormons, 62, 89
Morocco, 336*t*, 337–339, 337*i*
politics and government in, 330
Moscow (Russia), 280–282
settlement of, 252–253
**Mother country, 127**
Mount Fuji, 476, 476*i*
Mount Kenya, 356
Mount Kilimanjaro, 356
Mount Pinatubo, 578, 579
Mountains
formation of, 21–22, 22*i*
plate tectonics and, 21–22, 22*i*
rain shadows and, 25, 25*i*
Movement of Landless Rural Workers (MST), 144
Movie industry, in India, 467
Mozambique, 399*t*, 408–410, 408*i*
Mudumalai Wildlife Sanctuary, 450
Mugabe, Robert, 385, 409
**Mughals, 419–420,** 419*i*
*Mulattos,* 149
**Multiculturalism, 14.** *See also* **Cultural diversity**
Multimedia Super Corridor, 576
**Multinational corporations, 32.** *See also* **Globalization**
Mumbai (India), 429–430, 429*i*, 465–467
Murray River, 589
**Muslims, 305.** *See also* **Islam**
diversity among, 316
in East Asia, 500–501
in Europe, 211–212, 212*i*, 213, 213*i*, 230
cultural influences of, 189
in North Africa and Southwest Asia, 304–306
history of, 305–306
in Russia and the Newly Independent States, 272
Shi'ite, 314–315
in South Asia, 420, 423, 432, 433–434

in Southeast Asia, 540, 559–560, 579
spread of Islam and, 305–306
in sub-Saharan Africa, 389
Sunni, 314–315
vs. Arabs, 299
Myeka High School (South Africa), 371

Na Ilyinke, 245–246
**NAFTA (North American Free Trade Agreement),** 73, 75–77, 110, 112, 138*i*, **141**–142, 164–165
Namibia, 399*t*, 408–410, 408*i*
Narmada River, damming of, 413–414
**National identity, 49**
National parks, in sub-Saharan Africa, 395
**Nationalism, 193**
**Nations, 49**
**Nation-state, 49,** 193
Native Americans. *See also* **Indigenous peoples**
displacement of, 59, 62–63
extractive resources and, 100
origin of, 57–58
population patterns for, 105–106, 108
settlements of, 57–59
**NATO (North Atlantic Treaty Organization), 202,** 266
Natural gas. *See also* Oil and gas industry
in Southeast Asia, 572
Nauru, 596*t*, 614*t*
pollution in, 611
Nazca Plate, 119
Nefedova, Tatyana, 274
Nehru, Jawaharlal, 433
Nei Mongol, 516
**Neocolonialism, 137, 355**
Neolithic Revolution, 25
Nepal, 457–459, 457*i*
democratic movements in, 448
Dolpo-pa in, 458
human well-being in, 452–453, 453*t*
Rong-pa in, 458
trade in, 458
Netherlands, 222–226, 223*i*
agriculture in, 224, 225*i*, 226
colonies of, 362, 541, 556
human well-being in, 220*t*
land use in, 224–226, 224*i*, 225*i*
landforms of, 186
Muslims in, 213
population patterns in, 224
Netherlands Antilles, 156*t*, 158–160, 159*i*
New Caledonia, 596*t*, 614*t*
*New England and Atlantic Provinces* [NORTH AMERICA], 97–99, 98*i*
New Orleans, LA (USA), 102
**New Urbanism, 81**

New York, NY (USA), 61, 82
　terrorist attacks on, 49, 66–68
New Zealand, 582*i*, 583–615, 596*t*, 614*t*.
　　*See also* OCEANIA
　colonization of, 593–594
　environmental issues in, 608*i*, 610
　independence for, 594
News media
　in North Africa and Southwest Asia,
　　297–298, 298*i*
　in Russia and the Newly Independent
　　States, 267, 267*i*, 268*t*
Ngilu, Charity Kaluki, 384
Nicaragua, 157*t*, 166–168, 166*f*, 167*t*
　canals in, 168
　civil conflicts in, 168–169
　globalization and, 139
　politics and government in, 147
　population dynamics in, 167*t*
　wars and civil unrest in, 168–169
Nicholas II, 254
Niger, 398*t*, 400–402, 400*i*
Niger River, 401
Nigeria, 400–402, 400*i*
　corruption in, 382–383
　ethnic conflict in, 381–383
　history of, 381–382
　human well-being in, 398*t*
　oil industry in, 382
Nike, 550
Nile River, 301, 302*i*, 339, 340*i*
　in Egypt, 340, 340*i*, 342–343
　in Sudan, 340–341, 340*i*
*Nile: Sudan and Egypt* [NORTH AFRICA AND
　　SOUTHWEST ASIA], 336*t*, 339–343, 342*i*
9/11 attack, 49, 66–68, 73, 327
**Noble savages, 593**
*Nodi bhanga lok*, 462
**Nomad pastoralists, 251**
**Nongovernmental organizations (NGOs),**
　**49–50**
**Nonmaterial resources, 30**
**Nonpoint sources of pollution, 277**
Noriega, Manuel, 147
Norilsk (Siberia), 260*i*, 261, 284
**Norms, 14–15**
NORTH AFRICA AND SOUTHWEST ASIA, 297–350
　agriculture in, 301, 301*i*, 302–303, 302*i*,
　　332, 341–342, 341*i*
　Arab reformists in, 330–331
　children in, 320, 329
　climate of, 300, 300*i*
　in cold war, 306–308
　countries of, 314*i*, 337–350
　cultural diversity in, 320, 320*i*
　early civilizations in, 302–304
　economy of, 322–326
　education in, 320
　environmental issues in, 331–335

　desertification, 335
　　water resources, 331–335, 333*i*,
　　　341–342
　European colonization of, 306–308, 307*i*
　family values in, 316–317
　Fertile Crescent in, 302, 302*i*
　gender issues in, 303, 311, 317–320, 331,
　　335, 336*t*, 343–345
　globalization and, 321–322
　herding in, 324–325, 335
　history of, 302–309
　human well-being in, 327*t*, 335, 336*t*
　information technology in, 321
　Israel–Palestine dispute in, 308–309, 308*i*,
　　327–329
　land distribution in, 342
　landforms and vegetation of, 300–301
　languages of, 320, 320*i*
　migration in, 312–314, 338
　military spending in, 325, 325*i*
　news media in, 297–298, 298*i*, 321
　oil and gas industry in, 308, 322–323, 322*i*,
　　323*i*, 349–350
　Ottoman Empire in, 306
　physical geography of, 300–301
　politics and government in, 329–331
　polygyny in, 319–320
　population patterns in, 309–314, 309*i*
　public vs. private spaces in, 317–320
　refugees in, 313–314
　regional map of, 296*i*–297*i*
　religion in, 302, 303–305, 304*i*, 314–316
　sociocultural issues in, 314–322
　structural adjustment policies in, 325–326
　subregions of, 337–350
　　Arabian Peninsula, 342–345, 343*i*
　　Eastern Mediterranean, 345–346, 345*i*
　　Maghreb, 336*t*, 337–339, 337*i*
　　Nile (Sudan and Egypt), 339–342,
　　　340*i*
　　Northeast (Turkey, Iran, Iraq), 346–350,
　　　346*i*
　terminology for, 299
　terrorism in, 330
　urbanization in, 312–313, 312*i*
　wars and civil unrest in, 345
　　Gulf War, 326, 349–350
　　Iran-Iraq war, 326, 349
　　Israeli-Palestinian conflict, 308–309,
　　　308*i*, 327–329, 345
　　military spending and, 325, 325*i*
　　in Sudan, 340–341, 341*i*
　　Turkey-Kurdish conflict, 347
　　U.S.–Iraq war, 67–68, 326–327
　Western domination of, 306–308, 307*i*
NORTH AMERICA, 53–115
　agriculture in, 59–61, 70–72, 71*i*, 83, 96,
　　99, 104–106, 108, 108*i*, 109, 111
　central patterns in, 55

　climate of, 56–57, 56*i*
　economic core of, 60–61, 101–102, 101*i*,
　　102*i*
　economy of, 69–79
　　gender and, 79–80
　　health care and, 96–97, 96*t*
　　immigration and, 84–86
　　NAFTA and, 73, 75–77, 138*i*, 141–142,
　　　164–165
　　service and technology sectors, 74, 78
　　U.S.-Canada interdependence, 69, 70*i*,
　　　74*i*, 75–77, 75*i*
　environmental issues in, 92–96, 93*i*–95*i*,
　　108
　families in, 89–91, 89*i*
　gender issues in, 79–80, 89–91, 90*i*
　geopolitical issues in, 66–68
　human well-being in, 96–97, 96*t*
　　health care and, 78–79, 78*t*
　immigrants in, 53–55, 54*i*, 57, 61–66, 81,
　　82–86, 83*i*, 84*i*, 99, 103, 106, 110,
　　112, 212
　indigenous peoples of. *See* Native
　　Americans
　industrialization in, 60–61
　landforms of, 56–57, 56*i*
　migration in, 53–55, 54*i*, 57, 61–62, 63,
　　65–66, 81, 82–86, 83*i*, 84*i*
　permeable national borders in, 112
　physical geography of, 56–57, 56*i*
　politics and government in
　　federal system in, 79–80
　　gender and, 79–80
　　geopolitical issues and, 66–68
　population patterns in, 64–66, 64*i*, 65*i*, 91,
　　91*i*, 105
　regional map of, 52*i*–53*i*
　religion in, 87*i*, 88–89
　settlement of, 57–63
　　European, 57–61
　　pre-European, 57–59, 59*i*
　sociocultural issues in, 80–91
　subregions of, 97–112
　　American South (Southeast), 102–104,
　　　103*i*
　　Continental Interior, 107–108, 107*i*
　　Great Plains Breadbasket, 104–106, 105*i*
　　New England and Atlantic Provinces,
　　　97–99, 98*i*
　　Old Economic Core, 100–102, 101*i*
　　Pacific Northwest, 109–110, 109*i*
　　Québec, 99–100, 99*i*
　　Southern California and Southwest,
　　　110–112, 111*i*
　transportation in, 72–73
　urbanization in, 80–82
**North American Free Trade Agreement**
　**(NAFTA), 73, 75–77, 110, 112, 138*i*,**
　**141–142, 164–165**

North American Plate, 119
North Atlantic Drift, 187
**North Atlantic Treaty Organization (NATO),** **202**, 266
North China Plain, 509–510, 509*i*
*North Europe* [EUROPE], 185–186, 185*i*, 215–216, 216*i*, 233–237, 234*i*
  human well-being in, 220*t*, 221–222
North European Plain, 186, 248–249, 248*i*, 249*i*
North Korea, 525–527, 525*i*. *See also Korea* [EAST ASIA]
  human well-being in, 507–508, 507*t*
  partitioning of, 484
North Sea, pollution of, 218, 219*i*
*Northeast China* [EAST ASIA], 509–510, 509*i*
  Beijing and Tianjin, 511–512
  Far Northeast, 510–511
  Loess Plateau, 510
  North China Plain, 510
*Northeast: Turkey, Iran, Iraq* [*North Africa and Southwest Asia*], 336*t*, 346–350, 346*i*
*Northeastern South Asia* [SOUTH ASIA], 461–464, 462*i*
  Far Eastern India, 462*i*, 463–464
  Ganga-Bramaputra delta, 461–462, 462*i*
  village life in, gender equality in, 438
  West Bengal, 462, 462*i*
*Northern Andes and Caribbean Coast* [MIDDLE AND SOUTH AMERICA], 169–171, 169*i*
Northern Ireland, 228
Northern Mariana Islands, 614*t*
Northern Plateau, 476
*Northwest India* [SOUTH ASIA], 459–461, 460*i*, 461*i*
Norway, 233–235, 234*i*
  gender issues in, 235
  human well-being in, 220*t*
Nour, Nawal, 391
Novosibirsk (Siberia), 256, 284
**Nuclear family, 89**
Nuclear power. *See also* Energy production
  in Europe, 205
  in Russia and the Newly Independent States, 277
Nuclear weapons
  in Korea, 526
  in Oceania, 611
  in Russia and the Newly Independent States, 268

Obasanjo, Olusegun, 383
Obeah, 179
Occidental Petroleum, 115–116
OCEANIA, 583–615
  agriculture in, 604, 609–610
  Asian influence in, 594, 605, 606–608
  civil unrest in, 599–601

climate of, 587–589, 588*i*, 589*i*
colonization of, 593–594
cultural diversity in, 598–604, 598*i*
economy of, 604–609
education in, 613, 614*t*
environmental issues in, 608*i*, 609–613, 612*i*
European influence in, 592–594, 598–599
exploration of, 593
flora and fauna of, 589–591, 609–610
forging unity in, 601–604
gender issues in, 602–604, 613, 614*t*
global warming and, 612–613
history of, 591–594
human well-being in, 613, 614*t*
independence movements in, 594
indigenous peoples of, 591–592, 592*i*, 594, 598–601
international relations of, 594
island formation in, 587, 587*f*
languages of, 601
migration in, 591
Pacific Way and, 601
peopling of, 591–592
physical geography of, 584–590
population patterns in, 594–597, 595*i*, 596*t*
regional map of, 582*i*–583*i*
sociocultural issues in, 598–604
sports in, 601–602
tourism in, 604*i*, 605–606, 612
urbanization in, 596–597, 596*t*
Oceanic plates, 21–22, 21*i*
Oceans
  pollution of, 218–219, 219*i*
  U.N. Convention on the Law of the Sea and, 611–612
**Ogallala aquifer, 95**–96, 95*i*
Oil and gas industry
  in Europe, 205, 229, 234–235
  geopolitical aspects of, 69
  history of, 322–323
  in Middle and South America, 115–116, 164, 170, 178
  in North Africa and Southwest Asia, 308, 322–323, 322*i*, 323*i*, 343, 344, 349–350
  in North America, 108, 111
  OPEC and, 170, 322–323
  in Russia and the Newly Independent States, 262, 263, 264–266, 265*i*, 275
  in Southeast Asia, 572
  U.S. demand for oil and, 69
  water pollution from, 108, 116, 220–221, 275
Okapi Wildlife Reserve, 395–396
Okigbo, Bede, 376
*Old Economic Core* [NORTH AMERICA], 100–102, 101*i*

**Old-growth forests, 567**
**Oligarchs, 262**
Oman, 336*t*, 344
One-child policy, in China, 498
**OPEC, 322**–323
Opium Wars, 481
**Organization for Economic Co-operation and Development (OECD), 47**–48
**Organization of Petroleum Exporting Countries (OPEC), 170, 322**–323
**Orographic rainfall, 24**, 24*i*
Ortega, Daniel, 169
Ottoman Empire, 306
**Outsourcing, 78, 442**
Overfishing. *See* Fishing
Overseas Chinese, 515, 557–558
**Ozone layer, 613**

Pacific islands, 583–615. *See also* OCEANIA
  environmental issues in, 610–613
  formation of, 587, 587*f*
*Pacific Northwest* [NORTH AMERICA], 109–110, 109*i*
**Pacific Rim (Basin), 65**
Pacific Ring of Fire, 22, 537, 578, 587
**Pacific Way, 601**
**Paddy rice cultivation, 512, 547**–548
Pagan empire, 540
Pahlavi, Shah Reza, 348
*Pakistan* [SOUTH ASIA], 454–455, 454*i*, 456–457
  creation of, 423, 424, 446–447
  economy of, 457
  history of, 454
  human well-being in, 452–453, 453*t*
  income disparity in, 457
  India and, 423
  landforms of, 455–456
  languages of, 431
  military spending in, 457
  relations with India, 423, 424, 446–447, 446*i*
Palau, 596*t*, 614*t*
Palestine, 345–346, 345*i*
  conflict with Israel, 308–309, 308*i*, 327–329, 345
    water resources and, 332–335
  human well-being in, 327*t*, 335, 336*t*
  partitioning of, 308–309, 308*i*, 327, 328*i*
  refugees in, 313–314, 313*i*
  water resources of, 332–335
Palestine Liberation Organization (PLO), 327–328, 330
Pamir Mountains, 455, 455*i*
Panama, 147, 157*t*, 166–168, 166*i*, 167*t*
  population dynamics in, 167*t*
Panama Canal, 168
*Pancasila*, 556

*Panchayati raj*, 438
**Pangaea hypothesis, 20–21, 21***i*. See also **Plate tectonics**
Papua New Guinea, 596*t*, 614*t*
Paraguay, 157*t*, 174–175, 174*t*
    economy of, 174–176
Parallels, 5, 5*i*
*Pardo*, 149
Paris (France), 226
**Parsis, 431***i*, **432**
**Partition, 423**
**Pastoralism, 370**
Patagonia, 122
Paternostro, Silvana, 158
Pauoa Valley (Hawaii), 605–606, 605*i*
Pava, Will, 218
Pawson, Eric, 599
Pearl River (Zhu Jiang), 518*i*, 520
Peoples Republic of China, 481–482
**Perestroika, 255**
**Permafrost, 249**
**Permeable national borders, 112**
Perry, Matthew C., 482
Personal Vignettes
    Aguilar Busto Rosalino, 142–143
    bicycle rickshaw driver, 424–425
    Buaphet Kheunkaew, 560
    Cynthia Beall and Melvyn Goldstein, 527–528
    Fahreeden, 293–294
    Janis Neulans, 183–184
    Jesusa Ocampo, 579–580
    John and Terese Hart, 395–396
    Lee Xia, 495
    Lydia Pulsipher, 133
    Mak and Lin, 562
    Marta, 559–560
    Mary, 366
    Mau Piailug, 591
    Miguel Elbaje, 155–156
    Mohammed Abdel Wahaab Wad, 342
    Mosamad Shonabhan, 443–444
    Natasha, 282
    Nurbegum Sa'idi, 435–436
    Olivia, 29–30, 31
    Orbalin Hernandez, 165
    Paramgit Kumar, 264–265
    Pauoa Valley, 605–606
    Puerto Misahaulli, 154
    Raufa Hassan al-Sharki, 344–345
    Reyhan, 29–30, 31
    Samrith Phum and Choch, 543
    Sheikh Ahmed Abdulrahman Jahaf, 345
    Tanya, 30
    Tim Hermack, 109–110
    Valerii, 272
    Vera Kuzmic, 238–239
    Watsahah K., 552
Peru, 122, 140, 157*t*, 171–174, 172*f*

agriculture in, 144
climate of, 122
colonization of, 124, 125*i*
cultural diversity in, 144
economy of, 140, 140*i*
environmental issues in, 155
forestry in, 153
globalization and, 141
migration from, 129
poverty in, 174
wars and civil unrest in, 174
Peru Current, 122
Philadelphia, PA (USA), 61, 82
*Philippines* [SOUTHEAST ASIA], 534*i*, 536, 577–580, 578*i*
    civil unrest in, 578–579
    colonization of, 541
    economy of, 579–580
    ethnic groups in, 579
    human well-being in, 568*t*, 569
    income disparity in, 579
    independence movements in, 541
    urbanization in, 578–579, 579*i*
Phoenix, AZ (USA), 76, 82, 82*i*
Photographs, interpretation of, 6, 7*i*
**Physical geography, 3**
    of East Asia, 475–478, 476*i*
    of Europe, 186–188
    of Middle and South America, 119–122
    of North Africa and Southwest Asia, 300–301, 300*i*
    of North America, 300–301
    of Oceania, 584–590
    of Russia and the Newly Independent States, 248–250, 248*i*–250*i*
    of South Asia, 415–418, 416*i*–418*i*
    of Southeast Asia, 536–539
    of sub-Saharan Africa, 356–359, 357*i*
Piaguage, Colon, 115–116, 117
**Pidgin, 601**
Pinochet, Augusto, 147
Pittsford, NY (USA), 82
Pizarro, Francisco, 124
**Plantations, 136**
    in Middle and South America, 136
    in North America, 59–60
Plants
    domestication of, 25, 126*t*. See also Agriculture
    genetically modified, 72
    of Oceania, 589–590
**Plate tectonics, 21–22, 21***i*
    East Asia and, 475
    Europe and, 186
    Middle and South America and, 119
    North Africa and Southwest Asia and, 300–301
    North America and, 56
    Oceania and, 585

Russia and the Newly Independent States and, 249
South Asia and, 415–416
Southeast Asia and, 536–538
sub-Saharan Africa and, 356
Plateau of Tibet, 415, 476, 517–519
**Pluralistic states, 49**
Pogroms, 308
Pol Pot, 542
Poland, 237–242, 237*i*
    anti-Soviet uprisings in, 194
    economic transition in, 239–240
    environmental issues in, 218, 239–240
    human well-being in, 221*t*
    immigrants from, 210
Polar ice caps, melting of, 46
Polish Plumber, 210, 210*i*
**Political ecology, 41**
Politics and government, 48–50
    in East Asia, 481–482, 527, 529
    in Europe, 193, 230, 240–242
    in Middle and South America, 127–129, 127*i*, 145–147
    in North Africa and Southwest Asia, 329–331
    in North America, 66–68, 79–80, 89
    in Russia and the Newly Independent States, 254–255, 261–262, 266–269
    in South Asia, 444–448
    women in, 438
    in Southeast Asia, 555–557, 555*i*
    in sub-Saharan Africa, 380–385, 384*i*
Politkovskaya, Ana, 267, 267*i*
Pollution. *See also* Air pollution; Environmental issues; Water pollution
    nonpoint sources of, 277
Polo, Marco, 480
**Polygyny, 319**–320, 366, 390, 390*i*
**Polynesia, 592**
Pope Benedict, 213
Population control
    in East Asia, 486, 497–499
    education and, 427–428
    in North Africa and Southwest Asia, 310–311
    in South Asia, 425–428, 425*i*
    in Southeast Asia, 545, 545*i*
    in sub-Saharan Africa, 365–366
Population density. *See also* Population patterns
    average, 37
    global patterns of, 37–38, 37*i*
Population growth
    demographic transition and, 40, 40*i*
    global patterns of, 35–36, 36*i*–37*i*
    wealth and, 39–40, 40*i*
Population patterns, 35–40
    in East Asia, 485–486, 486*i*, 497–499
    in Europe, 195–198, 195*i*–197*i*

in Middle and South America, 128*f*, 129–134
in North Africa and Southwest Asia, 309–314, 309*i*
in North America, 64–66, 64*i*, 65*i*, 91, 91*i*
in Oceania, 594–597, 595*i*, 596*t*
in Russia and the Newly Independent States, 255–259, 256*i*, 258*i*
in South Asia, 424–428, 424*i*–426*i*
in Southeast Asia, 543–546, 543*t*, 544*i*–546*i*
in sub-Saharan Africa, 364–369
**Population pyramids, 38–39,** 38*i*
for East Asia, 498, 499*i*
for Europe, 197, 197*i*
for North Africa and Southwest Asia, 311, 311*i*
for North America, 91, 91*i*
for Russia and the Newly Independent States, 258*i*, 259
for South Asia, 426, 426*i*
for Southeast Asia, 545, 545*i*
for sub-Saharan Africa, 366, 366*i*
**Populist movements, 151**
Portugal, 221*t*
colonies of, 124, 541
economy of, 202
welfare system in, 215–216
Poverty. *See* **Income disparity**
Powers, Samantha, 341
Precipitation, 23–25, 23*i*, 24*i*. *See also* Climate; Rain/rainfall
frontal, 25
Prescott-Allen, Robert, 41
Press freedom
in North Africa and Southwest Asia, 297–298, 298*i*, 321
in Russia and the Newly Independent States, 267, 267*i*, 268*t*
**Primate cities, 131**–133
Buenos Aries, 176
London, 229
Paris, 226
in sub-Saharan Africa, 386–387
Prime meridian, 5, 5*i*
**Privatization, 261**
**Producer services, 205**
Projections, 5–6, 6*i*
Proselytization, 15
Prostitution. *See* Sex industry
Protectionism, 201
**Protectorates, 306**
**Protestant Reformation, 190**
Protestantism, evangelical, 151, 389
Public health. *See* Health and disease; HIV-AIDS
Pudong (China), 514
Puerto Rico, 156*t*, 159*i*, 161
**Pull factors, 562**

Pulsipher, Lydia, 133
Punjab, armed conflicts in, 445–446
**Purchasing power parity (PPP), 34**
**Purdah, 429, 436,** 436*i*
**Push factors, 562**
Putin, Vladimir, 255, 267, 275
**Pyroclastic flows,** 118*f*, 119

Qaidam Basin, 476
*Qanats,* **517**
Qatar, 343*i*, 344
politics and government in, 331
population pyramid for, 311*i*
Qin empire, 478
Qing Empire, 481
Qinghai Plateau (Plateau of Tibet), 415, 517–519
Qinghai Province, 516*i*, 517–518
*Québec* [NORTH AMERICA], 56, 84, 99–100, 99*i*
**Québecois, 56,** 84
**Quotas, import, 32**
**Qur'an, 304**

Rabari, 459, 460*i*
**Race, 19**–20
Racial/ethnic groups. *See also* Cultural diversity; **Indigenous peoples**
in East Asia, 500–502, 500*i*, 516–519
in Europe, 193
in Middle and South America, 149
in North Africa and Southwest Asia, 320, 320*i*
in North America, 86–88, 87–88, 87*i*, 99, 103
in Oceania, 591–592, 592*i*, 594, 598–601
in Russia and the Newly Independent States, 269–270, 269*i*
in South Asia, 430–431
of South Asia, 420, 455–456, 455*i*
in sub-Saharan Africa, 380–381, 381*i*
**Racism, 20**
Radio Sahar, 435–436
Railroads. *See* Transportation
Rain forest. *See also* Deforestation; Forests
in sub-Saharan Africa, 394
**Rain shadow, 24,** 24*i*
Rain/rainfall, 23–25, 23*i*, 24*i*. *See also* **Climate**
acid, 93, 93*i*
monsoon, 416–418, 416*i*, 418*i*
orographic, 24, 24*i*
Ram's Temple, 445
Ranching. *See* Agriculture; Animal husbandry
**Rate of natural increase, 38**
Rawanda, female politicians in, 384
**Recharge areas, 332**
Red Basin, 512–514
Red River, 534*i*, 536, 572, 573*i*

Refugees. *See also* Migration
in North Africa and Southwest Asia, 313–314, 313*i*
in Southeast Asia, 563–564
**Regional conflicts, 445**
**Regional geography, 4**
**Regional self-sufficiency, 490**
**Regional trade blocs, 32,** 75–77
**Regions, 6**–9
boundaries of, 7
characteristics of, 9
climate, 25, 26*i*–27*i*
contiguous, 9
overview of, 6–9
**Religion(s), 15.** *See also* **Cultural diversity** *and specific religions*
caste and, 433
communal conflict and, 433
in East Asia, 479–480, 500–501, 519
in Europe, 190
formal, 12, 12*i*
informal, 15
major world, 12*i*
in Middle and South America, 151, 179
monotheistic, 303–305
in North Africa and Southwest Asia, 302, 303–305, 304*i*
in North America, 87*i*, 88–89
in Russia and the Newly Independent States, 249, 271–272, 271*i*
in South Asia, 431–434, 431*i*–433*i*
nationalism and, 444–445
in Southeast Asia, 540, 541, 558–560
spread of, 15
in sub-Saharan Africa, 388–390, 388*i*
theocratic states and, 315
vs. secularism, 15
**Religious nationalism, 444**–445
**Remittances, 10**
**Renaissance, 190**
Republic of Ireland, 227–228. *See also* Ireland
Republic of Congo, 355, 398*t*, 402–404, 402*i*
Reservations, 63
**Resettlement schemes, 561**
**Resource base, 196**
**Resources, 30**
extraction/consumption of, 30. *See also* Logging; Mining; Oil and gas industry
changing patterns of, 45
in East Asia, 510–511
in Europe, 216–221, 236, 239–240
in Middle and South America, 115–116, 117, 124, 173
in North Africa and Southwest Asia, 331–335
in North America, 62, 98, 100, 108, 109, 110*i*, 112
in Oceania, 605, 610–611

Resources (continued)
    in Russia and the Newly Independent States, 275, 285
    in South Asia, 413–414, 448–452
    in Southeast Asia, 566–567
    in sub-Saharan Africa, 371–374, 372i, 373i
  extractive, 30
  nonmaterial, 30
**Responsibility system, 491–492**
Rhine River, 185i, 187
Rice cultivation, wet, 512, 547–548
**Ring of Fire, 22,** 22i, 537, 578, 587
Rio de Janeiro (Brazil), 176
**Roaring forties, 588**
Robinson projection, 5, 6i
Rocky Mountains, 56
**Roma, 193**
Roman Catholic Church
  in Europe, 190
  liberation theology and, 151
  in Middle and South America, 151
Romania, 221t, 237–242, 237i
  agriculture in, 207–208
Rome, ancient, 189
Root bridge, 464i
Rothermund, Dietmar, 421
Rousseau, Jean-Jacques, 49
Roy, Beth, 433–434
Ruedy, John, 338
Rushdie, Salman, 230
RUSSIA AND THE NEWLY INDEPENDENT STATES, 245–294
  Afghanistan war and, 255, 291
  agriculture in, 250, 251i, 254, 256, 257, 263
    irrigation in, 277, 278i
  anti-Soviet uprisings and, 194
  Aral Sea disaster in, 277, 278i
  Baltic states and, 233, 236–237
  Bolshevik Revolution in, 193
  climate of, 250, 250i
  in cold war, 193–194, 254–255, 255i
  communications industry in, 267, 268i, 268t
  Communism in, 254–255
  corporate raiders in, 245–246, 262
  corruption in, 274
  cultural diversity in, 269–270, 269i
  democratization in, 255
  economic transition in, 255, 259–266
    debt crisis in, 262–263
    foreign direct investment and, 262
    informal economy in, 262, 263i
    integration with Europe and U.S., 264
    obstacles to, 263–264
    price controls in, 262
    privatization in, 261
    trading partners and, 263
  environmental issues in, 218, 258, 274–277

  European, 249
  foreign direct investment in, 262
  gender issues in, 268–269, 269i, 272–273, 279–280, 279t
  global influence of, 255
  globalization and, 285
  history of, 251–255
  human impact in, 276i
  human well-being in, 257, 278–280, 279t
  immigration in, 285–286
  imperial expansion in, 253, 253i
  industrialization in, 254, 259–260, 284
  information technology in, 78, 267, 268i, 268t
  international relationships of, 247i
  landforms of, 248–249, 248i, 249i
  media in, 267
  military in, 267–269
  Muslims in, 272
  natural resources of, 275, 285–286
  nostalgia in, 274
  nuclear power in, 277
  oil and gas industry in, 262, 263, 264–266, 265i
  oligarchs in, 245–246, 262
  overview of, 245–247
  physical geography of, 248–250, 248i–250i
  political reforms in, 255, 261–262, 266–269
  population patterns in, 255–259, 256i, 258i
  in post-Soviet years, 255
  regional map of, 244i–245i
  religion in, 249, 271–272
  repression in, 254
  Russification and, 270, 271, 291
  settlement of, 251–252
  Slavic dominance in, 270, 271
  social instability in, 274
  social services in, 272
  sociocultural issues in, 270–274
  subregions of, 280–294
    Belarus, Moldova, and Ukraine, 246, 247i, 279t, 286–288, 287i
    Caucasia, 279t, 284–285, 288–290, 288i, 289i
    Central Asian Republics, 279t, 290–294, 290i
    European Russia, 279t, 280–285, 282i
    Russian Far East, 279t, 284–285, 285i
    Siberian Russia, 279t, 283–284, 283i
  terminology for, 248
  trade routes in, 251–252, 252i, 291
  transportation in, 260–261, 263–264
  underemployment in, 272
  urbanization in, 275–277, 281–283
  wars and civil unrest in, 193, 254, 255, 270, 272
Russia [RUSSIA AND THE NEWLY INDEPENDENT STATES], 280–285

  European, 249, 280–283, 281i
    history of, 251–255
    urbanization in, 275–277, 280–282
  history of, 251–255
  human well-being in, 278–280, 279t
  internal republics of, 248, 269–270, 272
  Siberia, 283–284, 283i
    climate of, 250, 250i
    conquest of, 253
    economy in, 261
    transportation in, 260–261
Russian Far East [RUSSIA AND THE NEWLY INDEPENDENT STATES], 284–285, 285i
**Russian Federation,** 255, **269–270,** 279t
**Russification,** 270, 271, 291
**Rust Belt, 101,** 102i
Rwanda, 392, 399t, 404–408, 405i

Sadek, Wafeya, 319
Sahara Desert, 300, 357, 394
**Sahel, 357,** 394
St. Basil's Cathedral, 252, 253
St. Kitts, 156t, 158–160, 159i
St. Louis, MO (USA), 105
St. Lucia, 156t, 158–160, 159i
St. Nevis, 156t, 158–160, 159i
St. Petersburg (Russia), 282i, 283
St. Vincent, 156t, 158–160, 159i
Salam, 331
Salaryman, 524
**Salinization, 323–324**
Salween River, 534i, 536
Samoa, 596t, 614t
San Francisco, CA (USA), 111
Sandinistas, 169
Sanitation, in urban areas, 44–45
Santeria, 179
Santiago (Chile), 131
São Paulo (Brazil), 176
São Tomé and Príncipe, 399t, 402–404, 403i
Saraswati River, 418
Sarawak (Borneo), 533–535, 534i
Sardar Sarovar Dam, 413–414, 414i
Saro-Wiwa, Ken, 382, 383
Saudi Arabia, 314, 342–345, 343i
  gender issues in, 343–345
  history of, 342–343
  Islamism in, 343–344
  news media in, 321
  oil industry in, 308, 322–323, 322i, 323i, 343, 344
  terrorism in, 330, 343
Sauer, Carl, 25
**Scale,** of maps, **2,** 4–5, 4i
Scales, geographical, 8–9, 9i
**Scandinavia, 185,** 233–237, 234i
  human well-being in, 220t, 222
  welfare systems in, 214–215
**Schengen Accord, 208**

Schistosomiasis, 367
Schlüter, Achim, 238
Schumacher, E. F., 45
Schurz, William, 179
Scotland, 229–230
Sea(s)
    pollution of, 218–219, 219*i*
    U.N. Convention on the Law of the Sea
        and, 611–612
Sea level changes, 46
    from global warming, 612–613
Seattle, WA (USA), 111
**Seclusion, 306**
    in North Africa and Southwest Asia, 306,
        317–318, 318*i*
    in South Asia, 429, 435, 436
Secoya, 115–116
**Secular states, 315**
**Secularism, 15**
Seko, Mobutu Sese, 404
Selective abortion, 427
**Self-reliant development, 379**
Senegal, 398*t*, 400–402, 400*i*
*Señorita Extraviada, La*, 164
September 11 attack, 49, 66–68, 73, 326
Serbia, 237–242, 237*i*
    ethnic tensions in, 195, 241–242
    human well-being in, 221*t*
Serbo-Croatian war, 195, 240–242
**Serfs, 253**
Sericulture, 513
**Service sector, 30, 73.** *See also* Tourism
    in Middle and South America, 154, 160,
        166
    in North America, 73
    in Oceania, 604–605
    in South Asia, 442
    in Southeast Asia, 551
Sex industry
    in East Asia, 524
    in Russia and the Newly Independent
        States, 273
    in Southeast Asia, 524, 552
**Sex tourism, 552**
Seychelles, 399*t*, 404–408, 405*i*
Shakr, Abdullah, 316
Shanghai, 514–516, 514*i*
Shantytowns, in Middle and South America,
    131–133, 131*i*
**Shari'a, 314,** 383
Sharoni, Simona, 329–330
**Sheikhs, 343**
**Shifting cultivation,** 123, **358**
**Shi'ite Muslims, 314–315,** 349
**Shoguns, 482**
Siakor, Silas, 353–354, 354*i*
*Siberian Russia* [RUSSIA AND THE NEWLY
    INDEPENDENT STATES], 279*t*, 283–284,
    283*i*

climate of, 250, 250*i*
    conquest of, 253
    economy in, 261
    transportation in, 260–261
Sichuan Basin, 512–514
Sichuan Province, 512–514, 513*i*
Sierra Leone, 398*t*, 400–402, 400*i*
    wars in, 401–402
Sierra Madre, 56, 119
**Sikhs, 431*i*, 432**
    Golden Temple attack and, 445–446, 446*i*
Silk industry, 513
Silk Road, 251–252, 252*i*, 291, 500, 509, 513
**Silt, 119**
Silva, Lula da, 144
*Singapore* [SOUTHEAST ASIA], 534*i*, 536,
    572–574, 573*i*
    cultural diversity in, 576
    economic development in, 550, 576
    human well-being in, 568*t*, 569
    politics and government in, 555–556, 555*i*
Singapore-Johor-Riau Growth Triangle, 550
Single-parent family, 89, 90
Sino-Japanese War, 481
Six Day War, 327
Slavery
    in Middle and South America, 149, 169,
        170
    in North America, 59–60, 360–361, 360*i*
    in sub-Saharan Africa, 360–361, 360*i*
**Slavs, 251**
Slovakia, 221*t*, 237–242, 237*i*
Slovenia, 17, 195, 221*t*, 237–242, 237*i*
    agriculture in, 208, 238
    in European Union, 203
*Small Is Beautiful* (Schumacher), 45
Smallholder horticulture, 372
Smallpox, 124
**Smog, 92**
Soccer, 211
**Social forestry movement, 451**
**Social welfare, 214**
**Socialism, 254**
    command economy in, 193, 254, 259–261,
        489–490
    in Middle and South America, 145–146
    in Russia. *See* Soviet Union
Sociocultural issues
    in East Asia, 497–502
    in Europe, 208–216
    in Middle and South America, 148–151
    in North Africa and Southwest Asia,
        314–322
    in North America, 80–91
    in Oceania, 598–604
    in Russia and the Newly Independent
        States, 270–274
    in South Asia, 428–438
    in Southeast Asia, 557–564

in sub-Saharan Africa, 385–392
**Soft power, 496**
Soil. *See also* Agriculture
    alluvial, 515
    in East Asia, 509–510
    in Southeast Asia, 539
Soil erosion, 22, 104. *See also* Deforestation
    in East Asia, 510, 515
Solar energy. *See also* Energy production
    in Europe, 205
Solomon Islands, 596*t*, 614*t*
Solzhenitsyn, Aleksandr, 285
Somalia, 399*t*, 402–404, 403*i*
    armed conflict in, 404–405
    U.S. intervention in, 404
*Some Trouble with Cows* (Roy), 433–434
Somoza regime, 169
Soufrière volcano, 118*i*, 119
South Africa, 408–410, 408*i*
    apartheid in, 362–363, 383, 410
    Boers in, 362–363
    development in, 409–410
    economy of, 374, 410
    ethnic groups of, 410
    human well-being in, 399*t*
    physical geography of, 409–410
**South America, 118.** *See also* MIDDLE AND
    SOUTH AMERICA
South American Plate, 119
SOUTH ASIA, 413–473
    agriculture in, 439–440
    bride burning in, 436–437
    caste in, 432, 434–435
        democracy and, 444
    child labor in, 427
    city life in, 429–430
    climate of, 416–418, 416*i*–418*i*
    colonial legacy in, 420–424
    dams in, 413–414, 414*i*, 415*i*
    early invaders of, 418–420, 419*i*, 430–431
    economy of, 423–424, 438–444
    education in, 437–438, 437*i*, 447, 453,
        453*t*
    environmental issues in, 413–414, 448–452
    ethnic groups in, 420, 430–431, 455–456,
        455*i*
    female infanticide in, 426*i*, 427, 436–437
    gender issues in, 426*i*, 427, 435–438, 453,
        453*t*
    globalization and, 465–467
    history of, 418–424
    HIV-AIDS in, 428
    human impact in, 448–450, 449*i*
    human well-being in, 452–453, 453*t*
    immigration and, 420, 422
    income disparity in, 425, 440–443, 441*i*,
        457, 466
    Indian subcontinent of, 414
    Indus Valley civilization of, 418

SOUTH ASIA (continued)
industrialization in, 440–443, 441i, 465
pollution from, 452
information technology in, 78, 423–424
landforms of, 415–416
languages of, 420, 430–431, 430i
marriage in, 436–437
Muslims in, 432, 433–434
physical geography of, 415–418, 416i–418i
politics and government in, 444–448, 448
women in, 438
population patterns in, 424i–426i, 425–428
regional conflicts in, 445–447, 446i
in Kashmir, 446–447, 447i
in Punjab, 445–446
regional map of, 412i–413i
religion in, 431–434, 431i–433i
religious nationalism in, 444–445
sociocultural issues in, 428–438
subregions of, 454–470, 454i
Afghanistan and Pakistan, 454–457,
454i–455i
Central India, 465–467, 465i
Himalayan Country, 457–459, 457i–459i
Northeastern South Asia, 461–464, 462i
Northwest India, 459–461, 460i, 461i
Southern South Asia, 467–470, 467i
terminology for, 415
tourism in, 442
trade in, in colonial era, 420–421
urbanization in, 456, 465
village life in, 428–429
wars and civil unrest in, 424, 469–470
water resources of, 413–414, 451–452
South Europe [EUROPE], 185, 185i, 222–230,
223i, 230–233, 231i
human well-being in, 221–222, 221t
welfare systems in, 215–216, 216i
South Korea, 525–527, 525i. See also Korea
[EAST ASIA]
human well-being in, 507–508, 507t
partitioning of, 484
South Pacific. See OCEANIA
SOUTHEAST ASIA, 533–580
agriculture in, 546–548, 547i, 548i
air pollution in, 567–568
climate of, 538–539, 538i
colonization of, 540i, 541
corruption in, 553
cultural diversity in, 539–541, 557–558
deforestation in, 533–535, 566–567
earthquakes in, 537
economy of, 543, 546–554
education in, 545, 561t, 562, 568t, 569
environmental issues in, 533–535, 548,
564–568
Export Processing Zones in, 549–550, 552
gender issues in, 552, 556, 560–561
globalization and, 552–554
growth triangles in, 550

history of, 539–543
HIV-AIDS in, 545–546
human impact on, 565i
human well-being in, 568–569, 568t
income disparity in, 549, 549t, 569,
575–576, 575i
independence movements in, 541–543
industrialization in, 548–552, 571–572
information technology in, 78
landforms of, 532i, 536–538, 537i
marriage in, 559–560
migration in, 561–564
mining in, 567
peopling of, 539
physical geography of, 536–539
politics and government in, 555–557, 555i
population patterns in, 543–546, 543t,
544i–546i
regional map of, 532i–533i
religion in, 540, 541, 558–560
sociocultural issues in, 557–564
subregions of, 569–580
Indonesia and Timor-Leste, 576–577,
577i
Island and Peninsular Southeast Asia:
Malaysia, Singapore, Brunei,
574–576, 575i
Mainland Southeast Asia: Burma and
Thailand, 569–572, 570i
Mainland Southeast Asia: Vietnam,
Laos, Cambodia, 572–574, 573i
Philippines, 577–580, 578i
terminology for, 536
terrorism in, 557–558
"tiger" economies in, 571
tourism in, 551–552
tsunamis in, 537, 537i
urbanization in, 561–562, 578–579, 579i
volcanoes in, 22, 537
wars and civil unrest in, 541–543
work patterns in, 561
working conditions in, 550–551
Southeastern Anatolia Project, 334
Southeastern China Coast, 519i, 520
Southeastern Europe, 186, 240–242, 241i
Southern Africa [SUB-SAHARAN AFRICA],
408–410, 408i
civil wars in, 409
development in, 409–410
economies of, 409–410
human well-being in, 399t
physical geography of, 408
Southern African Development Community
(SADC), 378
Southern California and Southwest [NORTH
AMERICA], 110–112
Southern China [EAST ASIA], 518i, 519–520
Hong Kong, 520
Macao, 520
Southeastern Coast, 520

Yunnan-Guizhou Plateau, 519–520
Southern Cone [MIDDLE AND SOUTH
AMERICA], 174–176, 175i–177i
economies of, 174–176
urbanization in, 176, 176i
Southern Growth Triangle, 550
Southern South Asia [SOUTH ASIA], 467–470,
469i
Bangalore, 467, 467i
Chennai, 467, 467i
Kerala, 467i, 468, 469
Sri Lanka, 467i, 468–470
Sovereignty, 49
Soviet Union (USSR), 246. See also
Communism; Russia
agriculture in, 207
Baltic states and, 233, 236–237
breakup of, 246
in cold war, 193–194, 254–255, 255i
collapse of, 194
history of, 254–255
Mongolia and, 529–530
uprisings against, 194
war with Afghanistan, 255, 291
Spain, 230–233, 231i
agriculture in, 231
civil unrest in, 231–232
colonies of, 124
economy of, 202, 231
human well-being in, 221t
immigrants in, 215–216
welfare system in, 215–216
Spatial analysis, 2
Special economic zones (SEZs), 494
Sports
Haka dance, 583–584
in Oceania, 601–602
soccer, 211
Sri Lanka, 452–453, 453t
infant mortality in, 427
Srivijaya empire, 540
Stalin, Joseph, 254
State-aided market economies, 487
Steel. See Industrialization; Manufacturing;
Metallurgy
Steppes, 26i, 249
Stock, Robert, 381
Strait, John B., 88
Structural adjustment policies (SAPs), 32,
139
currency devaluation and, 376
in Europe, 230, 242
in Middle and South America, 134,
139–140, 170, 172, 176
in North Africa and Southwest Asia,
325–326, 342
in South Asia, 440–442
in Southeast Asia, 552–553
in sub-Saharan Africa, 374–376
Subcontinents, 414

Subduction, 22
**Subduction zone, 119**
**Subregions, 8,** 9*i*
SUB-SAHARAN AFRICA, 353–411
   African Union and, 378–379
   agriculture in, 365, 370–374, 371*i*–373*i*
      colonization and, 361–362
      debt restructuring and, 376
      desertification and, 394
      development of, 359, 360
      irrigation in, 396–397
   carrying capacity of, 364–365
   climate of, 356–359, 357*i*
   cold war and, 383–384
   colonization of, 354–355, 361–364
      legacy of, 369, 380–383, 390
   corruption in, 353–354, 382–383, 385
   cultural diversity in, 380–381, 381*i*
   devolution in, 385, 404
   early humans in, 359
   economy of, 369–380
      grassroots development in, 379
      informal, 377–378, 377*i*
      regional integration in, 378–379
      restructuring of, 374–378
      self-reliant development in, 379
   education in, 371
   energy resources in, 394–395
   environmental issues in, 353–355, 358,
      371, 392–397, 401
      deforestation, 393
      desertification, 393
      loss of habitat, 395–396
   ethnic groups in, 380–385, 392
   exports from, 370–374, 372*i*, 373*i*
   female circumcision in, 391
   foreign investment in, 374–378, 376*i*
   gender issues in, 365–366, 379, 383, 384,
      390–391, 397, 398*t*–399*t*
   geopolitical shifts in, 384–385, 384*i*
   history of, 359–364
   HIV-AIDS in, 367–369, 368*i*, 374, 408, 409
   housing in, 385–386, 385*i*, 387
   human impact on, 393*i*
   human well-being in, 397, 398*t*–399*t*
   income disparity in, 374
   independence for, 363–364
   indigenous belief systems in, 388–389
   industrialization in, 359
      debt restructuring and, 376–377
   languages in, 391–392, 392*i*
   migration in, 386–387
   national parks in, 395–396
   natural resources of, 354–355
      foreign exploitation of, 355
   neocolonialism in, 355
   oil industry in, 382
   physical geography of, 356–359, 357*i*
   politics and government in, 384–385, 384*i*
   population patterns in, 364–369

   post-colonial era in, 363–364, 364*i*, 369
   regional map of, 352*i*–353*i*
   religion in, 388–390, 388*i*
   repression in, 384–385, 384*i*
   settlement patterns in, 385–387, 385*i*–387*i*
   slavery in, 360–361, 360*i*. *See also* Slavery
   sociocultural issues in, 385–392
   structural adjustment policies in, 374–376
   subregions of, 400–410, 403*i*
      Central Africa, 402–404, 403*i*
      East Africa, 404–408, 405*i*
      Southern Africa, 408–410, 408*i*
      West Africa, 400–402, 400*i*
   technological development in, 379–380
   terminology for, 355–356
   trade in, development of, 360
   transportation in, 379, 379*i*
   urbanization in, 386–387, 386*i*, 387*i*
   wars and civil unrest in, 355, 370, 380–385,
      392, 401–402
   water resources in, 357, 358*i*, 396–397
   wildlife sanctuaries in, 395–396
**Subsidence, 130**
**Subsidies, 206**
**Subsistence affluence, 607**
**Subsistence economy, 40**
Sudan, 339–341, 340*i*, 341*i*
**Sudras, 434**
Suharto, 553, 555
Sundaland, 537–538, 537*i*
**Sunni Muslims, 314**–315, 349
Suriname, 157*t*, 169–170, 169*i*
**Sustainable agriculture, 41**–43, 44*i*
**Sustainable development, 41**–45
   tourism in, 606
   urbanization and, 43–45
Suu Kyi, Aung San, 555–556
Swaziland, 399*t*, 408–410, 408*i*
Sweden, 233–235, 234*i*
   currency of, 200
   human well-being in, 220*t*
   social welfare in, 235
   welfare system in, 215–216, 216*i*
Switzerland, 220*t*
   politics and government in, 193
**Syncretism, 388**
Syria, 345–346, 345*i*

**Taiga, 108**
*Taiwan* [EAST ASIA], 524–525, 524*i*
   aboriginal peoples in, 502
   environmental issues in, 506–507
   history of, 481, 483, 484–485
   human well-being in, 507–508, 507*t*
Tajikistan, 246, 247*i*, 279*t*
Tajiks, 516
Taklimakan Desert, 476, 476*i*, 516*i*, 517
**Taliban, 435**–436, 447–448, 454, 456
Tamils, 468–469
Tanzania, 399*t*, 404–408, 405*i*

   Kenya and, 407–408
Taoism, in Southeast Asia, 559–560, 559*i*
**Tariffs, 32,** 74–75
   European Union and, 201, 206
Tarim Basin, 476, 476*i*, 477–478, 516*i*, 517
Tatarstan, 270, 272
Taylor, Charles, 353
Technocrats, 134
**Technology, 17.** *See also*
      **Internet/Information technology**
   sustainable development and, 41–45
Tectonic plates. *See* **Plate tectonics**
Television, in Bhutan, 443
**Temperate midlatitude climate, 187**–188,
   187*i*
Temperature, 23
   air pressure and, 23
**Temperature-altitude zones, 121**–122, 121*i*
Territorial disputes. *See* Contested space; Wars
   and civil unrest
**Terrorism, 49, 330**
   economic and political aims of, 330
   September 11 attack and, 49, 66–68, 73,
      326
   in Southeast Asia, 557–558
Texaco Corp., 115–116, 117*i*
Textile industry
   child labor in, 427
   in East Asia, 513
   in South Asia, 420, 464
*Thailand* [SOUTHEAST ASIA], 536, 569, 570*i*,
   571–572
   economy of, 571–572
   energy resources in, 572
   human well-being in, 568*t*, 569
Al-Thani, Hamad bin Khalifa, 331
Thar Desert, 459
Tharoor, Shashi, 430, 431
**Theocratic states, 315**
**Thermal inversion, 93**
Thinley, Lyonpo Jigma, 443
Three Gorges Dam, 502, 504*i*, 514
Tiananmen Square, 482
Tianjin (China), 511
Tibet, 501, 516*i*, 517–519
   China and, 501, 517–519
   railroads in, 501, 501*i*
   terminology for, 475
Tien Shan Mountains, 249
*Tierra caliente,* **122,** 129
*Tierra fria,* **122,** 130
*Tierra helada,* **122**
*Tierra templada,* **122,** 129
Tigre River, 119*i*
Tigris River, 301, 302*i*, 334
*Timor-Leste* [SOUTHEAST ASIA], 534*i*, 536, 556,
   568*t*, 569, 576–577, 577*i*
Tito, Josip Broz, 241
Togo, 398*t*, 400–402, 400*i*
Tonga, 596*t*, 614*t*

**Total fertility rate (TFR), 38**
Tourism
  ecotourism, 154
  environmental impact of, 154, 551, 606, 612
  in Europe, 197–198, 228, 240
  in Middle and South America, 154, 158–160, 161i
  in North America, 98
  in Oceania, 604i, 605–606, 612
  sex, 552
  in South Asia, 442
  in Southeast Asia, 551–552
  sustainable, 606
Towns, rise of, in Europe, 190
Toyota Corp., 77–78
Trade. *See also under* Economic
  agricultural development and, 25
  colonies and, 31–32. *See also* **Colonialism**
  development of, in sub-Saharan Africa, 359–360
  languages of, 16
**Trade blocs, 32,** 75–77. *See also*
  **Globalization**
**Trade deficit, 77**
**Trade winds, 122**
Trans-Alaska Pipeline, 108
Trans-Amazon Highway, 151
Transportation
  in East Asia, 501, 501i
  in Europe, 205, 206i
  in Middle and South America, 170
  in North America, 72–73, 103
  in Russia and the Newly Independent States, 260–261, 263–264
  in sub-Saharan Africa, 379, 379i
Trans-Siberian Highway, 260
Trans-Siberian Railroad, 260–261
Travel. *See also* Tourism; Transportation
  air, 72–73
Treaty of Tordesillas, 124
Trinidad and Tobago, 131, 156t, 158–160, 159i
**Tropical savanna, 26,** 26i
Tropical storms, 122
**Tsunamis, 476,** 537, 537i
  refugees from, 563–564
Tumen River, 525, 525i
**Tundra, 108,** 249
Tunisia, 336t, 337–339, 337i
Turfan Depression, 517
Turkey, 346–347, 346i
  as EU candidate, 346
  globalization and, 347
  history of, 346
  human well-being in, 335, 336t
  Kurds and, 347
  religion in, 316, 346
  strategic location of, 346–347, 346i
  water resources of, 334, 347
Turkic-speaking peoples, of China, 500–501

Turkmenistan, 246, 247i, 279t
Tutsi, 392
Tutu, Desmond, 363
Tuvalu, 596t, 614t
*Typee* (Melville), 603
**Typhoons, 507**

Uganda, 404–408, 405i
  HIV-AIDS in, 368
  human well-being in, 399t
*Ukraine* [Russia and the Newly Independent States], 246, 247i, 279t, 286–288, 287i
Ulan Bator (Mongolia), 528, 528i
Uluru, 586i
Uma Bawang longhouse, 533–535
Umbanda, 179
Underdevelopment, in Middle and South America, 125–129, 127i
**Underemployment, 272**
UNESCO World Heritage Sites, 548i, 551, 586
**Union of Soviet Socialist Republics, 246**
United Arab Emirates, 336t, 344
United Kingdom, 229–230
  agriculture in, 229
  climate of, 229
  colonies of. *See also* **Colonialism**
    in Oceania, 593–594
    in South Asia, 420–424, 421i
    in Southeast Asia, 541
  in creation of Israel, 308–309
  currency of, **200**
  Hong Kong and, 481, 520
  human well-being in, 220t
  Iraq and, 326, 349
  Nigeria and, 381–382
  welfare systems in, 215, 215i
**United Nations, 49**
  U.S. in, 74
United Nations, Convention on the Law of the Sea, 611–612
United Nations Food and Agriculture Organization (UNFAO), 41
United Nations Gender Development Index (GDI), 35, 335. *See also* Gender issues; Human well-being
United Nations Gender Empowerment Index (GEM), 35, 335. *See also* Gender issues; Human well-being
United Nations Human Development Index (HDI), 34–35, 335
United States. *See also* North America
  European Union and, 202
  federal system in, 79
  gross domestic product of, 96–97, 96t
  health care in, 78–79, 78t
  human well-being in, 96–97, 96t, 256t
  immigrants in, 53–55, 54i, 83–86, 83i, 84i

  intervention in Middle and South America, 147, 160–161, 168, 169
  population patterns in, 64, 64i, 65i
  relationship with Canada, 69, 70i
  terrorism and, 49, 66–68, 73, 326
  trade deficit of, 77
  trading partners of, 69, 70i, 73, 74, 74i, 75–78, 75i, 138i, 141–142
Untouchables, 434
Ural Mountains, 248–249, 250i
Urban gardens
  in Russia and the Newly Independent States, 257
  in sub-Saharan Africa, 370
**Urban growth poles, 178**
Urban planning, 132, 133i
**Urban sprawl, 82,** 82i
Urbanization. *See also* Migration
  in East Asia, 473–474, 523
  environmental impact of, 43–45
  in Europe, 191, 192i
  gentrification and, 81
  megalopolis and, 81–82, 81i
  in Middle and South America, 129, 131–134, 150–151, 165–166
  New Urbanism and, 81
  in North Africa and Southwest Asia, 312–313, 312i
  in North America, 54i, 57, 61–62, 63–64, 65–66, 80–86, 88
  in Oceania, 609
  in Russia and the Newly Independent States, 275–277, 281–283
  in South Asia, 429–430, 456, 465
  in Southeast Asia, 561–562, 578–579, 579i
  in sub-Saharan Africa, 386–387, 386i, 387i
Urdu, 431
Uribe, Alvaro, 171
Uruguay, 157t, 174–175, 174i
  economy of, 139, 141
  globalization and, 139, 141
  human well-being in, 157t
  population patterns in, 141
Uygurs, 500, 516–517
Uzbekistan, 246, 247i, 272, 279t
  agriculture in, 277
  Aral Sea disaster in, 277, 278i

**Vaishyas, 434**
Valledolid, Angel, 173
Vancouver, BC (Canada), 82
Vanuatu, 596t, 614t
*Varna,* **434**
*Vedas,* 431–432
Veddas, 468
**Veiling, 306,** 318–319, 319i
  in North Africa and Southwest Asia, 306, 318–319, 319i
  in South Asia, 435

Venezuela, 139–140, 145–146, 157t, 169, 169i, 170
  agriculture in, 144
  economy of, 137, 139, 141
  forestry in, 153
  globalization and, 137, 139–140, 141
  landforms of, 119
  politics and government in, 145, 147
  poverty in, 134
Vietnam [SOUTHEAST ASIA], 534i, 536, 572–574, 573i
  colonization of, 541
  economy of, 574
  human well-being in, 568t, 569
  independence movements in, 541–542
Vietnam War, 541–542
Vignettes. See Frontline World Video Vignettes; Personal Vignettes
Vogeler, Ingolf, 206
Volcanoes, 22
  in Middle and South America, 118i, 119
  pyroclastic flows from, 118i, 119
  in Ring of Fire, 22, 22i, 537, 578
  in Southeast Asia, 22, 537, 578
Volga River, 249
Voodoo, 179

Wadud, Nasiye, 316
Wales, 229–230
Wal-Mart, 75, 76
Walsh, Michael, 86
Wang Chunqiao, 492
War on Terror, 67–68
Wars and civil unrest
  cold war, 48–49, 193–194, 254–255, 255i
  communal conflicts in, 433
  drug trade and, 144, 146–147, 146i
  in East Asia, 481, 482
  in Europe, 193–194, 228
  Geneva Conventions in, 68
  genocide in, 49
  geopolitics and, 48–50
  Iraq invasion, 67–68
  in Middle and South America, 143–145, 168–169, 170–171, 172, 174, 176
  in North Africa and Southwest Asia, 308–309, 308i, 326–329
  regional conflicts and, 445–447
  in South Asia, 424, 445–448, 446i, 469–470
  in Southeast Asia, 541–543
  in sub-Saharan Africa, 355, 370, 380–385, 401–402
  terrorism and, 49, 66–68
Washington, D.C. (USA), 82
Water pollution. See also Environmental issues
  in East Asia, 506
  in Oceania, 611–612
  from oil spills, 108, 116, 220–221, 275
  in South Asia, 451–452

Water resources
  for agriculture. See Irrigation
  in Australia, 610
  in East Asia, 505–506
  global distribution of, 332–333, 333i
  India-Bangladesh conflict over, 451
  in Middle and South America, 155
  in North Africa and Southwest Asia, 323–324, 324i, 331–335, 347
  in North America, 95–97, 95i
  recharge areas and, 332
  in Russia and the Newly Independent States, 277, 278i
  in South Asia, 413–414, 451–452
  in sub-Saharan Africa, 357, 358i, 396–397, 401
  in urban areas, 44–45, 45i
Watts, Michael, 382
Wealth. See also under Economic; Income disparity
  population growth and, 39–40, 40i
Weather, 22. See also Climate
Weathering, 22
Wegener, Alfred, 20–21
Welfare states, 193, 214–216, 215f, 235
West Africa [SUB-SAHARAN AFRICA], 400–402, 400i
  cultural diversity in, 401
  human well-being in, 398, 398t
  physical geography of, 400–401, 401i
  wars and civil unrest in, 401–402
West African Economic and Monetary Union (UEMOA), 378
West Bank, of Jordan River, 328, 328i, 332–335
West Bengal, 461i, 462
West Europe [EUROPE], 222–230, 223i
  countries of, 222–230, 223i
  human well-being in, 220t, 221–222
  welfare system in, 215–216, 216i
West Germany, 227
West Pakistan, 423, 424
West Siberian Plain, 248–249, 250i
Western Europe, 186
Western Ghats, 415–416, 417
Wet rice cultivation, 512, 547–548
Wetlands, in sub-Saharan Africa, 401
When Work Disappears: The World of the New Urban Poor (Wilson), 101
Widows, in South Asia, 436
Wildlife sanctuaries
  in South Asia, 450
  in sub-Saharan Africa, 395–396
Wilson, William Julius, 101
Women. See Female(s); Gender
Wong, David, 81
Working women, 549–550
  in East Asia, 473–475, 499–500, 508
  in Europe, 212–214, 216

globalization and, 549–550
  in Middle and South America, 164–165
  in North Africa and Southwest Asia, 310, 310i
  in Southeast Asia, 549–550, 551, 552, 560, 561t
World Bank, 32, 174, 178
World cities, 191, 226, 229
World Heritage Sites, 548i, 551, 586
World regions, 8, 9i
World Trade Center attacks, 49, 66–68, 73, 326
World Trade Organization (WTO), 32, 141
  China in, 495–496
  United States in, 74–75
World War II
  Europe in, 193–194
  Holocaust in, 193–194, 308
  Japan in, 484
World Wide Web. See Internet/information technology

Xibe, 516, 517i
Xinjiang Uygur Autonomous Region, 516–517
Xizang Autonomous Region, 516i, 517–519
Xizang-Qinghai Plateau, 476

Yadana Project, 572
Yalu River (Amnok River), 525, 525i
Yangtze River (Chang Jiang), 475, 476, 512, 513i, 515
  damming of, 502, 504i, 514
Yarnal, Brent, 216–218
Yellow River (Huang He), 475, 476, 510, 511i
Yemen, 336t, 344–345
Yew, Lee Kuan, 555
Yom Kippur War, 327
Yoruba, 382, 383
Yugoslavia
  breakup of, 195
  wars and civil unrest in, 195, 240–242, 241i
Yunnan-Guizhou Plateau, 476, 519–520, 519i
Yunnis, Mohammed, 443
Yurts, 17i, 515–516

Zakat, 315
Zambia, 399t, 408–410, 408i
Zapatista movement, 143–144
Zheng He, 480
Zhu Jiang (Pearl River), 518i, 520
Zimbabwe, 408–410, 408i
  devolution in, 385
  history of, 359
  human well-being in, 399t
Zionism, 308
  terrorism and, 330
Zoroastrians, 432